W0079931

Space Sciences Series of ISSI

The Space Sciences Series of ISSI books are coherent reports of the findings, discussions, and ideas that result from international scientific workshops regularly held at the International Space Science Institute (ISSI) in Bern, Switzerland. ISSI's main task is to contribute to the achievement of a deeper understanding of the results from space-research missions, adding value to those results through multi-disciplinary research in an atmosphere of international cooperation. The books are reprints of special issues in the Space Science Reviews journal and occasionally of special issues in the Surveys in Geophysics journal.

More information about this series at https://link.springer.com/bookseries/6592

Thomas Widemann · Colin Wilson · Doris Breuer ·
Cédric Gillmann · Suzanne Smrekar · Tilman Spohn
Editors

Venus: Evolution Through Time

The book is a spin-off from the Topical Collection "Venus:
Evolution Through Time" of the journal Space Science Reviews

 Springer

Editors
Thomas Widemann
Paris Observatory/LIRA
Meudon Cedex, France

Colin Wilson
ESA
Noordwijk, ZH, The Netherlands

Doris Breuer
Institute of Planetary Research
   German Aerospace Center DLR
Berlin, Germany

Cédric Gillmann
Institut für Geophysik
   ETH Zürich
Zürich, Switzerland

Suzanne Smrekar
California Institute of Technology
   Jet Propulsion Laboratory
Pasadena, CA, USA

Tilman Spohn
Institute of Planetary Research
   German Aerospace Center DLR
Berlin, Germany

ISSN 1385-7525   Space Sciences Series of ISSI
ISBN 978-94-024-2269-6

Cover Image: Composite image showing the evolutionary pathways of Venus and Earth from formation in the Sun's proto-planetary disk via young planets of similar size and properties to the current habitable Earth and life-hostile Venus. Credit: Composite: Paul Byrne; protoplanetary disk, starfield: ESO/L. Calçada/S. Brunier; magma ocean world: ESA/HST/M. Kornmesser; Archaean Earth: A. Brenner; modern Earth: NASA/Apollo 11; ancient Venus: K. Llewellin; modern Venus: JAXA/ISAS/DARTS Akatsuki/UVI.

This Springer imprint is published by the registered company Springer Nature B.V.
The registered company address is: Van Godewijckstraat 30, 3311 GX Dordrecht, The Netherlands

Paper in this product is recyclable.

Contents

Space Science Reviews (2024) 220:39
https://doi.org/10.1007/s11214-024-01075-0

Venus: Evolution Through Time – Editorial

Thomas Widemann[1,2] · Colin Wilson[3,4] · Doris Breuer[5] · Cédric Gillmann[6] ·
Suzanne E. Smrekar[7] · Tilman Spohn[5,8]

Accepted: 6 May 2024 / Published online: 23 May 2024

1 Introduction

Why are the terrestrial planets so different from each other? Venus, our nearest neighbor, should be the most Earth-like of all our planetary siblings. Its size and presumed bulk composition are very similar to those of the Earth. Like Earth, the primordial atmosphere would have been H-dominated from the nebula. Then a secondary atmosphere would have outgassed, with expected large atmospheric abundances of carbon dioxide and water. The current state of Venus is the result of the cumulative effects of many processes from the planet's formation, and its magma ocean phase up to present-day. Venus's interior thermal evolution, volcanic, tectonic and outgassing history, as well as interaction with the atmosphere have left clues to piece back together what made Earth and Venus evolve differently.

Therefore, due to its position at the inner edge of the Solar System habitable zone (HZ), Venus evolution through time provides insights on surface habitability of rocky exoplanets. How and why did things go differently for Venus? What lessons can we learn about the history of life and habitability on Earth-sized planets and terrestrial planets in general, whether in our solar system or elsewhere? Such questions are examples of the major gaps in our understanding of how and when Venus's evolutionary path diverged from Earth's.

Now is a pivotal time in Venus exploration. A new fleet of Venus missions is being developed including the radar-equipped orbiters (ESA's EnVision and NASA's VERITAS missions) and an entry probe (NASA's DAVINCI). These missions will not only answer key questions about present day Venus, but also about Venus's long-term evolution and comparative planetary science. Their current definition phase is an ideal time to collate knowledge of Venus long-term evolution scenarios and the observations needed to discriminate between them. Future missions will greatly advance our knowledge of Venus. A broad range of research approaches including Earth-based observations, laboratory and modeling studies are needed to prepare to interpret future data.

2 Tectonics and Volcanism

The evolution of Venus' geodynamic state is poorly understood, despite the fact that it is our closest planetary neighbor. It is still unknown whether water has ever condensed on the

Venus: Evolution Through Time
Edited by Colin F. Wilson, Doris Breuer, Cédric Gillmann, Suzanne E. Smrekar, Tilman Spohn and
Thomas Widemann

Extended author information available on the last page of the article

surface of Venus. The current surface age is only ~150–1000 m.y. on average, and thus likely does not record Venus' distant past. This has led to suggestions of regional or global resurfacing events – but how this happened is an open question. Whether resurfacing was catastrophic or steady has important implications for interior and climate evolution. Because the evolution of a planetary body is strongly controlled by its thermal history, the strength of convection and the extent of partial melting of the mantle determine processes such as volcanism and tectonics—which in turn affect crustal production and the evolution of the atmosphere. Large-scale flood lava units would be indicative of episodic global or regional resurfacing events that are concentrated in time, with little activity between, whereas widespread, small-scale flow units would be indicative of a more equilibrium resurfacing style.

Another key question is how interior processes relate to surface deformation and volcanism. Venus is the only other Earth-sized planet in the Solar System but contrary to expectations, it does not show prominent evidence of plate tectonics, which dominates geological processes on Earth Venus lacks a network of connected, large scale plates, leaving the nature of Venus' dominant geodynamic process up for debate. The diverse range of volcanic and tectonic features on Venus' young surface provide ample evidence that mantle processes continue to drive deformation and volcanic resurfacing. Direct geological evidence of present-day volcanic activity has been observed on the surface of Venus based on the recent reanalysis of Magellan radar images. Thousands of volcanoes have been mapped on the surface, but how many of them may be active is unknown. At least a dozen sites have been interpreted as evidence of subduction. Such sites may offer critical insight into the initiation of subduction on early Earth, which is inferred to be the first step in initiating plate tectonics.

3 Volatiles

Characterizing the origin and evolution of Venus through time requires an accurate assessment of the history of its atmosphere. The atmosphere of Venus, like that of Earth and Mars, appears to have evolved substantially from its original composition. The extent to which the major processes that have shaped the atmospheres of Earth and Mars have also affected Venus is largely unknown.

Many fundamental questions remain about Venus, from the rate of current and past volcanism, to the dominant form of tectonism over time, to the nature of atmospheric and surface exchanges, and whether current rates of atmospheric loss are representative of the past. Detailed chemical measurements of the composition of the atmosphere – including noble gases and their isotopes – will help us understand how much of the modern (secondary) atmosphere is the result of interior degassing. Did the secondary atmosphere form during or after the magma ocean phase, was it delivered by impacts, or as part of volatile cycles with the interior?

Venus' enhanced D/H ratio suggests that it has lost a significant (but uncertain) amount of water. However, there is no consensus on how much water condensed at the surface, how much remains in the interior, how much has been outgassed over the planet's evolution, or how much was lost to space, e.g., during a steam atmosphere phase. This conundrum has important implications for the atmospheric evolution of Venus, and thus for its habitability through time. If Venus ever had a liquid water ocean, it could have been habitable for hundreds of thousands to billions of years.

What circumstances affect how volatiles shape habitable worlds? More than any other group, volatile elements have a strong influence on the evolutionary pathways of rocky bodies and are critical to understanding the evolution of the solar system. It is clear that Venus had a different volatile history than Earth. Exoplanet transit detection surveys have a bias toward detecting exoplanets close to their parent stars: the growing number of such Venus-like exoplanet discoveries underscores the importance of studying the long-term evolution of Venus in the search for habitable Earth-sized planets.

4 Overview of Collection Contents

The thirteen papers in this collection are the direct result of the effort of 73 authors reporting on the findings, discussions, and results from the ISSI Workshop "Venus: Evolution through Time" held September 13–17, 2021. They review our knowledge of Venus based on complementary approaches, the key open scientific questions and the observations needed to address these questions. Here we provide an overview of each group of topics: general/comparative planetology; initial conditions, accretion and early Venus; surface processes; interior regime throughout history, water and other volatiles, evidence for current activity; selected mission concepts and future investigations.

4.1 General/Comparative Planetology

O'Rourke et al. provide an overview of the planet Venus, from the dawn of the Space Age to the recent history of space missions, basic properties of the atmosphere, surface and interior; and recent advances in Venus-related theory and modeling; it explores big-picture hypotheses for how Venus may have evolved over time, and sets the stage for the rest of this topical collection—providing the general background needed to delve into detailed models for the evolution of Venus as a planetary system.

Way et al. address the comparison of Venus with exoplanets in a general context: Venus is the only Earth-sized planet, besides Earth, for which we have in-situ data here in our solar system. Thus, it is imperative that we collect improved information on Venus to aid in the modeling of planetary atmospheres, surfaces, and interiors. The Venus Zone is defined by Kane et al. (2014) as part of the habitable zone (HZ), where an Earth-sized planet is more likely to be a Venus analog than an Earth analog. Are there divergent evolutionary paths for Earth-sized exoplanets in the Venus Zone? If Venus had a habitable period, what constraints can interior, tectonic, and atmospheric escape models provide to understand the likelihood of long-term volatile cycling?

Westall et al. discuss more specifically the establishment of the conditions for the emergence of life: if liquid water once existed on the surface of the early planet, how long was the transition from habitable to uninhabitable? What was the water inventory of Venus, and was there liquid water on its surface at temperatures conducive to the emergence of life? The article addresses how terrestrial exoplanet habitability studies can inform Venus' evolutionary history and conversely, how Venus' early evolution can inform exoplanet studies regarding the importance of primordial & basal magma oceans, the divergent paths for planets in the Venus Zone, and the longevity of a habitable Venus.

4.2 Initial Conditions, Accretion, and Early Venus

Salvador et al. describe the earliest stages of planetary evolution; the outcomes of the accretion sequence, with particular emphasis on the sources and timing of water delivery and the initial thermal state of Venus and the so-called magma ocean phase. The magma ocean's molten surface and mantle allows for efficient thermal and chemical exchanges between the interior and the atmosphere. The resulting distribution of water between reservoirs is critical for the subsequent evolution of the planet. The early evolution has been suggested as a possible origin for the differences observed between Earth and Venus. This article reviews to what extent the magma ocean phase can affect Venus evolution, surface conditions and habitability.

Avice et al. discuss how the elemental and isotopic composition of volatile elements (H, C, N, O, S and noble gases) in the Venusian atmosphere provides clues to the origin and evolution of the entire planet. Noble gases are the best available geochemical tracers of such geophysical processes: they are inert, insensitive to silicate partial melting, and each of them (He, Ne, Ar, Kr, Xe) has at least one stable isotope that is non radiogenic; these are often referred to as primordial noble gas isotopes, and their budgets are established during accretion. Their measurement can thus provide key answers to fundamental questions such as the source of volatile elements from the protoplanetary solar nebula, asteroids, or comets, or a mixture of these sources, and whether the atmosphere of Venus was shaped by early impacts, outgassing and atmospheric escape. Future measurements such as those provided by instruments on DAVINCI's descent probe will provide important constraints on models that attempt to understand the late delivery of volatile-rich bodies originally formed in the outer Solar System to the terrestrial planets.

4.3 Surface Processes

The four articles by Carter et al., Ghail et al., Gilmore et al., and Herrick et al. address the geologic history of Venus to constrain the composition, thermal state and structure of the interior, the characterization of tectonic features, the determination of global mineralogy to distinguish major rock types and weathering regimes, and the nature, distribution and extent of sedimentary surface modification processes and their evolution through time. Carter et al. describe the formation of regolith, its transport and the possible lithification of regolith into sedimentary rock. Regolith formation processes may inform planetary climate and evolution. On Venus, a dominant source of sediments is impact cratering, with the apparent removal of fine-grained ejecta over time providing important information on surface changes. Other important sources are mass wasting, mechanical weathering and possible pyroclastic volcanism.

Our understanding of current surface processes and conditions effectively provides constraints for extrapolation back into Venus's geologic and convective history; The size, type, and distribution of volcanic features provide a window into interior processes such as upwelling/downwelling, decompression melting and mantle temperature; the wide variety of tectonic features such as rifts, extensional and dykes, tessera terrain, coronae, ridge belts and wrinkle ridges are discussed in Ghail et al. The overall pattern of surface features relative to inferred mantle convection and crustal thickness can be used to make some inferences regarding the evolution of the interior over time. Both the VERITAS and EnVision orbiter missions will search for evidence of ongoing volcanic and tectonic activity from space and, with an atmospheric entry probe, the DAVINCI mission will look for chemical signatures of volcanism in the atmosphere itself.

Gilmore et al. address more specifically the surface composition, weathering and oxidation mechanisms that may have important implications for the long-term evolution of Venus' atmosphere and climate: What is the composition and diversity of surface material? What is the weathering history of the surface rocks? Is the near-IR signature of the tesserae consistent with Fe-poor magmas, clay minerals, or primary sedimentary phases? To date, the only direct in situ measurements of Venusian surface materials are from the Venera and

VeGa landers, which collected data over a period from 1972 (Venera 8) to 1985 (VeGa 1,2). Mineralogy was not examined directly by these landers, but the geochemical data have been used to infer mineralogy. VERITAS and EnVision will measure iron content from orbit, providing proxy data for the SiO_2 content of the surface rocks, thus providing the first global maps of surface rock type. These data will address a critical question of whether or not the tessera are evolved crust.

Herrick et al. discuss how impact features and their associated deposits (ejecta, halos, parabolas) change through time, and how models of resurfacing designed to match some of the general observations can provide constraints and insights into the resurfacing history of the planet: has there been a systematic change in volcanic style? Are canali, for example, confined to a past regime or still active today? Were the plains formed from a few massive outpourings in a short period of time, or from many thousands of small flows over their entire history?

4.4 Interior Regime Throughout History, Water and Other Volatiles, Evidence for Current Activity

Three papers collectively address the composition and history of Venus' interior and the present-day couplings between the surface and the atmosphere.

Wilson et al. review possible effects of volcanic eruptions on the modern atmosphere, the signatures and potential detectability of present-day volcanically emitted material: which gas species would Venus' volcanoes emit today and which can we measure? What are the known constraints on volcanic processes, outgassing rates and vertical transport and mixing in the atmosphere? The article further discusses expectations for in-situ and remote measurements of volcanic plumes in the atmosphere with particular emphasis on the upcoming DAVINCI, EnVision and VERITAS missions, as well as possible future missions.

Rolf et al. discuss constraints on the evolution of Venus' mantle convection and surface tectonics, as well as the current state of our understanding of its mantle dynamics. Venus loses its interior heat by thermal conduction through the lithosphere and by volcanism, but how these fluxes vary through time and across the surface of Venus and how they are related to the various geological and tectonic features of the Venusian surface remains to be determined.

Gillmann et al. review the long-term evolution of the Venusian atmosphere and the modulation of its composition by the interior/exterior cycle. Moving from the deep interior to the top of the atmosphere, the authors describe sources and sinks of atmospheric volatile species, and the processes that are most likely to govern long-term changes: volcanic outgassing, surface-atmosphere interactions, and atmosphere escape. They explore how these various mechanisms are interrelated and have varied with time. The main current plausible scenarios for the evolutionary pathway that led to present-day Venus are explored and discussed in light of these volatile exchange processes.

4.5 Selected Mission Concepts and Future Investigations

In each article mentioned above, key open questions for the evolution of Venus' atmosphere, climate, surface, interior and habitability through time are identified, as well as the measurements or approaches that are needed to address them with particular emphasis on the upcoming missions. Widemann et al. discuss investigations and payload instrument concepts that support the science goals and open questions presented in the companion papers of this collection. Also included are their related investigations (observations & modeling) and discussion of what measurements and future data products are needed to better constrain Venus' evolution through time. Key advances will come from new types of data to better constrain the interior, surface and atmosphere, such as improved crustal thickness and structure, mantle viscosity/temperature from seismology, lithospheric thickness from electromagnetic sounding, in-situ heat flow to constrain thermal lithospheric thickness and radiogenic heat budget and distribution.

This *Space Science Reviews* topical collection is the first of its kind on Venus science since the new fleet of Venus missions was selected by ESA and NASA in 2021. Data from VERITAS, DAVINCI, and EnVision from the end of this decade will dramatically improve our understanding of the planet's long-term history, current activity and evolutionary path. Over the next 15 years, these three missions will work together to answer many of the outstanding questions of Venus science and rocky planet evolution described in this collection.

Other Venus missions have been considered in Russia, Japan, and China, and ISRO has announced a likely Indian space orbiter proposed for launch. The article also describes the Venera-D orbiter, descent module and lander mission concepts and the Shukrayaan-1 radar-equipped orbiter. The science observation strategy for all of these missions is being developed now and in the coming years; various concepts for detecting seismic activity, whether from landers, from aerial platforms, or from orbit are also being considered; therefore, now is an ideal time to gather knowledge of the Venus evolution scenarios and the observations needed to distinguish between them. The article by Widemann et al. also describes the synergies between these mission concepts, ground- and space-based observatories and facilities, laboratory measurements, and future algorithmic or modeling activities that pave the way for the development of a Venus program that extends into the 2040s.

Appendix

Venus: Evolution Through Time – Table of Contents

6. **Sedimentary Processes on Venus**
   Carter, Lynn, Martha Gilmore, Richard Ghail, Paul Byrne, Suzanne E. Smrekar, Terra M. Ganey, Noam Izenberg

7. **Volcanic and Tectonic Constraints on the Evolution of Venus**
   Ghail, Richard, Suzanne E. Smrekar, Thomas Widemann, Paul K. Byrne, Anna J.P. Gülcher, Joseph O'Rourke, Madison E. Borrelli, Martha S. Gilmore, Robert E. Herrick, Mikhail A. Ivanov, Ana-Catalina Plesa, Tobias Rolf, Leah Sabbeth, Joe W. Schools, J. Gregory Shellnutt

8. **Mineralogy of the Venus Surface**
   Gilmore, Martha, Jörn Helbert, Richard Ghail, Paul Byrne, Sue Smrekar, Noam Izenberg, Lynn Carter, Nils Mueller, Taras Gerya, Mikhail Ivanov, Darby Dyar, Justin Filiberto, Alison Santos, Jeremy Brossier

9. **Resurfacing History and Volcanic Activity of Venus**
   Herrick, Robert R., Evan T. Bjonnes, Lynn M. Carter, Taras Gerya, Richard C. Ghail, Cedric Gillmann, Martha Gilmore, Scott Hensley, Mikhail A. Ivanov, Noam R. Izenberg, Nils T. Mueller, Joseph G. O'Rourke, Tobias Rolf, Suzanne E. Smrekar, Matthew B. Weller

10. **Possible Effects of Volcanic Eruptions on the Modern Atmosphere of Venus**
    Wilson, Colin, Emmanuel Marcq, Cedric Gillmann, Thomas Widemann, Oleg Korablev, Nils Mueller, Maxence Lefevre, Paul Rimmer, Séverine Robert, Mikhail Zolotov

11. **Dynamics and Evolution of Venus' Mantle Through Time**
    Rolf, Tobias, Matt Weller, Anna Gülcher, Paul Byrne, Joseph G. O'Rourke, Robert Herrick, Evan Bjonnes, Anne Davaille, Richard Ghail, Cedric Gillmann, Ana-Catalina Plesa, Suzanne Smrekar,

12. **The Long-Term Evolution of the Atmosphere of Venus: Processes and Feedback Mechanisms**
    Gillmann, Cedric, Michael J. Way, Guillaume Avice, Doris Breuer, Gregor J. Golabek, Dennis Höning, Joshua Krissansen-Totton, Helmut Lammer, Joseph G. O'Rourke, Moa Persson, Ana-Catalina Plesa, Arnaud Salvador, Manuel Scherf, Mikhail Zolotov.

13. **Venus Evolution Through Time: Key Science Questions, Selected Mission Concepts and Future Investigations**
    Widemann, Thomas, Suzanne E. Smrekar, James B. Garvin, Anne Grete Straume-Lindner, Adriana C. Ocampo, Mitchell D. Schulte, Thomas Voirin, Scott Hensley, M. Darby Dyar, Jennifer L. Whitten, Daniel C. Nunes, Stephanie A. Getty, Giada N. Arney, Natasha M. Johnson, Erika Kohler, Tilman Spohn, Joseph G. O'Rourke, Colin Wilson, Michael J. Way, Colby Ostberg, Frances Westall, Dennis Höning, Seth Jacobson, Arnaud Salvador, Guillaume Avice, Doris Breuer, Lynn Carter, Martha S. Gilmore, Richard Ghail, Jörn Helbert, Paul Byrne, Alison R. Santos, Robert R. Herrick, Noam Izenberg, Emmanuel Marcq, Tobias Rolf, Matt Weller, Cedric Gillmann, Oleg Korablev, Lev Zelenyi, Ludmila Zasova, Dmitry Gorinov, Gaurav Seth, Ch. V. Narasimha Rao, Nilesh Desai

Acknowledgements The success of this topical collection and book is due first and foremost to the 73 authors, who undertook the perilous task of capturing the state of the art of planetary science in the perspective of Venus exploration in the coming decades; in addition, the expertise of dedicated and committed reviewers helped guide the authors to a complete collection of manuscripts. Equally important has been the ongoing support the project has continuously received from Springer Nature, at the complex interface between authors, reviewers, guest editors and publishers. The authors would like to thank the ISSI staff, in particular Silvia Wenger, Jenny Fankhauser, Andrea Fischer, Willy Wäfler and Mark Sargent for their competent support. Without their help, the workshop and the book would not have been possible. The financial support by ISSI is gratefully acknowledged.

Declarations

Competing Interests The authors declare no competing interests.

Publisher's Note Springer Nature remains neutral with regard to jurisdictional claims in published maps and institutional affiliations.

Authors and Affiliations

Thomas Widemann[1,2] · Colin Wilson[3,4] · Doris Breuer[5] · Cédric Gillmann[6] · Suzanne E. Smrekar[7] · Tilman Spohn[5,8]

✉ T. Widemann
thomas.widemann@obspm.fr

C. Wilson
colin.wilson@esa.int

D. Breuer
doris.breuer@dlr.de

C. Gillmann
cgillmann@ethz.ch

S.E. Smrekar
suzanne.e.smrekar@jpl.nasa.gov

T. Spohn
tilman.spohn@dlr.de

1 LESIA, Observatoire de Paris, Université PSL, CNRS, Sorbonne Université, Université Paris Cité, 5 place Jules Janssen, 92195 Meudon, France

2 Université Paris-Saclay, UVSQ, DYPAC, 78000 Versailles, France

3 ESA's European Space Research and Technology Centre, Keplerlaan 1, 2201 AZ Noordwijk, The Netherlands

4 Department of Atmospheric, Oceanic and Planetary Physics, Oxford University, Oxford OX1 3PU, UK

5 Deutsches Zentrum für Luft- und Raumfahrt, Institute of Planetary Research, Rutherfordstraße 2, 12489 Berlin, Germany

6 Institut für Geophysik, Geophysical Fluids Dynamics, ETH Zurich, Sonneggstrasse 5, 8092 Zürich, Switzerland

7 Jet Propulsion Laboratory, California Institute of Technology, 4800 Oak Grove Drive, Pasadena, CA 91109, USA

8 International Space Science Institute, Hallerstrasse 6, 3012 Bern, Switzerland

   Springer

Space Science Reviews (2023) 219:10
https://doi.org/10.1007/s11214-023-00956-0

Venus, the Planet: Introduction to the Evolution of Earth's Sister Planet

Joseph G. O'Rourke[1] · Colin F. Wilson[2,3] · Madison E. Borrelli[1] · Paul K. Byrne[4] · Caroline Dumoulin[5] · Richard Ghail[6] · Anna J.P. Gülcher[7] · Seth A. Jacobson[8] · Oleg Korablev[9] · Tilman Spohn[10,11] · M.J. Way[12,13] · Matt Weller[14,15] · Frances Westall[16]

Received: 10 July 2022 / Accepted: 5 January 2023 / Published online: 6 February 2023
© The Author(s) 2023

Abstract

Venus is the planet in the Solar System most similar to Earth in terms of size and (probably) bulk composition. Until the mid-20th century, scientists thought that Venus was a verdant world—inspiring science-fictional stories of heroes battling megafauna in sprawling jungles. At the start of the Space Age, people learned that Venus actually has a hellish surface, baked by the greenhouse effect under a thick, CO_2-rich atmosphere. In popular culture, Venus was demoted from a jungly playground to (at best) a metaphor for the redemptive potential of extreme adversity. However, whether Venus was much different in the past than it is today remains unknown. In this review, we show how now-popular models for the evolution of Venus mirror how the scientific understanding of modern Venus has changed over time. Billions of years ago, Venus could have had a clement surface with water oceans. Venus perhaps then underwent at least one dramatic transition in atmospheric, surface, and interior conditions before present day. This review kicks off a topical collection about all aspects of Venus's evolution and how understanding Venus can teach us about other planets, including exoplanets. Here we provide the general background and motivation required to delve into the other manuscripts in this collection. Finally, we discuss how our ignorance about the evolution of Venus motivated the prioritization of new spacecraft missions that will rediscover Earth's nearest planetary neighbor—beginning a new age of Venus exploration.

Keywords Venus · Planetary probes · Planetary climates · Planetary structure · Planetary dynamics · Planetary system formation

1 Introduction

Venus is so hot right now. Literally, its massive, CO_2-rich atmosphere creates a greenhouse effect that makes its surface the hottest in the Solar System on average. Figuratively, Venus is heating up in popularity following recent announcements that NASA and the European

---

Venus: Evolution Through Time
Edited by Colin F. Wilson, Doris Breuer, Cédric Gillmann, Suzanne E. Smrekar, Tilman Spohn and Thomas Widemann

Extended author information available on the last page of the article

Fig. 1 Venus and Earth perhaps resembled each other after their accretion but set off on divergent evolutionary paths after a few billion years. Broadly speaking, the goal of the three newly selected missions to Venus—ESA's EnVision and NASA's VERITAS and DAVINCI—is to determine if this "habitable hypothesis" for the evolution of Venus is correct

Space Agency (ESA) will send three new missions to Earth's sister planet. These capable missions—VERITAS, DAVINCI, and EnVision—will help end a thirty-year drought when visits to Venus were rare—a historical anomaly. Overall, more than forty missions have been launched with Venus on their itinerary (e.g., Taylor et al. 2018)—virtually the same total number that have been sent towards Mars. However, the last NASA-led mission to Venus (Magellan) death-spiraled into the atmosphere in 1994 after finishing its successful radar mapping campaign. Since then, Venus has been relatively lonely, hosting only three flybys (MESSENGER, Galileo, and Cassini) and two orbiters (Venus Express and Akatsuki, which is still operating as of 2023). In the last three decades, spacecraft data from Venus helped illuminate the workings of its modern atmosphere and hinted at its past. New missions will let us better address the profound question: How has Venus evolved over time?

Scientists can tell an alluring story about the evolution of Venus that is impossible to prove or disprove using available data (Fig. 1). According to this "habitable hypothesis," a time traveler could visit two clement worlds with oceans early in the Solar System's history. Venus and Earth likely accreted with similar bulk inventories of volatiles (e.g., Chambers 2001; Rubie et al. 2015). Both planets were probably born hot with a steam atmosphere above a magma ocean (e.g., Matsui and Abe 1986; Zahnle et al. 1988; Elkins-Tanton 2008). If Venus shed more heat to space than it absorbed from the Sun, then its magma ocean could solidify within ~ 10 Myr while its atmosphere cooled enough for the remaining steam to condense onto the surface (e.g., Hamano et al. 2013). Clouds on the dayside of Venus may have kept surface temperatures Earth-like even as the Sun brightened over time (e.g., Yang et al. 2014; Way et al. 2016; Way and Del Genio 2020). Eventually, perhaps as recently as half a billion years ago, increasing solar input and huge amounts of volcanism caused a climatic catastrophe that led to the current, caustic conditions (e.g., Strom et al. 1994; Weller and Kiefer 2020; Krissansen-Totton et al. 2021; Way et al. 2022a). This transition may have been rapid in geological terms but slow relative to biological generations. Life might have migrated from the increasingly inhospitable surface to the possibly last habitable niche in the clouds (e.g., Limaye et al. 2018, 2021; Seager et al. 2021). New missions can search for atmospheric signatures of a clement past, geological traces of ancient oceans, and evidence of active biology.

On the flip side, surface conditions may have been hellish since Venus accreted. If the early atmosphere could not radiate away all its absorbed solar radiation, then only the escape of water to space could cause cooling. This process would delay the solidification of the surficial magma ocean to ~ 100 Myr and could desiccate the interior and surface (e.g., Hamano et al. 2013; Lebrun et al. 2013). Frustratingly, the orbital distance of Venus is so close to the critical value(s) in models that conclusions about its early state are sensitive to assumptions about poorly understood processes. For example, clouds may preferentially exist on the nightside of Venus during the magma ocean phase (Turbet et al. 2021) instead of on the dayside (Way and Del Genio 2020)—in which case they would trap heat in the atmosphere rather than reflect it away. Venus's thick atmosphere could have degassed at early times from the primordial magma ocean, meaning that the total atmospheric mass has not changed much since the period of initial differentiation (e.g., Gillmann et al. 2009, 2020). Yet, at the moment there is little consensus on exactly what sorts of post-accretion atmospheres should be modeled (e.g., Gaillard et al. 2022; Bower et al. 2022; Salvador et al. 2023). Acidity and low water activity might make the present-day clouds uninhabitable despite their clement temperature and pressure conditions (e.g., Hallsworth et al. 2021). In any case, even if Venus were never habitable, the planetary system—including its atmosphere, crust, mantle, and core—should have evolved over geologic time.

The purpose of this review chapter is to explore big-picture hypotheses for how Venus may have evolved and their programmatic implications. We set the stage for the rest of this topical collection—providing the general background needed to delve into detailed models for the evolution of Venus as a planetary system. Section 2 describes the evolution of our understanding of present-day Venus, which was most rapid at the dawn of the Space Age. From the 1950s through the 1970s, more was learned about Venus in three decades than in the prior three centuries since the invention of the telescope. Interestingly, the now-popular story of how Venus evolved over geologic time mirrors the changes in our understanding of modern Venus. In the past, we thought Venus was habitable today. Today, we think that Venus was habitable in the past—but is advocacy of this idea only a coping mechanism for the disappointing discoveries of the Space Age? Sect. 3 presents the fundamental properties of Venus, especially those related to the evolution of its atmosphere, surface, and interior. This section also advertises the chapters in this topical collection that are most relevant to each aspect of Venus. Finally, Sect. 4 shows how the idea that Venus evolved over time motivates strategies for planetary exploration. For example, the new 2023–2032 Planetary Science and Astrobiology Decadal Survey from the National Academies in the United States highlighted the importance of exploring Venus—with and beyond the newly selected missions—to answering priority scientific questions about the origin and evolution of rocky (exo)planets.

2 Our Evolving Understanding of Venus's Modern State

Before scientists could construct informed models about the evolution of Venus, they needed to understand its present-day state. Venus has been an object of human fascination since prehistory (Fig. 2). Until recently, speculation about Venus was only anchored to the observation that Venus appears very bright in visible light. Assuming that the bright things in Venus's sky were H_2O-rich clouds, people thought that the surface of Venus was Earth-like, except with steady, planet-wide precipitation (Sect. 2.1). However, two key discoveries at

Fig. 2 Glyph related to Venus in a Mayan bas-relief from the museum of Copan, Honduras. Photo by C. Gillmann

the dawn of the space age—lots of CO_2 in the atmosphere and strong emission at radio wavelengths—challenged this fantasy. Models of Venus's atmosphere were most "up in the air" in the 1950s and early 1960s (Sect. 2.2). In the 1960s and 1970s, views of modern Venus completed a paradigm shift from habitable to hellish. By 1974, scientists had converged on the correct conception of the present-day atmosphere—it is massive and made almost entirely of gaseous CO_2 plus clouds formed from sulfuric acid droplets (Sect. 2.3). Building on the realization that the surface is scorched, the last few decades of spacecraft visits (Sect. 2.4) and advances in our theoretical understanding of rocky planets—including new numerical and laboratory tools with which to study them (Sect. 2.5)—have painted a detailed picture of Venus's modern state.

2.1 Pre-1920s Views of Modern Venus

Venus is typically the second-brightest object in the night sky—and surely has been noticed as such by people and non-human animals since time immemorial (Fig. 2). The first telescopic observations circa 1610 revealed that Venus always appears as a crescent from Earth, which served as strong evidence in support of the Copernican and Tychonic models of the Solar System. In 1761, a Russian scientist (Lomonosov) observed the refraction of solar rays during the transit of Venus across the Sun—thus discovering the atmosphere of Venus (e.g., Marov 2005). But little else was learned about Venus itself for hundreds of years. In 1891, a then-famous amateur astronomer wrote a guide for fellow enthusiasts titled *Telescopic Work for Starlight Evenings*. He declared Venus "the most attractive planet of our system" because "none of the other planets can compare with her in respect to brilliancy" (Denning 1891). However, he regretfully confessed that "when the telescope is directed to Venus it must be admitted that the result hardly justifies the anticipation" because "the lustre of Venus is so strong at night that her disk is rarely defined with satisfactory clearness" (Denning 1891). Close-up views of other planets at the time revealed fascinating details: craters on the Moon, polar caps on Mars, cloud bands on Jupiter, et cetera. However, Venus appears almost featureless in the visible wavelengths when viewed by eye through small telescopes (Fig. 3).

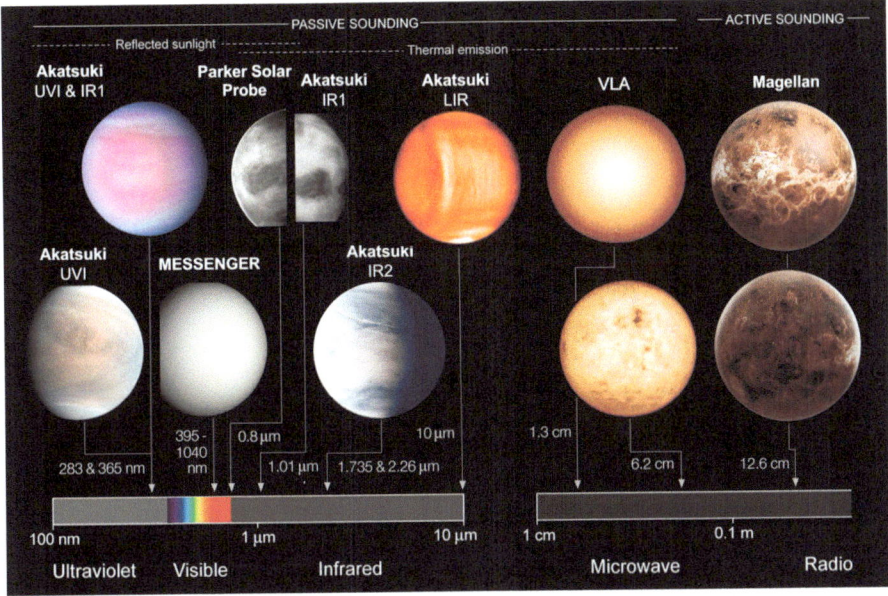

Fig. 3 Venus looks bland in nearly all the visible wavelengths—but other wavelengths reveal myriad details. From left to right: dayside false color image from Akatsuki's UVI instrument (PLANET-C Project); dayside false color image from Akatsuki's UVI and IR1 instruments (PLANET-C Project); composite of dayside imagery from MESSENGER that shows natural color (NASA/JHUAPL/CIW/ Gordon Ugarkovic); nightside image from PSP centered on Ovda Regio (Wood et al. 2022); nightside image of the same area from Akatsuki's IR1 instrument, but rotated (PLANET-C Project); nightside synthesized false color image from Akatsuki's IR2 instrument (PLANET-C Project); stack of five pseudo-color infrared images from Akatsuki's LIR instrument (PLANET-C Project); microwave observations from the Very Large Array (Butler et al. 2001); and a surface 3D model derived from Magellan radar imagery (NASA Visualization Technology Applications and Development). Images at shorter wavelengths are made using sunlight reflected from the dayside of Venus. Longer-wavelength images record thermal emission from the surface and/or atmosphere. Finally, the radar images show the power of active sounding to reveal surface features

Reflected light from the dayside crescent of Venus swamps optical observations made from Earth. Supposed sightings of a moon (named Neith by Cassini and Lagrange) starting in the late 1600s were discredited by the late 1700s. Attempts to track faint features to determine a rotation rate of the atmosphere were made inaccurately, and not widely accepted—in fact, rotation rate retrievals remained wrong into the 1960s (e.g., Sagan 1960). For example, Cassini and Bianchini proclaimed incorrect rotation periods of ~ 1 and 24 Earth-days in the 1660s and 1720s, respectively (Denning 1891). Starting in 1643, some observers reported faint emission from the nightside of Venus, called "ashen light." These reports were dismissed as optical illusions (e.g., Sheehan et al. 2014). However, glimpsing some thermal emission from the night side (Sheehan et al. 2014) or O_2 airglow at wavelengths of ~ 0.45–0.55 μm with the human eye is perhaps possible (Wood et al. 2022). Until the mid-20th century, observations of Venus had not advanced much from prehistory. To the naked eye, Venus seems bright. When magnified in an optical telescope, Venus looks bigger and brighter.

Planetary scientists never surrender to a lack of data. One secure fact about Venus—its brightness—is enough to tell a fantastic tale about its surface conditions. Specifically, we can calculate the temperature required for equilibrium between the thermal radiation from

Venus and the incident radiation from the Sun. First, if the planet radiates as a blackbody with a uniform temperature, then the total emitted flux (i.e., in units of Watts) is

$$F_{out} = 4\pi R^2 \left(\sigma T_{eq}^4\right), \tag{1}$$

where R is the planetary radius, σ is the Stefan-Boltzmann constant, and T_{eq} is the sought-after equilibrium temperature (e.g., Ingersoll 2013). In other words, the total outgoing radiation equals the product of the surface area of Venus and the flux per unit area from the Stefan-Boltzmann law. Second, any planet reflects a portion of the incident sunlight and absorbs the rest:

$$F_{abs} = \pi R^2 (1 - A) \left(\frac{1 \text{ AU}}{D}\right)^2 F_E, \tag{2}$$

where A is the Bond albedo, and D is the Sun-planet distance (in astronomical units, where 1 AU is roughly the Earth-Sun distance), and $F_E \sim 1361$ W/m^2 is the solar constant (i.e., the flux density of solar radiation) at 1 AU. In other words, the total absorbed radiation is proportional to the cross-sectional area of Venus (not its total surface area). The term $(1 \text{ AU}/D)^2 F_E$ equals the solar radiation at Venus's orbital distance. If Venus were rotating quickly as a whole with respect to its orbital period (false) or has efficient atmospheric circulation (true), then the incoming and outgoing energy can reach an equilibrium. Setting $F_{out} = F_{abs}$ and rearranging the various terms, the equilibrium temperature is

$$T_{eq} = \left[\frac{F_E}{4\sigma}(1 - A)\left(\frac{1 \text{ AU}}{D}\right)^2\right]^{\frac{1}{4}}. \tag{3}$$

Plugging in numerical values (Table 1), $T_{eq} \sim 225$ K for Venus today, which is almost 30 K colder than the equilibrium temperature of Earth. Venus orbits closer to the Sun and thus faces more incident radiation than Earth—however, most of that radiation is reflected away without being absorbed (i.e., Venus has a Bond albedo of 0.77 compared to only 0.31 for Earth). Many textbooks elaborate on this calculation and its implications in detail (e.g., Ingersoll 2013). The simplest (but incorrect) interpretation of this quick calculation is that Venus and Earth have similar climates.

Before the Space Age, scientists surmised correctly that Venus is hotter and cloudier than present-day Earth. However, they erred by hugely underestimating the atmospheric surface pressure on Venus, which underrated the climatic differences between Venus and Earth. They were also incorrect in assuming that the clouds of Venus were made of H_2O (as vapor, droplets, and/or ice), although H_2O clouds can exist on other hot planets in the upper layers of thick, H_2O/CO_2-dominated atmospheres (e.g., Pluriel et al. 2019). If the Venusian clouds and atmospheric surface pressure were Earth-like, then the climate of modern Venus would resemble that of Earth during, for example, the Carboniferous or Cretaceous Periods. Whereas roughly one third of Earth's skies are clear at any time on average (e.g., King et al. 2013), Venus is always completely shrouded. More clouds were thought to yield more rain, which would lead to a verdant surface. In his 1918 work of popular science, *The Destinies of the Stars*, Nobel laureate Svante Arrhenius declared that "Venus is no doubt covered with swamps" and thus with abundant life "belonging to the vegetable kingdom" (Arrhenius 1918). The uniformity of Venus's visual appearance also led to the idea that the surface climate was spatially consistent—a jungle-analogue from the equator to the poles. This supposition of homogeneity ultimately proved correct, albeit not in the sense that anyone anticipated in the early 20th century.

Fig. 4 Venus has played many roles in popular culture (photos by JGO). For example, *The Land of Crimson Clouds* (top left), published in 1959 by Boris and Arkady Strugatsky, described a trip to Venus in a progressive future when space exploration advanced economic prosperity and social harmony. *Old Venus* (top right) is a collection of short stories published in 2016 that pays homage to the sword and planet sub-genre of pre-1960s science fiction. *The Expanse* (bottom left) reflects the depression that prevailed in the immediate aftermath of Mariner 2—treating Venus as a boring scrap planet. *The House of Styx* (bottom right, published in 2020) is typical of recent fiction set on a Venus where protagonists are reforged in a crucible of pain

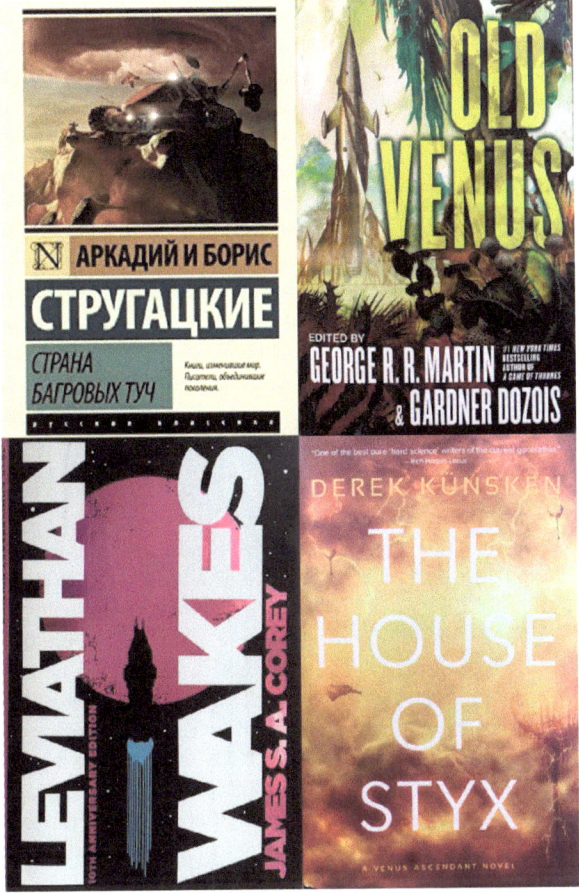

2.1.1 A Verdant Venus in Popular Culture

Pre-Space Age scientific views of Venus led to a delightful explosion of science fiction. As reviewed in a recent collection of short stories, *Old Venus*, which pays tribute to the classics, Venus was the stage for a subgenre dubbed "Planetary Romance" and/or "Sword and Planet" (Dozois 2015). Heroes tromped around the jungle, battling dinosaur-like beasts and other energetic megafauna. While Mars offered a sort of barren elegance, Venus had perhaps too much life. Exploration of Venus was seen as an optimistic endeavor—a path towards human progress. For example, the Strugatsky Brothers, perhaps the most famous Russian sci-fi authors at the time, published *The Land of Crimson Clouds* in 1959 at the dawn of the Space Age (Fig. 4). In this novel, a high casualty rate for the first crew to pierce the eponymous clouds was reckoned a fair trade for an interplanetary future for humanity. A Venus that was far away but maybe not too inhospitable seemed to offer risks and rewards that were relatable to the exploration (and exploitation) of Earth.

Some fictional works explored the downsides of life on a clement Venus. After all, humans are most comfortable on the planet that they evolved to inhabit—even the planet with the most Earth-like surface, Mars, would be a hard place to live (cf., Stirone 2021). In the novel *The Space Merchants*, published in 1952 by Frederik Pohl and Cyril M. Kornbluth, an advertising executive lures naive customers to new colonies on Venus. He sells Venus as a land of plenty, waiting to be seized, but economic activity is difficult and dangerous—and, of course, the plot demands dastardly deeds. At an intimate scale, Ray Bradbury portrayed Venus as a wet hell in his short story *The Long Rain*, published in 1950. Four characters compare their surroundings to "an immense cartoon nightmare" and the steady drops of rain to a torture technique. They search in vain for the only bearable habitat—a "Sun Dome... a yellow house, round and bright as the sun" filled with "warmth and quiet and hot food and freedom from rain" (Bradbury 1951). The weather drives the men to hallucination and insanity. They realize (in fleeting moments before the aforementioned insanity) that they cannot survive on Venus without what they left behind on Earth.

2.2 Competing Models to Explain New Observations (1920s to Early 1960s)

In the early 20th century, new observations set the stage for a paradigm shift about the evolution of Venus. First, scientists accidentally discovered that carbon dioxide was abundant in the atmosphere. Astronomers had tried but failed to find spectral signatures of oxygen and water vapor at infrared wavelengths from the atmosphere in order to prove that the clouds were Earth-like (e.g., St. John and Nicholson 1922; Adams and Dunham 1932). They instead found unexpected absorption bands that were matched to laboratory measurements of carbon dioxide with an equivalent path length of ~ 200–400 m at pressure/temperature conditions of 1 atm and 273 K (e.g., Adams and Dunham 1932; Adel 1937). That quantity of CO_2 (~ 2–4×10^{17} kg) was interpreted as the amount that existed in the atmosphere above the "reflecting layer" where the optical depth was approximately unity. Although roughly the same mass of CO_2 exists in Earth's entire atmosphere, we now know that this lower limit underestimated the true total for Venus by a factor of > 90. Still, Wildt (1940) realized that even the claimed amount of CO_2 would cause a greenhouse effect that could raise the surface temperatures to ~ 366–408 K, which would be incompatible with surface water.

A few decades later, another set of observations further challenged the fantasy of a jungly Venus. Radiometric measurements of Venus at various wavelengths provide "brightness temperatures" if blackbody spectra (Planck's law) are fit to the observed emission. Early studies found brightness temperatures of ~ 230–250 K in infrared wavelengths, which were near the equilibrium temperature predicted by Eq. (3) and (correctly) interpreted as the real temperatures at or near the cloud tops (e.g., Pettit and Nicholson 1955; Öpik 1956). Mayer et al. (1958) conducted the first observations of Venus (and any planet) at radio wavelengths, specifically at 3.15 and 9.4 cm. They measured unexpectedly strong emission with brightness temperatures of ~ 600 K (e.g., Mayer et al. 1958; Barrett 1961). A blackbody with a temperature of only ~ 260 K would emit less than half the measured radiation at those wavelengths. Subsequent observations at 10-cm wavelength also yielded high brightness temperatures and found little difference (~ 10s of degrees Celsius at most) between the effective temperatures of the dayside and nightside (Drake 1962). Ultimately, in the early 1960s, the ancient idea that Venus has roughly uniform surface conditions seemed correct—but, if the brightness temperatures from radio observations should indeed be interpreted as surface temperatures, those conditions were hellish, not humid.

When Mariner 2 was launched towards Venus in 1962, several models of its atmosphere and surface remained in contention (Fig. 5). Roughly speaking, in chronological order, they featured 1) the jungly fantasy with H_2O clouds (Arrhenius 1918), 2) a surface entirely covered with a carbonated ocean and H_2O clouds (Menzel and Whipple 1954), 3) a surface covered in hydrocarbons and clouds made of smog (Hoyle 1955), 4) an "aeolosphere" with the top of an unceasing, global dust storm at the clouds (Öpik 1961), and 5) a scorched surface below two cloud decks, the lower made of bright ice crystals and the higher made of

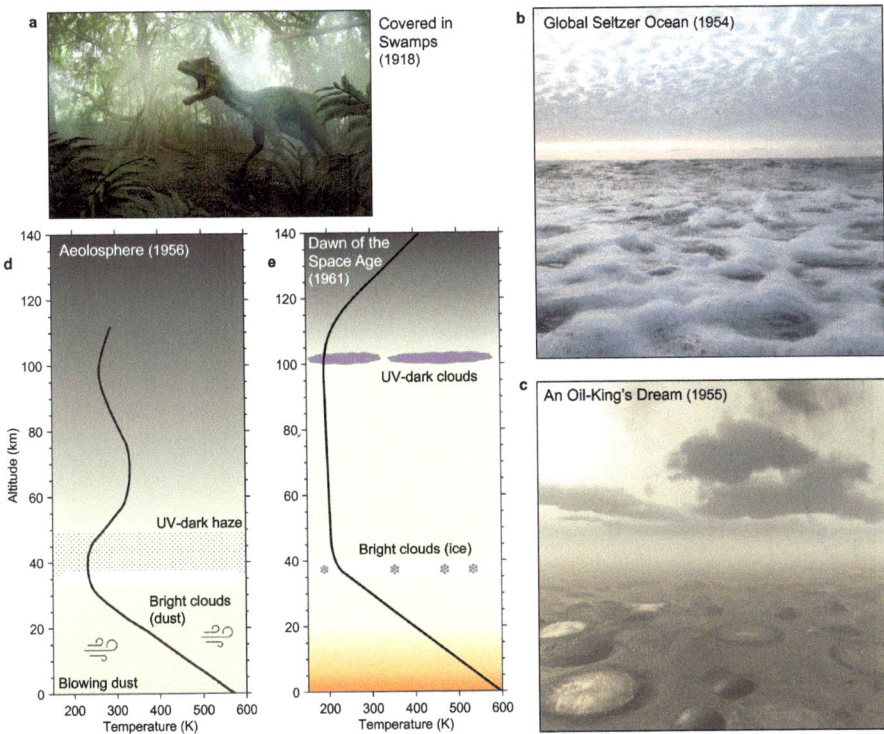

Fig. 5 Before the launch of Mariner 2 in 1962, scientific opinion was divided between several different models of the Venusian atmosphere and surface. The ancient notion of a jungly Venus (**a**) had its proponents but was becoming disfavored because the observed abundance of CO_2 gas was far above the value for Urey equilibrium. Some scientists proposed to avoid Urey equilibrium by drowning the entire surface (**b**) or coating it in oil (**c**). Die-hard devotees of these models invoked the ionosphere to explain the strong radio emission detected in the 1950s. Others (correctly) thought that the hot surface produced that radio emission. (**d**) A global dust storm (e.g., "model I" from Öpik 1961) or (**e**) an even greater abundance of CO_2 (e.g., Fig. 3 in Sagan 1961) were argued to produce bright clouds and the requisite greenhouse heating. Ultimately, all these models were different in key respects from the modern picture shown in Fig. 6. We used a stock image from Microsoft for (**a**) and DALL · E 2 to generate (**b**) and (**c**)

an unknown UV absorber (Sagan 1961). Only the last two models correctly predicted that the surface was far too hot for liquid water to survive. As discussed in Sect. 2.3, all these models were ultimately wrong about the composition of the clouds.

Imagine that you were a scientist studying Venus at the dawn of the Space Age. Picking your favorite model was a choose-your-own-adventure process with a few steps. First, you would need to decide if you believed that the "brightness temperature" inferred from radio and microwave observations was the real temperature of the surface (e.g., Mayer et al. 1958; Barrett 1961; Drake 1962). Some scientists argued that the ionosphere of Venus could emit in these wavelengths (e.g., Roberts 1963), meaning that the observed radiation may not be thermal emission from the surface. For example, Jones (1961) proposed that the solar wind could create high brightness temperatures from free-free transitions of electrons in an optically thick ionosphere. If you believed that liquid water could be stable on the surface, then you would next need to decide if you accepted the evidence that CO_2 was abundant in the atmosphere. If so, then you needed to explain why surface rocks had not drawn down that CO_2 gas. However, if you instead agreed that the surface was hellish, then you would

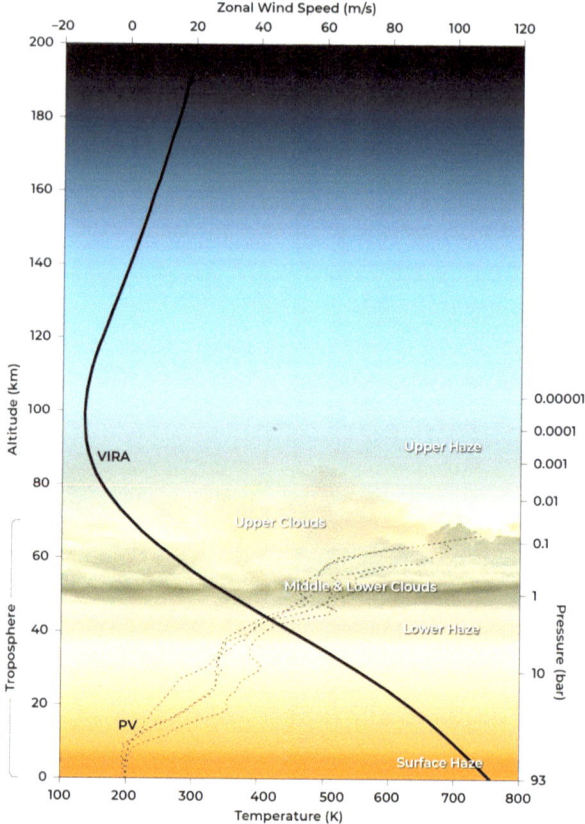

Fig. 6 The present-day atmosphere of Venus. Starting in the 1960s, several probes made in situ measurements of temperature and wind speeds in the Venusian atmosphere. Missions also determined the size distribution(s) of aerosol particles in the cloud and haze layers. Black curves show an equatorial temperature profile from the Venus International Reference Atmosphere (solid) and zonal wind speeds from four Pioneer Venus entry probes (dashed)

next consider if CO_2 alone—or another absorber—provides the opacity that creates a strong greenhouse effect. These branching choices led to the diverse pictures shown in Fig. 5.

Scientists struggled to reconcile models with both abundant CO_2 in the atmosphere and surface temperatures that were compatible with liquid water. Urey (1952) famously proposed that a series of reactions would maintain an equilibrium partial pressure of CO_2 in the atmosphere of a planet with both exposed silicates and liquid water on its surface. For example, enstatite ($Mg_2Si_2O_6$) could react with atmospheric CO_2 to produce magnesite ($MgCO_3$) and quartz (SiO_2). Similarly, wollastonite ($CaSiO_3$) could react with CO_2 to produce calcite ($CaCO_3$) and quartz. However, the inferred atmospheric abundance of CO_2 at Venus was much larger than calculated assuming this Urey equilibrium. Logically, Venus must lack either exposed silicates or liquid water. Menzel and Whipple (1954) proposed that Venus was covered with what Sagan (1961) called a "global Seltzer ocean." Drowning all the rocks could provide ample water vapor for clouds but prevent the Urey reactions. The high partial pressure of CO_2 would lead to carbonation of the ocean—fizzy! Hoyle (1955) argued instead that Venus lacked any surface water. He suggested that Venus accreted with an excess of hydrocarbons relative to water. The oxygen in water oxidized most of the hydrocarbons, producing atmospheric CO_2, while the hydrogen escaped to space. He predicted that the surface was still covered with hydrocarbons ("endowed beyond the dreams of the richest Texas oil-king") and that the clouds contained drops of oil. Petroleum geologists at the time pointed out that Hoyle's belief that hydrocarbons on Earth and Venus were primordial (i.e., delivered by meteorites and comets) conflicted with the oil-kings' notions that oil on Earth was a relatively recent byproduct of fossilized organic material (e.g., Pratt 1956).

Other scientists were quicker to accept that the surface of Venus was hundreds of Kelvins hotter than earlier believed. This paradigm shift eliminated any cognitive barrier to accepting that the atmosphere contained a huge mass of CO_2, which created a planetary greenhouse. However, tension still seemed to exist between the ~ 600 K brightness temperatures and the ~ 400 K surface temperatures that earlier greenhouse models predicted (Wildt 1940). Sagan (1961) argued that a CO_2-dominated atmosphere with quadruple the mass of Earth's atmosphere would provide the necessary heating. Based on his calculations of the adiabatic lapse rate, the temperature in the atmosphere would drop rapidly enough with altitude that H_2O could freeze at ~ 30–40 km to form ice-crystal clouds. Öpik (1961) claimed that CO_2 alone could not provide enough greenhouse heating, even if a minor contribution to the total opacity from water vapor was also considered. He proposed dust as an additional source of greenhouse heating and defined the "aeolosphere" as the region between the solid surface and the clouds. In his models, wind friction provided enough energy to keep the dust lofted. Öpik (1956) had previously argued that the atmosphere rotates at least once every ~ 10 Earth-days—fast enough to redistribute the required energy around the planet. Sagan (1961) criticized Öpik's aeolosphere models because they predicted a distribution of grain sizes for lofted dust that was inconsistent with the particle sizes derived from observations of their polarization. In contrast, Öpik (1961) argued that Sagan's proposed ice-crystal clouds were unlikely to form because the H_2O content of the atmosphere was below the saturation value. Both categories of cloud-centric criticism from these eminent scientists were soon proved correct. Mariner 2 and subsequent ground- and space-based observations revealed that the atmosphere and surface conditions were different than predicted by any previous study.

2.3 Convergence to the Modern Model of Modern Venus (1960s to 1970s)

Mariner 2 encountered Venus in December 1962 at a distance of $\sim 34{,}000$ km at closest approach—and became the last nail in the coffin for the dream of a swampy Venus. As the first successful interplanetary mission, Mariner 2 was designed to accomplish a broad range of scientific investigations, centered on understanding the atmosphere of Venus and the nearby particles and fields environment (e.g., Sonett 1963). One instrument—the microwave radiometer—was designed to test if the brightness temperatures at wavelengths ≥ 3 cm revealed the actual surface conditions. Arguments that the ionosphere could radiate intensely were the last gasp of the hypothesis that the surface of Venus is habitable today. However, many scientists considered this hypothesis unrealistic because it required huge electron densities (e.g., Roberts 1963). During the Venus flyby (called a "near-collision" by Sonett 1963), the microwave radiometer conducted three scans of the planetary disk at wavelengths of 13.5 and 19 mm to settle this debate (Barath et al. 1964).

Two competing hypotheses for the radio emission observed from Venus at Earth made opposing predictions. If the radio emission were an ionospheric and/or atmospheric phenomenon, then "limb brightening" would be observed with higher brightness temperatures near the edge of the disk, where the atmosphere appeared thickest from the instrument's point of view (e.g., Roberts 1963). If the emission originated from the surface, however, then "limb darkening" would result with the highest brightness temperatures measured at the center of the disk, where the atmospheric path length from the surface to the spacecraft was minimized (e.g., Sagan 1961). Measuring these phase effects from Earth required tracking Venus for its entire orbit and calibrating for the changing Earth-Venus distance, which was difficult. With a single flyby, Mariner 2 found limb darkening and proved that the hypothesis of surface emission was correct. Walker and Sagan (1966) published an "obituary" for the ionospheric hypothesis—and thus for the dream of a clement surface now. More recently, microwave observations of Venus from the Very Large Array (Fig. 3) resolved the limb darkening effect across the entire disk (Butler et al. 2001).

A flurry of spacecraft from the USSR provided in situ measurements of the composition and thermal conditions of the atmosphere in the 1960s and 70s. Reaching Venus was the goal of 16 early USSR launches (e.g., Avduevsky et al. 1983). Among these spacecraft, Venera 2 and 3 (1965) missed the planet by so little that the efforts were continued. The first spacecraft to hit Venus and successfully measure the atmospheric parameters during the entry and descent down to 24 km altitude was Venera 4 (1967). The measured temperature was 262 °C at 18 bar pressure. The gas analysis revealed $> 80\%$ CO_2 atmosphere with $< 2.5\%$ of nitrogen with an addition of O_2 and traces of H_2O—contrary to the expectation of $\geq 50\%$ N_2 (e.g., Avduevsky et al. 1983). The common understanding at that time was that the reached physical conditions were representative of the lower atmosphere down to the surface. However, the extrapolation of Venera 4 results already implied the surface temperature and pressure close to their actual values. Still, the next generation Venera 5 and 6 (1969) probes were designed to withstand the pressure of 25 bars only. They operated down to 18 and 22 km altitudes, largely confirming the Venera 4 results (e.g., Avduevsky et al. 1970, 1983). Venera 7 (1970) was the first probe to reach the surface of Venus. It collected a temperature profile from 55 km to the surface, where it measured 457–474 °C. Venera 8 (1972), designed to survive for ~ 90 minutes at the surface, was the precursor of all subsequent Soviet landers (e.g., Marov et al. 1973). It measured the atmospheric profile at altitudes of 0–100 km, including the first directly measured surface pressure of 93 ± 1.5 bar, detected three levels of clouds (including some that were not visible in IR and UV images), improved the knowledge of the atmospheric composition (97% CO_2, 2% N_2), and provided the first estimates of the surface composition (see Sect. 3.2).

The composition of the clouds remained a mystery even after doubts about the surface temperature dissipated. As reviewed in Sect. 2.2, previous studies suggested water vapor (e.g., Menzel and Whipple 1954), ice crystals (e.g., Sagan 1961), dust (Öpik 1961), oil droplets (e.g., Hoyle 1955), and several other possible candidates (e.g., Hansen and Hovenier 1974 and references therein). The decade after Mariner 2 featured several successful missions to Venus, including the Mariner 5 flyby and the Venera 4–7 atmospheric probes (e.g., Rea 1972; Avduevsky et al. 1970, 1977, 1983; Taylor et al. 2018 and references therein). However, early probes did not provide a convincing answer for the composition of the clouds. Scientists invoked observations of the polarization of reflected light from Venus to argue for and against models of particle size and composition (e.g., Sagan 1961; Rea 1972). Finally, Hansen and Hovenier (1974) developed high-quality models of scattering and matched them to polarization data. They showed that a concentrated solution of sulfuric acid (now estimated at ~ 80–99 wt% H_2SO_4, depending on altitude) was the best match to the properties of the cloud droplets. Parts of the upper atmosphere of Venus might be cold, but there are no large reservoirs of pure water. A few years later, Venera 9 and 10 provided the first in situ measurements of the clouds (e.g., Marov et al. 1980). The Venera missions (9–14) and Pioneer Venus confirmed that the atmosphere was very dry and the clouds were made of sulfuric acid droplets (e.g., Kawabata et al. 1980; Knollenberg and Hunten 1980; Moroz 1983; Esposito et al. 1983; Titov et al. 2018 and references therein).

Ground-based measurements also provided surprising information about the rotation of Venus. Scientists tracked surface features visible in radar images from the Arecibo and Goldstone installations. They determined that the solid body rotates very slowly—it takes ~ 243 Earth-days for the surface of Venus to spin 360° on its rotation axis (e.g., Pettengill et al.

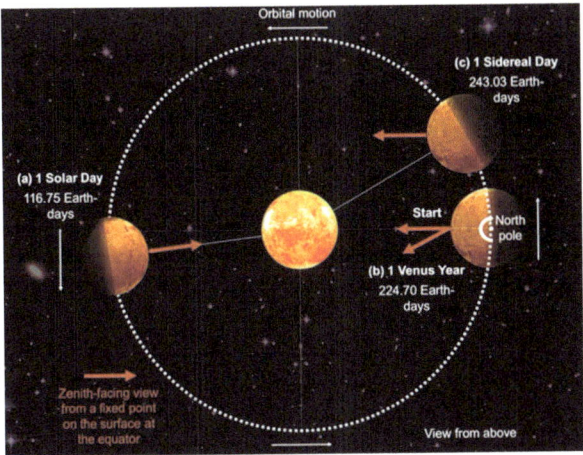

Fig. 7 The solid body of Venus rotates more slowly at present day than any terrestrial planet in our Solar System. This cartoon depicts the orbit of Venus around the Sun, viewed from above. The orbital angles are approximately to scale, but the relative sizes of Venus and the Sun are not. After one solar day (**a**), an observer at a fixed location on the surface of Venus would see the Sun return to its original position in the sky. In one year (**b**), Venus completes a single orbit around the Sun. In one sidereal day (**c**), Venus revolves once relative to the celestial sphere (e.g., the background stars). Earth's sidereal day is shorter than its solar day, which is much shorter than our year. In contrast, a Venus-year lasts less than two of its solar days—and one sidereal day on Venus is longer than a Venus-year

1962; Goldstein 1964; Carpenter 1966; Gold and Soter 1969). Venus also has retrograde rotation and a relatively small obliquity (e.g., Yoder 1997 and references therein). As shown in Fig. 7, a sidereal day on Venus is thus longer than a Venus year. In contrast, a sidereal day on Earth lasts a few minutes less than 24 hours. Many factors—including accretionary processes, giant impacts during or after accretion, and atmospheric and solar tides—can affect the spin dynamics of Venus over geologic time. Length-of-day variations on Earth are approximately a ms (or about 1 part in 10^{11}) whereas they may be up to about 20 min (or about 1 part in 10^5) on Venus due to solar tides and the coupling between the fluid atmosphere and the solid surface (e.g., Margot et al. 2021). Venus's slow rotation could have been established during its accretion—or may be a recent phenomenon. Altogether, the rotation state of Venus is strikingly unique in the Solar System.

2.3.1 A Hellish Venus in Popular Culture

Popular artists did not deny the new scientific consensus about present-day Venus. However, revealing the truth about Venus caused immediate depression. Dozois (2015) describes the angst caused by discovering that Venus "was just a ball of baking-hot rock and scalding poisonous gas, [allegedly] duller than a supermarket parking lot." Science-fiction writers committed to at least a smidgen of realism could no longer place heroes on Venus and expect them to have a good (or at least damp) time. New stories about Venus, such as those in the anthology *Farewell, Fantastic Venus!* (Aldiss and Harrison 1968) were "deliberately retro" and tinged with regret that reality ultimately failed to conform to optimistic expectations (Dozois 2015). When Bradbury's *The Long Rain* was adapted for television in the 1990s, it was stripped of any reference to Venus—the setting was shunted to an unnamed exoplanet. However, artistic work eventually reflected a transition from depression about the lost dream of jungly adventure to acceptance of the real Venus.

Many fictional works now treat Venus as less interesting than virtually all other planetary bodies in the Solar System. For example, in *Rendezvous with Rama* (Clarke 1973), human settlements stretch to Mercury but skip Venus. Likewise, *The Expanse* (Fig. 4) is a wildly popular series of novels, novellas, short stories, and television that concluded in early 2022 (Corey 2022). This space opera tours Earth, Mars, sundry asteroids, the outer solar system, and scores of exoplanetary systems—all home to diverse, memorable communities. In contrast, Venus is kept deserted with plans "to create a network of high-atmosphere floating cities" mired in "a labyrinth of lawsuits" (Corey 2011). Later, Venus is treated as a convenient dumping ground for a life-devouring horror—as a protagonist proclaims at a pivotal point, "Give [the life-devouring horror] Venus... it's an awful place" (Corey 2011). Inhabitants of *The Expanse* would place Venus at the bottom of any list of places to visit.

The best evocations of Venus in popular culture now embrace its superlative inhospitality. Venus provokes awe because its hostility is relatable—it is the Mr. Hyde to Earth's Dr. Jekyll. This duality echoes the symbolic tension between Venus's namesake (Aphrodite, the Greek goddess of love) and the malice of its surface. For example, a recent, award-winning novel, *Gideon the Ninth*, leans into this tension. This bestseller depicts a consequential gathering of representatives from nine "Houses", each situated on or near nine worlds: the Solar System's eight major planets plus Pluto. The representative from Venus (a necromancer, as it happens) remarks that "[her] House loves beauty... a kind of beauty in dying beautifully" (Muir 2019). A 2021 novel, *The House of Styx* (Fig. 4) is set on Venus as a proving ground for the protagonists to grow stronger through adversity. In this novel, bands of industrious, anarchist Quebecois live in atmospheric habitats, obliged to a constant "struggle to pit [their] cunning against Venus to stay alive and scrape some subsistence from the deep clouds" (Künsken 2020). Some of the characters worship Venus, although the planet demands "the same price as any goddess: she wants to be embraced," which hurts a lot (because of the acid) for the cultists among the cloud-dwellers who take "embraced" literally (Künsken 2020). Life on Venus demands sacrifice—the symbolism is potent but not subtle.

2.4 Recent History of Spacecraft Exploration

Back in real life, people never stopped launching spacecraft towards Venus, either as a primary target or as a waypoint on an interplanetary trajectory to another destination. Previous reviews contain comprehensive accounts of missions that targeted Venus. For example, Table 2 in Taylor et al. (2018) from the Venus III collection lists all Venus-related launches, including both successes and failures. Even missions that do not "care" about Venus, except as a convenient mass from which to steal momentum during a gravitational assist maneuver, provide snapshots of Venus's evolution (e.g., Gray et al. 2021). For example, MESSENGER made unique measurements of Venus's upper atmosphere during its flybys (e.g., Pérez-Hoyos et al. 2018; Peplowski et al. 2020)—and BepiColombo is executing similar observations on its way to Mercury (e.g., Mangano et al. 2021). Of course, missions that orbit Venus for years and/or perform in situ measurements make scientists rewrite textbooks. In this review, we do not aim to cover the full history of spacecraft exploration at Venus. Instead, here we highlight four missions from the last three decades that are foundational to our present understanding of Venus and exemplify how spacecraft can shed new light on Venus and possible models of its evolution.

2.4.1 Magellan

The NASA Magellan Mission entered orbit around Venus in August 1990 and operated until purposefully plunging into the atmosphere and burning up in October 1994. Magellan was a scaled-down version of a concept called Venus Orbiting Imaging Radar—to save money, Magellan was designed to re-use hardware from other flight programs as much as possible (e.g., Saunders et al. 1990). Magellan's primary instrument was a large radar sensor built on a high-gain antenna with a diameter of 3.7 m. For imaging, the antenna was operated as a synthetic aperture radar (SAR) instrument with a wavelength of 12.6 cm (S-band) and a look angle of $\sim 25°$ away from the vertical direction (i.e., off-nadir). A small horn antenna was nadir-pointed and collected altimetry data. In the burst-mode of data collection, the horn and high-gain antennas were operated in a careful sequence so their transmissions and the reflections from the Venus surface would not overlap (e.g., Pettengill et al. 1991). The spacecraft was placed in a polar orbit so the solid body would rotate underneath the orbital path every 243 Earth-days, which corresponded to a single "Cycle". During each orbit, Magellan alternated between collecting data and transmitting the data back to Earth with the high-gain antenna—and brief periods of spacecraft housekeeping (e.g., desaturating the reaction wheels and navigating).

Magellan achieved its primary requirements during its first Cycle—and achieved additional science during four subsequent Cycles. During Cycle 1 (1990–1991), Magellan succeeded at its primary objectives to acquire radar imagery of $> 70\%$ of the surface with a horizontal resolution of < 300 m (Saunders et al. 1992) and to determine the global topography with horizontal and vertical resolutions of ~ 10 km and ~ 80 m, respectively (Ford and Pettengill 1992). These SAR images were left-looking with incidence angles that varied from $\sim 45°$ at the equator to $\sim 16°$ near the poles. During Cycle 2 (1991–1992), the spacecraft was reoriented to a right-looking geometry with an incidence angle of $\sim 25°$ (slightly less toward the south pole). Because the spacecraft's electronic bays overheated during data transmission to Earth, only images of $\sim 55\%$ of the surface were returned during this phase. For Cycle 3 (1992), the spacecraft was reoriented back to a left-looking geometry—but with a smaller incidence angle to enable stereo imagery. Unfortunately, the spacecraft's transmitters experienced failures and only $\sim 21\%$ of the surface was imaged in this new geometry. After Cycles 1–3, however, over 98% of the surface had been imaged at least once with a spatial resolution of ~ 125 m or so—still the best global radar map of the surface to the date of writing. Towards the end of the mission, the spacecraft dedicated itself to acquiring gravity data via Doppler ranging in its original, elliptical orbit (Cycle 4) and after aerobraking that circularized the orbit (Cycle 5). Ultimately, the best gravity data have a resolution approaching spherical harmonic degree 180, equating to a horizontal resolution of > 250 km (Konopliv et al. 1999)—a thousand times worse than the image resolution.

Overstating the scientific importance of the Magellan datasets is likely impossible. Although the images, topography, and gravity data are now more than three decades old, they are still being mined for new scientific insights. Magellan revealed most of the properties of the surface and interior discussed below in Sects. 3.2 and 3.3. Globally, Magellan provided a snapshot of the planet's current geologic state, including a catalog of thousands of tectonic and volcanic features—some of which are analogous to those observed on other terrestrial planets, whereas others are superlative in the Solar System. In terms of the evolution of Venus, puzzling observations have spawned many debates. Because Magellan only operated for a few years, the rates of various volcanic and tectonic processes remain unknown—and attempts to detect changes to surface features between successive radar imaging cycles were inconclusive. Fortunately, the VERITAS and EnVision missions will serve as the spiritual successors to Magellan (e.g., Widemann et al. 2023), providing new geologic and geophysical data with orders-of-magnitude better image, topographic, and geodesic resolution over a temporal baseline of decades, which may well verify that Venus is geologically active in the present.

2.4.2 Venus Express

ESA's Venus Express (VEx) orbited Venus from 2006 until 2014 (e.g., Svedhem et al. 2007a, 2007b). Its scientific payload was mainly focused on characterizing the atmosphere, from the surface to the thermosphere, using a suite of spectrometers, imagers, and in situ instrumentation. While many of the investigations aboard VEx focused on how Venus works in the present day, much of their data informs our knowledge of how Venus has evolved.

VEx provided several indications which are indirectly suggestive of current or geologically recent volcanism. These observations build on the legacy of multiple prior missions to Venus. For example, an infrared atmospheric spectrometer (6–35 μm) onboard Venera 15 operated for two months, which demonstrated the power of such data to characterize the temperature structure (e.g., Oertel et al. 1985) as well as water and SO_2 content at the cloud tops (e.g., Zasova et al. 2004)—Pioneer Venus Orbiter also studied SO_2 in the ultraviolet (e.g., Stewart et al. 1979; Esposito 1984). Firstly, VEx's mapping of mesospheric SO_2 abundances showed a fourfold rise in the first year of observations followed by a tenfold fall over following years, a pattern which suggests episodic injection of SO_2 into the mesosphere, due to either volcanic activity or meteorological variability (e.g., Marcq et al. 2013, 2020). Secondly, mapping of surface emissivity at 1 μm wavelength, performed on the nightside of Venus by the VIRTIS instrument, found anomalously high emissivity surrounding some hotspot volcanoes, which may indicate fresh, as yet unweathered lava flows (Smrekar et al. 2010). Thirdly, some repeated 1 μm imagery from the Venus Monitoring Camera showed apparent changes in surface radiometric brightness, which could be a direct thermal signature of volcanic activity (Shalygin et al. 2015). However, this detection was only achieved at one location and at one wavelength without correction for cloud effects, so it cannot be considered a robust detection. Taken together, these three results from Venus Express support the case for active volcanism on Venus today—and have inspired follow-up observation on future Venus missions to search for new eruptions.

VEx also provided some first clues about compositional diversity on the surface of Venus through its 1 μm emissivity mapping—in particular, showing that tesserae highlands have low 1 μm emissivity, consistent with a felsic composition (e.g., Gilmore et al. 2015, 2017). If widespread felsic composition of highlands is confirmed, that would suggest a similarity to Earth's continental crust, whose formation required large water abundances, and would provide strong evidence of a water-rich past (see Sect. 3.2 below).

Escape of volatiles to space is another area in which VEx contributed to our understanding of Venus's evolution. Escape rates of hydrogen and oxygen were measured. While, at first, they seemed to be roughly in stoichiometric 2:1 ratio (i.e., for H_2O), subsequent analysis found that the ratio can be as low as 1:1 in times of solar maximum, with implications for the chemical evolution of the Venus atmosphere during water escape (Persson et al. 2018). The oxygen ion escape rates were found to be lower than those from Earth, although Venus is closer to the Sun and not shielded by a strong, global magnetic field at present day. This observation appears to contradict the commonly held belief that internal magnetic fields "protect" planets from atmospheric loss (e.g., Brain et al. 2016; Dong et al. 2020). Finally, Venus Express's SPICAV spectrometers showed that the HDO/H_2O ratio in the mesosphere of Venus is twice as highly enriched as in the troposphere (below the clouds), and that this enrichment factor rises by another order of magnitude above 100 km altitude (Bertaux et al. 2007; Fedorova et al. 2008; Vandaele et al. 2020). Venus's high D/H enrichment compared to that of Earth implies that Venus has lost vast amounts of water over its history (as will be discussed below in Sect. 3.1.1)—these measurements of its vertical distribution enable better understanding of D/H fractionation processes, and therefore of its implications for understanding the history of water on Venus.

2.4.3 Akatsuki

Akatsuki was almost a failure but turned into a spectacular success. Launched in May 2010 as Japan's third planetary mission (Planet-C project), Akatsuki was supposed to enter Venus's orbit in December 2010. However, the orbit insertion maneuver failed, leaving the spacecraft in a heliocentric orbit (Nakamura et al. 2011). Subsequent analysis found that fuel was unable to pass through a critical valve into the orbital maneuvering engine—but the spacecraft could use its reaction control system to perform an orbital maneuver (Nakamura et al. 2014, 2016). The team performed clever trajectory analyses and designed a new orbit insertion maneuver that placed the spacecraft into Venus's orbit in December 2015. Originally, the spacecraft planned to enter an equatorial orbit with a period of ~ 30 hours and periapsis and apoapsis altitudes of < 850 km and $\sim 80{,}000$ km, respectively (Nakamura et al. 2011). The final orbit is ~ 5–6 times further away from Venus than planned (e.g., with a periapsis altitude of $\sim 1{,}000$–8,000 km and an apoapsis altitude of 360,000 km) and has a period of ~ 10.5 Earth-days (Nakamura et al. 2011). However, the equatorial orbit (inclination of $3°$) still allows Akatsuki to track features in Venus's atmosphere for much longer than was previously possible using instruments on spacecraft in polar orbits (e.g., Nakamura et al. 2014, 2016).

Akatsuki's instrument payload was designed to make three-dimensional movies of the atmospheric dynamics. The spacecraft carries five photometric sensors (Nakamura et al. 2014): infrared cameras at 1 μm (IR1) and 2 μm (IR2), an ultraviolet imager (UVI), a long-wave infrared camera (LIR), and a lightning and airglow camera (LAC). Atmospheric gasses, clouds, and hazes absorb different wavelengths of light at different altitudes. For example, ground-based studies have used near-infrared spectral windows to study the lower atmosphere of Venus (e.g., Arney et al. 2014). Multispectral imaging by Akatsuki thus returns multiple "slices" of the atmosphere (Fig. 3). Wind speeds are inferred via tracking of morphological features (e.g., Peralta et al. 2017, 2019a, 2019b, 2020; Limaye et al. 2018). Images also constrain models of other cloud properties (e.g., thicker clouds block more thermal emission from the surface and thus appear darker in the near-infrared on the nightside). Akatsuki also performs radio sounding of the atmosphere, which provides vertical profiles of temperature and some molecular abundances (e.g., vertical profiles of H_2SO_4 vapor). These snapshots of the present-day atmosphere feed into models of its long-term evolution.

Akatsuki is still operating and continues to make notable discoveries. Characterizing the processes that drive the atmospheric superrotation (see Sect. 3.1.1) is a major goal. Based on studies of the cloud-level winds, the Akatsuki team discovered that thermal tides and large-scale turbulence promote and oppose superrotation at the cloud tops (Horinouchi et al. 2020). The overall dynamics are still uncertain because the winds at lower altitudes remain unknown, but Akatsuki images recently revealed a large-scale disruption in the lower cloud decks that propagates much faster than the prevailing winds (Peralta et al. 2020). Akatsuki also studies the coupling between the surface and the atmosphere. For example, new images revealed a large stationary gravity wave in the atmosphere, probably generated by mountain topography (Fukuhara et al. 2017)—meaning that the lower and upper atmosphere may interact more than previously believed (e.g., Brecht et al. 2021). Such waves may change the rotation rate of the solid body over time (e.g., Navarro et al. 2018). Overall, scientists need to understand how regions of the atmosphere interact with each other and with the surface. Understanding those dynamics at present day is a first step towards building models of how the entire planetary system evolved over geologic time.

2.4.4 Parker Solar Probe

NASA's Parker Solar Probe (PSP) launched in 2018 to study the solar wind and sample the low solar corona for the first time (Fox et al. 2016). PSP uses several Venus gravity assist (VGA) maneuvers to gradually lower its perihelion to < 10 solar radii from the Sun's center—seven VGA maneuvers are currently planned, the last scheduled for 2024. Although PSP was not designed to study Venus, many of its instruments operate during each gravity assist and make useful scientific measurements. The Solar Orbiter mission will behave similarly during its many Venus flybys (e.g., Allen et al. 2021). For example, PSP has yielded new insights into the Venus plasma and magnetospheric environment (e.g., Bowen et al. 2021; Malaspina et al. 2020; Collinson et al. 2022)—and discovered a circumsolar dust ring near Venus's orbit (Stenborg et al. 2021). During VGA2, PSP searched for but did not find radio signals from lighting on Venus (Pulupa et al. 2021)—supporting the result from Akatsuki that optical flashes from lightning (at least those visible from space) occur much less frequently, if at all, and/or more intermittently than terrestrial lightning (e.g., Lorenz et al. 2019). Finally, PSP returned some of the most striking images of Venus ever taken at visible wavelengths (Fig. 3). During flybys of Venus in 2020 and 2021, the Wide-Field Imager for Parker Solar Probe (WISPR) observed the nightside of Venus (Wood et al. 2022). WISPER was designed to study the solar wind at wavelengths from ~ 0.5–0.8 μm. Surprisingly, their images revealed thermal emission from the surface of Venus (mostly at ~ 0.7–0.8 μm) and O_2 nightglow emission at the limb (mostly at ~ 0.45–0.55 μm). The human eye is, in principle, sensitive enough to see the O_2 nightglow—and perhaps a lucky observer could catch a glimpse of the surface emission. Overall, multi-flyby missions such as PSP and Solar Orbiter help us understand Venus's evolution by better illustrating Venus's present-day state.

2.5 Recent Advances in Venus-Related Theory and Modeling

In parallel to new spacecraft launches, scientists leveraged advances in theories, techniques, and computational power—often first applied to Earth—to develop increasingly sophisticated models of Venus. In turn, exploration of Venus fed back into building a better understanding of all planets (e.g., Lapôtre et al. 2020). Here we provide a few examples of how efforts to understand Earth's tectonics (Sects. 2.5.1 and 2.5.2) and early habitability (Sects. 2.5.3 and 2.5.4) led to advances in our understanding of Venus's evolution.

2.5.1 Theory of Mantle Convection and Plate Tectonics on Earth

The hypothesis that Earth's mantle flows and circulates has slowly developed ever since the nineteenth century. General studies on the physics of thermal convection—not specifically applied to Earth's mantle—were gradually linked to observations (see, e.g., Bercovici 2015 for a detailed historical timeline). In particular, individual observations and concepts of continental drift, seafloor spreading, apparent polar wander, and subduction, together with the growing concept of a viscously deforming mantle on geological timescales, were combined into the plate tectonics theory that revolutionized geophysics in the mid-to-late 1960s. The plate tectonic model divides the solid, outer shell of the Earth (lithosphere) into a number of thin, rigid plates that move with respect to one another and that are continuously being created and consumed at their edges (e.g., Morgan 1968; McKenzie and Parker 1967; Le Pichon 1968). Turcotte and Oxburgh (1967) applied boundary layer theory for thermal convection to Earth's mantle, associating oceanic lithosphere with the cold, upper thermal boundary layer of mantle convection; ocean ridges with ascending convection; and ocean trenches with descending convection of the cold upper thermal boundary layer into the mantle. Finally, it was broadly accepted that both viscous (fluid-like) and elastic (solid-like) behavior, depending on the timescale of deformation, shape the Earth's interior and surface. Subsequent to this plate tectonics and mantle convection revolution, a wealth of fundamental studies explored key concepts such as nonlinear convection, mantle flow with increasingly complex variable rheologies, and convection in the presence of newly-established solid-solid phase transitions throughout Earth's mantle.

Once plate tectonics was established as the fundamental framework describing Earth's present-day dynamics, it was only a matter of time before scientists went beyond this framework and explored different regimes of mantle convection and tectonics, acknowledging planetary transitions over time. With increasingly improved studies on simplified mantle flow coupled with rigid plates, different relationships between surface kinematics and convective forces were soon established (e.g., Christensen 1985; Hager and O'Connell 1981; Ricard et al. 1993; Bunge and Grand 2000). Distinct "modes" of mantle convection were proposed that establish different wavelengths of convection and surface boundary mobility. These modes of mantle convection are highly dependent on, amongst other parameters, the thermal state of the convective system and the material properties (such as density and viscosity). These dependencies imply that during the thermal evolution of a planet (e.g., as it cools down), different mantle convection regimes may be encountered, with crucial implications for the planet's surface tectonics evolution (see Sect. 3.3.2 and Rolf et al. 2022).

Key questions related to planetary transitions that have been puzzling scientists are "When and how did plate tectonics start on Earth?" and, relatedly, "Why does Venus currently lack plate tectonics?" Roughly speaking, the fact that the surface of Venus is hot and dry—compared to the relatively cold and wet surface of Earth—is probably pivotal. However, even though we can study our own planet *in situ*, understanding of the initiation and evolution of plate tectonics on Earth is still wrapped in controversy. The lack of unambiguous data—such as pristine, unaltered, and completely contextualized rocks older than ~ 3.5 Ga sampling the deep interior and surface of the planet—is but one factor impeding our understanding. Despite this lack of direct evidence, more refined modeling combined with proxies for tectonic processes on the early Earth have helped us infer the nature of early tectonics on this planet. Examples include the formation of felsic rocks typical of (proto)continental crust, paired metamorphic zones typical of convergent tectonics (e.g., Hawkesworth et al. 2020), and strong, thickened crust that can support brittle breakage and the intrusion of dyke swarms (e.g., Van Kranendonk 2010; Hawkesworth et al. 2009; Cawood et al. 2013).

Suggestions for the timing of the onset of plate tectonics range from ~ 4–1 Ga (e.g., Van Kranendonk 2011; Hawkesworth et al. 2020). Indeed, the process appears to have been gradual—or perhaps episodic—with an initial transition from an earlier convection regime (possibly from a sluggish or more stagnant state, or already a plume-induced proto-plate tectonics) between ~ 3 and 4 Ga. Although the rock record shows evidence of major continental amalgamation by ~ 2.8 Ga (e.g., Evans 2013), there is earlier evidence of increased tectonic activity in the form of eroded continental crust (e.g., Belousova et al. 2010; Dhuime et al. 2012). Recent modeling studies on tectono-magmatic processes on Precambrian Earth (e.g., O'Neill et al. 2007; Gerya 2014; Rey et al. 2014; Bercovici and Ricard 2014; Fischer and Gerya 2016; Rozel et al. 2017; Sobolev and Brown 2019; Hawkesworth et al. 2020; Gerya 2022) enhanced our understanding of pre-plate tectonic regime with lid evolution driven by episodic tectono-magmatic activity in the absence of subduction (e.g., Sizova et al. 2015; Capitanio et al. 2019a, 2019b). Secular cooling of the mantle potential temperature during the Archean-Proterozoic period (~ 3 Ga and ~ 0.75 Ga) likely resulted in transitional tectonics on Earth, whereby a squishy- or plume-lid regime (see Rolf et al. 2022, for details) gradually, or episodically, evolved towards the modern plate tectonics regime by combining elements of different global tectonic styles in both space and time (e.g., Fischer and Gerya 2016; Chowdhury et al. 2017, 2020; Sobolev and Brown 2019; Perchuk et al. 2018, 2019, 2020).

2.5.2 Advances in Analogue Experiments and Numerical Techniques

Most—if not all—of the above-mentioned advances in understanding Earth's tectonics go hand-in-hand with developments in geodynamic and atmospheric modeling, mainly facilitated by improved theory, advanced laboratory experiments, numerical modeling techniques, and, importantly, computational power.

Analogue modeling is an experimental approach to investigate geological phenomena and geodynamic processes in a laboratory at convenient time- and length-scales. It has a long history starting over 200 years ago (see, e.g., Ranalli 2001 for a detailed timeline). While early analogue models mainly focused on individual geological structures, e.g., folds, thrust faults, and salt domes (e.g., Daubrée 1879; Cadell 1889; Escher and Kuenen 1928; Ramberg 1967), the focus shifted to plate tectonic processes as the theory of plate tectonics became well accepted in the 1960s. Another major step forward in analogue modeling came in the 1980s, when realistic models were built to simulate both brittle and viscous behavior, mimicking a rheologically stratified crust and mantle (e.g., Faugere and Brun 1984; Davy and Cobbold 1988). Analogue modeling underwent significant advances and proved itself an effective and relatively inexpensive tool for investigating tectonic and geodynamic processes. For example, analogue models were—and still are—key in describing mantle geodynamic regimes that can occur in rocky planets (e.g., Davaille 1999; Davaille and Limare 2007). Recently, analogue models have been applied to Venus to make important hypotheses on the style of mantle dynamics and the potential of plume-induced subduction on Venus (Davaille et al. 2017).

Numerical modeling developed from the mid-to-late 1970s onwards. The first 2D numerical model of subduction was presented in 1970 (Minear and Toksöz 1970), exactly during the start of the "Plate Tectonics Era," shortly followed by the first 2D mantle thermal convection model (Torrance and Turcotte 1971). It was not long before the first 2D mantle thermal-chemical convection models (Keondzhyan and Monin 1977, 1980) and the first 3D spherical mantle convection models (Baumgardner 1985; Machetel et al. 1986) were presented. Surprisingly, the first 3D models of mantle convection were in spherical geometry—not Cartesian as one might expect! Since the 1980s, the field of numerical geodynamic modeling developed very rapidly in terms of the applications and techniques. However, as most early models treated the mantle and the lithosphere with little to no feedback, the self-consistent generation of (plate) tectonics in these models was long an issue. Only at the end of the 1990s, the improved description of pseudo-plasticity allowed for numerical modeling of mantle convection that produced, in a self-consistent way, regions with little deformation (plates) bounded by regions of localized deformation (plate boundaries) (e.g., Moresi and Solomatov 1998; Tackley 1998; Trompert and Hansen 1998). These models opened novel perspectives on the exploration of a unified lithosphere-convective mantle system on Earth and, importantly, how the system operates on other rocky planets. Nowadays, computational power and ever-improving computational techniques (e.g., parallel high-performance computing, adaptive mesh refinement, solvers, inverse theory, etc.) allow us to obtain larger and—perhaps—higher-quality numerical data in less and less time. However, the exploration of high-resolution 3D global models of mantle convection and surface processes, potentially coupled with atmospheric dynamics, remain a computational frontier. Moreover, as numerical data gets more complex, it becomes even more important to thoroughly understand the physics behind the computations.

2.5.3 Theory and Modeling of a Runaway Greenhouse

In the last century, key developments were made in understanding the evolution of planetary climates and atmospheres that have applications to Venus. Simpson (1927) was the first to recognize that an atmosphere in radiative equilibrium under an increased solar insolation would lead to an excess of infrared radiation (IR). Plass (1961) demonstrated the role of increased anthropogenic CO_2 would have on the IR budget and the warming of the climate. Sagan (1960) was among the first to realize that the then estimated 600 K surface temperature on Venus made it "evident that a very efficient greenhouse effect is required" (see Sect. 2.3). Sagan (1960) and Gold (1964) both realized that this "efficient greenhouse effect" would prevent Venus from having surface liquid water, which Gold (1964) referred to as a "runaway process." Subsequent 1-D radiative-convective modeling work identified what we now term the "runaway greenhouse" (e.g., Komabayasi 1967; Komabayashi 1968; Ingersoll 1969; Pollack 1971; Kasting 1988; Abe and Matsui 1988; Nakajima et al. 1992).

   Ingersoll (1969) was likely the first to propose that a habitable planet with oceans at the orbit of Venus could switch to an uninhabitable state when greenhouse gasses in the atmosphere block thermal radiation from leaving the planet, preventing the planet's atmosphere from cooling, leading to a runaway greenhouse. Two years later, the first 1-D, non-grey radiative transfer simulations by Pollack (1971) demonstrated that ancient Venus could have had temperate conditions if the planet had 100% cloud cover, but with 50% it would be in a runaway state. Work by many authors over the subsequent decades discussed the possibility of an early temperate Venus that would warm up as the Sun increased in luminosity over the eons (Gough 1981), gradually increasing the atmospheric temperature and driving it into its present-day runaway greenhouse state (e.g., Bullock and Grinspoon 1996, 2001; Grinspoon and Bullock 2007). While the runaway greenhouse process has been successfully modeled in 1-D, radiative-convective models as mentioned above, it has proved to be devilishly difficult in 3-D general circulation models (GCMs) (e.g., Ding and Pierrehumbert 2020; Boukrouche et al. 2021; Chaverot et al. 2022). For example, most (but not all) Earth-derived GCMs used for planetary atmospheric modeling cannot handle multiple condensable species—or a variable atmospheric mass with more than one species—as the model moves forward in time (e.g., Fauchez et al. 2020). For the latter, as the atmosphere heats up, water becomes an ever-larger fraction of the atmosphere. This means the mean molecular weight of the atmosphere, which must be pre-set, becomes more and more inaccurate (e.g., Way et al. 2017; Appendix A). These factors will influence the accuracy of the atmospheric dynamics, including cloud convection processes. As well, most GCM parameterized radiative transfer schemes are limited in the temperature and pressure ranges allowed, although pressure is probably the easiest to accommodate.

2.5.4 Links Between Venus and (Early) Earth and Planetary Habitability

Earth has physical attributes that can be analyzed to provide information about its early habitability (i.e., rocks dating back to ~ 4.1 Ga, mantle zircon crystals dating back to ~ 4.3 Ga, and inherited geochemical signatures from erstwhile Hadean crust). However, the rarity of these attributes and the fact that the oldest rocks have been severely altered by metamorphism makes interpretation of the signatures they contain at times controversial (see Westall et al. 2023). Therefore, iteration of the rock and geochemical data with models of the geophysical and atmospheric evolution of the early Earth are essential to a better understanding of how Earth became habitable. Additionally, comparison with the early evolution of other terrestrial planets, especially Venus and Mars, is an important factor. Briefly, habitability on the early Earth during the Hadean and Eoarchean epochs (4.5–3.5 Ga) means the establishment of conditions for the emergence of life in the first place. Here, only the essential ingredients of water, organic molecules (C, H, N, O), other elements, such as P, S, and transition elements, as well as a source of energy are necessary—but only necessary for the time needed for life to emerge (which, of course, we do not know but is likely to have been relatively short, > 1–2 Ma). Important is also the "scenario" for the emergence of life, whether in submarine hydrothermal environments, subaerial ones, or any other geologic setting (see review in Westall et al. 2018). If exposed land mass is a prerequisite, it needs to be stable for the length of time for life to emerge, likewise submarine hydrothermal systems. On these timescales, the geophysical situation of a planet, whether it is one plate or not, whether the planet was dominated by plume tectonics or sluggish, shallow tectonics, is not critical. The tectonic regime and plate tectonics, specifically, become relevant once life is flourishing because of the necessity of recycling nutrients used up on the surface (e.g., Korenaga 2012; Foley and Driscoll 2016; Foley and Smye 2018). This cycle only comes into play after about a couple of billion years.

Thus, in terms of Venus, the nature of the tectonic regime that dominated the early history of the planet is irrelevant for the emergence of life, providing that the initial conditions were conducive to water at the surface (the other ingredients: organic molecules, essential elements, and energy sources would have been similar to those in early Earth). Tectonics only become critical if there was a flourishing (or, eventually, flailing) biosphere on the planet that needed to access renewable resources. Nevertheless, a better understanding of the physical mechanisms responsible for Venus's geologic history will greatly advance our understanding of what makes a rocky planet habitable and, ultimately, life emerge.

3 Fundamental Properties of Venus Relevant to Its Evolution

Understanding the evolution of Venus is, by definition, a more complex task than making direct observations of its modern properties. A detailed catalog of the fundamental properties of Venus is the foundation of attempts to study its past. Table 1 compares the basic properties of Venus and Earth. Because the bulk densities of these two planets are so similar, scientists often assume that Venus and Earth have similar bulk compositions (Sect. 1). However, measurements of key parameters for Venus are so uncertain that significant differences might await discovery. Here we describe the different parts of Venus as a planetary system and how they may have changed over time.

3.1 The Atmosphere of Venus

The atmosphere is the easiest part of Venus to study—yet many of its basic properties are still unknown or poorly understood. Virtually every planetary process affects the atmosphere. Equilibration (or lack thereof) with the early magma ocean set its initial conditions—volcanic degassing and reactions with the surface control its mass and composition over time. Any intrinsic magnetic field could have affected atmospheric escape processes. To understand the evolution of the atmosphere is thus to know the history of the entire planet. Crucially, the atmosphere also helps govern the evolution of the solid body. Surface temperature is the boundary condition for mantle convection—and controls the rheological properties of lavas and rocks that govern volcanic and tectonic processes now preserved in the geologic record.

Table 1 Basic properties of Venus and Earth. Unless otherwise given, data are extracted from the NASA Earth and Venus Fact Sheets (Williams 2022a, 2022b). Additional sources include: [1]Simon et al. (1994), [2]Konopliv et al. (1999), [3]Konopliv and Yoder (1996), [4]nominal Love number at degree and order 2 for an anelastic Earth (Petit and Luzum 2010), [5]Margot et al. (2021), [6]Von Zahn et al. (1983), Taylor et al. (1997), and De Bergh et al. (2006), [7]Lebonnois and Schubert (2017), [8]James et al. (2013), [9]Anderson and Smrekar (2006), [10]Jiménez-Díaz et al. (2015), [11]Dumoulin et al. (2017), [12]Kennett et al. (1995) (model ak135), [13]Tesauro et al. (2012)

Parameter [Units]	Venus	Earth
Orbital and Rotational Parameters		
Semimajor Axis [10^6 km]	108.210	149.598
Sidereal Orbital Period [days]	224.701	365.256
Orbit Inclination [deg]	3.395	0.000
[1]Orbit Eccentricity	0.006772	0.0167
Sidereal Rotation Period [hrs]	5832.6	23.9345
Obliquity to Orbit [deg]	177.36	23.44
Bulk Planetary Parameters		
[2]Mass [10^{24} kg]	4.8675	5.9722
Equatorial Radius [km]	6051.8	6378.1
Polar Radius [km]	6051.8	6356.8
Volumetric Mean Radius [km]	6051.8	6371.0
Mean Density [kg/m^3]	5243	5513
Equatorial Surface Gravity [m/s^2]	8.87	9.80
J_2 [$\times 10^{-6}$]	4.458	1082.63
[3,4]Tidal Love Number, k_2	0.295 ± 0.066	$0.30102 - i \cdot 0.00130$
[5]Moment of Inertia Factor	0.337 ± 0.024	0.3307
Surface and Atmosphere Parameters		
Solar Irradiance [W/m^2]	2601.3	1361.0
Average Surface Temperature [K]	737	288
Surface Pressure [10^5 Pa]	92	1.014
Mass of Atmosphere [10^{20} kg]	4.8	0.051
[6]Atmospheric constituents [by volume]	96.5% CO_2	78.1% N_2
	3.5% N_2	21.0% O_2
	20 ppm H_2O	$\sim$ 1% H_2O
	70 ppm Ar	9340 ppm Ar
	150 ppm SO_2	412 ppm CO_2 and rising
Fraction of angular momentum contained in the atmosphere[7]	1.6×10^{-3}	2.7×10^{-8}
Topographic Range [km]	13	20.4
Interior Structure Estimates		
[8,12]Thickness of the Crust [km]	8–25	35
[9,10,13]Thickness of the Elastic Lithosphere [km]	< 100	10–120
	< 20 for underneath 50% of the surface area	
[11,12]Radius of the Core [km]	2940–3425	3479.5

3.1.1 Basic Properties of the Atmosphere

Venus has the most massive atmosphere of any terrestrial planet in our Solar System. Its overhanging firmament comprises nearly 0.01% of the total planetary mass, compared to the factor of $\sim 8.5 \times 10^{-7}$ for Earth (Table 1). Figure 6 shows the vertical structure of the atmosphere, which is roughly consistent at low latitudes near the equator. In Earth's atmosphere, nitrogen is the most abundant gas—but Venus's atmosphere contains roughly three times as much nitrogen relative to the mass of each planet. As discussed in Sect. 2.2, carbon dioxide dominates the atmosphere of Venus. The total mass of gaseous CO_2 is estimated to equal or exceed the combined amounts of CO_2 present in Earth's atmosphere plus (as carbonates) in Earth's crust and mantle (e.g., Ingersoll 2013; Lécuyer et al. 2000; Donahue and Pollack 1983). However, we cannot conclude that Venus contains more carbon than Earth (or vice versa) because the carbon inventories of their metallic cores (e.g., Fischer et al. 2020) and Venus's solid body are uncertain and debated.

Not all atmospheric gasses are more plenteous at Venus than at Earth. The absolute amount of water vapor in the Venus atmosphere is about the same as on Earth, although it represents only about 30 parts per million of Venus's massive atmosphere. In Venusian water vapor, the ratio of deuterium to hydrogen (D/H) is ~ 157 times larger than D/H $\sim 1.5 \times 10^{-4}$ for Earth (e.g., Donahue et al. 1982; De Bergh et al. 2006), which may imply that large amounts of water vapor have been lost over Venus's history. Molecular oxygen, so important to us on Earth, is present on Venus at only 50 parts per million or less on Venus; this means that the absolute mass of molecular oxygen in the Venus atmosphere is at least two orders of magnitude less than on Earth. Beyond the bulk constituents of each atmosphere, scientists are quite interested in trace components such as the myriad isotopes of noble gasses (e.g., Baines et al. 2013; Chassefière et al. 2012; Avice et al. 2022). For example, atmospheric argon-40 is twice as prevalent at Earth compared to Venus at present day (e.g., Von Zahn et al. 1983; Kaula 1999; O'Rourke and Korenaga 2015).

Clouds and hazes are perhaps the most interesting features in the atmosphere of Venus. The Venus I, II, and III collections each include comprehensive reviews of these clouds and hazes (Esposito et al. 1983, 1997; Titov et al. 2018). Clouds on Venus are found between altitudes of ~ 48–70 km above the surface. Recent studies divide the cloud deck into three layers—all dominated by concentrated droplets of sulfuric acid with $< 25\%$ water by weight and water-activity values ≤ 0.004 (Hallsworth et al. 2021). As described in Titov et al. (2018), the upper clouds (~ 57–70 km) include both submicron- and micron-sized particles—and the mysterious UV absorber. A kilometer-thick gap separates the upper clouds from the middle and lower clouds. The boundary between the middle and lower clouds is (figuratively and literally) cloudy—both layers contain large particles (mean diameters of ~ 7–8 µm) that may be a separate "mode 3" population or simply the tail-end of a distribution of the ordinary particles. Hazes of fine aerosols are found both above and below the clouds up to altitudes of ~ 100 km and down to altitudes of ~ 33 km, respectively. A thin haze layer may also exist at the surface.

Atmospheric gases participate in complex chemical cycles that are most active in and near the clouds (e.g., Mills and Allen 2007; Marcq et al. 2018). Roughly speaking, photochemistry dominates the chemistry of the upper clouds where the temperatures are < 300 K (e.g., Titov et al. 2018). Photochemical processes depend on the amount of incident sunlight and thus vary spatially and temporally across the atmosphere based on the local time of day. The middle and lower clouds become more reliant on processes such as condensation and convective mixing with the lower atmosphere. Below the clouds, temperatures exceeding ~ 400 K enable relatively rapid reactions that enable gases to approach thermochemical equilibria. Sulfur and water (and many other species) cycle through all these regions. For example, photochemistry forms sulfuric acid (H_2SO_4) from sulfur oxides and water in the upper atmosphere, which condenses with water vapor to form the cloud droplets. At the base of the clouds, H_2SO_4 evaporates and, at lower altitudes, decomposes into water vapor and SO_3, which is highly reactive on its way to forming SO_2 (e.g., Dai et al. 2022). At the surface, volcanism may inject SO_2 into the lower atmosphere, explaining its elevated abundance relative to model predictions (e.g., Esposito 1984) and thus, perhaps, the existence of the clouds themselves (e.g., Bullock and Grinspoon 2001). However, existing models of these chemical cycles leave many questions unanswered (e.g., Bierson and Zhang 2020), including what surface reactions buffer the composition of the lower atmosphere (e.g., Gillmann et al. 2022) and why SO_2 and H_2O are depleted in and above the clouds of Venus relative to models that predict rapid fluxes of SO_2 through the clouds (e.g., Rimmer et al. 2021). Future missions that traverse or dwell in the clouds should help solve these puzzles.

Venus, like Earth, also features different atmospheric layers defined by vertical variations in temperature. The Venus International Reference Atmosphere (VIRA) provides temperature, density, pressure, and thermodynamic gas properties for the atmosphere at different altitudes and latitudes (Seiff et al. 1985). Roughly speaking, the atmosphere is hottest at the surface, above which temperatures decrease with altitude at nearly the dry adiabatic lapse rate of ~ 10 K/km. Convective equilibrium usually prevails in the troposphere, although stable stratification may exist at some altitudes. Recent work has focused on the (in)stability of the atmosphere near the surface, which is poorly understood because only the VeGa-2 probe provided a reliable temperature profile within < 12 km of the surface (e.g., Lebonnois and Schubert 2017). The base of the upper clouds typically defines the tropopause, above which the UV absorber (whatever it is) absorbs roughly half of the incoming solar energy. The layer above the troposphere is often called the stratosphere by analogy to Earth because radiative equilibrium sets the vertical thermal profile (e.g., Taylor et al. 2018)—although other studies prefer "mesosphere" instead because, unlike in Earth's stratosphere, temperature continues to decrease with altitude in this layer (e.g., Pätzold et al. 2007; Lebonnois and Schubert 2017). Temperature does increase with altitude from ~ 120–150 km in the thermosphere due to ionization and dissociation caused by solar radiation (e.g., Taylor et al. 2018). The outermost, ephemeral layers of the atmosphere are the exosphere, where collisions between molecules are so rare that they can easily escape, and the magnetosphere induced by the solar wind (but maybe not entirely, see Sect. 3.3.3).

Wind speeds vary dramatically with altitude in the Venusian atmosphere. Famously, Venus has one of the only super-rotating atmospheres in the Solar System—moving in the same direction as the solid body, but with a shorter period (e.g., Read and Lebonnois 2018). However, many exoplanets may have similar atmospheric dynamics (e.g., Imamura et al. 2020; Lee et al. 2020). In 1985, radio tracking of two balloons—a highlight of the VeGa mission—at altitudes near ~ 54 km provided the first in situ measurements of cloud-level wind speeds (Sagdeev et al. 1986; Crisp et al. 1990). Four Pioneer Venus entry probes measured zonal wind speeds (Fig. 6) that were near-zero below altitudes of ~ 10 km but rose pseudo-linearly to ~ 65–90 m/s at altitudes of ~ 60 km (e.g., Schubert et al. 1980). The rotation period of the atmosphere thus ranges from ~ 7 to 4 Earth-days from the bottom to top of the clouds, respectively. The angular momentum density of the atmosphere peaks at altitudes of ~ 20 km, although $> 70\%$ of the total atmospheric mass lies at lower altitudes. Sub-cloud altitudes contain $> 90\%$ of the total mass and angular momentum (e.g., Peralta et al. 2019a, 2019b; Schubert et al. 1980). In tandem with the zonal super-rotation, the atmosphere features Hadley cells at the cloud levels that extend north and south from the equator to cloud-top polar vortices. Because of the huge mass and thermal inertia of the atmosphere and the rapidity of zonal and meridional transport, changes in elevation cause the biggest changes in surface temperature with a > 100 K difference between Maxwell Montes and Diana Chasma—the highest and lowest elevations on Venus, respectively. In contrast, latitude, longitude, and local time cause temperature fluctuations of less than ±10 K. Solar insolation peaks at the equator and local noon, but excess heat is swiftly redistributed.

3.1.2 How the Atmosphere May Have Evolved

Before they learned the composition of the clouds (e.g., Sagan 1960), scientists speculated that the atmosphere of Venus underwent dramatic changes over time. Imagine that Venus once had an Earth-like climate with water oceans on the surface. Excess sunlight due to Venus's relative proximity to the Sun can drive the atmosphere into a "runaway greenhouse" state (Ingersoll 1969) as described in Sect. 2.5.3. Briefly, a post-accretion steam atmosphere above surface oceans can only radiate a certain amount of energy away to space. Because the saturation vapor pressure of water increases exponentially with temperature, a hotter troposphere has greater opacity. This feedback between temperature and opacity imposes an upper limit to the upward flux from a steam atmosphere in equilibrium. Only the evaporation of the oceans can provide an energy sink to balance any excess incoming radiation. With insolation above the critical value, thermal equilibrium is not achieved until the surface is dry and the relative humidity of the troposphere can decrease. One-dimensional models indicate that the radiation limit is ~ 280–300 W/m^2 (e.g., Ingersoll 1969; Abe and Matsui 1988; Kasting 1988; Nakajima et al. 1992; Marcq et al. 2017)—although different assumptions about the structure and composition of the atmosphere can change the exact value.

In any case, Venus probably absorbs less solar radiation than it can re-radiate away at present due to its bright clouds. However, the solar insolation would have exceeded that limit if Venus ever had fewer clouds and thus an Earth-like albedo. Clouds can help promote or prohibit the stability of surface oceans, depending on where they occur (e.g., Way and Del Genio 2020; Turbet et al. 2021). Clouds on the dayside reflect solar radiation, but clouds on the nightside act as a thermal blanket. Regardless, the Sun has brightened over time, so a simple story for the evolution of Venus features an Earth-like climate that was eventually forced into a runaway greenhouse state (Fig. 1), producing today's dry atmosphere as mentioned in Sects. 1 and 2.5.3.

Several chapters in this topical collection discuss how the atmosphere of Venus may have evolved—and how its evolution influences the rest of the planetary system. First, two chapters explore the possible histories and scientific value of trace gasses in the atmosphere. Salvador et al. (2023) investigate the role of water in the early atmosphere of Venus and the processes that may have caused volatile loss at early times. Avice et al. (2022) focus on how measurements of isotopes of volatiles and noble gasses can constrain models of Venus's evolution. Two other chapters discuss the myriad connections between the atmosphere and solid body. Wilson et al. (2023) tackle the hypothesis that magmatic degassing affects the composition of the present-day atmosphere. They also describe how mapping of anomalous plumes of gas (e.g., elevated concentrations of a wide array of possible species) or particulate matter (e.g., ash particles or sulphate aerosols) could help identify regions of the surface with active volcanism. Gillmann et al. (2022) present fully coupled models for the atmosphere, crust, mantle, and core of Venus that include feedbacks between surface temperature, the regime of mantle convection, and even the connections between an internal dynamo and atmospheric escape. Finally, the atmosphere is the easiest part of a Venus-like exoplanet to observe. Way et al. (2022b) review the prospects for characterizing such distant worlds—and how studies of exoplanets and the Earth/Venus dichotomy inform and feed into each other.

3.2 The Surface of Venus

Venus has a surface unlike any other world in the Solar System except, perhaps, parts of Earth. The Venus surface is relatively young, probably active, but not operating in an Earth-like regime of plate tectonics (likely because of its temperature and the lack of surface water now). Scientists vigorously debate how the surface has evolved recently and over geologic time. Does the surface preserve signs of a clement past? Or has recent, planet-wide volcanism erased the geologic "memory" of most of the history of Venus?

3.2.1 Basic Properties of the Surface

Scientists have used various techniques to peer through the atmosphere of Venus and unveil the surface. Since the 1960s, missions (e.g., Pioneer Venus Orbiter, Venera 15 and 16, and Magellan) and Earth-based facilities (e.g., Arecibo and Goldstone) have observed Venus with radar to which the atmosphere is transparent. So far, Magellan has provided the highest-quality, global imagery and topographic data so far from mapping cycles conducted over three Venus-years from 1990 to 1992 (detailed in Sect. 2.4.1 above). The radar images have a horizontal resolution of ~ 125 m per pixel (Saunders et al. 1992; Ford et al. 1993). The global topographic data has a horizontal resolution no better than ~ 10–20 km per pixel (Ford and Pettengill 1992). Recently, Herrick et al. (2012) processed stereo imagery acquired during Cycles 1 and 3 (Sect. 2.4.1) to create digital elevation models with horizontal and vertical resolutions of ~ 1 km and ~ 100 m, respectively, that cover $\sim 20\%$ of the surface. In comparison, the Venera 15 and 16 orbiters performed SAR mapping with a horizontal resolution of ~ 1 km and a swath width of ~ 10–40 km—and delivered some altimetry data with an accuracy of ~ 50 m. The two Soviet spacecraft operated up to spring of 1985 at coordinated orbits and mapped the northern hemisphere, corresponding to around 25% of the surface. Pioneer Venus Orbiter and then Venera 14 and 15 discovered many of the types of features described below (e.g., Barsukov et al. 1986), which Magellan revealed in sharper detail and found to be distributed across the entire surface.

Thermal radiation from the surface can penetrate the atmosphere in several "spectral windows" in the near infrared (e.g., Allen and Crawford 1984; Allen 1987; Carlson et al. 1991; Crisp et al. 1991). Although not all surface radiation is absorbed, escaping surface radiation is inevitably scattered with blurring at horizontal scales of > 50 km when observed remotely from above the cloud layer (i.e., from orbit). The VIRTIS (Visible InfraRed Thermal Imaging Spectrometer) instrument on Venus Express (see Sect. 2.4.2) observed Venus in three of these spectral windows (e.g., Drossart et al. 2007). Scientists used VIRTIS data to place coarse bounds on the infrared emissivity of surface units (e.g., Mueller et al. 2020), which provoke hypotheses about their rock type as discussed below (e.g., Gilmore et al. 2017, 2023). Ultimately, radar and near-infrared image data have shown that all major geologic processes—volcanism, tectonics, mass wasting, erosion, and impacts—have operated on Venus (Fig. 8). However, scientists debate the relative importance of these processes in shaping the present-day surface and the sequence(s) in which they may have occurred.

In the aftermath of the Venera, Pioneer Venus, and Magellan missions, mappers classified myriad types of named features on the surface of Venus. The United States Geological Survey (USGS) produced a handy guide to the official nomenclature (Tanaka et al. 1993). Some

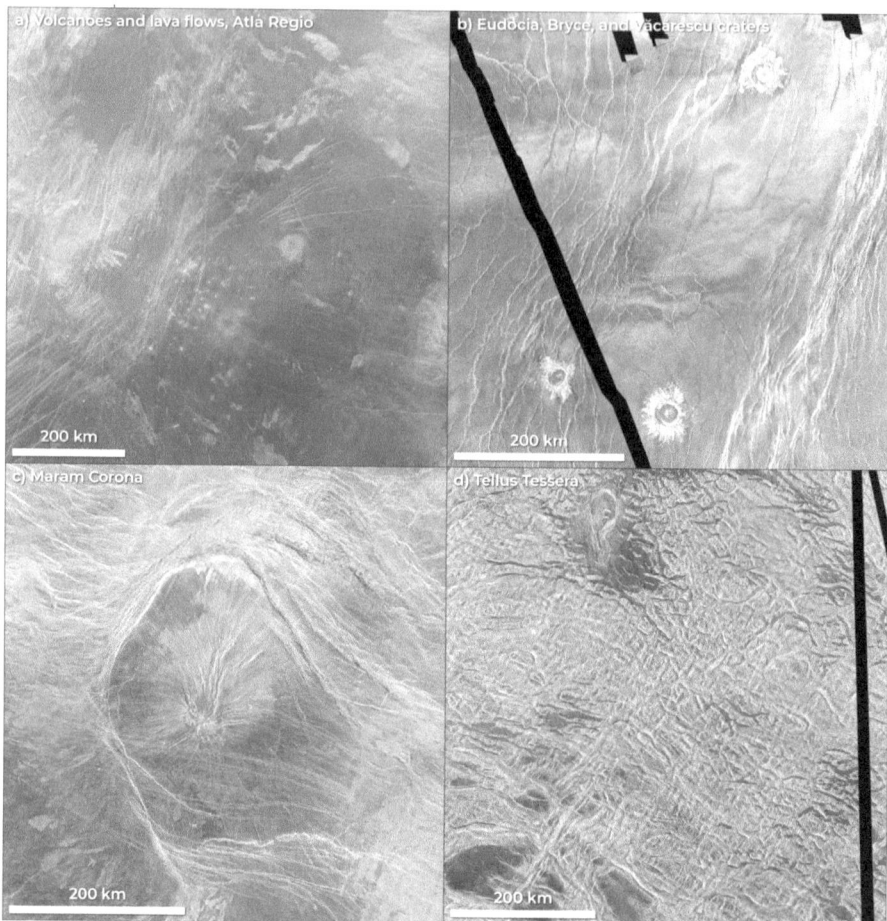

Fig. 8 Four types of geologic features—shown here in left-looking radar imagery overlain on inverted right-looking imagery from Magellan—exemplify how observations of Venus can constrain models of its evolution. (**a**) A field of volcanoes and lava flows near Atla Regio, in the planet's southern hemisphere. The morphology and distribution of volcanic landforms reflect the processes and compositions of magmas in the subsurface. (**b**) A cluster of three craters—Eudocia, Bryce, and Văcărescu—at high southern latitudes; impact craters can provide information on the age of the surface. (**c**) Coronae probe the properties of Venus's lithosphere, including its elastic thickness and heat flow. This example is Maram Corona, centered near 7.5°S, 221°E. (**d**) A portion of Tellus Tessera (centered at 37°N, 81°E), one of Venus' enigmatic and highly tectonically deformed surface units. North is to the top in all frames; the scale bars show 200 km. Black areas are gores (missing data) in the Magellan global radar image mosaic

features are huge—the first things someone would notice when presented with a map of the surface. For example, the most extensive land masses on Venus are called "terrae" (singular "terra"). These terrae are often compared to Earth's continents, although the hypsometry of Venus is unimodal, not bimodal as for Earth (and weakly on Mars). Near the north pole, Ishtar Terra hosts four mountain ranges (termed "montes"), including the superlative Maxwell Montes. The lower plain inside these ranges is Lakshmi Planum—"planum" being the general term for a plateau or high-standing plain. Near the equator, Aphrodite Terra is divided into two main "Regiones" (regions): Ovda Regio and Thetis Regio. Along the southeastern edge of Aphrodite Terra is Artemis Corona, the prime example of the more than 500 quasi-circular features that range from ~ 60 to over 1,000 km in diameter, and which are associated with a variety of tectonic and volcanic features (e.g., Barsukov et al. 1986; Smrekar and Stofan 1997; Stofan et al. 1992; McGovern et al. 2013). Artemis Corona is the largest with a diameter of $> 2,000$ km. Another superlative feature is Baltis Vallis—a thin (~ 1–3 km) channel, almost certainly volcanic, that is the longest ($\sim 6,800$ km) found anywhere in the Solar System. Not every named feature on Venus is gigantic. The surface is littered with smaller landforms both tectonic—e.g., "dorsae" (ridges), "fossae" (long, narrow depressions), "lineae" (elongated features)—and volcanic (Fig. 8), such as "tholi" (small domes or hills) and "fluctūs" (flow features). Merely cataloging the surface is the work of many lifetimes.

One of the few uncontroversial facts about the surface of Venus is that it is relatively (geologically) young on average. Venus hosts a unique and enigmatic population of impact craters (e.g., Herrick et al. 2023). Fewer than 1,000 craters have been identified on the surface (e.g., Phillips et al. 1991; Schaber et al. 1992). On airless bodies like the Moon, Mercury, and Mars, the size–frequency distributions of impact craters obey power laws. Smaller impactors (e.g., asteroids and comets) are more common than large ones, so smaller craters form more frequently than larger ones. However, the size–frequency distribution of impact craters on Venus is log-normal. One might suspect that available imagery prevents scientists from identifying small craters. However, the thick atmosphere is the real culprit behind their absence (e.g., Zahnle 1992)—as was predicted before any images were obtained (e.g., Tauber and Kirk 1976; Kahn 1982). Impactors that would otherwise form craters smaller than a kilometer or so across burn up or explode before reaching the surface. We can derive an approximate age for the surface on the basis of a production function for impactors that are large enough to plow through the atmosphere and actually reach the surface. Such impactors are expected to hit Venus every half a million years or so, meaning that 1,000 craters correspond to an age of ~ 0.5 Gyr. Careful calculations yield estimates for Venus ranging from ~ 240 Myr to ~ 1 Gyr (e.g., McKinnon et al. 1997; Le Feuvre and Wieczorek 2011). Crucially, this cratering age need not be the actual age of the surface. With so few craters, obtaining statistical constraints on the relative ages of different terrains is difficult (e.g., Hauck et al. 1998), if not impossible. Many areas of the surface could be many times older than the cratering age (e.g., Hansen and López 2010). Finally, craters are not always atop their local stratigraphic sequence (e.g., Herrick and Rumpf 2011).

The USSR delivered the only successful landers to the surface of Venus in the 1970s and 80s. Venera 7, the first probe to reach the surface, measured the surface temperature and the rigidity of the rocks—but did not survive for long. Venera 8 was the first design able to operate for more than an hour on the surface. Its gamma-ray K-Th-U measurements suggested a more evolved rock rather than the broadly basaltic composition found for all sites subsequently visited. Suggested explanations for the Venera 8 measurements include there being older terrain at the landing site or a specific kind of K-rich basalt (Surkov 1983; Treiman 2007). Their success and the failures of the 1973 Mars campaign pushed the USSR to put Mars on standby as they proceeded with sustained Venus exploration. From Venera 9 to VeGa, a heavier Proton launcher allowed for much higher complexity and capacity of Venus missions. Venera 9 and Venera 10 (1975) delivered the first panoramas of the surface (Florenskiy et al. 1983), supplemented with in situ composition (gamma-ray and photometry) measurements (Surkov 1983). The basaltic composition revealed by these missions confirmed the past differentiation of Venus into a mantle and crust, and presumably an iron-rich core. The density of the surface was measured at 2.7–2.9 $\mathrm{g\,cm^{-3}}$. Venera 11 and 12 (1978) attempted the analysis of the surface samples via X-ray fluorescence. However, the newly developed drilling device failed—and the panoramic cameras' openings remained closed. By 1981, having solved technical problems encountered by Venera 11 and 12, the USSR launched the Venera 13 and 14 landers (e.g., Moroz 1983). These later landers returned two color panoramas of the landing site—and analyzed two surface samples acquired by drilling from $\sim$ 3 cm depth and transferred into the protected lander volume. The samples from two geologic units, a hilly upland and a flat lowland near the eastern extension of Phoebe Regio in the planet's western hemisphere, were attributed to weakly differentiated alkaline gabbroids (Venera 13) and oceanic tholeiitic basalts (Venera 14). Overall, the legacy of Soviet Venera and VeGa landers consists of four panoramas and in-situ analysis of seven (Venera 8–10, 13, and 14 and VeGa 1 and 2) landing sites, five revealing a predominantly tholeiitic basaltic lavas and two (Venera 8 and 13) sites indicating more alkalic, lamprophyre-like lavas or ash beds (e.g., Weitz and Basilevsky 1993; Treiman 2007). In the context of Magellan radar images, the composition of Venus rocks compared with analogues on Earth suggested a different formation history of the two planets' crusts.

3.2.2 How the Surface May Have Evolved

Other chapters in this topical collection address three key questions related to the evolution of the surface: Are any geologic features relics of or otherwise related to the putative transition from clement to hellish conditions? Does the present surface record evidence of so-called "catastrophic resurfacing? And what are the feedbacks between the evolution of the surface and the rest of Venus as a planetary system?

Tesserae are units on Venus proposed to record evidence that water was once stable on the surface of Venus. Firstly, VIRTIS data provide tantalizing hints that at least some tesserae are felsic due to their low infrared emissivity (e.g., Mueller et al. 2009; Gilmore et al. 2015). Large volumes of felsic rock are most efficiently formed in the presence of oceans (e.g., Campbell and Taylor 1983). Secondly, tesserae have complex patterns of topography (Fig. 8), which has been compared with valley patterns that fluvial erosion can produce (e.g., Khawja et al. 2020). Gilmore et al. (2023) provide a general overview of available constraints on models of the composition of the surface. Carter et al. (2023) discuss how sedimentary processes, which are often assumed to play a minor role on Venus, may have shaped the surface—and perhaps confused our interpretations of many features. Finally, Westall et al. (2023) teach us what the present surface can tell us about the evolving habitability of Venus.

If temperate conditions existed in an earlier Venusian epoch, then any climatic catastrophe on Venus should probably have a geologic counterpart. A huge amount of the mantle would have needed to undergo partial melting to degass > 90 bars of carbon dioxide (e.g., Way and Del Genio 2020; Way et al. 2022a). Such massive melting would presumably cover the vast majority of the surface in thick lava flows. Scientists have long debated whether catastrophic resurfacing (i.e., an episode of planet-wide volcanism lasting < 100 Myr that covers > 80% of the surface in flows with thicknesses of $\sim$ 1 km or more) is the most parsimonious interpretation of the cratering record (e.g., Strom et al. 1994; Nimmo and McKenzie 1998; Ivanov and Head 2013). Alternatively, Venus could preserve a uniformitarian history where a variety of geologic processes have operated on a variety of scales at a variety of times (e.g., Guest and Stofan 1999). Herrick et al. (2023) delve into the facets of the impact record and the history of attempting to explain them with models both simple and complicated. Ghail et al. (2023) survey the volcanic and tectonic features on Venus as the foundation for building realistic models of their geologic evolution.

Ultimately, processes that originate in the deep interior of Venus govern the age and appearance of its surface. On Earth, plate tectonics started as a kinematic theory to describe the rotation and translation of the surface. Linking simple models of plate boundaries to a three-dimensional conception of mantle convection was a scientific revolution (see Sects. 2.5.1 and 2.5.2, above). Likewise, ongoing efforts attempt to link straightforward models for the resurfacing of Venus (e.g., Herrick et al. 2023) to theories about the long-term evolution of the lithosphere and mantle. Rolf et al. (2022) review different conceptions of Venus's mantle dynamics at present day (e.g., stagnant- versus episodic- versus squishy-lid). As noted in Sect. 3.1.2, Gillmann et al. (2022) discuss how atmospheric evolution can lead to changes in the tectonic regime on Venus over time via the feedbacks between volcanic degassing, surface temperature, and rock rheology.

3.3 The Interior of Venus

Little is known about the interior of Venus now—let alone how the interior has changed over time. Most models assumed that Venus, like all terrestrial planets, was initially hot due to the release of gravitational energy during accretion (e.g., Stevenson et al. 1983), even without invoking late energetic impacts. Venus, like Earth, is expected to start with a core that is fully molten and a mantle that was at least partially liquid. Radioactive decay of isotopes of uranium, thorium, and potassium provides additional heat over geologic time. Once Venus formed, it started losing heat to the void of space. Depending on the regime of mantle convection and atmospheric properties, the rate at which heat is lost to space might roughly exceed, equal, or pale in comparison to the rate of internal, radiogenic heat production. Consequently, the interior temperature of Venus can increase, stay roughly constant, or decrease at different periods during its history. Figure 9 shows how the present-day internal structure of Venus reflects its bulk temperature. Thus, the modern state of the core and mantle will tell us the extent to which Venus has cooled down from its putative hot start.

3.3.1 Basic Properties of the Interior

Venus is a differentiated planet, but the sizes, compositions, and physical states of its different layers are unknown. The Pioneer Venus Orbiter and Magellan missions used the Doppler tracking method to determine the planetocentric constant GM, where G is the gravitational constant (Konopliv et al. 1999). As is common for planetary bodies, these data are high-quality enough that only the fundamental uncertainty about the value of G limits the precision of our estimate of M for Venus. The radial distribution of Venus's internal mass is still largely putative because it is mainly based on models and comparisons with Earth (e.g., Shah et al. 2022; Smrekar et al. 2018). The first surface analyses from the Venera landers revealed that Venus has a crust that was derived from partial melting of its silicate mantle (e.g., Surkov 1983; Treiman 2007). However, available data do not discriminate between a wide range of models for the thicknesses of its crust, lithosphere, mantle, and core. The surface is obviously solid, but the state (liquid and/or solid) of the deep interior are largely unconstrained (Dumoulin et al. 2017). Future measurements of these parameters (e.g., Widemann et al. 2023) will be vital to constrain thermal evolution models.

At present, three tools provide access to the structure and internal properties of Venus: measurements of the moment of inertia, the tidal deformation, and the gravity field. The moment of inertia (MoI) is calculated by accurately determining the precession rate of the spin axis. If Venus were in hydrostatic equilibrium, the MoI could also be derived from its degree-2 gravity field. However, the polar flattening of the planet does not correspond to its

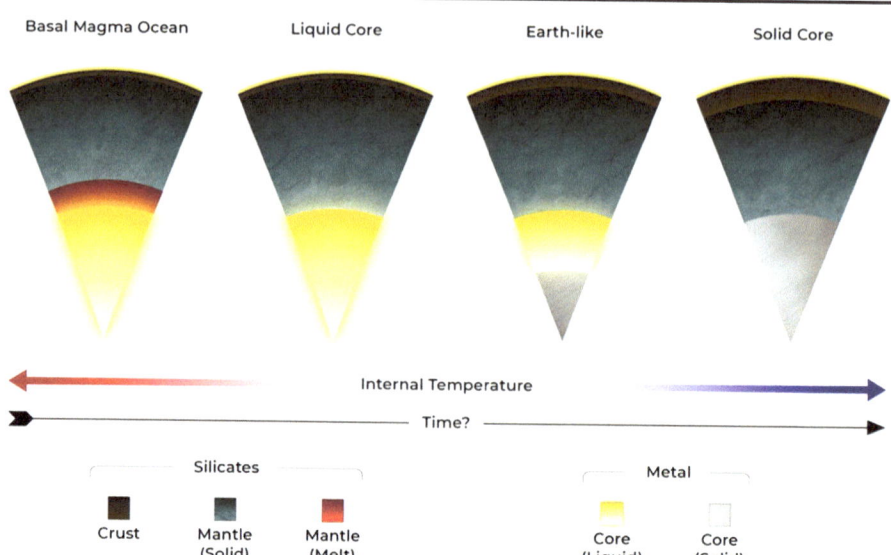

Fig. 9 The internal structure of Venus at present is unknown. Four possibilities are shown, corresponding to internal temperatures that decrease from left to right. First, Venus may be hot enough that the lowermost mantle is still molten. Next, the core could remain fully molten after the mantle completely solidifies. However, Venus may have an Earth-like internal structure with a partially frozen core. Finally, observations do not exclude the (unlikely) possibility that Venus is so cold that the core is entirely solid. Venus probably formed hot and cooled down over time (albeit the mantle may not cool monotonically due to radiogenic heating)—so the internal structure could evolve from left to right over geologic time

current rotation (Van Hoolst 2015), so hydrostatic equilibrium is not applicable. The long precession period of Venus ($\sim 29{,}000$ Earth-years) and the small obliquity of its rotation axis have long prevented the estimation of its MoI. A recent study by Margot et al. (2021) estimated for the first time the MoI with an uncertainty of $\sim 7\%$. They precisely measured the precession rate of the rotation axis through the analysis of radar echo speckles measured at ground-based antennas over the period 2006–2020. This result, although pathbreaking, does not allow us to determine the size of the core with useful accuracy (e.g., $\sim 3{,}500 \pm 500$ km). Fortunately, future Venus missions should improve this measurement of Venus's MoI thanks to radio science alone (e.g., Rosenblatt et al. 2021) or radio science coupled with radar imagery (e.g., Cascioli et al. 2021).

The deformation of the planet under the gravitational attraction of the Sun (solar solid body tide) is a source of information on the internal structure and also on its rheological parameters, such as the viscosity of the mantle or the state of the core. Indeed, the changes in the mass distribution during the tidal deformation generate a variable term in the gravity potential field, of degree 2. The proportional coefficient between the potential due to mass redistribution and the external potential due to solar tide is called the potential Love number and is denoted k_2. The first estimate of the potential Love number was made using tracking data from the PVO and Magellan probes by Konopliv and Yoder (1996). The value found ($k_2 = 0.295 \pm 0.066$) was then used to rule out the presence of a solid core based on elastic tidal deformation modeling results (Yoder 1995). However, if viscosity is taken into account, making solid interior layers more deformable, the current uncertainty on the value of k_2 does not rule out this possibility. Dumoulin et al. (2017) showed, for instance, that a model with a solid core having a low viscosity (i.e., less than 10^{17} Pa s) can account for the current estimate of k_2. Radio-science experiments onboard future missions orbiting Venus will also considerably reduce the uncertainty on k_2 (e.g., Rosenblatt et al. 2021; Cascioli et al. 2021). The imaginary part of the potential Love number, which is equal to k_2/Q, Q^{-1} being the global dissipation function of the planet, reflects the phase lag of the tidal bulge and therefore particularly provides a quantification of the mantle viscosity (cf., Dumoulin et al. 2017). Again, the estimation of this parameter by future space missions (Rosenblatt et al. 2021; Cascioli et al. 2021) could therefore allow an estimate of the mantle viscosity, and consequently of the mantle temperature, which is crucial to constrain thermal evolution models. These new constraints could also be useful for constraining the rotational evolution of the planet via solid body tidal dissipation (e.g., Way and Del Genio 2020, Sect. 5).

The planet's gravity field can also be used in conjunction with the global topography data to estimate crustal and lithospheric thickness. These studies require assumptions about crustal density and type of topographic support. Overall, recent work suggests that the crustal thickness of Venus is $\sim$ 5–70 km, depending on the region, with a mean of $\sim$ 15–30 km, depending on the study (e.g., Nimmo and McKenzie 1998; Anderson and Smrekar 2006; James et al. 2013; Jiménez-Díaz et al. 2015; Yang et al. 2016; Maia and Wieczorek 2022). Broadly speaking, exploitation of the low spherical harmonic degrees of the geoid and topography allows for estimation of global and/or regional lithospheric thickness. Scientists also apply models of lithospheric flexure to the topography of individual volcano-tectonic features (e.g., Johnson and Sandwell 1994; Russell and Johnson 2021; Borrelli et al. 2021; Smrekar et al. 2022a; Ghail et al. 2023) and impact craters (e.g., Ivanov et al. 1986; Grimm and Solomon 1988; Brown and Grimm 1996), which provide local estimates. Depending on the study, the lithospheric thickness of Venus has been reported as $\sim$ 0–600 km (but usually $<$ 100 km), depending on what isostatic equilibrium models are assumed and/or what types of features are studied (e.g., Anderson and Smrekar 2006; Moore and Schubert 1997; Orth and Solomatov 2011).

The bulk composition of the mantle has been estimated using accretion models of the protoplanetary disc. Because of the lack of constraints on these accretion scenarios, models of Venus's internal structure have been generally terrestrial models scaled to Venus to account for the slightly smaller radius and hence lower pressure (e.g., Zharkov 1983; Yoder 1995; Mocquet et al. 2011; Aitta 2012). However, accretion models can lead to different iron oxide contents in the mantle depending on the cosmochemical assumptions (e.g., Lewis 1972; Weidenschilling 1976; Ringwood and Anderson 1977; Morgan and Anders 1980; Rubie et al. 2015). Recent models of solar nebula condensation seem to favor a somewhat smaller Venusian mantle than that of the Earth, implying a larger core-to-mantle volume for Venus than for the Earth (e.g., Trønnes et al. 2019). Various radial structures have also been proposed to account for the variability of the mantle composition with respect to the accretion model hypothesis (e.g., Dumoulin et al. 2017; Zharkov and Gudkova 2019). If the FeO content of the mantle increases (or decreases), then the core should be smaller (or larger). In models, the size of the core varies by several hundred kilometers depending on the assumed interior composition (e.g., Shah et al. 2022). Depending on the core's exact radius, the mantle may or may not have a perovskite to post-perovskite phase transition at its base, similar to the one that can occur in the few hundred km above Earth's core (e.g., Dumoulin et al. 2017; Xiao et al. 2021; Margot et al. 2021). Once again, VERITAS and EnVision will sharpen the tools we use to study Venus's interior (e.g., Widemann et al. 2023), allowing us to make realistic models of its present structure.

Overall, the relative sizes of the core and mantle provide proxies for Venus's redox state. Redox state (or oxygen fugacity) is a central parameter in models of the evolution of Venus. In this collection, Gillmann et al. (2022) describe how the oxygen fugacity of the bulk mantle helps control the long-term evolution of the atmosphere. Because the properties of iron alloys change with pressure and temperature, the proportion and nature of light elements in Venus's core govern its eventual solidification history (e.g., Xiao et al. 2021; Shah et al. 2022). The DAVINCI mission aims to measure the oxygen fugacity of the lower atmosphere to constrain models of the surface mineralogy and atmosphere-surface interactions at present-day (e.g., Garvin et al. 2022). Broadly speaking, oxygen fugacity is also critical to measure for rocky exoplanets, since it is key to the formation and evolution of planetary interiors, surfaces, and atmospheres.

3.3.2 How the Lithosphere and Mantle May Have Evolved

The evolution of a planetary body is strongly controlled by its thermal history. The vigor of convection and extent of partial melting of the mantle determine processes like volcanism and tectonics—which translate in turn to crustal production and the evolution of the atmosphere. Hypotheses about how the lithosphere and mantle of Venus may have evolved developed in tandem with the advances in understanding plate tectonics on Earth, which were detailed in Sects. 2.5.1 and 2.5.2 above. Ultimately, recent studies establish that Venus may provide a modern example of the lithospheric and mantle dynamics of early Earth. However, many first-order questions about Venus's mantle dynamics, and how they have evolved over time, await answers from new missions and modeling studies.

Comparisons between the terrestrial planets in our Solar System suggest that Earth is currently unique as it operates within a plate tectonic regime. Perhaps consequently, Earth has had a habitable climate over geologic time scales. Plate tectonics (as defined on Earth) is characterized by a coherent network of fractured lithosphere, which self-organizes the surface into a series of rigid surface plates. Motion of these discrete plates is accommodated by localized failure along relatively narrow plate boundary zones. The cold surface plates of a plate tectonic regime participate in mantle overturn and in turn are associated with the cooling of the planetary interior. As a result, plate tectonics is considered to be a specific example of the mobile-lid style of mantle convection. Earth is the only body in the Solar System for which a large and robust dataset of its thermal, geologic, and tectonic evolution is accessible. However, our understanding of Earth's evolution remains contentious despite (or perhaps because of) this expansive dataset. For example, the onset time of plate tectonics and thus its duration—and the mechanisms that initiated it—are far from certain (e.g., O'Neill et al. 2007; Debaille et al. 2013; Gerya 2014; Foley et al. 2014; Weller and Lenardic 2018). The starting condition for Earth is uncertain. However, a long-standing consensus, from a thermal standpoint, is that plate tectonics will eventually wane as Earth cools—and Earth will eventually move into stagnant-lid regime, perhaps not dissimilar to Mars today.

In contrast to the current day Earth, observations of Mars suggest a planet operating within a stagnant-lid tectonic regime (e.g., Nimmo and Stevenson 2000). Within this regime, the cold and stiff outermost rock layer does not participate in mantle overturn. Unlike the fractured network of plates in a plate tectonic planet, the surface is made up of a single plate. Single-plate surfaces lack significant horizontal or vertical motions, which largely segregates the surface from the interior. The thick lithosphere of the single-plate, stagnant lid inhibits conductive heat loss, which in addition to the lack of chilling from downwelling slabs, leads to a warmer interior. Similar to Earth, there exists suggestions of an early mobile or plate tectonic phase for Mars (e.g., Sleep 2000; Zhang and O'Neill 2016). However, evidence of such a phase remains elusive. The tectonic history of Venus is even more elusive.

How Venus loses its heat remains a major unanswered question. Pure conduction through a stagnant lithosphere is unlikely to account for all the heat transfer from its interior (e.g., Reese et al. 1999). Several other tectonic regimes have been proposed for Venus, which are described in more detail in Rolf et al. (2022). A popular proposition is that the planet experiences episodical overturns of a lithosphere that is usually a stable stagnant lid, leading to an "episodic lid" regime (e.g., Turcotte 1993; Moresi and Solomatov 1998). This regime would account for global recycling of crust that is subducted during ephemeral bursts of activity. Such a crustal overturn event could produce a global resurfacing event favored by some, but certainly not all, models for the evolution of the surface discussed in Sect. 3.2.2 above (e.g., Strom et al. 1994; Nimmo and McKenzie 1998; Armann and Tackley 2012; Gillmann and Tackley 2014; Bercovici and Ricard 2014). The episodic-lid regime, however, does not account for intrusive or extrusive magmatism on the planet. The recently proposed "plume-lid" (Sizova et al. 2010; Fischer and Gerya 2016) or "plutonic-squishy lid" (Lourenço et al. 2018, 2020) regime emphasizes the importance of intrusive magmatic processes and is characterized by a set of strong plates separated by warm and weak regions generated by plutonism. Instead of lithospheric subduction, lithospheric material is recycled into the mantle by delamination and dripping. This squishy lid regime has also been applied to early Earth in the Archean Eon (4–2.5 Ga) (Fischer and Gerya 2016), when temperature conditions on Earth were thought to be similar to those on Venus today (e.g., Anderson 1981; Head et al. 2008; Van Kranendonk 2010; Harris and Bédard 2014).

The "kick-start" of subduction, a key element of modern-style plate tectonics, has been attributed to various factors. Perhaps most relevant for Earth—and Venus—is the hypothesis that the interaction of a buoyant mantle plume with oceanic lithosphere may have initiated subduction (e.g., Ueda et al. 2008). According to this scenario, under certain circumstances, a long-lived, buoyant mantle plume can overcome the strength of the lithosphere and penetrate through it, pushing the lithosphere downward into the asthenosphere, eventually initiating self-sustained subduction (e.g., Ueda et al. 2008; Stern and Gerya 2018). Numerical explorations of this theory in 3D were undertaken to investigate the initiation of subduction by a thermal plume on Archean Earth (Gerya et al. 2015), by a thermal-chemical plume on modern Earth (Baes et al. 2016), and on Venus (Gülcher et al. 2020). A combination of three key physical factors was proposed to be needed to trigger self-sustained plume-induced subduction (Gerya et al. 2015): 1) a strong, negatively buoyant lithosphere; 2) focused magmatic weakening and thinning of the lithosphere above the plume, and 3) lubrication of the slab interface by hydrated crust. The first and third factor may be (partially) absent on Venus (e.g., Smrekar et al. 2007; Huang et al. 2013). Furthermore, laboratory experiments undertaken by Davaille et al. (2017) advocate a limited plume-induced subduction regime in Venusian environments (see Sect. 2.5.2).

3.3.3 How Prospects for Intrinsic Magnetism May Have Evolved

Spacecraft missions have not yet provided clear evidence that Venus has any intrinsic magnetism. There are at least two types of intrinsic magnetism: 1) an active dynamo that relies on modern motions of electrically conductive fluid in the planetary interior and 2) crustal remanent magnetism that signals the past existence of a dynamo as found on, for example, Mars, Mercury, Earth, and Earth's Moon. Many relevant measurements have been made, but the detection limits for Venus are poor relative to those for other terrestrial planets. Specifically, magnetometers have been carried on at least 11 missions to Venus: Mariner 2 and 10; Venera 4 and 9–12; Pioneer Venus Orbiter; VeGa 1 and 2; and Venus Express (e.g., Russell 1993; Russell et al. 2007). Only Venera 4, PVO, and VEx made measurements at close enough distances to address the intrinsic magnetism of Venus (Fig. 10). The other missions established that Venus lacks an Earth-like magnetosphere and tackled science questions related to space physics and the solar wind.

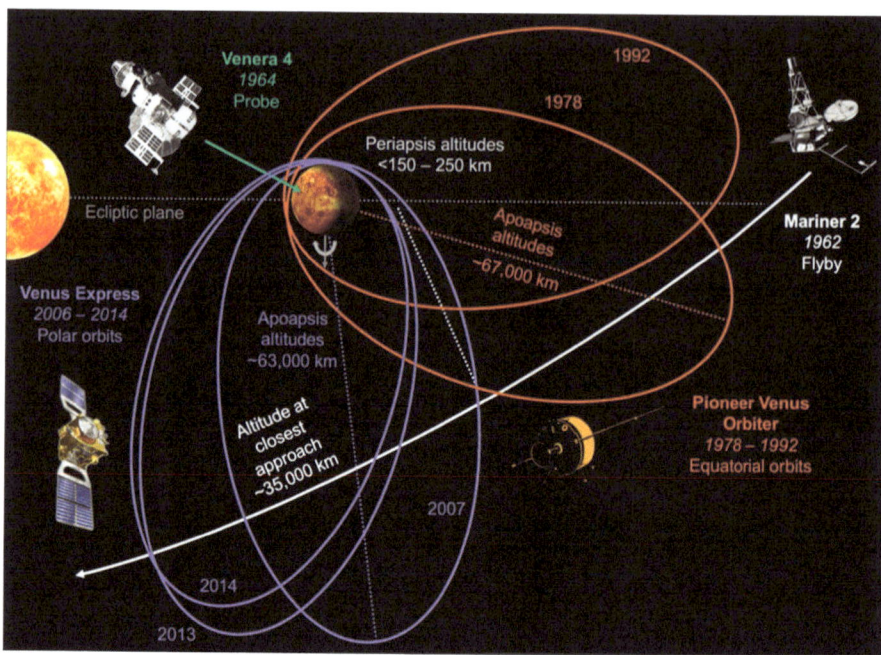

Fig. 10 No intrinsic magnetism has yet been discovered at Venus. This cartoon shows the orbits and trajectories (approximately to scale, following Futaana et al. 2017) of the first magnetometer flyby by Mariner 2 and of the three missions (Venera 4, Pioneer Venus Orbiter, and Venus Express) that supplied the most stringent constraints currently available on our models of Venus's intrinsic magnetism. Pioneer Venus Orbiter and Venus Express placed the most precise upper limit so far on the total magnetic moment of Venus, which is at least $\sim 10^5$ times lower than Earth's modern magnetic moment. Venera 4 collected magnetometer data until ~ 25 km above Eistla Regio (19°N 38°E)—but no other mission made relevant magnetometer measurements below the ionosphere. The southern hemisphere is unexplored magnetically—strong, horizontally coherent magnetization there is possible. Crustal remanent magnetization at horizontal scales smaller than orbital altitudes could exist almost anywhere on the surface

Pioneer Venus Orbiter provided the most stringent detection limit on (the putative lack of) Venus's dynamo. Measurements taken in orbit implied that the total magnetic moment of Venus is $< 10^{-5}$ times Earth's magnetic moment (Phillips and Russell 1987). Lowering this detection limit with a future orbiter is difficult because the thick atmosphere precludes operating at lower altitudes. Weaker magnetic fields would also interact with the solar wind and the ionosphere in complicated ways that require numerical simulations to interpret. In any case, the current limit from PVO is widely considered proof that no dynamo exists because most existing theories predict that a dynamo would be stronger if it existed (e.g., Stevenson 2003, 2010). For example, the "weak" dynamo of Mercury is still $\sim 10^3$ times stronger than the detection limit for Venus. However, some scientists argue that current models and observations are not stringent enough to exclude the possibility that a feeble dynamo exists (e.g., Luhmann et al. 2015). In fact, a very small intrinsic field, near the detection limit from PVO, has been proposed to explain observed structure in the nightside magnetosphere of Venus (e.g., Knudsen et al. 1982).

Crustal remnant magnetism could await discovery on Venus and explain any weak intrinsic magnetism. Intuitively, preserving crustal magnetism on Venus might seem difficult because Venus has the hottest surface on average of any terrestrial planet in the Solar System. However, temperatures in the top few kilometers of the crust should be low enough for common minerals such as magnetite and hematite to retain thermal remanent magnetism from a past dynamo for billions of years (O'Rourke et al. 2019a, 2019b and references therein) unless the surface was much hotter in the past (e.g., Bullock and Grinspoon 1996). In principle, the upper limit on the magnetic moment of Venus of PVO is consistent with the entire surface of Venus being magnetized with an intensity of ~ 1 A/m down to a depth of ~ 1 km (i.e., a total dipole moment of $\sim 5 \times 10^{17}$ A m^2 versus $\sim 8 \times 10^{22}$ A m^2 for Earth). For reference, the average magnetization intensity for magnetized regions in smoothed maps from orbital data of crustal magnetization on Mars is ~ 1 A/m (e.g., Langlais et al. 2019). Crustal magnetization on Venus could also be sparse with local regions having high magnetization intensities, as on Mars at the InSight landing site where the magnetization intensity could be ~ 10 A/m or higher locally (e.g., Johnson et al. 2020). In models, the structure of the nightside magnetosphere of Venus is governed by the total sum of the magnetization, not its spatial distribution (e.g., Knudsen et al. 1982).

New missions are required to search for weak intrinsic magnetism and, if successful, to characterize the source as either a weak dynamo or crustal remanence. As mentioned above, only Venera 4 measured magnetic fields below orbital altitudes—down to an altitude of ~ 25 km above Eistla Regio (Dolginov et al. 1969; Russell 1976). No crustal magnetism was detected at this particular location. Crustal magnetization is not apparent either in the PVO and VEx datasets. Orbital missions are only able to search for magnetization that is spatially coherent at length scales comparable to orbital altitudes (≥ 150 km above the equator and north pole). No mission has yet made magnetometer measurements at low orbital altitudes over the south pole. Magnetometer surveys below the ionosphere (e.g., from an aerial platform as in O'Rourke et al. 2021) are best suited to searching for intrinsic magnetism. Any low-altitude survey should be able to definitively test the dynamo hypothesis, which would produce global, albeit weak, fields. A non-detection of crustal magnetism is more equivocal because of the possible spatial variability. As an extreme analogy, the question of whether Earth's crust preserves a record of a Hadean dynamo is not yet settled (e.g., Borlina et al. 2020; Tarduno et al. 2020)—and will only be answered via small-scale, laboratory measurements of individual mineral crystals.

Because the magnetic history of Venus is so uncertain, the extent to which magnetism constrains models of the evolution of Venus is unclear. In other words, perhaps only a "small" difference between Earth and Venus can explain the absence of an Earth-like magnetosphere. Slow rotation of Venus is occasionally proposed as the culprit (e.g., Luhmann et al. 2015) but most modern scaling laws for the intensity of a dynamo-generated field do not depend on rotation rate above a critical threshold that Venus exceeds by several orders of magnitude (e.g., Stevenson 2003, 2010; Christensen 2010). Perhaps a more popular idea is that the deep interior of Venus cools slowly relative to Earth. Driving a dynamo with convection in an electrically conductive fluid—either in the metallic core or, perhaps, a basal magma ocean (e.g., O'Rourke 2020)—requires that fluid to cool at a certain rate. Slow cooling leads to stagnation and the rapid ($< 10^4$ years) dissipation of any magnetic fields (as happened on Mars). Because Venus has slightly lower pressure than Earth at its center, Venus may lack an inner core even if it has the same core-mantle boundary temperature as Earth. Thus, even if Earth and Venus were cooling at comparable rates, Venus would be less likely to host a dynamo (Stevenson et al. 1983). In general, the critical heat flow required to produce a dynamo in terrestrial planets is close to the actual heat flows expected in planetary interiors. This marginal criticality is why not all terrestrial planets have dynamos, whereas dynamos are basically universal in ice giants, gas giants, and stars. In the absence of plate tectonics, the operative mode of mantle convection may extract relatively less heat from the deep interior (e.g., Nimmo 2002; Driscoll and Bercovici 2013, 2014; O'Rourke et al. 2018).

Alternatively, a dramatic difference in the accretion of Earth and Venus would predict that Venus never had an intrinsic magnetic field. If Venus experienced a "gentle" accretion without any late energetic impacts, then primordial chemical stratification of the core would preclude convection and thus a dynamo from ever existing (Jacobson et al. 2017). At an extreme case, which would be contrary to all expectations, Venus could have accreted so gradually that internal temperatures were low enough for the deep interior to fully solidify. Ultimately, Venus is the only planet in the inner Solar System that lacks either an internal dynamo today (as in Earth and Mercury) or strong evidence that a dynamo existed in the past (as for Earth's Moon and Mars). Improving our understanding of magnetic fields at Venus would help us understand terrestrial planets in general.

4 How Understanding the Evolution of Venus Motivates the Future of Planetary Exploration

Nations invest in space exploration to advance scientific knowledge, to understand humanity's place in the universe, and to advance their prestige amongst the family of nations. National space agencies prioritize science missions based on the perceived importance and feasibility of answering different sets of questions. Historically, the popularity of Venus as a mission target has waxed and waned over the last several decades. As detailed in Sect. 2, interplanetary missions often targeted Venus at the start of the Space Age because it was relatively accessible from Earth and because scientists had long believed that it was habitable today. Other worlds—for example, Mars and myriad icy satellites—soon seemed more appealing once spacecraft became capable enough to reach and explore them—and when they were recognized as perhaps the most likely places to find life beyond Earth at present day. However, the popularity of Venus has recently rebounded from its post-Magellan nadir.

New missions to Venus happen when scientific desires converge with technological advances. The scientific desire to explore Venus has only grown in recent years. As detailed in Sect. 3, all the available observations of Venus are compatible with radically different stories about its evolution over time—a habitable past or a perpetual hell. Starting roughly in the 2000s, increasing awareness of anthropogenic climate change on Earth highlighted the importance of understanding the extreme greenhouse effect on Venus, motivating concepts such as the Venus Climate Mission (National Academies 2011). As studies of exoplanets exploded over the past decade, Venus gained prominence as the archetype of a hot, rocky planet. The field of noble gas isotope geochemistry developed over the past few decades and demonstrated the power to diagnose the formation and evolution of planetary systems (e.g., Baines et al. 2013; Chassefière et al. 2012). This broad-based interest in Venus recently aligned with mission proposals that had relatively low technical risk. Geophysical orbiters like VERITAS (e.g., Smrekar et al. 2022b) and EnVision are analogous to Magellan but will return orders-of-magnitude better data after thirty years of instrument development. The DAVINCI mission will use geochemistry instruments that would have been exotic in the 1990s but now have high heritage from recent Mars missions (Garvin et al. 2022). Future missions to Venus will benefit from technology development efforts that are happening now—focused on, for example, enabling long-lived aerial platforms and surface stations.

Here we review how open questions about the evolution of Venus fit into the strategic plans of space agencies in Europe and the United States. Ultimately, the next few decades of Venus exploration are guaranteed to be fruitful and exciting due to VERITAS, DAVINCI, and EnVision. However, both ESA and NASA are laying the groundwork for an even more ambitious sequence of follow-up missions to understand why Venus and Earth are so different—and what Venus can teach us about the evolution of terrestrial planets in general. Although not discussed here, space agencies in many other countries, including India, Japan, and Russia, also plan to explore Venus. We encourage interested readers to consult Widemann et al. (2023) in this collection to learn about the science goals and planned measurements for these upcoming and possible future missions.

4.1 Importance of Venus's Evolution to ESA

ESA's planetary exploration in the early 21st century has been based largely on its "Cosmic Vision" strategy (ESA 2005). This poses four big questions, the first two of which are "What are the conditions for planet formation and the emergence of life?" and "How does the Solar System work?" The first of these themes explicitly calls for a study of planetary formation and evolution, and the emergence of habitable environments, and then of life itself. These themes are clearly addressed by the EnVision orbiter that in 2021 was selected to be the next mission in ESA's space science program—and would also be addressed by cloud-level balloon missions like the European Venus Explorer concept proposed to ESA in 2007 and 2010 (Wilson et al. 2012). The second top-level question posed in the Cosmic Vision strategy deals with the role of the Sun in the solar system, and its interactions with the planetary systems. This question would support a mission focused on measurement of escape processes and solar wind interaction—or, equally, a mission focused on the radiation budget of Venus, monitoring ingoing and outgoing radiative fluxes and studying its greenhouse balance.

The two previous missions selected in ESA's space science program, PLATO and Ariel, both due for launch in the late 2020s, focus on exoplanet detection and characterization. Their search for, and study of, terrestrial exoplanets will inevitably lead them to confront questions of the diversity of evolutionary outcomes for terrestrial planets and refinement of habitable zones for terrestrial planets; in this their goals will be quite complementary to those of the solar system missions which allow study of a few nearby planets in great depth.

Cosmic Vision is now being succeeded as a strategy document by a new one called *Voyage 2050*. This new report identifies themes for both large- and medium-class missions for the next three decades. One of the large-class mission themes called for is the "characterization of temperate exoplanets." Venus's exploration will play an important role in informing what we know about planetary habitability and its evolution through time, and about what observations can be used to constrain these factors. *Voyage 2050* calls out "Venus Geology and Geophysics" as a key theme for understanding why Venus took an evolutionary path so different from that of Earth—a theme largely addressed by the EnVision mission. No further Venus-specific mission themes are singled out in the Voyage 2050 document. However, ESA's calls for medium-class mission ideas are open to mission concepts that would address Venus's evolution.

4.2 Importance of Venus's Evolution to NASA

In the United States, the National Academies of Sciences, Engineering, and Medicine provide strategic advice to NASA via the decadal survey process. For the last ten years, NASA has operated under the guidance of "Vision and Voyages for Planetary Science in the Decade 2013–2022" (National Academies 2011). Recommendations from the community of scientists interested in Venus, especially via the Venus Exploration Analysis Group (VEXAG), helped shape the recommendations in that survey. As described in a previous review (Glaze et al. 2018), Venus was central to the prioritized scientific objectives for the exploration of the inner solar system planets. At the end of the applicable decade, NASA decided to

Fig. 11 During the three years in the run-up to the new decadal survey in the United States, the Venus community produced several reports and mission concept studies that expressed a consensus vision of Venus exploration—highlighting the importance of Venus science to cross-cutting questions in planetary science and astrobiology. The top and bottom rows show the covers of the 2019 VEXAG strategic documents and three recent mission concept study reports, respectively

achieve many of these objectives by selecting the VERITAS and DAVINCI missions for the Discovery Program (Widemann et al. 2023).

Planning the next decade of Venus exploration began with a grassroots effort organized through VEXAG. In 2018 and 2019, VEXAG produced three strategic documents (Fig. 11): the Goals, Objectives, and Investigations (GOI) for Venus Exploration (O'Rourke et al. 2019a); the Venus Technology Plan (Hunter et al. 2019); and the Roadmap for Venus Exploration (Cutts et al. 2019). The GOI document identified three high-priority scientific goals (unprioritized):

1. Understand Venus's early evolution and potential habitability to constrain the evolution of Venus-sized (exo)planets.
2. Understand atmospheric composition and dynamics on Venus.
3. Understand the geologic history preserved on the surface of Venus and the present-day couplings between the surface and atmosphere.

The technology plan lauded the successes of existing programs such as the Heatshield for Extreme Entry Environment Technology (HEEET) and High Operating Temperature Technology (HOTTech)—and described the new technologies required to support ambitious missions in the extreme conditions found at Venus's atmosphere and surface. Finally, the Roadmap document recommended an ambitious program of Venus missions, starting with an orbiter and an atmospheric-entry probe in the near term—exactly what NASA selected in VERTIAS and DAVINCI. The new decadal survey considered other types of missions to Venus, including additional orbiters, probes, landers, and aerial platforms—all motivated by scientific questions that the triad of now-selected missions may not fully answer.

A new decade brings new opportunities to explore Venus to advance planetary science. Recently, the National Academies released "Origins, Worlds, Life: A Decadal Strategy for Planetary Science and Astrobiology 2023–2032" (National Academies 2022). This survey was organized around 12 priority science questions. These questions centered on three high-level scientific themes:

1. Origins: How did the solar system and Earth originate, and are systems like ours common or rare in the universe?
2. Worlds and processes: How did planetary bodies evolve from their primordial states to the diverse objects seen today?
3. Life and habitability: What conditions led to habitable environments and the emergence of life on Earth, and did life form elsewhere?

Future exploration of Venus is central to all three themes. Table 2 lists the priority science questions and sub-questions that are directly tied to Venus. The decadal survey lists the research activities that would help answer these (sub-)questions—and elaborates on the technological developments that would enable these activities.

NASA will likely select its next missions to Venus (after VERITAS and DAVINCI) through competitions. What NASA calls small-cost missions are commonly led by a single Principal Investigator (PI) and designed to achieve a focused set of science objectives. These missions are typically developed and launched within $\sim$ 3–5 years after their selection. The new decadal survey endorsed continuing three successful programs of competed, PI-led missions that include Venus as a possible destination. First, the Small Innovative Missions for Planetary Exploration (SIMPLEx) program includes very small, low-cost missions with a PI-managed mission cost cap of $55 million in 2018, which the decadal survey recommended raising to roughly $80 million (in fiscal year 2025 dollars, not including the launch). Similarly, ESA's Fast (F) mission opportunities support relatively small, low-cost spacecraft that launch along with a primary mission but then address any area of space science. NASA's Discovery Program enables more sophisticated (and less risky) missions with a PI-managed mission cost cap an order of magnitude above the cost cap for SIMPLEx (e.g., $\sim$$800 million or so), comparable to ESA's Medium (M-class) missions. Both the SIMPLEx and Discovery Programs are open to any mission that advances planetary science—obviously including Venus-targeting missions.

Higher-cost NASA missions are typically chosen from a restricted list. Venus is on the list of allowed targets for the New Frontiers program, which includes missions that are more scientifically ambitious (and thus technically challenging) than Discovery-class missions. Specifically, the new decadal survey described a New Frontiers mission called the Venus In Situ Explorer (VISE), which must address at least two of these three scientific objectives:

- Characterize past or present large-scale spatial and temporal (global, longitudinal and/or diurnal) processes within Venus's atmosphere.
- Investigate past or present surface-atmosphere interactions at Venus.
- Establish past or present physical and chemical properties of the Venus surface and/or interior.

Table 2 Priority questions from the new Planetary Science and Astrobiology Decadal Survey 2023–2032 (National Academies 2022) that relate to the evolution of Venus. Studying Venus also helps us understand the evolution of Earth (e.g., Q3.3) and rocky planets writ large (e.g., Q12). The VISE mission concept for the New Frontiers program would mostly address questions 3, 5, 6, 10, and 12. Venus-targeting missions in the Discovery and/or SIMPLEx programs could address any question(s)

Scientific theme	Priority science questions related to Venus	Sub-questions involving Venus
Origins	Q1. Evolution of the protoplanetary disk.	Q1.1 What were the initial conditions in the Solar System? Q1.2 How did distinct reservoirs of gas and solids form and evolve in the protoplanetary disk? Q1.3 What processes led to the production of planetary building blocks?
	Q3. Origin of Earth and inner solar system bodies.	Q3.1 How and when did asteroids and inner Solar System protoplanets form? Q3.2 Did giant planet formation and migration shape the formation of the inner solar system? Q3.3 How did the Earth-Moon system form? Q3.4 What processes yielded Mars, Venus, and Mercury and their varied initial states? Q3.5 How and when did the terrestrial planets and Moon differentiate? Q3.6 What established the primordial inventories of volatile elements and compounds in the inner Solar System?
Worlds and processes	Q4. Impacts and dynamics	Q4.2 How did impact bombardment vary with time and location in the Solar System? Q4.3 How did collisions affect the geological, geophysical, and geochemical evolution and properties of planetary bodies?
	Q5. Solid body interiors and surfaces.	Q5.1 How diverse are the compositions and internal structures within and among solid bodies? Q5.2 How have the interiors of solid bodies evolved? Q5.3 How have surface/near-surface characteristics and compositions of solid bodies been modified by, and recorded, interior processes? Q5.4 How have surface characteristics and compositions of solid bodies been modified by, and recorded, surface processes and atmospheric interactions? Q5.5 How have surface characteristics and compositions of solid bodies been modified by, and recorded, external processes? Q5.6 What drives active processes occurring in the interiors and on the surfaces of solid bodies?
	Q6. Solid body atmospheres, exospheres, magnetospheres, and climate evolution.	Q6.1 How do solid-body atmospheres form and what was their state during and shortly after accretion? Q6.2 What processes govern the evolution of planetary atmospheres and climates over geologic timescales? Q6.3 What processes drive the dynamics and energetics of atmospheres on solid bodies? Q6.4 How do planetary surfaces and interiors influence and interact with their host atmospheres? Q6.5 What processes govern atmospheric loss to space? Q6.6 What chemical and microphysical processes govern the clouds, hazes, chemistry, and trace gas composition of solid-body atmospheres?

Table 2 (*Continued*)

Scientific theme	Priority science questions related to Venus	Sub-questions involving Venus
Life and habitability	Q10. Dynamic habitability.	Q10.1 What is "habitability"? Q10.2 Where are or were the Solar System's past or present habitable environments? Q10.3 What controls the amount of available water on a body over time? Q10.4 Where and how are organic building blocks of life synthesized in the Solar System? Q10.5 What is the availability of nutrients and other inorganic ingredients to support life? Q10.6 What controls the energy available for life? Q10.7 What controls the continuity or sustainability of habitability?
	Q11. Search for life elsewhere	Q11.1 What is the extent and history of organic chemical evolution, potentially leading toward life, in habitable environments throughout the Solar System? Q11.2 What is the biosignature potential in habitable environments beyond Earth? What are the possible sources of false positives and false negatives? Q11.3 Is or was there life elsewhere in the Solar System?
Cross-cutting	Q12. Exoplanets	Studies of Venus's evolution would help address the exoplanetary analogues to all the priority science questions listed above.

VISE is anticipated to collect data that is not obtainable from an orbiter alone (e.g., VERITAS and EnVision) or a single descent probe (e.g., DAVINCI).

Innovative teams can design missions with a wide range of modalities that are responsive to the VISE concept in the decadal survey. For example, missions proposed to the New Frontiers program under a previous definition of VISE have included landers (targeting the plains or tesserae) and a combined orbiter plus descent probe. This decadal survey commissioned a study at NASA's Goddard Space Flight Center of a new concept called ADVENTS (Assessment and Discovery of Venus's Past Evolution and Near-Term Climatic and Geophysical State, Fig. 11), which included an orbiter, a variable-altitude aerobot, and a dropsonde (O'Rourke et al. 2021). The Steering Committee of the decadal survey ultimately decided that ADVENTS overlapped with the VISE concept—and thus an ADVENTS-like mission is eligible for the New Frontiers program to satisfy the listed VISE objectives. The Panel on Venus for the decadal survey suggested several other mission concepts, including Venus In Situ Seismic and Atmospheric Network, Venus Sub-Cloud Aerobot, the Venus Life Potential, and Venus Investigation of Dynamics From an Equatorial Orbit. These mission concepts were not studied in detail through the decadal survey process, but future teams could develop them under the umbrella of VISE (National Academies 2022, Appendix E).

The new decadal survey did not prioritize a Flagship mission to Venus. Flagship missions are directed by NASA—not led by a PI nor chosen by competition—and must have exceptional scientific merit. In preparation for the decadal survey, NASA commissioned a concept study for a Venus Flagship Mission (Fig. 11). This ambitious mission would deliver an orbiter, two small satellites, a lander, and a variable-altitude aerobot on a single launch to simultaneously study Venus with remote observations and in situ measurements

(Beauchamp et al. 2021). The full version of this mission was estimated to have a total cost above $7 billion and relatively high technical risk. A descoped version of the mission could include only an orbiter and a lander—but was still judged to have relatively high cost and technical risk (National Academies 2022). Mission teams could consider designing a lander that targeted the plains instead of a tessera to reduce risk and cost while still answering priority science questions. As technology matures, increased lander lifetime, mobility, and/or autonomy will enable more advanced scientific investigations (e.g., Kremic et al. 2021).

Beyond missions, the new decadal survey recognized that investments in facilities and technology are vital to Venus's exploration. NASA currently supports at least two facilities that can recreate some of the conditions at the surface of Venus. For example, scientists have used the Planetary Aeolian Laboratory at NASA Ames Research Center to study aeolian processes at extreme pressures relevant to Venus (e.g., Greeley and Iversen 1985). The NASA Glenn Extreme Environments Rig (GEER) is a newer, highly capable facility that can conduct scientific measurements and test equipment at the temperature, pressure, and chemical conditions applicable to Venus's surface (e.g., Lukco et al. 2018). The GEER facility is the staging ground for the team building the Long-Lived In Situ Solar System Explorer (LLISSE), which is a small lander that could serve as a technology demonstration on future Venus missions (e.g., Beauchamp et al. 2021). LLISSE uses electronics built with silicon-carbide integrated circuits that can survive near-indefinitely on Venus's surface (Kremic et al. 2021). Further work on LLISSE over the next ten years should enable the next decadal survey to treat a long-lived lander as technically feasible—capable of achieving the scientific promise intrinsic to such a mission modality.

Increased investment in other NASA infrastructure is also needed to support future exploration of Venus—and planetary science in general. Spacecraft missions return data to scientists on Earth via the Deep Space Network (DSN), the international collection of giant radio antennas. The decadal survey identified a clear need to expand the DSN to support the communications requirements of future missions, including human missions that involve huge amounts of video transmissions. Although one radio band (Ka) offers the best downlink speeds, the decadal survey endorsed maintaining the DSN's ability to transmit in other bands. In particular, S-band transmissions can penetrate Venus's atmosphere and are thus vital to any in situ Venus mission (National Academies 2022). Finally, the Goldstone Solar System Radar (GSSR) is a productive facility that works in tandem with the DSN. GSSR is key to Venus exploration via its ability to map the surface in the L- and S-bands—and to measure the spin state of Venus over a long temporal baseline (e.g., Margot et al. 2021).

5 Conclusions

Scientists have more questions than answers about the evolution of Venus. Unlike at the dawn of the Space Age, we now know that the surface of Venus is inhospitable—scorched under a thick blanket of CO_2 whose most visually striking feature is the global cloud layer made of concentrated sulfuric acid. National space agencies will soon spend billions of dollars to answer one question: Has Venus always been like this? Our uncertainty is the best motivation for continued exploration. Without understanding the Earth/Venus dichotomy, we cannot claim to understand rocky (exo)planets in general. Scientists can publish many articles supporting the "habitable hypothesis:" the idea that Venus once was as we once imagined it to be now. A time traveler might have found two "blue marbles" in our Solar System only a billion years ago. However, these optimistic models could be wrong. Perhaps Earth and its sister planet trod divergent evolutionary paths from the start. Venus may have toured only the gradations of hell over its lifetime. This review aimed to explain the key turning points in the evolution of our study of the evolution of Venus—and to arm the reader with enough background knowledge to tackle the other manuscripts in this topical collection. Ultimately, we will rediscover Venus over the next few decades. Some of our current ideas about Venus might seem as silly as the old "Seltzer ocean" hypothesis seems now. But one prophecy is secure: We will soon have satisfying answers to some of our most pressing questions—and be able to ask new questions that cannot yet congeal in our ignorant minds.

Acknowledgements JGO thanks the participants in the "Venus In Situ Sample Capture Mission" workshop organized by the W.M. Keck Institute for Space Studies, especially John Elliott and Francois Tissot, for sci-fi recommendations. Thanks also to all the participants of the "Venus: Evolution through Time" workshop and project through the International Space Science Institute for countless fruitful discussions. M.J.W. acknowledges support from the GSFC Sellers Exoplanet Environments Collaboration (SEEC) and ROCKE-3D: The evolution of solar system worlds through time, funded by the NASA Planetary and Earth Science Divisions Internal Scientist Funding Model. Thanks finally to two anonymous reviewers for providing helpful feedback.

Funding Note Open access funding provided by Swiss Federal Institute of Technology Zurich.

Declarations

Competing Interests The authors have no competing interests to declare that are relevant to the content of this article.

References

Abe Y, Matsui T (1988) Evolution of an impact-generated H_2O–CO_2 atmosphere and formation of a hot proto-ocean on Earth. J Atmos Sci 45:3081–3101. https://doi.org/10.1175/1520-0469(1988)045<3081:EOAIGH>2.0.CO;2

Adams WS, Dunham J (1932) Absorption bands in the infra-red spectrum of Venus. Publ Astron Soc Pac 44:243. https://doi.org/10.1086/124235

Adel A (1937) A determination of the amount of carbon dioxide above the reflecting layer in the atmosphere of the planet Venus. Astrophys J 85:345. https://doi.org/10.1086/143832

Aitta A (2012) Venus' internal structure, temperature and core composition. Icarus 218:967–974. https://doi.org/10.1016/j.icarus.2012.01.007

Aldiss BW, Harrison H (1968) Farewell, Fantastic Venus!: A History of the Planet Venus in Fact and Fiction. Macdonald, London

Allen DA (1987) The dark side of Venus. Icarus 69:221–229. https://doi.org/10.1016/0019-1035(87)90101-1

Allen DA, Crawford D (1984) Cloud structure on the dark side of Venus. Nature 307:222–224. https://doi.org/10.1038/307222a0

Allen RC, Cernuda I, Pacheco D, Berger L, Xu ZG, Freiherr von Forstner JL, Rodríguez-Pacheco J, Wimmer-Schweingruber RF, Ho GC, Mason GM, Vines SK, Khotyaintsev Y, Horbury T, Maksimovic M, Hadid LZ, Volwerk M, Dimmock AP, Sorriso-Valvo L, Stergiopoulou K, Andrews GB, Angelini V, Bale SD, Boden S, Böttcher SI, Chust T, Eldrum S, Espada PP, Espinosa Lara F, Evans V, Gómez-Herrero R, Hayes JR, Hellín AM, Kollhoff A, Krasnoselskikh V, Kretzschmar M, Kühl P, Kulkarni SR, Lees WJ, Lorfèvre E, Martin C, O'Brien H, Plettemeier D, Polo OR, Prieto M, Ravanbakhsh A, Sánchez-Prieto S, Schlemm CE, Seifert H, Souček J, Steller M, Štverák Š, Terasa JC, Trávníček P, Tyagi K, Vaivads A, Vecchio A, Yedla M (2021) Energetic ions in the Venusian system: insights from the first solar orbiter flyby. Astron Astrophys 656:A7. https://doi.org/10.1051/0004-6361/202140803

Anderson DL (1981) Plate tectonics on Venus. Geophys Res Lett 8:309–311. https://doi.org/10.1029/GL008i004p00309

Anderson FS, Smrekar SE (2006) Global mapping of crustal and lithospheric thickness on Venus. J Geophys Res, Planets 111:1–20. https://doi.org/10.1029/2004JE002395

Armann M, Tackley PJ (2012) Simulating the thermochemical magmatic and tectonic evolution of Venus's mantle and lithosphere: two-dimensional models. J Geophys Res, Planets 117:E12003. https://doi.org/10.1029/2012JE004231

Arney G, Meadows V, Crisp D, Schmidt SJ, Bailey J, Robinson T (2014) Spatially resolved measurements of H_2O, HCl, CO, OCS, SO_2, cloud opacity, and acid concentration in the Venus near-infrared spectral windows. J Geophys Res, Planets 119:1860–1891. https://doi.org/10.1002/2014JE004662

Arrhenius S (1918) The Destinies of the Stars. The Knickerbocker Press, New York

Avduevsky VS, Marov MY, Rozhdestvensky MK (1970) A tentative model of the Venus atmosphere based on the measurements of Veneras 5 and 6. J Atmos Sci 27:561–568. https://doi.org/10.1175/1520-0469(1970)027<0561:ATMOTV>2.0.CO;2

Avduevsky VS, Borodin NF, Burtsev VP, Malkov IV, Marov MI, Morozov SF, Rozhdestvenskii MK, Romanov RS, Sokolov SS, Fokin VG (1977) Automatic stations Venera 9 and Venera 10-functioning of descent vehicles and measurement of atmospheric parameters. Cosm Res 14:655–666

Avduevsky VS, Marov MY, Kulikov YN, Shari VP, Gorbachevskiy AY, Uspenskiy GR, Cheremukhina ZP (1983) Structure and parameters of the Venus atmosphere according to Venera probe data. In: Hunten DM, Colin L, Donahue TM, Moroz VI (eds) Venus. University of Arizona Press, Tucson, pp 280–298

Avice G, Parai R, Jacobson SA, Labidi J, Trainer MG, Petkov MP (2022) Noble gases and stable isotopes track the origin and early evolution of the Venus atmosphere. Space Sci Rev 218:60. https://doi.org/10.1007/s11214-022-00929-9

Baes M, Gerya T, Sobolev SV (2016) 3-D thermo-mechanical modeling of plume-induced subduction initiation. Earth Planet Sci Lett 453:193–203. https://doi.org/10.1016/j.epsl.2016.08.023

Baines KH, Atreya SK, Bullock MA, Grinspoon DH, Mahaffy P, Russell CT, Schubert G, Zahnle K (2013) The atmospheres of the terrestrial planets: clues to the origins and early evolution of Venus, Earth, and Mars. In: Comparative Climatology of Terrestrial Planets. University of Arizona Press, Tucson, pp 137–160. https://doi.org/10.2458/azu_uapress_9780816530595-ch006

Barath FT, Barrett AH, Copeland J, Jones DE, Lilley AE (1964) Symposium on radar and radiometric observations of Venus during the 1962 conjunction: Mariner 2 microwave radiometer experiment and results. Astron J 69:49. https://doi.org/10.1086/109227

Barrett AH (1961) Microwave absorption and emission in the atmosphere of Venus. Astrophys J 133:281. https://doi.org/10.1086/147024

Barsukov VL, Basilevsky AT, Burba GA, Bobinna NN, Kryuchkov VP, Kuzmin RO, Nikolaeva OV, Pronin AA, Ronca LB, Chernaya IM, Shashkina VP, Garanin AV, Kushky ER, Markov MS, Sukhanov AL, Kotelnikov VA, Rzhiga ON, Petrov GM, Alexandrov YN, Sidorenko AI, Bogomolov AF, Skrypnik GI, Bergman MY, Kudrin LV, Bokshtein IM, Kronrod MA, Chochia PA, Tyuflin YS, Kadnichansky SA, Akim EL (1986) The geology and geomorphology of the Venus surface as revealed by the radar images obtained by Veneras 15 and 16. J Geophys Res, Solid Earth 91:378–398. https://doi.org/10.1029/JB091iB04p0D378

Baumgardner JR (1985) Three-dimensional treatment of convective flow in the Earth's mantle. J Stat Phys 39:501–511. https://doi.org/10.1007/BF01008348

Beauchamp P, Gilmore MS, Lynch RJ, Sarli BV, Nicoletti A, Jones A, Ginyard A, Segura ME (2021) Venus flagship mission concept: a decadal survey study. IEEE Aerosp Conf Proc 2021:1–18. https://doi.org/10.1109/aero50100.2021.9438335

Belousova EA, Kostitsyn YA, Griffin WL, Begg GC, O'Reilly SY, Pearson NJ (2010) The growth of the continental crust: constraints from zircon Hf-isotope data. Lithos 119:457–466. https://doi.org/10.1016/j.lithos.2010.07.024

Bercovici D (2015) Mantle dynamics: an introduction and overview. In: Treatise on Geophysics: Second Edition. Elsevier, Oxford. https://doi.org/10.1016/B978-0-444-53802-4.00125-1

Bercovici D, Ricard Y (2014) Plate tectonics, damage and inheritance. Nature 508:513–516. https://doi.org/10.1038/nature13072

Bertaux JL, Nevejans D, Korablev O, Villard E, Quémerais E, Neefs E, Montmessin F, Leblanc F, Dubois JP, Dimarellis E, Hauchecorne A, Lefèvre F, Rannou P, Chaufray JY, Cabane M, Cernogora G, Souchon G, Semelin F, Reberac A, Van Ransbeek E, Berkenbosch S, Clairquin R, Muller C, Forget F, Hourdin F, Talagrand O, Rodin A, Fedorova A, Stepanov A, Vinogradov I, Kiselev A, Kalinnikov Y, Durry G, Sandel B, Stern A, Gérard JC (2007) SPICAV on Venus Express: three spectrometers to study the global structure and composition of the Venus atmosphere. Planet Space Sci 55:1673–1700. https://doi.org/10.1016/j.pss.2007.01.016

Bierson CJ, Zhang X (2020) Chemical cycling in the Venusian atmosphere: a full photochemical model from the surface to 110 km. J Geophys Res, Planets 125:e2019JE006159. https://doi.org/10.1029/2019JE006159

Borlina CS, Weiss BP, Lima EA, Tang F, Taylor RJM, Einsle JF, Harrison RJ, Fu RR, Bell EA, Alexander EW, Kirkpatrick HM, Wielicki MM, Mark Harrison T, Ramezani J, Maloof AC (2020) Reevaluating the evidence for a Hadean-Eoarchean dynamo. Sci Adv 6:eaav9634. https://doi.org/10.1126/sciadv.aav9634

Borrelli ME, O'Rourke JG, Smrekar SE, Ostberg CM (2021) A global survey of lithospheric flexure at steepsided domical volcanoes on Venus reveals intermediate elastic thicknesses. J Geophys Res, Planets 126:1–14. https://doi.org/10.1029/2020JE006756

Boukrouche R, Lichtenberg T, Pierrehumbert RT (2021) Beyond runaway: initiation of the post-runaway greenhouse state on rocky exoplanets. Astrophys J 919:130. https://doi.org/10.3847/1538-4357/ac1345

Bowen TA, Bale SD, Bandyopadhyay R, Bonnell JW, Case A, Chasapis A, Chen CHK, Curry S, Dudok de Wit T, Goetz K, Goodrich K, Gruesbeck J, Halekas J, Harvey PR, Howes GG, Kasper JC, Korreck K, Larson D, Livi R, MacDowall RJ, Malaspina DM, Mallet A, McManus MD, Page B, Pulupa M, Raouafi N, Stevens ML, Whittlesey P (2021) Kinetic-scale turbulence in the Venusian magnetosheath. Geophys Res Lett 48:1–12. https://doi.org/10.1029/2020GL090783

Bower DJ, Hakim K, Sossi PA, Sanan P (2022) Retention of water in terrestrial magma oceans and carbonrich early atmospheres. Planet Sci J 3:93. https://doi.org/10.3847/psj/ac5fb1

Bradbury R (1951) The Long Rain. In: The Illustrated Man. Doubleday & Company, Garden City

Brain DA, Bagenal F, Ma YJ, Nilsson H, Stenberg Wieser G (2016) Atmospheric escape from unmagnetized bodies. J Geophys Res, Planets 121:2364–2385. https://doi.org/10.1002/2016JE005162

Brecht A, Brecht S, Luhmann J, Bellan J, Jessup K-L, Navarro T, Lebonnois S, Bougher S, Ma Y, Parish H (2021) Closing the gap between theory and observations of Venus atmospheric dynamics with new measurements. Bull AAS 53. https://doi.org/10.3847/25c2cfeb.2c8c0bbc

Brown CD, Grimm RE (1996) Floor subsidence and rebound of large Venus craters. J Geophys Res, Planets 101:26057–26067. https://doi.org/10.1029/96JE02706

Bullock MA, Grinspoon DH (1996) The stability of climate on Venus. J Geophys Res, Planets 101:7521–7529. https://doi.org/10.1029/95JE03862

Bullock MA, Grinspoon DH (2001) The recent evolution of climate on Venus. Icarus 150:19–37. https://doi.org/10.1006/icar.2000.6570

Bunge H-P, Grand SP (2000) Mesozoic plate-motion history below the northeast Pacific Ocean from seismic images of the subducted Farallon slab. Nature 405:337–340. https://doi.org/10.1038/35012586

Butler BJ, Steffes PG, Suleiman SH, Kolodner MA, Jenkins JM (2001) Accurate and consistent microwave observations of Venus and their implications. Icarus 154:226–238. https://doi.org/10.1006/icar.2001.6710

Cadell HM (1889) VII.—Experimental researches in mountain building. Trans R Soc Edinb 35:337–357. https://doi.org/10.1017/S0080456800017658

Campbell IH, Taylor SR (1983) No water, no granites – no oceans, no continents. Geophys Res Lett 10:1061–1064. https://doi.org/10.1029/GL010i011p01061

Capitanio FA, Nebel O, Cawood PA, Weinberg RF, Chowdhury P (2019a) Reconciling thermal regimes and tectonics of the early Earth. Geology 47:923–927. https://doi.org/10.1130/G46239.1

Capitanio FA, Nebel O, Cawood PA, Weinberg RF, Clos F (2019b) Lithosphere differentiation in the early Earth controls Archean tectonics. Earth Planet Sci Lett 525:115755. https://doi.org/10.1016/j.epsl.2019.115755

Carlson R et al (1991) Galileo infrared imaging spectroscopy measurements at Venus. Science 253:1541–1548. https://doi.org/10.1126/science.253.5027.1541

Carpenter IRL (1966) Study of Venus by cw radar—1964 results. Astron J 71:142–151. https://doi.org/10.1086/109872

Carter L, Gilmore MS, Ghail RC, Byrne PK, Smrekar SE, Ganey TM, Izenberg NR (2023) Sedimentary processes on Venus. Space Sci Rev. In preparation

Cascioli G, Hensley S, De Marchi F, Breuer D, Durante D, Racioppa P, Iess L, Mazarico E, Smrekar SE (2021) The determination of the rotational state and interior structure of Venus with VERITAS. Planet Sci J 2:220. https://doi.org/10.3847/PSJ/ac26c0

Cawood PA, Hawkesworth CJ, Dhuime B (2013) The continental record and the generation of continental crust. Geol Soc Am Bull 125:14–32. https://doi.org/10.1130/B30722.1

Chambers JE (2001) Making more terrestrial planets. Icarus 152:205–224. https://doi.org/10.1006/icar.2001.6639

Chassefière E, Wieler R, Marty B, Leblanc F (2012) The evolution of Venus: present state of knowledge and future exploration. Planet Space Sci 63–64:15–23. https://doi.org/10.1016/j.pss.2011.04.007

Chaverot G, Bolmont E, Turbet M (2022) Study of the runaway greenhouse effect with a 3D global climate model. In: EGU General Assembly 2022, p EGU22-8125. https://doi.org/10.5194/egusphere-egu22-8125

Chowdhury P, Gerya T, Chakraborty S (2017) Emergence of silicic continents as the lower crust peels off on a hot plate-tectonic Earth. Nat Geosci 10:698–703. https://doi.org/10.1038/ngeo3010

Chowdhury P, Chakraborty S, Gerya TV, Cawood PA, Capitanio FA (2020) Peel-back controlled lithospheric convergence explains the secular transitions in Archean metamorphism and magmatism. Earth Planet Sci Lett 538:116224. https://doi.org/10.1016/j.epsl.2020.116224

Christensen UR (1985) Thermal evolution models for the Earth. J Geophys Res 90:2995. https://doi.org/10.1029/JB090iB04p02995

Christensen UR (2010) Dynamo scaling laws and applications to the planets. Space Sci Rev 152:565–590. https://doi.org/10.1007/s11214-009-9553-2

Clarke AC (1973) Rendezvous with Rama. Gollancz, London

Collinson GA, Ramstad R, Frahm R, Wilson L, Xu S, Whittlesey P, Brecht SH, Ledvina S (2022) A revised understanding of the structure of the Venusian magnetotail from a high-altitude intercept with a tail ray by Parker solar probe. Geophys Res Lett 49:1–11. https://doi.org/10.1029/2021GL096485

Corey JSA (2011) Leviathan Wakes. Orbit Books, London

Corey JSA (2022) Memory's Legion: The Complete Expanse Story Collection. Orbit, New York

Crisp D, Ingersoll AP, Hildebrand CE, Preston RA (1990) VEGA Balloon meteorological measurements. Adv Space Res 10:109–124. https://doi.org/10.1016/0273-1177(90)90172-V

Crisp D, Allen D, Grinspoon D, Pollack J (1991) The dark side of Venus: near-infrared images and spectra from the Anglo-Australian Observatory. Science 253:1263–1266. https://doi.org/10.1126/science.11538493

Cutts JA (VEXAG) (2019) Roadmap for Venus Exploration. https://www.lpi.usra.edu/vexag/documents/reports/VEXAG_Venus_Roadmap_2019.pdf

Dai L, Zhang X, Shao WD, Bierson CJ, Cui J (2022) A simple condensation model for the H_2SO_4-H_2O gas-cloud system on Venus. J Geophys Res, Planets 127:e2021JE007060. https://doi.org/10.1029/2021JE007060

Daubrée A (1879) Etudes Synthetiques de Geologie Experimentale. Paris

Davaille A (1999) Simultaneous generation of hotspots and superswells by convection in a heterogeneous planetary mantle. Nature 402:756–760. https://doi.org/10.1038/45461

Davaille A, Limare A (2007) Laboratory studies of mantle convection. In: Treatise on Geophysics. Elsevier, Amsterdam, pp 89–165. https://doi.org/10.1016/B978-044452748-6.00116-4

Davaille A, Smrekar SE, Tomlinson S (2017) Experimental and observational evidence for plume-induced subduction on Venus. Nat Geosci 10:349–355. https://doi.org/10.1038/ngeo2928

Davy P, Cobbold P (1988) Indentation tectonics in nature and experiments: experiments scaled for gravity. Bull Geol Inst Univ Uppsala 14:129–141

De Bergh C, Moroz VI, Taylor FW, Crisp D, Bézard B, Zasova LV (2006) The composition of the atmosphere of Venus below 100 km altitude: an overview. Planet Space Sci 54:1389–1397. https://doi.org/10.1016/j.pss.2006.04.020

Debaille V, O'Neill C, Brandon AD, Haenecour P, Yin QZ, Mattielli N, Treiman AH (2013) Stagnant-lid tectonics in early Earth revealed by [142]Nd variations in late Archean rocks. Earth Planet Sci Lett 373:83–92. https://doi.org/10.1016/j.epsl.2013.04.016

Denning WF (1891) Telescopic Work for Starlight Evenings. Taylor and Francis, London

Dhuime B, Hawkesworth CJ, Cawood PA, Storey CD (2012) A change in the geodynamics of continental growth 3 billion years ago. Science 335:1334–1336. https://doi.org/10.1126/science.1216066

Ding F, Pierrehumbert RT (2020) The phase-curve signature of condensable water-rich atmospheres on slowly rotating tidally locked exoplanets. Astrophys J 901:L33. https://doi.org/10.3847/2041-8213/abb941

Dolginov SS, Yeroshenko YG, Davis L (1969) On the nature of the magnetic field near Venus. Kosm Issled 7:747

Donahue TM, Hoffman JH, Hodges RR Jr, Watson AJ (1982) Venus was wet: a measurement of the ratio of deuterium to hydrogen. Science 216:630–633. https://doi.org/10.1126/science.216.4546.630

Donahue TM, Pollack JB (1983) Origin and evolution of the atmosphere of Venus. In: Hunten DM, Colin L, Donahue TM, Moroz VI (eds) Venus. University of Arizona Press, Tucson, pp 1003–1036

Dong C, Jin M, Lingam M (2020) Atmospheric escape from TOI-700 d: Venus versus Earth analogs. Astrophys J 896:L24. https://doi.org/10.3847/2041-8213/ab982f

Dozois G (2015) Introduction: return to Venusport. In: Martin GRR, Dozois G (eds) Old Venus: A Collection of Stories. Bantam Books, New York

Drake FD (1962) 10-cm observations of Venus near superior conjunction. Nature 195:894. https://doi.org/10.1038/195894a0

Driscoll P, Bercovici D (2013) Divergent evolution of Earth and Venus: influence of degassing, tectonics, and magnetic fields. Icarus 226:1447–1464. https://doi.org/10.1016/j.icarus.2013.07.025

Driscoll P, Bercovici D (2014) On the thermal and magnetic histories of Earth and Venus: influences of melting, radioactivity, and conductivity. Phys Earth Planet Inter 236:36–51. https://doi.org/10.1016/j.pepi.2014.08.004

Drossart P, Piccioni G, Gérard JC, Lopez-Valverde MA, Sanchez-Lavega A, Zasova L, Hueso R, Taylor FW, Bézard B, Adriani A, Angrilli F, Arnold G, Baines KH, Bellucci G, Benkhoff J, Bibring JP, Blanco A, Blecka MI, Carlson RW, Coradini A, Di Lellis A, Encrenaz T, Erard S, Fonti S, Formisano V, Fouchet T, Garcia R, Haus R, Helbert J, Ignatiev NI, Irwin P, Langevin Y, Lebonnois S, Luz D, Marinangeli L, Orofino V, Rodin AV, Roos-Serote MC, Saggin B, Stam DM, Titov D, Visconti G, Zambelli M, Tsang C (the VIRTIS-Venus Express Technical Team) (2007) A dynamic upper atmosphere of Venus as revealed by VIRTIS on Venus Express. Nature 450:641–645. https://doi.org/10.1038/nature06140

Dumoulin C, Tobie G, Verhoeven O, Rosenblatt P, Rambaux N (2017) Tidal constraints on the interior of Venus. J Geophys Res, Planets 122:1338–1352. https://doi.org/10.1002/2016JE005249

Elkins-Tanton LT (2008) Linked magma ocean solidification and atmospheric growth for Earth and Mars. Earth Planet Sci Lett 271:181–191. https://doi.org/10.1016/j.epsl.2008.03.062

ESA (2005) Cosmic vision: space science for Europe 2015–2025. In: ESA Broch BR-247, pp 1–111

Escher B, Kuenen PH (1928) Experiments in connection with salt domes. Leidse Geol Meded 3:151–182

Esposito LW (1984) Sulfur dioxide: episodic injection shows evidence for active Venus volcanism. Science 223:1072–1074. https://doi.org/10.1126/science.223.4640.1072

Esposito LW, Knollenberg RG, Marov MY, Toon OB, Turco RP (1983) The clouds and hazes of Venus. In: Hunten DM, Colin L, Donahue TM, Moroz VI (eds) Venus. University of Arizona Press, Tucson, pp 484–564

Esposito LW, Bertaux J-L, Krasnopolsky V, Moroz VI, Zasova LV (1997) Chemistry of lower atmosphere and clouds. In: Bougher SW, Hunten DM, Phillips RJ (eds) Venus II. University of Arizona Press, Tucson, pp 415–458

Evans DAD (2013) Reconstructing pre-Pangean supercontinents. Geol Soc Am Bull 125:1735–1751. https://doi.org/10.1130/B30950.1

Fauchez TJ, Turbet M, Wolf ET, Boutle I, Way MJ, Del Genio AD, Mayne NJ, Tsigaridis K, Kopparapu RK, Yang J, Forget F, Mandell A, Domagal Goldman SD (2020) TRAPPIST-1 Habitable Atmosphere Intercomparison (THAI): motivations and protocol version 1.0. Geosci Model Dev 13:707–716. https://doi.org/10.5194/gmd-13-707-2020

Faugere E, Brun J-P (1984) Modélisation Expérimentale de La Distention Continentale. C R Séances Acad Sci, Sér 2, Méc Phys, Chim Sci Univ, Sci Terre 299:365–370

Fedorova A, Korablev O, Vandaele AC, Bertaux JL, Belyaev D, Mahieux A, Neefs E, Wilquet WV, Drummond R, Montmessin F, Villard E (2008) HDO and H_2O vertical distributions and isotopic ratio in the Venus mesosphere by solar occultation at infrared spectrometer on board Venus Express. J Geophys Res, Planets 113:1–16. https://doi.org/10.1029/2008JE003146

Fischer R, Gerya T (2016) Regimes of subduction and lithospheric dynamics in the Precambrian: 3D thermomechanical modelling. Gondwana Res 37:53–70. https://doi.org/10.1016/j.gr.2016.06.002

Fischer RA, Cottrell E, Hauri E, Lee KKM, Le Voyer M (2020) The carbon content of Earth and its core. Proc Natl Acad Sci USA 117:8743–8749. https://doi.org/10.1073/pnas.1919930117

Florenskiy KP, Bazilevskiy AT, Burba GA, Nikolayeva OV, Pronin AA, Selivanov AS, Narayeva MK, Panfilov AS, Chemodanov VP (1983) Panorama of Venera 9 and 10 landing sites. In: Hunten DM, Colin L, Donahue TM, Moroz VI (eds) Venus. University of Arizona Press, Tucson, pp 137–153

Foley BJ, Driscoll PE (2016) Whole planet coupling between climate, mantle, and core: implications for rocky planet evolution. Geochem Geophys Geosyst 17:1885–1914. https://doi.org/10.1002/2015GC006210

Foley BJ, Smye AJ (2018) Carbon cycling and habitability of Earth-sized stagnant lid planets. Astrobiology 18:873–896. https://doi.org/10.1089/ast.2017.1695

Foley BJ, Bercovici D, Elkins-Tanton LT (2014) Initiation of plate tectonics from post-magma ocean thermochemical convection. J Geophys Res, Solid Earth 119:8538–8561. https://doi.org/10.1002/2014JB011121

Ford PG, Pettengill GH (1992) Venus topography and kilometer-scale slopes. J Geophys Res 97:13103. https://doi.org/10.1029/92JE01085

Ford JP, Plaut JJ, Weitz CM, Farr TG, Senske DA, Stofan ER, Michaels G, Parker TJ (1993) Guide to Magellan image interpretation. JPL Publ 93-24

Fox NJ, Velli MC, Bale SD, Decker R, Driesman A, Howard RA, Kasper JC, Kinnison J, Kusterer M, Lario D, Lockwood MK, McComas DJ, Raouafi NE, Szabo A (2016) The Solar Probe Plus mission: humanity's first visit to our star. Space Sci Rev 204:7–48. https://doi.org/10.1007/s11214-015-0211-6

Fukuhara T, Futaguchi M, Hashimoto GL, Horinouchi T, Imamura T, Iwagaimi N, Kouyama T, Murakami SY, Nakamura M, Ogohara K, Sato M, Sato TM, Suzuki M, Taguchi M, Takagi S, Ueno M, Watanabe S, Yamada M, Yamazaki A (2017) Large stationary gravity wave in the atmosphere of Venus. Nat Geosci 10:85–88. https://doi.org/10.1038/ngeo2873

Futaana Y, Wieser GS, Barabash S, Luhmann JG (2017) Solar wind interaction and impact on the Venus atmosphere. Space Sci Rev 212:1453–1509. https://doi.org/10.1007/s11214-017-0362-8

Gaillard F, Bernadou F, Roskosz M, Bouhifd MA, Marrocchi Y, Iacono-Marziano G, Moreira M, Scaillet B, Rogerie G (2022) Redox controls during magma ocean degassing. Earth Planet Sci Lett 577:117255. https://doi.org/10.1016/j.epsl.2021.117255

Garvin JB, Getty SA, Arney GN, Johnson NM, Kohler E, Schwer KO, Sekerak M, Bartels A, Saylor RS, Elliott VE, Goodloe CS, Garrison MB, Cottini V, Izenberg N, Lorenz R, Malespin CA, Ravine M, Webster CR, Atkinson DH, Aslam S, Atreya S, Bos BJ, Brinckerhoff WB, Campbell B, Crisp D, Filiberto JR, Forget F, Gilmore M, Gorius N, Grinspoon D, Hofmann AE, Kane SR, Kiefer W, Lebonnois S, Mahaffy PR, Pavlov A, Trainer M, Zahnle KJ, Zolotov M (2022) Revealing the mysteries of Venus: the DAVINCI mission. Planet Sci J 3:117. https://doi.org/10.3847/psj/ac63c2

Gerya T (2014) Precambrian geodynamics: concepts and models. Gondwana Res 25:442–463. https://doi.org/10.1016/j.gr.2012.11.008

Gerya T (2022) Numerical modeling of subduction: state of the art and future directions. Geosphere 18:503–561. https://doi.org/10.1130/GES02416.1

Gerya TV, Stern RJ, Baes M, Sobolev SV, Whattam SA (2015) Plate tectonics on the Earth triggered by plume-induced subduction initiation. Nature 527:221–225. https://doi.org/10.1038/nature15752

Ghail R, Smrekar SE, Byrne PK, Gilmore MS, Herrick RR, Ivanov MA, O'Rourke JG, Plesa I, Rolf T, Sabbeth L, Schools JW, Shellnutt JG (2023) Volcanic and tectonic constraints on the evolution of Venus. Space Sci Rev. In preparation

Gillmann C, Tackley P (2014) Atmosphere/mantle coupling and feedbacks on Venus. J Geophys Res, Planets 119:1189–1217. https://doi.org/10.1002/2013JE004505

Gillmann C, Chassefière E, Lognonné P (2009) A consistent picture of early hydrodynamic escape of Venus atmosphere explaining present Ne and Ar isotopic ratios and low oxygen atmospheric content. Earth Planet Sci Lett 286:503–513. https://doi.org/10.1016/j.epsl.2009.07.016

Gillmann C, Golabek GJ, Raymond SN, Schönbächler M, Tackley PJ, Dehant V, Debaille V (2020) Dry late accretion inferred from Venus's coupled atmosphere and internal evolution. Nat Geosci 13:265–269. https://doi.org/10.1038/s41561-020-0561-x

Gillmann C, Way MJ, Avice G, Breuer D, Golabek GJ, Höning D, Krissansen-Totton J, Lammer H, O'Rourke JG, Persson M, Plesa A-C, Salvador A, Scherf M, Zolotov MY (2022) The long-term evolution of the atmosphere of Venus: processes and feedback mechanisms. Space Sci Rev 218:56. https://doi.org/10.1007/s11214-022-00924-0

Gilmore MS, Mueller N, Helbert J (2015) VIRTIS emissivity of Alpha Regio, Venus, with implications for Tessera composition. Icarus 254:350–361. https://doi.org/10.1016/j.icarus.2015.04.008

Gilmore M, Treiman A, Helbert J, Smrekar S (2017) Venus surface composition constrained by observation and experiment. Space Sci Rev 212:1511–1540. https://doi.org/10.1007/s11214-017-0370-8

Gilmore MS, Helbert J, Ghail R, Byrne PK, Smrekar SE, Izenberg NR, Carter L, Mueller N, Gerya TV, Ivanov MA, Dyar D (2023) Surface composition and mineralogy of the Venus surface. Space Sci Rev. In preparation

Glaze LS, Wilson CF, Zasova LV, Nakamura M, Limaye S (2018) Future of Venus research and exploration. Space Sci Rev 214:89. https://doi.org/10.1007/s11214-018-0528-z

Gold T (1964) Outgassing processes on the Moon and Venus. In: The Origin and Evolution of Atmospheres and Oceans, p 249

Gold T, Soter S (1969) Atmospheric tides and the resonant rotation of Venus. Icarus 11:356–366. https://doi.org/10.1016/0019-1035(69)90068-2

Goldstein RM (1964) Venus characteristics by Earth-based radar. Astron J 69:12–18. https://doi.org/10.1086/109221

Gough DO (1981) Solar interior structure and luminosity variations. In: Domingo V (ed) Physics of Solar Variations. Springer, Dordrecht, pp 21–34. https://doi.org/10.1007/978-94-010-9633-1_4

Gray C, Byrne PK, Curry S, O'Rourke JG, Royer E (2021) Science on the fly! The importance of Venus flyby observations. Bull Am Astron Soc 53:1–9. https://doi.org/10.3847/25c2cfeb.d488434f

Greeley R, Iversen JD (1985) Wind as a Geologic Process. Cambridge University Press, Cambridge

Grimm RE, Solomon SC (1988) Viscous relaxation of impact crater relief on Venus: constraints on crustal thickness and thermal gradient. J Geophys Res, Solid Earth 93:11911–11929. https://doi.org/10.1029/JB093iB10p11911

Grinspoon DH, Bullock MA (2007) Astrobiology and Venus Exploration. Geophysical Monograph Series, pp 191–206. https://doi.org/10.1029/176GM12

Guest JE, Stofan ER (1999) A new view of the stratigraphic history of Venus. Icarus 139:55–66. https://doi.org/10.1006/icar.1999.6091

Gülcher AJP, Gerya TV, Montési LGJ, Munch J (2020) Corona structures driven by plume–lithosphere interactions and evidence for ongoing plume activity on Venus. Nat Geosci 13:547–554. https://doi.org/10.1038/s41561-020-0606-1

58

Hager BH, O'Connell RJ (1981) A simple global model of plate dynamics and mantle convection. J Geophys Res, Solid Earth 86:4843–4867. https://doi.org/10.1029/JB086iB06p04843

Hallsworth JE, Koop T, Dallas TD, Zorzano MP, Burkhardt J, Golyshina OV, Martín-Torres J, Dymond MK, Ball P, McKay CP (2021) Water activity in Venus's uninhabitable clouds and other planetary atmospheres. Nat Astron 5:665–675. https://doi.org/10.1038/s41550-021-01391-3

Hamano K, Abe Y, Genda H (2013) Emergence of two types of terrestrial planet on solidification of magma ocean. Nature 497:607–610. https://doi.org/10.1038/nature12163

Hansen JE, Hovenier JW (1974) Interpretation of the polarization of Venus. J Atmos Sci 31:1137–1160. https://doi.org/10.1175/1520-0469(1974)031<1137:IOTPOV>2.0.CO;2

Hansen VL, López I (2010) Venus records a rich early history. Geology 38:311–314. https://doi.org/10.1130/G30587.1

Harris LB, Bédard JH (2014) Crustal evolution and deformation in a non-plate-tectonic Archaean Earth: comparisons with Venus. In: Dilek Y, Furnes H (eds) Evolution of Archean Crust and Early Life. Modern Approaches in Solid Earth Sciences, vol 7. Springer, Dordrecht, pp 215–291. https://doi.org/10.1007/978-94-007-7615-9_9

Hauck SA, Phillips RJ, Price MH (1998) Venus: crater distribution and plains resurfacing models. J Geophys Res 103:13635–13642. https://doi.org/10.1029/98JE00400

Hawkesworth C, Cawood P, Kemp T, Storey C, Dhuime B (2009) A matter of preservation. Science 323:49–50. https://doi.org/10.1126/science.1168549

Hawkesworth CJ, Cawood PA, Dhuime B (2020) The evolution of the continental crust and the onset of plate tectonics. Front Earth Sci 8:326. https://doi.org/10.3389/feart.2020.00326

Head JW, Hurwitz DM, Ivanov MA, Basilevsky AT, Kumar PS (2008) Geological mapping of Fortuna Tessera (V-2): Venus and Earth's Archean process comparisons. In: Abstracts of the Annual Meeting of Planetary Geologic Mappers

Herrick RR, Rumpf ME (2011) Postimpact modification by volcanic or tectonic processes as the rule, not the exception, for Venusian craters. J Geophys Res, Planets 116:E02004. https://doi.org/10.1029/2010JE003722

Herrick RR, Stahlke DL, Sharpton VL (2012) Fine-scale Venusian topography from Magellan stereo data. Eos 93:125–126. https://doi.org/10.1029/2012EO120002

Herrick RR, Izenberg NR, Ghail R, Gulcher A, Weller MB, Bjonnes EE, O'Rourke JG, Rolf T, Smrekar SE, Carter L, Mueller N, Davaille A, Gillmann C, Hensley S, Gerya TV, Gilmore MS, Avice G, Ivanov MA (2023) Resurfacing history and volcanic activity of Venus. Space Sci Rev. In revision

Horinouchi T, Hayashi Y, Watanabe S, Yamada M, Yamazaki A, Kouyama T, Taguchi M, Fukuhara T, Takagi M, Ogohara K, Murakami S, Peralta J, Limaye SS, Imamura T, Nakamura M, Sato TM, Satoh T (2020) How waves and turbulence maintain the super-rotation of Venus' atmosphere. Science 368:405–409. https://doi.org/10.1126/science.aaz4439

Hoyle F (1955) Frontiers of Astronomy. Harper & Brothers, New York

Huang J, Yang A, Zhong S (2013) Constraints of the topography, gravity and volcanism on Venusian mantle dynamics and generation of plate tectonics. Earth Planet Sci Lett 362:207–214. https://doi.org/10.1016/j.epsl.2012.11.051

Hunter G (VEXAG) (2019) Venus Technology Plan. https://www.lpi.usra.edu/vexag/documents/reports/VEXAG_Venus_Techplan_2019.pdf

Imamura T, Mitchell J, Lebonnois S, Kaspi Y, Showman AP, Korablev O (2020) Superrotation in planetary atmospheres. Space Sci Rev 216:87. https://doi.org/10.1007/s11214-020-00703-9

Ingersoll AP (1969) The runaway greenhouse: a history of water on Venus. J Atmos Sci 26:1191–1198. https://doi.org/10.1175/1520-0469(1969)026<1191:TRGAHO>2.0.CO;2

Ingersoll AP (2013) Venus: atmospheric evolution. In: Planetary Climates. Princeton University Press, Princeton, pp 7–25

Ivanov MA, Head JW (2013) The history of volcanism on Venus. Planet Space Sci 84:66–92. https://doi.org/10.1016/j.pss.2013.04.018

Ivanov BA, Basilevsky AT, Kryuchokov VP, Chernaya IM (1986) Impact craters of Venus: analysis of Venera 15 and 16 data. J Geophys Res, Solid Earth 91:413–430. https://doi.org/10.1029/JB091iB04p0D413

Jacobson SA, Rubie DC, Hernlund J, Morbidelli A, Nakajima M (2017) Formation, stratification, and mixing of the cores of Earth and Venus. Earth Planet Sci Lett 474:375–386. https://doi.org/10.1016/j.epsl.2017.06.023

James PB, Zuber MT, Phillips RJ (2013) Crustal thickness and support of topography on Venus. J Geophys Res, Planets 118:859–875. https://doi.org/10.1029/2012JE004237

Jiménez-Díaz A, Ruiz J, Kirby JF, Romeo I, Tejero R, Capote R (2015) Lithospheric structure of Venus from gravity and topography. Icarus 260:215–231. https://doi.org/10.1016/j.icarus.2015.07.020

Johnson CL, Sandwell DT (1994) Lithospheric flexure on Venus. Geophys J Int 119:627–647. https://doi.org/10.1111/j.1365-246X.1994.tb00146.x

Johnson CL, Mittelholz A, Langlais B, Russell CT, Ansan V, Banfield D, Chi PJ, Fillingim MO, Forget F, Haviland HF, Golombek M, Joy S, Lognonné P, Liu X, Michaut C, Pan L, Quantin-Nataf C, Spiga A, Stanley S, Thorne SN, Wieczorek MA, Yu Y, Smrekar SE, Banerdt WB (2020) Crustal and time-varying magnetic fields at the InSight landing site on Mars. Nat Geosci 13:199–204. https://doi.org/10.1038/s41561-020-0537-x

Jones DE (1961) The microwave temperature of Venus. Planet Space Sci 5:166–167. https://doi.org/10.1016/0032-0633(61)90094-0

Kahn R (1982) Deducing the age of the dense Venus atmosphere. Icarus 49:71–85. https://doi.org/10.1016/0019-1035(82)90057-4

Kasting JF (1988) Runaway and moist greenhouse atmospheres and the evolution of Earth and Venus. Icarus 74:472–494. https://doi.org/10.1016/0019-1035(88)90116-9

Kaula WM (1999) Constraints on Venus evolution from radiogenic argon. Icarus 139:32–39. https://doi.org/10.1006/icar.1999.6082

Kawabata K, Coffeen DL, Hansen JE, Lane WA, Sato M, Travis LD (1980) Cloud and haze properties from Pioneer Venus polarimetry. J Geophys Res 85:8129. https://doi.org/10.1029/JA085iA13p08129

Kennett BLN, Engdahl ER, Buland R (1995) Constraints on seismic velocities in the Earth from traveltimes. Geophys J Int 122:108–124. https://doi.org/10.1111/j.1365-246X.1995.tb03540.x

Keondzhyan VP, Monin AS (1977) On pole wandering due to continental drift. Dokl Akad Nauk SSSR 233:316–319

Keondzhyan VP, Monin AS (1980) On the concentration convection in the Earth's mantle. Dokl Akad Nauk SSSR 253:78–81

Khawja S, Ernst RE, Samson C, Byrne PK, Ghail RC, MacLellan LM, (2020) Tesserae on Venus may preserve evidence of fluvial erosion. Nat Commun 11:5789. https://doi.org/10.1038/s41467-020-19336-1

King MD, Platnick S, Menzel WP, Ackerman SA, Hubanks PA (2013) Spatial and temporal distribution of clouds observed by MODIS onboard the Terra and Aqua satellites. IEEE Trans Geosci Remote Sens 51:3826–3852. https://doi.org/10.1109/TGRS.2012.2227333

Knollenberg RG, Hunten DM (1980) The microphysics of the clouds of Venus: results of the Pioneer Venus particle size spectrometer experiment. J Geophys Res 85:8039. https://doi.org/10.1029/JA085iA13p08039

Knudsen WC, Banks PM, Miller KL (1982) A new concept of plasma motion and planetary magnetic field for Venus. Geophys Res Lett 9:765–768. https://doi.org/10.1029/GL009i007p00765

Komabayashi M (1968) Conditions for the coexistence of the atmosphere and the oceans. Shizen 23:24–31

Komabayasi M (1967) Discrete equilibrium temperatures of a hypothetical planet with the atmosphere and the hydrosphere of one component-two phase system under constant solar radiation. J Meteorol Soc Japan Ser II 45:137–139. https://doi.org/10.2151/jmsj1965.45.1_137

Konopliv AS, Yoder CF (1996) Venusian K2 tidal love number from Magellan and PVO tracking data. Geophys Res Lett 23:1857–1860. https://doi.org/10.1029/96GL01589

Konopliv AS, Banerdt WB, Sjogren WL (1999) Venus gravity: 180th degree and order model. Icarus 139:3–18. https://doi.org/10.1006/icar.1999.6086

Korenaga J (2012) Plate tectonics and planetary habitability: current status and future challenges. Ann NY Acad Sci 1260:87–94. https://doi.org/10.1111/j.1749-6632.2011.06276.x

Kremic T, Amato M, Gilmore MS, Kiefer WS, Johnson N, Sauder J, Hunter G, Thompson T (2021). Venus surface platform study final report

Krissansen-Totton J, Fortney JJ, Nimmo F (2021) Was Venus ever habitable? Constraints from a coupled interior-atmosphere-redox evolution model. Planet Sci J 2:216. https://doi.org/10.3847/PSJ/ac2580

Künsken D (2020) The House of Styx. Solaris, Oxford

Langlais B, Thébault E, Houliez A, Purucker ME, Lillis RJ (2019) A new model of the crustal magnetic field of Mars using MGS and MAVEN. J Geophys Res, Planets 124:1542–1569. https://doi.org/10.1029/2018JE005854

Lapôtre MGA, O'Rourke JG, Schaefer LK, Siebach KL, Spalding C, Tikoo SM, Wordsworth RD (2020) Probing space to understand Earth. Nat Rev Earth Environ 1:170–181. https://doi.org/10.1038/s43017-020-0029-y

Le Feuvre M, Wieczorek MA (2011) Nonuniform cratering of the Moon and a revised crater chronology of the inner solar system. Icarus 214:1–20. https://doi.org/10.1016/j.icarus.2011.03.010

Le Pichon X (1968) Sea-floor spreading and continental drift. J Geophys Res 73:3661–3697. https://doi.org/10.1029/JB073i012p03661

Lebonnois S, Schubert G (2017) The deep atmosphere of Venus and the possible role of density-driven separation of CO_2 and N_2. Nat Geosci 10:473–477. https://doi.org/10.1038/ngeo2971

Lebrun T, Massol H, Chassefière E, Davaille A, Marcq E, Sarda P, Leblanc F, Brandeis G (2013) Thermal evolution of an early magma ocean in interaction with the atmosphere. J Geophys Res, Planets 118:1155–1176. https://doi.org/10.1002/jgre.20068

Lécuyer C, Simon L, Guyot F (2000) Comparison of carbon, nitrogen and water budgets on Venus and the Earth. Earth Planet Sci Lett 181:33–40. https://doi.org/10.1016/S0012-821X(00)00195-3

Lee YJ, García Muñoz A, Imamura T, Yamada M, Satoh T, Yamazaki A, Watanabe S (2020) Brightness modulations of our nearest terrestrial planet Venus reveal atmospheric super-rotation rather than surface features. Nat Commun 11:1–8. https://doi.org/10.1038/s41467-020-19385-6

Lewis JS (1972) Metal/silicate fractionation in the solar system. Earth Planet Sci Lett 15:286–290. https://doi.org/10.1016/0012-821X(72)90174-4

Limaye SS, Watanabe S, Yamazaki A, Yamada M, Satoh T, Sato TM, Nakamura M, Taguchi M, Fukuhara T, Imamura T, Kouyama T, Lee YJ, Horinouchi T, Peralta J, Iwagami N, Hashimoto GL, Takagi S, Ohtsuki S, Murakami S-y, Yamamoto Y, Ogohara K, Ando H, Sugiyama K-i, Ishii N, Abe T, Hirose C, Suzuki M, Hirata N, Young EF, Ocampo AC (2018) Venus looks different from day to night across wavelengths: morphology from Akatsuki multispectral images. Earth Planets Space 70:24. https://doi.org/10.1186/s40623-018-0789-5

Limaye SS, Mogul R, Baines KH, Bullock MA, Cockell C, Cutts JA, Gentry DiM, Grinspoon DH, Head JW, Jessup KL, Kompanichenko V, Lee YJ, Mathies R, Milojevic T, Pertzborn RA, Rothschild L, Sasaki S, Schulze-Makuch Di, Smith DJ, Way MJ (2021) Venus, an astrobiology target. Astrobiology 21:1163–1185. https://doi.org/10.1089/ast.2020.2268

Lorenz RD, Imai M, Takahashi Y, Sato M, Yamazaki A, Sato TM, Imamura T, Satoh T, Nakamura M (2019) Constraints on Venus lightning from Akatsuki's first 3 years in orbit. Geophys Res Lett 46:7955–7961. https://doi.org/10.1029/2019GL083311

Lourenço DL, Rozel AB, Gerya T, Tackley PJ (2018) Efficient cooling of rocky planets by intrusive magmatism. Nat Geosci 11:322–327. https://doi.org/10.1038/s41561-018-0094-8

Lourenço DL, Rozel AB, Ballmer MD, Tackley PJ (2020) Plutonic-squishy lid: a new global tectonic regime generated by intrusive magmatism on Earth-like planets. Geochem Geophys Geosyst 21:e2019GC008756. https://doi.org/10.1029/2019GC008756

Luhmann JG, Ma YJ, Villarreal MN, Wei HY, Zhang TL (2015) The Venus-solar wind interaction: is it purely ionospheric? Planet Space Sci 119:36–42. https://doi.org/10.1016/j.pss.2015.09.012

Lukco D, Spry DJ, Harvey RP, Costa GCC, Okojie RS, Avishai A, Nakley LM, Neudeck PG, Hunter GW (2018) Chemical analysis of materials exposed to Venus temperature and surface atmosphere. Earth Space Sci 5:270–284. https://doi.org/10.1029/2018EA000355

Machetel P, Rabinowicz M, Bernardet P (1986) Three-dimensional convection in spherical shells. Geophys Astrophys Fluid Dyn 37:57–84. https://doi.org/10.1080/03091928608210091

Maia JS, Wieczorek MA (2022) Lithospheric structure of Venusian crustal plateaus. J Geophys Res, Planets 127:1–27. https://doi.org/10.1029/2021JE007004

Malaspina DM, Goodrich K, Livi R, Halekas J, McManus M, Curry S, Bale SD, Bonnell JW, de Wit TD, Goetz K, Harvey PR, MacDowall RJ, Pulupa M, Case AW, Kasper JC, Korreck KE, Larson D, Stevens ML, Whittlesey P (2020) Plasma double layers at the boundary between Venus and the solar wind. Geophys Res Lett 47:1–9. https://doi.org/10.1029/2020GL090115

Mangano V, Dósa M, Fränz M, Milillo A, Oliveira JS, Lee YJ, McKenna-Lawlor S, Grassi D, Heyner D, Kozyrev AS, Peron R, Helbert J, Besse S, de la Fuente S, Montagnon E, Zender J, Volwerk M, Chaufray JY, Slavin JA, Krüger H, Maturilli A, Cornet T, Iwai K, Miyoshi Y, Lucente M, Massetti S, Schmidt CA, Dong C, Quarati F, Hirai T, Varsani A, Belyaev D, Zhong J, Kilpua EKJ, Jackson BV, Odstrcil D, Plaschke F, Vainio R, Jarvinen R, Ivanovski SL, Madár Á, Erdős G, Plainaki C, Alberti T, Aizawa S, Benkhoff J, Murakami G, Quemerais E, Hiesinger H, Mitrofanov IG, Iess L, Santoli F, Orsini S, Lichtenegger H, Laky G, Barabash S, Moissl R, Huovelin J, Kasaba Y, Saito Y, Kobayashi M, Baumjohann W (2021) BepiColombo science investigations during cruise and flybys at the Earth, Venus and Mercury. Space Sci Rev. https://doi.org/10.1007/s11214-021-00797-9

Marcq E, Bertaux JL, Montmessin F, Belyaev D (2013) Variations of sulphur dioxide at the cloud top of Venus's dynamic atmosphere. Nat Geosci 6:25–28. https://doi.org/10.1038/ngeo1650

Marcq E, Salvador A, Massol H, Davaille A (2017) Thermal radiation of magma ocean planets using a 1-D radiative-convective model of H_2O-CO_2 atmospheres. J Geophys Res, Planets 122:1539–1553. https://doi.org/10.1002/2016JE005224

Marcq E, Mills FP, Parkinson CD, Vandaele AC (2018) Composition and chemistry of the neutral atmosphere of Venus. Space Sci Rev 214:10. https://doi.org/10.1007/s11214-017-0438-5

Marcq E, Lea Jessup K, Baggio L, Encrenaz T, Lee YJ, Montmessin F, Belyaev D, Korablev O, Bertaux JL (2020) Climatology of SO_2 and UV absorber at Venus' cloud top from SPICAV-UV Nadir dataset. Icarus 335:113368. https://doi.org/10.1016/j.icarus.2019.07.002

Margot JL, Campbell DB, Giorgini JD, Jao JS, Snedeker LG, Ghigo FD, Bonsall A (2021) Spin state and moment of inertia of Venus. Nat Astron 5:676–683. https://doi.org/10.1038/s41550-021-01339-7

Marov MYA (2005) Mikhail Lomonosov and the discovery of the atmosphere of Venus during the 1761 transit. In: Proceedings of the International Astronomical Union. Cambridge University Press, Cambridge, pp 209–219. https://doi.org/10.1017/S1743921305001390. 2004(IAUC196)

Marov MYA, Lystsev VE, Lebedev VN, Lukashevich NL, Shari VP (1980) The structure and microphysical properties of the Venus clouds: Venera 9, 10, and 11 data. Icarus 44:608–639. https://doi.org/10.1016/0019-1035(80)90131-1

Marov MYA, Avduevsky VS, Kerzhanovich VV, Rozhdestevensky MK, Borodin NF, Ryabov OL (1973) Venera 8: measurements of temperature, pressure and wind velocity on the illuminated side of Venus. J Atmos Sci 30:1210–1214. https://doi.org/10.1175/1520-0469(1973)030<1210:VMOTPA>2.0.CO;2

Matsui T, Abe Y (1986) Evolution of an impact-induced atmosphere and magma ocean on the accreting Earth. Nature 319:303–305. https://doi.org/10.1038/319303a0

Mayer CH, McCullough TP, Sloanaker RM (1958) Observations of Venus at 3.15-cm wave length. Astrophys J 127:1. https://doi.org/10.1086/146433

McGovern PJ, Rumpf ME, Zimbelman JR (2013) The influence of lithospheric flexure on magma ascent at large volcanoes on Venus. J Geophys Res, Planets 118:2423–2437. https://doi.org/10.1002/2013JE004455

McKenzie DP, Parker RL (1967) The North Pacific: an example of tectonics on a sphere. Nature 216:1276–1280. https://doi.org/10.1038/2161276a0

McKinnon WB, Zahnle KJ, Ivanov BA, Melosh HJ (1997) Cratering on Venus: models and observations. In: Bougher SW, Hunten DM, Phillips RJ (eds) Venus II. University of Arizona Press, Tucson, pp 969–1014

Menzel DH, Whipple FL (1954) The case for H_2O clouds on Venus. Astron J 59:329. https://doi.org/10.1086/107037

Mills FP, Allen M (2007) A review of selected issues concerning the chemistry in Venus' middle atmosphere. Planet Space Sci 55:1729–1740. https://doi.org/10.1016/j.pss.2007.01.012

Minear JW, Toksöz MN (1970) Thermal regime of a downgoing slab and new global tectonics. J Geophys Res 75:1397–1419. https://doi.org/10.1029/JB075i008p01397

Mocquet A, Rosenblatt P, Dehant V, Verhoeven O (2011) The deep interior of Venus, Mars, and the Earth: a brief review and the need for planetary surface-based measurements. Planet Space Sci 59:1048–1061. https://doi.org/10.1016/j.pss.2010.02.002

Moore WB, Schubert G (1997) Venusian crustal and lithospheric properties from nonlinear regressions of highland geoid and topography. Icarus 128:415–428. https://doi.org/10.1006/icar.1997.5750

Moresi L, Solomatov V (1998) Mantle convection with a brittle lithosphere: thoughts on the global tectonic styles of the Earth and Venus. Geophys J Int 133:669–682. https://doi.org/10.1046/j.1365-246X.1998.00521.x

Morgan WJ (1968) Rises, trenches, great faults, and crustal blocks. J Geophys Res 73:1959–1982. https://doi.org/10.1029/JB073i006p01959

Morgan JW, Anders E (1980) Chemical composition of Earth, Venus, and Mercury. Proc Natl Acad Sci 77:6973–6977. https://doi.org/10.1073/pnas.77.12.6973

Moroz VI (1983) Summary of preliminary results of the Venera 13 and Venera 14 missions. In: Hunten DM, Colin L, Donahue TM, Moroz VI (eds) Venus. University of Arizona Press, Tucson, pp 45–68

Mueller N, Helbert J, Hashimoto GL, Tsang CCC, Erard S, Piccioni G, Drossart P (2009) Venus surface thermal emission at 1 μm in VIRTIS imaging observations: evidence for variation of crust and mantle differentiation conditions. J Geophys Res, Planets 114:1–21. https://doi.org/10.1029/2008JE003118

Mueller NT, Smrekar SE, Tsang CCC (2020) Multispectral surface emissivity from VIRTIS on Venus Express. Icarus 335:113400. https://doi.org/10.1016/j.icarus.2019.113400

Muir T (2019) Gideon the Ninth. Tom Doherty Associates, New York

Nakajima S, Hayashi Y-Y, Abe Y (1992) A study on the "Runaway Greenhouse Effect" with a one-dimensional radiative–convective equilibrium model. J Atmos Sci 49:2256–2266. https://doi.org/10.1175/1520-0469(1992)049<2256:ASOTGE>2.0.CO;2

Nakamura M, Imamura T, Ishii N, Abe T, Satoh T, Suzuki M, Ueno M, Yamazaki A, Iwagami N, Watanabe S, Taguchi M, Fukuhara T, Takahashi Y, Yamada M, Hoshino N, Ohtsuki S, Uemizu K, Hashimoto GL, Takagi M, Matsuda Y, Ogohara K, Sato N, Kasaba Y, Kouyama T, Hirata N, Nakamura R, Yamamoto Y, Okada N, Horinouchi T, Yamamoto M, Hayashi Y (2011) Overview of Venus orbiter Akatsuki. Earth Planets Space 63:443–457. https://doi.org/10.5047/eps.2011.02.009

Nakamura M, Kawakatsu Y, Hirose C, Imamura T, Ishii N, Abe T, Yamazaki A, Yamada M, Ogohara K, Uemizu K, Fukuhara T, Ohtsuki S, Satoh T, Suzuki M, Ueno M, Nakatsuka J, Iwagami N, Taguchi M, Watanabe S, Takahashi Y, Hashimoto GL, Yamamoto H (2014) Return to Venus of the Japanese Venus climate orbiter AKATSUKI. Acta Astronaut 93:384–389. https://doi.org/10.1016/j.actaastro.2013.07.027

Nakamura M, Imamura T, Ishii N, Abe T, Kawakatsu Y, Hirose C, Satoh T, Suzuki M, Ueno M, Yamazaki A, Iwagami N, Watanabe S, Taguchi M, Fukuhara T, Takahashi Y, Yamada M, Imai M, Ohtsuki S, Uemizu K, Hashimoto GL, Takagi M, Matsuda Y, Ogohara K, Sato N, Kasaba Y, Kouyama T, Hirata N, Nakamura R, Yamamoto Y, Horinouchi T, Yamamoto M, Hayashi YY, Kashimura H, Sugiyama KI, Sakanoi T, Ando H, Murakami SY, Sato TM, Takagi S, Nakajima K, Peralta J, Lee YJ, Nakatsuka J,

62

Ichikawa T, Inoue K, Toda T, Toyota H, Tachikawa S, Narita S, Hayashiyama T, Hasegawa A, Kamata Y (2016) AKATSUKI returns to Venus. Earth Planets Space 68:75. https://doi.org/10.1186/s40623-016-0457-6

National Academies of Sciences, Engineering, and Medicine (2011) Vision and Voyages for Planetary Science in the Decade 2013–2022. National Academies Press, Washington. https://doi.org/10.17226/13117

National Academies of Sciences, Engineering, and Medicine (2022) Origins, Worlds, and Life: A Decadal Strategy for Planetary Science and Astrobiology 2023–2032. National Academies Press, Washington. https://doi.org/10.17226/26522

Navarro T, Schubert G, Lebonnois S (2018) Atmospheric mountain wave generation on Venus and its influence on the solid planet's rotation rate. Nat Geosci 11:487–491. https://doi.org/10.1038/s41561-018-0157-x

Nimmo F (2002) Why does Venus lack a magnetic field? Geology 30:987. https://doi.org/10.1130/0091-7613(2002)030<0987:WDVLAM>2.0.CO;2

Nimmo F, McKenzie D (1998) Volcanism and tectonics on Venus. Annu Rev Earth Planet Sci 26:23–51. https://doi.org/10.1146/annurev.earth.26.1.23

Nimmo F, Stevenson DJ (2000) Influence of early plate tectonics on the thermal evolution and magnetic field of Mars. J Geophys Res, Planets 105:11969–11979. https://doi.org/10.1029/1999JE001216

Oertel D, Spänkuch D, Jahn H, Becker-Ross H, Stadthaus W, Nopirakowski J, Döhler W, Schäfer K, Güldner J, Dubois R, Moroz VI, Linkin VM, Kerzhanovich VV, Matsgorin IA, Lipatov AN, Shurupov AA, Zasova LV, Ustinov EA (1985) Infrared spectrometry of Venus from "Venera-15" and "Venera-16". Adv Space Res 5:25–36. https://doi.org/10.1016/0273-1177(85)90267-4

O'Neill C, Lenardic A, Moresi L, Torsvik TH, Lee CTA (2007) Episodic precambrian subduction. Earth Planet Sci Lett 262:552–562. https://doi.org/10.1016/j.epsl.2007.04.056

Öpik EJ (1956) The surface conditions on Venus. Ir Astron J 4:37–48.

Öpik EJ (1961) The aeolosphere and atmosphere of Venus. J Geophys Res 66:2807–2819. https://doi.org/10.1029/JZ066i009p02807

O'Rourke JG (2020) Venus: a thick basal magma ocean may exist today. Geophys Res Lett 47:e2019GL086126. https://doi.org/10.1029/2019GL086126

O'Rourke JG, Korenaga J (2015) Thermal evolution of Venus with argon degassing. Icarus 260:128–140. https://doi.org/10.1016/j.icarus.2015.07.009

O'Rourke JG, Gillmann C, Tackley P (2018) Prospects for an ancient dynamo and modern crustal remanent magnetism on Venus. Earth Planet Sci Lett 502:46–56. https://doi.org/10.1016/j.epsl.2018.08.055

O'Rourke JG (VEXAG) (2019a) Venus Goals, Objectives, and Investigations. https://www.lpi.usra.edu/vexag/documents/reports/VEXAG_Venus_GOI_2019.pdf

O'Rourke JG, Buz J, Fu RR, Lillis RJ (2019b) Detectability of remanent magnetism in the crust of Venus. Geophys Res Lett 46:5768–5777. https://doi.org/10.1029/2019GL082725

O'Rourke JG et al (2021) ADVENTS: assessment and discovery of Venus' past evolution and near-term climatic and geophysical state. Mission Concept Study Report to the NRC Planetary Science and Astrobiology Decadal Survey 2023–2032. NASA Goddard Space Flight Center, Green Bank, Maryland. https://tinyurl.com/2p88fx4f

Orth CP, Solomatov VS (2011) The isostatic stagnant lid approximation and global variations in the Venusian lithospheric thickness. Geochem Geophys Geosyst 12:1–17. https://doi.org/10.1029/2011GC003582

Pätzold M, Häusler B, Bird MK, Tellmann S, Mattei R, Asmar SW, Dehant V, Eidel W, Imamura T, Simpson RA, Tyler GL (2007) The structure of Venus' middle atmosphere and ionosphere. Nature 450:657–660. https://doi.org/10.1038/nature06239

Peplowski PN, Lawrence DJ, Wilson JT (2020) Chemically distinct regions of Venus's atmosphere revealed by measured N_2 concentrations. Nat Astron 4:947–950. https://doi.org/10.1038/s41550-020-1079-2

Peralta J, Hueso R, Sánchez-Lavega A, Lee YJ, Munõz AG, Kouyama T, Sagawa H, Sato TM, Piccioni G, Tellmann S, Imamura T, Satoh T (2017) Stationary waves and slowly moving features in the night upper clouds of Venus. Nat Astron 1:1–5. https://doi.org/10.1038/s41550-017-0187

Peralta J, Iwagami N, Sánchez-Lavega A, Lee YJ, Hueso R, Narita M, Imamura T, Miles P, Wesley A, Kardasis E, Takagi S (2019a) Morphology and dynamics of Venus's middle clouds with Akatsuki/IR1. Geophys Res Lett 46:2399–2407. https://doi.org/10.1029/2018GL081670

Peralta J, Sánchez-Lavega A, Horinouchi T, McGouldrick K, Garate-Lopez I, Young EF, Bullock MA, Lee YJ, Imamura T, Satoh T, Limaye SS (2019b) New cloud morphologies discovered on the Venus's night during Akatsuki. Icarus 333:177–182. https://doi.org/10.1016/j.icarus.2019.05.026

Peralta J, Navarro T, Vun CW, Sánchez-Lavega A, McGouldrick K, Horinouchi T, Imamura T, Hueso R, Boyd JP, Schubert G, Kouyama T, Satoh T, Iwagami N, Young EF, Bullock MA, Machado P, Lee YJ, Limaye SS, Nakamura M, Tellmann S, Wesley A, Miles P (2020) A long-lived sharp disruption on the lower clouds of Venus. Geophys Res Lett 47:1–10. https://doi.org/10.1029/2020GL087221

Perchuk AL, Safonov OG, Smit CA, van Reenen DD, Zakharov VS, Gerya TV (2018) Precambrian ultra-hot orogenic factory: making and reworking of continental crust. Tectonophysics 746:572–586. https://doi.org/10.1016/j.tecto.2016.11.041

Perchuk AL, Zakharov VS, Gerya TV, Brown M (2019) Hotter mantle but colder subduction in the Precambrian: what are the implications? Precambrian Res 330:20–34. https://doi.org/10.1016/j.precamres.2019.04.023

Perchuk AL, Gerya TV, Zakharov VS, Griffin WL (2020) Building cratonic keels in Precambrian plate tectonics. Nature 586:395–401. https://doi.org/10.1038/s41586-020-2806-7

Pérez-Hoyos S, Sánchez-Lavega A, García-Muñoz A, Irwin PGJ, Peralta J, Holsclaw G, McClintock WM, Sanz-Requena JF (2018) Venus upper clouds and the UV absorber from MESSENGER/MASCS observations. J Geophys Res, Planets 123:145–162. https://doi.org/10.1002/2017JE005406

Persson M, Futaana Y, Fedorov A, Nilsson H, Hamrin M, Barabash S (2018) H^+/O^+ escape rate ratio in the Venus magnetotail and its dependence on the solar cycle. Geophys Res Lett 45:10,805–10,811. https://doi.org/10.1029/2018GL079454

Petit G, Luzum B (2010) IERS convections. IERS Technical Note No. 36. Verlag des Bundesamts fur Kartographie und Geodasie, Frankfurt an Main, Germany

Pettengill GH, Briscoe HW, Evans JV, Gehrels E, Hyde GM, Kraft LG, Price R, Smith WB (1962) A radar investigation of Venus. Astron J 67:181–190. https://doi.org/10.1086/108692

Pettengill GH, Ford PG, Johnson WTK, Raney RK, Soderblom LA (1991) Magellan: radar performance and data products. Science 252:260–265. https://doi.org/10.1126/science.252.5003.260

Pettit E, Nicholson SB (1955) Temperatures on the bright and dark sides of Venus. Publ Astron Soc Pac 67:293. https://doi.org/10.1086/126823

Phillips JL, Russell CT (1987) Upper limit on the intrinsic magnetic field of Venus. J Geophys Res 92:2253. https://doi.org/10.1029/ja092ia03p02253

Phillips RJ, Arvidson RE, Boyce JM, Campbell DB, Guest JE, Schaber GG, Soderblom LA (1991) Impact craters on Venus: initial analysis from Magellan. Science 252:288–297. https://doi.org/10.1126/science.252.5003.288

Plass GN (1961) The influence of carbon dioxide variations on the atmospheric heat balance. Tellus 13:296–300

Pluriel W, Marcq E, Turbet M (2019) Modeling the albedo of Earth-like magma ocean planets with H_2O-CO_2 atmospheres. Icarus 317:583–590. https://doi.org/10.1016/j.Icarus.2018.08.023

Pollack JB (1971) A nongrey calculation of the runaway greenhouse: implications for Venus' past and present. Icarus 14:295–306. https://doi.org/10.1016/0019-1035(71)90001-7

Pratt WE (1956) Oil according to Hoyle. Am Assoc Pet Geol Bull 40:177–179. https://doi.org/10.1306/5ceae319-16bb-11d7-8645000102c1865d

Pulupa M, Bale SD, Curry SM, Farrell WM, Goodrich KA, Goetz K, Harvey PR, Malaspina DM, Raouafi NE (2021) Non-detection of lightning during the Second Parker Solar Probe Venus gravity assist. Geophys Res Lett 48:1–7. https://doi.org/10.1029/2020GL091751

Ramberg H (1967) Gravity, Deformation, and the Earth's Crust. Academic Press, London

Ranalli G (2001) Experimental tectonics: from Sir James Hall to the present. J Geodyn 32:65–76. https://doi.org/10.1016/S0264-3707(01)00023-0

Rea DG (1972) Composition of the upper clouds of Venus. Rev Geophys 10:369. https://doi.org/10.1029/RG010i001p00369

Read PL, Lebonnois S (2018) Superrotation on Venus, on Titan, and elsewhere. Annu Rev Earth Planet Sci 46:175–202. https://doi.org/10.1146/annurev-earth-082517-010137

Reese CC, Solomatov VS, Moresi LN (1999) Non-Newtonian stagnant lid convection and magmatic resurfacing on Venus. Icarus 139:67–80. https://doi.org/10.1006/icar.1999.6088

Rey PF, Coltice N, Flament N (2014) Spreading continents kick-started plate tectonics. Nature 513:405–408. https://doi.org/10.1038/nature13728

Ricard Y, Richards M, Lithgow-Bertelloni C, Le Stunff Y (1993) A geodynamic model of mantle density heterogeneity. J Geophys Res, Solid Earth 98:21895–21909. https://doi.org/10.1029/93JB02216

Rimmer PB, Jordan S, Constantinou T, Woitke P, Shorttle O, Hobbs R, Paschodimas A (2021) Hydroxide salts in the clouds of Venus: their effect on the sulfur cycle and cloud droplet pH. Planet Sci J 2:133. https://doi.org/10.3847/PSJ/ac0156

Ringwood AE, Anderson DL (1977) Earth and Venus: a comparative study. Icarus 30:243–253. https://doi.org/10.1016/0019-1035(77)90156-7

Roberts JA (1963) Radio emission from the planets. Planet Space Sci 11:221–259. https://doi.org/10.1016/0032-0633(63)90026-6

Rolf T, Weller MB, Gülcher A, Byrne PK, O'Rourke JG, Herrick RR, Bjonnes EE, Davaille A, Ghail R, Gillmann C, Plesa A-C, Smrekar SE (2022) Dynamics and evolution of Venus' mantle through time. Space Sci Rev 218:70. https://doi.org/10.1007/s11214-022-00937-9

64

 Springer

Rosenblatt P, Dumoulin C, Marty JC, Genova A (2021) Determination of Venus' interior structure with En-Vision. Remote Sens 13:1624. https://doi.org/10.3390/rs13091624

Rozel AB, Golabek GJ, Jain C, Tackley PJ, Gerya T (2017) Continental crust formation on early Earth controlled by intrusive magmatism. Nature 545:332–335. https://doi.org/10.1038/nature22042

Rubie DC, Jacobson SA, Morbidelli A, O'Brien DP, Young ED, de Vries J, Nimmo F, Palme H, Frost DJ (2015) Accretion and differentiation of the terrestrial planets with implications for the compositions of early-formed solar system bodies and accretion of water. Icarus 248:89–108. https://doi.org/10.1016/j.icarus.2014.10.015

Russell CT (1976) The magnetic moment of Venus: Venera-4 measurements reinterpreted. Geophys Res Lett 3:125–128. https://doi.org/10.1029/GL003i003p00125

Russell CT (1993) Magnetic fields of the terrestrial planets. J Geophys Res 98:18681–18695. https://doi.org/10.1029/93je00981

Russell MB, Johnson CL (2021) Evidence for a locally thinned lithosphere associated with recent volcanism at Aramaiti Corona, Venus. J Geophys Res, Planets 126:1–19. https://doi.org/10.1029/2020JE006783

Russell CT, Zhang TL, Delva M, Magnes W, Strangeway RJ, Wei HY (2007) Lightning on Venus inferred from whistler-mode waves in the ionosphere. Nature 450:661–662. https://doi.org/10.1038/nature05930

Sagan C (1960) The surface temperature of Venus. Astron J 65:352. https://doi.org/10.1086/108265

Sagan C (1961) The planet Venus. Science 133:849–858. https://doi.org/10.1126/science.133.3456.849

Sagdeev RZ, Linkin VM, Kerzhanovich VV, Lipatov AN, Shurupov AA, Blamont JE, Crisp D, Ingersoll AP, Elson LS, Preston RA, Hildebrand CE, Ragent B, Seiff A, Young RE, Petit G, Boloh L, Alexandrov YN, Armand NA, Bakitko RV, Selivanov AS (1986) Overview of VEGA Venus balloon in situ meteorological measurements. Science 231:1411–1414. https://doi.org/10.1126/science.231.4744.1411

Salvador A, Avice G, Breuer D, Gillmann C, Jacobson SA, Lammer H, Marcq E, Raymond SN, Sakuraba H, Scherf M, Way MJ (2023) Water and the early atmosphere of Venus. Space Sci Rev. In preparation

Saunders RS, Pettengill GH, Arvidson RE, Sjogren WL, Johnson WTK, Pieri L (1990) The Magellan Venus radar mapping mission. J Geophys Res 95:8339–8355. https://doi.org/10.1029/JB095iB06p08339

Saunders RS, Spear AJ, Allin PC, Austin RS, Berman AL, Chandlee RC, Clark J, Decharon AV, De Jong EM, Griffith DG, Gunn JM, Hensley S, Johnson WTK, Kirby CE, Leung KS, Lyons DT, Michaels GA, Miller J, Morris RB, Morrison AD, Piereson RG, Scott JF, Shaffer SJ, Slonski JP, Stofan ER, Thompson TW, Wall SD (1992) Magellan mission summary. J Geophys Res 97:13067. https://doi.org/10.1029/92JE01397

Schaber GG, Strom RG, Moore HJ, Soderblom LA, Kirk RL, Chadwick DJ, Dawson DD, Gaddis LR, Boyce JM, Russell J (1992) Geology and distribution of impact craters on Venus: what are they telling us? J Geophys Res 97:13257. https://doi.org/10.1029/92JE01246

Schubert G, Covey C, Del Genio A, Elson LS, Keating G, Seiff A, Young RE, Apt J, Counselman CC, Kliore AJ, Limaye SS, Revercomb HE, Sromovsky LA, Suomi VE, Taylor F, Woo R, von Zahn U (1980) Structure and circulation of the Venus atmosphere. J Geophys Res 85:8007. https://doi.org/10.1029/JA085iA13p08007

Seager S, Petkowski JJ, Gao P, Bains W, Bryan NC, Ranjan S, Greaves J (2021) The Venusian lower atmosphere haze as a depot for desiccated microbial life: a proposed life cycle for persistence of the Venusian aerial biosphere. Astrobiology 21:1206–1223. https://doi.org/10.1089/ast.2020.2244

Seiff A, Schofield JT, Kliore AJ, Taylor FW, Limaye SS, Revercomb HE, Sromovsky LA, Kerzhanovich VV, Moroz VI, Marov MY (1985) Models of the structure of the atmosphere of Venus from the surface to 100 kilometers altitude. Adv Space Res 5:3–58. https://doi.org/10.1016/0273-1177(85)90197-8

Shah O, Helled R, Alibert Y, Mezger K (2022) Possible chemical composition and interior structure models of Venus inferred from numerical modelling. Astrophys J 926:217. https://doi.org/10.3847/1538-4357/ac410d

Shalygin EV, Markiewicz WJ, Basilevsky AT, Titov DV, Ignatiev NI, Head JW (2015) Active volcanism on Venus in the Ganiki Chasma rift zone. Geophys Res Lett 42:4762–4769. https://doi.org/10.1002/2015GL064088

Sheehan W, Brasch K, Cruikshank D, Baum R (2014) The ashen light of Venus: the oldest unsolved solar system mystery. J Br Astron Assoc 124:209–215

Simon JL, Bretagnon P, Chapront J, Chapront-Touze M, Francou G, Laskar J (1994) Numerical expressions for precession formulae and mean elements for the Moon and the planets. Astron Astrophys 282:663–683

Simpson GC (1927) The mechanism of a thunderstorm. Proc R Soc Lond Ser A 114:376–401. https://doi.org/10.1098/rspa.1927.0048

Sizova E, Gerya T, Brown M, Perchuk LL (2010) Subduction styles in the Precambrian: insight from numerical experiments. Lithos 116:209–229. https://doi.org/10.1016/j.lithos.2009.05.028

Sizova E, Gerya T, Stüwe K, Brown M (2015) Generation of felsic crust in the Archean: a geodynamic modeling perspective. Precambrian Res 271:198–224. https://doi.org/10.1016/j.precamres.2015.10.005

Sleep NH (2000) Evolution of the mode of convection within terrestrial planets. J Geophys Res, Planets 105:17563–17578. https://doi.org/10.1029/2000JE001240

Smrekar SE, Stofan ER (1997) Corona formation and heat loss on Venus by coupled upwelling and delamination. Science 277:1289–1294. https://doi.org/10.1126/science.277.5330.1289

Smrekar SE, Elkins-Tanton L, Leitner JJ, Lenardic A, Mackwell S, Moresi L, Sotin C, Stofan ER (2007) Tectonic and thermal evolution of Venus and the role of volatiles: implications for understanding the terrestrial planets. In: Esposito LW, Stofan ER, Cravens TE (eds) Exploring Venus as a Terrestrial Planet, pp 45–71. https://doi.org/10.1029/176GM05

Smrekar SE, Stofan ER, Mueller N, Treiman A, Elkins-Tanton L, Helbert J, Piccioni G, Drossart P (2010) Recent hotspot volcanism on Venus from VIRTIS emissivity data. Science 328:605–608. https://doi.org/10.1126/science.1186785

Smrekar SE, Davaille A, Sotin C (2018) Venus interior structure and dynamics. Space Sci Rev 214:88. https://doi.org/10.1007/s11214-018-0518-1

Smrekar SE, Ostberg C, O'Rourke JG (2022a) Earth-like lithospheric thickness and heat flow on Venus consistent with active rifting. Nat Geosci. https://doi.org/10.1038/s41561-022-01068-0

Smrekar SE, Hensley S, Nybakken R, Wallace MS, Perkovic-Martin D, You T-H, Nunes D, Brophy J, Ely T, Burt E, Dyar MD, Helbert J, Miller B, Hartley J, Kallemeyn P, Whitten J, Iess L, Mastrogiuseppe M, Younis M, Prats P, Rodriguez M, Mazarico R (2022b) VERITAS (Venus Emissivity, Radio Science, InSAR, Topography, and Spectroscopy): a discovery mission. In: 2022 IEEE Aerospace Conference (AERO), pp 1–20. https://doi.org/10.1109/AERO53065.2022.9843269

Sobolev SV, Brown M (2019) Surface erosion events controlled the evolution of plate tectonics on Earth. Nature 570:52–57. https://doi.org/10.1038/s41586-019-1258-4

Sonett C (1963) A summary review of the scientific findings of the Mariner Venus mission. Space Sci Rev 2:751–777. https://doi.org/10.1007/BF00208814

St. John CE, Nicholson SB (1922) The absence of oxygen and water-vapor lines in the spectrum of Venus. Astrophys J 56:380. https://doi.org/10.1086/142712

Stenborg G, Gallagher B, Howard RA, Hess P, Raouafi NE (2021) Pristine PSP/WISPR observations of the circumsolar dust ring near Venus's orbit. Astrophys J 910:157. https://doi.org/10.3847/1538-4357/abe623

Stern RJ, Gerya T (2018) Subduction initiation in nature and models: a review. Tectonophysics 746:173–198. https://doi.org/10.1016/j.tecto.2017.10.014

Stevenson DJ (2003) Planetary magnetic fields. Earth Planet Sci Lett 208:1–11. https://doi.org/10.1016/S0012-821X(02)01126-3

Stevenson DJ (2010) Planetary magnetic fields: achievements and prospects. Space Sci Rev 152:651–664. https://doi.org/10.1007/s11214-009-9572-z

Stevenson DJ, Spohn T, Schubert G (1983) Magnetism and thermal evolution of the terrestrial planets. Icarus 54:466–489. https://doi.org/10.1016/0019-1035(83)90241-5

Stewart AI, Anderson DE, Esposito LW, Barth CA (1979) Ultraviolet spectroscopy of Venus: initial results from the Pioneer Venus orbiter. Science 203:777–779. https://doi.org/10.1126/science.203.4382.777

Stirone S (2021) Mars Is a Hellhole. The Atlantic

Stofan ER, Sharpton VL, Schubert G, Baer G, Bindschadler DL, Janes DM, Squyres SW (1992) Global distribution and characteristics of coronae and related features on Venus: implications for origin and relation to mantle processes. J Geophys Res 97:13347. https://doi.org/10.1029/92JE01314

Strom RG, Schaber GG, Dawson DD (1994) The global resurfacing of Venus. J Geophys Res 99:10899–10926. https://doi.org/10.1029/94je00388

Surkov YA (1983) Studies of Venus rocks by Veneras 8, 9, and 10. In: Hunten DM, Colin L, Donahue TM, Moroz VI (eds) Venus. University of Arizona Press, Tucson, pp 154–158

Svedhem H, Titov DV, McCoy D, Lebreton JP, Barabash S, Bertaux JL, Drossart P, Formisano V, Häusler B, Korablev O, Markiewicz WJ, Nevejans D, Pätzold M, Piccioni G, Zhang TL, Taylor FW, Lellouch E, Koschny D, Witasse O, Eggel H, Warhaut M, Accomazzo A, Rodriguez-Canabal J, Fabrega J, Schirmann T, Clochet A, Coradini M (2007a) Venus Express – the first European mission to Venus. Planet Space Sci 55:1636–1652. https://doi.org/10.1016/j.pss.2007.01.013

Svedhem H, Titov DV, Taylor FW, Witasse O (2007b) Venus as a more Earth-like planet. Nature 450:629–632. https://doi.org/10.1038/nature06432

Tackley PJ (1998) Self-consistent generation of tectonic plates in three-dimensional mantle convection. Earth Planet Sci Lett 157:9–22. https://doi.org/10.1016/S0012-821X(98)00029-6

Tanaka KL, Schaber GG, Chapman MG, Stofan ER, Campbell DB, Davis PA, Guest JE, McGill GE, Rogers PG, Saunders RS, Zimbelman JR (1993) The Venus Geologic Mappers' Handbook. USGS Open-File Report 93-516

Tarduno JA, Cottrell RD, Bono RK, Oda H, Davis WJ, Fayek M, van't Erve O, Nimmo F, Huang W, Thern ER, Fearn S, Mitra G, Smirnov AV, Blackman EG (2020) Paleomagnetism indicates that primary magnetite

66

in zircon records a strong Hadean geodynamo. Proc Natl Acad Sci USA 117:2309–2318. https://doi.org/10.1073/pnas.1916553117

Tauber ME, Kirk DB (1976) Impact craters on Venus. Icarus 28:351–357. https://doi.org/10.1016/0019-1035(76)90148-2

Taylor FW, Crisp D, Bezard B (1997) Near-infrared sounding of the lower atmosphere of Venus. In: Bougher SW, Hunten DM, Phillips RJ (eds) Venus II. University of Arizona Press, Tucson, pp 325–352

Taylor FW, Svedhem H, Head JW (2018) Venus: the atmosphere, climate, surface, interior and near-space environment of an Earth-like planet. Space Sci Rev 214:35. https://doi.org/10.1007/s11214-018-0467-8

Tesauro M, Kaban MK, Cloetingh SAPL (2012) Global strength and elastic thickness of the lithosphere. Glob Planet Change 90–91:51–57. https://doi.org/10.1016/j.gloplacha.2011.12.003

Titov DV, Ignatiev NI, McGouldrick K, Wilquet V, Wilson CF (2018) Clouds and hazes of Venus. Space Sci Rev 214:126. https://doi.org/10.1007/s11214-018-0552-z

Torrance KE, Turcotte DL (1971) Structure of convection cells in the mantle. J Geophys Res 76:1154–1161. https://doi.org/10.1029/JB076i005p01154

Treiman AH (2007) Geochemistry of Venus' surface: current limitations as future opportunities. In: Exploring Venus as a Terrestrial Planet, pp 7–22. https://doi.org/10.1029/176GM03

Trompert R, Hansen U (1998) Mantle convection simulations with rheologies that generate plate-like behaviour. Nature 395:686–689. https://doi.org/10.1038/27185

Trønnes RG, Baron MA, Eigenmann KR, Guren MG, Heyn BH, Løken A, Mohn CE (2019) Core formation, mantle differentiation and core-mantle interaction within Earth and the terrestrial planets. Tectonophysics 760:165–198. https://doi.org/10.1016/j.tecto.2018.10.021

Turbet M, Bolmont E, Chaverot G, Ehrenreich D, Leconte J, Marcq E (2021) Day–night cloud asymmetry prevents early oceans on Venus but not on Earth. Nature 598:276–280. https://doi.org/10.1038/s41586-021-03873-w

Turcotte DL (1993) An episodic hypothesis for Venusian tectonics. J Geophys Res 98:61–68. https://doi.org/10.1029/93je01775

Turcotte DL, Oxburgh ER (1967) Finite amplitude convective cells and continental drift. J Fluid Mech 28:29–42. https://doi.org/10.1017/S0022112067001880

Ueda K, Gerya T, Sobolev SV (2008) Subduction initiation by thermal-chemical plumes: numerical studies. Phys Earth Planet Inter 171:296–312. https://doi.org/10.1016/j.pepi.2008.06.032

Urey HC (1952) The Planets: Their Origin and Development. Yale University Press, New Haven

Van Hoolst T (2015) Rotation of the terrestrial planets. In: Treatise on Geophysics, 2nd edn. Elsevier, Amsterdam, pp 121–151. https://doi.org/10.1016/B978-0-444-53802-4.00168-8

Van Kranendonk MJ (2010) Two types of Archean continental crust: plume and plate tectonics on early Earth. Am J Sci 310:1187–1209. https://doi.org/10.2475/10.2010.01

Van Kranendonk MJ (2011) Onset of plate tectonics. Science 333:413–414. https://doi.org/10.1126/science.1208766

Vandaele AC, Mahieux A, Chamberlain S, Wilquet V, Robert S, Piccialli A, Thomas I, Trompet L (2020) Water vapor and hydrogen isotopic ratio at the Venus terminator from SOIR/VEX. In: 51st Lunar and Planetary Science Conference, p 1377

Von Zahn U, Kumar S, Niemann H, Prinn R (1983) Composition of the Venus atmosphere. In: Hunten DM, Colin L, Donahue TM, Moroz VI (eds) Venus. University of Arizona Press, Tucson, pp 299–430

Walker RG, Sagan C (1966) The ionospheric model of the Venus microwave emission: an obituary. Icarus 5:105–123. https://doi.org/10.1016/0019-1035(66)90014-5

Way MJ, Del Genio AD (2020) Venusian habitable climate scenarios: modeling Venus through time and applications to slowly rotating Venus-like exoplanets. J Geophys Res, Planets 125:e2019JE006276. https://doi.org/10.1029/2019JE006276

Way MJ, Del Genio AD, Kiang NY, Sohl LE, Grinspoon DH, Aleinov I, Kelley M, Clune T (2016) Was Venus the first habitable world of our solar system? Geophys Res Lett 43:8376–8383. https://doi.org/10.1002/2016GL069790

Way MJ, Aleinov I, Amundsen DS, Chandler MA, Clune TL, Del Genio AD, Fujii Y, Kelley M, Kiang NY, Sohl L, Tsigaridis K (2017) Resolving orbital and climate keys of Earth and extraterrestrial environments with dynamics (ROCKE-3D) 1.0: a general circulation model for simulating the climates of rocky planets. Astrophys J Suppl Ser 231:12. https://doi.org/10.3847/1538-4365/aa7a06

Way MJ, Ernst RE, Scargle JD (2022a) Large-scale volcanism and the heat death of terrestrial worlds. Planet Sci J 3:92. https://doi.org/10.3847/PSJ/ac6033

Way MJ, Ostberg CM, Foley BJ, Gillmann C, Höning D, Lammer H, O'Rourke JG, Persson M, Plesa A-C, Salvador A, Scherf M, Weller MB (2022b) Synergies between Venus and exoplanetary observations. Space Sci Rev. In revision

Weidenschilling SJ (1976) Accretion of the terrestrial planets. II. Icarus 27:161–170. https://doi.org/10.1016/0019-1035(76)90193-7

Weitz CM, Basilevsky AT (1993) Magellan observations of the Venera and Vega landing site regions. J Geophys Res 98:17069. https://doi.org/10.1029/93JE01776

Weller MB, Kiefer WS (2020) The physics of changing tectonic regimes: implications for the temporal evolution of mantle convection and the thermal history of Venus. J Geophys Res, Planets 125:1–22. https://doi.org/10.1029/2019JE005960

Weller MB, Lenardic A (2018) On the evolution of terrestrial planets: bi-stability, stochastic effects, and the non-uniqueness of tectonic states. Geosci Front 9:91–102. https://doi.org/10.1016/j.gsf.2017.03.001

Westall F, Hickman-Lewis K, Hinman N, Gautret P, Campbell KA, Bréhéret JG, Foucher F, Hubert A, Sorieul S, Dass AV, Kee TP, Georgelin T, Brack A (2018) A hydrothermal-sedimentary context for the origin of life. Astrobiology 18:259–293. https://doi.org/10.1089/ast.2017.1680

Westall F, Way MJ, Izenberg NR, Helbert J, Gilmore MS, Weller MB, Carter L, Gillmann C, Honing D, Gerya TV, Plesa A-C, Wilson CF, Selsis F (2023) The habitability of Venus. Space Sci Rev. In revision

Widemann T, Smrekar S, Garvin JB, Straume AG, Hensley S, Dyar D, Whitten J, Nunez D, Getty S, Arney G, Johnson NM, Kohler E, Way MJ, Westall F, Avice G, Gilmore M, Carter L, Ghail R, Helbert J, Izenberg N, Byrne P, Wilson C, Herrick R, Salvador A, Jacobson S, Breuer D, Höning D, Garcia RF, Plesa A, Gillmann C, Korablev O, Zelenyi L, Zasova L, Gorinov D (2023) Investigations to address Venus evolution through time: key science questions and selected mission concepts. Space Sci Rev. In review

Wildt R (1940) Note on the surface temperature of Venus. Astrophys J 91:266. https://doi.org/10.1086/144165

Williams DR (2022a) Venus fact sheet [WWW Document]. https://nssdc.gsfc.nasa.gov/planetary/factsheet/venusfact.html (accessed 6.15.22)

Williams DR (2022b) Earth fact sheet [WWW Document]. https://nssdc.gsfc.nasa.gov/planetary/factsheet/earthfact.html (accessed 6.15.22)

Wilson CF, Chassefière E, Hinglais E, Baines KH, Balint TS, Berthelier J-J, Blamont J, Durry G, Ferencz CS, Grimm RE, Imamura T, Josset J-L, Leblanc F, Lebonnois S, Leitner JJ, Limaye SS, Marty B, Palomba E, Pogrebenko SV, Rafkin SCR, Talboys DL, Wieler R, Zasova LV, Szopa C (2012) The 2010 European Venus Explorer (EVE) mission proposal. Exp Astron 33:305–335. https://doi.org/10.1007/s10686-011-9259-9

Wilson CF, Marcq E, Gillmann C, Widemann T, Korablev O, Mueller N, Lefevre M, Rimmer P, Robert S, Zolotov M (2023) Magmatic volatiles and effects on the modern atmosphere of Venus. Space Sci Rev. In preparation

Wood BE, Hess P, Lustig-Yaeger J, Gallagher B, Korwan D, Rich N, Stenborg G, Thernisien A, Qadri SN, Santiago F, Peralta J, Arney GN, Izenberg NR, Vourlidas A, Linton MG, Howard RA, Raouafi NE (2022) Parker solar probe imaging of the night side of Venus. Geophys Res Lett 49:1–8. https://doi.org/10.1029/2021gl096302

Xiao C, Li F, Yan J, Gregoire M, Hao W, Harada Y, Ye M, Barriot J (2021) Possible deep structure and composition of Venus with respect to the current knowledge from geodetic data. J Geophys Res, Planets 126:e2019JE006243. https://doi.org/10.1029/2019JE006243

Yang J, Boué G, Fabrycky DC, Abbot DS (2014) Strong dependence of the inner edge of the habitable zone on planetary rotation rate. Astrophys J Lett 787:L2. https://doi.org/10.1088/2041-8205/787/1/L2

Yang A, Huang J, Wei D (2016) Separation of dynamic and isostatic components of the Venusian gravity and topography and determination of the crustal thickness of Venus. Planet Space Sci 129:24–31. https://doi.org/10.1016/j.pss.2016.06.001

Yoder CF (1995) Venus' free obliquity. Icarus 117:250–286. https://doi.org/10.1006/icar.1995.1156

Yoder CF (1997) Venusian spin dynamics. In: Bougher SW, Hunten DM, Phillips RJ (eds) Venus II. University of Arizona Press, Tucson, pp 1087–1124

Zahnle KJ (1992) Airburst origin of dark shadows on Venus. J Geophys Res 97:10243. https://doi.org/10.1029/92JE00787

Zahnle KJ, Kasting JF, Pollack JB (1988) Evolution of a steam atmosphere during Earth's accretion. Icarus 74:62–97. https://doi.org/10.1016/0019-1035(88)90031-0

Zasova LV, Moroz VI, Formisano V, Ignatiev NI, Khatuntsev IV (2004) Infrared spectrometry of Venus: IR Fourier spectrometer on Venera 15 as a precursor of PFS for Venus Express. Adv Space Res 34:1655–1667. https://doi.org/10.1016/j.asr.2003.09.067

Zhang S, O'Neill C (2016) The early geodynamic evolution of Mars-type planets. Icarus 265:187–208. https://doi.org/10.1016/j.icarus.2015.10.019

Zharkov VN (1983) Models of the internal structure of Venus. Moon Planets 29:139–175. https://doi.org/10.1007/BF00928322

Zharkov VN, Gudkova TV (2019) On parameters of the Earth-like model of Venus. Sol Syst Res 53:1–4. https://doi.org/10.1134/S0038094618060084

68

 Springer

Publisher's Note Springer Nature remains neutral with regard to jurisdictional claims in published maps and institutional affiliations.

Authors and Affiliations

Joseph G. O'Rourke[1] · Colin F. Wilson[2,3] · Madison E. Borrelli[1] · Paul K. Byrne[4] · Caroline Dumoulin[5] · Richard Ghail[6] · Anna J.P. Gülcher[7] · Seth A. Jacobson[8] · Oleg Korablev[9] · Tilman Spohn[10,11] · M.J. Way[12,13] · Matt Weller[14,15] · Frances Westall[16]

✉ A.J.P. Gülcher
anna.guelcher@erdw.ethz.ch

J.G. O'Rourke
jgorourk@asu.edu

[1] School of Earth and Space Exploration, Arizona State University, Tempe, AZ, USA

[2] Department of Physics, Oxford University, Oxford, UK

[3] European Space Agency ESA/ESTEC, Noordwijk, The Netherlands

[4] Department of Earth and Planetary Sciences, Washington University in St. Louis, St. Louis, USA

[5] Laboratoire de Planétologie et Géosciences (UMR6112), Nantes Université, CNRS, Université d'Angers, Nantes, France

[6] Earth Sciences, Royal Holloway, University of London, Egham, UK

[7] Institute of Geophysics, Department of Earth Sciences, ETH Zürich, Zürich, Switzerland

[8] Department of Earth and Environmental Sciences, Michigan State University, East Lansing, MI 48824, USA

[9] Space Research Institute RAS (IKI), Profsoyuznaya 84/32, 117997 Moscow, Russia

[10] International Space Science Institute, Bern, Switzerland

[11] DLR Institute of Planetary Research, Berlin Germany

[12] NASA Goddard Institute for Space Studies, 2880 Broadway, New York, NY 10025, USA

[13] Theoretical Astrophysics, Department of Physics and Astronomy, Uppsala University, Uppsala, Sweden

[14] Department of Earth, Environmental & Planetary Sciences, Brown University, Providence, RI, USA

[15] Lunar and Planetary Institute/USRA, Houston, TX, USA

[16] CNRS-Centre de Biophysique Moléculaire, Orléans, France

Space Science Reviews (2023) 219:13
https://doi.org/10.1007/s11214-023-00953-3

Synergies Between Venus & Exoplanetary Observations

Venus and Its Extrasolar Siblings

M.J. Way[1,2] · Colby Ostberg[3] · Bradford J. Foley[4] · Cedric Gillmann[5] ·
Dennis Höning[6,7] · Helmut Lammer[8] · Joseph O'Rourke[9] · Moa Persson[10] ·
Ana-Catalina Plesa[11] · Arnaud Salvador[12,13,14] · Manuel Scherf[8,15,16] · Matthew Weller[17]

Received: 13 April 2022 / Accepted: 11 January 2023 / Published online: 9 February 2023
© The Author(s) 2023

Abstract
Here we examine how our knowledge of present day Venus can inform terrestrial exoplanetary science and how exoplanetary science can inform our study of Venus. In a superficial way the contrasts in knowledge appear stark. We have been looking at Venus for millennia and studying it via telescopic observations for centuries. Spacecraft observations began with Mariner 2 in 1962 when we confirmed that Venus was a hothouse planet, rather than the tropical paradise science fiction pictured. As long as our level of exploration and understanding of Venus remains far below that of Mars, major questions will endure. On the other hand, exoplanetary science has grown leaps and bounds since the discovery of Pegasus 51b in 1995, not too long after the golden years of Venus spacecraft missions came to an end with the Magellan Mission in 1994. Multi-million to billion dollar/euro exoplanet focused spacecraft missions such as JWST, and its successors will be flown in the coming decades. At the same time, excitement about Venus exploration is blooming again with a number of confirmed and proposed missions in the coming decades from India, Russia, Japan, the European Space Agency (ESA) and the National Aeronautics and Space Administration (NASA). Here we review what is known and what we may discover tomorrow in complementary studies of Venus and its exoplanetary cousins.

Keywords Exoplanets · Venus

1 Can Exoplanets Inform Venus' Evolutionary History?

It may sound preposterous to propose that terrestrial exoplanets, which are far from being explored in-situ, and which present challenges even to detection of their atmospheres, can in any way inform Venus' evolutionary history. Yet exoplanetary science has already provided a means to put ancient Venus 4.2 billion years ago within the habitable zone (Yang et al. 2014; Way et al. 2016). Initial studies of Venus' early climate by Ingersoll (1969),

Venus: Evolution Through Time
Edited by Colin F. Wilson, Doris Breuer, Cédric Gillmann, Suzanne E. Smrekar, Tilman Spohn and Thomas Widemann

Extended author information available on the last page of the article

Pollack (1971), Kasting et al. (1984), and others laid out the challenges for Venus having temperate surface conditions in its early history, given the $\sim 40\%$ higher incident solar radiation it received 4.2 Ga compared with modern-day Earth. However, Pollack (1971) demonstrated that temperate conditions were possible if Venus had 100% cloud cover, providing an albedo sufficiently high to block enough incoming sunlight to reduce surface temperatures to less than 300 K. Yet he provided no rationale for his choice of 100% cloud cover. Moving 40+ years into the future exoplanet researchers were beginning to look at large parameter sweeps using 3-D General Circulation Models (GCMs) to investigate how insolation and rotation rate influence climate (e.g. Yang et al. 2014). This effort was driven in part by the discovery of a large number of planets orbiting M-dwarf and K-dwarf stars – many in their habitable zones. One of the first of these exoplanet studies by Leconte et al. (2013) used the Laboratoire de Météorologie Dynamique (LMD)[1] GCM to demonstrate that temperate conditions were possible for the tidally locked world HD 85512 b, which orbits a K-dwarf star with a 58-day period. A year later, using the National Center for Atmospheric Research (NCAR)[2] Community Atmosphere Model (CAM) GCM, Yang et al. (2014) demonstrated that slowly rotating worlds (not necessarily tidally locked) with modern Earth-like atmospheres could in fact host temperate surface conditions with mean surface temperatures < 300 K at stellar insolations approaching 2.5 times what Earth receives today. This was due to large scale contiguous high albedo tropospheric clouds located in the substellar region. These were a byproduct of the extended single-hemisphere-sized Hadley cells from a weakened Coriolis force due to the slower rotation rate. This exoplanet related discovery had confirmed Pollack's proposed 100% cloud cover 43 years later. The Yang et al. (2014) work prompted a number of similar studies (Way et al. 2016, 2018) that confirmed the original result with a completely different 3-D GCM known as ROCKE-3D (Resolving Orbital Keys of Earth and Extraterrestrial Environments with Dynamics)[3] (Way et al. 2017). This research has had a profound effect on understanding the possible climate history of Venus and Venus-like worlds. Whereas earlier Venus focused studies claimed an early short-lived habitable period was possible (Grinspoon and Bullock 2007), these exoplanet studies demonstrated that Venus could have had quite long periods of habitability (Way and Del Genio 2020).

Thus far at least five different GCMs have produced the cloud-albedo feedback for slowly rotating worlds: ROCKE-3D, NCAR (Yang et al. 2014), the UK Met Office Unified Model (Walters et al. 2011), LMD, and Exocam.[4] While such coherence may appear definitive these model results must be verified with observations of planets within the canonical Venus Zone (e.g. Kane et al. 2014, hereafter VZ). At the same time, there is still great uncertainty related to the longevity of the early magma ocean atmospheres (see Sect. 1.4), in the composition of the atmospheres (e.g. Bower et al. 2022) and exactly what role clouds might play (Turbet et al. 2021). Are these atmospheres a mix of CO, CO_2, N_2, H_2O, CH_4, or H_2, and what sorts of clouds are involved, if any? Here again exoplanetary observations hold the keys to the kingdom, and are the only way to definitively test and refine our models and their underlying physics.

Planetary scientists recognize that the exploration of Venus can inform our understanding of exoplanets, and vice versa as discussed in this article. These linkages permeate the new decadal survey released by the United States of America's National Academies (National Academies of Sciences, Engineering, and Medicine 2021) as detailed in the introduction to

---

[1] https://www-planets.lmd.jussieu.fr/

[2] https://ncar.ucar.edu/

[3] https://simplex.giss.nasa.gov/gcm/ROCKE-3D/

[4] https://github.com/storyofthewolf/ExoCAM

this topical collection (O'Rourke et al. 2023, this collection). Table 1 pulls verbatim excerpts from this new report identifying some of the observations of Venus and exoplanets that scientists consider most important in the near term. We can study Venus as "the exoplanet in our backyard" and obtain measurements, including in situ data, that are not feasible at planets orbiting distant stars. We can also study a statistical sample of Venus-sized exoplanets to explore if a Venus-like evolutionary pathway is typical. These parallel approaches will promote synergies and strengthen ties between these oft-separated scientific communities.

1.1 Transiting Exoplanets in the Venus Zone and JWST

The Transiting Exoplanet Survey Satellite (TESS; Ricker et al. 2010) is currently observing our nearest and brightest stellar neighbors in search of exoplanets. Similar to the Kepler/K2 mission (e.g. Howell et al. 2014, and references within),[5] TESS is discovering exoplanets using the transit method. This method works by observing changes in the brightness of a star as a planet passes between the instrument and the star. The magnitude of the change in the star's brightness reveals the radius of the planet (assuming that one knows the radius of the star), while the periodicity of the brightness fluctuations is used to infer the planet's orbital period. The transit method is intrinsically biased towards planets with shorter orbital periods (Kane and von Braun 2008), since the probability of observing a planet transit is inversely proportional to the planet's orbital period. This observational bias has led to TESS discovering a large number of terrestrial planets in the Venus Zone (VZ; Kane et al. 2014). The VZ is defined as the area around a star where a planet is more likely to resemble a Venus analog than an Earth analog, but does not guarantee a planet will have Venus-like surface conditions. Temperate planets may also reside in the VZ, as recent works have highlighted the possibility of Venus sustaining a temperate climate in the past (Way et al. 2016; Way and Del Genio 2020). Ultimately, the VZ is a tool to guide target selection for follow-up observations of exoplanet atmospheres. These observations will provide information about the atmospheres of VZ planets, which helps infer information about their surface conditions and test the hypothesis of the VZ. Similar to the Habitable Zone (HZ; Kopparapu et al. 2013), the VZ is defined by two boundaries. The inner VZ boundary is defined, in terms of insolation flux, as 25x the flux received by Earth. This specific value was chosen as it is the flux needed to place Venus on the 'Cosmic Shoreline' (Zahnle and Catling 2017), which is an empirical relationship used to predict the insolation flux needed for a terrestrial body to lose the majority of its atmosphere via thermal escape processes. The outer VZ boundary is the runaway greenhouse boundary, which is the inner boundary of the HZ. This boundary is the insolation flux where an Earth-like planet is predicted to enter a runaway greenhouse state.

Unlike the Kepler/K2 mission, which observed stars nearly 1000 pc away, TESS is observing stars which are at a distance of ~ 60 pc. The closer vicinity of TESS stars makes them inherently brighter than Kepler/K2 stars, and therefore allows for more signal to be obtained from them. The increased number of photons from TESS stars creates an excellent opportunity to conduct follow-up observations of the atmospheres of TESS planets from ground and space based instruments. Planets detected by TESS are initially added to the TESS Object of Interest (TOI) list. However a TOI is required to be detected by additional observations in order for it to become a confirmed planet. All confirmed planets are listed on the NASA Exoplanet Archive.[6] At the time of writing, the NASA Exoplanet Archive and

---

[5]https://www.nasa.gov/mission_pages/kepler/overview/index.html.

[6]https://exoplanetarchive.ipac.caltech.edu/.

Table 1 Recently, the Planetary Science and Astrobiology Decadal Survey 2023–2032 highlighted many synergies between observations of Venus and exoplanets (National Academies of Sciences, Engineering, and Medicine 2021). This report prioritized scientific activities that would help answer two key questions: What does Venus teach us about the evolutionary pathways of exoplanets? Is the evolution of Venus typical of Venus-sized exoplanets? Below, we quoted priority questions, strategic research, and supportive activities from Chap. 15 ("Question 12: Exoplanets") that are related to many of the scientific connections between Venus and exoplanets discussed in this article and many others in this collection

Priority questions linking Venus and Exoplanets	
12.1	Evolution of the Protoplanetary Disk
12.3	Origin of Earth and Inner Solar System Bodies
12.4	Impacts and Dynamics
12.5	Solid Body Interiors and Surfaces
12.6	Atmosphere and Climate Evolution on Solid Bodies
12.10	Dynamic Habitability
12.11	Search for Life Elsewhere

Strategic Research to Benefit Exoplanetary Science	
Question(s)	Strategic Research
12.1, 12.3, 12.6	Measure abundances and isotopic compositions of noble gases and other key elements (in the atmosphere of Venus)
12.6	Determine the properties of the atmospheres of terrestrial planets (… Venus…) that would be observable on exoplanets
12.10	Constrain the inner edge of the habitable zone in the solar system by studying the surface geomorphology and geochemistry of Venus to assess whether it ever possessed oceans
12.11	Study methods to discriminate past and present false positive biosignatures on solar system bodies (e.g., abiotic O_2 on Venus…) from true biosignatures to inform false positives discrimination methods for exoplanets
	Devise metrics and frameworks to establish confidence in interpretation of biosignatures in the solar system and exoplanetary systems

Strategic Research on Exoplanets to Benefit Venusian Science	
Question(s)	Strategic Research
12.1	Characterize protoplanetary disks around young stars
12.3, 12.4, 12.5, 12.6, 12.10	Obtain an inventory of properties of solid body exoplanets (i.e., mass, composition, bulk Obtain an inventory of properties of solid body exoplanets (i.e., mass, composition, bulk atmospheric chemistry and abundance of clouds and hazes, potential biosignatures, rotation rates, relative distance from host star, type of host star)
12.4	Determine how impacts contribute volatiles to (or, in some cases, remove volatiles from) planetary bodies
12.5	Search for magnetospheric activity at exoplanets

Supportive Activities to Promote Synergy Between Venusian and Exoplanetary Science
Observations of [Venus] through transit spectroscopy and direct-imaging as analogs to exoplanet observations
Observations of particle and gas opacity in [Venus] as a function of phase angle to help determine the dependence of reflectivity and scattering on particles and clouds
Laboratory studies to understand the relationship between the bulk composition of a planet and its atmosphere, and to determine the optical properties of clouds and hazes
Increased interactions between the astronomy, planetary science, astrobiology communities

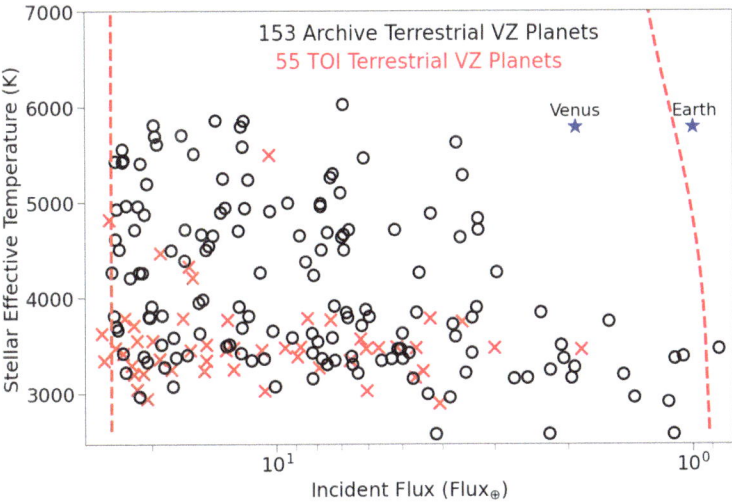

Fig. 1 The locations of terrestrial VZ planets ($R_p < 1.5\ R_\oplus$) from the NASA Exoplanet Archive and TOI list in reference to the VZ as a function of planetary insolation flux. Earth and Venus are shown for reference

TOI list contain 153 and 55 terrestrial planets ($R_p < 1.5\ R_\oplus$) that spend any portion of their orbit in the VZ, respectively (Fig. 1). A radius cutoff of 1.6 $R_\oplus$ is typically chosen as it may be the empirical upper size limit of terrestrial exoplanets (Fulton et al. 2017).

Determining that a planet resides in the VZ provides only a first-order estimate about the potential environment on that planet. In order to more accurately deduce possible surface conditions on a VZ planet, observations of its atmosphere will be required. JWST (launched in December 2021) may be humanity's first opportunity to peer into the atmospheres of terrestrial exoplanets via either transmission or secondary eclipse spectroscopy (e.g. Barstow et al. 2015; Batalha and Line 2017; Beichman et al. 2014; Belu et al. 2011; Clampin 2011; Crouzet et al. 2017; Deming et al. 2009; Greene et al. 2016; Howe et al. 2017; Mollière et al. 2017; Lustig-Yaeger et al. 2019b; Fauchez et al. 2019; Koll et al. 2019; Wunderlich et al. 2019).

1.2 Transmission and Secondary Eclipse Spectroscopy with JWST

Informed predictions of the surface conditions and climates on potential exo-Venuses will require observations of their atmospheres via transmission and secondary eclipse spectroscopy. Secondary eclipse spectroscopy is conducted by observing the appearance and disappearance of light reflected and/or emitted by the planet as it orbits its host star – there is no need to spatially resolve the light from the planet from that of the host star. Transmission spectroscopy involves observing starlight that passes through the atmosphere of a transiting exoplanet. Both techniques can be used to gather information about the composition and structure of an exoplanet atmosphere. The atmospheres of terrestrial exoplanets have been inaccessible to this point, but JWST may provide the light-gathering power necessary to retrieve information from terrestrial exoplanet atmospheres (e.g. Lustig-Yaeger et al. 2019b; Batalha et al. 2018; Morley et al. 2017; Lincowski et al. 2019; Fauchez et al. 2019; Turbet et al. 2016; Meadows et al. 2018).

The performance of JWST when observing exoplanets can be predicted using the Transmission Spectroscopy Metric (TSM; Kempton et al. 2018). The TSM provides a first-order

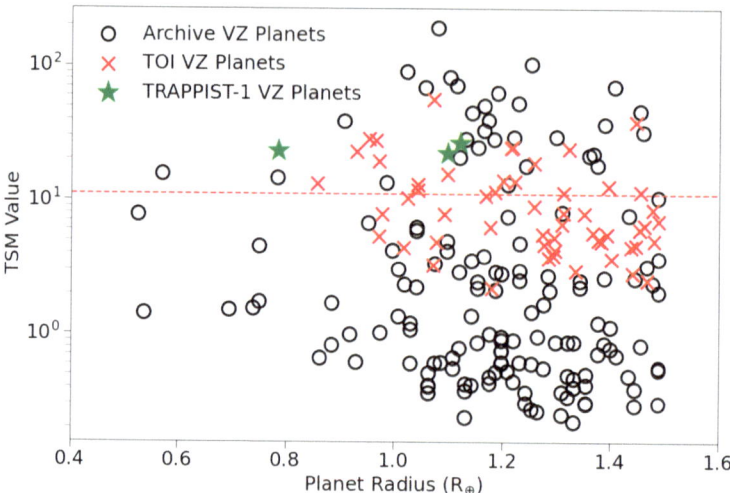

Fig. 2 Planetary radii versus associated TSM values for terrestrial planets ($R_p < 1.5$ $R_\oplus$) from the NASA Exoplanet Archive and TOI list. Planets with TSM values greater than 12 (red dotted line) are predicted to allow for a S/N of at least 12 from 10 hours of observations with JWST. The green stars denote the three TRAPPIST-1 planets in the VZ

approximation of the signal-to-noise ratio (S/N) of transmission spectra resolved from 10 hours of transit observations using the JWST NIRISS instrument (Louie et al. 2018) that can be used to prioritize targets that offer the best opportunity for JWST follow-up observations. Kempton et al. (2018) identified the top terrestrial targets as having TSM values greater than 12. Applying this threshold to known VZ planets shows there are 36 planets which qualify as top candidates for JWST observations (Fig. 2), including TRAPPIST-1b, c, and d (red stars in Fig. 2). Given that the TRAPPIST-1 system also has 3 planets in the HZ, observations of both the TRAPPIST-1 VZ and HZ planets could help us to discern whether the differences in climate between Earth and Venus is a common phenomena.

Here we simulate JWST observations of Kepler-1649b (Angelo et al. 2017) as an exo-Venus by modelling hypothetical JWST NIRSpec PRISM transmission spectra using the Planetary Spectrum Generator (PSG; Villanueva et al. 2018). NIRSpec PRISM has a wavelength range of 0.7–5.0 μm encompassing major H_2O and CO_2 features, and has been shown to be the optimal instrument for performing transmission spectroscopy in the NIR (Lustig-Yaeger et al. 2019b). PSG is a publicly available online interface that couples radiative transfer models, planetary databases, and spectral databases. Exo-Venus transmission or emission spectra can be produced with PSG by superimposing an atmosphere onto a terrestrial exoplanet in the VZ. Kepler-1649b is used as the hypothetical exo-Venus, as its size is similar to that of Venus, with a radius of 1.077 $R_\venus$ (1.017 $R_\oplus$), and has a incident insolation flux that is 2.21 times greater than that of Earth (Venus is 1.9), albeit orbiting a much redder M-dwarf star (Angelo et al. 2017). We used an atmosphere for the Kepler-1649b exo-Venus that uses data from a ROCKE-3D simulation of the planet documented in Kane et al. (2018). Specifically, we use data from simulation 10 in the previously mentioned work, which assumes an Earth-like input atmosphere (1 bar N_2 dominated with 376 ppmv CO_2), a lower insolation flux than Kepler 1649b of 1.4 and a mean surface temperature of 60 °C making it representative of a hypothetical temperate ancient-Venus. Note that using the actual insolation flux results in mean surface temperatures well over 100 °C as shown in simulations

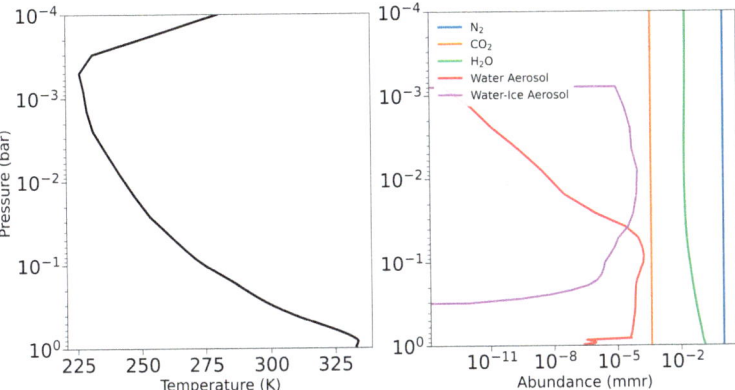

Fig. 3 Left: The globally averaged pressure-temperature profile of a Kepler-1649b Exo-Venus hypothetical atmosphere using data from a ROCKE-3D simulation of the planet. Right: Globally averaged Mean Mixing Ratio (mmr) composition versus Pressure. Note that the insolation for this exoplanet has been artificially reduced by a factor of 1.4, otherwise it would have most certainly entered a runaway greenhouse condition

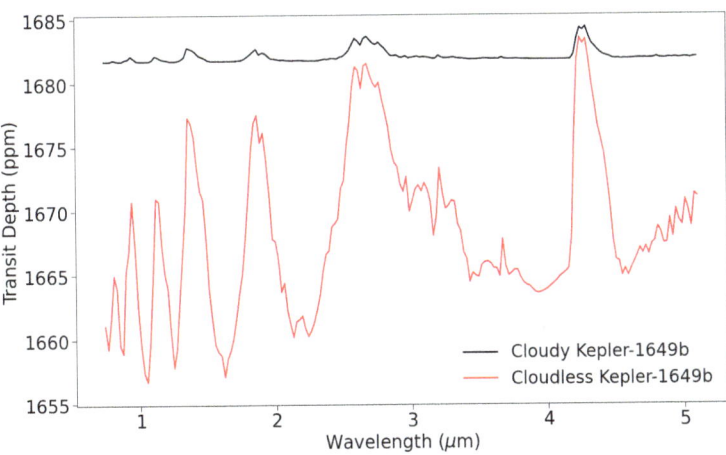

Fig. 4 Transmission spectra modelled with PSG for a temperate Kepler-1649b exo-Venus, assuming both a cloudy and cloudless atmosphere

1–3 in Kane et al. (2018) which is beyond the capabilities of the GCM used in this study (ROCKE-3D). Figure 3 illustrates the structure and chemical composition of the atmosphere from simulation 10.

Using the Kepler-1649b atmosphere from the ROCKE-3D simulation as an input for PSG, we modelled the transmission spectrum of Kepler-1649b from 0.6–5.3 μm, coinciding with the wavelength range of JWST NIRSpec PRISM. Since PSG is a 1-D radiative transfer model, the globally averaged pressure, temperature, and composition of the simulated Kepler-1649b atmosphere was used. Figure 4 displays the transmission spectra of the Kepler-1649b exo-Venus with and without water and water-ice aerosols, which is hereafter referred to as cloudy and cloudless, respectively. PSG determined that the atmosphere is opaque at elevations with higher aerosol densities, which had a significant affect on the absorption features in the transmission spectra. Prominent H_2O and CO_2 absorption features are visible in the cloudless spectrum, but are nearly completely truncated by the clouds in the modelled spectrum. The effect of clouds in the temperate Venus atmosphere will likely make it difficult for JWST to detect any absorption features, as shown in previous work (Fauchez et al. 2019).

The H_2SO_4 clouds in the atmosphere of present-day Venus have an equally significant effect on its transmission spectra (Ehrenreich et al. 2012). This was also demonstrated in Meadows et al. (2018) who simulated H_2SO_4 clouds and hazes in hypothetical modern Venus analogs. Hazes can form when the CH_4 to CO_2 ratio is greater than 0.1 and are an important contributor to the radiation budget and the detectability of Earth-like planets (Arney et al. 2016, 2017). Furthermore, Meadows et al. (2018) examined cloud and haze formation effects on the detectability of atmospheres on Proxima Centauri b using a "1-D coupled climate-photochemical models to generate self-consistent atmospheres for several evolutionary scenarios, including high-O_2, high-CO_2, and more Earth-like atmospheres, with both oxic and anoxic compositions." They also included the hydrocarbon hazes in instances when the CH_4/CO_2 ratio was greater than 0.1. Because their atmospheres were not cold enough they did not see any CO_2 clouds, but they have been shown to play an important role in the radiation budget in ancient Mars simulations (Colaprete and Toon 2003; Forget et al. 2013). However, it has long been postulated that the H_2SO_4 clouds on Venus are impermanent and require a regular supply of SO_2 from volcanism. As discussed in Sect. 2.1.3 the equilibrium level of SO_2 in the atmosphere is set by the volcanic outgassing rate versus the chemical reactions with surface materials (Zolotov 2018). The rate of present day volcanism on Venus is poorly constrained, although there are a number of studies from Venus Express demonstrating hot-spot volcanism (Shalygin et al. 2015; Smrekar et al. 2010). Other studies imply geologically recent volcanism due to the radar-dark floors of craters, presumably from volcanic fill-in (e.g. Herrick and Rumpf 2011) while others have demonstrated on-going plume activity (Gülcher et al. 2020). Recently, Byrne and Krishnamoorthy (2022) have used the recent Earth volcanic record as a proxy to derive estimates for Venus. If volcanism ceased today estimates of the lifetime of the clouds in different studies have ranged from $\sim$ 2–50 Myr (Fegley and Prinn 1989; Bullock and Grinspoon 1996, 2001) depending upon surface chemical reaction rates as mentioned above. Hence for some exo-Venus worlds H_2SO_4 clouds may not be an inhibitor to detection of major atmospheric species for a modern Venus-like atmosphere during periods of low volcanic sulfur outgassing.

It is important to note that the true nature and variety of environments on Venus-like worlds may be expansive, but will need to be investigated through atmospheric observations of exo-Venus candidates. Additionally, the atmospheric composition of an exo-Venus orbiting an M-dwarf star may differ from that of Venus. Placing Earth around Proxima-Centauri could enhance the abiotic production of CH_4 in its atmosphere (Meadows et al. 2018) which is often cited as an atmospheric biosignature (Thompson et al. 2022), and the atmospheric composition of Venus may be affected in a similar scenario. Furthermore, from an evolutionary point of view, the large energy deposition from stellar-winds produced by an M-dwarf could, over time, strip molecules from an exo-Venus atmosphere, which would affect the atmospheric composition as well (e.g. Airapetian et al. 2020), but was not accounted for when modelling the Kepler-1649b atmosphere.

The successful detection of transiting exo-Venus atmospheres with JWST remains uncertain, but models such as PandExo (Batalha et al. 2017) can provide insight into how JWST may perform. PandExo is an open-source code that allows users to simulate observations of exoplanets with JWST, and uses the Space Telescope Science Institute's Exposure Time Calculator, Pandeia (Pickering et al. 2016), to predict the S/N of observations. The performance of PandExo's simulated noise has been tested against noise simulations designed by

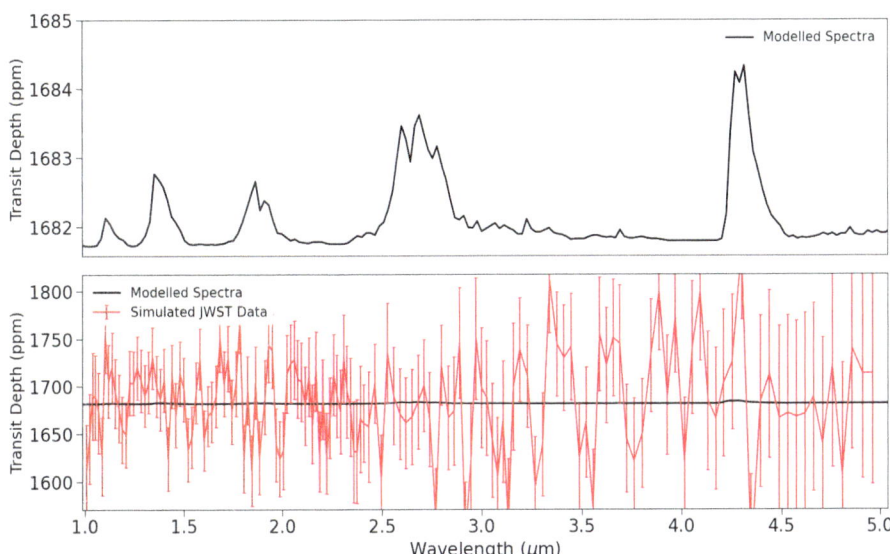

Fig. 5 PandExo simulated transmission spectrum of an exo-Venus Kepler-1649b from 30 transit observations using JWST NIRSpec PRISM. The upper figure displays the PSG modelled transmission spectrum with no noise, while the bottom figure compares data from JWST simulated observations of Kepler-1649b to that of the original spectrum. Note that the y-axes of the two plots are on different scales, illustrating the size of the uncertainties in comparison to the noise-less spectrum

the JWST instrument teams, and is within 10% agreement of their results (Batalha et al. 2017). Figure 5 shows a simulated transmission spectrum of the Kepler-1649b exo-Venus generated by PandExo, assuming 30 transit observations with JWST NIRSpec PRISM. The atmosphere used for the Pandexo simulated observations is the same as that used for Fig. 4. Given 30 transit observation of Kepler-1649b, the simulated JWST data is unable to resolve any of the major absorption features in the NIR. Furthermore, the large uncertainty in the data would make it difficult to differentiate the spectra from that of a flat-line, which may result in mistaking an exo-Venus as a planet with no atmosphere (Lustig-Yaeger et al. 2019a). Increasing the number of transit observations would decrease the uncertainty in the data, however acquiring the JWST time needed to conduct these observations will be a challenge. The features being less than 5 ppm make them smaller than the predicted 20 ppm noise floor of the NIRSpec instrument (Rustamkulov et al. 2022), making them potentially undetectable by JWST given any amount of observations and only accessible with future observatories.

Assuming that absorption features are detected in the atmosphere of an exoplanet, retrieval algorithms will then be used to estimate its atmospheric composition. Retrieval algorithms have been shown to experience difficulty differentiating Earth-like from Venus-like planets, since Venus' transmission spectra lacks unique absorption features that can be used to distinguish it from Earth (Barstow et al. 2016). The information gained from a retrieval model can then be applied to a GCM, which model the possible surface conditions of the planet based on the atmosphere estimated by the retrieval. The use of GCMs may play a critical role in constraining the potential climates of exoplanets (Turbet et al. 2016; Wolf et al. 2019) for the foreseeable future in coordination with JWST.

Emission spectroscopy will be attempted by JWST primarily using the Mid-Infrared Instrument (MIRI), which has a wavelength range between 5–29 μm. The emission spectra retrieved by MIRI will be useful for identifying the presence, or lack of an atmosphere on

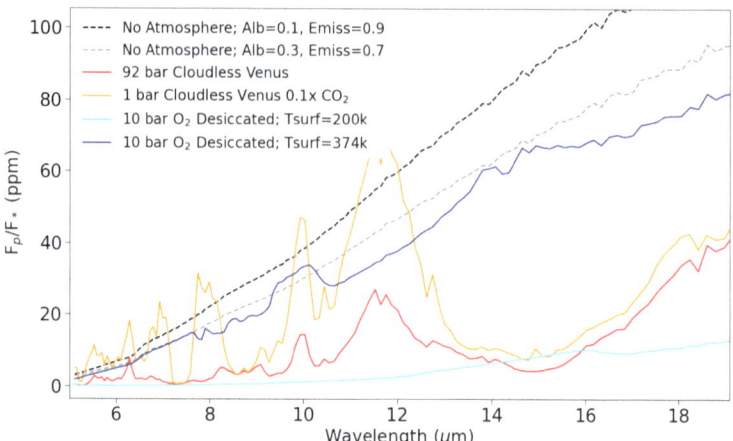

Fig. 6 A variety of emission spectra that could be potentially observed on exoplanets using the MIRI instrument aboard JWST. The planet-star flux ratio values are obtained by placing these atmospheres on the Venus-zone planet, L98-59d

a planet (Batalha et al. 2018; Meadows et al. 2018; Turbet et al. 2016). Figure 6 illustrates several hypothetical emission spectra that could be observed on the VZ planet, L98-59d. Included are the following atmospheres: cloudless 92 bar Venus analog (red); 1 bar cloudless Venus with $0.1\times$ the CO_2 of present-day Venus (yellow); 10 bar, O_2 dominated desiccated atmosphere with a surface temperature of 374 K; 10 bar, O_2 desiccated atmosphere with a surface temperature of 200 K; an atmosphere-less, black-body emission spectrum assuming bond albedo $= 0.1$ and emissivity $= 0.9$; an atmosphere-less, black-body emission spectrum assuming a bond albedo $= 0.3$ and emissivity $= 0.7$. All atmospheres assume no clouds to illustrate the dependence of emission spectra on atmospheric composition. It can be seen that the presence of CO_2 in the two Venus-like atmospheres causes the structure of their emission spectra to differ greatly from the other 4 spectra, particularly with the large CO_2 emission peaks at 10 and ~ 12 μm. The O_2 dominated desiccated atmospheres are included since many VZ planets orbit hyperactive M-dwarf stars, which could photodissociate any atmospheric H_2O in these planets over time (Wordsworth and Pierrehumbert 2013; Luger and Barnes 2015). In this scenario rapid hydrogen escape would ensue and an O_2 dominated, but H_2O desiccated, atmosphere would remain.

Coupling the PSG emission spectra with PandExo gives insight into the ability of JWST to detect an atmosphere on a hypothetical L98-59d, and whether JWST would be able to tell them apart (Fig. 7). Figure 7 displays simulated JWST data assuming both 5 and 15 secondary eclipse observations of an exo-Venus L98-59d with no atmosphere, and with a cloudless 92 bar Venus-like atmosphere. For 5 eclipse observations, the uncertainty in the simulated data for both cases make it difficult to determine whether there is an atmosphere. With 15 eclipse observations, the simulated data is a much better fit to the modelled spectra up to 11 μm. Retrieval models will also be used for JWST emission spectra to determine the likelihood of a planet having an atmosphere, but as earlier studies cited above have shown it is unlikely any individual atmospheric features will be discerned.

In summary, there are an abundance of VZ planets which are promising candidates for follow-up JWST observations, and the TESS mission will be discovering additional candidates throughout its lifetime. Of these candidates, the TRAPPIST-1 planets in the VZ are especially intriguing, as observations of their atmospheres, and the atmospheres of the

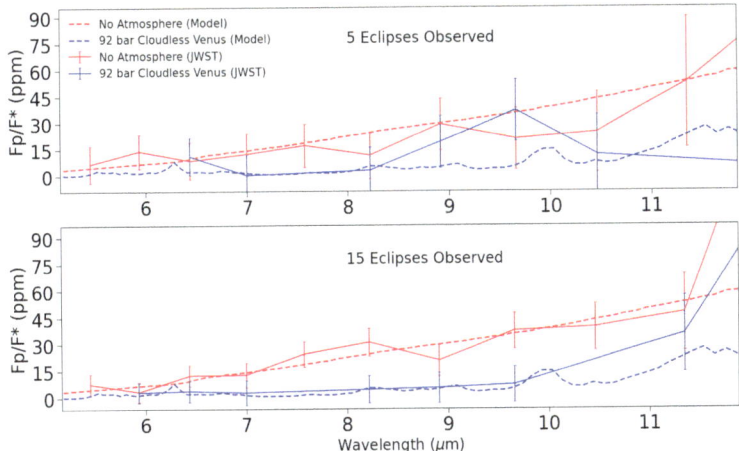

Fig. 7 Simulated JWST MIRI LRS data from 5 (top) and 30 (bottom) secondary eclipse observations of L98-59d assuming it has either no atmosphere, or a cloudless 92 bar Venus-like atmosphere. The dotted lines are the PSG modelled emission spectra, while the solid lines are PandExo simulated MIRI observations

TRAPPIST-1 HZ planets, will provide an opportunity to compare the differences between Earth and Venus to planets receiving similar insolation flux. JWST will be our first opportunity to obtain information about the atmospheres of terrestrial planets, including exo-Venuses. Simulated JWST data revealed that 15 transit observations with JWST NIRSpec PRISM would be insufficient for resolving the atmosphere of Kepler-1649b with both a temperate exo-Venus, and present-day Venus atmosphere. Venusian clouds and hazes severely truncate the absorption features in the present-day Venus spectrum, and will make it difficult to efficiently determine the atmospheric composition of an exo-Venus, or detect its atmosphere at all. The temperate exo-Venus atmosphere would be difficult to detect as well, despite the lack of Venus-like clouds. Even if significant JWST time is allotted for observations of exo-Venuses, it still may be the case that atmospheric information vital for understanding the climates of exo-Venuses may remain inaccessible during the JWST era. The inability to infer the surface conditions of exo-Venuses will inhibit exoplanets from being a resource to study Venus' evolution, and whether Venus could have sustained temperate surface conditions in its past.

1.3 Future Space and Ground Based Exo-Venus Observational Capabilities

There are at least three next generation ground-based (> 20 m in diameter) optical near-IR observatories currently under construction (circa 2022) or likely to be built in the near future. The European led Extremely Large Telescope (ELT) has a capable first generation set of instruments (Ramsay et al. 2020) and is the only next generation telescope both fully funded and under construction. The Magellan Giant Telescope (GMT) (Fanson et al. 2020) and the Thirty Meter Telescope (TMT) (Sanders 2013) are yet to be fully funded. The former two are currently under construction in Chile while the TMT is proposed for the northern hemisphere, although the exact location remains uncertain (Clery 2019). Once complete, these new observatories will offer the opportunity for a marked increase in collecting area and resolution. With increasing advances in adaptive optics, they will afford new opportunities to characterize the atmospheres of nearby exo-Venuses, as they are discovered by space observatories devoted to detecting such systems via the transit method (e.g. Kepler,[7] TESS,[8] CHEOPS,[9] PLATO[10]) complimented by ground based radial velocity instruments like that of the FLAMES facility at the VLT (e.g. Pasquini et al. 2002). In space, JWST has just launched. It may be able to detect atmospheres around a few nearby terrestrial planets in systems such as Trappist-1, although such observations will be challenging, as discussed above.

A mostly-US funded successor to The Hubble Space Telescope was recently recommended as a top priority in the US National Academy of Sciences (NAS) Decadal Survey (National Academies of Sciences, Engineering, and Medicine 2021, Sect. 7.4).[11] It is referred to as the "IR/O/UV Large Strategic Mission" (which we refer to as IROV, see Sect. 7.5.2 in the NAS report) and recently dubbed the Habitable Worlds Observatory (Clery 2023). It is "optimized for observing habitable exoplanets and general astrophysics", according to the report. The UV component is why IROV is more properly termed a successor to The Hubble Space Telescope rather than JWST – the latter being IR optimized. IROV is scheduled to launch in the early 2040s. IROV is expected to be some combination of The Large UV Optical Infrared Surveyor (LUVOIR) (The LUVOIR Team 2019) with 8 m diameter and HabEx (Martin et al. 2019) with a ~ 4 m diameter mirror, while including a coronagraph for direct imaging and spectroscopy of extrasolar planets. IROV would have a "light collecting area several times larger, 2–3 times sharper image quality, and instruments and detectors significantly more sensitive, providing 1–2 order-of-magnitude leaps in sensitivity and performance over HST." The report recommends a ~ 6 m sized mirror as a balance between a Habex 4 m, which would struggle to provide a "robust exoplanet census", and a LUVOIR 8 m, which would likely launch much later than IROV, in the late 2040s or early 2050s. As shown in the work of Checlair et al. (2020), the diameter of the mirror appears to be the critical factor in determining whether we will make the revolutionary discoveries intended. IROV will be capable of observing over 100 nearby Sun-like stars and would quantify the elements of any associated planetary systems, giving ample opportunity for the discovery of Venus-like worlds at various stages in their evolutionary history. For Proxima Centauri b Meadows et al. (2018) demonstrates the capabilities of a HabEx 6.5 m space telescope with coronagraph that could be similar to the capabilities of IROV. The inner working angle (IWA) is wavelength dependent and for the HabEx 6.5 m they calculate the optimal $IWA = 1\lambda/D = 1.17$ μm, but in fact the diffraction limit should be 1.22 instead of 1 and this gives 0.96 μm. Examining the estimated reflection spectra in Figs. 21–26 in Meadows et al. (2018) it is apparent that this instrument may be able to distinguish between 10 bar O_2 rich atmospheres, a 90 bar cloud covered Venus, Archean and modern Earth. Both Meadows et al. (2018) and Turbet et al. (2016) provide simulations for Proxima Centauri b as both temperate and Venus-like. Barnes et al. (2016) also demonstrated that it is possible for Proxima Centauri b to have a Venus-like evolutionary path, so our closest neighbor may be denuded, an exo-Earth or even an exo-Venus.

Finally, there is currently a mission proposal to ESA called LIFE (Konrad et al. 2021),[12] which would entail a space based nulling interferometer. This is more-or-less a scaled down

---

[7] https://www.nasa.gov/mission_pages/kepler/main/index.html.

[8] https://www.nasa.gov/tess-transiting-exoplanet-survey-satellite.

[9] https://sci.esa.int/web/cheops.

[10] https://platomission.com/2018/05/07/habitability-of-planets-around-solar-like-stars/.

[11] https://www.nationalacademies.org/our-work/decadal-survey-on-astronomy-and-astrophysics-2020-astro2020.

[12] https://www.life-space-mission.com.

Table 2 First generation ELT instruments relevant to exo-Venus characterization

Instrument	Main specifications			Exo-Venus science
	Field of view slit length pixel scale	Spectral resolution	Wavelength coverage (μm)	
METIS	Imager + coronagraph $\sim 10 \times 10''$ @ 5 mas/pix in L, M @ 7 mas/pix in N	L, M, N + narrowbands	3–19	Thermal Emission
	Single slit	R~1400 in L, 1900 in M, 400 in N.	3–13	
	IFU $0.6 \times 0.9''$ @ 8 mas/pix w/coronagraph	L, M Bands R~100 000	2.9–5.3	Transmission & Reflection Spectra
HARMONI	IFU 4 spaxel scales $0.8 \times 0.6''$ @ 4 mas/pix $6 \times 9''$ @ 30 × 60 mas/pix (w/coronagraph)	R~3200 R~7100 R~17 000	0.47–2.45	Reflection Spectra
ANDES/HIRES	Single Object IFU (SCAO)	R~100 000 R~100 000	0.4–1.8″	Transmission & Reflection Spectra

and more affordable version of one of the Terrestrial Planet Finder concept missions from nearly two decades ago (e.g. Coulter 2003).

As mentioned above, only one next generation large (> 30 m) optical ground based telescope is fully funded today, so we focus the rest of this section on what the ELT will deliver for exoplanetary investigations with applications to exo-Venuses.

There are presently seven different first generation instruments intended for use with the ELT.[13] Below we focus on three of the first generation instruments relevant to exo-Venus observations (see Table 2).

HARMONI (High Angular Resolution Monolithic Optical and Near-infrared Integral field spectrograph) (Rodrigues et al. 2018; Houllé et al. 2021) and METIS (Mid-infrared ELT Imager and Spectrograph) (Brandl et al. 2018) are funded via the telescope construction budget while HIRES (HIgh REsolution Spectrograph) (Marconi et al. 2018, 2021) is funded by a consortium. We note that HIRES has been renamed ANDES (ArmazoNes high Dispersion Echelle Spectrograph),[14] but the instrument architecture remains the same (we will use both names herein).

METIS will operate at 3–19 μm and will focus on high contrast imaging/spectroscopy, along with high spectral resolution integral field unit (IFU) observations. METIS is designed with a coronagraph which will reduce the brightness of an axially-symmetric source (star) by $\sim 10^{-5}$–10^{-7}. Low resolution spectra will be obtained with the remaining reflected light for attempted characterization of planets more than 3 Astronomical Units in distance. METIS' IFU mode will have a $1.0'' \times 0.5''$ field of view and will allow for 3 km s^{-1} spectral resolution over 2.9–5.3 μm with an angular resolution down to $0.02''$. METIS will also be capable of direct imaging in thermal emission which will be useful for detecting targets around Sun-

---

[13] https://elt.eso.org/instrument.

[14] https://elt.eso.org/instrument/ANDES/.

like stars where the contrast is less than that of M-dwarfs (mid-IR is 10^{-7} while 10^{-10} in the visible) although the yield estimates are at most a few such objects (Quanz et al. 2015; Bowens et al. 2021).

The near infrared arm of the HIRES instrument is a more capable version of the present day European Southern Observatory (ESO) Very Large Telescope (VLT) CRIRES+ (The CRyogenic InfraRed Echelle Spectrograph Upgrade Project) instrument[15] for transmission spectroscopy. Baseline wavelength coverage is expected to be 0.55–1.80 μm with a goal of 0.33–2.44 μm at a spectral resolution 100000–150000, the bigger mirror allowing higher resolution studies than with CRIRES+. With the Integral Field Unit (IFU) HIRES will observe reflection spectra of nearby exo-Venus candidates discovered via transits, and radial velocity (RV) surveys. Given the geometrical constraints of transiting candidates many more nearby candidates will be available via RV surveys. Figure 2 of Lovis et al. (2022) depicts the possible reflected light candidates for two different IWAs for ELT at 0.75 and 1.5 μm. Although the TRAPPIST-1 planets (Gillon et al. 2016) are beyond the reach of HIRES reflection spectroscopy because they are within the IWA, they will be accessible via transmission spectroscopy.

Given their capabilities for transmission, thermal and reflection spectra HIRES and METIS should allow us to disentangle the atmospheric chemical composition of exo-Venuses and exo-Earths within the habitable and possibly Venus zones (e.g. as shown for the Proxima Centauri b system by Turbet et al. 2016; Meadows et al. 2018) for nearby exoplanetary systems. They may be capable of catching a young exo-Venus in its magma ocean/steam atmosphere phase (e.g. Martins et al. 2013; Kawahara et al. 2014), possibly helping to constrain modelling studies (e.g. Matsui and Abe 1986; Elkins-Tanton 2008; Hamano et al. 2013; Lebrun et al. 2013; Salvador et al. 2017; Turbet et al. 2021).

HARMONI will leverage a combination of adaptive optics, a high-contrast imaging module, a medium resolution IFU (R up to 17 000) and a coronagraph to study exoplanets. The approach was first described by Sparks and Ford (2002) and in 2015 Snellen et al. (2015) demonstrated the potential for this combination for the ELT. Hoeijmakers et al. (2018) used a medium resolution IFS on the VLT SINFONI instrument (Eisenhauer et al. 2003) similar in many respects to HARMONI (but without a coronagraph) to characterize β Pic b. Hence the HARMONI instrument coupled to the ELT has tremendous potential for exo-Venus characterization. It is worth mentioning that a second generation high-contrast imager called PCS has been proposed for the ELT (Kasper et al. 2021). PCS would combine extreme adaptive optics with high spectral resolution exploiting the full potential of this technique on the ELT.

It may be possible to image accreting exoplanets in IR wavelengths (Mamajek and Meyer 2007; Miller-Ricci et al. 2009; Bonati et al. 2019). Miller-Ricci et al. (2009) predicted several near infrared windows that would allow detection of a magma ocean. However, if water vapor is a major component of the atmosphere (which is not a given, see work by e.g. Bower et al. 2022) Goldblatt et al. (2013, see Supplementary Information) has shown that the atmosphere may be opaque at most optical and IR wavelengths making characterization problematic. As mentioned above, the ELT HIRES & METIS instruments may have the capabilities to characterize not only the magma ocean and steam atmospheres (e.g. Lupu et al. 2014; Hamano et al. 2015; Bonati et al. 2019), but may also tell us if modelling studies of a temperate Venus (Way et al. 2016; Way and Del Genio 2020) are correct to place it in the habitable zone in its early history. The study by Bonati et al. (2019) points to a K-band window around 2.2 μm being optimal at ELT with the smallest inner working angle

---

[15] https://www.eso.org/sci/facilities/develop/instruments/crires_up.html.

of 24 milliarcseconds, but calculations by Turbet et al. (2021) could imply that the shorter wavelengths offered by HIRES may prove sufficient.

A number of studies have shown that it may be possible to detect the rotation rate, and other surface features such as ocean glint from single pixel images or low resolution spectroscopy of exoplanets (e.g. Pallé et al. 2008; Robinson et al. 2014; Fujii et al. 2014; Lustig-Yaeger et al. 2018; Jiang et al. 2018; Gómez-Leal et al. 2016; Mettler et al. 2020; Ryan and Robinson 2021; Li et al. 2021). Rotation rate in particular has direct application to Venusian studies. Venus' present day retrograde rotation rate and how it might have come about has been studied for decades (see Hoolst 2015, for a review). A variety of explanations have been put forward for its present-day obliquity and slow rotation rate, from impactors (e.g. McCord 1968), solid-body tidal dissipation (e.g. MacDonald 1964; Goldreich and Peale 1966; Way and Del Genio 2020), core-mantle friction (Goldreich and Peale 1970; Correia and Laskar 2001; Correia et al. 2003; Correia and Laskar 2003), oceanic tidal dissipation (Green et al. 2019), to atmospheric tides (Ingersoll and Dobrovolskis 1978; Dobrovolskis and Ingersoll 1980; Dobrovolskis 1980, 1983). Investigators have used Earth observation satellites, such as DSCOVR[16] (Jiang et al. 2018), and space missions such as EPOXI[17] (Robinson et al. 2014) for exoplanetary purposes. For example, DSCOVR has a charged coupled device array 2048 × 2048 pixels with sizes of 15 μm. Wavelength coverage is from 200 to 950 nanometers. Jiang et al. (2018) shrank the DSCOVR high-resolution 2-D images down to a single pixel and successfully extracted estimates of the land/ocean ratio and rotation rate. This implies that with a sufficient cadence, the same single pixel 'images' we obtain for exoplanets may allow us to constrain their rotation rate (Li et al. 2021) and possibly land/sea ratio. Robinson et al. (2010) also demonstrated that it may be possible to use JWST to detect ocean glint in single pixel images of extrasolar planets, but would require an external occulter which is not available. With similar techniques, we can hope to get better statistical constraints on exo-Venus rotation rates. We could also gain new insight on the causes behind Venus' present-day rotation rate and what it might have been in the distant past. The importance of discerning the rotation rate of planets in the VZ cannot be understated as it can be tied back to the slowly rotating cloud-albedo feedback seen in GCM models that may allow temperate climates under high insolations as discussed in Sect. 1. As well, observing glint in an planet in the VZ would also be an important discovery as it would show that VZ planets do exist in the liquid water habitable zone (Kasting et al. 1993; Kopparapu et al. 2013, e.g.). On the other hand if no glint nor cloud-albedo feedback is seen in slow rotators in the VZ then this would make a good case for Venus never having been in the habitable zone.

1.4 The Importance of Primordial & Basal Magma Oceans

Magma oceans are likely ubiquitous during the early history of terrestrial planets. During the accretion of Venus-sized planets, the gravitational energy released from gathering their mass is sufficient to melt their entire mantles (e.g. Elkins-Tanton 2012, and references therein). Giant impacts can provide additional energy. Early mantle melting is also favored by radiogenic heating of short-lived isotopes (Merk et al. 2002), the loss of potential energy during core formation (Sasaki and Nakazawa 1986; Samuel et al. 2010) and by tidal heating if one or several moons orbit the planet (Zahnle et al. 2007). Additional energy sources are available for planets that orbit close to their parent stars (e.g., in the Venus Zone around

---

[16]https://solarsystem.nasa.gov/missions/DSCOVR.

[17]https://www.nasa.gov/mission_pages/epoxi.

M dwarfs), including star-planet tidal heating (e.g. Driscoll and Barnes 2015) and, speculatively, magnetic induction (e.g. Kislyakova et al. 2017). Observations of young exoplanets can help test several hypotheses about the early atmosphere and magma ocean of Venus-like planets.

Salvador et al. (2023), Gillmann et al. (2022, this collection) contain a detailed discussion on Venus' primordial and basal magma oceans. Briefly stated, historical models assumed that Earth and Venus had primordial magma oceans that were overlain by an outgassed, dense atmosphere mostly consisting of H_2O and CO_2 (Arrhenius et al. 1974; Jakosky and Ahrens 1979). As reviewed in Massol et al. (2016), the idea of a steam & CO_2 magma ocean atmosphere continued to be the dominant hypothesis, although recent work has begun to question the simplicity of this formulation (Lichtenberg et al. 2021; Bower et al. 2022; Gaillard et al. 2022). Several 1-D models provide predictions about the longevity of the magma ocean in relation to the distance of Venus from its host-star (Matsui and Abe 1986; Elkins-Tanton 2008; Hamano et al. 2013; Lebrun et al. 2013; Salvador et al. 2017), but cannot conclusively constrain the timescale of the blanketing atmosphere. Either Venus' magma ocean was short-lived like that of Earth ($\sim$ 1 Myr), allowing water to condense on the surface, or so long ($\sim$ 100 Myr) that the steam atmosphere is photodissociated, with hydrogen loss via atmospheric escape and oxygen absorption by the magma ocean (see Westall et al. 2023; Salvador et al. 2023, this collection). Recent 3-D atmospheric modelling by Turbet et al. (2021) has shown that the steam atmosphere and subsequent magma ocean lifetime could be long, leading again to a desiccated atmosphere during the magma ocean phase. Their model examined N_2, H_2O and CO_2 constituents from 1–30 bar in partial pressure. While these results should be confirmed by another 3-D GCM, their importance cannot be overstated, as it may determine whether Venus kept most of its primordial water or not, and whether water ever condensed on the surface of Venus. See Salvador et al. (2023, this collection) for a more detailed discussion.

To inform studies of Venus, scientists should seek to determine how atmospheric properties vary with the intensity of incident starlight, especially for very young exoplanets. If models that feature an early steam atmosphere for Venus are correct, then we should expect to find steam atmospheres around Venus-like exoplanets that are < 100 Myr old (see Salvador et al. 2023, this collection). Under some critical threshold of stellar insolation, steam atmospheres may quickly condense into surface oceans. For example, Turbet et al. (2021) suggested that this threshold was 92% of Earth's present-day insolation, meaning that Earth narrowly escaped a Venusian fate. However, this critical value can vary depending on the details of the atmospheric model and uncertain parameters (Hamano et al. 2013; Lebrun et al. 2013; Goldblatt et al. 2013; Kopparapu et al. 2013). The predicted mass and composition of the magma ocean atmosphere results from the partitioning of volatile elements between the melt and the gas phase which is primarily controlled by their solubility within the melt and depends on the redox state of the magma ocean and thus the bulk composition of the exoplanet (e.g. Katyal et al. 2020; Barth et al. 2020). Observations of stellar composition can provide meaningful, but not exact, constraints on the compositions of terrestrial exoplanets (e.g. Hinkel and Unterborn 2018; Adibekyan et al. 2021). While magma ocean outgassing is generally thought to be efficient because of the vigorous convection and associated velocities, other mechanisms, such as interstitial trapping of volatile-rich melt (Hier-Majumder and Hirschmann 2017), could drastically alter this view and result in alternative outgassing scenarios (e.g., Ikoma et al. 2018). Furthermore, the convective dynamics and associated patterns might significantly increase the degassing timescales (Salvador and Samuel 2022). Then, magma ocean degassing efficiency would decrease with the planet size and increase with the initial water content. Because of its thermal blanketing effect, the outgassing rate of the atmosphere might strongly affect the cooling of the magma ocean and lead to divergent planetary evolution paths and resulting surface conditions. Many other parameters affecting mantle evolution and mixing such as the rotation rate or the crystallization sequence could significantly affect the volatile distribution and resulting outgassing with time. Yet, they have been poorly studied in the frame of volatile degassing. Thus a complete understanding of the interplay between magma ocean cooling rate, outgassing and their influence on post-MO mantle convection regime and surface conditions is still lacking. Ultimately, a large sample size of exoplanets is needed to derive statistical conclusions.

Detailed characterization of terrestrial exoplanets will remain difficult for at least the next decade. Schaefer and Parmentier (2021) provide a summary of some technical pitfalls. However, some hot, bright planets that orbit very close to their parent stars can be studied with modern technology. For example, observations of the infrared phase curve of the terrestrial exoplanet LHS 3844b, collected with the Spitzer Space Telescope, revealed that it does not have a substantial atmosphere (e.g. Kreidberg et al. 2019), which is consistent with a volatile-poor bulk composition (e.g. Kane et al. 2020) or with low outgassing rates. Future observatories could potentially use the direct imaging technique to detect superficial magma oceans for planets that also have thin or nonexistent atmospheres (Bonati et al. 2019). Alternatively, planets with huge amounts of outgassing from a magma ocean might have an atmosphere that is thick enough to affect mass-radius measurements (Bower et al. 2019). In the same way, the partition of water between the atmosphere and the magma ocean of water-rich exoplanets can affect their calculated radii by up to 16% in some cases (Dorn and Lichtenberg 2021), which would be enough to be tested for close-in bodies, and help understand the evolution of water budget in terrestrial planets. Furthermore, planets sustaining relatively long (~ 100 Myr) magma ocean states under a runaway greenhouse due to their proximity to the host star (Hamano et al. 2013, type-II planets) might also be distinguishable by a radius inflation effect (Turbet et al. 2019, 2020), thus providing additional constraints. In the history of exoplanetary studies, planets with extreme properties (e.g., hot Jupiters) were often the easiest and thus the earliest to be studied. Significant technical advances are needed to explore true exoplanetary analogues to Earth and Venus (see Sect. 1.3).

2 How Can Venus Inform Exoplanetary Studies

Our nearest planetary neighbor provides one of the end members of terrestrial habitability in our solar system. With its thick present-day atmosphere and inhospitable surface conditions, Venus is considered to be too close to our sun to be within the habitable zone, but was Venus ever within the habitable zone? The latter concept would be surprising to any modern-day climate scientist. How can a world that was receiving, 4.2 billion years ago, 1.4 times the incident solar radiation that Earth receives today be inside the habitable zone? As discussed above and in (e.g. Westall et al. 2023, this collection), an efficient cloud albedo feedback from a slowly rotating Venus may have kept ancient Venus temperate according to GCM modeling (Yang et al. 2014; Way et al. 2016) assuming sufficient surface liquid water and a short lived magma ocean phase (Hamano et al. 2013). If these GCM results are correct, we can expect to find habitable worlds well within the VZ around G-dwarf stars. For planets in the VZ of M-dwarfs, GCM results demonstrate severe limitations in the greater than modern-day Earth solar insolations (1361 W m^{-2}) allowed by the redder spectral energy distribution of such host stars (Kane et al. 2018). This is because Earth-like atmospheres are highly efficient at absorbing and trapping the infrared radiation of M-dwarfs, preventing the high insolations and temperate climates seen in GCM exoplanet modelling studies of VZ planets around G-dwarfs (Yang et al. 2014; Way et al. 2018). As well, the (likely tidally-locked) planets around low mass stars tend to "rotate" much faster (i.e. shorter orbital periods) than around more massive stars. This results in a reduced cloud albedo feedback at the substellar point (e.g. Kopparapu et al. 2017). Venus can also become a point of reference when it comes to the behaviour of its interior. For example, it is still debated if Venus' mantle convection is indeed in a stagnant lid regime at present-day, as has long been theorized (Solomatov 2004). However, Venus provides many more clues about the state of its mantle than any exoplanet, and can help discriminate between the multiple scenarios highlighted by numerical studies (Ballmer and Noack 2021). Finally, most mechanisms at work on Venus (or Earth), are likely to also affect exoplanets, in one form or another. Venus' ability to inform exoplanetary studies goes beyond providing us with an example of the atmospheric signature of a planet in a runaway greenhouse state with an inhospitable climate: Venus can also help us understand planetary evolution more generally. For these reasons it is important to understand how our present-day and near-future understanding of Venus can inform the study of exo-Venuses. In the rest of this article, we will provide an overview of our understanding of Venus through time.

2.1 Volatile Cycling and Weathering on Venus Through Time

In addition to a thick, CO_2-dominated atmosphere, resulting in an extremely hot climate, Venus also lacks modern Earth-style plate tectonics (e.g. Breuer and Moore 2007) and a strong, intrinsic magnetic field. The exact style of tectonics Venus currently exhibits is not well known, due, in large part, to the difficulty in mapping the Venusian surface in sufficient detail. Venus does not appear to fall neatly within either the plate-tectonic or stagnant-lid end-member regimes of tectonics. Although there is no evidence for a global network of plate boundaries and mobile plates, there are regions of the Venusian surface with features strikingly similar to subduction zones on Earth (e.g. Davaille et al. 2017; Gerya 2014b; Sandwell and Schubert 1992). Moreover, there is evidence for the motion of discrete crustal blocks on Venus, though it is difficult to constrain when this motion may have occurred during Venusian history (Byrne et al. 2021). Finally, Venus' lithosphere is estimated to be thinner than what would be expected if the planet were in a stagnant-lid state (Borrelli et al. 2021).

These significant differences in the magnetospheric, tectonic, and climatic state of Venus compared to Earth also possibly led to significant differences in atmospheric retention, surface weathering, and volatile cycling. Understanding these differences is crucial for interpreting future atmospheric observations from exoplanets, in particular those in the "Venus zone" (Kane et al. 2014) that are thus likely to also be in a runaway greenhouse state. In this section, we will explore how Venus' current state leads to different weathering, volatile cycling, and atmospheric retention processes and behavior than operate on Earth.

Like all rocky planets, Venus' climate is likely coupled to the interior (e.g. Gillmann and Tackley 2014) and the magnetosphere (e.g. Foley and Driscoll 2016). The hot, thick CO_2 greenhouse climate may be both a cause and a consequence of Venus' lack of plate tectonics. Likewise, the presence or absence of a magnetic field may be controlled by the style of tectonics the planet exhibits. Meanwhile, atmospheric evolution is influenced by the magnetosphere, which alters rates of atmospheric escape (see Sect. 2.3). Such atmospheric evolution then affects the climate, feeding back to interior processes (see Gillmann et al. 2022, this journal).

Coupling between surface and interior opens up further questions about the evolution of Venus and how it informs exoplanet studies. Do planets that experience a runaway greenhouse necessarily also lose plate tectonics and the operation of a core dynamo? Are runaway greenhouse climates, and their subsequent impact on a planet's interior always externally driven (e.g. due to changes in stellar luminosity), or can they be internally driven as well (e.g. due to changes in tectonics or rates of volatile outgassing via volcanism)? Are the current surface conditions inherited from the cooling of an early magma ocean stage or the results of the long-term evolution? Studying Venus' history can help shed light on these questions. We therefore structure this section as follows: first, we outline the weathering, and volatile cycling that operate on Venus today; next, we discuss how these processes might have evolved throughout Venusian history, and what constraints we have on this evolution; finally, we discuss how these processes are coupled to the interior evolution, and how this coupling could dictate rocky planet evolution in general.

2.1.1 Volatile Cycling and Weathering on Present-Day Venus

Volatile cycling on Earth is driven by volcanic outgassing from the interior and weathering processes, which reincorporate outgassed volatiles into rocks at the surface. The latter is typically facilitated by water-rock reactions, and ingassing of volatiles via the return of these volatilized surface rocks to the interior, typically through subduction. On Venus, the extremely hot climate, lack of liquid water at the surface, and lack of global-scale plate tectonics means volatile cycling, to the extent it can occur, must behave very differently than on Earth.

Some of the key volatiles for the evolution of Venus' atmosphere and surface environment are C, H, N, and S. Considering C & H first, there is a clear dichotomy in these species at the surface and in the atmosphere between Earth & Venus today: Venus' surface is dry and the atmosphere is dominated by ~ 90 bars of CO_2 (e.g. Mogul et al. 2022), while on Earth liquid water is abundant and CO_2 is only a trace gas in the atmosphere. This dichotomy leads to significant differences in weathering, but may also have been caused by differences in weathering.

2.1.2 Weathering

On Earth, the carbonate-silicate cycle operates to regulate the amount of CO_2 in the atmosphere, and maintain a temperate climate throughout most of Earth's history (e.g. Walker et al. 1981; Berner 1993; Kasting 1993). Silicate weathering is the primary mechanism for removing CO_2 from the atmosphere in this cycle, and the dependence of the rate of silicate weathering on climate state creates a negative feedback. Weathering on the modern Earth is driven by reactions between exposed rock on Earth's surface, as well as rock on the seafloor near mid-ocean ridges (e.g. Brady and Gíslason 1997; Coogan and Gillis 2013; Coogan and Dosso 2015; Krissansen-Totton et al. 2018), and CO_2 dissolved in rainwater and the oceans. Liquid water is therefore critical, and weathering will be severely limited on a planet lacking liquid water, like Venus. There is some chemical reaction between Venus' CO_2-rich atmosphere and surface rocks (see Gillmann et al. 2022, this journal for a detailed discussion), as evidenced by carbonate-rich coatings, which may form as an intermediate step in weathering of Venus' surface (Dyar et al. 2021). Nevertheless, the slow gas-solid reactions and the limited erosion in the absence of water prevents the efficient consumption of atmospheric CO_2 by the formation of carbonates (Zolotov 2019). In addition, carbonates are thermodynamically unstable at Venus' surface, where they react with sulfur species, in particular SO_2, from the atmosphere to form sulfates (Gilmore et al. 2017). Indeed, the elevated bulk sulfur content of 0.65 ± 0.40 wt% and 1.9 ± 0.6 wt% recorded at the Venera 13 and Vega 2 landing sites, respectively (Surkov et al. 1984, 1986) indicates net trapping of sulfur-bearing phases from the atmosphere into surface rocks (Zolotov 2019). All told, the lack of liquid water on Venus today means that weathering cannot act as an efficient removal process for atmospheric CO_2.

Such inefficient silicate weathering could in fact partly explain why Venus' present-day atmosphere is CO_2 dominated. Without weathering to remove it, CO_2 continuously accumulates in the atmosphere, as volcanic degassing from the interior proceeds. Earth contains a similar amount of CO_2 locked in carbonate rocks as exists in the Venusian atmosphere today (e.g. Ronov and Yaroshevsky 1969; Holland 1978; Lécuyer et al. 2000), thanks to active weathering processes on the Earth.

Another key factor is that weathering on Earth is also tied to tectonics. For weathering to be continuously active, erosion is needed to transport weathered rock away, and expose fresh rock. In the extreme case where there is no erosion whatsoever, weathering would cease entirely once a layer of weathered rock formed at the surface, as ground water would be unable to reach fresh, weatherable rock. A less extreme, and more common scenario, is when the rate of silicate weathering becomes limited by the supply of fresh rock brought to the near surface environment by erosion. In this case, all climate feedback involved in silicate weathering is lost; the weathering rate depends only on the erosion rate, as all fresh rock is weathered nearly instantly when brought into the weathering zone near the surface. Weathering reaching this state of being globally "supply limited" is another potential mechanism for forming a CO_2 dominated, hothouse climate, even if liquid water is still present on a planet's surface (e.g. Foley 2015; Kump 2018).

Silicate weathering is also linked to the land area of the planet: Wind and rainfall on emerged continents promote erosion and, in turn, the rate at which new surface is exposed. A large land area is however not vital for a stable climate: On a planet largely covered by oceans, seafloor-weathering dominates and can regulate the atmospheric CO_2 to some extent (e.g. Foley 2015; Höning et al. 2019; Krissansen-Totton et al. 2018).

As erosion rates are ultimately bounded by rates of tectonic uplift, it has been previously argued that plate tectonics might be essential for silicate weathering (e.g. Kasting and Catling 2003). As a result, another possible explanation for Venus' present-day atmospheric state could be that a lack of plate tectonics limits silicate weathering, allowing volcanically outgassed CO_2 to build up in the atmosphere. However, even without plate tectonics there are processes, such as volcanism, that act to supply weatherable rock to the surface. So whether a lack of plate tectonics leads to a hothouse climate depends on whether these other processes can supply enough fresh, weatherable rock to keep pace with CO_2 outgassing. Foley and Smye (2018) argue that even in a stagnant-lid regime, volcanism provides a sufficient supply of weatherable rock to sustain temperate climates. This study considered outgassing of CO_2 from the mantle and from decarbonation of crustal carbonate as it is buried by fresh lava flows, and found that a much higher concentration of CO_2 in erupted magma than on the modern Earth would be needed for a hothouse climate to form. However, the amount of CO_2 outgassed also depends on the types of materials through which magmas penetrate on their way to eruption (e.g. Henehan et al. 2016). If magmas erupt through C-rich crustal rocks, more CO_2 can be released than one would expect based on mantle CO_2 concentration alone. For example, in the case of the Siberian Traps, volatile release likely outweighed weathering as a result of magma interaction with crustal rocks (e.g. Svensen et al. 2009). However, such high CO_2 degassing rates may be anomalous and, geologically speaking, short-lived, as they require magmas to first hit regions where crustal rocks are C-rich, and then can only be maintained until these pockets of C-rich crustal rocks have been exhausted. Maintaining a permanent hothouse climate with liquid water present would require CO_2 degassing rates to continuously exceed silicate weathering rates through the planet's lifetime.

It therefore remains unclear exactly how the present atmosphere of Venus came about if there was an earlier temperate period (Head et al. 2021). A loss of water due to a runaway greenhouse climate would almost certainly lead to the buildup of a thick CO_2 atmosphere, as long as volcanism was still active. A lack of plate tectonics, with liquid water still present, could impede weathering to the point where a hothouse climate forms, but this would require either a CO_2-rich mantle or for magmas to interact with C-rich rocks as they erupt; without either of these two conditions weathering can still maintain a temperate climate even in a stagnant-lid regime of tectonics.

Whether the tectonic regime or the presence of liquid water is the more significant limitation on weathering processes has important implications for exoplanets. If weathering is not strongly affected by tectonic regime, then one does not need to know a planet's tectonic regime in order to assess whether a carbonate-silicate cycle, capable of sustaining habitable surface conditions, can operate. Estimating an exoplanet's tectonic state from remote observations will be a significant challenge, so testing whether habitability is possible without plate tectonics is critical for exoplanet studies. Future Venusian exploration can help test the importance of tectonics for weathering and habitability. If Venus is shown to have had active silicate weathering in the past, while also lacking plate tectonics, then we would have direct evidence that plate tectonics is not necessary for the carbonate-silicate cycle. On the other hand, if Venus' history indicates the loss of water through a runaway greenhouse was the primary causal factor for Venus' CO_2-rich atmosphere, then we'd expect exoplanets that have experienced runaway greenhouses to have similar atmospheric states. Such expectations can be tested with future observations, as outlined in Sect. 1. Going further, exploring when and why the carbonate-silicate cycle ultimately failed to regulate the climate on Venus, as must have happened at some point during Venus' history, would offer clues to the conditions for habitability of terrestrial planets (see also Westall et al. 2023, this collection).

2.1.3 Volcanism & Outgassing

Weathering is not the only aspect of the carbonate-silicate cycle that is essential for regulating atmospheric CO_2 levels. Volcanic outgassing is also necessary, at sufficiently high rates, to maintain enough CO_2 to prevent global glaciation (e.g. Walker et al. 1981; Kadoya and Tajika 2014; Foley and Smye 2018; Stewart et al. 2019). Venus today is of course near the other extreme limit, with a CO_2 dominated atmosphere, rather than a CO_2 poor one. However, the importance of volcanic outgassing to rocky planets in general highlights the question of whether Venus is actively outgassing today.

The variations of SO_2 in the atmosphere of Venus have been recorded by Venera 12 (Gelman et al. 1979), Pioneer Venus (Oyama et al. 1980; Esposito 1984) and Venus Express (Marcq et al. 2013). Combined with models these can give estimates of the column sulfur abundance (e.g. Schulze-Makuch et al. 2004; Krasnopolsky 2016). The variations of SO_2 and the maintenance of the H_2SO_4 cloud layer on Venus have been suggested to indicate volcanic activity. Since SO_2 reacts with calcite ($CaCO_3$) on the surface of Venus to form anhydrite ($CaSO_4$), it will be consumed unless replenished by volcanism. Following Gilmore et al. (2017) this can be written as $CaCO_3$(calcite)+1.5 SO_2(gas)$\rightarrow CaSO_4$(anhydrite)+CO_2(gas)+0.25 S_2(gas). Fegley and Prinn (1989) calculated a sulphur removal rate of 2.8×10^{13} g yr^{-1}. In order to maintain the global H_2SO_4 cloud layer, this removal rate needs to be balanced by a volcanic outgassing rate of 5.6×10^{13} g yr^{-1} or 1.1 Pa kyr^{-1} SO_2. Depending on the S/Si ratio of erupted material, Fegley and Prinn (1989) estimated the equivalent global volcanic eruption rate to 0.4–11 km^3/yr. This rate is lower than the total average output rates on Earth of about 26–34 km^3/yr, of which about 75% are contributed by ocean-ridge magmatism (Crisp 1984), while recent work by Byrne and Krishnamoorthy (2022) implies that Venusian volcanic rates should be similar to those on modern Earth. It should be noted, however, that atmospheric dynamics and chemistry may be responsible for the variability of sulfur species in the atmosphere of Venus (Hashimoto and Abe 2005; Marcq et al. 2013). The measurements mentioned above will be improved upon with mass spectrometer observations from the upcoming DAVINCI mission (Garvin et al. 2020)[18] which will help to better constrain column abundances of sulphur and a number of other species. As well, the DAVINCI in-situ infrared (IR) imaging camera should help connect surface observables to the orbiting IR and radar instruments on VERITAS and Envision (Widemann et al. 2022) to confirm or refute previous indications of on-going volcanism (e.g. Smrekar et al. 2010; Shalygin et al. 2015; Gilmore et al. 2017) as a possible sulfur source, and provide valuable insight to exoplanet studies.

Remote observations of H_2O and HDO have been made from Venus' orbit (e.g. Cottini et al. 2012), from Earth ground based instruments (e.g. Encrenaz et al. 1995; Sandor and Clancy 2005), and from in-situ instruments on the Pioneer Venus large probe and Venera 15 (Donahue et al. 1982; Koukouli et al. 2005). A compilation of H_2O measurements by De Bergh et al. (2006) gives atmospheric column values from 20–45 ppmv with one measurement at 200 ppmv. It is generally assumed that H_2O sources are volcanic like those of its sulphur counterparts (e.g. Fegley 2003, 2014; Truong and Lunine 2021).

Tying the abundances of N_2 in the upper atmosphere to lower atmosphere abundances remains challenging (e.g. Peplowski et al. 2020). N_2 as the second most abundant gas in the Venusian atmosphere is often overlooked, but it corresponds to nearly four times the atmospheric abundance on Earth when scaled by planetary mass. Here again the DAVINCI mission will give more accurate column abundances of N_2 and in combination with photochemical modelling (e.g. Krasnopolsky 2012) may help us to better understand the upper atmosphere abundances and how those tie to possible surface sources and the N_2 cycle in general. N_2 is certainly a challenging gas to detect in exoplanetary atmospheres, but Schwieterman et al. (2015) has shown that it may be possible.

Future atmospheric characterization of exoplanets can also help test models of volcanic outgassing, by potentially identifying ongoing volcanic activity on such planets. SO_2 has been proposed as a proxy for explosive volcanism (Kaltenegger et al. 2010), as well as sulfate aerosols (Misra et al. 2015). Sulfate aerosols are formed during volcanic eruptions and have a lifetime of months to years in the atmosphere; as such they may be detectable in transit transmission spectra (Misra et al. 2015). Venusian measurements are critical to helping us constrain the longevity and rate of volcanism on rocky exoplanets – a key question for interpreting future atmospheric observations performed by upcoming missions such as JWST and ELT (see Sect. 1.3). Additional modelling studies have investigated volcanism and outgassing of terrestrial exoplanets (Kite et al. 2009; Tosi et al. 2017; Noack et al. 2017; Dorn et al. 2018; Foley and Smye 2018; Foley 2019). These studies provide predictions for how long volcanism can last on planets in different tectonic regimes, with different sizes, heat budgets, and material properties. On Exo-Venus-like planets with an atmosphere similar to that of Venus, the signal of SO_2 and other volcanic gases needs to be detected above an optically thick lower atmosphere. However, volcanic gas plumes are less buoyant in a hot and dense atmosphere and may thus not reach high enough altitudes compared to altitudes reached in otherwise thinner and colder atmospheres (Henning et al. 2018).

In addition, analogs of present-day Venus may present a featureless spectra both in transit transmission and in direct imaging (see Sect. 1.2 and Fig. 4), making their characterization difficult (Arney and Kane 2018; Fauchez et al. 2019). Nevertheless, these challenges further

---

[18]https://www.nasa.gov/feature/goddard/2021/nasa-to-explore-divergent-fate-of-earth-s-mysterious-twin-with-goddard-s-davinci.

emphasize the necessity of additional Venus exploration. By studying Venus' present-day atmosphere, interaction with any present-day volcanism, and the evolution of the atmosphere over time, we could test these proposed proxies for exoplanetary volcanism, and perhaps develop more effective ones.

As mentioned above, studying Venus' evolution may help constrain further predictions from models of exoplanet outgassing and climate evolution. For example, in a study employing parameterized thermal evolution modelling and mantle outgassing, Tosi et al. (2017) investigated the habitability of a stagnant lid Earth (an Earth-like planet without plate tectonics) and found that depending on the mantle redox conditions, several hundreds bar of CO_2 may be outgassed. Moreover, models of mantle melting and volatile partitioning suggest that the chemical composition of the atmosphere and the dominant outgassed species are strongly controlled by the redox state of the mantle (Ortenzi et al. 2020). For sulfur species both fO_2 and water content are critical (Gaillard and Scaillet 2009, 2014). For a given water content, the outgassed sulfur increases for increasing fO_2. For oxidising conditions, SO_2 is the dominant sulfur species irrespective of the water content. For reduced conditions, SO_2 and S_2 are the dominant sulfur species for hydrated melts (Gaillard and Scaillet 2009). At the same time surface pressure also affects the final composition of the gases released into the atmosphere. For example, high surface pressures may limit outgassing of water, because the solubility of the latter in surface lava significantly increases for atmospheric pressures larger than 10 bar (Gaillard and Scaillet 2014). Under present-day Venus surface pressures, the most dominant outgassing species is CO_2, while only a small portion of SO_2 and water is expected to be outgassed, due to their high solubility in surface lava (Gaillard and Scaillet 2014). If constraints on Venus' interior oxidation state can be placed by measuring atmospheric H_2/H_2O and temperature (e.g. Sossi et al. 2020), then results from these models can potentially be tested by both the present-day atmospheric makeup, and whatever constraints on the long-term evolution of the atmospheric composition are developed from future missions. This ability to benchmark outgassing models against Venus will improve our predictions for the atmospheres of exoplanets. Future missions will be used to constrain the present-day atmospheric composition and perhaps surface water abundances. These are particularly interesting as they may be directly related to mantle water abundance which would help constrain the range of water content-dependent parameters associated with mantle melting (e.g., Hirschmann 2006; Ni et al. 2016) and convective dynamics such as viscosity and density (e.g., Lange 1994).

Venus may also be able to help us to predict the evolution and habitability of terrestrial exoplanets more generally. Since most exoplanets detected thus far are larger than Earth and Venus, a scaling of the main physics with planet size and mass is crucial. For Venus-like planets with a similar relative core mass fraction, the planet mass can be directly derived from its size (Valencia et al. 2006). When exploring the habitability of massive planets, it is important to attempt to quantify the volcanic outgassing rate which controls the atmospheric partial pressure of CO_2 regardless of their tectonic state. On the one hand, the mantle temperature generally increases with the size of a planet, which increases the strength of convection and the melting depth. This favours an increasing outgassing rate with planet size. On the other hand, the pressure gradient is higher in more massive planets, which reduces the strength of convection and the melting depth, favoring smaller outgassing rates of massive planets. The melting depth is particularly important for stagnant-lid planets, since on a planet with plate tectonics, mantle material can rise to the surface at mid-ocean ridges. An additional important factor to be considered for massive planets is the buoyancy of partial melt, which needs to be positively buoyant in order to rise to the surface. Since the density of melt increases more strongly with pressure than solid rock, only melt that forms below a certain pressure contributes to volcanic outgassing (Ohtani et al. 1995; Agee 1998). The above noted competing mechanisms typically lead to a higher degassing rate for planets between 2 and 4 Earth masses and a reduced outgassing rate for more massive planets (Noack et al. 2017; Dorn et al. 2018; Kruijver et al. 2021). Compared to smaller planets, high outgassing rates of large planets can last longer, since their larger ratio between volume and surface area implies a less efficient cooling. While for massive stagnant-lid planets, the above noted effects can even lead to a cessation of volcanism, (Noack et al. 2017; Dorn et al. 2018). This is not the case for planets with plate tectonics where the melting region is extended closer to the surface beneath mid-ocean ridges (Kite et al. 2009; Kruijver et al. 2021).

A recent study by Quick et al. (2020) finds that even massive exoplanets such as 55 Cancri e, an 8 M_E rocky exoplanet, might be volcanically active based on the estimated heat sources (radiogenic and tidal) available in their interior. Rocky exoplanets closely orbiting their parent star may experience volcanic activity focused only on one hemisphere, due to the strong surface temperature variations caused by their tidally locked orbit (Meier et al. 2021). Altogether, understanding physical processes that control volcanic outgassing of Venus throughout its evolution, and studying the sensitivity of these processes to planetary parameters such as size, bulk composition, and tectonic state, will greatly advance our estimates of the atmospheric composition of exoplanets.

2.1.4 Volatile Ingassing

As explained in Sect. 2.1.2 silicate weathering can regulate the amount of CO_2 in the atmosphere if liquid water is present on the surface. The carbon that is removed from the atmosphere eventually becomes stored in carbonate sediments, which are subsequently buried on the seafloor. The fate of these sediments on longer timescales is controlled by the tectonic regime of the planet. Plate tectonics allow for a relatively shallow temperature-depth gradient in subduction zones, which allows large parts of the carbonates to remain stable during subduction. On modern Earth, approximately half of the carbon that enters subduction zones is released at arc volcanoes, although this fraction strongly depends on the temperature-depth profile of the individual subduction zone (Sleep and Zahnle 2001; Dasgupta and Hirschmann 2010; Ague and Nicolescu 2014). The remaining carbon is subducted into the mantle, which closes the deep carbon cycle. On exoplanets with plate tectonics the fraction of subducted carbon that enters the mantle may differ significantly. On planets with higher plate speed, steeper angle of subduction and/or smaller mantle temperature, carbonates would not heat up as strongly during subduction and a larger fraction could remain stable. For example, cooling of the Earth's mantle during the past 3 Gyr could have enhanced the carbon fraction that enters the mantle by approximately 10% (Höning et al. 2019). On timescales of millions to billions of years, this variation can play a key role in the distribution of carbon between the mantle and the atmosphere.

Without plate tectonics, transporting carbon into the mantle is challenging. The slow sinking of carbonated crust, as it becomes buried by new lava flows, results in a thermal equilibrium with the surrounding rock. The bulk of the carbonates becomes unstable at a relatively narrow temperature interval (Foley and Smye 2018), which is usually exceeded within the stagnant lid. If the released CO_2 is transported with uprising lava or through cracks to the surface, recycling of carbon into the mantle is rare. As a result, the combined crust-atmosphere carbon reservoir on stagnant-lid planets would steadily increase with ongoing volcanic outgassing. Since the release rate of CO_2 from the crust into the atmosphere depends on the crustal carbon reservoir, an important consequence is that atmospheric CO_2 retains a memory of its initial value. The initial atmospheric CO_2 reservoir may be erased quickly, but if this then gets stored in the crust and is not recycled into the mantle, CO_2 release (and therefore atmospheric CO_2) in the subsequent evolution would still depend on the initial CO_2. However, on planets with plate tectonics, the initial carbon distribution becomes unimportant after some million years (Foley 2015), because of the recycling. Another important consequence is that weathering cessation could result in a dramatic rise of atmospheric CO_2, since all carbon that has been degassed during the entire history of the planet would accumulate in the atmosphere. In case of early Venus the atmospheric CO_2 concentration would have increased by approximately one order of magnitude within 100 Myr (Höning et al. 2021). Altogether, volatile ingassing strongly affects the long-term atmospheric evolution of a planet. Predicting volatile ingassing requires knowledge about the tectonic and thermal state of the planet and a precise understanding of the fate of released CO_2 in the crustal matrix.

As explained in Sect. 2.1, there maybe active subduction in localized regions of Venus today, possibly driven by lithospheric burial under plume-induced volcanism and subsequent rollback of the buried lithosphere (Gerya 2014b; Davaille et al. 2017). Although the Venusian crust is not highly volatilized today, due to the lack of liquid water and hence nearly non-existent weathering, this style of subduction could potentially drive volatile ingassing if it were active with liquid water present. Rates of ingassing possible with this style of limited subduction have not been well studied, but are likely much lower than ingassing rates seen with Earth-like plate tectonics. Venus exploration can thus potentially help constrain rates of volatile ingassing for planets that lie in between the end-member plate-tectonic and stagnant-lid regimes, and help inform the range of volatile cycling behavior that might be seen on exoplanets.

Bean et al. (2017) discussed a comparative planetology approach to test the habitable zone concept: If silicate weathering is generally temperature-dependent on exoplanets with liquid surface water, the atmospheric CO_2 concentration on the planet should decrease with increasing incident insolation, for example as a function of stellar type, age, distance between the star and the planet. When incident insolation exceeds a critical value, surface water would evaporate and weathering would cease. Therefore, we would expect to observe an abrupt increase of atmospheric CO_2 on planets at the inner edge of the habitable zone (Turbet 2019b; Graham and Pierrehumbert 2020). For stagnant-lid planets, this abrupt CO_2 increase might even be more pronounced, because volcanic degassing would be accompanied by a release of CO_2 from buried carbonates. From thermal evolution models coupled to a carbon cycle model for stagnant lid planets, Höning et al. (2021) predicted an increase of the CO_2 concentration on planets at the inner edge of the habitable zone of at least one order of magnitude.

2.1.5 Weathering and the Sulfur Cycle on Venus Today

The chemical interaction between the surface and atmosphere on Venus is particularly important as it can affect the sulfur cycle (see Gillmann et al. 2022, this collection). The latter plays a dominant role in the complex photochemistry and dynamics of Venus' atmosphere affecting sulphuric acid cloud formation (e.g. Fegley and Prinn 1989), the presence of an optically thick aerosol layer (Knollenberg and Hunten 1980) and variations of SO_2 atmospheric content (Esposito 1984; Marcq et al. 2013). While sulfur and other atmospheric species could be supplied to the atmosphere via volcanic activity, whose present-day level has large uncertainties (Mueller et al. 2017, and references therein), weathering processes act as a sink to remove these through complex multiphase chemistry. This is yet another area where exoplanet observations can play an important role in discerning not only the state of the atmosphere in a VZ planet, but may also provide some constraints on volcanic activity for a modern Venus-like world with measurable SO_2 abundances.

2.2 Venus' Magnetic Field

Venus lacks a global (i.e., strong) magnetic field today. As discussed in O'Rourke et al. (2023, this collection), any intrinsic magnetism in Venus must be relatively weak – specifically producing magnetic fields $\leq$ 5–10 times weaker at the surface than Earth's dynamo-generated field (Phillips and Russell 1987). However, we currently have no meaningful information about the magnetic history of Venus prior to the Mariner 2 flyby in 1962. Understanding why Venus has no global magnetic field now and whether one existed in the past is important for several reasons (e.g. Lapôtre et al. 2020; Laneuville et al. 2020). First, planetary magnetism is intrinsically interesting as a complex phenomenon (e.g. Stevenson 2003, 2010). Second, the absence (or presence) of a global magnetic field places constraints on models of planetary formation and thermal evolution. Finally, magnetic fields may play key roles in atmospheric escape processes over time (see Sect. 2.3 below). Studies of Venus provide clues about how magnetic fields will shape the evolution of exoplanets. At the same time, studies of exoplanets may elucidate if the magnetic aspect of the Earth/Venus dichotomy is a natural corollary to the differences in atmospheric conditions – that is, are the prospects for a long-lived, global magnetic field correlated with surface habitability?

Studying planetary magnetism is thus a "two-way street" between Venus and exoplanets (Lapôtre et al. 2020). Over the next few decades, we should advance our scientific understanding by both exploring Venus and searching for extrasolar magnetospheres. Various direct and indirect methods for detecting magnetic fields at exoplanets have been proposed. Space-based radio telescopes could search for direct radio emission (e.g. Driscoll and Olson 2011). Other ideas include searching for various types of auroral emission from exoplanets – or evidence of the interaction of stars and the stellar wind with magnetized exoplanets (e.g. Lazio et al. 2016; Vedantham et al. 2020; Pope et al. 2020). Brown dwarfs are the current frontier for direct detections of magnetic fields (e.g. Kao et al. 2018). Indirect evidence has been presented for the magnetic fields of hot Jupiters from stellar interactions (e.g. Cauley et al. 2019).

There are a number of geodynamic scenarios for Venus which may have implications for exoplanetary studies. Venus lacks a global magnetic field today because it does not have a strong dynamo operating in its deep interior. Although Venus rotates slowly compared to Earth, a dynamo would still exist if a large amount of electrically conductive liquid were churning vigorously. Such reservoirs (e.g., a metallic core that is at least partially liquid) might exist, but they are currently stagnant. Broadly speaking, two types of scenarios have been proposed to explain why no dynamo operates within Venus. These scenarios make different predictions about whether any crustal remnant magnetism might await detection on Venus. Moreover, these scenarios imply different predictions for what kinds of exoplanets will host global intrinsic magnetic fields.

The first type of story for Venus' magnetic history argues that the tectonic state of Venus prevents any dynamo from operating in the deep interior. As discussed in the previous section (and shown in Fig. 12), the interior of Venus is thought to cool more slowly than Earth's if its mantle operates in the episodic – and/or stagnant – lid regime. Venus could have a metallic core that has the same bulk composition and is chemically homogeneous, like Earth's core. However, iron alloys are thermally as well as electrically conductive (e.g. Williams 2018), so thermal conduction can transport all the heat from a slow-cooling core without any fluid motion. Earth's cooling rate is arguably only somewhat higher than the critical value required to sustain convection (e.g. Nimmo 2015; Davies et al. 2015; Labrosse 2015). Slow cooling is thus fatal to the chances for a dynamo in Venus at present-day (e.g. Nimmo 2002; Driscoll and Bercovici 2014; O'Rourke et al. 2018). This general conclusion also holds if Venus initially had a basal magma ocean (O'Rourke 2020). Critically, a dynamo seems more likely to have operated in the past. In this case, crustal remnant magnetism may provide a detectable record of an early dynamo (e.g. O'Rourke et al. 2019).

The second type of story proposes that the stochastic nature of the accretion of Venus doomed the chances for a dynamo from the start. Specifically, Jacobson et al. (2017) proposed that Venus did not suffer a late energetic impact. The absence of such an impact would mean that the core of Venus could have an onion-like structure where the outermost layers were added last. As proto-Venus grew, its interior grew hotter and had higher pressures. Core-forming material would thus equilibrate with silicates under progressively more extreme conditions, causing more light elements such as silicon and oxygen to partition into the iron alloy (e.g. Siebert et al. 2013; Fischer et al. 2015). This process would establish a stable density gradient in the core that prevents convection–material containing a few weight percent of extra light elements would need to cool by thousands of degrees (impossibly) to become negatively buoyant. This stable stratification would exist even if the core of Venus had the same bulk composition (and thus relative size) as Earth's. In this case, the subsequent thermal evolution of Venus is irrelevant to the prospects for a dynamo. No dynamo would exist even if the core cooled at Earth-like rates. Discovering any crustal remnant magnetism would thus probably disprove this scenario.

We can extrapolate predictions for exoplanets from these two types of stories about Venus. If tectonic state is the dominant factor, then Venus-like geodynamics should produce Venus-like magnetic histories. That is, a planet with a Venus-like atmosphere (and thus surface) would be less likely to have a long-lived global magnetic field (see Sect. 2.4) while modern Venus-like climates might be bad for plate tectonics (see Sect. 2.1). Planetary magnetism could thus serve as a probe of a planet's tectonic state, which is otherwise difficult to determine by observation. If planet-star distance controls atmospheric properties, then magnetospheres should be rare in the Venus Zone (VZ), but common in the habitable zone (HZ). In contrast, planet-star distance probably does not control the timing of giant impacts during planetary accretion (e.g. Rubie et al. 2015; Jacobson et al. 2017). If stochastic events are the dominant factor, then Venus-sized planets in both the VZ and HZ may or may not have magnetospheres. Hence the probability of a global magnetic field would not strongly depend on planet-star distance. Ultimately, exoplanets provide the large sample size necessary to tell us if Venus reflects general principles of planetary evolution, or if Venus trod an evolutionary pathway that is cosmically rare.

Planetary mass can also affect the prospects for a global magnetic field. The term "super-Earth" is often used for exoplanets with an Earth-like density but masses up to ~ 5 Earth-masses and ~ 1.5 Earth-radii (e.g. Rogers 2015; Weiss and Marcy 2014). However, this terminology may be misleading given the absence of definite facts about the surface of any super-Earth. Any massive planet, especially one in the VZ, could be a "super-Venus" with a Venus-like atmosphere and hellish surface conditions (e.g. Kane et al. 2013). All else being equal, larger planets are possibly more likely to host dynamos. Larger cores can have higher energy contents (e.g. Driscoll and Olson 2011) and, depending on their bulk composition, are still expected to grow solid inner cores that provide a strong power source for a dynamo (e.g. Boujibar et al. 2020; Bonati et al. 2021; van Summeren et al. 2013). Simple scaling laws predict that the actual cooling rate of the core would increase with planetary mass faster than the critical value required to drive convection (Blaske and O'Rourke 2021). Super-Venus (and super-Earth) planets are also likely to have basal magma oceans (Soubiran and Militzer 2018) made of liquid silicates that are electrically conductive enough to sustain a dynamo (e.g. Stixrude et al. 2020). Ultimately, a super-Venus could sustain a global magnetic field for much longer than Venus – meaning that tectonic state and dynamo occurrence might not correlate for massive exoplanets.

2.3 Atmospheric Escape and Importance of a Magnetic Field

Here, we discuss present-day observations of the terrestrial planets in our solar system with a focus on Venus, alongside simulations regarding the influence of a global magnetic field on atmospheric escape and habitability. These hold critical lessons for the longevity of exoplanetary atmospheres since the terrestrial worlds of our solar system hold the ground truth necessary to understand atmospheric evolution in general.

The lack of a global magnetic field at Venus today might lead one to believe that Venus' atmosphere is very vulnerable to the interaction with the solar wind, and thus to the loss of its atmosphere. The effect of the presence of a global magnetic field on atmospheric evolution via atmospheric escape has long been debated. The consensus was that a global magnetic field is important for protecting the atmosphere from being stripped by the solar wind (e.g. Lundin et al. 2007). However, recent spacecraft visiting the three terrestrial sibling planets, Venus, Earth, and Mars, have provided data to shed some new light on this question. Atmospheric escape rates for the three planets appear relatively similar (Strangeway et al. 2010). This new data is important in order to understand if a global magnetic field is necessary for terrestrial planets and exoplanets to retain their atmosphere despite loss caused by stellar radiation.

To understand the influence of solar wind on atmospheric evolution, we first have to compare the characteristics of the three planets. One of the major differences between them is that Venus and Mars do not have a global magnetic field, while Earth does. Secondly, the size of Venus and Earth is approximately the same, while the radius of Mars is about half of Venus' and Earth's. As a consequence, the mass of Mars is only a tenth of that of Venus or Earth. Third, while Earth's atmosphere is mainly composed of N_2 and O_2, Venus' and Mars' main atmospheric constituent is CO_2. Fourth, Mars has an atmospheric surface pressure of ≈ 6 mbar, Earth a comfortable 1 bar, and Venus a crushing 93 bar. Fifth, as Venus lies closer to the Sun, it resides in a harsher solar radiation and solar wind environment than Earth and Mars. Thus Venus receives about twice and five times more energy and solar wind particles from our host star than the other two planets. It may already be obvious that the solar wind cannot completely remove an atmosphere from a planet even when a global magnetic field is not present, as Venus has the thickest atmosphere of the three sibling planets.

However, we have no constraints on when Venus lost its magnetic field, nor the strength of any field it might have possessed (e.g. O'Rourke et al. 2018). Thus far, no crustal remnant field has been detected on Venus, as it has been on Mars (Acuna et al. 1999). The crustal remnant magnetic field on Mars tells us that Mars once had a magnetic field, and constraints on its strength can be approximated, even if it is vigorously debated (e.g. Langlais et al. 2019, and references therein). Many studies have asserted that remnant magnetism could not survive within the hot crust of Venus. However, at present-day, the surface is ~ 100 K below the Curie temperatures of common magnetic carriers such as magnetite and hematite. Therefore, crustal remnant magnetism could possibly have survived for billions of years, down to depths of a few kilometers (e.g. O'Rourke et al. 2019). A magnetometer survey below the ionosphere on a future mission could conduct the first capable search for crustal magnetization (O'Rourke et al. 2018).

A planet with a global magnetic field will interact with the solar wind and form a magnetosphere, such as at Earth. A planet without a global magnetic field will instead form an induced magnetosphere from the interaction between the solar wind and the ionosphere

(Luhmann et al. 2004), as at Venus and Mars. The difference is important for understanding how the solar wind can influence the escape rates from a planet, as different types of interactions cause different channels of escape to be important.

At Venus, the main escape channels are ion escape from ion pickup in the solar wind or ion acceleration in the magnetotail (for more details see the review of the main Venusian escape channels for O^+ and H^+ by Lammer et al. 2006 and in Gillmann et al. 2022, this collection). The O^+ ion escape rates at Venus have been estimated at $\sim 10^{24}$–10^{25} s^{-1} (Brace et al. 1987; McComas et al. 1986; Barabash et al. 2007; Fedorov et al. 2011; Persson et al. 2018, 2020; Masunaga et al. 2019). These escape rates were also found to be weakly dependent on the solar wind dynamic pressure and energy flux, but not so much with EUV flux (Edberg et al. 2011; Kollmann et al. 2016; Masunaga et al. 2019; Persson et al. 2020). In addition, extreme space weather, such as an Interplanetary Coronal Mass Ejection (ICME) events, may increase the escape rates by several orders of magnitude (e.g., Luhmann et al. 2007), for a time.

Mars' ion escape rates show a similar order of magnitude to Venus'. The O^+ escape rates lie in the range of 10^{24}–10^{25} s^{-1} (Bogdanov and Vaisberg 1975; Lundin et al. 1990; Nilsson et al. 2012; Ramstad et al. 2015; Brain et al. 2015; Dong et al. 2017; Nilsson et al. 2021; Scherf and Lammer 2021). In contrast with Venus, the O^+ escape rates at Mars were found to be inversely correlated with the solar wind dynamic pressure (Dubinin et al. 2017; Ramstad et al. 2018), but have a positive correlation with the EUV flux (Ramstad et al. 2015). Due to the lower gravity at Mars, and thus escape velocity, the ions need less acceleration in order to escape, compared to both Venus and Earth. A large part of escape at Mars is therefore the low energy ion escape, which also has a stronger correlation with upstream solar wind and solar XUV flux compared to their higher energy counterparts (Dubinin et al. 2017; Ramstad et al. 2017). The escaping ions of less than 50 eV were shown to contribute between 35–90% to the total ion escape (Ramstad et al. 2017). However, during space weather events it was shown that the high energy ion escape at Mars can increase as it does for Venus (Edberg et al. 2010; Jakosky et al. 2015). Hence even though Venus and Mars have the same type of interaction with the solar wind, the escape rates are not dependent on the same parameters.

Despite its strong global magnetic field, Earth displays escape rates of equal or even higher order of magnitude than both Venus and Mars. Several studies indicate average O^+ escape rates in the order of 10^{24}–10^{26} s^{-1} (e.g., Yau et al. 1985; Peterson et al. 2001; Andersson et al. 2005; Nilsson et al. 2012; Slapak et al. 2017; Schillings et al. 2019). The O^+ escape rates at Earth are closely related to geomagnetic activity, and increase with higher activity (e.g., Yau et al. 1985; Slapak et al. 2017). In addition, Schillings et al. (2019) showed that Earth's O^+ escape rate is strongly correlated with the solar wind dynamic pressure, but does not have a strong correlation with EUV flux.

A summary of the results from three studies on the average escape rates at Venus, Earth and Mars is shown in Fig. 8 as taken from Ramstad and Barabash (2021), where the heavy ion escape rates are presented as a function of the solar wind dynamic pressure. As is evident, the escape rates at Earth are higher and more dependent on the changes in the solar wind dynamic pressure than Venus and Mars. Gunell et al. (2018) went into the details on the effect of a global magnetic field on escape by running a set of simulations on how the H^+ and O^+ escape rates from a Venus-like, an Earth-like and a Mars-like planet would be affected by a change in the dipole magnetic moment of its core. The results of the simulations are shown in Fig. 9. They took into account the seven largest escape channels for magnetized and unmagnetized planets. The study gives us a similar picture to the recent measurements shown in Fig. 8: A magnetic field does not always protect the atmosphere, in some cases it can actually increase the escape rates. This conclusion was also supported by global MHD

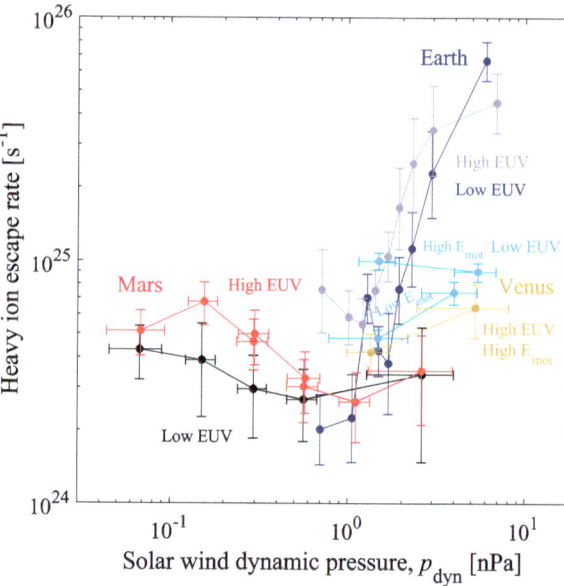

Fig. 8 Summary of measured heavy ion escape rates as a function of upstream solar wind dynamic pressure at Venus (blue and yellow, Masunaga et al. 2019), Earth (purple, Schillings et al. 2019) and Mars (black and red, Ramstad et al. 2018). Figure adapted from Ramstad and Barabash (2021)

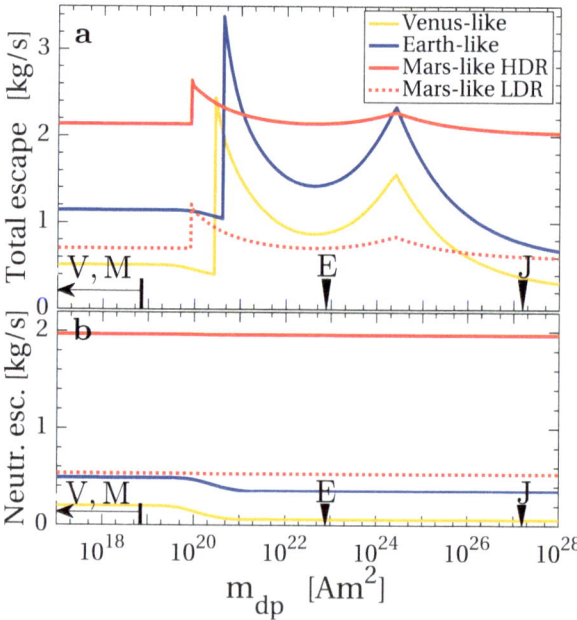

Fig. 9 Mass escape from Venus-, Earth- and Mars-like planets, for both neutral and ion (H^+ and O^+) escape, and how it varies with a change in the dipole magnetic moment of the planet. These are from model computations including seven of the most important escape channels. For Mars: LDR/HDR=Low/High Dissociative Recombination of molecular oxygen. Today's value of the magnetic moment of Venus (V), Mars (M), Earth (E), and Jupiter (J) is indicated. From Gunell et al. (2018)

simulations of Venus- and Earth-type exoplanets by Dong et al. (2020). This means that the global magnetic field is not the only characteristic that determines the escape rate from a planet, there are many other factors to consider.

One important factor to be considered is the composition of a planet's atmosphere, though it tends to be neglected within comparative studies of planetary escape. While CO_2, N_2, O_2, CO and O heat the upper atmosphere through photoionization by XUV radiation, O_2, and

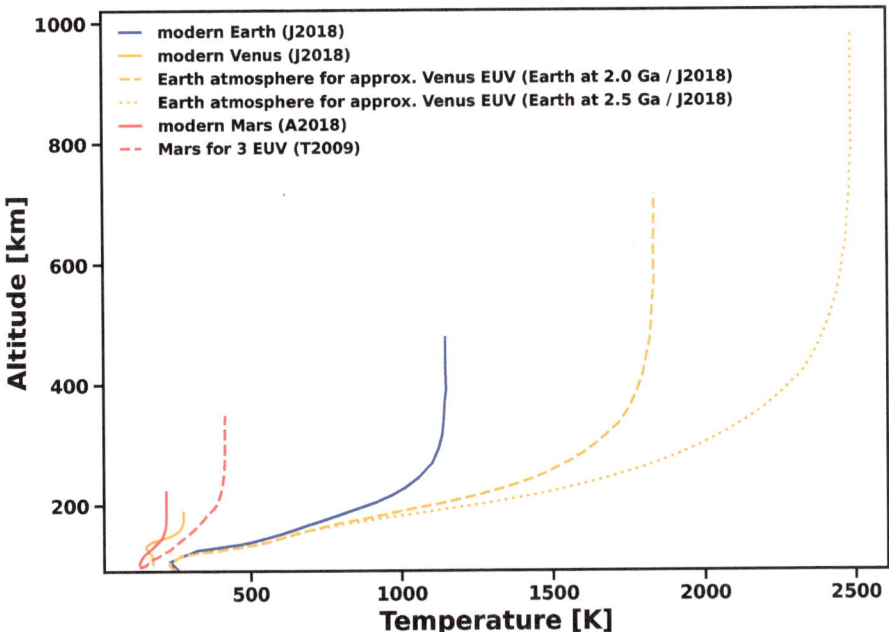

Fig. 10 The neutral upper atmosphere profiles for modern Earth (Johnstone et al. 2018), Venus (Johnstone et al. 2018), and Mars (Tian et al. 2009), and for three hypothetical planets (Tian et al. 2009; Johnstone et al. 2018) that resemble Earth's atmosphere approximately for Venus' EUV flux (dashed and dotted orange lines), and Mars closer to Venus' orbit (the EUV flux at Venus's orbit is about 5 times higher than for Mars, but this plot for 3 EUV is the closest profile available to this value)

O_3 through photodissociation by solar UV radiation, and O through exothermic three-body reactions (Kulikov et al. 2006), CO_2 molecules act as an infrared cooler in the thermosphere (e.g., Roble and Dickinson 1989; Roble 1995; Mlynczak et al. 2010; Cnossen 2020). It emits infrared radiation from the sun back into space, thereby reducing heat within the upper atmosphere. This not only leads to a decline of thermospheric temperature compared to admixtures with less CO_2, but also to a decrease of the exobase altitude (see also Gillmann et al. 2022, this collection). IR cooling through CO_2 might be the most important of the two effects (Kulikov et al. 2006).

This effect is exemplified through a comparison between the upper atmospheres of Venus and Earth, as can be seen in Fig. 10. Even though Venus receives twice as much energy from our Sun, the altitude of its exobase ($r_{exo,v} \approx 200$ km) is less than half that of the Earth ($r_{exo,e} \approx 500$ km). This is due to the main constituent of the Earth's atmosphere being 78% N_2 and 21% O_2, whereas CO_2 only constitutes a minor species (with a mixing ratio of $\approx 0.04\%$ CO_2), while Venus' atmosphere holds a mixing ratio of about 96% CO_2 and 4% N_2. Mars in turn has a similar atmospheric composition to Venus and a comparable exobase level of $r_{exo,m} \approx 200$ km. Thus its smaller mass is compensated by an EUV flux that is 5 times less intense than at Venus' orbital distance. In addition to the altitude, CO_2 also reduces the average exospheric temperature T_{exo} which varies for neutral particles from about 220 K and 250 K at Mars and Venus, respectively, to over 1000 K at Earth. Both characteristics might affect atmospheric escape.

Figure 10 shows simulated neutral upper atmosphere temperature profiles for present-day Venus (Johnstone et al. 2018), Earth (Johnstone et al. 2018), Mars (Amerstorfer et al. 2017), and three hypothetical planets. The dashed red line (Tian et al. 2009) is equivalent to a Martian atmosphere that is irradiated by an EUV flux that is three times as high as at present. For such an increase, exobase level and temperature rise towards $r_{exo} = 415$ km and $T_{exo} = 350$ K, respectively. If Mars resided at Venus' orbit, both values would be higher, since the EUV flux at Venus' orbit is about 5 times as high compared to the orbit of Mars. However, this profile is the closest analog to such a planet available in the literature. The dashed and dotted orange lines depict Earth's present-day atmosphere (Johnstone et al. 2018) for 2.0 and 2.5 Ga, respectively. This is the approximate time frame at which the EUV flux at Earth's orbit is believed to be about twice as high as at present day (see Tu et al. 2015, and Gillmann et al. 2022 this collection), i.e. comparable to the orbital location of Venus. For these two cases, the exobase levels and temperatures for an N_2-dominated atmosphere rise towards $r_{exo} = 700$ km and $T_{exo} = 1800$ K, and $r_{exo} = 980$ km and $T_{exo} = 2500$ K, respectively. If Venus would indeed have such an atmosphere, these levels would be even higher since this planet has a higher equilibrium temperature and about 80% of the Earth's mass. A nitrogen–oxygen dominated atmosphere around Venus instead of its present-day CO_2 atmosphere would, therefore, lead to a significantly different atmospheric structure, thereby illustrating that composition and orbital location indeed matters. But will this also affect the rates of atmospheric escape? Would they cease to be similar if the planets would change place and/or atmospheric composition?

As mentioned earlier, Gunell et al. (2018) derived a formalism to compare atmospheric escape at Venus-, Earth-, and Mars-like planets. Although they did not consider different atmospheric composition, even though this can affect the outcome significantly, as illustrated below. By way of example, these authors (Gunell et al. 2018, Equation (A.10)) semi-empirically parameterized the particle loss through ion pickup as,

$$Q_{pu,\alpha} = Q_{0,pu,\alpha} \frac{2h_a^3 r_b h_a^2 r_b h_a r_b^2}{2h_a^3 h_a^2 r_{exo} h_a r_{exo}^2} e^{\frac{\Delta r}{h_a}},$$ (1)

where $\Delta r = r_{exo} - r_b$ is the distance between r_{exo} and the outer boundary layer r_b, i.e., either the induced magnetosphere boundary r_{IMB} for an unmagnetized, or the magnetopause stand-off distance r_{sd} for a magnetized planet, $h_\alpha = (k_B T_{exo,\alpha} r_{exo}^2)/GM_{pl} m_\alpha$ is the scale height of species α, $T_{exo,\alpha}$ is the exospheric temperature of species α, k_B is the Boltzmann constant, G is the gravitational constant, and M_{pl} is the mass of the planet. The constant $Q_{0,pu,\alpha}$ is a scaling factor for retrieving today's escape rates in case r_{exo} and r_b resemble the present-day values of these planets. As one can see, r_{exo} and T_{exo} are important parameters within $Q_{pu,\alpha}$, and both values are affected by the composition of an atmosphere and the incident EUV flux it receives from its host star. Therefore our hypothetical planets – Mars with 3 times the present-day EUV flux, and the Venus-like planets with a nitrogen–oxygen dominated atmosphere – will end up with different values for $Q_{pu,\alpha}$.

With this formalism, it is thus in principle possible to directly compare atmospheric loss from Venus, Earth, and Mars with our hypothetical planets. However, it is not straight forward *since we do not know how r_{IMB} scales with the change of exobase level*. Moreover, it turns out that this equation is quite sensitive to the scaling factor $Q_{0,pu,\alpha}$ and the exobase temperature with which it was derived. This can be seen in Fig. 11, which illustrates how changes in T_{exo} (panel a), r_{exo} (panel b), and $Q_{0,pu,\alpha}$ (for Venus, both panels – see below) can affect the outcome of Equation (1) and mostly entail significant changes in ion-pickup escape rates at Mars and Venus. In all of the illustrated cases in Fig. 11 r_{IMB} was kept equal to the values employed in Gunell et al. (2018). Present-day O^+ escape rates for Mars and Venus are also shown within this figure; these are displayed for the same values of T_{exo} and

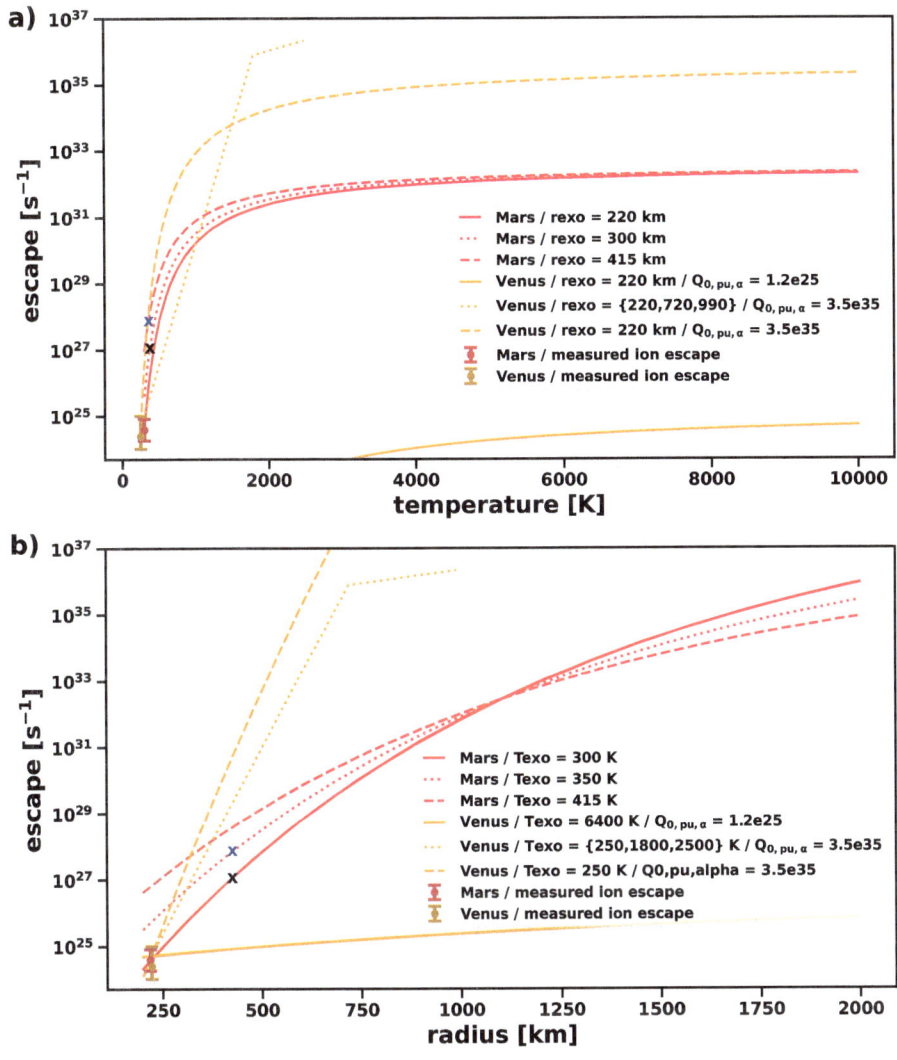

Fig. 11 Ion-pickup escape rates of Mars and Venus as calculated with Equation (1) vs. exobase temperature T_{exo} (panel **a**) and exobase radius r_{exo} (panel **b**). For Mars, the scaling factor $Q_{0,pu,\alpha}$ was kept at 2.6×10^{32} s^{-1} = const. for all displayed example cases; as one can see, escape rates change significantly for small changes in T_{exo} and r_{exo}. For Venus, changes in escape rates are more modest, if the same value for $Q_{0,pu,\alpha}$ is chosen as in Gunell et al. (2018). However, if one recalculates $Q_{0,pu,\alpha}$ by taking into account the exobase temperature of cold oxygen, small changes in T_{exo}, again, entail significant changes in escape rates (dashed orange lines). The dotted orange lines illustrate the 3 Venus cases discussed in the main text; here, T_{exo} and r_{exo} were changed simultaneously in both panels. The present-day ion escape rates of Mars and Venus are displayed for comparison; the blue and black crosses are Mars examples discussed in the main text

r_{exo} as used within Gunell et al. (2018) since there are no specific studies correlating ion escape rates at these planets with different exobase radii and temperatures. A few specific examples of Fig. 11 that are related to our hypothetical planets are discussed next.

For Mars, if we keep the scaling factor for oxygen loss at $Q_{0,pu,\alpha} = 2.6 \times 10^{32}$ s^{-1} and insert $r_{exo} = 415$ km of our hypothetical Martian planet but keep T_{exo} at 300 K as in

Gunell et al. (2018), the escape rate rises 3–46 times, depending on whether Δr or r_{IMB} is kept equal to Gunell et al. (2018) (Fig. 11, black 'x' with r_{IMB} kept equal). If we increase the temperature by 50 K to $T_{exo} = 350$ K, then the escape increases even further by about an order of magnitude (Fig. 11, blue 'x' with r_{IMB} kept equal).

For our hypothetical Venus-like planets with N_2-O_2dominated atmospheres, the change in escape rate is minimal between 1.2 and 4 times for both hypothetical cases and changes in Δr, if one keeps T_{exo} constant (Fig. 11b, solid orange line). However, Gunell et al. (2018) used the exospheric temperature of hot oxygen to retrieve their scaling factor of $Q_{0,pu,\alpha} = 1.2 \times 10^{25}$ s^{-1} for oxygen. If we instead scale with the neutral temperature of cold oxygen at the exobase (≈ 250 K), which is by far the main oxygen species at the exobase level (Lammer et al. 2006), and retrieve $Q_{0,pu,\alpha} \approx 10^{35}$ s^{-1}, then the loss of oxygen would rise by several orders of magnitude if we insert exobase temperatures of 1800 K and 2500 K for our 2.0 and 2.5 Ga cases, respectively (Fig. 11a and b, dotted orange lines). However, this might be above any reasonable escape for such an atmosphere even if it is significantly more expanded than Venus' real atmosphere.

From an exoplanet perspective this exercise illustrates that it is not trivial to scale the escape and compare different planets with different atmospheric compositions and to draw a definitive conclusion on the importance of intrinsic magnetic fields from the current state of research. Further investigation into atmospheric escape at magnetized and unmagnetized planets is therefore highly warranted. This uncertainty is even more critical if one goes back in time to higher EUV fluxes than at Venus' present-day orbit. As already illustrated in Fig. 10, Earth's nitrogen-dominated atmosphere starts to significantly expand for higher EUV fluxes (e.g., Tian et al. 2008; Johnstone et al. 2018, 2021). Crucially, even CO_2-dominated atmospheres will start to inflate for fluxes that are about 15 to 20 times higher than at present-day (Tian et al. 2009; Johnstone et al. 2021).

Given our present knowledge, it is difficult to estimate how these severely altered conditions (which also apply to young solar-like stars) will affect atmospheric escape, particularly at magnetized planets. Kislyakova et al. (2020) investigated polar escape at Earth for different EUV fluxes ranging back until the Archean eon. They found a significant increase in the polar loss of nitrogen and oxygen within their model from presently 2.1×10^{26} s^{-1} and 8.4×10^{21} s^{-1} for O^+ and N^+ to 1.6×10^{27} s^{-1} and 5.6×10^{26} s^{-1} at 2.5 Ga (or 7.6 and 66.7 times more respectively). This increase in escape of O^+ is more significant than in the case of unmagnetized Venus, for which it was recently extrapolated back in time by Persson et al. (2020). However, it is neither well established whether atmospheric escape would have been stronger at Earth without a magnetic field at 2.5 Ga, nor how escape at Venus would have evolved if it had a nitrogen-dominated atmosphere and/or if it had been "shielded" by an intrinsic magnetic field. Besides that, it seems probable that a Venus-like exoplanet with an Earth-like atmosphere would show larger escape rates than if it had a CO_2-dominated atmosphere, which is important for considering its potential habitability. Yet the early Earth atmosphere had very little O_2 and a higher pCO_2 (e.g. Catling and Zahnle 2020) which may have limited atmospheric escape (Lichtenegger et al. 2010). The same possibility exists for early Venus' atmospheric composition – its evolution would have changed the picture we see today in ways that are difficult to constrain without more information on the planet's distant past. However, whether an intrinsic magnetic field would diminish the escape remains poorly understood.

From these considerations, one finds that atmospheric composition is likely more important for defining atmospheric loss than the presence of an intrinsic magnetic field. However, even if Earth-like magnetospheres do not shield atmospheres from escape, they can separate particle fluxes according to their energy spectrum so that life forms on a planet's surface are protected from highly energetic primary and secondary solar cosmic rays. There are two sources of cosmic rays, the first originate from high energetic solar events (SCRs), while the second are called galactic cosmic rays (GCRs) that belong to energetic sources in the Milky Way or other galaxies. Upon impact with the Earth's atmosphere, cosmic rays produce showers of secondary particles, some of which reach the surface. SCRs can have global effects on life-forms that enhance mutation rates (Belisheva and Popov 1995; Belisheva et al. 2012; Dar et al. 1998; Brack et al. 2010).

Within Earth's magnetospheric cusp area over the Arctic it was found that secondary radiation produced by intense high energy SCR particle showers, like the October 1989 solar proton event (Reeves et al. 1992), caused various biological phenomena associated with DNA lesions on the cellular level (Belisheva and Popov 1995; Belisheva et al. 2012). These biological effects were detected during experiments with three cellular lines growing in culture during three events of ground level enhancements in the neutron count rate detected and correlated by ground-based neutron monitors, in October 1989 at Srednyi Island, in the White Sea of the Physical Research Institute of the St. Petersburg University, and at the Kola Science Centre of the Russian Academy of Sciences in Apatity, Murmansk region (e.g., Belisheva et al. 2012). Depending on the planetary magnetic field and atmospheric pressure, cosmic ray particles interact with the atmosphere where they generate secondary highly energetic particles of which some can reach the surface of planets for Earth-like pressure values or lower (e.g., Shea and Smart 1995).

The protection of Earth's surface against secondary high energy solar cosmic ray particles with a surface pressure of ≈ 1 bar atmosphere amounts to ≈ 1000 g cm^{-2}, whereas that of the thin Martian atmosphere with ≤ 10 mbar only results in ≈ 16 g cm^{-2} (e.g., Shea and Smart 1995; Brack et al. 2010). If the planetary atmosphere is dense enough, like that of Venus, these high-energy particles cannot penetrate to the surface. However, the atmospheric region on Venus that may be favourable for biology is located between and/or near the upper and lower bounds of the three Venusian cloud layers (Cockell 1999; Mogul et al. 2021; Kotsyurbenko et al. 2021) at ≈ 38–55 km (Marov and Greenspoon 1998), where the atmospheric pressure level is comparable to Earth's. Because Venus is not shielded by an intrinsic magnetosphere like the Earth, high-energy SCR particles will therefore precipitate into its atmosphere and are absorbed around the so-called thermally biological favourable atmospheric layers.

Finally, we point out that smaller magnetic moments that may originate due to tidally locking on terrestrial planets inside the habitable zones of low-mass M and K-type stars, and potentially also due to induced magnetospheres, would provide a weaker protection of planetary surfaces or biologically favourable atmospheric layers against GCRs (Grießmeier et al. 2005, 2009). However, in a follow-up study, Grießmeier et al. (2016) point out that for such planets, as well as for unmagnetized bodies, with atmospheric pressures similar or higher than the Earth's, the effects of the increased GCR radiation would be small. For thin atmospheres on the other hand, the shielding from GCRs would be entirely controlled by the magnetosphere, if present. If not, the surface radiation dose cannot be prevented from increasing up to several hundred times the background flux.

2.4 The Critical Dependence of and on Planetary Thermal History

The great divergence between Venus and Earth is critical to understanding potential exoplanetary evolution. Given comparable sizes, masses, and presumably chemical make-up, Venus is often thought of as Earth's twin. As such, one would naturally expect it to exhibit similar patterns of convection, heat loss, and tectonics. Venus, however, is strikingly different in its apparent convective, tectonic, and atmospheric conditions today. These observations lead to a key set of questions: given the broad similarities between Earth and Venus, (1) what led to the dramatic differences between the two planets; and (2) What can the divergence between Venus and Earth tell us about the thermal evolution of exoplanets? With significant attention (in both this article and others of this collection) devoted to the former, here, we will focus on the latter. To address this question in some detail, it is important to outline what we know about the thermal-tectonic regimes and evolution of the Earth. We will then extrapolate this knowledge to the Earth-Venus divergence, and outline potential implications for exoplanets.

The Earth is the only body in the Solar System for which significant information about its thermal, geologic, atmospheric, and tectonic evolution is readily accessible. Consequently, Earth derived data and observations are often used to inform general models of thermal evolution, which are then extrapolated to other bodies in our Solar System, and beyond. However, despite the Earth's large dataset, our knowledge and understanding of the Earth's thermal evolution remains largely opaque. For instance, while we know Earth is currently within a plate tectonics regime, its initiation and total life of activity are far from certain (e.g. O'Neill et al. 2007; Debaille et al. 2013; Gerya 2014a; Lu et al. 2021). These uncertain time frames have profound implications for understanding the long-term thermal and surface evolution of the Earth, let alone Venus, and extrasolar planets.

Critical to this discussion is the notion that the thermal and tectonic state of a planet are intimately connected, and tie into the long-term surface-interior geophysical cycles that influence and control both atmospheric and surface evolution (see Sect. 2.1; as well as Gillmann et al. 2022, this collection; Phillips et al. 2001; Lenardic et al. 2008; Driscoll and Bercovici 2014; Gillmann and Tackley 2014; O'Rourke et al. 2018; Krissansen-Totton et al. 2021). Consequently, a discussion of any one aspect of planetary thermal evolution inherently discusses the other aspects, even if only tacitly. As tectonic states have distinct characteristics, each affects planetary evolution and a planet's thermal state differently. For the purposes of this section, we will briefly outline three main tectonic end-members relative to their thermal implications (definitions of tectonic states are discussed in greater detail in 3.A).

Returning to the Earth, we can define plate tectonics as a subset of active (or mobile) lid convection (e.g. Schubert et al. 2001). This mode of tectonics is characterized by the outermost layer of cold and rigid rock participating in the mantle convective cycle. That outer layer is brought back into the interior along with the convective mantle. This leads to the cooling of the interior, a thin lithosphere, and generally efficient heat loss at the surface. In contrast to the mobile lid, the outermost cold and rigid surface layer of the stagnant lid regime resists convective motions (e.g. Schubert et al. 2001). As a consequence, this mode of tectonics has a thicker immobile surface that does not actively participate in mantle convection. The stagnant lid leads to inefficient heat loss and higher internal temperatures when compared to an active lid state. An additional regime considered is the episodic lid (Moresi and Solomatov 1998), sometimes identified as a transitional regime between active and stagnant lids (Weller et al. 2015; Weller and Lenardic 2018; Weller and Kiefer 2021). This regime is highly dynamic, characterized by periods of extreme quiescence punctuated with rapid episodes of surface-interior interaction (Armann and Tackley 2012). In a first order sense, an example of internal temperatures for each regime for an Earth or Venus sized body is indicated in Fig. 12. Critical to the discussion of planetary thermal evolution, each of these three states has been suggested to have once operated on the Earth in the past, to varying degrees, though the exact nature and expression of these tectonics, and indeed the thermal state the early Earth exhibited, is vigorously debated (e.g. Condie and Kröner 2008; Davies 1993; Debaille et al. 2013; Calvert et al. 1995; O'Neill et al. 2007, 2015, 2016;

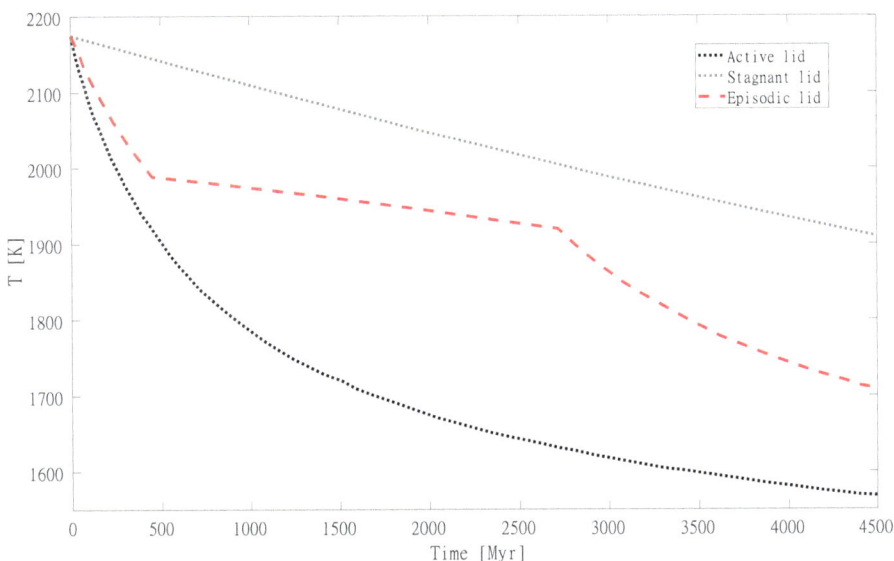

Fig. 12 Simple thermal history numerical models for an Earth/Venus sized Active lid (black dotted line) and a Stagnant lid (grey dotted line), taken from an identical initial thermal state (here taken as 2174 K) see Breuer and Moore (2007) (and references therein) for a detailed discussion of models. The Episodic lid thermal state is taken from O'Neill (2020), and shows three distinct evolutionary trends: early active episodic, middle quiescent-episodic, and final active lid. Here T[K] represents the average mantle temperature in Kelvin

Stern 2008; Moyen and van Hunen 2012; Moore and Webb 2013; Gerya 2014a; O'Neill and Debaille 2014). The list of citations is by no means meant to be exhaustive.

While the geologic record often is ambiguous, and as a consequence, the thermal evolution of the early Earth is passionately debated, it has long been agreed that, as the planet loses heat, the Earth will eventually cease operating in a plate tectonic regime and begin to move into a stagnant-lid regime, similar to observations for current day Mars (e.g. Nimmo and Stevenson 2000). While the time frame of this transition remains unclear, a key aspect of planetary tectonics and thermal evolution is highlighted here: the tectonic and thermal state of a planet may change significantly, and perhaps more than once, as the planet evolves. This idea, generally postulated to explain Earth observations, may be extended to other planetary bodies, as has been suggested by studies exploring the convective and tectonic sensitivities to changes in internal mantle temperatures over time, and surface temperature changes through planetary climatic evolution (e.g. O'Neill et al. 2007, 2016; Lenardic et al. 2008; Landuyt and Bercovici 2009; Foley et al. 2012; Lenardic and Crowley 2012; Stein et al. 2013; Gillmann and Tackley 2014; Weller et al. 2015; Weller and Lenardic 2018).

Earth and Venus can be seen as planetary end-members (in a bifurcation space). For the tectonic/thermal evolution of planets, there exist two main drivers of change: (1) Changes in internal temperatures from changes in heat loss and radiogenic heating rates; and (2) changes in surface temperatures from the long-term climate variations of the planet. First, we examine case (1) through the lens of secular cooling (loss of heat with time and depleting internal heat sources). Early in planetary thermal evolution, the internal temperatures are high due to leftover heat from accretion and high levels of radiogenic elements (e.g. Fig. 12). From both buoyancy and velocity/stress-scaling arguments (e.g. Lenardic et al. 2021, and references therein), these conditions tend to strongly favor early stagnant lid tectonic states (Weller et al. 2015; O'Neill et al. 2016; Weller and Lenardic 2018). However, as radiogenic heating, and consequently internal temperatures, decreases with time, this early stagnant state may yield, often through an intermediary episodic state, into an active lid regime. With further heat loss and decrease in radiogenic heating rates, the active lid may ultimately transition once again into a stagnant lid, potentially through an oscillatory episodic state. This stagnant → episodic → active lid pathway, as suggested for the Earth (e.g. O'Neill et al. 2007, 2016), can be thought of as the consequence of secular cooling and depletion of radiogenic heating. This then may be thought of as a system state driving force operating on (Earth or Venus sized) planetary bodies, moving the planetary system towards a specific evolutionary path over time, which then may be acted upon by other forces and processes.

While secular cooling (driver 1) serves to push the planet to an active lid state (and an eventual return to stagnant conditions as more heat is progressively lost), surface temperature changes (driver 2) can profoundly alter the expression of tectonics (e.g. Lenardic et al. 2008; Landuyt and Bercovici 2009; Foley et al. 2012; Gillmann and Tackley 2014; Weller et al. 2015). For a planet operating in active lid tectonics, an increase in surface temperatures on geologic time scales has been demonstrated to trigger a transition from active lid convection, into a significantly long-lived episodic lid regime (Gillmann and Tackley 2014), before eventually settling into stagnant lid behavior (Weller et al. 2015; Weller and Kiefer 2021). For an early stagnant lid thermal state, high surface temperatures can prevent the planet from transitioning states entirely. Conversely, a stagnant lid planet with high surface temperature could transition into a mobile lid state, if surface temperature dropped low enough (Lenardic et al. 2008; Gillmann and Tackley 2014). Therefore, surface temperatures may override the secular driven changes in tectonics for Venus/Earth sized bodies. Alternatively, it could enhance some of its effects, depending on the tectonic/thermal state of a planet at the time of surface temperature change.

For both early thermal states (hot, young, or enriched in radiogenic/tidal heating sources) and late thermal states (cold, old, or lacking significant radiogenic heating sources), there exists a strong thermal coupling that pushes the planet towards stagnant lid states (e.g. Weller et al. 2015; O'Neill et al. 2016; Weller and Lenardic 2018). However, a significant span of a planet's thermal evolution is controlled by competing and nonlinear forcing, both internal (e.g. heating and temperature) and external (e.g. surface temperature). As a result, the planetary thermal and tectonic state may be predominantly governed by the specific thermal history of the system, allowing stable and unstable active lids, episodic lids, stagnant lids, or all of the above. In fact, nonlinearity within the convective thermal system allows for a hysteresis of states and thermal evolutionary scenarios (Fig. 13). Within the hysteresis window, the specific evolutionary history of the system (e.g. the initial conditions, along with the specific thermal evolution) has been shown to play a significant control on the mode of tectonics and thermal state that a planet may operate within. This contrasts with a more traditional view, where a specific set of planetary parameters such as strength of the lithosphere, internal temperature, or surface conditions is directly associated with a specific tectonics/thermal state (Weller and Lenardic 2012; Lenardic and Crowley 2012; Weller et al. 2015; Weller and Lenardic 2018) (see Fig. 13 caption).

The hysteresis window is specifically a region of multiple stable tectonic/thermal solutions for otherwise similar planetary bodies. That is, otherwise identical planetary states (e.g. surface temperatures, heating rates, rock strength, volatile contents, etc.) can allow for entirely different tectonic and thermal regimes, depending on how the planet evolved toward this state. Interestingly, this window does not seem to be uniform in regard to system complexity or energetics. Figure 13 illustrates the hysteresis window conceptually as a function of system energy, or vigor of convection (traditionally considered by the Rayleigh number (Ra) or viscosity contrast). For simple systems with low energy (low convective vigor)

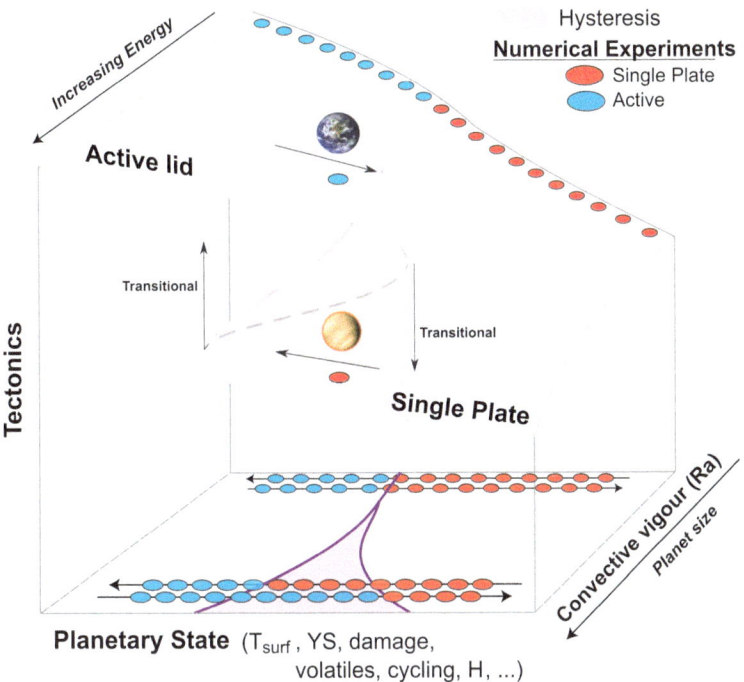

Fig. 13 Modified after Lenardic and Crowley (2012) (Tobias Rolf is credited with an earlier modification of this Figure). Schematic view of bifurcations in planetary tectonics. X-axis denotes changing planetary state variables, for example: Surface temperatures (Tsurf), global yield strength (YS), damage accumulation/healing, volatile abundances and cycling, radiogenic heating rates (H), etc.... Convective systems inherently allow for variations in tectonic stability space as a function of increasing convective vigor or energy (Y-axis, background to foreground). For systems with limited energy or low Ra, a single stability point exists (attractor) for a set combination of parameters (e.g. tectonic state has a functional relationship with planetary parameters). For these states, changing parameter paths, or the systems history (denoted by directional increasing/decreasing horizontal arrows with tectonic state indicators: active lid – blue circles, stagnant lid – red circles), has no effect on the final tectonic/thermal state (back projection on the phase space). As complexity increases, multiple attractors effect the stability space for a given set of planetary parameters. Instead on single attractor space (uni-tectonic space), multiple competing attractor wells ensure a path dependence on the final tectonic state. The system allows for rapid changes with parameter variations (direction transition arrows). Multiple solutions exist dependent on the initial conditions and history of the system (hysteresis space, purple shading) as indicated by both mobile and stagnant lid solution viable for the same parameters (foreground). Venus and Earth are plotted as possible endmembers in this hysteresis gap. Putative super-Earth's/Venus' would be projected to plot out of the page in ever widening hysteresis space

there exists a single coupled tectonic-thermal attractor space, or direction of evolution. To put it another way, there exists only one set of stable solutions for any combination of individual planetary states. However, we do not expect planets in general to operate at these low energy/low complexity system states (Lenardic and Crowley 2012; Weller and Lenardic 2012). As complexity and the energetics of the system increases (for example Ra and viscosity contrasts), the system is increasingly affected by competing stable tectonic/thermal solutions (Lenardic and Crowley 2012; Weller and Lenardic 2012). For conditions expected for real bodies, such as Earth or Venus, the hysteresis space may encompass most reasonable planetary parameters (Weller and Lenardic 2018), and consequently the thermal and tectonic evolution of a planet is almost entirely governed by the planet's specific geologic and climatic history. As system complexity and energy increase, as for example for so-called super-Earth's and super-Venus', this window may be expected to contain all real solutions. For the foreseeable future, the complexity of such systems make it computationally unfeasible to run in-depth (non-parameterized) numerical simulations to model them.

If we consider a putative proto-Venus/Earth type body, the hysteresis framework offers interesting insight into the coupled thermal tectonic evolution of terrestrial bodies. In this framework, both planetary states are equally possible, and dependent on the specific thermal evolution of each planet. In Fig. 13, these end-member states are indicated by the Earth evolving along a prior state that allowed active lid convection, whereas Venus' earlier evolution did not. However, that does not imply that Venus could not have been in an active lid state at some point, or that it fundamentally lacks the capacity to do so. In fact, there exists suggestive, but not unambiguous, evidence that Venus may have operated in some form of active lid mode of tectonics at one time in its past (see Rolf et al. 2022 this collection for discussion), or that tectonic state may exist as a continuum rather than just simple end-members.

In a general sense, the implications for exoplanets are that there may not exist a preferred tectonic or thermal state for any one planetary variable or type. Instead, the thermal and tectonic state of exoplanets may be much more strongly controlled by the planets' specific history, a history that we will not be able to sample or observe. As a corollary, this implies that tectonic regime may be vulnerable to change by random events, such as collisions with large impactors (Gillmann et al. 2016; O'Neill et al. 2017), given they occur at a favourable time to destabilize the current state. If the planetary tectonic/thermal state of extrasolar planets is non-unique, this suggests that we need to move towards considering tectonic and thermal states in a probability space, as opposed to known variable space (e.g. surface temperature, size, etc.). For example, water, if detected in planetary atmospheres, may not be indicative of an active lid state, as has been suggested as the requirement for plate tectonics on Earth (Hubbert and Rubey 1959; Bird 1978). These results further imply that finding both water and habitable surface conditions would not be an indicator of the tectonic or thermal state of a planet, nor its geologic and climatic history. On the other hand, this probability-oriented approach makes the characterization of exoplanets even more critical to bypass the Solar system assumptions that underpin our understanding of planetary evolution.

Planetary evolution in nonlinear space then is highly complex, but finding solutions is not insurmountable. Instead of focusing on key parameters that control tectonic or thermal states, we need to focus on and understand the probabilities of Venus type solutions relative to Earth (or even other) type solutions. If both Venus and Earth operated within an active lid mode of tectonics in our Solar System, then the potential for active lid modes may be common, but the systems could have strong temporal (e.g. O'Neill et al. 2016), stochastic (e.g. Weller and Lenardic 2018; Weller and Kiefer 2020), and reinforcing feedback (e.g. Lenardic et al. 2019) dependencies, that interface in extremely complex ways. The existence of the hysteresis window indicates that we need to understand the feedback effects between the evolution of the atmosphere, mantle, and surface tectonics in a more holistic and probabilistic way through suites of ensemble numerical simulations that focus on the interplay of planetary starting conditions, varying physical parameters and the physics they encompass, as well as stochastic fluctuations. Within our own Solar System, results from the InSight mission (Banerdt et al. 2020) have greatly improved our understanding of the interior structure of Mars (e.g. Knapmeyer-Endrun et al. 2021; Khan et al. 2021; Stähler et al. 2021). Compared to Mars, which is characterized by a stagnant lid regime throughout its thermal history, Venus tectonic evolution might have been significantly different. Though great care must be taken in extrapolating between dissimilar planets (e.g., Mars to Venus),

InSight's results demonstrate how geophysical measurements can provide valuable and detailed information about the interior of other planets. This type of data provides us with the ability to compare and contrast the differences in the interiors of terrestrial planets operating in different tectonic regimes.

The initial thermal state of the planet, which is intimately related to its accretion sequence, determines the amount of energy the planet will dissipate over its history and is thus of fundamental importance regarding its entire evolution. Despite the absence of direct evidence on the Earth and Venus, several heating mechanisms are thought to affect the earliest stages of planetary evolution (for a detailed discussion, see Salvador et al. 2023, this collection). The accretion process itself delivers a substantial amount of energy to the growing planets through the accumulation and burial of impact energy (e.g., Safronov 1978; Tonks and Melosh 1993). Radiogenic heating produced by the decay of short-lived isotopes (in particular ^{26}Al and ^{60}Fe) is responsible for substantial melting of early forming and growing planetary embryos, planetesimals, and proto-planets (e.g., Merk et al. 2002; Bhatia 2021). During the formation of the core, metal-silicate differentiation and metal downwards migration release gravitational energy dissipated by viscous heating which could increase the temperature of an entire Earth-sized planet by almost 2000 K (Tozer 1965; Flasar and Birch 1973). Due to the combination of these heat sources, terrestrial planets are generally thought to experience one or several episodes of early and large-scale mantle melting (e.g., Elkins-Tanton 2012). Without an atmosphere overlying the molten surface, the heat accumulated can be rapidly radiated to space but melting can be enhanced and sustained in the presence of a primordial atmosphere (e.g., Hayashi et al. 1979; Ikoma and Genda 2006) providing a thermal blanketing effect. On early Earth, the hypothetical Moon-forming giant impact is often referred to as being responsible for generating a last and global-scale magma ocean extending throughout the entire mantle (e.g. Benz et al. 1986; Canup 2004). From then on, its cooling, solidification, and associated chemical differentiation would then set the stage for the subsequent long-term evolution of the planet. On Venus, the absence of a moon cannot completely discard the likelihood of an early fully molten stage. Indeed, the orbital proximity of Earth and Venus implies similar bulk properties and suggests that they have experienced similar accretion sequences (e.g. Morbidelli et al. 2012; Raymond 2021) with similar endowments of radioactive elements so that the aforementioned heating mechanisms and resulting global-scale melting events would likely apply for both planets, although recent work may put some of this into question (e.g. Emsenhuber et al. 2021). While these energetic processes are inherent to the formation of terrestrial planets, the initial thermal state and the occurrence and timing of large scale melting events on exoplanets are critically related to the timescale of the accretion phase (see Salvador et al. 2023, and references therein). While the current orbital configuration might help put constraints on the tidal heating presently affecting an observed solidified exoplanet, inferring their initial thermal state is out of reach. However, observing a substantial number of young exoplanetary systems might help testing and informing planetary formation and early evolution models to draw more statistically robust trends, thus improving our understanding of early planetary pathways and associated thermal states.

Mantle viscosity is one of the most important parameters that controls the cooling behavior of the interior. This in turn affects magmatic and tectonic processes throughout the thermochemical evolution of the planet. The viscosity of silicate materials is strongly temperature and pressure dependent. The dependence of viscosity on temperature is given by the activation energy, which is the energy necessary to create vacancies in the crystal lattice and the barrier that atoms need to overcome in order to migrate into a vacant site. The activation

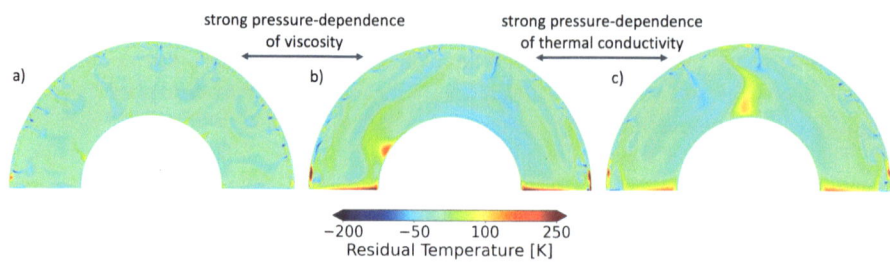

Fig. 14 Effects of pressure dependent parameters on the convection pattern for a Venus-like interior (Hirschberger et al. 2020): **a)** small pressure dependence of viscosity and thermal conductivity (i.e., viscosity increases with depth by a factor of 32 and thermal conductivity increases with depth by a factor of 1.7); **b)** strong pressure dependence of the viscosity but weak pressure dependence of thermal conductivity (i.e., viscosity increases with depth by about 4 orders of magnitude and thermal conductivity increases with depth the same as in panel **a)**; **c)** strong pressure dependence of both viscosity and thermal conductivity (i.e., viscosity increases with depth the same as in panel **b** and thermal conductivity increases with depth by a factor of about 6)

volume describes the pressure dependence of the viscosity and indicates that for higher pressure the energy necessary for the formation of vacancies and the barrier for atom migration increase. While rheological parameters have been measured in laboratory experiments (e.g., Hirth and Kohlstedt 2003), uncertainties in their values are large because such experiments need to be extrapolated to the conditions relevant for planetary interiors. In particular, the effects of the depth dependence of the viscosity has been highly debated for the deep interior of large rocky planets (super-Earths). Some authors suggest an almost isoviscous interior of large super-Earths indicating a fully convecting mantle (Si 2011), but others indicate that a strong pressure dependence of the viscosity will lead to the formation of a stagnant region in the lower mantle (the so-called CMB lid) (Stamenković et al. 2011).

While the pressure inside the mantles of Earth and Venus does not reach the range for which a CMB lid could form, a strong pressure dependence will affect the convection planform, as well as the number and shape of mantle plumes. Mantle convection models show that a strong pressure dependent viscosity will promote fewer and more prominent mantle plumes compared to cases where little or no pressure dependence is applied (Fig. 14). This in turn may affect the melt production in the interior and the geoid. A strong viscosity increase related to mineral phase transitions, as it is suggested to match the geoid on the Earth, has been found inconsistent with the gravity-topography correlation on Venus (Rolf et al. 2018). This suggests a more gradual increase of the viscosity with depth, possibly indicating a drier upper mantle than on Earth (Rolf et al. 2018). In addition to the viscosity, thermodynamic parameters such as thermal expansivity and thermal conductivity vary with temperature and pressure and can affect the dynamics of the mantle (Tosi et al. 2013). In particular, the increase of thermal conductivity with pressure promotes more diffuse plumes and downwellings thus decreasing the temperature variations in the mantle (Hirschberger et al. 2020). However the strongest effect on convection is expected for the pressure dependence of the viscosity as this increases by several orders of magnitude, compared to an increase by a factor of about 6 for the thermal conductivity (Armann and Tackley 2012).

3 Conclusions

The terrestrial worlds of our solar system are the benchmarks for exploring the exoplanetary realm of our galaxy. As shown herein there is a tremendous amount of knowledge from solar system objects that can be applied to exoplanetary observations of Venus analogs. Conversely with new ground and space based capabilities coming on-line in the coming decade we will also begin to take lessons from Venus' exoplanetary cousins to learn more about the evolutionary history of Venus and Earth. Yet there is a large imbalance in the knowledge each domain presents us today as reflected in the sizes of the exoplanet versus Venus sections of this article. The Venus sections are decidedly larger as one might expect of our nearest planetary neighbor whose atmosphere and surface has been studied intensely with spacecraft and ground based instruments for the past 60+ years, whereas exoplanetary science is still in its infancy. As noted throughout Sect. 2 Venus studies also benefit tremendously from the study of our home world Earth and our second closest neighbor Mars. For decades planetary scientists have struggled to understand how a possibly early habitable period on both Venus and Mars could result in their present apparently uninhabitable states. If Venus did evolve from an earlier temperate period with surface water reservoirs to it's present hothouse state exactly how did it occur, and what are the key processes involved? We still lack a full understanding of how such a catastrophic event could take place, but there is great anticipation that the study of planets in neighboring stellar systems will help inform our studies of Venus. Yet as shown in Sect. 1 we are at least two decades away from statistically characterizing the atmospheres of exo-Venus worlds. At the same time we are over a decade away until the data from the newly confirmed Venus missions from ESA and NASA begins to arrive. Even that data will take many years to process and understand, as we see today with the on-going studies of the Magellan Mission radar data from the 1990s (e.g. Byrne et al. 2021; Khawja et al. 2020; MacLellan et al. 2021; Brossier et al. 2021; Borrelli et al. 2021).

There are a number of takeaways to consider when looking at how Venus and exoplanetary studies might inform each other in the future as discussed within this article. Firstly, lets consider the key role that the early evolution of Venus' magma ocean plays in possibly deciding Venus' long-term H_2O budget and the possibility of surface liquid water. In this case exoplanetary observations of planets in the VZ can help us to constrain magma ocean lifetimes around a wide range of stellar hosts, including those explicitly resembling the G-dwarf that is our sun. This involves research programs explicitly looking for solar twins, defined as stellar hosts with chemical compositions or early XUV activity very similar to our sun (Gustafsson et al. 2010; Airapetian et al. 2021). Secondly, why did Earth and Venus take such divergent evolutionary paths when they otherwise appear to be so similar in size, density and possibly chemical composition (Lécuyer et al. 2000) in comparison with the other terrestrial planets within the solar system? Examining exoplanets in the VZ may tell us whether Venus ever had temperate surface conditions and whether rotation rate plays a role in stabilizing such conditions as demonstrated in GCM studies (Yang et al. 2014; Way et al. 2016). Unfortunately in the near term it could be that a modern Venus-like cloud and haze layer will prevent JWST from resolving atmospheric species that could give clues to exoplanetary atmospheric evolution histories. Clouds in general make observing even major species very challenging with JWST (Fauchez et al. 2019; Teinturier et al. 2022), although there may be some opportunities when observing more arid planets with fewer clouds (Ding and Wordsworth 2022). Thirdly, can we discern the longevity of any postulated climate state in Venus' history? For example, if Venus had a temperate period its longevity may be constrained from in-situ observations of the noble gas isotopes as described in (Avice et al. 2022, this collection) and in (Baines et al. 2013), while exoplanetary worlds in the VZ may also help us to bound the problem. There is an on-going debate as to the timescale of volcanic outgassing required to produce the basaltic plains that cover nearly 80% of Venus' surface (e.g. Phillips et al. 1992; Bullock et al. 1993; Herrick 1994; Strom et al. 1994; Basilevsky and Head 1996; Bjonnes et al. 2012; Ivanov and Head 2013; Kreslavsky et al. 2015). Then there is the nature of the 92 bar CO_2 atmosphere in place today. If there was a period of time with a lower atmospheric density (e.g. 1 bar) similar to that achieved by Earth throughout most of its history what mechanism or mechanisms occurred to emplace the present 92 bar atmosphere (e.g. Head et al. 2021)? In these last two cases observing a statistically relevant sample of VZ worlds in different evolutionary phases could help us bound the parameter space in ways we may only scarcely comprehend today.

Acknowledgements MJW would like to thank Nikolai Piskunov in Uppsala Astrophysics for useful discussions regarding the first generation instrumentation on the ELT. We would also like to thank the three anonymous referees who helped us to greatly improve our manuscript.

Funding Note Open Access funding enabled and organized by Projekt DEAL. MJW acknowledges support from the Goddard Space Flight Center Sellers Exoplanet Environments Collaboration (SEEC) and ROCKE-3D: The evolution of solar system worlds through time, funded by the NASA Planetary and Earth Science Divisions Internal Scientist Funding Model. CG acknowledges the support of Rice University and the CLEVER planets group (itself supported by NASA and part of NExSS). ACP gratefully acknowledges the financial support and endorsement from the DLR Management Board Young Research Group Leader Program and the Executive Board Member for Space Research and Technology.

Declarations

Competing Interests The authors declare no competing interests.

References

Acuna MH, Connerney JEP, Ness NF, Lin RP, Mitchell D, Carlson CW, McFadden J, Anderson KA, Reme H, Mazelle C, Vignes D, Wasilewski P, Cloutier P (1999) Global distribution of crustal magnetization discovered by the Mars Global Surveyor MAG/ER experiment. Science 284:790. https://doi.org/10.1126/science.284.5415.790

Adibekyan V, Dorn C, Sousa SG, Santos NC, Bitsch B, Israelian G, Mordasini C, Barros SC, Delgado Mena E, Demangeon OD et al (2021) A compositional link between rocky exoplanets and their host stars. Science 374(6565):330–332

Agee CB (1998) Crystal-liquid density inversions in terrestrial and lunar magmas. Phys Earth Planet Inter 107(1–3):63–74

Ague JJ, Nicolescu S (2014) Carbon dioxide released from subduction zones by fluid-mediated reactions. Nat Geosci 7(5):355–360

Airapetian VS, Barnes R, Cohen O, Collinson GA, Danchi WC, Dong CF, Del Genio AD, France K, Garcia-Sage K, Glocer A, Gopalswamy N, Grenfell JL, Gronoff G, Güdel M, Herbst K, Henning WG, Jackman CH, Jin M, Johnstone CP, Kaltenegger L, Kay CD, Kobayashi K, Kuang W, Li G, Lynch BJ, Lüftinger T, Luhmann JG, Maehara H, Mlynczak MG, Notsu Y, Osten RA, Ramirez RM, Rugheimer S, Scheucher M, Schlieder JE, Shibata K, Sousa-Silva C, Stamenković V, Strangeway RJ, Usmanov AV, Vergados P, Verkhoglyadova OP, Vidotto AA, Voytek M, Way MJ, Zank GP, Yamashiki Y (2020) Impact of space weather on climate and habitability of terrestrial-type exoplanets. Int J Astrobiol 19(2):136–194. https://doi.org/10.1017/S1473550419000132. arXiv:1905.05093

Airapetian VS, Jin M, Lueftinger T, Saikia SB, Kochukhov O, Guedel M, Van Der Holst B, Manchester W IV (2021) One year in the life of young suns: data constrained corona-wind model of kappa1 Ceti. ArXiv preprint. arXiv:2106.01284

Amerstorfer UV, Gröller H, Lichtenegger H, Lammer H, Tian F, Noack L, Scherf M, Johnstone C, Tu L, Güdel M (2017) Escape and evolution of Mars's CO_2 atmosphere: influence of suprathermal atoms. J Geophys Res, Planets 122(6):1321–1337. https://doi.org/10.1002/2016JE005175

Andersson E, Bauer P, Beljaars A, Chevallier F, Hólm E, Janisková M, Kållberg P, Kelly G, Lopez P, McNally A, Moreau E, Simmons AJ, Thépaut JN, Tompkins AM (2005) Assimilation and modeling of the atmospheric hydrological cycle in the ECMWF forecasting system. Bull Am Meteorol Soc 86(3):387–402. https://doi.org/10.1175/BAMS-86-3-387

Angelo I, Rowe JF, Howell SB, Quintana EV, Still M, Mann AW, Burningham B, Barclay T, Ciardi DR, Huber D, Kane SR (2017) Kepler-1649b: an exo-Venus in the solar neighborhood. Astron J 153(4):162. https://doi.org/10.3847/1538-3881/aa615f

Armann M, Tackley PJ (2012) Simulating the thermochemical magmatic and tectonic evolution of Venus's mantle and lithosphere: two-dimensional models. J Geophys Res, Planets 117(E12):E12003

Arney G, Kane S (2018) Venus as an analog for hot earths. arXiv:1804.05889

Arney G, Domagal-Goldman SD, Meadows VS, Wolf ET, Schwieterman E, Charnay B, Claire M, Hébrard E, Trainer MG (2016) The pale orange dot: the spectrum and habitability of hazy Archean Earth. Astrobiology 16(11):873–899. https://doi.org/10.1089/ast.2015.1422. arXiv:1610.04515

Arney GN, Meadows VS, Domagal-Goldman SD, Deming D, Robinson TD, Tovar G, Wolf ET, Schwieterman E (2017) Pale orange dots: the impact of organic haze on the habitability and detectability of Earthlike exoplanets. Astrophys J 836(1):49. https://doi.org/10.3847/1538-4357/836/1/49. arXiv:1702.02994

Arrhenius G, De BR, Alfvén H (1974) Origin of the ocean. The Sea 5:839–861

Avice G, Parai R, Jacobson S, Labidi J, Trainer M, Mikhail PP (2022) Noble gases and stable isotopes track the origin and early evolution of the Venus atmosphere. Space Sci Rev 218:60. https://doi.org/10.1007/s11214-022-00929-9

Baines KH, Atreya SK, Bullock MA, Grinspoon DH, Mahaffy P, Russell CT, Schubert G, Zahnle K (2013) The atmospheres of the terrestrial planets: clues to the origins and early evolution of Venus, Earth, and Mars. University of Arizona Press, Tucson, p 137. https://doi.org/10.2458/azu_uapress_9780816530595-ch006

Ballmer MD, Noack L (2021) The diversity of exoplanets: from interior dynamics to surface expressions. Elements 17(4):245–250

Banerdt WB, Smrekar SE, Banfield D, Giardini D, Golombek M, Johnson CL, Lognonné P, Spiga A, Spohn T, Perrin C et al (2020) Initial results from the InSight mission on Mars. Nat Geosci 13(3):183–189

Barabash S, Fedorov A, Sauvaud JJ, Lundin R, Russell CT, Futaana Y, Zhang TL, Andersson H, Brinkfeldt K, Grigoriev A, Holmström M, Yamauchi M, Asamura K, Baumjohann W, Lammer H, Coates AJ, Kataria DO, Linder DR, Curtis CC, Hsieh KC, Sandel BR, Grande M, Gunell H, Koskinen HEJ, Kallio E, Riihelä P, Säles T, Schmidt W, Kozyra J, Krupp N, Fränz M, Woch J, Luhmann J, McKenna-Lawlor S, Mazelle C, Thocaven JJ, Orsini S, Cerulli-Irelli R, Mura M, Milillo M, Maggi M, Roelof E, Brandt P, Szego K, Winningham JD, Frahm RA, Scherrer J, Sharber JR, Wurz P, Bochsler P (2007) The loss of ions from Venus through the plasma wake. Nature 450(7170):650–653. https://doi.org/10.1038/nature06434

Barnes R, Deitrick R, Luger R, Driscoll PE, Quinn TR, Fleming DP, Guyer B, McDonald DV, Meadows VS, Arney G, Crisp D, Domagal-Goldman SD, Foreman-Mackey D, Kaib NA, Lincowski A, Lustig-Yaeger J, Schwieterman E (2016) The habitability of Proxima Centauri b I: evolutionary scenarios. ArXiv e-prints. arXiv:1608.06919

Barstow JK, Aigrain S, Irwin PG, Kendrew S, Fletcher LN (2015) Transit spectroscopy with James Webb Space Telescope: systematics, starspots and stitching. Mon Not R Astron Soc 448(3):2546–2561

Barstow JK, Aigrain S, Irwin PG, Kendrew S, Fletcher LN (2016) Telling twins apart: exo-Earths and Venuses with transit spectroscopy. Mon Not R Astron Soc 458(3):2657–2666

Barth P, Carone L, Barnes R, Noack L, Mollière P, Henning T (2020) Magma ocean evolution of the TRAPPIST-1 planets. ArXiv e-prints. arXiv:2008.09599

Basilevsky AT, Head JW (1996) Evidence for rapid and widespread emplacement of volcanic plains on Venus: stratigraphic studies in the Baltis Vallis region. Geophys Res Lett 23:1497–1500. https://doi.org/10.1029/96GL00975

Batalha NE, Line MR (2017) Information content analysis for selection of optimal JWST observing modes for transiting exoplanet atmospheres. Astron J 153(4):151

Batalha NE, Mandell A, Pontoppidan K, Stevenson KB, Lewis NK, Kalirai J, Earl N, Greene T, Albert L, Nielsen LD (2017) PandExo: a community tool for transiting exoplanet science with *JWST* & *HST*. Publ Astron Soc Pac 129(976):064501. https://doi.org/10.1088/1538-3873/aa65b0

Batalha NE, Lewis NK, Line MR, Valenti J, Stevenson K (2018) Strategies for constraining the atmospheres of temperate terrestrial planets with JWST. Astrophys J Lett 856(2):L34

Bean JL, Abbot DS, Kempton EMR (2017) A statistical comparative planetology approach to the hunt for habitable exoplanets and life beyond the solar system. Astrophys J Lett 841(2):L24

Beichman C, Benneke B, Knutson H, Smith R, Lagage PO, Dressing C, Latham D, Lunine J, Birkmann S, Ferruit P et al (2014) Observations of transiting exoplanets with the James Webb Space Telescope (JWST). Publ Astron Soc Pac 126(946):1134

Belisheva NK, Popov A (1995) Dynamics of the morphofunctional state of cell cultures with variation in the geomagnetic field in high latitudes. Biophysics 40:737–745

Belisheva NK, Lammer H, Biernat HK, Vashenuyk EV (2012) The effect of cosmic rays on biological systems – an investigation during GLE events. Astrophys Space Sci Trans 8(1):7–17. https://doi.org/10.5194/astra-8-7-2012

Belu A, Selsis F, Morales JC, Ribas I, Cossou C, Rauer H (2011) Primary and secondary eclipse spectroscopy with JWST: exploring the exoplanet parameter space. Astron Astrophys 525:A83

Benz W, Slattery WL, Cameron AGW (1986) The origin of the Moon and the single-impact hypothesis I. Icarus 66(3):515–535. https://doi.org/10.1016/0019-1035(86)90088-6

Berner RA (1993) Paleozoic atmospheric CO_2: importance of solar radiation and plant evolution. Science 261(5117):68–70. https://doi.org/10.1126/science.261.5117.68

Bhatia GK (2021) Early thermal evolution of the embryos of Earth: role of [26]Al and impact-generated steam atmosphere. Planet Space Sci 207:105335. https://doi.org/10.1016/j.pss.2021.105335

Bird P (1978) Stress and temperature in subduction shear zones: Tonga and Mariana. Geophys J Int 55(2):411–434. https://doi.org/10.1111/j.1365-246X.1978.tb04280.x

Bjonnes EE, Hansen VL, James B, Swenson JB (2012) Equilibrium resurfacing of Venus: results from new Monte Carlo modeling and implications for Venus surface histories. Icarus 217:451–461. https://doi.org/10.1016/j.icarus.2011.03.033

Blaske CH, O'Rourke JG (2021) Energetic requirements for dynamos in the metallic cores of super-Earth and super-Venus exoplanets. J Geophys Res, Planets 126(7):e06739. https://doi.org/10.1029/2020JE006739

Bogdanov AV, Vaisberg OL (1975) Structure and variations of solar wind-Mars interaction region. J Geophys Res 80(4):487. https://doi.org/10.1029/JA080i004p00487

Bonati I, Lichtenberg T, Bower DJ, Timpe ML, Quanz SP (2019) Direct imaging of molten protoplanets in nearby young stellar associations. Astron Astrophys 621:A125

Bonati I, Lasbleis M, Noack L (2021) Structure and thermal evolution of exoplanetary cores. J Geophys Res, Planets 126(5):e06724. https://doi.org/10.1029/2020JE006724

Borrelli ME, O'Rourke JG, Smrekar SE, Ostberg CM (2021) A global survey of lithospheric flexure at steep-sided domical volcanoes on Venus reveals intermediate elastic thicknesses. J Geophys Res, Planets 126(7):e06756. https://doi.org/10.1029/2020JE006756

Boujibar A, Driscoll P, Fei Y (2020) Super-Earth internal structures and initial thermal states. J Geophys Res, Planets 125(5):e06124. https://doi.org/10.1029/2019JE006124

Bowens R, Meyer MR, Delacroix C, Absil O, van Boekel R, Quanz SP, Shinde M, Kenworthy M, Carlomagno B, Orban de Xivry G, Cantalloube F, Pathak P (2021) Exoplanets with ELT-METIS. I. Estimating the direct imaging exoplanet yield around stars within 6.5 parsecs. Astron Astrophys 653:A8. https://doi.org/10.1051/0004-6361/202141109. arXiv:2107.06375

Bower DJ, Kitzmann D, Wolf AS, Sanan P, Dorn C, Oza AV (2019) Linking the evolution of terrestrial interiors and an early outgassed atmosphere to astrophysical observations. Astron Astrophys 631:A103. https://doi.org/10.1051/0004-6361/201935710. arXiv:1904.08300

Bower DJ, Hakim K, Sossi PA, Sanan P (2022) Retention of water in terrestrial magma oceans and carbon-rich early atmospheres. Planet Sci J 3(4):93. https://doi.org/10.3847/PSJ/ac5fb1. arXiv:2110.08029

Brace LH, Kasprzak WT, Taylor HA, Theis RF, Russell CT, Barnes A, Mihalov JD, Hunten DM (1987) The ionotail of Venus: its configuration and evidence for ion escape. J Geophys Res 92(A1):15–26. https://doi.org/10.1029/JA092iA01p00015

Brack A, Horneck G, Cockell CS, Bérces A, Belisheva NK, Eiroa C, Henning T, Herbst T, Kaltenegger L, Léger A, Liseau R, Lammer H, Selsis F, Beichman C, Danchi W, Fridlund M, Lunine J, Paresce F, Penny A, Quirrenbach A, Röttgering H, Schneider J, Stam D, Tinetti G, White GJ (2010) Origin and evolution of life on terrestrial planets. Astrobiology 10(1):69–76. https://doi.org/10.1089/ast.2009.0374

Brady PV, Gíslason SR (1997) Seafloor weathering controls on atmospheric CO_2 and global climate. Geochim Cosmochim Acta 61(5):965–973. https://doi.org/10.1016/S0016-7037(96)00385-7

Brain DA, McFadden JP, Halekas JS, Connerney JEP, Bougher SW, Curry S, Dong CF, Dong Y, Eparvier F, Fang X, Fortier K, Hara T, Harada Y, Jakosky BM, Lillis RJ, Livi R, Luhmann JG, Ma Y, Modolo R, Seki K (2015) The spatial distribution of planetary ion fluxes near Mars observed by MAVEN. Geophys Res Lett 42(21):9142–9148. https://doi.org/10.1002/2015GL065293

Brandl BR, Absil O, Agócs T, Baccichet N, Bertram T, Bettonvil F, van Boekel R, Burtscher L, van Dishoeck E, Feldt M, Garcia PJV, Glasse A, Glauser A, Güdel M, Haupt C, Kenworthy MA, Labadie L, Laun W, Lesman D, Pantin E, Quanz SP, Snellen I, Siebenmorgen R, van Winckel H (2018) Status of the mid-IR

ELT imager and spectrograph (METIS). In: Evans CJ, Simard L, Takami H (eds) Ground-based and airborne instrumentation for astronomy VII. Society of photo-optical instrumentation engineers (SPIE) conference series, vol 10702, p 107021U. https://doi.org/10.1117/12.2311492

Breuer D, Moore W (2007) Dynamics and thermal history of the terrestrial planets, the Moon, and Io, vol 10. Am. Geophys. Union, Washington, pp 299–348. https://doi.org/10.1016/B978-044452748-6/00161-9

Brossier J, Gilmore M, Toner K, Stein A (2021) Distinct mineralogy and age of individual lava flows in Atla Regio, Venus derived from Magellan radar emissivity. J Geophys Res, Planets 126(3):e2020JE006722

Bullock MA, Grinspoon DH (1996) The stability of climate on Venus. J Geophys Res, Planets 101(E3):7521–7529

Bullock MA, Grinspoon DH (2001) The recent evolution of climate on Venus. Icarus 150(1):19–37

Bullock MA, Grinspoon DH, Head JW III (1993) Venus resurfacing rates: constraints provided by 3-D Monte Carlo simulations. Geophys Res Lett 20(19):2147–2150. https://doi.org/10.1029/93GL02505

Byrne PK, Krishnamoorthy S (2022) Estimates on the frequency of volcanic eruptions on Venus. J Geophys Res, Planets 127(1):e2021JE007040. https://doi.org/10.1029/2021JE007040

Byrne PK, Ghail RC, Şengör AMC, James PB, Klimczak C, Solomon SC (2021) A globally fragmented and mobile lithosphere on Venus. Proc Natl Acad Sci 118(26):e2025919118. https://doi.org/10.1073/pnas.2025919118

Calvert A, Sawyer E, Davis W, Ludden J (1995) Archaean subduction inferred from seismic images of a mantle suture in the superior province. Nature 375:670–674

Canup RM (2004) Simulations of a late lunar-forming impact. Icarus 168(2):433–456. https://doi.org/10.1016/j.icarus.2003.09.028

Catling DC, Zahnle KJ (2020) The Archean atmosphere. Sci Adv 6(9):eaax1420

Cauley PW, Shkolnik EL, Llama J, Lanza AF (2019) Magnetic field strengths of hot Jupiters from signals of star-planet interactions. Nat Astron 3:1128–1134. https://doi.org/10.1038/s41550-019-0840-x. arXiv:1907.09068

Checlair JH, Hayworth BPC, Olson SL, Komacek TD, Villanueva GL, Popović P, Yang H, Abbot DS (2020) Non-detection of O_2/O_3 informs frequency of Earth-like planets with LUVOIR but not HabEx. ArXiv e-prints. arXiv:2008.03952

Clampin M (2011) Overview of the James Webb Space Telescope Observatory. In: UV/optical/IR space telescopes and instruments: innovative technologies and concepts V. International society for optics and photonics, vol 8146, p 814605

Clery D (2019) No safe haven for the Thirty Meter Telescope. Science 365(6457):960–961. https://doi.org/10.1126/science.365.6457.960

Clery D (2023) Future NASA scope would find life on alien worlds. Science 379(6628):123–124. https://doi.org/10.1126/science.adg6273

Cnossen I (2020) Analysis and attribution of climate change in the upper atmosphere from 1950 to 2015 simulated by WACCM-X. J Geophys Res Space Phys 125(12):e28623. https://doi.org/10.1029/2020JA028623

Cockell CS (1999) Life on Venus. Planet Space Sci 47(12):1487–1501. https://doi.org/10.1016/S0032-0633(99)00036-7

Colaprete A, Toon OB (2003) Carbon dioxide clouds in an early dense Martian atmosphere. J Geophys Res, Planets 108(E4):5025

Condie KC, Kröner A (2008) When did plate tectonics begin? Evidence from the geologic record. In: When did plate tectonics begin on planet Earth? Geol. Soc. Am., Boulder. https://doi.org/10.1130/2008.2440(14)

Coogan LA, Dosso SE (2015) Alteration of ocean crust provides a strong temperature dependent feedback on the geological carbon cycle and is a primary driver of the Sr-isotopic composition of seawater. Earth Planet Sci Lett 415:38–46. https://doi.org/10.1016/j.epsl.2015.01.027

Coogan LA, Gillis KM (2013) Evidence that low-temperature oceanic hydrothermal systems play an important role in the silicate-carbonate weathering cycle and long-term climate regulation. Geochem Geophys Geosyst 14(6):1771–1786. https://doi.org/10.1002/ggge.20113

Correia ACM, Laskar J (2001) The four final rotation states of Venus. Nature 411:767–770. https://doi.org/10.1038/35081000

Correia ACM, Laskar J (2003) Long-term evolution of the spin of Venus. II. Numerical simulations. Icarus 163:24–45. https://doi.org/10.1016/S0019-1035(03)00043-5

Correia ACM, Laskar J, de Surgy ON (2003) Long-term evolution of the spin of Venus. I. Theory. Icarus 163:1–23. https://doi.org/10.1016/S0019-1035(03)00042-3

Cottini V, Ignatiev N, Piccioni G, Drossart P, Grassi D, Markiewicz W (2012) Water vapor near the cloud tops of Venus from Venus Express/VIRTIS dayside data. Icarus 217(2):561–569

Coulter DR (2003) NASA's Terrestrial Planet Finder mission: the search for habitable planets. In: Fridlund M, Henning T, Lacoste H (eds) Earths: DARWIN/TPF and the search for extrasolar terrestrial planets. ESA special publication, vol 539, pp 47–54

Crisp JA (1984) Rates of magma emplacement and volcanic output. J Volcanol Geotherm Res 20(3–4):177–211

Crouzet N, Bonfils X, Delfosse X, Boisse I, Hébrard G, Forveille T, Donati JF, Bouchy F, Moutou C, Doyon R et al (2017) Follow-up and characterization of the TESS exoplanets with SOPHIE, SPIRou, and JWST. ArXiv preprint. arXiv:1701.03539

Dar A, Laor A, Shaviv NJ (1998) Life extinctions by cosmic ray jets. Phys Rev Lett 80(26):5813–5816. https://doi.org/10.1103/PhysRevLett.80.5813. arXiv:astro-ph/9705008

Dasgupta R, Hirschmann MM (2010) The deep carbon cycle and melting in Earth's interior. Earth Planet Sci Lett 298(1–2):1–13

Davaille A, Smrekar S, Tomlinson S (2017) Experimental and observational evidence for plume-induced subduction on Venus. Nat Geosci 10(1):349–355. https://doi.org/10.1038/ngeo2928

Davies GF (1993) Conjectures on the thermal and tectonic evolution of the Earth. Lithos 30(3):281–289. https://doi.org/10.1016/0024-4937(93)90041-A

Davies C, Pozzo M, Gubbins D, Alfè D (2015) Constraints from material properties on the dynamics and evolution of Earth's core. Nat Geosci 8(9):678–685. https://doi.org/10.1038/ngeo2492

De Bergh C, Moroz V, Taylor F, Crisp D, Bézard B, Zasova L (2006) The composition of the atmosphere of Venus below 100 km altitude: an overview. Planet Space Sci 54(13–14):1389–1397

Debaille V, O'Neill C, Brandon AD, Haenecour P, Yin QZ, Mattielli N, Treiman AH (2013) Stagnant-lid tectonics in early Earth revealed by ^{142}Nd variations in late Archean rocks. Earth Planet Sci Lett 373:83–92. https://doi.org/10.1016/j.epsl.2013.04.016

Deming D, Seager S, Winn J, Miller-Ricci E, Clampin M, Lindler D, Greene T, Charbonneau D, Laughlin G, Ricker G et al (2009) Discovery and characterization of transiting super earths using an all-sky transit survey and follow-up by the James Webb Space Telescope. Publ Astron Soc Pac 121(883):952

Ding F, Wordsworth RD (2022) Prospects for water vapor detection in the atmospheres of temperate and arid rocky exoplanets around M-dwarf stars. ArXiv e-prints. arXiv:2201.08423

Dobrovolskis AR (1980) Atmospheric tides and the rotation of Venus. II – Spin evolution. Icarus 41:18–35. https://doi.org/10.1016/0019-1035(80)90157-8

Dobrovolskis AR (1983) Atmospheric tides on Venus. III – The planetary boundary layer. Icarus 56:165–175. https://doi.org/10.1016/0019-1035(83)90133-1

Dobrovolskis AR, Ingersoll AP (1980) Atmospheric tides and the rotation of Venus. I – Tidal theory and the balance of torques. Icarus 41:1–17. https://doi.org/10.1016/0019-1035(80)90156-6

Donahue TM, Hoffman JH, Hodges RR, Watson AJ (1982) Venus was wet – a measurement of the ratio of deuterium to hydrogen. Science 216:630–633. https://doi.org/10.1126/science.216.4546.630

Dong Y, Fang X, Brain DA, McFadden JP, Halekas JS, Connerney JEP, Eparvier F, Andersson L, Mitchell D, Jakosky BM (2017) Seasonal variability of Martian ion escape through the plume and tail from MAVEN observations. J Geophys Res Space Phys 122(4):4009–4022. https://doi.org/10.1002/2016JA023517

Dong C, Jin M, Lingam M (2020) Atmospheric escape from TOI-700 d: Venus versus Earth analogs. Astrophys J 896(2):L24. https://doi.org/10.3847/2041-8213/ab982f

Dorn C, Lichtenberg T (2021) Hidden water in magma ocean exoplanets. Astrophys J Lett 922(1):L4

Dorn C, Noack L, Rozel A (2018) Outgassing on stagnant-lid super-Earths. Astron Astrophys 614:A18

Driscoll PE, Barnes R (2015) Tidal heating of Earth-like exoplanets around M stars: thermal, magnetic, and orbital evolutions. Astrobiology 15(9):739–760. https://doi.org/10.1089/ast.2015.1325. arXiv:1509.07452

Driscoll P, Bercovici D (2014) On the thermal and magnetic histories of Earth and Venus: influences of melting, radioactivity, and conductivity. Phys Earth Planet Inter 236:36–51. https://doi.org/10.1016/j.pepi.2014.08.004

Driscoll P, Olson P (2011) Optimal dynamos in the cores of terrestrial exoplanets: magnetic field generation and detectability. Icarus 213(1):12–23. https://doi.org/10.1016/j.icarus.2011.02.010

Dubinin E, Fraenz M, Pätzold M, McFadden J, Halekas JS, DiBraccio GA, Connerney JEP, Eparvier F, Brain D, Jakosky BM, Vaisberg O, Zelenyi L (2017) The effect of solar wind variations on the escape of oxygen ions from Mars through different channels: MAVEN observations. J Geophys Res Space Phys 122(11):11,285–11,301. https://doi.org/10.1002/2017JA024741

Dyar MD, Helbert J, Cooper RF, Sklute EC, Maturilli A, Mueller NT, Kappel D, Smrekar SE (2021) Surface weathering on Venus: constraints from kinetic, spectroscopic, and geochemical data. Icarus 358:114139

Edberg NJT, Nilsson H, Williams AO, Lester M, Milan SE, Cowley SWH, Fränz M, Barabash S, Futaana Y (2010) Pumping out the atmosphere of Mars through solar wind pressure pulses. Geophys Res Lett 37(3):L03107. https://doi.org/10.1029/2009GL041814

Edberg NJT, Nilsson H, Futaana Y, Stenberg G, Lester M, Cowley SWH, Luhmann JG, McEnulty TR, Opgenoorth HJ, Fedorov A, Barabash S, Zhang TL (2011) Atmospheric erosion of Venus during stormy space weather. J Geophys Res Space Phys 116(A9):A09308. https://doi.org/10.1029/2011JA016749

118

Ehrenreich D, Vidal-Madjar A, Widemann T, Gronoff G, Tanga P, Barthélemy M, Lilensten J, Des Etangs AL, Arnold L (2012) Transmission spectrum of Venus as a transiting exoplanet. Astron Astrophys 537:L2

Eisenhauer F, Abuter R, Bickert K, Biancat-Marchet F, Bonnet H, Brynnel J, Conzelmann RD, Delabre B, Donaldson R, Farinato J, Fedrigo E, Genzel R, Hubin NN, Iserlohe C, Kasper ME, Kissler-Patig M, Monnet GJ, Roehrle C, Schreiber J, Stroebele S, Tecza M, Thatte NA, Weisz H (2003) SINFONI – integral field spectroscopy at 50 milli-arcsecond resolution with the ESO VLT. In: Iye M, Moorwood AFM (eds) Instrument design and performance for optical/infrared ground-based telescopes. Society of photo-optical instrumentation engineers (SPIE) conference series, vol 4841, pp 1548–1561. https://doi.org/10.1117/12.459468. arXiv:astro-ph/0306191

Elkins-Tanton LT (2008) Linked magma ocean solidification and atmospheric growth for Earth and Mars. Earth Planet Sci Lett 271(1–4):181–191. https://doi.org/10.1016/j.epsl.2008.03.062

Elkins-Tanton LT (2012) Magma oceans in the inner solar system. Annu Rev Earth Planet Sci 40(1):113–139. https://doi.org/10.1146/annurev-earth-042711-105503

Emsenhuber A, Asphaug E, Cambioni S, Gabriel TSJ, Schwartz SR (2021) Collision chains among the terrestrial planets. II. An asymmetry between Earth and Venus. Planet Sci J 2(5):199. https://doi.org/10.3847/PSJ/ac19b1. arXiv:2110.00221

Encrenaz T, Lellouch E, Cernicharo J, Paubert G, Gulkis S, Spilker T (1995) The thermal profile and water abundance in the Venus mesosphere from H_2O and HDO millimeter observations. Icarus 117(1):162–172

Esposito LW (1984) Sulfur dioxide: episodic injection shows evidence for active Venus volcanism. Science 223(4640):1072–1074

Fanson J, Bernstein R, Angeli G, Ashby D, Bigelow B, Brossus G, Bouchez A, Burgett W, Contos A, Demers R, Figueroa F, Fischer B, Groark F, Laskin R, Millan-Gabet R, Pi M, Wheeler N (2020) Overview and status of the Giant Magellan Telescope project. In: Ground-based and airborne telescopes VIII. Society of photo-optical instrumentation engineers (SPIE) conference series, vol 11445, p 114451F. https://doi.org/10.1117/12.2561852

Fauchez TJ, Turbet M, Villanueva GL, Wolf ET, Arney G, Kopparapu RK, Lincowski A, Mandell A, de Wit J, Pidhorodetska D, Domagal-Goldman SD, Stevenson KB (2019) Impact of clouds and hazes on the simulated JWST transmission spectra of habitable zone planets in the TRAPPIST-1 system. Astrophys J 887(2):194. https://doi.org/10.3847/1538-4357/ab5862

Fedorov A, Barabash S, Sauvaud JA, Futaana Y, Zhang TL, Lundin R, Ferrier C (2011) Measurements of the ion escape rates from Venus for solar minimum. J Geophys Res Space Phys 116(A7):A07220. https://doi.org/10.1029/2011JA016427

Fegley JB (2003) Venus. In: Treatise on geochemistry, vol 1, p 711. https://doi.org/10.1016/B0-08-043751-6/01150-6

Fegley JB (2014) Venus. In: Davis AM (ed) Planets, asteroids, comets and the solar system, vol 2. Elsevier, Amsterdam, pp 127–148

Fegley B, Prinn RG (1989) Estimation of the rate of volcanism on Venus from reaction rate measurements. Nature 337(6202):55–58

Fischer RA, Nakajima Y, Campbell AJ, Frost DJ, Harries D, Langenhorst F, Miyajima N, Pollok K, Rubie DC (2015) High pressure metal–silicate partitioning of Ni, Co, V, Cr, Si, and O. Geochim Cosmochim Acta 167:177–194. https://doi.org/10.1016/j.gca.2015.06.026

Flasar FM, Birch F (1973) Energetics of core formation: a correction. J Geophys Res 78(26):6101–6103. https://doi.org/10.1029/JB078i026p06101

Foley BJ (2015) The role of plate tectonic-climate coupling and exposed land area in the development of habitable climates on rocky planets. Astrophys J 812(1):36. https://doi.org/10.1088/0004-637X/812/1/36. arXiv:1509.00427

Foley BJ (2019) Habitability of Earth-like stagnant lid planets: climate evolution and recovery from snowball states. Astrophys J 875:72. https://doi.org/10.3847/1538-4357/ab0f31. arXiv:1903.12111

Foley BJ, Driscoll PE (2016) Whole planet coupling between climate, mantle, and core: implications for rocky planet evolution. Geochem Geophys Geosyst 17(5):1885–1914. https://doi.org/10.1002/2015GC006210. arXiv:1711.06801

Foley BJ, Smye AJ (2018) Carbon cycling and habitability of Earth-sized stagnant lid planets. Astrobiology 18(7):873–896. https://doi.org/10.1089/ast.2017.1695

Foley BJ, Bercovici D, Landuyt W (2012) The conditions for plate tectonics on super-Earths: inferences from convection models with damage. Earth Planet Sci Lett 331(332):281–290. https://doi.org/10.1016/j.epsl.2012.03.028

Forget F, Wordsworth R, Millour E, Madeleine JB, Kerber L, Leconte J, Marcq E, Haberle R (2013) 3D modelling of the early Martian climate under a denser CO_2 atmosphere: temperatures and CO_2 ice clouds. Icarus 222(1):81–99. https://doi.org/10.1016/j.icarus.2012.10.019

Fujii Y, Kimura J, Dohm J, Ohtake M (2014) Geology and photometric variation of solar system bodies with minor atmospheres: implications for solid exoplanets. Astrobiology 14(9):753–768. https://doi.org/10.1089/ast.2014.1165

Fulton BJ, Petigura EA, Howard AW, Isaacson H, Marcy GW, Cargile PA, Hebb L, Weiss LM, Johnson JA, Morton TD, Sinukoff E, Crossfield IJM, Hirsch LA (2017) The California-Kepler survey. III. A gap in the radius distribution of small planets. Astron J 154(3):109. https://doi.org/10.3847/1538-3881/aa80eb. arXiv:1703.10375

Gaillard F, Scaillet B (2009) The sulfur content of volcanic gases on Mars. Earth Planet Sci Lett 279(1–2):34–43

Gaillard F, Scaillet B (2014) A theoretical framework for volcanic degassing chemistry in a comparative planetology perspective and implications for planetary atmospheres. Earth Planet Sci Lett 403:307–316

Gaillard F, Bernadou F, Roskosz M, Bouhifd MA, Marrocchi Y, Iacono-Marziano G, Moreira M, Scaillet B, Rogerie G (2022) Redox controls during magma ocean degassing. Earth Planet Sci Lett 577:117255

Garvin J, Getty S, Arney G, Johnson N, Malespin C, Webster C, Ravine M, Lorenz R, Kiefer W, Atreya S et al (2020) Deep atmosphere of Venus investigation of noble gases, chemistry, and imaging plus (DAVINCI+): discovering a new Venus via a flyby, probe, orbiter mission. In: AGU fall meeting abstracts, vol 2020, p P026-0001

Gelman B, Zolotukhin V, Lamonov N, Levchuk B, Mukhin L, Nenarokov D, Khotnikov B, Rotin V, Lipatov A (1979) An analysis of the chemical composition of the atmosphere of Venus on an AMS of the Venera-12 using a gas chromatograph. NASA STI/Recon Technical Report N 79:25964

Gerya T (2014a) Precambrian geodynamics: concepts and models. Gondwana Res 25(2):442–463. https://doi.org/10.1016/j.gr.2012.11.008

Gerya TV (2014b) Plume-induced crustal convection: 3D thermomechanical model and implications for the origin of novae and coronae on Venus. Earth Planet Sci Lett 391:183–192. https://doi.org/10.1016/j.epsl.2014.02.005

Gillmann C, Tackley P (2014) Atmosphere/mantle coupling and feedbacks on Venus. J Geophys Res, Planets 119(6):1189–1217

Gillmann C, Golabek GJ, Tackley PJ (2016) Effect of a single large impact on the coupled atmosphere-interior evolution of Venus. Icarus 268:295–312. https://doi.org/10.1016/j.icarus.2015.12.024

Gillmann C, Way MJ, Avice G, Breuer D, Golabek GJ, Höning D, Krissansen-Totton J, Lammer H, Plesa AC, Persson M, O'Rourke JG, Salvador A, Scherf M, Zolotov MY (2022) The long-term evolution of the atmosphere of Venus: processes and feedback mechanisms. Space Sci Rev 218:56. https://doi.org/10.1007/s11214-022-00924-0

Gillon M, Jehin E, Lederer SM, Delrez L, de Wit J, Burdanov A, Grootel VV, Burgasser AJ, Triaud AHMJ, Opitom C, Demory BO, Sahu DK, Gagliuffi DCB, Magain P, Queloz D (2016) Temperate Earth-sized planets transiting a nearby ultracool dwarf star. Nature 533:221–224. https://doi.org/10.1038/nature17448

Gilmore M, Treiman A, Helbert J, Smrekar S (2017) Venus surface composition constrained by observation and experiment. Space Sci Rev 212(3):1511–1540. https://doi.org/10.1007/s11214-017-0370-8

Goldblatt C, Robinson TD, Zahnle KJ, Crisp D (2013) Low simulated radiation limit for runaway greenhouse climates. Nat Geosci 6(8):661–667. https://doi.org/10.1038/ngeo1892

Goldreich P, Peale SJ (1966) Resonant rotation for Venus? Nature 209:1117–1118. https://doi.org/10.1038/2091117a0

Goldreich P, Peale SJ (1970) The obliquity of Venus. Astron J 75:273. https://doi.org/10.1086/110975

Gómez-Leal I, Codron F, Selsis F (2016) Thermal light curves of Earth-like planets: 1. Varying surface and rotation on planets in a terrestrial orbit. Icarus 269:98–110. https://doi.org/10.1016/j.icarus.2015.12.050

Graham RJ, Pierrehumbert R (2020) Thermodynamic and energetic limits on continental silicate weathering strongly impact the climate and habitability of wet, rocky worlds. Astrophys J 896(2):115

Green JAM, Way MJ, Barnes R (2019) Consequences of tidal dissipation in a putative Venusian ocean. Astrophys J 876(2):L22. https://doi.org/10.3847/2041-8213/ab133b. arXiv:1903.07517

Greene TP, Line MR, Montero C, Fortney JJ, Lustig-Yaeger J, Luther K (2016) Characterizing transiting exoplanet atmospheres with JWST. Astrophys J 817(1):17

Grießmeier JM, Stadelmann A, Motschmann U, Belisheva NK, Lammer H, Biernat HK (2005) Cosmic ray impact on extrasolar Earth-like planets in close-in habitable zones. Astrobiology 5(5):587–603. https://doi.org/10.1089/ast.2005.5.587

Grießmeier JM, Stadelmann A, Grenfell JL, Lammer H, Motschmann U (2009) On the protection of extrasolar Earth-like planets around K/M stars against galactic cosmic rays. Icarus 199(2):526–535. https://doi.org/10.1016/j.icarus.2008.09.015. arXiv:0902.0952

Grießmeier JM, Tabataba-Vakili F, Stadelmann A, Grenfell JL, Atri D (2016) Galactic cosmic rays on extrasolar Earth-like planets. II. Atmospheric implications. Astron Astrophys 587:A159. https://doi.org/10.1051/0004-6361/201425452. arXiv:1603.06500

Grinspoon DH, Bullock MA (2007) Astrobiology and Venus exploration Am. Geophys. Union, Washington, pp 191–206. https://doi.org/10.1029/176GM12

Gülcher AJ, Gerya TV, Montési LG, Munch J (2020) Corona structures driven by plume–lithosphere interactions and evidence for ongoing plume activity on Venus. Nat Geosci 13(8):547–554. https://doi.org/10.1038/s41561-020-0606-1

Gunell H, Maggiolo R, Nilsson H, Stenberg Wieser G, Slapak R, Lindkvist J, Hamrin M, De Keyser J (2018) Why an intrinsic magnetic field does not protect a planet against atmospheric escape. Astron Astrophys 614:L3. https://doi.org/10.1051/0004-6361/201832934

Gustafsson B, Meléndez J, Asplund M, Yong D (2010) The chemical composition of solar-type stars in comparison with that of the Sun. Astrophys Space Sci 328(1):185–191. https://doi.org/10.1007/s10509-009-0257-6

Hamano K, Abe Y, Genda H (2013) Emergence of two types of terrestrial planet on solidification of magma ocean. Nature 497:607–610. https://doi.org/10.1038/nature12163

Hamano K, Kawahara H, Abe Y, Onishi M, Hashimoto GL (2015) Lifetime and spectral evolution of a magma ocean with a steam atmosphere: its detectability by future direct imaging. Astrophys J 806(2):216. https://doi.org/10.1088/0004-637X/806/2/216. arXiv:1505.03552

Hashimoto GL, Abe Y (2005) Climate control on Venus: comparison of the carbonate and pyrite models. Planet Space Sci 53(8):839–848

Hayashi C, Nakazawa K, Mizuno H (1979) Earth's melting due to the blanketing effect of the primordial dense atmosphere. Earth Planet Sci Lett 43(1):22–28. https://doi.org/10.1016/0012-821X(79)90152-3

Head J, Wilson L, Ivanov M, Wordsworth R (2021) Contributions of volatiles to the Venus atmosphere from the observed extrusive volcanic record: implications for the history of the Venus atmosphere. In: EGU general assembly conference abstracts, EGU general assembly conference abstracts, pp EGU21–13030

Henehan MJ, Hull PM, Penman DE, Rae JW, Schmidt DN (2016) Biogeochemical significance of pelagic ecosystem function: an end-Cretaceous case study. Philos Trans R Soc B, Biol Sci 371(1694):20150510

Henning WG, Renaud JP, Saxena P, Whelley PL, Mandell AM, Matsumura S, Glaze LS, Hurford TA, Livengood TA, Hamilton CW, Efroimsky M, Makarov VV, Berghea CT, Guzewich SD, Tsigaridis K, Arney GN, Cremons DR, Kane SR, Bleacher JE, Kopparapu RK, Kohler E, Lee Y, Rushby A, Kuang W, Barnes R, Richardson JA, Driscoll P, Schmerr NC, Genio ADD, Davies AG, Kaltenegger L, Elkins-Tanton L, Fujii Y, Schaefer L, Ranjan S, Quintana E, Barclay TS, Hamano K, Petro NE, Kendall JD, Lopez ED, Sasselov DD (2018) Highly volcanic exoplanets, lava worlds, and magma ocean worlds: an emerging class of dynamic exoplanets of significant scientific priority. National Academy of Sciences. White paper submitted in response to the National Academy of Sciences 2018 Exoplanet Science Strategy solicitation, from the NASA Sellers Exoplanet Environments Collaboration (SEEC) of the Goddard Space Flight Center. arXiv:1804.05110

Herrick RR (1994) Resurfacing history of Venus. Geology 22:703. https://doi.org/10.1130/0091-7613(1994)022<0703:RHOV>2.3.CO;2

Herrick RR, Rumpf ME (2011) Postimpact modification by volcanic or tectonic processes as the rule, not the exception, for Venusian craters. J Geophys Res, Planets 116(E2):E02004

Hier-Majumder S, Hirschmann MM (2017) The origin of volatiles in the Earth's mantle. Geochem Geophys Geosyst 18(8):3078–3092. https://doi.org/10.1002/2017GC006937. arXiv:1011.1669v3

Hinkel NR, Unterborn CT (2018) The star-planet connection. I. Using stellar composition to observationally constrain planetary mineralogy for the 10 closest stars. Astrophys J 853(1):83. https://doi.org/10.3847/1538-4357/aaa5b4. arXiv:1709.08630

Hirschberger P, Plesa AC, Breuer D (2020) Elastic lithosphere thickness calculations from numerical thermal evolution models of Venus' interior with a variable thermal conductivity. In: AGU fall meeting, p 690646

Hirschmann MM (2006) Water, melting, and the deep Earth H_2O cycle. Annu Rev Earth Planet Sci 34(1):629–653. https://doi.org/10.1146/annurev.earth.34.031405.125211

Hirth G, Kohlstedt D (2003) Rheology of the upper mantle and the mantle wedge: a view from the experimentalists. In: Geophysical monograph series, vol 138. Am. Geophys. Union, Washington, pp 83–106

Hoeijmakers HJ, Schwarz H, Snellen IAG, de Kok RJ, Bonnefoy M, Chauvin G, Lagrange AM, Girard JH (2018) Medium-resolution integral-field spectroscopy for high-contrast exoplanet imaging. Molecule maps of the β Pictoris system with SINFONI. Astron Astrophys 617:A144. https://doi.org/10.1051/0004-6361/201832902. arXiv:1802.09721

Holland HD (1978) The chemistry of the atmosphere and oceans. In: The chemistry of the atmosphere and oceans. Wiley, New York

Höning D, Tosi N, Spohn T (2019) Carbon cycling and interior evolution of water-covered plate tectonics and stagnant-lid planets. Astron Astrophys 627:A48

Höning D, Baumeister P, Grenfell JL, Tosi N, Way MJ (2021) Early habitability and crustal decarbonation of a stagnant-lid Venus. J Geophys Res, Planets 126(10):e2021JE006895. https://doi.org/10.1029/2021JE006895

Hoolst TV (2015) 10.04 – Rotation of the terrestrial planets. In: Schubert G (ed) Treatise on geophysics, second edn. Elsevier, Oxford, pp 121–151. https://doi.org/10.1016/B978-0-444-53802-4.00168-8

Houllé M, Vigan A, Carlotti A, Choquet É, Cantalloube F, Phillips MW, Sauvage JF, Schwartz N, Otten GPPL, Baraffe I, Emsenhuber A, Mordasini C (2021) Direct imaging and spectroscopy of exoplanets with the ELT/HARMONI high-contrast module. Astron Astrophys 652:A67. https://doi.org/10.1051/0004-6361/202140479. arXiv:2104.11251

Howe AR, Burrows A, Deming D (2017) An information-theoretic approach to optimize JWST observations and retrievals of transiting exoplanet atmospheres. Astrophys J 835(1):96

Howell SB, Sobeck C, Haas M, Still M, Barclay T, Mullally F, Troeltzsch J, Aigrain S, Bryson ST, Caldwell D, Chaplin WJ, Cochran WD, Huber D, Marcy GW, Miglio A, Najita JR, Smith M, Twicken JD, Fortney JJ (2014) The K2 mission: characterization and early results. Publ Astron Soc Pac 126(938):398. https://doi.org/10.1086/676406. arXiv:1402.5163

Hubbert MK, Rubey W (1959) Role of fluid pressure in mechanics of overthrust faulting, pts. I and II. Geol Soc Am Bull 70:115–205

Ikoma M, Genda H (2006) Constraints on the mass of a habitable planet with water of nebular origin. Astrophys J 648(1):696–706. https://doi.org/10.1086/505780

Ikoma M, Elkins-Tanton L, Hamano K, Suckale J (2018) Water partitioning in planetary embryos and protoplanets with magma oceans. Space Sci Rev 214(4):76. https://doi.org/10.1007/s11214-018-0508-3

Ingersoll AP (1969) The runaway greenhouse: a history of water on Venus. J Atmos Sci 26:1191–1198. https://doi.org/10.1175/1520-0469(1969)026<1191:TRGAHO>2.0.CO;2

Ingersoll AP, Dobrovolskis AR (1978) Venus' rotation and atmospheric tides. Nature 275:37. https://doi.org/10.1038/275037a0

Ivanov MA, Head JW (2013) The history of volcanism on Venus. Planet Space Sci 84:66–92. https://doi.org/10.1016/j.pss.2013.04.018

Jacobson SA, Rubie DC, Hernlund J, Morbidelli A, Nakajima M (2017) Formation, stratification, and mixing of the cores of Earth and Venus. Earth Planet Sci Lett 474:375–386. https://doi.org/10.1016/j.epsl.2017.06.023

Jakosky BM, Ahrens TJ (1979) The history of an atmosphere of impact origin. In: Lunar and planetary science conference proceedings, vol 3, pp 2727–2739

Jakosky BM, Grebowsky JM, Luhmann JG, Connerney J, Eparvier F, Ergun R, Halekas J, Larson D, Mahaffy P, McFadden J, Mitchell DF, Schneider N, Zurek R, Bougher S, Brain D, Ma YJ, Mazelle C, Andersson L, Andrews D, Baird D, Baker D, Bell JM, Benna M, Chaffin M, Chamberlin P, Chaufray YY, Clarke J, Collinson G, Combi M, Crary F, Cravens T, Crismani M, Curry S, Curtis D, Deighan J, Delory G, Dewey R, DiBraccio G, Dong C, Dong Y, Dunn P, Elrod M, England S, Eriksson A, Espley J, Evans S, Fang X, Fillingim M, Fortier K, Fowler CM, Fox J, Gröller H, Guzewich S, Hara T, Harada Y, Holsclaw G, Jain SK, Jolitz R, Leblanc F, Lee CO, Lee Y, Lefevre F, Lillis R, Livi R, Lo D, Mayyasi M, McClintock W, McEnulty T, Modolo R, Montmessin F, Morooka M, Nagy A, Olsen K, Peterson W, Rahmati A, Ruhunusiri S, Russell CT, Sakai S, Sauvaud JA, Seki K, Steckiewicz M, Stevens M, Stewart AIF, Stiepen A, Stone S, Tenishev V, Thiemann E, Tolson R, Toublanc D, Vogt M, Weber T, Withers P, Woods T, Yelle R (2015) MAVEN observations of the response of Mars to an interplanetary coronal mass ejection. Science 350(6261):0210. https://doi.org/10.1126/science.aad0210

Jiang JH, Zhai AJ, Herman J, Zhai C, Hu R, Su H, Natraj V, Li J, Xu F, Yung YL (2018) Using deep space climate observatory measurements to study the Earth as an exoplanet. Astron J 156(1):26. https://doi.org/10.3847/1538-3881/aac6e2. arXiv:1805.05834

Johnstone CP, Güdel M, Lammer H, Kislyakova KG (2018) Upper atmospheres of terrestrial planets: carbon dioxide cooling and the Earth's thermospheric evolution. Astron Astrophys 617:A107. https://doi.org/10.1051/0004-6361/201832776. arXiv:1806.06897

Johnstone CP, Lammer H, Kislyakova K, Scherf M, Guedel M (2021) The young Sun's XUV-activity as a constraint for lower CO$_2$-limits in the Earth's Archean atmosphere. Earth Planet Sci 576:117197

Kadoya S, Tajika E (2014) Conditions for oceans on Earth-like planets orbiting within the habitable zone: importance of volcanic CO$_2$ degassing. Astrophys J 790:107. https://doi.org/10.1088/0004-637X/790/2/107

Kaltenegger L, Henning W, Sasselov D (2010) Detecting volcanism on extrasolar planets. Astron J 140(5):1370

Kane SR, von Braun K (2008) Constraining orbital parameters through planetary transit monitoring. Astrophys J 689(1):492

Kane SR, Barclay T, Gelino DM (2013) A potential super-Venus in the Kepler-69 system. Astrophys J 770(2):L20. https://doi.org/10.1088/2041-8205/770/2/L20. arXiv:1305.2933

Kane SR, Kopparapu RK, Domagal-Goldman SD (2014) On the frequency of potential Venus analogs from Kepler data. Astrophys J 794(1):L5. https://doi.org/10.1088/2041-8205/794/1/l5

Kane SR, Ceja AY, Way MJ, Quintana EV (2018) Climate modeling of a potential ExoVenus. Astrophys J 869(1):46. https://doi.org/10.3847/1538-4357/aaec68. arXiv:1810.10072

Kane SR, Roettenbacher RM, Unterborn CT, Foley BJ, Hill ML (2020) A volatile-poor formation of LHS 3844b based on its lack of significant atmosphere. Planet Sci J 1(2):36. https://doi.org/10.3847/PSJ/abaab5. arXiv:2007.14493

Kao MM, Hallinan G, Pineda JS, Stevenson D, Burgasser A (2018) The strongest magnetic fields on the coolest brown dwarfs. Astrophys J Suppl Ser 237(2):25. https://doi.org/10.3847/1538-4365/aac2d5. arXiv:1808.02485

Kasper M, Cerpa Urra N, Pathak P, Bonse M, Nousiainen J, Engler B, Heritier CT, Kammerer J, Leveratto S, Rajani C, Bristow P, Le Louarn M, Madec PY, Ströbele S, Verinaud C, Glauser A, Quanz SP, Helin T, Keller C, Snik F, Boccaletti A, Chauvin G, Mouillet D, Kulcsár C, Raynaud HF (2021) PCS – a roadmap for exoearth imaging with the ELT. Messenger 182:38

Kasting J (1993) Earth's early atmosphere. Science 259(5097):920–926. https://doi.org/10.1126/science.11536547

Kasting JF, Catling D (2003) Evolution of a habitable planet. Annu Rev Astron Astrophys 41(1):429–463. https://doi.org/10.1146/annurev.astro.41.071601.170049

Kasting JF, Pollack JB, Ackerman TP (1984) Response of Earth's atmosphere to increases in solar flux and implications for loss of water from Venus. Icarus 57:335–355. https://doi.org/10.1016/0019-1035(84)90122-2

Kasting JF, Whitmire DP, Reynolds RT (1993) Habitable zones around main sequence stars. Icarus 101(1):108–128. https://doi.org/10.1006/icar.1993.1010

Katyal N, Ortenzi G, Lee Grenfell J, Noack L, Sohl F, Godolt M, García Muñoz A, Schreier F, Wunderlich F, Rauer H (2020) Effect of mantle oxidation state and escape upon the evolution of Earth's magma ocean atmosphere. Astron Astrophys 643:A81. https://doi.org/10.1051/0004-6361/202038779. arXiv:2009.14599

Kawahara H, Murakami N, Matsuo T, Kotani T (2014) Spectroscopic coronagraphy for planetary radial velocimetry of exoplanets. Astrophys J Suppl Ser 212(2):27. https://doi.org/10.1088/0067-0049/212/2/27. arXiv:1404.5712

Kempton EMR, Bean JL, Louie DR, Deming D, Koll DD, Mansfield M, Christiansen JL, López-Morales M, Swain MR, Zellem RT et al (2018) A framework for prioritizing the TESS planetary candidates most amenable to atmospheric characterization. Publ Astron Soc Pac 130(993):114401

Khan A, Ceylan S, van Driel M, Giardini D, Lognonné P, Samuel H, Schmerr NC, Stähler SC, Duran AC, Huang Q, Kim D, Broquet A, Charalambous C, Clinton JF, Davis PM, Drilleau M, Karakostas F, Lekic V, McLennan SM, Maguire RR, Michaut C, Panning MP, Pike WT, Pinot B, Plasman M, Scholz JR, Widmer-Schnidrig R, Spohn T, Smrekar SE, Banerdt WB (2021) Upper mantle structure of Mars from insight seismic data. Science 373(6553):434–438. https://doi.org/10.1126/science.abf2966

Khawja S, Ernst R, Samson C, Byrne P, Ghail R, MacLellan L (2020) Tesserae on Venus may preserve evidence of fluvial erosion. Nat Commun 11(1):1–8

Kislyakova KG, Noack L, Johnstone CP, Zaitsev VV, Fossati L, Lammer H, Khodachenko ML, Odert P, Güdel M (2017) Magma oceans and enhanced volcanism on TRAPPIST-1 planets due to induction heating. Nat Astron 1:878–885. https://doi.org/10.1038/s41550-017-0284-0. arXiv:1710.08761

Kislyakova KG, Johnstone CP, Scherf M, Holmström M, Alexeev II, Lammer H, Khodachenko ML, Güdel M (2020) Evolution of the Earth's polar outflow from mid-Archean to present. J Geophys Res Space Phys 125(8):e27837. https://doi.org/10.1029/2020JA027837. arXiv:2008.10337

Kite ES, Manga M, Gaidos E (2009) Geodynamics and rate of volcanism on massive Earth-like planets. Astrophys J 700(2):1732

Knapmeyer-Endrun B, Panning MP, Bissig F, Joshi R, Khan A, Kim D, Lekić V, Tauzin B, Tharimena S, Plasman M et al (2021) Thickness and structure of the Martian crust from InSight seismic data. Science 373(6553):438–443

Knollenberg R, Hunten D (1980) The microphysics of the clouds of Venus: results of the Pioneer Venus particle size spectrometer experiment. J Geophys Res Space Phys 85(A13):8039–8058

Koll DDB, Malik M, Mansfield M, Kempton EMR, Kite E, Abbot D, Bean JL (2019) Identifying candidate atmospheres on rocky M dwarf planets via eclipse photometry. Astrophys J 886(2):140. https://doi.org/10.3847/1538-4357/ab4c91. arXiv:1907.13138

Kollmann P, Brandt PC, Collinson G, Rong ZJ, Futaana Y, Zhang TL (2016) Properties of planetward ion flows in Venus' magnetotail. Icarus 274:73–82. https://doi.org/10.1016/j.icarus.2016.02.053

Konrad BS, Alei E, Angerhausen D, Carrión-González Ó, Fortney JJ, Grenfell JL, Kitzmann D, Mollière P, Rugheimer S, Wunderlich F, Quanz SP (the LIFE Collaboration) (2021) Large Interferometer for Exoplanets (LIFE): III. Spectral resolution, wavelength range and sensitivity requirements based on atmospheric retrieval analyses of an exo-Earth. ArXiv e-prints. arXiv:2112.02054

Kopparapu RK, Ramirez R, Kasting JF, Eymet V, Robinson TD, Mahadevan S, Terrien RC, Domagal-Goldman S, Meadows V, Deshpande R (2013) Habitable zones around main-sequence stars: new estimates. Astrophys J 765(2):131

Kopparapu Rk, Wolf ET, Arney G, Batalha NE, Haqq-Misra J, Grimm SL, Heng K (2017) Habitable moist atmospheres on terrestrial planets near the inner edge of the habitable zone around M dwarfs. Astrophys J 845(1):5. https://doi.org/10.3847/1538-4357/aa7cf9. arXiv:1705.10362

Kotsyurbenko OR, Cordova JA, Belov AA, Cheptsov VS, Kölbl D, Khrunyk YY, Kryuchkova MO, Milojevic T, Mogul R, Sasaki S, Słowik GP, Snytnikov V, Vorobyova EA (2021) Exobiology of the Venusian clouds: new insights into habitability through terrestrial models and methods of detection. Astrobiology 21(10):1186–1205. https://doi.org/10.1089/ast.2020.2296

Koukouli M, Irwin P, Taylor F (2005) Water vapor abundance in Venus' middle atmosphere from Pioneer Venus OIR and Venera 15 FTS measurements. Icarus 173(1):84–99

Krasnopolsky VA (2012) A photochemical model for the Venus atmosphere at 47–112 km. Icarus 218(1):230–246

Krasnopolsky VA (2016) Sulfur aerosol in the clouds of Venus. Icarus 274:33–36

Kreidberg L, Koll DDB, Morley C, Hu R, Schaefer L, Deming D, Stevenson KB, Dittmann J, Vanderburg A, Berardo D, Guo X, Stassun K, Crossfield I, Charbonneau D, Latham DW, Loeb A, Ricker G, Seager S, Vanderspek R (2019) Absence of a thick atmosphere on the terrestrial exoplanet LHS 3844b. Nature 573(7772):87–90. https://doi.org/10.1038/s41586-019-1497-4. arXiv:1908.06834

Kreslavsky MA, Ivanov MA, Head JW (2015) The resurfacing history of Venus: constraints from buffered crater densities. Icarus 250:438–450. https://doi.org/10.1016/j.icarus.2014.12.024

Krissansen-Totton J, Arney GN, Catling DC (2018) Constraining the climate and ocean pH of the early Earth with a geological carbon cycle model. Proc Natl Acad Sci 115(16):4105–4110. https://doi.org/10.1073/pnas.1721296115

Krissansen-Totton J, Fortney JJ, Nimmo F (2021) Was Venus ever habitable? Constraints from a coupled interior–atmosphere–redox evolution model. Planet Sci J 2(5):216

Kruijver A, Höning D, van Westrenen W (2021) Carbon cycling and habitability of massive Earth-like exoplanets. Planet Sci J 2(5):208

Kulikov YN, Lammer H, Lichtenegger H, Terada N, Ribas I, Kolb C, Langmayr D, Lundin R, Guinan E, Barabash S et al (2006) Atmospheric and water loss from early Venus. Planet Space Sci 54(13–14):1425–1444

Kump LR (2018) Prolonged late Permian–early Triassic hyperthermal: failure of climate regulation? Philos Trans R Soc A, Math Phys Eng Sci 376(2130):20170078. https://doi.org/10.1098/rsta.2017.0078

Labrosse S (2015) Thermal evolution of the core with a high thermal conductivity. Phys Earth Planet Inter 247:36–55. https://doi.org/10.1016/j.pepi.2015.02.002. Transport Properties of the Earth's Core

Lammer H, Lichtenegger HIM, Biernat HK, Erkaev NV, Arshukova IL, Kolb C, Gunell H, Lukyanov A, Holmstrom M, Barabash S, Zhang TL, Baumjohann W (2006) Loss of hydrogen and oxygen from the upper atmosphere of Venus. Planet Space Sci 54(13–14):1445–1456. https://doi.org/10.1016/j.pss.2006.04.022

Landuyt W, Bercovici D (2009) Variations in planetary convection via the effect of climate on damage. Earth Planet Sci Lett 277(1):29–37. https://doi.org/10.1016/j.epsl.2008.09.034

Laneuville M, Dong C, O'Rourke JG, Schneider AC (2020) Magnetic fields on rocky planets. In: Planetary diversity, vol 2514–3433. IOP Publishing, Bristol, pp 3-1–3-47. https://doi.org/10.1088/2514-3433/abb4d9ch3

Lange RA (1994) Chap. 9. The effect of H_2O, CO_2 and F on the density and viscosity of silicate melts. In: Carroll MR, Holloway JR (eds) Volatiles in magmas, vol 9. Mineralogical Society of America, Washington, pp 331–370. https://doi.org/10.1515/9781501509674-015

Langlais B, Thébault E, Houliez A, Purucker ME, Lillis RJ (2019) A new model of the crustal magnetic field of Mars using MGS and MAVEN. J Geophys Res, Planets 124(6):1542–1569. https://doi.org/10.1029/2018JE005854

Lapôtre MGA, O'Rourke JG, Schaefer LK, Siebach KL, Spalding C, Tikoo SM, Wordsworth RD (2020) Probing space to understand Earth. Nat Rev Earth Environ 1(3):170–181. https://doi.org/10.1038/s43017-020-0029-y

Lazio J, Shkolnik E, Hallinan G (2016) Planetary magnetic fields – planetary interiors and habitability – final report. Study Report prepared for the Keck Institute for Space Studies, KISS. https://kiss.caltech.edu/final_reports/Magnetic_final_report.pdf

Lebrun T, Massol H, ChassefièRe E, Davaille A, Marcq E, Sarda P, Leblanc F, Brandeis G (2013) Thermal evolution of an early magma ocean in interaction with the atmosphere. J Geophys Res, Planets 118:1155–1176. https://doi.org/10.1002/jgre.20068

Leconte J, Forget F, Charnay B, Wordsworth R, Selsis F, Millour E, Spiga A (2013) 3D climate modeling of close-in land planets: circulation patterns, climate moist bistability, and habitability. Astron Astrophys 554:A69. https://doi.org/10.1051/0004-6361/201321042. arXiv:1303.7079

Lécuyer C, Simon L, Guyot F (2000) Comparison of carbon, nitrogen and water budgets on Venus and the Earth. Earth Planet Sci Lett 181(1):33–40. https://doi.org/10.1016/S0012-821X(00)00195-3

Lenardic A, Crowley JW (2012) On the notion of well-defined tectonic regimes for terrestrial planets in this solar system and others. Astrophys J 755(2):132. https://doi.org/10.1088/0004-637X/755/2/132

Lenardic A, Jellinek M, Moresi L (2008) A climate change induced transition in the tectonic style of a terrestrial planet. Earth Planet Sci Lett 271:34–42

Lenardic A, Weller M, Höink T, Seales J (2019) Toward a boot strap hypothesis of plate tectonics: feedbacks between plates, the asthenosphere, and the wavelength of mantle convection. Phys Earth Planet Inter 296:106299. https://doi.org/10.1016/j.pepi.2019.106299

Lenardic A, Seales J, Moore W, Weller M (2021) Convective and tectonic plate velocities in a mixed heating mantle. Geochem Geophys Geosyst 22(2):e2020GC009278

Li J, Jiang JH, Yang H, Abbot DS, Hu R, Komacek TD, Bartlett SJ, Yung YL (2021) Rotation period detection for Earth-like exoplanets. Astron J 163(1):27. https://doi.org/10.3847/1538-3881/ac36ce

Lichtenberg T, Bower DJ, Hammond M, Boukrouche R, Sanan P, Tsai SM, Pierrehumbert RT (2021) Vertically resolved magma ocean–protoatmosphere evolution: H_2, H_2O, CO_2, CH_4, CO, O_2, and N_2 as primary absorbers. J Geophys Res, Planets 126(2):e2020JE006711. https://doi.org/10.1029/2020JE006711

Lichtenegger HIM, Lammer H, Grießmeier JM, Kulikov YN, von Paris P, Hausleitner W, Krauss S, Rauer H (2010) Aeronomical evidence for higher CO_2 levels during Earth's Hadean epoch. Icarus 210(1):1–7. https://doi.org/10.1016/j.icarus.2010.06.042

Lincowski AP, Lustig-Yaeger J, Meadows VS (2019) Observing isotopologue bands in terrestrial exoplanet atmospheres with the James Webb Space Telescope: implications for identifying past atmospheric and ocean loss. Astron J 158(1):26

Louie DR, Deming D, Albert L, Bouma LG, Bean J, Lopez-Morales M (2018) Simulated JWST/NIRISS transit spectroscopy of anticipated TESS planets compared to select discoveries from space-based and ground-based surveys. Publ Astron Soc Pac 130(986):044401. https://doi.org/10.1088/1538-3873/aaa87b. arXiv:1711.02098

Lovis C, Blind N, Chazelas B, Kühn JG, Genolet L, Hughes I, Sordet M, Schnell R, Turbet M, Fusco T, Sauvage JF, Bugatti M, Billot N, Hagelberg J, Hocini E, Guyon O (2022) RISTRETTO: high-resolution spectroscopy at the diffraction limit of the VLT. In: Evans CJ, Bryant JJ, Motohara K (eds) Ground-based and airborne instrumentation for astronomy IX. Society of photo-optical instrumentation engineers (SPIE) conference series, vol 12184, p 121841Q. https://doi.org/10.1117/12.2627923. arXiv:2208.14838

Lu G, Zhao L, Chen L, Wan B, Wu F (2021) Reviewing subduction initiation and the origin of plate tectonics: what do we learn from present-day Earth? Earth Planet Phys 5(2):123–140

Luger R, Barnes R (2015) Extreme water loss and abiotic O_2 buildup on planets throughout the habitable zones of M dwarfs. Astrobiology 15(2):119–143. https://doi.org/10.1089/ast.2014.1231. arXiv:1411.7412

Luhmann J, Ledvina S, Russell C (2004) Induced magnetospheres. Adv Space Res 33(11):1905–1912

Luhmann JG, Kasprzak WT, Russell CT (2007) Space weather at Venus and its potential consequences for atmosphere evolution. J Geophys Res, Planets 112(E4):E04S10. https://doi.org/10.1029/2006JE002820

Lundin R, Zakharov A, Pellinen R, Barabasj SW, Borg H, Dubinin EM, Hultqvist B, Koskinen H, Liede I, Pissarenko N (1990) Aspera/Phobos measurements of the ion outflow from the MARTIAN ionosphere. Geophys Res Lett 17(6):873–876. https://doi.org/10.1029/GL017i006p00873

Lundin R, Lammer H, Ribas I (2007) Planetary magnetic fields and solar forcing: implications for atmospheric evolution. Space Sci Rev 129(1–3):245–278. https://doi.org/10.1007/s11214-007-9176-4

Lupu RE, Zahnle K, Marley MS, Schaefer L, Fegley B, Morley C, Cahoy K, Freedman R, Fortney JJ (2014) The atmospheres of Earthlike planets after giant impact events. Astrophys J 784(1):27. https://doi.org/10.1088/0004-637X/784/1/27. arXiv:1401.1499

Lustig-Yaeger J, Meadows VS, Tovar Mendoza G, Schwieterman EW, Fujii Y, Luger R, Robinson TD (2018) Detecting ocean glint on exoplanets using multiphase mapping. Astron J 156(6):301. https://doi.org/10.3847/1538-3881/aaed3a. arXiv:1901.05011

Lustig-Yaeger J, Meadows VS, Lincowski AP (2019a) A mirage of the cosmic shoreline: Venus-like clouds as a statistical false positive for exoplanet atmospheric erosion. Astrophys J 887(1):L11. https://doi.org/10.3847/2041-8213/ab5965. arXiv:1911.09132

Lustig-Yaeger J, Meadows VS, Lincowski AP (2019b) The detectability and characterization of the TRAPPIST-1 exoplanet atmospheres with JWST. Astron J 158(1):27. https://doi.org/10.3847/1538-3881/ab21e0. arXiv:1905.07070

MacDonald GJF (1964) Tidal friction. Rev Geophys Space Phys 2:467–541. https://doi.org/10.1029/RG002i003p00467

MacLellan L, Ernst R, El Bilali H, Ghail R, Bethell E (2021) Volcanic history of the Derceto large igneous province, Astkhik Planum, Venus. Earth-Sci Rev 220:103619

Mamajek EE, Meyer MR (2007) An improbable solution to the underluminosity of 2M1207B: a hot proto-planet collision afterglow. Astrophys J 668(2):L175–L178. https://doi.org/10.1086/522957. arXiv:0709.0456

Marconi A, Allende Prieto C, Amado PJ, Amate M, Augusto SR, Becerril S, Bezawada N, Boisse I, Bouchy F, Cabral A, Chazelas B, Cirami R, Coretti I, Cristiani S, Cupani G., de Castro Leão I, de Medeiros JR, de Souza MAF, Di Marcantonio P, Di Varano I, D'Odorico V, Drass H, Figueira P, Fragoso AB, Fynbo JPU, Genoni M, González Hernández JI, Haehnelt M, Hughes I, Huke P, Kjeldsen H, Korn AJ, Landoni M, Liske J, Lovis C, Maiolino R, Marquart T, Martins CJAP, Mason E, Monteiro MA, Morris T, Murray G, Niedzielski A, Oliva E, Origlia L, Pallé E, Parr-Burman P, Parro VC, Pepe F, Piskunov N, Rasilla JL, Rees P, Rebolo R, Riva M, Rousseau S, Sanna N, Santos NC, Shen TC, Sortino F, Sosnowska D, Sousa S, Stempels E, Strassmeier K, Tenegi F, Tozzi A, Udry S, Valenziano L, Vanzi L, Weber M, Woche M, Xompero M Zackrisson E (2018) ELT-HIRES, the high resolution spectrograph for the ELT: results from the Phase A study. In: Evans CJ, Simard L, Takami H (eds) Ground-based and airborne instrumentation for astronomy VII. Society of photo-optical instrumentation engineers (SPIE) conference series, vol 10702, p 107021Y. https://doi.org/10.1117/12.2311664

Marconi A, Abreu M, Adibekyan V, Aliverti M, Allende Prieto C, Amado P, Amate M, Artigau E, Augusto S, Barros S, Becerril S, Benneke B, Bergin E, Berio P, Bezawada N, Boisse I, Bonfils X, Bouchy F, Broeg C, Cabral A, Calvo-Ortega R, Canto Martins BL, Chazelas B, Chiavassa A, Christensen L, Cirami R, Coretti I, Covino S, Cresci G, Cristiani S, Cunha Parro V, Cupani G, de Castro Leão I, Renan de Medeiros J, Furlande Souza MA, Di Marcantonio P, Di Varano I, D'Odorico V, Doyon R, Drass H, Figueira P, Belen Fragoso A, Uldall Fynbo JP, Gallo E, Genoni M, González Hernández J, Haehnelt M, Hlavacek-Larrondo J, Hughes I, Huke P, Humphrey A, Kjeldsen H, Korn A, Kouach D, Landoni M, Liske J, Lovis C, Lunney D, Maiolino R, Malo L, Marquart T, Martins C, Mason E, Molaro P, Monnier J, Monteiro M, Mordasini C, Morris T, Mucciarelli A, Murray G, Niedzielski A, Nunes N, Oliva E, Origlia L, Pallé E, Pariani G, Parr-Burman P, Peñate J, Pepe F, Pinna E, Piskunov N Rasilla Piñeiro JL, Rebolo R, Rees P, Reiners A, Riva M, Romano D, Rousseau S, Sanna N, Santos N, Sarajlic M, Shen TC, Sortino F, Sosnowska D, Sousa S, Stempels E, Strassmeier K, Tenegi F, Tozzi A, Udry S, Valenziano L, Vanzi L, Weber M, Woche M, Xompero M, Zackrisson E, Zapatero Osorio MR (2021) HIRES, the high-resolution spectrograph for the ELT. Messenger 182:27

Marcq E, Bertaux JL, Montmessin F, Belyaev D (2013) Variations of sulphur dioxide at the cloud top of Venus's dynamic atmosphere. Nat Geosci 6(1):25–28

Marov MK, Greenspoon DH (1998) The planet Venus. Yale University Press, New Haven

Martin S, Kuan G, Stern P, Scowen P, Krist J, Mawet D, Ruane G (2019) Habitable Exoplanet Observatory (HabEx) telescope and optical instruments. In: Techniques and instrumentation for detection of exoplanets IX. Society of photo-optical instrumentation engineers (SPIE) conference series, vol 11117, p 1111704. https://doi.org/10.1117/12.2530737

Martins JHC, Figueira P, Santos NC, Lovis C (2013) Spectroscopic direct detection of reflected light from extrasolar planets. Mon Not R Astron Soc 436(2):1215–1224. https://doi.org/10.1093/mnras/stt1642. arXiv:1308.6516

Massol H, Hamano K, Tian F, Ikoma M, Abe Y, Chassefière E, Davaille A, Genda H, Güdel M, Hori Y et al (2016) Formation and evolution of protoatmospheres. Space Sci Rev 205(1):153–211. https://doi.org/10.1007/s11214-016-0280-1

Masunaga K, Futaana Y, Persson M, Barabash S, Zhang TL, Rong ZJ, Fedorov A (2019) Effects of the solar wind and the solar EUV flux on O^+ escape rates from Venus. Icarus 321:379–387. https://doi.org/10.1016/j.icarus.2018.11.017

Matsui T, Abe Y (1986) Impact-induced atmospheres and oceans on Earth and Venus. Nature 322(6079):526–528. https://doi.org/10.1038/322526a0

McComas DJ, Spence HE, Russell CT, Saunders MA (1986) The average magnetic field draping and consistent plasma properties of the Venus magnetotail. J Geophys Res 91(A7):7939–7953. https://doi.org/10.1029/JA091iA07p07939

McCord TB (1968) The loss of retrograde satellites in the solar system. J Geophys Res 73(4):1497–1500. https://doi.org/10.1029/JB073i004p01497

Meadows VS, Arney GN, Schwieterman EW, Lustig-Yaeger J, Lincowski AP, Robinson T, Domagal-Goldman SD, Deitrick R, Barnes RK, Fleming DP et al (2018) The habitability of Proxima Centauri b: environmental states and observational discriminants. Astrobiology 18(2):133–189

Meier TG, Bower DJ, Lichtenberg T, Tackley PJ, Demory BO (2021) Hemispheric tectonics on LHS 3844b. Astrophys J Lett 908(2):L48

Merk R, Breuer D, Spohn T (2002) Numerical modeling of ^{26}al-induced radioactive melting of asteroids considering accretion. Icarus. https://doi.org/10.1006/icar.2002.6872

Mettler JN, Quanz SP, Helled R (2020) Earth as an exoplanet. I. Time variable thermal emission using spatially resolved moderate imaging spectroradiometer data. Astron J 160(6):246. https://doi.org/10.3847/1538-3881/abbc15. arXiv:2010.02589

Miller-Ricci E, Meyer MR, Seager S, Elkins-Tanton L (2009) On the emergent spectra of hot protoplanet collision afterglows. Astrophys J 704(1):770–780. https://doi.org/10.1088/0004-637X/704/1/770. arXiv: 0907.2931

Misra A, Krissansen-Totton J, Koehler MC, Sholes S (2015) Transient sulfate aerosols as a signature of exoplanet volcanism. Astrobiology 15(6):462–477

Mlynczak MG, Hunt LA, Thomas Marshall B, Martin-Torres FJ, Mertens CJ, Russell JM, Remsberg EE, López-Puertas M, Picard R, Winick J, Wintersteiner P, Thompson RE, Gordley LL (2010) Observations of infrared radiative cooling in the thermosphere on daily to multiyear timescales from the TIMED/SABER instrument. J Geophys Res Space Phys 115(A3):A03309. https://doi.org/10.1029/2009JA014713

Mogul R, Limaye SS, Lee YJ, Pasillas M (2021) Potential for phototrophy in Venus' clouds. Astrobiology 21(10):1237–1249. https://doi.org/10.1089/ast.2021.0032

Mogul R, Limaye SS, Way MJ (2022) The CO_2 profile and analytical model for the Pioneer Venus Large Probe neutral mass spectrometer. Icarus 392(115374):1–21. https://doi.org/10.1016/j.icarus.2022.115374

Mollière P, van Boekel R, Bouwman J, Henning T, Lagage PO, Min M (2017) Observing transiting planets with JWST-prime targets and their synthetic spectral observations. Astron Astrophys 600:A10

Moore W, Webb AA (2013) Heat-pipe Earth. Nature 501:501–505. https://doi.org/10.1038/nature12473

Morbidelli A, Lunine JI, O'Brien DP, Raymond SN, Walsh KJ (2012) Building terrestrial planets. Annu Rev Earth Planet Sci 40:251–275. https://doi.org/10.1146/annurev-earth-042711-105319

Moresi L, Solomatov V (1998) Mantle convection with a brittle lithosphere: thoughts on the global tectonic styles of the Earth and Venus. Geophys J Int 133(3):669–682. https://doi.org/10.1046/j.1365-246X.1998.00521.x

Morley CV, Kreidberg L, Rustamkulov Z, Robinson T, Fortney JJ (2017) Observing the atmospheres of known temperate Earth-sized planets with JWST. Astrophys J 850(2):121

Moyen JF, van Hunen J (2012) Short-term episodicity of Archaean plate tectonics. Geology 40(5):451–454. https://doi.org/10.1130/G322894.1

Mueller N, Smrekar S, Helbert J, Stofan E, Piccioni G, Drossart P (2017) Search for active lava flows with VIRTIS on Venus Express. J Geophys Res, Planets 122(5):1021–1045

National Academies of Sciences, Engineering, and Medicine (2021) Pathways to discovery in astronomy and astrophysics for the 2020s. The National Academies Press, Washington. https://doi.org/10.17226/26141

Ni H, Zhang L, Guo X (2016) Water and partial melting of Earth's mantle. Sci China Earth Sci 59(4):720–730. https://doi.org/10.1007/s11430-015-5254-8

Nilsson H, Stenberg G, Futaana Y, Holmström M, Barabash S, Lundin R, Edberg NJT, Fedorov A (2012) Ion distributions in the vicinity of Mars: signatures of heating and acceleration processes. Earth Planets Space 64(2):135–148. https://doi.org/10.5047/eps.2011.04.011

Nilsson H, Zhang Q, Stenberg Wieser G, Holmström M, Barabash S, Futaana Y, Fedorov A, Persson M, Wieser M (2021) Solar cycle variation of ion escape from Mars. Icarus, 114610. https://doi.org/10.1016/j.icarus.2021.114610

Nimmo F (2002) Why does Venus lack a magnetic field? Geology 30(11):987–990. https://pubs.geoscienceworld.org/gsa/geology/article-pdf/30/11/987/3524399/i0091-7613-30-11-987.pdf

Nimmo F (2015) Energetics of the core Elsevier, Amsterdam, pp 31–65. https://doi.org/10.1016/B978-0-444-53802-4.00167-6

Nimmo F, Stevenson DJ (2000) Influence of early plate tectonics on the thermal evolution and magnetic field of Mars. J Geophys Res 105(E5):11969–11980. https://doi.org/10.1029/1999JE001216

Noack L, Rivoldini A, Van Hoolst T (2017) Volcanism and outgassing of stagnant-lid planets: implications for the habitable zone. Phys Earth Planet Inter 269:40–57

Ohtani E, Nagata Y, Suzuki A, Kato T (1995) Melting relations of peridotite and the density crossover in planetary mantles. Chem Geol 120(3–4):207–221

O'Neill C (2020) Planetary thermal evolution models with tectonic transitions. Planet Space Sci 192:105059. https://doi.org/10.1016/j.pss.2020.105059

O'Neill C, Debaille V (2014) The evolution of Hadean–Eoarchaean geodynamics. Earth Planet Sci Lett 406:49–58. https://doi.org/10.1016/j.epsl.2014.08.034

O'Neill C, Lenardic A, Moresi L, Torsvik T, Lee CT (2007) Episodic Precambrian subduction. Earth Planet Sci Lett 262(3):552–562. https://doi.org/10.1016/j.epsl.2007.04.056

O'Neill C, Lenardic A, Condie KC (2015) Earth's punctuated tectonic evolution: cause and effect. Geol Soc (Lond) Spec Publ 389(1):17–40. https://doi.org/10.1144/SP389.4. https://sp.lyellcollection.org/content/389/1/17.full.pdf

127

O'Neill C, Lenardic A, Weller M, Moresi L, Quenette S, Zhang S (2016) A window for plate tectonics in terrestrial planet evolution? Phys Earth Planet Inter 255:80–92. https://doi.org/10.1016/j.pepi.2016.04.002

O'Neill C, Marchi S, Zhang S, Bottke W (2017) Impact-driven subduction on the Hadean Earth. Nat Geosci 10:793–797. https://doi.org/10.1038/ngeo3029

O'Rourke JG (2020) Venus: a thick basal magma ocean may exist today. Geophys Res Lett 47(4):e86126. https://doi.org/10.1029/2019GL086126

O'Rourke JG, Gillmann C, Tackley P (2018) Prospects for an ancient dynamo and modern crustal remanent magnetism on Venus. Earth Planet Sci Lett 502:46–56. https://doi.org/10.1016/j.epsl.2018.08.055

O'Rourke JG, Buz J, Fu RR, Lillis RJ (2019) Detectability of remanent magnetism in the crust of Venus. Geophys Res Lett 46(11):5768–5777. https://doi.org/10.1029/2019GL082725

O'Rourke JG, Wilson C, Borrelli M, Byrne P, Dumoulin C, Ghail R, Gülcher A, Jacobson S, Spohn T, Weller M, Westall F (2023) Venus, the planet: introduction to Earth's sister planet. Space Sci Rev 219:10. https://doi.org/10.1007/s11214-023-00956-0

Ortenzi G, Noack L, Sohl F, Guimond C, Grenfell J, Dorn C, Schmidt J, Vulpius S, Katyal N, Kitzmann D et al (2020) Mantle redox state drives outgassing chemistry and atmospheric composition of rocky planets. Sci Rep 10(1):1–14. https://doi.org/10.1038/s41598-020-67751-7

Oyama V, Carle G, Woeller F, Pollack J, Reynolds R, Craig R (1980) Pioneer Venus gas chromatography of the lower atmosphere of Venus. J Geophys Res Space Phys 85(A13):7891–7902

Pallé E, Ford EB, Seager S, Montañés-Rodríguez P, Vazquez M (2008) Identifying the rotation rate and the presence of dynamic weather on extrasolar Earth-like planets from photometric observations. Astrophys J 676(2):1319–1329. https://doi.org/10.1086/528677. arXiv:0802.1836

Pasquini L, Avila G, Blecha A, Cacciari C, Cayatte V, Colless M, Damiani F, de Propris R, Dekker H, di Marcantonio P, Farrell T, Gillingham P, Guinouard I, Hammer F, Kaufer A, Hill V, Marteaud M, Modigliani A, Mulas G, North P, Popovic D, Rossetti E, Royer F, Santin P, Schmutzer R, Simond G, Vola P, Waller L, Zoccali M (2002) Installation and commissioning of FLAMES, the VLT multifibre facility. Messenger 110:1–9

Peplowski PN, Lawrence DJ, Wilson JT (2020) Chemically distinct regions of Venus's atmosphere revealed by measured N_2 concentrations. Nat Astron 4(10):947–950

Persson M, Futaana Y, Fedorov A, Nilsson H, Hamrin M, Barabash S (2018) H^+/O^+ escape rate ratio in the Venus magnetotail and its dependence on the solar cycle. Geophys Res Lett 45(20):10,805–10,811. https://doi.org/10.1029/2018GL079454

Persson M, Futaana Y, Ramstad R, Masunaga K, Nilsson H, Hamrin M, Fedorov A, Barabash S (2020) The Venusian atmospheric oxygen ion escape: extrapolation to the early solar system. J Geophys Res, Planets 125(3):e06336. https://doi.org/10.1029/2019JE006336

Peterson WK, Collin HL, Yau AW, Lennartsson OW (2001) Polar/Toroidal Imaging Mass-Angle Spectrograph observations of suprathermal ion outflow during solar minimum conditions. J Geophys Res 106(A4):6059–6066. https://doi.org/10.1029/2000JA003006

Phillips JL, Russell CT (1987) Upper limit on the intrinsic magnetic field of Venus. J Geophys Res Space Phys 92(A3):2253–2263. https://doi.org/10.1029/JA092iA03p02253

Phillips RJ, Raubertas RF, Arvidson RE, Sarkar IC, Herrick RR, Izenberg N, Grimm RE (1992) Impact craters and Venus resurfacing history. J Geophys Res, Planets 97(E10):15923–15948

Phillips RJ, Bullock MA, Hauck SA (2001) Climate and interior coupled evolution on Venus. Geophys Res Lett 28(9):1779–1782

Pickering TE, Pontoppidan KM, Laidler VG, Sontag CD, Robberto M, Karakla DM, Hanley C, Gilbert K, Slocum C, Earl NM et al (2016) Pandeia: a multi-mission exposure time calculator for JWST and WFIRST. In: Observatory operations: strategies, processes, and systems VI. https://doi.org/10.1117/12.2231768

Pollack JB (1971) A nongrey calculation of the runaway greenhouse: implications for Venus' past and present. Icarus 14:295–306. https://doi.org/10.1016/0019-1035(71)90001-7

Pope BJS, Bedell M, Callingham JR, Vedantham HK, Snellen IAG, Price-Whelan AM, Shimwell TW (2020) No massive companion to the coherent radio-emitting M dwarf GJ 1151. Astrophys J 890(2):L19. https://doi.org/10.3847/2041-8213/ab5b99

Quanz SP, Crossfield I, Meyer MR, Schmalzl E, Held J (2015) Direct detection of exoplanets in the 3–10 μm range with E-ELT/METIS. Int J Astrobiol 14(2):279–289. https://doi.org/10.1017/S1473550414000135. arXiv:1404.0831

Quick LC, Roberge A, Mlinar AB, Hedman MM (2020) Forecasting rates of volcanic activity on terrestrial exoplanets and implications for cryovolcanic activity on extrasolar ocean worlds. Publ Astron Soc Pac 132(1014):084402

Ramsay S, Amico P, Bezawada N, Cirasuolo M, Derie F, Egner S, George E, Gonté F, González Herrera JC, Hammersley P, Haupt C, Heijmans J, Ives D, Jakob G, Kerber F, Koehler B, Mainieri V, Manescau A,

128

Oberti S, Padovani P, Peroux C, Siebenmorgen R, Tamai R, Vernet J (2020) The ESO Extremely Large Telescope instrumentation programme. In: Advances in optical astronomical instrumentation 2019. Society of photo-optical instrumentation engineers (SPIE) conference series, vol 11203, p 1120303. https://doi.org/10.1117/12.2541400

Ramstad R, Barabash S (2021) Do intrinsic magnetic fields protect planetary atmospheres from stellar winds? Space Sci Rev 217(2):36. https://doi.org/10.1007/s11214-021-00791-1

Ramstad R, Barabash S, Futaana Y, Nilsson H, Wang XD, Holmström M (2015) The Martian atmospheric ion escape rate dependence on solar wind and solar EUV conditions: 1. Seven years of Mars Express observations. J Geophys Res, Planets 120(7):1298–1309. https://doi.org/10.1002/2015JE004816

Ramstad R, Barabash S, Futaana Y, Nilsson H, Holmström M (2017) Global Mars-solar wind coupling and ion escape. J Geophys Res Space Phys 122(8):8051–8062. https://doi.org/10.1002/2017JA024306

Ramstad R, Barabash S, Futaana Y, Nilsson H, Holmström M (2018) Ion escape from Mars through time: an extrapolation of atmospheric loss based on 10 years of Mars Express measurements. J Geophys Res, Planets 123(11):3051–3060. https://doi.org/10.1029/2018JE005727

Raymond SN (2021) A terrestrial convergence. Nat Astron 5:875–876. https://doi.org/10.1038/s41550-021-01488-9

Reeves GD, Cayton TE, Gary SP, Belian RD (1992) The great solar energetic particle events of 1989 observed from geosynchronous orbit. J Geophys Res 97(A5):6219–6226. https://doi.org/10.1029/91JA03102

Ricker GR, Latham D, Vanderspek R, Ennico K, Bakos G, Brown T, Burgasser A, Charbonneau D, Clampin M, Deming L et al (2010) Transiting Exoplanet Survey Satellite (TESS). In: American astronomical society meeting abstracts, vol 215, p 450-06

Robinson TD, Meadows VS, Crisp D (2010) Detecting oceans on extrasolar planets using the glint effect. Astrophys J 721(1):L67–L71. https://doi.org/10.1088/2041-8205/721/1/L67. arXiv:1008.3864

Robinson TD, Ennico K, Meadows VS, Sparks W, Bussey DBJ, Schwieterman EW, Breiner J (2014) Detection of ocean glint and ozone absorption using LCROSS Earth observations. Astrophys J 787(2):171. https://doi.org/10.1088/0004-637X/787/2/171. arXiv:1405.4557

Roble RG (1995) Major greenhouse cooling (yes, cooling): the upper atmosphere response to increased CO_2. Rev Geophys 33(S1):539–546. https://doi.org/10.1029/95RG00118

Roble RG, Dickinson RE (1989) How will changes in carbon dioxide and methane modify the mean structure of the mesosphere and thermosphere? Geophys Res Lett 16(12):1441–1444. https://doi.org/10.1029/GL016i012p01441

Rodrigues M, Capone J, Earle A, Foster T, Hidalgo A, Lewis I, Lynn J, O'brien K, Tosh I, George EM, Accardo M, Alvarez D, Conzelmann R, Hopgood J, Clarke F, Schnetler H, Tecza M, Thatte N (2018) The HARMONI/ELT spectrographs. In: Evans CJ, Simard L, Takami H (eds) Ground-based and airborne instrumentation for astronomy VII. Society of photo-optical instrumentation engineers (SPIE) conference series, vol 10702, p 107029M. https://doi.org/10.1117/12.2313396

Rogers LA (2015) Most 1.6 Earth-radius planets are not rocky. Astrophys J 801(1):41

Rolf T, Steinberger B, Sruthi U, Werner SC (2018) Inferences on the mantle viscosity structure and the post-overturn evolutionary state of Venus. Icarus 313:107–123

Rolf T, Weller M, Gülcher A, Byrne P, O'Rourke JG, Herrick R, Bjonnes E, Davaille A, Ghail R, Gillmann C, Plesa AC (2022) Dynamics and evolution of Venus' mantle through time. Space Sci Rev 218:70. https://doi.org/10.1007/s11214-022-00937-9

Ronov AB, Yaroshevsky AA (1969) Chemical composition of the Earth's crust. In: The Earth's crust and upper mantle. American Geophysical Union, Washington, pp 37–57. https://doi.org/10.1029/GM013p0037

Rubie D, Jacobson S, Morbidelli A, O'Brien D, de Young E, Vries J, Nimmo F, Palme H, Frost D (2015) Accretion and differentiation of the terrestrial planets with implications for the compositions of early-formed solar system bodies and accretion of water. Icarus 248:89–108. https://doi.org/10.1016/j.icarus.2014.10.015

Rustamkulov Z, Sing DK, Liu R, Wang A (2022) Analysis of a JWST NIRSpec lab time series: characterizing systematics, recovering exoplanet transit spectroscopy, and constraining a noise floor. Astrophys J Lett 928(1):L7

Ryan DJ, Robinson TD (2021) Detecting oceans on exoplanets with phase-dependent spectral principal component analysis. ArXiv e-prints. arXiv:2109.11062

Safronov VS (1978) The heating of the Earth during its formation. Icarus 33(1):3–12. https://doi.org/10.1016/0019-1035(78)90019-2

Salvador A, Samuel H (2022) Convective outgassing efficiency in planetary magma oceans: insights from computational fluid dynamics. Icarus 390:115265. https://doi.org/10.1016/j.icarus.2022.115265

Salvador A, Massol H, Davaille A, Marcq E, Sarda P, Chassefière E (2017) The relative influence of H_2O and CO_2 on the primitive surface conditions and evolution of rocky planets. J Geophys Res, Planets 122(7):1458–1486. https://doi.org/10.1002/2017JE005286

Salvador A, Avice G, Breuer D, Gillmann C, Jacobson S, Lammer H, Marcq E, Raymond SN, Sakuraba H, Scherf M, Way M (2023) Magma ocean, water, and the early atmosphere of Venus. Space Sci Rev

Samuel H, Tackley PJ, Evonuk M (2010) Heat partitioning during core formation by negative diapirism in terrestrial planets. Earth Planet Sci Lett 290:13–19

Sanders GH (2013) The Thirty Meter Telescope (TMT): an international observatory. J Astrophys Astron 34(2):81–86. https://doi.org/10.1007/s12036-013-9169-5

Sandor BJ, Clancy RT (2005) Water vapor variations in the Venus mesosphere from microwave spectra. Icarus 177(1):129–143

Sandwell DT, Schubert G (1992) Evidence for retrograde lithospheric subduction on Venus. Science 257(5071):766–770. https://doi.org/10.1126/science.257.5071.766

Sasaki S, Nakazawa K (1986) Metal-silicate fractionation in the growing Earth: energy source for the terrestrial magma ocean. J Geophys Res, Solid Earth 91(B9):9231–9238. https://doi.org/10.1029/JB091iB09p09231

Schaefer L, Parmentier V (2021) The air over there: exploring exoplanet atmospheres. ArXiv e-prints. arXiv:2108.08387

Scherf M, Lammer H (2021) Did Mars possess a dense atmosphere during the first ~400 million years? Space Sci Rev 217(1):2. https://doi.org/10.1007/s11214-020-00779-3. arXiv:2102.05976

Schillings A, Slapak R, Nilsson H, Yamauchi M, Dandouras I, Westerberg LG (2019) Earth atmospheric loss through the plasma mantle and its dependence on solar wind parameters. Earth Planets Space 71(1):70. https://doi.org/10.1186/s40623-019-1048-0

Schubert G, Turcotte D, Olson P (2001) Mantle convection in the Earth and planets. Cambridge University Press, Cambridge. https://books.google.com/books?id=2lwnV2xCMmoC

Schulze-Makuch D, Grinspoon DH, Abbas O, Irwin LN, Bullock MA (2004) A sulfur-based survival strategy for putative phototrophic life in the Venusian atmosphere. Astrobiology 4(1):11–18

Schwieterman EW, Robinson TD, Meadows VS, Misra A, Domagal-Goldman S (2015) Detecting and constraining N_2 abundances in planetary atmospheres using collisional pairs. Astrophys J 810(1):57

Shalygin EV, Markiewicz WJ, Basilevsky AT, Titov DV, Ignatiev NI, Head JW (2015) Active volcanism on Venus in the Ganiki Chasma rift zone. Geophys Res Lett 42(12):4762–4769

Shea MA, Smart DF (1995) History of solar proton event observations. Nucl Phys B, Proc Suppl 39(1):16–25. https://doi.org/10.1016/0920-5632(95)00003-R

Si K (2011) Rheological structure of the mantle of a super-Earth: some insights from mineral physics. Icarus 212(1):14–23

Siebert J, Badro J, Antonangeli D, Ryerson FJ (2013) Terrestrial accretion under oxidizing conditions. Science 339(6124):1194–1197. https://doi.org/10.1126/science.1227923

Slapak R, Schillings A, Nilsson H, Yamauchi M, Westerberg LG, Dandouras I (2017) Atmospheric loss from the dayside open polar region and its dependence on geomagnetic activity: implications for atmospheric escape on evolutionary timescales. Ann Geophys 35(3):721–731. https://doi.org/10.5194/angeo-35-721-2017

Sleep NH, Zahnle K (2001) Carbon dioxide cycling and implications for climate on ancient Earth. J Geophys Res, Planets 106(E1):1373–1399

Smrekar SE, Stofan ER, Mueller N, Treiman A, Elkins-Tanton L, Helbert J, Piccioni G, Drossart P (2010) Recent hotspot volcanism on Venus from VIRTIS emissivity data. Science 328(5978):605–608

Snellen I, de Kok R, Birkby JL, Brandl B, Brogi M, Keller C, Kenworthy M, Schwarz H, Stuik R (2015) Combining high-dispersion spectroscopy with high contrast imaging: probing rocky planets around our nearest neighbors. Astron Astrophys 576:A59. https://doi.org/10.1051/0004-6361/201425018. arXiv:1503.01136

Solomatov VS (2004) Initiation of subduction by small-scale convection. J Geophys Res, Solid Earth 109(B1):B01412. https://doi.org/10.1029/2003JB002628

Sossi PA, Burnham AD, Badro J, Lanzirotti A, Newville M, O'neill HSC (2020) Redox state of Earth's magma ocean and its Venus-like early atmosphere. Sci Adv 6(48):eabd1387

Soubiran F, Militzer B (2018) Electrical conductivity and magnetic dynamos in magma oceans of Super-Earths. Nat Commun 9:3883. https://doi.org/10.1038/s41467-018-06432-6

Sparks WB, Ford HC (2002) Imaging spectroscopy for extrasolar planet detection. Astrophys J 578(1):543–564. https://doi.org/10.1086/342401. arXiv:astro-ph/0209078

Stähler SC, Khan A, Banerdt WB, Lognonné P, Giardini D, Ceylan S, Drilleau M, Duran AC, Garcia RF, Huang Q, Kim D, Lekic V, Samuel H, Schimmel M, Schmerr N, Sollberger D, Stutzmann É, Xu Z, Antonangeli D, Charalambous C, Davis PM, Irving JCE, Kawamura T, Knapmeyer M, Maguire R, Marusiak AG, Panning MP, Perrin C, Plesa AC, Rivoldini A, Schmelzbach C, Zenhäusern G, Beucler É, Clinton J, Dahmen N, van Driel M, Gudkova T, Horleston A, Pike WT, Plasman M, Smrekar SE (2021) Seismic detection of the Martian core. Science 373(6553):443–448. https://doi.org/10.1126/science.abi7730

Stamenković V, Breuer D, Spohn T (2011) Thermal and transport properties of mantle rock at high pressure: applications to super-Earths. Icarus 216(2):572–596

Stein C, Lowman J, Hansen U (2013) The influence of mantle internal heating on lithospheric mobility: implications for super-Earths. Earth Planet Sci Lett 361:448–459. https://doi.org/10.1016/j.epsl.2012.11.011

Stern RJ (2008) Modern-style plate tectonics began in Neoproterozoic time: an alternative interpretation of Earth's tectonic history. In: When did plate tectonics begin on planet Earth? Geol. Soc. Am., Boulder. https://doi.org/10.1130/2008.2440(13)

Stevenson DJ (2003) Planetary magnetic fields. Earth Planet Sci Lett 208(1):1–11. https://doi.org/10.1016/S0012-821X(02)01126-3

Stevenson DJ (2010) Planetary magnetic fields: achievements and prospects. Space Sci Rev 152(1–4):651–664. https://doi.org/10.1007/s11214-009-9572-z

Stewart E, Ague JJ, Ferry JM, Schiffries CM, Tao RB, Isson TT, Planavsky NJ (2019) Carbonation and de-carbonation reactions: implications for planetary habitability. Am Mineral 104(10):1369–1380. https://doi.org/10.2138/am-2019-6884

Stixrude L, Scipioni R, Desjarlais M (2020) A silicate dynamo in the early Earth. Nat Commun 11:935. https://doi.org/10.1038/s41467-020-14773-4

Strangeway R, Russell C, Luhmann J, Moore T, Foster J, Barabash S, Nilsson H (2010) Does a planetary-scale magnetic field enhance or inhibit ionospheric plasma outflows? In: Agu fall meeting abstracts, vol 2010, pp SM33B–1893

Strom RG, Schaber GG, Dawson DD (1994) The global resurfacing of Venus. J Geophys Res, Planets 99(E5):10899–10926

Surkov YA, Barsukov V, Moskalyeva L, Kharyukova V, Kemurdzhian A (1984) New data on the composition, structure, and properties of Venus rock obtained by Venera 13 and Venera 14. J Geophys Res, Solid Earth 89(S02):B393–B402

Surkov YA, Moskalyova L, Kharyukova V, Dudin A, Smirnov G, Zaitseva SY (1986) Venus rock composition at the Vega 2 landing site. J Geophys Res, Solid Earth 91(B13):E215–E218

Svensen H, Planke S, Polozov AG, Schmidbauer N, Corfu F, Podladchikov YY, Jamtveit B (2009) Siberian gas venting and the end-Permian environmental crisis. Earth Planet Sci Lett 277(3):490–500. https://doi.org/10.1016/j.epsl.2008.11.015

Teinturier L, Vieira N, Jacquet E, Geoffrion J, Bestavros Y, Keating D, Cowan NB (2022) Mapping the surface of partially cloudy exoplanets is hard. Mon Not R Astron Soc. https://doi.org/10.1093/mnras/stac030. arXiv:2201.00825

The LUVOIR Team (2019) The LUVOIR mission concept study final report. ArXiv e-prints. arXiv:1912.06219

Thompson MA, Krissansen-Totton J, Wogan N, Telus M, Fortney JJ (2022) The case and context for atmospheric methane as an exoplanet biosignature. Proc Natl Acad Sci 119(14):e2117933119. https://doi.org/10.1073/pnas.2117933119

Tian F, Kasting JF, Liu HL, Roble RG (2008) Hydrodynamic planetary thermosphere model: 1. Response of the Earth's thermosphere to extreme solar EUV conditions and the significance of adiabatic cooling. J Geophys Res, Planets 113(E5):E05008. https://doi.org/10.1029/2007JE002946

Tian F, Kasting JF, Solomon SC (2009) Thermal escape of carbon from the early Martian atmosphere. Geophys Res Lett 36(2):L02205. https://doi.org/10.1029/2008GL036513

Tonks WB, Melosh HJ (1993) Magma ocean formation due to giant impacts. J Geophys Res, Planets 98(E3):5319–5333. https://doi.org/10.1029/92JE02726

Tosi N, Yuen DA, de Koker N, Wentzcovitch RM (2013) Mantle dynamics with pressure-and temperature-dependent thermal expansivity and conductivity. Phys Earth Planet Inter 217:48–58

Tosi N, Godolt M, Stracke B, Ruedas T, Grenfell JL, Höning D, Nikolaou A, Plesa AC, Breuer D, Spohn T (2017) The habitability of a stagnant-lid Earth. Astron Astrophys 605:A71

Tozer DC (1965) Thermal history of the Earth: I. The formation of the core. Geophys J Int 9(2–3):95–112. https://doi.org/10.1111/j.1365-246X.1965.tb02064.x

Truong N, Lunine JI (2021) Volcanically extruded phosphides as an abiotic source of Venusian phosphine. Proc Natl Acad Sci 118(29):e2021689118. https://doi.org/10.1073/pnas.2021689118

Tu L, Johnstone CP, Güdel M, Lammer H (2015) The extreme ultraviolet and X-ray Sun in Time: high-energy evolutionary tracks of a solar-like star. Astron Astrophys 577:L3

Turbet M (2019b) Two examples of how to use observations of terrestrial planets orbiting in temperate orbits around low mass stars to test key concepts of planetary habitability. In: Di Matteo P, Creevey O, Crida A, Kordopatis G, Malzac J, Marquette JB, N'Diaye M, Venot O (eds) SF2A-2019: proceedings of the annual meeting of the French society of astronomy and astrophysics

Turbet M, Leconte J, Selsis F, Bolmont E, Forget F, Ribas I, Raymond SN, Anglada-Escudé G (2016) The habitability of Proxima Centauri b. II. Possible climates and observability. Astron Astrophys 596:A112. https://doi.org/10.1051/0004-6361/201629577. arXiv:1608.06827

Turbet M, Ehrenreich D, Lovis C, Bolmont E, Fauchez T (2019) The runaway greenhouse radius inflation effect – an observational diagnostic to probe water on Earth-sized planets and test the habitable zone concept. Astron Astrophys 628:A12. https://doi.org/10.1051/0004-6361/201935585

Turbet M, Bolmont E, Ehrenreich D, Gratier P, Leconte J, Selsis F, Hara N, Lovis C (2020) Revised mass-radius relationships for water-rich rocky planets more irradiated than the runaway greenhouse limit. Astron Astrophys 638:A41. https://doi.org/10.1051/0004-6361/201937151

Turbet M, Bolmont E, Chaverot G, Ehrenreich D, Leconte J, Marcq E (2021) Day–night cloud asymmetry prevents early oceans on Venus but not on Earth. Nature 598:276–280. https://doi.org/10.1038/s41586-021-03873-w

Valencia D, O'Connell RJ, Sasselov D (2006) Internal structure of massive terrestrial planets. Icarus 181(2):545–554

van Summeren J, Gaidos E, Conrad CP (2013) Magnetodynamo lifetimes for rocky, Earth-mass exoplanets with contrasting mantle convection regimes. J Geophys Res, Planets 118(5):938–951. https://doi.org/10.1002/jgre.20077. arXiv:1304.2437

Vedantham H, Callingham J, Shimwell T, Tasse C, Pope B, Bedell M, Snellen I, Best P, Hardcastle M, Haverkorn M et al (2020) Coherent radio emission from a quiescent red dwarf indicative of star–planet interaction. Nat Astron 4(6):577–583

Villanueva GL, Smith MD, Protopapa S, Faggi S, Mandell AM (2018) Planetary spectrum generator: an accurate online radiative transfer suite for atmospheres, comets, small bodies and exoplanets. J Quant Spectrosc Radiat Transf 217:86–104

Walker JCG, Hays PB, Kasting JF (1981) A negative feedback mechanism for the long-term stabilization of the Earth's surface temperature. J Geophys Res 86:9776–9782. https://doi.org/10.1029/JC086iC10p09776

Walters DN, Best MJ, Bushell AC, Copsey D, Edwards JM, Falloon PD, Harris CM, Lock AP, Manners JC, Morcrette CJ, Roberts MJ, Stratton RA, Webster S, Wilkinson JM, Willett MR, Boutle IA, Earnshaw PD, Hill PG, MacLachlan C, Martin GM, Moufouma-Okia W, Palmer MD, Petch JC, Rooney GG, Scaife AA, Williams KD (2011) The Met Office Unified Model Global Atmosphere 3.0/3.1 and JULES Global Land 3.0/3.1 configurations. Geosci Model Dev 4(4):919–941. https://doi.org/10.5194/gmd-4-919-2011

Way MJ, Del Genio AD (2020) Venusian habitable climate scenarios: modeling Venus through time and applications to slowly rotating Venus-like exoplanets. J Geophys Res, Planets 125(5):e2019JE006276

Way MJ, Del Genio AD, Kiang NY, Sohl LE, Grinspoon DH, Aleinov I, Kelley M, Clune T (2016) Was Venus the first habitable world of our solar system? Geophys Res Lett 43:8376

Way MJ, Aleinov I, Amundsen DS, Chandler MA, Clune TL, Del Genio AD, Fujii Y, Kelley M, Kiang NY, Sohl L, Tsigaridis K (2017) Resolving orbital and climate keys of Earth and extraterrestrial environments with dynamics (ROCKE-3D) 1.0: a general circulation model for simulating the climates of rocky planets. Astrophys J Suppl Ser 231:12. https://doi.org/10.3847/1538-4365/aa7a06. arXiv:1701.02360

Way MJ, Del Genio AD, Aleinov I, Clune TL, Kelley M, Kiang NY (2018) Climates of warm Earth-like planets I: 3-D model simulations. Astrophys J Suppl 239:24. https://doi.org/10.3847/1538-4365/aae9e1. arXiv:1808.06480

Weiss LM, Marcy GW (2014) The mass-radius relation for 65 exoplanets smaller than 4 Earth radii. Astrophys J Lett 783(1):L6

Weller MB, Kiefer WS (2020) The physics of changing tectonic regimes: implications for the temporal evolution of mantle convection and the thermal history of Venus. J Geophys Res, Planets 125(1):e2019JE005960. https://doi.org/10.1029/2019JE005960

Weller MB, Kiefer WS (2021) Punctuated evolution of the Venusian atmosphere from mantle outgassing. In: 52nd Lunar and planetary science conference, p 1555

Weller MB, Lenardic A (2012) Hysteresis in mantle convection: plate tectonics systems. Geophys Res Lett 39(10):L10202. https://doi.org/10.1029/2012GL051232

Weller MB, Lenardic A (2018) On the evolution of terrestrial planets: bi-stability, stochastic effects, and the non-uniqueness of tectonic states. Geosci Front 9(1):91–102. Lid Tectonics. https://doi.org/10.1016/j.gsf.2017.03.001

Weller M, Lenardic A, O'Neill C (2015) The effects of internal heating and large scale climate variations on tectonic bi-stability in terrestrial planets. Earth Planet Sci Lett 420:85–94. https://doi.org/10.1016/j.epsl.2015.03.021

Westall F, Höning D, Gillmann C, Way M (2023) The habitability of Venus. Space Sci Rev

Widemann T, Avice G, Breuer D, Gillmann C, Salvador A, Way M (2022) Future observations and missions. Space Sci Rev

Williams Q (2018) The thermal conductivity of Earth's core: a key geophysical parameter's constraints and uncertainties. Annu Rev Earth Planet Sci 46(1):47–66. https://doi.org/10.1146/annurev-earth-082517-010154

Wolf ET, Kopparapu R, Airapetian V, Fauchez T, Guzewich SD, Kane SR, Pidhorodetska D, Way MJ, Abbot DS, Checlair JH et al (2019) The importance of 3D general circulation models for characterizing the climate and habitability of terrestrial extrasolar planets. ArXiv preprint. arXiv:1903.05012

Wordsworth R, Pierrehumbert R (2013) Water loss from terrestrial planets with CO_2-rich atmospheres. Astrophys J 778(2):154

Wunderlich F, Godolt M, Grenfell JL, Städt S, Smith AMS, Gebauer S, Schreier F, Hedelt P, Rauer H (2019) Detectability of atmospheric features of Earth-like planets in the habitable zone around M dwarfs. Astron Astrophys 624:A49. https://doi.org/10.1051/0004-6361/201834504. arXiv:1905.02560

Yang J, Boué G, Fabrycky DC, Abbot DS (2014) Strong dependence of the inner edge of the habitable zone on planetary rotation rate. Astrophys J 787:L2. https://doi.org/10.1088/2041-8205/787/1/L2. arXiv:1404.4992

Yau AW, Shelley EG, Peterson WK, Lenchyshyn L (1985) Energetic auroral and polar ion outflow at DE 1 altitudes: magnitude, composition, magnetic activity dependence, and long-term variations. J Geophys Res 90(A9):8417–8432. https://doi.org/10.1029/JA090iA09p08417

Zahnle KJ, Catling DC (2017) The cosmic shoreline: the evidence that escape determines which planets have atmospheres, and what this may mean for Proxima Centauri b. Astrophys J 843(2):122

Zahnle K, Arndt N, Cockell C, Halliday A, Nisbet E, Selsis F, Sleep NH (2007) Emergence of a habitable planet. Space Sci Rev 129:35–78. https://doi.org/10.1007/s11214-007-9225-z

Zolotov MY (2018) Gas–solid interactions on Venus and other solar system bodies. Rev Mineral Geochem 84(1):351–392. https://doi.org/10.2138/rmg.2018.84.10

Zolotov M (2019) Chemical weathering on Venus. In: Oxford research encyclopedia of planetary science. Oxford University Press, London. https://doi.org/10.1093/acrefore/9780190647926.013.146

Publisher's Note Springer Nature remains neutral with regard to jurisdictional claims in published maps and institutional affiliations.

Authors and Affiliations

M.J. Way[1,2] ⓘ · Colby Ostberg[3] · Bradford J. Foley[4] · Cedric Gillmann[5] · Dennis Höning[6,7] · Helmut Lammer[8] · Joseph O'Rourke[9] · Moa Persson[10] · Ana-Catalina Plesa[11] · Arnaud Salvador[12,13,14] · Manuel Scherf[8,15,16] · Matthew Weller[17]

✉ A.-C. Plesa
Ana.Plesa@dlr.de

M.J. Way
Michael.J.Way@nasa.gov

[1] NASA Goddard Institute for Space Studies, 2880 Broadway, New York, NY 10025, USA

[2] Theoretical Astrophysics, Department of Physics and Astronomy, Uppsala University, Uppsala, Sweden

[3] Department of Earth and Planetary Sciences, University of California, Riverside, CA 92521, USA

[4] Department of Geosciences, Pennsylvania State University, University Park, PA, USA

[5] Department of Earth, Environmental and Planetary Sciences, Rice University, Houston, TX 77005, USA

[6] Potsdam Institute for Climate Impact Research, Potsdam, Germany

[7] Department of Earth Sciences, Vrije Universiteit Amsterdam, Amsterdam, The Netherlands

[8] Space Research Institute, Austrian Academy of Sciences, Schmiedlstr. 6, 8042, Graz, Austria

[9] School of Earth and Space Exploration, Arizona State University, Tempe, AZ, USA

[10] Institut de Recherche en Astrophysique et Planétologie, Centre National de la Recherche Scientifique, Université Paul Sabatier – Toulouse III, Centre National d'Etudes Spatiales, Toulouse, France

11   Institute of Planetary Research, DLR, Berlin, Germany

12   Department of Astronomy and Planetary Science, Northern Arizona University, Box 6010, Flagstaff, AZ 86011, USA

13   Habitability, Atmospheres, and Biosignatures Laboratory, University of Arizona, Tucson, AZ, USA

14   Lunar and Planetary Laboratory, University of Arizona, Tucson, AZ, USA

15   Institute of Physics, University of Graz, Graz, Austria

16   Institute for Geodesy, Technical University, Graz, Austria

17   Lunar and Planetary Institute, 3600 Bay Area Blvd., Houston, TX 77058, USA

Space Science Reviews (2023) 219:17
https://doi.org/10.1007/s11214-023-00960-4

The Habitability of Venus

F. Westall[1] · D. Höning[2] · G. Avice[3] · D. Gentry[4] · T. Gerya[5] · C. Gillmann[6] · N. Izenberg[7] · M.J. Way[8,9] · C. Wilson[10,11]

Received: 5 June 2022 / Accepted: 30 January 2023 / Published online: 22 February 2023
© The Author(s) 2023

Abstract
Venus today is inhospitable at the surface, its average temperature of 750 K being incompatible to the existence of life as we know it. However, the potential for past surface habitability and upper atmosphere (cloud) habitability at the present day is hotly debated, as the ongoing discussion regarding a possible phosphine signature coming from the clouds shows. We review current understanding about the evolution of Venus with special attention to scenarios where the planet may have been capable of hosting microbial life. We compare the possibility of past habitability on Venus to the case of Earth by reviewing the various hypotheses put forth concerning the origin of habitable conditions and the emergence and evolution of plate tectonics on both planets. Life emerged on Earth during the Hadean when the planet was dominated by higher mantle temperatures (by about 200 °C), an uncertain tectonic regime that likely included squishy lid/plume-lid and plate tectonics, and proto continents. Despite the lack of well-preserved crust dating from the Hadean and Paleoarchean, we attempt to review current understanding of the environmental conditions during this critical period based on zircon crystals and geochemical signatures from this period, as well as studies of younger, relatively well-preserved rocks from the Paleoarchean. For these early, primitive life forms, the tectonic regime was not critical but it became an important means of nutrient recycling, with possible consequences on the global environment in the long-term, that was essential to the continuation of habitability and the evolution of life. For early Venus, the question of stable surface water is closely related to tectonics. We discuss potential transitions between stagnant lid and (episodic) tectonics with crustal recycling, as well as consequences for volatile cycling between Venus' interior and atmosphere. In particular, we review insights into Venus' early climate and examine critical questions about early rotation speed, reflective clouds, and silicate weathering, and summarize implications for Venus' long-term habitability. Finally, the state of knowledge of the Venusian clouds and the proposed detection of phosphine is covered.

1 Introduction

With an average temperature of ~ 750 K, today the surface of Venus is far from environmental conditions suitable for life as we know it. However, billions of years ago, when the

---

Venus: Evolution Through Time
Edited by Colin F. Wilson, Doris Breuer, Cédric Gillmann, Suzanne E. Smrekar, Tilman Spohn and Thomas Widemann

Extended author information available on the last page of the article

Sun was much fainter (Sagan and Mullen 1972; Hart 1979; Gough 1981; Claire et al. 2012), Venus was located in the middle of the classical habitable zone around our sun (Kasting 1993; Kopparapu et al. 2013), thus fueling speculations about the early habitability of the planet (Pollack 1971; Grinspoon and Bullock 2007; Way et al. 2016). Important preconditions for habitability at the surface of Venus include a temperature range allowing the existence of liquid water; surface geochemistry with available chemical energy and appropriate elemental and molecular constituents, such as active water/rock interfaces; and protection from lethal solar radiation. The latter could have been provided by liquid water, as well as by a large reflective cloud cover resulting from the presence of liquid water. Surface thermochemical conditions would have ultimately been controlled by Venus' tectonic activity, although different models suggest different scenarios. Venus' convection regime may have changed over the course of the planet's history and at least some models suggest that Venus could have maintained temperate surface conditions until as recently as 0.7 Ga (Way et al. 2016). Lastly, although the early Sun was fainter, Venus' more sunward position means that solar incident insolation during its early history was still 40% higher than for present-day Earth (Lammer et al. 2008). Even if the overall radiation environment at the surface (as determined by absorption in Venus' early atmosphere) was clement, it may have affected conditions for early life potentially inhabiting water on exposed landmasses or subsea environments, such as hydrothermal vents. Increasing solar luminosity, continuous degassing of CO_2 from Venus' mantle into the atmosphere, or large-scale volcanic eruptions (Way et al. 2022) may have brought an end to a potential early habitable period.

In order to gain insight into whether these preconditions existed in the early history of Venus, we address the mechanisms relevant to Earth's early planetary and biological evolution in Sect. 2. The main challenge with this approach is the loss of the earliest records of Earth's ancient surface due to tectonic recycling. Understanding of the early environmental conditions on Earth can be approached through a combination of modelling, inherited geochemical signatures in younger rocks, and comparison with well-preserved, younger crustal rocks formed about 1 billion years (Gyr) after solidification of the planet. We will also elaborate on tectonic processes and interior-atmosphere volatile exchange relevant to the evolution of Venus and a potential early habitable period in Sect. 3. Based on these constraints, the climate throughout Venus' history from general circulation models will be discussed in Sect. 4.

Speculation on present-day habitability is primarily focused on Venus' cloud aerosols. Although larger reservoirs of water are likely to be dissolved in the mantle and as atmospheric vapor, cloud droplets are the only known place where liquid water is found on Venus. This liquid water is dissolved in sulfuric acid, the aerosols' primary constituent. Section 5 discusses what is known about the requirements for Venus cloud habitability by comparison with Earth's aerobiosphere, as well as discussion of suggested Venusian biosignatures, such as UV absorption and the controversial report of phosphine detection. We conclude with an evaluation of mission requirements necessary to improve our constraints on Venus' past and present habitability in Sect. 6.

2 Early Earth History

2.1 Water on the Earth

The history of initial habitability on any planet is first and foremost the history of water, although long-lived habitability is also controlled by tectonics. Temperature and pressure, as well as the various gaseous species that comprise the atmosphere, determine the possibility of liquid water at the surface. A large part of the conditions that govern the onset (or lack thereof) of a habitable era is therefore set by the composition of the atmosphere and its interaction with factors, such as planetary characteristics, solar energy input, or material delivery.

Recent investigation of calcium-aluminium-rich inclusions (CAIs) in some of the most primitive meteorites suggests the admixture of a significant amount of interstellar water during the early evolution of the protosolar cloud. This, in turn, implies very early formation of planetary reservoirs of volatile elements (Aléon et al. 2022; cf. Grossman and Larimer 1974). Thus, early volatile-containing materials in the Solar System would have contributed to the building blocks (pebbles, e.g. Morbidelli et al. 2012; Raymond 2021; Johansen et al. 2021 and/or planetesimals, Chambers and Wetherill 1998; Levison et al. 2015; Burkhardt et al. 2021) of the inner rocky planets. Some of the early water (and other volatiles) would have been degassed and lost during the Moon forming impact (Benz et al. 1986; Canup 2004) with a Mars-sized planet (named Theia cf. Halliday 2000) that occurred approximately 4.51 Ga (Barboni et al. 2017). Although (Connelly and Bizzarro 2016), also using Pb isotope data, suggest slighter younger dates between 4.426–4.417 Ga. Recent calculations suggest that all of Earth's water and other volatiles may have been delivered by volatile-rich carbonaceous chondrites, initially formed outside the orbit of Jupiter but displaced inwards by the planet's growth and migration (Kleine et al. 2020). However, timing of the accretion of the volatiles to Earth is still an active area of research (Avice et al. 2022; Salvador et al. 2023, this journal). Note that Marty (2012) suggests that the isotope signatures of terrestrial H, N, Ne and Ar may be the result of mixing between two end-members of solar and chondritic compositions, with the N and H isotopic compositions suggesting a primitive meteoritic origin.

Liquid water is critical to magmatic processes on the Earth, including partial melting of the mantle and crustal recycling. Indeed, water is essential for the production of significant amounts of granitic melts formed by melting of pre-existing crustal rocks (Campbell and Taylor 1983; Jacob et al. 2021; Turcotte and Schubert 2002; Korenaga 2018). Although non-hydrous fractionation will form feldspathoids, as testified by the lunar anorthosites (Norman et al. 2003). These granitic melts, in turn, formed the early, buoyant, less dense granitoid rocks that were the cores of early continents. Thus, evidence of any of these phenomena can be used as proxies for the presence of liquid water.

Physical evidence for the existence of early granitoid crust, however, is restricted to: (1) zircon crystals formed by crustal fractionation during the Hadean (4.5–4.0 Ga) and Eoarchean (4.0–3.6 Ga) that were eroded from the initial crustal rocks and then sedimented. These ancient zircons have re-emerged in Palaeoarchean (3.5–3.3 Ga) rocks in Western Australia (Wilde et al. 2001; Mojzsis et al. 2001). (2) Small enclaves of granitoid rocks from this period still exist and are occasionally associated with metamorphosed sediments (metasediments), such as the 4.3 (O'Neil et al. 2008) to 3.8 (Cates and Mojzsis 2007) Nuvvuagittuq Supracrustal Belt and the 4.02 Ga Acasta Gneiss (Bowring and Williams 1999) in Canada, the 3.7–3.8 Ga Isua terrane in West Greenland (Moorbath et al. 1973), and the 3.5–3.2 Ga greenstone belts of the Pilbara in W. Australia (Nelson et al. 1999) and Barberton in South Africa (Lowe and Byerly 1999b). Finally, (3) inherited uranium-lead and hafnium isotope signatures in the reworked zircon crystals provide a certain amount of information pertaining to the pre-existing Hadean crust (Mulder et al. 2021).

Oxygen isotopic signatures preserved in zircon crystals, and dated at up to 4.4 Ga (Wilde et al. 2001; Mojzsis et al. 2001; Valley et al. 2014) (but possibly younger in age, Whitehouse et al. 2017) have been interpreted to suggest the exposure of the crust from which the crystals formed via hydrothermal processing, implying the presence of water recycled into the crust from the surface of the Earth by 4.4 Ga. Indeed, recent combined oxygen and silicon isotope measurements of zircons from the Hadean support the existence of significant quantities of siliceous sediments during the Hadean (Trail et al. 2018).

Another proxy for the presence of water is the existence of sediments; they imply erosion by and/or deposition in a body of water. Sediments are associated with the most ancient terranes preserved, the 4.3–3.8 Ga Nuvvuagittuq (Canada) and the 3.7–3.8 Ga Isua (West Greenland) supracrustal terranes, the latter of which includes also metamorphosed pillow basalts, undeniable structures produced under water. Further evidence of hydrosphere-crustal interactions comes from extremely high $\delta^{18}O$ values of up to $+9‰$ measured in metamorphic zircons formed about 3.5 Ga by reworking of metamorphic crust in the ca. 3.86 Ga, Eoarchean Saglek Block (North Atlantic Craton) (Vezinet et al. 2019).

2.2 Interior and Tectonic Processes on Earth

2.2.1 Brief Overview

The interior and tectonic processes on the early Earth had important implications for the building of a habitable planet (e.g. Schubert et al. 1989; Korenaga 2012; Höning et al. 2019a). Indeed, our present-day Solar System provides a perfect correlation between the occurrence of plate tectonics and planetary habitability, although with a sample size of one. A possible reason for this is the increased exchanges between the interior and the atmosphere of planets with plate tectonics, compared, for example with stagnant lid convection (Foley and Smye 2018; Höning et al. 2019b; Rolf et al. 2022; Gillmann et al. 2022, this journal). It has also long been suggested that plate tectonics and the presence of surface liquid water were entwined (Campbell and Taylor 1983, for example) and favoured volatile cycles, and possibly stabilizing feedback process for surface conditions. Moreover, while Venus (and Mars) appear to be operating under stagnant-lid-like convection today, it is possible that their convection regime changed over their past respective histories (e.g. Sleep 1994; Gillmann and Tackley 2014; Smrekar et al. 2018).

In a review of the evolution of continental crust and the onset of plate tectonics, Hawkesworth et al. (2020) note the paucity of early crustal preservation and reiterate the fact that inferences based on the few preserved remnants, represent only a part of the geological history of this early time. This fact is all the more important because it is apparent that tectonic signatures varied in time and place, and that a form of subduction may have been catalysed, at least temporarily, by impacts and mantle plumes, as well as by plate tectonics (Gillmann et al. 2016; O'Neill et al. 2017; Gerya et al. 2015a). A recent study of Paleoarchean zircons ages recording submantle $\delta^{18}O$ relates their production to impact induced crustal recycling (Johnson et al. 2022).

The timing of the onset of plate tectonics is still debated and ranges from ca. 4 Ga to 1 Ga (as reviewed by Lammer et al. 2018; Dehant et al. 2019; Korenaga 2021). Most estimates place the transition from an earlier convection regime (possibly from a more stagnant state, or already a plume-induced proto-plate tectonics, see also Fig. 1) between 3 and 4 Ga, with the process taking place gradually at different places and at different times. Prior to about 3.0 Ga, xenon isotopes suggest little recycling of volatiles in the crust (Péron and Moreira 2018). However, the recent review by Korenaga (2021) hypothesises an early start to plate tectonics during the Hadean, as soon as there was water at the surface of the planet. The existence of plate tectonics has numerous implications: firstly that the crust was sufficiently rigid as to allow crustal breakup under stress caused by vigorous mantle convection, as well

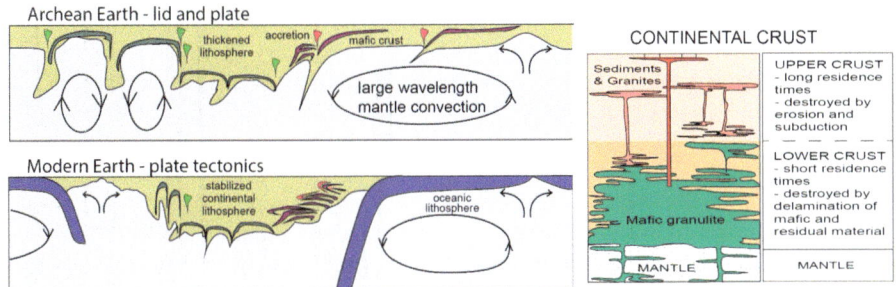

Fig. 1 Comparison of two styles of tectonics, lid and plate tectonics during the Archean epoch and modern-style Wilson plate tectonics (after Hawkesworth et al. 2020)

Fig. 2 Overview of changes in crustal growth, crustal thickness, crustal reworking, lithospheric stabilisation and the formation of supercontinents due to lateral accretion, the appearance of dykes swarms indicating rigid crust, changes in the oxygen isotope composition reflecting increasing continental sediment incorporated into the mantle/crust with time, and the appearance of blueschists indicating high temperature metamorphism, all signs of and influenced by the emergence of plate tectonics (after Hawkesworth et al. 2020)

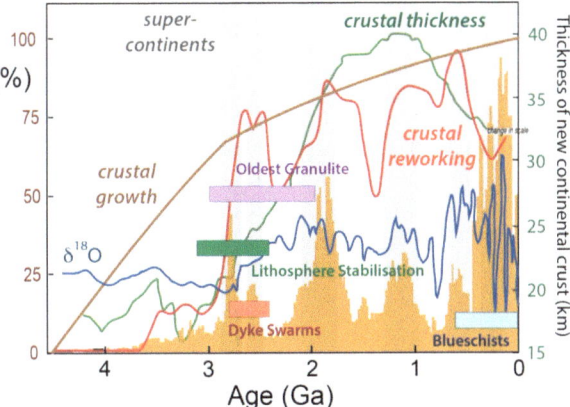

as to allow the intrusion of dyke swarms (Cawood et al. 2018), and secondly, that it was dense enough (i.e. mafic in composition) to subduct (Van Kranendonk 2010; Hawkesworth et al. 2009; Cawood et al. 2013). The paired metamorphic zones so typical of convergent tectonics, and recognised by Th/Nb ratios, suggest that magmas, both related to subduction (suites of high Th/Nb magmas) and not related to subduction (low Th/Nb magmas), were concomitant in different locations of the planet (Hawkesworth et al. 2020).

In parallel to the initiation of plate tectonics, there was a change in the composition of juvenile continental crust from mafic to intermediate andesitic compositions (Dhuime et al. 2015), the latter characterising the upper continental crust (Chowdhury et al. 2017; Perchuk et al. 2018). Increasing crustal thickness and more acidic compositions of the granitic cores of the continents led to landmasses with higher relief, which influenced erosion and sedimentation, and hence the composition of the oceans and the atmospheres.

Major continental amalgamation to form super continents started at least by 2.8 Ga (Evans 2013). Rates of continental reworking (estimated from Hf isotope ratios, Belousova and Kostitsyn 2010; Dhuime et al. 2012) and destruction linked to tectonic processes started increasing from about 3.0 Ga, an indication of efficient recycling of the older, less buoyant, mafic continental crust. There was also a change in the global oxygen isotope ratios in zircons indicating incorporation of eroded sediments and, therefore, the presence of exposed landmasses (Valley et al. 2005; Spencer et al. 2014). Figure 2 compares the evolution of crustal growth and thickness with factors such as lithospheric stabilisation, change in style of metamorphism, as well as dyke swarm frequency. Korenaga (2018) has compiled a list of published models for continental growth through Earth's history which underlines the wide range of estimates for the initiation of plate tectonics from the Hadean to the Archean.

Continental landmass is an important source of phosphorus, one of the rate limiting nutrients for biomass development. On the early Earth, relatively low weathering rates of exposed land masses (because of their low relief) led to a relatively low influx of P in the form of apatite (Hao et al. 2020). New experimental and analytical work suggests that phosphate (HPO_4^{2-}) in the form of apatite (insoluble) can be reduced to phosphite (HPO_3^{2-}) by concurrent oxidation of Fe^{2+}. Phosphite is much more soluble and therefore would have been available for biomass development.

2.2.2 Evidence for Interior Processes and Tectonics, the Results of Modelling Studies

Recently, tectono-magmatic processes on pre-Phanerozoic Earth have been the subject of growing numerical geodynamic modelling efforts (e.g. Gerya 2022, and references therein). The resulting holistic modelling- and observation-based view of the global Precambrian tectono-magmatic evolution that has emerged (Gerya 2014; Rey et al. 2014; Bercovici and Ricard 2014; Rozel et al. 2017; Sobolev and Brown 2019; Gerya 2019; Hawkesworth et al. 2020; Gerya 2022) is briefly summarized below.

As envisaged in these models, Hadean-Archean plutonic squishy-lid/plume-lid/lid-and-plate tectonics before about 3 Ga were characterised by mantle potential temperatures 250–200 K higher than present day, which resulted in the widespread development of mantle-derived magmatism and rheologically-weak crust (Richter 1985; Gerya 2014; Rozel et al. 2017; Hawkesworth et al. 2020, see Figs. 1 and 2). Models suggest that the global tectono-magmatic style was dominated by plume- and drip-induced tectono-magmatic processes under conditions of an internally deformable (squishy, non-stagnant, non-rigid) lithospheric lid that is often compared to conditions on present-day Venus (e.g. Van Kranendonk 2010; Gerya et al. 2015b; Rozel et al. 2017; Harris and Bédard 2014; Hansen 2018). In this hypothesised pre-plate tectonics regime, both proto-oceanic and proto-continental lithospheres were formed by a combination of several tectono-magmatic differentiation processes (e.g. Sizova et al. 2015; Capitanio et al. 2020). Lid evolution was driven by episodic tectono-magmatic activity (e.g. Moore and Webb 2013; Johnson et al. 2014; Piccolo et al. 2019, 2020) controlling crustal and lithospheric growth and removal with a periodicity of ~ 100 Myr (Sizova et al. 2015; Fischer and Gerya 2016a), which is comparable to the geological-geochemical record from some major Archean greenstone belts, e.g. East Pilbara in Western Australia and Kaapvaal in South Africa (cf. discussions in Fischer and Gerya 2016a, and references therein). Thermal regimes of crustal reworking produced by this non-plate tectonic environment are also broadly consistent with the metamorphic record (cf. discussions in Capitanio et al. 2019, and references therein). (Ultra)-slow rifting and oceanic spreading with intense decompression melting and thick mafic crust were capable of developing in the absence of subduction (e.g. Sizova et al. 2015; Capitanio et al. 2019, 2020). Due to the elevated mantle potential temperature (e.g., Herzberg et al. 2010), (ultra)slow spreading is associated with intense mantle decompression melting leading to thick mafic crust formation (e.g., Sizova et al. 2015) and thereby to high heat fluxes at ridges. In addition, heat fluxes were also higher in the continental crust, which was much hotter than present day due to widespread intrusions of mantle-derived magma (e.g., Sizova et al. 2015; Rozel et al. 2017; Piccolo et al. 2019). Importantly, the modern-style global mosaic of rigid plates separated by narrow, rheologically weak, plate boundaries did not exist in this pre-plate tectonics period (Bercovici and Ricard 2014). Voluminous melting of the upper mantle caused the formation of both cold lithospheric and hot sub-lithospheric, highly depleted, proto-cratonic mantles with lowered density and increased viscosity (e.g. Sizova et al. 2015; Capitanio et al. 2020; Perchuk et al. 2020, 2021). Note that this scenario is in direct contrast to the scenario proposed by Korenaga (2021), in which the Hadean was characterised by a vigorous plate tectonic regime and recycling of the earlier, thinner crust. This regime then slowed down during the Archean as a result of increasing mantle temperatures and therefore thicker crust that would have been more difficult to subduct.

Subsequently, during the period of protracted Archean-Proterozoic transitional tectonics between about 3 Ga and 0.75 Ga, notable secular cooling of the mantle potential temperature occurred (to 200–100 K above present day). As a result, squishy-lid/plume-lid/lid-and-plate tectonics may have gradually evolved towards the modern plate tectonics regime by combining elements of these two contrasting global styles in both space and time (e.g. Fischer and Gerya 2016b; Chowdhury et al. 2017, 2020; Sobolev and Brown 2019; Perchuk et al. 2018, 2019, 2020). The transitional tectonic regime was controlled by gradual stabilization of rheologically-strong continental and oceanic plate interiors (e.g. Sizova et al. 2010; Fischer and Gerya 2016a). Plume-induced subduction was likely common in the beginning, and triggered the onset of this transitional tectonic regime (Gerya et al. 2015a). Due to the hot mantle temperature and weak lithospheric plates subjected to bending-induced segmentation near trenches (Gerya et al. 2021), shallow slab break-off would have been very frequent, causing intermittent rather than continued subduction (e.g. van Hunen and van den Berg 2008; Perchuk et al. 2019, 2020; Gerya et al. 2021).

Elements of squishy-lid/plume-lid/lid-and-plate tectonics were also locally present and controlled continued development of granite-greenstone belts in (proto)continental domains (Fischer and Gerya 2016b). As noted above, different elements of modern plate tectonics likely emerged at different geological times and oceanic subduction likely became widespread earlier than modern-style (cold) continental collision (e.g. Sizova et al. 2010, 2014; Perchuk et al. 2018). Delamination of the mantle lithosphere in long-lived accretionary orogens controlled gradual changes of continental crust composition from mafic to more felsic components with related rising of the continents due to efficient recycling of lower continental, mafic crust and tectono-magmatic reworking and thickening of more felsic upper continental crust (Chowdhury et al. 2017; Perchuk et al. 2018). The intermittent subduction was likely initially inefficient in creating large volumes of silicic continental crust and, associated with massive decompression melting of the mantle, resulted in the formation of oceanic plateau-basalts (Perchuk et al. 2019). The presence of low-density, highly depleted, hot, ductile mantle under oceanic plates contributed to the formation of chemically layered cratonic keels through a viscous emplacement mechanism driven by oceanic subduction (Perchuk et al. 2020). This peculiar mechanism of cratonic growth deactivated after about 2 Ga due to a decrease in mantle temperature (Perchuk et al. 2020).

Finally, the establishment of modern plate tectonics after about 0.75 Ga followed cooling of mantle potential temperatures to less than 100 K above present day values. This process was attained gradually by a combination of four interrelated factors (Bercovici and Ricard 2014; Gerya 2014; Gerya et al. 2015b; Sobolev and Brown 2019; Gerya et al. 2021): (1) cooling and strengthening of the oceanic lithosphere that stabilized continued long-lived subduction, (2) emergence of a global mosaic of rigid plates divided by strongly localized, long-lived, rheologically-weak boundaries, (3) stabilisation and cooling of thick, rheologically strong continental lithospheres and the rise of the continents above the sea level, and (4) the growing intensity of surface erosion providing rheologically weak sediments deposited in the oceans that increasingly lubricated subduction in trenches. The transition to modern plate tectonics followed a long period of reduced tectono-magmatic activity – the boring billion, 1.7 to 0.75 Ga (Sobolev and Brown 2019).

Fig. 3 Evolution of the composition of the Earth's atmosphere through geological time (Catling and Zahnle 2020)

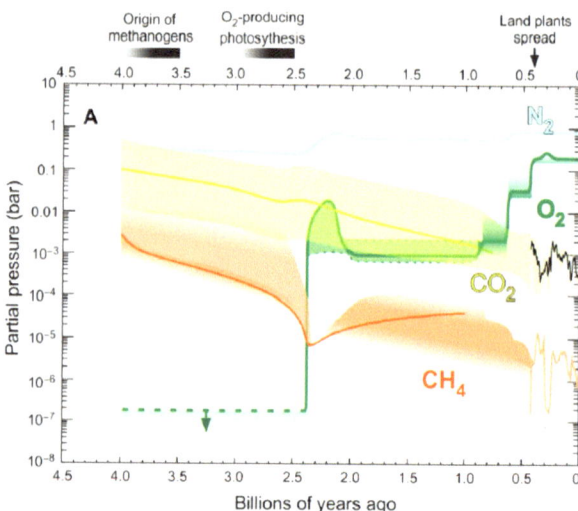

2.2.3 Establishment of the Conditions for the Emergence of Life

Understanding the internal, dynamic processes of the early Earth is certainly essential for appreciating the building of habitable conditions on a global scale. However, the emergence of life and its early evolution were events that occurred on local scales, although perhaps combining the results of different prebiotic reactions occurring in different microenvironments (Stüeken et al. 2013). In this section we will review our present understanding of the environmental conditions reigning on early Earth that were of immediate importance for the emergence of life.

The primary requirement for establishing an environment conducive to the emergence of life is the presence of liquid water. We noted above various proxies indicating liquid water on the Hadean-Eoarchean Earth. One of the main constraints for liquid water at the surface is the composition and partial pressure of the atmosphere (Table 1 in Catling and Zahnle 2020, and references therein). After the Moon-forming impact about 4.5 Ga (e.g. Barboni et al. 2017) that effectively vaporised the surface of the Earth as well as the impactor, the Si-rich vapor recondensed and a thick CO_2 plus water greenhouse atmosphere formed (Zahnle et al. 2015; Sleep et al. 2014; Sossi et al. 2020). Removal of much of the CO_2 though crustal recycling during the Hadean would have resulted in an atmosphere containing approximately 1 bar CO_2 atmosphere and temperatures permitting oceans to form (at $\sim$ 500 K, Zahnle et al. 2015; Sleep et al. 2014).

In contrast to Sleep et al. (2014), Catling and Zahnle (2020) conclude that the early atmosphere could not have been very thick and that it was compensated by the presence of greenhouse gases (Fig. 3). These interpretations are based on geochemical investigations of nitrogen contained in fluid inclusions in quartz crystals of Paleoarchean age (Marty et al. 2013; Avice et al. 2018) and on physical phenomena, such as the sizes of gas bubbles in submarine lavas of similar age indicating hydrostatic pressures of not more than 0.5 bars (Som et al. 2012).

To date, we have no hard and fast evidence of when oceans formed but have listed the different proxies in Sect. 2.1, which suggest an early appearance of water (Catling and Zahnle 2020). Indeed, the Hadean Earth would have been more of an ocean planet and its primitive continents being characterised by submerged plateaus with emergent volcanic edifices and their surrounding land masses, similar to those characteristic of the Paleoarchean, as we will see below.

2.2.4 Early Habitable Environments

There are only a few exposed locations where Paleoarchean terranes are well-preserved (namely the $\sim$ 3.5–3.3 Ga Barberton, South Africa, and Pilbara, Australia, Greenstone Belts), most of which are subaqueous deposits. Indeed, until about 3.2 Ga, very little sub-aerial material from this time period exists. There are reports of quartzites and quartz-biotite schists from the Isua and Nuvvuagittuq terranes that are interpreted to be the metamorphosed remnants of sandstone and conglomerate protoliths (Bolhar et al. 2005; Cates and Mojzsis 2007; O'Neil et al. 2011), respectively, as well as some horizons of pebble conglomerates and sands attesting to deposition in a terrestrial setting in the 3.48 Ga Hooggenoeg Formation. Subaerial deposits are far more common in the younger, $<$ 3.2 Ga Moodies Group in the Barberton Greenstone Belt (Lowe and Byerly 1999b,a; Heubeck 2009; Hofmann et al. 1999), while subaerial spring deposits associated with a caldera have been described in the 3.48 Ga Dresser Formation in the Pilbara (Djokic et al. 2021). All the preserved subaqueous sedimentary deposits in the Barberton and Pilbara Greenstone Belts formed at relatively shallow water depths in depositional basins on top of the plateau-like protocontinents (i.e. at water depths ranging from littoral to below wave base, which could have been some tens to a few 100s m) (Lowe and Byerly 1999b). Nijman et al. (2017) compared the Paleoarchean depositional basins to collapse basins on Venus or Mars, forming on softened crust atop mantle plumes, although it has been argued that the early Earth's crust was not thick enough to support such a tectonic situation (comment by an anonymous reviewer). Nevertheless, the thickness of the Archean Earth's crust is modelled to have been greater than that of the present day owing to hotter mantle temperatures and magmas (Hawkesworth et al. 2020). The group of van Kranendonk (Djokic et al. 2021) proposes an alternative caldera-like scenario for at least some of the shallow basins. Although we have geochemical evidence for the existence of open ocean via a positive Eu anomaly reflecting a global, background hydrothermal signature (Jacobsen and Pimentel-Klose 1988; Hofmann and Wilson 2007; Hofmann and Harris 2008; Hickman-Lewis et al. 2020b), there is no morphological preservation of deep oceanic crust, which was probably removed (together with much of the early proto-continental crust) by a combination of tectonic overturn and possibly the high rate of impacts on the Hadean Earth (Melosh and Vickery 1989; Abramov et al. 2013; Kemp et al. 2010; Kamber 2015; Griffin et al. 2014; Maher and Stevenson 1988).

The early basins and emergent landmasses likely hosted a variety of habitable environments including subaqueous, littoral (i.e. tidal therefore partially subaerial), subaerial and hydrothermal settings (note, however, that hydrothermal settings were ubiquitous). These sedimentary environmental settings are attested by sedimentary structures, pillow lavas and geochemical signatures. Various proxies are used to infer the environmental conditions. Estimates of water temperatures on the early Earth from oxygen, silicon, and hydrogen isotopic signatures preserved in chert sediments are wide-ranging, from a cool 26 °C (Hren et al. 2009; Blake et al. 2010) to $\sim$ 50 °C and up to $\sim$ 70 °C (Robert and Chaussidon 2006; Van den Boorn et al. 2010; Marin-Carbonne et al. 2012; Tartèse et al. 2017). The latter studies especially noted the strong influence of the early Earth's abundant hydrothermal activity on the temperature signatures, as evidenced also by the aforementioned REE signatures (positive Eu and Y anomalies, and Y/Ho ratio Hofmann and Harris 2008; Hickman-Lewis et al. 2020b). It should be borne in mind that the sediments analysed were formed at the interface between the relatively shallow column of water and above warm or hot rock that could be easily heated; this may not have been the case in the "deep" ocean, and we do not know how deep the Earth's oceans were outside the shallow plateau areas. However, in the scenario where there was not much exposed continental landmass, the average ocean depths would have been about 2 km. The pH of the early oceans would have been variable, with alkaline conditions enhanced by aqueous alteration of the predominantly ultramafic and mafic crust (Kempe and Degens 1985; García-Ruiz et al. 2020), while more acidic conditions were the consequence of boiling and hydrogen-rich hydrothermal fluids as well as the CO_2-rich atmosphere (Morse and Mackenzie 1998; Catling and Zahnle 2020). Intermediate values were estimated by Friend et al. (2008), who interpreted circum-neutral pH from geochemical analyses of Eoarchean rocks from West Greenland. On a local scale, variations in pH could have been readily maintained around hydrothermal vents, where exiting fluids of a certain pH mix with seawater of different pH, or where the pH is changed by interaction with adjacent rock/sediment materials, e.g. acidic fluids flowing through mafic or ultramafic rocks and sediments becomes initially alkaline before returning to and slightly acidic pH if that is the ambient condition (Dass et al. 2018; Westall et al. 2018). Estimations of salinity for the early oceans vary, but fluid inclusion studies suggest that they range from present day values to about double these values (Marty et al. 2018; see also Knauth 2011; Catling and Zahnle 2020). These estimations will be relevant for the shallow water basins on top of the submerged protocontinents but environmental conditions in the open ocean may have differed.

While all environmental parameters indicate an anaerobic early Earth, extremely small amounts of oxygen would have formed locally by EUV photodissociation of water vapour in the atmosphere and at the surface of the seawater (Kasting et al. 1979). Oxygenated species could have resulted also from dissociation of boiling, pressurised hydrothermal fluids as they exited vents in shallow waters (pers. comm. C. Ramboz, 2009). Note also that the early oceans were more enriched in the transition metals essential for Earth-like life (Fe, V, Ni, As and Co) than today, owing to leaching of the early ultramafic and mafic rocks characteristic of the early volcanic crust (Hickman-Lewis et al. 2020a). Indeed, the early oceans could be considered iron-rich environments.

2.3 Early Life on Earth

2.3.1 Scenarios for the Emergence of Life

There is strong evidence for diversified life forms comprising chemotrophic and phototrophic microorganisms already in the Paleoarchean (3.6 to 3.2 Ga) based on morphological structures, as well as geochemical and organic biosignatures (Hofmann et al. 1999; Hassenkam et al. 2017; Djokic et al. 2017; Hickman-Lewis and Westall 2021). This suggests that life must have emerged during the Hadean. Or, if life appeared very rapidly (and we have no idea how long it took for life to emerge) at the latest, in the very early Eoarchean. Microbial fossils have been interpreted from the 4.28–3.8 Ga Nuvvuagittuq rocks of Canada (Dodd et al. 2017; Papineau et al. 2022), where jaspilite deposits probably represent hydrothermal chemical sediments. Relatively large filaments (about 16.5 μm in diameter and up to 1000s of μm in length) were hypothesised to be of microbial origin. Associated multiple sulfur isotopes are consistent with a microbial signature. However, Greer et al. (2020) and Lan et al. (2022) suggested that these and other Fe-rich microbial filaments in the highly metamorphosed, Nuvvuagittuq rocks are abiotic mineral features. Microbial life is strongly associated with hydrothermal deposits in later Paleoarchean sediments 3.33 Ga (Westall et al. 2015; Hickman-Lewis et al. 2020b), and confirmation of such traces in more ancient deposits is actively being sought.

There are many scenarios suggested for the emergence of life on Earth, including hydrothermal environments undersea (Baross and Hoffman 1985; Russell and Hall 1997; Martin and Russell 2003) and on land (Damer and Deamer 2020; Van Kranendonk et al. 2021); associated with impact craters (Sasselov et al. 2020); pumice rafts (Brasier et al. 2011); deep seated faults (Schreiber et al. 2012); and mixing of chemical precursors produced in combinations of these and other environments (Stüeken et al. 2013). Each of the scenarios has relative merits and some disadvantages, as reviewed by Westall et al. (2018). We will briefly summarise the different scenarios below.

An important point in addressing the scenarios for the origin of life is that the environmental requirements for this are not necessarily the same as those for flourishing, more evolved life forms. This will become evident also later in this chapter during the discussion of possible life forms in the clouds of Venus today. For life as we know it, based on organic carbon molecules and liquid water, the basic ingredients include the six essential elements, C, H, O, N, P, S, as well as transition metals (especially Fe), liquid water, an energy source (e.g., chemical, photonic, heat), and a suitable geological context. According to our current understanding of prebiotic chemistry processes, life as we know it could not emerge in an environment with free oxygen, thus anaerobic conditions are also important. According to some researchers (Pascal et al. 2013; Pross and Pascal 2013), the initial energy for pushing prebiotic reactions past the required activation level needs to be very high and can only be provided by UV radiation. Conversely, the complex compounds necessary for biological functions, such as peptides, information transferring molecules (RNA, DNA), or the lipids of cell membranes (Kminek and Bada 2006; Reisz et al. 2014), would rapidly break down under UV radiation. Others (Adam 2007; Adam et al. 2018) have hypothesized that beach sands enriched in uranium could have provided the radiation necessary for activating prebiotic processes. Although the existence of uranium placer deposits during the Hadean is highly unlikely owing to the small quantities of uranium in the early terrestrial rocks and the limited availability of oxygenated environments for leaching and concentrating it out of the rocks. If correct, the necessity of UV radiation for early prebiotic reactions would place serious constraints on where life could have emerged, i.e. life could only have emerged where there was exposed land and not, for example, on an ice covered ocean planet.

Another important condition for the emergence of life is the presence of natural gradients: in temperature, pH, ionic concentrations, water and osmotic potential, and energy (Westall et al. 2018). Gradients drive the diffusion of essential components for prebiotic chemistry and primitive metabolisms, via hydrothermal fluids, seawater, pore waters (in porous materials), river water, or (impact) lakes. As one commonly hypothesized requirement for life is compartmentalization, chemical constituents would have needed to be transferred into and out of the micro-scale compartments (e.g., pores in rocks and minerals, naturally-forming gels, vesicles, or micelles) in which prebiotic reactions would have taken place. In terms of the emergence of life, three key factors are critical: (1) the concentration of the various molecular building blocks of life, (2) their stabilisation and structural conformation, and (3) chemical evolution (as summarised from previous works by Westall et al. 2018).

In a manner that is *a priori* counter intuitive for non-prebiotic chemists, there are stages during prebiotic reactions when water is a hindrance. This is when it is necessary to concentrate the ingredients of life. Darwin's dilute, warm little pond will not work. Concentration allows basic prebiotic molecules to interact sufficiently with each other to create additional, more complex conformations. For example, Russell and Hall (1997), Russell (2021), Martin et al. (2008) view the reactive mineral-rich walls of pores in deep sea hydrothermal vents as a likely location for concentration and condensation of organic molecules. Porous silica gel was suggested by Westall et al. (2018) and Dass et al. (2018) because of its ubiquity

Fig. 4 Model for the emergence of cellular life in porous hydrothermal vent systems (after Martin and Russell 2003)

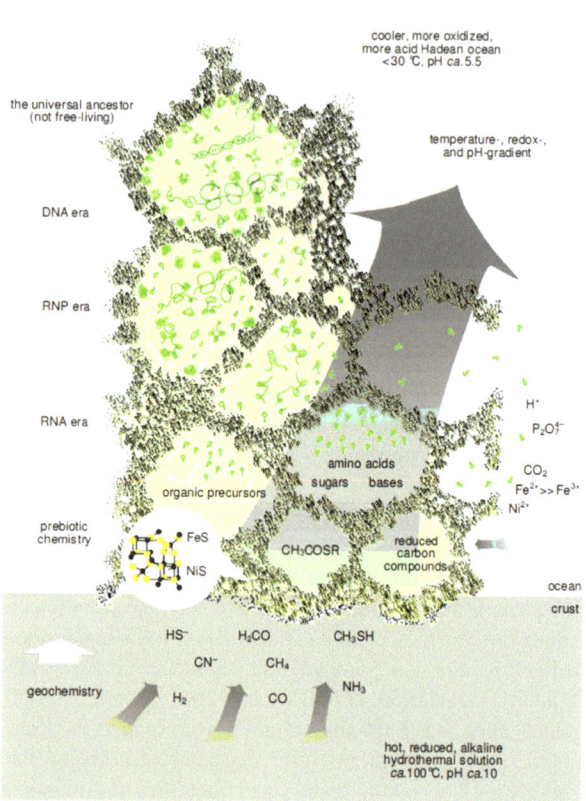

in the early terrestrial oceans, and its association with hydrothermal environments. Organic molecules chelate to the surface of the pores in the gel, which is permeable, letting through nutrients, molecules and enabling gradients. Other researchers prefer wetting-drying cycles that imply exposure of the organic molecules to the early atmosphere, either in a beach environment (Deamer 1997), or on land (Damer and Deamer 2020; Marshall 2020; Sasselov et al. 2020).

Deep sea hydrothermal vents were suggested as a suitable location for the origin of life by Baross and Hoffman (1985). This idea was further developed in great detail by Russell and various colleagues since the mid 1990s. Russell et al. (2010) noted the particular importance of alkaline vents for the emergence of life (Fig. 4). These were environments from which metal-rich fluids and small organic molecules formed during serpentinising reactions in the crust, including hydrogen, methane, minor formate, and ammonia, as well as calcium and traces of acetate, molybdenum and tungsten. Chemiosmotic energy would have been provided by proton and redox gradients across the porous vent walls. According to this hypothesis, prebiotic chemical reactions in the porous, reactive mineral constructs of the vents would concentrate molecules, helping them to form new structures and combinations. Eventually, all the constituents of life, except cell membranes, would be found within the pores, thus forming the first living entities (i.e. non membrane-bound cells). Finally membranes would form around the edges of the pores to enclose the proteins and RNA molecules, allowing the protocells to be expelled into the ocean. In this scenario, UV radiation is not necessary to surmount the activation energy barrier.

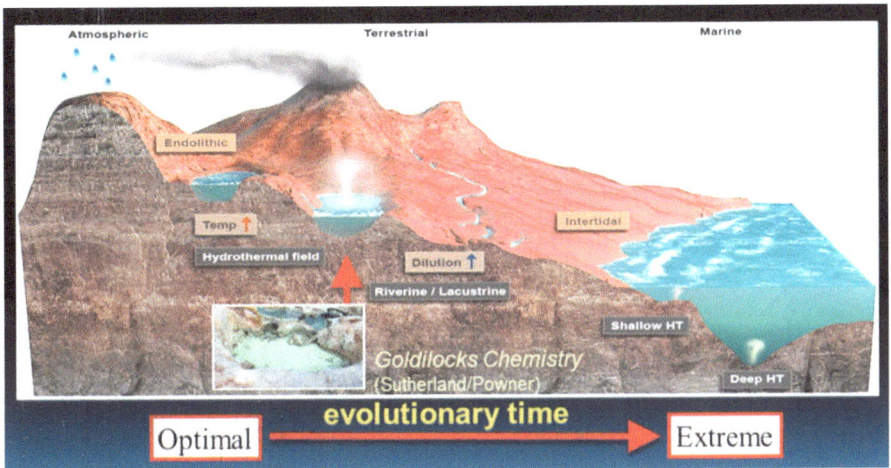

Fig. 5 Hypothetical emergence of life in subaerial hydrothermal springs. After Damer and Deamer (2020)

In support of the deep-sea hydrothermal vent scenario, Ménez et al. (2018) note that serpentinite-bearing hydrothermal environments requiring exhumation of mantle rocks to the surface are common for (ultra)slow spreading mid-ocean ridges and/or oceanic rifts. Such tectonic settings require lithospheric extension and were likely present since very early stages of lithospheric evolution and crustal differentiation (e.g. Sizova et al. 2015). Models show that their existence does not require global plate tectonics and/or subduction and can associate with several other styles of mantle convection and surface dynamics such as ridge-only convection or plutonic squishy lid that might be common styles for young Venus/Earth-sized terrestrial planets (Rozel et al. 2017; Sizova et al. 2015; Lourenço et al. 2018).

In a variant on the deep-sea hydrothermal scenario and based on their studies of well-preserved hydrothermal sediments from the Paleoarchean, Westall et al. (2018) suggested that volcanic sediments in the vicinity of hydrothermal vents may have hosted prebiotic reactions leading to the emergence of life. The scenario is very similar in principal to that of Russell (a porous medium comprised of reactive minerals), with the exception of the inclusion of porous silica gel, as noted above, a ubiquitous by-product of the early silica-rich seawater. The presence of these sediments around hydrothermal effluent extends the available environments for the emergence of life. Moreover, such environments existed at all water depths, from tidally-influenced littoral environments down to the deep sea. In this case, if UV radiation were an essential factor in prebiotic reactions, shallow water systems in the tidal zone would offer exposure to UV radiation, as well as protection of the more complex molecules under water and within the subaqueous sediments.

Another popular scenario suggests subaerial hydrothermal environments for the emergence of life. This is largely because of the findings that (1) UV radiation can contribute to the neoformation of prebiotic molecules (Pascal et al. 2013; Pross and Pascal 2013), (2) hydrophobic conditions are necessary at certain stages of prebiotic reactions to concentrate molecules (Damer and Deamer 2020; Deamer 1997; Marshall 2020; Sasselov et al. 2020), and (3) subaerial vents have been interpreted in ancient Paleoarchean terranes (Van Kranendonk et al. 2021). Hydrothermal vents on land (Fig. 5) would have provided a suitable environment for prebiotic processes as they are exposed to UV radiation, when necessary, as well as protected by water. The porous sediments and vent walls would have served the same function as hypothesized in the deep-sea vent scenario, mini-reactors localizing and supporting the prebiotic chemical reactions. Fully-formed microbial cells would have been transported to the oceans by rivers (UV avoidance being a necessity in this scenario). One could also envisage the transport of microbes attached to each other or to dust particles in small clumps that can travel significant distances through the air. In this scenario, despite exposure to UV and desiccation, cells in the interior of the clump will remain shielded and wet for a certain time even though cells on the surface of the clump die (Madronich et al. 2018). This scenario works on the Earth today but the higher UV flux on the early Earth may have been a considerable constraint.

The common denominator in the most popular of the above origin-of-life scenarios is hydrothermal environments, either undersea or on land. Here, the contact of hot water with reactive mineral surfaces would have provided the chemical energy for prebiotic reactions. The early volcanic rocks were more ultramafic than today, comprising predominantly iron and magnesium-rich basalts and komatiites. There would have been the possibility of exposure to UV radiation in beach environments, shallow water vents, or subaerial vents, at significant moments (if it was indeed essential). The mineral surfaces, perhaps assisted by the presence of ubiquitous silica gel, would have facilitated increasing molecular concentration, conformation and complexity in mineral and silica pores. The necessary gradients would have been provided by the through flow of fluids and nutrients from the vent effluent to the immediately surrounding environment (including sediments) and protocells would have formed.

We have only considered here in detail a few of the wide variety of environments suggested for the emergence of life, but we noted above some of the other hypotheses. However, we may never know exactly where life originated. It is clear that there were numerous possibilities for prebiotic chemistry in different scenarios. It is possible that important components of living cells were formed in different types of environments and eventually concentrated together in one location. Although certain prebiotic chemists do not endorse this idea, considering that the processes leading to life needed to have occurred in one location (N. Lane, pers. comm,, 2022). It is also possible that life emerged in more than one place during the course of the Hadean, possibly even under different scenarios in different times and places. Large impacts or other more localized environmental changes could have wiped out life in some regions while in others it continued to flourish. Given the biochemical, genetic, and other evidence we have today that all modern life shares a common ancestor (Weiss et al. 2018, and references therein), eventually, the early world ocean must have been dominated by one form of life, presumably the metabolically most effective, using the molecular machinery that we know today.

2.3.2 Scenarios for the Emergence of Life and the Problem of Prebiotic Chemistry in the Laboratory

The origin of life has traditionally been addressed through experiments in prebiotic chemistry in carefully controlled laboratory conditions. This has led to a significant amount of confusion in the origins of life community because the realities of the early terrestrial environment were, and are still, rarely taken into account. A prime example of this situation is the stabilisation of ribose (sugar), one of the essential ingredients of RNA. The element boron has been suggested to have been critical to the stabilisation of the sugar (Benner et al. 2010; Scorei 2012). Boron is a constituent of tourmaline, a mineral present in the sediments of Eoarchean terranes and was certainly present on the Hadean Earth (Grew et al. 2011) but some researchers have suggested its presence as boron salts on exposed landmasses, thereby inferring that life could have emerged in subaerial rivers (Benner et al. 2010). Another solution for the stabilisation of ribose hypothesises is exposure to ice. Szostak (2016) invokes seasonal changes in a subaerial hydrothermal setting (similar to Yellowstone), whereby temperature changes could have induced a temporally icy setting. Trinks et al. (2005) suggest sea ice as an important setting for prebiotic chemistry in terms of concentration of organic molecules and for providing optimal conditions for the early replication of nucleic acids. Support for this scenario was based on models suggesting that the early Earth had a cold start (Catling and Zahnle 2020). They are predicated on the lower luminosity of the Sun and the necessity of either high partial pressure for a CO_2 atmosphere or a large amount of greenhouse gases, such as CH_4 (Sagan and Mullen 1972).

Is such a scenario realistic? While there is no evidence during the Archean of glacial conditions (beaches were bathed by tidal waves and hosted evaporite mineral precipitation), heat flow from the mantle during the Hadean is modelled to have been lower than during the Archean (Ruiz 2017; Korenaga 2018). Radiogenic heating of the mantle continued up to about 3.0–2.5 Ga when it reached a maximum (1500–1600 °C compared with 1350 °C today) and then decreased thereafter (Herzberg et al. 2010). Although Foley et al. (2014) propose mantle temperatures more than 2000 °C for post magma ocean times.

The aforementioned examples underlines two important points regarding the origin of life on Earth. In the first place, the difficulties in interpreting early habitable environmental conditions are based on the relatively distorted prism of the relatively rare occurrences of metamorphosed and altered rocks from the Eoarchean. Secondly, the fact that experiments in prebiotic chemistry often do not take into account realistic early Earth scenarios. There is also the point that, while life on Earth may have emerged in one particular environment (or several), this does not mean that life on another terrestrial planet, such as Venus, could not have originated in an alternative scenario.

2.3.3 Evidence for Early Life

After life on Earth became established, the variety of environments on the early Earth seem to have provided a plethora of habitats, each characterised by different and, likely, time-variable characteristics. Given the anaerobic conditions prevailing on the early Earth and the negative effects of oxygen on modelled prebiotic chemistry, early terrestrial organisms had to have inhabited anaerobic environments. The earliest ecosystems would have supported chemoautotrophs, *i.e.* microorganisms obtaining their energy from oxidation of organic substances, such as methyl compounds (chemoorganotrophs), or inorganic substances, such as ferrous iron, hydrogen sulfide, elemental sulfur, thiosulfate, or ammonia (*i.e.* chemolithotrophs). These were the key substrates or electron donors that were immediately available to support life on the early Earth. Heterotrophs, or organisms that depend on carbon for their nutrient supply (by consuming them, their debris, or their waste products), may have emerged as early as the first autotrophs.

Organisms that developed the ability to use sunlight as a more powerful and effective source of energy, *i.e.* phototrophs, emerged after the chemotrophs. Initially (3.8 to 3.5 Ga), phototrophs may have used hydrogen and/or sulfur as a reductant. A decrease in the availability of these compounds in Earth's atmosphere, possibly related to the global production of methane by the early biosphere (methanogenesis), may have led to the development of ferrous iron-based phototrophy at or before 3.0 Ga (Olson 2006). The efficiency of even these early forms of photosynthesis, combined with the abundant availability of sunlight, appears to have conferred an immense metabolic advantage to the phototrophs; they were already relatively widespread by about 3.5 Ga (Noffke et al. 2013; Hickman-Lewis et al. 2018a), which indicates their relatively rapid evolution and spread (*i.e.* biomass) in environments with sufficient insolation. Liu et al. (2020) considers this an additional point in favour of an origin of life during the Hadean and not after the (now) controversial Late Heavy Bombardment (LHB) of 3.9 Ga. Given the implication for the timing of the origin of life it is important to understand why the LHB is currently out of favor. The hypothesised LHB was originally tied to dating of the returned lunar samples by Turner and Cadogan (1975) and Tera et al. (1973) that showed a preponderance of ca. 3.96 Ga ages that they hypothesised could have been produced by a "lunar cataclysm". Modelling has shown that such an event could have been created by instability in the orbits of the giant planets, particularly Jupiter (Walsh et al. 2011; Deienno et al. 2016). Recent analysis of lunar impact glass ages (Zellner 2017) demonstrates the unlikeliness of a late bombardment peak. Observations of a binary Jupiter Trojan (Nesvorný 2018), coupled with studies based on cratering statistics, geochronological databases tied to closure temperature, and resolved ages using orbital dynamics and thermal modeling (Mojzsis et al. 2019; Clement et al. 2019), show that any giant planet instability would have occurred before about 4.45 Ga.

We noted above purported microbial fossils associated with hydrothermal activity from the 3.8 Ga Nuvvuagittuq Supracrustal terrane (Dodd et al. 2017; Papineau et al. 2022) that have since been reinterpreted as of abiotic origin (McMahon 2019; Greer et al. 2020; Lan et al. 2022). Furthermore, while organic molecules in garnets from the 3.7 Ga Isua Greenstone Belt may be remnants of microbial life (Hassenkam et al. 2017), purported microbial stromatolites described from the same rocks (Nutman et al. 2016, 2019) are apparently of abiotic origin (Allwood et al. 2018; Zawaski et al. 2020). Nevertheless, by 3.5 Ga, the Barberton and Pilbara Greenstone Belts, the two main locations with well-preserved crustal rocks, document abundant evidence of microbial life. Most readily visible are small, domical stromatolites ∼several cm in height occurring in shallow water environments in the Pilbara (Hofmann et al. 1999; Allwood et al. 2006), that represent the macroscopic evidence of phototrophic microbial mat formation. However, most phototrophic biofilms and mats from the Paleoarchean in Barberton and the Pilbara are represented by tabular mats (Byerly et al. 1986; Westall et al. 2011a, 2006a; Noffke et al. 2013; Hickman-Lewis et al. 2018b). Evidently, these phototrophic biosignatures occur in very shallow water environments, the organisms relying on access to sunlight to obtain their energy.

The shallow water environment, together with warm seawater, would have led to relatively high salt concentrations. Westall et al. (2006a) describe microcrystalline, silica-pseudomorphed evaporate mineral sequences in 3.33 Ga coastal sediments in the Barberton Greenstone Belt, South Africa, while (Lowe and Byerly 1999a) document an horizon of nacholite crystals in 3.42 Ga sediments from Barberton (Knauth 2011). Thus, given the abundant evidence for microbial life in these shallow water environments, it must have been at least partially halophilic (Westall et al. 2015; Hickman-Lewis and Westall 2021). Moreover, the volcanic sedimentary environment with its associated hydrothermal activity, hosted chemotrophic life, including chemolithotrophs, as well as chemoorganotrophs, the latter in the direct vicinity of hydrothermal vents (Westall et al. 2006b, 2011b; Hickman-Lewis et al. 2020a). Colonies inhabiting hydrothermal environments would have comprised thermophiles and probably hyperthermophiles.

(Hickman-Lewis and Westall 2021) review the widespread distribution of early life in the Barberton Greenstone Belt through the Archean (3.5–2.6 Ga), showing how its nature and distribution throughout this early period of Earth's history were controlled by both the gradual evolution of the environment, as well as the rise of oxygenic phototrophs at about 3.0 Ga. Figure 6 illustrates a variety of early biogenic remains from the Paleoarchean Barberton Greenstone Belt (South Africa) and Strelley Pool Chert (Pilbara Greenstone Belt, Australia) sediments, including macroscopic stromatolites from the 3.43 Ga Strelley Pool (Fig. 6A, B) (e.g. Hofmann et al. 1999; Allwood et al. 2006), tufted tabular stromatolites

Fig. 6 Early terrestrial microorganisms. (**A**, **B**) 3.44 Ga old stromatolites from the Pilbara, Australia in plan view and cross section. (**C**) Tabular phototrophic mats from the 3.472 Ga old Middle Marker Formation, Barberton, South Africa. (**D**) Layers of carbonaceous clots representing chemotrophic colonies in the vicinity of hydrothermal vents

from the 3.72 Ga Middle Marker Horizon, Barberton (Hickman-Lewis et al. 2018a), and clotted probable chemolithic microbial colonies from the 3.33 Ga Josefsdal Chert, Barberton (Hickman-Lewis et al. 2020a).

3 Tectono-Magmatic Processes on Venus vs. Hadean-Archean Earth

3.1 Tectonics

High surface temperatures that prevail on present-day Venus may strongly affect the interior and the surface (Phillips et al. 2001; Noack et al. 2012; Gillmann and Tackley 2014). Extensive outgassing and a greenhouse effect such as the one observed on Venus directly affect surface mobilization (horizontal velocity and inclusion of the lithosphere in the convective cell). Several studies that coupled 1D, 2D and 3D interior dynamics models with atmospheric evolution models have investigated the effects of an evolving atmosphere formed through mantle degassing of H_2O and CO_2 (Noack et al. 2012; Gillmann and Tackley 2014). However, the feedback between the atmosphere and the mantle can be quite complex. Some models, in which digitized atmospheric temperature values from a non-grey (wavelength-dependent) radiative-convective atmospheric model by Bullock and Grinspoon (2001) were used, suggest that high surface temperatures lead to surface mobilization (Noack et al. 2012).

Higher surface temperatures translate into lower surface viscosity, reducing the viscosity contrast between the surface and the mantle, and allowing the surface layer to be mobilized by mantle convection.

On the other hand, models that consider plastic yielding of the lithosphere and couple the interior evolution to a grey atmosphere (thermal opacity uses a single value, independent of wavelength), find that a high surface temperature stops surface recycling and promotes a stagnant-lid regime (Gillmann and Tackley 2014). Instead, a lower surface temperature will lead to higher viscosities and higher convective stresses that, in turn, may promote plastic yielding and surface mobilization (Lenardic et al. 2008). Venus may have experienced lower surface temperatures during its early thermal history, due to the efficient removal of water by escape processes (Gillmann and Tackley 2014). During this time more moderate conditions may have existed at its surface and allowed the sequestration of atmospheric CO_2, preventing its accumulation in the atmosphere. Global circulation models even suggest that a temperate Venus could have been maintained under habitable surface conditions until as recently as 0.7 Ga (Way et al. 2016). The difference between the results from numerical models is mostly due to the various rheologies and mechanisms considered, and may indicate possible competition between multiple processes on real planets. As such, there may be a sweet spot when rheology remains stiff enough for the lid to break and convective stress to be transmitted to the lid, but soft enough to allow vigorous convection and prevent the lid from growing too static. Additionally, the specifics of the regime may depend on the history of the planet (for instance Weller et al. 2015; Weller and Kiefer 2020) and in particular the transition between the magma ocean solidification and the solid mantle convection (see Salvador et al. 2017, 2023). This topic is discussed further in (Gillmann et al. 2022; Rolf et al. 2022, this journal).

The exact style of resurfacing on Venus is still debated (Rolf et al. 2022, this journal). Different scenarios that have been discussed in various studies (Armann and Tackley 2012; Gillmann and Tackley 2014; Karlsson et al. 2020; Lourenço et al. 2020) could have operated at various times throughout Venus' history: (1) stagnant lid: the surface was continuously renewed by volcanic activity without any kind of surface mobilization (Armann and Tackley 2012; Gillmann and Tackley 2014; Karlsson et al. 2020); (2) episodic lid: at periods plate tectonic like surface mobilization takes place with more quiescent periods in between (Armann and Tackley 2012; Gillmann and Tackley 2014; Uppalapati et al. 2020); (3) plutonic squishy lid (Lourenço et al. 2020): recycling of the lithosphere by eclogitic dripping and delamination, with strong plates separated by hot magmatic intrusions. These scenarios are different in terms of the efficiency of volatile recycling, which, in turn, has important implications for mantle dynamics and thermochemical history. In addition to this, the exact convection regime is probably not static and could have changed throughout the history of Venus (Gillmann and Tackley 2014; Weller et al. 2015; Weller and Kiefer 2020).

Understanding the tectonic regime throughout Venus' history is important as it is closely linked to its volatile history. First, outgassing, and thus the atmosphere thickness and bulk composition, are directly governed by the mantle dynamics and volatile release by volcanism. More detailed overviews of the mantle based outgassing processes are proposed in Rolf et al. (2022) and Gillmann et al. (2022), respectively. It has therefore been postulated that one could infer the tectonic style of a planet based on its volatile history and atmosphere characteristics. In particular, Venus' [40]Ar measurements have been used to suggest that the planet only outgassed 10 to 34% of its total [40]Ar inventory (Kaula 1999; O'Rourke and Korenaga 2015; Namiki and Solomon 1998; Volkov and Frenkel 1993), compared to Earth's 50%. That would imply Venus outgassing was limited during most of its evolution or was only important during its early history (when [40]Ar had not formed yet). Therefore such measurements argue against an Earth-like tectonic regime for all but primitive Venus at least. One should note that the thick CO_2 atmosphere and large N_2 inventory could point toward strong outgassing at some point in the history of Venus; this question is not yet solved. It has been suggested that such a period may have been very ancient, possibly dating back to the magma ocean phase (Gaillard et al. 2022) or the Late Accretion (Gillmann et al. 2020). CO_2 could also have been released by different processes depending on surface conditions (Höning et al. 2021).

Planetary tectonics also ties into the possibility of a carbonate-silicate cycle and thereby on the potential of long-term habitability. On Earth, the climate is stabilized as CO_2, outgassed at mid-ocean ridges and other volcanic units, is consumed by silicate weathering processes, precipitated as carbonates on the seafloor, and recycled back into the mantle at subduction zones (Walker et al. 1981; Kasting and Catling 2003). The land fraction is an important parameter in this cycle, since silicate weathering is particularly efficient on land that is emerged over sea-level. Higher rates of seafloor weathering at mid-ocean ridges can partly compensate for a smaller land fraction (e.g., Foley 2015), but the global mean surface temperature would nevertheless be expected to be higher if most of the planet's surface is covered by oceans.

In the stagnant lid scenario, recycling of volatiles is the least efficient because the thick static lid is not part of the convection. Water, carbonates (if formed during moderate atmospheric conditions), and sulfates may have never been recycled into the mantle. The observed tessera terrains, that represent around 8% of the crust on Venus, have been suggested to resemble continental crust on the Earth (Gilmore et al. 2015). If so, they may be difficult to form in this scenario, as their formation would require some kind of crustal recycling in the presence of water. However, models (Karlsson et al. 2020) have not yet been able to simulate their behaviour satisfactorily. While delamination of the lower crust may take place in this scenario, if the crust grows thicker than the basalt to eclogite transition depth (Sizova et al. 2015; Fischer and Gerya 2016a, e.g.,), water recycling remains unlikely. On Earth, felsic material can form without water. However, such a mechanism would struggle to produce enough felsic material to account for the total volume of present-day Venus tesserae (Smrekar et al. 2018, and references therein). Future missions may place a constraint on how much of the tessera terrains can actually be considered felsic.

Volatile recycling would be efficient during plate tectonic periods in the episodic lid scenario, while in the plutonic squishy lid case, the efficiency would presumably be lower than in the episodic case but still notably higher (Sizova et al. 2015; Fischer and Gerya 2016a, e.g.,) than in the stagnant lid scenario. Volatiles that are introduced back into the mantle would have major consequences for mantle dynamics and subsequent magmatic evolution. Recycled crust will become negatively buoyant when undergoing the phase transition from basalt to eclogite. The recycled crust is rich in incompatible elements, such as heat producing elements and volatiles, which can significantly affect subsequent melting of the mantle. The subducted crustal material will refertilize the mantle and promote partial melting, both by increasing the amount of heat producing elements in the mantle and recycling of volatiles that would locally decrease the melting temperature. In addition to decreasing the local solidus, recycled volatiles will also decrease the viscosity of the mantle material, thus affecting the interior dynamics and the cooling behavior of the mantle, and consequently the subsequent outgassing.

Constraints on the style of recycling on Venus may be derived from the inferred crustal thickness, and variations in the crustal age and geoid (Kiefer and Hager 1991; Armann and Tackley 2012; King 2018). The episodic lid models and the plutonic squishy lid, with a low reference mantle viscosity and a low eruption rate, seem to produce a crustal thickness that is closer to the inferred crustal thickness of Venus compared to stagnant lid cases (Rolf et al. 2018; Lourenço et al. 2020). Surface age variations indicated by Venus' cratering record may be easier to reconcile with the episodic lid scenario (Uppalapati et al. 2020) and the long wavelength of the gravity spectrum can be matched well if the last resurfacing event ended a few hundred Myr ago (Rolf et al. 2018). Whether these observations are consistent with the plutonic squishy lid regime remains to be tested in future models. It should be noted again, however, that the stagnant lid, episodic lid, and plutonic squishy lid scenarios are not mutually exclusive, but may have been active at different times during Venus' history, as seems to have been the case on early Earth. Additionally, they constitute a continuum of behaviours rather than distinct, clear-cut end-members. Finally, local variations are to be expected, and different crust deformation processes could occur at different locations of the surface of Venus at a given time (see Rolf et al. 2022). Thus, recycling of volatiles may have significantly changed during the thermal evolution of Venus, and its present-day state is the result of complex feedback mechanisms between the interior, surface and atmosphere (Gillmann et al. 2022). Future work assessing volatile exchange associated with changes in the tectonic regime throughout Venus' history is necessary in order to advance our knowledge of stable surface water in the past.

3.2 Outgassing

Outgassing is an important source of secondary volatiles for the atmosphere of a terrestrial-type planet. It therefore directly affects surface conditions and the surface habitability of a planet. Broadly speaking, three processes can lead to significant outgassing and affects on the atmosphere (including the fluid envelope) in the long term: (i) magma ocean solidification, (ii) collision with impactors, especially large ones at an early stage, and (iii) volcanism.

Magma ocean evolution, solidification, outgassing and its consequences on the atmosphere and surface conditions on rocky planets, in particular on Venus, are discussed in detail in Salvador et al. (2023, this journal). After accretion and the capture of a possible primordial hydrogen atmosphere, it is the source of the early volatiles and a secondary atmosphere. It is generally understood that CO_2 outgasses early and in large quantities, while water should be released near the end of the magma ocean phase, if at all (Salvador et al. 2017). It has been proposed that a freezing magma ocean could retain a large portion of its water (Solomatova and Caracas 2021). Outgassing from magma oceans is an active research topic, and it has been highlighted that the actual species outgassed to form this early secondary atmosphere could heavily depend on the magma ocean's redox state (e.g. Lichtenberg et al. 2021; Gaillard et al. 2022) It has been suggested that this phase could already set the planet on an habitable or uninhabitable evolutionary path, depending on the duration of the magma ocean, and ability of the planet to cool down fast enough to allow liquid water to condense on its surface (Gillmann et al. 2009; Lebrun et al. 2013; Hamano et al. 2013; Salvador et al. 2017; Turbet et al. 2021) before it is lost to space.

Large impacts and their consequences are discussed in Gillmann et al. (2022), Salvador et al. (2023, this journal). The velocity and mass of impactors means that large amounts of kinetic energy are transferred to the planet as thermal energy during a collisional event. For bodies that are large enough (or fast enough), that is, above a few tens of kilometers in radius, impacts can cause large-scale melting of the crust and the mantle of the planet. They may create magma ponds/seas, and lead to the release of volatiles into the atmosphere (e.g. for Venus, Gillmann et al. 2016). Such events are more frequent and important during early evolution, especially during the accretion phase and are accompanied by the additional release of volatiles contained in the impactors (Sakuraba et al. 2019; Gillmann et al. 2020;

Sakuraba et al. 2021). Volatile release by impactors can substantially modify the composition of the atmosphere and the state of the surface. If the impactor is large enough there could be implications for habitability ranging from the short term (earthquakes, tsunamis, storms, lava ponds/oceans) to the long term (increased greenhouse effect, increase in CO_2 concentrations).

Volcanic outgassing is discussed in Gillmann et al. (2022, this journal). Its causes are explained in Rolf et al. (2022, this journal) and include partial melting of the mantle due to local pressure-temperature conditions exceeding the local solidus of the mantle material. Its surface expressions are addressed in Smerkar et al. (2022, this journal) and Herrick et al. (2022, this journal). For Venus, recent volcanic production can be roughly estimated, but with large uncertainties. Observation of the surface and atmosphere can provide some constraints for modelling efforts, but recent volcanic production rates are debated. They could be very low, $\ll 1$ km^3, in agreement with the minimal effect of volcanism on randomly distributed impact craters (Basilevsky and Head 1997; Schaber et al. 1992); moderately lower than on Earth (~ 1 km^3/yr) (Head et al. 1991; Phillips et al. 1992); or similar (~ 10 km^3/yr; Fegley and Prinn 1989; Bullock and Grinspoon 2001) to Earth's production rates (e.g. Byrne and Krishnamoorthy 2021). Long-term numerical modelling of mantle dynamics offers a reasonable solution for estimating a range of possible outgassing rates from volcanic production, keeping in mind the lack of hard constraints on values before 1 Ga, or on the state of Venus' mantle. However, little evidence exists for volatile concentration in Venus' lava, leading to further uncertainties, as volatile output strongly depends on the redox state (oxygen fugacity) of the mantle. In addition, it has been suggested that the high surface pressure in the Venusian atmosphere could suppress water outgassing compared to CO_2 (Gaillard and Scaillet 2014). It is currently debated if the present-day atmosphere is geologically (<1Ga) recent (as suggested by Way and Del Genio 2020) or a fossil atmosphere (Head et al. 2021).

Extrapolating Venus' outgassing rates back in time is subject to even greater uncertainties since the tectonic regime may have changed (discussed in Rolf et al. 2022, this journal). Magmatic outgassing is a consequence of partial melting of hot, uprising mantle material. The shallower the depth to which the mantle material rises, the smaller the lithostatic pressure and therefore the lower the solidus temperature. A thick, insulating lid on top of the mantle would usually create a barrier to hot, uprising plumes and therefore a relatively small melting region (O'Neill et al. 2014). In contrast, on planets with plate tectonics, the melting region beneath mid-ocean ridges is extended close to the surface. On Earth, this mechanism causes a basaltic crust production rate of at least ~ 19 km^3/yr (Cogné and Humler 2006). An additional effect associated with plate tectonics is the subduction of water. In subduction zones at depths of 100–200 km, subducted hydrous minerals become unstable and release their water (Rüpke et al. 2004; Stern 2011). Since water reduces the solidus temperature of the surrounding rock, partial melt is produced that rises to the surface, which is also accompanied by outgassing. Altogether, the tectonic regime has a tremendous effect on the outgassing rate. If early Venus possessed plate tectonics, its outgassing rate would likely have been much higher than it is today. The nature of outgassed species is also important for surface conditions. It has been highlighted that the mantle redox state (the mantle oxygen fugacity) could greatly affect the speciation in the atmosphere, with oxidised mantles (such as the Earth's) leading to the outgassing of CO_2 and water. On the other hand, a reduced mantle could rather favor CO or H_2 (e.g. Kasting et al. 1993a; Gaillard et al. 2021; Frost and McCammon 2008; Hirschmann 2012).

On Venus, not only is CO_2 the major current component of the atmosphere, but it is also responsible in large part for the high surface temperatures. Large rates of CO_2 outgassing can inhibit global glaciation due to this species' role as a greenhouse gas. This is particularly important if a carbonate-silicate cycle is active on the planet, where silicate weathering serves as a sink to atmospheric CO_2 (e.g., Kadoya and Tajika 2014). On the other hand, high rates of CO_2 degassing can enhance the greenhouse effect and ultimately lead to the evaporation of water. Whereas an active carbonate silicate cycle would balance high rates of outgassing to some extent, recycling of CO_2 into the mantle on planets without plate tectonics is rare (e.g., Foley and Smye 2018; Höning et al. 2019b). The solar flux that Venus receives today, and has received in its history, is substantially higher than that the present-day Earth receives, and the threshold towards surface water evaporation is the most relevant bottleneck to Venus' habitability (see also Gillmann et al. 2022, this journal). Small outgassing rates during Venus' early evolution, in combination with an active carbonate-silicate cycle, would increase the likelihood of an early, habitable Venus, while significant early outgassing would go against habitable conditions.

On Earth, water is a major component of volcanic outgassing (on the order of 1%). It is possible that Venus is much drier, due to loss during the magma ocean phase and the inability to condense water early on (Gillmann et al. 2009; Hamano et al. 2013), or that high surface pressure stifles water outgassing (Gaillard and Scaillet 2014), but the planet's current state and modelling seem consistent with marginal water outgassing during recent history at least (Gillmann and Tackley 2014). Beyond its availability for a possible liquid layer, water also affects surface conditions by being a strong greenhouse gas and, despite its low abundance in Venus' current atmosphere, is the second highest contributor to high surface temperatures on the planet.

Initial analysis by the Pioneer Venus Large Probe Neutral Mass Spectrometer (PV-LNMS) indicated the possible presence of CH_4 (Donahue and Hodges 1992). It was speculated that CH_4 was likely not well mixed in the atmosphere, given the measured variations in abundance. Later analysis of the same data by the same team indicated that the detection was unlikely (Donahue and Hodges 1993) and was due to contamination from terrestrial CH_4 brought along in the instrument and hence "was generated by a reaction between an unidentified highly deuterated atmospheric constituent and a poorly deuterated instrumental contaminant." However, the instrumental contaminant has never been identified and hence the detection of CH_4 in the Venusian atmosphere remains an open question.

3.3 D/H Ratio

Venus' atmosphere today contains only about 30 ppm H_2O (Fegley 2014, and refs. therein). The first in-situ D/H measurement by the PV-LNMS demonstrated a ratio ~ 150 times that of Earth (Donahue et al. 1982). Upper atmosphere measurements of D/H by Venus Express (Fedorova et al. 2008) documented much higher values, which are inconsistent with those of Donahue et al. (1982). Fedorova et al. (2008) attributed these differences to "a lower photodissociation of HDO and/or a lower escape rate of D atoms versus H atoms." Ground based measurements do not always concord with both space and in-situ measurement, although in some cases they are consistent (Matsui et al. 2012). Various measurements have been suggested to imply a vertical variation of HDO/H_2O (see Marcq et al. 2018, and references therein) at odds with current chemical models. Despite these discrepancies, the general view is that hydrogen in Venus' atmosphere has a D/H much higher than any other reservoirs of hydrogen in the solar system, which implies that Venus lost hydrogen to space. This may indicate that after an early wet, possibly habitable, time (Donahue et al. 1982; Way and Del Genio 2020), water dissociated and the hydrogen was removed from the atmosphere to space. Dissociation of water molecules and the escape of hydrogen probably had a strong influence on the entire geodynamical history of the planet (Baines et al. 2013)

However, while recent studies propose scenarios for the evolution of the D/H of Earth's water (Pahlevan et al. 2019; Kurokawa et al. 2018), estimating the amount of water lost from

Venus over the last 4.5 Gyr remains extremely challenging for several reasons. Firstly, the history of hydrogen escape cannot be easily re-constructed since it depends on numerous factors (solar irradiation history, atmospheric composition and vertical structure, regime of escape etc.). Secondly, the starting D/H ratio for hydrogen in the Venus atmosphere remains unknown. An important contribution from solar gases would imply a low starting D/H ratio, while contributions from comets could have increased this ratio up to 4–5 times that of Earth (Altwegg et al. 2015). New investigations of the elemental and isotopic composition of noble gases in the Venus atmosphere, especially of xenon, would help shed further light on the history of hydrogen (and water) escape from Venus (see Avice and Marty 2020 and Avice et al. 2022, this journal).

4 The Origin and Persistence of Habitability on Venus

4.1 Climate History

The initial stages of Venus' habitable state are more shrouded in mystery than those of Earth's. We begin our analysis of Venusian habitability with the longevity of the post-accretion magma ocean. Early work by Hamano et al. (2013) and Lebrun et al. (2013) showed that, if the time needed for the crystallization of Venus' magma ocean is ~ 100 million years or longer, there is the risk of dissociation and loss of its primordial H_2O steam atmosphere: the hydrogen, as well as some of the oxygen, escapes during this time while any leftover oxygen is absorbed into the magma ocean. After the cooling and crystallization of the magma ocean, the planet (denoted by Hamano as a Type II world) may inherit a thick CO_2 dominated atmosphere not that different from what we observe today on Venus. In this scenario, the D/H ratio measured by Donahue et al. (1982) is a possible remnant of the primordial CO_2+Steam(H_2O) dominated atmosphere. In an alternative scenario for Venus (which Hamano et al. termed a Type I world), the magma ocean crystallization takes place over ~ 1 million years (similar to that of Earth: Katyal et al. 2020; Nikolaou et al. 2019). This scenario avoids the loss of the steam atmosphere, which may then condense out onto the surface, possibly allowing for a period of habitability of undetermined length.

More recent work by Turbet et al. (2021) has expanded the 1-D models of Hamano et al. (2013), Lebrun et al. (2013) to a 3-D GCM where cloud effects can be modeled and their importance quantified. The Turbet et al. (2021) models of steam+CO_2 and steam+N_2 atmospheres demonstrate that there are little to no day side clouds at the substellar point to shield the planet from high solar insolation (as will be seen in the cold-start cases below). Their model also demonstrates the presence of high clouds at the polar and night side, which are effective at trapping outgoing infrared radiation preventing the cooling of the planet. The Turbet study supports the Type-II outcome modeled in Hamano et al. (2013), where the magma ocean steam atmosphere is never able to condense out on the surface, and once again the H_2 escapes and most of the oxygen is absorbed by the magma ocean. One major shortcoming in all of the models above is the inability to provide better constraints on exactly what the constituents were of the outgassed magma ocean atmosphere, which presently is an active area of research (e.g. Bower et al. 2022; Gaillard et al. 2022). Alongside these unknowns can be added the inability to constrain the albedo (Salvador et al. 2017, 2023), which would again influence whether Venus becomes a Type I or Type II world. The lack of definitive evidence for one scenario or the other implies that the question of the existence of Venus' past habitability remains open. More details on magma ocean and atmospheric evolution is provided in Salvador et al. (2023, this journal). One method of testing which hypothesis for Venus' evolution is correct is by examining data from the upcoming DAVINCI mission (Garvin et al. 2022), which should provide better constraints on when Venus lost its water and the timescale over which it happened by examining a number of noble gas isotopes (see Avice et al. 2022, this journal). Another more indirect method would be possible by looking at exoplanet demographics – if we observe planets in the Venus Zone (Kane et al. 2014) that have temperate conditions then at least we know it is possible. See Way et al. (2023, this journal) for how exoplanet research may inform Venus' history.

To date, very little work has been done to examine how Venus (or the Earth for that matter) moves from a post-magma ocean state to a period of habitability with moderate surface temperatures and oceans, despite this transition being a cornerstone of the onset of habitability. As mentioned above, the first step would be to provide better constraints on exactly what the constituents were of the magma ocean atmosphere (e.g. Bower et al. 2022; Gaillard et al. 2022). One also needs to account for large impacts occurring in the first few hundred million years of the planet's evolution (see Salvador et al. 2023; Gillmann et al. 2022, this journal), which could have major consequences on the atmospheric mass and composition, as large amounts of water, CO_2, N_2, and other species could be delivered or removed (e.g. Schlichting and Mukhopadhyay 2018; Gillmann et al. 2020). Thus, it is unlikely that surface conditions would remain consistent throughout that early time. Some works have attempted to examine the first 100s of million years (e.g. Harrison 2020), as described above, but many unknowns still remain.

There is another problem for those interested in Venus' habitability: How could Venus ever have been habitable like early (or modern) Earth when Venus at 4.5 Ga received ~ 1.5 times the incident solar flux that Earth receives today? Most studies resulting in temperate conditions have been made assuming a cold start in the post-magma ocean phase, which hypothesizes that the early magma ocean cooled quickly (a few million years) and that water was able to condense out on the surface (Hamano Type I discussed above). The first to successfully model such temperate conditions in the Pre-Fortunian (Hiesinger and Tanaka 2020) was Pollack (1971), who used a 1-D radiative convective non-grey model. He presented two options at 4.5 Ga when the solar luminosity was $\sim 30\%$ lower than today. The first was an early Venus with a 50% cloud cover – the motivation for 50% was that he believed modern Earth has roughly this amount (modern measurements indicate $70 \pm 10\%$; Holdaway and Yang 2016) – and in that scenario early Venus had temperatures (depending upon the atmosphere assumed) ranging from ~ 320–500 K. The second choice was a 100% cloud cover model, yet the motivation for such a model was not disclosed. In this scenario Pollack discovered that the planet could host moderate surface temperatures below 300 K. Pollack also demonstrated that, even at today's insolation (~ 1.9 times Earth's), the surface temperature could have remained below 300 K. This 100% cloud cover assumption was the basis for all subsequent Venus habitability studies (e.g. Grinspoon and Bullock 2007) yet no mechanism for producing 100% cloud cover mechanism was ever provided. It would not be until the exoplanet work of Yang et al. (2014) that such a mechanism was discovered. Yang et al. (2014) used the NCAR CAM General Circulation Model (GCM)[1] and discovered that, for slowly rotating planetary atmospheres, an expanded Hadley cell would provide the 100% cloud cover at the subsolar point. Modern Earth actually contains three Hadley cells in the north and three in the south. The reason Earth does not have a single Hadley cell in each hemisphere is because its 'fast' rotation generates a strong Coriolis force deflecting the north-south overturning cells. In a slowly rotating world, the Coriolis force is very weak and hence a single north and south Hadley cell is present. Subsequent Venus-focused work

---

[1] https://www.cesm.ucar.edu/models/atm-cam/.

by Way et al. (2016), using the ROCKE-3D (Way et al. 2017) GCM with a fully coupled dynamic ocean, confirmed Yang's work which utilzied a simplified single mixed-layer/slab ocean without any horizontal heat transport. Later ROCKE-3D GCM work by Way et al. (2018) utilized both fully coupled dynamic oceans and mixed-layer/slab oceans to confirm Yang's general conclusions over a large range of rotation rates and insolation. The fact that two independent GCMs observe the same behavior is encouraging, but cannot be considered conclusive until these effects are observed in exoplanetary systems in the future. However, these models require at least some surface water. Some tens of cm in soil would be sufficient according to more recent work in Way and Del Genio 2020. Whether or not water was able to condense on the surface at all after the magma ocean phase depends on the atmosphere at this time whose composition is a matter of on-going debate as mentioned above (e.g. Bower et al. 2022). It should be noted that we have no constraints on what early Venus' rotation rate was, but an early slow rotation rate can be achieved via a number of mechanisms including solid body tidal dissipation (e.g. MacDonald 1964; Goldreich and Peale 1966; Way and Del Genio 2020), Core-Mantle friction (Goldreich and Peale 1970; Correia and Laskar 2001; Correia et al. 2003; Correia and Laskar 2003), and oceanic tidal dissipation (Green et al. 2019). The possible role of impactors in Venus' rotational evolution goes back at least to the work of McCord (1968), although no detailed hydrodynamical simulations have ever been performed to examine the impactor parameters and lack of an observable moon as we have for Earth. Our moon is likely the remnant of an impactor (e.g. Benz et al. 1986; Canup 2004; Lock and Stewart 2017), see section on Archean Earth above. Moreover, given the youthful age (200–750 Myr) of the surface of Venus (e.g. McKinnon et al. 1997; Bottke et al. 2016), there is little chance of observing cratered remains of any such ancient impactors. If such an impactor did collide with the planet in Venus' past, it may be possible to detect it isotopically if it was sufficiently different from the bulk composition of Venus, but measuring this would be challenging. To paraphrase Way and Del Genio (2020) "it is clear that Brasser et al. (2016) and Mojzsis et al. (2019) prefer the hypothesis that the Earth's late veneer was mainly delivered by a single Charon- or Ceres-sized impactor. For that reason, if a larger object was involved in the late evolution of Venus' spin or obliquity, it may be possible to detect its geochemical fingerprints in a future in situ mission." Thus, all discussions of the habitability of Venus over any time scale discussed herein assumes that the planet was rotating slowly enough to generate $\sim 100\%$ cloud cover at the subsolar point and extending across most of the sunlit hemisphere of the planet (see Fig. 7).

It should be noted that the persistence of Venus' habitability was originally predicated upon the notion that the faint young Sun's increase in brightness through time (e.g. Gough 1981; Claire et al. 2012) would take some hundreds of million years to subsequently increase surface temperatures, driving the planet into a runaway greenhouse. Regardless, some form of volatile cycling would be required to keep the planet's climate stable over geological time (Höning et al. 2021; Krissansen-Totton et al. 2021; Gillmann et al. 2022; Way et al. 2023, this journal), as for Earth, normally through some form of weathering (e.g. Walker et al. 1981; Kasting and Catling 2003; Krissansen-Totton et al. 2018; Höning 2020; Graham and Pierrehumbert 2020). At the same time, the work of Yang et al. (2014), Way and Del Genio (2020) has definitively shown that, if the cloud albedo feedback for slowly rotating worlds is correct, then increases in solar insolation through time cannot be the deciding factor in the evolution of Venus from a temperate planet to a hothouse as long as volatile cycling takes place.

From this broad overview of the possible conditions at the onset, persistence, and loss of habitability, it appears that the question of transitioning from one state or era to another is a major challenge that will need to be addressed by future models. The period toward the

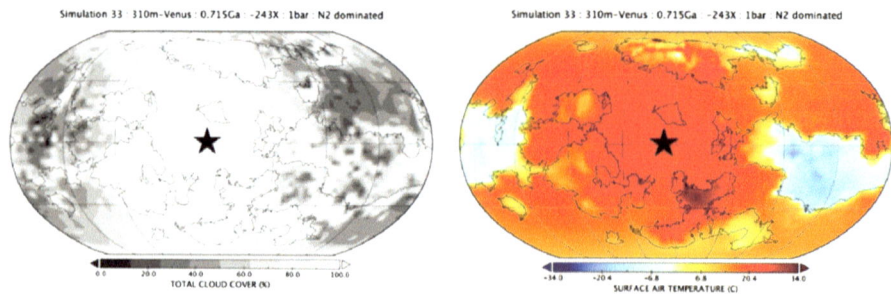

Fig. 7 Generated from General Circulation Model simulation 33 in Way and Del Genio (2020). This is a 1 bar N_2 Dominated atmosphere including 400 ppmv CO_2 and 1 ppmv CH_4. The sidereal rotation rate is the same as modern Venus ($-243 \times$ Earth). Insolation is set to the value that Venus received 715 Ma (1.7 $\times$ Modern Earth or 2358.9 W/m^2). This is a snapshot of 1/12 of a Venusian solar year. Left: percentage total cloud cover. The black star represents the location of the subsolar point. Right: surface temperature map. Note that the highest temperature regions (dark red) are located near the subsolar point and on the southern landmass, while the coldest regions are on the anti-solar continental landmasses (blue). The fully coupled dynamic ocean, which includes horizontal heat transport, keeps the oceans warm (red/orange colors) even in the anti-solar regions of the simulation

end of the magma ocean phase has been highlighted as an important criterion for subsequent evolution and needs to be studied more intensely before any definitive conclusions can be drawn. In the same way, much more work needs to be done to explore how a planet may go from a temperate to a moist and then a runaway greenhouse state (e.g. Kasting 1988), and 3D GCMs will be needed (e.g. Boukrouche et al. 2021).

4.2 Linking Possible Past Habitable States to Present-Day Observations

Present-day Venus looks nothing like what habitable models suggest Venus could have been like in the past. Therefore, we should first attempt to understand how the planet could have radically changed from an hypothesized temperate climate with a relatively thin atmosphere to the dense hothouse we observe today. Then, we take a look at what signs of a previous habitable time interval or of the stages required to bring Venus to its present state could be observable today.

The evolution of Venus from an hypothetical habitable time interval to the present-day must bring its atmosphere to the current inventory of major species ($11 \cdot 10^{18}$ kg of N_2, 10^{16} kg of H_2O and $4.69 \cdot 10^{20}$ kg of CO_2). It must also remove any molecular oxygen. Ideally, it would also bring the D/H ratio to its present value (Donahue et al. 1982), but due to uncertainties in the loss mechanisms, and varying isotopic ratios for the volcanic and meteoritic sources, this is challenging (Grinspoon 1987, 1992; Gurwell 1995). Likewise, the stable isotope ratios of noble gases have been suggested to derive from early hydrodynamic escape but cannot be modelled by a unique self-consistent scenario due to the lack of constraints on early atmospheric conditions, structure and composition, as well as solar energy input (see Avice et al. 2022, this journal).

Nitrogen evolution has long been assumed to be relatively straightforward once the magma ocean crystallized, since, in the absence of fixation by living organisms, it was not expected to be part of complex cycles (e.g. Stüeken et al. 2016). However, the understanding of the nitrogen cycle, even on Earth, is much less advanced than that of CO_2 (Stüeken et al. 2020). On Earth (Marty and Dauphas 2003), it is thought to be approximately in balance between sources (half volcanic outgassing and half oxidative weathering) and the sink (burial with a touch of subducted flux). In the past, though, nitrogen fluxes are likely to have significantly changed (Goldblatt 2018). This possibly affected surface pressure, despite variations that are much lower than those expected on Venus for CO_2, for instance. Some reasons behind these variations include possible volcanic production changes with time (from mantle conditions and composition evolution), and changes in the composition of the atmosphere (e.g. presence/absence of oxygen) (e.g. Som et al. 2016; Catling and Zahnle 2020), leading to changes in the chemical reactions between the atmosphere and the surface (i.e. weathering). For example, some results suggest maximum atmosphere pressures of about 0.5 bar on Earth (probably much less), 2.7 Gyr ago, using barometric calculations from fossilized raindrops and gas bubbles in basaltic lava (Som et al. 2012, 2016). However, the past N_2 abundance in the atmosphere is still poorly constrained and generally thought to have possibly varied by a factor 2–3 relative to present-day (Goldblatt et al. 2009; Johnson and Goldblatt 2015; Goldblatt 2018). In such a scenario, considerable build-up of nitrogen in the atmosphere of Earth over its history may be expected. What this could mean for Venus is still uncertain, given the differences between the two planets, the lack of data relative to Venus' past and the dependence of the nitrogen fluxes on surface conditions. Comparing the nitrogen abundances on the two planets, one should also consider that some nitrogen is stored in Earth's continental crust. Still, a better understanding of the nitrogen exchanges applied to Venus will provide valuable insight on the planet's evolution.

The greater abundance of nitrogen in Venus' atmosphere compared to Earth's ($4 \cdot 10^{18}$ kg) could imply it has escaped even less than on Earth and was thus protected from losses (Lammer et al. 2018), despite the fact that the ^{40}Ar value suggests that Venus' mantle is less degassed than Earth's. Some early temperate Venus models use an atmospheric nitrogen content similar to Earth's (e.g. Way et al. 2016). While it is likely that, by the end of the magma ocean phase, the primordial nitrogen-based species would have been trapped by the hot surface and removed from the atmosphere, collisions with large impactors would have delivered additional nitrogen over the first few hundred million years. Gillmann et al. (2020) proposed that about $5 \cdot 10^{18}$ kg N_2 could have been brought to the atmosphere this way, despite this number being highly dependent on impactor vaporization and composition. The rest (about half) of the present-day inventory of N_2 could realistically be released into the atmosphere over the following 4 Gyr by volcanic activity, but actual fluxes depend on volcanic production rates, mantle composition and surface conditions (Gillmann et al. 2020; Gaillard and Scaillet 2014), as well as burial fluxes, which are all poorly constrained. Confirmation of the volatile composition of the lava and the volcanic plumes could help refine these estimates and better assess the feasibility of long term outgassing of the current nitrogen content of Venus' atmosphere.

CO_2 evolution is a more complex issue, since it can interact more easily with the surface. The main question is how it was possible to evolve from very low CO_2 abundances, in a temperate atmosphere, to a full-fledged atmosphere with 88 bar CO_2. Volcanic outgassing has been proposed to be responsible for Venus' atmospheric CO_2 inventory (see Gillmann et al. 2022, this journal), despite the possible high surface pressure (Gaillard and Scaillet 2014). This implies that, if CO_2 is available for outgassing (i.e. present in the mantle and transferred into the melt; see Gillmann et al. 2022 this journal), it will be released into the atmosphere. However, it has been shown that, with Earth-like outgassing (Earth-like composition of the gases released into the atmosphere during a volcanic eruption, indicating probable Earth-like oxidation of the Venusian mantle), at least the equivalent of ~ 10 global resurfacing events is needed to build up Venus' CO_2 atmosphere (Lopez et al. 1998). More recent work (Head et al. 2021), estimates that the number of equivalent global resurfacing events needed to obtain the amount of CO_2 in the present atmosphere of Venus is about 100. This would indicate that most of the present-day atmosphere would have originated from the period before the present geological record, which is in line with the interpretation of ^{40}Ar in the atmosphere of Venus that suggests that the bulk of the outgassing occurred rather early during its evolution. Numerical modeling of the mantle of Venus (e.g. Armann and Tackley 2012; Gillmann and Tackley 2014) also implies that global volcanic events are unlikely to occur with such a high frequency due to the massive internal heat dissipation they cause. The mantle requires time for heat to accumulate again before a new event is triggered. Therefore, volcanism is unlikely to have been the cause for the full atmosphere build up, or even for more than a fraction of the build-up. It does not preclude volcanism from triggering a transition, though (e.g. Way et al. 2022). Instead, whatever outgassing was due to volcanism took place on the long term, but without allowing us to be more specific about a precise age for a possible transition.

Weller and Kiefer (2021) present an alternative picture of how Venus' atmosphere could go from a very low CO_2 abundance to 20–60 bar. Weller and Kiefer (2021) demonstrate that a significant fraction of the present-day atmosphere can be produced with a single overturn (early hot planet, about 5 bar CO_2 per overturn), no overturns, or multiple overturns (later cold planet). Interestingly, these overturns do not need to be global and can occur on geologically short timescales. Weller et al. (2022) also suggests that a significant portion of Venus' atmosphere present-day N_2 and CO_2 inventory could be best produced under an early plate tectonics regime and may be reached without initial magma ocean contribution.

An alternative solution that has been suggested is that a global volcanic event, possibly akin to Earth-like Large Igneous Provinces (LIPs), could have both outgassed CO_2 from mantle reservoirs and destabilized carbon crustal reservoirs (such as carbonates and other carbon rich sediments, see Retallack et al. 2006; Svensen et al. 2009; Ganino and Arndt 2009; Nabelek et al. 2014), leading to the accumulation of CO_2 in the atmosphere on a short timescale (Way and Del Genio 2020; Krissansen-Totton et al. 2021; Höning et al. 2021; Way et al. 2022) at an undefined date. Such a mechanism and its feasibility are rather difficult to assess in the absence of observation. While Earth has experienced several LIPs during its evolution, no trace of CO_2 partial pressure increase on the order of tens of bars has been recorded (Schaller et al. 2011, for an example of estimate in the order of tens of thousands ppm CO_2). However, such events have been associated with dramatic climate change and global extinction events (e.g. Wignall 2001), making them important enough to affect life on a global scale and planetary habitability. They could possibly trigger a climate transition by overwhelming any volatile cycling in effect hence driving the planet into a moist and then runaway greenhouse (Way and Del Genio 2020; Way et al. 2022).

The remaining issue with this habitable scenario can simply be stated as "where is the oxygen?" If there were once substantial surface reservoirs of water on the surface of Venus and they were driven into the atmosphere as the planet warmed up, then why does the atmosphere not contain many bars of oxygen? In essence, as a planet enters the moist greenhouse state, water is transported into the stratosphere. Over time this water is photodissociated, the hydrogen can escape via diffusion and the oxygen should be left over (Kasting 1988). Studies have shown that it is difficult for oxygen to escape in substantial quantities in the present day Venusian atmosphere (e.g. Persson et al. 2020) where the $H^+{:}O^+$ ratio is $\sim 2{:}1$ over the solar cycle (e.g. Barabash et al. 2007; Persson et al. 2018). Assuming Venus' atmosphere has not changed over geological time, Persson et al. (2020) demonstrated that it would have lost between 0.02–0.6 meters of a global equivalent layer of water over the past 3.9 billion years via atmospheric escape. Additionally, non-thermal escape is a slow, ongoing process that declines with time, as the solar extreme UV input decreases. This implies that it takes a long time to remove any significant amount of oxygen from Venus' atmosphere. In turn, if atmospheric escape alone is considered, progressive loss after an early habitable time billions of years ago would be favored. In fact, it has been speculated that exoplanetary worlds

162

with multiple bars of oxygen may indicate a former temperate period with oceans (Luger and Barnes 2015; Wordsworth and Pierrehumbert 2013). It has been suggested that surface interaction and oxidation could have been a major sink of oxygen during the magma ocean phase (Kasting et al. 1993b; Gillmann et al. 2009), but this can be an efficient way to suppress oxygen accumulation only until the magma ocean solidifies.

Way and Del Genio (2020) speculated that the resurfacing we see on Venus today could have been the means to sequester the leftover oxygen. Gillmann et al. (2020) have simulated ongoing oxidation of the fresh, solid, basaltic crust and found it able to extract oxygen from the atmosphere at a maximum rate slightly higher than atmospheric escape, at most. Pieters et al. (1986), Lécuyer et al. (2000) have calculated that a hypothetical equivalent layer of approximately 50 km of hematite would be necessary to account for the oxidation of the content of an Earth ocean on Venus. More recent work by Warren and Kite (2021) has suggested that, for this hypothesis to be valid volcanic ash produced by explosive volcanism needs to be oxidized. In their model oxidation efficiency was increased by the larger free surface of the material (they therefore assume a 100% oxidation efficiency). However, such a mechanism still requires layers of kilometers to tens of kilometers of oxidized material to be emplaced onto the surface of the planet. This hypothesis also needs to consider that only very limited pyroclastic activity has been identified on Venus today (Campbell and Clark 2006; Ghail and Wilson 2015; Grosfils et al. 2000, 2011; Keddie and Head 1995; McGill 2000), as explosive volcanism requires volatile contents > 3–5 wt%, several wt% higher than typical Earth magmas (< 1 wt%) (Head et al. 2021). As a result, it is possible that such a mechanism might have actually played a role in the more distant past of Venus, rather than relatively recently. Again, better understanding of the nature and composition of the surface layers of Venus would be a tremendous help to understanding its history. Gillmann et al. (2022, this journal) expands on this topic and surface-atmosphere interaction.

Section 2 describes what to look for in the atmosphere today (D/H and noble gases, see Avice et al. 2022 this journal, for more details on noble gases) and what to look for on the surface in terms of felsic materials (similar to material from Earth's continents, formed at subduction factories) that may have a connection to surface water-rock interactions. On the other hand, if tesserae prove to be mainly basaltic, they formed without the need for liquid water, which would support a dryer evolution at least at the time they were formed.

5 Present-Day Habitability

5.1 The Clouds of Venus

The question of Venus' present-day habitability has been discussed for decades (Morowitz and Sagan 1967; Cockell 1999; Grinspoon and Bullock 2007). As covered in prior sections, it is often reasoned that if conditions on early Venus were similar to conditions on early Earth during the period in which Earth life arose – carbon molecules, surface water and rock-water interactions, and N, P, S, and transition metals, as well as suitable surface geology, volcanism and hydrothermal activity – this indicates the potential for an Earth-like biochemistry to have arisen on early Venus. However, modern-day Venus' surface is too hot for liquid water to be present, which rules out such biochemical reactions. Speculative alternative biochemistries compatible with the modern Venus surface have been proposed, such as the use of supercritical carbon dioxide as a polar solvent (Budisa and Schulze-Makuch 2014); the possibility for water-based life to have retreated to underground high-pressure water refugia

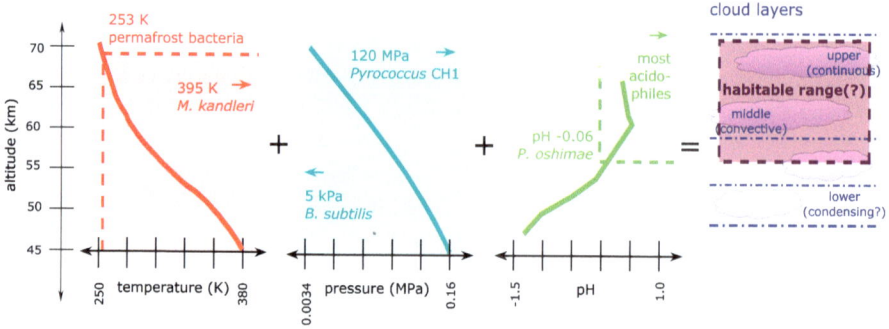

Fig. 8 The calculated temperature, pressure, and pH prevailing in Venus cloud aerosols in the height range 45–70 km from the surface (solid lines) and respective observed limits for terrestrial life (dashed lines and solid fill). The limits of terrestrial organisms may or may not reflect the possibilities for Venus

has also been discussed (Schulze-Makuch and Irwin 2002). However, the only liquid water known to exist today on Venus is that dissolved within its sulphuric acid clouds. There has therefore been a great deal of interest in the potential for present-day habitability of the Venus cloud aerosols, and whether such a habitat could have existed contemporaneously with surface water for long enough for life to have made the transition.

Figure 8 shows the calculated temperature, pressure, and pH in the middle Venus atmosphere (Grinspoon and Bullock 2007; Dartnell et al. 2015), juxtaposed with the respective observed limits for terrestrial life. The resulting discussion of a potential 'habitable range', and its relation to the limits of terrestrial cloud- and airborne microorganisms, is the subject of the following section. Future missions may help to constrain additional major variables affecting habitability in this altitude range, such as water availability and better constraining ultraviolet radiation flux (e.g. Mogul et al. 2021b).

5.2 Life in the Clouds

Earth's biosphere is a dynamic system of organisms and their interactions with the physical environment, including both transient and enduring habitats as well as short-term transport pathways (wind, rain) and long-term dormant refugia (polar ice, the deep subsurface). It includes a significant atmospheric component, the aerobiosphere. Tropospheric cloud water – the warmest and wettest airborne habitat – carries between 10^3 and 10^5 viable cells per mL (Amato et al. 2007), some of which is metabolically active (Amato et al. 2017). In addition to liquid cloud water, viable microbes are transported both regionally and globally as dust (Schuerger et al. 2018), ranging from 10^1 to 10^6 cells per cubic meter of air (Bowers et al. 2011; Burrows et al. 2009). These viable, dry bioaerosols extend throughout the troposphere and into the stratosphere (Bryan et al. 2019). However, most and possibly all of these desiccated microbes are inactive. Airborne microbial reproduction has not yet been directly observed in the field, although there have been indirect laboratory demonstrations (e.g. Sattler et al. 2001); this is likely in part because microbes in the field do not stay airborne for very long compared to typical generation times (Gentry et al. 2021).

At its most abstract, the requirements needed to support life (as we know it) were summarized by Hoehler (2007) as: a solvent (water); nutrients (C, H, N, O, P, S, and trace elements like Fe); energy for primary producers (autotrophs, chemical or photonic); and a stable environment (temperature, pH, radiation, etc.). The limits of habitability are often reasoned by analogy to therefore be the limits of life with respect to these requirements; the limits for the emergence of life are not fully understood, but may be different from or more constrained than established life which has had time to adapt and diversify, as noted above in Sect. 2.3.

By these metrics, hypothetical life in Venusian aerosols may be within the bounds of temperature, pressure, and pH (Grinspoon and Bullock 2007; Nicholson et al. 2010); radiation (Schulze-Makuch and Irwin 2002; Dartnell et al. 2015; Mogul et al. 2021b); C, H, N, O, and S, with some evidence for P (Limaye et al. 2021; Milojevic et al. 2021; Mogul et al. 2021a); and energy sources (photonic and/or oxidation-reduction potential, Limaye et al. 2018; Mogul et al. 2021b). Seager et al. (2020) argue that because the pH scale becomes highly compressed at extreme acid (or base) concentrations, the Hammett acidity function (H_0) is a more representative metric for the Venus aerosols' acid activity. H_0 for Venus' aerosols is poorly constrained. It has been estimated from as low as -11 (Seager et al. 2020) based on the current understanding of bulk aerosol composition, to ≥ -1.5 by Mogul et al. (2021b) with favorable assumptions regarding trace aerosol composition, the latter of which has some support in recent modeling by Rimmer et al. (2021) speculating the presence of ammonium or other hydroxide salts; however, even under the most favorable conditions, the results are at or below the acidity of any known Earth habitat, a substantial challenge for the hypothesis of an Earth-like biochemistry. The previously discussed alternative biochemistry hypotheses, such as a theoretical biochemistry based on sulphuric acid as a polar solvent instead of water, are not sufficiently detailed to be constrained in the same way.

An airborne ecosystem faces the additional unique requirement that its organisms must be able to stay aloft long enough to reproduce in a suitable, microbial-scale environment. Otherwise, the aerobiosphere will eventually settle out to extinction (if the planetary surface is uninhabitable, as with Venus), or be limited to transportation of a continual flux of organisms from the surface (as appears to be the case on Earth). A stable microbial aerobiosphere – using the term 'microbe' generally, without implied similarity to terrestrial microbiology – therefore has much stricter constraints than initially apparent. Microbes in a long-lived aerobiosphere (i.e., an atmospheric habitat) cannot rely on the common survival strategy of dormancy, i.e., 'waiting it out' to grow and reproduce during brief influxes of water, light, heat, etc. as is observed in microbes from Earth's deserts, poles, and other extreme environments. In effect, the 'soft' constraints of surviving versus thriving (activity, growth, and reproduction) become converted to hard habitability constraints when assessing potential atmospheric habitability, and are further related to the typical particle residence time determined by the large- and small-scale atmospheric dynamics.

On Earth, residence time for liquid water cloud particles, the most clement airborne microenvironment, ranges from hours to days; this is roughly on order with typical microbial generation times for common surface soil- or water-dwelling microbes. Smaller and lighter particles in drier and colder parts of Earth's atmosphere, such stratospheric aerosols, may be resident for as much as a few years; however, extremophilic microbes observed capable of withstanding similar conditions in other terrestrial habitats reproduce far more slowly, with an example of a 60-day mean generation time reported for Siberian permafrost at $-10\,°C$ (Bakermans et al. 2003). Another survival strategy often found in extremophilic environments with highly dynamic conditions – for example, a desert which might receive all of its rainfall on one or two days a year – is adaptation to long periods of dormancy followed by brief periods of repair and growth (e.g., Friedmann et al. 1993). Microbes have been observed to survive decades and perhaps far longer of complete desiccation, freezing, or other extreme conditions in the field (see Schulze-Makuch et al. 2018; Lowenstein et al. 2011; Knowlton et al. 2013 and references therein), but it should be emphasized that they do not reproduce during these periods and thus this phenomenon does not necessarily extend

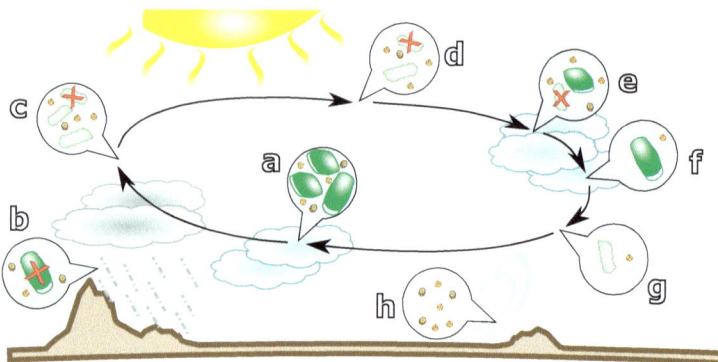

Fig. 9 Life cycle within a notional aerobiosphere. [**a**] Microbes (green) accumulate enough nutrients (brown) in a warm, wet cloud to divide. [**b**] Loss with precipitation. [**c**] Encounter with drier region; some transition to desiccated, inactive forms. [**d**] Dry forms accumulate damage, e.g., radiation. [**e**] Encounter with high-humidity region; some rehydrate and repair. [**f**] Survivors grow, potentially exhausting available nutrients. [**g**] Wet/dry cycles may repeat, depending on cloud dynamics. [**h**] Nutrients (e.g., surface minerals), energy, and wet periods become sufficient to allow division, beginning the cycle anew

the criteria for long-term aerobiosphere habitability. This is an important point: on Earth, airborne life has so far been observed to originate within at most a few generations from surface habitats.

Our knowledge of the microenvironments and typical residence times of Venus cloud droplets is limited, though they are likely longer-lasting than Earth's clouds. Seager et al. (2020) implemented a model that suggests coagulation rates constrain 3 μm-diameter cloud aerosols to 6 months aloft; Grinspoon and Bullock (2007) note that Hadley circulation may impose an overall 70–90 day upper bound.

Conditions favorable to metabolic activity and reproduction must also have sufficient continuity. A generalization of the constraints that shape Earth's aerobiosphere is shown in Fig. 9: cloud formation, precipitation, particle trajectories, cycles of dehydration and rehydration, and nutrient and radiation flux, among others. The hydrated periods of metabolic activity must align with the availability of nutrients and energy to allow growth to reproduce before the particle containing the microbe(s) rains or settles below the surface, or habitable altitude range.

Given the above, the most significant question for the potential present-day habitability of Venus' clouds is whether sufficient water exists in the cloud aerosols to allow occasional microbial growth. Both Seager et al. (2020) and Hallsworth et al. (2021a) estimate the water activity (a_w) as ≤ 0.004, far below the microbial activity limit of ~ 0.6. Limaye et al. (2021) calculated a higher but still prohibitive estimate of 0.02; as with calculations of H_0 above, the speculative models of Mogul et al. (2021b) and Rimmer et al. (2021) could allow currently unmeasured trace aerosol constituents to raise this to within known limits.

Understanding the habitability of Venus' clouds will require both future missions and modeling, with close coordination between experts in atmospheric dynamics, aerosol properties, cloud microphysics, and aerobiology. Key parameters to constrain include detailed measurements of Venus' aerosol composition, including trace constituents that could be nutrients or provide acid neutralization; typical residence times for particles with microbe-like properties in the Venusian atmosphere; and typical generation times for potential Earth analogue microorganisms, especially primary producers able to survive repeated desiccation and high acidity.

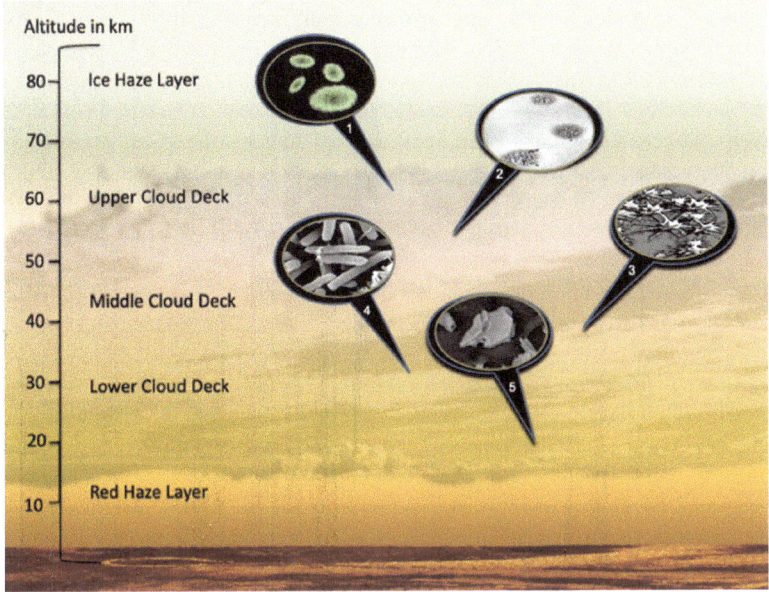

Fig. 10 Notional particles potentially to be encountered in the Venus cloud decks, inspired by terrestrial atmospheric sampling, to guide future instrument and analysis selection: (1) complex shapes with fluorescent properties, (2) particulate aggregates of sulfates and related compounds, (3) unidentified group of complex shapes adhered to an aerosol particle, (4) objects that resemble Earth bacteria or archaea, and (5) volcanic ash particles

Given the importance of microenvironments within cloud droplets to habitability, it is relevant to note several lines of evidence pointing to the existence of multiple cloud aerosol constituents beyond sulphuric acid and water: (1) UV absorption in the upper clouds of Venus is caused by an as-yet-unidentified "unknown UV absorber"; (2) VEx/VMC imager's analysis of the phase functions of light reflected from the upper clouds show more variation in refractive index than can be explained by H_2SO_4:H_2O mixtures alone; (3) particulates are observed to exist at altitudes below the main cloud base, where temperatures are too high for H_2SO_4:H_2O droplets to persist in liquid form; (4) X-ray Fluorescence analysis of collected droplets conducted from Venera and Vega descent probes found evidence of iron, chlorine and phosphorus in cloud droplets; and (5) the recent reanalysis of the Pioneer Venus LNMS data by Mogul et al. (2021a) which may provide further evidence for phosphorus in the cloud layer. The latter two results have not yet been reconciled with other in situ measurements and therefore remain something of an enigma (see review in Titov et al. 2018). The identification of cloud particle composition, down to the trace level, is clearly of great importance for assessing the present day habitability of the cloud deck (Fig. 10). In this respect, new investigations including measurement of the abundance and isotope ratios of volatile elements would likely shed light on the past and present dynamics of the cloud region of the Venus atmosphere (Avice et al. 2022).

5.3 Suggested Venusian Biosignatures

Interest in the habitability of Venus' clouds is furthered by several currently unexplained observations of the Venusian atmosphere that bear some similarities to known terrestrial biosignatures; if the clouds can be shown to bear equivalent similarities to the corresponding terrestrial habitats, the case for dedicated life detection investigation strengthens, and vice versa.

The Venus cloud layers have significant spectral absorption features not currently explained by what is known about the bulk aerosol composition, most notably in the UV but also at some longer wavelengths. Limaye et al. (2018) and Mogul et al. (2021b) suggested that this could be caused by phototrophy and/or 'sunscreen' pigments similar to carotenoids – in other words, analogous to the green 'color' of Earth resulting from the global presence of chlorophyll.

There are also discontinuities or unexplained variances in atmospheric sulfur and other chemical cycling (Bierson and Zhang 2020; Shao et al. 2020). Spacek and Benner (2021) suggested that these result from the presence of organic carbon, while Limaye et al. (2018) suggested that redox-based metabolic processes could play a role.

Recently, there have been controversial claims for the presence of a biosignature, the molecule phosphine (PH_3), in Venusian clouds. We summarize the controversy below. In September of 2020, Greaves et al. (2021b) published an analysis of JCMT (James Clerk Maxwell Telescope) and ALMA (Atacama Large Millimeter Array) spectra of the Venusian atmosphere that demonstrated that phosphine (PH_3) may have been detected. At the same time another paper by Bains et al. (2021), with many of the same authors as in the Greaves et al. (2021b) work, was submitted to arXiv with the title "Phosphine on Venus Cannot be Explained by Conventional Processes."

Subsequently a series of papers were submitted (posted to arXiv) and eventually published that put into question the veracity of the original JCMT and ALMA observations (e.g. Snellen et al. 2020; Thompson 2021; Villanueva et al. 2021; Akins et al. 2021; Lincowski et al. 2021). Additional papers placed upper limits via other space and ground based measurements that further questioned the PH_3 detection (Encrenaz et al. 2020; Trompet et al. 2021). Greaves et al. (2021a,c,d) offered a response to such criticisms, and more back-and-forth rebuttals continue in the literature today. At the same time another paper may have offered support to the Greaves et al. (2021b) ground based observations (Mogul et al. 2021a) by looking at in-situ archival data from the Pioneer Venus Large Probe Neutral Mass Spectrometer (PV-LNMS). Subsequently, a few papers have been published that look into the possible origins of PH_3 in planetary atmospheres in addition to the Bains 2020 paper (e.g. Bains et al. 2019a,b, 2021; Sousa-Silva et al. 2020; Omran et al. 2021; Cockell et al. 2021; Truong and Lunine 2021; Limaye et al. 2021) and whether factors, such as water activity, pH, etc. play a role in PH_3 production (e.g. Hallsworth et al. 2021a; Rimmer et al. 2021). Other work has considered the effects of Cosmic Rays on PH_3 production, but found it difficult to produce as much as 20 ppb (McTaggart 2022). While there is yet no consensus on the detection of phosphine in the atmosphere of Venus, its potential discovery has initiated many efforts including a mission to search for life in the clouds of Venus (Seager et al. 2021). This demonstrates that detection of potential biosignatures in planetary atmospheres is a high priority goal for investigations targeting Venus and its exoplanet cousins.

5.4 The Venus Life Equation

Izenberg et al. (2021) proposed a general framework for assessing the probability of extant life on modern-day Venus. This 'Venus Life Equation' breaks down the qualitative factors affecting the probability of the *origination* of life (O), the *robustness* (size and diversity) of the supportable biosphere (R), and whether habitable conditions could have persisted *continuously* between the origin of life and the current day (C). The factors supporting a high value for O by analogy between early Venus and Earth are discussed above. R, by contrast, is very low for the atmospheric habitat hypothesis, as a result of both limited substrate (the total liquid volume of Venus' aerosols is at least five orders of magnitude less than, say, Earth's surface and ground water) and the typical low biodiversity of ecosystems highly constrained by water availability. The value C is affected by both the potential for global extinction events, such as asteroid strikes and coronal mass ejections, and overall planetary climate history, as affected by volcanism, stellar evolution, and many other factors. The former is relatively similar for Venus and Earth; the latter depends primarily on the water history of Venus as discussed above. The Venus Life Equation thus suggests a non-zero value for the probability of extant Venusian life. It also confirms that continuity (spatial and temporal) of conditions amenable to life is one of the most important unknowns that can be quantitatively constrained by direct in situ observation, through robust improvement of understanding of atmospheric zones and geologic/hydrologic history.

This latter point is of particular importance where the Venusian aerosols are concerned. Unlike on Earth, where localized extinction-level events occurred but conditions for life persisted in other habitats (and cf. the subsurface punctuated habitability suggested for Mars by Melosh and Vickery 1989), this may not have been a possibility on Venus.

6 Investigation Priorities

There are two investigation priorities concerning the habitability of Venus:

1. To study past habitability. This can be addressed both through orbital observations of the surface and crust, in order to understand the geodynamic regime through time, and through noble gas and light element isotope measurements, to obtain insights into the history of volatiles through formation and evolution.
2. To characterize the present cloud-level environment including searching for molecular biosignatures of past or present-day life. This can be partially addressed by descent probes, but a more comprehensive investigation would require sustained presence in the clouds as from a balloon platform.

1. Studying the past habitability of Venus

It is very difficult to access Venus' history. Modelling and comparative planetology with other planets of our Solar system alongside exoplanets (for example, their age in conjunction with their rotation rate) will provide insight into the past conditions on Venus but models are only as good as the data initially used. Both the onset of, and exit from, a potential habitable phase need to be modelled. The period toward the end of the magma ocean phase has been highlighted as an important criterion for subsequent evolution and needs to be studied more intensely before any definitive conclusions can be drawn as to whether Venus ever hosted liquid water at its surface (e.g. Salvador et al. 2023, this journal). Similarly, much more work needs to be done to explore how a planet may go from a temperate to a moist and then a runaway greenhouse state (e.g. Kasting 1988), and coupled interior-atmosphere models as well as 3D GCMs will be needed (e.g. Boukrouche et al. 2021). Future Venus missions will address habitability in a range of different investigations. One approach to reconstructing Venus' history is to study its geologic record, as preserved in its surface and crust; this provides a record of the last billion years or so of surface evolution.

Of particular interest is the possibility that the tessera highlands show emissivity signatures consistent with widespread (continental-scale) granitic composition, like that found in Earth's continental crust; such a detection would suggest that large volumes of liquid water were present during their formation (Gilmore et al. 2017). However, non-detection of this felsic signature would not be conclusive, as such continental crust might have been covered by aeolian or other deposits, or otherwise not detectable from orbit. Determining Venus' current geodynamic regime – through gravity mapping and through searches for recent or ongoing geological activity – will help to constrain estimates of current heat and volatile loss from the interior, important factors in modelling the evolution of Venus' climate and habitability. The EnVision and VERITAS orbiters both will provide extensive datasets to address these investigations, as will DAVINCI descent imaging of tesserae, as will discussed in far more detail in companion publications in this journal (Chaps. 2–4).

Another approach, isotope geochemistry, allows one to constrain Venus' evolution in the distant past, right back to its formation and early evolution. The isotopic abundances of noble gases and light elements provide constraints on acquisition and loss processes of volatiles, and about their exchanges between mantle and atmosphere, so are particularly important for reconstructing the history of water. For example, measuring the magnitude of radiogenic/fissiogenic excesses of ^{4}He and $^{129, 131-136}$Xe produced at different times over Venus' history, will help to distinguish between scenarios for the geological evolution of Venus (stagnant lid, episodic plate tectonics episodes etc. see Gillmann et al. 2022, this journal). Although some noble gas isotope measurements were already obtained from Pioneer Venus and Venera in the 1970s and early 1980s, the upcoming DAVINCI entry probe mission will measure a greater variety of these isotopes with much greater precision, including the first measurements of krypton and xenon isotopes, permitting much better constraints on formation and evolution scenarios than are currently possible. Venus atmospheric sample return missions, though technically demanding, would allow isotopic ratio measurement to even higher precision and thus would offer correspondingly greater constraints on evolution scenarios. These investigations, and their implications for determining the history of water, are reviewed in detail in (Avice et al. 2022, this journal).

Conducting these investigations will not only give us a better understanding of Venus' evolution and potential habitability through time, but also will help us to assess habitability in terrestrial worlds in other planetary systems; these parallels are explored in much more detail in (Way et al. 2023, this journal).

2. The search for biosignatures in Venusian clouds

If Venus was habitable in the past (meaning, it had liquid water on its surface, the other ingredients of life being a given on a rocky planet such as Venus, and similar to early Earth), and life emerged, could it have survived to the present day in atmospheric aerosols? With regard to the habitability of the Venusian cloud deck, high priority in situ investigations include the structure of the atmosphere and variables, such as temperature, pressure, pH, UV radiation flux (cf. Grinspoon and Bullock 2007; Dartnell et al. 2015), and above all, detailed composition of the cloud aerosols, including water activity, acid activity, and trace constituents such as organics and ammonia. Comparison of these variables with those on Earth would substantially improve our ability to provide a 'habitable range'. Moreover, analysis of the aerosols will permit detailed study of the micro-environmental conditions within them. Could they be conducive to hosting an airborne biosphere, permitting life to thrive (not just survive), even if they are "extreme" by terrestrial standards? Future descent probes, like the upcoming DAVINCI (Garvin et al. 2022) probe, or the descent phase of lander missions such as Venera-D (Zasova et al. 2019) will be essential for this investigation by providing vertical profiles of atmospheric composition with far greater sensitivity and vertical resolution than is available from past missions. Far more spatially and temporally extensive investigations will require cloud-level aerial platforms (e.g. Cutts et al. 2018; Baines et al. 2018; Arredondo et al. 2021) offering sustained presence in the clouds. Instrumentation on such platforms should, in particular, seek to measure the composition, size, and lifetime of cloud and aerosol particles, as these are the most habitable environmental niches of astrobiological interest.

In situ investigations should also look for signs of extant life (biomolecules, metabolites). A staged approach to detection of biosignatures was developed by the Venus Life Finder Missions team (Seager et al. 2022), consisting of missions increasing in size and complexity. In a first mission, a small entry probe would descend through the clouds carrying an autofluorescence backscatter nephelometer, which would characterize the shape and composition of cloud particles and search for the fluorescence in UV light as a biomarker (Baumgardner et al. 2022). Such a mission is in development, at the time of writing, and may launch as soon as 2023 (French et al. 2022). A second proposed mission would put more capable chemical and environmental detection instrumentation on a long-lived balloon floating in Venus' clouds; such instrumentation could eventually include an aerosol mass spectrometer (Baines et al. 2021) and/or a fluorescent microscope which provides particular sensitivity to biomolecules (Sasaki et al. 2022). These could eventually lead to a third mission which would bring back a sample of Venus cloud material to Earth, so that it could be examined with the highly sensitive instrumentation available in terrestrial laboratories. Even if no metabolically active life forms are detected, information from these investigations will inform the models used to determine the history of the planet and its potential for having been habitable and seen the independent emergence of life.

One final note is that any information from Venus *in situ* and any possible, future sample return mission would be extremely valuable for studying the habitability of exoplanets.

Acknowledgements The comments of two anonymous reviews are gratefully acknowledged. ISSI is acknowledged for hosting the Venus workshop and also for providing FW with quiet conditions for rewriting parts of the manuscript.

Funding Note Open Access funding enabled and organized by Projekt DEAL. M.J.W. acknowledges support from the Goddard Space Flight Center Sellers Exoplanet Environments Collaboration (SEEC) and ROCKE-3D: The evolution of solar system worlds through time, funded by the NASA Planetary and Earth Science Divisions Internal Scientist Funding Model.

Declarations

Competing Interests The authors declare that they have no conflict of interest.

References

Abramov O, Kring DA, Mojzsis SJ (2013) The impact environment of the Hadean earth. Geochemistry 73(3):227–248
Adam Z (2007) Actinides and life's origins. Astrobiology 7:852–872
Adam ZR, Hongo Y, Cleaves HJ, Yi R, Fahrenbach AC, Yoda I, Aono M (2018) Estimating the capacity for production of formamide by radioactive minerals on the prebiotic Earth. Sci Rep 8(1):1–8

Akins AB, Lincowski AP, Meadows VS, Steffes PG (2021) Complications in the ALMA detection of phosphine at Venus. Astrophys J Lett 907(2):L27

Aléon J, Lévy D, Aléon-Toppani A, Bureau H, Khodja H, Brisset F (2022) Determination of the initial hydrogen isotopic composition of the solar system. Nat Astron 6(4):458–463

Allwood AC, Walter MR, Kamber BS, Marshall CP, Burch IW (2006) Stromatolite reef from the early Archaean era of Australia. Nature 441(7094):714–718

Allwood AC, Rosing MT, Flannery DT, Hurowitz JA, Heirwegh CM (2018) Reassessing evidence of life in 3,700-million-year-old rocks of Greenland. Nature 563(7730):241–244

Altwegg K, Balsiger H, Bar-Nun A, Berthelier JJ, Bieler A, Bochsler P, Briois C, Calmonte U, Combi M, De Keyser J, Eberhardt P, Fiethe B, Fuselier S, Gasc S, Gombosi TI, Hansen KC, Hässig M, Jäckel A, Kopp E, Korth A, LeRoy L, Mall U, Marty B, Mousis O, Neefs E, Owen T, Reme H, Rubin M, Semon T, Tzou CY, Waite H, Wurz P (2015) 67P/Churyumov-Gerasimenko, a Jupiter family comet with a high D/H ratio. Science 347(6220):3. https://doi.org/10.1126/science.1261952

Amato P, Parazols M, Sancelme M, Laj P, Mailhot G, Delort AM (2007) Microorganisms isolated from the water phase of tropospheric clouds at the Puy de Dôme: major groups and growth abilities at low temperatures: microorganisms from the water phase of tropospheric clouds. FEMS Microbiol Ecol 59(2):242–254. https://doi.org/10.1111/j.1574-6941.2006.00199.x

Amato P, Joly M, Besaury L, Oudart A, Taib N, Moné AI, Deguillaume L, Delort AM, Debroas D (2017) Active microorganisms thrive among extremely diverse communities in cloud water. PLoS ONE 12(8):e0182–869. https://doi.org/10.1371/journal.pone.0182869

Armann M, Tackley PJ (2012) Simulating the thermochemical magmatic and tectonic evolution of Venus's mantle and lithosphere: two-dimensional models. J Geophys Res, Planets 117(E12):E12003

Arredondo A, Hodges A, Abrahams JNH, Bedford CC, Boatwright BD, Buz J, Cantrall C, Clark J, Erwin A, Krishnamoorthy S, Magana L, McCabe RM, McIntosh EC, Noviello JL, Pellegrino M, Ray C, Styczinski M, Weigel P (2021) VALENTiNE: a concept for a new frontiers class long duration in-situ balloon mission to Venus. In: 52nd lunar and planetary science conference, Lunar and Planetary Institute, Virtual, p 1526. https://ui.adsabs.harvard.edu/abs/2021LPI....52.1526A

Avice G, Marty B (2020) Perspectives on atmospheric evolution from noble gas and nitrogen isotopes on Earth, Mars & Venus. Space Sci Rev 216(3):36. https://doi.org/10.1007/s11214-020-00655-0

Avice G, Marty B, Burgess R, Hofmann A, Philippot P, Zahnle K, Zakharov D (2018) Evolution of atmospheric xenon and other noble gases inferred from Archean to Paleoproterozoic rocks. Geochim Cosmochim Acta 232:82–100. https://doi.org/10.1016/j.gca.2018.04.018

Avice G, Parai R, Jacobson S, Labidi J, Trainer M, Mikhail PP (2022) Noble gases and stable isotopes track the originand early evolution of the Venus atmosphere. Space Sci Rev 218:60. https://doi.org/10.1007/s11214-022-00929-9

Baines KH, Atreya SK, Bullock MA, Grinspoon DH, Mahaffy P, Russell CT, Schubert G, Zahnle K (2013) The atmospheres of the terrestrial planets: clues to the origins and early evolution of Venus, Earth, and Mars. In: Comparative climatology of terrestrial planets. University of Arizona Press, Tucson, pp 1–28. https://doi.org/10.2458/azu_uapress_9780816530595-ch006

Baines KH, Cutts JA, Nikolić D, Madzunkov SM, Renard J, Mousis O, Barge LM, Limaye SS (2018) An aerosol instrument package for characterizing the Venus cloud habitability zone. Astrobiology 18(9):1181–1198. https://www.liebertpub.com/doi/10.1089/ast.2017.1783

Baines KH, Nikolić D, Cutts JA, Delitsky ML, Renard JB, Madzunkov SM, Barge LM, Mousis O, Wilson C, Limaye SS et al (2021) Investigation of Venus cloud aerosol and gas composition including potential biogenic materials via an aerosol-sampling instrument package. Astrobiology 21(10):1316–1323

Bains W, Petkowski JJ, Sousa-Silva C, Seager S (2019a) Trivalent phosphorus and phosphines as components of biochemistry in anoxic environments. Astrobiology 19(7):885–902

Bains W, Petkowski JJ, Sousa-Silva C, Seager S (2019b) New environmental model for thermodynamic ecology of biological phosphine production. Sci Total Environ 658:521–536

Bains W, Petkowski JJ, Seager S, Ranjan S, Sousa-Silva C, Rimmer PB, Zhan Z, Greaves JS, Richards AM (2021) Phosphine on Venus cannot be explained by conventional processes. Astrobiology 21(10):1277–1304. https://doi.org/10.1089/ast.2020.2352

Bakermans C, Tsapin AI, Souza-Egipsy V, Gilichinsky DA, Nealson KH (2003) Reproduction and metabolism at −10 °C of bacteria isolated from Siberian permafrost. Environ Microbiol 5:321–326. https://doi.org/10.1046/j.1462-2920.2003.00419.x

Barabash S, Fedorov A, Sauvaud J, Lundin R, Russell C, Futaana Y, Zhang T, Andersson H, Brinkfeldt K, Grigoriev A et al (2007) The loss of ions from Venus through the plasma wake. Nature 450(7170):650–653

Barboni M, Boehnke P, Keller B, Kohl IE, Schoene B, Young ED, McKeegan KD (2017) Early formation of the Moon 4.51 billion years ago. Sci Adv 3(1):e1602,365

172

Baross JA, Hoffman SE (1985) Submarine hydrothermal vents and associated gradient environments as sites for the origin and evolution of life. Orig Life Evol Biosph 15(4):327–345

Basilevsky A, Head J (1997) Onset time and duration of corona activity on Venus: stratigraphy and history from photogeologic study of stereo images. Earth Moon Planets 76(1):67–115

Baumgardner D, Fisher T, Newton R, Roden C, Zmarzly P, Seager S, Petkowski JJ, Carr CE, Špaček J, Benner SA et al (2022) Deducing the composition of Venus cloud particles with the autofluorescence nephelometer (AFN). Aerospace 9(9):492

Belousova E, Kostitsyn Y (2010). Griffi n, wl, begg, gc

Benner SA, Kim HJ, Kim MJ, Ricardo A (2010) Planetary organic chemistry and the origins of biomolecules. Cold Spring Harb Perspect Biol 2(7):a003467

Benz W, Slattery WL, Cameron AGW (1986) The origin of the Moon and the single-impact hypothesis I. Icarus 66(3):515–535. https://doi.org/10.1016/0019-1035(86)90088-6

Bercovici D, Ricard Y (2014) Plate tectonics, damage and inheritance. Nature 508(7497):513–516

Bierson CJ, Zhang X (2020) Chemical cycling in the Venusian atmosphere: a full photochemical model from the surface to 110 km. J Geophys Res, Planets 125(7):e2019JE006,159. https://doi.org/10.1029/2019JE006159

Blake RE, Chang SJ, Lepland A (2010) Phosphate oxygen isotopic evidence for a temperate and biologically active Archaean ocean. Nature 464(7291):1029–1032

Bolhar R, Kamber BS, Moorbath S, Whitehouse MJ, Collerson KD (2005) Chemical characterization of Earth's most ancient clastic metasediments from the Isua Greenstone Belt, southern West Greenland. Geochim Cosmochim Acta 69(6):1555–1573

Bottke WF, Vokrouhlicky D, Ghent B, Mazrouei S, Robbins S, Marchi S (2016) On asteroid impacts, crater scaling laws, and a proposed younger surface age for Venus. In: Lunar and planetary science conference, p 2036

Boukrouche R, Lichtenberg T, Pierrehumbert RT (2021) Beyond runaway: initiation of the post-runaway greenhouse state on rocky exoplanets. Astrophys J 919(2):130. https://doi.org/10.3847/1538-4357/ac1345. arXiv:2107.14150

Bower D, Hakim K, Sossi P, Sanan P (2022) Retention of water in terrestrial magma oceans and carbon-rich early atmospheres. Planet Sci J 3(93):1–28

Bowers RM, McLetchie S, Knight R, Fierer N (2011) Spatial variability in airborne bacterial communities across land-use types and their relationship to the bacterial communities of potential source environments. ISME J 5(4):601–612. https://doi.org/10.1038/ismej.2010.167

Bowring SA, Williams IS (1999) Priscoan (4.00–4.03 Ga) orthogneisses from northwestern Canada. Contrib Mineral Petrol 134(1):3–16

Brasier MD, Matthewman R, McMahon S, Wacey D (2011) Pumice as a remarkable substrate for the origin of life. Astrobiology 11(7):725–735

Brasser R, Mojzsis S, Werner S, Matsumura S, Ida S (2016) Late veneer and late accretion to the terrestrial planets. Earth Planet Sci Lett 455:85–93. https://doi.org/10.1016/j.epsl.2016.09.013

Bryan NC, Christner BC, Guzik TG, Granger DJ, Stewart MF (2019) Abundance and survival of microbial aerosols in the troposphere and stratosphere. ISME J 13(1111):2789–2799. https://doi.org/10.1038/s41396-019-0474-0

Budisa N, Schulze-Makuch D (2014) Supercritical carbon dioxide and its potential as a life-sustaining solvent in a planetary environment. Life 4(33):331–340. https://doi.org/10.3390/life4030331

Bullock MA, Grinspoon DH (2001) The recent evolution of climate on Venus. Icarus 150(1):19–37

Burkhardt C, Spitzer F, Morbidelli A, Budde G, Render JH, Kruijer TS, Kleine T (2021) Terrestrial planet formation from lost inner solar system material. Sci Adv 7(52):eabj7601

Burrows SM, Elbert W, Lawrence MG, Pöschl U (2009) Bacteria in the global atmosphere – part 1: review and synthesis of literature data for different ecosystems. Atmos Chem Phys 9(23):9263–9280. https://doi.org/10.5194/acp-9-9263-2009

Byerly GR, Lower DR, Walsh MM (1986) Stromatolites from the 3,300–3,500-Myr Swaziland Supergroup, Barberton mountain land, South Africa. Nature 319(6053):489–491

Byrne PK, Krishnamoorthy S (2021) Estimates on the frequency of volcanic eruptions on Venus. J Geophys Res, Planets 127(1):e2021JE007040

Campbell BA, Clark DA (2006) Geologic map of the Mead quadrangle (V-21), Venus, vol 2897. US Geological Survey, Denver

Campbell IH, Taylor SR (1983) No water, no granites-no oceans, no continents. Geophys Res Lett 10(11):1061–1064

Canup RM (2004) Simulations of a late lunar-forming impact. Icarus 168(2):433–456. https://doi.org/10.1016/j.icarus.2003.09.028

Capitanio F, Nebel O, Cawood P, Weinberg R, Chowdhury P (2019) Reconciling thermal regimes and tectonics of the early Earth. Geology 47(10):923–927

Capitanio FA, Nebel O, Cawood PA (2020) Thermochemical lithosphere differentiation and the origin of cratonic mantle. Nature 588(7836):89–94

Cates N, Mojzsis S (2007) Pre-3750 Ma supracrustal rocks from the Nuvvuagittuq supracrustal belt, northern Québec. Earth Planet Sci Lett 255(1–2):9–21

Catling DC, Zahnle KJ (2020) The Archean atmosphere. Sci Adv 6(9):eaax1420

Cawood PA, Hawkesworth C, Dhuime B (2013) The continental record and the generation of continental crust. Geol Soc Am Bull 125(1–2):14–32

Cawood PA, Hawkesworth CJ, Pisarevsky SA, Dhuime B, Capitanio FA, Nebel O (2018) Geological archive of the onset of plate tectonics. Philos Trans R Soc A, Math Phys Eng Sci 376(2132):20170,405

Chambers J, Wetherill G (1998) Making the terrestrial planets: N-body integrations of planetary embryos in three dimensions. Icarus 136(2):304–327

Chowdhury P, Gerya T, Chakraborty S (2017) Emergence of silicic continents as the lower crust peels off on a hot plate-tectonic Earth. Nat Geosci 10(9):698–703

Chowdhury P, Chakraborty S, Gerya TV, Cawood PA, Capitanio FA (2020) Peel-back controlled lithospheric convergence explains the secular transitions in Archean metamorphism and magmatism. Earth Planet Sci Lett 538:116,224

Claire MW, Sheets J, Cohen M, Ribas I, Meadows VS, Catling DC (2012) The evolution of solar flux from 0.1 nm to 160 µm: quantitative estimates for planetary studies. Astrophys J 757:95. https://doi.org/10.1088/0004-637X/757/1/95

Clement MS, Morbidelli A, Raymond SN, Kaib NA (2019) A record of the final phase of giant planet migration fossilized in the asteroid belt's orbital structure. Mon Not R Astron Soc Lett 492(1):L56–L60. https://doi.org/10.1093/mnrasl/slz184

Cockell CS (1999) Life on Venus. Planet Space Sci 47(12):1487–1501. https://doi.org/10.1016/S0032-0633(99)00036-7

Cockell CS, McMahon S, Biddle JF (2021) When is life a viable hypothesis? The case of Venusian phosphine. Astrobiology 21(3):261–264

Cogné JP, Humler E (2006) Trends and rhythms in global seafloor generation rate. Geochem Geophys Geosyst 7(3):Q03011

Connelly J, Bizzarro M (2016) Lead isotope evidence for a young formation age of the Earth–Moon system. Earth Planet Sci Lett 452:36–43

Correia ACM, Laskar J (2001) The four final rotation states of Venus. Nature 411:767–770. https://doi.org/10.1038/35081000

Correia ACM, Laskar J (2003) Long-term evolution of the spin of Venus. II. Numerical simulations. Icarus 163:24–45. https://doi.org/10.1016/S0019-1035(03)00043-5

Correia ACM, Laskar J, de Surgy ON (2003) Long-term evolution of the spin of Venus. I. Theory. Icarus 163:1–23. https://doi.org/10.1016/S0019-1035(03)00042-3

Cutts JA, Baines K, Grimm R, Matthies L, Hall JL, Limaye S, Thompson TW (2018) Aerial platforms for scientific investigation of Venus. https://trs.jpl.nasa.gov/bitstream/handle/2014/49516/CL%2318-7416.pdf?sequence=1

Damer B, Deamer D (2020) The hot spring hypothesis for an origin of life. Astrobiology 20(4):429–452

Dartnell LR, Nordheim TA, Patel MR, Mason JP, Coates AJ, Jones GH (2015) Constraints on a potential aerial biosphere on Venus: I. Cosmic rays. Icarus 257:396–405. https://doi.org/10.1016/j.icarus.2015.05.006

Dass AV, Jaber M, Brack A, Foucher F, Kee TP, Georgelin T, Westall F (2018) Potential role of inorganic confined environments in prebiotic phosphorylation. Life 8(1):7

Deamer DW (1997) The first living systems: a bioenergetic perspective. Microbiol Mol Biol Rev 61(2):239–261

Dehant V, Debaille V, Dobos V, Gaillard F, Gillmann C, Goderis S, Grenfell JL, Höning D, Javaux EJ, Karatekin Ö et al (2019) Geoscience for understanding habitability in the solar system and beyond. Space Sci Rev 215(6):1–48

Deienno R, Gomes RS, Walsh KJ, Morbidelli A, Nesvornỳ D (2016) Is the grand tack model compatible with the orbital distribution of main belt asteroids? Icarus 272:114–124

Dhuime B, Hawkesworth CJ, Cawood PA, Storey CD (2012) A change in the geodynamics of continental growth 3 billion years ago. Science 335(6074):1334–1336

Dhuime B, Wuestefeld A, Hawkesworth CJ (2015) Emergence of modern continental crust about 3 billion years ago. Nat Geosci 8(7):552–555

Djokic T, Van Kranendonk MJ, Campbell KA, Walter MR, Ward CR (2017) Earliest signs of life on land preserved in ca. 3.5 Ga hot spring deposits. Nat Commun 8(1):1–9

Djokic T, Van Kranendonk MJ, Campbell KA, Havig JR, Walter MR, Guido DM (2021) A reconstructed sub-aerial hot spring field in the 3.5 billion-year-old Dresser Formation, North Pole Dome, Pilbara Craton, Western Australia. Astrobiology 21(1):1–38

Dodd MS, Papineau D, Grenne T, Slack JF, Rittner M, Pirajno F, O'Neil J, Little CT (2017) Evidence for early life in Earth's oldest hydrothermal vent precipitates. Nature 543(7643):60–64

Donahue T, Hodges R Jr (1992) Past and present water budget of Venus. J Geophys Res, Planets 97(E4):6083–6091

Donahue TM, Hodges RR Jr (1993) Venus methane and water. Geophys Res Lett 20(7):591–594

Donahue T, Hoffman J, Hodges R, Watson A (1982) Venus was wet: a measurement of the ratio of deuterium to hydrogen. Science 216(4546):630–633

Encrenaz T, Greathouse TK, Marcq E, Widemann T, Bézard B, Fouchet T, Giles R, Sagawa H, Greaves J, Sousa-Silva C (2020) A stringent upper limit of the PH_3 abundance at the cloud top of Venus. Astron Astrophys 643:L5

Evans DA (2013) Reconstructing pre-Pangean supercontinents. Geol Soc Am Bull 125(11–12):1735–1751

Fedorova A, Korablev O, Vandaele AC, Bertaux JL, Belyaev D, Mahieux A, Neefs E, Wilquet W, Drummond R, Montmessin F et al (2008) HDO and H_2O vertical distributions and isotopic ratio in the Venus mesosphere by Solar Occultation at Infrared spectrometer on board Venus Express. J Geophys Res, Planets 113(E5):E00B22

Fegley B (2014) Venus. In: Treatise on geochemistry. Elsevier, Amsterdam, pp 127–148. https://doi.org/10.1016/B978-0-08-095975-7.00122-4

Fegley B, Prinn RG (1989) Estimation of the rate of volcanism on Venus from reaction rate measurements. Nature 337(6202):55–58

Fischer R, Gerya T (2016a) Early Earth plume-lid tectonics: a high-resolution 3D numerical modelling approach. J Geodyn 100:198–214

Fischer R, Gerya T (2016b) Regimes of subduction and lithospheric dynamics in the Precambrian: 3D thermomechanical modelling. Gondwana Res 37:53–70

Foley BJ (2015) The role of plate tectonic–climate coupling and exposed land area in the development of habitable climates on rocky planets. Astrophys J 812(1):36

Foley BJ, Smye AJ (2018) Carbon cycling and habitability of Earth-sized stagnant lid planets. Astrobiology 18(7):873–896

Foley BJ, Bercovici D, Elkins-Tanton LT (2014) Initiation of plate tectonics from post-magma ocean thermochemical convection. J Geophys Res, Solid Earth 119(11):8538–8561. https://doi.org/10.1002/2014JB011121

French R, Mandy C, Hunter R, Mosleh E, Sinclair D, Beck P, Seager S, Petkowski JJ, Carr CE, Grinspoon DH et al (2022) Rocket lab mission to Venus. Aerospace 9(8):445

Friedmann EI, Kappen L, Meyer MA, Nienow JA (1993) Long-term productivity in the cryptoendolithic microbial community of the Ross Desert, Antarctica. Microb Ecol 25(1):51–69

Friend CR, Nutman AP, Bennett VC, Norman M (2008) Seawater-like trace element signatures (REE + Y) of Eoarchaean chemical sedimentary rocks from southern West Greenland, and their corruption during high-grade metamorphism. Contrib Mineral Petrol 155(2):229–246

Frost DJ, McCammon CA (2008) The redox state of Earth's mantle. Annu Rev Earth Planet Sci 36:389–420

Gaillard F, Scaillet B (2014) A theoretical framework for volcanic degassing chemistry in a comparative planetology perspective and implications for planetary atmospheres. Earth Planet Sci Lett 403:307–316

Gaillard F, Bouhifd MA, Füri E, Malavergne V, Marrocchi Y, Noack L, Ortenzi G, Roskosz M, Vulpius S (2021) The diverse planetary ingassing/outgassing paths produced over billions of years of magmatic activity. Space Sci Rev 217(1):1–54

Gaillard F, Bernadou F, Roskosz M, Bouhifd MA, Marrocchi Y, Iacono-Marziano G, Moreira M, Scaillet B, Rogerie G (2022) Redox controls during magma ocean degassing. Earth Planet Sci Lett 577:117,255

Ganino C, Arndt NT (2009) Climate changes caused by degassing of sediments during the emplacement of large igneous provinces. Geology 37(4):323–326

García-Ruiz JM, Van Zuilen MA, Bach W (2020) Mineral self-organization on a lifeless planet. Phys Life Rev 34:62–82

Garvin JB, Getty SA, Arney GN, Johnson NM, Kohler E, Schwer KO, Sekerak M, Bartels A, Saylor RS, Elliott VE et al (2022) Revealing the mysteries of Venus: the DAVINCI mission. Planet Sci J 3(5):117

Gentry DM, Iraci LT, Barth E, McGouldrick K, Jessup KL (2021) Habitability of cloudy worlds: intersecting constraints and unknowns. In: Proceedings of LPSC LII. Lunar and Planetary Institute, vol 52, p 2691. https://www.hou.usra.edu/meetings/lpsc2021/pdf/2691.pdf. Abstract #2691

Gerya T (2014) Precambrian geodynamics: concepts and models. Gondwana Res 25(2):442–463

Gerya T (2019) Geodynamics of the early Earth: quest for the missing paradigm. Geology 47(10):1006–1007

Gerya T (2022) Numerical modeling of subduction: state of the art and future directions. Geosphere 18(2):503–561

Gerya T, Stern R, Baes M, Sobolev S, Whattam S (2015a) Plume-induced subduction initiation triggered plate tectonics on Earth. Nature 527(7577):221–225

Gerya TV, Stern RJ, Baes M, Sobolev SV, Whattam SA (2015b) Plate tectonics on the Earth triggered by plume-induced subduction initiation. Nature 527(7577):221–225

Gerya TV, Bercovici D, Becker TW (2021) Dynamic slab segmentation due to brittle–ductile damage in the outer rise. Nature 599(7884):245–250

Ghail RC, Wilson L (2015) A pyroclastic flow deposit on Venus. Geol Soc (Lond) Spec Publ 401(1):97–106

Gillmann C, Tackley P (2014) Atmosphere/mantle coupling and feedbacks on Venus. J Geophys Res, Planets 119(6):1189–1217

Gillmann C, Chassefière E, Lognonné P (2009) A consistent picture of early hydrodynamic escape of Venus atmosphere explaining present Ne and Ar isotopic ratios and low oxygen atmospheric content. Earth Planet Sci Lett 286(3–4):503–513. https://doi.org/10.1016/j.epsl.2009.07.016

Gillmann C, Golabek GJ, Tackley PJ (2016) Effect of a single large impact on the coupled atmosphere-interior evolution of Venus. Icarus 268:295–312

Gillmann C, Golabek GJ, Raymond SN, Schönbächler M, Tackley P, Dehant V, Debaille V (2020) Dry late accretion inferred from Venus's coupled atmosphere and internal evolution. Nat Geosci 13(4):265–269

Gillmann C, Way MJ, Avice G, Breuer D, Golabek GJ, Höning D, Krissansen-Totton J, Lammer H, Plesa AC, Persson M, O'Rourke JG, Salvador A, Scherf M, Zolotov MY (2022) The long-term evolution of the atmosphere of Venus: processes and feedback mechanisms. Space Sci Rev 218:56. https://doi.org/10.1007/s11214-022-00924-0

Gilmore MS, Mueller N, Helbert J (2015) VIRTIS emissivity of Alpha Regio, Venus, with implications for tessera composition. Icarus 254:350–361. https://doi.org/10.1016/j.icarus.2015.04.008

Gilmore M, Treiman A, Helbert J, Smrekar S (2017) Venus surface composition constrained by observation and experiment. Space Sci Rev 212(3):1511–1540

Goldblatt C (2018) Atmospheric evolution Springer, Cham, pp 62–76. https://doi.org/10.1007/978-3-319-39312-4_107

Goldblatt C, Claire MW, Lenton TM, Matthews AJ, Watson AJ, Zahnle KJ (2009) Nitrogen-enhanced greenhouse warming on early Earth. Nat Geosci 2(12):891–896. https://doi.org/10.1038/ngeo692

Goldreich P, Peale SJ (1966) Resonant rotation for Venus? Nature 209:1117–1118. https://doi.org/10.1038/2091117a0

Goldreich P, Peale SJ (1970) The obliquity of Venus. Astron J 75:273. https://doi.org/10.1086/110975

Gough DO (1981) Solar interior structure and luminosity variations. Sol Phys 74(1):21–34. https://doi.org/10.1007/BF00151270

Graham RJ, Pierrehumbert R (2020) Thermodynamic and energetic limits on continental silicate weathering strongly impact the climate and habitability of wet, rocky worlds. Astrophys J 896(2):115. https://doi.org/10.3847/1538-4357/ab9362. arXiv:2004.14058

Greaves JS, Richards A, Bains W, Rimmer PB, Clements DL, Seager S, Petkowski JJ, Sousa-Silva C, Ranjan S, Fraser HJ (2021a) Reply to: no evidence of phosphine in the atmosphere of Venus from independent analyses. Nat Astron 5(7):636–639

Greaves JS, Richards AM, Bains W, Rimmer PB, Sagawa H, Clements DL, Seager S, Petkowski JJ, Sousa-Silva C, Ranjan S et al (2021b) Phosphine gas in the cloud decks of Venus. Nat Astron 5(7):655–664

Greaves JS, Richards A, Bains W, Rimmer PB, Sagawa H, Clements DL, Seager S, Petkowski JJ, Sousa-Silva C, Ranjan S et al (2021c) Addendum: phosphine gas in the cloud deck of Venus. Nat Astron 5(7):726–728

Greaves JS, Rimmer PB, Richards A, Petkowski JJ, Bains W, Ranjan S, Seager S, Clements DL, Silva CS, Fraser HJ (2021d) Low levels of sulphur dioxide contamination of phosphine spectra from Venus' atmosphere. ArXiv preprint. arXiv:2108.08393

Green JAM, Way MJ, Barnes R (2019) Consequences of tidal dissipation in a putative Venusian ocean. Astrophys J Lett 876(2):L22. https://doi.org/10.3847/2041-8213/ab133b. arXiv:1903.07517

Greer J, Caro G, Cates NL, Tropper P, Bleeker W, Kelly NM, Mojzsis SJ (2020) Widespread polymetamorphosed Archean granitoid gneisses and supracrustal enclaves of the southern Inukjuak Domain, Québec (Canada). Lithos 364:105,520

Grew ES, Bada JL, Hazen RM (2011) Borate minerals and origin of the RNA world. Orig Life Evol Biosph 41(4):307–316

Griffin W, Belousova E, O'Neill C, O'Reilly SY, Malkovets V, Pearson N, Spetsius S, Wilde S (2014) The world turns over: Hadean–Archean crust–mantle evolution. Lithos 189:2–15

Grinspoon DH (1987) Was Venus wet? Deuterium reconsidered. Science 238(4834):1702–1704. https://doi.org/10.1126/science.238.4834.1702

Grinspoon DH (1992) Venusian hydrology: steady state reconsidered. In: International colloquium on Venus. LPI contributions, vol 789, p 36

Grinspoon DH, Bullock MA (2007) Astrobiology and Venus exploration Am. Geophys Union, Washington, pp 191–206. https://doi.org/10.1029/176GM12

Grosfils EB, Aubele J, Crumpler L, Gregg TK, Sakimoto S (2000) Volcanism on Earth's seafloor and Venus. In: Environmental effects on volcanic eruptions. Springer, Berlin, pp 113–142

Grosfils EB, Long SM, Venechuk EM, Hurwitz DM, Richards JW, Kastl B, Drury DE, Hardin JS (2011) Geologic map of the Ganiki Planitia quadrangle (V-14), Venus. Venus: US Geological Survey Scientific Investigations Map 3121

Grossman L, Larimer JW (1974) Early chemical history of the solar system. Rev Geophys 12(1):71–101

Gurwell MA (1995) Evolution of deuterium on Venus. Nature 378(6552):22–23. https://doi.org/10.1038/378022b0

Halliday AN (2000) Terrestrial accretion rates and the origin of the Moon. Earth Planet Sci Lett 176(1):17–30

Hallsworth JE, Koop T, Dallas TD, Zorzano MP, Burkhardt J, Golyshina OV, Martín-Torres J, Dymond MK, Ball P, McKay CP (2021a) Water activity in Venus's uninhabitable clouds and other planetary atmospheres. Nat Astron 5(7):665–675

Hamano K, Abe Y, Genda H (2013) Emergence of two types of terrestrial planet on solidification of magma ocean. Nature 497:607–610. https://doi.org/10.1038/nature12163

Hansen VL (2018) Global tectonic evolution of Venus, from exogenic to endogenic over time, and implications for early Earth processes. Philos Trans R Soc A, Math Phys Eng Sci 376(2132):20170,412

Hao J, Knoll AH, Huang F, Hazen RM, Daniel I (2020) Cycling phosphorus on the Archean Earth: Part I. Continental weathering and riverine transport of phosphorus. Geochim Cosmochim Acta 273:70–84

Harris LB, Bédard JH (2014) Crustal evolution and deformation in a non-plate-tectonic Archaean Earth: comparisons with Venus. In: Evolution of Archean crust and early life. Springer, Berlin, pp 215–291

Harrison T (2020) Hadean Earth. Springer, Berlin. https://books.google.se/books?id=LFfsDwAAQBAJ

Hart MH (1979) Habitable zones about main sequence stars. Icarus 37(1):351–357

Hassenkam T, Andersson M, Dalby K, Mackenzie D, Rosing M (2017) Elements of Eoarchean life trapped in mineral inclusions. Nature 548(7665):78–81

Hawkesworth C, Cawood P, Kemp T, Storey C, Dhuime B (2009) A matter of preservation. Science 323(5910):49–50

Hawkesworth CJ, Cawood PA, Dhuime B (2020) The evolution of the continental crust and the onset of plate tectonics. Front Earth Sci 8:326

Head JW, Campbell DB, Elachi C, Guest JE, McKenzie DP, Saunders RS, Schaber GG, Schubert G (1991) Venus volcanism: initial analysis from Magellan data. Science 252(5003):276–288

Head J, Wilson L, Ivanov M, Wordsworth R (2021) Contributions of volatiles to the Venus atmosphere from the observed extrusive volcanic record: implications for the history of the Venus atmosphere. In: EGU general assembly conference abstracts. EGU general assembly conference abstracts, EGU21, p 13,030

Herrick R, Izenberg N, Ghail R (2022) Resurfacing history and volcanic activity of Venus. Space Sci Rev

Herzberg C, Condie K, Korenaga J (2010) Thermal history of the Earth and its petrological expression. Earth Planet Sci Lett 292(1–2):79–88. https://doi.org/10.1016/j.epsl.2010.01.022

Heubeck C (2009) An early ecosystem of Archean tidal microbial mats (Moodies Group, South Africa, ca. 3.2 Ga). Geology 37(10):931–934

Hickman-Lewis K, Westall F (2021) A southern African perspective on the co-evolution of early life and environments. S Afr J Geol 124(1):225–252

Hickman-Lewis K, Cavalazzi B, Foucher F, Westall F (2018a) Most ancient evidence for life in the Barberton greenstone belt: microbial mats and biofabrics of the 3.47 Ga Middle Marker horizon. Precambrian Res 312:45–67

Hickman-Lewis K, Westall F, Cavalazzi B (2018b). Trace of early life in the Barberton greenstone belt

Hickman-Lewis K, Cavalazzi B, Sorieul S, Gautret P, Foucher F, Whitehouse MJ, Jeon H, Georgelin T, Cockell CS, Westall F (2020a) Metallomics in deep time and the influence of ocean chemistry on the metabolic landscapes of Earth's earliest ecosystems. Sci Rep 10(1):1–16

Hickman-Lewis K, Gourcerol B, Westall F, Manzini D, Cavalazzi B (2020b) Reconstructing Palaeoarchaean microbial biomes flourishing in the presence of emergent landmasses using trace and rare earth element systematics. Precambrian Res 342:105,689

Hiesinger H, Tanaka K (2020) Chap. 15 – The planetary time scale. In: Gradstein FM, Ogg JG, Schmitz MD, Ogg GM (eds) Geologic time scale 2020. Elsevier, Amsterdam, pp 443–480. https://doi.org/10.1016/B978-0-12-824360-2.00015-2

Hirschmann MM (2012) Magma ocean influence on early atmosphere mass and composition. Earth Planet Sci Lett 341:48–57

Hoehler TM (2007) An energy balance concept for habitability. Astrobiology 7(6):824–838. https://doi.org/10.1089/ast.2006.0095

Hofmann A, Harris C (2008) Silica alteration zones in the Barberton greenstone belt: a window into subseafloor processes 3.5–3.3 Ga ago. Chem Geol 257(3–4):221–239

Hofmann A, Wilson AH (2007) Silicified basalts, bedded cherts and other sea floor alteration phenomena of the 3.4 Nondweni greenstone belt, South Africa. Dev Precambrian Geol 15:571–605

Hofmann H, Grey K, Hickman A, Thorpe R (1999) Origin of 3.45 Ga coniform stromatolites in Warrawoona Group, Western Australia. Geol Soc Am Bull 111(8):1256–1262

Holdaway D, Yang Y (2016) Study of the effect of temporal sampling frequency on DSCOVR observations using the GEOS-5 nature run results (part II): cloud coverage. Remote Sens 8(5):431. https://doi.org/10.3390/rs8050431

Höning D (2020) The impact of life on climate stabilization over different timescales. Geochem Geophys Geosyst 21(9):e2020GC009,105

Höning D, Tosi N, Hansen-Goos H, Spohn T (2019a) Bifurcation in the growth of continental crust. Phys Earth Planet Inter 287:37–50

Höning D, Tosi N, Spohn T (2019b) Carbon cycling and interior evolution of water-covered plate tectonics and stagnant-lid planets. Astron Astrophys 627:A48

Höning D, Baumeister P, Grenfell JL, Tosi N, Way MJ (2021) Early habitability and crustal decarbonation of a stagnant-lid Venus. J Geophys Res, Planets 126(10):e2021JE006,895

Hren M, Tice M, Chamberlain C (2009) Oxygen and hydrogen isotope evidence for a temperate climate 3.42 billion years ago. Nature 462(7270):205–208

Izenberg NR, Gentry DM, Smith DJ, Gilmore MS, Grinspoon DH, Bullock MA, Boston PJ, Słowik GP (2021) The Venus life equation. Astrobiology 21(10):1305–1315. https://doi.org/10.1089/ast.2020.2326

Jacob JB, Moyen JF, Fiannacca P, Laurent O, Bachmann O, Janoušek V, Farina F, Villaros A (2021) Crustal melting vs. fractionation of basaltic magmas: part 2, attempting to quantify mantle and crustal contributions in granitoids. Lithos 402:106,292

Jacobsen SB, Pimentel-Klose MR (1988) Nd isotopic variations in Precambrian banded iron formations. Geophys Res Lett 15(4):393–396

Johansen A, Ronnet T, Bizzarro M, Schiller M, Lambrechts M, Nordlund Å, Lammer H (2021) A pebble accretion model for the formation of the terrestrial planets in the Solar System. Sci Adv 7(8):eabc0444. https://doi.org/10.1126/sciadv.abc0444. arXiv:2102.08611

Johnson B, Goldblatt C (2015) The nitrogen budget of Earth. Earth-Sci Rev 148:150–173

Johnson TE, Brown M, Kaus BJ, VanTongeren JA (2014) Delamination and recycling of Archaean crust caused by gravitational instabilities. Nat Geosci 7(1):47–52

Johnson TE, Kirkland CL, Lu Y, Smithies RH, Brown M, Hartnady MI (2022) Giant impacts and the origin and evolution of continents. Nature 608(7922):330–335

Kadoya S, Tajika E (2014) Conditions for oceans on Earth-like planets orbiting within the habitable zone: importance of volcanic CO_2 degassing. Astrophys J 790(2):107

Kamber BS (2015) The evolving nature of terrestrial crust from the Hadean, through the Archaean, into the Proterozoic. Precambrian Res 258:48–82. https://doi.org/10.1016/j.precamres.2014.12.007

Kane SR, Kopparapu RK, Domagal-Goldman SD (2014) On the frequency of potential Venus analogs from Kepler data. Astrophys J 794(1):L5. https://doi.org/10.1088/2041-8205/794/1/l5

Karlsson R, Cheng KW, Crameri F, Rolf T, Uppalapati S, Werner SC (2020) Implications of anomalous crustal provinces for Venus' resurfacing history. J Geophys Res, Planets 125(10):e2019JE006,340

Kasting JF (1988) Runaway and moist greenhouse atmospheres and the evolution of Earth and Venus. Icarus 74:472–494. https://doi.org/10.1016/0019-1035(88)90116-9

Kasting J (1993) Earth's early atmosphere. Science 259(5097):920–926. https://doi.org/10.1126/science.11536547

Kasting JF, Catling D (2003) Evolution of a habitable planet. Annu Rev Astron Astrophys 41(1):429–463. https://doi.org/10.1146/annurev.astro.41.071601.170049

Kasting JF, Liu S, Donahue T (1979) Oxygen levels in the prebiological atmosphere. J Geophys Res, Oceans 84(C6):3097–3107

Kasting JF, Eggler DH, Raeburn SP (1993a) Mantle redox evolution and the oxidation state of the Archean atmosphere. J Geol 101(2):245–257

Kasting JF, Whitmire DP, Reynolds RT (1993b) Habitable zones around main sequence stars. Icarus 101(1):108–128. https://doi.org/10.1006/icar.1993.1010

Katyal N, Ortenzi G, Lee Grenfell J, Noack L, Sohl F, Godolt M, García Muñoz A, Schreier F, Wunderlich F, Rauer H (2020) Effect of mantle oxidation state and escape upon the evolution of Earth's magma ocean atmosphere. Astron Astrophys 643:A81. https://doi.org/10.1051/0004-6361/202038779. arXiv:2009.14599

Kaula WM (1999) Constraints on Venus evolution from radiogenic argon. Icarus 139(1):32–39

Keddie ST, Head JW (1995) Formation and evolution of volcanic edifices on the Dione Regio rise, Venus. J Geophys Res, Planets 100(E6):11,729–11,754

Kemp A, Wilde S, Hawkesworth C, Coath C, Nemchin A, Pidgeon R, Vervoort J, DuFrane S (2010) Hadean crustal evolution revisited: new constraints from Pb–Hf isotope systematics of the Jack Hills zircons. Earth Planet Sci Lett 296(1–2):45–56

Kempe S, Degens ET (1985) An early soda ocean? Chem Geol 53(1–2):95–108

Kiefer WS, Hager BH (1991) A mantle plume model for the equatorial highlands of Venus. J Geophys Res, Planets 96(E4):20,947–20,966

King SD (2018) Venus resurfacing constrained by geoid and topography. J Geophys Res, Planets 123(5):1041–1060

Kleine T, Budde G, Burkhardt C, Kruijer T, Worsham E, Morbidelli A, Nimmo F (2020) The non-carbonaceous–carbonaceous meteorite dichotomy. Space Sci Rev 216(4):1–27

Kminek G, Bada JL (2006) The effect of ionizing radiation on the preservation of amino acids on Mars. Earth Planet Sci Lett 245(1–2):1–5

Knauth LP (2011) Salinity history of the Earth's ocean. In: Encyclopedia of geobiology. Springer, Netherlands

Knowlton C, Veerapaneni R, D'Elia T, Rogers SO (2013) Microbial analyses of ancient ice core sections from Greenland and Antarctica. Biology 2(11):206–232. https://doi.org/10.3390/biology2010206

Kopparapu RK, Ramirez R, Kasting JF, Eymet V, Robinson TD, Mahadevan S, Terrien RC, Domagal-Goldman S, Meadows V, Deshpande R (2013) Habitable zones around main-sequence stars: new estimates. Astrophys J 765:131. https://doi.org/10.1088/0004-637X/765/2/131. arXiv:1301.6674

Korenaga J (2012) Plate tectonics and planetary habitability: current status and future challenges. Ann NY Acad Sci 1260(1):87–94

Korenaga J (2018) Crustal evolution and mantle dynamics through Earth history. Philos Trans R Soc A, Math Phys Eng Sci 376(2132):20170,408

Korenaga J (2021) Hadean geodynamics and the nature of early continental crust. Precambrian Res 359:106,178

Krissansen-Totton J, Arney GN, Catling DC (2018) Constraining the climate and ocean pH of the early Earth with a geological carbon cycle model. Proc Natl Acad Sci 115:4105–4110. https://doi.org/10.1073/pnas.1721296115. arXiv:1804.00763

Krissansen-Totton J, Fortney JJ, Nimmo F (2021) Was Venus ever habitable? Constraints from a coupled interior-atmosphere-redox evolution model. Planet Sci J 2(5):216. https://doi.org/10.3847/PSJ/ac2580. arXiv:2111.00033

Kurokawa H, Foriel J, Laneuville M, Houser C, Usui T (2018) Subduction and atmospheric escape of Earth's seawater constrained by hydrogen isotopes. Earth Planet Sci Lett 497:149–160. https://doi.org/10.1016/j.epsl.2018.06.016

Lammer H, Kasting JF, Chassefière E, Johnson RE, Kulikov YN, Tian F (2008) Atmospheric escape and evolution of terrestrial planets and satellites. Space Sci Rev 139(1):399–436

Lammer H, Zerkle AL, Gebauer S, Tosi N, Noack L, Scherf M, Pilat-Lohinger E, Güdel M, Grenfell JL, Godolt M et al (2018) Origin and evolution of the atmospheres of early Venus, Earth and Mars. Astron Astrophys Rev 26(1):1–72

Lan Z, Kamo SL, Roberts NM, Sano Y, Li XH (2022) A Neoarchean (ca. 2500 Ma) age for jaspilite-carbonate BIF hosting purported micro-fossils from the Eoarchean ($\geq$ 3750 Ma) Nuvvuagittuq supracrustal belt (Québec, Canada). Precambrian Res 377:106,728

Lebrun T, Massol H, Chassefière E, Davaille A, Marcq E, Sarda P, Leblanc F, Brandeis G (2013) Thermal evolution of an early magma ocean in interaction with the atmosphere. J Geophys Res, Planets 118:1155–1176. https://doi.org/10.1002/jgre.20068

Lécuyer C, Simon L, Guyot F (2000) Comparison of carbon, nitrogen and water budgets on Venus and the Earth. Earth Planet Sci Lett 181(1–2):33–40

Lenardic A, Jellinek A, Moresi LN (2008) A climate induced transition in the tectonic style of a terrestrial planet. Earth Planet Sci Lett 271(1–4):34–42

Levison HF, Kretke KA, Duncan MJ (2015) Growing the gas-giant planets by the gradual accumulation of pebbles. Nature 524(7565):322–324

Lichtenberg T, Bower DJ, Hammond M, Boukrouche R, Sanan P, Tsai SM, Pierrehumbert RT (2021) Vertically resolved magma ocean–protoatmosphere evolution: H_2, H_2O, CO_2, CH_4, CO, O_2, and N_2 as primary absorbers. J Geophys Res, Planets 126(2):e2020JE006,711

Limaye SS, Mogul R, Smith DJ, Ansari AH, Słowik GP, Vaishampayan P (2018) Venus' spectral signatures and the potential for life in the clouds. Astrobiology 18(9):1181–1198. https://doi.org/10.1089/ast.2017.1783

Limaye SS, Mogul R, Baines KH, Bullock MA, Cockell C, Cutts JA, Gentry DM, Grinspoon DH, Head JW, Jessup KL, Kompanichenko V, Lee YJ, Mathies R, Milojevic T, Pertzborn RA, Rothschild L, Sasaki S, Schulze-Makuch D, Smith DJ, Way MJ (2021) Venus, an astrobiology target. Astrobiology 21(10):1163–1185. https://doi.org/10.1089/ast.2020.2268

Lincowski AP, Meadows VS, Crisp D, Akins AB, Schwieterman EW, Arney GN, Wong ML, Steffes PG, Parenteau MN, Domagal-Goldman S (2021) Claimed detection of PH_3 in the clouds of Venus is consistent with mesospheric SO_2. Astrophys J Lett 908(2):L44

Liu B, Raymond SN, Jacobson SA (2020) Early solar system instability triggered by dispersal of the gaseous disk. Nature 604(7907):643–646. https://doi.org/10.1038/s41586-022-04535-1

Lock SJ, Stewart ST (2017) The structure of terrestrial bodies: impact heating, corotation limits, and synestias. J Geophys Res, Planets 122(5):950–982. https://doi.org/10.1002/2016JE005239

Lopez I, Oyarzun R, Marquez A, Doblas-Reyes F, Laurrieta A (1998) Progressive build up of CO_2 in the atmosphere of Venus through multiple volcanic resurfacing events. Earth Moon Planets 81(3):187–192. https://doi.org/10.1023/A:1006369831384

Lourenço DL, Rozel AB, Gerya T, Tackley PJ (2018) Efficient cooling of rocky planets by intrusive magmatism. Nat Geosci 11(5):322–327

Lourenço DL, Rozel AB, Ballmer MD, Tackley PJ (2020) Plutonic-squishy lid: a new global tectonic regime generated by intrusive magmatism on Earth-like planets. Geochem Geophys Geosyst C008:756

Lowe DR, Byerly G (1999a) Petrology and sedimentology of cherts and related silicified sedimentary rocks in the Swaziland Supergroup. In: Geologic evolution of the Barberton greenstone belt, South Africa. Geological Society of America Special Paper, vol 329, pp 83–114

Lowe DR, Byerly GR (1999b) Geologic evolution of the Barberton greenstone belt, South Africa, vol 329. Geol. Soc. Am., Boulder

Lowenstein TK, Schubert BA, Timofeeff MN (2011) Microbial communities in fluid inclusions and long-term survival in halite. GSA Today 21(1):4–9. https://doi.org/10.1130/GSATG81A.1

Luger R, Barnes R (2015) Extreme water loss and abiotic O_2 buildup on planets throughout the habitable zones of M dwarfs. Astrobiology 15(2):119–143. https://doi.org/10.1089/ast.2014.1231. arXiv:1411.7412

MacDonald GJF (1964) Tidal friction. Rev Geophys Space Phys 2:467–541. https://doi.org/10.1029/RG002i003p00467

Madronich S, Björn LO, McKenzie RL (2018) Solar UV radiation and microbial life in the atmosphere. Photochem Photobiol Sci 17(12):1918–1931

Maher KA, Stevenson DJ (1988) Impact frustration of the origin of life. Nature 331(6157):612–614

Marcq E, Mills FP, Parkinson CD, Vandaele AC (2018) Composition and chemistry of the neutral atmosphere of Venus. Space Sci Rev 214(1):1–55

Marin-Carbonne J, Chaussidon M, Robert F (2012) Micrometer-scale chemical and isotopic criteria (O and Si) on the origin and history of Precambrian cherts: implications for paleo-temperature reconstructions. Geochim Cosmochim Acta 92:129–147

Marshall M (2020) The water paradox and the origins of life. Nature 588(7837):210–213

Martin W, Russell MJ (2003) On the origins of cells: a hypothesis for the evolutionary transitions from abiotic geochemistry to chemoautotrophic prokaryotes, and from prokaryotes to nucleated cells. Philos Trans R Soc Lond B, Biol Sci 358(1429):59–85

Martin W, Baross J, Kelley D, Russell MJ (2008) Hydrothermal vents and the origin of life. Nat Rev Microbiol 6(11):805–814

Marty B (2012) The origins and concentrations of water, carbon, nitrogen and noble gases on Earth. Earth Planet Sci Lett 313:56–66

Marty B, Dauphas N (2003) The nitrogen record of crust–mantle interaction and mantle convection from Archean to present. Earth Planet Sci Lett 206(3–4):397–410

Marty B, Zimmermann L, Pujol M, Burgess R, Philippot P (2013) Nitrogen isotopic composition and density of the Archean atmosphere. Science 342(6154):101–104. https://doi.org/10.1126/science.1240971

Marty B, Avice G, Bekaert DV, Broadley MW (2018) Salinity of the Archaean oceans from analysis of fluid inclusions in quartz. C R Géosci 350(4):154–163

Matsui H, Iwagami N, Hosouchi M, Ohtsuki S, Hashimoto G (2012) Latitudinal distribution of HDO abundance above Venus' clouds by ground-based 2.3 μm spectroscopy. Icarus 217(2):610–614

McCord TB (1968) The loss of retrograde satellites in the solar system (1896–1977). J Geophys Res 73(4):1497–1500. https://doi.org/10.1029/JB073i004p01497

McGill GE (2000) Geologic map of the Sappho Patera quadrangle (V-20), Venus. The Survey

McKinnon WB, Zahnle KJ, Ivanov BA, Melosh HJ (1997) Cratering on Venus: models and observations. In: Bougher SW, Hunten DM, Phillips RJ (eds) Venus II: geology, geophysics, atmosphere, and solar wind environment, p 969

McMahon S (2019) Earth's earliest and deepest purported fossils may be iron-mineralized chemical gardens. Proc R Soc B 286(1916):20192,410

McTaggart R (2022) The cosmogenic production of phosphorus in the atmosphere of Venus. Icarus 374:114,791

Melosh H, Vickery A (1989) Impact erosion of the primordial atmosphere of Mars. Nature 338(6215):487–489

Ménez B, Pisapia C, Andreani M, Jamme F, Vanbellingen QP, Brunelle A, Richard L, Dumas P, Réfrégiers M (2018) Abiotic synthesis of amino acids in the recesses of the oceanic lithosphere. Nature 564(7734):59–63

Milojevic T, Treiman AH, Limaye S (2021) Phosphorus in the clouds of Venus: potential for bioavailability. Astrobiology 21(10):1250–1263. https://doi.org/10.1089/ast.2020.2267

Mogul R, Limaye SS, Way MJ, Cordova JA (2021a) Venus' mass spectra show signs of disequilibria in the middle clouds. Geophys Res Lett 48(7):e2020GL091,327. https://doi.org/10.1029/2020GL091327

Mogul R, Limaye SS, Lee YJ, Pasillas M (2021b) Potential for phototrophy in Venus' clouds. Astrobiology 21(10):1237–1249

Mojzsis SJ, Harrison TM, Pidgeon RT (2001) Oxygen-isotope evidence from ancient zircons for liquid water at the Earth's surface 4,300 Myr ago. Nature 409(6817):178–181

Mojzsis SJ, Brasser R, Kelly NM, Abramov O, Werner SC (2019) Onset of giant planet migration before 4480 million years ago. Astrophys J 881(1):44

Moorbath S, O'nions R, Pankhurst R (1973) Early Archaean age for the Isua iron formation, West Greenland. Nature 245(5421):138–139

Moore WB, Webb AAG (2013) Heat-pipe Earth. Nature 501(7468):501–505

Morbidelli A, Lunine JI, O'Brien DP, Raymond SN, Walsh KJ (2012) Building terrestrial planets. Annu Rev Earth Planet Sci 40:251–275. https://doi.org/10.1146/annurev-earth-042711-105319

Morowitz H, Sagan C (1967) Life in the clouds of Venus? Nature 215(5107):1259–1260. https://doi.org/10.1038/2151259a0

Morse JW, Mackenzie FT (1998) Hadean ocean carbonate geochemistry. Aquat Geochem 4(3):301–319

Mulder JA, Nebel O, Gardiner NJ, Cawood PA, Wainwright AN, Ivanic TJ (2021) Crustal rejuvenation stabilised Earth's first cratons. Nat Commun 12(1):1–7

Nabelek PI, Bédard JH, Rainbird RH (2014) Numerical constraints on degassing of metamorphic CO_2 during the Neoproterozoic Franklin large igneous event, Arctic Canada. Geol Soc Am Bull 126(5–6):759–772

Namiki N, Solomon SC (1998) Volcanic degassing of argon and helium and the history of crustal production on Venus. J Geophys Res, Planets 103(E2):3655–3677

Nelson D, Trendall A, Altermann W (1999) Chronological correlations between the Pilbara and Kaapvaal cratons. Precambrian Res 97(3–4):165–189

Nesvorný D (2018) Dynamical evolution of the early solar system. Annu Rev Astron Astrophys 56(1):137–174. https://doi.org/10.1146/annurev-astro-081817-052028

Nicholson WL, Fajardo-Cavazos P, Fedenko J, Ortíz-Lugo JL, Rivas-Castillo A, Waters SM, Schuerger AC (2010) Exploring the low-pressure growth limit: evolution of Bacillus subtilis in the laboratory to enhanced growth at 5 kilopascals. Appl Environ Microbiol 76(22):7559–7565. https://doi.org/10.1128/AEM.01126-10

Nijman W, Kloppenburg A, de Vries ST (2017) Archaean basin margin geology and crustal evolution: an East Pilbara traverse. J Geol Soc 174(6):1090–1112

Nikolaou A, Katyal N, Tosi N, Godolt M, Grenfell JL, Rauer H (2019) What factors affect the duration and outgassing of the terrestrial magma ocean? Astrophys J 875(1):11. https://doi.org/10.3847/1538-4357/ab08ed

Noack L, Breuer D, Spohn T (2012) Coupling the atmosphere with interior dynamics: implications for the resurfacing of Venus. Icarus 217(2):484–498

Noffke N, Christian D, Wacey D, Hazen RM (2013) Microbially induced sedimentary structures recording an ancient ecosystem in the ca. 3.48 billion-year-old Dresser Formation, Pilbara, Western Australia. Astrobiology 13(12):1103–1124

Norman MD, Borg LE, Nyquist LE, Bogard DD (2003) Chronology, geochemistry, and petrology of a ferroan noritic anorthosite clast from Descartes breccia 67215: clues to the age, origin, structure, and impact history of the lunar crust. Meteorit Planet Sci 38(4):645–661

Nutman AP, Bennett VC, Friend CR, Van Kranendonk MJ, Chivas AR (2016) Rapid emergence of life shown by discovery of 3,700-million-year-old microbial structures. Nature 537(7621):535–538

Nutman AP, Bennett VC, Friend CR, Van Kranendonk MJ, Rothacker L, Chivas AR (2019) Cross-examining Earth's oldest stromatolites: seeing through the effects of heterogeneous deformation, metamorphism and metasomatism affecting Isua (Greenland) 3700 Ma sedimentary rocks. Precambrian Res 331:105,347

Olson JM (2006) Photosynthesis in the Archean era. Photosynth Res 88(2):109–117. https://doi.org/10.1007/s11120-006-9040-5

Omran A, Oze C, Jackson B, Mehta C, Barge LM, Bada J, Pasek MA (2021) Phosphine generation pathways on rocky planets. Astrobiology 21(10):1264–1276

O'Neil J, Carlson RW, Francis D, Stevenson RK (2008) Neodymium-142 evidence for Hadean mafic crust. Science 321(5897):1828–1831

O'Neil J, Francis D, Carlson RW (2011) Implications of the Nuvvuagittuq greenstone belt for the formation of Earth's early crust. J Petrol 52(5):985–1009

O'Neill C, Lenardic A, Höink T, Coltice N (2014) Mantle convection and outgassing on terrestrial planets. In: Comparative climatology of terrestrial planets, pp 473–486

O'Neill C, Marchi S, Zhang S, Bottke W (2017) Impact-driven subduction on the Hadean Earth. Nat Geosci 10(10):793–797

O'Rourke JG, Korenaga J (2015) Thermal evolution of Venus with argon degassing. Icarus 260:128–140

Pahlevan K, Schaefer L, Hirschmann MM (2019) Hydrogen isotopic evidence for early oxidation of silicate Earth. Earth Planet Sci Lett 526:115,770. https://doi.org/10.1016/j.epsl.2019.115770

Papineau D, She Z, Dodd MS, Iacoviello F, Slack JF, Hauri E, Shearing P, Little CT (2022) Metabolically diverse primordial microbial communities in Earth's oldest seafloor-hydrothermal jasper. Sci Adv 8(15):eabm2296

Pascal R, Pross A, Sutherland JD (2013) Towards an evolutionary theory of the origin of life based on kinetics and thermodynamics. Open Biol 130:156

Perchuk AL, Safonov OG, Smit CA, van Reenen DD, Zakharov VS, Gerya T (2018) Precambrian ultra-hot orogenic factory: making and reworking of continental crust. Tectonophysics 746:572–586

Perchuk AL, Zakharov VS, Gerya T, Brown M (2019) Hotter mantle but colder subduction in the Precambrian: what are the implications? Precambrian Res 330:20–34

Perchuk AL, Gerya TV, Zakharov VS, Griffin WL (2020) Building cratonic keels in Precambrian plate tectonics. Nature 586(7829):395–401

Perchuk AL, Gerya TV, Zakharov VS, Griffin WL (2021) Depletion of the upper mantle by convergent tectonics in the early Earth. Sci Rep 11(1):1–12

Péron S, Moreira M (2018) Onset of volatile recycling into the mantle determined by xenon anomalies. Geochem Perspect Lett 9:21–25

Persson M, Futaana Y, Fedorov A, Nilsson H, Hamrin M, Barabash S (2018) H^+/O^+ escape rate ratio in the Venus magnetotail and its dependence on the solar cycle. Geophys Res Lett 45(20):10,805–10,811. https://doi.org/10.1029/2018GL079454

Persson M, Futaana Y, Ramstad R, Masunaga K, Nilsson H, Hamrin M, Fedorov A, Barabash S (2020) The Venusian atmospheric oxygen ion escape: extrapolation to the early solar system. J Geophys Res, Planets 125(3):e06336. https://doi.org/10.1029/2019JE006336

Phillips RJ, Raubertas RF, Arvidson RE, Sarkar IC, Herrick RR, Izenberg N, Grimm RE (1992) Impact craters and Venus resurfacing history. J Geophys Res, Planets 97(E10):15,923–15,948

Phillips RJ, Bullock MA, Hauck SA (2001) Climate and interior coupled evolution on Venus. Geophys Res Lett 28(9):1779–1782

Piccolo A, Palin RM, Kaus BJ, White RW (2019) Generation of Earth's early continents from a relatively cool Archean mantle. Geochem Geophys Geosyst 20(4):1679–1697

Piccolo A, Kaus BJ, White RW, Palin RM, Reuber GS (2020) Plume – lid interactions during the Archean and implications for the generation of early continental terranes. Gondwana Res 88:150–168

Pieters CM, Head JW, Pratt S, Patterson W, Garvin J, Barsukov VL, Basilevksy AT, Khodakovsky IL, Selivanov AS, Panfilov AS, Gektin YM, Narayeva YM (1986) The color of the surface of Venus. Science 234(4782):1379–1383. https://doi.org/10.1126/science.234.4782.1379

Pollack JB (1971) A nongrey calculation of the runaway greenhouse: implications for Venus' past and present. Icarus 14(3):295–306

Pross A, Pascal R (2013) The origin of life: what we know, what we can know and what we will never know. Open Biol 120:190

Raymond SN (2021) A terrestrial convergence. Nat Astron 5:875–876. https://doi.org/10.1038/s41550-021-01488-9

Reisz JA, Bansal N, Qian J, Zhao W, Furdui CM (2014) Effects of ionizing radiation on biological molecules—mechanisms of damage and emerging methods of detection. Antioxid Redox Signal 21(2):260–292

Retallack GJ, Metzger CA, Greaver T, Jahren AH, Smith RM, Sheldon ND (2006) Middle-late Permian mass extinction on land. Geol Soc Am Bull 118(11–12):1398–1411

Rey PF, Coltice N, Flament N (2014) Spreading continents kick-started plate tectonics. Nature 513(7518):405–408

Richter FM (1985) Models for the Archean thermal regime. Earth Planet Sci Lett 73(2–4):350–360

Rimmer PB, Jordan S, Constantinou T, Woitke P, Shorttle O, Hobbs R, Paschodimas A (2021) Hydroxide salts in the clouds of Venus: their effect on the sulfur cycle and cloud droplet pH. Planet Sci J 2(4):133. https://doi.org/10.3847/psj/ac0156

Robert F, Chaussidon M (2006) A palaeotemperature curve for the Precambrian oceans based on silicon isotopes in cherts. Nature 443(7114):969–972

Rolf T, Capitanio FA, Tackley PJ (2018) Constraints on mantle viscosity structure from continental drift histories in spherical mantle convection models. Tectonophysics 746:339–351

Rolf T, Weller M, Gülcher A, Byrne P, O'Rourke JG, Herrick R, Bjonnes E, Davaille A, Ghail R, Gillmann C, Plesa AC (2022) Dynamics and evolution of Venus' mantle through time. Space Sci Rev 218:70. https://doi.org/10.1007/s11214-022-00937-9

182

Rozel A, Golabek GJ, Jain C, Tackley PJ, Gerya T (2017) Continental crust formation on early Earth controlled by intrusive magmatism. Nature 545(7654):332–335

Ruiz J (2017) Heat flow evolution of the Earth from paleomantle temperatures: evidence for increasing heat loss since 2.5 Ga. Phys Earth Planet Inter 269:165–171

Rüpke LH, Morgan JP, Hort M, Connolly JA (2004) Serpentine and the subduction zone water cycle. Earth Planet Sci Lett 223(1–2):17–34

Russell MJ (2021) The "water problem"(sic), the illusory pond and life's submarine emergence—a review. Life 11(5):429

Russell MJ, Hall A (1997) The emergence of life from iron monosulphide bubbles at a submarine hydrothermal redox and pH front. J Geol Soc 154(3):377–402

Russell M, Hall A, Martin W (2010) Serpentinization as a source of energy at the origin of life. Geobiology 8(5):355–371

Sagan C, Mullen G (1972) Earth and Mars: evolution of atmospheres and surface temperatures. Science 177(4043):52–56

Sakuraba H, Kurokawa H, Genda H (2019) Impact degassing and atmospheric erosion on Venus, Earth, and Mars during the late accretion. Icarus 317:48–58

Sakuraba H, Kurokawa H, Genda H, Ohta K (2021) Numerous chondritic impactors and oxidized magma ocean set Earth's volatile depletion. Sci Rep 11(1):1–14

Salvador A, Massol H, Davaille A, Marcq E, Sarda P, Chassefière E (2017) The relative influence of H_2O and CO_2 on the primitive surface conditions and evolution of rocky planets. J Geophys Res, Planets 122(7):1458–1486

Salvador A, Avice G, Breuer D, Gillmann C, Lammer H, Marcq E, Raymond SN, Sakuraba H, Scherf M, Way MJ (2023) Magma ocean, water, and the early atmosphere of Venus. Space Sci Rev

Sasaki S, Yamagishi A, Yoshimura Y, Enya K, Miyakawa A, Ohno S, Fujita K, Usui T, Limaye SS (2022) In situ biochemical characterization of Venus cloud particles using a life-signature detection microscope. Can J Microbiol 68(6):413–425

Sasselov DD, Grotzinger JP, Sutherland JD (2020) The origin of life as a planetary phenomenon. Sci Adv 6(6):eaax3419

Sattler B, Puxbaum H, Psenner R (2001) Bacterial growth in supercooled cloud droplets. Geophys Res Lett 28(2):239–242. https://doi.org/10.1029/2000GL011684

Schaber G, Strom R, Moore H, Soderblom LA, Kirk RL, Chadwick D, Dawson D, Gaddis L, Boyce J, Russell J (1992) Geology and distribution of impact craters on Venus: what are they telling us? J Geophys Res, Planets 97(E8):13,257–13,301

Schaller MF, Wright JD, Kent DV (2011) Atmospheric P_{CO_2} perturbations associated with the Central Atlantic Magmatic Province. Science 331(6023):1404–1409

Schlichting HE, Mukhopadhyay S (2018) Atmosphere impact losses. Space Sci Rev 214(1):1–31

Schreiber U, Locker-Grütjen O, Mayer C (2012) Hypothesis: origin of life in deep-reaching tectonic faults. Orig Life Evol Biosph 42(1):47–54

Schubert G, Turcotte D, Solomon S, Sleep N (1989) Coupled evolution of the atmospheres and interiors of planets and satellites. In: Origin and evolution of planetary and satellite atmospheres, pp 450–483

Schuerger AC, Smith DJ, Griffin DW, Jaffe DA, Wawrik B, Burrows SM, Christner BC, Gonzalez-Martin C, Lipp EK, Schmale DG III, Yu H (2018) Science questions and knowledge gaps to study microbial transport and survival in Asian and African dust plumes reaching North America. Aerobiologia 34(4):425–435. https://doi.org/10.1007/s10453-018-9541-7

Schulze-Makuch D, Irwin LN (2002) Reassessing the possibility of life on Venus: proposal for an astrobiology mission. Astrobiology 2(2):197–202. https://doi.org/10.1089/15311070260192264

Schulze-Makuch D, Wagner D, Kounaves SP, Mangelsdorf K, Devine KG, de Vera JP, Schmitt-Kopplin P, Grossart HP, Parro V, Kaupenjohann M, Galy A, Schneider B, Airo A, Frösler J, Davila AF, Arens FL, Cáceres L, Cornejo FS, Carrizo D, Dartnell L, DiRuggiero J, Flury M, Ganzert L, Gessner MO, Grathwohl P, Guan L, Heinz J, Hess M, Keppler F, Maus D, McKay CP, Meckenstock RU, Montgomery W, Oberlin EA, Probst AJ, Sáenz JS, Sattler T, Schirmack J, Sephton MA, Schloter M, Uhl J, Valenzuela B, Vestergaard G, Wörmer L, Zamorano P (2018) Transitory microbial habitat in the hyperarid Atacama desert. Proc Natl Acad Sci 115(11):2670–2675. https://doi.org/10.1073/pnas.1714341115

Scorei R (2012) Is boron a prebiotic element? A mini-review of the essentiality of boron for the appearance of life on Earth. Orig Life Evol Biosph 42(1):3–17

Seager S, Petkowski JJ, Gao P, Bains W, Bryan NC, Ranjan S, Greaves J (2020) The Venusian lower atmosphere haze as a depot for desiccated microbial life: a proposed life cycle for persistence of the Venusian aerial biosphere. Astrobiology. https://doi.org/10.1089/ast.2020.2244

Seager S, Petkowski JJ, Carr CE, Grinspoon D, Ehlmann B, Saikia SJ, Agrawal R, Buchanan W, Weber MU, French R, Klupar P, Worden SP (2021) Venus life finder mission study. ArXiv e-prints. arXiv:2112.05153

183

Seager S, Petkowski JJ, Carr CE, Grinspoon DH, Ehlmann BL, Saikia SJ, Agrawal R, Buchanan WP, Weber MU, French R et al (2022) Venus life finder missions motivation and summary. Aerospace 9(7):385

Shao WD, Zhang X, Bierson CJ, Encrenaz T (2020) Revisiting the sulfur-water chemical system in the middle atmosphere of Venus. J Geophys Res, Planets 125(8):e2019JE006,195. https://doi.org/10.1029/2019JE006195

Sizova E, Gerya T, Brown M, Perchuk L (2010) Subduction styles in the Precambrian: insight from numerical experiments. Lithos 116(3–4):209–229

Sizova E, Gerya T, Brown M (2014) Contrasting styles of Phanerozoic and Precambrian continental collision. Gondwana Res 25(2):522–545

Sizova E, Gerya T, Stüwe K, Brown M (2015) Generation of felsic crust in the Archean: a geodynamic modeling perspective. Precambrian Res 271:198–224

Sleep NH (1994) Martian plate tectonics. J Geophys Res, Planets 99(E3):5639–5655

Sleep NH, Zahnle KJ, Lupu RE (2014) Terrestrial aftermath of the Moon-forming impact. Philos Trans R Soc A, Math Phys Eng Sci 372(2024):20130,172. https://doi.org/10.1098/rsta.2013.0172

Smerkar S, Ghail R, Byrne P (2022) Volcano-tectonic processes on Venus. Space Sci Rev

Smrekar SE, Davaille A, Sotin C (2018) Venus interior structure and dynamics. Space Sci Rev 214(5):88. https://doi.org/10.1007/s11214-018-0518-1

Snellen I, Guzman-Ramirez L, Hogerheijde M, Hygate A, Van der Tak F (2020) Re-analysis of the 267 GHz ALMA observations of Venus – no statistically significant detection of phosphine. Astron Astrophys 644:L2

Sobolev SV, Brown M (2019) Surface erosion events controlled the evolution of plate tectonics on Earth. Nature 570(7759):52–57

Solomatova NV, Caracas R (2021) Genesis of a CO_2-rich and H_2O-depleted atmosphere from Earth's early global magma ocean. Sci Adv 7(41):eabj0406

Som D, Catling S, Harnmeijer J, Polivka P, Buick R (2012) Air density 2.7 billion years ago limited to less than twice modern levels by fossil raindrop imprints. Nature 484:359. https://doi.org/10.1038/nature10890

Som S, Buick R, Hagadorn J, Blake T, Perreault J, Harnmeijer J, Catling D (2016) Earth's air pressure 2.7 billion years ago constrained to less than half of modern levels. NatGeo 9:448. https://doi.org/10.1038/ngeo2713

Sossi PA, Burnham AD, Badro J, Lanzirotti A, Newville M, O'Neill HS (2020) Redox state of Earth's magma ocean and its Venus-like early atmosphere. Sci Adv 6(48):eabd1387. https://doi.org/10.1126/sciadv.abd1387

Sousa-Silva C, Seager S, Ranjan S, Petkowski JJ, Zhan Z, Hu R, Bains W (2020) Phosphine as a biosignature gas in exoplanet atmospheres. Astrobiology 20(2):235–268

Spacek J, Benner SA (2021) The organic carbon cycle in the atmosphere of Venus and evolving red oil. In: LPI contributions, Lunar and Planetary Institute. https://ui.adsabs.harvard.edu/abs/2021LPICo2629.10520. Abstract #4052

Spencer CJ, Cawood PA, Hawkesworth CJ, Raub TD, Prave AR, Roberts NM (2014) Proterozoic onset of crustal reworking and collisional tectonics: reappraisal of the zircon oxygen isotope record. Geology 42(5):451–454

Stern CR (2011) Subduction erosion: rates, mechanisms, and its role in arc magmatism and the evolution of the continental crust and mantle. Gondwana Res 20(2–3):284–308

Stüeken EE, Anderson RE, Bowman JS, Brazelton WJ, Colangelo-Lillis J, Goldman AD, Som SM, Baross JA (2013) Did life originate from a global chemical reactor? Geobiology 11(2):101–126. https://doi.org/10.1111/gbi.12025

Stüeken E, Kipp M, Koehler M, Schwieterman E, Johnson B, Buick R (2016) Modeling pN_2 through geological time: implications for planetary climates and atmospheric biosignatures. Astrobiology 16(12):949–963. https://doi.org/10.1089/ast.2016.1537

Stüeken EE, Som SM, Claire M, Rugheimer S, Scherf M, Sproß L, Tosi N, Ueno Y, Lammer H (2020) Mission to planet Earth: the first two billion years. Space Sci Rev 216(2):31. https://doi.org/10.1007/s11214-020-00652-3

Svensen H, Planke S, Polozov AG, Schmidbauer N, Corfu F, Podladchikov YY, Jamtveit B (2009) Siberian gas venting and the end-Permian environmental crisis. Earth Planet Sci Lett 277(3–4):490–500

Szostak JW (2016) On the origin of life. Medicina 76(4):199–203

Tartèse R, Chaussidon M, Gurenko A, Delarue F, Robert F (2017) Warm Archean oceans reconstructed from oxygen isotope composition of early-life remnants. Geochem Perspect Lett 3:55–65

Tera F, Papanastassiou D, Wasserburg G (1973) A lunar cataclysm at $\sim$ 3.95 AE and the structure of the lunar crust. In: Lunar and planetary science conference, vol 4

Thompson MA (2021) The statistical reliability of 267-GHz JCMT observations of Venus: no significant evidence for phosphine absorption. Mon Not R Astron Soc Lett 501(1):L18–L22

Titov DV, Ignatiev NI, McGouldrick K, Wilquet V, Wilson CF (2018) Clouds and hazes of Venus. Space Sci Rev 214(8):126. https://doi.org/10.1007/s11214-018-0552-z

Trail D, Boehnke P, Savage PS, Liu MC, Miller ML, Bindeman I (2018) Origin and significance of Si and O isotope heterogeneities in Phanerozoic, Archean, and Hadean zircon. Proc Natl Acad Sci 115(41):10,287–10,292

Trinks H, Schröder W, Bierbricher C (2005) Sea ice as a promoter of the emergence of first life. Orig Life Evol Biosph 35:429–445

Trompet L, Robert S, Mahieux A, Schmidt F, Erwin J, Vandaele A (2021) Phosphine in Venus' atmosphere: detection attempts and upper limits above the cloud top assessed from the SOIR/VEx spectra. Astron Astrophys 645:L4

Truong N, Lunine J (2021) Volcanically extruded phosphides as an abiotic source of Venusian phosphine. Proc Natl Acad Sci 118(29):e2021689118

Turbet M, Bolmont E, Chaverot G, Ehrenreich D, Leconte J, Marcq E (2021) Day–night cloud asymmetry prevents early oceans on Venus but not on Earth. Nature 598(7880):276–280

Turcotte DL, Schubert G (2002) Geodynamics. Cambridge University Press, Cambridge

Turner G, Cadogan P (1975) The history of lunar bombardment inferred from ^{40}Ar-^{39}Ar dating of highland rocks. In: Lunar and planetary science conference proceedings, vol 6, pp 1509–1538

Uppalapati S, Rolf T, Crameri F, Werner S (2020) Dynamics of lithospheric overturns and implications for Venus's surface. J Geophys Res, Planets 125(11):e2019JE006,258

Valley J, Lackey J, Cavosie A, Clechenko C, Spicuzza M, Basei M, Bindeman I, Ferreira V, Sial A, King E et al (2005) 4.4 billion years of crustal maturation: oxygen isotope ratios of magmatic zircon. Contrib Mineral Petrol 150(6):561–580

Valley JW, Cavosie AJ, Ushikubo T, Reinhard DA, Lawrence DF, Larson DJ, Clifton PH, Kelly TF, Wilde SA, Moser DE et al (2014) Hadean age for a post-magma-ocean zircon confirmed by atom-probe tomography. Nat Geosci 7(3):219–223

Van den Boorn S, Van Bergen M, Vroon P, De Vries S, Nijman W (2010) Silicon isotope and trace element constraints on the origin of 3.5 Ga cherts: implications for early Archaean marine environments. Geochim Cosmochim Acta 74(3):1077–1103

van Hunen J, van den Berg AP (2008) Plate tectonics on the early Earth: limitations imposed by strength and buoyancy of subducted lithosphere. Lithos 103(1–2):217–235

Van Kranendonk MJ (2010) Two types of Archean continental crust: plume and plate tectonics on early Earth. Am J Sci 310(10):1187–1209

Van Kranendonk MJ, Baumgartner R, Djokic T, Ota T, Steller L, Garbe U, Nakamura E (2021) Elements for the origin of life on land: a deep-time perspective from the Pilbara Craton of Western Australia. Astrobiology 21(1):39–59

Vezinet A, Thomassot E, Pearson DG, Stern RA, Luo Y, Sarkar C (2019) Extreme δ^{18}O signatures in zircon from the Saglek Block (North Atlantic Craton) document reworking of mature supracrustal rocks as early as 3.5 Ga. Geology 47(7):605–608

Villanueva G, Cordiner M, Irwin P, de Pater I, Butler B, Gurwell M, Milam S, Nixon C, Luszcz-Cook S, Wilson C et al (2021) No evidence of phosphine in the atmosphere of Venus from independent analyses. Nat Astron 5(7):631–635

Volkov V, Frenkel MY (1993) The modeling of Venus' degassing in terms of K-Ar system. Earth Moon Planets 62(2):117–129

Walker JCG, Hays PB, Kasting JF (1981) A negative feedback mechanism for the long-term stabilization of the Earth's surface temperature. J Geophys Res 86:9776–9782. https://doi.org/10.1029/JC086iC10p09776

Walsh KJ, Morbidelli A, Raymond SN, O'Brien DP, Mandell AM (2011) A low mass for Mars from Jupiter's early gas-driven migration. Nature 475(7355):206–209

Warren AO, Kite ES (2021) Degassing, decarbonation, and dehydration: investigating the likelihood of a habitable era on Venus. In: 52nd lunar and planetary science conference. Lunar and planetary science conference, p 1253

Way MJ, Del Genio AD (2020) Venusian habitable climate scenarios: modeling Venus through time and applications to slowly rotating Venus-like exoplanets. J Geophys Res, Planets 125(5):e2019JE006,276

Way MJ, Del Genio AD, Kiang NY, Sohl LE, Grinspoon DH, Aleinov I, Kelley M, Clune T (2016) Was Venus the first habitable world of our solar system? Geophys Res Lett 43(16):8376–8383

Way MJ, Aleinov I, Amundsen DS, Chandler MA, Clune TL, Del Genio AD, Fujii Y, Kelley M, Kiang NY, Sohl L, Tsigaridis K (2017) Resolving orbital and climate keys of Earth and extraterrestrial environments with dynamics (ROCKE-3D) 1.0: a general circulation model for simulating the climates of rocky planets. Astrophys J Suppl Ser 231:12. https://doi.org/10.3847/1538-4365/aa7a06. arXiv:1701.02360

Way MJ, Del Genio AD, Aleinov I, Clune TL, Kelley M, Kiang NY (2018) Climates of warm Earth-like planets I: 3-D model simulations. ArXiv e-prints. arXiv:1808.06480

Way MJ, Ernst RE, Scargle JD (2022) Large-scale volcanism and the heat death of terrestrial worlds. Planet Sci J 3(4):92. https://doi.org/10.3847/psj/ac6033

Way MJ, Ostberg C, Foley BJ, Gillmann C, Höning D, Lammer H, O'Rourke J, Persson M, Plesa I, Salvador A, Scherf M, Weller M (2023) Synergies between Venus & exoplanetary observations. Space Sci Rev. 219:13. https://doi.org/10.1007/s11214-023-00953-3

Weiss MC, Preiner M, Xavier JC, Zimorski V, Martin WF (2018) The last universal common ancestor between ancient Earth chemistry and the onset of genetics. PLoS Genet 14(8):e1007,518. https://doi.org/10.1371/journal.pgen.1007518

Weller MB, Kiefer WS (2020) The physics of changing tectonic regimes: implications for the temporal evolution of mantle convection and the thermal history of Venus. J Geophys Res, Planets 125(1):e05960. https://doi.org/10.1029/2019JE005960

Weller MB, Kiefer WS (2021) Punctuated evolution of the Venusian atmosphere from mantle outgassing. In: 52nd lunar and planetary science conference. Lunar and planetary science conference, p 1555

Weller M, Lenardic A, O'Neill C (2015) The effects of internal heating and large scale climate variations on tectonic bi-stability in terrestrial planets. Earth Planet Sci Lett 420:85–94

Weller M, Evans A, Ibarra D, Johnson A, Kukla T (2022) Atmospheric evidence of early plate tectonics on Venus. LPI Contrib 2678:2328

Westall F, De Ronde CE, Southam G, Grassineau N, Colas M, Cockell C, Lammer H (2006a) Implications of a 3.472–3.333 Gyr-old subaerial microbial mat from the Barberton greenstone belt, South Africa for the UV environmental conditions on the early Earth. Philos Trans R Soc B, Biol Sci 361(1474):1857–1876

Westall F, de Vries ST, Nijman W, Rouchon V, Orberger B, Pearson V, Watson J, Verchovsky A, Wright I, Rouzaud JN, Marchesini D, Severine A (2006b) The 3.466 Ga "Kitty's Gap Chert," an early Archean microbial ecosystem. In: Processes on the early Earth. Geol. Soc. Am., Boulder. https://doi.org/10.1130/2006.2405(07)

Westall F, Cavalazzi B, Lemelle L, Marrocchi Y, Rouzaud JN, Simionovici A, Salomé M, Mostefaoui S, Andreazza C, Foucher F et al (2011a) Implications of in situ calcification for photosynthesis in a ∼ 3.3 Ga-old microbial biofilm from the Barberton greenstone belt, South Africa. Earth Planet Sci Lett 310(3–4):468–479

Westall F, Foucher F, Cavalazzi B, de Vries ST, Nijman W, Pearson V, Watson J, Verchovsky A, Wright I, Rouzaud JN et al (2011b) Volcaniclastic habitats for early life on Earth and Mars: a case study from 3.5 Ga-old rocks from the Pilbara, Australia. Planet Space Sci 59(10):1093–1106

Westall F, Campbell KA, Bréhéret JG, Foucher F, Gautret P, Hubert A, Sorieul S, Grassineau N, Guido DM (2015) Archean (3.33 Ga) microbe-sediment systems were diverse and flourished in a hydrothermal context. Geology 43(7):615–618

Westall F, Hickman-Lewis K, Hinman N, Gautret P, Campbell K, Bréhéret JG, Foucher F, Hubert A, Sorieul S, Dass AV et al (2018) A hydrothermal-sedimentary context for the origin of life. Astrobiology 18(3):259–293

Whitehouse MJ, Nemchin AA, Pidgeon RT (2017) What can Hadean detrital zircon really tell us? A critical evaluation of their geochronology with implications for the interpretation of oxygen and hafnium isotopes. Gondwana Res 51:78–91

Wignall P (2001) Large igneous provinces and mass extinctions. Earth-Sci Rev 53(1–2):1–33. https://doi.org/10.1016/S0012-8252(00)00037-4

Wilde SA, Valley JW, Peck WH, Graham CM (2001) Evidence from detrital zircons for the existence of continental crust and oceans on the Earth 4.4 Gyr ago. Nature 409(6817):175–178

Wordsworth R, Pierrehumbert R (2013) Water loss from terrestrial planets with CO_2-rich atmospheres. Astrophys J 778(2):154

Yang J, Boué G, Fabrycky DC, Abbot DS (2014) Strong dependence of the inner edge of the habitable zone on planetary rotation rate. Astrophys J 787:L2. https://doi.org/10.1088/2041-8205/787/1/L2. arXiv:1404.4992

Zahnle KJ, Lupu R, Dobrovolskis A, Sleep NH (2015) The tethered Moon. Earth Planet Sci Lett 427:74–82

Zasova L, Gorinov D, Eismont N, Kovalenko I, Abbakumov A, Bober S (2019) Venera-d: a design of an automatic space station for Venus exploration. Sol Syst Res 53(7):506–510

Zawaski MJ, Kelly NM, Orlandini OF, Nichols CI, Allwood AC, Mojzsis SJ (2020) Reappraisal of purported ca. 3.7 Ga stromatolites from the Isua supracrustal belt (West Greenland) from detailed chemical and structural analysis. Earth Planet Sci Lett 545:116,409

Zellner NEB (2017) Cataclysm no more: new views on the timing and delivery of lunar impactors. Orig Life Evol Biosph 47(3):261–280. https://doi.org/10.1007/s11084-017-9536-3

Publisher's Note Springer Nature remains neutral with regard to jurisdictional claims in published maps and institutional affiliations.

  Springer

Authors and Affiliations

F. Westall[1] · D. Höning[2] · G. Avice[3] · D. Gentry[4] · T. Gerya[5] · C. Gillmann[6] · N. Izenberg[7] · M.J. Way[8,9] · C. Wilson[10,11]

✉ D. Höning
dennis.hoening@pik-potsdam.de

[1] CNRS-Centre de Biophysique Moléculaire, Orléans, France

[2] Potsdam Institute for Climate Impact Research, Potsdam, Germany

[3] Institut de physique du globe de Paris, CNRS, Université Paris Cité, 75005 Paris, France

[4] NASA Ames Research Center, Moffett Field, CA, USA

[5] Dep. of Earth Sciences, ETH Zürich, Zürich, Switzerland

[6] Department of Earth, Environmental and Planetary Sciences, Rice University, Houston, TX 77005, USA

[7] The Applied Geophysics Laboratory, The Johns Hopkins University, 3400 North Charles Street, Malone 140, Baltimore, MD 21218, USA

[8] NASA Goddard Institute for Space Studies, 2880 Broadway, New York, NY 10025, USA

[9] Theoretical Astrophysics, Department of Physics and Astronomy, Uppsala University, Uppsala, Sweden

[10] Department of Physics, University of Oxford, Sherrington Road, Oxford OX1 3PU, UK

[11] European Space Agency, Noordwijk, The Netherlands

Space Science Reviews (2023) 219:51
https://doi.org/10.1007/s11214-023-00995-7

Magma Ocean, Water, and the Early Atmosphere of Venus

Arnaud Salvador[1,2,3] · Guillaume Avice[4] · Doris Breuer[5] · Cédric Gillmann[6] ·
Helmut Lammer[7] · Emmanuel Marcq[8] · Sean N. Raymond[9] ·
Haruka Sakuraba[10] · Manuel Scherf[7,11,12] · M.J. Way[13,14]

Received: 22 February 2023 / Accepted: 10 August 2023 / Published online: 20 September 2023
© The Author(s) 2023

Abstract

The current state and surface conditions of the Earth and its twin planet Venus are drastically
different. Whether these differences are directly inherited from the earliest stages of planetary evolution, when the interior was molten, or arose later during the long-term evolution is
still unclear. Yet, it is clear that water, its abundance, state, and distribution between the different planetary reservoirs, which are intimately related to the solidification and outgassing
of the early magma ocean, are key components regarding past and present-day habitability,
planetary evolution, and the different pathways leading to various surface conditions.

In this chapter we start by reviewing the outcomes of the accretion sequence, with particular emphasis on the sources and timing of water delivery in light of available constraints,
and the initial thermal state of Venus at the end of the main accretion. Then, we detail the processes at play during the early thermo-chemical evolution of molten terrestrial planets, and
how they can affect the abundance and distribution of water within the different planetary
reservoirs. Namely, we focus on the magma ocean cooling, solidification, and concurrent
formation of the outgassed atmosphere. Accounting for the possible range of parameters for
early Venus and based on the mechanisms and feedbacks described, we provide an overview
of the likely evolutionary pathways leading to diverse surface conditions, from a temperate
to a hellish early Venus. The implications of the resulting surface conditions and habitability
are discussed in the context of the subsequent long-term interior and atmospheric evolution.
Future research directions and observations are proposed to constrain the different scenarios
in order to reconcile Venus' early evolution with its current state, while deciphering which
path it followed.

Keywords Venus · Interior evolution · Atmosphere · Degassing

1 Introduction

Water is not only an essential element for the chemical and biological processes responsible
for the emergence and development of life and living organisms (e.g., Westall and Brack
2018; Hoehler et al. 2018). Its abundance and distribution within and between the different

Venus: Evolution Through Time
Edited by Colin F. Wilson, Doris Breuer, Cédric Gillmann, Suzanne E. Smrekar, Tilman Spohn and
Thomas Widemann

Extended author information available on the last page of the article

reservoirs of a rocky planet, from the deep interior to the exosphere, is also of fundamental importance in controlling their properties and thus the physical and chemical processes at play within and between these reservoirs, making it a crucial factor in setting the conditions suitable for the development of life and for the overall planetary evolution (e.g., Kasting and Catling 2003; Hirschmann 2006; Karato 2015; Foley and Driscoll 2016; Korenaga et al. 2017; Ohtani 2020). Understanding the mechanisms shaping the water budget within the planets and how they evolve through time is thus required to build a consistent picture of planetary evolution and characterize the potential habitability of terrestrial planets (classically defined as the ability of a planet to sustain liquid water at its surface; e.g., Kasting et al. 1993b).

From that perspective, the early stages of planetary evolution appear to be of first importance. Indeed, during this relatively short but intense period of time, the combination of different heat sources, in particular the energy delivered by successive collisions of planetary building-blocks during planetary accretion and growth, is believed to substantially melt the planetary body (e.g., Safronov 1978; Kaula 1979; Tonks and Melosh 1993; Nakajima et al. 2021), resulting in the creation of a so-called "magma ocean" ("MO"): a possibly global, from mantle-thick to near-surface shell of magma of low, water-like viscosity (hence "ocean"; e.g., Warren 1985). During this magma ocean stage, because of the molten surface, the absence of a thick, long-lasting boundary layer between the planetary interior and the atmosphere allows for free and extremely efficient thermal and chemical exchanges between these reservoirs.

The modern concept of magma ocean originally took root in the study of the mineralogy of the Moon's surface from the lunar samples brought back to Earth in the seventies. The formation of the bright anorthositic crust, making up the ancient lunar highlands covering most of the surface (e.g., Ji et al. 2022), was proposed to result from the flotation of light Ca- and Al-rich anorthite minerals (plagioclase feldspar) over a crystallizing, denser molten silicate layer, hence requiring the upper part of the lunar mantle to be initially molten (e.g., Smith et al. 1970; Wood et al. 1970; Warren 1985; Binder 1986). Early and extensive melting of the surface and subsurface was confirmed by trace element analysis and experimental petrology (Schnetzler and Philpotts 1971; Taylor 1986). The origin of the lunar magma ocean was then tied to the formation process of the Moon itself and in particular to the energetics of collisional accretion (e.g., Wetherill 1976; Taylor 1986; Matsui and Abe 1986c; Hartmann 1986; Cameron 1986), where conversion of kinetic energy into heat from impacts could increase the temperature up to the melting point of silicates and potentially melt the surface and the interior of the impacted body to an extent depending primarily on the impactor size and resulting heat burial.

Reciprocally, the Earth was thought to be significantly heated and molten as a result of the Moon-forming event(s) and more generally out of the accretion sequence (Safronov 1978; Kaula 1979; Wetherill 1985), thought to be of increasing violence with time, with the increasing size of impactors. By considering each plausible combination of heating mechanisms involved in melting the Moon, Hostetler and Drake (1980) extended the analysis on the early thermal state of the Earth and other terrestrial planets and suggested that early global melting (and differentiation) was also likely for all terrestrial planets (e.g., Drake 2000). The hypothesis of a single moon-forming giant impact gained popularity and became the overarching paradigm for lunar origin (e.g., Hartmann and Davis 1975; Cameron and Ward 1976; Benz et al. 1986; Boss 1986; Taylor 1986; Newsom and Ross Taylor 1989; Drake 2000; Canup and Asphaug 2001). The reciprocal idea that the Earth experienced a planetary-scale global magma ocean, possibly melting the entire mantle molten (Wetherill 1985), as a result of such collision (or from violent accretion) gained interest (Melosh 1990;

Tonks and Melosh 1993). The existence of a molten stage on rocky planets was also independently confirmed by early melting in planetesimals triggered by radiogenic heating from the decay of radioactive elements (e.g., Urey 1955; Fish et al. 1960; Lee et al. 1976). It is now generally accepted that all terrestrial planets, including Venus, experience at least a transient magma ocean stage early in history but the number of occurrences, timing, timescale and depth of the molten layer may significantly vary from one planet to another (see Elkins-Tanton 2012; Schaefer and Elkins-Tanton 2018, for reviews).

In this phase, some of the major processes shaping the conditions at the surface and in the interior of the planet take place, and set the initial conditions for the later long-term state and evolution of the planet (e.g., Schaefer and Elkins-Tanton 2018). The volatile and "atmophile" (i.e., air-loving) species dissolved within the molten mantle are outgassed and build-up the atmosphere, the rapid segregation of heavy metallic iron (and associated "siderophile", i.e., metal-loving, elements) and lighter silicate (and "lithophile", i.e., stone-loving, elements) favored by the low viscosity of the melt is responsible for the chemical differentiation of the planet. The distribution of water between the different reservoirs, the amount reaching the surface as well as the early climate and associated liquid water sustainability at the surface are direct outcomes of the magma ocean solidification. The resulting internal structure and elemental distribution thus set the starting point for the onset of mantle convection and for the long-term evolution of the planet. These early stages of planetary evolution are thus of fundamental importance in controlling the potential past and present-day habitability of a planet and in understanding the pathways to the present-day surface conditions.

For a planet to be habitable, two *sine qua non* conditions must be achieved. First, water needs to reach and be abundant enough at the surface of the planet, and second, the surface conditions (pressure and temperature) must allow for the water to condense. These two conditions are often addressed separately, while implicitly assuming that the other condition is satisfied. One positive condition does not imply the other, yet they are intricately linked. As mentioned above, these two requirements strongly rely on the outcomes of the magma ocean phase. On the other hand, maintaining a temperate and stable climate with sustainable water oceans long enough for life to emerge then relies on additional processes such as atmospheric escape and geochemical cycles, which affect the planetary environment over longer timescales, throughout the entire evolution of the planet, including when the internal heat is no longer climatologically significant.

Given the fact that planetary surface conditions are essentially controlled by atmospheric and climate dynamics, habitable zone and habitability-related studies have been historically conducted from an atmospheric point-of-view, using atmospheric models as main tools to assess the surface temperature and pressure and infer the presence of liquid water at the planetary surface (e.g., Kasting et al. 1993b; Selsis et al. 2007; Kopparapu et al. 2013; Ramirez 2018). Despite being highly influential, the early evolution stages have been neglected in habitable zone studies, thus possibly missing important pieces of the puzzle. Only recently has the importance of interior–atmosphere feedback on habitability been recognized (e.g., Zahnle et al. 2007; Noack et al. 2014; Foley and Driscoll 2016; Ramirez et al. 2018; Dehant et al. 2019).

The case of Venus is still puzzling and these neglected components might hold the key. Given their vicinity in the inner Solar System, the Earth and Venus likely experienced many similarities during their formation, including analogous accretion sequences, chemical elements endowments and volatile delivery. Their nearly identical sizes and bulk densities suggesting similar bulk compositions (e.g., Smrekar et al. 2018) are also indicative of a common past. Yet, Earth's present-day liquid water- and life-sustaining temperate climate differs strikingly from the mostly dry and hellish Venusian surface conditions. Whether these differences are directly inherited from the tumultuous magma ocean stage or appeared later in history remains unclear. Studying the planetary evolution and in particular this common molten past, at the earliest ages of the Solar System, may be one of the keys to resolving this paradox and unveil the evolutionary pathway(s) leading to a habitable and inhabited planet. This is the object of the present chapter, with particular emphasis on the role and feedback of and between water and interior-atmosphere interaction when the planet was molten. Throughout the chapter, we will then address the following key questions:

- What are the initial conditions and water content of the molten interior of Venus? (Sect. 2)
- What processes affect the thermal evolution of the magma ocean and the concurrent atmospheric formation? (Sect. 3)
- How is water distributed between the solidifying mantle, the molten interior, and the atmosphere during magma ocean evolution? (Sect. 3)
- What are the outcomes of the magma ocean phase and the corresponding implications for the habitability and long-term evolution of the planet? (Sect. 4)
- When and how did the Earth and Venus surface conditions diverge? (Sect. 4)
- What are the next steps to achieve towards a better understanding of early Venusian evolution? (Sect. 5)

To do so, the chapter is organized as follows: Sect. 2 describes the initial conditions of the molten stage evolution, by focusing on its initial thermo-chemical state resulting from the accretion sequence, and on the initial amount of water accreted as a function of the different sources and processes of volatiles delivery. Section 3 focuses on the processes at play during magma ocean thermal evolution and concurrent formation of the atmosphere, including the feedback between the interior and the atmospheric reservoirs. Section 4 examines the possible outcomes of the magma ocean cooling sequence and how the different evolutionary paths could influence the subsequent surface conditions and long-term evolution of the planet. Finally, Sect. 5 discusses how further research directions combined with future observations and measurements could provide clues to decipher the early evolution of Venus and choose between the different scenarios to eventually reconstruct a consistent picture of the history of the planet.

2 Water Delivery and Initial Conditions for the Molten Mantle Evolution: What do We Start with?

Near the end of planetary accretion, the planet's water inventory and thermal state play the two key roles in its subsequent evolution. Yet, even for Earth, these parameters are highly unconstrained. In this section, we provide an overview of the processes that could be responsible for water delivery on early Venus. We estimate the amount of water delivered, and describe the early energy sources that control the initial heat budget of the planet. In Sect. 2.1 we discuss the possibility for Venus to produce water from the nebular gas gravitationally captured within the protoplanetary disk. We then describe the outcomes of the accretion sequence as a function of the different planetary formation scenarios in Sect. 2.2. While the sources, timing, and amount of water delivered to Earth are still debated, in Sect. 2.2.1 we discuss how the different planetary formation scenarios and their implications for water delivery apply to Venus. We also describe the different heat sources at play and the corresponding thermal state of the mantle resulting from the accretion sequence (Sect. 2.2.2). We then examine scenarios of water delivery, budget, and fate in light of available constraints from isotopic ratios in Sect. 2.3.

2.1 Proto-Venus and H₂/He Envelope: Did Early Venus Produce Water from the Protoplanetary Nebula?

If proto-Venus accreted a sufficient mass within the protoplanetary disk lifetime, the growing planet could have captured an H_2-dominated primordial atmosphere by accreting nebular gas from the circumstellar disk, which could have then been quickly lost by Extreme UltraViolet-driven (EUV) hydrodynamic escape after the disk dissipated (e.g., Hayashi et al. 1979; Stökl et al. 2016; Sharp 2017). If this was the case, the efficient hydrodynamic hydrogen flow dragged heavier elements such as noble gas isotopes with it, leading to modifications of their initial isotope ratios (Pepin 1991).

Recently, Lammer et al. (2020b) applied a hydrodynamic upper atmosphere and a Smooth Particle Hydrodynamics (SPH) impact model for the loss calculations of captured H_2-dominated envelopes. This was done for various protoplanetary masses and a wide range of possible EUV-activity evolution tracks of the young Sun and initial atmospheric compositions based on mixtures of captured nebula gas, outgassed and delivered materials. As shown in Fig. 1, these authors could reproduce the present Venus atmospheric $^{36}Ar/^{38}Ar$, $^{20}Ne/^{22}Ne$ isotope ratios if proto-Venus accreted a mass of $\sim$0.84–1.0 M_{Venus} ($\sim$0.68–0.81 $M_{\oplus}$) during the disk lifetime of $\sim$3–4.5 Myr (Bollard et al. 2017; Wang et al. 2017) after the origin of the Sun.

Lammer et al. (2020b) also showed that proto-Venus could have captured primordial H_2-dominated envelopes with hydrogen surface partial pressures between $\sim$40–1000 bar (Fig. 1). The blanketing effect of such dense and opaque atmospheres would be responsible

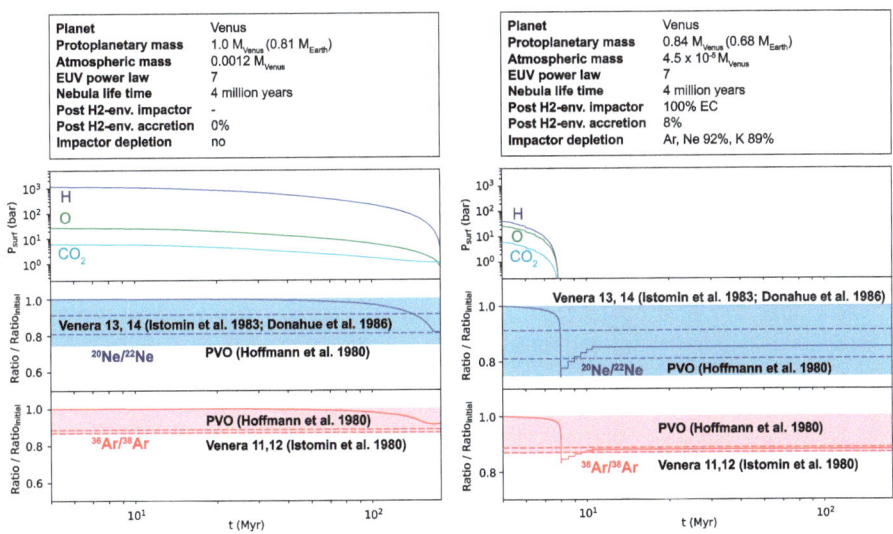

Fig. 1 a) Successful reproduction attempt of the Pioneer Venus, Venera 11 and 12 $^{36}Ar/^{38}Ar$, and Venera 13 and 14 $^{20}Ne/^{22}Ne$ noble gas isotope ratios within the measured error bars if Venus accreted its final mass during the disk lifetime. In such a case the planet would have lost, a hydrogen-dominated primordial atmosphere with $\sim$1000 bar partial pressure after $\geq$ 200 Myr if the young Sun was a weakly to moderately active young G-star. b) Successful reproduction attempt of the Venera 11 and 12 $^{36}Ar/^{38}Ar$, and Venera 13 and 14 $^{20}Ne/^{22}Ne$ noble gas ratios within the measured error bars for a proto-Venus with $\sim$0.84 M_{Venus} ($\sim$0.68 $M_{\oplus}$). We use a 100% EC-like building block composition that accreted about 17% of the planet mass after a primordial atmosphere with a hydrogen partial pressure of $\sim$40 bar was lost until $\sim$7 Myr after the origin of the Sun and $\sim$3 Myr after the disk evaporated (after Lammer et al. 2020b)

for extensive mantle melting (e.g., Hayashi et al. 1979; Olson and Sharp 2019). Then, if proto-Venus grew partly within the gas disk (Ikoma and Genda 2006; Ikoma et al. 2018), the primordial atmospheric H_2 can be oxidized by interactions between the gas and the underlying molten surface. This requires surface temperatures above the melting point of silicate ($\sim$1400 K), thereby producing water that originates from the protoplanetary nebula. Hydrogen and water would thus be ingassed into the molten interior while some other fraction of hydrogen would possibly be sequestered into the core (e.g., Sharp 2017; Wu et al. 2018; Olson and Sharp 2018, 2019; Young et al. 2023). How much water can be produced from the nebula depends both on the mass of the surrounding H_2-envelope and on the oxidation state of the magma ocean, characterized by the available iron oxides (e.g., wüstite and magnetite) and fayalite (Fe_2SiO_4) in the magma ocean (Ikoma and Genda 2006). Logically, the amount of water produced from primordial H_2 increases with the increasing chemical potential of oxygen (i.e., oxygen fugacity; see Sect. 3.2.5), as more oxygen atoms available combine with hydrogen atoms to produce more water. Indeed, according to Ikoma and Genda (2006), at 1500 K the mass ratio of produced H_2O to the primordial atmospheric H_2, M_{H_2O}/M_{H_2}, increases from approximately 0.49 to 0.88, and to 24.02 for increasingly oxidizing environments, corresponding to the buffer assemblages of quartz-iron-fayalite (QIF):

$$\underset{\text{iron}}{2\,Fe} + \underset{\text{quartz}}{SiO_2} + O_2 \rightleftarrows \underset{\text{fayalite}}{Fe_2SiO_4}, \tag{R1}$$

iron-wüstite (IW):

$$\underset{\text{iron}}{2\,x\,Fe} + O_2 \rightleftarrows \underset{\text{wüstite}}{2\,Fe_xO}, \tag{R2}$$

and wüstite-magnetite (WM):

$$\underset{\text{wüstite}}{6/(4x-3)\,Fe_xO} + O_2 \rightleftarrows \underset{\text{magnetite}}{2x/(4x-3)\,Fe_3O_4} \tag{R3}$$

respectively (e.g., Frost 1991). Note that the side of the reaction that has greater entropy,[1] that is the side with oxygen, is stable at higher temperature than the other side (e.g., Fig. 1 in Frost 1991). Ikoma and Genda (2006) used values of $x = 0.947$ for IW and $x = 0.974$ for WM taken from Robie et al. (1978).

Note that the oxidation state of the magma ocean, and therefore water production efficiency, relates to the timing and location of metal–silicate separation. If metallic iron has not yet segregated from the magma ocean and coexists with the molten silicate, as believed for the earliest evolution steps, the magma ocean is thought to be reducing. It would be at the QIF or IW redox buffer rather than at the WM buffer, thus producing less water than if the metallic iron has already segregated into the core (see Sect. 3.2.5 for definitions of mineral redox buffers and discussion regarding the magma ocean oxidation state). However, the picture might be more complicated, making the influence of the magma ocean oxidation state on the process of hydrogen ingassing and water production less straightforward. For instance, the ingassing hydrogen may cause Fe reduction, followed by iron removal to the core. In this case, Sharp (2017) suggested that far larger amounts of H_2 could be incorporated into the melt, which could in turn enhance water production even with abundant metallic iron in a reducing environment.

---

[1]By convention reactions are written with the high entropy side on the right. Here we kept it on the left, so that the oxidation reads from left to right.

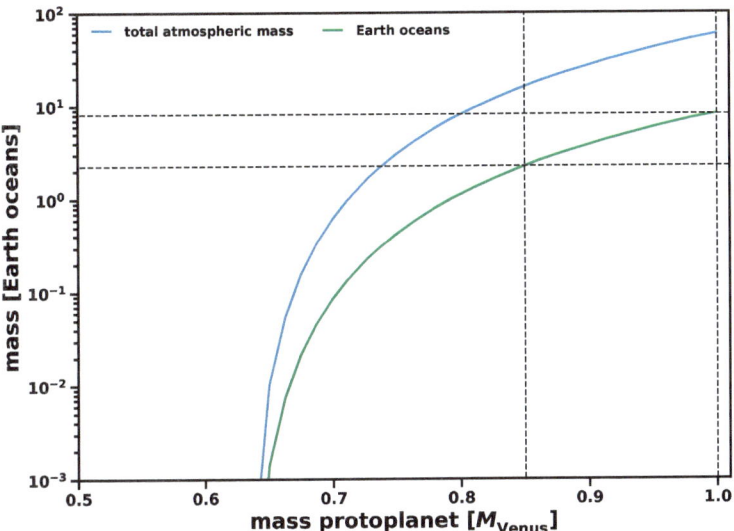

Fig. 2 The amount of Earth oceans (green line) that may be produced by a captured H_2-dominated primordial atmosphere (blue line) due to gas–melt interactions between the atmosphere and a magma ocean according to Ikoma and Genda (2006). The vertical and horizontal dashed-lines correspond to the possible proto-Venus mass and corresponding primordial atmospheric mass ranges where Venus' present-day Ar and Ne isotope ratios can be reproduced within the error bars

If one applies the same assumption as Ikoma and Genda (2006) for the hydrogen envelopes as simulated by Lammer et al. (2020b), shown in Fig. 1, and assumes that such oxygen buffers (i.e., QIF, IW, and WM) are available, a proto-Venus with a mass of ~ 0.84 M_{Venus} could produce a water reservoir on early Venus from the captured H_2-envelope of at least ~ 2 Earth oceans, as shown in Fig. 2. In case proto-Venus accreted its final mass during the disk lifetime, then a nebular-based H_2O amount of at least ~ 8 Earth oceans could have been produced (see Fig. 2). The probability of terrestrial planet formation within the nebular gas is still debated. This is because the dissipation mechanisms remain uncertain. Yet, the capture of other nebular gases such as Ne followed by their incorporation into molten planetary interiors has further been inferred from mantle isotopic ratios (Williams and Mukhopadhyay 2019). It supports the idea that terrestrial planets may partly form before the nebular gas dissipated. One should note that molten silicate–metallic core separation may not be complete by the time the nebular gas fully vanishes, suggesting a rather reducing magma ocean and associated relatively low water production (e.g., at the water production rates imposed by the QIF or IW buffers).

In comparison, Lammer et al. (2020a) found that Earth's present-day atmospheric $^{36}Ar/^{38}Ar$, $^{20}Ne/^{22}Ne$, and $^{36}Ar/^{22}Ne$ isotope ratios could only be reproduced if the protoplanet grew a mass of ≤ 0.6 $M_{\oplus}$ within the solar nebula. It implies that, contrary to Venus, only $\sim 2\%$ of the current value of Earth's seawater could have been produced from the protoplanetary nebula. This low value for the Earth lies within the error bars of the D/H seawater-carbonaceous chondrite "match" of $150\pm 10 \times 10^{-6}$ (Robert et al. 2000; Pahlevan et al. 2019), and is in agreement with measurements of the D/H ratios in glassy melt inclusions in two deep mantle basaltic rock samples (Hallis et al. 2015). This suggests that only a small fraction of Earth's water was derived from nebular accretion (e.g., Wu et al. 2018). Nevertheless, Young et al. (2023) recently emphasized that primordial H_2 atmospheres remain an influential and promising mechanism to explain fundamental chemical features of terrestrial planets, such as Earth's water content, core density, and oxidation state.

If proto-Venus grew partly within the gas disk and produced nebula-based water, its initial D/H ratio of $21\pm5 \times 10^{-6}$ could then resemble that of nebular gas which is ~7 times lower than Earth's seawater (Geiss and Gloeckler 1998; Robert et al. 2000) and CCs (e.g., Marty 2012) measured D/H ratios. If this was the case, then Venus' present-day D/H ratio of $160\pm20 \times 10^{-4}$ (Donahue et al. 1982) would be ~762 times higher than initially instead of ~106 times, if the planets initial H_2O inventory were CC-based as on Earth. In such a case atmospheric escape must have fractionated D/H much more significantly than expected in previous studies (Donahue 1999).

Note that many stars are surrounded by gaseous and dusty debris disks. Compared to the protoplanetary disks surrounding young newly formed stars, those long-lasting (possibly for hundreds of millions of years; Wyatt 2008; Matrà et al. 2017) debris disks can be found and remain after planets have already formed. They result from the collisions and destruction of asteroids, comets, planetesimals, and dwarf planets (rather than being of primordial origin as the younger protoplanetary disks; e.g., Dent et al. 2014; Hughes et al. 2018; Matrà et al. 2019). Rather than being dominated by hydrogen and helium, the gas of those disks is mainly composed of CO, carbon, and oxygen (e.g., Cataldi et al. 2014; Kral et al. 2017). Terrestrial planets could form massive atmospheres from the capture of the gas contained in those debris disks (Kral et al. 2020). Yet, the amount of water produced and incorporated into the planet from such accreted atmospheres has not been assessed. It may be limited compared to protoplanetary disks due to the absence of hydrogen in their carbon and oxygen rich compositions (e.g., Cataldi et al. 2014; Kral et al. 2017).

2.2 Outcomes of Planetary Formation Scenarios

2.2.1 Water Delivery

Understanding the growth of the rocky planets is essential because it sets their volatile inventories. Yet our picture of the growth of the rocky planets remains incomplete. This comes in large part from a lack of hard constraints during the era of accretion. The initial conditions for Solar System formation are likely represented in the circumstellar disks that are directly observed around nearby young stars, for instance with the ALMA telescope (e.g., ALMAPartnership et al. 2015; Andrews et al. 2018). The population of known exoplanets represent thousands of outcomes of planet formation (e.g., Winn and Fabrycky 2015). At intermediate size- and time scales between young circumstellar disks and mature planetary systems, empirical constraints are hard to find.

The current paradigm of Solar System formation involves a series of different stages of growth (for a review, see Raymond and Morbidelli 2022). First, micron-sized dust grains collide and coagulate until they reach one of several growth barriers at roughly mm-sizes (Blum and Wurm 2008; Zsom et al. 2010). These large dust grains are commonly called "pebbles", and have sufficient inertia that their orbits are mainly Keplerian (see Johansen and Lambrechts 2017). However, the gas orbits slightly slower than the Keplerian speed because of partial pressure support. Pebbles therefore feel a headwind from the gas, which causes them to lose orbital energy and spiral inward – or rather, along the local pressure gradient – on short timescales (Weidenschilling 1977; Haghighipour and Boss 2003). Pebbles generally drift inward, but can become concentrated in regions in pressure bumps within the gas disk (Birnstiel et al. 2012). When pebbles are sufficiently concentrated in any given region, they can be further concentrated by the streaming instability and directly clump into

10–100 km-scale planetesimals (Youdin and Goodman 2005; Johansen et al. 2009, 2014). These are generally considered the "building blocks" of the planets. While the classical model assumes that planetesimals formed uniformly across the disk (e.g., Raymond et al. 2014), new models propose that planetesimals form in different regions at different times, often at (moving) condensation fronts (Armitage et al. 2016; Drążkowska and Alibert 2017; Drążkowska and Dullemond 2018; Morbidelli et al. 2022; Izidoro et al. 2022). Planetesimals continue to grow by both collisions between planetesimals and by accreting pebbles that continue to drift within the disk (Johansen and Lambrechts 2017; Ormel 2017). The giant planet cores are thought to have rapidly accreted by accreting pebbles (Lambrechts and Johansen 2014; Levison et al. 2015). In contrast, the terrestrial planets are generally thought to have mainly grown by planetesimal accretion (Raymond et al. 2009; Chambers 2016; Izidoro et al. 2021b), although a new class of models proposes that pebble accretion may have played a role (Johansen et al. 2021). Next, planetesimals grow into terrestrial planet embryos and giant planet cores. In the terrestrial planet-forming region, planetary embryos were perhaps a Mars-mass ($\sim$10% of an Earth-mass; Morbidelli et al. 2012), whereas the giant planets' cores are thought to have been $\sim$10–20 $M_\oplus$. The final assembly of the terrestrial planets is thought to have involved giant impacts between planetary embryos as well as a sweep-up of leftover planetesimals (for reviews, see Morbidelli et al. 2012; Raymond et al. 2014; Jacobson and Walsh 2015; Raymond and Morbidelli 2022).

Because it is smaller and faster orbiting, Venus is significantly more likely to have experienced more "hit-and-run" (bouncing; e.g., Asphaug et al. 2006; Reufer et al. 2012) collisions than Earth (Emsenhuber et al. 2021). In addition, Emsenhuber et al. (2021) showed that Venus is expected to eventually accrete most of its runners while Earth loses about half, with many of those ending up at Venus. Generally, Venus serves as a sink of runners emerging from farther out collisions and is more likely to have accreted a massive outer Solar System body. This means that despite their proximity, the collisional histories of the two planets were not identical.

The first quantitative model for terrestrial planet formation, developed from the 1970s to the 2000s, is commonly referred to as the "classical model", and was built upon the work of several pioneers in the field, notably that of George Wetherill (key papers include Safronov 1972; Greenberg et al. 1978; Wetherill 1978, 1985, 1990). Yet, the classical model – which assumes that a smooth disk of planetesimals existed between the present-day orbits of Venus and Jupiter upon dissipation of the Sun's gaseous disk – systematically fails to reproduce the planets' orbital architecture, forming Mars analogs that are as massive as Earth (Wetherill 1991; Raymond et al. 2009; Lammer et al. 2020a). In recent years, three planetesimal accretion-based solutions have been proposed to this "small Mars" problem and are illustrated in Fig. 3 (for details, see Raymond et al. 2020). The Low-mass asteroid belt model proposes that Venus and Earth grew within a narrow ring/annulus of planetesimals, but few planetesimals ever formed in the primordial asteroid belt. Mars was stranded outside this ring and its growth was stunted (Hansen 2009; Drążkowska et al. 2016; Raymond and Izidoro 2017b; Izidoro et al. 2022). The Grand Tack model starts from the same extended planetesimal disk assumed in the classical model, but Mars' feeding zone was strongly depleted by Jupiter's gas-driven orbital migration, which first drove the gas giant inward to 1.5–2 AU, then back outward past 5 AU (Walsh et al. 2011; Jacobson and Morbidelli 2014; Raymond and Morbidelli 2014; Brasser et al. 2016). The third model is based on the giant planets' instability, which has been invoked by many studies to explain the giant planets' orbits and populations of small Solar System bodies (for a review, see Nesvorný 2018). While it was originally invoked as a late instability (Gomes et al. 2005), re-analysis of the dynamics and constraints suggest that it instead took place within 100 million years

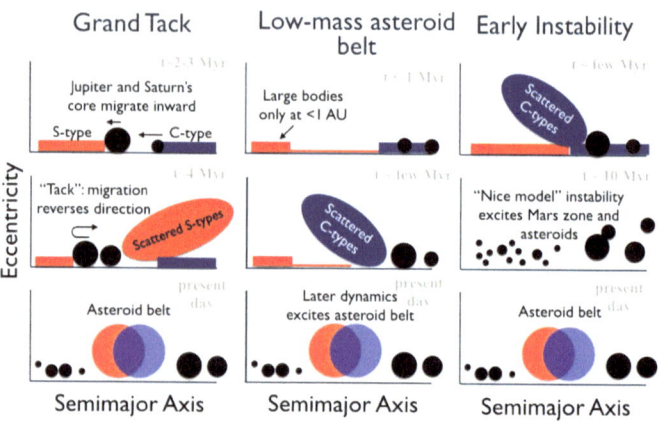

Fig. 3 Cartoon depiction of three global models of terrestrial planet formation that can match the inner Solar System (adapted from Raymond et al. 2020)

of the start of planet formation (Zellner 2017; Nesvorný et al. 2018; Morbidelli et al. 2018; Mojzsis et al. 2019), and perhaps during the dispersal of the gas disk itself (Liu et al. 2022b). An early instability naturally depletes the planetesimal disk exterior to Earth's present-day orbit and can match the terrestrial planets and asteroid belt (Clement et al. 2018, 2019,b, 2021). Finally, two recent models have proposed that pebble accretion and gas-driven migration may have been the dominant processes in terrestrial planet formation (Johansen et al. 2021; Brož et al. 2021), with just a single late giant impact to explain the formation of the Moon (Ćuk and Stewart 2012; Canup 2012).

Recent measurements have uncovered a dichotomy in the isotopic signatures of meteorites (Warren 2011; Kleine and Walker 2017; Kruijer et al. 2020). These signatures are apparent in a number of different elements with different chemical characteristics. This has led to a change in paradigm regarding the source material that built the planets: rather than a smooth gradient of chemical characteristics across a broad planetesimal disk, it is thought that two distinct reservoirs of planetesimals formed concurrently during the gas disk lifetime in different regions within the Solar System (e.g., Budde et al. 2016). In fact, recent analysis of samples of the asteroid Ryugu, returned from JAXA's Hayabusa2 mission, indicates that the meteorite isotopes actually represent a "trichotomy," with three distinct clusters such that there may have been three distinct planetesimal reservoirs in the Sun's planet-forming disk (Hopp et al. 2022). Exactly how the different reservoirs were kept apart is debated: the flux of inward-drifting pebbles may have been cut off by Jupiter's growing core (Kruijer et al. 2017) or by a pressure bump in the disk (Brasser and Mojzsis 2020; Izidoro et al. 2021a) or by something else (Lichtenberg et al. 2021b; Liu et al. 2022). The two main types of meteorites are called non-carbonaceous and carbonaceous. The parent bodies of non-carbonaceous meteorites are thought to have originated in the inner Solar System, whereas carbonaceous meteorites likely originated beyond Jupiter's orbit. The present-day orbital distributions of the different types of meteorites overlap in the asteroid belt (DeMeo and Carry 2014), but they almost certainly formed far apart (Raymond and Izidoro 2017b; Raymond and Nesvorný 2022).

Meteorite isotopes constrain the origin of Earth's water, and although we do not have comparison samples from Venus, we can use dynamical arguments to extrapolate to Venus. It has long been known that bulk Earth's D/H and $^{15/14}$N ratios are well-matched by carbonaceous chondrite meteorites (Marty and Yokochi 2006; Marty 2012). For the past 20 years it has been generally accepted that most of Earth's water was delivered from a carbonaceous source (Morbidelli et al. 2000; Meech and Raymond 2020). To match the isotopic signatures of a range of meteorites, Earth must have accreted almost entirely from non-carbonaceous (Enstatite chondrite-like) bodies indigenous to the inner Solar System, but with a ~5% contribution from carbonaceous material, assumed to have originated beyond the orbit of Jupiter (Burkhardt et al. 2021). This small "pollution" of carbonaceous chondrite-like planetesimals represents the source of Earth's water in this context. However, recent measurements of Enstatite chondrites revealed that these objects contain far more water than previously thought, and with D/H and $^{15/14}$N ratios consistent with inner Earth (Piani et al. 2020). Measurements and analysis of zinc isotopes in different types of meteorites can concurrently match the isotope ratios of hydrogen, nitrogen and zinc if roughly two-thirds of Earth's volatiles were sourced locally and one-third was delivered from a carbonaceous source (Steller et al. 2022; Savage et al. 2022). Future isotopic measurements of other volatiles will certainly refine this analysis and give stronger constraints on the relative fraction of Earth's water that was sourced locally vs. delivered.

The question remains: what was the source of the carbonaceous component of Earth's water? In the classical model, Earth's feeding zone was broad enough that it extended past 2.5 AU, where it is assumed that water-rich planetesimals and planetary embryos originated (Morbidelli et al. 2000; Raymond et al. 2006). This mechanism delivered a modestly-smaller amount of water to Venus than Earth, simply because the tails of feeding zones of closer-in planets are less extended (Raymond et al. 2004). However, the initial conditions of the classical model are suspect, as it assumes that carbonaceous and non-carbonaceous planetesimals formed as close neighbors, which would predict a gradient in isotopic compositions rather than the observed dichotomy. Indeed, the present-day orbits of water-rich asteroids cannot represent the correct initial conditions for Solar System formation. Rather, carbonaceous asteroids were likely implanted from beyond Jupiter's orbit during the gas giants' growth (Raymond and Izidoro 2017) or migration (Walsh et al. 2011; Raymond and Izidoro 2017). During this implantation, a large population of scattered carbonaceous planetesimals would also have crossed the orbits of the terrestrial planets (Walsh et al. 2011; O'Brien et al. 2014; Raymond and Izidoro 2017; O'Brien et al. 2018). Given their source beyond Jupiter, scattered planetesimals would more efficiently deliver water to rocky planets at larger orbital radii. Earth (or its building blocks) would have accreted modestly more scattered planetesimals than Venus.

In the pebble accretion framework, water can be delivered by water-rich pebbles but only to planets orbiting beyond the snow line (Sato et al. 2016; Ida et al. 2019). As the disk cools in time, the snow line moves inward, and most models find a snow line interior to 1 AU during the late phases of the disk (Lecar et al. 2006; Kennedy and Kenyon 2008; Martin and Livio 2012; Bitsch et al. 2015). If the flux of carbonaceous pebbles was not blocked at this time, then Earth could have accreted water from drifting pebbles (as in the models of Johansen et al. 2021; Brož et al. 2021). Earth and Venus (or their building blocks) would have accreted similar amounts of water unless the snow line happened to lie in between the planets' orbits, in which case Earth could have accreted much wetter than Venus. While this is a narrow range of orbital radii, that range falls within the potential values for the snow line's position for reasonable disk parameters (Martin and Livio 2012; Bitsch et al. 2015). However, the relative contribution of carbonaceous material could not have been more than ~5% according to meteorite isotopic studies that include a number of different elements (Burkhardt et al. 2021). This makes it difficult to reconcile the pebble accretion scenario with the terrestrial planets' growth and water delivery.

We can make an educated guess as to Venus' initial water content. Let us first assume that two-thirds of Earth's water was indeed sourced locally, as suggested by recent zinc isotopic analyses (Savage et al. 2022; Steller et al. 2022), and that the rest was delivered by scattered carbonaceous planetesimals (Walsh et al. 2011; Raymond and Izidoro 2017b). Given their near-identical feeding zones (Izidoro et al. 2022), Venus would presumably have the same concentration of locally-sourced water as Earth. However, water delivery from scattered planetesimals has a radial gradient in efficiency such that Venus was likely delivered somewhat less water from carbonaceous sources than Earth (O'Brien et al. 2014; Raymond and Izidoro 2017b), implying an overall amount of water delivery within a factor of a few between the two planets, with Earth being modestly more water-rich. While there is still some variation in water contents in different Enstatite chondrites (Piani et al. 2020), it is hard to imagine that Venus could have formed completely dry.

The relative amount of incorporated water within Earth and Venus sensitively depends on the delivery mechanism, as well as their relative growth rates and the behavior of the underlying gaseous disk. Models for the source of Earth and Venus' water are clearly heavily influenced by isotopic measurements of different types of meteorites. We expect that new measurements that include a wider range of elements – in particular, more and more volatile elements and elements with different chemical properties – will guide our future thinking.

2.2.2 Energy Delivery, Heat Sources, and Mantle Initial Thermal State

In addition to water and other volatiles, the accretion process delivers a substantial amount of energy to the growing planets through impacts. Whether due to the accumulation of accretional energy after successive impacts (e.g., Safronov 1978; Tonks and Melosh 1993) or resulting from the thermal blanketing effect of a proto-atmosphere (e.g., Hayashi et al. 1979; Abe and Matsui 1985), the early stages of planetary evolution have long been thought to be punctuated by one or several events of surface and mantle melting (see Elkins-Tanton 2012, for a review). Yet, the extent and timing of this molten phase remains unclear. We review below the early heating mechanisms at play during planetary formation and discuss how they may affect the initial thermal state of the Venusian mantle.

Accretional Heating Several processes likely contribute significantly to increasing the protoplanet temperature and thus favor melting. During the accretion phase, planetary growth by impacts is thought to provide a large amount of kinetic energy. This is dissipated in the form of heat upon impact and responsible for local to global melting of the surface, extending throughout the mantle (Safronov 1978). The size/mass of colliding bodies, the impact velocity, angle, and frequency are key parameters controlling the time- and spatial-extent of melting (e.g., Nakajima et al. 2021).

During the so-called runaway accretion phase (e.g., Kokubo and Ida 1996; Kortenkamp et al. 2001), frequent successive/continuous collisions and associated heat accumulation are thought to provide an important heat source leading possibly to global-scale melting events. At the end of the accretion phase, planetary system formation models (e.g., N-body simulations) predict that the size of impactors increases so that collisions, although less frequent and well separated in time, become more and more energetic (e.g., Kokubo and Ida 1998, 2000; Chambers 2010; Morbidelli et al. 2012). Several tens of Mars-sized protoplanets are thought to form from planetesimal accretion, thus leading to several giant impacts between these protoplanets in the late accretion stages (e.g., Quintana et al. 2016). This ultimately implies melting episodes of greater extent (Kaula 1979; Tonks and Melosh 1993; Nakajima and Stevenson 2015), but likely more isolated in time. Giant impacts, followed by isostatic readjustement are even expected to entirely melt the mantle (Canup 2008; Nakajima and Stevenson 2015). However, in these late accretion stages, it has been suggested that the energy of the giant impacts could be quickly (~1 Myr) radiated to space because of their low frequency (Zahnle et al. 2007). Other studies suggest that the accretional heat generation rate does not significantly change with time, i.e., planetary heating is almost constant with/insensitive to planetary growth time, as longer time implies higher relative velocities, larger planetesimals and thus deeper heat burial (Kaula 1979). Furthermore, a remnant or outgassing steam atmosphere could sustain a runaway greenhouse state (e.g., Hamano et al. 2013) and significantly slow down the cooling of the protoplanet, and thus possibly maintaining a molten surface and mantle between successive giant collisions.

On early Earth, the hypothesis of a moon-forming single giant impact is often referred to as being responsible for the last global-scale mantle melting event (Canup 2004; Nakajima and Stevenson 2015). On Venus, the apparent absence of a moon and associated late giant impact may challenge the occurrence of such a deep and global magma ocean. Yet, it should be noted that the standard model of a single giant moon-forming impact is still debated for the Earth (e.g., Asphaug 2014). Alternative scenarios, involving smaller multiple impacts have been proposed and may better explain the compositional similarity of the Earth and Moon (Rufu et al. 2017). Regardless, the absence of a giant moon-forming impact cannot completely discard the likelihood of a fully or large-scale molten mantle stage, neither on Earth nor on Venus. The frequency of impacts and thus the accretion time scale must be considered. The collisional histories of the two planets may not have been identical, with Venus experiencing more hit-and-run collisions (that could explain its lack of satellite; Emsenhuber et al. 2021). Yet, giant impacts and associated tremendous energy delivery remain ubiquitous in inner Solar System planetary accretion (Quintana et al. 2016), thus suggesting that magma oceans were common events in the early evolution of terrestrial planets (e.g., Elkins-Tanton 2012; Schaefer and Elkins-Tanton 2018).

However, the absence of a Venusian moon certainly rules out the possibility of substantial tidal heating (solid-body tidal dissipation from the Sun alone could still slow Venus' rotation rate and induce tidal heating; Way and Del Genio 2020). This additional heat source affects solid materials more than the melt and thus is likely concentrated at the bottom of the crystallizing mantle (Zahnle et al. 2007). As mentioned previously, bulk similarities between the Earth and Venus, in terms of sizes and densities, suggest that the two planets have similar bulk compositions (e.g., Smrekar et al. 2018). Combined with their vicinity in the Solar System, and although no moon orbits Venus, it seems reasonable to think that they are composed of similar materials. As discussed above they likely accreted in a relatively similar way with associated accretional heating and therefore analogous thermal histories including substantial interior melting.

Radiogenic Heating The presence of radioactive elements, generating heat with decay, is also thought to be a significant *internal* heat source of rocky planets, thus contributing to melting of the planetary interior (e.g., Elkins-Tanton 2012; Solomatov 2015). Active on a timescale of about one million years and thus affecting the earliest evolutionary stages (within the first few million years after CAI formation[2]), radioactive decay of short-lived radionuclides, such as now extinct ^{26}Al and ^{60}Fe isotopes (half-lives of 0.717 Myr and 2.61 Myr, respectively), contributed to the heating and differentiation of early forming and growing bodies: planetary embryos, planetesimals, and proto-planets, starting before the end of the accretion phase (Urey 1955; Merk et al. 2002; Yoshino et al. 2003; Elkins-Tanton et al. 2011; Šrámek et al. 2012; Fu et al. 2017; Bhatia 2021). ^{26}Al decay dominates the heat budget while the decay of other radionuclides such as ^{10}Be, ^{53}Mn, ^{146}Sm, ^{182}Hf, or ^{244}Pu have

---

[2]Calcium-Aluminum-rich Inclusions are believed to be the earliest formed materials in the Solar System and thus define its formation age, ~4.568 billion years ago (Bouvier et al. 2007; Burkhardt et al. 2008).

a minor contribution (Lee et al. 1976; Chaussidon and Gounelle 2007; Elkins-Tanton 2012; Fu et al. 2017; Lugaro et al. 2018). To a lesser extent but on longer timescales and still active today (e.g., Schubert et al. 2001; Jaupart et al. 2015), long-lived radioactive elements ^{238}U, ^{235}U, ^{40}K, and ^{232}Th (half-lives ranging between 0.7 and 15 Gyr) were more abundant early in the Solar System and also took part in the early heat budget (Urey 1955; Lebrun et al. 2013; Nikolaou et al. 2019).

The amount of radiogenic heat produced relates directly to the abundance of the radioactive elements. As mentioned above the proximity of Earth and Venus combined with their similar bulk compositions (e.g., Smrekar et al. 2018), also suggest that they have accreted roughly similar materials with similar endowments of radioactive elements. Both planets have thus likely experienced similar amounts of radiogenic heating and resulting extent of interior melting.

Thermal Blanketing Effect of a Primordial Atmosphere Whether the energy sources primarily heat up the surface and extend downward through the mantle (accretional heating) or primarily affect the interior (radiogenic heating), as soon as the planet is heated, the excess energy must escape to reach thermal equilibrium with the relatively cold surroundings. Then, whole planet cooling involves heat being radiated to space. This is particularly important when the surface is molten.

Without an atmosphere, the planet radiates directly into interplanetary space as a black body with the temperature of the planet's surface, making radiation of energy very efficient (dashed purple line, Fig. 4). In such a case, the cooling stops when the planet reaches thermal equilibrium, i.e., when the surface temperature reaches the equilibrium temperature of the planet (that is the temperature at which the power supplied by the star is equal to the power emitted by the planet).

In the presence of an atmosphere, radiative dissipation of energy from the surface is buffered by the thermal blanketing effect of the surrounding atmosphere and heat is retained. This can maintain the surface temperature above the rock melting point. Several works have reviewed the different stages of the evolution, corresponding to different types of early/primordial atmosphere (i.e., not outgassed from the interior during the solidification process) and related forming-mechanisms.

In the earliest disk-embedded stages of planetary evolution (if the proto-planet accreted enough mass within the disk lifetime), the gravitational capture of gas from the surrounding solar nebula (see Sect. 2.1) and the blanketing effect of the accreted opaque atmosphere is itself thought to keep the surface temperature of the proto-planet above the silicate melting point (e.g., Hayashi et al. 1979; Ikoma and Genda 2006; Olson and Sharp 2019). Ikoma and Genda (2006) have shown that starting from 0.5 Earth masses, proto-planets surrounded by such nebula-captured atmospheres are expected to have a molten surface. If the surface is already molten because of the previously discussed energy sources, such an atmosphere would undoubtedly favor a long-lasting molten layer. Depending on the thermal opacity of the primordial atmosphere and on the prior thermal state of the interior, the molten shell is thought to extend more or less deeply into the mantle. The interior could thus be maintained in a partially or fully molten state throughout the protoplanetary disk lifetime, until dissipation of the solar nebula and the primordial atmosphere.

In the eighties, Abe & Matsui pioneering magma ocean-related studies addressed the evolution of an impact-generated steam atmosphere and its implications for the early thermal history of the Earth (Matsui and Abe 1984; Abe and Matsui 1985; Matsui and Abe 1986a,b; Abe and Matsui 1988). Such atmospheres result from the devolatilization of incoming planetesimals upon impact, thus releasing volatile species, such as H_2O (due to shock dehydration; e.g., Benlow and Meadows 1977; Boslough et al. 1980; Lange and Ahrens 1982) and

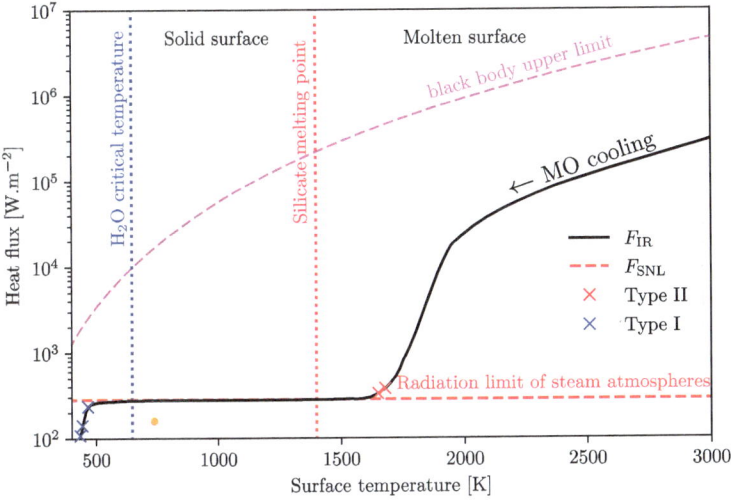

Fig. 4 Outgoing thermal flux emitted at the top of the atmosphere (F_{IR}, black line) as a function of surface temperature for a steam (H_2O-dominated) atmosphere (see Fig. 13 for CO_2-dominated atmospheres). Global radiative equilibrium and associated climate stability are reached on either sides of the radiation limit of steam atmospheres: the asymptotic Simpson–Nakajima limit ($F_{SNL} \approx 280$ W.m^{-2}, dashed red line), when the thermal emissions balance the absorbed insolation ($F_{IR} \approx F_{abs}$). This radiation limit is responsible for a runaway greenhouse. At the ERCS (i.e., when the heat flux from the interior of the planet becomes negligible compared to the net absorbed stellar flux; see Sect. 3.3), climate stability can be reached either with temperate surface conditions and a water ocean (Type I planets, e.g., blue crosses), for planets whose net absorbed flux is below the radiation limit ($F_{abs} < F_{SNL}$), or with a molten surface (Type II planets, e.g., red crosses), for planets whose absorbed flux exceeds the radiation limit ($F_{abs} > F_{SNL}$). Crosses indicate the surface conditions reached at the ERCS for early Venus (i.e., at Venus' orbital distance around the young Sun) and considering different Bond albedo values ($\alpha = [0.2, 0.3, 0.5, 0.7, 0.77]$ from right to left). These crosses correspond to the ones shown in the lower part of Fig. 12. Depending on its Bond albedo at the end of the rapid cooling stage, early Venus can either be habitable or not. Reciprocally, these surface conditions would be reached on a planet absorbing (and emitting) the corresponding amount of solar radiation, i.e., located at various orbital distances, while considering a constant Bond albedo. The orange dot denotes present-day Venus

CO_2, directly *into* the atmosphere. Alongside other early studies (Zahnle et al. 1988; Kasting 1988), they showed that once such an impact-generated atmosphere is formed, its thermal blanketing effect is responsible for heating the surface of the accreting Earth up to its melting point (~1400 K), thus forming a magma ocean. This may also apply to Venus. Matsui and Abe (1986b) emphasized that the distance from the star and the heat flux equilibria at the end of the magma ocean stage might explain the current differences between the Earth and Venus, thus demonstrating the importance of the early evolution of terrestrial planets in setting their potential habitability. These results were revisited almost 30 years later in the light of a more sophisticated molten interior–atmosphere coupled model (Hamano et al. 2013, see Sects. 3 and 4). Note that a hybrid-type proto-atmosphere, made of a mixture of both solar nebula and impact degassing products is also thought to be thermally opaque enough to produce a deep magma ocean, and hence sustain an already existing one (Abe et al. 2000; Saito and Kuramoto 2017).

The increase of atmospheric water vapor concentration, or the increase of heat to be radiated through a steam atmosphere after an energy delivery (such as after an impact), results in high surface temperatures triggered by the runaway greenhouse effect of steam dominated atmospheres (e.g., Simpson 1927; Kasting 1988; Nakajima et al. 1992). This runaway be-

Fig. 5 Schematic cross section of a planet in the magma ocean stage surrounded by an atmosphere and associated heat fluxes. The vigorous convection of the molten mantle provides a substantial heat flux at the surface of the planet, F_{conv} (dotted red arrows). A fraction α, the Bond albedo, of the total solar radiation received by the planet $F_{\odot}$ (plain lines at the top of the atmosphere) is reflected back into space. The remaining flux is absorbed by the atmosphere (F_{abs}, dashed black arrows entering the atmosphere). These two incoming heat fluxes balance the outgoing infrared/thermal flux radiated at the top of the atmosphere, F_{IR} (blue arrows), that ultimately control the cooling of the whole planet

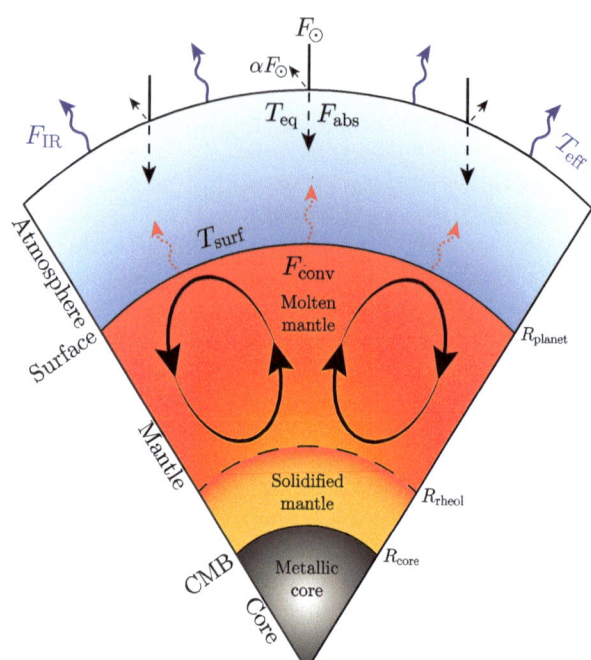

havior is due to the existence of a nearly-asymptotic radiation limit in the thermal (infrared) radiation a planet can emit as a function of its temperature (Fig. 4). This asymptotic radiation limit is often referred to as the Simpson–Nakajima limit (see Goldblatt and Watson 2012, for a review), whose corresponding flux value is $F_{SNL} \approx 280$ W.m^{-2}. In planetary atmospheres, steady-state global radiative balance, and thus climate stability, is achieved by the equilibrium between the net, absorbed incoming stellar radiation, F_{abs}, and the outgoing thermal radiation (F_{IR}) emitted by the atmosphere (also referred to as the outgoing longwave radiation, OLR) (Fig. 5), as a response to the heating of the surface and atmosphere resulting from their absorption of the stellar flux (sunlight). The net, absorbed incoming stellar radiation is:

$$F_{abs} = (1 - \alpha) \times F_{\odot} \ , \tag{1}$$

where α is the fraction of the total incident solar radiation scattered back out into space in all directions and integrated spatially over the whole globe (spherically) and spectrally over all wavelengths, i.e., the global bolometric or Bond albedo (e.g., Hanel et al. 2003). $F_{\odot}$ is the total incident stellar flux received at the top of the atmosphere (TOA) of a planet orbiting the host star at a distance D (in AU) and averaged over the planetary surface area:

$$F_{\odot} = \frac{1}{4} \frac{S_{\odot}^{N}}{D^2} \ . \tag{2}$$

$S_{\odot}^{N} = 1361$ W.m^{-2} is the total solar irradiance, i.e., the spectrally integrated, mean total electromagnetic energy received from the Sun at a distance of 1 AU per unit area (Kopp and Lean 2011; Prša et al. 2016). The factor 1/4 comes from the ratio of the planet's cross-sectional area to its surface area: the planet intercepts the incident solar beam as a disk but radiates as a sphere over its entire surface (e.g., Pierrehumbert 2010). Thus, if there is a heat supply, through an increase of the absorbed stellar flux or after an impact, the planet warms and consequently emits more thermal radiation to space to maintain energy balance. In the steam atmosphere case, most of this emission occurs through the water vapor infrared atmospheric "window," between $\sim$8–13 µm. It thus matches the Wien peak in surface thermal emission at 10 µm; the spectral region essentially transparent (with relatively little absorption) to infrared radiation. However, when the heat flux reaches the critical radiation limit, or when atmospheric water vapor becomes too abundant as a result of oceans evaporation or following impact-degassing, the moist atmosphere becomes optically too thick for heat to be radiated. The outgoing thermal flux can no longer increase and lose heat despite an increasing temperature, i.e., surface warming no longer leads to more thermal emission. This inevitably leads to an increase in the surface temperature T_{surf}, hence a runaway greenhouse. This remains until the planet is warm enough to radiate heat in the 4 µm water vapor window, when T_{surf} reaches about 1600 K (which is above the surface melting point), thus restoring radiative balance and climate stability (e.g., Goldblatt and Watson 2012; Goldblatt et al. 2013, see Fig. 4, Sects. 3 and 4 for more details and implications for habitability). This explains the abrupt transition between these extreme stable climatic states on either side of the radiation limit (Fig. 4): temperate or relatively cold surface conditions on one side ($F_{IR} < F_{SNL}$), and a molten surface on the other side ($F_{IR} > F_{SNL}$). Conversely, a molten surface state can only transition to a temperate climate by two means. Through removing water vapor from the atmosphere by photo-dissociation and hydrogen escape, thus decreasing atmospheric opacity. Alternatively it is possible via a decrease of the thermal radiation: if the additional heat supply stops and the absorbed stellar flux remains below the radiation limit for instance.

However, it is important to note that the runaway greenhouse effect and associated abrupt climate transition occurs only for H_2O-dominated atmospheres, as their main component is both a condensable and thermal absorber species. Its triggering radiation threshold value may be sensitive to the atmospheric relative humidity and planetary mass (e.g., Ishiwatari et al. 2002; Goldblatt 2015). For other atmospheric compositions, the radiation limit, if relevant, is not that steep and radiative balance and thus stable climates can be reached for intermediate conditions where the surface is not molten (e.g., Type III planets in Salvador et al. 2017, see Fig. 13 and Sect. 4 for a detailed discussion). In the presence of such "moderately-blanketing" primordial atmospheres, the atmosphere may not melt the surface by itself, but would still buffer the dissipation of heat resulting from other processes, thus contributing to maintaining a molten surface.

Overall, the blanketing effect of a proto-atmosphere, regardless if it is nebula-captured, impact-degassed, or a mixture of both, can melt the planetary surface. Consequently, if the surface is already molten, it will buffer radiative dissipation and maintain the surface temperature above the melting point. If not, it will enhance the thermal effect of any heating mechanism, significantly increasing the likelihood and subsistence of a magma ocean. Yet, it is important to note that any pre-existing primordial atmosphere is subject to depletion, either through the disk dissipation for nebula-captured atmospheres, or through impact erosion. In the case of a nebula-captured atmosphere, the timescale of disk dissipation and following atmospheric escape must be compared to the energy delivery timescale (accretion + radioactive decay timescales) to evaluate the thermal contribution of the atmosphere in maintaining the molten surface.

For an impact-generated atmosphere, if the atmospheric mass added by impact degassing overcomes or balances atmospheric losses due to impact erosion, the atmosphere keeps being replenished after successive collisions or keeps a roughly constant mass and its thermal blanketing effect is maintained over time. In such a scenario, the atmosphere helps to maintain the molten surface between successive impacts and the heat generated keeps accumulating. This likely keeps the planet molten during the entire accretion sequence. Conversely, if the atmosphere loses mass after each collision, its blanketing power would decrease according to the accretion time scale and resulting erosion rate until it becomes too thin to retain heat. Then, heat would more efficiently dissipate and the atmosphere would not be responsible for melting the surface or keeping it molten during the entire accretion sequence. If the heating rate due to the other mechanisms is lower than the dissipation rate, the planet would only experience transient and local magma oceans/ponds. Depending on the dissipation timescale of the proto-solar nebula relative to Venus' accretion timescale, and in case a protoatmosphere survives atmospheric erosion by late impacts, the blanketing effect provides an additional heat source to extend and enhance the existing ones.

Based on element partitioning analysis, Sakuraba et al. (2019) suggested that the survival of the primordial atmosphere through the late accretion may partially account for the present-day atmosphere of Venus. This implies that such an early and long-lasting atmosphere maintained a strong greenhouse effect during the accretion of Venus. Thus it would significantly retain any heat produced – strongly supporting the idea that Venus' evolution started with a molten surface and an extensively molten mantle.

Metal–Silicate Differentiation and Core Formation The early thermal state and melting extent of the mantle is tightly linked to metal–silicate segregation and metal downwards migration that relate to core formation. Due to the strong fractionation of lithophile (silicate-loving) Hf from siderophile (metal-loving) W during core formation, the ^{182}Hf–^{182}W system (^{182}Hf decays to ^{182}W with a half life of 8.9 Ma, which can be used to date processes in the first $\sim$60 Myr) is particularly well suited to trace and date metal–silicate differentiation (Lee and Halliday 1995; Harper and Jacobsen 1996; Lee and Halliday 1997; Horan et al. 1998; Quitté et al. 2000; Jacobsen 2005; Rubie et al. 2015b).

Early work on ^{182}Hf–^{182}W chronometry of meteorites indicate that an idealized, instant core formation in the terrestrial planets must have been completed during the first $\sim$30 Myr of the Solar System history (Kleine et al. 2002; Yin et al. 2002). Yet, this simplistic instantaneous core formation model only provides a lower time constraint and does not apply for larger, Earth-sized bodies, where core formation did not occur as a single event but rather as a combination of discrete delivery of metallic components to the deep interior (e.g., Kleine et al. 2009; Deguen et al. 2014; Rubie et al. 2015b; Nimmo and Kleine 2015).

Accounting for the different mechanisms of core formation in more realistic scenarios leads to estimates ranging between 30 to 200 Myr (e.g., Chambers 2001) at the very most (see Rubie et al. 2015b; Nimmo and Kleine 2015, for extensive reviews). Similarly, the more radiogenic W isotopic composition of the silicate Earth compared to that of chondritic meteorites demonstrates that Earth has differentiated into a mantle and core within the lifetime of ^{182}Hf ($\sim$30–50 Myr; e.g., Jacobsen 2005). Rapid terrestrial planet formation and early metal–silicate separation are also consistent with rapid planetesimals formation and differentiation informed from ^{53}Mn–^{53}Cr (Lugmair and Shukolyukov 1998) and ^{26}Al–^{26}Mg (Srinivasan et al. 2000) systematics (Nyquist et al. 2001; Wadhwa et al. 2006; Nyquist et al. 2009), and observations of other planetary systems (Briceño et al. 2001; Bodenheimer and Lin 2002).

Core formation is a complex, multi-stage phenomenon most likely resulting from the combination of different processes operating over different timescales such as percolation of liquid metal through a compacting solid silicate matrix, iron droplets sinking through a magma ocean, negative diapirism generated by Rayleigh-Taylor instability, or hydraulic fracturing (e.g., Ricard et al. 2009; Deguen et al. 2014; Rubie et al. 2015b; Landeau et al. 2016; Wacheul and Le Bars 2018; Clesi et al. 2020; Landeau et al. 2021). Because the relative contribution and timing of each process is still unclear, large uncertainties remain. In any case, mechanisms of core formation depend on the thermal state of the planetary body and require some extent of partial melting (e.g., Rubie et al. 2015b). Given its suggested relatively short timescale, core formation is also generally thought to require a mechanism for efficiently segregating metal (Stevenson 1990; Yoshino et al. 2003). The low viscosity of the melt and high temperatures facilitate material deformation needed for mass redistribution (e.g., Šrámek et al. 2010, 2012; Nimmo and Kleine 2015). Therefore, large extent of mantle melting is thought to facilitate metal–silicate segregation and metal migration. Thus, a great extent of mantle melting is believed to favor rapid core formation suggested by geochemical evidence. This view is also supported by the mantellic abundance of moderately siderophile elements (MSEs; e.g., Fe, Ni, Co, W, and Mo) that may be explained by a final metal–silicate equilibration at very high P and T, i.e., at the base of a deep magma ocean (e.g., Li and Agee 1996; Righter and Drake 1997, 1999; Rubie et al. 2003, 2015b). Note however that alternative mechanisms, such as percolation of iron-rich melt through solid silicate, have been proven to result in rapid core formation without necessarily involving a substantial amount of mantle melting (Ghanbarzadeh et al. 2017; Berg et al. 2018). The respective contribution and timing of the different mechanisms involved in core formation need to be further considered to draw a realistic picture of this complex, multi-stage phenomenon, accounting for both geochemical and geophysical constraints (e.g., Badro et al. 2015).

Other terrestrial isotopic data analysis suggest that the mantle experienced early and global/major chemical differentiation ($^{146}Sm–^{142}Nd$ system; Caro et al. 2003, Boyet et al. 2003, Boyet and Carlson 2005), and that the growth of continental crust began early ($^{176}Lu–^{176}Hf$ and $^{147}Sm–^{143}Nd$; e.g., Bennett et al. 1993, Amelin et al. 1999, Bizzarro et al. 2003, Harrison et al. 2005), within 30 Myr and in the first 200 Myr of evolution, respectively, which both might require a global, deep magma ocean (Harrison 2009; Caro 2011; Solomatov 2015). Yet, one should note that these results could reflect multiple possible scenarios given the uncertainties affecting the isotopic measurements themselves, their representativeness and the various fractionation processes. One promising way to overcome individual limitations of each isotopic system and to provide tight constraints on the timescale of accretion and early differentiation is to build scenarios that satisfy several different isotopic systems simultaneously and that are consistent with the results of dynamical models (e.g., Halliday 2004; Wood and Halliday 2005; Nimmo and Kleine 2015).

Overall, geochemical constraints may indicate that core formation and large-scale mantle differentiation occurred at similar timescales. These are likewise comparable to the accretion timescales constrained by planetary formation models. If planetary accretion, core formation, and mantle differentiation occurred concurrently and on short timescales, the rapidity of the processes themselves would be indicative of a significant degree of mantle melting. The amount of energy involved would induce large-scale melting, thus likely suggesting the existence of a global and deep magma ocean (Solomatov 2015).

Core formation is not only linked to the degree of mantle melting through the ease of metal–silicate segregation and iron migration but has also important energetic implications for the mantle thermal state (e.g., Jaupart et al. 2015; Solomatov 2015). Indeed, the release of gravitational energy from the gravitational differentiation between metallic and silicate materials during core formation, is dissipated by viscous heating in both the iron and silicate phases. This supplies an additional and substantial source of heat supporting a whole molten silicate layer scenario (e.g., Solomon 1979; Sasaki and Nakazawa 1986; Ricard et al. 2009; Monteux et al. 2009; Šrámek et al. 2010; Samuel et al. 2010; Rubie et al. 2015b; Landeau et al. 2016). The energy released by rapid core formation, itself favored by extensive mantle melting, could be responsible for heating the entire Earth to $\sim$1700 K (Tozer 1965; Flasar and Birch 1973). Additional release of gravitational energy due to the redistribution of chemical and thermal heterogeneities in the mantle and segregation of leftover iron (density heterogeneities) following core formation is also thought to increase the temperature by several hundred degrees (Tonks and Melosh 1992; Solomatov 2015).

Numerical simulations of giant impacts emphasize that during collisions of large differentiated bodies the iron core of the impactor is the most heated material (Canup 2004). After its presumably rapid segregation into the impacted protoplanetary core, a substantial fraction of the associated gravitational potential energy might be retained within the segregated iron, resulting in an early superheated core (Stevenson 2001; Solomatov 2015). The heat transferred to the mantle may induce additional mantle melting or sustain its molten state and extend the magma ocean lifetime (Ke and Solomatov 2006, 2009).

Similarly and over longer timescales, once the core is formed, the release of heat from the core at the core–mantle boundary (CMB) constitutes an additional heat source. While little is known about early magma ocean–core interactions, they certainly affect each other's evolution. Indeed, the core constitutes an evolving boundary condition of the mantle and contributes to the heat budget of the interior. It thus certainly affects the cooling timescale and crystallization pattern of the magma ocean. In turn, the lower mantle conditions resulting from the magma ocean evolution may buffer the heat loss from the core and provide additional heat. This could thus affect the thermal state and dynamics of the core and influence the resulting dynamo and potential magnetic field strength with time (e.g., Stevenson 2001; Ke and Solomatov 2009). However, in the absence of evidence of a present or past magnetic field on Venus, that could inform the thermal and dynamical state of its core, the core–mantle interactions and their effects on the thermal history of the planet remain highly unconstrained. For the early Earth, estimates of the initial CMB temperature are above the mantle melting temperature and thus indicate extensive lower mantle melting at that time, therefore implying a deep magma ocean (e.g., Andrault et al. 2016, see Fiquet 2018 for a review). Relics of this early magma ocean retained at the CMB may exist today as a hypothetical deep molten zone on top of the Earth's core that might explain the current zones/patches of extremely low seismic velocities (ultra-low velocity zones, ULVZs) shown by seismology. On Venus, the supposedly less efficient cooling of the interior relative to Earth due to the absence of plate tectonics might favor the existence of a thick basal magma ocean today (O'Rourke 2020), inherited from the primordial heat accumulated. Testing this hypothesis might provide constraints on Venus early state and accretion vigor.

Young Sun Stars are expected to be significantly more active during their early evolution, and as a result, stellar evolution models predict that both the X-ray–EUV emissions and the magnetic field strength of the Sun were larger in the past (e.g., Bahcall et al. 2001; Ribas et al. 2005; Tu et al. 2015; Johnstone et al. 2021). On top of affecting planets' atmospheric properties and habitability (e.g., Gallet et al. 2017), the energetic environment in which forming planets and planetesimals are embedded can be responsible for significantly heating and possibly melting their interiors. In particular, the strong magnetic field and enhanced solar wind of young stars generates electromagnetic induction heating that could substantially melt planetary mantles and result in the formation of magma oceans (Sonett et al. 1968; Herbert et al. 1977; Kislyakova et al. 2017; Kislyakova and Noack 2020; Noack et al. 2021). While this energy source was likely effective during the early stages of planetary formation and certainly affected proto-Venus, its contribution remains unconstrained. Finally, despite a relatively weaker contribution compared to the other aforementioned mechanisms, the enhanced high energy X-ray and EUV emissions of the young Sun (e.g., Ribas et al.

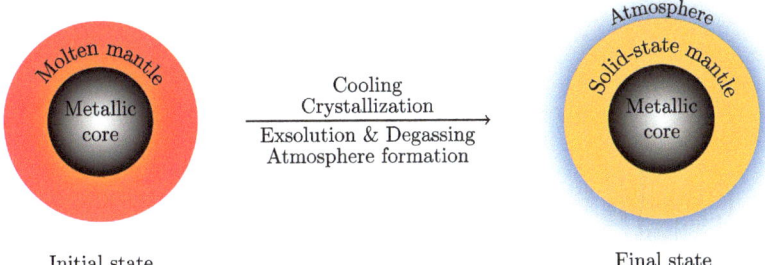

Fig. 6 Sketches of the hypothetical initial and final states usually considered in magma ocean–atmosphere thermal evolution numerical studies. Initially, as the aftermath of a hypothetical last giant impact, the silicate mantle is assumed to be fully molten, meaning that the magma ocean extends throughout the entire mantle. Any primary, captured atmosphere (if one existed at all) has been lost. All volatile species are then assumed to be initially dissolved within the molten mantle. Heavier, core-forming elements, such as metallic iron, are thought to have already fully segregated to the core, meaning that the metallic core is already formed. During the cooling and crystallization of the magma ocean, volatile species initially dissolved within the melt are exsolved and outgassed, thus progressively building-up the atmosphere. One should distinguish the end of the molten stage, that is when the mantle has fully solidified, from the end of the rapid cooling stage (ERCS; see Sect. 3.3). The latter is reached when the heat flux from the interior of the planet becomes negligible compared to the net absorbed stellar flux (F_{abs}) that thereafter balances the thermal emissions at the top of the atmosphere, F_{IR} (i.e., the planet has reached global radiative equilibrium), and controls the surface conditions. Depending on the net absorbed flux, the planet can achieved global radiative balance with a fully or partially solidified mantle at the ERCS. Once the ERCS is reached, the evolution is no longer controlled by the molten mantle (if still molten) but rather by the balance between the absorbed stellar flux and the thermal emissions

2005; Johnstone et al. 2021) may also induce additional heating of the planetary surface and atmosphere, and should then be accounted for in the thermal budget.

Initial Thermal State To conclude, the respective contribution and timing of each individual heating source to the warming and melting of the terrestrial planets' mantle remain unclear. Their combination and temporal overlap imply that generated temperatures were high enough to melt a large part – if not all – of the silicate mantle. Furthermore, the different heat sources involved are likely common and inherent to the early history of rocky planets. Thus, it seems reasonable to assume that, like Earth, Venus experienced at least one global magma ocean phase extending throughout the entire mantle during its early evolution.

For the sake of simplicity, thermal evolution studies of the early magma ocean phase generally model the cooling of a hypothetical last episode of mantle global-scale melting. By cooling down and crystallizing, the molten, liquid-state mantle evolves toward a solid-state mantle, which represents the starting point for the long-term evolution of the planet (Fig. 6; see Sects. 3 and 4). Note that depending on the absorbed stellar flux, global radiative equilibrium can be reached with a molten or solidified surface (see Sect. 3.3). For the Earth, such a global melting event often refers to the result of the late, giant moon-forming impact scenario. Yet, as discussed above, it seems reasonable to follow the same approach to address the early evolution of Venus (and other Earth-sized rocky planets) even in the absence of a Venusian moon. It should be noted that subsequent events such as late giant impacts may substantially affect, at least locally, the heat budget, interior stratification, and volatile distribution of the solidifying planet (e.g., Gabriel and Cambioni 2023). The classically described early thermal evolution sequence remains a simple and theoretical view.

As mentioned above, isotopic signatures can provide more specific constraints on the timescales and energetic of the accretion phase and early evolution and thus on the resulting early thermal state of rocky planets. Such future measurements on Venus would provide helpful constraints to decipher its evolutionary pathway and understand why it ended up being so different from Earth (see Sect. 5 and Widemann et al. 2023, in this collection).

2.3 Constraints from Isotopes

The very high D/H ratio of hydrogen in the Venus atmosphere (157 times higher than terrestrial water; Von Zahn et al. 1983) has often been interpreted as evidence that Venus underwent intense episodes of hydrogen escape (Donahue et al. 1982). These escape episodes would be responsible for the low abundance of water on Venus and would have influenced the entire geological evolution of the planet (Baines et al. 2013). However, fractionation of D from H isotopes during atmospheric escape of hydrogen depends on the regime and parameters of atmospheric escape (Hunten et al. 1987). Some escape regimes of hydrogen create only a limited isotopic fraction of the remaining hydrogen (Zahnle et al. 1990). Thus the D/H ratio gives only partial information and it remains unclear how much water really escaped from Venus. One could imagine a scenario in which considerable amounts of water escaped but that the isotopic fractionation was relatively inefficient. Conversely, the planet could have accreted almost dry and lost only a small amount of water but with a regime of escape leading to intense isotopic fractionation.

The initial D/H ratio of water is an important unknown in models of hydrogen escape from Venus. This ratio was probably lower than it is today but by how much remains unclear. It cannot be excluded that Venus' water was initially sourced by comets which can show elevated D/H ratios (e.g., 3 times the terrestrial value for water emitted from comet 67P/Churyumov–Gerasimenko; Altwegg et al. 2015). Interestingly, there is a high abundance of neon in the atmosphere of Venus and its isotopic composition seems intermediate between solar and meteoritic components (see Avice et al. 2022, this collection). Comets are likely to be devoid or highly depleted in neon (whose $T_{condensation} \approx 20$ K). The presence of neon in its atmosphere likely indicates that comets cannot be the sole contributors to volatile elements on Venus. Cometary water could have been brought to Venus and added to an original solar- or chondritic-derived atmosphere, or primordial neon could have been degassed from the mantle through time. The value of the Ar/Ne elemental ratio of Venus' atmosphere seems high compared to what would be expected from a pure mixture between solar and chondritic neon. While the large error bars on the Ar/Ne ratio does not allow one to draw any firm conclusions, a high Ar/Ne would be compatible with a cometary contribution to the Venus atmosphere. Such cometary contribution has been identified for Earth, based on the isotopic composition of primordial Earth's atmospheric xenon (Pepin 1991; Marty et al. 2017; Avice et al. 2017). However, the D/H ratio of Earth's water is close to a chondritic value, meaning that comets probably contributed less than 1 percent of Earth's total water budget (Fig. 7; Marty et al. 2016, Bekaert et al. 2020). Yet, this estimate is only valid if data collected for 67P/C–G are representative of cometary diversity. The possibility of an asteroid–comet continuum makes definitive statements difficult given the few available measurements and their uncertainties (Gounelle 2011). Furthermore, D/H ratio interpretations remain problematic given the number, uncertainties, and complexity of the processes affecting them (Stephant et al. 2016, 2018; Piani et al. 2020; Stephant et al. 2020).

It must be noted that elements such as noble gases can also escape from planetary atmospheres during episodes of hydrogen escape and that such escape would lead to elemental and isotopic fractionation of these elements (Hunten et al. 1987; Zahnle et al. 2019). Future data on the elemental and isotopic composition of noble gases collected in the atmosphere of Venus, for example during the recently selected DAVINCI (NASA) mission (Garvin et al.

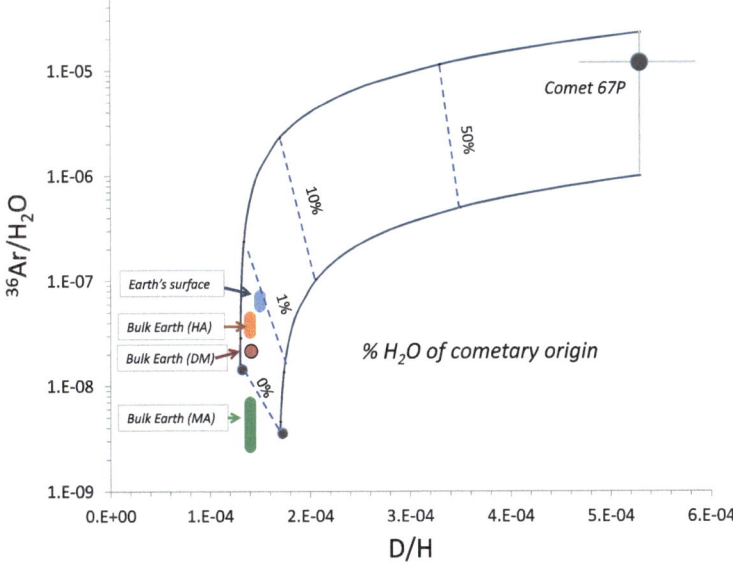

Fig. 7 ^{36}Ar/H$_2$O versus D/H mixing diagram between cometary (represented by 67P/C–G) and chondritic end-members. The mixing curves depend on the elemental and isotopic composition of the end-members. Comets have high Ar/H$_2$O ratios implying that more than 90% of Ar could of cometary origin while only about 1% of Earth's water would have been brought by comets (without altering the initial chondritic-like D/H ratio). HA, DM and MA are the bulk Earth estimates from Halliday (2013), Dauphas and Morbidelli (2014) and Marty (2012), respectively. Figure modified after Marty et al. (2016, see also refs. therein)

2020, 2022), will certainly help to put constraints on models attempting to track the joint history of water and noble gases on Venus and to compare the case of Venus with other terrestrial planets (Avice and Marty 2020).

3 Magma Ocean Evolution and the Formation of the Secondary Atmosphere

3.1 Magma Ocean Cooling and Solidification

There is no strict definition of what is referred to as "magma ocean". The term "magma" refers to a multiphase mixture of liquid (melts), solid (crystals), and gas (bubbles) composing a molten rock (e.g., Lesher and Spera 2015) and "ocean" was used to reflect the low, water-like viscosity of the mixture and the fact that the system is virtually 100% liquid (e.g., Warren 1985; Solomatov 2015). The magma ocean must then behave rheologically as a liquid, with a crystal fraction small enough that crystals are suspended within the melt without being interconnected. In addition, previous definitions include a criterion regarding the extent of melt, such that the magma must encompass a substantial fraction (more than 10%) of the body (Taylor and Norman 1992).

This general definition can be confusing as it may encompass a large variety of cases. These include drastically different scenarios such as a periodically molten planet as a result of tidal heating in an eccentric orbit, and a planet *ad-aeternum* molten owing to its proximity to the star. A fundamental distinction can be made compared to the early transient molten stage. The latter is likely inherent to the early evolution of rocky planets. This is in contrast to the long-lasting, steady-state molten stage experienced by planets in close-in orbits around their star; sometimes referred to as "lava worlds". Here we focus on the former, which is likely experienced by most terrestrial planets due to the combination of the heat sources detailed in Sect. 2.2.2. Compared to the steady-state, sustained molten stage of lava worlds, early transient magma oceans are not in thermal equilibrium with their colder stellar environment. They are cooling down, losing excess heat to reach equilibrium. As a fluid-like material, the magma ocean can efficiently transport and lose heat by advecting matter. Its dynamics and cooling are thus controlled by thermal convection. More precisely, as the molten mantle loses its internal heat while being cooled from above, it corresponds to the classical problem of Rayleigh-Bénard convection.

In such a case, the Navier–Stokes equations are used to describe the motion and evolution of the fluid. Using appropriate characteristic scales, two key dimensionless parameters regarding the fluid convective dynamics arise from the non-dimensional form of the momentum conservation equation (itself originating from Newton's second law): the Rayleigh number, Ra, and the Prandtl number, Pr. The thermal Rayleigh number,

$$Ra = \frac{\theta g \Delta T L^3}{\kappa \nu} \ , \tag{3}$$

compares the thermal buoyancy force that drives convection to the advection-resisting effects of thermal and momentum diffusion (viscous dissipation). The motion-driving parameters are the thermal expansion coefficient, θ, gravity, g, the superadiabatic temperature difference ΔT across the convective layer of thickness L (here being the magma ocean thickness). The parameters buffering advection are the thermal diffusivity, κ, and the kinematic viscosity, $\nu = \eta/\rho$, where η is the dynamic viscosity, and ρ is the fluid density. The Rayleigh number is thus a measure of the convective vigor. It is estimated to be up to $Ra = 10^{31}$ for a typical magma ocean encompassing an Earth-sized mantle (e.g., Solomatov 2000). It decreases with magma ocean crystallization (the present-day Earth's mantle is estimated to be between 10^6 and 10^8; e.g., Ricard 2015). The Prandtl number is

$$Pr = \frac{\nu}{\kappa} \ , \tag{4}$$

and expresses the ratio of momentum to heat diffusion. For the present-day solid-state mantle of Solar System terrestrial planets $Pr \gg 1$ ($Pr > 10^{23}$ on Earth where mantle convection is essentially a buoyancy-driven Stokes flow; e.g., Schubert et al. 2001, Ricard 2015), which implies that fluid motions stop as soon as the heat source disappears because of negligible inertial effects compared to viscous effects. Conversely, inertia cannot be neglected in the frame of vigorously convective, low viscosity magma oceans. In that case $Pr \approx 1$ for a global Earth-sized mantle magma ocean and progressively increases with solidification. These two parameters characterize the convective dynamics.

The heat loss is a function of the convective vigor so that the dimensionless heat flux, i.e., the Nusselt number, Nu, which is the ratio of convective to conductive heat transfer ($Nu = 1$ when heat is dissipated by conduction only, and exceeds unity as soon as convection starts), is representative of the heat flux (F_{conv}) leaving the magma ocean through the top boundary of the convective layer (i.e., the surface). F_{conv} scales as a power-law of the Rayleigh number (e.g., Jaupart et al. 2015):

$$Nu = \frac{F_{conv} L}{k \Delta T} = a_{Nu} Ra^{\gamma_{Nu}} \ , \tag{5}$$

where the scaling law pre-factor a_{Nu} and exponent γ_{Nu} values depend on the convective regime and patterns (e.g., Malkus 1954; Grossmann and Lohse 2000; Chillà and Schumacher 2012; Stevens et al. 2013, 2018). For extremely vigorous and turbulent convection relevant to magma ocean parameters, the convective heat flux is usually parameterized using $a_{Nu} = 0.089$ and $\gamma_{Nu} = 1/3$ (e.g., Siggia 1994, Solomatov 2015; regime IV_u in Grossmann and Lohse 2000). This implies that the heat loss is governed solely by local instabilities of the upper thermal boundary layer and is independent of the magma ocean thickness (e.g., Lebrun et al. 2013; Salvador et al. 2017; Nikolaou et al. 2019).

Given its low viscosity and the large temperature difference between the base of the MO and the surface, the convection is highly vigorous. The temperature across the molten layer is nearly adiabatic and isentropic (e.g., Abe 1997; Solomatov 2015). For a one-phase system, such as for the completely molten or completely solid and homogeneous layer, it can be expressed as:

$$\frac{dT}{dP} = \frac{\alpha T}{\rho c_p} \, , \tag{6}$$

where P is the pressure and c_p is the isobaric specific heat. In two-phase systems, i.e., where the melt and crystals co-exist, the equation for the adiabat is modified to account for the effects of phase changes (e.g., Abe 1997; Solomatov and Stevenson 1993; Solomatov 2015). The secular cooling of the magma ocean is then often computed by integrating the equation of conservation of energy over the evolving magma ocean thickness (e.g., Abe 1997; Lebrun et al. 2013; Jaupart et al. 2015).

Starting from a magma ocean extending throughout the entire mantle, the depth at which crystallization starts is the depth at which the adiabat first intersects the liquidus (above which temperature the material is completely liquid). Where it starts and how it proceeds with cooling thus firstly depends on the respective slopes of the melting curves (in particular that of the liquidus) and the adiabat (Fig. 8e). While the liquidus is well constrained over the first 1000 km of Earth's mantle, it remains uncertain at lower depth (e.g., Andrault et al. 2017; Fiquet 2018; Andrault 2019, for reviews). If the $P - T$ slope of the adiabatic temperature of the magma ocean is flatter than the liquidus (such as for a chondritic composition; Andrault et al. 2011), crystallization would start at the base of the mantle and proceeds from the bottom upward, as generally assumed in magma ocean studies (Fig. 8a–c, e and f). But this may vary if the liquidus profile presents a strong curvature at mid-lower mantle depths (such as for a fertile peridotite; Fiquet et al. 2010; Thomas et al. 2012), resulting in a crystallization starting at mid-mantle depths, where a growing shell of solid mantle material would form. Such a scenario would induce the formation of a persistent molten layer at the base of the mantle, a so-called "basal magma ocean" (Labrosse et al. 2007), and an early thermochemical separation between the upper and lower parts of the molten mantle. Whether the differences between these two crystallization scenarios arise from the different compositions considered or from experimental uncertainties remains unclear.

The depth at which crystallization starts is insufficient information to describe the subsequent solidification pattern of the magma ocean. Of equal importance is the relative buoyancy and thus the density difference between crystals and melts, which controls the way solid–liquid segregation proceeds (e.g., Boukaré and Ricard 2017; Caracas et al. 2019; see Andrault et al. 2017; Asimow 2018, for detailed reviews). At low pressures, forming crystals are typically assumed to be denser than the surrounding melt and thus sink to the bottom. Yet, the larger compressibility of liquids compared to that of crystals may induce the latter to be buoyant with respect to silicate melts in the lower mantle (Nomura et al. 2011;

Andrault et al. 2012; Thomas et al. 2012). Changes of Fe melt–crystal partition coefficient may also result in denser (and sinking) Fe-rich melts than co-existing crystals (e.g., Tateno et al. 2014; Andrault et al. 2017). This scenario contrasts with the classical view of crystals settling, compaction, and melt ejection in a crystallizing magma ocean (Solomatov 2007; Boukaré et al. 2015), and could also produce an early, isolated basal magma ocean above the CMB. Accounting for the effect of convective dynamics on the settling behavior of crystals (resulting from the competition between crystals settling and entrainment within the convective flow; e.g., Maas and Hansen 2015, 2019; Patočka et al. 2020, 2022) as a function of their evolving properties such as their shape, size, density, and distribution may also significantly affect the solidification and differentiation pattern (e.g., Solomatov 2015). The location of initial crystallization as well as the subsequent solidification pattern have critical implications regarding both the chemical and thermal stratification of the mantle, and may significantly affect the distribution of volatile species within the planetary interior and their outgassing. Despite being an open question whose influence is still poorly constrained, it might profoundly affect the early and long-term evolution of terrestrial planets.

As a result of cooling and crystallization, the melt fraction of the magma, ϕ_{melt}, decreases. Conversely its crystal fraction, $\phi_{\text{crystal}} = 1 - \phi_{\text{melt}}$, increases. It depends on the temperature at which crystals start to form, i.e., the liquidus T_{liq}, and on the temperature below which the material is completely solid, i.e., the solidus T_{sol}. In a linear and simplified form it can be written (Abe 1997):

$$\phi_{\text{melt}} = \frac{T - T_{\text{sol}}}{T_{\text{liq}} - T_{\text{sol}}} \ . \tag{7}$$

Because the temperature, T, is a function of depth and time, ϕ_{melt} varies accordingly and globally decreases with magma ocean cooling from $\phi_{\text{melt}} = 1$ ($T \geqslant T_{\text{liq}}$), when no crystals are present ($\phi_{\text{crystal}} = 0$), to $\phi_{\text{melt}} = 0$ ($T \leqslant T_{\text{sol}}$), when the magma is fully crystallized ($\phi_{\text{crystal}} = 1$).

The viscosity of the magma and associated rheological behavior greatly varies as a function of the temperature/crystal content and controls virtually all dynamic processes in the magma ocean. For instance, it strongly affects the convective vigor, the associated heat redistribution, and thus, the amount of heat the mantle can release. Throughout the magma ocean lifetime, viscosity changes over several orders of magnitude (Fig. 8d). Where and when the magma temperature is above liquidus, the dynamic viscosity of the silicate melt can be as low as the viscosity of water at ambient conditions ($\eta \approx 10^{-3}$ Pa.s; e.g., Urbain et al. 1982, Solomatov 2007, Kono 2018). A rheological transition associated with an abrupt viscosity increase occurs when the crystal fraction becomes large enough such that the connectivity of the crystals hampers efficient, liquid-like motion (e.g., Lejeune and Richet 1995; Abe 1993b, 1997). The mantle depth at which the rheological transition occurs is called the rheological front and separates the upper, low viscosity, vigorously convective, melt-dominated-motion mantle from the lower, rheologically solid-like mantle. The critical melt fraction at which this transition occurs is classically taken to be $\phi_{\text{melt,c}} \approx 0.4$ (Marsh 1981; Abe 1993b), but other mechanisms, such as diffusion creep and melt contiguity (Faul and Jackson 2007; Takei and Holtzman 2009a), may keep the viscosity low and fluid-like even at significantly lower melt fractions (from $\phi_{\text{melt,c}} \approx 0.019$ for 15 μm grain size and down to $\phi_{\text{melt,c}} \approx 10^{-4}$ for typical mantle grain size of 3 mm; Takei and Holtzman 2009b). Assuming that the solidification of the magma ocean proceeds from the bottom upward, with crystals efficiently segregating from the melt and accumulating downward, two layers of very distinct rheological behaviors exist from either side of the rheological front (Fig. 8b).

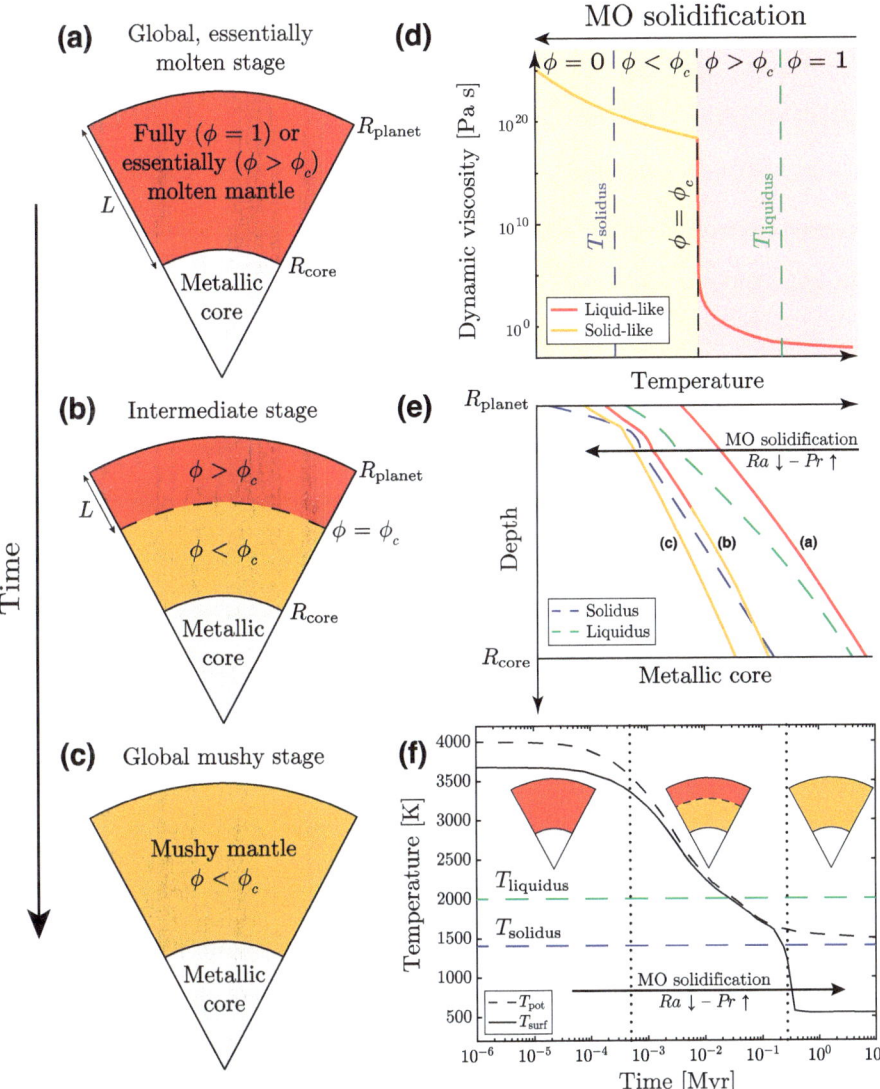

Fig. 8 (a–c) Magma ocean schematic evolution sequence from (a) a fully ($\phi = 1$)/essentially ($\phi_c < \phi < 1$) molten stage of global extent ($L = R_{\text{planet}} - R_{\text{core}}$), to (b) an intermediate stage where an essentially molten upper unit coexists with a lower mushy unit, to (c) a globally mushy stage where the liquid fraction is below the critical value of the rheological transition everywhere. (d) Viscosity and (e) temperature profile of the silicate mantle as a function of its evolving temperature/melt fraction. (f) Typical thermal evolution of an Earth-like magma ocean overlaid with an outgassing H_2O–CO_2–N_2 atmosphere (modified from Salvador et al. 2017)

Here we consider that the magma ocean is rheologically liquid-like, either fully ($\phi_{\text{melt}} = 1$) or essentially molten, with a melt fraction above $\phi_{\text{melt,c}}$ ($\phi_{\text{melt,c}} < \phi_{\text{melt}} \leqslant 1$), while the "mushy" layer below ($0 \leqslant \phi_{\text{melt}} \leqslant \phi_{\text{melt,c}}$) convects less vigorously and behaves rheologically more like the present-day solidified terrestrial mantle, with a melt fraction-dependent viscosity. Starting from a fully molten mantle, the magma ocean thus extends from the surface to the rheological front (Figs. 5 and 8a and b) and with time, it shrinks until the rheological front reaches the surface (Fig. 8c). From then on, the mantle dynamics and mixing is much less vigorous and the associated convective flux is much lower: the end of the transient magma ocean stage is reached.

Note that global thermal equilibrium can be reached independently of the magma ocean end: the former does not necessarily imply that the magma ocean phase is over. Indeed, global thermal equilibrium and climate stability depend on the radiative balance between the net absorbed insolation (F_{abs}) and the outgoing radiation (F_{IR}), and can thus be achieved at surface temperatures higher than the surface melting point. This can be the case for close-in planets that can keep their planetary surface molten indefinitely (e.g., lava worlds). For early Venus, depending on the atmospheric composition and on the absorbed solar flux resulting in part from the cloud–albedo feedback, climate stability may have been achieved with a molten or solidified surface (see Sects. 3.3 and 4, and Fig. 12).

3.2 Water/Volatiles Partitioning and Outgassing

At a given magma ocean crystallization stage, water and other volatile species are distributed between the different planetary reservoirs, from the deep interior to the atmosphere. These volatile species are chemical compounds with a low boiling point that thus vaporize readily and partition preferentially into a fluid or gas phase (e.g., Chap. 7 in White 2013). We first describe how this distribution proceeds at a given crystallization stage and then how it evolves with magma ocean cooling and solidification.

3.2.1 Melt–Crystal Volatile Partitioning

Within the molten mantle, the partitioning of a species between the co-existing melt (silicate liquid) and solid phase (crystals) is determined through partition coefficients that dictate how much of a given volatile species dissolved in the melt can be incorporated into a given mineral phase at given thermodynamic conditions. This is sometimes referred to as the storage capacity (e.g., Hirschmann et al. 2005). "Water" is not incorporated as an integral part of the crystal structure. It occurs mostly in the form of hydroxyl groups (–OH) in hydrous minerals, or as hydrogen atoms decorating point defects of nominally anhydrous minerals, the so-called NAMs, and bonded to structural oxygen, but rarely as molecular water. For extensive reviews regarding hydrogen/"water" in the solid mantle, refer to Hirschmann et al. (2005), Hirschmann (2006), Keppler and Bolfan-Casanova (2006), Demouchy and Bolfan-Casanova (2016), Peslier et al. (2017). Most coupled magma ocean–atmosphere thermal evolution studies consider a single solid–melt partition coefficient (ψ_{vol}) throughout the entire mantle (Hamano et al. 2013, 2015; Schaefer et al. 2016; Bower et al. 2019; Barth et al. 2021), usually taken as the partition coefficient of perovskite ($\psi_{H_2O} = 10^{-4}$ for water; except in Schaefer et al. (2016) and Barth et al. (2021), where $\psi_{H_2O} = 10^{-2}$ is close from the partition coefficient of lherzolite). Or they consider the partition coefficient of lherzolite for the upper mantle ($\psi_{H_2O} = 1.1 \times 10^{-2}$), and of perovskite for the lower mantle ($\psi_{H_2O} = 10^{-4}$; after Bolfan-Casanova et al. 2003), thus accounting for differences due to pressure and associated stable mineral phases (Elkins-Tanton 2008; Lebrun et al. 2013; Salvador et al. 2017; Nikolaou et al. 2019).

In addition to water partitioning between silicate melts and cumulates (i.e., crystals accumulated by gravitational settling in a magma), another process responsible for the incorporation of water into solidifying silicates is the interstitial trapping of water-enriched melt (Elkins-Tanton 2008; Hamano et al. 2013, 2015; Ikoma et al. 2018). Some models assume that 1% of the melt mass is incorporated into cumulates through interstitial trapping throughout solidification (Elkins-Tanton 2008; Hamano et al. 2013, 2015). Yet, as discussed in Hamano et al. (2013), estimates for magma ocean solidification remain highly unconstrained. They are based on the degree of melting at which melt separation starts for abyssal peridotites (Johnson and Dick 1992) and likely overestimates the mass fraction of interstitially trapped melt. Furthermore, interstitial trapping would only store significant amounts of water into cumulates in the presence of water-rich melts. The latter correspond to evolved magmas found at late solidification stages where water exsolution and outgassing are highly efficient (e.g., Elkins-Tanton and Grove 2011; Lebrun et al. 2013; Salvador et al. 2017; Nikolaou et al. 2019; Bower et al. 2019) and thought to be the water-draining dominant processes. Yet, it has been recently suggested that this interstitial trapping and melt retention process might be highly effective at the magma ocean freezing front (i.e., at the rheological transition) and may retain a substantial fraction of volatile-rich melt at the end of magma ocean evolution (Hier-Majumder and Hirschmann 2017; Miyazaki and Korenaga 2021, 2022). Volatile species that are highly soluble in the melt, such as water, would be very sensitive to this process and largely incorporated into the primordial solidified mantle. As this relates to volatile partitioning during the entire magma ocean evolution and crystallization sequence rather than being specific to a given crystallization stage, we discuss it in more detail in Sect. 3.2.4.

If one does not take into account volatile trapping in the porous melt phase, the incorporation of volatile species into the solidifying mantle is negligible compared to its partitioning into the melt during the magma ocean crystallization. Volatile trapping in the solidified mantle might thus be ignored to first order when addressing magma ocean and atmospheric thermal evolution (Lichtenberg et al. 2021). Yet this very process is of major importance when the magma ocean is eventually fully solidified, and when addressing the long term evolution of the planet and the dynamics of the solidified mantle.

Water has a strong effect on solid-state mantle creep strength (Hirth and Kohlstedt 1995; Mei and Kohlstedt 2000; Hirth and Kohlstedt 2004; Demouchy and Bolfan-Casanova 2016), melting relations (Hirth and Kohlstedt 1996; Asimow and Langmuir 2003), shear zones development (e.g., Skemer et al. 2013), viscosity and density (e.g., Lange 1994), and phase transitions (Wood 1995). Its partition between the melt and the solidifying mantle during the crystallization of the magma ocean will be of fundamental importance for the style and strength of convective flow during the following stages of planetary evolution (e.g., Schaefer and Elkins-Tanton 2018). Namely, the distribution of water within the planet has been proposed to resolve the distinct tectonic behaviors between the Earth's wet mantle and that supposedly dry of Venus (Mackwell et al. 1998; Richards et al. 2001). This water could be directly inherited from the magma ocean stage in the case of a gravitationally stable stratified crystallization sequence, or acquired later after mantle global overturn (e.g., Tikoo and Elkins-Tanton 2017). Furthermore, by decreasing the solidus temperature and thus facilitating melting at greater depth and greater extent (because water lowers the Gibbs free energy of the melt), water abundance in the solid mantle is a crucial parameter for Earth's mantle melting and volcanism. Particularly at subduction zones, mid-ocean ridges and for intraplate hotspot volcanism (e.g., Hirschmann 2006; Litasov and Ohtani 2007; Ni et al. 2016). The resulting interior melting and volcanism can in turn affect the atmospheric composition, climate, volatile cycling, and consequently, the surface conditions and habitability on rocky planets (e.g., Foley and Driscoll 2016; Dehant et al. 2019).

3.2.2 Melt–Gas Volatile Partitioning

Compared to crystals (i.e., magma's solid phase), the melt (i.e., magma's liquid phase) can contain a large amount of volatile species, including water, that are dissolved within the latter. Water is sometimes said to be more soluble in the melt than in the crystals/solid phase. This even though, technically, water is rather incorporated in the solid phase than dissolved within it. Thus, compared to the crystallizing solid mantle, the molten mantle has a larger volatile storage capacity. Unsurprisingly, the thicker the magma ocean, the more water can be dissolved into it.

Yet, while water is highly soluble in the silicate melt, the molten mantle still has a limited maximum storage capacity for dissolved water. This is also referred to as the so-called water solubility (see Moore 2008; Baker and Alletti 2012, for reviews). The solubility of volatile species in melts has been constrained both experimentally (e.g., Holloway and Blank 1994) and theoretically (e.g., Ottonello et al. 2018). It depends primarily on the volatile species considered, on the melt composition, and pressure, but also changes according to temperature, and volatile species mixture (H_2O only or $H_2O + CO_2$ for instance; see Papale 1997, 1999, Papale et al. 2006).

It is important to note that rather than giving the actual water content, solubility studies constrain the storage capacity of the mantle. This is the maximum amount of water that may be stored, which is the mantle saturation limit (e.g., Keppler and Bolfan-Casanova 2006). It does not give any constraint on its saturation state, i.e., if it is actually volatile-saturated or not. It means that the solubility of a given volatile species into the melt does not provide any information regarding the amount of this very species delivered to the planet. It is rather a bulk property of the molten mantle, so that one should always distinguish the actual water content of the magma ocean, inherited from the planetary formation, from its intrinsic storage capacity.

For a given melt composition and mass fraction of water, the higher the pressure, the larger amount of a volatile species can be maintained dissolved within the melt (Fig. 9). In the deep molten mantle, i.e., at high pressure, the silicate melt can retain more than 10 wt% of water and carbon dioxide in solution (e.g., Papale 1997). Thus, the decrease of ambient overburden pressure experienced by a fluid parcel rising up through convective motions will be responsible for a decrease of the melt storage capacity (Fig. 9), similarly to magma ascent in volcanic systems (e.g., Chap. 5 in Parfitt and Wilson 2008; Sparks et al. 1994; Gonnermann 2015; Wallace et al. 2015). This overburden pressure is the lithostatic load, that is the pressure at a given depth below the surface due to the weight of the overlying layers of silicate material. When the water-dissolved maximum storage capacity is reached, the molten mantle is water-saturated, meaning that all supplementary water exceeding the saturation limit cannot remain dissolved within the liquid phase and is thus exsolved out of the melt to form gas bubbles (i.e., decompression-induced degassing; Figs. 9 and 10; e.g., Edmonds and Woods 2018). Due to their low density compared to the ambient liquid, gas bubbles naturally tend to rise up through the fluid towards the surface, unless they are entrained downwards by the convective currents. Mantle degassing and atmosphere formation and replenishment then occur when gas bubbles reach and burst at the surface of the magma ocean, thus transferring volatile species from the gas phase of the magma to the atmosphere.

In volcanic systems, ascent of gas bubbles is a complex process. Whether they are able to reach and burst at the surface involves the interplay between diffusion, bubbles coalescence, growth, and re-entrainment, which are influenced by several parameters such as magma viscosity and melt–bubble relative density, themselves controlled by the amount of dissolved/exsolved volatiles (e.g., Berlo et al. 2011). Thus far this has been poorly investigated in the context of magma ocean outgassing. For volcanic eruptions, it is believed that rapid growth and ascent of gas bubbles is favored by the high temperatures and low viscosity of the magma, which allows gas to easily segregate from the melt in the form of small bubbles that can themselves easily merge leading to larger bubbles with a faster ascending speed (Sparks 1978; Sparks et al. 1994; Lesher and Spera 2015). The dynamics of a magma ocean may significantly differs from that of a volcanic erupting system. Yet, it is generally assumed that as soon as bubbles are formed, they also rapidly reach the surface without being re-entrained downwards by convective current, albeit vigorous, thus assuming optimistic efficient outgassing.

Instead of outgassing after bubbles formation and ascent to the surface, water can also be released to the atmosphere by molecular diffusion through the upper thermal boundary layer (e.g., Hamano et al. 2013, 2015; Massol et al. 2016; Ikoma et al. 2018). This degassing process dominates if bubbles are too small to rise up and reach the surface before being transported back downwards by vigorous convective flow. Such a diffusion-limited volatile flux provides a minimum estimate for the degassing rate. Using an experimentally-derived molecular diffusion coefficient for basaltic melts (Zhang et al. 2007), Hamano et al. (2013) found that even with a modest degree of supersaturation, this mechanism can still be responsible for the formation of a massive steam atmosphere, by keeping its water content in solution equilibrium with that of the silicate melt.

3.2.3 The Core as an Additional Reservoir of Hydrogen/Water

The metallic core may also be an additional reservoir of water (in the form of hydrogen) ($M_{H_2O}^{core}$ in Eq. (8)), potentially storing several oceans' mass equivalents (e.g., Genda 2016; Peslier et al. 2017). The core density deficit and sound velocity excess (compared to those of pure iron) inferred from seismology (e.g., Birch 1952; Poirier 1994) suggest that the inner core contains a few percents of light elements (see Hirose et al. 2021, for a recent review). Similarly, the density of Earth's outer core is estimated to be 5–10 wt% lower than that of pure iron. Approximately 10 wt% light alloys in the core have been suggested (e.g., Birch 1952; Hirose et al. 2013). Furthermore, some outer core temperature estimates, about 400 K lower than previously thought, would also support a similar fraction of elements lighter than iron to maintain its liquid state (Nomura et al. 2014).

During the differentiation, core-forming metal transports hydrogen and other volatile species from the magma ocean to the core (e.g., Rubie et al. 2015b; Dasgupta and Grewal 2019). Iron-loving elements are prone to be incorporated with metal droplets suspended in the magma ocean, and then partitioned into the core. A large amount of H may be partitioned into the metallic iron at high pressure and be incorporated to the core this way (e.g., Kuramoto and Matsui 1996; Okuchi 1997; Abe et al. 2000; Shibazaki et al. 2009; Li et al. 2020). Umemoto and Hirose (2015) found that $\sim$1 wt% of H in the outer core (equivalent to the hydrogen content of $\sim$130 oceans of water) can reproduce its lower density and faster bulk compressional sound velocity. More conservative estimates suggest an equivalent of 0.2 to up to 90 Earth oceans mass if H is the only volatile in the core (see Peslier et al. 2017, and references therein). The partitioning process is controlled by the chemical equilibration between metal liquids and silicate melt. The partitioning coefficients depend on the $P - T$ conditions at which silicate melt–metal equilibration and core sequestration occur, and on the redox state of the magma ocean (e.g., Li and Agee 1996; Okuchi 1997; Badro et al. 2015).

H has been proposed as a strong candidate because of its abundance in the Solar System and in planetary interiors and its high incompatibility in Fe liquid metal (Okuchi 1997; Abe et al. 2000; Iizuka-Oku et al. 2017). Recent ab initio molecular dynamics calculations have suggested that at the $P - T$ conditions of both magma ocean and the current Earth's CMB, H_2 and H_2O are both more compatible in the core than in silicate melts (Yuan and Steinle-Neumann 2020; Li et al. 2020). This implies that the Earth's core can potentially act as a large reservoir of water. This is also supported by recent experimentally determined partitioning of hydrogen between liquid Fe and molten silicate under high pressures and temperatures relevant to the conditions of core-forming metal segregation from silicate (Tagawa et al. 2021).

However, other recent experimental results contradict H strong siderophile behavior (Clesi et al. 2018; Malavergne et al. 2019), favoring alternative light elements such as O, S, C, or Si (e.g., Helffrich and Kaneshima 2004; Badro et al. 2015; Shahar et al. 2016). The long-standing question of hydrogen in the Earth's core still remains an unresolved topic. Determining if this discussion applies to Venus would first require estimates of its core structure and density, that may be assessed through future ionospheric and/or ground-based seismology and would certainly improve our understanding of current and early interior processes. As a first approximation it seems reasonable to think that such process may also apply to Venus. This relies on Earth and Venus having similar bulk density and composition, and if one considers that they experienced roughly similar accretion sequences due to their vicinity (which might be challenged; see Emsenhuber et al. 2021), Although little is know about Venus' interior today, future exploration missions, such as NASA's VERITAS (Smrekar et al. 2022) and ESA's EnVision are promising to provide some constraints on its physical properties (see Widemann et al. 2023, this collection).

3.2.4 Evolution During Magma Ocean Solidification

Distribution Between the Different Reservoirs As the magma ocean crystallizes, each volatile species is distributed between the different and evolving reservoirs of the magma ocean–atmosphere system: the solid mantle, the remaining melt (molten mantle), and the atmosphere. As discussed above, some amount of volatile species may also be incorporated into the core during the segregation of metallic iron to the core if metal–silicate separation is not completed. This distribution can be described by the mass conservation equation. For water, it is (following Lebrun et al. 2013; Schaefer et al. 2016; Salvador et al. 2017):

$$M_{H_2O}^{MO,t=0} = M_{H_2O}^{melt} + M_{H_2O}^{solid} + M_{H_2O}^{atm} + M_{H_2O}^{core} , \qquad (8)$$

where the initial and total mass of water of the system, dissolved within the molten mantle at $t = 0$, $M_{H_2O}^{MO,t=0}$, is distributed between the molten part of the mantle, the solidifying mantle, and the atmosphere, whose water masses are $M_{H_2O}^{melt}$, $M_{H_2O}^{solid}$, and $M_{H_2O}^{atm}$, respectively. The mass of water transferred to the core, $M_{H_2O}^{core}$, is often neglected as it is generally assumed that metallic iron has fully segregated into the core by the time of the last hypothetical global magma ocean, and that chemical exchanges between the mantle and the core are then limited. Before metallic iron has fully segregated to the core, a substantial amount of hydrogen may be incorporated into the core that could thus possibly contain a significant amount of water initially accreted by the planet (Sect. 3.2.3). Here, $t = 0$ represents the starting point of the cooling of the last and global magma ocean stage following a giant impact, whose timing is uncertain (e.g., Elkins-Tanton 2012), and implicitly relies on the accretion timeline.

The magma ocean initial water content is usually expressed as a mass fraction of water, $X_{H_2O}^{MO,t=0}$, relative to the initial magma ocean mass, $M_{MO,t=0}$:

$$M_{H_2O}^{MO,t=0} = X_{H_2O}^{MO,t=0} \times M_{MO,t=0} . \qquad (9)$$

This initial water content delivered to the planet is highly unconstrained. For Earth, it is estimated to be 18_{-15}^{+81} times the equivalent mass of the oceans (Peslier et al. 2017), and implicitly relies on assumptions regarding the accretion sequence (see Sect. 2). This includes, but is not limited to the timing and dynamics of planetary formation, both affecting the sources and the processes of water delivery (e.g., Morbidelli et al. 2012; Marty 2012; O'Brien et al. 2018). It is generally assumed that any pre-existing atmosphere has been blown away due to the energetics of the giant impact such that water and other volatile species are initially dissolved within the molten interior. Yet, atmospheric erosion may not be as efficient (e.g., Genda and Abe 2003; Pham et al. 2011; Schlichting et al. 2015) and a fraction of accreted volatiles may be released directly into the atmosphere upon impact (e.g., Lange and Ahrens 1982; Matsui and Abe 1986b). $M_{H_2O}^{melt}$ is the mass of water dissolved within the molten part of the mantle and staying in the residual liquid as solidification proceeds. It is a mass fraction of the rheologically liquid-like magma ocean, $X_{H_2O}^{melt}$, of mass M_{melt}, and is given by:

$$M_{H_2O}^{melt} = X_{H_2O}^{melt} \times M_{melt} \; . \tag{10}$$

The mass of water stored within the solidifying mantle depends both on the melt–crystals partition coefficient, ψ_{H_2O}, and on the mass fraction of water incorporated through interstitial melt trapping, $X_{H_2O}^{inter}$, following:

$$M_{H_2O}^{solid} = \psi_{H_2O} X_{H_2O}^{melt} M_{solid} + X_{H_2O}^{inter} M_{melt} \iota \; , \tag{11}$$

where ι is the mass fraction of melt incorporated into the solid mantle through interstitial trapping. It is relative to the melt mass, M_{melt}, taken as $\iota = 1\%$ of M_{melt} in Elkins-Tanton 2008, Hamano et al. 2013, 2015. Note that most coupled magma ocean–atmosphere studies do not consider any incorporation of water through interstitial trapping of water-enriched melt (i.e., $\iota = 0$), so that the corresponding last term in the right-hand side of Eq. (11) is neglected.

The partition coefficient is either computed as a mantle-averaged value and taken as such (e.g., Hamano et al. 2013, 2015; Schaefer et al. 2016; Bower et al. 2019; Barth et al. 2021), or written as (Lebrun et al. 2013; Salvador et al. 2017):

$$\psi_{H_2O} = \frac{M_{perov}^{solid} \psi_{H_2O}^{melt-perov} + M_{lherz}^{solid} \psi_{H_2O}^{melt-lherz}}{M_{perov}^{solid} + M_{lherz}^{solid}} \; , \tag{12}$$

when considering distinct and specific melt–solid partition coefficients for the lower and upper mantle mineral phases, such as perovskite and lherzolite (e.g., Elkins-Tanton 2008; Lebrun et al. 2013; Salvador et al. 2017; Nikolaou et al. 2019). In such a case, M_{perov}^{solid} and M_{lherz}^{solid} stand for the evolving masses of the perovskite and lherzolite mineral phases, and $\psi_{H_2O}^{melt-perov}$ and $\psi_{H_2O}^{melt-lherz}$ their melt–solid partition coefficients. Because of its dependence on the mass of crystallized phases, ψ_{H_2O} changes with solidification.

Finally, the mass of outgassed atmospheric water ($M_{H_2O}^{atm}$) when considering a pure steam atmosphere is given as (e.g., Elkins-Tanton 2008; Schaefer et al. 2016):

$$M_{H_2O}^{atm} = 4\pi R_p^2 \frac{P_{H_2O}^{atm}}{g} \; , \tag{13}$$

where R_p is the planetary radius, $P_{H_2O}^{atm}$ is the atmospheric partial pressure of water (at the planetary surface), and g is the gravity. Importantly, one should note that when considering multi-species atmospheres, the molar mass needs to be considered to relate the partial pressure of each volatile species to its atmospheric mass (e.g., Pierrehumbert 2010; Bower et al.

2019):

$$M_{\text{vol}}^{\text{atm}} = 4\pi R_{\text{p}}^2 \left(\frac{\mu_{\text{vol}}}{\bar{\mu}} \right) \frac{P_{\text{vol}}}{g} \quad , \tag{14}$$

where $M_{\text{vol}}^{\text{atm}}$ is the mass of the volatile species vol in the atmosphere, μ_{vol} is its molar mass, P_{vol} its atmospheric partial pressure at the planetary surface, and $\bar{\mu}$ is the mean molar mass of the atmosphere. As discussed above, the atmospheric partial pressure of a volatile species at the surface comes from its outgassing from the interior. Its efficiency depends mainly on its abundance and solubility within the melt, as well as on the convective style and crystallization sequence of the molten mantle.

Because water is an incompatible element, it partitions preferentially into the remaining silicate melt relative to the silicate solid when the magma crystallizes. Therefore, as magma ocean crystallization proceeds, the residual and thinning melt layer becomes more and more enriched in water. Its mass fraction of dissolved water (relative to the magma ocean mass), $X_{\text{H}_2\text{O}}^{\text{melt}} = M_{\text{H}_2\text{O}}/M_{\text{MO}}$, increases with time (e.g., Elkins-Tanton 2008; Massol et al. 2016; Ikoma et al. 2018). The magma ocean mass, M_{MO}, decreases with mantle solidification while the mass of dissolved water $M_{\text{H}_2\text{O}}$ remains similar due its preferential partitioning into the melt (unless a significant amount of water goes into the solid phase). This facilitates the water saturation of the melt with magma ocean solidification (Fig. 9). Yet, a parcel of magma still needs to reach shallow depths and pressures low enough for saturation and exsolution to occur (Figs. 9 and 10).

Given its low, liquid-like viscosity ($\sim$0.1 Pa.s) and associated vigorous convection, magma ocean large convective velocities (v_{conv}, on the order of 0.5 m.s^{-1}; see Solomatov 2000, 2015; Elkins-Tanton 2008) are expected to rapidly carry volatile-bearing melt close to the surface (black arrows in Fig. 10). There exsolution can occur at the exsolution depth, d_{exsol}; blue dashed line in Fig. 10. The complete circulation of a deep magma ocean would be achieved in about 1 to 3 weeks (Elkins-Tanton 2008). Compared to the solidification timescale, this is fast enough for the entire magma ocean volume to reach the exsolution depth and outgas its volatiles before the solidification front substantially rises-up. Its velocity is v_{sol}; see yellow arrow in Fig. 10. Furthermore, this 1–3 weeks timescale being smaller than computational time-steps, magma ocean–atmosphere models generally assume that at each time-step, all the magma ocean volume has "seen" the surface, reached volatile saturation, and has thus virtually instantaneously outgassed all its volatiles in excess of saturation at each modeling time-step (Zahnle et al. 1988; Elkins-Tanton 2008; Hamano et al. 2013; Lebrun et al. 2013; Hamano et al. 2015; Schaefer et al. 2016; Salvador et al. 2017; Nikolaou et al. 2019; Bower et al. 2019; Lichtenberg et al. 2021; Barth et al. 2021; Krissansen-Totton et al. 2021). Based on this optimistic and efficient outgassing assumption, these models simulate volatile degassing by considering the pressure equilibrium of the gas phase between the magma and the atmosphere (i.e., on either side of the planetary surface). All volatiles in excess of saturation are thus directly transferred to the atmosphere. Solubility laws are then used to relate the atmospheric partial pressure of each volatile, $P_{\text{vol}}^{\text{atm}}$, to its abundance in the melt, $X_{\text{vol}}^{\text{melt}}$, according to (e.g., Elkins-Tanton 2008; Lebrun et al. 2013; Massol et al. 2016):

$$P_{\text{vol}}^{\text{atm}} = \left(\frac{X_{\text{vol}}^{\text{melt}}}{\zeta_{\text{vol}}} \right)^{\beta_{\text{vol}}} \quad , \tag{15}$$

where ζ_{vol} and β_{vol} are volatile-specific solubility constants taken from experimental data (such as Pan et al. 1991; Papale 1997). In magma ocean–atmosphere models, the atmospheric pressures are generally computed out of the volatile species abundances in the melt based on their solubility.

Typical Outgassing Sequence Under this assumption and outgassing treatment, typical coupled magma ocean–atmosphere models describe the formation of an H_2O–CO_2 atmosphere through magma ocean degassing as follows[3] (e.g., Elkins-Tanton 2008; Lebrun et al. 2013; Salvador et al. 2017; Nikolaou et al. 2019; Bower et al. 2019). Given the low solubility of CO_2 in silicate melts, it exsolves from magma at much greater depths beneath the surface than H_2O and outgasses early. Most of the carbon dioxide is outgassed to the atmosphere in the very first time-steps, resulting in high CO_2 atmospheric partial pressures, P_{CO_2}, at the earliest evolutionary stages.

Water being much more soluble in the melt, its outgassing occurs later on (starting after $\sim 10^4$ years of cooling), as it requires a more advanced crystallization stage for its concentration in the melt to be high enough to reach saturation. Once it starts, H_2O outgassing and resulting atmosphere build-up reflects the progressive replacement of melt mantle volume with solid mantle volume that has a smaller volatile storage capacity (e.g., Lebrun et al. 2013; Massol et al. 2016; Nikolaou et al. 2019). As an example, Fig. 11 shows the progressive partitioning of water between the different reservoirs during a typical magma ocean outgassing sequence (derived from Salvador et al. 2017 simulations). Interestingly, because the exsolution of a given volatile species depends on its concentration within the melt, lower initial abundances of water within the magma ocean may result in larger relative amounts of water stored in the planet's interior after the magma ocean ends (Nikolaou et al. 2019).

Compared to the H_2O–CO_2 atmospheres typically considered, Lichtenberg et al. (2021) has comprehensively studied how alternative atmospheric components may individually affect the radiative transfer of the atmosphere and the cooling of the magma ocean. How the combination of these individual components in more complex and realistic multi-species atmospheres would affect the heat transfer and magma ocean cooling rate has recently been investigated in a model framework considering variable oxidation states of the magma ocean (e.g., Katyal et al. 2020; Bower et al. 2022; Gaillard et al. 2022a; Maurice et al. 2023, see Sect. 3.2.5 for a detailed discussion).

Solubility Laws The implicit and model-dependent choice of solubility law has strong implications regarding magma ocean outgassing efficiency. That is because it ultimately dictates the outgassing rate, timescale, and the volatile amounts that can be dissolved in/exsolved out of the melt. Yet, this is poorly emphasized in the literature. Solubility laws obtained for basaltic systems are generally used (e.g., Elkins-Tanton 2008; Lebrun et al. 2013; Massol et al. 2016) as they are more appropriate for magma oceans since basalts, i.e., relatively low silica content melts, are high temperature and low viscosity melts (e.g., Sparks et al. 1994; Lesher and Spera 2015). Yet, they also have high gas diffusivity and tend to have more modest volatile contents than more viscous, silicic melts such as rhyolitic (as considered in Lichtenberg et al. 2021) or dacitic melts. In the case of water, at a given depth, a silicate liquid of basaltic composition can contain less dissolved water than one of rhyolitic composition (Fig. 9). Reciprocally, for a given water content, basaltic melts become saturated and exsolve H_2O at greater depths, i.e., at larger confining pressures, compared to rhyolitic melts. This facilitates exsolution and outgassing in basalt compared to rhyolite as the magma reaches saturation deeper and does not have to travel to shallower depths, that are closer to the surface (Fig. 9).

---

[3]Recall that coupled magma ocean–atmosphere models that consider the outgassing from the molten interior (in contrast to impact-generated atmosphere) generally assume that any prior atmosphere has dissipated after a giant impact responsible for melting the entire mantle, so that no pre-existing atmosphere is considered at the beginning of the simulations.

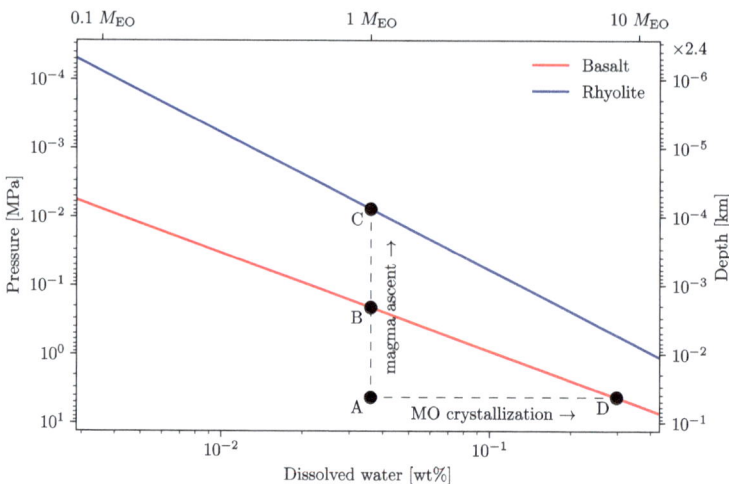

Fig. 9 Water solubility in basalt (red) and rhyolite (blue) as a function of the pressure acting on the magma (i.e., confining pressure) and corresponding depth beneath the surface. The top axis indicates the amount of water dissolved within a fully molten Earth-sized mantle corresponding to the mass fractions of dissolved water of the bottom axis. For a given water content, the saturation curve gives the location of the exsolution depth, d_{exsol}, at which gas bubbles start to form (Fig. 10, blue dashed line). At a given cooling stage, a parcel of magma ascending from a depth corresponding to point A undergoes decompression and would become saturated in H_2O at point B if it were a basalt and would start exsolving the excess of water, if any. If it were a rhyolite, it would not become saturated until it had ascended to the lower pressure marked by point C. Reciprocally, the saturation curve indicates the maximum amount of water that can be dissolved within the melt at a given pressure. As solidification proceeds, water partitions readily into the silicate melt, whose volume decreases owing to crystallization, such that evolving magma ocean liquids become more and more enriched in water. With magma ocean crystallization, a melt starting at a mass fraction of dissolved water corresponding to point A would see its water concentration increase and would become saturated in H_2O at point D. Once saturation is reached, the melt's mass fraction of dissolved water would evolve along the saturation curve. Solubility laws taken from Parfitt and Wilson (2008) and Carroll and Holloway (1994)

However, it should be noted that additional complexities may arise. For instance, the viscosity of mafic and ultramafic melts (such as basalt and komatiite, respectively) is comparatively much less affected by the dissolved water content than that of polymerized, low temperatures rhyolitic or dacitic melts (e.g., Fig. 5.5 in Lesher and Spera 2015, and references therein). In turn, viscosity influences magma motion and transport and may affect the amount of melt reaching the exsolution depth.

Furthermore, even for a given silicate melt composition, several solubility laws are available and their differences will affect outgassing efficiency as described before. These differences become more and more important when decreasing the mass fraction of dissolved water and are thus larger for low water contents and increase with planetary size for a given absolute water content (see Appendix A in Salvador and Samuel 2023, for a detailed discussion). Predictions regarding magma ocean outgassing, both in terms of degassing timescale and amount of outgassed volatiles, are thus ultimately influenced by the generally implicit and model-dependent choice of solubility law. In turn, the solubility law choice will alter the cooling rate of the magma ocean through the amount of radiative energy that can be dissipated to space, which is dictated by the outgassed atmospheric composition.

In addition, while the solubility of each volatile species is often treated separately when several volatile species are present in the magma ocean (for instance, H_2O and CO_2), solubility functions are actually more complex in such a case. This is in fact virtually always

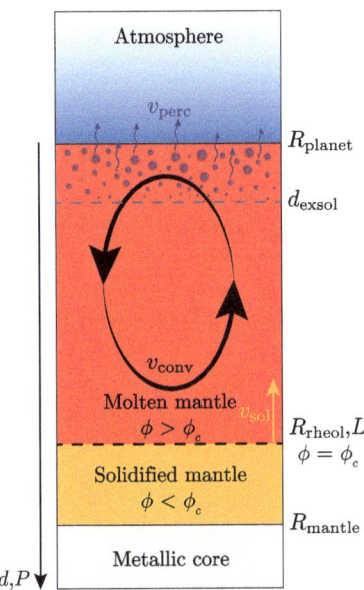

Fig. 10 Schematic of the magma ocean degassing process and associated velocities. v_{sol} is the upward solidification velocity associated with the uprising rheological front where $\phi = \phi_c$ (assuming that solidification proceeds from bottom-up). v_{conv} is the liquid-like convection velocity and v_{perc} is the percolation velocity of the gas bubbles rising up to the surface. d_{exsol} is the exsolution depth above which oversaturated volatiles are exsolved out of the melt and start forming gas bubbles. In a rapidly convecting magma ocean, volatile-bearing melt is assumed to reach the exsolution depth much faster than solidification proceeds and loses its volatiles in excess of saturation on that occasion. The outgassing is thus often assumed to be efficient. Adapted from Salvador and Samuel (2023)

the case: the gas phase is never just pure H_2O or CO_2. Then, chemical interactions between the volatiles themselves and the magma affect the solubility (e.g., Papale 1997; Papale et al. 2006; Massol et al. 2016; Edmonds and Woods 2018). For instance, as soon as the least soluble volatile starts to exsolve and form gas bubbles, some small amounts of most of the other species present will also diffuse into those bubbles (e.g., Parfitt and Wilson 2008). Such migration of the molecules of volatiles into the gas bubbles will increase the concentration gradient in the surrounding liquid, which will enhance the diffusion of more molecules toward the growing bubble. The non-ideal mixing behavior of the volatile species in the vapor phase is also responsible for the curvature of the saturation curves (see Papale et al. 2006; Moore 2008; Berlo et al. 2011, for detailed discussions).

Limits of the Efficient Degassing Assumption Stemming from the vigorous convection of the magma ocean, the efficient degassing scenario, relying on one a priori robust – yet debatable – assumption is mainly considered. Yet, despite high convective velocities, several mechanisms may buffer the outgassing efficiency of the magma ocean (see Table 1 for a summary). Alternatively, minimal to no degassing scenarios may be considered (e.g., Ikoma et al. 2018), where a substantial fraction of volatiles is retained in the melt on longer timescales and possibly ends up being trapped in the solidified mantle.

Although vigorous convection is prone to efficient and rapid magma ocean mixing, the fluid motions at relevant magma ocean dynamics can be organized according to large-scale circulations (e.g., Krishnamurti and Howard 1981; Castaing et al. 1989; Siggia 1994). This

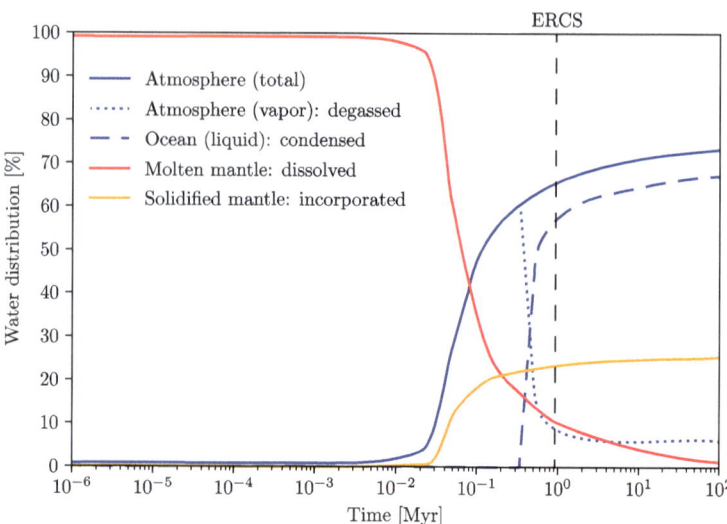

Fig. 11 Evolution of the distribution of water between the molten (red) and solidified (orange) mantle, the atmosphere (plain blue line for the total amount of water outgassed, including both vapor and liquid phase) and the ocean (dashed blue) during magma ocean solidification of an Earth-sized planet, while assuming efficient outgassing. The vertical dashed line indicates the time at which the End of the Rapid Cooling Stage is reached, i.e., when the planet has lost its excess of internal heat and reached global radiative balance with the net absorbed solar flux. Derived from Salvador et al. (2017) simulations

may isolate a significant part of the magma ocean from the surface in the center of large-scale convection rolls (e.g., Fig. 1 in Banzhaf 2003), thereby limiting the amount of silicate melt reaching low pressures at which volatiles are exsolved, and thus, tamping down magma outgassing. Using numerical fluid dynamics experiments, a recent study investigated how the detailed convective patterns may influence the time-dependent ability of the convecting fluid to reach shallow depths at which volatile exsolution and degassing proceeds (Salvador and Samuel 2023). They showed that vigorous convection and associated mixing does not ensure magma ocean chemical homogeneity. Rather, vortex-like small-scale features entrained in large-scale convective currents might generate local chemical heterogeneities. Then, despite high convective velocities and effective mixing, additional convective cycles and longer timescales would be required to erase those small-scale heterogeneities and to bring the volatile-rich melt blobs close to the surface where exsolution and degassing can occur. In some cases, these timescales may be comparable to or even larger than the magma ocean time spent in a given dynamic regime. Accumulated over the whole magma ocean lifetime, the amount of outgassed water may substantially decrease. There may be cases resulting in no outgassing of water initially dissolved deeper than the exsolution depth.

The crystallization sequence of the magma ocean may also significantly affect volatile distribution and outgassing efficiency. If the crystallization of the magma ocean starts from mid-mantle depths (Fiquet et al. 2010), rather than proceeding from bottom-up as it is commonly assumed in coupled magma ocean–atmosphere models, two distinct and separate molten layers may exist on either side of the crystallization front (e.g., Labrosse et al. 2007). While the upper molten layer would remain in contact with the surface where exsolution and outgassing may proceed as described above, the lower, deep basal magma ocean may keep its volatile species dissolved thanks to the large pressures at depth. The two molten layers would have diverging compositions, with the upper one being progressively depleted in volatiles and dehydrated through outgassing, while the primordial lower one would become more and more volatile-rich with crystallization due to volatiles' melt affinity. Depending on the volume of this lower molten layer, dictated by the depth at which crystallization would start, a substantial amount of volatile species might remain isolated from the surface and thus unable to be outgassed.

In a similar fashion, melt–crystals relative density and resulting segregation (e.g., Boukaré and Ricard 2017; Caracas et al. 2019) may prevent possibly volatile-rich melt to reach the surface. If the melt becomes more dense than coexisting minerals at pressures near the base of the mantle (Mosenfelder et al. 2007; Asimow 2018), denser melts and their dissolved volatiles would be entrained downwards. Solidification would then proceed downward from a septum, and the dissolved volatiles would be trapped in the deep interior rather than rising-up throughout the melt and being released in the atmosphere (Elkins-Tanton 2011). Finally, even within a given reservoir of magma, the mode of crystallization, and more specifically melt–crystals segregation is believed to affect the amount of exsolved volatiles. In particular, the fraction of exsolved volatiles from a given magma is enhanced if the evolved liquids and exsolved volatiles are efficiently segregated from their crystalline products (e.g., Edmonds and Woods 2018).

If the magma ocean crystallization results in a gravitationally stable mantle stratification (e.g., Solomatov 2015), volatile species that have not been outgassed may remain in the solidified mantle and mantle chemical heterogeneities may persist until the present-day. On the other hand, if the crystallization leads to an unstable cumulate density stratification, the solidified mantle may overturn to a stable configuration (e.g., Elkins-Tanton et al. 2003). Mantle overturn may induce late episodes of outgassing through the melting of remnant volatile-rich mantle reservoirs, thus possibly releasing to the atmosphere volatile species that have not been outgassed during magma ocean solidification.

The planetary rotation rate may also affect the chemical structure of the crystallizing magma ocean and favor the development of chemical heterogeneities (Maas and Hansen 2015, 2019). In particular, for moderate and high rotation rates, the resulting crystallization pattern may induce an inhomogeneous magma ocean solidification with changes in depth and latitude. Major chemical segregation in distinct reservoirs might isolate a substantial magma ocean fraction from efficiently and rapidly reaching shallow depths where volatile exsolution occurs. Similarly to a basal magma ocean, these molten layers would likely retain their dissolved volatiles and outgassing may not be as efficient as commonly modeled. Yet, if the rotation rate of Venus during magma ocean crystallization was similar to its present-day value, this process would most likely not induce significant latitudinal crystal dynamics variations and would thus be negligible.

Another mechanism, that may significantly alter magma ocean outgassing efficiency and contradicts the classical efficient degassing assumption, is the efficient retention of interstitial, volatile-rich melt at the freezing front of a rapidly crystallizing magma ocean (Hier-Majumder and Hirschmann 2017; Miyazaki and Korenaga 2021, 2022). It stems from the fact that, if crystal–melt separation, matrix compaction, and melt upwards percolation are not efficient, a substantial fraction of melt and dissolved volatiles could be trapped within the crystallizing magma ocean.

The crystallizing melt is a solid-dominated, yet porous matrix with interstitial melt. McKenzie (2011) emphasized that depending on the competition between compaction within the freezing front and the rate of freezing, a substantial volume of the melt could be trapped during melt crystallization in a magma chamber, given no horizontal variations. Compaction within the freezing front expels out trapped melt, while rapid crystallization of the front freezes the melt in place and thus inhibits expulsion. Using a two-phase flow (of melt and matrix) model, Hier-Majumder and Hirschmann (2017) showed that for magma ocean crystallization characteristic timescales, compaction within the moving freezing front is inefficient. Hence melt keeps being trapped at the rheological front while solidification proceeds. Then, if melt percolates upward through the solidified mantle insufficiently fast, Rayleigh-Taylor instabilities would continuously transport the newly formed solid matrix downwards (Miyazaki and Korenaga 2021, 2022), thus entraining the trapped melt and its dissolved volatiles at depth. Volatiles would thus also be trapped within the solidifying mantle and then sequestered in the deep mantle, rather than enriching the remaining melt. Volatile trapping would be enhanced by the fact that their solubility is greater at depth, where magma ocean crystallization is often thought to start, thus trapping melts with high volatile storage capacity and potential high water content. This way, up to 77% H_2O and 12% CO_2 of the total accretion-delivered volatile content could have been trapped in the mantle during magma ocean crystallization (Hier-Majumder and Hirschmann 2017). Yet, the low volatile storage capacity of solid mantle mineral phases (e.g., Bolfan-Casanova et al. 2003; Shcheka et al. 2006) do not support such large amounts of remnants water and carbon. To reconcile their results with such a low volatile storage capacity, Hier-Majumder and Hirschmann (2017) suggest different storage sites for the magma ocean crystallization leftover volatiles. Mantle carbon would be stored in solid carbonates and carbonate-rich melts in the upper mantle, and in diamond and metal carbides in the mid and lower mantle. Volatile-rich accessory minerals and water-enriched melt remaining along grain boundaries or as hydrous fluids or melts would act as storage sites of mantle water. They argue that such a buffered magma ocean outgassing reconciles the fact that the Earth's deep interior contains significant reservoirs of volatiles despite their incompatible nature. This process would be more effective in the case of equilibrium crystallization, where crystals and liquid remain in equilibrium (i.e., in contact) throughout crystallization. As opposed to fractional crystallization where melt–crystal separation is efficient. Assuming that 1% of the melt mass is interstitially trapped, lower estimates of the initial amount of water retained in the solid mantle range from 3% (Hamano et al. 2013) to 13% (Elkins-Tanton 2008; Hier-Majumder and Hirschmann 2017).

Note that in several cases, the amount of remnant water in the solidified mantle would mainly depend on the crystal–melt partition coefficients and on its solubility in the melt. This would include cases of high compaction and resulting efficient expulsion of melt from the porous matrix, or cases of slow magma ocean solidification relative to the compaction rate. It would also apply to cases of efficient crystal–melt separation at the rheological front. The latter owing to effective fractional crystallization or even tearing off and re-entrainment of the newly formed solid matrix within the molten mantle by vigorous convection currents and subsequent settle down of individual crystals.

Miyazaki and Korenaga (2021, 2022) have studied how such hampered magma ocean degassing and subsequent solidified mantle evolution and degassing patterns might affect the formation of early water oceans on rocky planets. Hamano et al. (2013) suggested that the rapidly-outgassed, steam atmosphere of Venus had prevented its climate from being temperate owing to the strong thermal blanketing effect. Miyazaki and Korenaga (2021) have suggested that, even for inefficient water degassing during the magma ocean phase, Venus may have continued to lack oceans throughout its history whereas Earth was able to form water oceans since the solidification of the mantle surface (Miyazaki and Korenaga 2022). In particular, if no ocean were formed upon solidification of the surface on early Venus, comparable rates of subsequent solid-state mantle degassing and water loss (through hydrogen escape) would not supply enough water to allow for later ocean formation.

Overall, any process that significantly reduces the amount of melt reaching the low pressures at which volatile exsolution (i.e., melt–gas partitioning) occurs during magma ocean

Table 1 Mechanisms possibly stifling magma ocean outgassing by isolating volatile-rich melt from the surface and contrasting with the classical efficient outgassing scenario

Mechanism	Reference(s)
Magma ocean crystallization starting at mid-mantle depth	Labrosse et al. (2007), Fiquet et al. (2010)
Planetary rotation and asymmetric crystal settling	Maas and Hansen (2015, 2019)
Melt–crystal density-driven segregation	Boukaré and Ricard (2017), Caracas et al. (2019)
Interstitial melt retention at magma ocean freezing front	Hier-Majumder and Hirschmann (2017), Miyazaki and Korenaga (2021)
Magma ocean convective dynamics	Salvador and Samuel (2023)

crystallization would alter outgassing efficiency and contrast with the classical view (Table 1). Then, if the melt exsolves volatiles, gas bubbles further need to reach the surface. If not, this would also stifle magma ocean outgassing. The global contribution of these combined mechanisms is still poorly constrained but the amount of volatile species remaining in the solidified mantle during the magma ocean solidification may have strong implications regarding the early and long-term evolution of the planet and should be accounted for in the future generation of models.

3.2.5 Influence of the Mantle Redox State

Generalities Molten mantle chemistry plays an important role in magma ocean outgassing. This is less by affecting its efficiency and the *amount* of outgassed species, such as the physical mechanisms mentioned above, than by controlling the *type* of outgassed species (see Table 2 for a summary of the different early outgassed atmospheres and underlying assumptions/constraints). In particular, the oxidation (redox) state of the mantle is critical. It has been interpreted in different ways but can be fundamentally defined by its availability of electrons and the resulting oxidation/valence state of its constituent elements (e.g., McCammon 2005a,b).

Oxidation–reduction reactions are chemical reactions generally[4] involving a transfer of electrons between two species. One is an electron donor (a chemical reductant or reducing agent, which readily gives up electrons) that becomes oxidized and the other an electron acceptor (a chemical oxidant or oxidizing agent, which possesses a strong affinity for electrons) that becomes reduced (e.g., Chap. 3 in White 2013; Kasting 2014). Oxidation thus refers to the loss of one or more electrons and the increase in the oxidation state/number. Reduction[5] is the gain of one or more electrons.

To maintain conservation of electrons (i.e., electrical neutrality), the oxidation of one compound necessarily implies the reduction of another, as electrons are transferred from one species to another. While originally used to refer to the reaction of a compound with oxygen (oxygen being the most common electron acceptor), hence oxidation, the term oxidation has

---

[4]While the transfer of electrons will always cause a change in oxidation state, there are many reactions that are classified as redox reactions even though no electron transfer occurs, such as those involving covalent bonds.

[5]The term reduction originates from the loss in weight upon heating a metallic ore, such as a metal oxide, to extract the metal: the ore was then "reduced" to the metal. Its meaning has then been generalized to any processes involving a gain of electrons.

been generalized to any electron-transfer reactions, without necessarily involving oxygen, after realizing that electron loss was the core of a substance being oxidized.

Yet, as it is the most common electron acceptor, the abundance of oxygen usually controls the oxidation state of a system, which is the case for the mantle, but this need not be the case. The oxidation state of the mantle is controlled by its chemical potential of oxygen, commonly referred to as the "oxygen fugacity", f_{O_2}. It is the equivalent of the partial pressure of oxygen in a particular environment corrected for the non-ideal character of the gas. It describes the formal abundance of O_2 gas in units of atmospheres (although often not representing an actual gas phase). Given its strong oxidizing power, a high oxygen fugacity/abundance corresponds to an oxidized environment while a low oxygen fugacity is associated with abundant reduced species in a reduced environment.

While most major elements constituting the mantle have only one stable oxidation state (e.g., Si, Al, Mg, O, Ca, Na, Ti), others, like Fe and several minor elements (e.g., C, S, Ni, Co, V) exhibit several stable oxidation states. Their geochemical behavior is thus affected by oxidation–reduction processes and reciprocally, their distribution, behavior, and abundance affect the oxidation state of their mantellic environment (e.g., Wood et al. 1990; Mallmann and O'Neill 2009). Because of their abundance, iron- and carbon-bearing solid species play an important role in controlling the oxygen fugacity.

Oxidation States of Iron and Redox Buffers From its most reduced to most oxidized state, iron most commonly occurs as metallic iron (oxidation state 0, noted Fe^0, Fe(O), or simply Fe), ferrous (divalent) iron (oxidation state of $+2$, meaning that it lost 2 electrons in the oxidation process; noted Fe^{2+} or Fe(II)), and ferric (trivalent) iron (oxidation state of $+3$ after the loss of 3 electrons; noted Fe^{3+} or Fe(III)). Iron is, by mass, the most abundant element on Earth (followed by oxygen). It is the most abundant element with a variable oxidation state. Iron's relative abundance in different oxidation states is also indicative of that of the mantle.

Furthermore, some specific iron-bearing (but not only) mineral assemblages, are in chemical equilibrium at a certain oxygen fugacity. For these two reasons, such mineral assemblages, called *redox buffers* (e.g., Reactions (R1), (R2), and (R3)), are used as references to constrain the oxygen fugacity of the mantle, hence its oxidation state (e.g., Frost 1991; Hirschmann 2021). For instance, in a given set of physicochemical conditions, at a given temperature, the oxidation of fayalite (F, which contains two ferrous ions Fe^{2+}) to magnetite (M, which contains one Fe^{2+} and two Fe^{3+}) and quartz (Q) imposes or "buffers" a specific value of f_{O_2} at equilibrium when all three mineral species are present. The reaction is (e.g., Frost 1991):

$$3\ Fe_2SiO_4 + O_2 \rightleftarrows 2\ Fe_3O_4 + 3\ SiO_2. \tag{R4}$$
$$\underset{\text{fayalite}}{} \qquad \underset{\text{magnetite}}{} \quad \underset{\text{quartz}}{}$$

This particular quartz-fayalite-magnetite (QFM) redox buffer is representative of relatively oxidized conditions that are typical of modern Earth's volcanic degassing and presumably of Earth's upper mantle. It is important to understand that the oxygen fugacity is fixed at a given value in a given oxygen fugacity–pressure–temperature space (e.g., Wood et al. 1990). This is simply by the equilibrium coexistence of fayalite, magnetite, and quartz. The oxygen fugacity does not depend on the proportion of these minerals. This is why it is appropriately called a buffer. This can be illustrated by considering a magma containing some amount of fayalite, magnetite, quartz, and oxygen. If the oxygen fugacity is increased by adding oxygen to the system, equilibrium in Reaction (R4) is driven to the right until the log of the oxygen fugacity returns to the given value (which is temperature and pressure dependent).

Only when all fayalite is converted to magnetite (and quartz) can the oxygen fugacity rise (e.g., Chap. 3 in White 2013). A decrease of oxygen fugacity would be buffered in exactly the opposite way until all magnetite and quartz were gone.

Another common buffer of oxygen fugacity is the iron-wüstite (IW) buffer (Reaction (R2)), which describes the coexistence of metallic (Fe) and ferrous (Fe^{2+}, which combines with O^{2-} to form FeO) iron and indicates a highly reducing environment. Here, an increase in f_{O_2} favors the reaction from left to right, which stabilises iron in its higher oxidation state (wüstite), whereas low f_{O_2} stabilises iron in its metallic, lower oxidation state.

Geochemical Implications, Speciation, and Outgassed Composition Like minerals, gas species can exist in various oxidation/valence states depending on the oxygen fugacity. The redox state of gases in equilibrium with rocks is often described by the mineral buffers that govern the capacity of the rock to consume or release oxygen (e.g., Zahnle et al. 2020). As mentioned above, oxidation–reduction processes affect the geochemical behavior of elements of several stable oxidation states, and consequently, that of the other elements interacting and having some chemical affinities with them (e.g., siderophile elements for iron). Overall, the redox state of the mantle (and its oxygen fugacity) controls the molecular speciation of C–H–O–N–S-bearing fluids and melts (French 1966; Frost 1979; Holloway 1981; Wood et al. 1990; Holloway and Blank 1994; Hirschmann et al. 2012; Hirschmann 2012; Grewal et al. 2020). It influences the solid–melt–gas partitioning of volatile species (Holloway and Blank 1994; Libourel et al. 2003; Miyazaki et al. 2004; Hirschmann 2012; Armstrong et al. 2015; Boulliung et al. 2020; Gaillard et al. 2021) and trace elements in the silicate magma – and thus, the composition of the atmosphere –, their partitioning among silicate melt and molten or solid metallic iron (Drake et al. 1989; Hillgren 1991; Holzheid et al. 1994; Righter et al. 1997; Gaillard et al. 2022b), as well as the mineral saturation and bulk mineralogy (e.g., Feig et al. 2010, and references therein). Indeed, given an elemental inventory of C–H–O–N–S, the abundance of oxygen, f_{O_2}, critically affects the identity of molecular species outgassed and controls the relative abundances of the oxidized and reduced form of major gases (e.g., H_2/H_2O, CO/CO_2).

The question of atmospheric oxidation state (hence, of its composition) is thus inextricably interlinked to the mantle redox evolution (e.g., Kasting et al. 1993a; Lécuyer and Ricard 1999). Essentially, an oxidized magma, i.e., of high oxygen fugacity (for instance at the QFM buffer, consistently with present-day upper-mantle conditions), induces outgassing of oxidized volatile species, such as H_2O, CO_2, SO_2, and N_2. Conversely, a magma of reduced composition (of low oxygen fugacity) releases reduced volatile species, such as CO, H_2, CH_4, NH_3, and H_2S (e.g., Holland 1962; Frost 1979; Wood et al. 1990; Kasting 1993; Hirschmann 2012). This speciation process can be explained by the fact that an environment of high oxygen fugacity/abundance makes it readily available to bond with C–H–O–N–S fluids and produce oxygen-bearing compounds. Note however that redox changes during magma ascent (Sato 1978; Burgisser and Scaillet 2007), mixing (Ueki et al. 2020), and degassing (Mathez 1984; Humphreys et al. 2015) may occur, making the link between the redox state of magmas and their outgassed products less straightforward than commonly assumed.

Present-Day Redox State The Earth's present-day atmosphere is highly oxidized as its second most abundant constituent is O_2, which is a powerful electron acceptor. However, early magma ocean and subsequent long-term volcanic degassing only cannot explain this current characteristic feature, which certainly results from the complex interplay between geochemical and biotic processes (e.g., Holland 2006; Kasting 2013; Catling 2014; Catling and Zahnle

2020). Yet, at present-day, H_2O and CO_2 are the two most abundant gases released by volcanic degassing, which indicates that the Earth's present-day upper mantle is in an oxidized state (e.g., Frost and McCammon 2008). With CO_2 its most abundant constituent, Venus' atmosphere is currently relatively oxidized, despite being considered possibly weakly reducing in its lower part (e.g., Catling 2015). Surface–atmosphere interactions likely result in an oxidizing surface with a high oxygen fugacity at about the Magnetite-Hematite (MH) redox buffer (e.g., Fegley et al. 1997; Zolotov 2018). The mean FeO/MnO ratio measured at Venus' surface by Venera 13, 14, and Vega 2 is slightly less than that of the bulk silicate Earth (~52 for Venus' surface, Surkov et al. 1984, 1986; Volkov et al. 1986; and ~60 for the BSE). Used as a proxy of the planet's oxidation state (Wänke et al. 1973), it also suggests a rather oxidized surface and interior. If the present-day Venusian SO_2 atmospheric inventory originates from volcanic outgassing, it could indicate that at least some parts of the Venusian mantle are highly oxidized, at about the QFM buffer.

Early Oxidation State While present-day volcanic outgassing can be related to the current magma sources redox state, the earliest oxidation state of planetary mantles is much more speculative and little is known about the magma ocean initial and evolving redox conditions. The oldest clues available on Earth come from measurements on Hadean zircons, recording oxygen fugacity of Hadean magmatic melts identical to those in the present-day mantle, thus indicative of an oxidized upper mantle (at about QFM buffer) as early as 4.35 Gyr ago (Trail et al. 2011). Earth's oldest igneous rocks also suggest that the mantle redox state was only weakly reduced and has not varied significantly over at least the past 3.9 Gyr of Earth's history (Canil 1997; Delano 2001; Canil 2002; Lee et al. 2003; Li and Lee 2004; Rollinson et al. 2017; Nicklas et al. 2018). As today, the redox state was then likely set by the QFM buffer, so that the resulting volcanically-outgassed atmosphere would have been mostly made of H_2O, CO_2, and N_2, with small amounts of CO and H_2 (Holland 1962; Abelson 1966). There is no such records available for Venus and whether these conditions are systematic outcomes of planetary early evolution and can be generalized to early Venus or are Earth-specific is unclear.

Constraints from Prebiotic Chemistry Prebiotic chemistry can provide independent but indirect constraints on the environment at the surface of early Earth. Yet, the prevalence of an early oxidized environment suggested by geochemical evidence is hard to reconcile with the reduced conditions requirements inferred from chemical studies on the origin of life. Indeed, prebiotic chemistry experiments, starting from the famous Miller–Urey experiments in the early 1950s (Miller 1953, 1955; Miller and Urey 1959), have shown that amino acids and a wide number of other building blocks of living organisms can be synthesized from electrical discharges in highly reduced conditions with the presence of methane and ammonia and where free oxygen is absent (see Oro et al. 1990; McCollom 2013; Benner et al. 2020, for reviews). The fact that such conditions result in efficient abiotic synthesis of biologically important organic compounds suggested that the origin of life required a very reducing environment. Hence hinting that this was the case for the earliest terrestrial atmosphere overlying a solidified surface.

It should be noted that bridging the gap between the non-agnostic conditions under which prebiotic chemistry experiments are successful in producing life-essential compounds and actual early Earth's conditions where life emerged remains challenging and is still an extensive area of study (e.g., McCollom 2013). For instance, Cleaves et al. (2008) have suggested that yields of amino acids observed in more neutral (N_2/CO_2-dominated) or oxidized environment experiments may have been biased low/underestimated owing to oxidation of the acids during sample processing. Then, using oxidation inhibitors during sample processing or adding $CaCO_3$ to buffer the pH of the aqueous phase would significantly increase the yields.

Energy sources other than electrical discharges/lightnings have also been invoked to synthesized complex biological molecules without necessarily requiring highly reducing environments (e.g., Chyba and Sagan 1992; Kurosawa et al. 2013). In particular, coronal mass ejection events from the active young Sun could trigger nitrogen fixation and ignite reactive chemistry yielding complex molecules in less reducing environments (Airapetian et al. 2016, 2020).

Based on the abundance of organic molecules in the Solar System, alternative scenarios have invoked exogenic sources of organics for the origin of life rather than an in-situ production (e.g., Anders 1989; Chyba et al. 1990; Chyba and Sagan 1992; Whittet 1997; Matthews and Minard 2006). This abundance of organic molecules may imply that they are not that hard to make (see Zahnle et al. 2010 and references therein for a thorough discussion). Yet, the extraterrestrial delivery of complex and organic materials by comets, asteroids, and interplanetary dust particles and their implications for the emergence of life is challenged by their survival upon impact (e.g., Whittet 1997; Pierazzo and Chyba 1999; Pasek and Lauretta 2008).

The apparent contradiction between the reducing conditions required by prebiotic chemistry and the likelihood of an oxidized early atmosphere (and mantle) might also be resolved by alternative scenarios involving an oxidizing surface environment. Such an environment may result from magma ocean degassing. But the oxidizing environment might experience a perturbation making it temporarily and/or locally reducing (Airapetian et al. 2016; Itcovitz et al. 2022). In fact, such transient conditions may actually be the most favorable for prebiotic chemistry (Benner et al. 2020).

Impact-Induced Degassing and Transient Reducing Environments To produce such a perturbation and reconcile geological evidence of an oxidizing past with the requirements for life's emergence, impact-induced degassing has been proposed. Gases are then directly released into the atmosphere on impact (Benlow and Meadows 1977; Jakosky and Ahrens 1979; Lange and Ahrens 1982; Matsui and Abe 1984, 1986b). It is an efficient mechanism to release strongly reduced species into the early atmosphere, and thus to provide favorable conditions for the origin of life despite a rather oxidized, volcanically-produced, environment (e.g., Zahnle et al. 2010, 2020). It is thought to proceed as follows: due to the energetics of impact, all of the volatiles carried by the impactors are expected to be vaporized, to equilibrate chemically with the other materials of the impactor, and to directly enter the atmosphere. Atmospheres resulting from impact degassing would then tend to have a composition reflective of the impacting bodies (rather than the proto-planet mantle). These are mostly volatile-rich and much more reduced than the mantle (e.g., Zahnle et al. 2010). Note that additional and more complex reactions between the impactor components and the impacted body's atmosphere or crust can increase the production rates of reduced species (e.g., Kasting 1990).

Models of impact-generated atmospheres generally suggest that the outgassing of reducing species dominates for a variety of meteoritic materials (e.g., carbonaceous, ordinary, and enstatite chondrites; Kasting 1990; Schaefer and Fegley 2007; Hashimoto et al. 2007; Sugita and Schultz 2009; Schaefer and Fegley 2010). If the gases of Earth's earliest atmosphere had equilibrated with material like that of primitive meteorites, the resulting atmosphere would be much more reduced than present-day volcanic gases (e.g., Zahnle et al. 2010). This is due to the reducing power of the substantial amounts of metallic iron and iron sulfides contained in many meteorites, including ordinary and enstatite chondrites. Even for CI carbonaceous chondrites the resulting gases are rather strongly reducing owing to the reducing power of the abundant carbon contained in CI chondrites and present in reduced form (Hashimoto et al. 2007; Zahnle et al. 2010). These are the most highly oxidized meteorites (e.g., Urey and Craig 1953, Rubin et al. 1988, Krot et al. 2014; and Fig. 5.2 in Trieloff et al. 2006) and thus expected to generate the most highly oxidized gases among the different meteoritic sources. Note however that Schaefer and Fegley (2010) found that carbonaceous chondritic material such as CI, CM, and CV chondritic materials would rather produce oxidized atmospheres.

Yet, mixtures of solid meteoritic material (from highly oxidized carbonaceous chondritic material to highly reduced enstatite chondritic or eucritic materials) may also produce more oxidizing compositions (Schaefer and Fegley 2017). They are relevant to the mixing of materials from different locations in the solar nebula and thus representative of the primitive solar nebular materials. Indeed, because of variations in the abundance of buffering phases (such as metallic iron) and volatiles within the solid materials, gases produced out of their mixtures are not a straightforward linear mixture of the gases produced by individual materials. As a result, adding only a few percent of carbonaceous material to a differentiated body may be sufficient to raise the f_{O_2} of the mixture by several log units. Given the stochastic nature of accretion (Sossi et al. 2022) and since the oxygen fugacities differ between individual materials and their mixtures, the behavior in mixtures remains uncertain (Schaefer and Fegley 2017). Schaefer and Fegley (2017) also emphasized that other processes affecting the redox state of the atmosphere, such as hydrogen escape, photochemistry, or surface-atmosphere interactions, would certainly induce additional complexities and should be further investigated. For instance, they have found that oxidized materials become more oxidized by hydrogen loss whereas reduced materials do not. While NH_3- and methane-rich atmospheres are prone to efficient photochemical dissociation by solar ultraviolet radiation (e.g., Leighton and Steiner 1936; Lasaga et al. 1971; Kuhn and Atreya 1979; Kasting et al. 1983) and thus unlikely stable as steady states, impact-induced outgassing of infalling meteoritic material could repeatedly create transients and life emergence-favorable conditions many times.

Initial Redox State of the Magma Ocean and Its Evolution Based on the aforementioned direct and indirect constraints and processes, a consensual view may emerge. It suggests that the mantle oxidation state was relatively oxidized and close to its current state early in history, once the surface was solidified (i.e., post-magma ocean). The overlying atmosphere was also likely oxidized and composed dominantly of H_2O and CO_2. In such an oxidizing environment, impact-induced degassing transiently produced reduced conditions favorable for the development of life on Earth. Before the surface was solid, prior to the available geological records and to the setting of life-sustainable conditions on a solidified planet, the oxidation state of the earlier evolution stages, such as the magma ocean phase itself and its evolution upon cooling, is highly unconstrained. In particular, whether oxidized conditions are a direct/intrinsic consequence of the magma ocean evolution or inherited from other mechanisms related to planetary formation remains unclear. To address this issue, we discuss below the processes thought to affect the redox state of the earliest stages of planetary evolution, with particular emphasis on the magma ocean stage, and how they may be reconciled with the available constraints of the subsequent stages.

Hydrogen-Rich and Reducing Early Atmospheres A primordial H_2–He atmosphere captured from the solar nebula (see Sect. 2.1) would induce a reduced start. Before being lost after the dissipation of the solar nebula (e.g., Zahnle et al. 2007), its reducing power could have affected both the surface and the mantle of the proto-planet. Similarly the H_2–He solar nebula could have reduced the planetesimals that subsequently encountered the planet, thus being a source of reduced materials (Zahnle et al. 2010). Tian et al. (2005) have suggested that in a hydrogen-rich early atmosphere, the escape rate of hydrogen would be energy-limited due to lower exospheric temperatures rather than diffusion-limited and thus operating much slower. In such a case, a primordial hydrogen-rich atmosphere and its reducing power would have been sustained longer and maintained after dissipation of the solar nebula. Without necessarily involving the solar nebula, on longer timescales, the balance between slow hydrogen escape and hydrogen outgassing (requiring a reducing upper mantle) could have preserved/induced an atmospheric hydrogen mixing ratio of up to 30% (Tian et al. 2005; Liggins et al. 2020) and sustained a reducing environment. Note however that Tian et al. (2005) hydrogen escape rate estimates may be biased low and are controversial (Catling 2006; Tian et al. 2006; Kuramoto et al. 2013; Tian 2015).

Link with Core Formation and Evidence for a Reduced Start The depth, timing, and timescale of metal–silicate equilibration, separation, and resulting core formation, play a critical role in determining the oxidation state of the magma ocean and crystallizing mantle. These processes are likely occurring throughout the accretion sequence and possibly extend through the different magma ocean stages. Because of the reducing power of metallic iron, metal–silicate equilibration necessarily implies a reducing environment (of low oxygen fugacity, near the IW buffer) imposed by the coexistence of metallic iron in equilibrium with silicate (Reaction (R2)), thought to significantly reduce the silicate melt (O'Neill 1991; Wade and Wood 2005; Frost and McCammon 2008; Rubie et al. 2011). One may then expect that the magma ocean was initially in such a reduced state, prior to the complete metal segregation to the core. Importantly, the redox state of the upper part of the mantle, where the gases originate, would mainly matter for the oxidation state of contemporary outgassed species.

The segregation of metallic iron through the mantle, as well as the pressure–temperature dependence of metal–silicate partition coefficients (e.g., Wade and Wood 2005) could induce a vertical zonation/gradient of the redox state. However, vigorous convection and mixing of the molten mantle could also buffer metallic iron sinking through Rayleigh-Taylor instabilities and could repeatedly bring reducing iron to shallow depths. In addition, effective convective re-entrainment could slow down iron accumulation at the base of the molten layer (alike crystals re-entrainment; e.g., Solomatov 2015). It could also buffer the formation of metal diapirs large enough to sink to the core, therefore delaying metal–silicate separation. Thus, the upper mantle may have remained in such a reduced state prior to at least advanced planetary differentiation and core formation (e.g., Catling and Claire 2005). Here the metallic iron would have sunk deep enough to no longer affect the upper mantle redox state, and every time the metallic iron of accreting planetesimals would merge into the mantle of the growing proto-planet. Yet, the latter effect strongly relies on the details of metallic iron distribution upon impact, itself influenced by impact parameters and mantle mixing (Deguen et al. 2014; Landeau et al. 2016, 2021; Maas et al. 2021; Itcovitz et al. 2022).

Keppler and Golabek (2019) suggested that, in the case of a highly reduced mantle resulting from metal–silicate equilibration, most of the carbon could be reduced to graphite, which is less dense than a peridotite melt. In such a case, regardless of metal–silicate equilibration depth, as long as such highly reduced conditions prevail somewhere within the magma ocean, graphite flotation and accumulation on top of the magma ocean may have imposed low f_{O_2} conditions in its shallow parts given the CCO buffer ($C + O_2 = CO_2$).

These mechanisms favor a reduced start, but predicting the early oxidation state of the shallow part of the magma ocean and the resulting atmospheric composition remains challenging. In part because linking core formation to the mantle oxidation state is not straightforward. Yet also because core formation itself is complex, and likely results from the interplay of different processes with lots of remaining uncertainties (e.g., Rubie et al. 2015b).

Changes of partition coefficients, phase separation and phase changes (and resulting density contrasts; e.g., Boukaré and Ricard 2017) under extreme and varying $P/T/f_{O_2}$ conditions relevant to magma ocean cooling could induce unknown yet possibly significant feedback that need more investigations. Furthermore, additional complexities regarding the fate of iron in the post-impact mantle and the details of metal–silicate equilibration in a turbulent molten mantle, as well as their influence on the mantle redox state have been poorly studied. Deguen et al. (2014) have shown that the turbulent mixing of a vigorously convecting magma ocean enhances equilibration. By considering the effect of rotation in the convective magma ocean in spherical geometry, Maas et al. (2021) emphasized that metal dispersion and settling also depend on the impactor's target latitude. Citron and Stewart (2022) and Itcovitz et al. (2022) recently showed that the impact parameters strongly influence the distribution, mixing, and availability of the reduced phases of the impactor (e.g., metallic iron) to potentially reduce the atmosphere. They also investigated the interactions between the post-impact atmosphere and the impact-generated melt phase. Importantly, the manner in which iron is accreted by and mixed with the target, and the timing and depth at which metal–silicate equilibration occurs requires further investigation to constrain the oxidation state of the shallow magma ocean layers upon impacts. It would also determine the oxidation state of the outgassed species, and hence the composition of the resulting atmosphere. Because most of the projectile iron (more than 70%) is deposited in the crust and upper mantle, where it is not available to reduce surface water, reducing environments could be less likely to arise after large impacts than previously suggested.

In any case, the fact that the molten mantle must have been in equilibrium with the metallic iron now forming the core, and hence, characterized by low oxygen fugacity, is hard to reconcile with the prevailing early oxidized conditions of the solidified surface, by the time minerals and rocks were formed as described before. From a reduced start, some mantle oxidation mechanisms must then have been in place to resolve this paradox.

Oxidation Mechanisms from a Reduced Start To explain the formation of the core in agreement with the abundance of siderophile elements of the mantle and their experimentally-determined partition coefficients, several models have been suggested. These have implications for the oxidation state of the mantle (e.g., Li and Agee 1996; Wood et al. 2006; Fischer et al. 2015). From a reduced start, several oxidation mechanisms have been proposed to increase the oxygen fugacity of the mantle to the QFM buffer (e.g., McCammon 2005b; Wood et al. 2006; Frost et al. 2008; Frost and McCammon 2008; Wood et al. 2009; Schaefer and Elkins-Tanton 2018; Armstrong et al. 2019).

Photo-dissociation of H_2O followed by the escape of H_2 from the atmosphere (Hunten 1993) may have caused oxidation of both the atmosphere and mantle by the leftover oxygen (Kasting et al. 1993a; Sharp et al. 2013). This mechanism likely occurred to some extent. The composition of martian meteorites (Dreibus and Wänke 1987) and the elevated abundance of light elements in its core (Khan et al. 2022) suggest that Mars was possibly more volatile-rich than Earth. If hydrogen escape was the main oxidizing mechanism, it is unclear why Mars produced basalts more reduced than MORB and indicative of a relatively more reduced (Herd et al. 2002; Wadhwa 2008) and FeO-rich (Wänke et al. 1988) mantle (at about the IW buffer), despite efficient atmospheric water loss through hydrodynamic escape

(Pepin 1991; Jakosky and Phillips 2001). Furthermore, while Venus likely experienced massive loss of hydrogen, its silicate portion is not significantly more oxidized than the Earth (Schaefer and Fegley 2017).

An alternative solution involves the increase of oxidation state of accreting bodies during late accretion (Wänke et al. 1988; O'Neill 1991; Wood et al. 2009; Rubie et al. 2011, 2015; Fischer et al. 2015; Shi et al. 2022), also referred to as heterogeneous accretion. This allows trace element partitioning models to match the present-day mantle FeO abundance, but fails to predict the current redox state-governing $Fe^{3+}/\Sigma Fe$ (which is the proportion of the Fe that is ferric) and the final oxygen fugacities of these models is generally still too low (typically at IW-2, which is about 5 $\log f_{O_2}$ units below the Earth's present-day value; e.g., Schaefer and Fegley 2017; Schaefer and Elkins-Tanton 2018).

Another possibility is that the mantle self-oxidized as a result of perovskite crystallization (Mao and Bell 1977; Frost et al. 2004; Wade and Wood 2005; Galimov 2005; Frost et al. 2008; Frost and McCammon 2008; Hirschmann 2012; Armstrong et al. 2019; Hirschmann 2022). At depth, the growth of silicate perovskite (bridgmanite), the dominant lower-mantle mineral, forced the disproportionation[6] of ferrous iron (Fe^{2+}) into ferric iron (Fe^{3+}) (dissolved in perovskite; McCammon 1997) plus iron metal (segregated to the core; Frost et al. 2004): $3FeO \longrightarrow Fe^{0} + Fe_2O_3$ (or $4FeO \longrightarrow Fe^{0} + Fe_3O_4$). Repeated crystallisation and dissolution of Fe^{3+}-containing perovskite at the base of the magma ocean acted as an "oxygen pump" (Wade and Wood 2005). This raised the oxygen fugacity of the growing mantle by releasing ferric (Fe^{3+}) iron to the magma ocean, while the produced reduced metallic iron segregated to the core (e.g., Armstrong et al. 2019). Infalling metallic iron from subsequent impactors would have been oxidized by the ferric iron, therefore raising the FeO content of the silicate magma and the oxygen fugacity of core separation. Mantle convection and homogenization would further raise the oxygen fugacity of the upper mantle to its present-day value toward the end of core formation and by the time of Hadean zircon formation (e.g., McCammon 2005a; Frost and McCammon 2008). This self-oxidation process relies on the ability of Earth's mantle to stabilize substantial amounts of perovskite at depth. It is essentially planetary size-dependent and could have thus operated similarly on Venus (assuming it experienced a deep enough magma ocean) but not on Mars (e.g., Wood et al. 2006). Note that Schaefer and Elkins-Tanton (2018) proposed that the disproportionation of Fe^{2+} would occur directly in the silicate melt phase in the magma ocean during equilibration with the sinking core-forming metal delivered by impactors, rather than during crystallization (see also Hirschmann 2022). This alternative scenario allows immediate separation of the metallic liquid, leaving the oxidized mantle material behind without requiring additional mixing nor complex scenarios of repeated crystallisation and melting of bridgmanite.

Thus, despite a favored early reduced start for the magma ocean, these oxidation mechanisms may have brought the molten mantle to an oxidized state before its complete solidification. If the planet is still molten once core formation is complete and the planet has a differentiated Fe metal and FeS core, Lupu et al. (2014) showed that the equilibrium of a molten surface of either bulk silicate Earth or continental crust composition results in the formation of H_2O–CO_2-dominated (i.e., oxidized) atmospheres. This corresponds to a case where metal–silicate segregation occurs before the last magma ocean episode or rapidly compared to its solidification.

Finally, an increasing amount of evidence has suggested that magma ocean evolution itself might be the determining factor in the evolution of the oxidation state of the magma

---

[6] A disproportionation reaction is a redox reaction in which the initial material of intermediate oxidation state undergoes oxidation as well as reduction and converts simultaneously to two compounds, one more oxidized and one more reduced.

 Springer

ocean–atmosphere system (Armstrong et al. 2019; Solomatova and Caracas 2021). It may also reconcile the two favored end-members: a reduced start, owing to metal–silicate equilibration, and an oxidized end, as demonstrated by geological records. For instance, Maurice et al. (2023) suggested that the crystallization sequence of the magma ocean itself could be an oxidizing process, because of the incompatibility of ferric iron. This would imply that the shift towards an oxidized atmosphere would be a systematic outcome of the magma ocean evolution.

Indirect Evidence of Magma Ocean Redox State Different attempts recently aimed at assessing the oxidation state of the molten mantle prior to the aforementioned available direct or indirect constraints (Pahlevan et al. 2019; Deng et al. 2020; Thompson et al. 2021; Solomatova and Caracas 2021). Thompson et al. (2021) recently conducted laboratory investigations of meteorite outgassing on CM-type carbonaceous chondrites (i.e., undifferentiated), as representative samples of the bulk composition of material in the solar nebula. This was in order to assess the composition of resulting atmospheres. However, one should note that their experimental setup and procedure is more representative of outgassing from the molten interior of a planet (of bulk chondritic composition) rather than impact-induced degassing as modeled in the aforementioned studies. Indeed, by simply heating chondrites and measuring the abundances of released volatiles, they reproduce the conditions expected for the outgassing from a molten planet and resulting secondary atmosphere formation. Here shock-induced devolatilization does not occur. Yet, devolatilization along with atmospheric entry and structural changes upon impact most likely altered the chondrite that was sampled and its overall representativeness. They found that water would makeup most of such an outgassed atmosphere (66%), followed by CO (18%) and CO_2 (16%), thus producing a rather oxidizing atmosphere (Thompson et al. 2021).

Pahlevan et al. (2019) used hydrogen isotopic records (the deuterium content) of Earth's hydrosphere to independently assess the oxygen fugacity of terrestrial magma ocean outgassing. They emphasized that the preservation of a carbonaceous chondritic (i.e., undifferentiated) D/H signature in the terrestrial oceans requires a hydrogen-poor (relative to water, i.e., $H_2/H_2O < 0.3$) Earth's outgassed envelope. This is indicative of oxidizing conditions ($\log f_{O_2} >$ IW + 1) for last atmospheric equilibration with the magma ocean. In agreement with the earliest geological record (Trail et al. 2011), they concluded that oxidation of the silicate Earth occurred during the crystallization of the (last) magma ocean, and may not require later geological processes such as subduction (and associated oxidation through slab dehydration for instance; e.g., Wood et al. 1990; Kasting et al. 1993a) to reach the oxidized present-day state.

Earth and chondrites similar volatile isotopic compositions suggest that the source of terrestrial volatiles is chiefly chondritic (Marty 2012). Yet, the elemental composition of the bulk silicate Earth is depleted in C, N, and H relative to chondritic material. Accretion of some non-chondritic materials has thus been proposed to account for this depletion (Hirschmann 2016). Sakuraba et al. (2021) modeled the evolution of volatile abundances during the accretion by considering elemental partitioning, including the effect of the magma ocean redox state on volatile solubility and metal–silicate partition coefficient, and impact erosion. They found that the BSE depletion pattern can be reproduced from continuous accretion of chondritic bodies only, without the need of non-chondritic materials, but requires a relatively oxidized magma ocean ($\log f_{O_2} \approx$ IW + 1). In good agreement with Pahlevan et al. (2019), their results indicate a relatively oxidizing magma ocean at the end of the accretion stage.

Based on first-principles molecular dynamics calculations, Solomatova and Caracas (2021) showed that during the first stages of magma ocean evolution, C is rapidly devolatilized (in the form of CO_2). Yet H remains dissolved in the magma for longer and is volatilized (as H_2O) during later magma ocean stages. This would have resulted in the outgassing of a CO_2-rich and H_2O-depleted atmosphere during the early stages of the magma ocean. H_2O would have been outgassed later on with solidification, thus supporting the classically modeled outgassing sequence of coupled magma ocean–multi-species atmosphere models (e.g., Elkins-Tanton 2008; Lebrun et al. 2013; Salvador et al. 2017; Nikolaou et al. 2019; Bower et al. 2019). They also suggested that, despite the uncertain oxidation state after a giant impact, the amount of oxygen available in the system influences the speciation of the vaporized carbon. Hence more oxidized systems favor the release of more oxidized vapor species, i.e., CO_2 outgassing rather than CO. The composition of the melt itself also affects the composition of the vapor phase. In addition to increasing with the oxidation state, they demonstrated that the relative proportion of CO_2 to CO increases with decreasing density and temperature, thus suggesting that the abundance of atmospheric CO_2 increased with time and magma ocean cooling.

Deng et al. (2020) combined first-principle molecular dynamics simulations with thermodynamic modeling at relevant P/T conditions. They studied the redox controlling reactions in magma oceans. In agreement with Armstrong et al. (2019) high pressure experiments, they found that compared to ferrous iron (Fe^{2+}), ferric iron (Fe^{3+}) becomes increasingly energetically favorable with pressure because of its small partial molar volume in silicate melts under large compression (Fe^{3+}-bearing melts are more compressible). In a convectively homogenized magma ocean, this would produce a vertical gradient in oxygen fugacity (as proposed by Hirschmann 2012). Here the upper part being relatively more oxidized than the deeper part where redox values were taken to be representative of local equilibrium with reduced molten iron ponds, i.e., assuming that core–mantle equilibration occurred at the bottom of a relatively deep magma ocean (consistent with metal–silicate partitioning experiments; e.g., Li and Agee 1996; Fischer et al. 2015). Yet, before gradually decreasing downwards throughout the rest of the mantle, the relative oxygen fugacity first increases slightly with pressure in the uppermost mantle. The increasing trend was also reported previously in experiments establishing the effect of pressure on the $Fe^{3+}/\Sigma Fe$ ratio of silicate melts (Zhang et al. 2017). Yet, as Zhang et al. (2017) experiments were restricted to relatively shallow depths (up to 7 GPa), they concluded that oxygen fugacities at the surface of shallow magma oceans are more reduced than at depth. They also found that when extrapolated to higher pressures relevant for magma oceans on Mars- or Earth-/Venus-sized planets, atmospheres overlying magma oceans should be highly reduced and rich in H_2 and CO. Importantly, Deng et al. (2020) showed that this trend actually reverses at larger depth and emphasized that an oxidized upper mantle is a natural consequence of a magma ocean due to the increasing stability of ferric (oxidized) iron with depth, even at relatively reduced conditions, which raises the $Fe^{3+}/\Sigma Fe$ ratio of silicate melts in equilibrium with metal alloy. By calculating the speciation of volatiles to determine the composition of an outgassed atmosphere at chemical equilibrium with the underlying magma ocean, they found that the resulting early atmosphere of Earth-/Venus-sized planets (i.e., with deep magma oceans) should be enriched in H_2O (~ 70 mol%) and CO_2 (~ 15 mol%), and therefore, be highly oxidized.

However, Righter and Ghiorso (2012) used a different approach to estimate f_{O_2} and despite agreeing with a similar vertical zonation in magma oceans (from shallow oxidized portions to deep reduced portions), they argued that metal–silicate experiments would bias f_{O_2} estimations such that none of the magma ocean part would be more oxidized than the

IW buffer. Conversely to what is assumed in accretion models described previously, increasing the FeO content of planetary mantles during accretion may not lead to oxidation. It would rather lead to significant reduction relative to the IW buffer, therefore decreasing f_{O_2} from high to low during accretion. This view indicates that further investigations are needed but remains challenged by the apparent prevalence of oxidized conditions indicated in rock records. It also contrasts with the studies mentioned above (Pahlevan et al. 2019; Deng et al. 2020; Thompson et al. 2021; Solomatova and Caracas 2021), all suggesting that the degassing of the upper magma ocean produced oxidized species and was thus already in a relatively oxidized state early on.

Influence of the Magma Ocean Redox State on the Thermo-Chemical Evolution Because of the uncertainties regarding the magma ocean redox state, it is important to assess its influence on the thermo-chemical evolution of the magma ocean–atmosphere system. This has been the focus of recent studies that simulated the solidification and/or outgassing of the magma ocean while considering f_{O_2} as a free parameter, thus considering both reduced and oxidized conditions (Katyal et al. 2020; Bower et al. 2022; Gaillard et al. 2022a; Maurice et al. 2023).

Katyal et al. (2020) investigated the effect of mantle oxidation state on volatile outgassing and chemical speciation at the surface at specific conditions relevant to different steps of magma ocean evolution for various redox states. They showed that the oxidation state of the magma ocean, through the composition of the outgassed atmosphere, affects the atmospheric thermal opacity, surface pressure, and scale height, and thus, ultimately, the cooling rate of the magma ocean. Because of its relatively low mean molecular weight, a H_2/CO-dominated reduced atmosphere outgassed from a reduced magma ocean has a larger scale height than a H_2O/CO_2-dominated oxidized atmosphere (outgassed from a oxidized mantle, which is consistent with experimental determinations; e.g., Grewal et al. 2020) containing heavier species. The higher thermal blanketing effect of the latter dense and small scale height atmosphere made of efficient greenhouse gases is expected to slow down the cooling of the magma ocean. Conversely, a reduced mantle surrounded by reduced outgassed species is expected to cool down faster. Yet, the greenhouse effect of reduced species is far from being negligible. In particular for dense atmospheres, the collision-induced absorption (CIA) of reduced species such as H_2 and CH_4 can extend the outer edge of the habitable zone (Pierrehumbert and Gaidos 2011; Ramirez and Kaltenegger 2017, 2018) and maintain ancient Mars temperate (Ramirez et al. 2014; Wordsworth et al. 2017; Turbet et al. 2019, 2020; Godin et al. 2020). Note that both the atmospheric thermal opacity and scale height affect the thermal emission and transmission spectra that could be used to infer the oxidation state of an extrasolar planetary interior and constrain its evolution with time.

Bower et al. (2022) reached similar conclusions while considering additional C/H ratios and hydrogen budgets. As shown in earlier coupled magma ocean–atmosphere studies for oxidized species (Salvador et al. 2017; Nikolaou et al. 2019; Bower et al. 2019), because of their larger solubility in silicate melts the H-bearing species are outgassed later than C-bearing ones. This also applies to reduced conditions. Because of the combined influence of volatile solubility and redox reactions, Bower et al. (2022) emphasized that an atmosphere may evolve from dominantly CO-rich and reducing to H_2O-rich and oxidizing with cooling. They further suggested that the mode of crystallization (i.e., equilibrium versus fractional) would induce additional complexity. For example, the formation of an early surface lid (at around 30% of remaining melt), as a result of faster magma ocean freezing under equilibrium crystallization, could delay or even prevent subsequent outgassing of highly soluble volatile species (such as water).

Maurice et al. (2023) also reported that the redox state of the crystallizing magma ocean evolves systematically towards oxidizing conditions. This owing to the incompatibility of Fe^{3+} in the minerals resulting in its enrichment in the melt and in the increase of the oxygen fugacity throughout magma ocean crystallization. They showed that the outgassed atmosphere at the final stages of crystallization is consistently oxidized and dominated by H_2O and CO_2. Yet, unlike other studies, they found that the cooling and solidification of initially reduced magma oceans is slower than that of initially oxidized magma oceans. Indeed, despite the stronger greenhouse effect of oxidized atmospheric species, they emphasized that initially reduced atmospheres are not able to radiate as efficiently as oxidized atmospheres owing to their steeper lapse rate, so that the molten phase is protracted.

The Gaillard et al. (2022a) magma ocean static outgassing model also supports an enhanced outgassing of C-bearing species compared to H-bearing ones in all oxidation states. They further included N- and S-bearing species and predicted that an oxidized magma ocean would mainly outgas C–N–S. They therefore form atmospheres made of CO_2–SO_2–N_2–H_2O with relative proportions varying according to the degree of oxidation, while a reduced magma ocean would mainly outgas CO–H_2 with some CH_4 for the most reduced cases and N_2 otherwise.

Besides affecting the chemical composition of the outgassed atmosphere, and thus indirectly its thermal evolution through atmospheric opacity, the redox state may affect intrinsic properties of the magma ocean itself. Using high-temperature experiments, Lin et al. (2021) showed that oxygen fugacity has a direct effect on rock melting properties by affecting the liquidus. In particular, they reported that a higher oxygen abundance lowers the liquidus temperature of iron-free basalt at 1 atm, thus facilitating melting. How these findings apply to larger pressures is not yet clear, but they could have significant implications for magma ocean evolution. Indeed, these results suggest that oxidized interiors would melt to a greater extent compared to reduced mantles, and would thus sustain magma ocean and associated outgassing for longer periods of time.

Summary Overall, a number of lines of evidence suggest that the magma ocean was initially in a reducing state, because of the coexistence of metal and silicate melt, before the former fully segregated to the core. From then on, whether as a result of self-oxidation, metal segregation to the core, hydrogen escape, or oxidized material endowment (or a combination of all), the crystallizing magma ocean likely became more oxidized with time. Crystallization of the magma ocean is also required to concentrate enough incompatible species in the melt to reach melt-saturation of highly soluble volatile species and their extensive outgassing. By this time, the oxidation state of the magma ocean may have been high enough to reach the present-day QFM buffer. It would have produced an oxidized atmosphere, predominantly composed of H_2O–CO_2–N_2. This is consistent with the earliest terrestrial rock records (see Fig. 1 in Scaillet and Gaillard 2011, for a schematic reconstruction of the mantle redox state evolution). Given their similar size, density, and their vicinity in the Solar System, such a scenario may apply for both early Earth and Venus. Yet, the absence of measurements for the latter makes any statement even more speculative than for Earth. In such an early oxidizing environment after the surface solidified, late impact-induced degassing may have created local and transient reducing conditions. These would be favorable for prebiotic chemistry and life emergence on Earth, while other surface conditions may have compromised its development or sustainability on Venus.

While this scenario might tentatively reconcile the different available pieces of evidence (see Table 2 for a summary of the different early outgassed atmospheres and underlying assumptions/constraints), one should keep in mind that early atmospheric formation was most

Table 2 Early outgassed atmospheres and underlying constraints or assumptions

Associated mechanism	Reference(s)
Oxidized atmospheres: dominated by H_2O–CO_2–SO_2–N_2	
——————————————— Geologic constraints ———————————————	
Crystallization of terrestrial Hadean zircons 4.35 Gyr ago under oxidizing conditions	Trail et al. (2011)
Oxidation state of terrestrial Archean (up to 3.9 Gyr old) igneous rocks at-or-near current oxidizing state	Canil (1997), Delano (2001), Canil (2002), Lee et al. (2003), Li and Lee (2004), Rollinson et al. (2017), Nicklas et al. (2018)
——————————————— Prebiotic chemistry ———————————————	
Production of reduced organics triggered by coronal mass ejection events from the young Sun or impact energy	Kurosawa et al. (2013), Airapetian et al. (2016)
Exogenous delivery of reduced organic materials	Anders (1989), Chyba et al. (1990), Chyba and Sagan (1992), Whittet (1997), Matthews and Minard (2006)
Production of amino acids and other organic compounds using oxidation inhibitors	Cleaves et al. (2008)
——————————————— Accretion ———————————————	
Accreting bodies of increasing oxidation state (heterogeneous accretion) inferred from trace element abundances	Wänke et al. (1988), O'Neill (1991), Wood et al. (2009), Rubie et al. (2011, 2015), Fischer et al. (2015), Shi et al. (2022)
Elemental partitioning and impact erosion during continuous accretion of chondritic bodies	Sakuraba et al. (2021)
Impact-generated atmosphere from carbonaceous (CI, CM, CV) chondritic material	Schaefer and Fegley (2010, 2017)
Heating and outgassing experiments of carbonaceous chondrites	Thompson et al. (2021)
——————————————— Core formation ———————————————	
Core formation (rapidly) completed with metallic iron fully segregated from the silicate	Holland (1962), Kleine et al. (2002), Frost and McCammon (2008)
Deep metal–silicate separation	Li and Agee (1996), Fischer et al. (2015)
Chemical equilibrium with molten surface of bulk silicate Earth or molten Earth's continental crust composition	Lupu et al. (2014)
Impactor's iron not available to reduce the upper molten layers of the target or the atmosphere	Itcovitz et al. (2022), Citron and Stewart (2022)
——————————————— Magma ocean evolution ———————————————	
Gradient of the magma ocean redox state with oxidized shallow parts	Hirschmann (2012), Deng et al. (2020)
Disproportionation of FeO at depth to Fe metal and Fe_2O_3 components (in post-spinel and perovskite form)	Mao and Bell (1977), Frost et al. (2004), Galimov (2005), Wade and Wood (2005), Wood and Halliday (2005), Frost et al. (2008), Schaefer and Elkins-Tanton (2018), Armstrong et al. (2019), Hirschmann (2022)

likely a continuous, dynamic, and evolving process. It would have overlapped with stochastic planetary accretion, core formation, and cooling of possibly several episodes of more or less deep magma oceans (e.g., Tucker and Mukhopadhyay 2014). The magma oceans would have been influenced by several mechanisms at play during these early stages, where out-

Table 2 (*Continued*)

Associated mechanism	Reference(s)
Increasing outgassed CO_2/CO ratio with magma ocean cooling from ab initio molecular dynamics simulations	Solomatova and Caracas (2021)
Self-oxidation of the magma ocean with crystallization owing to Fe^{3+} incompatibility	Maurice et al. (2023)
———————————— Atmospheric evolution ————————————	
Preservation of carbonaceous chondritic D/H signature in the terrestrial oceans	Pahlevan et al. (2019)
Photodissociation of H_2O and(/or) atmospheric escape of H_2	Hunten (1993), Kasting et al. (1993a), Sharp et al. (2013)
Reduced atmospheres: dominated by H_2–CO–CH_4–NH_3–H_2S	
———————————— Prebiotic chemistry ————————————	
Production of amino acids and other organic compounds from electrical discharges	Miller (1953, 1955), Miller and Urey (1959)
———————————— Accretion ————————————	
Impact-induced degassing	Kasting (1990), Hashimoto et al. (2007), Sugita and Schultz (2009), Zahnle et al. (2010, 2020)
Impact-generated atmosphere from ordinary (H, L, LL) and enstatite (EH, EL) chondritic material	Schaefer and Fegley (2007, 2010, 2017)
Reduction of the magma ocean during accretion	Righter and Ghiorso (2012)
———————————— Core formation ————————————	
Incomplete metallic iron–silicate separation in the shallow part of the magma ocean	Holland (1962), O'Neill (1991), Frost and McCammon (2008)
Impactor's iron available to reduce the upper molten layers of the target and the atmosphere	Itcovitz et al. (2022), Citron and Stewart (2022)
———————————— Magma ocean evolution ————————————	
Graphite precipitation (requires a highly reducing magma ocean) and flotation at the surface of the magma ocean	Keppler and Golabek (2019)
———————————— Atmospheric evolution ————————————	
Reducing power of the primordial H_2-dominated atmosphere acting on proto-planets and constituting planetesimals	Zahnle et al. (2010)
Outgassing and slow (energy limited) escape of hydrogen	Tian et al. (2005), Liggins et al. (2020)

gassing of the growing planets' interior, impact degassing of accreting planetesimals, impact erosion, and atmospheric self-escape all affected both its physical and chemical properties over different time scales. A comprehensive theory explaining such a complex process requires further investigation of each of these contributions and would remain elusive without considering further inputs from incoming Venus exploration and ongoing exoplanets characterization.

3.3 Magma Ocean–Atmosphere Interactions

In the absence of a thick, long-lasting, viscous boundary layer at the surface of the magma ocean, thermal and volatile exchanges proceed efficiently between the molten interior and the overlying atmosphere. This is the case for most of the magma ocean evolution sequence. Note that a crust-like, thin, viscous boundary layer might form at the surface, either as observed on lava lakes or by flotation of crystals less dense than the surrounding melt, but would be quickly, i.e., relative to the magma ocean cooling rate, destabilized and broken apart by bursting bubbles and strong convective currents. After having described the partitioning of volatile species between the different reservoirs of the planet in the previous section, we discuss below the interaction between the solidifying magma ocean and the outgassed atmosphere and how they can affect the evolution of the coupled system. In particular, we focus on the heat transfer and exchanges from the magma ocean to space and how they control the cooling of the molten interior.

In its most simple form, the conservation of energy at the top of a planet's atmosphere can be described schematically as the equilibrium between all the incoming, $F_\downarrow$, and outgoing, $F_\uparrow$, fluxes:

$$F_\uparrow = F_\downarrow \ , \tag{16}$$

where the total outgoing flux $F_\uparrow = F_{\mathrm{IR}} + \alpha F_\odot$ comprises the thermal, infrared flux, F_{IR}, representing the heat loss at the top of the atmosphere, and the reflected fraction α (i.e., the Bond albedo) of the gross incident solar radiation $F_\odot$ that peaks at shorter wavelengths given the higher temperature of the star (the so-called Wien peak, that is inversely proportional to the temperature), in the visible spectral range, hence referred to as the Outgoing Shortwave Radiation (OSR). The total incoming flux is the sum of the heat sources:

$$F_\downarrow = F_{\mathrm{abs}} + F_{\mathrm{conv}} \ , \tag{17}$$

and includes the contribution of the net (incident minus reflected) solar flux absorbed by the planet, F_{abs} (Eq. (1)), and the heat flux coming from the planetary interior and generated by the convection of the magma ocean, F_{conv} (or F_{MO}). Regarding thermal exchanges more specifically, the magma ocean–atmosphere system has then two incoming heat sources: the absorbed solar flux and the convective flux from the magma ocean, that must balance the thermal radiation F_{IR} emitted to space at the top of the atmosphere (Fig. 5; e.g., Zahnle et al. 1988; Abe 1993a):

$$F_{\mathrm{IR}} = F_{\mathrm{abs}} + F_{\mathrm{conv}} \ . \tag{18}$$

During the magma ocean stage, the temperature of the mantle is so hot that the heat flux coming from the boiling convective mantle far exceeds the incoming solar flux ($F_{\mathrm{conv}} \gg F_{\mathrm{abs}}$). This implies that the outgoing thermal flux is dominated by the heat flux from the magma ocean: $F_{\mathrm{IR}} \approx F_{\mathrm{conv}}$. Because the heat flux received from the star is substantially lower than the planetary interior heat flux, i.e., the equilibrium temperature is much lower than the molten surface temperature, the planet has an excess of heat. To evacuate the excess of internal heat and reach thermal and radiative equilibrium with the stellar environment, the magma ocean cools down through convection. The generated heat flux F_{conv} is initially high and decreases with time. As $F_{\mathrm{IR}} \approx F_{\mathrm{conv}}$, the outgoing thermal flux also decreases with time.

The outgoing heat flux depends on the surface temperature, and the surface temperature itself results from the heat flux balance (Eq. (18)). It can be written $F_{\mathrm{IR}} = \sigma T_{\mathrm{eff}}^4 = \varepsilon \sigma T_{\mathrm{surf}}^4$, where σ is the Stefan–Boltzmann constant, T_{eff} the effective temperature of the planet,[7] and

---

[7]The effective temperature of the planet, T_{eff}, is the temperature a black body would have to emit the same total amount of radiation (F_{IR}). It should not be confused with the equilibrium temperature, T_{eq}, which is the

ε is the emissivity of the atmosphere, which measures its effectiveness in emitting energy as thermal radiation.

In the absence of an atmosphere, as usually initially assumed as a result of giant impact in magma ocean–atmosphere models, Eq. (18) is evaluated at the surface of the planet from which thermal radiation is directly emitted to space. In such a case, the planet efficiently radiates the energy from the magma ocean and emits all incoming absorbed radiation ($\varepsilon = 1$). It loses heat to space as a black-body with a temperature of the planet's surface (dashed purple line, Fig. 4): the surface temperature equals the effective temperature, $T_{surf} = T_{eff}$. F_{IR} then only depends on T_{surf}, which is directly set by the magma ocean, and F_{IR} and F_{conv} decrease rapidly. In the absence of an atmosphere, complete solidification of a fully molten mantle and global magma ocean would theoretically proceed in less than 5000 years (Lebrun et al. 2013; Nikolaou et al. 2019).

Once the volatile species are exsolved out of the melt and reach the surface of the magma ocean, they progressively build-up the overlying atmosphere alongside magma ocean cooling and solidification. When an atmosphere is present, thermal emission to space is buffered by the blanketing effect of the atmosphere heat-trapping greenhouse gases that absorb part of the incoming (net solar plus interior) radiation ($\varepsilon < 1$). The surface temperature is then larger than the effective temperature ($T_{surf} > T_{eff}$). ε is inversely correlated with the atmospheric greenhouse effect: the smaller ε, the less radiation is emitted at the top of the atmosphere (F_{IR}), i.e., the stronger the blanketing effect. By buffering the amount of heat that can be radiated/dissipated to space through the atmosphere (F_{IR}), its blanketing effect buffers the magma ocean cooling rate. The surface temperature is then set by the equilibrium between the heat flux transported by convection in the magma ocean to the surface, and the heat flux that can be radiated through the atmosphere, from the surface of the magma ocean to space (e.g., Solomatov 2015).

The thermal opacity of the atmosphere is due to the absorption of radiation by greenhouse gases and is thus primarily a function of the atmospheric mass and composition, which are direct outcomes of magma ocean outgassing (Sects. 3.2.4 and 3.2.5). For instance, H_2O and CO_2 are two powerful greenhouse gases that can delay the complete magma ocean solidification for a few millions to hundred millions of years, depending on their atmospheric concentration (e.g., Hamano et al. 2013; Lebrun et al. 2013; Salvador et al. 2017; Nikolaou et al. 2019). During this phase where the outgoing emissions at the top of the atmosphere are dominated by the convective flux from the molten mantle (i.e., $F_{IR} \approx F_{conv}$), the evolution of the magma ocean and of the atmosphere are thus inextricably coupled and cannot be addressed separately.

Based on the heat fluxes equilibrium, the thermal evolution can be divided in two main phases: the first one, where thermal emissions going out of the atmosphere are primarily controlled by the convective flux coming from the magma ocean (i.e., $F_{IR} \approx F_{conv}$), as described above, and a second phase starting when F_{conv} decreases below the absorbed solar radiation and becomes negligible in front of it, where F_{IR} is dominated by F_{abs} (i.e., $F_{IR} \approx F_{abs} \gg F_{conv}$). The transition between these two phases may be defined as the time where the convective heat flux from the mantle becomes an order of magnitude lower than the absorbed solar flux ($F_{conv} \leq 0.1 \times F_{abs}$) and is referred to as the end of the rapid cooling stage ("ERCS"; Salvador et al. 2017).

From this moment, the planet has lost its excess internal heat and has reached global radiative equilibrium with its stellar environment. The molten interior and associated heat

---

temperature the planet would be at if only heated by the host star, i.e., if the planet was at equilibrium with the absorbed radiation.

 Springer

is no longer dominantly driving the thermal evolution. Both the surface temperature and the thermal emissions at the top of the atmosphere are then set by the radiative equilibrium between the net incoming absorbed solar radiation and the outgoing thermal radiation (e.g., Leconte et al. 2013), and are thus mainly controlled by atmospheric-related parameters: the absorbed solar flux and the atmospheric opacity. Note that the surface temperature reached at global radiative equilibrium may be above (e.g., if the planet absorbs a large amount of sunlight) or below the melting point of the surface. The complete solidification of the molten mantle, i.e., the end of the magma ocean phase, is independent of and should be distinguished from the ERCS/global radiative balance.

For a planet located at a given distance from its host star, because of the relatively slow variation of the total incoming solar flux with time (if stellar properties and orbital distance remain relatively stable), all F_{abs}-governing parameters are relatively constant and unrelated to the thermal evolution, besides the Bond albedo (Eq. (1)). Thus, the formation of a water ocean at the surface critically depends on two main atmospheric properties: its Bond albedo and greenhouse effect, both ultimately relying on the surface temperature and pressure prevailing at the ERCS, that are directly inherited from the magma ocean cooling and outgassing sequences.

In that respect, clouds play a vital role. Indeed, they both regulate the amount of solar energy reaching the surface through their albedo (Pluriel et al. 2019), and affect the amount of heat radiated back into space through their greenhouse effect. Because of their potential cooling or warming effect, their presence, properties, location, and distribution may drastically impact the surface conditions reached at the ERCS. However, they remain the major source of uncertainties in current climate models and in future Earth's climate predictions (e.g., Kasting and Catling 2003; Siebesma et al. 2020). In particular, accounting for their time-dependent, 3D spatial distribution, dynamics and microphysical properties to constrain their optical properties and global contribution is highly challenging. Modelling cloud formation and behavior throughout magma ocean cooling and assessing their significance for the early surface conditions suffers from additional complexities due to the prevailing extreme conditions. Different cloud configurations and their implications regarding the surface conditions of Venus at the ERCS and early water ocean formation are further discussed in the following section

When discussing the outcomes of the magma ocean phase in the next Section, we focus on the aftermath of the evolutionary phase affected by the mechanisms described throughout Sect. 3. This involves the conditions reached at the ERCS (Fig. 12), regardless of whether the mantle has fully solidified or not at the time global radiative balance is reached. Indeed, the subsequent surface conditions and evolution of both the interior and of the atmosphere rely on other processes. Even if the mantle is still molten at the ERCS, its dynamics may significantly differ from that of the pre-ERCS magma ocean convection, since the excess internal energy would have then been lost and a thermal steady-state reached.

4 Outcomes of the Magma Ocean Phase and Implications for the Long-Term Evolution of Venus

4.1 Surface Conditions and Potential Early Habitability

At the end of the rapid cooling stage, the conditions may allow for liquid water to condense at the surface and enable ocean formation. Water ocean formation essentially requires both the atmospheric water vapor pressure and the temperature at the surface to lay between

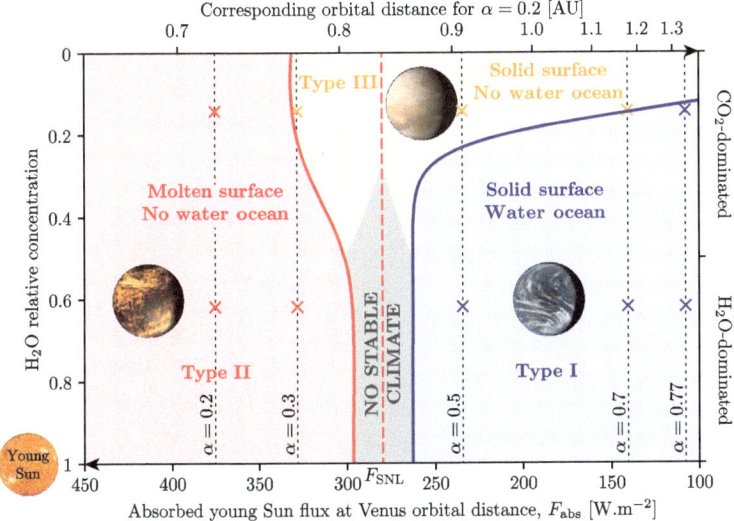

Fig. 12 Schematic summary of the various surface conditions reached at the End of the Rapid Cooling Stage (ERCS), i.e., when the heat flux coming from the interior becomes negligible compared to the net absorbed stellar flux. From then on, the latter balances the outgoing thermal radiation and this balance mainly controls the climate. The solid lines represent the transitions between the different types of surface conditions, but note that these are not hard limits. The crosses correspond to the ones shown in Figs. 4 and 13 for H_2O- and CO_2-dominated atmospheres, respectively. Note that alternative atmospheric compositions (e.g., reduced atmospheres or desert worlds) would alter the location and shape of the limits separating the different types of planets (see text). From any surface conditions reached at the ERCS and resulting from the magma ocean evolution, Venus may have transitioned to any other states before reaching its present-day conditions (see Fig. 14)

the triple and critical points of water. We describe below which magma ocean evolution scenarios would favor water ocean formation, both early on at the ERCS and later, and scenarios where water ocean formation is unlikely.

Insufficient Accreted or Outgassed Water Prior to magma ocean evolution, the amount of water delivered to the planet during the accretion phase may be the first limiting factor for ocean formation. If the planet accreted dry or did not get enough bulk water, water ocean formation at the end of the rapid cooling stage is compromised (uppermost part of Fig. 12). Unless a substantial amount of water is delivered later in history, through a "late veneer" for instance, and temperate conditions are met, water formation is unlikely over the entire planet lifetime (Fig. 14). Yet, as discussed in Sect. 2.2.1, this dry accretion scenario is currently not favored by planetary formation models for Venus. In addition, as discussed in Sect. 2, based on Venus proximity with Earth, their similar density, and its high atmospheric D/H ratio, a substantial, likely Earth-like, early water endowment is the prevailing scenario for Venus today.

As described in Sect. 3, the amount of atmospheric water vapor is a direct outcome of magma ocean cooling and outgassing. If the water vapor mixing ratio of the outgassed atmosphere is too low (i.e., the relative amount of outgassed water is low), insufficient vapor pressure would prevent water saturation pressure to be reached at the surface. This would happen if magma ocean outgassing is not efficient for water or produces a reduced atmosphere. Early ocean formation at the ERCS would then be impossible (uppermost part of Fig. 12). Later ocean formation could still be possible if enough mantellic remnant water is volcanically outgassed during the long-term evolution of the planet, if atmospheric water is delivered by late impacts, and/or if the redox state of the mantle or atmosphere evolves and becomes more oxidized (Fig. 14). Such late oxidation could be the result of volatile cycling and geo-/photo-chemical evolution, involving atmospheric hydrogen escape and surface oxidation processes.

H₂O-Dominated Atmospheres Provided there is sufficient atmospheric water at the ERCS, the surface temperature determines whether water vapor saturation pressure can be reached and a water ocean forms. Recall that surface temperature at the ERCS results from the global radiative balance between the net incoming absorbed solar flux, F_{abs}, and the thermal outgoing flux the planet is able to radiate to space, F_{IR}. Both fluxes rely on atmospheric state and composition. These are direct outcomes of the magma ocean outgassing and evolution: atmospheric mass, composition, thermal profile, resulting clouds (and distribution), aerosols types and locations.

For steam, H₂O-dominated atmospheres, the existence of the outgoing radiation limit with respect to surface temperature, at about 280 W.m⁻² (dashed red line in Fig. 4), creates two drastically different types of surface conditions at the ERCS (Fig. 12; Hamano et al. 2013; Salvador et al. 2017). If the planet's outgoing thermal flux is lower than the radiation limit (i.e., $F_{IR} \approx F_{abs} < F_{SNL}$), the resulting climate should be temperate. Surface temperatures would be below the water critical temperature and allow water ocean formation (e.g., blue crosses in Figs. 4 and 12). Such planets are expected to have a solidified surface at the ERCS and are referred to as "Type I" planets following Hamano et al. (2013) classification (blue area in Fig. 12). For Venus, an early habitable state could be reconciled with the present-day conditions via water loss resulting from later on atmospheric escape (Fig. 14; see Sect. 4.2). Yet, ocean water would first have to be vaporized, reach the upper atmosphere, and photodissociate. The hydrogen would be lost to space, while most of the oxygen would remain. If temperate conditions and oceans are stable, water loss would be less efficient than if water were found mainly in the vapor phase.

Conversely, at the ERCS if the planet absorbs more solar energy than the Simpson–Nakajima radiation limit ($F_{IR} \approx F_{abs} > F_{SNL}$), climate stability can only be achieved at a surface temperature above $T_{surf} \approx 1500$ K (e.g., red crosses in Figs. 4 and 12; Abe and Matsui 1988; Zahnle et al. 1988; Kasting 1988; Hamano et al. 2013; Goldblatt et al. 2013; Marcq et al. 2017; Salvador et al. 2017). At this temperature, which is above the solidus temperature of silicates, the planet surface is expected to be molten and able to radiate in the visible and near-infrared (NIR). It would have an emission peak around 5 μm (Wien peak), thus allowing emission through the 4 μm water vapor absorption window to balance the absorbed solar flux and achieve radiative equilibrium. Such molten surface planets at the ERCS are referred to as "Type II" planets (red area in Fig. 12; Hamano et al. 2013). Because the planet is at global radiative balance, the surface of type II planets are expected to remain molten and the magma ocean phase is sustained until there is a decrease of the net absorbed solar flux or a decrease of the atmospheric water abundance (Fig. 14). Atmospheric water could be lost via escape processes. The absorbed flux could then decrease owing to the weakening greenhouse effect or via the formation of reflective clouds triggered by a change of rotation rate or water condensation in the upper atmosphere.

The critical distance separating Type I and Type II planets is actually equivalent to the inner edge of the circumstellar habitable zone (IHZ). The latter is defined in relation to the radiation limit of the runaway greenhouse state and is classically calculated using atmospheric models (e.g., Kasting et al. 1993b; Selsis et al. 2007; Kopparapu et al. 2013; Leconte et al. 2013; Ramirez 2018). Such models typically start with temperate Earth-like surface conditions, with a surface water ocean and an H_2O-dominated atmosphere. From this base-line configuration, they vary the surface temperature (or the insolation; e.g., Leconte et al. 2013) to evaluate the atmospheric/climate response. They then calculate the corresponding solar flux needed to sustain the specified surface temperature. Note that this surface temperature variation could result from a change of insolation, absorbed stellar flux, or greenhouse gases concentration. They proceed until the ocean evaporates and a runaway greenhouse state is reached. The corresponding net absorbed and/or emitted critical flux is translated into a critical insolation threshold and orbital distance – this defines the inner edge of the habitable zone.

Compared to those studies, early evolution magma ocean models start with molten surfaces, where the water is outgassed from the interior and first found in the atmosphere in the vapor phase rather than being condensed into an ocean. As soon as enough water is outgassed, these atmospheres are in a runaway-greenhouse-like state. Coupled models then simulate the cooling and solidification of the magma ocean. As the interior flux decreases, their atmospheres evolve toward thermal equilibrium (reached at the ERCS, when the interior flux becomes negligible compared to the absorbed stellar flux), where their outgoing thermal emission balances the absorbed part of the incoming sunlight. The habitability of the planet is then assessed via the likelihood of water ocean formation given the surface conditions reached at the ERCS. Here, the location of the inner edge of the habitable zone is thus evaluated the other way round. They start from hot, uninhabitable conditions. They next evaluate the insolation threshold and corresponding orbital distance at which a stable climatic state (i.e., $F_{IR} \approx F_{abs}$) is able to produce a water ocean at the surface of the solidified planet.

It is important to emphasize that while the habitable zone boundaries are generally expressed in terms of insolation thresholds and orbital distances from the star, these two parameters only consider the incoming stellar flux $F_{\downarrow}$. Yet, the parameter that determines the climate state and associated surface conditions is the net absorbed stellar flux F_{abs}, where the albedo is critical. Accordingly, a planet of high Bond albedo (e.g., due to a highly reflective cloud cover) may sustain temperate surface conditions and have an early water ocean (type I planet). This is despite being closer to the star than a type II planet of lower albedo and with an incident flux larger than the predicted insolation threshold. Any parameter affecting the planetary albedo may thus dramatically alter the fate of the planet especially if it is located near the edge of the habitable zone, as is the case for Venus (e.g., Hamano et al. 2013). Provided that Venus' orbital parameters have remained stable and accounting for the lower luminosity of the young Sun (e.g., Gough 1981; Bahcall et al. 2001), the amount of absorbed stellar radiation F_{abs} and associated surface conditions depend upon the Bond albedo (Fig. 12).

CO_2-Dominated Atmospheres Similar to IHZ studies, Hamano et al. (2013) type I and II planet classification is based on steam, water-dominated atmospheres. Indeed, the runaway greenhouse occurs in such atmospheres because their main component is both a condensable and thermal absorber species. For atmospheres that are more diluted with respect to their water vapor concentrations (i.e., those whose composition is not water-dominated but non-condensable[8] species-dominated), the strong runaway greenhouse effect of H_2O-dominated atmospheres is buffered. For instance, as the longwave radiative forcing, or thermal absorption, of CO_2 is lower than that of H_2O (e.g., Kiehl and Trenberth 1997), CO_2-dominated atmospheres are more transparent/less opaque to thermal radiation as a result of atmospheric

---

[8] At least for conditions where water would be able to condense.

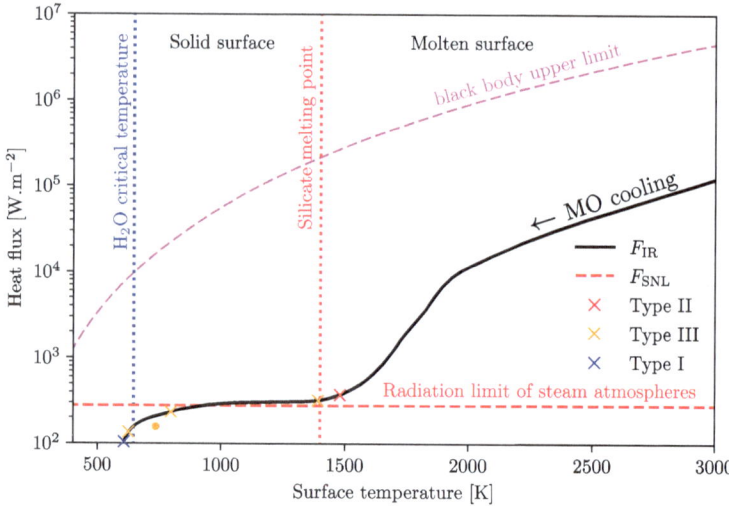

Fig. 13 Outgoing thermal flux emitted at the top of the atmosphere (F_{IR}, black line) as a function of surface temperature for a CO_2-dominated atmosphere. Compared to H_2O-dominated atmospheres (Fig. 4), global radiative equilibrium and stable climatic states can be reached at intermediate surface conditions between temperate Type I (e.g., blue cross, where $\alpha = 0.77$) and molten surface Type II (e.g., red cross, where $\alpha = 0.2$) planets. These Type III planets (e.g., orange crosses, where $\alpha = [0.3, 0.5, 0.7]$ from right to left) have solidified at the ERCS ($T_{surf} < T_{solidus}$) but water does not condense at the surface. This either because the atmosphere is too dry and the saturation vapor pressure of water is not reached at the surface or because the temperature is too high and above the critical temperature of water ($T_{surf} > 647$ K). The crosses correspond to the ones shown in the upper part of Fig. 12. The orange dot denotes present-day Venus

water vapor dilution and associated lower specific humidity (Salvador et al. 2017). Owing to their increased thermal transparency, CO_2-dominated atmospheres are able to radiate thermally (F_{IR}) even above the Simpson–Nakajima limit in response to an increase of temperature or net absorbed solar flux. These CO_2-dominated atmospheres thus maintain radiative balance at intermediate surface temperatures in contrast to H_2O-dominated atmospheres which are thermally opaque. In order to reach radiative balance, H_2O-dominated atmospheres have to reach higher temperatures to radiate again and stabilize their climate. Here the radiation limit of steam H_2O-dominated atmospheres is no longer relevant. The sharp climatic transition and surface temperature jump between type I and type II planets are soften and additional intermediate stable climate states can be achieved (Fig. 13). These climate states can have intermediate surface temperatures, including temperatures greater than that of the critical point of water, where water ocean formation is not possible (orange crosses in Figs. 13 and 12). In addition to a decrease in atmospheric opacity, the decrease of water vapor partial pressure implied by the atmospheric water vapor dilution effect can prevent water ocean formation even below the critical temperature of water. This is the case if the water vapor saturation pressure is not reached at the surface (e.g., lowermost orange cross in Fig. 13). Thus, for CO_2-dominated atmospheres, water ocean formation becomes highly sensitive to the atmospheric volatile concentration. The water vapor mixing ratio can prevent water ocean formation at any orbital distance if it is too low (Fig. 12). At the ERCS, intermediate and stable surface conditions between that of type I and II planets can be achieved where the surface has completely solidified thanks to surface temperatures below the solidus temperature but where no water ocean is formed (Figs. 12, 13 and 14). Planets reaching this state are referred to as "Type III" planets (orange area in Fig. 12; Salvador et al. 2017).

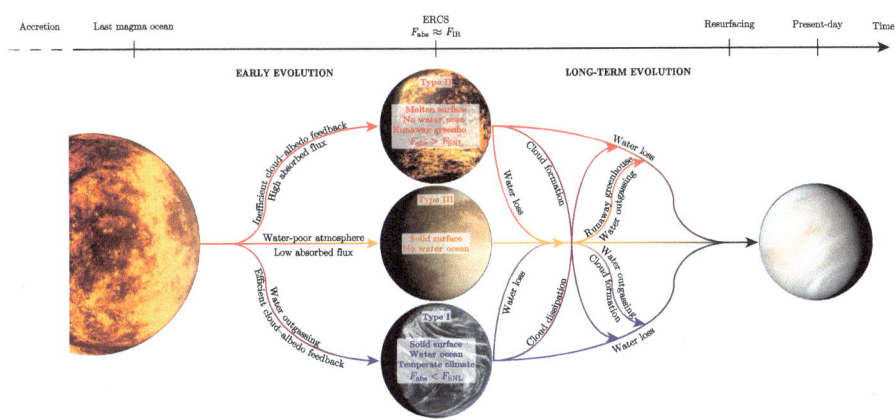

Fig. 14 Possible evolution pathways taken by Venus from the last major magma ocean episode to the present day (not to scale). The End of the Rapid Cooling Stage (ERCS) separates the early and long-term evolution and is reached when the heat flux coming from the interior becomes negligible compared to the net absorbed stellar flux (F_{abs}; Eq. (1)). From then on, the latter balances the outgoing thermal radiation (F_{IR}) and this balance mainly controls the climate. F_{SNL} is the radiation limit of steam atmospheres. Early evolution: Relatively dry accretion or inefficient magma ocean outgassing of water owing to its retention in the interior or a reduced upper mantle would induce low concentrations of water vapor in the atmosphere. This would prevent the formation of water oceans even if the surface has solidified at the ERCS owing to relatively low absorbed fluxes (MO to Type III; orange area of Fig. 12). Water outgassing from the magma ocean would result in temperate conditions at the ERCS if an efficient cloud–albedo feedback operates (MO to Type I; blue area of Fig. 12). If not, regardless of the water content, high absorbed fluxes would induce a molten surface at the ERCS (MO to Type II; red area of Fig. 12). Long-term evolution: Changes of the amount of water vapor in the atmosphere and of the absorbed stellar flux may trigger transitions from one type to another. Atmospheric water loss could be due to atmospheric escape (Sect. 4.2) or to its incorporation into the interior. It could induce the evaporation of the ocean (Type I to III) or the solidification of the surface by weakening the greenhouse effect (Type I to III). Although less likely, volcanic outgassing of CO_2 would have the same effect. Conversely, an increase of the amount of water vapor in the atmosphere through volcanic outgassing or late accretion of water-rich bodies could trigger a late runaway greenhouse (Type III to II) or the formation of water ocean (Type III to I) depending on the absorbed stellar flux. A temperate climate would thus be favored by synchronous formation of reflective clouds on the dayside of the planet, possibly helped by rotation rate changes, but threatened by the brightening Sun. If the atmosphere is already water-dominated, cloud formation would also initiate a temperate climate era (Type II to I) but their dissipation would end it (Type I to II). Note that these are just examples of the possible mechanisms and that the location of the transitions between the different branches are not indicative of their timing

Type III planets may experience post-ERCS habitable periods if the atmospheric water vapor mixing ratio increases enough, through long-term volcanic outgassing for instance, and provided that the absorbed flux remains/decreases below the runaway greenhouse threshold (Fig. 14). If not, a runaway greenhouse would prevent the formation of water oceans and the planet would be desiccated via atmospheric escape.

Desert Worlds Abe et al. (2011) modeled the climate of desert worlds with globally limited but locally abundant surface water, also referred to as "land planets", to assess their habitability. This is an intermediate case between H_2O- and CO_2-dominated atmospheres. Because the tropics of such land planets have very low relative humidities, the air is unsaturated in those regions where longwave radiation can be efficiently emitted above the traditional runaway greenhouse limit. Compared to planets with globally abundant liquid water, land planets thus have a wider habitable zone. They can sustain moderate climates where liquid water remains stable at the poles despite being closer to their host stars. Abe et al. (2011) suggested that if Venus were such a land planet with a 1 bar N_2-dominated atmosphere with 345 ppm CO_2, it could have remained habitable until as recently as 1 billion years ago.

Zsom et al. (2013) studied a more extreme case of hot desert/dry worlds representative of greenhouse gas-poor atmospheres. Here the atmospheric composition was dominated by a non-greenhouse gas (e.g., N_2) with low relative humidity ($\sim$1%), and low CO_2 mixing ratio (10^{-4}). They showed that the reduced greenhouse effect may bring the inner edge of the habitable zone as close as 0.38 AU from a Sun-like star, for a surface albedo of 0.8. For a more moderate surface albedo of 0.2, it would be 0.59 AU. Note that for such arid planets where the water budget is limited (e.g., Abe et al. 2011), their definition of habitability is not restricted to the formation of a global water ocean. Despite having a water reservoir limited to cold areas such as the poles or the night side (Abe et al. 2011; Menou 2013; Leconte et al. 2013) they also consider a planet habitable when: (i) water is able to precipitate mostly in the form of rain on a large fraction of the planet's surface, and (ii) where the water reservoir is stable on a multi-billion-year timescale. This way, most of the surface water is in liquid form, a large fraction of the planet remains habitable, and the water cycle is not broken (Zsom et al. 2013). While both early and present-day Venus could be located within this extended habitable zone, this particular atmospheric configuration is hard to relate to any early Venusian evolution scenario. While a low relative humidity could be explained by low water delivery, an inefficient or highly reduced magma ocean outgassing, a low CO_2 mixing ratio is hard to reconcile with the present-day atmospheric CO_2 inventory. Indeed, CO_2 is poorly soluble in the magma ocean and likely outgassed early. Otherwise, if C is not outgassed as CO_2, the outgassing of other carbon-bearing reduced species would challenge the restricted atmospheric composition and the absence of other greenhouse gases required in the Zsom et al. (2013) and Abe et al. (2011) calculations.

Critical Albedo Value The key to form early oceans under temperate climates at the ERCS is an optimal magma ocean cooling and outgassing sequence, producing enough water vapor in an atmosphere absorbing the right range of solar flux (e.g., Fig. 12). Abe and Matsui (1988) and Kasting (1988) used 1D radiative-convective, cloud-free atmospheric models to simulate the climates and surface conditions associated with hot H_2O–CO_2 atmospheres. They found that if Venus initially accreted a similar amount of water to Earth, a proto-ocean could form on early Venus. With Pollack (1971), they suggested that clouds could play a crucial role in producing early oceans on Venus by increasing the albedo and deflecting part of the incoming sunlight to space.

Salvador et al. (2017) estimated that less than 0.1 Earth water ocean initially dissolved within the early Venusian magma ocean and efficiently outgassed into the atmosphere would be sufficient to form a water ocean at the ERCS. They used the insolation of the young Sun ($\sim$70% of its current value; e.g., Gough 1981; Bahcall et al. 2001) while using Venus' present-day geometric albedo ($\sim$0.7) and atmospheric CO_2 content ($\sim$90 bar). They also found that using an albedo of 0.45 would reflect enough sunlight to form a water ocean at the ERCS if only 0.3 Earth's ocean of water were initially dissolved within the magma ocean and efficiently outgassed (Fig. 12). This is in good agreement with Abe (1993a), who showed that for an albedo of 0.35, early Venus would lie at the critical flux between a habitable and a runaway greenhouse state, indicating that a water ocean would be stable at Venus' surface for a larger albedo (their Fig. 3). More recent modeling by Krissansen-Totton et al. (2021) showed that all transient habitable solutions for early Venus have a Bond albedo larger than 0.4.

The Importance of Clouds As noted above clouds might be the critical factor in producing an albedo suitable for temperate surface conditions, by reducing the amount of absorbed sunlight at Venus' distance. Venus' present-day thick sulfuric acid (H_2SO_4) cloud cover is responsible for its high albedo, reflecting more than 75% of incoming stellar flux back to space (Moroz et al. 1985; see Titov et al. 2013, for a review). Despite being 30% closer from the Sun, Venus presently absorbs less solar energy than the Earth does: $F_{abs}^{\venus} \approx 157 \ W \cdot m^{-2}$ (Moroz et al. 1985) versus $F_{abs}^{\oplus} \approx 240 \ W.m^{-2}$ on average. The corresponding equilibrium temperatures are $\sim$231 K for Venus and 255 K for the Earth (assuming an albedo of 0.75 and 0.29, respectively). Note that the thick cloud layer also constitutes a second infrared opacity source in the Venusian atmosphere (the first being the 92 bars of CO_2), and contributes about 140 K to the greenhouse effect (Titov et al. 2013).

Given the flux received from the young Sun at Venus' orbital distance, if early Venus had enough atmospheric water at the ERCS and a high albedo, water ocean formation would have been likely (Fig. 12; Salvador et al. 2017). Several physical mechanisms and associated cloud effects could produce a high enough albedo to allow water oceans to form on early Venus. While maximizing the cooling effect of clouds by neglecting their warming effect, i.e., by assuming that clouds are transparent to IR outgoing radiation, Selsis et al. (2007) estimated that covering 100% and 50% of a planet's dayside with highly reflecting clouds can produce albedos as high as 0.8 and 0.6. This would shift the runaway greenhouse limit and associated inner edge of the habitable zone to an orbital distance of 0.46 AU and 0.68 AU for the current Sun. Such a highly optimistic habitable zone would encompass both present-day and early Venus.

Using a 3D general circulation model with a modern cloud scheme and Earth-like atmospheric compositions, Yang et al. (2014) showed that Venus' slow planetary rotation rate (of about 243 Earth days; e.g., Carpenter 1970; Campbell et al. 2019; Margot et al. 2021) would induce an albedo high enough to support temperate conditions. Indeed, by weakening the Coriolis force and extending daytime illumination compared to faster rotating planets, slow rotation rates promote strong convergence and convection in the substellar region. This produces a large area of optically thick clouds in the most illuminated part of the atmosphere, which greatly increases the albedo and decreases surface temperatures (as first evidenced for tidally locked planets; see Yang et al. 2013). Even at Venus' present-day orbital distance and insolation, this mechanism would sustain low enough surface temperatures to support surface liquid water.

Way et al. (2016) used early Venus' insolation and an Earth-like atmospheric composition to model the climatic history of Venus via 3D GCM simulations. They included a slow rotation rate like Yang et al. (2014), its present-day orbital parameters, modern topography from the Magellan mission, and an ocean volume consistent with the Pioneer Venus measured D/H ratio (Donahue et al. 1982). By modeling the climate dynamics and resulting surface conditions for several hypothetical scenarios, Way et al. (2016) showed that early Venus could have sustained stable water oceans and a temperate climate until as recently as 0.715 Ga ago. The 0.715 Ga was chosen as the approximate time of Venus' global resurfacing event (e.g., McKinnon et al. 1997). Following Yang et al. (2014) and Abe et al. (2011) they demonstrated that the rotation rate, through the cloud–albedo feedback, and the availability/distribution of surface water play a key role in climate circulation and heat redistribution resulting in the possible habitability of ancient Venus (Fig. 14). Extending their previous analysis to a significantly larger parameter space, Way and Del Genio (2020) showed that a shallow water ocean and habitable conditions could have persisted nearly 3 billion years. They speculated that this clement period ended via large igneous provinces formation (LIPs) before the global resurfacing period (lowermost branch of Fig. 14). However, if liquid water is not initially present at the surface, subsequent habitability is rather unlikely. Hence the importance of the early history of Venus and its magma ocean evolution cannot be understated.

Another important finding regarding the early history and relevant to magma ocean outcomes is that, while early water oceans and temperate climates would be confidently sustained for a few billion years, if liquid water is not initially present at the surface, subsequent habitability and formation of water ocean is likely compromised. It would require to both replenish the atmosphere with water through volcanic outgassing or late accretion and form reflective clouds on the planet's dayside (from type III to type I branch in Fig. 14) while reconciling such scenarios with existing constraints (e.g., Way and Del Genio 2020; Krissansen-Totton et al. 2021). An early enough slow rotation rate and associated highly reflective clouds feedback might then be critical.

At present we have no constraints on early Venus' rotation rate. Its current slow retrograde rotation rate could be a result of a number of processes working together or separately over Venus' lifetime (e.g., Cottereau et al. 2011; Hoolst 2015):

- core–mantle friction (e.g., Goldreich and Peale 1970; Lago and Cazenave 1979; Dobrovolskis 1980);
- solid-body tides (e.g., Correia and Laskar 2003; Hoolst 2015; Way and Del Genio 2020);
- spin–orbit resonance with the Earth (e.g., Goldreich and Peale 1967; Gold and Soter 1969; Bills 2005);
- thermal atmospheric tides (e.g., Gold and Soter 1969; Ingersoll and Dobrovolskis 1978; Dobrovolskis and Ingersoll 1980; Leconte et al. 2015) specific to its dense atmosphere (e.g., Correia and Laskar 2001);
- an impactor (e.g., McCord 1968; Alemi and Stevenson 2006).

In particular, most of these mechanisms depend on the atmospheric thickness, interior structure, or insolation. While insolation was negligible compared to the magma ocean convective flux, the atmosphere was building up and the interior structure was evolving throughout the magma ocean rapid cooling stage. This makes any estimation of the rotation rate at that time and post-ERCS highly challenging.

The cloud–albedo feedback for slowly rotating planets appears to extend the habitable zone to Venus' orbital distance (Yang et al. 2014; Way et al. 2016). Yet such temperate conditions are only possible with a pre-existing water ocean condensed at the surface and an Earth-like, water-poor atmosphere. However, it may not apply to planets with hot and steamy atmospheres. Here water does not condense into water oceans at the surface and is found entirely in the atmosphere, as expected for an efficiently outgassing magma ocean. Under such conditions, 3D GCM simulations by Turbet et al. (2021) demonstrated that water clouds form preferentially on the nightside, while being absent on the dayside. Their result is independent of the planetary rotation speed or day length. Such a day–night cloud asymmetry appears to prevent the efficient cloud–albedo effect at the sub-stellar point found by Yang et al. (2014). The Turbet et al. (2021) result induces strong solar absorption by atmospheric water vapor on the dayside and strong greenhouse effect by clouds (i.e., a net warming effect of clouds) on the nightside. Without the cloud–albedo effect at the substellar point, Turbet et al. (2021) suggested that if water is not first condensed into oceans (as discussed above) Venus would inevitably maintain a runaway greenhouse state from early on (uppermost branch of Fig. 14). High surface temperatures would prevent water oceans from forming at the surface, which would be in agreement with Hamano et al. (2013) cloud-free predictions identifying Venus as a type II planet.

Limitations and Perspectives The modeling of Turbet et al. (2021) is more realistic than previous 1D studies in assessing early water ocean formation on Venus under magma ocean-related, hot and steamy conditions. Extensive work yet remains to provide a conclusive statement. Firstly, cloud and atmospheric circulation feedback can vary non-linearly and non-monotonically with the rotation period (Jansen et al. 2019). Turbet et al. (2021) state that the applicability of their results to a wide range of rotation periods and to a slowly rotating Venus should be quantitatively confirmed by a comprehensive sensitivity study. This would also allow for a full characterization of day–night heat transport patterns on hot planets where water is found in the vapor phase in the atmosphere. It would also quantify the atmospheric circulation, climate dynamics, and resulting cloud effects (Noda et al. 2017).

As discussed in Sect. 3, the hot surface temperatures found in the early evolution stages are not ad-hoc conditions: the atmosphere is not an isolated system but instead, its bulk properties result from and are embedded in the coupled magma ocean–atmosphere evolution. Depending on the processes considered, water outgassing efficiency may vary significantly, and the differential outgassing of CO_2 and H_2O may induce additional complexities. To model a realistic early evolution and predict accurate associated surface conditions, such sophisticated 3D GCMs need to be coupled to molten interior models to consider self-consistent outgassing and cooling sequences. Such coupled models could thus account for evolving atmospheric properties, such as composition, surface pressure, and thermal surface boundary conditions, while considering interior–atmosphere feedback and possible non-linearities such as climatic hysteresis and multi-stability (e.g., Abe et al. 2011; Goldblatt 2015; Ishiwatari et al. 2021).

However, if the magma ocean covers the entire surface of the planet its heat flux is likely to be isotropic for most of the rapid cooling stage and would not depend on local time or latitude. The atmosphere would be uniformly and dominantly heated from below, such that atmospheric circulation and associated heat redistribution may not be affected by day–night or latitude-dependent insolation gradients. How the interior flux-controlled climate dynamics would look like in an outgassing atmosphere and how it would transition to a more classic (solar flux-dominated) climate circulation at the ERCS is unknown. In the latter the sunlight becomes the dominant energy source, and the implications for atmospheric circulation patterns and planet surface conditions again remain unknown. These questions will certainly benefit from the ongoing efforts and future improvements of both interior, atmosphere, and coupled numerical models (Sect. 5).

Krissansen-Totton et al. (2021) have developed a sophisticated coupled interior–atmosphere framework accounting for an extensive number of processes. These include geochemical cycling, atmospheric escape, and consistent atmospheric redox evolution, to self-consistently model the thermo-chemical evolution of Venus from the magma ocean stage to present-day. In light of all presently available atmospheric and surface constraints, they show that both habitable and never-habitable histories are possible, and that none can be conclusively ruled out. Indeed, both cases can successfully produce the current dense CO_2-dominated atmosphere with low water and oxygen content. They get consistent [40]Ar and [4]He outgassed abundances, and inferred surface heat flow. Importantly, they emphasize that Venus' habitability relies critically on the value of the Bond albedo, which appears to be the main controlling factor. Hence they demonstrate yet again the importance of the cloud–albedo feedback described above.

As it is generally the case for many processes affecting Venus history and evolution, there is a pressing need to update the precision of available constraints to disentangle Venus past. Upcoming missions may provide key in-situ measurements that may significantly help in constraining the likelihood of the different scenarios discussed herein (see Widemann et al. 2023, for an extensive discussion).

4.2 Atmospheric Escape and the Fate of Water

A high water vapor content in the early atmosphere of Venus has to be reconciled with its present-day low concentrations of water and oxygen (e.g., Johnson and de Oliveira 2019). It has long been hypothesized that Venus underwent large scale atmospheric escape during its evolution, providing a way to lose its water (e.g., Ingersoll 1969; Rasool and de Bergh 1970; Pollack 1971; Kasting and Pollack 1983). Several mechanisms can be responsible for atmospheric escape and may have operated to various extent throughout Venus evolution (Fig. 14). They include thermal hydrodynamic escape, non-thermal (suprathermal) escape, and mechanical impact erosion that are all discussed below. Note that high energies and temperatures resulting from extreme post-impact conditions may also be responsible for early intense atmospheric escape (Modirrousta-Galian and Korenaga 2023).

4.2.1 Hydrodynamic Escape

Part of the water loss could have occurred early during the evolution of the planet (Lammer et al. 2018). At that time, the EUV flux from the Sun was at its highest (Tu et al. 2015), powering hydrodynamic escape of water and possibly heavier species like noble gases and CO_2. Strong EUV heated the upper atmosphere, exciting and photo-dissociating the gas molecules (Lammer 2013). Water being easily dissociated implies that the upper atmosphere was likely dominated by hydrogen (in terms of number of atoms) and oxygen (in terms of mass) (Kasting and Pollack 1983). This upper atmosphere expanded hydrodynamically and "flowed" outward to escape into space. H loss is very efficient (O loss potentially less so; e.g., Chassefière 1996) and can also drag off heavier species (Watson et al. 1981; Erkaev et al. 2014). Hydrodynamic losses are usually estimated by using a simple energy-limited approximation (Watson et al. 1981; Odert et al. 2018).

Mass fractionation can occur as a result of hydrodynamic escape (Hunten 1973; Hunten et al. 1987). Numerous studies have tried to explain the isotopic profile of noble gases through hydrodynamic escape (Sekiya et al. 1980; Hunten et al. 1987; Sasaki and Nakazawa 1988; Zahnle et al. 1990; Pepin 1991; Gillmann et al. 2009; Odert et al. 2018). Simulations have identified scenarios that could reproduce fractionation consistent with current observations. However, the scarcity of data regarding the structure, temperature, and composition of the atmosphere, the energy deposition, the solar wind, and the solar nebula dissipation time implies that the problem is under-constrained, with multiple solutions explaining the available data. In addition, large error bars on the measurements allow for a wide range of acceptable evolutionary scenarios. Finally, the building blocks of the primitive atmosphere themselves may have already been subject to outgassing (Lichtenberg et al. 2019) and escape (Odert et al. 2018) before accretion and thus already fractionated.

The rotation rate of the star is one of the main factors governing EUV emissions (Johnstone et al. 2021). The EUV flux decreases with time as rotation rates decrease (Güdel et al. 1997). During the first 200 Myr, a fast rotating Sun could emit about 500 times higher EUV fluxes than at present-day. A slowly rotating Sun could reach around 30 times present-day EUV fluxes. Most G-stars EUV fluxes converge to a common evolutionary path after 1.5 Gyr. Whether the Sun was a slow, moderate or fast rotator 4+ billion years ago is uncertain. While some clues such as the overall escape intensity or lunar regolith Na and K depletion suggest a slow to moderate rotation rate, the question remains unsolved (Odert et al. 2018; Saxena et al. 2019; Johnstone et al. 2021). Since atmospheric thermal escape depends on the EUV flux heating the upper atmosphere of a planet, uncertainties on EUV fluxes translate into uncertainties in loss rates, with fast stellar rotators leading to more efficient loss.

Loss of water from hydrodynamic escape can be massive but is strongly parameter dependent. The energy available for escape and photodissociation is defined by the EUV flux and therefore, cumulative losses would vary by more than an order of magnitude whether the Sun was a slow or fast rotator. For example, over the first 500 Myr evolution, one Earth ocean equivalent could escape an Earth-like planet orbiting a slow-rotating Sun-like star, compared to 45 Earth oceans for a fast-rotating star (Johnstone 2020). Estimations for Venus are similar, with a total loss ranging from a few (Gillmann et al. 2009) to 15–20 Earth oceans equivalent (Odert et al. 2018). CO_2 can also be lost this way, on the order of 10 bar, for moderate stellar rotators, and up to 100 s of bars for fast rotating stars (Odert et al. 2018), in which case CO_2 is probably photodissociated too. Finally, oxygen escapes together with hydrogen. In the case of fast-rotating stars, the loss of O is efficient and can be so high that most of the initial constituents of water could be lost to space (requiring the loss of tens of Earth oceans; Odert et al. 2018). However, for lower EUV fluxes oxygen loss is less efficient than H escape, which means that oxygen would accumulate in the atmosphere (Gillmann et al. 2009; Wordsworth and Pierrehumbert 2013; Luger and Barnes 2015; Johnstone 2020). The overall escape for planets orbiting slow stellar rotators is weaker but mass fractionation between H and O is more efficient. Considering both aspects, the final remaining oxygen after a hydrodynamic escape phase is of similar order regardless of the rotation rate: around 300–400 bar (Johnstone 2020).

Oxygen accumulation is an interesting problem on Venus, as free oxygen is practically absent from the present-day atmosphere (Johnson and de Oliveira 2019). During the magma ocean phase and the possible extended runaway greenhouse phase, it could have been efficiently trapped and re-injected into the mantle through interaction with the hot surface (Hamano et al. 2013; Wordsworth 2016; Wordsworth et al. 2018), thus oxidizing the interior. If it became the main species in the upper atmosphere, it could also have escaped (Tian 2015). However, those sinks stop with the solidification of the magma ocean and the rise of a volcanically-produced, CO_2-dominated atmosphere (Gillmann et al. 2020).

Later in the evolution, enhanced magmatic production causing extensive resurfacing combined with crustal oxidation could also provide a substantial oxygen sink (Way and Del Genio 2020; Krissansen-Totton et al. 2021), but other later sinks such as non-thermal escape and solid surface oxidation are likely inefficient. It is therefore likely that most of any build-up of oxygen caused by hydrodynamic escape would have been removed by the end of the magma ocean phase, by the time the molten surface completely solidified (Gillmann et al. 2020) or shortly after that (Warren and Kite 2023). While several sink mechanisms can be effective in depleting atmospheric oxygen throughout Venus' early and long-term evolution to match the present-day atmospheric composition, removing water from the atmosphere appears to be more difficult than removing oxygen (Krissansen-Totton et al. 2021). This may confine the end of the habitable period to 1–3 Gyr ago or limit the amount of surface liquid water during that period.

In summary, hydrodynamic escape acts as a potential massive sink of volatiles that could desiccate a planet's atmosphere; but the magnitude of early escape and its consequences for planetary evolution are not governed solely by that sink. Instead they depend on the availability and state of water in the system, which ultimately rely on the interactions with the solidifying magma ocean.

4.2.2 Non-thermal Escape

Along with further possible episodes of hydrodynamic escape in a H-rich atmosphere, even after magma ocean solidification, long-term, non-thermal escape will lead to the loss of water over the last 4 billion years. Over this timescale spanning the entire planet lifetime, it is unlikely that the existence of water oceans limits volatile and water loss, given the relatively fast exchanges between the ocean and the atmosphere. However, the structure of the atmosphere can have large consequences on loss mechanisms. For example, without a functioning cold trap an atmosphere risks a wet stratosphere and is thus prone to efficient water photodissociation and H escape. Kasting et al. (1993b) demonstrated that once an Earth-like planet enters a moist greenhouse state where the stratospheric water vapor mixing ratio is at least 3×10^{-3} v/v one may lose an entire Earth's ocean worth of H over $\sim$4.5 Gyr via photodissociation of H_2O and hydrogen diffusion-limited escape.

The efficiency of the escape is governed by the atmospheric composition, itself inherited from atmosphere–interior interaction during the magma ocean phase. As an example, a thick CO_2 atmosphere outgassed from the magma ocean can reduce the exosphere temperature, and thus lead to slower escape (e.g., Way et al. 2023). This could have buffered the escape of N_2 in the CO_2-rich atmosphere of Venus and has been suggested to explain the larger amount of N_2 in Venus' atmosphere compared to Earth (Lammer et al. 2018).

Interaction between surface minerals and the atmosphere can further affect atmospheric composition both during the magma ocean phase and afterward (see Gillmann et al. 2022, for more details). Fresh surface basalt can trap oxygen by oxidizing iron (e.g., Gillmann et al. 2009, 2020). It has also been suggested that N_2 could have been trapped by the magma ocean before being released later on by volcanism (Wordsworth et al. 2018). The specific effects of this process depends on the composition and structure of the atmosphere, as determined by its primordial evolution.

4.2.3 Impact Erosion

On growing planets, impact erosion is another possible – yet hard to estimate – atmospheric loss process (e.g., Cameron 1983; Melosh and Vickery 1989; Griffith and Zahnle 1995). In this case, atmospheric escape results from the kinetic energy imparted to the atmosphere by the impact of large objects. When large asteroids or comets collide onto protoplanets with atmospheres, hot vapor plumes and high-speed ejecta resulting from the impact may also provide sufficient energy for part of the atmosphere to escape and be lost to space (e.g., Melosh and Vickery 1989; Zahnle 1990; Ahrens 1993). For giant impacts atmospheric escape can also be caused by subsequent global ground motion (e.g., Genda and Abe 2003, 2005; Schlichting et al. 2015).

Quantitative estimates for impact-induced atmospheric escape have been proposed by Vickery and Melosh (1990), Newman et al. (1999), Shuvalov and Artemieva (2002). Further realistic numerical simulations have been developed by Svetsov (2007) and Shuvalov (2009), who considered the cratering flow induced by impacts on Earth and Mars. Escaping efficiency generally depend on properties such as impactor sizes, impact velocities and angles of impact. These rely on planetary accretion scenarios. The size of impactors typically increases with time as planetesimals grow and would favor impact-induced atmospheric erosion. Yet the frequency of impact decreases with time, which buffers net escaping efficiency.

Atmospheric properties, and in particular density, also affect impact-induced erosion efficiency. For a given impact low density tenuous atmospheres are prone to efficient impact-induced erosion. However, when impactors enter dense atmospheres, such as the ones typically expected for type II planets, aerial burst and projectile fragmentation upon entry may enhance atmospheric erosion (Shuvalov et al. 2014).

Overall, impact-induced atmospheric erosion efficiency and consequences on atmospheric evolution result from a subtle balance between erosion and atmospheric replenishment through impact-induced degassing. It strongly depends on the timing and timescale of giant impacts relative to that of magma ocean solidification and outgassing. Any conclusive estimation assessing its importance is challenged by the large uncertainties regarding how multiple magma ocean events may be embedded into a bombardment history.

While EUV-driven hydrodynamic escape on early Venus would have (weakly; Kasting and Pollack 1983; Johnstone 2020) contributed to the D/H fractionation, impact erosion does not cause isotopic fractionation because the atmosphere escapes *en masse* in this mechanism. Compared to other flux-triggered escape mechanisms, impact erosion is not sensitive to the presence of a magnetic field. Erosion processes could have been important to set absolute atmospheric abundances (Sakuraba et al. 2019; Gillmann et al. 2020). However, in the presence of condensed reservoirs like liquid water, impact erosion during the late accretion could have affected the atmospheric elemental composition by favoring the escape of inert gases from the atmosphere (Sakuraba et al. 2019). Thus the effect of impact erosion on isotopic composition and elemental abundance is tightly linked to both the accretional history and thermal evolution, and therefore remains hard to assess.

4.3 Initial Conditions for the Long-Term Evolution of the Interior

4.3.1 Water, Mantle Properties and Evolution

Magma ocean outgassing efficiency and oxidation state are responsible for the water inventory left inside the planet. Atmospheric radiative balance and resulting circulation, escape processes, atmosphere–surface interactions, and volcanic outgassing govern the surface water inventory once the surface has solidified (see Gillmann et al. 2022, for an extensive discussion). The different magma ocean evolution pathways and outcomes discussed above are associated with a broad diversity of scenarios regarding water distribution between the planetary interior, the atmosphere, and the amount lost to space (Fig. 14). Whether Venus' mantle is completely water-depleted at the ERCS and left dry for its long-term evolution, partially depleted, or water-rich remains unclear. Yet, the amount of mantellic water inherited from magma ocean evolution and its distribution would strongly influence the long-term evolution and surface conditions of the planet, by affecting both the atmosphere and the mantle.

For instance, water in the mantle lowers the melting temperature (e.g., Hirschmann 2006; Ni et al. 2016), thus possibly favoring mantle melting and triggering volcanism and volcanic outgassing. If a substantial amount of water is supplied to the atmosphere through post-ERCS volcanic outgassing while the planet is habitable at ERCS, it may trigger a runaway greenhouse. This could lead to complete vaporization of the ocean followed by water photodissociation and hydrogen loss in the upper atmosphere, thus sterilizing the planet and making it uninhabitable. If the magma ocean outgassing efficiency of water were low (e.g., Hier-Majumder and Hirschmann 2017; Miyazaki and Korenaga 2021; Salvador and Samuel 2023), there may not have been enough water to form an ocean at the ERCS. Yet post-ERCS volcanic outgassing may supply enough water for clouds to form in the upper atmosphere, increase the albedo, and possibly create transient temperate conditions suitable for water ocean formation (from type III to type I branch in Fig. 14). If Venus were a faster rotator early on, the decrease of its planetary rotation rate could have helped in the formation of reflective clouds. On the other hand, a decrease in water outgassing rate associated with a steady loss of atmospheric water would slow down atmospheric replenishment and could result in thinner and dissipating clouds (Bullock and Grinspoon 2001). Cloud dissipation would induce a decrease in the Bond albedo and an increase in the absorbed solar flux and surface temperature. This could trigger a runaway greenhouse and vaporize any existing water ocean, making Venus exit a temperate climate (from type I to type II branch in Fig. 14).

Water also decreases mantle viscosity (e.g., Lange 1994) which in turn controls the rheology and convective dynamics of the interior (e.g., Karato 2015). Seismic wave propagation and mantle electrical conductivity are also affected by the mantellic water content (see Karato 2015, for a review). The spatial distribution of mantellic water and potential vertical and horizontal variations could then induce local changes of mantle properties and possibly lead to very different rheological behaviour and tectonic regimes (e.g., Korenaga 2011). The low viscosity layer under the stiff lithosphere, i.e., the asthenosphere, is thought to result from the presence of localized volatiles in Earth's mantle. This could be sustained by long-term volatile recycling (Richards and Lenardic 2018) and may promote plate tectonics on Earth (Bunge et al. 1996; Richards et al. 2001). Its absence on modern Venus (inferred from gravity/topography; e.g., Kiefer et al. 1986) together with the dry conditions at the surface have been historically invoked to explain the lack of modern Earth-style plate tectonics (Bercovici et al. 2000).

Indeed, the state of water at the planetary surface has crucial implications for the mantle convective regime. The presence or absence of liquid water at the planetary surface affects the rheological behavior of the lithosphere, lithosphere–mantle lubrication, upper mantle hydration, and plate bending (Grevemeyer et al. 2018). These may all influence subduction initiation, sustainability, and the prevailing tectonic regime (Regenauer-Lieb et al. 2001; Gerya et al. 2008; Bercovici and Ricard 2014; Korenaga et al. 2017; Stern 2018; Westall et al. 2023). In turn, those differences that are inextricably inherited from the magma ocean evolution affect magmatic and tectonic processes throughout the entire thermochemical evolution of the planet (e.g., Smrekar et al. 2007; Stamenković and Breuer 2014). This has implications for the mantle convection regime (e.g., Weller and Kiefer 2020), thermal history, volatile redistribution (e.g., Korenaga 2017) and outgassing (e.g., Driscoll and Bercovici 2013), that must all ultimately match present-day conditions (see Rolf et al. 2022; Gillmann et al. 2022, for reviews).

4.3.2 Late Impacts

After the end of the hypothetical last global magma ocean phase caused by the final giant impact (or Moon-forming impact in the case of Earth), a number of subsequent and additional large impacts may have occurred (Rubie et al. 2015), during the so-called "late accretion". The highly siderophile elements in Earth's mantle could be explained by such collisions (Chyba 1991). If they occurred, they also certainly affected water distribution (Sakuraba et al. 2019, 2021). As discussed previously, large impacts could have opposite effects on the planet's atmosphere depending on the balance between atmospheric erosion and replenishment (e.g., Pham et al. 2011). Large collisions could also cause widespread melting of the mantle of the target body and deplete its shallow layers through subsequent degassing of the molten parts (see Gillmann et al. 2022).

In cases where the impactor is not fully vaporized upon impact, the collision could also inject water-bearing material from the impactor into the impacted mantle. The nature of the material, as well as the portion of the total water inventory delivered to Venus during this late accretion era are still debated. It has been suggested that significant amounts of water could be delivered to a planet's surface and/or mantle through these late impacts (Albarede 2009). On the other hand, recent studies and Earth-based isotopic data imply that most of a terrestrial planet's water inventory is already accreted by the end of its main accretion phase (e.g., Gillmann et al. 2020), and that late accretion is mostly composed of volatile-poor material (e.g., Fischer-Gödde and Kleine 2017; Dauphas 2017). Depending on the composition of the impactors and on how they merge with the target (e.g., Itcovitz et al.

2022; Citron and Stewart 2022), the oxidation state of both the atmosphere and mantle could be significantly affected, with implications for prebiotic chemistry (Sect. 3.2.5 and Table 2). Late collisions could also have affected the mantellic convection style of the planet, since impacts could favor lithospheric damage and mobility, and thus trigger subduction (Gillmann et al. 2016; O'Neill et al. 2017; Borgeat and Tackley 2022).

4.3.3 Interior–Atmosphere Volatile Exchanges: The Carbonate–Silicate Cycle

After the magma ocean has fully solidified, solid-state convection takes over in mixing the entire mantle. It is most likely already in place in the solid-state part of the mantle while the upper molten mantle continues to cool down (e.g., Maurice et al. 2017; Ballmer et al. 2017). It produces partial melting, and ultimately volcanism and outgassing, thereby bringing water and other volatiles to the surface while locally depleting the mantle. The convection regime of the mantle thus determines the specifics of the volatile exchanges (see Rolf et al. 2022; Gillmann et al. 2022) and is therefore thought to play a major role in how young planets could branch in or out of a habitable state (Fig. 14). Long-term climate stability is required to maintain habitability throughout eons and is believed to rely on specific interior–atmosphere volatile cycling such as the critical carbon cycle and associated CO_2-climate feedback (e.g., Walker et al. 1981; Kasting et al. 1993b; Kasting and Catling 2003; Ramirez 2018).

On Earth, to compensate for the increase in surface temperature due to the brightening Sun, long-term habitability may have required quick removal of most of the early atmospheric CO_2 outgassed during the magma ocean phase (or delivered to the atmosphere by other means), thus reducing the greenhouse effect. Interaction with the surface and interior of the planet (and convection regime) would be involved in the process. The mechanism is not yet understood, given the significant differences between the conditions at that time and at present-day. Notably, the higher temperatures of a thicker atmosphere or the lack of modern-day plate tectonics. The latter plays a major role in the present-day carbon cycle and volatile return flux from the atmosphere into the Earth's interior.

The carbon cycle regulates the atmospheric concentration of CO_2 and has several different parts. The most important one in terms of long-term climate ($t > 10^6$ years) is the inorganic carbonate–silicate cycle. It results from slow interactions of atmospheric CO_2 with the crustal rock reservoir (e.g., Holland 1978; Walker et al. 1981; Kasting et al. 1993b; Kasting and Catling 2003). Dissolution of atmospheric CO_2 into liquid water drives the cycle. CO_2 is first dissolved into rainwater to form carbonic acid that dissolves continental silicate rocks over time. The products of such silicate weathering, including calcium (Ca^{2+}), magnesium (Mg^{2+}), and bicarbonate (HCO_3^-) ions, flow within streams and rivers into the ocean. There they react with more ocean-dissolved atmospheric CO_2 to precipitate as carbonate minerals. Those carbonates deposit and accumulate on the seafloor where they are buried in sediments. Their subduction, with that of continental carbonates and carbonatized oceanic basalt, injects carbon into the mantle (e.g., Sleep and Zahnle 2001). At depth, high temperatures and pressures force carbonate minerals to recombine with SiO_2 to reform silicate minerals while releasing CO_2. Following this so-called carbonate metamorphism, the released CO_2 is eventually volcanically outgassed and thus re-enters the atmosphere, therefore closing the cycle.

Because evaporation (and hence, precipitation) and chemical reaction rates rise with surface temperature, silicate weathering rates increase with increasing surface temperatures. This consumes more atmospheric CO_2, and buffers the rise of surface temperature. Conversely, atmospheric CO_2 consumption by silicate weathering decreases when surface temperatures fall, while volcanically-outgassed atmospheric CO_2 concentrations increase and counteract falling temperatures. This carbonate–silicate cycle thus contains a negative feedback regulating surface temperatures and atmospheric CO_2 concentrations. It is responsible for long-term climate stability (e.g., Walker et al. 1981; Sleep and Zahnle 2001; Kasting 2019).

If ancient Venus had liquid water, emerged lands (to allow silicate weathering), and an initially CO_2-dominated atmosphere, the carbonate–silicate cycle may have been in place, therefore stabilizing a long-term temperate climate (Way and Del Genio 2020). Via this cycle, Venus' atmosphere could have evolved toward a more Earth-like atmospheric composition (i.e., CO_2-poor) and maintained a stable temperate climate against the brightening Sun. In addition, temporarily sequestering CO_2 in carbonates through an effective carbonate–silicate cycle permits elevated H_2O loss before CO_2 is volcanically outgassed and returned to the atmosphere, which is key to recover the modern Venusian atmospheric composition from habitable scenarios (Krissansen-Totton et al. 2021).

Yet, the existence and efficiency of an early carbonate–silicate cycle on Venus strongly relies on its mantle dynamic regime and associated convection/tectonic style. Subduction from an active/mobile-lid regime (possibly plate tectonics) would allow for the complete cycle to operate. Other non-classical tectonic signatures of convective regimes, such as a post-magma ocean, non-plate tectonic, sluggish subduction may be alternative ways of injecting CO_2 into the mantle and removing it from the atmosphere (Foley et al. 2014). On the other hand, a stagnant lid regime, possibly operating today on Venus (e.g., Smrekar et al. 2007), limits the magnitude of any recycling of atmosphere/surface species into the mantle. However, even in this regime, progressive burial of material by successive lava flows could ultimately allow temporary storage of volatiles, or even a limited return flux of material into the mantle, provided the water-bearing minerals are not destabilized by the increasing pressure and temperature conditions before delamination (Höning et al. 2021).

Thus, the present-day stagnant lid regime likely prevents significant incorporation of CO_2 into the mantle and could even be responsible for its accumulation into the atmosphere through long-term volcanism (e.g., Way and Del Genio 2020; Krissansen-Totton et al. 2021; Gillmann et al. 2022). However, it is clear that terrestrial planets, including Earth itself, experience different tectonic styles through the eons (e.g., Stern 2018; Spencer et al. 2021), indicating that other mantle convective regimes certainly prevailed in Venus' past. This could allow for a transient effective carbonate–silicate cycle under a habitable climate. Yet, for both planets, the various convective regimes that have been in place, their relative timing and transition time scales and the magnitude of associated volatile exchanges remain unclear (e.g., Stern 2018).

5 Perspectives and Future Research Directions

Many parameters and processes with direct or indirect implications for the magma ocean phase and evolution remain highly unconstrained. The influence of some of them has not even been assessed yet. They concern both the molten interior and the outgassed atmosphere themselves, but also include external and indirect factors such as the planet's early environment and history.

Accretion Sequence and Stellar Environment The magma ocean phase is not an isolated process: it is inextricably embedded into the accretion sequence and related to the evolution of the stellar environment. To provide a more accurate understanding of the early evolution of rocky planets and implications for their habitability and long-term evolution, magma ocean models should be coupled to stellar evolution and planetary formation models.

In particular, planetary formation models would provide self-consistent constraints regarding volatile delivery timing, type, and budget (e.g., Morbidelli et al. 2012; Li et al. 2016; Burger et al. 2020; Venturini et al. 2020). They would also provide the frequency and energetics of impacts, and avoid making implicit assumptions regarding number occurrence of magma oceans, timing, initial depth, thermal state (e.g., Nakajima et al. 2021), and evolving chemical constituents abundance and distribution (e.g., Maas et al. 2021; Itcovitz et al. 2022; Citron and Stewart 2022). In particular, the effect of late giant impacts, that may affect volatile species and melt redistribution, atmospheric and mantle oxidation states, atmospheric erosion or replenishment (e.g., Pham et al. 2011; Modirrousta-Galian and Korenaga 2023), would be worth systematically assessing.

Stellar evolution models would allow for a more realistic treatment of upper atmospheric properties, including photochemistry and escape (e.g., Airapetian et al. 2020). This may in turn affect the heat flux balance and resulting climate stability, dynamics, and circulation, as well as atmospheric volatile abundance and oxidation state.

Molten Interior Regarding the mechanisms at play within the molten interior, many fluid dynamics aspects of terrestrial magma oceans emphasized in Solomatov (2000) remain poorly understood. They include the mode of crystallization (fractional vs equilibrium) resulting from the competition between crystals settling and re-entrainment, the convective regime or the effect of rotation (Maas and Hansen 2015, 2019). Additional complexities may arise from the convection in a multi-phase system (Boukaré and Ricard 2017), or spatial and time variations of melt and crystal properties such as density, viscosity, crystals size and shape, or chemical affinity. In that respect, magma ocean numerical models could significantly benefit from advances in characterization of material properties and behaviors at magma ocean extreme conditions performed by laboratory experiments.

The oxidation state of the molten mantle is also highly unconstrained. It is influenced by the timing, location, extent, and fashion of metal–silicate separation and how it may evolves in space and time. Yet it is highly influential regarding the partitioning behavior of chemical elements and the identity of outgassed volatile species.

Some possibly critical physical aspects of magma ocean outgassing have been poorly studied. For example, the processes of bubble formation, nucleation, and ascent; and the crystal–melt–gas partitioning and chemical affinity of many volatile species at magma ocean relevant conditions remain uncertain. Ingassing processes are generally not considered in magma ocean studies. Yet, how, when, which, and how much volatile species could be re-incorporated into the melt and then into the mantle may alter many aspects of the early and long-term evolution.

The thermo-chemical conditions at the base of the molten mantle, the interaction with the underlying metallic core (Foley and Driscoll 2016), and the potential existence of a basal magma ocean (Labrosse et al. 2007; O'Rourke 2020) have also received little attention in the framework of magma ocean–atmosphere evolution modeling. Yet, all of these processes may significantly affect the convective dynamics, outgassing, heat transfer, dynamo generation, and chemical differentiation of the solidifying mantle, and their relative influence remains to be quantified.

Long-Term Evolution of the Interior Addressing the subsequent long-term evolution of the planet requires considering the outcomes of the early molten evolution as potential initial states. Indeed, considering the surface conditions, the heat budget, and the distribution of volatiles resulting from the magma ocean evolution would help constrain the solid-state mantle convective regime, and the sustainability of possible habitable climates. Furthermore, the transition from the magma ocean to the solidified mantle may significantly affect the post-magma ocean mantellic convective regime and the long-term evolution of the planet (Maurice et al. 2017; Ballmer et al. 2017; Agrusta et al. 2019; Bolrão et al. 2021). In particular the transition to a global solid-state convecting mantle can be affected by the coexistence of convective layers of different rheological behavior, i.e., one melt-dominated and another crystal-dominated, and how they interact. Coupling the highly influential magma ocean evolution with subsequent solid-state mantle convection models would thus help reconcile the modelled evolution paths with the present-day observed constraints, allowing one to assess the likelihood of different evolutionary pathways (e.g., Gillmann et al. 2022).

Atmospheric State and Properties Planets are complex systems and their evolution results from the strong interaction between the different planetary reservoirs. Understanding their evolving surface conditions and determining their potential past habitability demand to account for both atmospheric and interior processes.

As discussed earlier, the evolving Bond albedo of Venus and resulting absorbed sunlight is one of the most important parameters to solve its potential habitability. Yet, the contribution of clouds remains poorly constrained and assessing the effect of realistic clouds remains a key challenge for atmospheric modeling. In particular, microphysical and optical properties of aerosols, including photochemical hazes and others cloud condensation nuclei, as well as their time-dependent 3D spatial distribution are poorly constrained, and even more uncertain under extreme magma ocean conditions.

The occurrence of silicate vapor atmospheres after giant impacts is expected but their behavior and how they affect the thermo-chemical state of the planet is unknown. The resulting cloud behaviour, heat redistribution, climate circulation and escape post-impacts, changes in atmospheric composition, oxidation state throughout the magma ocean cooling, as well as the climatic transition from an interior- to a sunlit-heated atmosphere are all poorly understood. Further coupling between sophisticated interior outgassing and thermal evolution models with 3D climate models are required to provide a self-consistent scenario. Yet the development of such coupled models is hobbled by our understanding of these processes and by current computational capabilities. For now, a feasible and promising avenue would be to use available parameterized models to conduct sensitivity studies and assess the relative influence of each of these processes.

Observational Constraints and Future Missions In addition to these modeling shortcomings, our understanding of Venus history and present state is challenged by the lack of in-situ measurements as well as by important uncertainties in existing observations. Along with the aforementioned future research directions, forthcoming planned Venus missions are expected to significantly improve our comprehension of Venus' past evolution and present state. How these missions are expected to address key science questions regarding Venus evolution is discussed in Widemann et al. (2023, this issue).

Crucial measurements include elemental abundances and isotopic composition of noble gases (He, Ne, Ar, Kr, Xe), and stable isotopes (C, H, O, N, S) in the atmosphere. These are well suited to track outgassing throughout the coupled interior–atmosphere evolution (see Table 1 in Baines et al. 2013). For a thorough discussion see Avice et al. (2022, this collection). As an example, hydrogen isotopic measurements may provide valuable insight into the primordial atmospheric composition and associated early mantle oxidation state (Pahlevan et al. 2019).

Regarding planetary formation, tungsten isotopic composition can be used to constrain the timing of core formation and planetary growth rate (e.g., Zahnle et al. 2007). In particular, because W is moderately siderophile and Hf is lithophile, Hf/W ratios are strongly fractionated by core formation. The ^{182}Hf–^{182}W chronometer can then be used to assess planetary formation time-scales that ultimately relate to the energetic of accretion and thus have direct implications regarding the timing of the magma ocean phase, its initial depth and spatial extent. Constraints on core formation timing can be used to infer the depth and timing of metal–silicate separation. They would also infer the resulting molten mantle oxidation state and hence, that of the outgassed volatile species.

As discussed throughout the chapter, water (and its distribution) is a key element for early planetary evolution, but it is also highly influential over the entire planet's lifetime. A present-day signature of the evolution path taken could be the amount of remaining water in the mantle. The water content of the mantle affects its properties, the resulting mantle dynamics, rheology, tectonic style, melting behavior, and outgassing. It also affects the surface composition and mineralogy. Any constraint on these parameters could be in-turn used to inform the amount of mantellic water associated with the interior state, thus providing indirect hints of the past and of the evolutionary pathway. Surface mineralogy measurements could similarly indicate what conditions prevailed by the time ancient rocks or mineral inclusions have been put in place.

Finally, together with future missions dedicated to Venus (e.g., Widemann et al. 2023, this collection), the expanding field of rocky exoplanets detection and characterization is key to improve our general understanding of planetary formation, evolution, and habitability, that will in turn inform Venus' evolutionary history and help us unveil the potential habitable past of our closest planetary neighbor (see Way et al. 2023, this collection).

6 Conclusion

Given the strong coupling between all planetary reservoirs, the magma ocean stage has critical implications for terrestrial planet evolution and habitability. Importantly, it determines the initial chemical and thermal state that will dictate the pathways taken by the planet during its entire history. In addition to being a required ingredient for the emergence and development of life, water abundance and distribution between the planetary layers influences many of the processes at play. Its partitioning is the result of magma ocean crystallization and is of primary importance. The present-day dichotomy between Earth and Venus surface conditions and habitable state might be directly inherited from the cooling and solidification of the coupled magma ocean–atmosphere system. If not, the latter remains the determining factor by setting the initial conditions for the long-term evolution of the planet.

Many uncertainties, unconstrained processes and associated parameters still exist and need further investigation in the coming years to build a consistent picture of the earliest stages of planetary evolution. Combined with interdisciplinary modeling efforts, future measurements of Earth's twin planet may provide significant clues to the early mechanisms at play. They could also shed light not only on Venus' and Earth's distant pasts, but also improve our understanding of what makes a planet habitable and how to search for life elsewhere in the Universe.

Acknowledgements The authors thank the anonymous reviewer for their comments who helped us to greatly improve the manuscript. The authors also thank Tilman Spohn and team members of the International Space Science Institute in Bern for hosting the "Venus: Evolution through Time" workshop. AS also warmly thanks them for inviting him as a visiting scientist. AS gratefully acknowledges support from NASA's Habitable

Worlds Program (No. 80NSSC20K0226). MJW acknowledges support from the GSFC Sellers Exoplanet Environments Collaboration (SEEC) and ROCKE-3D: The evolution of solar system worlds through time, funded by the NASA Planetary and Earth Science Divisions Internal Scientist Funding Model. Work by C.G. was carried out within the framework of the NCCR PlanetS.

Funding Open Access funding enabled and organized by Projekt DEAL.

Declarations

Competing Interests The authors have no conflict of interest to declare that are relevant to the content of this article.

References

Abe Y (1993a) Physical state of the very early Earth. Lithos 30(3–4):223–235. https://doi.org/10.1016/0024-4937(93)90037-D

Abe Y (1993b) Thermal evolution and chemical differentiation of the terrestrial magma ocean. In: Takahashi E, Jeanloz R, Rubie D (eds) Evolution of the Earth and planets. AGU, Washington DC, pp 41–54. https://doi.org/10.1029/GM074p0041

Abe Y (1997) Thermal and chemical evolution of the terrestrial magma ocean. Phys Earth Planet Inter 100(1–4):27–39. https://doi.org/10.1016/S0031-9201(96)03229-3.

Abe Y, Matsui T (1985) The formation of an impact-generated H_2O atmosphere and its implications for the early thermal history of the Earth. J Geophys Res, Solid Earth 90(S02):545–559. https://doi.org/10.1029/JB090iS02p0C545

Abe Y, Matsui T (1988) Evolution of an impact-generated H_2O–CO_2 atmosphere and formation of a hot proto-ocean on Earth. J Atmos Sci 45(21):3081–3101. https://doi.org/10.1175/1520-0469(1988)045<3081:EOAIGH>2.0.CO;2

Abe Y, Ohtani E, Okuchi T, Righter K, Drake M (2000) Water in the early Earth. In: Canup R, Righter K et al (eds) Origin of the Earth and Moon. University of Arizona Press, Tucson, pp 413–433

Abe Y, Abe-Ouchi A, Sleep NH, Zahnle KJ (2011) Habitable zone limits for dry planets. Astrobiology 11(5):443–460. https://doi.org/10.1089/ast.2010.0545

Abelson PH (1966) Chemical events on the primitive Earth. Proc Natl Acad Sci 55(6):1365–1372. https://doi.org/10.1073/pnas.55.6.1365

Agrusta R, Morison A, Labrosse S, Deguen R, Alboussiére T, Tackley PJ, Dubuffet F (2019) Mantle convection interacting with magma oceans. Geophys J Int 220(3):1878–1892. https://doi.org/10.1093/gji/ggz549

Ahrens TJ (1993) Impact erosion of terrestrial planetary atmospheres. Annu Rev Earth Planet Sci 21(1):525–555

Airapetian VS, Glocer A, Gronoff G, Hébrard E, Danchi W (2016) Prebiotic chemistry and atmospheric warming of early Earth by an active young Sun. Nat Geosci 9(6):452–455. https://doi.org/10.1038/ngeo2719

Airapetian VS, Barnes R, Cohen O, Collinson GA, Danchi WC, Dong CF, Del Genio AD, France K, Garcia-Sage K, Glocer A, Gopalswamy N, Grenfell JL, Gronoff G, Güdel M, Herbst K, Henning WG, Jackman CH, Jin M, Johnstone CP, Kaltenegger L, Kay CD, Kobayashi K, Kuang W, Li G, Lynch BJ, Lüftinger T, Luhmann JG, Maehara H, Mlynczak MG, Notsu Y, Osten RA, Ramirez RM, Rugheimer S, Scheucher M, Schlieder JE, Shibata K, Sousa-Silva C, Stamenković V, Strangeway RJ, Usmanov AV, Vergados P, Verkhoglyadova OP, Vidotto AA, Voytek M, Way MJ, Zank GP, Yamashiki Y (2020) Impact of space weather on climate and habitability of terrestrial-type exoplanets. Int J Astrobiol 19(2):136–194. https://doi.org/10.1017/S1473550419000132

Albarede F (2009) Volatile accretion history of the terrestrial planets and dynamic implications. Nature 461(7268):1227–1233

Alemi A, Stevenson D (2006) Why Venus has no Moon. In: AAS/division for planetary sciences meeting abstracts #38, AAS/division for planetary sciences meeting abstracts. p 07.03

Altwegg K, Balsiger H, Bar-Nun A, Berthelier JJ, Bieler A, Bochsler P, Briois C, Calmonte U, Combi M, De Keyser J, Eberhardt P, Fiethe B, Fuselier S, Gasc S, Gombosi TI, Hansen KC, Hässig M, Jackel A, Kopp E, Korth A, LeRoy L, Mall U, Marty B, Mousis O, Neefs E, Owen T, Reme H, Rubin M, Semon T, Tzou CY, Waite H, Wurz P (2015) 67P/Churyumov-Gerasimenko, a Jupiter family comet with a high D/H ratio. Science 347(6220):3. https://doi.org/10.1126/science.1261952

Amelin Y, Lee DC, Halliday AN, Pidgeon RT (1999) Nature of the Earth's earliest crust from hafnium isotopes in single detrital zircons. Nature 399:252 EP. https://doi.org/10.1038/20426

Anders E (1989) Pre-biotic organic matter from comets and asteroids. Nature 342(6247):255–257. https://doi.org/10.1038/342255a0

Andrault D (2019) Thermodynamical constraints on the crystallization of a deep magma-ocean on Earth. C R Géosci. https://doi.org/10.1016/j.crte.2018.06.003

Andrault D, Bolfan-Casanova N, Nigro GL, Bouhifd MA, Garbarino G, Mezouar M (2011) Solidus and liquidus profiles of chondritic mantle: implication for melting of the Earth across its history. Earth Planet Sci Lett 304(1–2):251–259. https://doi.org/10.1016/j.epsl.2011.02.006

Andrault D, Petitgirard S, Lo Nigro G, Devidal JL, Veronesi G, Garbarino G, Mezouar M (2012) Solid-liquid iron partitioning in Earth's deep mantle. Nature 487(7407):354–357. https://doi.org/10.1038/nature11294

Andrault D, Monteux J, Bars ML, Samuel H (2016) The deep Earth may not be cooling down. Earth Planet Sci Lett 443:195–203. https://doi.org/10.1016/j.epsl.2016.03.020

Andrault D, Bolfan-Casanova N, Bouhifd MA, Boujibar A, Garbarino G, Manthilake G, Mezouar M, Monteux J, Parisiades P, Pesce G (2017) Toward a coherent model for the melting behavior of the deep Earth's mantle. Phys Earth Planet Inter 265:67–81. https://doi.org/10.1016/j.pepi.2017.02.009

Andrews SM, Huang J, Pérez LM, Isella A, Dullemond CP, Kurtovic NT, Guzmán VV, Carpenter JM, Wilner DJ, Zhang S, Zhu Z, Birnstiel T, Bai XN, Benisty M, Hughes AM, Öberg KI, Ricci L (2018) The disk substructures at high angular resolution project (DSHARP). I. Motivation, sample, calibration, and overview. Astrophys J Lett 869(2):L41. https://doi.org/10.3847/2041-8213/aaf741

Armitage PJ, Eisner JA, Simon JB (2016) Prompt planetesimal formation beyond the snow line. Astrophys J Lett 828(1):L2. https://doi.org/10.3847/2041-8205/828/1/L2

Armstrong LS, Hirschmann MM, Stanley BD, Falksen EG, Jacobsen SD (2015) Speciation and solubility of reduced C–O–H–N volatiles in mafic melt: Implications for volcanism, atmospheric evolution, and deep volatile cycles in the terrestrial planets. Geochim Cosmochim Acta 171:283–302. https://doi.org/10.1016/j.gca.2015.07.007

Armstrong K, Frost DJ, McCammon CA, Rubie DC, Ballaran TB (2019) Deep magma ocean formation set the oxidation state of Earth's mantle. Science 365(6456):903–906. https://doi.org/10.1126/science.aax8376

Asimow PD (2018) Chap. 15 - Melts under extreme conditions from shock experiments. In: Kono Y, Sanloup C (eds) Magmas under pressure, vol 15. Elsevier, Amsterdam, pp 387–418. https://doi.org/10.1016/B978-0-12-811301-1.00015-0

Asimow PD, Langmuir CH (2003) The importance of water to oceanic mantle melting regimes. Nature 421(6925):815–820. https://doi.org/10.1038/nature01429

Asphaug E (2014) Impact origin of the moon? Annu Rev Earth Planet Sci 42(1):551–578. https://doi.org/10.1146/annurev-earth-050212-124057

Asphaug E, Agnor CB, Williams Q (2006) Hit-and-run planetary collisions. Nature 439(7073):155–160. https://doi.org/10.1038/nature04311

Avice G, Marty B (2020) Perspectives on atmospheric evolution from noble gas and nitrogen isotopes on Earth, Mars & Venus. Space Sci Rev 216(3):36. https://doi.org/10.1007/s11214-020-00655-0

Avice G, Marty B, Burgess R (2017) The origin and degassing history of the Earth's atmosphere revealed by Archean xenon. Nat Commun 8:15455. https://doi.org/10.1038/ncomms15455

Avice G, Parai R, Jacobson S, Labidi J, Trainer MG, Petkov MP (2022) Noble gases and stable isotopes track the origin and early evolution of the Venus atmosphere. Space Sci Rev 218(8):60. https://doi.org/10.1007/s11214-022-00929-9

Badro J, Brodholt JP, Piet H, Siebert J, Ryerson FJ (2015) Core formation and core composition from coupled geochemical and geophysical constraints. Proc Natl Acad Sci 112(40):12310–12314. https://doi.org/10.1073/pnas.1505672112

Bahcall JN, Pinsonneault MH, Basu S (2001) Solar models: current epoch and time dependences, neutrinos, and helioseismological properties. Astrophys J 555(2):990. https://doi.org/10.1086/321493

Baines KH, Atreya SK, Bullock MA, Grinspoon DH, Mahaffy P, Russell CT, Schubert G, Zahnle K (2013) The atmospheres of the terrestrial planets: clues to the origins and early evolution of Venus, Earth, and Mars. In: Mackwell SJ, Simon-Miller AA, Harder JW, Bullock MA (eds) Comparative climatology of terrestrial planets. University of Arizona Press, Tucson, pp 137–160. https://doi.org/10.2458/azu_uapress_9780816530595-ch006

Baker DR, Alletti M (2012) Fluid saturation and volatile partitioning between melts and hydrous fluids in crustal magmatic systems: the contribution of experimental measurements and solubility models. Earth-Sci Rev 114:298–324. https://doi.org/10.1016/j.earscirev.2012.06.005

Ballmer MD, Lourenço DL, Hirose K, Caracas R, Nomura R (2017) Reconciling magma-ocean crystallization models with the present-day structure of the Earth's mantle. Geochem Geophys Geosyst 18(7):2785–2806. https://doi.org/10.1002/2017GC006917

Banzhaf W (2003) Self-organizing systems. In: Meyers RA (ed) Encyclopedia of physical science and technology, 3rd edn. Academic Press, New York, pp 589–598. https://doi.org/10.1016/B0-12-227410-5/00681-5

Barth P, Carone L, Barnes R, Noack L, Mollière P, Henning T (2021) Magma ocean evolution of the TRAPPIST-1 planets. Astrobiology 21(11):1325–1349. https://doi.org/10.1089/ast.2020.2277

Bekaert DV, Broadley MW, Marty B (2020) The origin and fate of volatile elements on Earth revisited in light of noble gas data obtained from comet 67P/Churyumov-Gerasimenko. Sci Rep 10(1):5796. https://doi.org/10.1038/s41598-020-62650-3

Benlow A, Meadows AJ (1977) The formation of the atmospheres of the terrestrial planets by impact. Astrophys Space Sci 46(2):293–300. https://doi.org/10.1007/BF00644376

Benner SA, Bell EA, Biondi E, Brasser R, Carell T, Kim HJ, Mojzsis SJ, Omran A, Pasek MA, Trail D (2020) When did life likely emerge on Earth in an RNA-first process? ChemSystemsChem 2(2):e1900035. https://doi.org/10.1002/syst.201900035

Bennett VC, Nutman AP, McCulloch MT (1993) Nd isotopic evidence for transient, highly depleted mantle reservoirs in the early history of the Earth. Earth Planet Sci Lett 119(3):299–317. https://doi.org/10.1016/0012-821X(93)90140-5

Benz W, Slattery WL, Cameron AGW (1986) The origin of the moon and the single-impact hypothesis I. Icarus 66(3):515–535. https://doi.org/10.1016/0019-1035(86)90088-6

Bercovici D, Ricard Y (2014) Plate tectonics, damage and inheritance. Nature 508:513 EP. https://doi.org/10.1038/nature13072

Bercovici D, Ricard Y, Richards MA (2000) The relation between mantle dynamics and plate tectonics: a primer. In: Richards MA, Gordon RG, Van Der Hilst RD (eds) The history and dynamics of global plate motions. AGU, Washington DC, pp 5–46. https://doi.org/10.1029/GM121p0005

Berg MTL, Bromiley GD, Le Godec Y, Philippe J, Mezouar M, Perrillat JP, Potts NJ (2018) Rapid Core Formation in Terrestrial Planets by Percolative Flow: in-Situ Imaging of Metallic Melt Migration Under High Pressure/Temperature Conditions. Front Earth Sci 6(77). https://doi.org/10.3389/feart.2018.00077

Berlo K, Gardner JE, Blundy JD (2011) Timescales of magma degassing. In: Timescales of magmatic processes: from core to atmosphere, vol 11. Wiley, New York. https://doi.org/10.1002/9781444328509.ch11

Bhatia GK (2021) Early thermal evolution of the embryos of Earth: role of ^{26}Al and impact-generated steam atmosphere. Planet Space Sci 207:105335. https://doi.org/10.1016/j.pss.2021.105335

Bills BG (2005) Variations in the rotation rate of Venus due to orbital eccentricity modulation of solar tidal torques. J Geophys Res, Planets 110(E11). https://doi.org/10.1029/2003JE002190

Binder AB (1986) The initial thermal state of the Moon. In: Hartmann W, Phillips R, Taylor G (eds) Origin of the Moon. Lunar and Planetary Institute, p 425. https://ui.adsabs.harvard.edu/abs/1986ormo.conf..425B/abstract

Birch F (1952) Elasticity and constitution of the Earth's interior. J Geophys Res 57(2):227–286

Birnstiel T, Klahr H, Ercolano B (2012) A simple model for the evolution of the dust population in protoplanetary disks. Astron Astrophys 539:A148. https://doi.org/10.1051/0004-6361/201118136

Bitsch B, Johansen A, Lambrechts M, Morbidelli A (2015) The structure of protoplanetary discs around evolving young stars. Astron Astrophys 575:A28. https://doi.org/10.1051/0004-6361/201424964

Bizzarro M, Baker JA, Haack H, Ulfbeck D, Rosing M (2003) Early history of Earth's crust–mantle system inferred from hafnium isotopes in chondrites. Nature 421(6926):931–933. https://doi.org/10.1038/nature01421

Blum J, Wurm G (2008) The growth mechanisms of macroscopic bodies in protoplanetary disks. Annu Rev Astron Astrophys 46(1):21–56. https://doi.org/10.1146/annurev.astro.46.060407.145152

Bodenheimer P, Lin DNC (2002) Implications of extrasolar planets for understanding planet formation. Annu Rev Earth Planet Sci 30(1):113–148. https://doi.org/10.1146/annurev.earth.30.091201.140357

Bolfan-Casanova N, Keppler H, Rubie DC (2003) Water partitioning at 660 km depth and evidence for very low water solubility in magnesium silicate perovskite. Geophys Res Lett 30(17). https://doi.org/10.1029/2003GL017182

268

Bollard J, Connelly JN, Whitehouse MJ, Pringle EA, Bonal L, Jørgensen JK, Nordlund Å, Moynier F, Bizzarro M (2017) Early formation of planetary building blocks inferred from Pb isotopic ages of chondrules. Sci Adv 3(8):e1700407. https://doi.org/10.1126/sciadv.1700407

Bolrão DP, Ballmer MD, Morison A, Rozel AB, Sanan P, Labrosse S, Tackley PJ (2021) Timescales of chemical equilibrium between the convecting solid mantle and over- and underlying magma oceans. Solid Earth 12(2):421–437. https://doi.org/10.5194/se-12-421-2021

Borgeat X, Tackley PJ (2022) Hadean/Eoarchean tectonics and mantle mixing induced by impacts: a three-dimensional study. Prog Earth Planet Sci 9(1):38. https://doi.org/10.1186/s40645-022-00497-0

Boslough M, Weldon R, Ahrens T (1980) Impact-induced water loss from serpentine, nontronite and kernite. In: Lunar and planetary science conference proceedings, vol 3, pp 2145–2158. https://ui.adsabs.harvard.edu/abs/1980LPSC...11.2145B/abstract

Boss AP (1986) The origin of the moon. Science 231(4736):341–345. https://doi.org/10.1126/science.231.4736.341

Boukaré CE, Ricard Y (2017) Modeling phase separation and phase change for magma ocean solidification dynamics. Geochem Geophys Geosyst. https://doi.org/10.1002/2017GC006902

Boukaré CE, Ricard Y, Fiquet G (2015) Thermodynamics of the $MgO-FeO-SiO_2$ system up to 140 GPa: application to the crystallization of Earth's magma ocean. J Geophys Res, Solid Earth 120(9):6085–6101. https://doi.org/10.1002/2015JB011929

Boulliung J, Füri E, Dalou C, Tissandier L, Zimmermann L, Marrocchi Y (2020) Oxygen fugacity and melt composition controls on nitrogen solubility in silicate melts. Geochim Cosmochim Acta 284:120–133. https://doi.org/10.1016/j.gca.2020.06.020

Bouvier A, Blichert-Toft J, Moynier F, Vervoort JD, Albarède F (2007) Pb–Pb dating constraints on the accretion and cooling history of chondrites. Geochim Cosmochim Acta 71(6):1583–1604. https://doi.org/10.1016/j.gca.2006.12.005

Bower DJ, Kitzmann D, Wolf AS, Sanan P, Dorn C, Oza AV (2019) Linking the evolution of terrestrial interiors and an early outgassed atmosphere to astrophysical observations. Astron Astrophys 631:A103. https://doi.org/10.1051/0004-6361/201935710

Bower DJ, Hakim K, Sossi PA, Sanan P (2022) Retention of water in terrestrial magma oceans and carbon-rich early atmospheres. Planet Sci. J. 3(4):93. https://doi.org/10.3847/psj/ac5fb1

Boyet M, Carlson RW (2005) [142]Nd evidence for early (>4.53 Ga) global differentiation of the silicate Earth. Science 309(5734):576–581. https://doi.org/10.1126/science.1113634

Boyet M, Blichert-Toft J, Rosing M, Storey M, Télouk P, Albarède F (2003) [142]Nd evidence for early Earth differentiation. Earth Planet Sci Lett 214(3):427–442. https://doi.org/10.1016/S0012-821X(03)00423-0

Brasser R, Mojzsis SJ (2020) The partitioning of the inner and outer Solar System by a structured protoplanetary disk. Nat Astron 4(5):492–499. https://doi.org/10.1038/s41550-019-0978-6

Brasser R, Matsumura S, Ida S, Mojzsis SJ, Werner SC (2016) Analysis of terrestrial planet formation by the grand tack model: system architecture and tack location. Astrophys J 821(2):75. https://doi.org/10.3847/0004-637X/821/2/75

Briceño C, Vivas AK, Calvet N, Hartmann L, Pacheco R, Herrera D, Romero L, Berlind P, Sánchez G, Snyder JA, Andrews P (2001) The CIDA-QUEST large-scale survey of orion OB1: evidence for rapid disk dissipation in a dispersed stellar population. Science 291(5501):93–96. https://doi.org/10.1126/science.291.5501.93

ALMA Partnership, Brogan CL, Pérez LM, Hunter TR, Dent WRF, Hales AS, Hills RE, Corder S, Fomalont EB, Vlahakis C, Asaki Y, Barkats D, Hirota A, Hodge JA, Impellizzeri CMV, Kneissl R, Liuzzo E, Lucas R, Marcelino N, Matsushita S, Nakanishi K, Phillips N, Richards AMS, Toledo I, Aladro R, Broguiere D, Cortes JR, Cortes PC, Espada D, Galarza F, Appadoo DG, Ramirez LG, Humphreys EM, Jung T, Kameno S, Laing RA, Leon S, Marconi G, Mignano A, Nikolic B, Nyman LA, Radiszcz M, Remijan A, Rodón JA, Sawada T, Takahashi S, Tilanus RPJ, Vilaro BV, Watson LC, Wiklind T, Akiyama E, Chapillon E, de Gregorio-Monsalvo I, Francesco JD, Gueth F, Kawamura A, Lee CF, Luong QN, Mangum J, Pietu V, Sanhueza P, Saigo K, Takakuwa S, Ubach C, van Kempen T, Wootten A, Carrizo AC, Francke H, Gallardo J, Garcia J, Gonzalez S, Hill T, Kaminski T, Kurono Y, Liu HY, Lopez C, Morales F, Plarre K, Schieven G, Testi L, Videla L, Villard E, Andreani P, Hibbard JE, Tatematsu K (2015) The 2014 ALMA long baseline campaign: first results from high angular resolution observations toward the HL tau region. Astrophys J Lett 808(1):L3. https://doi.org/10.1088/2041-8205/808/1/L3

Brož M, Chrenko O, Nesvorný D, Dauphas N (2021) Early terrestrial planet formation by torque-driven convergent migration of planetary embryos. Nat Astron 5(9):898–902. https://doi.org/10.1038/s41550-021-01383-3

Budde G, Burkhardt C, Brennecka GA, Fischer-Gödde M, Kruijer TS, Kleine T (2016) Molybdenum isotopic evidence for the origin of chondrules and a distinct genetic heritage of carbonaceous and non-carbonaceous meteorites. Earth Planet Sci Lett 454:293–303. https://doi.org/10.1016/j.epsl.2016.09.020

Bullock MA, Grinspoon DH (2001) The recent evolution of climate on Venus. Icarus 150(1):19–37. https://doi.org/10.1006/icar.2000.6570

Bunge HP, Richards MA, Baumgardner JR (1996) Effect of depth-dependent viscosity on the planform of mantle convection. Nature 379(6564):436–438. https://doi.org/10.1038/379436a0

Burger C, Bazsó Á, Schäfer CM (2020) Realistic collisional water transport during terrestrial planet formation - self-consistent modeling by an N-body-SPH hybrid code. Astron Astrophys 634:A76. https://doi.org/10.1051/0004-6361/201936366

Burgisser A, Scaillet B (2007) Redox evolution of a degassing magma rising to the surface. Nature 445(7124):194–197. https://doi.org/10.1038/nature05509

Burkhardt C, Kleine T, Bourdon B, Palme H, Zipfel J, Friedrich JM, Ebel DS (2008) Hf–W mineral isochron for Ca, Al-rich inclusions: age of the solar system and the timing of core formation in planetesimals. Geochim Cosmochim Acta 72(24):6177–6197. https://doi.org/10.1016/j.gca.2008.10.023

Burkhardt C, Spitzer F, Morbidelli A, Budde G, Render JH, Kruijer TS, Kleine T (2021) Terrestrial planet formation from lost inner solar system material. Sci Adv 7(52):eabj7601. https://doi.org/10.1126/sciadv.abj7601

Cameron AG (1983) Origin of the atmospheres of the terrestrial planets. Icarus 56(2):195–201

Cameron A (1986) The impact theory for origin of the moon. In: Hartmann W, Phillips R, Taylor G (eds) Origin of the Moon, Lunar and Planetary Institute, p 609. https://ui.adsabs.harvard.edu/abs/1986ormo.conf..609C/abstract

Cameron AGW, Ward WR (1976) The origin of the moon. In: Lunar and planetary science conference, vol 7, pp 120–122. https://ui.adsabs.harvard.edu/abs/1976LPI.....7..120C/abstract

Campbell BA, Campbell DB, Carter LM, Chandler JF, Giorgini JD, Margot JL, Morgan GA, Nolan MC, Perillat PJ, Whitten JL (2019) The mean rotation rate of Venus from 29 years of Earth-based radar observations. Icarus 332:19–23. https://doi.org/10.1016/j.icarus.2019.06.019

Canil D (1997) Vanadium partitioning and the oxidation state of Archaean komatiite magmas. Nature 389(6653):842–845. https://doi.org/10.1038/39860

Canil D (2002) Vanadium in peridotites, mantle redox and tectonic environments: Archean to present. Earth Planet Sci Lett 195(1):75–90. https://doi.org/10.1016/S0012-821X(01)00582-9

Canup RM (2004) Simulations of a late lunar-forming impact. Icarus 168(2):433–456. https://doi.org/10.1016/j.icarus.2003.09.028

Canup RM (2008) Accretion of the Earth. Philos Trans R Soc, Math Phys Eng Sci 366(1883):4061–4075. https://doi.org/10.1098/rsta.2008.0101

Canup RM (2012) Forming a moon with an Earth-like composition via a giant impact. Science 338(6110):1052–1055. https://doi.org/10.1126/science.1226073

Canup RM, Asphaug E (2001) Origin of the moon in a giant impact near the end of the Earth's formation. Nature 412:708 EP. https://doi.org/10.1038/35089010

Caracas R, Hirose K, Nomura R, Ballmer MD (2019) Melt–crystal density crossover in a deep magma ocean. Earth Planet Sci Lett 516:202–211. https://doi.org/10.1016/J.EPSL.2019.03.031

Caro G (2011) Early silicate Earth differentiation. Annu Rev Earth Planet Sci 39(1):31–58. https://doi.org/10.1146/annurev-earth-040610-133400

Caro G, Bourdon B, Birck JL, Moorbath S (2003) ^{146}Sm–^{142}Nd evidence from Isua metamorphosed sediments for early differentiation of the Earth's mantle. Nature 423(6938):428–432. https://doi.org/10.1038/nature01668

Carpenter R (1970) A radar determination of the rotation of Venus. Astron J 75:61. https://doi.org/10.1086/110941

Carroll MR, Holloway JR (eds) (1994) Volatiles in magmas, vol 30. Mineralogical Society of America, Washington

Castaing B, Gunaratne G, Heslot F, Kadanoff L, Libchaber A, Thomae S, Wu XZ, Zaleski S, Zanetti G (1989) Scaling of hard thermal turbulence in Rayleigh-Bénard convection. J Fluid Mech 204:1–30. https://doi.org/10.1017/S0022112089001643

Cataldi G, Brandeker A, Olofsson G, Larsson B, Liseau R, Blommaert J, Fridlund M, Ivison R, Pantin E, Sibthorpe B, Vandenbussche B, Wu Y (2014) Herschel/HIFI observations of ionised carbon in the Pictoris debris disk. Astron Astrophys 563:A66. https://doi.org/10.1051/0004-6361/201323126

Catling DC (2006) Comment on "A hydrogen-rich early Earth atmosphere". Science 311(5757):38. https://doi.org/10.1126/science.1117827

Catling DC (2014) 6.7 - The great oxidation event transition. In: Holland HD, Turekian KK (eds) Treatise on geochemistry, 2nd edn. Elsevier, Oxford, pp 177–195. https://doi.org/10.1016/B978-0-08-095975-7.01307-3

Catling DC (2015) 10.13 - Planetary atmospheres. In: Schubert G (ed) Treatise on geophysics, 2nd edn. Elsevier, Oxford, pp 429–472. https://doi.org/10.1016/B978-0-444-53802-4.00185-8

Catling DC, Claire MW (2005) How Earth's atmosphere evolved to an oxic state: a status report. Earth Planet Sci Lett 237(1):1–20. https://doi.org/10.1016/j.epsl.2005.06.013

Catling DC, Zahnle KJ (2020) The Archean atmosphere. Sci Adv 6(9). https://doi.org/10.1126/sciadv.aax1420

Chambers JE (2001) Making more terrestrial planets. Icarus 152(2):205–224. https://doi.org/10.1006/icar.2001.6639

Chambers JE (2010) Terrestrial planet formation. In: Seager S (ed) Exoplanets. University of Arizona Press, Tucson, pp 297–317

Chambers JE (2016) Pebble accretion and the diversity of planetary systems. Astrophys J 825(1):63. https://doi.org/10.3847/0004-637X/825/1/63

Chassefière E (1996) Hydrodynamic escape of oxygen from primitive atmospheres: applications to the cases of Venus and Mars. Icarus 124(2):537–552. https://doi.org/10.1006/icar.1996.0229

Chaussidon M, Gounelle M (2007) Short-lived radioactive nuclides in meteorites and early solar system processes. C R Géosci 339(14):872–884. https://doi.org/10.1016/j.crte.2007.09.005

Chillà F, Schumacher J (2012) New perspectives in turbulent Rayleigh-Bénard convection. Eur Phys J E 35(7):58. https://doi.org/10.1140/epje/i2012-12058-1

Chyba CF (1991) Terrestrial mantle siderophiles and the lunar impact record. Icarus 92(2):217–233

Chyba C, Sagan C (1992) Endogenous production, exogenous delivery and impact-shock synthesis of organic molecules: an inventory for the origins of life. Nature 355(6356):125–132. https://doi.org/10.1038/355125a0

Chyba CF, Thomas PJ, Brookshaw L, Sagan C (1990) Cometary delivery of organic molecules to the early Earth. Science 249(4967):366–373. https://doi.org/10.1126/science.11538074

Citron RI, Stewart ST (2022) Large impacts onto the early Earth: planetary sterilization and iron delivery. Planet Sci. J. 3(5):116. https://doi.org/10.3847/psj/ac66e8

Cleaves HJ, Chalmers JH, Lazcano A, Miller SL, Bada JL (2008) A reassessment of prebiotic organic synthesis in neutral planetary atmospheres. Orig Life Evol Biosph 38(2):105–115. https://doi.org/10.1007/s11084-007-9120-3

Clement MS, Kaib NA, Raymond SN, Walsh KJ (2018) Mars' growth stunted by an early giant planet instability. Icarus 311:340–356. https://doi.org/10.1016/j.icarus.2018.04.008

Clement MS, Kaib NA, Raymond SN, Chambers JE, Walsh KJ (2019) The early instability scenario: terrestrial planet formation during the giant planet instability, and the effect of collisional fragmentation. Icarus 321(7):78–790. https://doi.org/10.1016/j.icarus.2018.12.033

Clement MS, Raymond SN, Kaib NA (2019b) Excitation and depletion of the asteroid belt in the early instability scenario. Astron J 157(1):38. https://doi.org/10.3847/1538-3881/aaf21e

Clement MS, Kaib NA, Raymond SN, Chambers JE (2021) The early instability scenario: Mars' mass explained by Jupiter's orbit. Icarus 367:114585. https://doi.org/10.1016/j.icarus.2021.114585

Clesi V, Bouhifd MA, Bolfan-Casanova N, Manthilake G, Schiavi F, Raepsaet C, Bureau H, Khodja H, Andrault D (2018) Low hydrogen contents in the cores of terrestrial planets. Sci Adv 4(3). https://doi.org/10.1126/sciadv.1701876

Clesi V, Monteux J, Qaddah B, Le Bars M, Wacheul JB, Bouhifd MA (2020) Dynamics of core-mantle separation: influence of viscosity contrast and metal/silicate partition coefficients on the chemical equilibrium. Phys Earth Planet Inter 306:106547. https://doi.org/10.1016/j.pepi.2020.106547

Correia ACM, Laskar J (2001) The four final rotation states of Venus. Nature 411:767 EP. https://doi.org/10.1038/35081000

Correia ACM, Laskar J (2003) Different tidal torques on a planet with a dense atmosphere and consequences to the spin dynamics. J Geophys Res, Planets 108(E11). https://doi.org/10.1029/2003JE002059

Cottereau L, Rambaux N, Lebonnois S, Souchay J (2011) The various contributions in Venus rotation rate and LOD. Astron Astrophys 531:A45. https://doi.org/10.1051/0004-6361/201116606

Ćuk M, Stewart ST (2012) Making the moon from a fast-spinning Earth: a giant impact followed by resonant despinning. Science 338(6110):1047–1052. https://doi.org/10.1126/science.1225542

Dasgupta R, Grewal DS (2019) Origin and early differentiation of carbon and associated life-essential volatile elements on Earth. In: Deep carbon. Cambridge University Press, Cambridge, pp 4–39

Dauphas N (2017) The isotopic nature of the Earth's accreting material through time. Nature 541(7638):521–524

Dauphas N, Morbidelli A (2014) 6.1 - Geochemical and planetary dynamical views on the origin of Earth's atmosphere and oceans. In: Holland HD, Turekian KK (eds) Treatise on geochemistry, 2nd edn. Elsevier, Oxford, pp 1–35. https://doi.org/10.1016/B978-0-08-095975-7.01301-2

Deguen R, Landeau M, Olson P (2014) Turbulent metal–silicate mixing, fragmentation, and equilibration in magma oceans. Earth Planet Sci Lett 391:274–287. https://doi.org/10.1016/j.epsl.2014.02.007

Dehant V, Debaille V, Dobos V, Gaillard F, Gillmann C, Goderis S, Grenfell JL, Höning D, Javaux EJ, Karatekin Ö, Morbidelli A, Noack L, Rauer H, Scherf M, Spohn T, Tackley P, Van Hoolst T, Wünnemann

K (2019) Geoscience for understanding habitability in the solar system and beyond. Space Sci Rev. https://doi.org/10.1007/s11214-019-0608-8

Delano JW (2001) Redox history of the Earth's interior since ~3900 Ma: implications for prebiotic molecules. Orig Life Evol Biosph 31(4):311–341. https://doi.org/10.1023/A:1011895600380

DeMeo FE Carry B (2014) Solar system evolution from compositional mapping of the asteroid belt. Nature 505(7485):629–634. https://doi.org/10.1038/nature12908

Demouchy S, Bolfan-Casanova N (2016) Distribution and transport of hydrogen in the lithospheric mantle: a review. Lithos 240–243:402–425. https://doi.org/10.1016/j.lithos.2015.11.012

Deng J, Du Z, Karki BB, Ghosh DB, Lee KKM (2020) A magma ocean origin to divergent redox evolutions of rocky planetary bodies and early atmospheres. Nat Commun 11(1):2007. https://doi.org/10.1038/s41467-020-15757-0

Dent WRF, Wyatt MC, Roberge A, Augereau JC, Casassus S, Corder S, Greaves JS, de Gregorio-Monsalvo I, Hales A, Jackson AP, Hughes AM, Lagrange AM, Matthews B, Wilner D (2014) Molecular gas clumps from the destruction of icy bodies in the β pictoris debris disk. Science 343(6178):1490–1492. https://doi.org/10.1126/science.1248726

Dobrovolskis AR (1980) Atmospheric tides and the rotation of Venus II. Spin evolution. Icarus 41(1):18–35. https://doi.org/10.1016/0019-1035(80)90157-8

Dobrovolskis AR, Ingersoll AP (1980) Atmospheric tides and the rotation of Venus I. Tidal theory and the balance of torques. Icarus 41(1):1–17. https://doi.org/10.1016/0019-1035(80)90156-6

Donahue TM (1999) New analysis of hydrogen and deuterium escape from Venus. Icarus 141(2):226–235. https://doi.org/10.1006/icar.1999.6186

Donahue TM, Hoffman JH, Hodges RR, Watson AJ (1982) Venus was wet: a measurement of the ratio of deuterium to hydrogen. Science 216(4546):630–633. https://doi.org/10.1126/science.216.4546.630

Drake MJ (2000) Accretion and primary differentiation of the Earth: a personal journey. Geochim Cosmochim Acta 64(14):2363–2369. https://doi.org/10.1016/S0016-7037(00)00372-0

Drake MJ, Newsom HE, Capobianco CJ (1989) V, Cr, and Mn in the Earth, Moon, EPB, and SPB and the origin of the Moon: experimental studies. Geochim Cosmochim Acta 53(8):2101–2111. https://doi.org/10.1016/0016-7037(89)90328-1

Drążkowska J, Alibert Y (2017) Planetesimal formation starts at the snow line. Astron Astrophys 608:A92. https://doi.org/10.1051/0004-6361/201731491

Drążkowska J, Dullemond CP (2018) Planetesimal formation during protoplanetary disk buildup. Astron Astrophys 614:A62. https://doi.org/10.1051/0004-6361/201732221

Drążkowska J, Alibert Y, Moore B (2016) Close-in planetesimal formation by pile-up of drifting pebbles. Astron Astrophys 594:A105. https://doi.org/10.1051/0004-6361/201628983

Dreibus G, Wänke H (1987) Volatiles on Earth and Mars: a comparison. Icarus 71(2):225–240. https://doi.org/10.1016/0019-1035(87)90148-5

Driscoll P, Bercovici D (2013) Divergent evolution of Earth and Venus: influence of degassing, tectonics, and magnetic fields. Icarus 226(2):1447–1464. https://doi.org/10.1016/j.icarus.2013.07.025

Edmonds M, Woods AW (2018) Exsolved volatiles in magma reservoirs. J Volcanol Geotherm Res 368:13–30. https://doi.org/10.1016/j.jvolgeores.2018.10.018

Elkins-Tanton LT (2008) Linked magma ocean solidification and atmospheric growth for Earth and Mars. Earth Planet Sci Lett 271(1–4):181–191. https://doi.org/10.1016/j.epsl.2008.03.062

Elkins-Tanton LT (2011) Formation of early water oceans on rocky planets. Astrophys Space Sci 332(2):359–364. https://doi.org/10.1007/s10509-010-0535-3

Elkins-Tanton LT (2012) Magma oceans in the inner solar system. Annu Rev Earth Planet Sci 40:113–139. https://doi.org/10.1146/annurev-earth-042711-105503

Elkins-Tanton LT, Grove TL (2011) Water (hydrogen) in the lunar mantle: results from petrology and magma ocean modeling. Earth Planet Sci Lett 307(1–2):173–179. https://doi.org/10.1016/j.epsl.2011.04.027

Elkins-Tanton LT, Parmentier EM, Hess PC (2003) Magma ocean fractional crystallization and cumulate overturn in terrestrial planets: implications for Mars. Meteorit Planet Sci 38(12):1753–1771. https://doi.org/10.1111/j.1945-5100.2003.tb00013.x

Elkins-Tanton LT, Weiss BP, Zuber MT (2011) Chondrites as samples of differentiated planetesimals. Earth Planet Sci Lett 305(1):1–10. https://doi.org/10.1016/j.epsl.2011.03.010

Emsenhuber A, Asphaug E, Cambioni S, Gabriel TSJ, Schwartz SR (2021) Collision chains among the terrestrial planets. II. An asymmetry between Earth and Venus. Planet Sci. J. 2(5):199. https://doi.org/10.3847/psj/ac19b1

Erkaev NV, Lammer H, Elkins-Tanton LT, Stökl A, Odert P, Marcq E, Dorfi EA, Kislyakova KG, Kulikov Y, Leitzinger M, Güdel M (2014) Escape of the Martian protoatmosphere and initial water inventory. Planet Space Sci 98:106–119. https://doi.org/10.1016/j.pss.2013.09.008

Faul UH, Jackson I (2007) Diffusion creep of dry, melt-free olivine. J Geophys Res, Solid Earth 112(B4). https://doi.org/10.1029/2006JB004586

  Springer

Fegley B, Zolotov MY, Lodders K (1997) The oxidation state of the lower atmosphere and surface of Venus. Icarus 125(2):416–439. https://doi.org/10.1006/icar.1996.5628

Feig ST, Koepke J, Snow JE (2010) Effect of oxygen fugacity and water on phase equilibria of a hydrous tholeiitic basalt. Contrib Mineral Petrol 160(4):551–568. https://doi.org/10.1007/s00410-010-0493-3

Fiquet G (2018) Chap. 4 - Melting in the Earth's deep interior. In: Kono Y, Sanloup C (eds) Magmas under pressure. Elsevier, Amsterdam, pp 115–134. https://doi.org/10.1016/B978-0-12-811301-1.00004-6

Fiquet G, Auzende AL, Siebert J, Corgne A, Bureau H, Ozawa H, Garbarino G (2010) Melting of peridotite to 140 gigapascals. Science 329(5998):1516–1518. https://doi.org/10.1126/science.1192448

Fischer RA, Nakajima Y, Campbell AJ, Frost DJ, Harries D, Langenhorst F, Miyajima N, Pollok K, Rubie DC (2015) High pressure metal–silicate partitioning of Ni, Co, V, Cr, Si, and O. Geochim Cosmochim Acta 167:177–194. https://doi.org/10.1016/j.gca.2015.06.026

Fischer-Gödde M, Kleine T (2017) Ruthenium isotopic evidence for an inner solar system origin of the late veneer. Nature 541(7638):525–527

Fish RA, Goles GG, Anders E (1960) The record in the meteorites. III. On the development of meteorites in asteroidal bodies. Astrophys J 132:243. https://doi.org/10.1086/146918

Flasar FM, Birch F (1973) Energetics of core formation: a correction. J Geophys Res 78(26):6101–6103. https://doi.org/10.1029/JB078i026p06101

Foley BJ, Driscoll PE (2016) Whole planet coupling between climate, mantle, and core: implications for rocky planet evolution. Geochem Geophys Geosyst 17(5):1885–1914. https://doi.org/10.1002/2015GC006210

Foley BJ, Bercovici D, Elkins-Tanton LT (2014) Initiation of plate tectonics from post-magma ocean thermochemical convection. J Geophys Res, Solid Earth 119(11):8538–8561. https://doi.org/10.1002/2014JB011121

French BM (1966) Some geological implications of equilibrium between graphite and a C-H-O gas phase at high temperatures and pressures. Rev Geophys 4(2):223–253. https://doi.org/10.1029/RG004i002p00223

Frost BR (1979) Mineral equilibria involving mixed-volatiles in a C-O-H fluid phase; the stabilities of graphite and siderite. Am J Sci 279(9):1033–1059. https://doi.org/10.2475/ajs.279.9.1033

Frost BR (1991) Chap. 1. Introduction to oxygen fugacity and its petrologic importance. Rev Mineral Geochem 25(1):1–9. https://doi.org/10.1515/9781501508684-004

Frost DJ, McCammon CA (2008) The redox state of Earth's mantle. Annu Rev Earth Planet Sci 36(1):389–420. https://doi.org/10.1146/annurev.earth.36.031207.124322

Frost DJ, Liebske C, Langenhorst F, McCammon CA, Trønnes RG, Rubie DC (2004) Experimental evidence for the existence of iron-rich metal in the Earth's lower mantle. Nature 428(6981):409–412. https://doi.org/10.1038/nature02413

Frost DJ, Mann U, Asahara Y, Rubie DC (2008) The redox state of the mantle during and just after core formation. Philos Trans R Soc, Math Phys Eng Sci 366(1883):4315–4337. http://www.jstor.org/stable/25197401

Fu RR, Young ED, Greenwood RC, Elkins-Tanton LT (2017) Silicate melting and volatile loss during differentiation in planetesimals. In: Planetesimals: early differentiation and consequences for planets. Cambridge University Press, Cambridge. https://doi.org/10.1017/9781316339794.006

Gabriel TSJ, Cambioni S (2023) The role of giant impacts in planet formation. Annu Rev Earth Planet Sci 51(1):671–695. https://doi.org/10.1146/annurev-earth-031621-055545

Gaillard F, Bouhifd MA, Füri E, Malavergne V, Marrocchi Y, Noack L, Ortenzi G, Roskosz M, Vulpius S (2021) The diverse planetary ingassing/outgassing paths produced over billions of years of magmatic activity. Space Sci Rev 217(1):22. https://doi.org/10.1007/s11214-021-00802-1

Gaillard F, Bernadou F, Roskosz M, Bouhifd MA, Marrocchi Y, Iacono-Marziano G, Moreira M, Scaillet B, Rogerie G (2022a) Redox controls during magma ocean degassing. Earth Planet Sci Lett 577:117255. https://doi.org/10.1016/j.epsl.2021.117255

Gaillard F, Malavergne V, Bouhifd MA, Rogerie G (2022b) A speciation model linking the fate of carbon and hydrogen during core – magma ocean equilibration. Earth Planet Sci Lett 577:117266. https://doi.org/10.1016/j.epsl.2021.117266

Galimov EM (2005) Redox evolution of the Earth caused by a multi-stage formation of its core. Earth Planet Sci Lett 233(3):263–276. https://doi.org/10.1016/j.epsl.2005.01.026

Gallet F, Charbonnel C, Amard L, Brun S, Palacios A, Mathis S (2017) Impacts of stellar evolution and dynamics on the habitable zone: the role of rotation and magnetic activity. Astron Astrophys 597:A14. https://doi.org/10.1051/0004-6361/201629034

Garvin J, Getty S, Arney G, Johnson N, Malespin C, Webster C, Ravine M, Lorenz R, Kiefer W, Atreya S et al (2020) Deep atmosphere of Venus investigation of noble gases, chemistry, and imaging plus (DAVINCI+): discovering a new Venus via a flyby, probe, orbiter mission. In: AGU fall meeting abstracts, vol 2020, p P026–0001

Garvin JB, Getty SA, Arney GN, Johnson NM, Kohler E, Schwer KO, Sekerak M, Bartels A, Saylor RS, Elliott VE, Goodloe CS, Garrison MB, Cottini V, Izenberg N, Lorenz R, Malespin CA, Ravine M, Webster CR, Atkinson DH, Aslam S, Atreya S, Bos BJ, Brinckerhoff WB, Campbell B, Crisp D, Filiberto JR, Forget F, Gilmore M, Gorius N, Grinspoon D, Hofmann AE, Kane SR, Kiefer W, Lebonnois S, Mahaffy PR, Pavlov A, Trainer M, Zahnle KJ, Zolotov M (2022) Revealing the mysteries of Venus: the DAVINCI mission. Planet Sci. J. 3(5):117. https://doi.org/10.3847/PSJ/ac63c2

Geiss J, Gloeckler G (1998) Abundances of deuterium and helium-3 in the protosolar cloud. Space Sci Rev 84(1):239–250. https://doi.org/10.1023/A:1005039822524

Genda H (2016) Origin of Earth's oceans: an assessment of the total amount, history and supply of water. Geochem J 50(1):27–42. https://doi.org/10.2343/geochemj.2.0398

Genda H, Abe Y (2003) Survival of a proto-atmosphere through the stage of giant impacts: the mechanical aspects. Icarus 164(1):149–162. https://doi.org/10.1016/S0019-1035(03)00101-5

Genda H, Abe Y (2005) Enhanced atmospheric loss on protoplanets at the giant impact phase in the presence of oceans. Nature 433(7028):842–844

Gerya TV, Connolly JAD, Yuen DA (2008) Why is terrestrial subduction one-sided? Geology 36(1):43–46. https://doi.org/10.1130/G24060A.1

Ghanbarzadeh S, Hesse MA, Prodanović M (2017) Percolative core formation in planetesimals enabled by hysteresis in metal connectivity. Proc Natl Acad Sci 114(51):13406–13411. https://doi.org/10.1073/pnas.1707580114

Gillmann C, Chassefière E, Lognonné P (2009) A consistent picture of early hydrodynamic escape of Venus atmosphere explaining present Ne and Ar isotopic ratios and low oxygen atmospheric content. Earth Planet Sci Lett 286(3–4):503–513. https://doi.org/10.1016/j.epsl.2009.07.016

Gillmann C, Golabek GJ, Tackley PJ (2016) Effect of a single large impact on the coupled atmosphere-interior evolution of Venus. Icarus 268:295–312. https://doi.org/10.1016/j.icarus.2015.12.024

Gillmann C, Golabek G, Raymond S, Schönbächler M, Tackley P, Dehant V, Vinciane D (2020) Dry late accretion inferred from Venus's coupled atmosphere and internal evolution. Nat Geosci 13:1–5. https://doi.org/10.1038/s41561-020-0561-x

Gillmann C, Way MJ, Avice G, Breuer D, Golabek GJ, Höning D, Krissansen-Totton J, Lammer H, O'Rourke JG, Persson M, Plesa AC, Salvador A, Scherf M, Zolotov MY (2022) The long-term evolution of the atmosphere of Venus: processes and feedback mechanisms. Space Sci Rev 218(7):56. https://doi.org/10.1007/s11214-022-00924-0

Godin PJ, Ramirez RM, Campbell CL, Wizenberg T, Nguyen TG, Strong K, Moores JE (2020) Collision-induced absorption of CH_4-CO_2 and H_2-CO_2 complexes and their effect on the ancient Martian atmosphere. J Geophys Res, Planets 125(12):e2019JE006357. https://doi.org/10.1016/10.1029/2019JE006357

Gold T, Soter S (1969) Atmospheric tides and the resonant rotation of Venus. Icarus 11(3):356–366. https://doi.org/10.1016/0019-1035(69)90068-2

Goldblatt C (2015) Habitability of waterworlds: runaway greenhouses, atmospheric expansion, and multiple climate states of pure water atmospheres. Astrobiology 15. https://doi.org/10.1089/ast.2014.1260

Goldblatt C, Watson AJ (2012) The runaway greenhouse: implications for future climate change, geoengineering and planetary atmospheres. Philos Trans R Soc Lond Ser A 370(1974):4197–4216. https://doi.org/10.1098/rsta.2012.0004. arXiv:1201.1593

Goldblatt C, Robinson TD, Zahnle KJ, Crisp D (2013) Low simulated radiation limit for runaway greenhouse climates. Nat Geosci 6(8):661–667. https://doi.org/10.1038/ngeo1892

Goldreich P, Peale S (1967) Spin-orbit coupling in the solar system. II. The resonant rotation of Venus. Astron J 72:662–666. https://doi.org/10.1086/110239

Goldreich P, Peale S (1970) The obliquity of Venus. Astron J 75:273. https://doi.org/10.1086/110975

Gomes R, Levison HF, Tsiganis K, Morbidelli A (2005) Origin of the cataclysmic late heavy bombardment period of the terrestrial planets. Nature 435:466–469. https://doi.org/10.1038/nature03676

Gonnermann HM (2015) Magma fragmentation. Annu Rev Earth Planet Sci 43(1):431–458. https://doi.org/10.1146/annurev-earth-060614-105206

Gough DO (1981) Solar interior structure and luminosity variations. Sol Phys 74(1):21–34. https://doi.org/10.1007/BF00151270

Gounelle M (2011) The asteroid–comet continuum: in search of lost primitivity. Elements 7(1):29–34. https://doi.org/10.2113/gselements.7.1.29

Greenberg R, Wacker JF, Hartmann WK, Chapman CR (1978) Planetesimals to planets: numerical simulation of collisional evolution. Icarus 35(1):1–26. https://doi.org/10.1016/0019-1035(78)90057-X

Grevemeyer I, Ranero CR, Ivandic M (2018) Structure of oceanic crust and serpentinization at subduction trenches. Geosphere 14(2):395–418. https://doi.org/10.1130/GES01537.1

Grewal DS, Dasgupta R, Farnell A (2020) The speciation of carbon, nitrogen, and water in magma oceans and its effect on volatile partitioning between major reservoirs of the Solar System rocky bodies. Geochim Cosmochim Acta 280:281–301. https://doi.org/10.1016/j.gca.2020.04.023

Griffith CA, Zahnle K (1995) Influx of cometary volatiles to planetary moons: the atmospheres of 1000 possible titans. J Geophys Res, Planets 100(E8):16907–16922

Grossmann S, Lohse D (2000) Scaling in thermal convection: a unifying theory. J Fluid Mech 407:27–56. https://doi.org/10.1017/S0022112099007545

Güdel M, Guinan EF, Skinner SL (1997) The X-ray sun in time: a study of the long-term evolution of coronae of solar-type stars. Astrophys J 483(2):947–960. https://doi.org/10.1086/304264

Haghighipour N, Boss AP (2003) On pressure gradients and rapid migration of solids in a nonuniform solar nebula. Astrophys J 583(2):996. https://doi.org/10.1086/345472

Halliday AN (2004) Mixing, volatile loss and compositional change during impact-driven accretion of the Earth. Nature 427(6974):505–509. https://doi.org/10.1038/nature02275

Halliday AN (2013) The origins of volatiles in the terrestrial planets. Geochim Cosmochim Acta 105:146–171. https://doi.org/10.1016/j.gca.2012.11.015

Hallis LJ, Huss GR, Nagashima K, Taylor GJ, Halldórsson SA, Hilton DR, Mottl MJ, Meech KJ (2015) Evidence for primordial water in Earth's deep mantle. Science 350(6262):795–797. https://doi.org/10.1126/science.aac4834

Hamano K, Abe Y, Genda H (2013) Emergence of two types of terrestrial planet on solidification of magma ocean. Nature 497(7451):607–610. https://doi.org/10.1038/nature12163

Hamano K, Kawahara H, Abe Y, Onishi M, Hashimoto GL (2015) Lifetime and spectral evolution of a magma ocean with a steam atmosphere: its detectability by future direct imaging. Astrophys J 806(2):216. https://doi.org/10.1088/0004-637X/806/2/216

Hanel RA, Conrath BJ, Jennings DE, Samuelson RE (2003) Retrieval of physical parameters from measurements. In: Exploration of the solar system by infrared remote sensing, 2nd edn. Cambridge University Press, Cambridge, pp 352–404. https://doi.org/10.1017/CBO9780511536106.010

Hansen BMS (2009) Formation of the terrestrial planets from a narrow annulus. Astrophys J 703(1):1131–1140. https://doi.org/10.1088/0004-637X/703/1/1131

Harper CL, Jacobsen SB (1996) Evidence for 182Hf in the early Solar System and constraints on the timescale of terrestrial accretion and core formation. Geochim Cosmochim Acta 60(7):1131–1153. https://doi.org/10.1016/0016-7037(96)00027-0

Harrison TM (2009) The hadean crust: evidence from >4 Ga zircons. Annu Rev Earth Planet Sci 37(1):479–505. https://doi.org/10.1146/annurev.earth.031208.100151

Harrison TM, Blichert-Toft J, Müller W, Albarede F, Holden P, Mojzsis SJ (2005) Heterogeneous hadean hafnium: evidence of continental crust at 4.4 to 4.5 Ga. Science 310(5756):1947–1950. https://doi.org/10.1126/science.1117926

Hartmann WK (1986) Moon origin: the impact-trigger hypothesis. In: Hartmann W, Phillips R, Taylor G (eds) Origin of the Moon, Lunar and Planetary Institute, pp 579–608. https://ui.adsabs.harvard.edu/abs/1986ormo.conf..579H/abstract

Hartmann WK, Davis DR (1975) Satellite-sized planetesimals and lunar origin. Icarus 24(4):504–515. https://doi.org/10.1016/0019-1035(75)90070-6

Hashimoto GL, Abe Y, Sugita S (2007) The chemical composition of the early terrestrial atmosphere: Formation of a reducing atmosphere from CI-like material. J Geophys Res, Planets 112(E5). https://doi.org/10.1029/2006JE002844

Hayashi C, Nakazawa K, Mizuno H (1979) Earth's melting due to the blanketing effect of the primordial dense atmosphere. Earth Planet Sci Lett 43(1):22–28. https://doi.org/10.1016/0012-821X(79)90152-3

Helffrich G, Kaneshima S (2004) Seismological constraints on core composition from Fe-O-S liquid immiscibility. Science 306(5705):2239–2242. https://doi.org/10.1126/science.1101109

Herbert F, Sonett CP, Wiskerchen MJ (1977) Model 'zero-age' lunar thermal profiles resulting from electrical induction. J Geophys Res 82(14):2054–2060. https://doi.org/10.1029/JB082i014p02054

Herd CDK, Borg LE, Jones JH, Papike JJ (2002) Oxygen fugacity and geochemical variations in the Martian basalts: implications for Martian basalt petrogenesis and the oxidation state of the upper mantle of Mars. Geochim Cosmochim Acta 66(11):2025–2036. https://doi.org/10.1016/S0016-7037(02)00828-1

Hier-Majumder S, Hirschmann MM (2017) The origin of volatiles in the Earth's mantle. Geochem Geophys Geosyst 18(8). https://doi.org/10.1002/2017GC006937. arXiv:1011.1669v3

Hillgren VJ (1991) Partitioning behavior of NI, CO, MO, and W between basaltic liquid and Ni-rich metal: implications for the origin of the moon and lunar core formation. Geophys Res Lett 18(11):2077–2080. https://doi.org/10.1029/91GL02534

Hirose K, Labrosse S, Hernlund J (2013) Composition and state of the core. Annu Rev Earth Planet Sci 41(1):657–691. https://doi.org/10.1146/annurev-earth-050212-124007

Hirose K, Wood B, Vočadlo L (2021) Light elements in the Earth's core. Nat. Rev. Earth Environ. 2(9):645–658. https://doi.org/10.1038/s43017-021-00203-6

Hirschmann MM (2006) Water, melting, and the deep Earth H_2O cycle. Annu Rev Earth Planet Sci 34(1):629–653. https://doi.org/10.1146/annurev.earth.34.031405.125211

Hirschmann MM (2012) Magma ocean influence on early atmosphere mass and composition. Earth Planet Sci Lett 341–344:48–57. https://doi.org/10.1016/j.epsl.2012.06.015

Hirschmann MM (2016) Constraints on the early delivery and fractionation of Earth's major volatiles from C/H, C/N, and C/S ratios. Am Mineral 101(3):540–553. https://doi.org/10.2138/am-2016-5452

Hirschmann MM (2021) Iron-wüstite revisited: a revised calibration accounting for variable stoichiometry and the effects of pressure. Geochim Cosmochim Acta 313:74–84. https://doi.org/10.1016/j.gca.2021.08.039

Hirschmann MM (2022) Magma oceans, iron and chromium redox, and the origin of comparatively oxidized planetary mantles. Geochim Cosmochim Acta 328:221–241. https://doi.org/10.1016/j.gca.2022.04.005

Hirschmann MM, Aubaud C, Withers AC (2005) Storage capacity of H_2O in nominally anhydrous minerals in the upper mantle. Earth Planet Sci Lett 236(1):167–181. https://doi.org/10.1016/j.epsl.2005.04.022

Hirschmann MM, Withers AC, Ardia P, Foley NT (2012) Solubility of molecular hydrogen in silicate melts and consequences for volatile evolution of terrestrial planets. Earth Planet Sci Lett 345–348:38–48. https://doi.org/10.1016/j.epsl.2012.06.031

Hirth G, Kohlstedt DL (1995) Experimental constraints on the dynamics of the partially Molten upper mantle: 2. Deformation in the dislocation creep regime. J Geophys Res, Solid Earth 100(B8):15441–15449. https://doi.org/10.1029/95JB01292

Hirth G, Kohlstedt DL (1996) Water in the oceanic upper mantle: implications for rheology, melt extraction and the evolution of the lithosphere. Earth Planet Sci Lett 144(1):93–108. https://doi.org/10.1016/0012-821X(96)00154-9

Hirth G, Kohlstedt DL (2004) Rheology of the upper mantle and the mantle wedge: a view from the experimentalists. In: Eiler J (ed) Inside the subduction factory. AGU, Washington DC, pp 83–105. https://doi.org/10.1029/138GM06

Hoehler TM, Som SM, Kiang NY (2018) Life's requirements. In: Deeg HJ, Belmonte JA (eds) Handbook of exoplanets. Springer, Berlin, pp 2795–2816. https://doi.org/10.1007/978-3-319-55333-7_74

Holland HD (1962) Model for the evolution of the Earth's atmosphere. In: Engel AEJ, James HL, Leonard BF (eds) Petrologic studies: a volume in honor of A. F. Buddington. The Geological Society of America, New York, pp 447–477. https://doi.org/10.1130/Petrologic.1962.447

Holland HD (1978) The chemistry of the atmosphere and oceans. Wiley, New York

Holland HD (2006) The oxygenation of the atmosphere and oceans. Philos Trans R Soc B, Biol Sci 361(1470):903–915. https://doi.org/10.1098/rstb.2006.1838

Holloway JR (1981) Volatile interactions in magmas. In: Thermodynamics of minerals and melts. Springer, New York, pp 273–293. https://doi.org/10.1007/978-1-4612-5871-1_13

Holloway JR, Blank JG (1994) Application of experimental results to C-O-H species in natural melts. In: Carroll MR, Holloway JR (eds) Volatiles in magmas, vol 6. Mineralogical Society of America, Washington. https://doi.org/10.1515/9781501509674-012

Holzheid A, Borisov A, Palme H (1994) The effect of oxygen fugacity and temperature on solubilities of nickel, cobalt, and molybdenum in silicate melts. Geochim Cosmochim Acta 58(8):1975–1981. https://doi.org/10.1016/0016-7037(94)90429-4

Höning D, Baumeister P, Grenfell JL, Tosi N, Way MJ (2021) Early habitability and crustal decarbonation of a stagnant-lid Venus. J Geophys Res, Planets 126(10):e2021JE006895

Hoolst TV (2015) 10.04 - Rotation of the terrestrial planets. In: Schubert G (ed) Treatise on geophysics, 2nd edn. Elsevier, Oxford, pp 121–151. https://doi.org/10.1016/B978-0-444-53802-4.00168-8

Hopp T, Dauphas N, Abe Y, Aléon J, Alexander CMO, Amari S, Amelin Y, Bajo K, Bizzarro M, Bouvier A, Carlson RW, Chaussidon M, Choi BG, Davis AM, Rocco TD, Fujiya W, Fukai R, Gautam I, Haba MK, Hibiya Y, Hidaka H, Homma H, Hoppe P, Huss GR, Ichida K, Iizuka T, Ireland TR, Ishikawa A, Ito M, Itoh S, Kawasaki N, Kita NT, Kitajima K, Kleine T, Komatani S, Krot AN, Liu MC, Masuda Y, McKeegan KD, Morita M, Motomura K, Moynier F, Nakai I, Nagashima K, Nesvorný D, Nguyen A, Nittler L, Onose M, Pack A, Park C, Piani L, Qin L, Russell SS, Sakamoto N, Schönbächler M, Tafla L, Tang H, Terada K, Terada Y, Usui T, Wada S, Wadhwa M, Walker RJ, Yamashita K, Yin QZ, Yokoyama T, Yoneda S, Young ED, Yui H, Zhang AC, Nakamura T, Naraoka H, Noguchi T, Okazaki R, Sakamoto K, Yabuta H, Abe M, Miyazaki A, Nakato A, Nishimura M, Okada T, Yada T, Yogata K, Nakazawa S, Saiki T, Tanaka S, Terui F, Tsuda Y, Watanabe S, Yoshikawa M, Tachibana S, Yurimoto H (2022) Ryugu's nucleosynthetic heritage from the outskirts of the Solar System. Sci Adv 8(46):eadd8141. https://doi.org/10.1126/sciadv.add8141

Horan MF, Smoliar MI, Walker RJ (1998) 182W and 187Re-187Os systematics of iron meteorites: chronology for melting, differentiation, and crystallization in asteroids. Geochim Cosmochim Acta 62(3):545–554. https://doi.org/10.1016/S0016-7037(97)00368-2

Hostetler C, Drake M (1980) On the early global melting of the terrestrial planets. In: Lunar and planetary science conference proceedings, vol 11, pp 1915–1929

276

🌀 Springer

Hughes AM, Duchêne G, Matthews BC (2018) Debris disks: structure, composition, and variability. Annu Rev Astron Astrophys 56(1):541–591. https://doi.org/10.1146/annurev-astro-081817-052035

Humphreys MCS, Brooker RA, Fraser DG, Burgisser A, Mangan MT, McCammon C (2015) Coupled interactions between volatile activity and Fe oxidation state during Arc crustal processes. J Petrol 56(4):795–814. https://doi.org/10.1093/petrology/egv017

Hunten DM (1973) The escape of light gases from planetary atmospheres. J Atmos Sci 30(8):1481–1494. https://doi.org/10.1175/1520-0469(1973)030<1481:TEOLGF>2.0.CO;2

Hunten DM (1993) Atmospheric evolution of the terrestrial planets. Science 259(5097):915–920. https://doi.org/10.1126/science.259.5097.915

Hunten DM, Pepin RO, Walker JCG (1987) Mass fractionation in hydrodynamic escape. Icarus 69(3):532–549. https://doi.org/10.1016/0019-1035(87)90022-4

Ida S, Yamamura T, Okuzumi S (2019) Water delivery by pebble accretion to rocky planets in habitable zones in evolving disks. Astron Astrophys 624:A28. https://doi.org/10.1051/0004-6361/201834556

Iizuka-Oku R, Yagi T, Gotou H, Okuchi T, Hattori T, Sano-Furukawa A (2017) Hydrogenation of iron in the early stage of Earth's evolution. Nat Commun 8(1):14096. https://doi.org/10.1038/ncomms14096

Ikoma M, Genda H (2006) Constraints on the mass of a habitable planet with water of nebular origin. Astrophys J 648(1):696–706. https://doi.org/10.1086/505780

Ikoma M, Elkins-Tanton L, Hamano K, Suckale J (2018) Water partitioning in planetary embryos and protoplanets with magma oceans. Space Sci Rev 214(4):76. https://doi.org/10.1007/s11214-018-0508-3. arXiv:1804.09294

Ingersoll AP (1969) The runaway greenhouse: a history of water on Venus. J Atmos Sci 26(6):1191–1198. https://doi.org/10.1175/1520-0469(1969)026<1191:TRGAHO>2.0.CO;2

Ingersoll AP, Dobrovolskis AR (1978) Venus' rotation and atmospheric tides. Nature 275(5675):37–38. https://doi.org/10.1038/275037a0

Ishiwatari M, Takehiro S, Nakajima K, Hayashi YY (2002) A numerical study on appearance of the runaway greenhouse state of a three-dimensional gray atmosphere. J Atmos Sci 59(22):3223–3238. https://doi.org/10.1175/1520-0469(2002)059<3223:ANSOAO>2.0.CO;2

Ishiwatari M, Nakajima K, Takehiro S, Hayashi YY, Kawai Y, Takahashi YO (2021) Revision of "Dependence of climate states of gray atmosphere on solar constant: from the runaway greenhouse to the snowball states" by Ishiwatari et al. (2007). J Geophys Res, Atmos 126(11):e2019JD031761. https://doi.org/10.1029/2019JD031761

Itcovitz JP, Rae ASP, Citron RI, Stewart ST, Sinclair CA, Rimmer PB, Shorttle O (2022) Reduced atmospheres of post-impact worlds: the early Earth. Planet Sci. J. 3(5):115. https://doi.org/10.3847/psj/ac67a9

Izidoro A, Bitsch B, Dasgupta R (2021a) The effect of a strong pressure bump in the Sun's natal disk: terrestrial planet formation via planetesimal accretion rather than pebble accretion. Astrophys J 915(1):62. https://doi.org/10.3847/1538-4357/abfe0b

Izidoro A, Bitsch B, Raymond SN, Johansen A, Morbidelli A, Lambrechts M, Jacobson SA (2021b) Formation of planetary systems by pebble accretion and migration - Hot super-Earth systems from breaking compact resonant chains. Astron Astrophys 650:A152. https://doi.org/10.1051/0004-6361/201935336

Izidoro A, Dasgupta R, Raymond SN, Deienno R, Bitsch B, Isella A (2022) Planetesimal rings as the cause of the Solar System's planetary architecture. Nat Astron 6(3):357–366. https://doi.org/10.1038/s41550-021-01557-z

Jacobsen SB (2005) The Hf-W isotopic system and the origin of the Earth and Moon. Annu Rev Earth Planet Sci 33(1):531–570. https://doi.org/10.1146/annurev.earth.33.092203.122614

Jacobson SA, Morbidelli A (2014) Lunar and terrestrial planet formation in the Grand Tack scenario. Philos Trans R Soc A, Math Phys Eng Sci 372(2024):20130174. https://doi.org/10.1098/rsta.2013.0174

Jacobson SA, Walsh KJ (2015) Earth and terrestrial planet formation. In: The early Earth. Am. Geophys. Union, Washington, pp 49–70. https://doi.org/10.1002/9781118860359.ch3

Jakosky BM, Ahrens T (1979) The history of an atmosphere of impact origin. In: Lunar and planetary science conference proceedings, vol 3, pp 2727–2739. https://ui.adsabs.harvard.edu/abs/1979LPSC...10.2727J

Jakosky BM, Phillips RJ (2001) Mars' volatile and climate history. Nature 412(6843):237–244. https://doi.org/10.1038/35084184

Jansen T, Scharf C, Way M, Genio AD (2019) Climates of warm Earth-like planets. II. Rotational "Goldilocks" zones for fractional habitability and silicate weathering. Astrophys J 875(2):79. https://doi.org/10.3847/1538-4357/ab113d

Jaupart C, Labrosse S, Lucazeau F, Mareschal JC (2015) 7.06 - Temperatures, heat, and energy in the mantle of the Earth. In: Schubert G (ed) Treatise on geophysics, 2nd edn. Elsevier, Oxford, pp 223–270. https://doi.org/10.1016/B978-0-444-53802-4.00126-3

Ji J, Guo D, Liu J, Chen S, Ling Z, Ding X, Han K, Chen J, Cheng W, Zhu K, Liu J, Wang J, Chen J, Ouyang Z (2022) The 1:2,500,000-scale geologic map of the global Moon. Sci Bull. https://doi.org/10.1016/j.scib.2022.05.021

Johansen A, Lambrechts M (2017) Forming planets via pebble accretion. Annu Rev Earth Planet Sci 45(1):359–387. https://doi.org/10.1146/annurev-earth-063016-020226

Johansen A, Youdin A, Low MMM (2009) Particle clumping and planetesimal formation depend strongly on metallicity. Astrophys J 704(2):L75–L79. https://doi.org/10.1088/0004-637X/704/2/L75

Johansen A, Blum J, Tanaka H, Ormel C, Bizzarro M, Rickman H (2014) The multifaceted planetesimal formation process. In: Beuther H et al (eds) Protostars and planets VI. University of Arizona Press, Tucson, p 547. https://doi.org/10.2458/azu_uapress_9780816531240-ch024

Johansen A, Ronnet T, Bizzarro M, Schiller M, Lambrechts M, Nordlund Å, Lammer H (2021) A pebble accretion model for the formation of the terrestrial planets in the Solar System. Sci Adv 7(8):eabc0444. https://doi.org/10.1126/sciadv.abc0444.

Johnson NM de Oliveira MRR (2019) Venus atmospheric composition in situ data: a compilation. Earth Space Sci 6(7):1299–1318. https://doi.org/10.1029/2018EA000536

Johnson KTM, Dick HJB (1992) Open system melting and temporal and spatial variation of peridotite and basalt at the Atlantis II Fracture Zone. J Geophys Res, Solid Earth 97(B6):9219–9241. https://doi.org/10.1029/92JB00701

Johnstone CP (2020) Hydrodynamic escape of water vapor atmospheres near very active stars. Astrophys J 890(1):79. https://doi.org/10.3847/1538-4357/ab6224

Johnstone CP, Bartel M, Güdel M (2021) The active lives of stars: a complete description of the rotation and XUV evolution of F, G, K, and M dwarfs. Astron Astrophys 649:A96. https://doi.org/10.1051/0004-6361/202038407. arXiv:2009.07695

Karato S (2015) 9.05 - Water in the evolution of the Earth and other terrestrial planets. In: Schubert G (ed) Treatise on geophysics, 2nd edn. Elsevier, Oxford, pp 105–144. https://doi.org/10.1016/B978-0-444-53802-4.00156-1

Kasting JF (1988) Runaway and moist greenhouse atmospheres and the evolution of Earth and Venus. Icarus 74(3):472–494. https://doi.org/10.1016/0019-1035(88)90116-9

Kasting JF (1990) Bolide impacts and the oxidation state of carbon in the Earth's early atmosphere. Orig Life Evol Biosph 20(3):199–231. https://doi.org/10.1007/BF01808105

Kasting JF (1993) Earth's early atmosphere. Science 259(5097):920–926. https://doi.org/10.1126/science.11536547

Kasting JF (2013) What caused the rise of atmospheric O_2? Chem Geol 362:13–25. https://doi.org/10.1016/j.chemgeo.2013.05.039

Kasting JF (2014) 6.6 - Modeling the archean atmosphere and climate. In: Holland HD, Turekian KK (eds) Treatise on geochemistry, 2nd edn. Elsevier, Oxford, pp 157–175. https://doi.org/10.1016/B978-0-08-095975-7.01306-1

Kasting JF (2019) The Goldilocks planet? How silicate weathering maintains Earth "just right". Elements 15(4):235–240. https://doi.org/10.2138/gselements.15.4.235

Kasting JF, Catling D (2003) Evolution of a habitable planet. Annu Rev Astron Astrophys 41(1):429–463. https://doi.org/10.1146/annurev.astro.41.071601.170049

Kasting JF, Pollack JB (1983) Loss of water from Venus. I. Hydrodynamic escape of hydrogen. Icarus 53(3):479–508. https://doi.org/10.1016/0019-1035(83)90212-9

Kasting JF, Zahnle KJ, Walker JCG (1983) Photochemistry of methane in the Earth's early atmosphere. Precambrian Res 20(2):121–148. https://doi.org/10.1016/0301-9268(83)90069-4

Kasting JF, Eggler DH, Raeburn SP (1993a) Mantle redox evolution and the oxidation state of the archean atmosphere. J Geol 101(2):245–257. https://doi.org/10.1086/648219

Kasting JF, Whitmire DP, Reynolds RT (1993b) Habitable zones around main sequence stars. Icarus 101(1):108–128. https://doi.org/10.1006/icar.1993.1010

Katyal N, Ortenzi G, Lee Grenfell J, Noack L, Sohl F, Godolt M, García Muñoz A, Schreier F, Wunderlich F, Rauer H (2020) Effect of mantle oxidation state and escape upon the evolution of Earth's magma ocean atmosphere. Astron Astrophys 643:A81. https://doi.org/10.1051/0004-6361/202038779

Kaula WM (1979) Thermal evolution of Earth and Moon growing by planetesimal impacts. J Geophys Res, Solid Earth 84:999–1008. https://doi.org/10.1029/JB084iB03p00999

Ke Y, Solomatov VS (2006) Early transient superplumes and the origin of the Martian crustal dichotomy. J Geophys Res, Planets 111(E10). https://doi.org/10.1029/2005JE002631

Ke Y, Solomatov VS (2009) Coupled core-mantle thermal evolution of early Mars. J Geophys Res, Planets 114(E7). https://doi.org/10.1029/2008JE003291

Kennedy GM, Kenyon SJ (2008) Planet formation around stars of various masses: the snow line and the frequency of giant planets. Astrophys J 673(1):502. https://doi.org/10.1086/524130

Keppler H, Bolfan-Casanova N (2006) Thermodynamics of water solubility and partitioning. Rev Mineral Geochem 62(1):193–230. https://doi.org/10.2138/rmg.2006.62.9

Keppler H, Golabek G (2019) Graphite floatation on a magma ocean and the fate of carbon during core formation. Geochem Perspect Lett 11:12–17. https://doi.org/10.7185/geochemlet.1918

Khan A, Sossi PA, Liebske C, Rivoldini A, Giardini D (2022) Geophysical and cosmochemical evidence for a volatile-rich Mars. Earth Planet Sci Lett 578:117330. https://doi.org/10.1016/j.epsl.2021.117330

Kiefer WS, Richards MA, Hager BH, Bills BG (1986) A dynamic model of Venus's gravity field. Geophys Res Lett 13(1):14–17. https://doi.org/10.1029/GL013i001p00014

Kiehl JT, Trenberth KE (1997) Earth's annual global mean energy budget. Bull Am Meteorol Soc 78(2):197–208. https://doi.org/10.1175/1520-0477(1997)078<0197:EAGMEB>2.0.CO;2

Kislyakova K, Noack L (2020) Electromagnetic induction heating as a driver of volcanic activity on massive rocky planets. Astron Astrophys 636:L10. https://doi.org/10.1051/0004-6361/202037924

Kislyakova KG, Noack L, Johnstone CP, Zaitsev VV, Fossati L, Lammer H, Khodachenko ML, Odert P, Güdel M (2017) Magma oceans and enhanced volcanism on TRAPPIST-1 planets due to induction heating. Nat Astron 1(12):878–885. https://doi.org/10.1038/s41550-017-0284-0

Kleine T, Walker RJ (2017) Tungsten isotopes in planets. Annu Rev Earth Planet Sci 45(1):389–417. https://doi.org/10.1146/annurev-earth-063016-020037

Kleine T, Münker C, Mezger K, Palme H (2002) Rapid accretion and early core formation on asteroids and the terrestrial planets from Hf–W chronometry. Nature 418(6901):952–955. https://doi.org/10.1038/nature00982

Kleine T, Touboul M, Bourdon B, Nimmo F, Mezger K, Palme H, Jacobsen SB, Yin QZ, Halliday AN (2009) Hf–W chronology of the accretion and early evolution of asteroids and terrestrial planets. Geochim Cosmochim Acta 73(17):5150–5188. https://doi.org/10.1016/j.gca.2008.11.047

Kokubo E, Ida S (1996) On runaway growth of planetesimals. Icarus 123(1):180–191. https://doi.org/10.1006/icar.1996.0148

Kokubo E, Ida S (1998) Oligarchic growth of protoplanets. Icarus 131(1):171–178. https://doi.org/10.1006/icar.1997.5840

Kokubo E, Ida S (2000) Formation of protoplanets from planetesimals in the solar nebula. Icarus 143(1):15–27. https://doi.org/10.1006/icar.1999.6237

Kono Y (2018) Chap. 10 - Viscosity measurement. In: Kono Y, Sanloup C (eds) Magmas under pressure. Elsevier, Amsterdam, pp 261–280. https://doi.org/10.1016/B978-0-12-811301-1.00010-1

Kopp G, Lean JL (2011) A new, lower value of total solar irradiance: Evidence and climate significance. Geophys Res Lett 38(1). https://doi.org/10.1029/2010GL045777

Kopparapu RK, Ramirez R, Kasting JF, Eymet V, Robinson TD, Mahadevan S, Terrien RC, Domagal-Goldman S, Meadows V, Deshpande R (2013) Habitable zones around main-sequence stars: new estimates. Astrophys J 770(1). https://doi.org/10.1088/0004-637X/770/1/82

Korenaga J (2011) Thermal evolution with a hydrating mantle and the initiation of plate tectonics in the early Earth. J Geophys Res, Solid Earth 116(12):1–20. https://doi.org/10.1029/2011JB008410

Korenaga J (2017) On the extent of mantle hydration caused by plate bending. Earth Planet Sci Lett 457:1–9. https://doi.org/10.1016/j.epsl.2016.10.011

Korenaga J, Planavsky NJ, Evans DA (2017) Global water cycle and the coevolution of the Earth's interior and surface environment. Philos Trans R Soc A, Math Phys Eng Sci 375(2094). https://doi.org/10.1098/rsta.2015.0393

Kortenkamp SJ, Wetherill GW, Inaba S (2001) Runaway growth of planetary embryos facilitated by massive bodies in a protoplanetary disk. Science 293(5532):1127–1129. https://doi.org/10.1126/science.1062391

Kral Q, Matrà L, Wyatt MC, Kennedy GM (2017) Predictions for the secondary CO, C and O gas content of debris discs from the destruction of volatile-rich planetesimals. Mon Not R Astron Soc 469(1):521–550. https://doi.org/10.1093/mnras/stx730

Kral Q, Davoult J, Charnay B (2020) Formation of secondary atmospheres on terrestrial planets by late disk accretion. Nat Astron 4(8):769–775. https://doi.org/10.1038/s41550-020-1050-2

Krishnamurti R, Howard LN (1981) Large-scale flow generation in turbulent convection. Proc Natl Acad Sci. https://doi.org/10.1073/pnas.78.4.1981

Krissansen-Totton J, Fortney JJ, Nimmo F (2021) Was Venus ever habitable? Constraints from a coupled interior–atmosphere–redox evolution model. Planet Sci. J. 2(5):216. https://doi.org/10.3847/psj/ac2580

Krot AN, Keil K, Scott ERD, Goodrich CA, Weisberg MK (2014) 1.1 - Classification of meteorites and their genetic relationships. In: Holland HD, Turekian KK (eds) Treatise on geochemistry, 2nd edn. Elsevier, Oxford, pp 1–63. https://doi.org/10.1016/B978-0-08-095975-7.00102-9

Kruijer TS, Burkhardt C, Budde G, Kleine T (2017) Age of Jupiter inferred from the distinct genetics and formation times of meteorites. Proc Natl Acad Sci 114(26):6712–6716. https://doi.org/10.1073/pnas.1704461114

Kruijer TS, Kleine T, Borg LE (2020) The great isotopic dichotomy of the early Solar System. Nat Astron 4(1):32–40. https://doi.org/10.1038/s41550-019-0959-9

Kuhn WR, Atreya SK (1979) Ammonia photolysis and the greenhouse effect in the primordial atmosphere of the Earth. Icarus 37(1):207–213. https://doi.org/10.1016/0019-1035(79)90126-X

Kuramoto K, Matsui T (1996) Partitioning of H and C between the mantle and core during the core formation in the Earth: its implications for the atmospheric evolution and redox state of early mantle. J Geophys Res, Planets 101(E6):14909–14932. https://doi.org/10.1029/96JE00940

Kuramoto K, Umemoto T, Ishiwatari M (2013) Effective hydrodynamic hydrogen escape from an early Earth atmosphere inferred from high-accuracy numerical simulation. Earth Planet Sci Lett 375:312–318. https://doi.org/10.1016/j.epsl.2013.05.050

Kurosawa K, Sugita S, Ishibashi K, Hasegawa S, Sekine Y, Ogawa NO, Kadono T, Ohno S, Ohkouchi N, Nagaoka Y Matsui T (2013) Hydrogen cyanide production due to mid-size impacts in a redox-neutral N2-rich atmosphere. Orig Life Evol Biosph 43(3):221–245. https://doi.org/10.1007/s11084-013-9339-0

Labrosse S, Hernlund JW, Coltice N (2007) A crystallizing dense magma ocean at the base of the Earth's mantle. Nature 450(7171):866–869. https://doi.org/10.1038/nature06355

Lago B, Cazenave A (1979) Possible dynamical evolution of the rotation of Venus since formation. Moon Planets 21(2):127–154. https://doi.org/10.1007/BF00897084

Lambrechts M, Johansen A (2014) Forming the cores of giant planets from the radial pebble flux in proto-planetary discs. Astron Astrophys 572:A107. https://doi.org/10.1051/0004-6361/201424343

Lammer H (2013) Origin and evolution of planetary atmospheres. Springer, Berlin. https://doi.org/10.1007/978-3-642-32087-3

Lammer H, Zerkle AL, Gebauer S, Tosi N, Noack L, Scherf M, Pilat-Lohinger E, Güdel M, Grenfell JL, Godolt M, Nikolaou A (2018) Origin and evolution of the atmospheres of early Venus, Earth and Mars. Astron. Astrophys. Rev. 26:2. https://doi.org/10.1007/s00159-018-0108-y

Lammer H, Brasser R, Johansen A, Scherf M, Leitzinger M (2020a) Formation of Venus, Earth and Mars: constrained by isotopes. Space Sci Rev 217(1):7. https://doi.org/10.1007/s11214-020-00778-4

Lammer H, Leitzinger M, Scherf M, Odert P, Burger C, Kubyshkina D, Johnstone C, Maindl T, Schäfer CM, Güdel M, Tosi N, Nikolaou A, Marcq E, Erkaev NV, Noack L, Kislyakova KG, Fossati L, Pilat-Lohinger E, Ragossnig F, Dorfi EA (2020b) Constraining the early evolution of Venus and Earth through atmospheric Ar, Ne isotope and bulk K/U ratios. Icarus 339:113551. https://doi.org/10.1016/j.icarus.2019.113551

Landeau M, Olson P, Deguen R, Hirsh BH (2016) Core merging and stratification following giant impact. Nat Geosci 9(1). https://doi.org/10.1038/ngeo2808

Landeau M, Deguen R, Phillips D, Neufeld JA, Lherm V, Dalziel SB (2021) Metal-silicate mixing by large Earth-forming impacts. Earth Planet Sci Lett 564:116888. https://doi.org/10.1016/j.epsl.2021.116888

Lange RA (1994) Chap. 9. The effect of H2O, CO2 and F on the density and viscosity of silicate melts. In: Carroll MR, Holloway JR (eds) Volatiles in magmas, vol 9. Mineralogical Society of America, Washington, pp 331–370. https://doi.org/10.1515/9781501509674-015

Lange MA, Ahrens TJ (1982) The evolution of an impact-generated atmosphere. Icarus 51(1):96–120. https://doi.org/10.1016/0019-1035(82)90031-8

Lasaga AC, Holland HD, Dwyer MJ (1971) Primordial oil slick. Science 174(4004):53–55. https://doi.org/10.1126/science.174.4004.53

Lebrun T, Massol H, Chassefière E, Davaille A, Marcq E, Sarda P, Leblanc F, Brandeis G (2013) Thermal evolution of an early magma ocean in interaction with the atmosphere. J Geophys Res, Planets 118(6):1155–1176. https://doi.org/10.1002/jgre.20068

Lecar M, Podolak M, Sasselov D, Chiang E (2006) On the location of the snow line in a protoplanetary disk. Astrophys J 640(2):1115–1118. https://doi.org/10.1086/500287

Leconte J, Forget F, Charnay B, Wordsworth R, Pottier A (2013) Increased insolation threshold for runaway greenhouse processes on Earth-like planets. Nature 504(7479):268. https://doi.org/10.1038/nature12827

Leconte J, Forget F, Charnay B, Wordsworth R, Selsis F, Millour E, Spiga A (2013) 3D climate modeling of close-in land planets: circulation patterns, climate moist bistability, and habitability. Astron Astrophys 554:A69. https://doi.org/10.1051/0004-6361/201321042

Leconte J, Wu H, Menou K, Murray N (2015) Asynchronous rotation of Earth-mass planets in the habitable zone of lower-mass stars. Science 347(6222):632–635. https://doi.org/10.1126/science.1258686

Lécuyer C, Ricard Y (1999) Long-term fluxes and budget of ferric iron: implication for the redox states of the Earth's mantle and atmosphere. Earth Planet Sci Lett 165(2):197–211. https://doi.org/10.1016/S0012-821X(98)00267-2

Lee DC, Halliday AN (1995) Hafnium–tungsten chronometry and the timing of terrestrial core formation. Nature 378(6559):771–774. https://doi.org/10.1038/378771a0

Lee DC, Halliday AN (1997) Core formation on Mars and differentiated asteroids. Nature 388(6645):854–857. https://doi.org/10.1038/42206

Lee T, Papanastassiou DA, Wasserburg GJ (1976) Demonstration of ^{26}Mg excess in Allende and evidence for ^{26}Al. Geophys Res Lett 3(1):41–44. https://doi.org/10.1029/GL003i001p00041

Lee CTA, Brandon AD, Norman M (2003) Vanadium in peridotites as a proxy for paleo-fO_2 during partial melting: prospects, limitations, and implications. Geochim Cosmochim Acta 67(16):3045–3064. https://doi.org/10.1016/S0016-7037(03)00268-0

Leighton PA, Steiner AB (1936) The photochemical decomposition of methane. J Am Chem Soc 58(9):1823. https://doi.org/10.1021/ja01300a512

Lejeune AM, Richet P (1995) Rheology of crystal-bearing silicate melts: an experimental study at high viscosities. J Geophys Res, Solid Earth 100(B3):4215–4229. https://doi.org/10.1029/94JB02985

Lesher CE, Spera FJ (2015) Chap. 5 - Thermodynamic and transport properties of silicate melts and magma. In: Sigurdsson H (ed) The encyclopedia of volcanoes, 2nd edn. Academic Press, Amsterdam, pp 113–141. https://doi.org/10.1016/B978-0-12-385938-9.00005-5

Levison HF, Kretke KA, Duncan MJ (2015) Growing the gas-giant planets by the gradual accumulation of pebbles. Nature 524(7565):322–324. https://doi.org/10.1038/nature14675

Li J, Agee CB (1996) Geochemistry of mantle-core differentiation at high pressure. Nature 381(6584):686–689. https://doi.org/10.1038/381686a0

Li ZXA, Lee CTA (2004) The constancy of upper mantle fO_2 through time inferred from V/Sc ratios in basalts. Earth Planet Sci Lett 228(3):483–493. https://doi.org/10.1016/j.epsl.2004.10.006

Li Y, Dasgupta R, Tsuno K, Monteleone B, Shimizu N (2016) Carbon and sulfur budget of the silicate Earth explained by accretion of differentiated planetary embryos. Nat Geosci. https://doi.org/10.1038/ngeo2801

Li Y, Vočadlo L, Sun T, Brodholt JP (2020) The Earth's core as a reservoir of water. Nat Geosci 13(6):453–458. https://doi.org/10.1038/s41561-020-0578-1

Libourel G, Marty B, Humbert F (2003) Nitrogen solubility in basaltic melt. Part I. Effect of oxygen fugacity. Geochim Cosmochim Acta 67(21):4123–4135. https://doi.org/10.1016/S0016-7037(03)00259-X

Lichtenberg T, Golabek G, Burn R, Meyer M, Alibert Y, Gerya T, Mordasini C (2019) A water budget dichotomy of rocky protoplanets from 26al-heating. Nat Astron 307. https://doi.org/10.1038/s41550-018-0688-5

Lichtenberg T, Bower DJ, Hammond M, Boukrouche R, Sanan P, Tsai SM, Pierrehumbert RT (2021) Vertically resolved magma ocean–protoatmosphere evolution: H_2, H_2O, CO_2, CH_4, CO, O_2, and N_2 as primary absorbers. J Geophys Res, Planets 126(2):e2020JE006711. https://doi.org/10.1029/2020JE006711

Lichtenberg T, Drążkowska J, Schönbächler M, Golabek GJ, Hands TO (2021b) Bifurcation of planetary building blocks during Solar System formation. Science 371(6527):365–370. https://doi.org/10.1126/science.abb3091

Liggins P, Shorttle O, Rimmer PB (2020) Can volcanism build hydrogen-rich early atmospheres? Earth Planet Sci Lett 550:116546. https://doi.org/10.1016/j.epsl.2020.116546

Lin Y, van Westrenen W, Mao HK (2021) Oxygen controls on magmatism in rocky exoplanets. Proc Natl Acad Sci 118(45). https://doi.org/10.1073/pnas.2110427118. https://www.pnas.org/content/118/45/e2110427118

Litasov KD, Ohtani E (2007) Effect of water on the phase relations in Earth's mantle and deep water cycle. In: Ohtani E (ed) Special paper of the geological society of America, vol 421. Geol. Soc. Am., Boulder, pp 115–156. https://doi.org/10.1130/2007.2421(08)

Liu B, Johansen A, Lambrechts M, Bizzarro M, Haugbølle T (2022) Natural separation of two primordial planetary reservoirs in an expanding solar protoplanetary disk. Sci Adv 8(16):eabm3045. https://doi.org/10.1126/sciadv.abm3045

Liu B, Raymond SN, Jacobson SA (2022b) Early solar system instability triggered by dispersal of the gaseous disk. Nature 604(7907):643–646. https://doi.org/10.1038/s41586-022-04535-1

Lugaro M, Ott U, Keresztúri Á (2018) Radioactive nuclei from cosmochronology to habitability. Prog Part Nucl Phys 102:1–47. https://doi.org/10.1016/j.ppnp.2018.05.002

Luger R, Barnes R (2015) Extreme water loss and abiotic O_2 buildup on planets throughout the habitable zones of M dwarfs. Astrobiology 15(2):119–143. https://doi.org/10.1089/ast.2014.1231

Lugmair GW, Shukolyukov A (1998) Early solar system timescales according to 53Mn-53Cr systematics. Geochim Cosmochim Acta 62(16):2863–2886. https://doi.org/10.1016/S0016-7037(98)00189-6

Lupu RE, Zahnle K, Marley MS, Schaefer L, Fegley B, Morley C, Cahoy K, Freedman R, Fortney JJ (2014) The atmospheres of earthlike planets after giant impact events. Astrophys J 784(1):27. https://doi.org/10.1088/0004-637X/784/1/27

Maas C, Hansen U (2015) Effects of Earth's rotation on the early differentiation of a terrestrial magma ocean. J Geophys Res, Solid Earth 120(11):7508–7525. https://doi.org/10.1002/2015JB012053

Maas C, Hansen U (2019) Dynamics of a terrestrial magma ocean under planetary rotation: a study in spherical geometry. Earth Planet Sci Lett 513:81–94. https://doi.org/10.1016/j.epsl.2019.02.016

Maas C, Manske L, Wünnemann K, Hansen U (2021) On the fate of impact-delivered metal in a terrestrial magma ocean. Earth Planet Sci Lett 554:116680. https://doi.org/10.1016/j.epsl.2020.116680

Mackwell SJ, Zimmerman ME, Kohlstedt DL (1998) High-temperature deformation of dry diabase with application to tectonics on Venus. J Geophys Res, Solid Earth 103(B1):975–984. https://doi.org/10.1029/97JB02671

Malavergne V, Bureau H, Raepsaet C, Gaillard F, Poncet M, Surblé S, Sifré D, Shcheka S, Fourdrin C, Deldicque D, Khodja H (2019) Experimental constraints on the fate of H and C during planetary core-mantle differentiation. Implications for the Earth. Icarus 321:473–485. https://doi.org/10.1016/j.icarus.2018.11.027

Malkus WVR (1954) Discrete transitions in turbulent convection. Proc R Soc Lond, Ser A, Math Phys Eng Sci 225(1161):185–195. https://doi.org/10.1098/rspa.1954.0196

Mallmann G, O'Neill HSC (2009) The crystal/melt partitioning of V during mantle melting as a function of oxygen fugacity compared with some other elements (Al, P, Ca, Sc, Ti, Cr, Fe, Ga, Y, Zr and Nb). J Petrol 50(9):1765–1794. https://doi.org/10.1093/petrology/egp053

Mao HK, Bell PM (1977) Disproportionation equilibrium in iron-bearing systems at pressures above 100 Kbar with applications to chemistry of the Earth's mantle. In: Saxena SK, Bhattacharji S, Annersten H, Stephansson O (eds) Energetics of geological processes. Springer, Berlin, pp 236–249. https://doi.org/10.1007/978-3-642-86574-9_12

Marcq E, Salvador A, Massol H, Davaille A (2017) Thermal radiation of magma ocean planets using a 1-D radiative-convective model of H_2O-CO_2 atmospheres. J Geophys Res, Planets 122(7):1539–1553. https://doi.org/10.1002/2016JE005224

Margot JL, Campbell DB, Giorgini JD, Jao JS, Snedeker LG, Ghigo FD, Bonsall A (2021) Spin state and moment of inertia of Venus. Nat Astron 5(7):676–683. https://doi.org/10.1038/s41550-021-01339-7

Marsh BD (1981) On the crystallinity, probability of occurrence, and rheology of lava and magma. Contrib Mineral Petrol 78(1):85–98. https://doi.org/10.1007/BF00371146

Martin RG, Livio M (2012) On the evolution of the snow line in protoplanetary discs. Mon Not R Astron Soc Lett 425(1):L6–L9. https://doi.org/10.1111/j.1745-3933.2012.01290.x

Marty B (2012) The origins and concentrations of water, carbon, nitrogen and noble gases on Earth. Earth Planet Sci Lett 313–314:56–66. https://doi.org/10.1016/j.epsl.2011.10.040

Marty B, Yokochi R (2006) Water in the early Earth. Rev Mineral Geochem 62(1):421–450. https://doi.org/10.2138/rmg.2006.62.18

Marty B, Avice G, Sano Y, Altwegg K, Balsiger H, Hässig M, Morbidelli A, Mousis O, Rubin M (2016) Origins of volatile elements (H, C, N, noble gases) on Earth and Mars in light of recent results from the Rosetta cometary mission. Earth Planet Sci Lett 441:91–102. https://doi.org/10.1016/j.epsl.2016.02.031

Marty B, Altwegg K, Balsiger H, Bar-Nun A, Bekaert DV, Berthelier JJ, Bieler A, Briois C, Calmonte U, Combi M, De Keyser J, Fiethe B, Fuselier SA, Gasc S, Gombosi TI, Hansen KC, Hässig M, Jäckel A, Kopp E, Korth A, Le Roy L, Mall U, Mousis O, Rème H, Owen T, Rubin M, Sémon T, Tzou CY, Waite JH, Wurz P (2017) Xenon isotopes in 67P/Churyumov-Gerasimenko show that comets contributed to Earth's atmosphere. Science 356(6342):1069–1072. https://doi.org/10.1126/science.aal3496

Massol H, Hamano K, Tian F, Ikoma M, Abe Y, Chassefière E, Davaille A, Genda H, Güdel M, Hori Y, Leblanc F, Marcq E, Sarda P, Shematovich VI, Stökl A, Lammer H (2016) Formation and evolution of protoatmospheres. Space Sci Rev 205:153–211. https://doi.org/10.1007/s11214-016-0280-1

Mathez EA (1984) Influence of degassing on oxidation states of basaltic magmas. Nature 310(5976):371–375. https://doi.org/10.1038/310371a0

Matrà L, MacGregor MA, Kalas P, Wyatt MC, Kennedy GM, Wilner DJ, Duchene G, Hughes AM, Pan M, Shannon A, Clampin M, Fitzgerald MP, Graham JR, Holland WS, Panić O, Su KYL (2017) Detection of exocometary CO within the 440 Myr old fomalhaut belt: a similar CO + CO_2 ice abundance in exocomets and solar system comets. Astrophys J 842(1):9. https://doi.org/10.3847/1538-4357/aa71b4

Matrà L, Öberg KI, Wilner DJ, Olofsson J, Bayo A (2019) On the ubiquity and stellar luminosity dependence of exocometary CO gas: detection around M dwarf TWA 7. Astron J 157(3):117. https://doi.org/10.3847/1538-3881/aaff5b

Matsui T, Abe Y (1984) The formation of an impact-generated H_2O atmosphere and its implications for the early thermal history of the Earth. In: Lunar and planetary science XV, pp 517–518. https://ui.adsabs.harvard.edu/abs/1984LPI....15..517M

Matsui T, Abe Y (1986a) Formation of a 'magma ocean' on the terrestrial planets due to the blanketing effect of an impact-induced atmosphere. Earth Moon Planets 34(3):223–230. https://doi.org/10.1007/BF00145081

Matsui T, Abe Y (1986b) Impact-induced atmospheres and oceans on Earth and Venus. Nature 322(6079):526–528. https://doi.org/10.1038/322526a0

Matsui T, Abe Y (1986c) Origin of the Moon and its early thermal evolution. In: Hartmann W, Phillips R, Taylor G (eds) Origin of the Moon, Lunar and Planetary Institute, pp 453–468. https://ui.adsabs.harvard.edu/abs/1986ormo.conf..453M/abstract

282

Matthews CN, Minard RD (2006) Hydrogen cyanide polymers, comets and the origin of life. Faraday Discuss 133:393–401. https://doi.org/10.1039/B516791D

Maurice M, Tosi N, Samuel H, Plesa AC, Hüttig C, Breuer D (2017) Onset of solid-state mantle convection and mixing during magma ocean solidification. J Geophys Res, Planets 122(3):577–598. https://doi.org/10.1002/2016JE005250

Maurice M, Dasgupta R, Hassanzadeh P (2023) Redox evolution of the crystallizing terrestrial magma ocean and its influence on the outgassed atmosphere. Planet Sci. J. 4(2):31. https://doi.org/10.3847/PSJ/acb2ca

McCammon C (1997) Perovskite as a possible sink for ferric iron in the lower mantle. Nature 387(6634):694–696. https://doi.org/10.1038/42685

McCammon C (2005a) The paradox of mantle redox. Science 308(5723):807–808. https://doi.org/10.1126/science.1110532

McCammon CA (2005b) Mantle oxidation state and oxygen fugacity: constraints on mantle chemistry, structure, and dynamics. In: Earth's deep mantle: structure, composition, and evolution. Am. Geophys. Union, Washington, pp 219–240. https://doi.org/10.1029/160GM14

McCollom TM (2013) Miller-urey and beyond: what have we learned about prebiotic organic synthesis reactions in the past 60 years? Annu Rev Earth Planet Sci 41(1):207–229. https://doi.org/10.1146/annurev-earth-040610-133457

McCord TB (1968) The loss of retrograde satellites in the solar system. J Geophys Res 73(4):1497–1500. https://doi.org/10.1029/JB073i004p01497

McKenzie D (2011) Compaction and crystallization in magma chambers: towards a model of the skaergaard intrusion. J Petrol 52(5):905–930. https://doi.org/10.1093/petrology/egr009

McKinnon WB, Zahnle KJ, Ivanov BA, Melosh HJ (1997) Cratering on Venus: models and observations. In: Bougher SW, Hunten DM, Phillips RJ (eds) Venus II: geology, geophysics, atmosphere, and solar wind environment, p 969

Meech K, Raymond S (2020) Origin of Earth's water: sources and constraints. In: Meadows VS, Arney GN, Schmidt BE, Des Marais DJ (eds) Planetary astrobiology. University of Arizona Press, Tucson, p 325. https://doi.org/10.2458/azu_uapress_9780816540068

Mei S, Kohlstedt DL (2000) Influence of water on plastic deformation of olivine aggregates: 2. Dislocation creep regime. J Geophys Res, Solid Earth 105(B9):21471–21481. https://doi.org/10.1029/2000JB900180

Melosh HJ (1990) Giant impacts and the thermal state of the early Earth. In: Newsom HE, Jones JH (eds) Origin of the Earth, Lunar and Planetary Institute, pp 69–83

Melosh H, Vickery A (1989) Impact erosion of the primordial atmosphere of Mars. Nature 338(6215):487–489

Menou K (2013) Water-trapped worlds. Astrophys J 774(1):51. https://doi.org/10.1088/0004-637X/774/1/51

Merk R, Breuer D, Spohn T (2002) Numerical modeling of ^{26}Al-induced radioactive melting of asteroids considering accretion. Icarus https://doi.org/10.1006/icar.2002.6872

Miller SL (1953) A production of amino acids under possible primitive Earth conditions. Science 117(3046):528–529. https://doi.org/10.1126/science.117.3046.528

Miller SL (1955) Production of some organic compounds under possible primitive Earth Conditions1. J Am Chem Soc 77(9):2351–2361. https://doi.org/10.1021/ja01614a001

Miller SL, Urey HC (1959) Organic compound synthesis on the primitive Earth. Science 130(3370):245–251. https://doi.org/10.1126/science.130.3370.245

Miyazaki Y, Korenaga J (2021) Inefficient Water Degassing Inhibits Ocean Formation on Rocky Planets: an Insight from Self-Consistent Mantle Degassing Models. Astrobiology. https://doi.org/10.1089/ast.2021.0126

Miyazaki Y, Korenaga J (2022) A wet heterogeneous mantle creates a habitable world in the Hadean. Nature 603(7899):86–90. https://doi.org/10.1038/s41586-021-04371-9

Miyazaki A, Hiyagon H, Sugiura N, Hirose K, Takahashi E (2004) Solubilities of nitrogen and noble gases in silicate melts under various oxygen fugacities: implications for the origin and degassing history of nitrogen and noble gases in the Earth. Geochim Cosmochim Acta 68(2):387–401. https://doi.org/10.1016/S0016-7037(03)00484-8

Modirrousta-Galian D, Korenaga J (2023) The three regimes of atmospheric evaporation for super-earths and sub-neptunes. Astrophys J 943(1):11. https://doi.org/10.3847/1538-4357/ac9d34

Mojzsis SJ, Brasser R, Kelly NM, Abramov O, Werner SC (2019) Onset of giant planet migration before 4480 million years ago. Astrophys J 881(1):44. https://doi.org/10.3847/1538-4357/ab2c03

Monteux J, Ricard Y, Coltice N, Dubuffet F, Ulvrova M (2009) A model of metal-silicate separation on growing planets. Earth Planet Sci Lett 287(3–4):353–362. https://doi.org/10.1016/j.epsl.2009.08.020

Moore G (2008) Interpreting H_2O and CO_2 contents in melt inclusions: constraints from solubility experiments and modeling. Rev Mineral Geochem 69:333–361. https://doi.org/10.2138/rmg.2008.69.9

Morbidelli A, Chambers J, Lunine JI, Petit JM, Robert F, Valsecchi GB, Cyr KE (2000) Source regions and timescales for the delivery of water to the Earth. Meteorit Planet Sci 35(6):1309–1320. https://doi.org/10.1111/j.1945-5100.2000.tb01518.x

Morbidelli A, Lunine JI, O'Brien DP, Raymond SN, Walsh KJ (2012) Building terrestrial planets. Annu Rev Earth Planet Sci 40:251–275. https://doi.org/10.1146/annurev-earth-042711-105319

Morbidelli A, Nesvorny D, Laurenz V, Marchi S, Rubie DC, Elkins-Tanton L, Wieczorek M, Jacobson S (2018) The timeline of the lunar bombardment: revisited. Icarus 305:262–276. https://doi.org/10.1016/j.icarus.2017.12.046

Morbidelli A, Baillié K, Batygin K, Charnoz S, Guillot T, Rubie DC, Kleine T (2022) Contemporary formation of early Solar System planetesimals at two distinct radial locations. Nat Astron 6(1):72–79. https://doi.org/10.1038/s41550-021-01517-7

Moroz VI, Ekonomov AP, Moshkin BE, Revercomb HE, Sromovsky LA, Schofield JT, Spänkuch D, Taylor FW, Tomasko MG (1985) Solar and thermal radiation in the Venus atmosphere. Adv Space Res 5(11):197–232. https://doi.org/10.1016/0273-1177(85)90202-9

Mosenfelder JL, Asimow PD, Ahrens TJ (2007) Thermodynamic properties of Mg_2SiO_4 liquid at ultra-high pressures from shock measurements to 200 GPa on forsterite and wadsleyite. J Geophys Res, Solid Earth 112(B6). https://doi.org/10.1029/2006JB004364

Nakajima M, Stevenson DJ (2015) Melting and mixing states of the Earth's mantle after the Moon-forming impact. Earth Planet Sci Lett 427:286–295. https://doi.org/10.1016/j.epsl.2015.06.023

Nakajima S, Hayashi YY, Abe Y (1992) A study on the "runaway greenhouse effect" with a one-dimensional radiative–convective equilibrium model. J Atmos Sci 49(23):2256–2266. https://doi.org/10.1175/1520-0469(1992)049<2256:ASOTGE>2.0.CO;2

Nakajima M, Golabek GJ, Wünnemann K, Rubie DC, Burger C, Melosh HJ, Jacobson SA, Manske L, Hull SD (2021) Scaling laws for the geometry of an impact-induced magma ocean. Earth Planet Sci Lett 568:116983. https://doi.org/10.1016/j.epsl.2021.116983

Nesvorný D (2018) Dynamical evolution of the early solar system. Annu Rev Astron Astrophys 56(1):137–174. https://doi.org/10.1146/annurev-astro-081817-052028

Nesvorný D, Vokrouhlický D, Bottke WF, Levison HF (2018) Evidence for very early migration of the Solar System planets from the Patroclus–Menoetius binary Jupiter Trojan. Nat Astron 2(11):878–882. https://doi.org/10.1038/s41550-018-0564-3

Newman WI, Symbalisty EM, Ahrens TJ, Jones EM (1999) Impact erosion of planetary atmospheres: some surprising results. Icarus 138(2):224–240

Newsom HE, Ross Taylor S (1989) Geochemical implications of the formation of the moon by a single giant impact. Nature 338(6210):29–34. https://doi.org/10.1038/338029a0

Ni H, Zhang L, Guo X (2016) Water and partial melting of Earth's mantle. Sci China Earth Sci 59(4):720–730. https://doi.org/10.1007/s11430-015-5254-8

Nicklas RW, Puchtel IS, Ash RD (2018) Redox state of the Archean mantle: evidence from V partitioning in 3.5–2.4 Ga komatiites. Geochim Cosmochim Acta 222:447–466. https://doi.org/10.1016/j.gca.2017.11.002

Nikolaou A, Katyal N, Tosi N, Godolt M, Lee Grenfell J, Rauer H (2019) What factors affect the duration and outgassing of the terrestrial magma ocean? Astrophys J 875:11. https://doi.org/10.3847/1538-4357/ab08ed

Nimmo F, Kleine T (2015) Early differentiation and core formation. In: Badro J, Walter M (eds) The early Earth: accretion and differentiation. AGU, Washington DC, pp 83–102. https://doi.org/10.1002/9781118860359.ch5

Noack L, Godolt M, Von Paris P, Plesa AC, Stracke B, Breuer D, Rauer H (2014) Can the interior structure influence the habitability of a rocky planet? Planet Space Sci 98:14–29. https://doi.org/10.1016/j.pss.2014.01.003

Noack L, Kislyakova KG, Johnstone CP, Güdel M, Fossati L (2021) Interior heating and outgassing of proxima centauri b: identifying critical parameters. Astron Astrophys 651:1–17. https://doi.org/10.1051/0004-6361/202040176

Noda S, Ishiwatari M, Nakajima K, Takahashi YO, Takehiro S, Onishi M, Hashimoto GL, Kuramoto K, Hayashi YY (2017) The circulation pattern and day-night heat transport in the atmosphere of a synchronously rotating aquaplanet: dependence on planetary rotation rate. Icarus 282:1–18. https://doi.org/10.1016/j.icarus.2016.09.004

Nomura R, Ozawa H, Tateno S, Hirose K, Hernlund J, Muto S, Ishii H, Hiraoka N (2011) Spin crossover and iron-rich silicate melt in the Earth/'s deep mantle. Nature 473(7346):199–202. https://doi.org/10.1038/nature09940

Nomura R, Hirose K, Uesugi K, Ohishi Y, Tsuchiyama A, Miyake A, Ueno Y (2014) Low core-mantle boundary temperature inferred from the solidus of pyrolite. Science 343(6170):522–525. https://doi.org/10.1126/science.1248186

Nyquist LE, Reese YD, Wiesmann H, Shih CY, Takeda H (2001) Live ^{53}Mn and ^{26}Al in an unique cumulate eucrite with very calcic feldspar (An∼ 98). Meteorit. Planet. Sci. Suppl. 36:A151–A152

Nyquist LE, Kleine T, Shih CY, Reese YD (2009) The distribution of short-lived radioisotopes in the early solar system and the chronology of asteroid accretion, differentiation, and secondary mineralization. Geochim Cosmochim Acta 73(17):5115–5136. https://doi.org/10.1016/j.gca.2008.12.031

O'Brien DP, Walsh KJ, Morbidelli A, Raymond SN, Mandell AM (2014) Water delivery and giant impacts in the 'Grand Tack' scenario. Icarus 239:74–84. https://doi.org/10.1016/j.icarus.2014.05.009

O'Brien DP, Izidoro A, Jacobson SA, Raymond SN, Rubie DC (2018) The delivery of water during terrestrial planet formation. Space Sci Rev 214(1):47. https://doi.org/10.1007/s11214-018-0475-8

Odert P, Lammer H, Erkaev N, Nikolaou A, Lichtenegger H, Johnstone C, Kislyakova K, Leitzinger M, Tosi N (2018) Escape and fractionation of volatiles and noble gases from Mars-sized planetary embryos and growing protoplanets. Icarus 307:327–346. https://doi.org/10.1016/j.icarus.2017.10.031

Ohtani E (2020) The role of water in Earth's mantle. Nat Sci Rev 7(1):224–232. https://doi.org/10.1093/nsr/nwz071

Okuchi T (1997) Hydrogen partitioning into Molten iron at high pressure: implications for Earth's core. Science 278(5344):1781–1784

Olson P, Sharp ZD (2018) Hydrogen and helium ingassing during terrestrial planet accretion. Earth Planet Sci Lett 498:418–426. https://doi.org/10.1016/j.epsl.2018.07.006

Olson PL, Sharp ZD (2019) Nebular atmosphere to magma ocean: a model for volatile capture during Earth accretion. Phys Earth Planet Inter 294. https://doi.org/10.1016/j.pepi.2019.106294

O'Neill H (1991) The origin of the moon and the early history of the Earth—a chemical model. Part 2: the Earth. Geochim Cosmochim Acta 55(4):1159–1172. https://doi.org/10.1016/0016-7037(91)90169-6

O'Neill C, Marchi S, Zhang S, Bottke W (2017) Impact-driven subduction on the Hadean Earth. Nat Geosci 10(10):793–797. https://doi.org/10.1038/ngeo3029

Ormel CW (2017) The emerging paradigm of pebble accretion. In: Pessah M, Gressel O (eds) Formation, evolution, and dynamics of young solar systems. Springer, Cham, pp 197–228. https://doi.org/10.1007/978-3-319-60609-5_7

Oro J, Miller SL, Lazcano A (1990) The origin and early evolution of life on Earth. Annu Rev Earth Planet Sci 18(1):317–356. https://doi.org/10.1146/annurev.ea.18.050190.001533

O'Rourke JG (2020) Venus: a thick basal magma ocean may exist today. Geophys Res Lett 47(4):e86126. https://doi.org/10.1029/2019GL086126

Ottonello G, Richet P, Papale P (2018) Bulk solubility and speciation of H_2O in silicate melts. Chem Geol 479:176–187. https://doi.org/10.1016/J.CHEMGEO.2018.01.008

Pahlevan K, Schaefer L, Hirschmann MM (2019) Hydrogen isotopic evidence for early oxidation of silicate Earth. Earth Planet Sci Lett 526:115770. https://doi.org/10.1016/j.epsl.2019.115770

Pan V, Holloway JR, Hervig RL (1991) The pressure and temperature dependence of carbon dioxide solubility in tholeiitic basalt melts. Geochim Cosmochim Acta 55(6):1587–1595. https://doi.org/10.1016/0016-7037(91)90130-W

Papale P (1997) Modeling of the solubility of a one-component H_2O or CO_2 fluid in silicate liquids. Contrib Mineral Petrol 126(3):237–251. https://doi.org/10.1007/s004100050247

Papale P (1999) Modeling of the solubility of a two-component $H_2O + CO_2$ fluid in silicate liquids. Am Mineral 84(4):477–492. https://doi.org/10.1007/s004100050247

Papale P, Moretti R, Barbato D (2006) The compositional dependence of the saturation surface of $H_2O + CO_2$ fluids in silicate melts. Chem Geol https://doi.org/10.1016/j.chemgeo.2006.01.013

Parfitt EA, Wilson L (2008) Fundamentals of physical volcanology. Blackwell, Malden. https://www.wiley.com/en-us/Fundamentals+of+Physical+Volcanology-p-9780632054435

Pasek M, Lauretta D (2008) Extraterrestrial flux of potentially prebiotic C, N, and P to the early Earth. Orig Life Evol Biosph 38(1):5–21. https://doi.org/10.1007/s11084-007-9110-5

Patočka V, Calzavarini E, Tosi N (2020) Settling of inertial particles in turbulent Rayleigh-Bénard convection. Phys Rev Fluids 5(11):114304. https://doi.org/10.1103/PhysRevFluids.5.114304

Patočka V, Tosi N, Calzavarini E (2022) Residence time of inertial particles in 3D thermal convection: implications for magma reservoirs. Earth Planet Sci Lett 591:117622. https://doi.org/10.1016/j.epsl.2022.117622

Pepin RO (1991) On the origin and early evolution of terrestrial planet atmospheres and meteoritic volatiles. Icarus 92(1):2–79. https://doi.org/10.1016/0019-1035(91)90036-S

Peslier AH, Schönbächler M, Busemann H, Karato SI (2017) Water in the Earth's interior: distribution and origin. Space Sci Rev 212(1–2):743–810. https://doi.org/10.1007/s11214-017-0387-z

Pham LBS, Karatekin Ö Dehant V (2011) Effects of impacts on the atmospheric evolution: comparison between Mars, Earth, and Venus. Planet Space Sci 59(10):1087–1092. https://doi.org/10.1016/j.pss.2010.11.010

Piani L, Marrocchi Y, Rigaudier T, Vacher LG, Thomassin D, Marty B (2020) Earth's water may have been inherited from material similar to enstatite chondrite meteorites. Science 369(6507):1110–1113. https://doi.org/10.1126/science.aba1948

Pierazzo E, Chyba CF (1999) Amino acid survival in large cometary impacts. Meteorit Planet Sci 34(6):909–918. https://doi.org/10.1111/j.1945-5100.1999.tb01409.x

Pierrehumbert RT (2010) Principles of planetary climate. Cambridge University Press, Cambridge. https://doi.org/10.1017/CBO9780511780783

Pierrehumbert R, Gaidos E (2011) Hydrogen greenhouse planets beyond the habitable zone. Astrophys J Lett 734(1):L13. https://doi.org/10.1088/2041-8205/734/1/L13

Pluriel W, Marcq E, Turbet M (2019) Modeling the albedo of Earth-like magma ocean planets with H_2O-CO_2 atmospheres. Icarus 317:583–590. https://doi.org/10.1016/j.icarus.2018.08.023. arXiv:1809.02036

Poirier JP (1994) Light elements in the Earth's outer core: a critical review. Phys Earth Planet Inter 85(3):319–337. https://doi.org/10.1016/0031-9201(94)90120-1

Pollack JB (1971) A nongrey calculation of the runaway greenhouse: implications for Venus' past and present. Icarus 14(3):295–306. https://doi.org/10.1016/0019-1035(71)90001-7

Prša A, Harmanec P, Torres G, Mamajek E, Asplund M, Capitaine N, Christensen-Dalsgaard J, Depagne É, Haberreiter M, Hekker S, Hilton J, Kopp G, Kostov V, Kurtz DW, Laskar J, Mason BD, Milone EF, Montgomery M, Richards M, Schmutz W, Schou J, Stewart SG (2016) Nominal values for selected solar and planetary quantities: IAU 2015 resolution B3. Astron J 152(2):41. https://doi.org/10.3847/0004-6256/152/2/41

Quintana EV, Barclay T, Borucki WJ, Rowe JF, Chambers JE (2016) The frequency of giant impacts on Earth-like worlds. Astrophys J 821(2):126. https://doi.org/10.3847/0004-637X/821/2/126

Quitté G, Birck JL, Allègre CJ (2000) 182Hf–182W systematics in eucrites: the puzzle of iron segregation in the early solar system. Earth Planet Sci Lett 184(1):83–94. https://doi.org/10.1016/S0012-821X(00)00303-4

Ramirez RM (2018) A more comprehensive habitable zone for finding life on other planets. Geosciences 8(280). https://doi.org/10.3390/geosciences8080280. https://www.mdpi.com/2076-3263/8/8/280

Ramirez RM, Kaltenegger L (2017) A volcanic hydrogen habitable zone. Astrophys J Lett 837(1):L4. https://doi.org/10.3847/2041-8213/aa60c8

Ramirez RM, Kaltenegger L (2018) A methane extension to the classical habitable zone. Astrophys J 858(2):72. https://doi.org/10.3847/1538-4357/aab8fa

Ramirez RM, Kopparapu R, Zugger ME, Robinson TD, Freedman R, Kasting JF (2014) Warming early Mars with CO_2 and H_2. Nat Geosci 7(1):59–63. https://doi.org/10.1038/ngeo2000

Ramirez RM, Abbot DS, Airapetian V, Fujii Y, Hamano K, Levi A, Robinson TD, Schaefer L, Wolf ET, Wordsworth RD (2018) The continued importance of habitability studies. In: National academies of sciences engineering and medicine. Exoplanet science strategy - white papers. National Academies Press. https://doi.org/10.48550/arXiv.1803.00215

Rasool SI de Bergh C (1970) The runaway greenhouse and the accumulation of CO_2 in the Venus atmosphere. Nature 226(5250):1037–1039. https://doi.org/10.1038/2261037a0

Raymond SN, Izidoro A (2017) Origin of water in the inner solar system: planetesimals scattered inward during Jupiter and Saturn's rapid gas accretion. Icarus 297:134–148. https://doi.org/10.1016/j.icarus.2017.06.030

Raymond SN, Izidoro A (2017b) The empty primordial asteroid belt. Sci Adv 3(9):e1701138. https://doi.org/10.1126/sciadv.1701138

Raymond SN, Morbidelli A (2014) The grand tack model: a critical review. In: Complex planetary systems. Proceedings of the international astronomical union, vol 310, pp 194–203. https://doi.org/10.1017/S1743921314008254. arXiv:1409.6340

Raymond SN, Morbidelli A (2022) Planet formation: key mechanisms and global models. In: Astrophysics and space science library, vol 466. Springer, Cham, pp 3–82. https://doi.org/10.1007/978-3-030-88124-5_1

Raymond SN, Nesvorný D (2022) Origin and dynamical evolution of the asteroid belt. In: Marchi S, Raymond CA, Russell CT (eds) Vesta and Ceres: insights from the Dawn Mission for the origin of the solar system. Cambridge planetary science. Cambridge University Press, Cambridge, pp 227–249. https://doi.org/10.1017/9781108856324.019

Raymond SN, Quinn T, Lunine JI (2004) Making other earths: dynamical simulations of terrestrial planet formation and water delivery. Icarus 168(1):1–17. https://doi.org/10.1016/j.icarus.2003.11.019

Raymond SN, Quinn T, Lunine JI (2006) High-resolution simulations of the final assembly of Earth-like planets I. Terrestrial accretion and dynamics. Icarus 183(2):265–282. https://doi.org/10.1016/j.icarus.2006.03.011

Raymond SN, O'Brien DP, Morbidelli A, Kaib NA (2009) Building the terrestrial planets: constrained accretion in the inner solar system. Icarus 203(2):644–662. https://doi.org/10.1016/j.icarus.2009.05.016

Raymond SN, Kokubo E, Morbidelli A, Morishima R, Walsh KJ (2014) Terrestrial planet formation at home and abroad. In: Protostars and planets VI. University of Arizona Press, Tucson, pp 595–618. https://doi.org/10.2458/azu_uapress_9780816531240-ch026

Raymond SN, Izidoro A, Morbidelli A (2020) Solar system formation in the context of extrasolar planets. In: Meadows VS, Arney GN, Schmidt BE, Des Marais DJ (eds) Planetary astrobiology. University of Arizona Press, Tucson, p 287. https://doi.org/10.2458/azu_uapress_9780816540068

Regenauer-Lieb K, Yuen DA, Branlund J (2001) The initiation of subduction: criticality by addition of water? Science 294(5542):578–580. https://doi.org/10.1126/science.1063891

Reufer A, Meier MMM, Benz W, Wieler R (2012) A hit-and-run giant impact scenario. Icarus 221(1):296–299. https://doi.org/10.1016/j.icarus.2012.07.021

Ribas I, Guinan EF, Güdel M, Audard M (2005) Evolution of the solar activity over time and effects on planetary atmospheres. I. High-energy irradiances (1–1700 Å). Astrophys J 622(1):680–694. https://doi.org/10.1086/427977

Ricard Y (2015) 7.02 - Physics of mantle convection. In: Schubert G (ed) Treatise on geophysics, 2nd edn. Elsevier, Oxford, pp 23–71. https://doi.org/10.1016/B978-0-444-53802-4.00127-5

Ricard Y, Šrámek O, Dubuffet F (2009) A multi-phase model of runaway core–mantle segregation in planetary embryos. Earth Planet Sci Lett 284(1):144–150. https://doi.org/10.1016/j.epsl.2009.04.021

Richards MA, Lenardic A (2018) The cathles parameter (Ct): a geodynamic definition of the asthenosphere and implications for the nature of plate tectonics. Geochem Geophys Geosyst 19(12):4858–4875. https://doi.org/10.1029/2018GC007664

Richards MA, Yang WS, Baumgardner JR, Bunge HP (2001) Role of a low-viscosity zone in stabilizing plate tectonics: Implications for comparative terrestrial planetology. Geochem Geophys Geosyst 2(8). https://doi.org/10.1029/2000GC000115

Righter K, Drake MJ (1997) Metal-silicate equilibrium in a homogeneously accreting Earth: new results for Re. Earth Planet Sci Lett 146(3):541–553. https://doi.org/10.1016/S0012-821X(96)00243-9

Righter K, Drake MJ (1999) Effect of water on metal–silicate partitioning of siderophile elements: a high pressure and temperature terrestrial magma ocean and core formation. Earth Planet Sci Lett 171(3):383–399. https://doi.org/10.1016/S0012-821X(99)00156-9

Righter K, Ghiorso MS (2012) Redox systematics of a magma ocean with variable pressure-temperature gradients and composition. Proc Natl Acad Sci 109(30):11955–11960. https://doi.org/10.1073/pnas.1202754109

Righter K, Drake MJ, Yaxley G (1997) Prediction of siderophile element metal-silicate partition coefficients to 20 GPa and 2800 °C: the effects of pressure, temperature, oxygen fugacity, and silicate and metallic melt compositions. Phys Earth Planet Inter 100(1):115–134. https://doi.org/10.1016/S0031-9201(96)03235-9

Robert F, Gautier D, Dubrulle B (2000) The solar system D/H ratio: observations and theories. Space Sci Rev 92(1):201–224. https://doi.org/10.1023/A:1005291127595

Robie RA, Hemingway BS, Fisher JR (1978) Thermodynamic properties of minerals and related substances at 298.15 K and 1 bar (10^5 pascals) pressure and at higher temperatures. https://doi.org/10.3133/b1452. Tech. Rep. http://pubs.er.usgs.gov/publication/b1452

Rolf T, Weller M, Gülcher A, Byrne P, O'Rourke JG, Herrick R, Bjonnes E, Davaille A, Ghail R, Gillmann C, Plesa AC, Smrekar S (2022) Dynamics and evolution of Venus' mantle through time. Space Sci Rev 218(8):70. https://doi.org/10.1007/s11214-022-00937-9

Rollinson H, Adetunji J, Lenaz D, Szilas K (2017) Archaean chromitites show constant $Fe3+/\Sigma Fe$ in Earth's asthenospheric mantle since 3.8 Ga. Lithos 282–283:316–325. https://doi.org/10.1016/j.lithos.2017.03.020

Rubie DC, Melosh HJ, Reid JE, Liebske C, Righter K (2003) Mechanisms of metal–silicate equilibration in the terrestrial magma ocean. Earth Planet Sci Lett 205(3):239–255. https://doi.org/10.1016/S0012-821X(02)01044-0

Rubie DC, Frost DJ, Mann U, Asahara Y, Nimmo F, Tsuno K, Kegler P, Holzheid A, Palme H (2011) Heterogeneous accretion, composition and core–mantle differentiation of the Earth. Earth Planet Sci Lett 301(1):31–42. https://doi.org/10.1016/j.epsl.2010.11.030

Rubie DC, Jacobson SA, Morbidelli A, O'Brien DP, Young ED, de Vries J, Nimmo F, Palme H, Frost DJ (2015) Accretion and differentiation of the terrestrial planets with implications for the compositions of early-formed Solar System bodies and accretion of water. Icarus 248:89–108. https://doi.org/10.1016/j.icarus.2014.10.015

Rubie DC, Nimmo F, Melosh HJ (2015b) 9.03 - Formation of the Earth's core. In: Schubert G (ed) Treatise on geophysics, 2nd edn. Elsevier, Oxford, pp 43–79. https://doi.org/10.1016/B978-0-444-53802-4.00154-8

Rubin AE, Fegley B, Brett R (1988) Oxidation state in chondrites. In: Kerridge JF, Matthews MS (eds) Meteorites and the early solar system. University of Arizona Press, Tucson, pp 488–511. https://ui.adsabs.harvard.edu/abs/1988mess.book..488R

Rufu R, Aharonson O, Perets HB (2017) A multiple-impact origin for the Moon. Nat Geosci 10. https://doi. org/10.1038/ngeo2866

Safronov VS (1972) Evolution of the protoplanetary cloud and formation of the Earth and planets. Keter Publishing House. https://www.semanticscholar.org/paper/Evolution-of-the-protoplanetary-cloud-and-formation-Safronov/97c82912840accc7096404cbea0b59d5f02d8c21

Safronov VS (1978) The heating of the Earth during its formation. Icarus 33(1):3–12. https://doi.org/10.1016/0019-1035(78)90019-2

Saito H, Kuramoto K (2017) Formation of a hybrid-type proto-atmosphere on Mars accreting in the solar nebula. Mon Not R Astron Soc 475(1):1274–1287. https://doi.org/10.1093/mnras/stx3176

Sakuraba H, Kurokawa H, Genda H (2019) Impact degassing and atmospheric erosion on Venus, Earth, and Mars during the late accretion. Icarus 317:48–58. https://doi.org/10.1016/J.ICARUS.2018.05.035

Sakuraba H, Kurokawa H, Genda H, Ohta K (2021) Numerous chondritic impactors and oxidized magma ocean set Earth's volatile depletion. Sci Rep 11(1):20894. https://doi.org/10.1038/s41598-021-99240-w

Salvador A, Samuel H (2023) Convective outgassing efficiency in planetary magma oceans: insights from computational fluid dynamics. Icarus 390:115265. https://doi.org/10.1016/j.icarus.2022.115265

Salvador A, Massol H, Davaille A, Marcq E, Sarda P, Chassefière E (2017) The relative influence of H_2O and CO_2 on the primitive surface conditions and evolution of rocky planets. J Geophys Res, Planets 122(7):1458–1486. https://doi.org/10.1002/2017JE005286

Samuel H, Tackley PJ, Evonuk M (2010) Heat partitioning in terrestrial planets during core formation by negative diapirism. Earth Planet Sci Lett 290(1):13–19. https://doi.org/10.1016/j.epsl.2009.11.050

Sasaki S, Nakazawa K (1986) Metal-silicate fractionation in the growing Earth: energy source for the terrestrial magma ocean. J Geophys Res, Solid Earth 91(B9):9231–9238. https://doi.org/10.1029/JB091iB09p09231

Sasaki S, Nakazawa K (1988) Origin of isotopic fractionation of terrestrial Xe: hydrodynamic fractionation during escape of the primordial H_2-He atmosphere. Earth Planet Sci Lett 89(3):323–334. https://doi.org/10.1016/0012-821X(88)90120-3

Sato M (1978) Oxygen fugacity of basaltic magmas and the role of gas-forming elements. Geophys Res Lett 5(6):447–449. https://doi.org/10.1029/GL005i006p00447

Sato T, Okuzumi S, Ida S (2016) On the water delivery to terrestrial embryos by ice pebble accretion. Astron Astrophys 589:A15. https://doi.org/10.1051/0004-6361/201527069

Savage PS, Moynier F, Boyet M (2022) Zinc isotope anomalies in primitive meteorites identify the outer solar system as an important source of Earth's volatile inventory. Icarus 386:115172. https://doi.org/10.1016/j.icarus.2022.115172

Saxena P, Killen RM, Airapetian V, Petro NE, Curran NM, Mandell AM (2019) Was the sun a slow rotator? Sodium and potassium constraints from the lunar regolith. Astrophys J Lett 876(1):L16. https://doi.org/10.3847/2041-8213/ab18fb

Scaillet B, Gaillard F (2011) Redox state of early magmas. Nature 480(7375):48–49. https://doi.org/10.1038/480048a

Schaefer L, Elkins-Tanton LT (2018) Magma oceans as a critical stage in the tectonic development of rocky planets. Philos Trans R Soc A, Math Phys Eng Sci 376(2132). https://doi.org/10.1098/rsta.2018.0109

Schaefer L, Fegley B (2007) Outgassing of ordinary chondritic material and some of its implications for the chemistry of asteroids, planets, and satellites. Icarus 186(2):462–483. https://doi.org/10.1016/j.icarus.2006.09.002

Schaefer L, Fegley BJ (2010) Chemistry of atmospheres formed during accretion of the Earth and other terrestrial planets. Icarus 208(1):438–448. https://doi.org/10.1016/j.icarus.2010.01.026

Schaefer L, Fegley B (2017) Redox states of initial atmospheres outgassed on rocky planets and planetesimals. Astrophys J 843:120. https://doi.org/10.3847/1538-4357/aa784f

Schaefer L, Wordsworth RD, Berta-Thompson Z, Sasselov D (2016) Predictions of the atmospheric composition of GJ 1132b. Astrophys J 829(2):63. https://doi.org/10.3847/0004-637X/829/2/63

Schlichting HE, Sari R, Yalinewich A (2015) Atmospheric mass loss during planet formation: the importance of planetesimal impacts. Icarus 247:81–94. https://doi.org/10.1016/j.icarus.2014.09.053

Schnetzler C, Philpotts J (1971) Alkali, alkaline Earth and rare-Earth element concentrations in some Apollo 12 soils, rocks, and separated phases. In: Lunar and planetary science conference proceedings, vol 2, p 1101. https://ui.adsabs.harvard.edu/abs/1971LPSC....2.1101S/abstract

Schubert G, Turcotte DL, Olson P (2001) Mantle convection in the Earth and planets. Cambridge University Press, Cambridge. https://doi.org/10.1017/CBO9780511612879

Sekiya M, Nakazawa K, Hayashi C (1980) Dissipation of the rare gases contained in the primordial Earth's atmosphere. Earth Planet Sci Lett 50(1):197–201. https://doi.org/10.1016/0012-821X(80)90130-2

Selsis F, Kating JF, Levrard B, Paillet J, Ribas I, Delfosse X (2007) Habitable planets around the star Gliese 581? Astron Astrophys 476(3):1373–1387. https://doi.org/10.1051/0004-6361:20078091

Shahar A, Schauble EA, Caracas R, Gleason AE, Reagan MM, Xiao Y, Shu J, Mao W (2016) Pressure-dependent isotopic composition of iron alloys. Science 352(6285):580–582. https://doi.org/10.1126/science.aad9945

Sharp ZD (2017) Nebular ingassing as a source of volatiles to the Terrestrial planets. Chem Geol 448:137–150. https://doi.org/10.1016/j.chemgeo.2016.11.018

Sharp ZD, McCubbin FM, Shearer CK (2013) A hydrogen-based oxidation mechanism relevant to planetary formation. Earth Planet Sci Lett 380:88–97. https://doi.org/10.1016/j.epsl.2013.08.015

Shcheka SS, Wiedenbeck M, Frost DJ, Keppler H (2006) Carbon solubility in mantle minerals. Earth Planet Sci Lett 245(3):730–742. https://doi.org/10.1016/j.epsl.2006.03.036

Shi L, Lu W, Kagoshima T, Sano Y, Gao Z, Du Z, Liu Y, Fei Y, Li Y (2022) Nitrogen isotope evidence for Earth's heterogeneous accretion of volatiles. Nat Commun 13(1):4769. https://doi.org/10.1038/s41467-022-32516-5

Shibazaki Y, Ohtani E, Terasaki H, Suzuki A, Ki F (2009) Hydrogen partitioning between iron and ringwoodite: implications for water transport into the Martian core. Earth Planet Sci Lett 287(3):463–470. https://doi.org/10.1016/j.epsl.2009.08.034

Shuvalov V (2009) Atmospheric erosion induced by oblique impacts. Meteorit Planet Sci 44(8):1095–1105. https://doi.org/10.1111/j.1945-5100.2009.tb01209.x

Shuvalov VV, Artemieva NA (2002) Atmospheric erosion and radiation impulse induced by impacts. In: Catastrophic events and mass extinctions: impacts and beyond. Geological society of America. https://doi.org/10.1130/0-8137-2356-6.695

Shuvalov V, Kührt E, de Niem D, Wünnemann K (2014) Impact induced erosion of hot and dense atmospheres. Planet Space Sci 98:120–127

Siebesma A, Bony S, Jakob C, Stevens B (eds) (2020) Clouds and climate: climate science's greatest challenge Cambridge University Press, Cambridge. https://doi.org/10.1017/9781107447738

Siggia E (1994) High Rayleigh number convection. Annu Rev Fluid Mech 26(1):137–168. https://doi.org/10.1146/annurev.fluid.26.1.137

Simpson GC (1927) Some studies in terrestrial radiation. Memoirs of the Royal Meteorological Society 11(16):69–95

Skemer P, Warren JM, Hansen LN, Hirth G, Kelemen PB (2013) The influence of water and LPO on the initiation and evolution of mantle shear zones. Earth Planet Sci Lett 375:222–233. https://doi.org/10.1016/j.epsl.2013.05.034

Sleep NH, Zahnle KJ (2001) Carbon dioxide cycling and implications for climate on ancient Earth. J Geophys Res, Planets 106(E1):1373–1399. https://doi.org/10.1029/2000JE001247

Smith J, Anderson A, Newton R, Olsen E, Crewe A, Isaacson M, Johnson D, Wyllie P (1970) Petrologic history of the moon inferred from petrography, mineralogy and petrogenesis of Apollo 11 rocks. Geochim Cosmochim Acta, Suppl 11:897–925. Proceedings of the Apollo 11 Lunar Science Conference, https://ui.adsabs.harvard.edu/abs/1970GeCAS...1..897S

Smrekar SE, Elkins-Tanton L, Leitner JJ, Lenardic A, Mackwell S, Moresi L, Sotin C, Stofan ER (2007) Tectonic and thermal evolution of Venus and the role of volatiles: implications for understanding the terrestrial planets. In: Exploring Venus as a terrestrial planet. Am. Geophys. Union, Washington, pp 45–71. https://doi.org/10.1029/176GM05

Smrekar SE, Davaille A, Sotin C (2018) Venus interior structure and dynamics. Space Sci Rev 214(5):88. https://doi.org/10.1007/s11214-018-0518-1

Smrekar S, Hensley S, Nybakken R, Wallace MS, Perkovic-Martin D, You TH, Nunes D, Brophy J, Ely T, Burt E, Dyar MD, Helbert J, Miller B, Hartley J, Kallemeyn P, Whitten J, Iess L, Mastrogiuseppe M, Younis M, Prats P, Rodriguez M, Mazarico E (2022) VERITAS (Venus emissivity, radio science, InSAR, topography, and spectroscopy): a discovery mission. In: 2022 IEEE aerospace conference (AERO), pp 1–20. https://doi.org/10.1109/AERO53065.2022.9843269

Solomatov V (2000) Fluid dynamics of a terrestrial magma ocean. In: Origin of the Earth and Moon, pp 323–338

Solomatov V (2007) 9.04 - Magma oceans and primordial mantle differentiation. In: Schubert G (ed) Treatise on geophysics. Elsevier, Amsterdam, pp 91–119. https://doi.org/10.1016/B978-044452748-6.00141-3

Solomatov V (2015) 9.04 - Magma oceans and primordial mantle differentiation. In: Schubert G (ed) Treatise on geophysics, 2nd edn. Elsevier, Oxford, pp 81–104. https://doi.org/10.1016/B978-0-444-53802-4.00155-X

Solomatov VS, Stevenson DJ (1993) Nonfractional crystallization of a terrestrial magma ocean. J Geophys Res, Planets 98(E3):5391–5406. https://doi.org/10.1029/92JE02579

Solomatova NV, Caracas R (2021) Genesis of a CO_2-rich and H_2O-depleted atmosphere from Earth's early global magma ocean. Sci Adv 7(41):eabj0406. https://doi.org/10.1126/sciadv.abj0406

Solomon SC (1979) Formation, history and energetics of cores in the terrestrial planets. Phys Earth Planet Inter 19(2):168–182. https://doi.org/10.1016/0031-9201(79)90081-5

Sonett CP, Colburn DS, Schwartz K (1968) Electrical heating of meteorite parent bodies and planets by dynamo induction from a pre-main sequence T tauri "solar wind". Nature 219(5157):924–926. https://doi.org/10.1038/219924a0

Sossi PA, Stotz IL, Jacobson SA, Morbidelli A, O'Neill HSC (2022) Stochastic accretion of the Earth. Nat Astron 6(8):951–960. https://doi.org/10.1038/s41550-022-01702-2

Sparks RSJ (1978) The dynamics of bubble formation and growth in magmas: a review and analysis. J Volcanol Geotherm Res 3(1):1–37. https://doi.org/10.1016/0377-0273(78)90002-1

Sparks RSJ, Barclay J, Jaupart C, Mader HM, Phillips JC (1994) Chapter 11a. Physical aspects of magma degassing I. Experimental and theoretical constraints on vesiculation. In: Carroll MR, Holloway JR (eds) Volatiles in magmas. Mineralogical Society of America, Washington, pp 413–446. https://doi.org/10.1515/9781501509674-017

Spencer CJ, Mitchell RN, Brown M (2021) Enigmatic mid-proterozoic orogens: hot, thin, and low. Geophys Res Lett 48(16):e2021GL093312. https://doi.org/10.1029/2021GL093312.

Šrámek O, Ricard Y, Dubuffet F (2010) A multiphase model of core formation. Geophys J Int 181(1):198–220. https://doi.org/10.1111/j.1365-246X.2010.04528.x

Šrámek O, Milelli L, Ricard Y, Labrosse S (2012) Thermal evolution and differentiation of planetesimals and planetary embryos. Icarus 217(1):339–354. https://doi.org/10.1016/j.icarus.2011.11.021

Srinivasan G, Papanastassiou DA, Wasserburg GJ, Bhandari N, Goswami JN (2000) Re-examination of ^{26}Al-^{26}Mg systematics in the Piplia Kalan eucrite. Lunar and Planetary Science XXXI, 1795

Stamenković V, Breuer D (2014) The tectonic mode of rocky planets: part 1 – driving factors, models & parameters. Icarus 234:174–193. https://doi.org/10.1016/j.icarus.2014.01.042

Steller T, Burkhardt C, Yang C, Kleine T (2022) Nucleosynthetic zinc isotope anomalies reveal a dual origin of terrestrial volatiles. Icarus 386:115171. https://doi.org/10.1016/j.icarus.2022.115171

Stephant A, Remusat L Robert F (2016) Water in type I chondrules of Paris CM chondrite. Geochim Cosmochim Acta 199:75–90. https://doi.org/10.1016/j.gca.2016.11.031

Stephant A, Garvie LAJ, Mane P, Hervig R, Wadhwa M (2018) Terrestrial exposure of a fresh Martian meteorite causes rapid changes in hydrogen isotopes and water concentrations. Sci Rep 8(1):12385. https://doi.org/10.1038/s41598-018-30807-w

Stephant A, Anand M, Tartèse R, Zhao X, Degli-Alessandrini G, Franchi IA (2020) The hydrogen isotopic composition of lunar melt inclusions: an interplay of complex magmatic and secondary processes. Geochim Cosmochim Acta 284:196–221. https://doi.org/10.1016/j.gca.2020.06.017

Stern RJ (2018) The evolution of plate tectonics. Philos Trans R Soc A, Math Phys Eng Sci 376(2132):20170406. https://doi.org/10.1098/rsta.2017.0406

Stevens RJAM, van der Poel EP, Grossmann S, Lohse D (2013) The unifying theory of scaling in thermal convection: the updated prefactors. J Fluid Mech 730:295–308. https://doi.org/10.1017/jfm.2013.298

Stevens RJAM, Blass A, Zhu X, Verzicco R, Lohse D (2018) Turbulent thermal superstructures in Rayleigh-Bénard convection. Phys Rev Fluids 3(4):041501. https://doi.org/10.1103/PhysRevFluids.3.041501

Stevenson D (1990) Fluid dynamics of core formation. In: Newsom H, Jones J (eds) Origin of the Earth. Oxford University Press, London, pp 231–249

Stevenson DJ (2001) Mars' core and magnetism. Nature 412(6843):214–219. https://doi.org/10.1038/35084155

Stökl A, Dorfi EA, Johnstone CP, Lammer H (2016) Dynamical accretion of primordial atmospheres around planets with masses between 0.1 and 5 $M_\oplus$ in the habitable zone. Astrophys J 825(2):86. https://doi.org/10.3847/0004-637X/825/2/86

Sugita S, Schultz PH (2009) Efficient cyanide formation due to impacts of carbonaceous bodies on a planet with a nitrogen-rich atmosphere. Geophys Res Lett 36(20). https://doi.org/10.1029/2009GL040252

Surkov YA, Barsukov VL, Moskalyeva LP, Kharyukova VP, Kemurdzhian AL (1984) New data on the composition, structure, and properties of Venus rock obtained by Venera 13 and Venera 14. J Geophys Res, Solid Earth 89(S02):B393–B402. https://doi.org/10.1029/JB089iS02p0B393

Surkov YA, Moskalyova LP, Kharyukova VP, Dudin AD, Smirnov GG, Zaitseva SY (1986) Venus rock composition at the Vega 2 landing site. J Geophys Res, Solid Earth 91(B13):E215–E218. https://doi.org/10.1029/JB091iB13p0E215

Svetsov V (2007) Atmospheric erosion and replenishment induced by impacts of cosmic bodies upon the Earth and Mars. Sol Syst Res 41(1):28–41

Tagawa S, Sakamoto N, Hirose K, Yokoo S, Hernlund J, Ohishi Y, Yurimoto H (2021) Experimental evidence for hydrogen incorporation into Earth's core. Nat Commun 12(1):2588. https://doi.org/10.1038/s41467-021-22035-0

Takei Y, Holtzman BK (2009a) Viscous constitutive relations of solid-liquid composites in terms of grain boundary contiguity: 1. Grain boundary diffusion control model. J Geophys Res, Solid Earth 114(B6). https://doi.org/10.1029/2008JB005850

290

 Springer

Takei Y, Holtzman BK (2009b) Viscous constitutive relations of solid-liquid composites in terms of grain boundary contiguity: 2. Compositional model for small melt fractions. J Geophys Res, Solid Earth 114(B6). https://doi.org/10.1029/2008JB005851

Tateno S, Hirose K, Ohishi Y (2014) Melting experiments on peridotite to lowermost mantle conditions. J Geophys Res, Solid Earth 119(6):4684–4694. https://doi.org/10.1002/2013JB010616

Taylor SR (1986) The origin of the Moon: geochemical considerations. In: Hartmann W, Phillips R, Taylor G (eds) Origin of the Moon, Lunar and Planetary Institute, pp 125–143. https://ui.adsabs.harvard.edu/abs/1986ormo.conf..125T/abstract

Taylor G, Norman M (1992) Evidence for magma oceans on asteroids, the Moon, and Earth. Houston: Lunar Planet Inst pp 58–75. https://adsabs.harvard.edu/full/1992pcmo.work...58T

Thomas CW, Liu Q, Agee CB, Asimow PD, Lange RA (2012) Multi-technique equation of state for Fe_2SiO_4 melt and the density of Fe-bearing silicate melts from 0 to 161 GPa. J Geophys Res, Solid Earth 117(10):1–18. https://doi.org/10.1029/2012JB009403

Thompson MA, Telus M, Schaefer L, Fortney JJ, Joshi T, Lederman D (2021) Composition of terrestrial exoplanet atmospheres from meteorite outgassing experiments. Nat Astron 5(6):575–585. https://doi.org/10.1038/s41550-021-01338-8

Tian F (2015) Atmospheric escape from solar system terrestrial planets and exoplanets. Annu Rev Earth Planet Sci 43(1):459–476. https://doi.org/10.1146/annurev-earth-060313-054834

Tian F, Toon OB, Pavlov AA, De Sterck H (2005) A hydrogen-rich early Earth atmosphere. Science 308(5724):1014–1017. https://doi.org/10.1126/science.1106983

Tian F, Toon OB, Pavlov AA (2006) Response to comment on "A hydrogen-rich early Earth atmosphere". Science 311(5757):38. https://doi.org/10.1126/science.1118412

Tikoo SM, Elkins-Tanton LT (2017) The fate of water within Earth and super-earths and implications for plate tectonics. Philos Trans R Soc A, Math Phys Eng Sci 375(2094):20150394. https://doi.org/10.1098/rsta.2015.0394

Titov DV, Piccioni G, Drossart P, Markiewicz WJ (2013) Radiative energy balance in the Venus atmosphere. In: Bengtsson L, Bonnet R-M, Grinspoon D, Koumoutsaris S, Lebonnois S, Titov D (eds) Towards understanding the climate of Venus: applications of terrestrial models to our sister planet. Springer, New York, pp 23–53. https://doi.org/10.1007/978-1-4614-5064-1_4

Tonks WB, Melosh HJ (1992) Core formation by giant impacts. Icarus 100(2):326–346. https://doi.org/10.1016/0019-1035(92)90104-F

Tonks WB, Melosh HJ (1993) Magma ocean formation due to giant impacts. J Geophys Res, Planets 98(E3):5319–5333. https://doi.org/10.1029/92JE02726

Tozer DC (1965) Thermal history of the Earth: I. The formation of the core. Geophys J Int 9(2–3):95–112. https://doi.org/10.1111/j.1365-246X.1965.tb02064.x

Trail D, Watson EB, Tailby ND (2011) The oxidation state of Hadean magmas and implications for early Earth's atmosphere. Nature 480(7375):79–82. https://doi.org/10.1038/nature10655

Trieloff M, Palme H, Brandner W (2006) The origin of solids in the early Solar System. In: Klahr H, Brandner W (eds) Planet formation: theory, observations, and experiments. Cambridge astrobiology. Cambridge University Press, Cambridge, pp 64–89. https://doi.org/10.1017/CBO9780511536571.006

Tu L, Johnstone CP, Güdel M, Lammer H (2015) The extreme ultraviolet and X-ray sun in time: high-energy evolutionary tracks of a solar-like star. Astron Astrophys 577:L3. https://doi.org/10.1051/0004-6361/201526146

Tucker JM, Mukhopadhyay S (2014) Evidence for multiple magma ocean outgassing and atmospheric loss episodes from mantle noble gases. Earth Planet Sci Lett 393:254–265. https://doi.org/10.1016/j.epsl.2014.02.050

Turbet M, Tran H, Pirali O, Forget F, Boulet C, Hartmann JM (2019) Far infrared measurements of absorptions by $CH_4 + CO_2$ and $H_2 + CO_2$ mixtures and implications for greenhouse warming on early Mars. Icarus 321:189–199. https://doi.org/10.1016/j.icarus.2018.11.021

Turbet M, Boulet C, Karman T (2020) Measurements and semi-empirical calculations of $CO_2 + CH_4$ and $CO_2 + H_2$ collision-induced absorption across a wide range of wavelengths and temperatures. Application for the prediction of early Mars surface temperature. Icarus 346:113762. https://doi.org/10.1016/j.icarus.2020.113762

Turbet M, Bolmont E, Chaverot G, Ehrenreich D, Leconte J, Marcq E (2021) Day–night cloud asymmetry prevents early oceans on Venus but not on Earth. Nature 598(7880):276–280. https://doi.org/10.1038/s41586-021-03873-w

Ueki K, Inui M, Matsunaga K, Okamoto N, Oshio K (2020) Oxidation during magma mixing recorded by symplectites at Kusatsu–Shirane Volcano, Central Japan. Earth Planets Space 72(1):68. https://doi.org/10.1186/s40623-020-01192-4

Umemoto K, Hirose K (2015) Liquid iron-hydrogen alloys at outer core conditions by first-principles calculations. Geophys Res Lett 42(18):7513–7520. https://doi.org/10.1002/2015GL065899·

Urbain G, Bottinga Y, Richet P (1982) Viscosity of liquid silica, silicates and alumino-silicates. Geochim Cosmochim Acta 46(6):1061–1072. https://doi.org/10.1016/0016-7037(82)90059-X

Urey HC (1955) The cosmic abundances of potassium, uranium, and thorium and the heat balances of the Earth, the Moon, and Mars. Proc Natl Acad Sci USA 41(3):127–144. https://doi.org/10.1073/pnas.41.3.127

Urey HC, Craig H (1953) The composition of the stone meteorites and the origin of the meteorites. Geochim Cosmochim Acta 4(1):36–82. https://doi.org/10.1016/0016-7037(53)90064-7

Venturini J, Ronco MP, Guilera OM (2020) Setting the stage: planet formation and volatile delivery. Space Sci Rev 216(5):86. https://doi.org/10.1007/s11214-020-00700-y

Vickery AM, Melosh HJ (1990) Atmospheric erosion and impactor retention in large impacts, with application to mass extinctions. Spec Pap, Geol Soc Am 247:289–300

Volkov VP, Zolotov MY, Khodakovsky IL (1986) Lithospheric-atmospheric interaction on Venus. In: Saxena SK (ed) Chemistry and physics of terrestrial planets. Springer, New York, pp 136–190. https://doi.org/10.1007/978-1-4612-4928-3_4

Von Zahn U, Kumar S, Niemann H, Prinn R (1983) Composition of the Venus atmosphere. In: Venus. The University of Arizona Press, Tucson

Wacheul JB Le Bars M (2018) Experiments on fragmentation and thermo-chemical exchanges during planetary core formation. Phys Earth Planet Inter 276:134–144. https://doi.org/10.1016/j.pepi.2017.05.018

Wade J, Wood BJ (2005) Core formation and the oxidation state of the Earth. Earth Planet Sci Lett 236(1):78–95. https://doi.org/10.1016/j.epsl.2005.05.017

Wadhwa M (2008) Redox conditions on small bodies, the Moon and Mars. Rev Mineral Geochem 68(1):493–510. https://doi.org/10.2138/rmg.2008.68.17

Wadhwa M, Srinivasan G, Carlson R (2006) Timescales of planetesimal differentiation in the early solar system. In: Lauretta DS, McSween HY (eds) Meteorites and the early solar system II. University of Arizona Press, Tucson, p 715

Walker JCG, Hays PB, Kasting JF (1981) A negative feedback mechanism for the long-term stabilization of Earth's surface temperature. J Geophys Res, Oceans 86(C10):9776–9782. https://doi.org/10.1029/JC086iC10p09776

Wallace PJ, Plank T, Edmonds M, Hauri EH (2015) Chap. 7 – Volatiles in magmas. In: Sigurdsson H (ed) The encyclopedia of volcanoes, 2nd edn. Academic Press, San Diego, pp 163–183. https://doi.org/10.1016/B978-0-12-385938-9.00007-9

Walsh KJ, Morbidelli A, Raymond SN, O'Brien DP, Mandell AM (2011) A low mass for Mars from Jupiter's early gas-driven migration. Nature 475(7355):206–209. https://doi.org/10.1038/nature10201

Wang H, Weiss BP, Bai XN, Downey BG, Wang J, Wang J, Suavet C, Fu RR, Zucolotto ME (2017) Lifetime of the solar nebula constrained by meteorite paleomagnetism. Science 355(6325):623–627. https://doi.org/10.1126/science.aaf5043

Wänke H, Baddenhausen H, Dreibus G, Jagoutz E, Kruse H, Palme H, Spettel B, Teschke F (1973) Multielement analyses of Apollo 15, 16, and 17 samples and the bulk composition of the moon. In: Lunar and planetary science conference proceedings, vol 4, p 1461. https://ui.adsabs.harvard.edu/abs/1973LPSC....4.1461W/abstract

Wänke H, Dreibus G, Runcorn SK, Turner G, Woolfson MM (1988) Chemical composition and accretion history of terrestrial planets. Philos Trans R Soc Lond A, Math Phys Eng Sci 325(1587):545–557. https://doi.org/10.1098/rsta.1988.0067

Warren PH (1985) The magma ocean concept and lunar evolution. Annu Rev Earth Planet Sci 13(1):201–240. https://doi.org/10.1146/annurev.ea.13.050185.001221

Warren PH (2011) Stable-isotopic anomalies and the accretionary assemblage of the Earth and Mars: a subordinate role for carbonaceous chondrites. Earth Planet Sci Lett 311(1):93–100. https://doi.org/10.1016/j.epsl.2011.08.047

Warren AO, Kite ES (2023) Narrow range of early habitable Venus scenarios permitted by modeling of oxygen loss and radiogenic argon degassing. Proc Natl Acad Sci 120(11):e2209751120. https://doi.org/10.1073/pnas.2209751120

Watson AJ, Donahue TM, Walker JCG (1981) The dynamics of a rapidly escaping atmosphere: applications to the evolution of Earth and Venus. Icarus 48(2):150–166. https://doi.org/10.1016/0019-1035(81)90101-9

Way M, Del Genio AD (2020) Venusian habitable climate scenarios: modeling venus through time and applications to slowly rotating Venus-like exoplanets. J Geophys Res, Planets 125(5). https://doi.org/10.1029/2019je006276

Way MJ, Del Genio AD, Kiang NY, Sohl LE, Grinspoon DH, Aleinov I, Kelley M, Clune T (2016) Was Venus the first habitable world of our solar system? Geophys Res Lett 43(16):8376–8383. https://doi.org/10.1002/2016GL069790

Way MJ, Ostberg C, Foley BJ, Gillmann C, Höning D, Lammer H, O'Rourke J, Persson M, Plesa AC, Salvador A, Scherf M, Weller M (2023) Synergies between Venus & exoplanetary observations. Space Sci Rev 219(13):64. https://doi.org/10.1007/s11214-023-00953-3

Weidenschilling SJ (1977) Aerodynamics of solid bodies in the solar nebula. Mon Not R Astron Soc 180(2):57–70. https://doi.org/10.1093/mnras/180.2.57

Weller MB, Kiefer WS (2020) The physics of changing tectonic regimes: implications for the temporal evolution of mantle convection and the thermal history of Venus. J Geophys Res, Planets 125(1):e2019JE005960. https://doi.org/10.1029/2019je005960

Westall F, Brack A (2018) The importance of water for life. Space Sci Rev 214(2):50. https://doi.org/10.1007/s11214-018-0476-7

Westall F, Höning D, Avice G, Gentry D, Gerya T, Gillmann C, Izenberg N, Way MJ, Wilson C (2023) The habitability of Venus and a comparison to early Earth. Space Sci Rev 219:17. https://doi.org/10.1007/s11214-023-00960-4

Wetherill GW (1976) The role of large bodies in the formation of the Earth and moon. In: Lunar and planetary science conference proceedings, vol 3, pp 3245–3257. https://ui.adsabs.harvard.edu/abs/1976LPSC....7.3245W/abstract

Wetherill GW (1978) Accumulation of the terrestrial planets. In: Gehrels T, Matthews MS (eds) IAU colloq. 52: protostars and planets, p p 565. https://ui.adsabs.harvard.edu/abs/1978prpl.conf..565W

Wetherill GW (1985) Occurrence of giant impacts during the growth of the terrestrial planets. Science 228(4701):877–879. https://doi.org/10.1126/science.228.4701.877

Wetherill GW (1990) Formation of the Earth. Annu Rev Earth Planet Sci 18(1):205–256. https://doi.org/10.1146/annurev.ea.18.050190.001225

Wetherill GW (1991) Why isn't Mars as big as Earth? In: Lunar and planetary science conference, vol 22, p 1495. https://ui.adsabs.harvard.edu/abs/1991LPI....22.1495W

White WM (2013) Geochemistry. Wiley, New York

Whittet DCB (1997) Is extraterrestrial organic matter relevant to the origin of life on Earth?. In: Whittet DCB (ed) Planetary and interstellar processes relevant to the origins of life Springer, Dordrecht, pp 249–262. https://doi.org/10.1007/978-94-015-8907-9_13

Widemann T, Smrekar SE, Garvin JB, Straume-Lindner AG, Ocampo AC, Voirin T, Hensley S, Dyar MD, Whitten JL, Nunes D, Getty S, Arney GN, Johnson NM, Kohler E, Spohn T, O'Rourke JR, Wilson C, Way MJ, Ostberg C, Westall F, Höning D, Jacobson S, Salvador A, Avice G, Breuer D, Carter L, Gilmore MS, Ghail R, Helbert J, Byrne P, Santos AR, Herrick RR, Izenberg N, Marcq E, Rolf T, Weller M, Gillmann C, Korablev O, Zelenyi LM, Zasova L, Gorinov D, Seth G, Rao CVN, Desai N (2023) Venus evolution through time: Key science questions, selected mission concepts and future investigations. Space Sci Rev 219. https://doi.org/10.1007/s11214-023-00992-w

Williams CD, Mukhopadhyay S (2019) Capture of nebular gases during Earth's accretion is preserved in deep-mantle neon. Nature 565(7737). https://doi.org/10.1038/s41586-018-0771-1

Winn JN, Fabrycky DC (2015) The occurrence and architecture of exoplanetary systems. Annu Rev Astron Astrophys 53(1):409–447. https://doi.org/10.1146/annurev-astro-082214-122246

Wood BJ (1995) The effect of H_2O on the 410-kilometer seismic discontinuity. Science 268(5207):74–76. https://doi.org/10.1126/science.268.5207.74

Wood BJ, Halliday AN (2005) Cooling of the Earth and core formation after the giant impact. Nature 437(7063):1345–1348. https://doi.org/10.1038/nature04129

Wood J, Dickey J Jr, Marvin U, Powell B (1970) Lunar anorthosites and a geophysical model of the moon. Geochim Cosmochim Acta, Suppl, 1:965–988. Proceeding of the Apollo 11 Lunar Science Conference

Wood BJ, Bryndzia LT, Johnson KE (1990) Mantle oxidation state and its relationship to tectonic environment and fluid speciation. Science 248(4953):337–345. https://doi.org/10.1126/science.248.4953.337

Wood BJ, Walter MJ, Wade J (2006) Accretion of the Earth and segregation of its core. Nature 441:825–833. https://doi.org/10.1038/nature04763

Wood BJ, Wade J, Kilburn MR (2009) Core formation and the oxidation state of the Earth: additional constraints from Nb, V and Cr partitioning. Geochim Cosmochim Acta 72(5):1415–1426. https://doi.org/10.1016/j.gca.2007.11.036

Wordsworth RD (2016) Atmospheric nitrogen evolution on Earth and Venus. Earth Planet Sci Lett 447:103–111. https://doi.org/10.1016/j.epsl.2016.04.002

Wordsworth RD, Pierrehumbert RT (2013) Water loss from terrestrial planets with CO_2-rich atmospheres. Astrophys J 778(2):154. https://doi.org/10.1088/0004-637X/778/2/154

Wordsworth R, Kalugina Y, Lokshtanov S, Vigasin A, Ehlmann B, Head J, Sanders C, Wang H (2017) Transient reducing greenhouse warming on early Mars. Geophys Res Lett 44(2):665–671. https://doi.org/10.1002/2016GL071766

Wordsworth RD, Schaefer LK, Fischer RA (2018) Redox evolution via gravitational differentiation on low-mass planets: implications for abiotic oxygen, water loss, and habitability. Astron J 155(5):195. https://doi.org/10.3847/1538-3881/aab608

Wu J, Desch SJ, Schaefer L, Elkins-Tanton LT, Pahlevan K, Buseck PR (2018) Origin of Earth's water: chondritic inheritance plus nebular ingassing and storage of hydrogen in the core. J Geophys Res, Planets 123(10):2691–2712. https://doi.org/10.1029/2018JE005698

Wyatt MC (2008) Evolution of debris disks. Annu Rev Astron Astrophys 46(1):339–383. https://doi.org/10.1146/annurev.astro.45.051806.110525

Yang J, Cowan NB, Abbot DS (2013) Stabilizing cloud feedback dramatically expands the habitable zone of tidally locked planets. Astrophys J Lett 771. https://doi.org/10.1088/2041-8205/771/2/L45

Yang J, Boué G, Fabrycky DC, Abbot DS (2014) Strong dependence of the inner edge of the habitable zone on planetary rotation rate. Astrophys J Lett. https://doi.org/10.1088/2041-8205/787/1/L2. arXiv:1404.4992

Yin Q, Jacobsen SB, Yamashita K, Blichert-Toft J, Télouk P, Albarède F (2002) A short timescale for terrestrial planet formation from Hf–W chronometry of meteorites. Nature 418(6901):949–952. https://doi.org/10.1038/nature00995

Yoshino T, Walter MJ, Katsura T (2003) Core formation in planetesimals triggered by permeable flow. Nature 422(6928):154–157. https://doi.org/10.1038/nature01459

Youdin AN, Goodman J (2005) Streaming instabilities in protoplanetary disks. Astrophys J 620(1):459–469. https://doi.org/10.1086/426895

Young ED, Shahar A, Schlichting HE (2023) Earth shaped by primordial H_2 atmospheres. Nature 616(7956):306–311. https://doi.org/10.1038/s41586-023-05823-0

Yuan L, Steinle-Neumann G (2020) Strong sequestration of hydrogen into the Earth's core during planetary differentiation. Geophys Res Lett 47(15):e2020GL088303. https://doi.org/10.1029/2020GL088303.

Zahnle KJ (1990) Atmospheric chemistry by large impacts. In: Global catastrophes in Earth history; an interdisciplinary conference on impacts, volcanism, and mass mortality. Geol. Soc. Am., Boulder, pp 271–288. https://doi.org/10.1130/SPE247-p271

Zahnle KJ, Kasting JF, Pollack JB (1988) Evolution of a steam atmosphere during Earth's accretion. Icarus 74(1):62–97

Zahnle K, Kasting JF, Pollack JB (1990) Mass fractionation of noble gases in diffusion-limited hydrodynamic hydrogen escape. Icarus 84(2):502–527. https://doi.org/10.1016/0019-1035(90)90050-J

Zahnle K, Arndt N, Cockell C, Halliday A, Nisbet E, Selsis F, Sleep NH (2007) Emergence of a habitable planet. Space Sci Rev 129:35–78. https://doi.org/10.1007/s11214-007-9225-z

Zahnle K, Schaefer L, Fegley B (2010) Earth's earliest atmospheres. Cold Spring Harb Perspect Biol 2(10):a004895–a004895. https://doi.org/10.1101/cshperspect.a004895

Zahnle KJ, Gacesa M, Catling DC (2019) Strange messenger: a new history of hydrogen on Earth, as told by Xenon. Geochim Cosmochim Acta 244:56–85. https://doi.org/10.1016/j.gca.2018.09.017

Zahnle KJ, Lupu R, Catling DC, Wogan N (2020) Creation and evolution of impact-generated reduced atmospheres of early Earth. Planet. Sci. J. 1(1):11. https://doi.org/10.3847/PSJ/ab7e2c

Zellner NEB (2017) Cataclysm no more: new views on the timing and delivery of lunar impactors. Orig Life Evol Biosph 47(3):261–280. https://doi.org/10.1007/s11084-017-9536-3

Zhang Y, Xu Z, Zhu M, Wang H (2007) Silicate melt properties and volcanic eruptions. Rev Geophys 45(4). https://doi.org/10.1029/2006RG000216

Zhang HL, Hirschmann MM, Cottrell E, Withers AC (2017) Effect of pressure on Fe3+/ΣFe ratio in a mafic magma and consequences for magma ocean redox gradients. Geochim Cosmochim Acta 204:83–103. https://doi.org/10.1016/j.gca.2017.01.023

Zolotov MY (2018) Gas-solid interactions on Venus and other solar system bodies. Rev Mineral Geochem 84:351–392. https://doi.org/10.2138/rmg.2018.84.10

Zsom A, Ormel CW, Güttler C, Blum J, Dullemond CP (2010) The outcome of protoplanetary dust growth: pebbles, boulders, or planetesimals? - II. Introducing the bouncing barrier. Astron Astrophys 513:A57. https://doi.org/10.1051/0004-6361/200912976

Zsom A, Seager S, de Wit J, Stamenković V (2013) Toward the minimum inner edge distance of the habitable zone. Astrophys J 778(2):109. https://doi.org/10.1088/0004-637X/778/2/109

Publisher's Note Springer Nature remains neutral with regard to jurisdictional claims in published maps and institutional affiliations.

Authors and Affiliations

Arnaud Salvador[1,2,3] ⓘ · Guillaume Avice[4] ⓘ · Doris Breuer[5] ⓘ · Cédric Gillmann[6] ⓘ · Helmut Lammer[7] ⓘ · Emmanuel Marcq[8] ⓘ · Sean N. Raymond[9] ⓘ · Haruka Sakuraba[10] ⓘ · Manuel Scherf[7,11,12] · M.J. Way[13,14] ⓘ

✉ D. Breuer
   doris.breuer@dlr.de

  Springer

A. Salvador
arnaudsalvador@arizona.edu

G. Avice
avice@ipgp.fr

C. Gillmann
cgillmann@ethz.ch

H. Lammer
helmut.lammer@oeaw.ac.at

E. Marcq
emmanuel.marcq@latmos.ipsl.fr

S.N. Raymond
sean.raymond@u-bordeaux.fr

H. Sakuraba
sakuraba@eps.sci.titech.ac.jp

M. Scherf
manuel.scherf@oeaw.ac.at

M.J. Way
Michael.J.Way@nasa.gov

1   Department of Astronomy and Planetary Science, Northern Arizona University, Box 6010,
    Flagstaff, AZ 86011, USA

2   Habitability, Atmospheres, and Biosignatures Laboratory, University of Arizona, Tucson, AZ, USA

3   Present address: Lunar and Planetary Laboratory, University of Arizona, Tucson, AZ, USA

4   Université Paris Cité, Institut de physique du globe de Paris, CNRS, F-75005 Paris, France

5   DLR, Institute of Planetary Research, 12489 Berlin, Germany

6   Department of Earth Sciences, Institute of Geophysics, Geophysical Fluid Dynamics, ETH Zurich,
    Switzerland

7   Space Research Institute, Austrian Academy of Sciences, Graz, Austria

8   LATMOS/IPSL, UVSQ Université Paris-Saclay, Sorbonne Université, CNRS, Guyancourt, France

9   Laboratoire d'Astrophysique de Bordeaux, CNRS and Université de Bordeaux, Pessac, France

10 Department of Earth and Planetary Sciences, Tokyo Institute of Technology, Ookayama,
    Meguro-ku, Tokyo 152-8551, Japan

11 Institute of Physics, University of Graz, Graz, Austria

12 Institute for Geodesy, Technical University, Graz, Austria

13 NASA Goddard Institute for Space Studies, 2880 Broadway, New York, NY 10025, USA

14 Theoretical Astrophysics, Department of Physics and Astronomy, Uppsala University, Uppsala,
    Sweden

Space Science Reviews (2022) 218:60
https://doi.org/10.1007/s11214-022-00929-9

Noble Gases and Stable Isotopes Track the Origin and Early Evolution of the Venus Atmosphere

Guillaume Avice[1] · Rita Parai[2] · Seth Jacobson[3] · Jabrane Labidi[1] ·
Melissa G. Trainer[4] · Mihail P. Petkov[5]

Received: 15 March 2022 / Accepted: 10 October 2022 / Published online: 26 October 2022
© The Author(s) 2022

Abstract

The composition the atmosphere of Venus results from the integration of many processes entering into play over the entire geological history of the planet. Determining the elemental abundances and isotopic ratios of noble gases (He, Ne, Ar, Kr, Xe) and stable isotopes (H, C, N, O, S) in the Venus atmosphere is a high priority scientific target since it could open a window on the origin and early evolution of the entire planet. This chapter provides an overview of the existing dataset on noble gases and stable isotopes in the Venus atmosphere. The current state of knowledge on the origin and early and long-term evolution of the Venus atmosphere deduced from this dataset is summarized. A list of persistent and new unsolved scientific questions stemming from recent studies of planetary atmospheres (Venus, Earth and Mars) are described. Important mission requirements pertaining to the measurement of volatile elements in the atmosphere of Venus as well as potential technical difficulties are outlined.

Keywords Noble gases · Stable isotopes · Venus · Atmosphere

1 Introduction and Overview

Like all planetary atmospheres, the composition of the atmosphere of Venus informs us about the entire geological history of Earth's sister planet (*e.g.* Catling and Kasting 2017). Past space missions have already demonstrated that the elemental and isotopic compositions of volatile elements in the atmosphere of Venus differ greatly from those of other terrestrial planets or reservoirs of volatile elements (solar gas, meteorites, comets) in the Solar System (*e.g.* Pepin and Porcelli 2002; Baines et al. 2013; Wieler 2002; Chassefière et al. 2012; Avice and Marty 2020). Recent discoveries about the detailed origins and evolution of volatile inventories on Earth, Mars, and in other Solar System reservoirs have raised compelling new questions that motivate the pursuit of improved noble gas and stable isotope measurements in the atmosphere of Venus. Future space missions targeting atmospheric measurements are needed to make critical comparisons of volatile accretion and transport between terrestrial planets.

Venus: Evolution Through Time
Edited by Colin F. Wilson, Doris Breuer, Cédric Gillmann, Suzanne E. Smrekar, Tilman Spohn and Thomas Widemann

Extended author information available on the last page of the article

Noble gases are the best available geochemical tracers of geophysical processes. Noble gases are inert, incompatible during silicate partial melting, and atmophile (Ozima and Podosek 2002). Accordingly, noble gases act as tracers of planetary differentiation, including the coupled evolution of planetary interiors and atmospheres through outgassing. Each noble gas element has at least one stable isotope that is non-radiogenic; these are often referred to as primordial noble gas isotopes, and their budgets are established during accretion. Ratios of primordial noble gas isotopes serve as fingerprints of volatile origins, but these ratios may also be affected by mass-fractionating loss. Each noble gas element also has at least one stable isotope that is produced by nuclear reactions (including radioactive decay) involving non-atmophile reactants. Noble gas abundances are typically so low in planetary solids that their isotopic compositions are very strongly affected by nuclear reactions, even when the reactants are themselves rare in planetary materials. Thus, radioactive decay and other nuclear reactions generate large variations (from a few percent to orders of magnitude) that track planetary differentiation, which fractionates non-atmophile parent to atmophile daughter ratios, on a variety of timescales. Taken together, noble gas elemental abundances and isotopic compositions provide a rich record of planetary volatile origins, and the timing and mechanisms of volatile transport between planetary reservoirs.

Stable isotope ratios of relatively light elements (C, H, O, N, S) have been widely used to put constraints on the budget of volatile elements of terrestrial planets and on the origin and evolution of planetary atmospheres (*e.g.* Marty 2012; Halliday 2013). These elements do not show the strong nuclear effects seen for noble gases. Nonetheless, various reservoirs in the Solar System show variable mass-dependant and mass-independent isotopic composition. In solar-system materials, isotope variations may have been caused by physico-chemical processes such as photochemistry, partial melting, and partial evaporation. Planetary formation and differentiation may lead to further isotopic variations (Young et al. 2002). The specific case of atmospheric loss may also lead to isotopic fractionation (Jakosky and Pepin 1994). In other words, the isotope ratios of light elements can help track the origin of planetary materials but also place constraints on the evolution of planetary mantles and atmospheres.

The possible sources of Venusian atmophile elements (noble gases and light elements) can be categorized roughly into two types: direct or indirect. Venus could have accreted a significant primary atmosphere *directly* from the nebula if it accreted as quickly as Mars (Dauphas and Pourmand 2011) and reached near its final size while gas was likely still present in the protoplanetary disk (Weiss et al. 2021). In this case, the isotopic composition of atmophile elements sourced directly from the solar nebula would presumably match those of gases of the proto-Sun. While hydrodynamic escape assisted by later bombardment from small bodies could have eroded away hydrogen and helium, other gases directly accreted from the nebula may have been in-gassed into the mantle of Venus and protected or may have survived in the atmosphere (Olson and Sharp 2019). However, if Venus accreted more slowly and was Mars-sized or smaller only growing to larger sizes after the gaseous protoplanetary disk had dissipated, one may anticipate that a primary atmosphere would have been insignificant. Regardless of the size of its primary atmosphere, Venus also probably accreted volatile-rich building blocks during its growth providing an *indirect* source of atmophile elements. The isotopic composition of atmophile elements sourced indirectly from accreted building blocks would vary in composition according to the parent-body source region as well as fractionating processes taking place within the parent-body. During the dynamical evolution of the solar system, different reservoirs of planetary building blocks contributed to the growth of the terrestrial planets (O'Brien et al. 2018). Initial accretion in the inner disk near the current location of Venus is presumed to be from volatile-depleted material, although even enstatite and ordinary chondrites, examples of primordial inner solar system material, contain substantial amounts of volatile species like water (e.g. Piani et al. 2020). If not supplied in Venus's building blocks, most volatiles within Venus could have been delivered from more distant parts of the protoplanetary disk, where volatile abundances in solids were significantly higher. However, this region is vast, encompassing primitive asteroidal and cometary compositions with clear evidence for strong isotopic gradients throughout (Marty et al. 2016). Only with a detailed understanding of the isotopic composition of the Venusian atmosphere will we be able to disentangle the different contributions to the atmosphere of Venus.

Apart from remote observations, and compared to missions involving landing on the surface of Venus, an exhaustive measurement of the elemental and isotopic composition of volatile elements in a sample of Venus's atmosphere is one of the less risky types of investigations with an astounding scientific outcome. Measuring the elemental and isotopic composition of noble gases and stable isotopes in the atmosphere of Venus is for example one of the main scientific goals of the recently selected NASA DAVINCI mission (Glaze et al. 2017).

This chapter presents an overview of the current state of knowledge and the outlook for future discoveries brought from the study of noble gases and of stable isotopes of C, H, O, N and S regarding the origin and early evolution of the atmosphere of Venus and on the implications for the geological history of the entire planet. Note that other contributions in this journal will cover in detail the role of water in Venus's history (Salvador et al. 2022) and the complex interactions between the interior and the surface of Venus (Gillmann et al. 2022).

2 Existing Dataset and Current State of Knowledge

The current state of knowledge on the elemental and isotopic compositions of noble gases and stable isotopes of H, C, N and O in the Venus atmosphere has already been reviewed in details elsewhere (von Zahn et al. 1983; Donahue and Pollack 1983; Wieler 2002; Johnson and de Oliveira 2019; Chassefière et al. 2012). This section describes the most important features of the atmosphere of Venus and highlights which scientific data are severely lacking.

2.1 Noble Gases (He, Ne, Ar, Kr, Xe)

Existing data on the elemental abundance and isotopic composition of noble gases in the atmosphere of Venus are reported in Table 1.

2.1.1 Elemental Abundances

Elemental abundances of Ne, Ar, Kr and Xe in the atmosphere of Venus and in other reservoirs of volatile elements in the Solar System are plotted in Fig. 1 (Dauphas and Morbidelli 2014; Avice and Marty 2020). Chondritic meteorites, especially carbonaceous chondrites, are enriched in heavy noble gases relative to light ones (Porcelli and Ballentine 2002; Wieler 2002). Although Mars is depleted in noble gases relative to Earth by about two orders of magnitude, note that noble gases in these two atmospheres follow globally the same abundance pattern with a similar magnitude of enrichment of Kr relative to Ne relative to Ar. Atmospheric xenon on Earth and Mars presents a depletion relative to a chondritic-like abundance pattern (this is the "missing xenon problem," see Avice and Marty (2020) for a recent review). For Venus, abundances of Ne and Ar are high compared to Mars and Earth

Table 1 Noble gases in the atmosphere of Venus. Abundances are reported as molar mixing ratios in the atmosphere. See details in the text for estimates for the abundance of xenon. Values in blue font give the percentages for errors. Recommended precisions for future investigations are taken from Chassefière et al. (2012). Table modified after Wieler (2002)

Non-radiogenic isotopes (mixing ratios)		Recommended precision	Radiogenic isotopes (mixing ratios)	
^{20}Ne (ppm)	7	5-10%	^{4}He (ppm)	12 (+24,−6) (+200%,−50%)
^{36}Ar (ppm)	30 (+20,−10) (+67%,−33%)	5-10%	^{40}Ar (ppm)	31 (+20,−10) (+64%,−32%)
^{84}Kr (ppb)[a]	25 (+3,−18) (+12%,−72%)	5-10%	^{129}Xe (ppb)	<9.5
^{84}Kr (ppb)[b]	600±200 (±33%)	5-10%		
^{132}Xe (ppb)[c]	1-10 (see text)	5-10%		

Isotopic ratios		Recommended precision
^{3}He/^{4}He	<3x10^{-4}	10%
^{20}Ne/^{22}Ne	11.8±0.6 (±5%)	5%
^{21}Ne/^{22}Ne	<0.067	5%
^{38}Ar/^{36}Ar[d]	0.183±0.003 (±1.6%)	1%
^{38}Ar/^{36}Ar[e]	0.18±0.02 (±11%)	1%*
^{40}Ar/^{36}Ar	1.11±0.02 (±1.8%)	1%*
$^{78-80}$Kr/^{84}Kr	n.d.	5%
$^{82-86}$Kr/^{84}Kr	n.d.	1%
$^{124-128}$Xe/^{130}Xe	n.d.	5%
$^{129-136}$Xe/^{130}Xe	n.d.	1%

Note: Table modified after Wieler (2002). See refs therein. a: Pioneer results reported by Donahue and Pollack (1983), b: Venera results taken from Donahue and Pollack (1983), c: lower limit deduced from Venera's results (see text), d: data from Istomin et al. (1983), e: data from Donahue and Pollack (1983). Recommended precisions are taken from Chassefière et al. (2012). n.d. = not determined. "*" denotes a new recommendation.

and are similar to those measured in meteorites. Two ranges of estimates exist for the abundance of krypton (see Table 1, Donahue and Pollack (1983) and refs. therein). If there is 25 ppb of ^{84}Kr in the Venus's atmosphere (Pioneer's results), krypton would be depleted relative to a chondritic-like abundance pattern, with almost a solar-like ^{84}Kr/^{36}Ar ratio. Conversely, if the abundance is closer to 600 ppb (Venera's results), the ^{84}Kr/^{36}Ar ratio is close to the chondritic ratio (Fig. 1), suggesting that Ne, Ar and Kr in the Venus's atmosphere could have been sourced by meteorites. Measuring precisely the abundance of ^{84}Kr would be decisive for understanding the origin of volatile elements in Venus. For xenon, an upper limit of 10 ppb for the abundance of ^{132}Xe has often been cited based on results obtained during the Pioneer mission (e.g. Wieler 2002). However, Istomin et al. (1982, 1983) argued that xenon was detected during the Russian Venera 14 mission and that, according to the sensitivity of the instrument, a detection of $^{131+132}$Xe isotopes above background implies a minimum abundance of Xe of 1-2 ppb (Fig. 2). A new precise determination of the abundance of Xe is still pending but considering a 1-10 ppb range suggests that Xe is not more depleted in the Venus atmosphere than atmospheric Xe on Earth and Mars. The fact that Xe in the atmospheres of Earth, Mars and Venus could be depleted in a similar fashion compared to chondritic abundances is striking, especially if the underabundance is due to a selective escape mechanism of xenon ions coupled to hydrogen ions during hydrodynamic escape of hydrogen (Zahnle et al. 2019; Avice and Marty 2020).

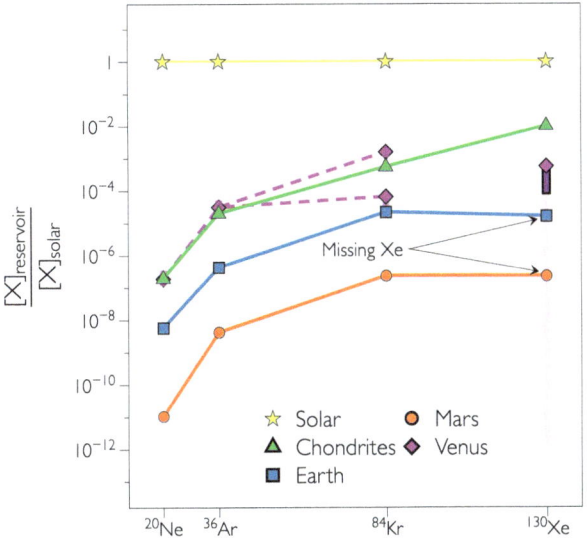

Fig. 1 Elemental abundances of noble gases in the atmosphere of Venus and in other reservoirs of the Solar System normalized to the abundance of silicium and to the Solar abundances (=1). For Venus, uncertainties on the abundances of Kr and Xe appear with a light purple range (this is an upper limit for Xe). The dark purple range for ^{130}Xe corresponds to another estimate with a minimum of 1 ppb of Xe (see text). The dotted purple line depicts the expected abundances of ^{130}Xe for an Earth-like/Mars-like ^{130}Xe/^{84}Kr ratio. Figure and data modified after Dauphas and Morbidelli (2014). See also references therein

Fig. 2 Signal of $^{131+132}$Xe detected during the Venera 14 mission. Mass and signal are given without units and the signal at the top of the ^{136}Xe peak is divided by 100. The very high ^{136}Xe peak is due to the presence of calibrant gas (99.99% pure ^{136}Xe). The $^{131+132}$Xe could correspond to a minimum abundance of 1-2 ppb of Xe in the atmosphere of Venus. Figure modified after Istomin et al. (1982)

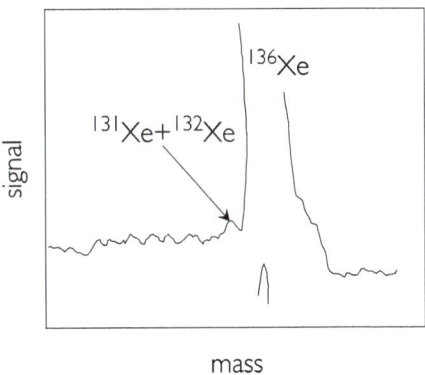

2.1.2 Isotope Ratios

Helium – An upper limit for the ^{3}He/^{4}He ratio of atmospheric helium has been estimated at 3×10^{-4} (Hoffman et al. 1980). This leaves a wide range of potential helium isotopic compositions viable for the atmosphere of Venus. The rather imprecise determination could be due to the scarcity of ^{3}He relative to ^{4}He, to the presence of an isobaric interference due to the presence of ^{1}H^{2}D, or could be explained by the fact that droplets of sulfuric acid clogged the inlet port of the Pioneer probe, which interfered with measurements of noble gases (Hoffman et al. 1980). This upper limit is compared to the ^{3}He/^{4}He ratio of other reservoirs in the Solar System in Fig. 3. The atmospheric ^{3}He/^{4}He ratio for Venus

Fig. 3 ^{3}He/^{4}He ratio of the atmosphere of Venus and of other reservoirs in the Solar System. Atmospheric escape of helium or degassing of radiogenic ^{4}He leads to a decrease of the ^{3}He/^{4}He ratio. Values are from Porcelli et al. (2002) and refs. therein

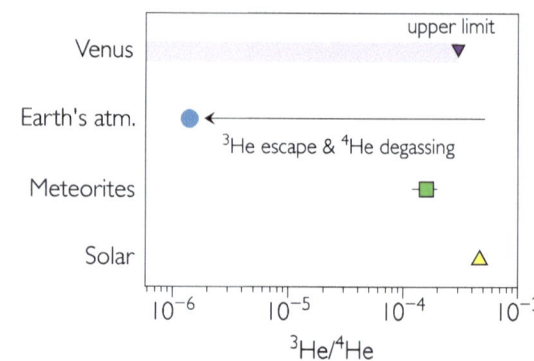

Fig. 4 Three-isotope diagram of neon isotopes (^{20,21,22}Ne). The isotopic composition of Earth atmospheric neon lies on a mixing range between Solar and Meteoritic end-members. The current estimate for Venus atmospheric neon suggests that, similarly to Mars atmospheric neon, Venus could lie outside of this mixing range although the uncertainty on the Ar/Ne ratio in the Venus atmosphere is high. Figure modified after Marty (2012)

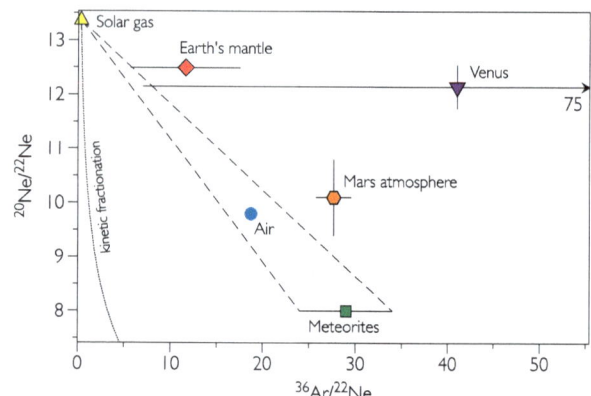

seems only marginally lower than the solar value, but since even the order of magnitude of the exact value remains unknown, it is impossible to evaluate the roles of relatively recent (hundreds of millions of years) outgassing of radiogenic ^{4}He and of preferential escape of ^{3}He relative to ^{4}He from the atmosphere of Venus to space.

Neon – The ^{20}Ne/^{22}Ne ratio in the Venus atmosphere is close to 12 (Table 1, Fig. 4). This ratio is higher than that measured for terrestrial atmospheric neon (9.8, Ozima and Podosek (2002)), which suggests that solar-wind irradiated material ($^{20}Ne/^{22}Ne = 12.8$, Heber et al. (2009)) or solar gas ($^{20}Ne/^{22}Ne = 13.4$, Heber et al. (2012)) contributed volatile elements in greater proportion to the atmosphere of Venus relative to Earth. For the ^{21}Ne/^{22}Ne ratio, only an upper limit of 0.067 has been proposed by Istomin et al. (1983). This value is too imprecise to provide any further constraint on the origin of neon on Venus.

Argon – There are two estimates for the ^{38}Ar/^{36}Ar ratio, one with a value of 0.183±0.003 (Istomin et al. 1983), and another with a much less precise value of 0.18±0.02 (Donahue and Pollack 1983). These two values are plotted in Fig. 5. If the conservative error estimate is considered, the ^{38}Ar/^{36}Ar ratio could correspond to any primordial end-member of the Solar System. However, if the value taken for the ^{38}Ar/^{36}Ar ratio is 0.183±0.003, then the primordial isotopes of argon are similar to Solar argon. The ^{40}Ar/^{36}Ar ratio has been measured at 1.11±0.02 (Istomin et al. 1983). Such a low value compared to the isotopic ratio of Earth's atmospheric argon (≈300, Ozima and Podosek (2002)) has been interpreted as evidence that Venus outgassed only about 25% of the radiogenic ^{40}Ar produced by the radioactive

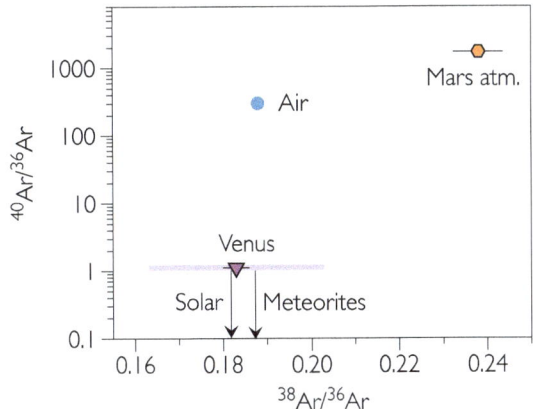

Fig. 5 Three-isotope diagram of argon isotopes. The thin black line corresponds to the error on the $^{38}Ar/^{36}Ar$ ratio of Venus reported by Istomin et al. (1983). The purple range represents the uncertainty given by Donahue and Pollack (1983). Values for air are from Ozima and Podosek (2002). Values for Mars are taken from Avice et al. (2018a) and refs. therein. Error bars at 1σ

decay of ^{40}K ($T_{1/2}=1.25$ Ga) during 4.56 Ga (Kaula 1999). This estimate for the amount of degassed radiogenic argon is about half the value estimated for Earth (Allègre et al. 1987). Note that estimates for the total outgassing of ^{40}Ar over Venus's history strongly rely on the assumed K/U ratio for bulk Venus, which remains debated (see the discussions by Lammer et al. (2020) and by Gillmann et al. (2022)). Limited outgassing of radiogenic Ar over 4.56 Ga is the basis for the common view that Venus remained quiescent, in a stagnant-lid tectonic regime, during most of its history (O'Rourke and Korenaga 2015). The high abundance of ^{36}Ar means that a large fraction of argon in the Venus atmosphere (Fig. 1) could be derived from primordial cosmochemical sources (solar gas, meteorites, comets). However, the relative abundances and isotope ratios of Ne and Ar do not provide a simple view of the source of light noble gases to Venus. The primordial Ar/Ne ratio is similar to values measured in meteorites, but the $^{20}Ne/^{22}Ne$ ratio may be solar-like (see above). The large uncertainty on the determination of the elemental Ar/Ne ratio does not permit evaluation of whether Venus plots on the mixing line between solar and meteoritic end-members (Fig. 4), or if the delivery of gas with an extremely high Ar/Ne ratio (*e.g.* via comets, Marty et al. (2016)) induced an excess of primordial Ar relative to Ne without changing the $^{20}Ne/^{22}Ne$ ratio.

Krypton and Xenon – Today, there is no available measurement of the isotopic composition of krypton or xenon in the atmosphere of Venus. The isotopic compositions of Kr and Xe in different volatile element reservoirs of the Solar system are plotted in Fig. 6 and briefly discussed here to illustrate the high potential of Kr and Xe isotopic measurements to shed light on volatile origins and loss processes on Venus.

The isotopic composition of terrestrial atmospheric krypton is close to the Chondritic end-member. Note that the latter two components are not simply related to one another by mass-dependent fractionation (which would appear as a rotation of the isotopic spectrum centered on ^{84}Kr in Fig. 6). On Earth, mantle Kr is showing a chondritic affinity with a deficit in neutron-rich ^{86}Kr relative to the chondritic composition whose origin remains poorly understood (Péron et al. 2021). The isotopic composition of Mars atmospheric Kr has been measured in-situ (Conrad et al. 2016). Mars Kr might present important excesses in ^{80}Kr and ^{82}Kr isotopes (Fig. 6) which have been attributed to the degassing from the Martian regolith of Kr isotopes produced by neutron capture reactions on bromine (Rao 2002). After correction for these excesses, Mars atmospheric Kr appears to be solar-like (Conrad et al. 2016). However, Mars atmospheric Kr trapped in glass phases of Martian meteorites does not exhibit the excesses measured in situ, and this discrepancy remains unexplained (Conrad

Fig. 6 Isotopic composition of krypton (**a**) and xenon (**b**) in important reservoirs of the Solar system. Isotopic ratios are expressed in delta values, which correspond to deviations in permil relative to the isotopic composition of the Solar Wind (Meshik et al. 2014, 2020). Data for the Earth's atmosphere are from Ozima and Podosek (2002). Data for Mars atmospheric Kr are from Conrad et al. (2016) for in-situ measurements and from Swindle et al. (1986) for measurements made on Martian meteorites. Data for Mars atmospheric Xe are from Conrad et al. (2016) for in-situ measurements and from Mathew et al. (1998) for Xe measured in Martian meteorites. Note that the exact values, especially for light Kr and Xe isotopes, remain debated at this time (Avice et al. 2018a). Data for Kr and Xe in the phase Q, the major carrier of heavy noble gases in carbonaceous chondrites, are from Busemann et al. (2000). Error bars at 1σ

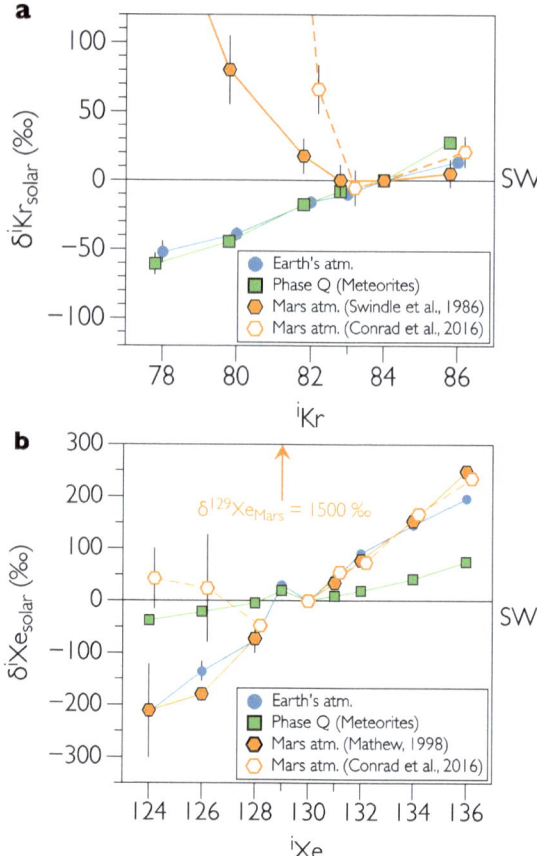

et al. 2016; Swindle et al. 1986; Avice et al. 2018a). Note that recent measurements of krypton contained in the Chassigny meteorite reveal that chondritic Kr is present in Mars's interior (Péron and Mukhopadhyay 2022). This suggests that, contrary to the common view, delivery of chondritic volatile elements can happen *before* the accretion of solar nebular gases to the Mars atmosphere.

For xenon, the isotopic composition of the Xe in the phase Q component (Q-Xe, Wieler et al. (1992), Busemann et al. (2000)), which is the main carrier of noble gases in carbonaceous chondrites, is enriched in heavy isotopes and depleted in light isotopes, corresponding to a mass-dependent isotopic fractionation of about 1 % per atomic mass unit (or %.amu^{-1}) compared to the solar composition (Meshik et al. 2020). The origin of phase Q and of the isotopic difference between Q-Xe and Solar-Xe remains debated, but common explanations invoke ionization processes, causing the mass-dependent isotopic fractionation, (e.g. Kuga et al. 2017) accompanied by addition of Xe isotopes either derived from presolar components (Crowther and Gilmour 2013) or produced during radioactive decay of I, U and Pu (Marrocchi et al. 2015).

Xenon on Earth and Mars presents a strong positive (i.e. enrichment in heavy isotopes) isotopic fractionation (3-4 %.u^{-1}) relative to primordial end-members in the Solar system. Previous studies argued that Xe had been lost from Earth and Mars during early episodes of hydrodynamic escape (Pepin 1991; Pepin and Porcelli 2002). However, recent studies of paleo-atmospheric xenon trapped in terrestrial rocks of Archean age revealed that the isotopic fractionation of atmospheric xenon was progressive over time (Pujol et al. 2011, 2009; Avice et al. 2017, 2018b; Bekaert et al. 2018; Almayrac et al. 2021) and ceased around the time of the Great Oxidation Event (Avice et al. 2018b; Ardoin et al. 2022). Zahnle et al. (2019) proposed that xenon ions escaped from the Earth's atmosphere through Coulomb interactions in an ionized hydrogen wind, which caused a progressive depletion and isotopic fractionation of atmospheric Xe on Earth. A similar process was probably operating on Mars (Cassata et al. 2022) although existing data suggest that escape and isotopic fractionation of Xe ceased at least 4.2 Ga ago (Cassata 2017).

Xenon in the Earth's atmosphere also presents interesting mass-independent features. Even after correction for the mass-dependent fractionation, Earth atmospheric Xe cannot be simply attributed to Solar Xe, Chondritic Xe or a mixture of these two end-member compositions present in the Solar system. Primordial unfractionated xenon is depleted in ^{134}Xe and ^{136}Xe isotopes by several percents relative to Solar or Chondritic Xe. This feature has been recognized decades ago (Takaoka 1972). Interestingly, the severe ^{134}Xe and ^{136}Xe anomalies on Earth relative to other bodies of the solar system correspond to a deficit in r-process Xe isotopes several orders of magnitude higher than other nucleosynthetic deviations reported for refractory nuclides measured in meteorites (*e.g.*, Kleine et al. 2020). Primordial Xe for Earth atmospheric Xe has been named "U-Xe" (Pepin 1991) and has resisted decades of investigation (Pepin 1994). Recent measurements of gases emitted by the comet 67P/Churyumov-Gerasimenko revealed that cometary xenon is also depleted in ^{134}Xe and ^{136}Xe relative to Solar xenon (Marty et al. 2017). This depletion is even more pronounced than for U-Xe. Mass balance calculations suggest that U-Xe could result from a mixing between 78% of Chondritic and 22% of Cometary xenon (Marty et al. 2017). The depletion in ^{134}Xe and ^{136}Xe suggests that comets could be depleted in r-process nuclides and that large-scale nucleosynthetic heterogeneities persisted across the Solar system (Avice et al. 2020).

Radiogenic ^{129}Xe produced by the decay of the short-lived radionuclide ^{129}I ($T_{1/2} = 15.7$ Myr) sheds light on early volatile loss from planetary reservoirs. Radiogenic ^{129}Xe excesses compared to Solar or Chondritic compositions are only generated by I/Xe fractionation within the first $\approx$100 Myr of Solar System history. On Earth, the mantle exhibits large ^{129}Xe/^{130}Xe excesses compared to chondrites, generally interpreted as evidence for significant early outgassing to generate high I/Xe ratios (*e.g.* Parai et al. 2019), as the solubility of iodine in silicate melts exceeds that of Xe (Musselwhite et al. 1991). Earth's atmosphere has a low ^{129}Xe/^{130}Xe ratio relative to the mantle, but exhibits ^{129}Xe excesses compared to a mass-fractionated non-radiogenic composition; an estimated 6.8 +/- 0.3 % of Earth's atmospheric ^{129}Xe budget is radiogenic (Porcelli and Ballentine 2002). Part of the radiogenic ^{129}Xe in Earth's atmosphere is derived from mantle outgassing over time: samples of Archean atmosphere show a deficit in ^{129}Xe/^{130}Xe compared to the modern composition after accounting for mass-dependent fractionation, providing a constraint on the rate of mantle Xe outgassing to the atmosphere, which, at the end of the Archean, could have been a least one order of magnitude than today (Avice et al. 2017; Marty et al. 2019). On Mars, the atmospheric radiogenic ^{129}Xe excess relative to a fractionated non-radiogenic composition is much greater than on Earth (Swindle et al. 1986; Garrison and Bogard 1998) (Fig. 6). However, the mantle composition determined from the martian meteorite Chassigny exhibits almost no excess in ^{129}Xe relative to chondrites (Ott 1988; Mathew and Marti 2001); outgassing of the mantle reservoir sampled by Chassigny cannot explain the large radiogenic excess in ^{129}Xe in the martian atmosphere. A distinct reservoir that experienced very strong

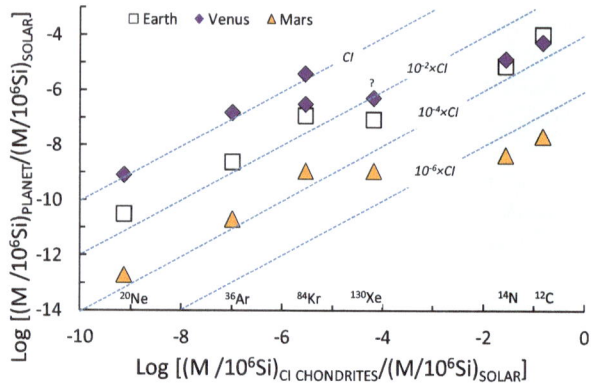

Fig. 7 Elemental abundances of noble gases, nitrogen and carbon on bulk Earth, Venus and Mars compared to abundances in CI chondrites. Elemental abundances are normalized to the abundance of silicon and to the Solar composition. Dashed lines depict different levels of depletion compared to volatile elements in CI chondrites. The two very different estimates for the abundance of krypton are also represented (see text). See Halliday (2013) for details. Figure modified after Halliday (2013)

early degassing (to generate high I/Xe within the lifetime of ^{129}I) must exist on Mars (Musselwhite et al. 1991), and outgassing of this reservoir has added radiogenic ^{129}Xe to the martian atmosphere over time (Mathew and Marti 2001; Cassata 2017).

Although there is no clear determination of the abundance of Xe in the Venus atmosphere, the fact that ^{131,132}Xe isotopes may have been detected during the Venera mission (Fig. 2, Istomin et al. (1983)) but that no ^{129}Xe peak had been identified suggests that the ^{129}Xe/^{132}Xe ratio is not extremely high and is likely lower than on Mars. Taken together, Xe isotopic signatures tell a complex story of the history of a planetary atmosphere. Bulk abundance and relative ratios of primordial isotopes can be particularly diagnostic regarding the nature of accreted volatiles and the mechanism and extent of atmospheric loss to space, such as giant impact erosion or hydrodynamic escape. The presence and abundances of radiogenic isotopes provide further insight regarding the timing and the degree of early volatile loss from mantle reservoirs that have outgassed to the atmosphere. Measurements of Xe in the atmosphere of Venus would thus provide potent insights into volatile origins, major evolutionary events, and transport for both the atmosphere and interior of the planet. These all have important implications for the inventory and history of other volatiles on Venus.

2.2 Stable Isotopes (C, H, O, N, S)

2.2.1 Elemental Abundances

Carbon, hydrogen, oxygen, nitrogen and sulfur in the Venus atmosphere are mainly stored in CO_2 (96.5±0.8% by vol.) for carbon and oxygen, H_2O (30±15 ppm) for hydrogen, N_2 (3.5±0.8%) for nitrogen and SO_2 (150±30 ppm) for sulfur (Fegley 2014, and refs. therein).

Estimates for the elemental abundances of carbon and nitrogen in the Venus atmosphere have been reported and discussed by Halliday (2013). Earth and Venus have similar abundances of carbon and nitrogen and the C/N ratio is close to chondritic proportions (Fig. 7). Comparatively, C and N are depleted in abundances on Mars relative to chondrites, Earth and Venus, but the C/N ratio is close to the chondritic value in all three terrestrial planets.

Fig. 8 Isotopic composition of hydrogen and nitrogen in reservoirs of the solar system. (**a**) Venus shows an extreme δD value compared to all reservoirs of hydrogen in the solar system. The two dashed lines show the upper and lower bound for the $\delta^{15}N$ value. The blue rectangle represents the zoom region for sub-panel (**b**). Isotope ratios are expressed with the delta notation, which corresponds to the deviation in permil relative to a standard isotope composition (mean Earth's ocean water for hydrogen, and Earth's air for nitrogen). Figure adapted from Marty (2012), Füri and Marty (2015), Marty et al. (2016). Data are from McCubbin and Barnes (2019), Marty et al. (2016) and refs. therein

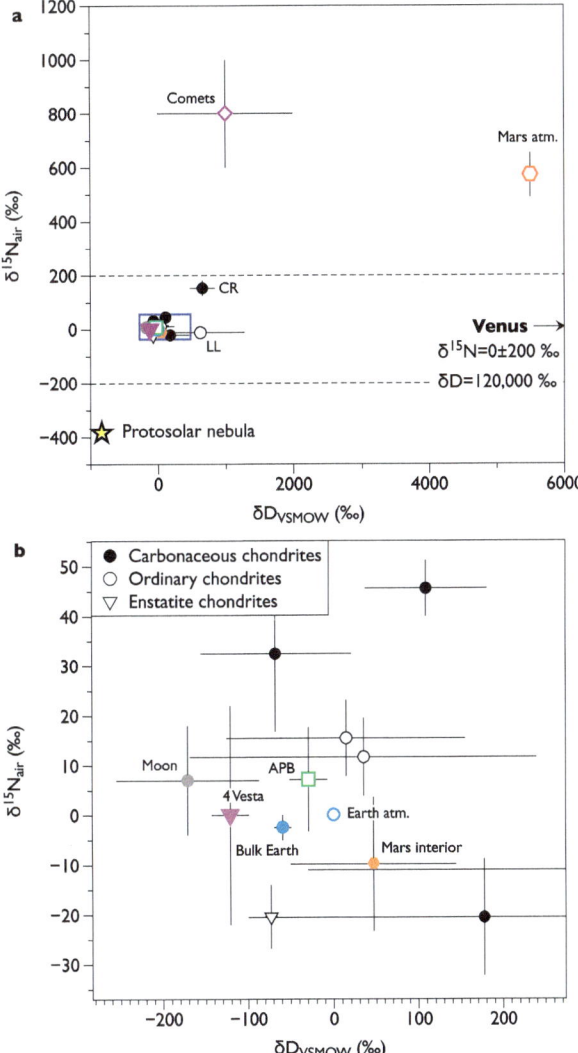

However, a near-chondritic C/N ratio should be interpreted with caution considering that strong isotopic fractionation of nitrogen isotopes in the Mars atmosphere demonstrates that Mars suffered from atmospheric escape of nitrogen (Wong et al. 2013, and refs. therein). The low abundance of water in the Venus atmosphere probably results from intense episodes of hydrogen escape (see Sect. 2.2.2 and Salvador et al. (2022)).

2.2.2 Isotope Ratios

The isotopic composition of hydrogen and nitrogen for Venus are compared to other Solar System reservoirs in Fig. 8. One of the most important feature of the Venus atmosphere is that the D/H ratio (where D=^{2}H) of hydrogen is extremely high compared to all other reservoirs in the Solar System. The δD_{VSMOW} value reaches 120,000‰. This very high ratio has been interpreted as evidence that Venus suffered from hydrogen escape episodes during its geological history (Donahue et al. 1982). When hydrogen escapes from an atmosphere, deuterium is more likely to be left behind, as it is twice as heavy (Hunten et al. 1987). The fraction of hydrogen remaining in the atmosphere becomes enriched in deuterium (D) relative to hydrogen (^{1}H). Atmospheric escape was thus an important process in the history of hydrogen on Venus. It must be noted that, since the starting D/H ratio of Venus remains unknown, an elevated δD_{SMOW} also leaves room for some cometary contribution to the delivery mix of volatile elements to Venus. For nitrogen, the δ^{15}N value has been estimated at $0\pm200‰$. This values is far too imprecise compared to typical values measured for nitrogen in the Solar system (Fig. 8, Füri and Marty (2015)) to draw any conclusion on the origin of Venus nitrogen and to evaluate whether the original ratio has been increased by atmospheric escape processes, as on Mars (McElroy et al. 1976).

The ^{12}C/^{13}C ratio of carbon in CO_2 has been measured at 89.3 ± 1.6. This value is identical, within errors, to the isotopic composition of carbon on Earth (Istomin et al. 1980). For oxygen, the ^{17}O/^{18}O ratio remains unknown and the ^{18}O/^{16}O value is 0.0020 ± 0.0001 (Hoffman et al. 1980). This estimate is far too imprecise to evaluate if the isotopic composition of oxygen on Venus differs from other oxygen reservoirs in the Solar system (Greenwood and Anand 2020). Finally, there is no isotopic data for sulfur, an important constituent of the atmosphere of Venus.

3 Unsolved and New Questions

There have been no new data acquired in-situ on the elemental abundances and isotopic compositions of noble gases and stable isotopes in the atmosphere of Venus for the last 40 years. However, recent discoveries regarding the origin and evolution of volatile elements Earth and Mars and in other Solar System reservoirs have raised new and critical scientific questions pertaining to Venus, such that previous efforts to define the future of scientific investigations on the atmosphere of Venus (e.g. Chassefière et al. 2012) must be updated. This section presents the most pressing outstanding scientific questions on the origin and early evolution of the atmosphere of Venus.

3.1 Origin of Venus and of Its Atmosphere

Noble gases and stable isotopes are often used to estimate the contributions of volatile elements from different accreted components to planetary atmospheres. In the case of Venus, current estimates of the elemental abundances and isotope ratios of those elements remain too imprecise to draw any firm conclusions, but the existing dataset allows one to pose scientific questions and frame testable hypotheses:

– *To what extent did solar-derived gas contribute to the Venus atmosphere?* Although the elemental Ne/Ar ratio of the Venus atmosphere is close to the chondritic value, the ^{20}Ne/^{22}Ne ratio seems higher than the average ratio measured in carbonaceous chondrites and closer to values measured in material irradiated by the Solar wind (e.g. Péron et al. 2018, for a recent review on SW noble gases) or to the ratio measured for solar Ne (Heber et al. 2009). One possible explanation would be that Venus Ne is of solar origin and that some of this Ne escaped from the Venus atmosphere. This would lead to a increase in the Ar/Ne ratio together with a decrease of the ^{20}Ne/^{22}Ne ratio. However, the existing dataset remains imprecise and a mixture between Solar Ne and Meteoritic Ne could also be envisaged (Fig. 4). Better measurements of the elemental abundances and the isotope ratios of Ne and Ar could shed new light on this problem. Precise measurements of the isotopic composition of Kr and Xe, down to the percent level, could also provide answers since the different volatile reservoirs of the Solar system (Solar gas, meteorites and even comets) have distinct Kr and Xe isotopic signatures. These end-members are not simply related by mass-dependent isotopic fractionation, which means that distinction between cosmochemical sources is possible even if the original signature has been modified by atmospheric escape leading to mass-dependent isotopic fractionation. A precise measurement of the $\delta^{15}N$ value (to the 5-10‰ precision level) is also desirable since the different classes of meteorites have distinct $^{15}N/^{14}N$ ratios (Füri and Marty 2015);

- *Did comets contribute volatile elements to Venus?* Recent studies of cometary noble gases and comparisons with Earth's atmospheric noble gases revealed that comets may have contributed a significant portion of Earth's noble gases, while the contribution to the budget of water would have stayed relatively minor ($\sim 1\%$) (Marty et al. 2016, 2017; Bekaert et al. 2020). This cometary contribution might have left its fingerprint in the isotopic composition of Earth atmospheric xenon (Avice et al. 2017; Marty et al. 2017) with a marked depletion in ^{134}Xe and ^{136}Xe isotopes relative to Solar or Meteoritic end-members. Contrary to the case of Earth, Mars Xe could be purely of solar origin (Ott 1988). Measuring precisely the isotopic composition of Venus atmospheric Xe could thus help to evaluate if comets contributed volatile elements to Venus. The Ar/Ne and $^{20}Ne/^{22}Ne$ ratios are also of interest here. Neon condenses only at very low temperatures ($<20K$) and comets are thus probably devoid of this element (Bar-Nun and Owen 1998). A significant cometary contribution to the atmosphere of Venus would result in a significant shift to the right in the $^{20}Ne/^{22}Ne$ vs $^{36}Ar/^{22}Ne$ space (Fig. 4);

- *Was the early inner Solar system efficiently homogenized for a major element like oxygen?* Families of primitive meteorites have widely distinct O isotope compositions. Although the Sun represents more than 99% of the total mass of the Solar System, meteorites and planets such as Earth and Mars are distinct from the Sun in terms of O isotope composition. Earth and Mars show enrichments in ^{17}O up to 70‰ at a given $^{18}O/^{16}O$ ratio, relative to the sun (McKeegan et al. 2011). This ground-breaking observation, pioneered in the 1970's (Clayton et al. 1973), spurred entire fields of research. Although conflicting interpretations still exist, the current paradigm suggests the starting $\delta^{17}O$-$\delta^{18}O$ composition of solids were solar, reflecting an inheritance from the average ^{16}O-rich molecular cloud. Solids subsequently evolved towards non-solar $\delta^{17}O$-$\delta^{18}O$ values via interaction with ^{17}O-rich water: they developed positive $\Delta^{17}O$ signatures relative to the Sun. Oxygen is a major mineral-forming element and makes up more than 45 wt.% of silicate planets (Javoy et al. 2010). Scientific reasons justifying sample return from Venus for O isotope measurements were recently reviewed (Greenwood and Anand 2020). Briefly, the later stages of terrestrial planet formation are thought to involve collisions with Moon-to-Mars sized planetesimals (Kaib and Cowan 2015). The final $\Delta^{17}O$ signature of a planet is the result of the weighted average of the various parent bodies (Young et al. 2016). The Moon has essentially the same $\Delta^{17}O$ composition as the Earth for reasons which most likely reflect its mode of formation in a giant impact event (Young et al. 2016). In contrast, Mars and Earth have distinct oxygen isotope $\Delta^{17}O$ compositions by +0.3‰ (Clayton and Mayeda 1983). A variable accretionary make-up may explain differences between these two planets. Large random $\Delta^{17}O$ variations may have existed among accreting bodies. Mars is only $\approx 10\%$ of the mass of Earth and sampled a relatively small number of accreting bodies whose weighted average $\Delta^{17}O$ may have deviated from the average inner Solar System composition, whereas Earth may have inherited a composition close to the average inner Solar System, with $\Delta^{17}O$ differences homogenized away by prolonged planetary growth. Alternatively, the composition of Mars may reflect an inner Solar System poorly homogenized for $\Delta^{17}O$ signatures, with a radial gradient in composition. Venus is much larger than Mars, as the second most massive terrestrial planet (after Earth) in our solar system. Whether the Earth and Venus are different in terms of $\Delta^{17}O$ would be a test of the potential inhomogeneity of the inner solar system, conceivably recorded by Mars's composition.

3.2 The Problem of Photochemistry

Photochemistry could prevent a straightforward interpretation of future O and S isotope data in terms of the origin of Venus. Processes associated with photochemistry are known to redistribute ^{17}O between the various O-bearing molecules in the Earth's atmosphere. Ozone carries a $> 100‰$ ^{17}O anomaly in air (Thiemens 2006). Isotope exchange between terrestrial CO_2 and ozone results in atmospheric CO2 with an anomalous $\Delta^{17}O$ relative to its source (Thiemens 2006). In fact, oxygen isotope exchange in modern air, between stratospheric ozone and any O-bearing molecule (SO_2, NO_x, CO_2, etc...) results in ^{17}O signatures skewed towards anomalously positive values for all those oxygen carriers. As a result, stratospheric CO_2 on Earth show ^{17}O anomalies up to 15‰ (Thiemens 2006), far from tracing the composition of bulk Earth. Photochemistry is known to occur in the modern atmosphere of Venus (Yung and Demore 1982, and refs. therein), which is a concern for how representative Venusian CO_2 may be to the interior of the planet. However, the Earth mechanism described in Thiemens (2006) is unlikely to be translated to Venus. First, a glaring difference is that there is no ozone in the Venusian atmosphere. Second, the Venusian atmospheric composition is crushingly dominated by CO_2, in stark contrast with Earth. For instance, the second most abundant O-bearing molecule on Venus is SO_2 (150 ppm), followed by water vapor (20 ppm) and carbon monoxide (17 ppm). We suggest that the mass balance is favorable for CO_2 to remain mostly unaffected by exotic unidentified chemistry. In other words, it is unclear how any unidentified chemistry involving third-party molecules could alter the ^{17}O of Venus's CO_2 atmosphere, as occurs on Earth.

Sulfur may help to understand the role of photochemistry. It occurs in the nebular gas as H_2S (Lauretta et al. 1997; Lodders 2003), which absorbs light and photo-dissociates when irradiated by the deep UV at wavelengths between 160 and 60 nm (Okabe 1978). In theory, photodissociation of H_2S produces S_0 with ^{33}S and ^{36}S isotopic anomalies (Chakraborty et al. 2013), like what is seen for ^{17}O. However, differentiated meteorites show almost no ^{33}S and ^{36}S variations (Antonelli et al. 2014; Dottin et al. 2018a). This is a major difference with oxygen: only extremely small ^{33}S and ^{36}S variations are anticipated for bulk planets. The similarity for ^{33}S and ^{36}S of Earth and Mars support this suggestion (Franz et al. 2014; Labidi et al. 2013). However, sulfur-bearing molecules such as SO_2 are readily photolyzed by the modern Sun's light, in planetary atmospheres. Volcanic SO_2 on both Earth and Mars is known to develop ^{33}S and ^{36}S anomalies, when sent flying to optically thin regions of planetary atmospheres (Baroni et al. 2007; Gautier et al. 2019). ^{33}S and ^{36}S measurements of Venusian SO_2 would provide first order constraints on whether any photochemistry is able to modify the composition of Venusian gases, like it does on the surface of Mars (Dottin et al. 2018b; Franz et al. 2014). These measurements will help discussing whether ^{17}O in Venusian CO_2 is a pristine measurement of the bulk planet composition, or if atmospheric samples are inevitably skewed by the occurrence of photochemistry.

3.3 Early Evolution of the Planet

3.3.1 Atmospheric Escape and Xenon

The high D/H ratio of the Venus atmosphere has often been interpreted as evidence that Venus lost its original water (Donahue et al. 1982). Dissociation of water molecules in the upper layers of the Venus atmosphere followed by escape of hydrogen to space, both powered by the strong irradiation from the early Sun, would have resulted in a depletion of water and an increase in the D/H ratio. However, both the initial water content and starting D/H ratio of water on Venus remain unknown. It remains possible that small amounts of water with a high D/H ratio (Altwegg et al. 2015) have been delivered to Venus by cometary bodies. Interestingly, a recent study combined outputs from several modeling approaches (thermochemical model of convection on Venus, atmospheric escape model and N-body simulations of the formation of the solar system) to point out that atmospheric escape on Venus was not able to remove large quantities of water over Venus history (Gillmann et al. 2020). This could imply that Venus was never affected by late accretion of water-rich material. Note however that existing models proposing scenarios for the evolution of water on Venus rely on rather poorly constrained parameters such as the initial water content or the EUV flux from the young Sun. Also, most of the isotopic fractionation of hydrogen, leading to an increase of the D/H ratio, would probably have happened only during the late stages of atmospheric escape of hydrogen. Indeed, when large amounts of hydrogen are lost during intense episodes of escape, deuterium also leaves the atmosphere efficiently and the D/H ratio stays relatively constant (Zahnle et al. 1990). In other words, when the planet suffered from different atmospheric escape regimes in its history, the D/H ratio the present atmosphere brings information on only a modest part of the history of atmospheric escape.

Noble gases hold clues on the extent of mass-fractionating escape processes suffered by Venus. Several models attempted to reproduce the elemental and isotopic composition of neon and argon measured in the atmosphere of Venus (e.g. Pepin 1991; Gillmann et al. 2009; Lammer et al. 2020, 2021). Most models involving atmospheric escape manage to reproduce the data in a consistent way but new measurements are required to really draw firm conclusions on the history of atmospheric escape on Venus. For the time being, the existing dataset only allows to conclude that the early Sun powering atmospheric escape was either a moderate or a slow rotator (Lammer et al. 2020).

Recent studies of ancient terrestrial samples containing paleo-atmospheric gases revealed that the isotopic composition of terrestrial atmospheric Xe evolved during at least 2 Ga after the Earth formed (Pujol et al. 2011, 2009; Avice et al. 2017, 2018b; Bekaert et al. 2018; Ardoin et al. 2022) (Fig. 9). This evolution corresponds to a progressive mass-dependent fractionation of Xe isotopes probably starting from a composition corresponding to a mixture between cometary and meteoritic Xe (Marty et al. 2017) and reaching the modern-like isotopic composition around the time of the Great Oxidation Event ca. 2.3 Ga ago (Avice et al. 2018b; Ardoin et al. 2022). The current favored explanation for explaining this protracted evolution is an escape of xenon ions together with hydrogen ions from the Archean atmosphere to outer space (Zahnle et al. 2019; Catling and Zahnle 2020). The isotopic composition of atmospheric Xe could thus be a tracer of hydrogen escape from planetary atmospheres. Compared to Earth, xenon in the atmosphere of Mars is likely derived from a different source, *i.e.* solar gas (Garrison and Bogard 1998; Ott 1988) but also presents a mass-dependent fractionation of 3-4 $\%.u^{-1}$ relative to the starting isotopic composition (Swindle et al. 1986). The magnitude of fractionation is similar to the case of Earth atmospheric xenon. Assuming that atmospheric escape of Xe is responsible for both the depletion and the isotopic fractionation of the remaining fraction, this similarity in the extent of isotopic fractionation is intriguing given that the escape process will be governed by parameters (abundance of total H_2, irradiation from the Sun, magnetic field etc.) which are likely to have been very distinct for early Earth and Mars. Note that, although present-day isotopic fractionation of atmospheric xenon is similar for Earth and Mars, the fractionation (and maybe escape) of xenon on Mars probably ceased much earlier than for Earth, around

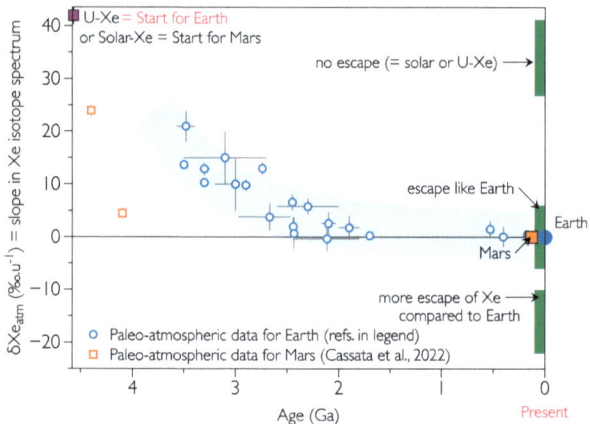

Fig. 9 Evolution of the isotopic composition of atmospheric xenon on Earth and Mars. Isotopic fractionation is expressed relative the isotopic composition of atmospheric xenon of modern Earth and Mars. Three theoretical scenarios are displayed for Venus Xe: No escape of xenon and an isotopic composition similar to primordial xenon from Solar, Chondritic or Cometary sources; Same fractionation as for Earth and Mars; More escape (and isotopic fractionation) of xenon. Paleo-atmospheric data for Earth are from Avice and Marty (2020) and refs. therein and from Ardoin et al. (2022) and Broadley et al. (2022). Paleo-atmospheric data for Mars are taken from Cassata et al. (2022). Error bars are at 1σ

4.2 Ga ago (Cassata 2017; Cassata et al. 2022) (Fig. 9). Collecting data on the abundance and isotopic composition of Xe in the Venus atmosphere is a high priority to determine to what extent "missing and isotopically fractionated xenon" is a common feature of planetary atmospheres of the inner Solar system and to further elucidate if atmospheric escape played a role in shaping the elemental and isotopic composition of atmospheric Xe. Note that potential fissiogenic additions to the budget of atmospheric $^{131-136}$Xe isotopes pose a challenge when one wants to determine the fractionation of atmospheric xenon (*e.g.*, Avice et al. 2017). A precise determination of the abundances of less abundant light isotopes such as ^{124,126,128,130}Xe plays then an important role to estimate the degree of fractionation of atmospheric xenon.

Further, the isotopic composition of nitrogen may also shed light on atmospheric loss in Venus' history. It is a reasonable hypothesis that the terrestrial planets formed with a similar nitrogen source, and the ^{15}N/^{14}N ratio places an important constraint on if and when Venus could have lost its putative primordial ocean and on the timing of the runaway greenhouse onset. Nitrogen could have been preferentially lost from Venus' atmosphere largely during its early geologic history when the atmosphere was presumably relatively thin (Baines et al. 2013, and refs. therein). Once the atmosphere became massive and CO_2-rich, further fractionation of isotopes in bulk nitrogen may be less evident due to dilution of N_2 by CO_2 at the exobase. Therefore, if the ^{15}N/^{14}N ratio is significantly greater than the terrestrial value (thus closer to Mars), this would suggest that intense hydrodynamic escape must have occurred prior to the rise of CO_2.

3.3.2 Contributions from Mantle Outgassing and Characterization of Mantle Reservoirs

It is of vital importance to understand the role of mantle outgassing in shaping surface conditions on Venus (Gillmann et al. 2022). Identification of radiogenic ^{129}Xe excesses in the atmosphere of Venus would suggest that at least some portion of the interior experienced early degassing to fractionate I/Xe during the lifetime of ^{129}I, and has subsequently out-gassed to affect the Venus atmospheric composition. The magnitude of the radiogenic excess, if any, would provide a lower limit constraint on early outgassing of a Venus mantle reservoir for comparison with terrestrial and martian mantle reservoirs, yielding insights into comparative geodynamics during accretion, and could allow comparisons of the degree of outgassing to the atmosphere. If no radiogenic ^{129}Xe excess is observed in the atmosphere, this would suggest either that the Venus mantle did not experience early outgassing (similar to the Chassigny mantle source on Mars), or that outgassing of the mantle has been very limited over the past 4.45 Ga, pushing any catastrophic early outgassing to explain low atmospheric ^{40}Ar/^{36}Ar to the earliest stages of Venus history. Xe isotopic measurements of the Venus atmosphere are needed to understand the timing and extent of mantle outgassing, and the relationship between early outgassing and atmospheric loss to space (Cassata 2017; Cassata et al. 2022). Note that atmospheric erosion by impacts could also have altered the budget of radiogenic xenon, depending on the timings of outgassing relative to the impacts. Combined, abundances of the radiogenic noble gases ^{129}Xe, ^{40}Ar, and ^{4}He will contribute to the understanding of the relative rates and significance of early, long-term, and recent outgassing, respectively.

4 Recommendations for Future Investigations

The existing dataset on the abundance and isotopic composition of noble gases and stable isotopes in the atmosphere of Venus is partial and imprecise and new investigations are urgent. A list of key measurements of noble gases and of their associated maximal uncertainties required to answer the scientific questions described in Sect. 3 is summarized in Table 1.

Two broad types of science investigation could be envisaged for gathering data on the elemental and isotopic compositions of noble gases and stable isotopes (H, C, N, O, S) in the atmosphere of Venus. One type would be an in-situ mission carrying a scientific payload able to measure the abundances and isotope ratios of the chemical elements of interest. Another one would be a sample return mission during which a portion of the Venus atmosphere would be sampled. The collected sample(s) would then be returned to Earth for characterization with state-of-the-art technologies available in international laboratories. The pros and cons regarding measurements of noble gases and stable isotopes are presented and briefly discussed in the next sections. See also Widemann et al. (2022) of this issue for a detailed discussion of future missions as well as new concepts for exploring Venus.

4.1 In Situ Measurements

To date, almost all data on the abundance and isotopic composition of volatile elements in the atmosphere of Venus (Johnson and de Oliveira 2019) have been collected during in situ investigations by probes plunging through the atmosphere of Venus and carrying mass spectrometers (*e.g.* Mahaffy 1999). This is the approach that will be used by the recently-selected DAVINCI mission from NASA (Garvin et al. 2022, Fig. 10).

One major advantage of in situ investigations is that the sample of interest is taken directly from the well-mixed atmosphere around the probe. In situ investigations at low velocity and well below the homopause ensure that the measured gas is representative of the entire atmosphere and that detected abundances and ratios are unambiguous (Tian 2015). Deep atmosphere sampling also avoids any uncertainties introduced by temporal or spatial

Fig. 10 The NASA DAVINCI mission (Garvin et al. 2022) will conduct in situ sampling of the Venus atmosphere during its ≈hour-long descent to the surface, providing critical measurements of noble gases and stable isotopes

variability of the homopause location by species or other upper atmosphere stratification (von Zahn et al. 1980; Mahieux et al. 2012; Gruchola et al. 2019; Peplowski et al. 2020).

Further, in situ sampling greatly minimizes complications of terrestrial contamination, as instrumentation can be cleaned and sealed off prior to launch, particularly for the volatiles. The captured in situ sample will thus greatly exceed any residual terrestrial species and dominate the measured signal. Finally, measurement of the atmosphere during a descent allows for sampling at multiple altitudes for species where there is the potential for vertical distribution that would otherwise complicate interpretation. Variable measurements of the D/H ratio at Venus suggest there may be a gradient (Bertaux et al. 2007), potentially driven by photolysis-induced isotopic fractionation (Liang and Yung 2009), preferential escape, or selective condensation, a process that has been found to be important for fractionating D and H on Mars and Earth (Bertaux and Montmessin 2001).

In situ investigations do present important challenges to be mitigated by each mission. Mass spectrometers on board atmospheric probes are inherently limited compared to ground-based analytical abilities, due to necessary constraints in terms of power consumption, size, mass, measurement duration, and robustness against spaceflight environments

(Arevalo et al. 2020). Three key considerations for mass spectrometric measurements of noble gases in particular are sensitivity, selectivity, and dynamic range. Sensitivity and resolution (as a proxy for selectivity) are two commonly limiting factors for flight mass spectrometers and become extremely important when targeted species are rare isotopes (D, ^{3}He, ^{78,80}Kr, $^{124-128}$Xe, ^{33}S etc.). The instrument must be sufficiently sensitive to detect a signal corresponding to these rare species. The dynamic range must also be sufficiently large to accurately measure ratios of species whose abundances span orders of magnitude (i.e., Xe vs. Ar). Another complication stems from the fact that isobaric interferences are always present, either in the residual background of the instrument itself or in the sample, even after purification steps (Wieler 2014). For example, for some ionization energies and instruments, doubly ionized CO_2^{++} ions are detected at mass 21.9949 while ^{22}Ne$^+$ ions are detected at mass 21.9914. The mass resolving power of the instrument required to simultaneously measure pure signals of each species if both are present in the instrument is higher than 6,000. Similarly, H^{35}Cl and H^{37}Cl signals are isobaric interferences of ^{36}Ar and ^{38}Ar, respectively (mass resolving power of 3,000). In the absence of high resolution for in situ instruments, the ionization energy can be adjusted to minimize the prevalence of doubly-charged ions. For other interferences, additional preprocessing and gas-cleaning techniques are required.

One way these three challenges have been addressed in previous in situ investigations has been to include an enrichment system to process samples prior to introduction into the instrument (Niemann et al. 1992; Mahaffy et al. 2012). Isobaric interferences can be minimized by significantly increasing the noble gas to interference ratio by working in ultra-clean conditions and by employing efficient purification methods to remove the bulk atmospheric components. Gas enrichment units have been employed (e.g. the SAM experiment onboard the Curiosity rover, Franz et al. (2017)) to artificially increase the partial pressure of noble gases, successfully enabling the measurement of Xe isotopes during a descent at Jupiter (Mahaffy et al. 2000) as well as recently in situ Ar, N$_2$, Kr, and Xe isotopic ratios from the surface of Mars (Atreya et al. 2013; Wong et al. 2013). Note that for Mars data, the N$_2$/Ar ratio was subsequently corrected after taking into account results obtained with calibration cells (Franz et al. 2017). The corrected N$_2$/Ar ratio is now in agreement with results obtained by previous space missions (Owen et al. 1977) and by analyses of SNC meteorites (Avice et al. 2018a). Performing enrichments in a step-wise manner can also address isobaric interferences across targeted species as well expand the effective dynamic range. The improved capabilities are necessarily traded against increases in complexity, size, power consumption, and measurement duration of the experiments.

Measurements of stable isotopes are also challenged by the presence of isobaric interferences within a mass spectrometer (e.g. ^{12}C^{16}O signal interfering with ^{28}N$_2$ signal). Additional steps of chromatography can partially solve this problem since it allows to introduce the chemical species at distinct times into the mass spectrometer. Alternatively, unlike noble gases, many of the volatile species that carry isotopes of interest have strong spectral features and thus can be measured with high sensitivity and specificity using Tunable Laser Spectroscopy (TLS) (Tarsitano and Webster 2007). Isotopic ratios of carbon, oxygen, and hydrogen have been measured using TLS on Mars (Franz et al. 2020), and similar technique could be adapted to in situ measurements at Venus even on a descent probe.

Finally, given the harsh conditions encountered by a probe plunging in the Venus atmosphere, all the scientific tasks must be achieved very rapidly, on the order of the hour, and the scientific data must be transferred either to an orbiting spacecraft or directly to Earth as soon as possible. Long-lived surface platforms, or aerial platforms at more temperate altitudes, could enable repeatable measurements with longer processing and integration times. However, these are less technically mature, more costly and the discussion of their inherent challenges is beyond the scope of this contribution.

4.2 Sample Return

There are many scientific motivations for mission concepts proposing to return a sample from the Venus atmosphere and several mission scenarios have already been envisaged (*e.g.* Rodgers et al. 2000; Sweetser et al. 2003; Greenwood et al. 2018). In most cases, the sequence of atmospheric sampling is part of a larger complex mission involving a descent stage, ground operations by a lander etc. One technical challenge is then to find a method to launch the atmospheric samples from Venus ground to orbit (*e.g.* Rodgers et al. 2000). One alternative scenario for a sample return mission scenario would be a probe launched from Earth, plunging at high velocity through the Venus atmosphere, sampling the atmosphere (Sotin et al. 2018) and returning the sample back to Earth. Such a mission can be achieved with a free-return ballistic trajectory (Sweetser et al. 2003). One important advantage of this type of scenario is that the scientific payload for this type of mission could be very simple with only few instruments to characterize the sampled gas such as pressure gauges. Most of the payload would consist in the sampling system (pipes, valves and sampling cylinders). This simple approach would certainly save mission costs and reduce the risk of this type of mission. Importantly, there would be no stringent constraint on the delay between sampling and transfer in new containers followed by scientific analyses if the sampling cylinders present very low leak rates and very low degassing rates of unwanted background gases (*e.g.* H_2) that could compromise the original chemical composition of the sample. Finally, and this is probably the most relevant argument here, sampled gas could be measured on Earth with state-of-the-art mass spectrometry & spectroscopy techniques, some of them being non-destructive such as Tunable Laser Spectroscopy (TLS) (*e.g.* Crosson et al. 2002; Tarsitano and Webster 2007). This non-destructive technique would allow to make a thoughtful characterization of the sample with, for example, measurements of the isotope ratio of H, O, C. High precision noble gas mass spectrometry could be done on very small amounts of the collected gas (Wieler 2014), especially if high-sensitive methods are used such as resonance ionization mass spectrometry (Gilmour et al. 1994). Although many of the scientific questions raised in Sect. 3 could be, at least partially, answered with an *in-situ* sampling and characterization mission, returning an atmospheric sample from Venus would bring the scientific output of any mission to Venus to another level. For example, determination of the $\Delta^{15}N^{15}N$ value of Venus atmospheric nitrogen, which quantifies the excess in the $^{15}N^{15}N$ molecule relative to a random distribution of N atoms in N_2, would allow to investigate the photochemistry of atmospheric nitrogen on Venus and to evaluate the amount of exchanges between Venus interior and its atmosphere (Yeung et al. 2017; Labidi et al. 2020; Gillmann et al. 2022). Similarly, applying very sensitive isotope determination techniques to a sample from the Venus atmosphere would allow detection and precise measurement of the abundances of minor isotopes of noble gases (^3He but also ^{78}Kr and $^{124,126}Xe$) in order to better evaluate the delivered mix of volatile elements to the Venus atmosphere but also to put constraints on atmospheric escape processes or on interior-surface interactions.

However, a sample return mission faces important challenges. In the case of a free-return ballistic trajectory, the sample will be collected at high velocity ($> 10\,\mathrm{km\,s^{-1}}$) and the gas in the front of the probe will be shocked, brought to high temperatures, and eventually turned into a plasma (Sweetser et al. 2003). Gas will then be fed into the sampling system via an inlet port and will have to travel through pipes and valves until a final expansion step into the sampling cylinder. All the steps described above could alter the chemical and isotopic integrity of the sample. The presence of a plasma implies that molecules will be dissociated. Cooling of the plasma will eventually lead to a recombination of the atoms into new molecules not originally present in the atmosphere of Venus. Noble gas atoms will also be ionized in the plasma and will be implanted on the surfaces of the probe. Numerous studies pointed out that the isotopic composition of noble gas trapped in the upper layers of solid exposed to ion implantation is mass-dependently fractionated relative to the starting composition (Bernatowicz and Hagee 1987; Kuga et al. 2017). Transfer of the gas from the atmosphere to the cylinder could also induce chemical and/or isotopic fractionation of gaseous species depending on the parameters of the gas flow. The effects of high velocity sampling on the chemical and isotopic composition of the sampled gas are currently studied via numerical simulations using a Direct Simulation Monte Carlo (DSMC) approach (Rabinovitch et al. 2019) and analog experiments for high velocity sampling are currently under development.

Returning an atmospheric sample from Venus requires important technical developments for the sampling system and the sample containers, and a full curation procedure for the sample(s) must be carefully prepared. Technologies for storing gas samples have already been employed in previous space missions (Allen et al. 2011; Moeller et al. 2021). A sample container must fit important criteria including: (i) a proper seal with a helium leak-rate on the order of 10^{-10} scc/s or below (Moeller et al. 2021); (ii) minimal interactions between sampled gas and container material; (iii) a sufficiently robust mechanical structure to resist Earth return methods (capture on orbit or direct re-entry). Sample curation is an extremely complex and costly task (*e.g.* McCubbin et al. 2019). Furthermore, curation of an atmospheric sample from another planet would be a novelty and would rest on very different principles compared to most existing setups for curating extraterrestrial rocks. For example, part of the modern sample curation process for samples recovered from asteroid Itokawa by the Japanese *Hayabusa* mission (JAXA) involves high-vacuum chambers and the use of pure gases in order to preserve the sample from any terrestrial contamination (Yada et al. 2014). Preserving and handling a gas sample from the Venus atmosphere would, of course, rely on different principles (no direct pumping nor gas injection), which remain to be defined by the international community.

Interestingly, a mission to Venus returning atmospheric samples to Earth would likely fit in the "Unrestricted Category V" of the classification scheme established by the Committee on Space Research (COSPAR 2020). If the classification is not revisited in the future, it would mean that such a sample would be subject to less stringent restrictions regarding its potential bearing of traces of life compared to samples from Mars, Europa or Enceladus. This would certainly relax the constraints for sample handling, preliminary characterization and preparation for long-term curation. However, some studies pointed out that Venus may have presented habitable surface conditions until as little as 700 Ma ago (Way et al. 2016) and that life forms could be hosted in the lower cloud layer on present-day Venus (Limaye et al. 2018). See also Westall et al. (this journal). It is plausible that, in the near future, Venus will join Mars, Europa and Enceladus in the list of objects for which a special care must be taken if a sample return is planned.

5 Conclusions

Knowing the elemental abundances and isotopic compositions of noble gases and H, C, N, O, S in the Venus atmosphere enables setting important constraints on the origin and early evolution of Earth's sister planet. Existing data are incomplete, but important differences compared to Earth and Mars have already been pointed out by previous studies. Venus is enriched in light noble gases compared to Earth and Mars. The Ar/Ne ratio is close to chondritic values whereas the isotopic composition of neon may be solar-like. There is little radiogenic ^{40}Ar in the atmosphere of Venus compared to Earth, suggesting that Venus is less degassed compared to Earth. Data on the abundance and isotope composition of krypton and xenon are cruelly lacking; these data are needed to put new constraints on the delivery mix of volatile elements to the entire planet and on the age and history of the atmosphere. For stable isotopes, the very high D/H ratio suggests that Venus suffered from hydrogen escape although the starting abundance and isotopic composition of Venus water remains unknown. A precise determination of the isotopic composition of nitrogen could help constrain the extent of elemental and isotopic fractionation of hydrogen and nitrogen by thermal and non-thermal atmospheric escape processes. Measurements of the isotopic composition of xenon will determine if, like Earth and Mars, Venus suffered from coupled hydrogen-xenon escape processes. About future investigations, the scientific community and institutional decision makers can opt for two types of investigations offering interesting complimentary perspectives: in-situ measurements or sample return. There is a strong and useful heritage of space technologies designed for in-situ measurements with an impressive list of outstanding scientific results obtained on diverse objects in the Solar System. Base on this approach, the recently selected DAVINCI (NASA) mission will provide a wealth of new data on the elemental and isotopic composition of noble gases and stable isotopes contained in the atmosphere of Venus. Sample return concepts present important but surmountable technical, scientific and organisational challenges. It may be within reach in the coming decades. Returning a sample from the atmosphere of Venus would provide, with a careful curation of the returned sample, enough analysable material to achieve high-precision measurements and would certainly lead to scientific breakthroughs.

Acknowledgements G.A. acknowledges the French Centre National d'Etudes Spatiales (CNES) for its funding support for Venus related studies. We warmly thank Tilman Spohn and members of the team at ISSI (International Space Science Institute) in Bern for their hospitality. We also thank the two anonymous reviewers for their comments and the editors for their careful handling of the manuscript.

Declarations

References

Allègre CJ, Staudacher T, Sarda P (1987) Rare gas systematics: formation of the atmosphere, evolution and structure of the Earth's mantle. Earth Planet Sci Lett 81(2):127–150. https://doi.org/10.1016/0012-821X(87)90151-8

Allen C, Allton J, Lofgren G, Righter K, Zolensky M (2011) Curating NASA's extraterrestrial samples—past, present, and future. Geochemistry 71(1):1–20. https://doi.org/10.1016/j.chemer.2010.12.003

Almayrac MG, Broadley MW, Bekaert DV, Hofmann A, Marty B (2021) Possible discontinuous evolution of atmospheric xenon suggested by Archean barites. Chem Geol 581:120,405. https://doi.org/10.1016/j.chemgeo.2021.120405

Altwegg K, Balsiger H, Bar-Nun A, Berthelier JJ, Bieler A, Bochsler P, Briois C, Calmonte U, Combi M, De Keyser J, Eberhardt P, Fiethe B, Fuselier S, Gasc S, Gombosi TI, Hansen KC, Hässig M, Jackel A, Kopp E, Korth A, LeRoy L, Mall U, Marty B, Mousis O, Neefs E, Owen T, Reme H, Rubin M, Semon T, Tzou CY, Waite H, Wurz P (2015) 67P/Churyumov-Gerasimenko, a Jupiter family comet with a high D/H ratio. Science 347(6220):3. https://doi.org/10.1126/science.1261952

Antonelli MA, Kim ST, Peters M, Labidi J, Cartigny P, Walker RJ, Lyons JR, Hoek J, Farquhar J (2014) Early inner solar system origin for anomalous sulfur isotopes in differentiated protoplanets. Proc Natl Acad Sci 111(50):17,749–17,754. https://doi.org/10.1073/pnas.1418907111

Ardoin L, Broadley M, Almayrac M, Avice G, Byrne D, Tarantola A, Lepland A, Saito T, Komiya T, Shibuya T, Marty B (2022) The end of the isotopic evolution of atmospheric xenon. Geochem Perspect Lett 20:43–47. https://doi.org/10.7185/geochemlet.2207

Arevalo R, Ni Z, Danell RM (2020) Mass spectrometry and planetary exploration: a brief review and future projection. J Mass Spectrom 55(1):e4454. https://doi.org/10.1002/jms.4454

Atreya SK, Trainer MG, Franz HB, Wong MH, Manning HLK, Malespin CA, Mahaffy PR, Conrad PG, Brunner AE, Leshin LA, Jones JH, Webster CR, Owen TC, Pepin RO, Navarro-Gonzalez R (2013) Primordial argon isotope fractionation in the atmosphere of Mars measured by the SAM instrument on Curiosity and implications for atmospheric loss. Geophys Res Lett 40(21):5605–5609. https://doi.org/10.1002/2013GL057763

Avice G, Marty B (2020) Perspectives on atmospheric evolution from noble gas and nitrogen isotopes on Earth, Mars & Venus. Space Sci Rev 216(3):36. https://doi.org/10.1007/s11214-020-00655-0.

Avice G, Marty B, Burgess R (2017) The origin and degassing history of the Earth's atmosphere revealed by Archean xenon. Nat Commun 8:15,455. https://doi.org/10.1038/ncomms15455

Avice G, Bekaert DV, Chennaoui Aoudjehane H, Marty B (2018a) Noble gases and nitrogen in Tissint reveal the composition of the Mars atmosphere. Geochem Perspect Lett 6:11–16. https://doi.org/10.7185/geochemlet.1802

Avice G, Marty B, Burgess R, Hofmann A, Philippot P, Zahnle K, Zakharov D (2018b) Evolution of atmospheric xenon and other noble gases inferred from Archean to Paleoproterozoic rocks. Geochim Cosmochim Acta 232:82–100. https://doi.org/10.1016/j.gca.2018.04.018

Avice G, Moreira M, Gilmour JD (2020) Xenon isotopes identify large-scale nucleosynthetic heterogeneities across the solar system. Astrophys J 889(1):68. https://doi.org/10.3847/1538-4357/ab5f0c

Baines KH, Atreya SK, Bullock MA, Grinspoon DH, Mahaffy P, Russell CT, Schubert G, Zahnle K (2013) The atmospheres of the terrestrial planets: clues to the origins and early evolution of Venus, Earth, and Mars. In: Comparative climatology of terrestrial planets. University of Arizona Press, Tucson, pp 1–28. http://muse.jhu.edu/books/9780816599752/9780816599752-13.pdf

Bar-Nun A, Owen T (1998) Trapping of gases in water ice and consequences to comets and the atmospheres of the inner planets. In: Solar system ices. Springer, Dordrecht, pp 353–366. https://doi.org/10.1007/978-94-011-5252-5_15

Baroni M, Thiemens MH, Delmas RJ, Savarino J (2007) Mass-independent sulfur isotopic compositions in stratospheric volcanic eruptions. Science 315(5808):84–87. https://doi.org/10.1126/science.1131754

Bekaert DV, Broadley MW, Delarue F, Avice G, Robert F Marty B (2018) Archean kerogen as a new tracer of atmospheric evolution: implications for dating the widespread nature of early life. Sci Adv 4(2):eaar2091. https://doi.org/10.1126/sciadv.aar2091

Bekaert DV, Broadley MW, Marty B (2020) The origin and fate of volatile elements on Earth revisited in light of noble gas data obtained from comet 67P/Churyumov-Gerasimenko. Sci Rep 10(1):5796. https://doi.org/10.1038/s41598-020-62650-3

Bernatowicz TJ, Hagee BE (1987) Isotopic fractionation of Kr and Xe implanted in solids at very low energies. Geochim Cosmochim Acta 51(6):1599–1611. https://doi.org/10.1016/0016-7037(87)90341-3

Bertaux JL, Montmessin F (2001) Isotopic fractionation through water vapor condensation: the Deuteropause, a cold trap for deuterium in the atmosphere of Mars. J Geophys Res, Planets 106(E12):32,879–32,884. https://doi.org/10.1029/2000JE001358

Bertaux JL, Vandaele AC, Korablev O, Villard E, Fedorova A, Fussen D, Quémerais E, Belyaev D, Mahieux A, Montmessin F, Muller C, Neefs E, Nevejans D, Wilquet V, Dubois JP, Haucheronе A, Stepanov A, Vinogradov I, Rodin A, the SPICAV/SOIR team (2007) A warm layer in Venus' cryosphere and high-altitude measurements of HF, HCl, H2O and HDO. Nature 450(7170):646–649. https://doi.org/10.1038/nature05974

Broadley M, Byrne D, Ardoin L, Almayrac M, Bekaert D, Marty B (2022) High precision noble gas measurements of hydrothermal quartz reveal variable loss rate of Xe from the Archean atmosphere. Earth Planet Sci Lett 588:117,577. https://doi.org/10.1016/j.epsl.2022.117577

Busemann H, Baur H, Wieler R (2000) Primordial noble gases in "phase Q" in carbonaceous and ordinary chondrites studied by closed-system stepped etching. Meteorit Planet Sci 35(5):949–973. https://doi.org/10.1111/j.1945-5100.2000.tb01485.x

Cassata WS (2017) Meteorite constraints on Martian atmospheric loss and paleoclimate. Earth Planet Sci Lett 479:322–329. https://doi.org/10.1016/j.epsl.2017.09.034

Cassata WS, Zahnle KJ, Samperton KM, Stephenson PC, Wimpenny J (2022) Xenon isotope constraints on ancient Martian atmospheric escape. Earth Planet Sci Lett 580:117,349. https://doi.org/10.1016/j.epsl.2021.117349

Catling DC, Kasting JF (2017) Atmospheric evolution on inhabited and lifeless worlds. Cambridge University Press, Cambridge. https://doi.org/10.1017/9781139020558

Catling DC, Zahnle KJ (2020) The Archean atmosphere. Sci Adv 6(9):eaax1420. https://doi.org/10.1126/sciadv.aax1420

Chakraborty S, Jackson TL, Ahmed M, Thiemens MH (2013) Sulfur isotopic fractionation in vacuum UV photodissociation of hydrogen sulfide and its potential relevance to meteorite analysis. Proc Natl Acad Sci 110(44):17,650–17,655. https://doi.org/10.1073/pnas.1213150110

Chassefière E, Wieler R, Marty B, Leblanc F (2012) The evolution of Venus present state of knowledge and future exploration. Planet Space Sci 63–64(C):15–23. https://doi.org/10.1016/j.pss.2011.04.007

Clayton RN, Mayeda TK (1983) Oxygen isotopes in eucrites, shergottites, and chassignites. Earth Planet Sci Lett 62(1):1–6. https://doi.org/10.1016/0012-821X(83)90066-3

Clayton RN, Grossman L, Mayeda TK (1973) A component of primitive nuclear composition in carbonaceous meteorites. Science 182(4111):485–488. https://doi.org/10.1126/science.182.4111.485

Conrad PG, Malespin CA, Franz HB, Pepin RO, Trainer MG, Schwenzer SP, Atreya SK, Freissinet C, Jones JH, Manning H, Owen T, Pavlov AA, Wiens RC, Wong MH, Mahaffy PR (2016) In situ measurement of atmospheric krypton and xenon on Mars with Mars Science Laboratory. Earth Planet Sci Lett 454:1–9. https://doi.org/10.1016/j.epsl.2016.08.028

COSPAR (2020) COSPAR policy on planetary protection. Space Res Today 208:10–22. https://doi.org/10.1016/j.srt.2020.07.009

Crosson ER, Ricci KN, Richman BA, Chilese FC, Owano TG, Provencal RA, Todd MW, Glasser J, Kachanov AA, Paldus BA, Spence TG, Zare RN (2002) Stable isotope ratios using cavity ring-down spectroscopy: determination of 13C/12C for carbon dioxide in human breath. Anal Chem 74(9):2003–2007. https://doi.org/10.1021/ac025511d

Crowther SA, Gilmour JD (2013) The Genesis solar xenon composition and its relationship to planetary xenon signatures. Geochim Cosmochim Acta 123(C):17–34. https://doi.org/10.1016/j.gca.2013.09.007

Dauphas N, Morbidelli A (2014) Geochemical and planetary dynamical views on the origin of Earth's atmosphere and oceans. In: Treatise on geochemistry, 2nd edn. Elsevier, Oxford, pp 1–35. http://www.sciencedirect.com/science/article/pii/B9780080959757013012

Dauphas N, Pourmand A (2011) Hf–W–Th evidence for rapid growth of Mars and its status as a planetary embryo. Nature 473(7348):489–492. https://doi.org/10.1038/nature10077

Donahue TM, Pollack JB (1983) Origin and evolution of the atmosphere of Venus. In: Venus. University of Arizona Press, Tucson, pp 1003–1036

Donahue TM, Hoffman JH, Hodges RR, Watson AJ (1982) Venus was wet: a measurement of the ratio of deuterium to hydrogen. Science 216(4546):630–633. https://doi.org/10.1126/science.216.4546.630

Dottin JW, Farquhar J, Labidi J (2018a) Multiple sulfur isotopic composition of main group pallasites support genetic links to IIIAB iron meteorites. Geochim Cosmochim Acta 224:276–281. https://doi.org/10.1016/j.gca.2018.01.013

Dottin JW, Labidi J, Farquhar J, Piccoli P, Liu MC, McKeegan KD (2018b) Evidence for oxidation at the base of the nakhlite pile by reduction of sulfate salts at the time of lava emplacement. Geochim Cosmochim Acta 239:186–197. https://doi.org/10.1016/j.gca.2018.07.029

Fegley B (2014) Venus. In: Treatise on geochemistry. Elsevier, Amsterdam, pp 127–148. https://doi.org/10.1016/B978-0-08-095975-7.00122-4

Franz HB, Kim ST, Farquhar J, Day JMD, Economos RC, McKeegan KD, Schmitt AK, Irving AJ, Hoek J, Iii JD (2014) Isotopic links between atmospheric chemistry and the deep sulphur cycle on Mars. Nature 508(7496):364–368. https://doi.org/10.1038/nature13175

Franz HB, Trainer MG, Malespin CA, Mahaffy PR, Atreya SK, Becker RH, Benna M, Conrad PG, Eigenbrode JL, Freissinet C, Manning HLK, Prats BD, Raaen E, Wong MH (2017) Initial SAM calibration gas experiments on Mars: quadrupole mass spectrometer results and implications. Planet Space Sci 138:44–54. https://doi.org/10.1016/j.pss.2017.01.014

Franz HB, Mahaffy PR, Webster CR, Flesch GJ, Raaen E, Freissinet C, Atreya SK, House CH, McAdam AC, Knudson CA, Archer PD, Stern JC, Steele A, Sutter B, Eigenbrode JL, Glavin DP, Lewis JMT, Malespin CA, Millan M, Ming DW, Navarro-González R, Summons RE (2020) Indigenous and exogenous organics and surface–atmosphere cycling inferred from carbon and oxygen isotopes at Gale crater. Nat Astron 4(5):526–532. https://doi.org/10.1038/s41550-019-0990-x

Füri E, Marty B (2015) Nitrogen isotope variations in the Solar System. Nat Geosci 8(7):1–8. https://doi.org/10.1038/ngeo2451

Garrison DH, Bogard DD (1998) Isotopic composition of trapped and cosmogenic noble gases in several Martian meteorites. Meteorit Planet Sci 33(4):721–736. https://doi.org/10.1111/j.1945-5100.1998.tb01678.x

Garvin JB, Getty SA, Arney GN, Johnson NM, Kohler E, Schwer KO, Sekerak M, Bartels A, Saylor RS, Elliott VE, Goodloe CS, Garrison MB, Cottini V, Izenberg N, Lorenz R, Malespin CA, Ravine M, Webster CR, Atkinson DH, Aslam S, Atreya S, Bos BJ, Brinckerhoff WB, Campbell B, Crisp D, Filiberto JR, Forget F, Gilmore M, Gorius N, Grinspoon D, Hofmann AE, Kane SR, Kiefer W, Lebonnois S, Mahaffy PR, Pavlov A, Trainer M, Zahnle KJ, Zolotov M (2022) Revealing the mysteries of Venus: the DAVINCI mission. Planet Sci J 3(5):117. https://doi.org/10.3847/PSJ/ac63c2

Gautier E, Savarino J, Hoek J, Erbland J, Caillon N, Hattori S, Yoshida N, Albalat E, Albarede F, Farquhar J (2019) 2600-years of stratospheric volcanism through sulfate isotopes. Nat Commun 10(1):127. https://doi.org/10.1038/s41467-019-08357-0

Gillmann C, Chassefière E, Lognonné P (2009) A consistent picture of early hydrodynamic escape of Venus atmosphere explaining present Ne and Ar isotopic ratios and low oxygen atmospheric content. Earth Planet Sci Lett 286(3–4):503–513. https://doi.org/10.1016/j.epsl.2009.07.016

Gillmann C, Golabek GJ, Raymond SN, Schönbächler M, Tackley PJ, Dehant V, Debaille V (2020) Dry late accretion inferred from Venus's coupled atmosphere and internal evolution. Nat Geosci 13:265–269. https://doi.org/10.1038/s41561-020-0561-x

Gillmann C, Way MJ, Avice G, Golabek GJ, Höning D, Breuer D, Krissansen-Totton J, Lammer H, Plesa AC, Persson M, O'Rourke JG, Scherf M Zolotov MY (2022, this journal) The long-term evolution of the atmosphere of Venus: processes and feedback mechanisms. Space Sci Rev https://doi.org/10.1007/s11214-022-00924-0

Gilmour JD, Lyon IC, Johnston WA, Turner G (1994) RELAX: an ultrasensitive, resonance ionization mass spectrometer for xenon. Rev Sci Instrum 65(3):617–625. https://doi.org/10.1063/1.1145127

Glaze LS, Garvin JB, Robertson B, Johnson NM, Amato MJ, Thompson J, Goodloe C, Everett D (2017) DAVINCI: deep atmosphere Venus investigation of noble gases, chemistry, and imaging. In: 2017 IEEE aerospace conference. IEEE, Big Sky, pp 1–5. https://doi.org/10.1109/AERO.2017.7943923

Greenwood RC, Anand M (2020) What is the oxygen isotope composition of Venus? The scientific case for sample return from Earth's "Sister" planet. Space Sci Rev 216(4):52. https://doi.org/10.1007/s11214-020-00669-8

Greenwood RC, Barrat JA, Miller MF, Anand M, Dauphas N, Franchi IA, Sillard P, Starkey NA (2018) Oxygen isotopic evidence for accretion of Earth's water before a high-energy Moon-forming giant impact. Sci Adv 4(3):eaao5928. https://doi.org/10.1126/sciadv.aao5928

Gruchola S, Galli A, Vorburger A, Wurz P (2019) The upper atmosphere of Venus: model predictions for mass spectrometry measurements. Planet Space Sci 170:29–41. https://doi.org/10.1016/j.pss.2019.03.006

Halliday AN (2013) The origins of volatiles in the terrestrial planets. Geochim Cosmochim Acta 105(C):146–171. https://doi.org/10.1016/j.gca.2012.11.015

Heber VS, Wieler R, Baur H, Olinger C, Friedmann TA, Burnett DS (2009) Noble gas composition of the solar wind as collected by the Genesis mission. Geochim Cosmochim Acta 73(24):7414–7432. https://doi.org/10.1016/j.gca.2009.09.013

Heber VS, Baur H, Bochsler P, McKeegan KD, Neugebauer M, Reisenfeld DB, Wieler R, Wiens RC (2012) Isotopic mass fractionation of solar wind: evidence from fast and slow solar wind collected by the Genesis mission. Astrophys J 759(121):1–13. https://doi.org/10.1088/0004-637X/759/2/121

Hoffman JH, Hodges RR, Donahue TM, McElroy MB (1980) Composition of the Venus lower atmosphere from the Pioneer Venus Mass Spectrometer. J Geophys Res Space Phys 85(A13):7882–7890. https://doi.org/10.1029/JA085iA13p07882

Hunten DM, Pepin RO, Walker JC (1987) Mass fractionation in hydrodynamic escape. Icarus 69(3):532–549. https://doi.org/10.1016/0019-1035(87)90022-4

Istomin V, Grechnev K, Kotchnev V (1980) Mass spectrometer measurements of the composition of the lower atmosphere of Venus. In: COSPAR colloquia series, vol 20. Elsevier, Amsterdam, pp 215–218. https://doi.org/10.1016/S0964-2749(13)60044-X

Istomin VG, Grechnev KV, Kochnev VA (1982) Preliminary results of mass-spectrometric measurements on the descent "Venera 13" and "Venera 14" space probes. Pis'ma Astron Zh 8(7):391–398

Istomin VG, Grechnev KV, Kochnev VA (1983) Venera 13 and Venera 14: mass spectrometry of the atmosphere. Kosm Issled 21:410–420

Jakosky BM, Pepin RO (1994) Mars atmospheric loss and isotopic fractionation by solar-wind-induced sputtering and photochemical escape. Icarus 111:271–288

Javoy M, Kaminski E, Guyot F, Andrault D, Sanloup C, Moreira M, Labrosse S, Jambon A, Agrinier P, Davaille A, Jaupart C (2010) The chemical composition of the Earth: enstatite chondrite models. Earth Planet Sci Lett 293(3–4):259–268. https://doi.org/10.1016/j.epsl.2010.02.033

Johnson NM, de Oliveira MRR (2019) Venus atmospheric composition in situData: a compilation. Earth Space Sci 6(7):1299–1318. https://doi.org/10.1029/2018EA000536

Kaib NA, Cowan NB (2015) The feeding zones of terrestrial planets and insights into Moon formation. Icarus 252(C):161–174. https://doi.org/10.1016/j.icarus.2015.01.013

Kaula WM (1999) Constraints on Venus evolution from radiogenic argon. Icarus 139(1):32–39. https://doi.org/10.1006/icar.1999.6082

Kleine T, Budde G, Burkhardt C, Kruijer TS, Worsham EA, Morbidelli A, Nimmo F (2020) The non-carbonaceous–carbonaceous meteorite dichotomy. Space Sci Rev 216(4):55. https://doi.org/10.1007/s11214-020-00675-w

Kuga M, Cernogora G, Marrocchi Y, Tissandier L, Marty B (2017) Processes of noble gas elemental and isotopic fractionations in plasma-produced organic solids: cosmochemical implications. Geochim Cosmochim Acta 217:219–230. https://doi.org/10.1016/j.gca.2017.08.031

Labidi J, Cartigny P, Moreira M (2013) Non-chondritic sulphur isotope composition of the terrestrial mantle. Nature 501(7466):208–211. https://doi.org/10.1038/nature12490

Labidi J, Barry PH, Bekaert DV, Broadley MW, Marty B, Giunta T, Warr O, Sherwood Lollar B, Fischer TP, Avice G, Caracausi A, Ballentine CJ, Halldórsson SA, Stefánsson A, Kurz MD, Kohl IE, Young ED (2020) Hydrothermal 15N15N abundances constrain the origins of mantle nitrogen. Nature 580(7803):367–371. https://doi.org/10.1038/s41586-020-2173-4

Lammer H, Leitzinger M, Scherf M, Odert P, Burger C, Kubyshkina D, Johnstone C, Maindl T, Schäfer C, Güdel M, Tosi N, Nikolaou A, Marcq E, Erkaev N, Noack L, Kislyakova K, Fossati L, Pilat-Lohinger E, Ragossnig F, Dorfi E (2020) Constraining the early evolution of Venus and Earth through atmospheric Ar, Ne isotope and bulk K/U ratios. Icarus 339:113,551. https://doi.org/10.1016/j.icarus.2019.113551

Lammer H, Brasser R, Johansen A, Scherf M, Leitzinger M (2021) Formation of Venus, Earth and Mars: constrained by isotopes. Space Sci Rev 217(1):7. https://doi.org/10.1007/s11214-020-00778-4

Lauretta DS, Lodders K, Fegley B (1997) Experimental simulations of sulfide formation in the solar nebula. Science 277(5324):358–360. https://doi.org/10.1126/science.277.5324.358

Liang MC, Yung YL (2009) Modeling the distribution of H 2 O and HDO in the upper atmosphere of Venus. J Geophys Res 114:E00B28. https://doi.org/10.1029/2008JE003095

Limaye SS, Mogul R, Smith DJ, Ansari AH, Słowik GP, Vaishampayan P (2018) Venus' spectral signatures and the potential for life in the clouds. Astrobiology 18(9):1181–1198. https://doi.org/10.1089/ast.2017.1783

Lodders K (2003) Solar system abundances and condensation temperatures of the elements. Astrophys J 591(2):1220. https://doi.org/10.1086/375492

Mahaffy P (1999) Mass spectrometers developed for planetary missions. In: Ehrenfreund P, Krafft C, Kochan H, Pirronello V (eds) Laboratory astrophysics and space research. Astrophysics and space science library, vol 236. Springer, Dordrecht, pp 355–376. https://doi.org/10.1007/978-94-011-4728-6_13

Mahaffy PR, Niemann HB, Alpert A, Atreya SK, Demick J, Donahue TM, Harpold DN, Owen TC (2000) Noble gas abundance and isotope ratios in the atmosphere of Jupiter from the Galileo Probe Mass Spectrometer. J Geophys Res, Planets 105(E6):15,061–15,071. https://doi.org/10.1029/1999JE001224

Mahaffy PR, Webster CR, Cabane M, Conrad PG, Coll P, Atreya SK, Arvey R, Barciniak M, Benna M, Bleacher L, Brinckerhoff WB, Eigenbrode JL, Carignan D, Cascia M, Chalmers RA, Dworkin JP, Errigo T, Everson P, Franz H, Farley K, Feng S, Frazier G, Freissinet C, Glavin DP, Harpold DN, Hawk D, Holmes V, Johnson CS, Jones A, Jordan P, Kellogg J, Lewis J, Lyness E, Malespin CA, Martin DK, Maurer J, McAdam AC, McLennan D, Nolan TJ, Noriega M, Pavlov AA, Prats B, Raaen E, Sheinman O, Sheppard D, Smith J, Stern JC, Tan F, Trainer M, Ming DW, Morris RV, Jones J, Gundersen C, Steele A, Wray J, Botta O, Leshin LA, Owen T, Battel S, Jakosky BM, Manning H, Squyres S, Navarro-González R, McKay CP, Raulin F, Sternberg R, Buch A, Sorensen P, Kline-Schoder R, Coscia D, Szopa C, Teinturier S, Baffes C, Feldman J, Flesch G, Forouhar S, Garcia R, Keymeulen D, Woodward S, Block BP, Arnett K, Miller R, Edmonson C, Gorevan S, Mumm E (2012) The sample analysis at Mars investigation and instrument suite. Space Sci Rev 170(1–4):401–478. https://doi.org/10.1007/s11214-012-9879-z

Mahieux A, Vandaele AC, Robert S, Wilquet V, Drummond R, Montmessin F, Bertaux JL (2012) Densities and temperatures in the Venus mesosphere and lower thermosphere retrieved from SOIR on board Venus Express: carbon dioxide measurements at the Venus terminator. J Geophys Res, Planets 117(E07001):1–15. https://doi.org/10.1029/2012JE004058

Marrocchi Y, Avice G, Estrade N (2015) Multiple carriers of Q noble gases in primitive meteorites. Geophys Res Lett 42:2093–2099. https://doi.org/10.1002/(ISSN)1944-8007

Marty B (2012) The origins and concentrations of water, carbon, nitrogen and noble gases on Earth. Earth Planet Sci Lett 313–314:56–66. https://doi.org/10.1016/j.epsl.2011.10.040

Marty B, Avice G, Sano Y, Altwegg K, Balsiger H, Hässig M, Morbidelli A, Mousis O, Rubin M (2016) Origins of volatile elements (H, C, N, noble gases) on Earth and Mars in light of recent results from the ROSETTA cometary mission. Earth Planet Sci Lett 441:91–102

322

Marty B, Altwegg K, Balsiger H, Bar-Nun A, Bekaert DV, Berthelier JJ, Bieler A, Briois C, Calmonte U, Combi M, De Keyser J, Fiethe B, Fuselier SA, Gasc S, Gombosi TI, Hansen KC, Hässig M, Jackel A, Kopp E, Korth A, Le Roy L, Mall U, Mousis O, Owen T, Reme H, Rubin M, Semon T, Tzou CY, Waite JH, Wurz P (2017) Xenon isotopes in 67P/Churyumov-Gerasimenko show that comets contributed to Earth's atmosphere. Science 356(6342):1069–1072. https://doi.org/10.1126/science.aal3496

Marty B, Bekaert DV, Broadley MW, Jaupart C (2019) Geochemical evidence for high volatile fluxes from the mantle at the end of the Archaean. Nature 575(7783):485–488. https://doi.org/10.1038/s41586-019-1745-7

Mathew KJ, Marti K (2001) Early evolution of Martian volatiles: nitrogen and noble gas components in ALH84001 and Chassigny. J Geophys Res 106(E1):1401–1422. https://doi.org/10.1029/2000JE001255

Mathew KJ, Kim JS, Marti K (1998) Martian atmospheric and indigenous components of xenon and nitrogen in SNC meteorites. Meteorit Planet Sci 33:655–664

McCubbin FM, Barnes JJ (2019) Origin and abundances of H_2O in the terrestrial planets, Moon, and asteroids. Earth Planet Sci Lett 526:115,771. https://doi.org/10.1016/j.epsl.2019.115771

McCubbin FM, Herd CDK, Yada T, Hutzler A, Calaway MJ, Allton JH, Corrigan CM, Fries MD, Harrington AD, McCoy TJ, Mitchell JL, Regberg AB, Righter K, Snead CJ, Tait KT, Zolensky ME, Zeigler RA (2019) Advanced curation of astromaterials for planetary science. Space Sci Rev 215(8):48. https://doi.org/10.1007/s11214-019-0615-9

McElroy MB, Yung YL, Nier AO (1976) Isotopic composition of nitrogen: implications for the past history of Mars' atmosphere. Science 194(4260):70–72. https://doi.org/10.1126/science.194.4260.70

McKeegan KD, Kallio APA, Heber VS, Jarzebinski G, Mao PH, Coath CD, Kunihiro T, Wiens RC, Nordholt JE, Moses RW, Reisenfeld DB, Jurewicz AJG, Burnett DS (2011) The oxygen isotopic composition of the Sun inferred from captured solar wind. Science 332(6037):1528–1532. https://doi.org/10.1126/science.1204636

Meshik A, Hohenberg C, Pravdivtseva O, Burnett D (2014) Heavy noble gases in solar wind delivered by Genesis mission. Geochim Cosmochim Acta 127(C):326–347. https://doi.org/10.1016/j.gca.2013.11.030

Meshik A, Pravdivtseva O, Burnett D (2020) Refined composition of Solar Wind xenon delivered by Genesis NASA mission: comparison with xenon captured by extraterrestrial regolith soils. Geochim Cosmochim Acta 276:289–298. https://doi.org/10.1016/j.gca.2020.03.001

Moeller RC, Jandura L, Rosette K, Robinson M, Samuels J, Silverman M, Brown K, Duffy E, Yazzie A, Jens E, Brockie I, White L, Goreva Y, Zorn T, Okon A, Lin J, Frost M, Collins C, Williams JB, Steltzner A, Chen F, Biesiadecki J (2021) The sampling and caching subsystem (SCS) for the scientific exploration of Jezero crater by the Mars 2020 perseverance rover. Space Sci Rev 217(1):5. https://doi.org/10.1007/s11214-020-00783-7

Musselwhite DS, Drake MJ, Swindle TD (1991) Early outgassing of Mars supported by differential water solubility of iodine and xenon. Nature 352(6337):697–699. https://doi.org/10.1038/352697a0

Niemann HB, Harpold DN, Atreya SK, Carignan GR, Hunten DM (1992) Galileo probe mass spectrometer experiment. Space Sci Rev 60:111–142

O'Brien DP, Izidoro A, Jacobson SA, Raymond SN, Rubie DC (2018) The delivery of water during terrestrial planet formation. Space Sci Rev 214(1):47. https://doi.org/10.1007/s11214-018-0475-8

Okabe H (1978) Photochemistry of small molecules. Wiley, New York

Olson PL, Sharp ZD (2019) Nebular atmosphere to magma ocean: a model for volatile capture during Earth accretion. Phys Earth Planet Inter 294:106294. https://doi.org/10.1016/j.pepi.2019.106294

O'Rourke JG, Korenaga J (2015) Thermal evolution of Venus with argon degassing. Icarus 260:128–140. https://doi.org/10.1016/j.icarus.2015.07.009

Ott U (1988) Noble gases in SNC meteorites: Shergotty, Nakhla, Chassigny. Geochim Cosmochim Acta 52:1937–1948. https://doi.org/10.1016/0016-7037(88)90017-8

Owen T, Biemann K, Rushneck DR, Biller JE, Howarth DW, Lafleur AL (1977) The composition of the atmosphere at the surface of Mars. J Geophys Res 82(28):4635–4639. https://doi.org/10.1029/JS082i028p04635

Ozima M, Podosek FA (2002) Noble gas geochemistry, 2nd edn. Cambridge University Press, Cambridge

Parai R, Mukhopadhyay S, Tucker JM, Petö MK (2019) The emerging portrait of an ancient, heterogeneous and continuously evolving mantle plume source. Lithos 346–347:105,153. https://doi.org/10.1016/j.lithos.2019.105153

Pepin RO (1991) On the origin and early evolution of terrestrial planet atmospheres and meteoritic volatiles. Icarus 92(1):2–79. https://doi.org/10.1016/0019-1035(91)90036-s

Pepin RO (1994) The hunt for U-xenon. Meteoritics 29:568–569

Pepin RO, Porcelli D (2002) Origin of noble gases in the terrestrial planets. Rev Mineral Geochem 47(1):191–246. https://doi.org/10.2138/rmg.2002.47.7

Peplowski PN, Lawrence DJ, Wilson JT (2020) Chemically distinct regions of Venus's atmosphere revealed by measured N2 concentrations. Nat Astron 4:947–950. https://doi.org/10.1038/s41550-020-1079-2

Péron S, Mukhopadhyay S (2022) Krypton in the Chassigny meteorite shows Mars accreted chondritic volatiles before nebular gases. Science 377:320–324. https://doi.org/10.1126/science.abk1175

Péron S, Moreira M, Agranier A (2018) Origin of light noble gases (He, Ne, and Ar) on Earth: a review. Geochem Geophys Geosyst 461(4):1227. https://doi.org/10.1002/2017GC007388

Péron S, Mukhopadhyay S, Kurz MD, Graham DW (2021) Deep-mantle krypton reveals Earth's early accretion of carbonaceous matter. Nature 600(7889):462–467. https://doi.org/10.1038/s41586-021-04092-z

Piani L, Marrocchi Y, Rigaudier T, Vacher LG, Thomassin D, Marty B (2020) Earth's water may have been inherited from material similar to enstatite chondrite meteorites. Science 369(6507):1110–1113. https://doi.org/10.1126/science.aba1948

Porcelli D, Ballentine CJ (2002) Models for distribution of terrestrial noble gases and evolution of the atmosphere. Rev Mineral Geochem 47(1):411–480. https://doi.org/10.2138/rmg.2002.47.11

Porcelli D, Ballentine CJ, Wieler R (2002) An overview of noble gas geochemistry and cosmochemistry. Rev Mineral Geochem 47(1):1–19. https://doi.org/10.2138/rmg.2002.47.1

Pujol M, Marty B, Burnard P, Philippot P (2009) Xenon in Archean barite: weak decay of 130Ba, mass-dependent isotopic fractionation and implication for barite formation. Geochim Cosmochim Acta 73(22):6834–6846. https://doi.org/10.1016/j.gca.2009.08.002

Pujol M, Marty B, Burgess R (2011) Chondritic-like xenon trapped in Archean rocks: a possible signature of the ancient atmosphere. Earth Planet Sci Lett 308(3–4):298–306. https://doi.org/10.1016/j.epsl.2011.05.053

Rabinovitch J, Borner A, Gallis MA, Sotin C (2019) Hypervelocity noble gas sampling in the upper atmosphere of Venus. In: AIAA aviation 2019 forum. American Institute of Aeronautics and Astronautics, Dallas. https://doi.org/10.2514/6.2019-3223

Rao M (2002) Neutron capture isotopes in the Martian regolith and implications for Martian atmospheric noble gases. Icarus 156(2):352–372. https://doi.org/10.1006/icar.2001.6809

Rodgers D, Gilmore M, Sweetser T, Cameron J, Chen G-S, Cutts J, Gershman R, Hall J, Kerzhanovich V, McRonald A, Nilsen E, Petrick W, Sauer C, Wilcox B, Yavrouian A, Zimmerman W (2000) Venus sample return. A hot topic. In: 2000 IEEE aerospace conference. Proceedings (Cat. No. 00TH8484), vol 7. IEEE, Big Sky, pp 473–484. https://doi.org/10.1109/AERO.2000.879315

Salvador A, Avice G, Breuer D, Gillmann C, Jacobson SA, Lammer H, Marcq E, Raymond SN, Sakuraba H, Scherf M, Way MJ (2022, this journal) Magma ocean, water, and the early atmosphere of Venus. Space Sci Rev

Sotin C, Avice G, Baker J, Freeman T, Madzunkov SM, Stevenson T, Arora N, Darrach MR, Lightsey G, Marty B (2018) Cupid's arrow: a small satellite to measure noble gases in Venus atmosphere. In: 49th lunar and planetary science conference

Sweetser T, Peterson C, Nilsen E, Gershman B (2003) Venus sample return missions—a range of science, a range of costs. Acta Astronaut 52(2–6):165–172. https://doi.org/10.1016/S0094-5765(02)00153-4

Swindle TD, Caffee MW, Hohenberg CM (1986) Xenon and other noble gases in shergottites. Geochim Cosmochim Acta 50(6):1001–1015. https://doi.org/10.1016/0016-7037(86)90381-9

Takaoka N (1972) An interpretation of general anomalies of xenon and the isotopic composition of primitive xenon. J Mass Spectrom Soc Jpn 20(4):287–302. https://doi.org/10.5702/massspec1953.20.287

Tarsitano CG, Webster CR (2007) Multilaser Herriott cell for planetary tunable laser spectrometers. Appl Opt 46(28):6923. https://doi.org/10.1364/AO.46.006923

Thiemens MH (2006) History and applications of mass-independent isotope effects. Annu Rev Earth Planet Sci 34(1):217–262. https://doi.org/10.1146/annurev.earth.34.031405.125026

Tian F (2015) Atmospheric escape from solar system terrestrial planets and exoplanets. Annu Rev Earth Planet Sci 43:459–476. https://doi.org/10.1146/annurev-earth-060313-054834

von Zahn U, Fricke KH, Hunten DM, Krankowsky D, Mauersberger K, Nier AO (1980) The upper atmosphere of Venus during morning conditions. J Geophys Res 85(A13):7829–7840

von Zahn U, Kumar S, Niemann H, Prinn R (1983) Composition of the Venus atmosphere. In: Venus. The University of Arizona Press, Tucson, p 299

Way MJ, Del Genio AD, Kiang NY, Sohl LE, Grinspoon DH, Aleinov I, Kelley M, Clune T (2016) Was Venus the first habitable world of our solar system? Geophys Res Lett 43:8376–8383. https://doi.org/10.1002/(ISSN)1944-8007

Weiss BP, Bai XN, Fu RR (2021) History of the solar nebula from meteorite paleomagnetism. Sci Adv 7(1):eaba5967. https://doi.org/10.1126/sciadv.aba5967

Widemann T, Smrekar SE, Garvin JB, Straume-Lindner AG, Ocampo AC, Voirin T, Hensley S, Darby Dyar M, Whitten JL, Nunes DC, Getty SA, Arney GN, Johnson NM, Kohler E, Spohn T, O'Rourke JG, Wilson C, Way MJ, Ostberg C, Westall F, Höning D, Jacobson S, Salvador A, Avice G, Carter L, Gilmore M, Ghail R, Helbert J, Byrne P, Herrick RR, Izenberg N, Marcq E, Rolf T, Weller M, Gillmann C, Korablev

O, Zelenyi L, Zasova L, Gorinov D (2022) Venus evolution through time: key science questions, selected mission concepts and future investigations. Space Sci Rev

Wieler R (2002) Noble gases in the solar system. Rev Mineral Geochem 47(1):21–70. https://doi.org/10.2138/rmg.2002.47.2

Wieler R (2014) Noble gas mass spectrometry. In: Treatise on geochemistry, 2nd edn. Elsevier, Oxford, pp 355–373. https://doi.org/10.1016/B978-0-08-095975-7.01428-5

Wieler R, Anders E, Baur H, Lewis RS, Signer P (1992) Characterisation of Q-gases and other noble gas components in the Murchison meteorite. Geochim Cosmochim Acta 56(7):2907–2921. https://doi.org/10.1016/0016-7037(92)90367-R

Wong MH, Atreya SK, Mahaffy PN, Franz HB, Malespin C, Trainer MG, Stern JC, Conrad PG, Manning HLK, Pepin RO, Becker RH, McKay CP, Owen TC, Navarro-González R, Jones JH, Jakosky BM, Steele A (2013) Isotopes of nitrogen on Mars: atmospheric measurements by Curiosity's mass spectrometer. Geophys Res Lett 40(23):6033–6037. https://doi.org/10.1002/2013GL057840

Yada T, Fujimura A, Abe M, Nakamura T, Noguchi T, Okazaki R, Nagao K, Ishibashi Y, Shirai K, Zolensky ME, Sandford S, Okada T, Uesugi M, Karouji Y, Ogawa M, Yakame S, Ueno M, Mukai T, Yoshikawa M, Kawaguchi J (2014) Hayabusa-returned sample curation in the Planetary Material Sample Curation Facility of JAXA: Hayabusa return sample curation in JAXA. Meteorit Planet Sci 49(2):135–153. https://doi.org/10.1111/maps.12027

Yeung LY, Li S, Kohl IE, Haslun JA, Ostrom NE, Hu H, Fischer TP, Schauble EA, Young ED (2017) Extreme enrichment in atmospheric 15N15N. Sci Adv 3(11):eaao6741. https://doi.org/10.1126/sciadv.aao6741

Young ED, Galy A, Nagahara H (2002) Kinetic and equilibrium mass-dependent isotope fractionation laws in nature and their geochemical and cosmochemical significance. Geochim Cosmochim Acta 66(6):1095–1104. https://doi.org/10.1016/S0016-7037(01)00832-8

Young ED, Kohl IE, Warren PH, Rubie DC, Jacobson SA, Morbidelli A (2016) Oxygen isotopic evidence for vigorous mixing during the Moon-forming giant impact. Science 351(6272):493–496. https://doi.org/10.1126/science.aad0525

Yung YL, Demore W (1982) Photochemistry of the stratosphere of Venus: implications for atmospheric evolution. Icarus 51(2):199–247. https://doi.org/10.1016/0019-1035(82)90080-X

Zahnle K, Kasting JF, Pollack JB (1990) Mass fractionation of noble gases in diffusion-limited hydrodynamic hydrogen escape. Icarus 84(2):502–527. https://doi.org/10.1016/0019-1035(90)90050-J

Zahnle KJ, Gacesa M, Catling DC (2019) Strange messenger: a new history of hydrogen on Earth, as told by xenon. Geochim Cosmochim Acta 244:56–85. https://doi.org/10.1016/j.gca.2018.09.017

Publisher's Note Springer Nature remains neutral with regard to jurisdictional claims in published maps and institutional affiliations.

Authors and Affiliations

Guillaume Avice[1]  **· Rita Parai[2] · Seth Jacobson[3] · Jabrane Labidi[1] · Melissa G. Trainer[4] · Mihail P. Petkov[5]**

✉ G. Avice
avice@ipgp.fr

R. Parai
parai@wustl.edu

S. Jacobson
seth@msu.edu

J. Labidi
labidi@ipgp.fr

M.G. Trainer
melissa.trainer@nasa.gov

M.P. Petkov
mihail.p.petkov@jpl.nasa.gov

[1]   Université Paris Cité, Institut de physique du globe de Paris, CNRS, 75005 Paris, France

2   Department of Earth and Planetary Sciences and McDonnell Center for Space Sciences,
    Washington University in St. Louis, St. Louis, MO, USA

3   Department of Earth and Environmental Sciences, Michigan State University, East Lansing, MI
    4886, USA

4   NASA Goddard Space Flight Center, Greenbelt, MD 20771, USA

5   NASA Jet Propulsion Laboratory, California Institute of Technology, Pasadena, CA 91109, USA

326   ⓐ Springer

Space Science Reviews (2023) 219:85
https://doi.org/10.1007/s11214-023-01033-2

Sedimentary Processes on Venus

Lynn M. Carter[1] · Martha S. Gilmore[2] · Richard C. Ghail[3] · Paul K. Byrne[4] ·
Suzanne E. Smrekar[5] · Terra M. Ganey[6] · Noam Izenberg[7]

Received: 18 January 2023 / Accepted: 17 October 2023 / Published online: 6 December 2023
© The Author(s) 2023

Abstract

The sedimentary cycle, including the processes of erosion, transport, and lithification, is a key part of how planets evolve over time. Early images of Venus's vast volcanic plains, numerous volcanoes, and rugged tectonic regions led to the interpretation that Venus is a volcanic planet with little sediment cover and perhaps few processes for generating sedimentary rocks. However, in the years since the Magellan mission in the 1990s we have developed a better understanding of sedimentary process on Venus. Impact craters are the largest present-day source of sediments, with estimates from the current crater population suggesting an average sediment layer 8–63 cm in thickness if distributed globally. There is clear evidence of fine-grained material in volcanic summit regions that is likely produced through volcanism, and dune fields and yardangs indicate transport of sediments and erosion of rocks through wind. Landslides and fine-grained materials in highland tessera regions demonstrate erosive processes that move sediment downhill. It is clear that sediments are an important part of Venus's geology, and it is especially important to realize that they mantle features that may be of interest to future landed or low-altitude imaging missions. The sinks of sediments are less well known, as it has been difficult to identify sedimentary rocks with current data. Layering observed in Venera images and in Magellan images of some tessera regions, as well as calculated rock densities, suggest that sedimentary rocks are present on Venus. New data is needed to fully understand and quantify the present-day sedimentary cycle and establish with certainty whether sedimentary rock packages do, in fact, exist on Venus. These data sets will need to include higher-resolution optical and radar imaging, experimental and geochemical measurements to determine how chemical weathering and lithification can occur, and topography to better model mesospheric winds. Sediments and sedimentary rocks are critical to understanding how Venus works today, but are also extremely important for determining how Venus's climate has changed through time and whether it was once a habitable planet.

Keywords Venus · Sediments · Sedimentology · Regolith · Aeolian · Volcanism

Venus: Evolution Through Time
Edited by Colin F. Wilson, Doris Breuer, Cédric Gillmann, Suzanne E. Smrekar, Tilman Spohn and Thomas Widemann

Extended author information available on the last page of the article

1 Introduction

The production of regolith and the transport and lithification of sediments are key processes that are intimately tied to planetary climate and evolution. On Earth and Mars, erosion by water, wind, and ice lead to the production and re-working of sedimentary rocks, and outcrops of sedimentary rocks can record evidence of changing climates. At face value, Venus—currently a hot, dry planet with a thick atmosphere—is expected to have a considerably different history of erosion and sedimentation to that of modern Earth. However, the processes that create sediments and the nature of Venusian sedimentary rocks are still largely unknown.

The Magellan mission to Venus, with its 150 m-resolution radar imaging at a wavelength of 12.6 cm, provided the first detailed look at sediments on a global scale. At this resolution, it was possible to identify dunes, yardangs, and wind streaks (Greeley et al. 1992, 1995), as well as mass-wasting features in the tesserae (Bindschadler et al. 1992; Solomon et al. 1992; Malin 1992). Early analysis suggested there were few dune fields and possibly a lack of sand-sized particles that could be transported (Greeley et al. 1992). Radar-dark and -bright crater ejecta and parabolas suggested sediment production associated with impacts (Campbell et al. 1992; Schultz 1992), as well as a possible erosive process for these deposits (Izenberg et al. 1994; Basilevsky et al. 2003). But there was little evidence seen for widespread regolith cover in the synthetic aperture radar (SAR) imaging. Radar is capable of penetrating into the surface and imaging through mantling deposits, however, so the Magellan SAR data could possibly be missing the detection of deposits that would otherwise be visible in optical or higher-spatial-resolution radar data.

It was also clear from crater statistics that the Venus surface is mostly very young, with an average surface crater retention age of $\sim$500 Mya (Phillips et al. 1991, 1992; Herrick et al. 2023). The presence of extensive volcanic plains led to the assumption that most of the former sedimentary history of Venus, provided it ever existed, is buried and inaccessible. In the absence of water, chemical weathering on Venus is thought to be a relatively slow process that may not produce considerable volumes of fines (e.g., sand size or smaller particles; Dyar et al. 2021). Images of the Venera landing sites show a variety of rock distributions and rock sizes (Florensky 1977; Florensky et al. 1983; Surkov and Barsukov 1985), but the rocks were largely interpreted as volcanic in origin, commensurate with their basaltic chemistry (Garvin et al. 1984). The Venera data do not show an abundance of fine sediments or landforms that would indicate the presence of extensive deposits of sand-size and smaller materials (Garvin et al. 1984). These observations together led to the general consensus that mantling deposits on Venus are localized, sediment transport is limited, and, aside from the highlands, the surface is composed of basaltic volcanic rocks.

Work in the decades since Magellan, including both new data and the application of new analysis techniques, have led to a different view of sediment cover and sedimentary rocks on Venus. These studies, including the identification of possible pyroclastic deposits, mapping of layering within tesserae, re-analyses of landing site data, and ground-based polarimetric radar imaging, reveal that Venus has a more complex rock record than was previously thought. The following sections present a summary of what we currently know about sediment sources and sinks on Venus, and provide a discussion of possible sedimentary rock sequences and their implications. These observations set the stage for future missions to Venus that can provide answers to the many questions that remain about Venus sedimentology, and what sediments there can reveal about the planet's surface processes, surface–atmosphere interactions, and climate evolution.

2 Sources of Sediment

The main processes leading to sediments on Venus today are impact cratering, volcanism, mass wasting and mechanical weathering, and the chemical breakdown of surface rocks. With our current data, it is difficult to completely assess how much sediment is generated through these various processes. Radar can image through thin deposits, and the currently available low-resolution orbital images likely miss smaller such deposits. In addition, there is usually little direct information available about the thickness of those deposits that have been recognized. However, there is still ample evidence that fine-grained (<1 cm) material covers a substantial part of the Venus surface and provides a material source for lithification and the potential generation of new sedimentary rocks.

2.1 Impact Craters

Impact cratering is likely to be one of the primary, if not the dominant, means of generating sediments on Venus today. The largest impact-related features are parabola-shaped deposits associated with some craters. Most of these parabolas are visible in SAR images, but there are several parabolas visible in microwave emissivity that are not always commensurately visible in SAR data (Campbell et al. 1992; Schaller and Melosh 1998). The parabola features likely form when debris from the impact is lofted into the atmosphere and then later carried downstream by the prevailing westward winds; models suggest that the settling time of cm-sized particles is a few hours (Campbell et al. 1992).

Individual parabolas can cover up to 2 million km^2, and the 58 parabolas currently recognized in the Magellan data cover cumulatively at least 40 million km^2 (or almost 9%) of the Venus surface (Campbell et al. 1992; Schaller and Melosh 1998). A two-layer radar backscattering model suggests that the parabola features may be at least 1–3 m thick (Campbell et al. 1992). In some cases, these parabolas appear to thin out as underlying plains features become visible in the radar data near the edges, supporting the idea that the deposits may be centimetres thick at the distal parts and metres thick in the central portions, consistent with the predictions of analytical models (Vervack and Melosh 1992). Wind streaks are associated with many of the parabolas, supporting the hypothesis of the presence of mobile fine materials (Greeley et al. 1992). In several cases, the parabolas clearly extend into nearby regions of tesserae, darkening the surface slightly and changing the local radar polarization signature of that terrain type (Campbell et al. 2015; Whitten and Campbell 2016).

The parabolas have a variety of morphologies and backscatter properties. In some cases, the parabolas are mostly radar bright, in others mostly radar dark, and in some instances, they consist of alternating bright and dark sections (Fig. 1). The radar-bright portions are likely rough at wavelength scales (Campbell et al. 1992). Low backscatter power values indicate smooth surfaces, either stripped of material or mantled in fines. These bright–dark patterns may indicate deposits with different particle sizes, differences in stripping of fine material, differences in composition, or possibly rippled surface textures.

Radar imaging polarimetry from the Arecibo observatory was used to measure the degree of linear polarization, an indicator of penetration of the radar wave and hence of the depths of possible surface mantling layers (Carter et al. 2004). The circular polarization ratio (CPR) is the ratio of the same-sense circular polarization that was transmitted to the opposite-sense circular polarization received. CPR is used as measure of surface roughness, with higher values indicating more rugged surface textures at the wavelength scale. The degree of linear polarization (values between 0 and 1) is the fraction of the received radar wave that is linearly polarized. When a circularly polarized wave is transmitted, such as with

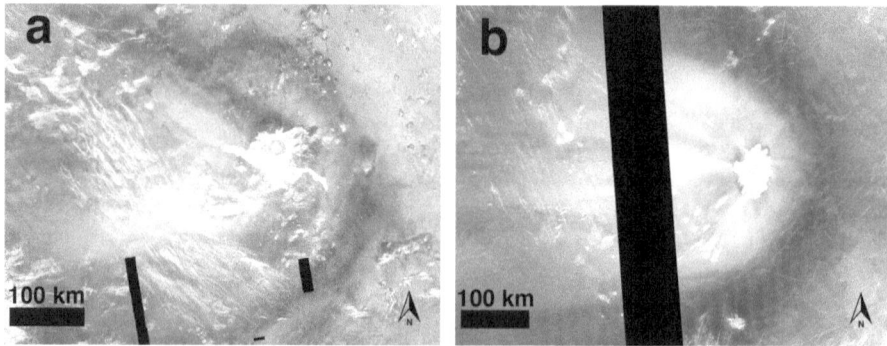

Fig. 1 Impacts are likely the dominant source of sediments on Venus. Parabola craters Carson (a) and Adivar (b) have both radar-bright and -dark markings that are probably caused by changes both in surface roughness and ejecta thickness (Campbell et al. 1992; Carter et al. 2004). These parabolas are both seemingly pristine, but over time parabola features erode (Izenberg et al. 1994), indicative of sediment removal by winds and/or physical or chemical modification occurring on timescales of tens of millions of years (Arvidson et al. 1992; Izenberg et al. 1994)

the Arecibo radar system, a linearly polarized component is introduced through interaction with the surface. For natural rock surfaces, this linearly polarized signal is most often generated by penetration into the surface and reflection from buried rocks or structures (Carter et al. 2011). These data reveal that high degree-of-linear polarization values are common near impact deposits, with both radar-bright and -dark deposits showing evidence of surface mantling (Carter et al. 2004). This finding suggests that there is fine material (i.e., low-density mantling deposits easily penetrable by radar) distributed across the parabolas, and that the observed variations in backscatter correspond to surface changes such as rock exposures or mantling deposits.

In addition to the parabolas, radar-dark halos around some craters are evidence of sediment deposition (possibly from pulverized rock) near the impact site (Izenberg et al. 1994). Similar features have also been seen around lunar craters, where they are interpreted to be areas of crushed and finer rock near the point of impact (Ghent et al. 2005). Small, dark haloes just outside the primary ejecta blanket are common on Venus, but there are some dark haloes that are tens to hundreds of kilometres in diameter (Campbell et al. 1992; Schaller and Melosh 1998). For example, Stanton crater (23.4°S, 199.90°E) has a 110 km-diameter halo (Schaller and Melosh 1998; Schaber et al. 1992). Radar polarimetric images of the largest haloes visible with the Arecibo telescope reveal polarimetric signatures (high degrees of linear polarization and low circular polarization ratios) indicative of fine-grained surface mantling and/or low-density rock (Carter et al. 2004).

A positive relationship between crater diameter and parabola size supports the hypotheses that the parabolas are generated in the impact process, and that all craters above ~11 km diameter—large enough to loft material into the upper atmosphere—will generate a parabola deposit during the impact process (Schaller and Melosh 1998; Campbell et al. 1992). Global maps of impact deposits on Venus (Fig. 2; Ganey et al. 2023) illustrate that up to ~60% of the planet's surface area was at one time blanketed with an ejecta deposit. This coverage includes the areal extent of all parabola deposits (both observed and modelled; see Fig. 2) associated with craters ≥11 km in diameter.

However, intact parabolas are associated with ~10% of impact all craters. Other craters are observed to have possible remnants of impact deposits, or have no associated impact deposits at all, suggesting that parabolas, once created, are altered and erased through time

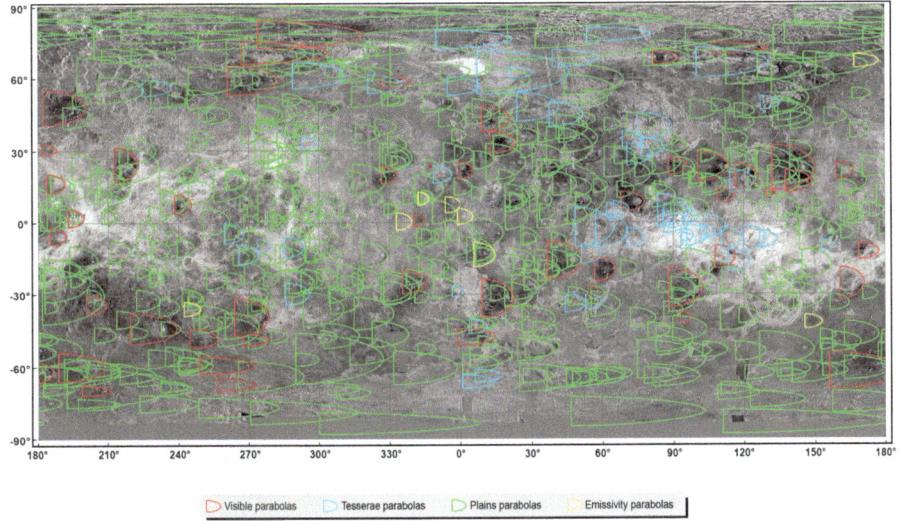

Fig. 2 Global map of Venus with parabola deposit outlines. The visible parabolas (red outlines) are the 49 radar-dark deposits with dimensions reported by Campbell et al. (1992). The tesserae and plains parabolas (blue and green outlines, respectively) are those parabolas modelled from the method from Basilevsky et al. (2004), with the name indicating the terrain where the impact crater is located. The emissivity parabolas (Campbell et al. 1992) are shown in yellow

(Izenberg et al. 1994; Schaller and Melosh 1998). Alternatively, some of the non-parabolic impact features may have formed during an earlier epoch when there was no atmospheric super-rotation (Kreslavsly and Bondarenko 2017). Crater ejecta can be erased via several processes, including volcanic resurfacing or aeolian erosion, both of which are observed on Venus. Chemical or physical weathering may also play a role (Izenberg et al. 1994). Whatever the mechanism(s) of removal, these observations suggest intact crater parabolas have a lifetime of approximately 10% of the average age of the surface and are eroded on timescales of several tens of millions of years (Campbell et al. 1992; Arvidson et al. 1992). Yet impact ejecta may persist longer as an identifiable unit in tesserae than in plains. For example, there is evidence for some fine-grained mantling deposits in tessera terrain with no corresponding impact crater visible on adjacent plains regions, suggesting that volcanic resurfacing may have erased a crater and ejecta in the plains, but not that portion of the ejecta deposit that mantled the higher-elevation tessera (Whitten and Campbell 2016).

Estimates vary for the total volume of impact deposits on Venus. Garvin (1990) estimated a cumulative impact deposit volume of 5.5×10^6 km^3 by applying the crater excavation models of Maxwell (1977) to the population of craters imaged by Venera (25% of the surface), extrapolated over the remainder of the planet. After Magellan, this estimate was revised using several methods. Schaller and Melosh (1998) modelled the distribution of particles produced by each impact crater as a function a diameter, and then tracked the distribution of sand-size (60–2000 µm) particles as they were lofted into space, re-entered the Venus atmosphere, were transported by wind, and then settled to the surface. Applying this model to all craters >2.5 km, Schaller and Melosh estimated a cumulative sediment volume of 3.5×10^4 km^3. Lorenz (1999) derived an equation for sediment volume as a function of crater diameter based on the methodology of McGetchin et al. (1973); this equation applied to all craters with diameters greater than 11 km yields a volume of 1.2×10^5 km^3. Basilevsky et al. (2004) used an empirical fit between crater diameter and the size of parabolas visible in the Magellan data and reported by Campbell et al. (1992), and assumed that all craters greater than 11 km in diameter would have produced a parabola. Basilevsky et al. (2004) modelled each parabola as a half ellipse, and then assumed that parabolas have an average thickness of 10 cm, which together yielded a total volume of 2.8×10^4 km^3.

Ganey et al. (2023) mapped the location of known and inferred parabolas of craters >11 km in diameter (Fig. 2) and used the McGetchin et al. (1973) method to calculate a revised minimum global impact deposit volume of 2.9×10^5 km^3. All of these estimates yield an average sediment layer 8–63 cm in thickness if distributed globally (Ganey et al. 2023). Several of the summarized approaches for estimating sediment volume rely on the McGetchin et al. (1973) relationship between ejecta thickness and distance from the host crater (Ganey et al. 2023; Lorenz 1999; Schaller and Melosh 1998; Vervack and Melosh 1992). Although this relationship was originally derived from lunar cratering models, laboratory experiments, and meteorite impacts, it has been verified against ejecta deposits on Earth and found to describe Venusian deposits reasonably well (Lorenz 1999). The McGetchin method may even underpredict deposit thickness by up to a factor of two, which suggests that some global volume estimates (e.g., Ganey et al. 2023) are most useful as minimum endmembers.

Other impact processes contribute to the production of sediments on a smaller scale. Radar-bright or -dark patches or haloes, called "splotches", have been interpreted as evidence of airburst meteoroids (Kirk and Chadwick 1994; Schaber et al. 1992). These bursts may cause crushing and breakup of surface rocks and deposition of ablated impactor material (Schultz 1992). They may also be responsible for some sediment transport, for example by creating small dunes or blowing away fine materials. In addition, many craters on Venus have long runout flows with lobate margins and sometimes complex structures such as interior channels (Phillips et al. 1991; Schultz 1992). These features have typically been interpreted as impact melt deposits (Chadwick and Schaber 1993), but some have also been interpreted as ground-hugging debris flow deposits (Schultz 1992). Most models of these flows incorporate multiple phases of deposition, including both melted rock and ejected country rock admixed with vapor (Herrick et al. 1997). Although not as spatially extensive as parabolas, these impact features also contribute to local sediment deposits.

2.2 Volcanism

Pyroclastic volcanism is a source of sedimentary deposits on Earth, Mercury, Mars, the Moon, and Io, providing varying fractions of the sediment supply based on the nature of volcanism and (where applicable) prevailing atmospheric conditions. On Venus, the dense atmosphere likely inhibits pyroclastic volcanism unless there is a substantial volatile content (Garvin et al. 1982; Airey et al. 2015). Nevertheless, there have been numerous candidate pyroclastic deposits identified across the planet (Campbell et al. 2017; Ghail and Wilson 2013; Carter et al. 2006; Campbell and Rogers 1994; Guest et al. 1992).

These putative Venus pyroclastic deposits have a range of radar properties. For example, both radar-bright and radar-dark features have been proposed as pyroclastic deposits (Fig. 3), and these features can have irregular outlines or clearly delimited unit boundaries. These deposits sometimes also have higher or lower microwave emissivities compared to their surroundings, as mapped and calculated by the Magellan radiometer (Brossier et al. 2020; Carter et al. 2006; Campbell and Rogers 1994). On Earth, the Moon, and Mars, pyroclastic deposits are typically radar-dark because the deposits are smooth (e.g., Solikhin et al. 2015; Gaddis et al. 1985; Carter et al. 2009; Harmon et al. 2012). Such deposits also typically have low values of CPR due to the fine-grained nature of the deposits and lack of embedded rocks in the upper centimetres of the flow (e.g., Carter et al. 2009).

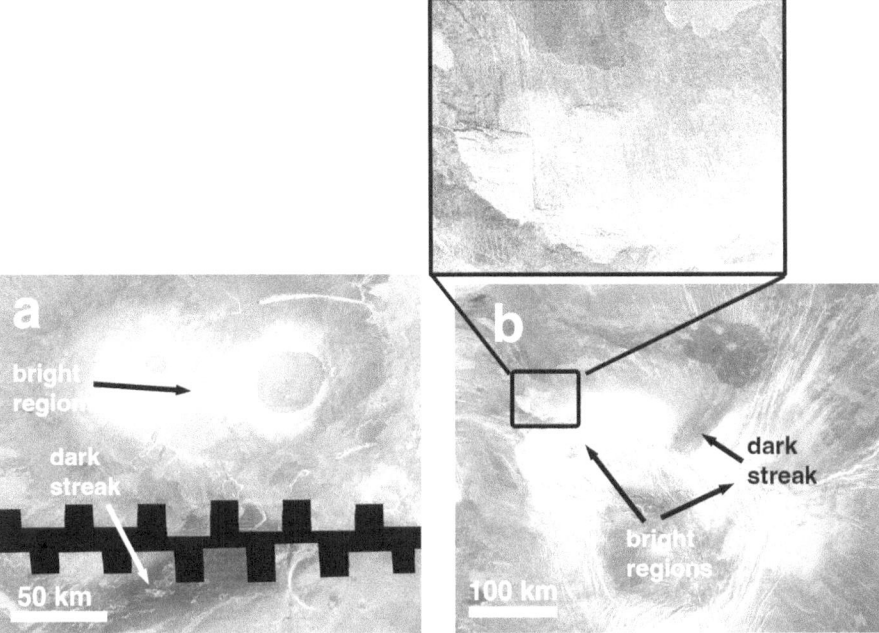

Fig. 3 The summits and flanks of volcanoes and coronae on Venus have diffuse deposits that are likely pyroclastic sediments (Magellan SAR images). A) The Tepev Mons summit has radar-bright, diffuse material between the two calderas, as well as a dark streak thought to be a fine-grained volcanic deposit (Campbell and Rogers 1994). B.) The summit of Irnini Mons has diffuse, radar-bright deposits downslope of the corona ring fractures that are likely ponded pyroclastic flows (Campbell et al. 2017; Ganesh et al. 2021). There is also a dark streak that is probably a relatively smoother tephra deposit. An inset shows detail of the edges of the diffuse bright material: fine, linear, dark streaks that cross the radar-bright lava flow and blend into the surrounding terrain imply that aeolian erosion is stripping or moving the fine material

Radar-dark streaks at Tepev and Irnini Montes on Venus are examples of features that have been proposed as pyroclastic deposits (Campbell and Rogers 1994; Carter et al. 2006; Ganesh et al. 2022), possibly as the result of an ash column blown by local winds or travelling downhill as pyroclastic flows (Fig. 3). Their radar-dark signatures suggest that the particles making up these deposits are small, and that their surfaces are smooth. The radar-dark streaks have low CPR in Arecibo data and there is no evidence of an increased degree of linear polarization, although these flows are almost too small to detect reliably given the Arecibo polarimetry resolution of 12-16 km/pixel (Carter et al. 2006).

In contrast, radar-bright deposits near some summits (e.g., Irnini, Anala, and Innini Montes, Pavlova Corona) have high CPR values suggestive of a rugged surface that was deposited as a column collapse or pyroclastic flow (Fig. 3, Campbell et al. 2017). These features have diffuse edges that may have been caused by some amount of erosion of fine material near the distal edges (Fig. 3, Campbell et al. 2017). Modelling of the radar-bright flows at Irnini Mons demonstrates that they can be deposited by pyroclastic density currents (Ganesh et al. 2021), and their backscatter and emissivity values suggest that they could be rough deposits consistent with a debris flow with cm-sized clasts (Ganesh et al. 2022). It is not clear if these deposits have been welded or partially welded, or consist solely of rock fragments, but the diffuse edges and dark streaks suggest that they may be relatively unconsolidated (Fig. 3).

Fig. 4 A shield field in the Venusian plains (22.08° N, 331.94° E) has wind-blown sediments and likely downstream scouring (Magellan SAR image). The origin of the sediments in uncertain. This small shield field is located north of the parabola crater Aurelia, so the debris may have come from the impact or directly from volcanic tephra. Multiple shield fields have radar polarimetric signatures indicative of sediment cover (Carter et al. 2006)

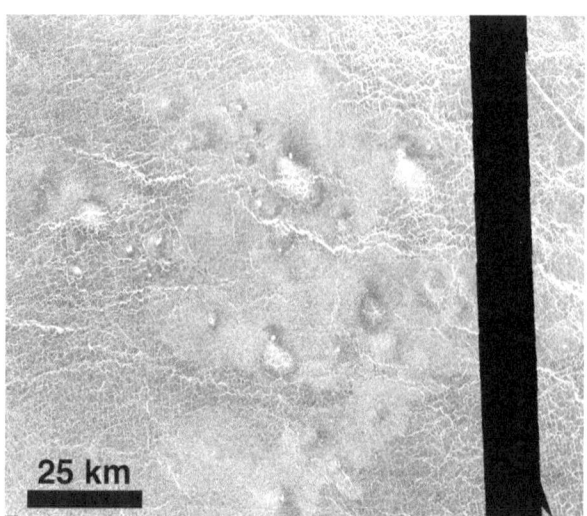

25 km

The summit regions of some volcanic edifices are also associated with wind streaks, providing evidence of sediment motion there. For example, plains shield fields sometimes have tails indicative of aeolian deposition and/or scouring of material (Fig. 4). Degree of linear polarization data for some of these volcanic fields (including that shown in Fig. 4) suggest that there is likely increased mantling material present relative to nearby plains regions (Carter et al. 2006). It is not possible to establish with certainty the source of the sediments; for example, the material could be generated by volcanism or be simply wind-blown sediment from elsewhere trapped by local topography. In some examples, the material appears to be sourced from a nearby parabola crater, such as in the case of dome fields near the crater Aurelia. Nevertheless, the increased concentrations of sediments in these dome fields suggest these deposits may in at least some instances be related to primary tephra deposition.

Even if pyroclastic volcanism is not widespread today on Venus, the features discussed above show that deposits of seemingly sedimentary material 10 s of km^2 in spatial extent can be generated. Ganesh et al. (2021) estimated the bulk volumes for pyroclastic deposits located in Eistla Regio: 310–780 km^3 at Irnini Mons, and 140–280 km^3 for the smaller deposits at Anala Mons and at Pavlova and Didilia coronae. Individual pyroclastic flow volumes are small, but an abundance of such deposits could potentially produce significant volumes of sediment. These deposits are likely later eroded by wind and could also potentially be lithified into volcaniclastic rocks (Byrne et al. 2021). It is also possible for pyroclastic flows to form lobate margins similar to lava flows (e.g., Ghail and Wilson 2013), and so it may be difficult to differentiate lava and pyroclastic deposits from radar backscatter data alone. Identifying pyroclastic deposits with future radar and microwave emissivity data is critical for the search for regions where lavas may have a high volatile content, and to understand the prevalence, erosion, and redistribution of volcanic sediments on Venus generally.

2.3 Mass Wasting and Mechanical Weathering

Mass wasting can contribute to the global breakdown of rocks and sediment production. New (i.e., present-day) landslides and avalanches have been observed with orbiting spacecraft on both Mars and the Moon (McEwen et al. 2021; Speyerer et al. 2015), and these images provide key information on current surface changes, outcrop structural stability, mechanical rock breakdown, and seismic shaking.

2.3.1 Evidence for Mass Wasting Deposits

There is some evidence for landslides on Venus with current data. Venera 9 is inferred to have landed on a talus slope. Small-scale steep slopes are very common on Venus, most often associated with normal faulting and graben, and mass wasting appears to be a common occurrence in these steep areas (Malin 1992). Rock slumps, rock and/or block slides, rock avalanches, debris avalanches, and a possible example of a debris flow have all been identified in Magellan imagery, suggesting a range of rock failure mechanisms (slides, slumps, flows) and material types (regolith versus rock) (Malin 1992; Bulmer and Guest 1996). Although landslides have not yet been identified on the flanks of Venus shield volcanoes, likely due to shallow slopes, debris avalanches and rock slides have been observed on steep sided volcanic domes (Bulmer and Guest 1996). The lengths of the landslide runouts fall in between the typical sizes of Earth and Mars landslides and appear, based on radar imagery, to have a range of surface debris sizes ranging from metre-scale rubble to very fine (radar-dark) dust deposits at some of the distal edges (Malin 1992). More so than on Earth, Venusian landslides are found in groups, and most of them occur in areas where neighbouring escarpments are as steep or steeper than the one that failed. Therefore, either landslides occur at roughly the same time because of one triggering event, or rock failure occurs frequently in a given location due to steep slopes and frequent triggering events (Malin 1992).

Some of the largest landslides are located in Diana Chasma (Fig. 5), where there is evidence for ongoing tectonic activity on the basis of the proportion of highly fractured craters and the disruption of a parabolic dark halo (Ghail 2002). However, this example also clearly demonstrates that, although mass wasting does occur, the retention of steep slopes several kilometres high implies that mass wasting is likely slower on and thus modifies the landscape to a much lesser extent on Venus than on Earth. The lesser degree of thermal cycling, combined with the lack of surface water and ice, could in addition slow the fracturing and movement of rock relative to Earth. A reduction in joint formation would then help prevent slope failure, as well as inhibit further weathering by limiting access of the atmosphere to interior rock faces (Balme et al. 2004). A reduction in jointing could also be caused by the high atmospheric pressure (equivalent to 900 m water depth on Earth) (Balme et al. 2004). The net effect is that the styles, and probably rates, of mass-wasting-induced surface modification on Venus are different from those on Earth and Mars. Mass wasting on Venus may appear more similar to Earth's submarine than subaerial surfaces (Iversen et al. 1976)—and so thinking of the dense, supercritical fluid atmosphere as an ocean may be helpful when trying to understand its behaviour (Malin 1992). In particular, the entrainment of atmospheric gas may provide the fluidization needed for putative debris flows (Malin 1992).

As on Earth, fresh normal fault scarps are expected to form at angles of 60–70°, although mechanical weathering will serve to reduce those slopes to the angle of repose ($\sim$35°) over time. Radargrammetric measurements of 170 normal faults in Venus plains regions yield an average slope of 36.4 $\pm$1.2° (Connors and Suppe 2001). These observations are consistent with the formation of talus slopes across Venus. However, the resolution of Magellan images, as well as unfavourable look angles in rugged regions, makes it difficult to discern all but the largest and most obvious mass-wasting structures (Malin 1992).

The lack of small impactors on Venus may preclude some of the sources of seismic shaking that can cause mass wasting on the Moon and other planets, but Venus is likely

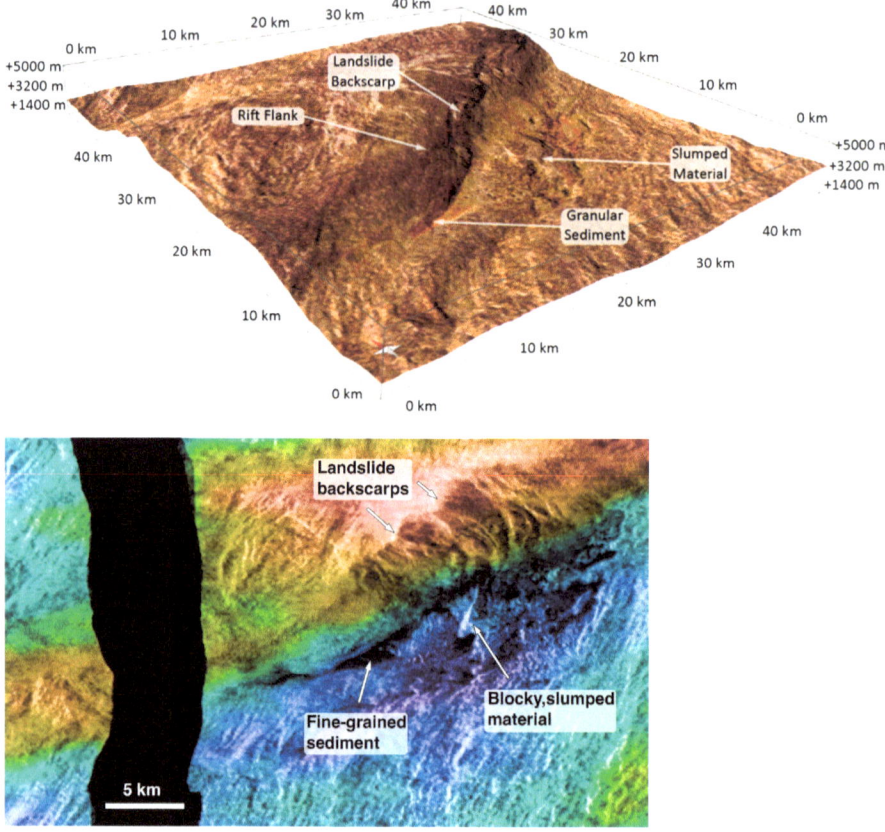

Fig. 5 One of a number of large landslides in Diana Chasma, this example is located at 19.09°S, 142.19°E. Top image: Topography is derived from stereo image pairs; the colour and shading is from right- and left-looking radar images, in which fine granular material is orange–red and bare rock is pinkish red, as seen on the rift flank and backscarp. The yellowish material is consistent with lithified weathered sediments. Bottom: Another, 2D, view of the landslide. The colour topography ranges from 400 m elevation in blue at the foot of the scarp to 3700 m in red at the top. Blocky slumped material is visible below the scarp; fine-grained (dark) material is visible along the scarp floor for tens of km. The fresh appearance of both the scarp and the slumped material implies either a recent origin or minimal erosion and transport, similar to submarine slides on Earth

to be tectonically and volcanically active seismically. Indeed, a primary source for mass movements on Venus is likely to be seismicity. If Venus has a level of seismic activity similar to intraplate levels on Earth, then we should expect 10 s to 100 s of events per year (Lognonné 2005; Lorenz 2012) Additionally, the Venera 14 vertical seismometer may have detected a ground-motion event that was unrelated to wind measurements (Ksanfomality et al. 1982; Lorenz and Panning 2018). A comprehensive mapping of landslides with future higher-resolution data could be used to identify areas more prone to seismicity (or possibly with weaker rock types or internal structures).

In summary, although landslides have been identified and contribute to local sediment budgets, it has been difficult to discern widespread evidence for mass wasting at the scale of the Magellan data. It has also been nearly impossible to search for active landslides with the differing look angles of the SAR images in the steep and rugged areas where landslides form. New, considerably higher-resolution imaging and topographic data and monitoring over time are needed to reach any firm conclusions about the recent history of mass wasting on Venus. This requirement is especially relevant when it comes to detecting the smaller landslides ($\sim$500 m^3 volume) that are typically more common on Earth and Mars than larger mass-wasting events (de Haas et al. 2014).

2.3.2 Upland Erosion

Higher-elevation regions on Venus are cooler and experience different wind regimes than at low elevations (Lefèvre 2022), which could lead to differences in the production of sediments in uplands relative to the plains. Upland erosion via mass wasting from steeper surfaces and/or aeolian scouring from updraft or downdraft winds can produce materials that fill in depressions in mountainous regions and result in deposition of material at lower altitudes. At higher elevations, crustal materials appear to be strongly affected by surface winds (Fig. 6), but even these areas have dielectric constants consistent with lithified sediments, not loose accumulations. The low resolution of the Magellan altimeter nadir reflectivity data, from which the dielectric permittivity (Ford and Pettengill 1992) and thus density of surface materials is derived (Ulaby et al. 1988), returns a mixed signal; in Fig. 6, the blue areas (dielectric constant 4-8) may well be a mixture of high dielectric (highland) material and loose sediments with lower permittivity (such as talus), whereas the green areas to the east (dielectric constant 2-4) may likewise be a mixture of talus and volcanic rock. Even so, the relatively large areas of streamlined material have the dielectric constant signature of lithified sedimentary rock rather than loose regolith.

This area is tectonically active, or at least has been in the geologically recent past (Ghail 2002), so fresh sedimentary material, primarily talus from over-steepened slopes, may be continually generated. Mean wind speeds are strongly altitude dependent and are able to erode and transport material throughout the highland regions (e.g., Greeley et al. 1984; see below). In some areas, this upland material is deposited on adjacent plains—leading to the possibility that the largely featureless plains covering several tens of percent of the planet may be at least partly sedimentary, rather than purely igneous in origin. The settling out of sedimentary material is consistent with observed crater degradation, in which low-lying crater floors are infilled with some combination of sediments and lava flows first (Herrick and Rumpf 2011). In the Fig. 6 example, coarser material (i.e. centimetre-sized) from mass wasting is apparently mobilised locally by the wind (see windstreaks/yardangs label), to form streamlines that then at least partially lithify into rock with the observed somewhat higher dielectric constant than the unconsolidated sediments.

2.4 Chemical Weathering

The Venusian surface is also eroded via chemical weathering. Interactions between new surface materials, such as recent lava flows, and the Venus atmosphere produce a series of chemical reactions. Some of these reactions may drive volume changes, as well as a change in the surface composition through the development of coatings. This process may alter surface compositions on short timescales; for example, Dyar et al. (2021) extrapolated a weathering rate for basaltic glass from the experiments of Teffeteller et al. (2022), predicting a surface alteration coating of $\sim$30 μm over 500,000 years. This coating is not expected to lead to the kind of blocky surface rocks seen at the Venera sites (Dyar et al. 2021). Models based on current Venus data have led to possible chemical alteration pathways for a variety basaltic minerals, but the relative roles and rates of CO_2, SO_2, and H_2O in weathering processes are still unclear (Zolotov 2018). The role of supercritical CO_2 and mixtures of gasses

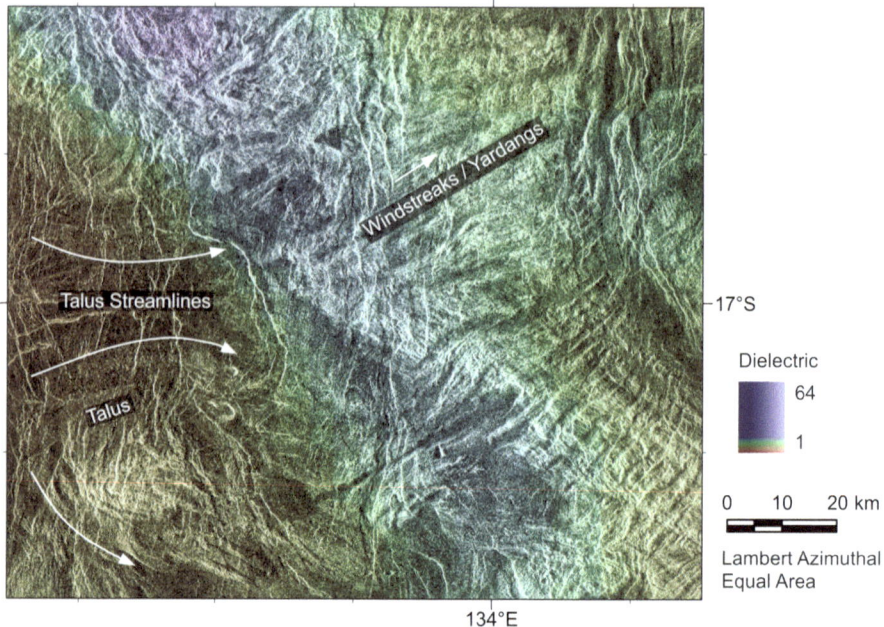

Fig. 6 Windblown sediments in south-east Thetis Regio shown with derived surface materials. Dielectric permittivity is calculated from Magellan altimeter nadir reflectivity; lower dielectric permittivity values indicate a relatively lower surface density. Radar-dark features ending in lobate shapes are interpreted to be talus (western side of image). A radar-bright hill in the southwest corner (above the arrow) has a streamlined shape, suggesting aeolian weathering. In the northeast part of the image, fine radar-dark and radar-bright alternating linear structures with a southwest to northeast orientation are interpreted as windstreaks or yardangs overprinting pre-existing topography. These features have dielectric constants consistent with lithified rock

and glass/solid rock are also currently unknown and could lead to unexpected weathering products or timescales (Zolotov 2018). Major sources of uncertainty remain in understanding the chemistry of Venus's surface–atmosphere interactions, and more laboratory experiments and measurements of surface chemistry and mineralogy are needed to fill these gaps (Filiberto et al. 2020; Dyar et al. 2021; Zolotov 2018). Our current understanding of surface weathering suggests that, over long timescales, it is possible that chemical weathering could contribute some volume to the sediment budget, but not at the scale of impacts, mass wasting, and volcanism (Dyar et al. 2021). For further information about surface mineralogy and weathering, see Gilmore et al. (2023, this collection).

In summary, sediment production on Venus is substantially different from sediment production on the Earth and Mars. Impact crater deposits cover larger areas on Venus and are thought to represent the major component of current sediment production, with volcanism as a possible secondary source. Compared to Earth and Mars, temperatures on Venus are remarkably constant, which reduces thermal cycling and resultant fatigue of rocks (which is also a major process on airless objects). The low surface wind speeds, thick atmosphere, and corresponding low saltation velocity mean that aeolian abrasion, an important source of erosion on Mars and in deserts on Earth, is considerably less effective on Venus (Kreslavsly and Bondarenko 2017). The lack of flowing and freezing water also dramatically reduces the weathering and erosion rate on Venus relative to Earth and prior warmer and wetter epochs on Mars. Water is also responsible for major chemical erosion rates and on Mars and

Earth. These aqueous processes will not occur on Venus, where chemical weathering is restricted to surface–atmosphere interactions or possibly volcanism-related interactions. The volume of sediments produced through each process and their timescales are poorly known for Venus, and require further data, laboratory studies, and modelling to reach a quantitative understanding.

3 Aeolian Transport

Once sediments are generated, they are transported from their place of origin by winds. Greeley et al. (1984) developed a Venus Wind Tunnel and demonstrated that sand grains of 60 to 2000 μm in diameter can become entrained by surface winds at Venus and are transportable by saltation. The empirical saltation threshold is $0.63 \, \mathrm{m \, s^{-1}}$, and bedforms (microdunes, ripples, etc.) will form below velocities of $1.5 \, \mathrm{m \, s^{-1}}$. Dust grains (below 40 μm diameter) move primarily in suspension, whereas coarser grains (2–16 mm diameter) move by creep, saltating at wind velocities over $2.5 \, \mathrm{m \, s^{-1}}$ (Greeley et al. 1984; Greeley and Arvidson 1990).

In order to assess how particles are transported, it is important to measure and model boundary-level wind velocities on Venus. Venus landers measured winds $< 2 \, \mathrm{m \, s^{-1}}$, and detected direct evidence of sediment motion (Lorenz 2016; Selivanov et al. 1982). Lorenz (2016) used a probability distribution (Weibull function) to estimate that surface wind speeds around $1 \, \mathrm{m \, s^{-1}}$ are common and may keep sand-sized particles in constant saltation. The upper limit of surface winds estimated by the Lorenz (2016) probabilistic technique is $\sim 2.2 \, \mathrm{m \, s^{-1}}$, which is similar to an upper limit of $1.5 \, \mathrm{m \, s^{-1}}$ derived through Global Circulation Model (GCM) simulations (Lebonnois et al. 2018). These findings suggest that course-grained materials probably do not saltate except during extreme weather events.

In addition, planetary boundary-layer models suggest that there are spatial and diurnal changes in winds. GCM models show slope-induced winds that change during the diurnal cycle, with downward katabatic winds at night and upward anabatic winds during the day that occur along the westward slopes of high-elevation terrains (Lebonnois et al. 2018). A turbulent-resolving boundary layer model by Lefèvre (2022) shows that at noon, for both high and low terrains (modelled as 1030 m and -230 m elevation, respectively), the friction velocity is above the saltation threshold, so turbulence can transport dust particles. At night, the velocities are lower and no transport occurs. In the low plains, the horizontal wind velocities are in the range where dunes can form, but at higher elevations the wind speed is too low to support dust transport. The Lefèvre (2022) model also produced convective vortices over the highlands; if present on Venus, such vortices could lift dust and produce dust devils.

The possibility of the movement of sediment on Venus is supported by a variety of independent observations (see also Kreslavsly and Bondarenko 2017). Firstly, the removal of crater parabolas over the timescale of 10 s to 100 s Myr requires redistribution by aeolian processes. A positive correlation between dark-floored craters and those with degraded or missing parabolas (Izenberg et al. 1994) is consistent with volcanic or aeolian infill of craters over time. On Mars, many impact craters are filled with layered sediments that may be the result of trapping of aeolian materials or wind-carried tephra deposits (e.g., Day and Catling 2019; Anderson et al. 2018), and a similar process could occur on Venus as sediment settles in basins. In addition, almost 6000 wind streaks have been identified on Venus, with various morphologies consistent with deposition and erosion of sedimentary materials (Greeley et al. 1995). These wind steaks have been used to infer local wind directions through mapping of their orientations (Greeley et al. 1994). The wind streaks are predominantly oriented westward and towards the equator, but when local effects from impact cratering are removed, the equatorward component is dominant, as expected for a Hadley circulation of the lower atmosphere (Greeley et al. 1994).

In some regions of Venus, the radar backscatter of stratigraphically older volcanic flows is lower than younger flows in a manner consistent with the smoothing of lava flows on Earth by weathering (Arvidson et al. 1992). By comparison to radar images of flows on Earth, Arvidson et al. (1992) estimated that the degree of smoothing on Venus corresponds to $\sim$1 m of surface modification, which translates to a vertical modification rate of $\sim$2 $\times$ 10^{-3} μm $\cdot$ yr^{-1}, assuming a mean surface age of 0.6 Gyr. This sediment production rate is orders of magnitude lower than that on Earth, consistent with a lack of fluvial erosion. However, age dating of terrains on Venus, even relative relationships, is uncertain and the surface texture of emplaced flows on Venus are likely considerably different than that of lava flows on Earth. In situ imaging is likely needed to understand how lava flows are smoothed through aeolian weathering and deposition on Venus.

Bondarenko and Kreslavsky (2018) conducted a principal-component analysis of various surface properties related to Fresnel reflectivity and surface roughness to evaluate the hypothesis that small aeolian features may in fact be common on Venus. Their analysis showed that aeolian features below the resolution of Magellan (microdunes, ripples, sand sheets, etc.) could be ubiquitous, covering much ($\sim$40%) of the surface of the planet. Remaining unknown is the relative contributions of impact cratering, volcanism, aeolian erosion, or chemical erosion to the production, timing, and distribution of sediment on the planet generally.

Only two dune fields have been unequivocally identified in SAR imagery: Menat Undae (25°S, 339°E; also named the Aglaonice dune field) and Al-Uzza Undae (68°N, 90°E; also named the Fortuna–Meskhent dune field) (Greeley et al. 1992). Both fields are interpreted to contain transverse dunes because they demonstrate bright backscatter in Magellan images from slip faces oriented near-normal to the radar illumination, and their crests are perpendicular to the orientation of associated wind streaks (Greeley et al. 1992). Microdune fields have been suggested to lie near Stowe (47°S, 230°E; diameter, $D = 75.3$ km), Guan-Daosheng (61°S, 178°E; $D = 43$ km) and Eudocia (58°S, 160°E; $D = 25.9$ km) craters based on differences in radar backscatter in left- and right-looking images of the regions (Weitz et al. 1994). Although too small to be directly resolved in Magellan SAR images, dunes with asymmetric slopes can produce the observed strong radar brightness if the slip faces are oriented near-normal to the radar illumination (Weitz et al. 1994). Venus also bears erosional aeolian landforms called yardangs; a field of such features has been identified southeast of Mead crater (Greeley et al. 1992). Our present information on dunes and yardangs may be limited by the available look angles and resolution of the Magellan data set. It will be important to search for evidence of dunes in new data from upcoming Venus missions to better quantify the sediment budget, and the spatial and temporal links between sediment sources and sinks, across the planet.

4 Sediments and Sedimentary Rocks

Venus sedimentary processes have been less studied then for other planets, largely from the assumption that initial Magellan data analysis showed little evidence for erosion and sediments—but expectations prior to Magellan were very different. The Venera landers, in

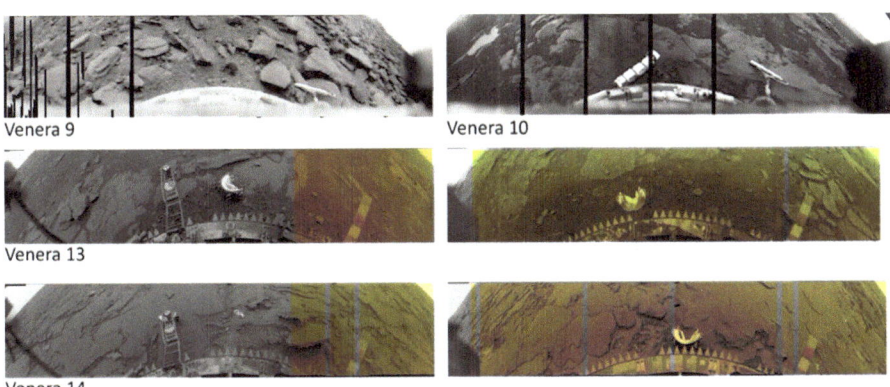

Fig. 7 Images of the Venus surface recorded by Venera Landers (reprocessed images © Don P Mitchell, used with permission). Venera 9 landed on a 15°–20° slope, the top of which is visible to the upper right; downhill is behind the camera view at approximately the 8 o'clock position (Florensky 1977). The left images of Venera 13 and 14 are from the spacecraft Camera 1 (rear camera), and the right images are from Camera 2 (front camera). The color images had lower signal-to-noise, and here the color portions have been combined with the clear filter images (Don P Mitchell per processing description at http://mentallandscape.com/V_DigitalImages.htm)

particular, show a landscape originally interpreted to be dominated by sedimentary deposits (Florensky et al. 1983) that only later were reinterpreted as lava flows (Garvin et al. 1984).

Sedimentary rocks are formed through reworking and lithification of eroded sediments, through deposition and lithification of volcanic or landslide deposits, or (on Earth) through biological processes. They are a sink for sediments. These rocks record the history of planets and are a key part of determining past climate, so it is important to understand how sedimentary rocks form on Venus and where deposits of sedimentary rocks are located.

4.1 Evidence of Plains Regolith

Four Venera landers returned panoramic views of the Venus surface: Veneras 9, 10, 13, and 14 (Fig. 7). The lander images have a resolution of 10 mm per line pair for Venera 9 and 10, and 4–5 mm for Venera 13 and 14 (Garvin et al. 1984), so pebble and larger sized rocks could be resolved.

Although the surfaces in these images may be interpreted as thin, platy pahoehoe-type lava flows (Garvin et al. 1984), aerially extensive, thin flows are unlikely under Venus conditions, where rapid early cooling and a thick lava crust are expected (Head and Wilson 1986). Morphologically, the rocks in the lander images most closely resemble partially eroded, lithified clastic sediments (Florensky et al. 1983), with either expanses of fractured layered strata or areas of fragmented strata broken into cobbles and fine-grained sediment (Surkov and Barsukov 1985). This sediment always occurs as deposits in the localized topographic lows between bedrock outcrops, cobbles, and in the troughs between what appear to be ripples. Active aeolian transport is demonstrated by the removal of sediment from the lander ring of Venera 13 over a period of about an hour (Selivanov et al. 1982). In addition, photometric measurements at the Venera 9 and 10 landing sites suggest that dust clouds (particle sizes <100 μm) were kicked up by the landing and then dissipated (Garvin et al. 1984; Moshkin et al. 1979).

Venera 9 recorded images of rock materials similar in appearance to the bedrock elsewhere, but in the form of subangular boulders up to 60 cm wide and 20 cm tall within a coarse sediment or gravel on a 15° to 20° slope interpreted as talus (Florensky 1977). The gravels are subangular to subrounded and with a distinct anisotropy in sizes, with some rocks falling in the diameter range of 1–10 cm and some in the range of 30–70 cm (Garvin et al. 1984). The distribution of these two size groups is spatially continuous around the landing site (Garvin et al. 1984), suggesting the anisotropy may arise from weathering of local layered materials. Finer-sized materials are not apparent in those images; their absence may be a result of removal by wind, although the resolution of the images precludes detecting sand-sized particles. Even the sluggish 0.2–2 m s^{-1} surface winds recorded by the landers are readily able to transport particles that are sand sized and smaller (Warner 1983; Lorenz 2016). More in-situ imaging of the Venus surface in multiple locations and at high resolution is needed to fully assess the detailed sizes and morphologies of sediment particles.

The bedrock imaged at the Venera 10, 13, and 14 sites consists of laminated or thinly bedded sheets a few centimetres thick, with varying degrees of coarse sediment or regolith in between the slabs (Garvin et al. 1984). The 5 cm-distance between notches on the lander rings of Venera 13 and 14 give an indication of scale of materials close to the landers. The Venera 14 images, which recorded the lowest sediment fraction of all the Venera landing sites, show that the bedrock consists of a series of interlocked, subangular to subrounded plates that are possibly jointed. Indeed, Florensky et al. (1983) described the Venera 14 layers as either sedimentary bedding or possibly basaltic volcanic tuffs formed by cycles of air fall or ground flow. Penetrometer measurements and landing stress–strain profiles from Venera 13 and 14 demonstrated that the surface has a low bearing strength and low density, properties that are consistent with porous materials such as ash or other sediments (Surkov et al. 1984; Basilevsky et al. 1985).

There is also evidence from Earth-based radar data that Venus may have widespread regolith cover, even in plains regions. Degree-of-linear polarization maps created with Arecibo Observatory radar data show some amount of surface penetration of the radar wave across much of the surface (Carter et al. 2011). In plains regions, the degree of linear polarization is ∼7–10%—substantially lower than the values of 10–40% observed near impact craters and volcanoes, but still above the expected noise level of ∼1–2% (Carter et al. 2004). At the relatively short wavelengths used by Arecibo (12.6 cm), these degree-of-linear polarization values suggest at least partial covering by low-density material such as fine regolith or a low-density upper crust on lithified sediments or volcanic rocks. The precise thickness of the surface covering is not easily determinable from current data sets, but must be at least a substantial fraction of the radar wavelengths to be detectable (i.e., at least a couple of centimetres in thickness). The origin of the surface regolith or coating is unknown, but it could be caused by rock weathering and/or transport of fine materials by winds.

Additional evidence for widespread sediments comes from quantitative analysis of Magellan radar backscatter. Modelling of Magellan altimeter radar backscatter profiles revealed a widespread anisotropy in the Doppler shifts of the echoes (Tyler et al. 1992). Mapping these Doppler shifts across the surface shows a latitude-dependent "striping", a phenomenon most likely caused by sloping surfaces (Tyler et al. 1992) and interpreted as possible evidence for widespread, small aeolian bedforms (Bondarenko et al. 2006).

Impacts also produce deposits that correspond to the local environment and may provide evidence of relative sediment cover. For example, the meteoroid airburst splotches discussed in Section 2.1 have a non-uniform distribution on the surface, which has been interpreted as evidence for the presence of increased regolith in some regions relative to others (Bondarenko and Kreslavsky 2018). The emissivity parabolas, which do not have a strong radar backscatter parabola, are also strongly clustered, suggesting a possible difference in surface materials in these areas (Campbell et al. 1992). These parabolas are thought to have thinner deposits than other SAR-visible parabolas (Campbell et al. 1992), but it is still not certain how these features form.

Finally, apparent aging of the surface also suggests that mantling deposits accumulate over time. Arvidson et al. (1992) noted a progression from fresh-looking to degraded lava flow morphology in Sedna Planitia. Younger flows, as determined from cross-cutting relationships, are brighter with clear lobate margins, whereas older flows have degraded or difficult-to-discern edges and darker centers (Arvidson et al. 1992). This progression appears similar to processes on other planets where lava flows are filled in and covered with sediment over time (Arvidson et al. 1992). Bondarenko et al. (2003) compared the radiophysical properties of lava flows with their stratigraphic position and found that locally older units usually had a lower permittivity, which is consistent with either the deposition of sedimentary material and/or weathering of the surface that produces a lower density material.

The current data suggest there is more substantial surface cover than what is observed in the Venera landing site panoramas, but equally those data make it difficult to determine the thickness and distribution of these materials. With a spatial resolution of 12–16 km/pixel, the Arecibo polarimetry radar data offer limited opportunity to map local changes in surface cover. Similarly, the low resolution of the Magellan emissivity data also makes it challenging to measure dielectric constants for small features. Observations of the surface with high-resolution(tens-of-metres) polarimetric radar and microwave emissivity at tens-of-kilometres spatial scales would help improve our identification and geological interpretation of present-day Venus surface cover.

4.2 Lithification

Weathering is an important process on Venus, but lithification may be equally important, rapid, and mask the sedimentary cycle in Magellan SAR data. Oxidation and the growth of sulphates and scapolites (a carbonate–sulphate mineral) likely rapidly cement loose material into a weakly lithified rock, akin to a marl (Berger et al. 2019; Filiberto et al. 2020). Sintering can also lead to lithification and is thought to be a very effective process on Venus due to the high temperatures and corresponding higher diffusion coefficients. Kreslavsly and Bondarenko (2017) estimate diffusion smoothing of particle shapes based on temperatures of 450 °C and diffusion coefficients of $\sim 10^{-7}$ cm/s, and found that 1 μm particles can be smoothed in ~ 2 days and 100 μm particles can be smoothed in $\sim 5 \times 10^5$ years. This process can readily lead to the rapid lithification of sand-sized particles.

Lithified rocks could appear similar to those observed in Venera lander images, and large deposits of loose materials by themselves may be uncommon. The simplest conclusion is that, although the surface materials are geochemically basaltic in composition, primary igneous rocks are weathered, transported, and deposited as wind-blown sand that rapidly lithifies into a sandstone. These rocks may then become jointed, perhaps by tectonic processes (e.g., unroofing), weathered, and disaggregated into gravels, with the fines removed by wind and deposited as the shale-like layered rocks seen in the plains. The sedimentary cycle may therefore have an important role in the exchange of materials, particularly volatiles, between the near-surface interior and atmosphere. More in-situ data as well as laboratory measurements are needed to understand the lithification process and timescales on Venus.

4.3 Evidence of Layered Deposits That Could Be Sedimentary

On the basis of their morphological appearance, the Venus uplands are far from pristine. In numerous places, highland tesserae show evidence for being composed of layers of some

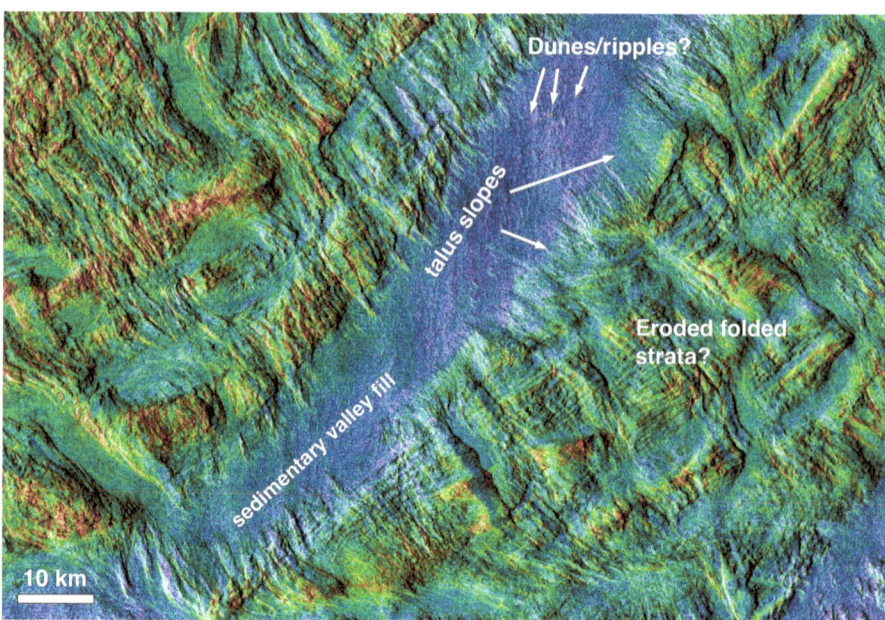

Fig. 8 Magellan left- and inverted right-look relief image, colour-coded with left- and right-look roughness, of tessera upland located at 1°S, 81°E. Blue corresponds to surfaces smoother than 2 cm (gravel or finer) and red to surfaces rougher than 8 cm (pebbles or coarser). Mass wasting (talus) material eroded from layered and possibly sedimentary rock layers lie on the steep slopes and fill the valley floor, possibly transported along the valley in dunes (north-south trending ridges indicated by arrows). The image is based on the magic airbrush technique of Kirk (1993), which combines normalised left- and inverted-right-look images to cancel out roughness variations and generate a shaded relief image. By also combining the normalised left and right look images, the opposite effect is achieved, primarily a roughness image (albeit with dielectric variations included). That image is colour-coded with a spectrum and the shaded relief image colourised with it to create a pseudo-full colour image that contains all the information of the left and right look pairs. This works perfectly on Earth, where the viewing geometries are almost identical and the topography is very well known, but not so well on Venus, where neither is true, so some warping of the right image is required to match it to the left (Ghail 1996)

material—presumably sediments, lithified volcanoclastics, lava flows, or some combination of the three (Byrne et al. 2021). Yet these layers are exposed along the flanks of locally high-standing terrain such as ridges and domes, and have outcrop patterns that seem to follow undulations in topography. By analogy with Earth, those units may have experienced considerable erosion (preceded by folding) to expose their tilted component layers. Given the current erosion rates, such a large amount of rock removal would likely need to have happened gradually over hundreds of millions of years or more, or occurred during an earlier epoch with higher erosion rates (Byrne et al. 2021). Slopes proximal to these exposed layers are characterized by likely mass wasting deposits (Fig. 8), which fill the valleys floors with material consistent with being sedimentary (e.g. smoother) and derived from those slopes. Those sediments, in turn, seem to be modified by the wind into linear features that may dunes. The nature of the layered strata, such as their age and composition, is still largely unknown and will be an important subject of research with upcoming missions.

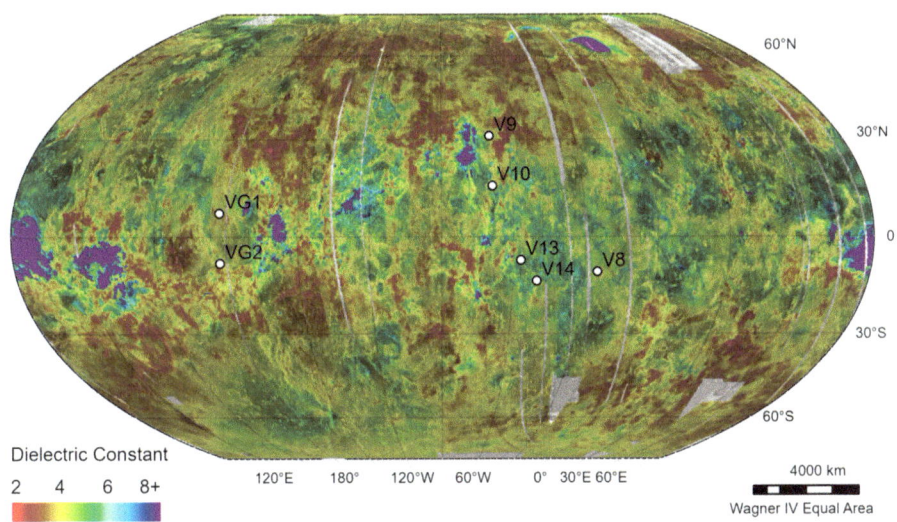

Fig. 9 Surface dielectric constant inferred from Magellan emissivity data, and using SAR backscatter as an approximation for roughness to correct for its effect on the dielectric constant. A dielectric constant of 2 to 3 is consistent with unconsolidated sediment or pyroclastic deposits; between 3 and 5 to lithified sedimentary rock or weathered volcanic rock; and >5 to slightly weathered or fresh volcanic rock. Blue-to-violet regions have anomalously high dielectric constants as well as high radar reflectivity (see Gilmore et al. 2023, this collection). The approximate locations of the Soviet Venera and Vega landing sites are indicated with small circles

4.4 A Global View of Rock Densities

Evidence for a global sedimentary cycle is apparent in the Magellan emissivity data (Pettengill et al. 1992), when converted dielectric constant (Fig. 9) and used to infer densities (Ulaby et al. 1988). A few areas of loose, low-density sediment with low dielectric constant (2-3) are evident, but nearly half the planet has surface characteristics consistent with partially lithified sediments (yellow and green in Fig. 9), i.e., having dielectric constants <5 and densities $<2500 \, \text{kg m}^{-3}$. Because the porosity and composition of the sediments and lithified rocks are not known, it is not possible to distinguish between sediments and rocks based solely on density. However, the denser surface materials are more likely to be lithified. Rocks with densities greater than $2500 \, \text{kg m}^{-3}$ (green-cyan in Fig. 9) are either fully lithified sediments, fresh volcanic rock, or welded pyroclastic flows. Although the 150 km-diameter landing circles of all the Venera and Vega landers include signatures from partially and fully lithified sediments and fresh rock, the regions in which the Venera 14 and Vega 2 sites are situated have the highest proportion of fully lithified and potentially fresh volcanic material—consistent with those landers' elemental analyses and panoramic images (Fig. 7).

Lithified sedimentary rock is consistent with the actual densities ($\sim 1500 \, \text{kg m}^{-3}$) and porosities ($\sim 50\%$) inferred from mechanical data recorded by the Venera 13 and 14 probes (Florensky et al. 1983). Based on its measurements of sulphate content, Venera 14 sampled the least weathered material, in agreement with the morphological evidence of a smooth surface with little fine material (Garvin et al. 1984), but the modal mineralogy recorded by the lander nonetheless implies considerable loss of alkali feldspar, very possibly to scapolite. The likely more altered and relatively poorly bound grains imaged at the Venera 13 site may be easily transported by surface winds and therefore still mobile; it is not clear whether these deposits were derived by erosion of the local bedrock, or were transported a long distance from elsewhere. The modal mineralogy, unconfined compressive strength, low porosity ($\sim$20%), and high density ($\sim$2600 $\mathrm{kg\,m^{-3}}$) of the Vega 2 site implies a fully weathered and cemented clastic rock, but no images were acquired to confirm this inference (Marov and Grinspoon (1998) and references therein).

5 Summary and Future Measurements

5.1 Summary of Current Understanding of Sediments and Sedimentary Rocks

Our understanding of sediments and sedimentary deposits on Venus has advanced considerably since the Magellan mission, but there is still much to learn about the extent and formation of these deposits. The distribution of sediments is clearly non-uniform over the surface, with impact-related materials forming the bulk of these deposits and pointing to estimates of hundreds of thousands of cubic kilometres of impact-derived sediments on Venus overall. The parabolas and radar-dark haloes erode over time, indicating that wind transport and possibly lithification are happening at the surface today. Although there are only a few probable dune fields on Venus, the several hundred identified areas of wind streaks show that there is substantial transport of material. Pyroclastic deposits and mass wasting are also likely sources of sediment in regional to local areas, but it is impossible to estimate the contribution of these materials from our current data sets. The locations where these deposits are observed, however, may lead us to sites with recent volcanism and seismicity, assuming that sediments are dispersed by wind, buried, or otherwise altered over time.

Whereas the sources of sediment are rapidly becoming better understood, determining the timescales for their formation and the likely sediment sinks is much more challenging and will require additional study. Additional surface geochemical information is needed, coupled with surface optical and perhaps radar observations, to understand how new sedimentary rocks are being created on Venus today. Higher-resolution imagery, topography, and mesospheric wind data could help us determine how material is transported to form aeolian deposits.

Finally, our knowledge of sedimentary rocks on Venus is in the early stages but is already intriguing. Layering in the tesserae suggests that a complex history is available for us to decipher if we have higher-resolution radar data or surface optical imagery of this enigmatic terrain type. Tesserae are likely one of our best options for determining whether Venus was ever a habitable planet, so it is important that we understand how this terrain type formed and whether tesserae represent ancient sedimentary layers deposited in fluvial environments, or feature volcanic layering, or are the result of some yet-to-be recognized formational process.

5.2 Future Measurements of Sediments and Surface Sedimentary Rocks

Mantling deposits may have very different radar properties based on thickness, texture, and dielectric properties, and the radar backscatter is also subject to look angle and radar wavelength. To better understand the present sediment cycle on Venus it is critical to acquire a range of data products, including high-resolution orbital radar imagery (at scales of metres per pixel), radar polarimetry, and radiometry, as well as decent imagers, and imaging systems mounted on aerial platforms below the cloud deck. Differing radar look angles or wavelengths are important for identifying dune fields. In-situ ground-penetrating radar data would also be extremely useful for assessing the depth of sediment layers. Long-term monitoring of areas such as cliff faces, volcanic centres, and aeolian fields at high resolution would also allow us to search for ongoing surface changes. Radar interferometry can also be used to detect surface changes; for example, repeat-pass interferometry can detect subtle aeolian movement through measurement of decorrelation signatures (Zebker et al. 1996). Laboratory measurements of chemical reactions using relevant compositions, temperatures, and densities are needed to determine how weathering and lithification progress on Venus at different altitudes and for different materials. And finally, more sensitive mineralogic measurements, high-resolution surface lander optical images able to differentiate transported grains, matrix and cement, and mechanical measurements of material properties to estimate density are needed in as many distinct locations as possible to fully understand the nature, properties, and origins of the surface rocks present today on Venus.

The planet Venus has had a unique climate evolution that may be recorded in surface sedimentary rocks from prior eras, such as in the tesserae. Mapping the sediments at the surface today can point us toward the story that tesserae hold, indicate active volcanoes and areas of possible seismic activity, and provide a planet-wide view of the rock cycle on a planet with a dense CO_2 atmosphere.

Acknowledgements The authors thank M.A. Kreslavsky and J.L. Whitten for reviews that substantially improved this article.

Funding The authors did not receive support from any organization for the submitted work.

Declarations

Competing Interests The authors have no competing interests to declare that are relevant to the content of this article.

References

Airey MW, Mather TA, Pyle DM, Glaze LS, Ghail RC, Wilson CF (2015) Explosive volcanic activity on Venus: the roles of volatile contribution, degassing, and external environment. Planet Space Sci 113–114:33–48

Anderson R, Edgar LA, Rubin DM, Lewis KW, Newman C (2018) Complex bedding geometry in the upper portion of Aeolis Mons, Gale crater, Mars. Icarus 314:246–264. https://doi.org/10.1016/j.icarus.2018.06.009

Arvidson RE, Greeley R, Malin MC, Saunders RS, Izenberg N, Plaut JJ, Stofan ER, Shepard MK (1992) Surface modification of Venus as inferred from Magellan observations of plains. J Geophys Res 97:13303

Balme MR, Sammonds PR, Vita-Finzi C, Couchman JP (2004) Experimental and theoretical fracture mechanics applied to fracture of the crust of Venus. J Geophys Res 109:E03005. https://doi.org/10.1029/2002JE001992

Basilevsky A, Kuzmin R, Nikolaeva O, Pronin A, Ronca L, Avduevsky V, Uspensky G, Cheremukhina Z, Semenchenko V, Ladygin V (1985) The surface of Venus as revealed by the Venera landings: part II. Geol Soc Am Bull 96(1):137–144

Basilevsky AT, Head JW, Setyaeva IV (2003) Venus: estimation of age of impact craters on the basis of degree of preservation of associated radar-dark deposits. Geophys Res Lett 30(18):1950. https://doi.org/10.1029/2003GL017504

Basilevsky AT, Head JW, Abdrakhimov AM (2004) Impact crater air fall deposits on the surface of Venus: areal distribution, estimated thickness, recognition in surface panoramas, and implications for provenance of sampled surface materials. J Geophys Res 109:E12003

Berger G, Cathala A, Fabre S, Borisova AY, Pages A, Aigouy T, Esvan J, Pinet P (2019) Experimental exploration of volcanic rocks-atmosphere interaction under Venus surface conditions. Icarus 329:8–23

Bindschadler DL, DeCharon A, Beratan KK, Smrekar SE, Head JW (1992) Magellan observations of Alpha Regio: implications for formation of complex ridged terrains on Venus. J Geophys Res 97:13563–13577

Bondarenko NV, Kreslavsky MA (2018) Surface properties and surficial deposits on Venus: new results from Magellan Radar Altimeter data analysis. Icarus 309:162–176

Bondarenko NV, Kreslavsky MA, Raitala J (2003) Correlation of dielectric permittivity of volcanic units on Venus with age. J Geophys Res 108:5013. https://doi.org/10.1029/2002JE001929

Bondarenko NV, Kreslavsky MA, Head JW (2006) North-south roughness anisotropy on Venus from the Magellan Radar Altimeter: correlation with geology. J Geophys Res 111:E06S12. https://doi.org/10.1029/2005JE002599

Brossier JF, Gilmore MS, Toner K (2020) Low radar emissivity signatures on Venus volcanoes and coronae: new insights on relative composition and age. Icarus 343:113693. https://doi.org/10.1016/j.icarus.2020.113693

Bulmer MH, Guest JE (1996) Modified volcanic domes and associated debris aprons on Venus. In: McGuire WJ, Jones AP, Neuberg J (eds) Volcano instability on the Earth and other planets. Geological society special publication, vol 110, pp 349–371

Byrne PK, Ghail RC, Gilmore MS, Şengör AMC, Klimczak C, Senske DA, Whitten JL, Khawja S, Ernst RE, Solomon SC (2021) Venus tesserae feature layered, folded, and eroded rocks. Geology 49:81–85

Campbell BA, Rogers PG (1994) Bell Regio, Venus: investigation of remote sensing data and terrestrial analogs for geologic analysis. J Geophys Res 99:21153–21171

Campbell DB, Stacy NJS, Newman WI, Arvidson RE, Jones EM, Musser GS, Roper AY, Schaller C (1992) Magellan observations of extended impact crater related features on the surface of Venus. J Geophys Res 97:16,249–16,277

Campbell BA, Campbell DB, Morgan GA, Carter LM, Nolan MC, Chandler JF (2015) Evidence for crater ejecta on Venus tessera terrain from Earth-based radar images. Icarus 250:123–130

Campbell BA, Morgan GA, Whitten JL, Carter LM, Glaze LS, Campbell DB (2017) Pyroclastic flow deposits on Venus as evidence of renewed magmatic activity. J Geophys Res 122:1580–1596

Carter LM, Campbell DB, Campbell BA (2004) Impact crater related surficial deposits on Venus: multi-polarization radar observations with arecibo. J Geophys Res 109:E06009. https://doi.org/10.1029/2003JE002227

Carter LM, Campbell DB, Campbell BA (2006) Volcanic deposits in shield fields and highland regions on Venus: surface properties from radar polarimetry. J Geophys Res 111:E06005. https://doi.org/10.1029/2005JE002519

Carter LM, Campbell BA, Hawke BR, Campbell DB, Nolan MC (2009) Radar remote sensing of Pyroclastic Deposits in the Mare Serenitatis and Mare Vaporum Regions of the Moon. J Geophys Res 114:E11004. https://doi.org/10.1029/2009JE003406

Carter LM, Campbell DB, Campbell BA (2011) Geologic studies of planetary surfaces using radar polarimetric imaging. In: Proc. IEEE, vol 99. https://doi.org/10.1109/JPROC.2010.2099090

Chadwick DJ, Schaber GG (1993) Impact crater outflows on Venus: morphology and emplacement mechanisms. J Geophys Res 98:20891–20902

Connors C, Suppe J (2001) Constraints on magnitudes of extension on Venus from slope measurements. J Geophys Res, Planets 106:3237–3260

Day MD, Catling DC (2019) Potential aeolian deposition of intra-crater layering: a case study of Henry crater, Mars. Geol Soc Am Bull 132(3/4):608–616. https://doi.org/10.1130/B35230.1

de Haas T, Ventra D, Carbonneau PE, Kleinhans MG (2014) Debris-flow dominance of alluvial fans masked by runoff reworking and weathering. Geomorphology 217:165–181

Dyar MD, Helbert J, Cooper FRF, Sklute EC, Maturilli A, Mueller NT, Kappel D, Smrekar SE (2021) Surface weathering on Venus: constraints from kinetic, spectroscopic, and geochemical data. Icarus 358:114139

Filiberto J, Trang D, Treiman AH, Gilmore MS (2020) Present-day volcanism on Venus as evidenced from weathering rates of olivine. Sci Adv 6(1):eaax7445

Florensky KP (1977) Surface of Venus as revealed by Soviet Venera 9 and 10. Geol Soc Am Bull 88:1537

Florensky KP, Bazilevsky AT, Burba CA, Nicolayeva OV, Pronin AA, Selivanov AS, Narayeva MK, Panfilov AS, Chemondanov VP (1983) Panorama of Venera 9 and 10 landing sites. In: Hunten PM et al (eds) Venus. Univ. of Ariz. Press, Tucson, pp 137–153

Ford PG, Pettengill GH (1992) Venus topography and kilometer-scale slopes. J Geophys Res 97:13103–13144

Gaddis LR, Pieters CM, Hawke BR (1985) Remote sensing of lunar pyroclastic mantling deposits. Icarus 61:461–489

Ganesh I, McGuire LA, Carter LM (2021) Modeling the dynamics of dense pyroclastic flows on Venus: insights into pyroclastic eruptions. J Geophys Res. https://doi.org/10.1029/2021JE006943

Ganesh I, Carter LM, Henz TN (2022) Radar backscatter and emissivity models of proposed pyroclastic density current deposits on Venus. J Geophys Res, Planets 127:e2022JE007318. https://doi.org/10.1029/2022JE007318

Ganey TM, Gilmore MS, Brossier J (2023) Reassessment of the volumes of sediment sources and sinks on Venus. Plan Sci J 4. https://doi.org/10.3847/PSJ/aca521

Garvin JB (1990) The global budget of impact-derived sediments on Venus. Earth Moon Planets 50/51:175–190. https://doi.org/10.1007/BF00142394

Garvin JB, Head JW, Wilson L (1982) Magma vesiculation and pyroclastic volcanism on Venus. Icarus 52(2):365–372. https://doi.org/10.1016/0019-1035(82)90119-1

Garvin JB, Head JW, Zuber MT, Helfenstein P (1984) Venus: the nature of the surface from Venera panoramas. J Geophys Res Solid Plan 89:3381–3399

Ghail RC (1996) An investigation of regional tectonic processes on Venus. PhD thesis, Lancaster University

Ghail RC (2002) Structure and evolution of southwest Thetis Regio, vol 107 p 5060. https://doi.org/10.1029/2001JE001514

Ghail RC, Wilson L (2013) A pyroclastic flow deposit on Venus. Geol Soc (Lond) Spec Publ 401:97–106

Ghent RR, Leverington DW, Campbell BA, Hawke BR, Campbell DB (2005) Earth-based observations of radar-dark crater haloes on the Moon: implications for regolith properties. J Geophys Res 110:E02005. https://doi.org/10.1029/2004JE002366

Gilmore MS Darby Dyar M, Mueller N et al (2023) Mineralogy of the Venus surface. Space Sci Rev 219:52. https://doi.org/10.1007/s11214-023-00988-6

Greeley R, Arvidson RE (1990) In: Aeolian processes on Venus, Earth, Moon and Planet, vol 50, pp 127–157

Greeley R, Iversen J, Leach R, Marshall J, White B, Williams S (1984) Windblown sand on Venus: preliminary results from laboratory simulations. Icarus 57:112–124

Greeley R, Arvidson RE, Elachi C, Geringer MA, Plaut JJ, Saunders RS, Schubert G, Stofan ER, Thouvenot EJP, Wall SD, Weitz CM (1992) Aeolian features on Venus: preliminary Magellan results. J Geophys Res 97:13319–13345

Greeley R, Schubert G, Limonadi D, Bender KC, Newman WI, Thomas PE, Weitz CM, Wall SD (1994) Wind streaks on Venus: clues to atmospheric circulation. Science 263:358–361

Greeley R, Bender K, Thomas PE, Schubert G, Limonadi D, Weitz CM (1995) Wind-related features and processes on Venus: summary of Magellan results. Icarus 115:399–420

Guest JE, Bulmer MH, Aubele J, Beratan K, Greeley R, Head JW, Michaels G, Weitz C, Wiles C (1992) Small volcanic edifices and volcanism in the plains of Venus. J Geophys Res 97:15,949–15,966

Harmon JK, Nolan MC, Campbell BA (2012) Arecibo radar imagery of Mars: the major volcanic provinces. Icarus 220:990–1030

Head JW, Wilson L (1986) Volcanic process and landforms on Venus: theory, predictions, and observations. J Geophys Res, Solid Earth 91:9407–9446

Herrick RR, Rumpf ME (2011) The resurfacing histories of Venusian impact craters. J Geophys Res 116:E02004. https://doi.org/10.1029/2010JE003722

Herrick RR, Sharpton VL, Malin MC, Lyons SN, Feely K (1997) Morphology and morphometry of impact craters. In: Bougher SW, Hunten DM, Phillips RJ (eds) Venus 2. University of Arizona Press, Tucson, pp 1015–1046

Herrick RR, Bjonnes E, Carter L et al (2023) Resurfacing history and volcanic evolution of Venus. Space Sci Rev 219:29. https://doi.org/10.1007/s11214-023-00966-y

Iversen J, Greeley R, Pollack JB (1976) Windblown dust on Earth, Mars and Venus. J Atmos Sci 33:2425–2429

Izenberg NR, Arvidson RE, Phillips RJ (1994) Impact crater degradation on Venusian plains. Geophys Res Lett 21:289–292

Kirk RL (1993) Separation of topographic and intrinsic backscatter variations in biscopic radar images: a "Magic airbrush". In: 24th Lunar and planetary science conference, pp 803–804

Kirk RL, Chadwick DJ (1994) Splotches on Venus: distribution, properties and classification. In: 25th Lunar and Planetary Science Conference, pp 705–706

Kreslavsly MA, Bondarenko NV (2017) Aeolian sand transport and aeolian deposits on Venus: a review. Aeolian Res 26:29–46

Ksanfomality LV, Zubkova VM, Morozov NA, Petrova EV (1982) Microseisms at the VENERA-13 and VENERA-14 landing sites. Sov Astron Lett 8:241

Lebonnois S, Schubert G, Forget F, Spiga A (2018) Planetary boundary layer and slope winds on Venus. Icarus 314:149–158. https://doi.org/10.1016/j.icarus.2018.06.006

Lefèvre M (2022) Venus boundary layer dynamics: Eolian transport and convective vortex. Icarus 387:115167. https://doi.org/10.1016/j.icarus.2022.115167

Lognonné P (2005) Planetary seismology. Annu Rev Earth Planet Sci 33:571–604

Lorenz RD (1999) Microtektites on Mars: volume and texture of distal impact ejecta deposits. Icarus 144:353–366. https://doi.org/10.1006/icar.1999.6303

Lorenz RD (2012) Planetary seismology – expectations for lander and wind noise with application to Venus. Planet Space Sci 62:86–96

Lorenz RD (2016) Surface winds on Venus: probability distribution from in-situ measurements. Icarus 264:311–315

Lorenz RD, Panning M (2018) Empirical recurrence rates for ground motion signals on planetary surfaces. Icarus 303:273–279

Malin MC (1992) Mass movements on Venus: preliminary results from Magellan Cycle 1 observations. J Geophys Res 97:16337–16352

Marov MYa, Grinspoon DH (1998) The planet Venus. Yale University Press, New Haven

Maxwell DE (1977) Simple z model of cratering, ejection, and the overturned flap. In: Roddy DJ et al (eds) Impact and explosion cratering. Pergamon, New York, pp 1003–1008

McEwen AS, Schaefer EI, Dundas CM, Sutton SS, Tamppari LK, Chojnacki M (2021) Mars: abundant recurring slope lineae (RSL) following the planet-encircling dust event (PEDE) of 2018. J Geophys Res 126:e2020JE006575

McGetchin TR, Settle M, Head JW (1973) Radial thickness variation in impact crater ejecta: implications for lunar basin deposits. Earth Planet Sci Lett 20:226–236. https://doi.org/10.1016/0012-821X(73)90162-3

Moshkin BE, Ekonomov AP, Golovin YM (1979) Dust on the surface of Venus. Cosmic Res Engl Transl 17:232

Pettengill GH, Ford PG, Wilt RJ (1992) Venus surface radiothermal emission as observed by Magellan. J Geophys Res 97:13091–13102

Phillips RJ, Arvidson RE, Boyce JM, Campbell DB, Guest JE, Schaber GG, Soderblom LA (1991) Impact craters on Venus: initial analysis from Magellan. Science 252(5003):288–297

Phillips RJ, Raubertas RF, Arvidson RE, Sarkar IC, Herrick RH, Izenberg N, Grimm RE (1992) Impact craters and Venus resurfacing history. J Geophys Res 97:15,923–15,948

Schaber GG, Strom RG, Moore HJ, Soderblom LA, Kirk RL, Chadwick DJ, Lawson DD, Gaddis LR, Boyce JM, Russell J (1992) Geology and distribution of impact craters on Venus: what are they telling us? J Geophys Res 97:13257–13301

Schaller CJ, Melosh HJ (1998) Venusian ejecta parabolas: comparing theory with observations. Icarus 131:123–137. https://doi.org/10.1006/icar.1997.5855

Schultz P (1992) Atmospheric effects on ejecta emplacement and crater formation on Venus from Magellan. J Geophys Res 97:16,183–16,248

Selivanov AS, Gektin YM, Naraeva MK, Panfilov AS, Fokin AB (1982) Evolution of the Venera 13 imagery. Sov Astron Lett 8:433–436

Solikhin A, Pinel V, Vandemeulebrouck J, Thouret J-C, Hendrasto M (2015) Mapping the 2010 Merapi pyroclastic deposits using dual-polarization Synthetic Aperture Radar (SAR) data. Remote Sens Environ 158:180–192

Solomon SC, Smrekar SE, Bindschadler DL, Grimm RE, Kaula WM, McGill GE, Phillips RJ, Saunders RS, Schubert G, Squyres SW, Stofan ER (1992) Venus tectonics: an overview of Magellan observations. J Geophys Res 91:13199–13255

Speyerer EJ, Robinson MS, Povilaitis RZ, Wagner RV (2015) Dynamic Moon revealed with high resolution temporal imaging. In: 46th Lunar and Planetary Science Conference, Abstract #2325

Surkov YuA, Barsukov VL (1985) Composition, structure and properties of Venus rocks. Adv Space Res 5:17–29

Surkov YA, Barsukov VL, Moskalyeva LP, Kharyukova VP, Kemurdzhian AL (1984) New data on the compositions, structure, and properties of Venus rock obtained by Venera 13 and Venera 14. J Geophys Res 89(suppl):B393–B402

Teffeteller H, Filiberto JM, McCanta C, Treiman AH, Keller L, Cherniak D, Rutherford M, Cooper RF (2022) An experimental study of the alteration of basalt on the surface of Venus. Icarus 115085

Tyler GL, Simpson RA, Maurer MJ, Holmann E (1992) Scattering properties of the Venusian surface: preliminary results from Magellan. J Geophys Res 97:13115–13139

Ulaby FT, Bengal T, East J, Dobson MC, Garvin J, Evans D (1988) Microwave dielectric spectrum of rocks. (Report 23817-1-T), University of Michigan Radiation Laboratory

Vervack RJ, Melosh HJ (1992) Wind interaction with falling ejecta: origin of the parabolic features on Venus. Geophys Res Lett 19:525–528

Warner JL (1983) Sedimentary processes and crustal cycling on Venus. J Geophys Res 88:A495–A500

Weitz CM, Plaut JJ, Greeley R, Saunders RS (1994) Dunes and microdunes on Venus: why were so few found in the Magellan data? Icarus 112:282–295

Whitten JL, Campbell BA (2016) Recent volcanic resurfacing of Venusian craters. Geology 44(7):519–522

Zebker HA, Rosen P, Hensley S, Mouginis-Mark PJ (1996) Analysis of active lava flows on Kilauea volcano, Hawaii, using SIR-C radar correlation measurements. Geology 24(6):495–498. https://doi.org/10.1130/00917613

Zolotov MY (2018) Gas-solid interactions on Venus and other Solar System Bodies. Rev Mineral Geochem 84:351–392. https://doi.org/10.2138/rmg.2018.84.10

Publisher's Note Springer Nature remains neutral with regard to jurisdictional claims in published maps and institutional affiliations.

Authors and Affiliations

Lynn M. Carter[1] · **Martha S. Gilmore**[2] · **Richard C. Ghail**[3] · **Paul K. Byrne**[4] · **Suzanne E. Smrekar**[5] · **Terra M. Ganey**[6] · **Noam Izenberg**[7]

✉ L.M. Carter
lmcarter@arizona.edu

✉ R.C. Ghail
richard.ghail@rhul.ac.uk

M.S. Gilmore
mgilmore@wesleyan.edu

P.K. Byrne
paul.byrne@wustl.edu

S.E. Smrekar
suzanne.e.smrekar@jpl.nasa.gov

T.M. Ganey
tganey@ucsc.edu

N. Izenberg
noam.izenberg@jhuapl.edu

[1] Lunar and Planetary Laboratory, University of Arizona, Tucson, AZ 85721, USA

[2] Dept. of Earth and Environmental Sciences, Wesleyan University, Middletown CT 06459, USA

[3] Department of Earth Sciences, Royal Holloway, University of London, Egham, Surrey, TW20 0EX, UK

[4] Department of Earth, Environmental, and Planetary Sciences, Washington University, St. Louis, MO 63130, USA

[5] Jet Propulsion Laboratory, California Institute of Technology, 4800 Oak Grove Drive, Pasadena, CA 91109, USA

[6] Department of Earth and Planetary Sciences, University of California Santa Cruz Santa Cruz, CA 95064, USA

[7] Applied Physics Laboratory, Johns Hopkins University, Laurel, MD 20723, USA

Space Science Reviews (2024) 220:36
https://doi.org/10.1007/s11214-024-01065-2

Volcanic and Tectonic Constraints on the Evolution of Venus

Richard C. Ghail[1] · Suzanne E. Smrekar[2] · Thomas Widemann[3,4] ·
Paul K. Byrne[5] · Anna J.P. Gülcher[2] · Joseph G. O'Rourke[6] ·
Madison E. Borrelli[6] · Martha S. Gilmore[7] · Robert R. Herrick[8] ·
Mikhail A. Ivanov[9] · Ana-Catalina Plesa[10] · Tobias Rolf[11] · Leah Sabbeth[2] ·
Joe W. Schools[12] · J. Gregory Shellnutt[13]

Received: 6 March 2023 / Accepted: 27 March 2024 / Published online: 29 April 2024
© The Author(s) 2024

Abstract

Surface geologic features form a detailed record of Venus' evolution. Venus displays a profusion of volcanic and tectonics features, including both familiar and exotic forms. One challenge to assessing the role of these features in Venus' evolution is that there are too few impact craters to permit age dates for specific features or regions. Similarly, without surface water, erosion is limited and cannot be used to evaluate age. These same observations indicate Venus has, on average, a very young surface (150–1000 Ma), with the most recent surface deformation and volcanism largely preserved on the surface except where covered by limited impact ejecta. In contrast, most geologic activity on Mars, the Moon, and Mercury occurred in the 1st billion years. Earth's geologic processes are almost all a result of plate tectonics. Venus' lacks such a network of connected, large scale plates, leaving the nature of Venus' dominant geodynamic process up for debate. In this review article, we describe Venus' key volcanic and tectonic features, models for their origin, and possible links to evolution. We also present current knowledge of the composition and thickness of the crust, lithospheric thickness, and heat flow given their critical role in shaping surface geology and interior evolution. Given Venus' hot lithosphere, abundant activity and potential analogues of continents, roll-back subduction, and microplates, it may provide insights into early Earth, prior to the onset of true plate tectonics. We explore similarities and differences between Venus and the Proterozoic or Archean Earth. Finally, we describe the future measurements needed to advance our understanding of volcanism, tectonism, and the evolution of Venus.

Keywords Venus · Formation of volcanic and tectonic features · Mantle dynamics · Subduction and plate tectonics · Topography · Hotspot volcanism · Plume-induced subduction · Crust and surface composition · Lithospheric thickness · Coronae · Lava channels · Venus resurfacing history

Venus: Evolution Through Time
Edited by Colin F. Wilson, Doris Breuer, Cédric Gillmann, Suzanne E. Smrekar, Tilman Spohn and Thomas Widemann

Extended author information available on the last page of the article

1 Introduction

This article describes current thinking on Venus' tectonic and volcanic processes, which provide essential insights on Venus' evolution through time. Given the often-noted gross similarities between Earth and Venus, an essential comparative planetology question is to know which processes differ or are the same between the two planets, and why? As the role of plate tectonics in Earth's long-term habitability becomes more evident, many exoplanet models have focussed on predicting the likelihood of this process. As the only other Earth-sized planet in the Solar System, Venus is the logical planet to explore this question (see Rolf et al. 2022 and many references therein). Venera measurements of radiogenic elements in Venus' crust show Earth-like concentrations, implying a similar internal heat budget. However, plate tectonics has not always been the dominant process on Earth either. Exactly when plate tectonics started and continents formed is debated, but certainly processes in the Archean (much of the first half of Earth's history) appear to have been very different to the modern Earth. The high surface temperature on Venus creates lithospheric temperatures similar to the elevated heat flow in the Archean (e.g., Van Kranendonk, 2010; Harris and Bédard 2014, 2015; Gerya et al. 2015), although with a transition to eclogite at a shallower depth than in Earth's present-day crust (e.g., Namiki and Solomon 1993). Much of Earth's early record has been erased by both plate tectonics and erosion. Tectonic and volcanic processes on Venus today may therefore inform our understanding of Archean Earth and vice versa. Some features on Venus appear consistent with localized subduction of the lithosphere (e.g., Schubert and Sandwell 1995; Harris and Bédard 2014, 2015). As subduction is believed to be the first step in creating a plate tectonic system, understanding the conditions necessary to permit this process to occur, is the absence of a plate tectonic framework. Other key issues include the formation of Venus' tesserae and whether or not Earth's Archean cratons formed in an analogous manner.

A key constraint on interpreting geologic processes on another planet is the size and spatial distribution of impact craters. Venus has only $\sim$1000 impact craters and their distribution cannot be distinguished from a random one. The average surface age is very young, relative to other terrestrial planets, but poorly constrained (estimates span at least $\sim$0.15–1 Ga). The processes that removed older terrains thereby removing craters are controversial (Herrick et al. 2023, this collection). The absence of large regions of conspicuously different ages on Venus has been used to infer a very different tectonic style from Earth (e.g., Herrick et al. 2023, this collection). Furthermore, only $\sim$10% of the craters are widely regarded as modified by subsequent geologic activity. However, many more craters might have suffered modification by sedimentary (Carter et al. 2023, this collection) and/or volcanic processes (e.g., Herrick and Rumpf 2011; Herrick et al. 2023, this collection). The random spatial distribution and ambiguous modification state of the impact craters has led to competing interpretations of Venus' impact record. In the "catastrophic resurfacing" model, the entire surface was wiped clean of impact craters, either by burial under a thick layer of lavas (e.g., Schaber et al. 1992; Head et al. 1994; Strom et al. 1994) and/or by recycling of the surface via overturn of the lithosphere (e.g., Parmentier and Hess 1992). Subsequently, geologic activity declined, allowing the surface to accumulate impact craters at random. The globally ancient surfaces of Mars, Mercury, and the Moon are in a stagnant lid convective regime, with interior heat lost primarily through conduction across a thick, immobile lithosphere. However, for Venus, a fully stagnant lid does not provide sufficient heat loss, leading to the idea that Venus has cycled between a stagnant lid regime today and a past plate tectonic regime, allowing for greater heat loss over time (Turcotte et al. 1999), consistent with the inference of a past catastrophic resurfacing event.

The opposite end-member is the 'equilibrium resurfacing' hypothesis (e.g., Phillips et al. 1992). Under this scenario, impact craters are removed at a roughly constant rate by geologic activity that has no spatial organization, in contrast to Earth. Perhaps more likely is an intermediate hypothesis of 'regional equilibrium resurfacing' that allows for spatial and temporal variability of regions less than ~1000 km in diameter, consistent with the impact crater distribution (e.g., Phillips et al. 1992). Herrick et al. (2023, this collection) provides a thorough discussion of these models and their implications. We introduce these interpretations here because of the fundamental role that both the mean surface age and local variability play in framing the geologic evolution of Venus and the interpretation of interior and atmospheric processes (e.g., Rolf et al. 2022, this collection). Recently, the concept of a partially mobile 'squishy' lid regime has been proposed, in which intrusive volcanism allows for plate-tectonic-like levels of heat loss (Lourenço et al. 2020). Below, we discuss a range of geologic features observed on Venus, along with implications for lithospheric structure and convective regime.

In parallel with the notion of 'episodic vs. equilibrium' resurfacing is the concept of 'directional vs non-directional' geologic history (Guest and Stofan 1999). Does the variety of geologic features observed on Venus occur in a specific sequence everywhere, consistent with an episodic (or 'catastrophic') resurfacing event? Ivanov and Head (2011) mapped the whole Venus surface at 1:10 million scale and identified a global stratigraphic sequence with 13 distinct members, with tesserae identified everywhere as the oldest unit, then various deformed plains, 'groove' (extensional) and mountain (compressional) belts, through undeformed plains, to 'lobate' (lava flow) plains and rift zones. The authors argued that since this sequence is never found next to, and offset from, an adjacent sequence, the sequence must represent a global change through time, i.e., a directional geologic history. Arguments against this sequence (Guest and Stofan 1999) take two principal forms: first, that the stratigraphic methods and interpretation of some authors (e.g., Ivanov and Head 2011) mix material (rock) units with tectonic (structural) events; and second, that radar is sensitive to surface texture and relatively insensitive to rock type. For example, the regional plains, which cover a third of the planet, may have very different ages but appear texturally the same, and hence cannot be used as a stratigraphic marker. A further problem for interpretation is that since radar is very sensitive to surface roughness near the radar wavelength, apparent boundaries and onlap relationships, such as the 'flooding' of tesserae by regional plains, may be a result of gradational downslope grain size changes, indicating the opposite age relationship (Carter et al. 2023, this collection, and references therein). Further, others argue that this sequence is not observed everywhere and that in places, tesserae may be younger than the plains (e.g., Ghail 2002; Ivanov and Head 1996; Gilmore and Head 2000). The scarcity of impact craters and the lack of any other independent dating technique hamper a resolution to this question.

Our understanding of current processes and conditions effectively provide boundary conditions for extrapolation backward into Venus' geologic and convective history (see Rolf et al. 2022, this collection). The size, type, distribution, and (where available) chemistry of volcanic features offers a window into interior processes such as upwelling/downwelling, decompression melting and mantle temperature. Venus hosts a wide array of volcanic features, from those seen on Earth including volcanic edifices and lava flows, to those unique (at least in abundance) to Venus, such as novae and coronae (e.g. Glaze et al. 2002; Krassilnikov and Head 2003). The overlap between volcanoes, coronae, and novae is an interesting puzzle, as discussed below. Tectonic features provide constraints on the rheology and thickness of the lithosphere and the origin and magnitude of driving stress fields. Gravity data provide further information on the thickness of the crust, the lithosphere, and subsurface density variations due to processes such as mantle upwellings. The low resolution of current datasets often results in multiple interpretations of a given feature or process. The goal here is not to provide a single view, but to represent a range of possible interpretations, the constraints available from current data, and to outline specific needs for future datasets.

A rocky body's crustal layer provides a complex archive for the planet's evolution. The crust may be primary, a direct consequence of a magma ocean freezing, secondary, a result of active magmatism and volcanism in the solidified interior, or tertiary, which invokes remelting of primary and/or secondary crust (e.g., like the Earth's continental crust). In the case of Venus, crustal composition is dominantly basaltic (Hess and Head 1990) and its young surface age dictates that most of its volume is likely secondary crust. Tessera terrains may represent tertiary crust (see Gilmore et al., 2023, this collection). The formation of eclogite from basalt may play a key role on Venus in terms of both mantle dynamics and the formation of specific volcanic and tectonic features. Delamination driven by eclogite formation has been proposed to drive the formation of both coronae (e.g., Piskorz et al. 2014; Gülcher et al. 2023) and tessera plateaus (e.g., Bindschadler and Parmentier 1990). Delamination would also result in the formation of secondary or tertiary crust (Elkins-Tanton et al. 2007). Whether or not delamination leads to mobilisation of an otherwise stagnant lid depends on the details of the mantle convection models (e.g., Armann and Tackley 2012; Rolf et al. 2018; Adams et al. 2022, see Rolf et al. 2022, this collection). Mantle convection models are useful tools to simulate the growth history of Venus' crust through its long-term evolution and how this relates to the planet's thermal state and geodynamic regime. A stagnant-lid scenario may lead to a relatively hot mantle, promoting magmatism and crustal growth. In contrast, episodically or partially mobile (plutonic squishy lid) lithosphere may reset crustal thickness and therefore limit its effective thickness (see Rolf et al. 2022, this collection, for details).

2 Overview of Volcanic Features

The surface of Venus is replete with volcanoes—that is, discrete edifices, although there is evidence for non-constructional volcanoes such as fissure vents, too. Early post-Magellan surveys (e.g., Head et al. 1992) reported on the positional details of almost 1,700 volcanic edifices across the planet, finding that roughly half the large volcanoes on Venus are located in the Beta, Atla, and Themis region, which covers less than a quarter of the surface (Crumpler et al. 1993). The largest volcanoes are well in excess of 100 km across: some, such as Maat (Fig. 1), Sif, and Tepev Montes, are several hundred kilometres in diameter (e.g., Janle et al. 1988; Senske et al. 1992), and lava flows as long as $\sim$500–1000 km are observed (Head et al. 1992). Indeed, vast lava flows are a common feature of the largest volcanoes on Venus, which are typically located atop topographic rises (e.g., Ivanov and Head 2013). These edifices are classic shield volcanoes, being far broader than they are tall; generally, Venus' largest volcanoes are up to 9 km tall, comparable to the largest volcanoes on Earth (Fig. 1).

Venus also exhibits $\sim$200 large flow fields, with areal extents of $\sim$39,000 to 744,000 km^2 (e.g., Magee and Head 2001). The majority of these flow fields emanate from either central volcanos or coronae, and are thus covered under the discussion of these features below. In addition, another $\sim$40 large flow fields occur in association with chasmata or fracture zones (e.g., Magee and Head 2001; Lancaster et al. 1995). These tectonic features are discussed below as well.

Volcanoes less than 100 km in diameter are distributed widely across the planet and are often characterized by radial lava flows and conical or shield-like shapes; some have flat-topped summits (Head et al. 1992). By far the greatest number of individual edifices are

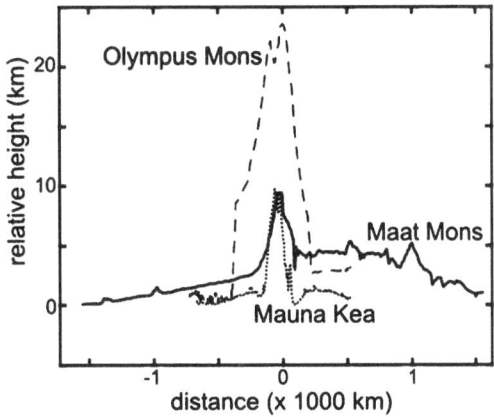

Fig. 1 Maat Mons and Mauna Kea (Hawaii) both reach ~9 km of prominence, but Maat Mons is several hundred km across. For comparison, Olympus Mons (Mars) reaches ~24 km of prominence, but still is narrower than Maat Mons

those less than 20 km in diameter (e.g., Crumpler et al. 1997; Grindrod et al. 2010), with recent mapping suggesting that there are at least tens of thousands of volcanoes <5 km across (Hahn and Byrne, 2022). Crumpler et al. (1997), classified volcanic edifices as "large", "intermediate" and "small" on the basis of Venera 15 and 16 results; large volcanoes are those equal to or larger than 100 km, intermediate volcanoes are those $\geq$ 20 km and < 100 km, and small volcanoes are those < 20 km in diameter, regardless of their geologic significance. Establishing the exact shape of these small edifices is difficult with the limited topographic resolution of Magellan and so determining whether they are conical or truly shield-like in appearance requires new data from future missions. Given their prevalence, volcanic processes likely have a key role in forming the young surface through the burial of impact craters and other old terrains (see Herrick et al. 2023, this collection). As on Earth, some volcanism can be linked directly to decompression melting above mantle plumes, while other features may be associated with lithospheric extension, delamination or dripping, or other processes (e.g. Head et al. 1992; Martin et al. 2007). The features that can most clearly be linked to mantle plumes are the ten or so volcanic rises, which share many key characteristics with Hawaii and other terrestrial hotspots (Stofan and Smrekar 2005).

The ~500 enigmatic coronae (mean diameter ~250 km) are widely distributed across the surface (e.g., Stofan et al. 1992), although ~62% occur in association with rifts or fracture belts (e.g., Glaze et al. 2002). These features likely have multiple formation mechanisms, but many probably formed above small-scale mantle plumes (e.g., Gülcher et al. 2020). Some proposed mechanisms have terrestrial analogues, such as large igneous provinces (LIPs) and plume-induced subduction. However, these processes are not active on Earth today. Two key questions for the formation of coronae are: (i) why do coronae form on Venus but not on present-day Earth, and (ii) why does Venus appear to have two largely distinct scales of mantle upwellings? For the latter question, one hypothesis is that different volcanic features, such as volcanic rises and coronae, formed at different times in Venus' evolution; another is that layered (or otherwise variable scales of) mantle convection generates different features (see Rolf et al. 2022, this collection).

2.1 Spatial Distribution of Volcanism

Given the seeming lack of obvious fluvial or aeolian sedimentary processes (although see Carter et al. 2023, this collection), volcanism is the major contributor to the formation of surface material on Venus (Head 1990) at least during its observable part of geologic history.

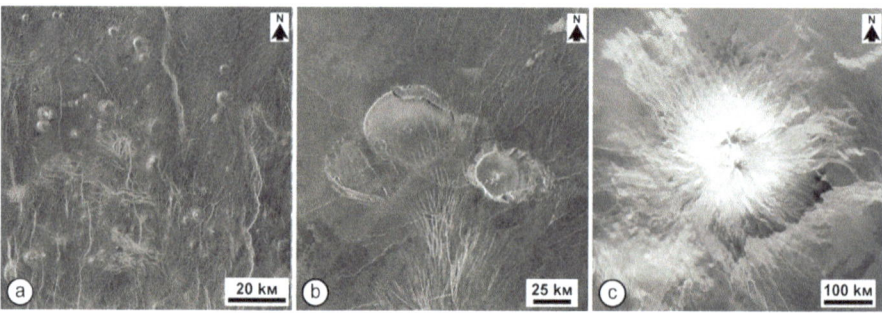

Fig. 2 - Examples of volcanic edifices on Venus: (a) a cluster of small edifices (< 20 km in diameter) that populate an exposure of shield plains; (b) a cluster of steep-sided domes that represent the intermediate volcanoes (from Crumpler and Aubele 2000); (c) a large volcano (> 100 km) of Sapas Mons (from Crumpler and Aubele 2000)

Indeed, Venus is a "volcanic" planet on which volcanic materials compose ∼92% of the surface (e.g., Ivanov and Head 2011). The remaining surface consists of tessera (as discussed below in Sect. 3.1).

The most areally extensive volcanic landforms on Venus are volcanic plains, some of which are as large as a few million square kilometres (e.g., Ivanov and Head 2011, 2013).

Small edifices (Fig. 2a) populate the surface of shield plains and are the most abundant volcanic form (Addington 2001; Crumpler et al. 1993; Hahn and Byrne, 2022). The exact number of these volcanoes is unknown but it may exceed the tens of thousands. These edifices are relatively small with the decile diameter range (central 80% of the population) from ∼2.5 to ∼5.5 km (Hahn and Byrne, 2022). In an early global catalogue of volcanic landforms of Venus (Head et al. 1992), the sub-population of small volcanoes on the shield plains were assigned to a class of "shield fields," which may represent isolated outcrops of shield plains embayed by younger materials (e.g., Ivanov and Head 2004).

A recently compiled, new global volcano catalogue for Venus indicates the presence of more than 85,000 edifices, the vast majority of which (∼52,000) are barely resolvable with available Magellan radar image data (Hahn and Byrne, 2022) (Fig. 3a). Of those volcanoes that are more readily identifiable, ∼32,500 are <5 km in diameter, 729 are 5–100 km in diameter, and 118 have diameters greater than 100 km. Previous studies had further grouped volcanoes into "fields" (e.g., Grosfils et al. 2000; Pavri et al. 1992); Hahn and Byrne (2022) found that (for specific criteria, including 25 edifices as a minimum number) the global volcano population could be grouped into 566 fields comprising about 65% of all mapped volcanoes; the remainder did not fall within clusters of $\geq$25 edifices.

The spatial distribution of volcanoes on Venus (Fig. 3a) appears to be drastically different from that on Earth, where the active volcanoes are mostly occurring within zones along convergent and divergent boundaries and along hotspot tracks (Fig. 3b) (Byrne and Krishnamoorthy 2022), although it should be noted that the distribution of (presumably) active volcanoes on Venus is currently unknown. Even so, the difference in spatial distribution suggests that Venus possesses a mechanism of the internal heat leakage which is dissimilar to terrestrial plate tectonics (e.g., Solomon and Head 1982; Solomon et al. 1992; Stofan and Saunders 1990). Grossly speaking, volcanoes on Venus show slightly greater spatial concentration in the Beta–Atla–Themis region and in a region north of Aphrodite Terra relative to the rest of the planet surface (Hahn & Byrne, 2022) (Fig. 3a).

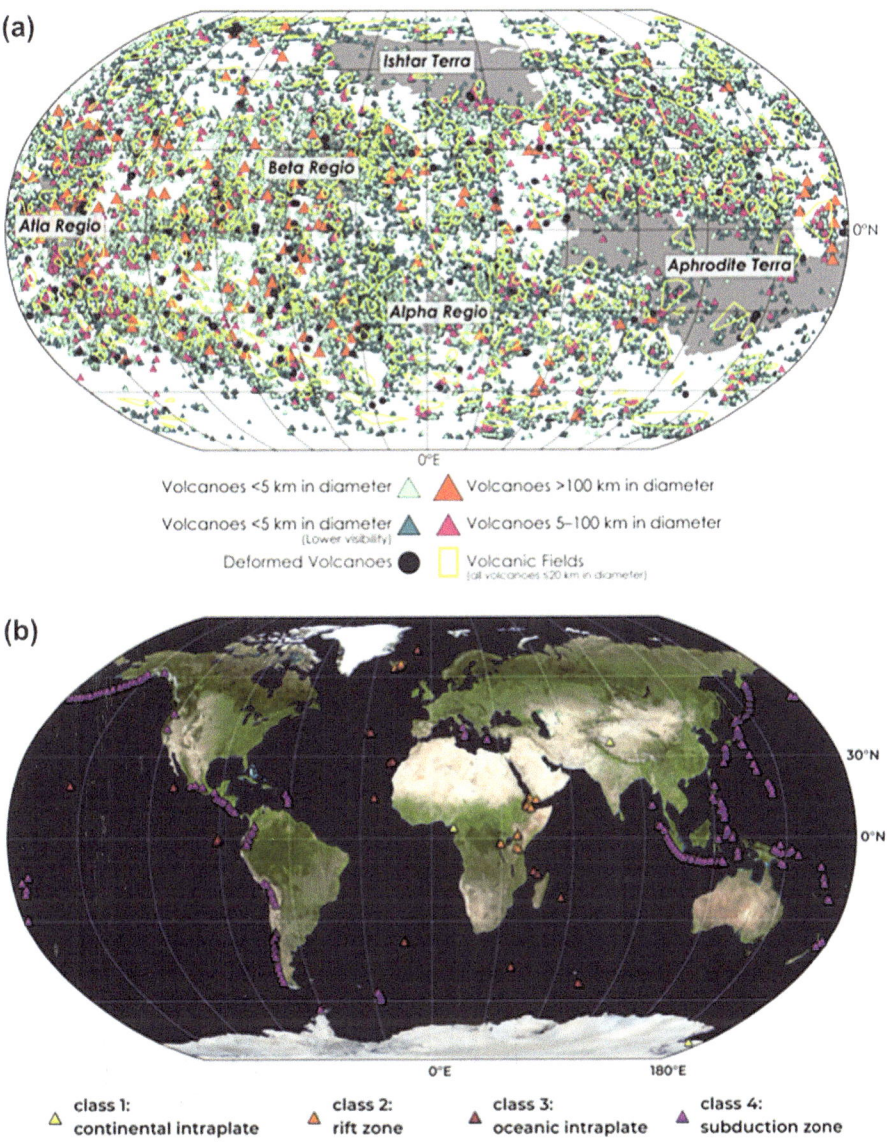

Fig. 3 The spatial distribution of volcanoes on: (a) Venus, and (b) Earth. The databases of Venusian volcanic edifices are from Hahn and Byrne (2022). Terrestrial volcanoes are those that are thought to have been active during the Holocene volcanoes, which are mostly located at plate boundaries (Global Volcanism Program 2023). This database from the Smithsonian also has a list of terrestrial volcanoes that may have been active during the Pleistocene

2.1.1 Lava Channels

Lava channels are widespread on Venus. Almost 200 such features were mapped using Magellan data, divided into six different types (e.g., Komatsu et al. 1993): sinuous rilles; canali; simple channels with associated flow margins; complex channels with flow margins; complex channels without flow margins; and compound channels. Some of these channels resemble features seen on Mars and Earth's Moon. For example, sinuous rilles have large depth-to-width ratios and morphologies similar to those of lunar and Martian rilles (e.g., Komatsu and Baker 1994; Oshigami et al. 2009). Usually, no lateral flow deposits are observed near rilles, but flow deposits are sometimes found at their termini. Channel sources are sometimes unidentifiable, but they are often found to originate in depressions. Mappers have identified ~59 rilles with average lengths of ~10–300 km and widths up to several kilometres (e.g., Baker et al. 1992; Komatsu et al. 1993). More than half of the rilles are associated with coronae or coronae-like features (e.g., arachnoids), while the remainder are almost evenly distributed between the plains and highlands (e.g., Komatsu et al. 1993). Valley networks on Venus are also typically associated with coronae and sinuous rilles—and they seem to be subject to tectonic control than rilles (Komatsu et al. 2001). Overall, both rilles and valley networks appear to have formed via incision into the surrounding terrain. Other types of simple, complex, and compound channels appear shallower and often have flow margins—suggesting a constructional (i.e., volcanic) origin (e.g., Komatsu et al. 1993; Komatsu and Baker 1994; Komatsu et al. 2001).

Canali are long, narrow channels that superficially resemble meandering rivers on Earth. They are typically more than 500 km long, ~1 km wide, and ~24 m deep on average (Baker et al. 1992; Komatsu et al. 1993; Williams-Jones et al. 1998). Baltis Vallis is ~6,800 km long—the longest canali on Venus and the longest channel of any sort so far found in the solar system (e.g., Baker et al. 1992). This extreme length makes Baltis Vallis useful as a stratigraphic marker (e.g., Basilevsky and Head 1996) and as a tool with which to study the evolution of topography (e.g., Conrad and Nimmo 2023). River-like features such as cut-off meander bends, abandoned channel segments, and terminal deposits are often observed at canali. Narrow levees are sometimes seen, but only at the limit of the resolution of the radar imagery. Canali are concentrated on the plains—although identifying their sources and/or termini is often difficult or impossible. Unusual types of lava are often invoked to explain the apparently unique properties of canali (e.g., Kargel et al. 1994). For example, some studies favored carbonatite-rich (or carbonatite-sulphate-rich, with a eutectic composition) lavas that have low viscosity (to explain river-like meanders) and low melting temperatures (to explain the extreme lengths). For example, one carbonatite lava from the Oldoinyo Lengai volcano on Earth was observed to have a liquidus temperature of only 763 K, although ~1000 K is a more typical value of this type of lava (e.g., Dawson et al. 1990; Keller and Krafft 1990). Mafic and ultramafic lavas that are observed on Earth and other rocky planets, such as basalt or komatiite, have relatively high melting temperatures (>1300–1600 K). They could create channels via thermal erosion, but they might tend to solidify after travelling tens (or, at most, hundreds) of km across the surface (e.g., Flynn et al. 2023), rather than the thousands of km needed for some canali (e.g., Head and Wilson 1986).

Speculatively, the formation of canali may be important to models of the evolution of Venus. Carbonatite lavas on Earth typically contain >30–50 volume percent of carbonate minerals (Dawson 1962). Some studies suggest that canali formed with discharge rates >10^5 m^3 s^{-1} for up to ~100 years (Kargel et al. 1994). Lava that formed each canale seen on the surface today thus could have degassed enough CO_2 to produce a small portion of the modern atmosphere (e.g., Kargel et al. 1994). Such rapid resurfacing with carbon-rich lava may lead to a coupled evolution of the surface and atmosphere—perhaps burying old impact craters (e.g., Komatsu et al. 1993; Strom et al. 1994) and creating a greenhouse effect that drives the global climate from clement to hellish (e.g., Way and Del Genio, 2020; Way et al., 2016). However, many questions about canali and carbonatite volcanism on Venus await answers.

2.1.2 Hotspots and Hotspot Volcanoes

Some of the large volcanoes on Venus can be classified as 'hotspot' volcanoes. On Earth, hotspots are defined as the surface manifestation of hot mantle that diapirically rises from the lower mantle or core–mantle boundary, generating volcanoes via pressure-release melting (Griffiths and Campbell 1990; Hoggard et al. 2020; Jellinek and Manga 2004; Morgan 1972). Mantle plumes and their tails are thought to be responsible for flood basalt volcanism and the formation of hotspots (White and McKenzie 1995); specifically, the bulbous head of a mantle plume impinges to the base of the lithosphere and generates mafic and ultramafic melts for a short time. On Earth, plumes have mantle potential temperatures that are 150 K to 350 K above ambient conditions. Seismic data indicates these mantle upwellings are generated at specific regions of the Earth's lower mantle that are known as Large Low Shear Velocity Provinces (LLSVPs) (Torsvik et al., 2014).

Nine hotspots are inferred to have originated from or near the core–mantle boundary on Venus (McGill 1994; Smrekar et al. 1997). They have been identified on the basis of their great topographic swells ($\sim$1000–2000 km diameter, $\sim$1–2 km high), large volcanic edifices, gravity anomalies that indicate compensation by low-density material at the base of the lithosphere (the plume head) (e.g., Smrekar and Phillips 1991), and, in some cases, rifts and coronae (Stofan et al. 1995). No hotspot tracks are observed in association with mantle plumes, consistent with a lack of plate tectonics operating at Venus. However, the large interior volcanoes and ring of wrinkle ridges surrounding Laufey Regio were interpreted as evidence of a past plume at that location (Brian et al. 2004). Once that rising mantle plume has cooled, the thermal isostasy from it no longer supports the topographic rise proposed to account for the circumferential wrinkle ridges.

Many coronae likely formed above small plumes but are typically too small to be caused by core–mantle boundary-sourced plumes. Artemis is larger (at $\sim$2600 km in diameter) than many of the hotspot features discussed above, and has been proposed to be a site of plume-induced subduction (Davaille et al. 2017; McKenzie et al. 1992; Sandwell and Schubert 1992b). Other large coronae could have formed over deep mantle plumes: despite their relatively low resolution, gravity data at Quetzelpetlatl corona ($\sim$900 km diameter) are indicative of an active plume (Davaille et al. 2017). Heng-o ($\sim$1100 km diameter) is another large corona, but neither its topography nor gravity suggests activity today. The nine large volcanic rises (Stofan and Smrekar 2005) on Venus (ten, if Laufey is included) exclude these few very large coronae. This is based on both on the smaller size of all features exclusive of Artemis and the analogy to Earth's hotspots, which do not have the evidence of plume-induced subduction seen at Quetzelpetlatl and Artemis Coronae (Ivanov and Head, 2011; Davaille et al. 2017). However, plume-induced subduction is proposed to have occurred earlier in Earth's history (e.g. Whattam and Stern, 2015), and thus excluding Artemis is perhaps artificial. Additionally, the larger features can be resolved in the Magellan gravity data, which provides evidence of an active plume at these features (e.g. Smrekar and Phillips 1991). As discussed below, there are multiple lines of evidence supporting these hotspots as likely active. Other (non-hotspot) volcanoes may also be active, but available data are inadequate to assess this possibility.

The classic volcanic edifices at terrestrial oceanic hotspots (e.g., Hawaii, Iceland) form shield volcanoes that are composed of a thick sequence (sometimes more than 10 km) of many individual lava flows (typically $\sim$2–20 m thick). These flows in turn form a broad base, on the order of 100 km across, with relatively steep lower slopes ($\sim$10°) that grade to more gentle upper slopes (2° to 3°). The morphology of oceanic hotspot volcanoes is due to the sequential accumulation of low viscosity lava flows that erupt over a period of 1 million years; volcanic activity can last up to 200 million years for a single hotspot (Heaman and Kjarsgaard 2000; Kasbohm and Schoene 2018; Wei et al. 2020). The lava compositions are 90 to 95% basaltic, with minor volumes of silicic volcanic rocks (e.g., rhyolite, trachyte, dacite) present within the youngest flow sequences (Jeffery and Gertisser 2018). After the initial effusive flows, the preponderance of subsequent flows during the shield-building stage are tholeiitic basalt, but minor volumes of alkali basalt may erupt during waning or rejuvenation stages (Thordarson and Garcia 2018).

On Venus, many of the hotspots described above have large shield volcanoes (e.g., Stofan et al. 1995), as well as other types of volcanism. Large shield volcanoes occur in many locations and geologic settings across Venus, as is the case on Earth (e.g., Hahn and Byrne, 2022). Without high-resolution geophysical and detailed geochemical data, the deep mantle origin and general eruption history of Venusian hotspot volcanism must be inferred from the geologic setting, gravity and topography, morphology and by analogy with Earth. Moreover, the quality of Magellan-based altimetric and stereo-derived topographic data are not sufficient to fully characterise volcano shape and slopes to the extent possible for Earth (and Mars). Nonetheless, on the basis of Magellan radar imagery, shield volcanoes on Venus appear to have broadly comparable morphologies and, presumably, slope values.

2.1.3 Coronae

Coronae are oblong to circular volcano-tectonic features ranging in diameter from 60 km to up to 2600 km, with a median size on the order of 200 km (Glaze et al. 2002). A ring of closely spaced concentric fractures and/or ridges is the defining characteristic of coronae. Stofan et al. (2001) divided coronae into two categories: Type 1, with >180° of a fracture annulus (Fig. 4c), and Type 2, or stealth coronae (Fig. 4b), with <180° (Stofan et al. 2001). Glaze et al. (2002) identified 406 coronae of Type 1 and 107 of Type 2. Coronae have also been categorized into five classes based on their fracture annuli: concentric, concentric double-ring, radial/concentric, asymmetric, and multiple (Stofan et al. 1992). Coronae often have circumferential rims or trenches that can extend beyond the annulus, which are not necessarily correlated radially or azimuthally with fracture annuli and can be equally incomplete and irregular in shape. They can also have interior domes, plateaus, and depressions. Based on these topographic elements, Smrekar and Stofan (1997) defined nine topographic classes. Three coronae that fall into a common topographic class (3a, rim surrounding interior high) with varying fracture patterns are illustrated in Fig. 4.

Coronae are found in every geologic environment on Venus (e.g., Stofan et al. 1992). The majority (62%) occur along fracture/rift belts (chasmata) (Glaze et al. 2002). Coronae appear randomly located in these extensional environments, suggesting that the detailed locations of coronae are not influenced by rifts and vice versa (Martin et al., 2007). Nonetheless, the concentration of coronae in extension environments implies a genetic relationship with these tectonic structures (e.g., Piskorz et al. 2014), although the remaining coronae occur in the plains (25%), at large topographic rises (11%), and tessera (1%) (Glaze et al. 2002).

Large volcanoes (see below) and smaller coronae overlap in scale. Morphologies can be transitional as well, leading some to propose that (some) coronae form via volcanic construction (Dombard et al. 2007; Lang and López 2015; McGovern et al. 2013). Others have pointed to the similarity between giant radiating dyke swarms on Earth and the radiating lineaments that occur at some coronae (e.g., Ernst et al. 1995).

The margins of some larger coronae, such as Artemis and Quetzalpetlal, have been proposed to be sites of retrograde lithospheric subduction (e.g., McKenzie et al. 1992; Sandwell

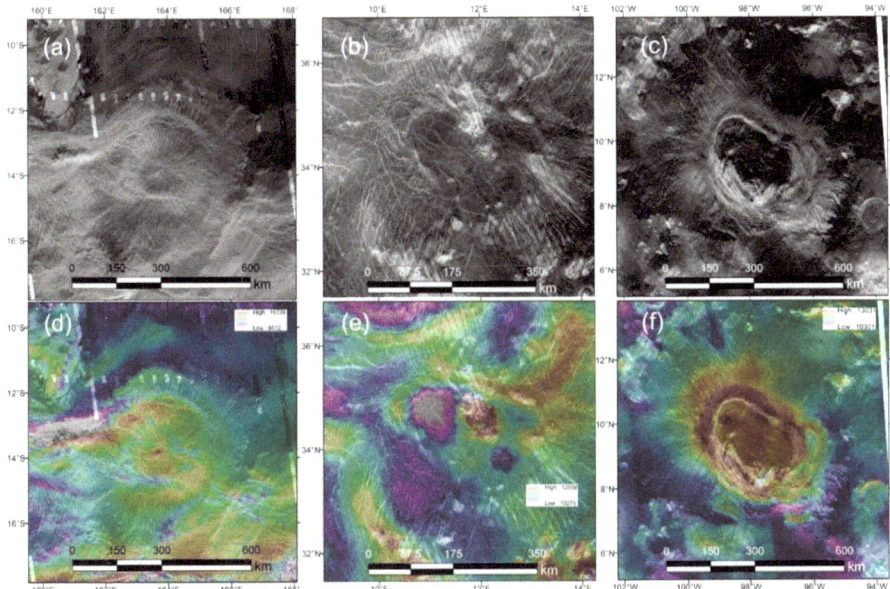

Fig. 4 Magellan left-look SAR images of (a) Miralaidji,, (b) Ponmakya, and (c) Aruru coronae, with respective stereo-derived topography (Herrick et al., 2012) in (d) and (e), and Magellan topography (f). Miralaidji Corona, near Latona and Dali Chasma, has radial fractures, and thus a morphology similar to that of novae (Krassilnikov and Head 2003). Ponmakya Corona, a stealth corona, has <180° concentric fractures, with a short NE-SW axis of ∼175 km and long NW-SE axis of ∼300 km, and >180° of a topographic rim. The appearance both in SAR and in topography is obscured by regional strain and interior volcanism, and orthogonal fractures obscure its appearance in SAR, in particular. Aruru Corona, a large volcanic centre (Tucker and Dombard 2023), has interior lava flooding, shield fields in the SW, and lava flows to the ESE

and Schubert 1992b; Schubert and Sandwell 1995). These coronae show topographic flexural signatures, beginning in the fracture annulus and extending outward, that resemble flexural signatures caused by retrograde subduction observed on Earth (Johnson and Sandwell 1994; O'Rourke and Smrekar 2018; Sandwell and Schubert 1992b). Moreover, interior extensional features displayed by Artemis corona are reminiscent of back-arc spreading centres (e.g., Brown and Grimm 1995; Sandwell and Schubert 1992b), contributing to the interpretation of retreating lithospheric subduction occurring at the south-eastern margin of the corona (McKenzie et al. 1992; Sandwell and Schubert 1992b), although others suggest Artemis Chasma is a site of convergence (Brown and Grimm 1995; Suppe and Connors 1992).

The diverse morphologies and sizes displayed by coronae may indicate that they form via a range of mechanisms. Modelling studies of the various mechanisms proposed to be responsible for corona formation are elaborated in more detail in Sect. 5.1.4 below. Regardless of the specific processes involved in their formation, coronae likely contribute to heat loss (Smrekar and Stofan 1997; Turcotte 1993). Moreover, coronae offer the opportunity to estimate the elastic lithospheric thickness through flexural modelling (e.g., O'Rourke and Smrekar 2018; Russell and Johnson 2021; Smrekar et al. 2023 and refer to Sect. 3 below and Rolf et al. 2022, this collection). As such, coronae offer important clues on the tectonic and resurfacing history of the planet.

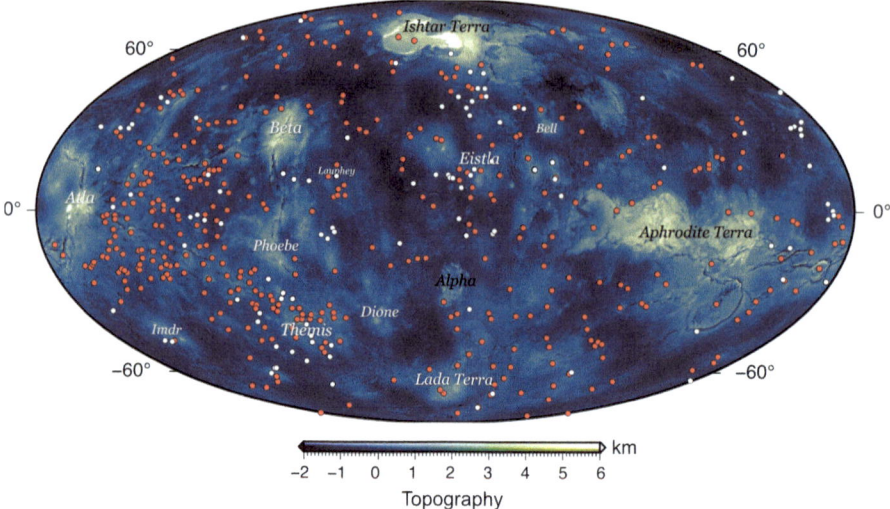

Fig. 5 - Magellan topography data superposed by locations of topographic rises (labelled with white names), major crustal plateaus (labelled with black names), Type 1 coronae (red dots), and Type 2 coronae (white dots). Blues and purple are topographically low regions; yellow and white are topographically high. Coronae locations and classifications from Stofan et al. (2001); Glaze et al. (2002)

2.1.4 Novae

Novae (singular: nova) are structures with prominent, stellate fracture patterns consisting of a mesh of graben centered on a central summit (Head et al. 1992; Krassilnikov and Head 2003). They are usually 100–300 km in diameter, have a prominent upraised topography (refer to Fig. 4a), and are associated with lava flows. Apparently distinct from volcanoes, novae show a dominance of tectonic over volcanic activity. Sixty-four novae have been identified on the Venusian surface (Krassilnikov and Head 2003), mainly located in areas of regional rises and rift zones, with a small number in the lowlands.

It is generally assumed that novae are formed by volcanic updoming and fracturing due to impingement of hot mantle diapirs on the upper part of the lithosphere (Janes et al. 1992; Koch and Manga 1996). Novae have been proposed to be the initial stages of corona development (e.g., Janes et al. 1992), or even perhaps "failed coronae" if the source of dynamic support were retracted early in the formation of what would go on to become a corona. Some studies suggested that novae within coronae represent reactivation of the corona location (Aittola and Kostama 2002), while others pointed out that the lengths of the radial structures often exceed the anticipated distance for diapiric uplift and fracturing, and instead linked novae to radial fracturing caused by dyke emplacement processes from a central magma source (e.g., Parfitt and Head 1993). However, it is important to note that <5% of coronae have radial fractures similar to novae, calling into question the hypothesis that novae are the first stage of corona formation.

2.2 Evidence for Recent Volcanism and Likely Active Features

One of the biggest scientific debates about Venus concerns its style of volcanism throughout its geologic history. As discussed above and in Herrick et al. (2023, this collection), both catastrophic and regional resurfacing models are viable for Venus. Rolf et al. (2022, this collection) discuss the importance of understanding the history of volcanism for internal evolution models. Over the last ~15 years, numerous studies have shown evidence for current volcanic activity, suggesting Venus is likely more active today than previously understood. Many studies have argued for recent volcanism based on signatures of chemical, thermal, atmospheric gas, and surface change. Analysis of geophysical data supports the location of activity at rifts and hotspots.

Chemical signatures of recent volcanism are based on interpreting surface emissivity data as an indication that the weathering of new lava flows has not proceeded to completion. Evidence of 'recent' volcanism was first noted in Magellan emissivity data. The relative youth of Idunn and Maat Montes as well as some coronae is further corroborated by high values of Magellan radar emissivity at their summits, consistent with relatively little weathering (Klose et al. 1992; Robinson and Wood 1993). Brossier et al. (2020) further investigated these and other regions, suggesting additional minerals that could explain these signatures. Measurements made by the Visible and Infrared Thermal Imaging Spectrometer (VIRTIS) instrument aboard Venus Express were used to derive a map of thermal emission at 1.02 microns of much of Venus' southern hemisphere added further evidence of recent volcanism. Higher thermal emissivity displayed by some lava flows were interpreted as the chemical weathering signature of geologically recent lava flows (Smrekar et al. 2010; Stofan et al. 2016; D'Incecco et al. 2017). The presence of volcanic activity was found at Idunn Mons (in Imdr Regio), Hathor and Innini Montes (in Dione Regio), and Mielikki Mons (in Themis Regio). Prior interpretation of gravity and topography data in these areas provided evidence of mantle plumes beneath these regions, providing further evidence of activity in these areas (Smrekar et al. 2010 and references therein). The age of these incompletely weathered flows is debated. Filiberto et al. (2020) suggest that the flows are as young as a few years based on weathering experiments done using terrestrial air. Others argue for much slower weathering (up to ~500,000 yr) based on kinetic calculations applied to Venus spectral data (Dyar et al. 2020). Much work remains to fully interpret the age of 'recent' flows.

Other studies have provided evidence of a thermal signature of active flows. Transient bright spots in data from the Venus Monitoring Camera (VMC) were suggested to arise from volcanic activity associated with the Ganis Chasma rift zone in the Atla Regio (Shalygin et al. 2015). Some transient features could also be artefacts of cloud motion or instrumental corrections for background temperature. However, the areas with the bright spots also have high radar emissivity associated with relative geologic youthfulness (Brossier et al. 2020). Bondarenko et al. (2010) argued for subsurface thermal anomalies in Bereghinia Planitia using Magellan microwave radiometer data. This area, along with ~50% of the planet, is not covered by VIRTIS data, which cover most of the southern hemisphere. Additionally, the Magellan radar emissivity signatures, which may be due to incomplete weathering, are found only above ~0.7 km altitude.

Together with Earth-based radar maps, Campbell et al. (2017) interpreted several radar-bright deposits to reflect the early stage of recently renewed magmatic activity in eastern Eistla, western Eistla, Phoebe, and Dione Regiones. Herrick and Hensley (2023) found a volcanic vent near the summit of Maat Mons that seems to have changed in shape during the Magellan mission, providing the first direct observational evidence for active surface change driven by volcanism. Both studies offer evidence for volcanism using radar imaging in the same areas where Magellan radar and/or VIRTIS thermal emissivity offer evidence of incomplete weathering.

Interpretation of the gravity and topography data (See Sect. 4) is also consistent with the locations of many of these recent and perhaps currently active centers of volcanism by revealing regions of upwelling and thin lithosphere. The long-wavelength geoid data suggest that both rifts and many areas that exhibit wrinkle ridges are likely supported by dynamic compensation and thermal anomalies at depth (Sandwell et al., 1997). The strong spatial correlation of the transient bright spots with rift zones suggests a correlation between deep and broad thermal upwellings and surface volcanism (Piskorz et al. 2014; Shalygin et al. 2015). Smrekar et al. (2010) provide an overview of several studies pointing to mantle plumes beneath the volcanic centers where emissivity anomalies are located. Other studies estimate elastic thickness from topographic flexure or admittance, and then derive heat flow as discussed in Sect. 3.2. High heat flow is a strong indication of activity, and is found at most coronae and nearby features where flexure is observed, including those coronae concentrated along Parga Chasma (e.g., O'Rourke and Smrekar 2018, Russell and Johnson 2021, Smrekar et al. 2023). Additionally, Gülcher et al. (2020) argued that 37 coronae with outer trench and rise morphology require ongoing suction above downwards-moving lithospheric material and an elevated interior supported by upwelling buoyancy.

Atmospheric data also offer a whiff of recent volcanism at unspecified sites. Varying rates of atmospheric SO_2 were interpreted as evidence of volcanic activity (e.g., Marcq et al. 2013). Analysis of the Soviet VeGa-2 probe (1985) data indicated, in the last few kilometres above the ground level, a lapse rate that would be unstable in the absence of a stabilising chemical gradient, which has been suggested to arise from a vertical gradient in nitrogen caused by density-driven separation of N_2 from CO_2 (Lebonnois and Schubert 2017). Cordier et al. (2019) found that producing a stabilising gradient with only CO_2 released from the Venusian crust would require an unrealistically huge rate of volcanic degassing. See also Wilson et al. (2024, this collection) for a discussion of effects of volcanic eruptions on the modern atmosphere of Venus.

3 Overview of Tectonic Features

3.1 Tessera Terrain

Approximately 8% of the surface of Venus is classified as tessera terrain (Ivanov and Head 1996), defined as having relatively complex patterns of deformation (e.g., Barsukov et al. 1986; Ivanov and Head; 1996). Tesserae are found as large (~1000 s km) plateaus standing ~1–4 km above the mean planetary radius, and as small outcrops (~10 s –100 s km) scattered in the plains (Gilmore et al., 2023, this collection). Global mapping of the tessera shows that most tessera margins are embayed by plains, thus establishing the tessera as the oldest materials in a proposed global stratigraphic column (e.g., Ivanov and Head 2011, 2013). This interpretation is supported by the total crater population of tessera terrain, which with ~80 craters is up to ~1.4 times the global average crater age (Gilmore et al. 1997; Ivanov and Basilevsky 1993), and may be older if deformed craters are recognized with future, higher-resolution imaging and topography data. The crater record, while sparse, does indicate that the observable phases of tessera deformation occurred at the time of, or just before, the global average age, although the rock age of the tesserae is unknown. Some tessera boundaries show interaction with plains units. Margin-parallel tessera structures are observed to uplift and deformed later plains units, e.g., in the southeast part of Thetis Regio (Ghail 2002), the western boundary of Alpha Regio (Gilmore and Head 2000), and the northern margin of Ovda Regio (Romeo and Capote 2011), suggesting that tessera deformation continued in places after the emplacement of marginal plains.

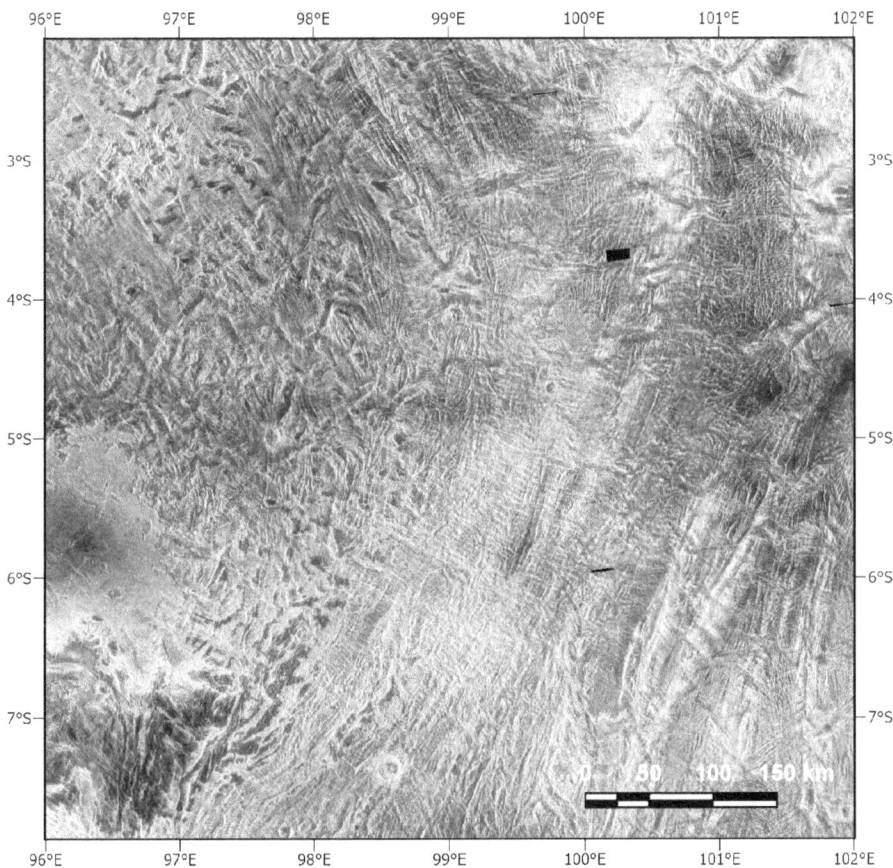

Fig. 6 - Margin-parallel ridge belts adjacent to interior facies of Ovda Regio

Shortening, extensional and transpressional structures are recognized in the tesserae (e.g., Ghent and Hansen 1999; Gilmore et al. 1998; Tuckwell and Ghail 2003). The style of extensional structures affects the interpretation of plateau origin and stratigraphy (i.e., whether their formation started with tectonic compression versus extension). First, these extensional features may be graben resulting from crustal relaxation that deformed older compression structures that, in turn, originated presumably from an earlier compressional, plateau-building phase (e.g., Bindschadler et al. 1992). Second, an alternative interpretation holds that the extensional features are steep (∼90°), open tensile fractures (called ribbon terrain). This hypothesis requires that they formed during an early extensional phase that precedes later compression and wide graben formation (e.g., Ghent and Tibuleac 2002; Hansen and Willis 1996). These endmember models call for the formation of tessera plateaus during mantle downwelling or upwelling (e.g., Phillips and Hansen 1998; Gilmore et al. 1998; Hansen et al. 2000), respectively, which can be tested with high-resolution images of extensional features. In several large plateaus (e.g., Ovda Regio, Tellus Regio), ridge belts are found parallel to the plateau margins, surrounding an interior with a different structural fabric (Fig. 6). At Tellus Regio, specifically, lateral accretion of the margins possibly occurred after the formation of the structures in the facies of the plateau interior (Gilmore and Head

2018; Resor et al. 2021). Therefore, at least some tessera facies possibly formed prior to their incorporation via lateral accretion—and subsequent deformation during plateau formation.

Several lines of evidence show that tessera formation is linked to an extinct geodynamic regime. Buckle analysis of the wavelength of tesserae folds show that they require both elevated heat flows and strain rates relative to structures preserved in the plains, coronae, and volcanoes (e.g., Brown and Grimm, 1997; 1999). This interpretation is valid for dry rocks of basaltic through felsic compositions if the rocks are not quartz-dominated (Resor et al. 2021). Multiple wavelengths of deformation might alternatively reflect physical layering in the crust (e.g., Romeo and Capote 2011), although this explanation may require layers that are thicker than observed (e.g., Byrne et al. 2021). The tesserae (average crustal thickness of ∼20 km) are also currently isostatically compensated, again consistent with a lack of recent mantle contribution (e.g., Anderson and Smrekar, 2006; Maia and Wieczorek, 2022).

The composition of the tesserae has not yet been measured directly, but that of some tessera exposures can be inferred from both NIR and radar observations. Observations at ∼1–1.2 μm from the Galileo flyby (Hashimoto et al. 2008) and VEx VIRTIS (Mueller et al. 2008; Gilmore et al. 2015) show that the Alpha Regio tessera has a lower emissivity than the global average, which is dominated by plains of presumably basaltic composition. Since emissivity near 1 μm is inversely correlated to Fe^{2+} content in minerals (e.g., Hunt and Salisbury 1970; Dyar et al. 2020), the signature of the tessera is consistent with rocks with lower Fe content, or more felsic compositions, assuming they are igneous rocks (e.g., Gilmore et al. 2023). The generation of substantial amounts of felsic melt (i.e., on the order of the volume of the tesserae) could require both water and a lithospheric recycling mechanism for formation (e.g., Campbell and Taylor 1983), and thus possibly must have formed during a more habitable era if they are ultimately shown to be chemically equivalent to Earth's continental crust (e.g., Campbell and Taylor 1983; Gilmore et al. 2023).

The interpretations above generally assume that the tesserae are igneous in origin. The complex structural fabric of tesserae makes it difficult to document the morphology of the original surfaces at the scale of the Magellan imagery, although Ivanov (2001) asserted from a survey that tessera precursor terrain is smooth and plains-like. However, layered beds have been identified in Tellus Regio, which resemble sedimentary beds on Earth and Mars exposed by erosion (Byrne et al., 2021). Additionally, Khawja et al. (2020) proposed that the topography of some tesserae is similar to fluvial drainage basins. It also should be recognized that the tesserae need not be structurally, morphologically, compositionally, or stratigraphically homogeneous: e.g., Brossier et al. (2020) showed that the radar emissivity properties of tesserae vary geographically, which may be evidence for differences in rock composition. The tesserae are also likely to contain sedimentary deposits from nearby, upwind craters (e.g., Campbell et al. 2015; Whitten and Campbell 2016) and internal mass wasting (e.g., Carter et al. 2023, this collection). The imaging, topography, and gravity data provided from VERITAS, DAVINCI and EnVision (Widemann et al. 2023, this collection; see also Sect. 6) will substantially advance our understanding of this important terrain type.

3.2 Rifts, Extension and Dykes

Long systems of extensional structures on Venus have variously been termed "fracture belts" or "groove belts" in the literature (e.g., Squyres et al. 1992; Ivanov and Head 2011) but, by analogy with Earth, are sites of crustal extension or rifting. These systems host normal faults that form graben and half graben with abundant evidence for fault linkage. Many of these systems are tens of kilometres across and hundreds of kilometres long. At larger scales, belts of extensional structures up to ten thousand kilometres long (Fig. 6) have been

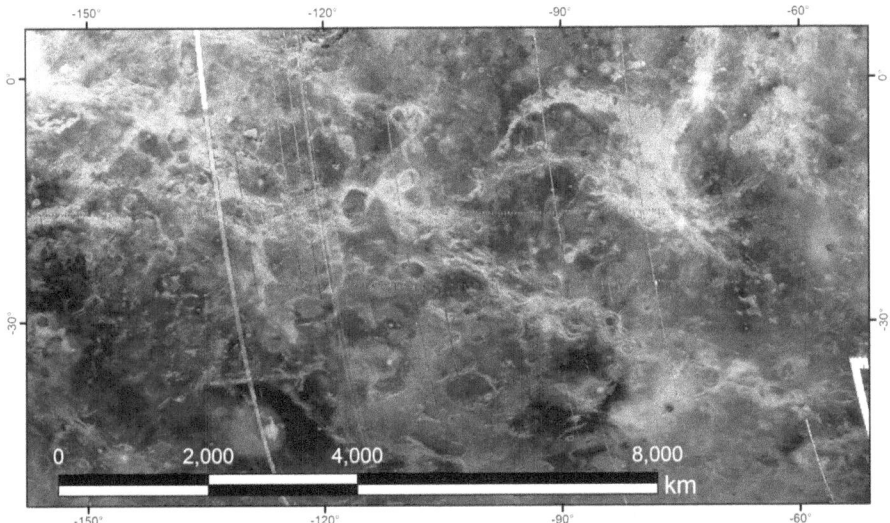

Fig. 7 - Parga Chasma, trending from NW to SE, or Atla Regio to Themis Regio, appears bright in Magellan SAR imagery. Note dozens of coronae in proximity to the rift

assigned the term chasmata or "rift zone" based on analogies to features on Earth and Mars (e.g., Solomon et al. 1991, 1992; Ivanov and Head 2011). These rifts form deep, elongated troughs, generally accompanied by broad expanses of effusive volcanic deposits, but are themselves typically located within regional topographic highs.

A large fraction of coronae are situated in proximity to the major rifts of Parga (Fig. 7), Hecate, and Dali/Diana (Hamilton and Stofan, 1996; Martin et al., 2007); many more occur in association with smaller rift zones (Glaze et al. 2002) (Sect. 2.1.3). Coronae at large rifts were found to have various age relationships with the rifts, such as coronae younger than the rift, synchronous formation with the rift, and post-tectonic (Hamilton and Stofan, 1996; Martin et al. 2007). At least one rift zone, Ganis Chasma, may be the site of active volcanism today (Shalygin et al. 2015). The collocation of rifts on Venus with elevated topography, together with considerable depths of compensation given by gravity–topography admittance functions, suggests that they are supported at least in part by mantle upwelling (e.g., Solomon et al. 1991; Smrekar et al. 2010). There are numerous examples of radiating extensional fracture systems on Venus (Grosfils and Head, 1994b) (Sect. 2.1.4), with at least 100 morphologically consistent with being associated with, and perhaps formed by, dyke emplacement (Ernst et al., 2001). Several of these (putative) radiating dyke swarms on Venus are associated with broad, domical swells comparable in size to mantle upwelling-induced uplifts on Earth (Grosfils and Head, 1994b).

3.3 Ridge Belts and Wrinkle Ridges

Where crustal shortening is spatially concentrated on Venus, such deformation is typically manifest as bands of structures interpreted as thrust faults and folds, and which have been termed "ridge belts" in the literature (e.g., Barsukov et al. 1986; Squyres et al., 1992, see Fig. 8). This term is not an established geologic one. Although it may be useful in certain circumstances, using more general terms such as "mountain ranges" may better align the study of Venus geology with the other terrestrial worlds (Klimczak et al., 2019). In any

Fig. 8 - Left-look Magellan SAR showing radar-bright wrinkle ridges trending ~E-W with ~20 km spacing

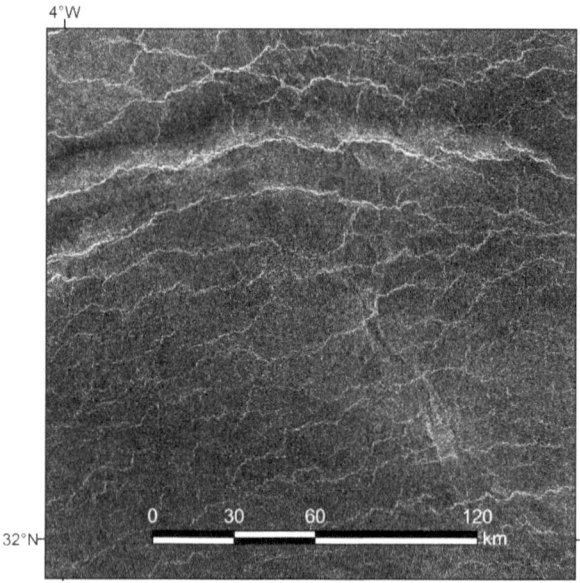

case, these belts are by any measure orogenic, in that they represent the (local) convergence of crustal material. Generally appearing as broad, linear rises a few hundred metres in relief, tens of kilometres in width, and many hundreds of kilometres long (Squyres et al., 1992), these systems typically show multiple anastomosing secondary arches and ridges super-posed on the larger rise that superficially resemble the "wrinkle ridge" anticlinal folds commonly observed elsewhere on Venus and on the Moon, Mars, and Mercury. Larger mountain ranges have long been recognized on Venus, and given their sizes are generally referred to as such (e.g., Barsukov et al. 1986). Those wrinkle ridges are typically tens to hundreds of kilometres long but only a few pixels across (corresponding to a width of a few hundred metres) when seen in Magellan data. Venus's wrinkle ridges are likely thrust-fault-related folds, given their morphological similarity to such features on other planetary bodies, which individually denote low amounts of shortening strain. These structures are ubiquitous on the planet, deforming the two most dominant surface units (the so-called upper and lower "ridged plains") mapped by Bilotti and Suppe (1999) and Ivanov and Head (2011).

3.4 Polygons and Distributed Deformation

In addition to wrinkle ridges, numerous other tectonic features occur in the plains, comprising polygonal fractures and small-scale fracture patterns (Fig. 9). Large fields of polygonal fractures with spacing ranging from a ~1–2 km (essentially the limit of resolution) up to 25 km have been identified in over 200 regions (Moreels and Smrekar, 2003). These features have been proposed to form as a result of cooling of lava flows, analogous to columnar fractures in basaltic lava flows, although extremely thick flows would be required for these Venus features (Johnson and Sandwell, 1992). In contrast, the largest columnar joint patterns on Earth are up to 30 m. For this reason, as well as their frequent association with volcanic features, Johnson and Sandwell (1994) favoured a model with thermal stresses from either cooling lava flows or heating at the base of the lithosphere. An alternative explanation is the propagation of climate-change driven cooling into the subsurface (e.g., Anderson

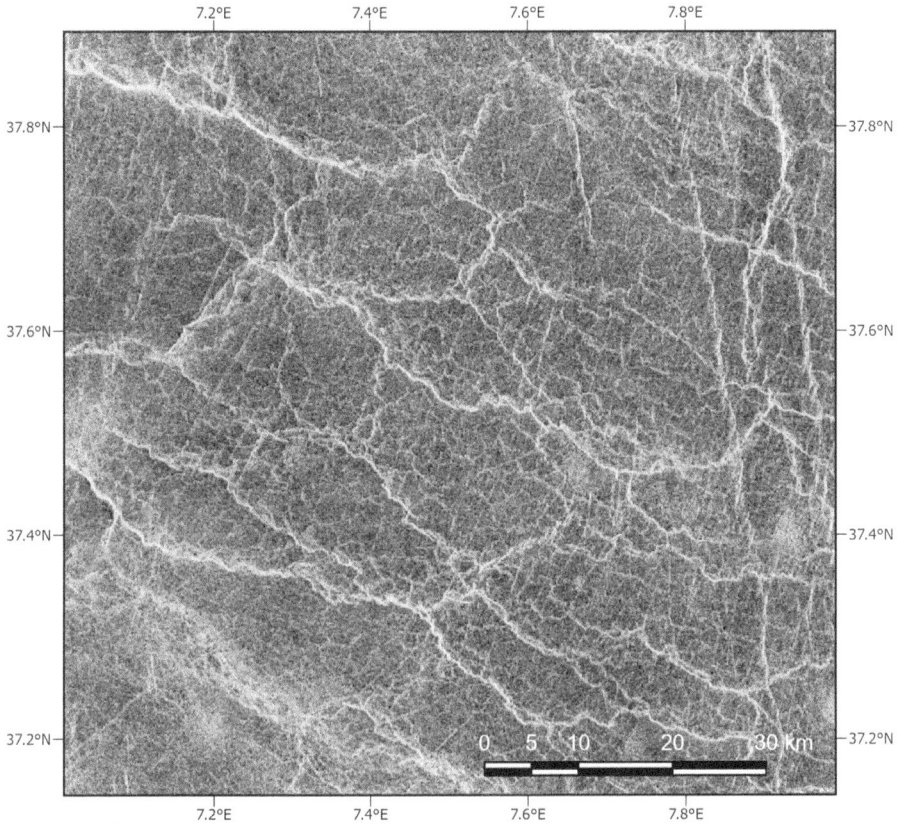

Fig. 9 - This section of FMAP 37N007 (Fig. 4 from Smrekar et al. 2002), shows two scales of polygons. The larger ones are controlled by intersections of NW-SE-trending wrinkle ridges

and Smrekar, 1999; Solomon et al., 1999; Smrekar et al. 2002). This hypothesis relies on huge volumes of volcanism, as predicted by the catastrophic resurfacing hypothesis, and associated outgassing as a driver for hundreds of degrees of climate change (Bullock and Grinspoon 1996; 2001). These two particular models differ in detail: Anderson and Smrekar (1999) used a continuous-plate model of deformation due to thermal stresses to assess the effect of large predicted positive and negative temperature changes, and found that thermal cooling stresses can predict the observed polygonal features, but that heating is insufficient to produce deformation. In contrast, Solomon et al. (1999) employed a broken-plate model, as had been used for mid-ocean ridge studies, and found that, in that configuration, heating can produce wrinkle ridges and cooling can produce polygonal fractures. The origin of these features may be tested by determining the minimum size of the polygons with higher-resolution imaging, and whether or not they occur as flows or other distinct topographic features using high resolution topography.

3.5 Subduction and Terrestrial-Style Plate Tectonics

Subduction is believed to be a necessary first step for plate tectonics (e.g., Whattam and Stern, 2015; Lithgow-Bertelloni and Richards, 1995). Since plate tectonics began long ago on Earth ($\sim$4–1 Ga, still debated), neither the mechanism for initiation of subduction nor for plate tectonics is clearly identifiable. Thus, understanding the nature of subduction on Venus and the conditions that enable it to form would provide important insights into the initiation of terrestrial subduction. In addition to creating the slab-pull force that helps drive terrestrial plate tectonics, subduction plays a key role in recycling volatiles into the interior. These recycled volatiles can later be released back into the atmosphere, stabilizing Earth's climate. On Venus, subduction provides links to interior processes and contributes to heat loss (e.g., Gülcher et al. 2020, 2023), and a means to estimate local elastic and lithospheric thickness.

Numerous locations (e.g., Artemis Corona, Fig. 10) on Venus appear to exhibit 'rollback' subduction, a process in which a plate sinks into the mantle under some combination of surface loading and negative plate buoyancy, without requiring lateral plate motion. The plate 'rolls back' as the downward-flexed plate advances away from the original subduction location as the plate sinks further. By analogy with terrestrial subduction zones such as the arcuate trenches of the South Sandwich Islands, the Aleutian trench, and north Fiji Basin, McKenzie et al. (1992), Sandwell and Schubert (1992a, 1992b), and Schubert and Sandwell (1995) identified a dozen sites of possible subduction totalling 10,000 km in length. Their criteria included a deep, narrow trench with arcuate or linear planform (Fig. 10), a curvature $>10^{-7}$ m^{-1} on the 'outer rise' side of the trench (the side attached to the subducting plate, e.g., on the southeast, outer side of Artemis), fractures parallel to the strike of the trench on both sides, and the absence of fractures cutting across the trenches. Hansen and Phillips (1993) presented an alternative interpretation of some of these features as resulting from mantle upwellings. This interpretation was based on the arcuate shape of many of these features, abundant volcanism in many locations, and fractures crossing the trench in one location.

Even with those examples of possible roll-back subduction associated with coronae, there is no evidence for a global network of tectonic plates diving into the planet's interior or being produced anew at spreading centres (e.g., Solomon et al. 1992). Harris and Bédard (2014, 2015) provided a general overview of strike-slip faulting on Venus. They argued for substantial lateral motions and indenter-like escape tectonics in Ishtar Terra, hypothesising that Lakshmi Planum collided with Ishtar Terra. These authors further suggested the presence of lateral displacements of hundreds to thousands of kilometres within shear zones in Ovda and Thetis Regiones. Harris and Bédard (2014, 2015) proposed that mantle flow tractions could have acted on the deep lithospheric keels of Lakshmi Planum and Ovda and Thetis Regiones, driving them across the planet in a manner akin to how large continental blocks move on Earth. Importantly, such motions produce the kinds of shortening and transpressional structures, and at the same approximate scales, as are recorded in the Archean rocks. If this hypothesis is correct, then the large-scale "drift" of major terranes on Venus may be a contemporary analogue of Archean Earth, when cratonic nuclei were mobile but modern plate tectonics, including subduction, had yet to take hold (Harris and Bédard 2014, 2015).

4 Crust, Lithosphere and Heat Flow

The surface geologic processes described above are a result of the interior heat loss mechanisms of convection and conduction. The crust, a compositional layer, and the lithosphere, a mechanical layer, strongly control the specific manifestation of these geologic processes. Models of both geologic processes and interior evolution require assumptions about crustal and lithospheric properties. In addition, viscous-deformation models require knowledge of

372

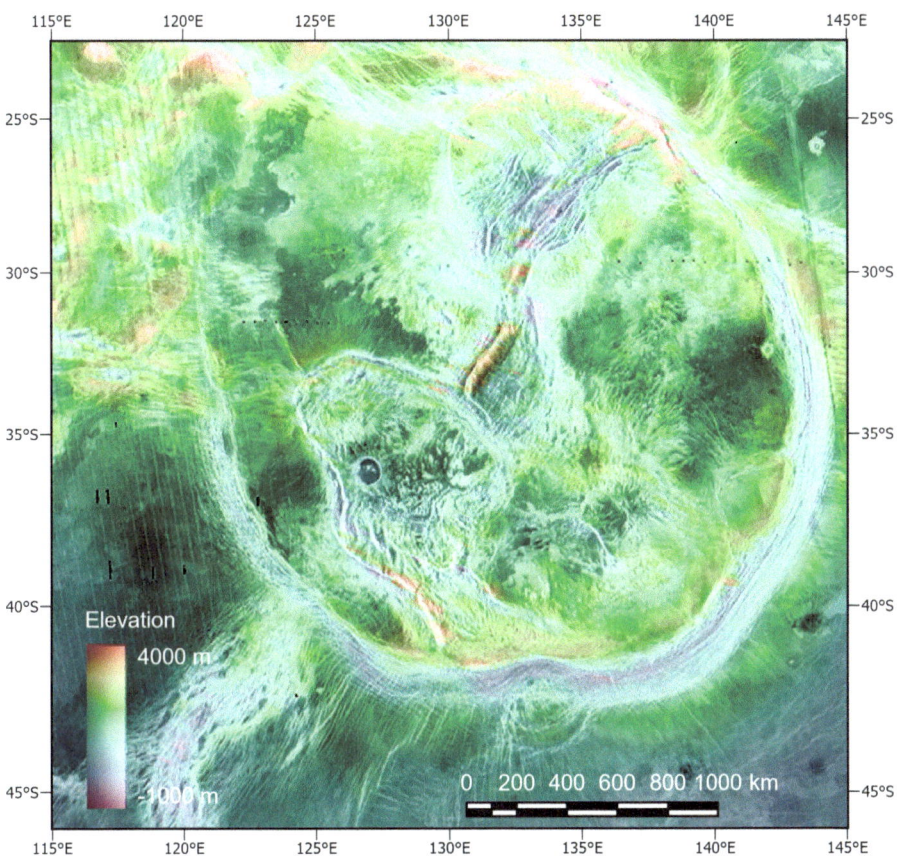

Fig. 10 - Magellan image data overlain on global topography of Artemis Corona. Artemis is the largest of all coronae, with a diameter of ∼2600 km. Its southeast is a likely site of subduction. Relative topography is in m

strain rate since the material strength is a function of the rate at which stress is applied. Models of deformation require laboratory-derived relationships between stress and strain. Even for Earth studies, there are substantial open questions about how to apply these laboratory-derived rheological laws to geologic processes, with laboratory predictions, done at very high strain rates of necessity, apparently overpredicting the strength of rocks deforming at the lower but more geologically realistic strain rates. In this section, we describe the available data to constrain estimates of Venus' crustal and lithospheric rheology, and review what is inferred about the composition and thickness of the crust, the temperature and thickness of the lithosphere, and strain rates based on terrestrial analogy.

4.1 Crust and Upper Mantle

4.1.1 Crust and Surface Composition from Geochemical Constraints

The similar bulk densities of Venus and Earth imply that they have broadly similar elemental contents (e.g., Taylor and McLennan, 1995; Lee et al. 2009; Rubie et al., 2015; Weller and

Duncan 2015; Shellnutt 2016). However, the crusts of Venus and Earth could have quite different, and variable, compositions. In-situ data on crustal composition is limited to data from several Soviet landers in plains locations with instrumentation that measured elemental geochemistry (see Treiman 2007 for a review). Measured compositions in these locations were primarily mafic and likely basaltic (Surkov et al. 1984, 1986; Kargel et al. 1993). The rocks for which near-complete major elemental chemistry measurements were made are terrestrial equivalents of tholeiitic (Venera 14) and alkalic (Venera 13) basalt, whereas the Venera 8 rock could be similar (e.g., Kargel et al. 1993) or notably different, with anomalously high K, Th, and U concentrations similar to terrestrial granodiorite or dacite (e.g., Nikolayeva 1990). Trace elemental (K, Th, U) analyses at other landing sites (Vega 1, Venera 9, Venera 10) also indicated the presence of terrestrial basalt. On the basis of morphology, much of Venus is inferred to be covered in basaltic material, similar to the measured plains locations.

There is some evidence from flyby and orbital spectroscopy for felsic rock in the massive tessera plateaus (e.g., Hashimoto et al. 2008; Helbert et al., 2008; Mueller et al. 2008; Gilmore et al. 2015; Gilmore et al., 2023). A small suite of features such as 'festoons' also have morphologies arguably consistent with high viscosity silicious lavas (e.g., Pavri et al. 1992; Bridges 1997), but perhaps also with basaltic lavas (e.g., Stofan et al., 2000; Wroblewski et al. 2019). However, morphology is a poor indicator of composition and in no way a substitute for remote or in-situ chemical data. Thus, the degree and scale of crustal differentiation that has taken place on Venus is debated, with some arguing for less than Earth's continents even in the tesserae and festoon flows (e.g., Kargel et al. 1993; Grimm and Hess, 1997; Treiman 2007; Treiman et al. 2016; Wrobleski et al., 2019).

Magma differentiation processes in terrestrial intraplate settings, thought to be the closest analogue for Venus tectonic settings (e.g., rifts, hotspots), are primarily related to crystallisation (i.e., equilibrium or fractional) and partial melting. In contrast, mixing or mingling is of secondary influence in comparison to convergent margin settings (Bachman and Bergantz, 2008; Christiansen and McCurry, 2008). Thus, forward petrological modelling using a range of magmatic conditions (i.e., water content, oxygen fugacity, and pressure), together with the rock compositions at the Venera 13 and Venera 14 landing sites and to a lesser extent the Vega 2 compositional measurements, can offer useful insight into the possible igneous rocks that could comprise a portion of the upper to middle crust. Fractional crystallisation modelling demonstrates that melt compositions similar to within-plate intermediate and silicic rocks (e.g., dacite, granodiorite, andesite, diorite, rhyolite, granite, trachyte, syenite), including the theoretical whole-rock composition of the Venera 8 landing site rocks and their K, Th, and U concentrations, can be produced at middle-to-upper crustal pressures (i.e., $\sim$0.1–0.5 GPa) under reducing or oxidizing, and hydrous or anhydrous, conditions (Shellnutt 2013, 2018, 2019). The residual solid after fractionation is mostly composed of cumulus olivine, orthopyroxene, clinopyroxene, plagioclase, and Fe–Ti oxide minerals (i.e., titanomagnetite, ilmenite) and are analogous to the cumulate rocks of layered mafic–ultramafic intrusions of Earth, such as the Bushveld Complex, Stillwater Intrusion, and Skaergaard Intrusion. On Earth, rocks similar to the modelled residual silicic liquids and their mafic/ultramafic cumulate rocks are commonly found at intraplate settings associated with rifting and hotspot volcanism. Therefore, they are likely present within the crust of Venus as layered igneous complexes, and their silicic volcanic rocks may be associated with volcanic rises, coronae, pancake domes, and shield volcanoes. The lower viscosity of intraplate silicic volcanic rocks compared with convergent margin silicic volcanic rocks suggests that their volcanic edifices may not necessarily have meaningfully different morphologies from their mafic counterparts (Christiansen, 2005). As discussed above, volcanic morphology is *suggestive* of composition (e.g., Wroblewski et al. 2019), but not necessarily *definitively* so.

Partial melting of the mafic crust of Venus can also yield magmas with intermediate-to-silicic compositions under high pressure and temperature ($\sim$1 GPa, >1250 K) conditions expected for the lower crust of Venus (James et al. 2013; Shellnutt 2013). Of particular interest is the Venera 14 landing site, which could be very similar to basalts of terrestrial Archean greenstone belts. It is interesting because rocks of similar composition to those near Venera 14 can yield silicic liquids that are indistinguishable from the tonalite–trondhjemite–granodiorite (TTG) series of rocks found in Archean cratons (Shellnutt 2013, 2018; Johnson et al., 2017). The formation of some TTGs may be related to thickening of mafic crust and the subsequent partial melting of the lower parts of the thickened crust either by ambient geothermal conditions or by anomalously hot mantle plumes (Smithies, 2000; Moyen and Martin, 2012). Venus has regions of tectonically thickened crust and evidence for hotspot volcanism and high temperature, and low viscosity (perhaps ultramafic) lava flows (e.g., Lenardic et al., 1991; Komatsu and Baker 1994; Hansen et al., 1999) that may be experiencing similar processes. If TTG suites are generated on Venus, they might form within regions of thickened crust such as Ishtar Terra and Ovda Regio. Additionally, gravitational instabilities, in the form of rollback subduction, delamination, or lithospheric dripping (see Sect. 3.5. above), may occur on Venus. Such instabilities would lead to remelting of the crust, recycling of volatiles, and potentially to the production of more evolved lavas erupting at the surface, including even low-viscosity flows (Elkins-Tanton et al. 2007).

The basalt at the Venera 14 site is broadly similar to olivine tholeiite and possibly normal-mid-ocean ridge basalt (N-MORB) and thus may have originated from a mantle thermal regime similar to Earth (McKenzie et al., 1992; Kargel et al. 1993; Fegly Jr., 2003; Filiberto, 2014). The material measured at the Vega 2 landing site appears to be a mixture of soil or weathered material and rock, since the composition is broadly basaltic but the SO_3 content is extremely high ($\sim$4.7 $\pm$ 1.5 wt%). If those measurements were accurate, such rocks would not have a direct terrestrial equivalent, with the exception of a mixture of sulphate rock (e.g., gypsum, anhydrite) or sulphide ore with a mafic (basalt, gabbro) host rock that is typical of some basalt country rock relationships, layered intrusions, and/or volcanogenic mass-sulphide deposits (VMS). Sulphate rocks near Vega 2 might be related to the carbonate sulphate volcanism that was proposed to form the canali (e.g., Kargel et al. 1994).

4.1.2 Crustal Thickness and Upper Mantle Viscosity

Gravity and topography measurements are currently the only datasets available with which we can investigate the interior structure of Venus. Without more direct measurements, derivation of interior structure is inherently non-unique and subject to the assumptions made about such parameters as crust and mantle density, thermal gradient, and strain rate. A key challenge for Venus is the resolution of the measured gravity field, which ranges from $\sim$400 to $\sim$1000 km (Konopliv et al., 1999). The spherical harmonic field is expanded to degree and order 180 to prevent aliasing of the signal. The actual resolution of the field is represented by degree strength, which is a maximum of degree and order $\sim$100 (Konopliv et al., 1999).

One approach to interpreting gravity and topography is to invert their spherical harmonics representation to provide an estimate of crustal thickness and upper-mantle viscosity and temperature (e.g., Banerdt, 1986; Herrick and Phillips, 1992; James et al. 2013). The correlations between thick/thin crust and hot/cold upper mantle with geologic surface features provides insight into the relationship between mantle convection and volcanism and tectonics. Different lithospheric properties and upper-mantle viscosity structures can be assumed in the inversions and the results then evaluated in terms of realism (e.g., are the temperature anomalies consistent with expected mantle behaviour?). The relationship between gravity highs and lows and stress patterns has also been used to infer areas of present or recent activity (e.g. Smrekar, 1994). The long-wavelength geoid (the integral of the gravity field) can be interpreted as an anomalous density distribution that gives rise to lithospheric stresses (Sandwell et al., 1997).

Two-layer modelling with a low-order, pre-Magellan gravity field indicated that the mantle temperatures required to reproduce the observed gravity and topography become unrealistic if the viscosity structure includes an Earth-like asthenosphere (Herrick and Phillips, 1992). Other studies also suggested that Venus lacks a narrow, low-viscosity zone (e.g., Kiefer and Hager, 1991; Smrekar and Phillips 1991; Steinberger et al. 2010). However, different parameterizations (Pauer et al. 2006) and an approach of first removing large areas of thick crust (Maia et al. 2023) produced results that favor a $\sim$200 km-thick zone below the lithosphere with a viscosity $\sim$10–100 times lower than the mantle beneath, i.e., more akin to Earth's asthenosphere.

A low-viscosity asthenosphere on Venus may indicate that the interior of Venus is wet. Importantly, however, alternative explanations are possible. A low-viscosity zone can be produced by CO_2 rather than water (Sifré et al., 2014; Ghail 2015; Chantel et al., 2016) and by a specific dry olivine activation volume (Armann and Tackley 2012).

The associations between crustal thickness and inferred mantle convection with geologic features (e.g., James et al. 2013) clearly show that areas of thickened crust are associated with tessera plateaus, while significant mantle upwellings are associated with large volcanic rises (Fig. 11), some of which are connected by major rift systems.

The overall pattern of surface features relative to calculated mantle convection and crustal thickness can be used to make some inferences regarding the evolution of the interior over time. The correlation of the present gravity field with the global volcano–rift–corona pattern in many regions suggest that the large-scale mantle convection pattern has not changed dramatically during the time period over which these features were emplaced. The global wrinkle ridge pattern generally, but not everywhere, roughly parallels the geoid gradient, consistent with these features forming in response to current mantle patterns (Bilotti and Suppe, 1999; Sandwell et al., 1997). Laufey Regio, which hosts large volcanoes and concentric wrinkle ridges, appears to have formed in response to a mantle upwelling that has waned or become inactive (Brian et al. 2004). That there is no clear association of tessera plateaus with the inferred mantle convection pattern is consistent with most, but not necessarily all – Ishtar Terra is a notable exception, of these areas being inactive. While they may no longer be forming, high topography is likely to be gravitationally relaxing, depending on its crustal strength (e.g., Smrekar and Phillips, 1988; Nunes et al. 2004; Romeo and Turcotte 2008; Maia and Wieczorek, 2022; Nimmo and Mackwell 2023).

Most estimates of Venus' average crustal thickness are tens of km, with regional maxima up to $\sim$70–90 km. Anderson and Smrekar (2006) and Jimenez-Daiz et al. (2015) both compiled global maps of crustal thickness using spectral admittance methods to solve for both local crustal and elastic thickness, without a contribution from mantle dynamics. Using somewhat different approaches, Anderson and Smrekar (2006) find an average crustal thickness on the order of $\sim$10–20 km and Jimenez-Daiz et al. (2015) find $\sim$20–25 km. James et al. (2013) employed a different method: analysing geoid to topography ratios and performing a two-layer layer inversion of crustal thicknesses and mantle forces to derive smoothed maps of global crustal thickness and dynamic mantle forces. They assumed a uniform elastic thickness of 20 km, which is reasonable (see the next subsection). They estimated a mean crustal thickness of $\sim$8–25 km with the assumption that crustal thickness is non-zero everywhere and is limited by the transition of basalt to eclogite at $\sim$60–70 km.

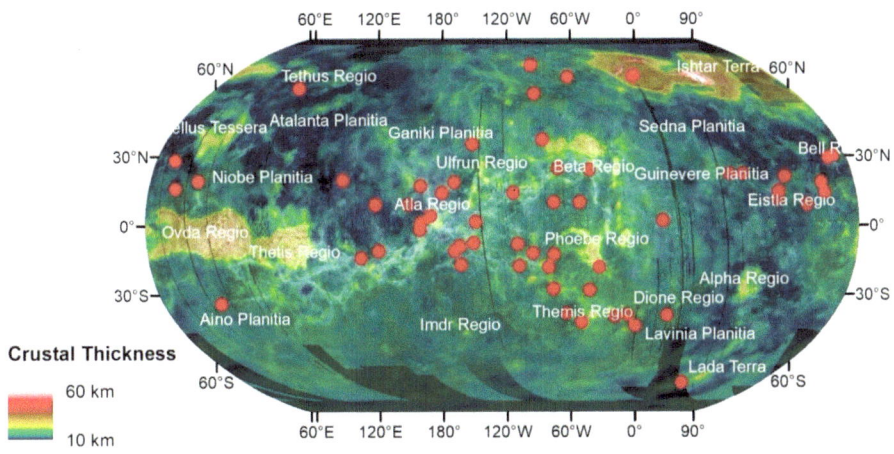

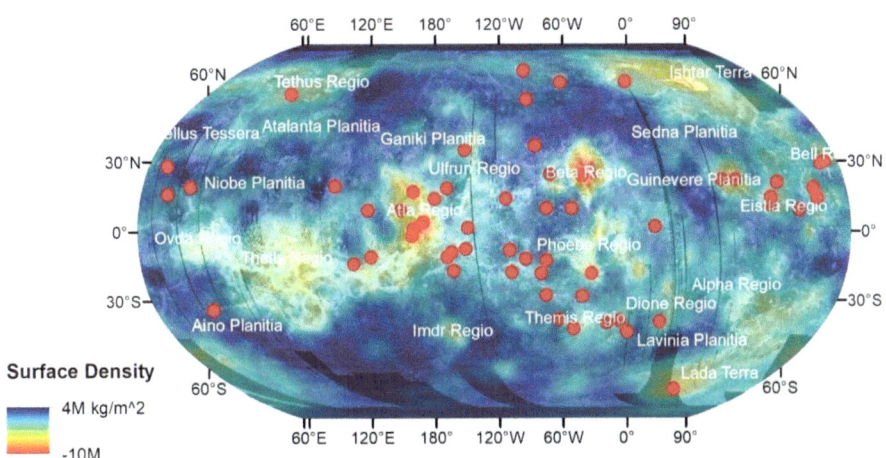

Fig. 11 Two-layer solution of James et al. (2013) superposed on Cycle 1 Magellan imaging of Venus. Top panel shows crustal thickness in km, and bottom panel density contrast in the uppermost mantle in kg m^{-2}. In the bottom panel, these Regios have been identified as hotspots: Imdr, Atla, Beta, Eistla (Western, Central, Eastern), Themis, Bell, Dione and Lada Terra (Stofan et al. 1995). Red dots show locations of large volcanoes in the database of Crumpler et al. (1997) with diameters >500 km. Among the many relationships between gravity, topography, and geology observable in these maps, large volcanoes are most often found in association with mantle upwellings or major rift zones and mostly avoid the areas of thickest crust, particularly tessera regions

Their calculated mean would increase to ~45 km if the Maxwell Montes formed tectonically in relatively recent times—meaning that they can be excluded from the assumption that crustal thickness cannot exceed ~70 km because of the basalt-eclogite phase transition has not reached completion, and/or the eclogite has not delaminated (e.g., Namiki and Solomon 1993; Jull and Arkani-Hamed, 1995).

Relatedly, convection models use gravity and topography to estimate a combination of crustal thickness and mantle properties (see Rolf et al. 2022, this collection, for more details). Such models typically focus on the longest-wavelength portion of the gravity spectrum, and tend to yield larger estimates of crustal thickness. Steinberger et al. (2010) found an average thickness of ~60 km using the range of the observed gravity spectrum that is likely dominated by crustal sources (i.e., above spherical harmonic degree 40). Wei et al. (2014) used models of Venus' mantle convection to include dynamical (mantle flow-sourced) contributions to derive a crustal thickness range of ~28–70 km, with the greatest values under highland regions such as Ishtar Terra and Aphrodite Terra.

4.2 Lithospheric Thickness and Heat Flow

A planet's lithosphere is the outermost part of the solid body. We can distinguish between three definitions of the lithosphere. First, the elastic lithosphere is the calculated thickness of a layer that is assumed to exhibit only elastic behavior. Second, the mechanical lithosphere represents a rheological layer that is too strong to convect even over long timescales. It encompasses the elastic lithosphere, and can additionally support loads via viscous deformation. Third, the thermal lithosphere contains the mechanical lithosphere and is defined as the layer in which heat is transported by thermal conduction, not convection. The thermal lithospheric thickness is usually defined by the depth to the mantle convection temperature, often taken as ~1575 K. In contrast, the crust and the uppermost mantle layers are defined by their composition. Composition implies a specific rheology, which has implications for the strength of the viscous portion of the mechanical lithosphere. The boundary between the crust and the uppermost mantle (or lithospheric mantle) may fall anywhere within these three layers, depending on parameters such as thermal gradient, strain rate, and rheology of the crust versus uppermost mantle.

The thermal lithosphere is essential to models of Venus's evolution. As the boundary layer, or 'lid,' of the convecting mantle, it couples the thermal state of the convecting interior to the surface. The heat transport across the thermal lithosphere limits the amount of heat leaving the mantle, determined by the thermal conductivity of mantle rocks, as well as the thickness and the temperature contrast across the lithosphere (i.e., Fourier's law). Both the temperature contrast and the thickness strongly depend on Venus' thermal history and its geodynamic regime (e.g., Rolf et al. 2022, this collection). The present thermal lithospheric thickness thus provides useful constraints on estimates of the thermal state of Venus' interior and how it may have evolved to this state.

Although convection models directly predict the thickness of the thermal lithosphere, it is the elastic lithospheric thickness that we can estimate more accurately. The wavelength at which the elastic lithosphere deforms in response to loads depends on its thickness. As discussed above, observations of gravity and topography thus yield quantitative estimates of the thickness of the elastic lithosphere. In idealized elastic models, lithospheric bending gives rise to internal stresses that increase linearly from zero at the mid-point of the elastic lithosphere and reach maximal values at its upper and lower boundaries. However, in reality, brittle failure and ductile flow limit the maximum possible stresses at the top and bottom of the mechanical lithosphere, respectively. Although the bottom of the mechanical lithosphere is viscous, it deforms only slowly and thus can also support surface topography or subsurface loads.

Specifically, the mechanical lithosphere encompasses both elastically and viscously strong parts, as defined by a yield-stress envelope. For a given rheological model of the lithosphere, the elastic thickness can be used to determine the mechanical thickness, which in turn is tied to the lithosphere's thermal state (e.g., McNutt 1984). The mechanical lithosphere is always thicker than the elastic lithosphere—but the total bending moments of both

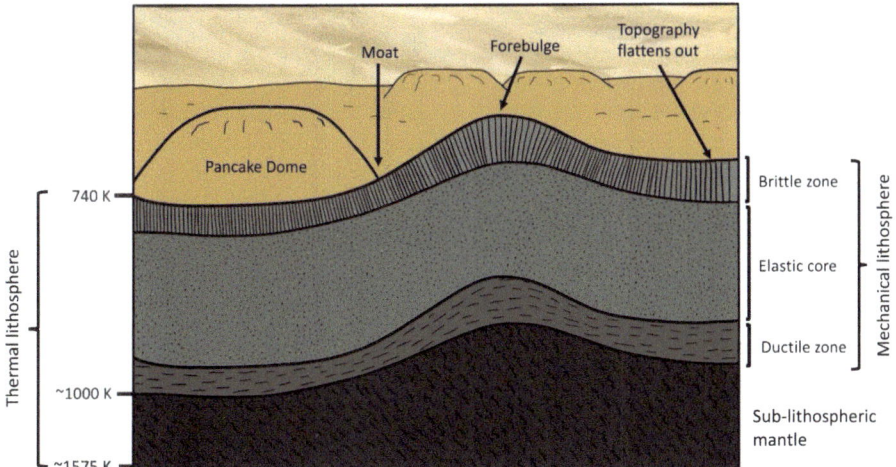

Fig. 12 - Volcanic features on Venus (here, a pancake dome) impose loads on the lithosphere. Even in the absence of gravity data, the topographic response of the lithosphere to these loads places constraints on models of lithospheric thickness. Rheological models enable conversions between inferred elastic and mechanical thicknesses—and then from mechanical thickness to the vertical thermal gradient and thus heat flow. Modified from Borrelli et al. (2021) with illustration by Joanna Wendel

(i.e., integrated over their yield strength envelopes) are equal. As shown in Fig. 12, the bottom of the mechanical lithosphere is associated with the temperature above which the ductile strength drops below a critical value (e.g., ~50 MPa). The corresponding critical temperature depends on the assumed rheology and strain rate but is typically ~1000 K for models with rheologies of dry olivine or diabase (e.g., Molnar, 2020). Stresses in the viscous layer can relax with time, such that the thickness of the mechanical layer can decrease with time. The thickness of the mechanical lithosphere can be used to estimate the thermal gradient due to the strong dependence of rock strength on temperature. Finally, the surface heat flow equals the product of the thermal gradient and the thermal conductivity of the lithosphere. Past papers have used a range of thermal conductivity values, from ~2–4 W m^{-1} K^{-1} (e.g., Brown and Grimm 1996; Johnson and Sandwell 1994; Sandwell and Schubert 1992b; Solomon and Head 1982; Reese et al. 1998; O'Rourke and Smrekar 2018; Russell and Johnson 2021; Borrelli et al., 2021; Smrekar et al. 2023). Thus, comparing heat flow values from different studies requires normalizing them to a given thermal conductivity. In this paper, any reference to a heat flow value is normalized to a value of 3 W m^{-1} K^{-1}.

Strain rate is an essential parameter for calculating the thickness of the mechanical lithosphere and thus thermal gradient and heat flow. Similarly, strain rate directly influences calculated yield stress values used to approximate the complex deformation of the lithosphere in models of convective regime, as it determines whether an active or stagnant lid is predicted. At present there are no direct measurements of strain rates for Venus. Currently the only potential method of accurately constraining local strain rates under consideration for Venus is repeat-pass interferometry (RPI), which could measure active deformation on the timescale of years (Hensley et al., 2022). The strain rate for terrestrial intraplate deformation of ~10^{-16} s^{-1} is often applied to Venus as a best guess (e.g. Johnson and Sandwell 1994). However, numerous Earth studies have shown that, in actively deforming regions, and especially volcanically active regions, strain rates can be orders of magnitude higher (e.g., Fagereng and Biggs, 2019; Molnar, 2020) and these locally higher rates should also be considered for the Venus lithosphere. The implication is that the strength of the lithosphere is likely higher during deformation than previously recognized. Ultimately, based on the effect of strain rate, lithospheric yield strength may be frequently underestimated.

Models of topographic profiles exhibiting flexural bending have been applied to deformation at a variety of features on Venus, including volcanic domes, coronae, and ridge belts. Interestingly, no flexural signatures have been found at larger volcanoes, possibly due to volcanic flooding of the moats (e.g., McGovern and Solomon, 1997). Most coronae have an elastic thickness of 20 km or less, with a handful of greater values mostly at larger coronae (Johnson and Sandwell 1994; Schubert and Sandwell 1995; O'Rourke and Smrekar 2018). Borrelli et al. (2021) examined topographic flexure at steep-sided volcanic domes (diameters of ~20–40 km). Topographic profiles at 20% of the ~60 domes on Venus show evidence of flexure, with most values of elastic thickness in the range from ~15–40 km. For some domes and small-scale features, topographic resolution may be a factor. McGovern et al. (2013) used magma-ascent models to predict that volcanoes may tend to form on thicker lithosphere than typical coronae. The lithospheric thicknesses estimated by Borrelli et al. (2021) generally support this hypothesis. Russell and Johnson (2021) estimated an elastic thickness of ~3–9 km for a small dome that sits on the fracture annulus of Aramaiti Corona. They advocated for the use of a lower Young's modulus and the resulting relatively low elastic thickness due to the presence of major fracturing in the annulus, as well as the best location of flexure per the location of fractures surrounding the dome. Borrelli et al. (2021) also noted that steep-sided domes in the vicinity of coronae tend to have lower elastic thickness than domes elsewhere.

Collectively, these studies found a wide range of heat flow values. The estimates of heat flow associated with coronae are typically $\sim >50$–75 mW m^{-2}, with a range of ~15 to >200 mW m^{-2} (Johnson and Sandwell 1994; Schubert and Sandwell 1995; O'Rourke and Smrekar 2018; Smrekar et al. 2023). Steep-sided domes typically have a lower heat flow than common values for coronae (e.g., ~50 mW m^{-2} for domes that are not near coronae; see Borrelli et al., 2021; Russell and Johnson, 2021), unless they are themselves near coronae. Additionally, most of the coronae identified as likely subduction zones (Schubert and Sandwell 1995) generally have greater elastic thicknesses and lower heat flows than average coronae values. Finally, the modelling of tectonic and impact features can also provide estimates of thermal gradient and heat flow. In a recent example, Bjonnes et al. (2021) simulated the multi-ring structure of Mead crater and found that the spacing of the ring faults can be explained by a thermal gradient of up to ~14 mK m^{-1} (or a heat flow of ~42 mW m^{-2}, for a conductivity of 3 W m^{-1} K^{-1}), indicating a locally thick lithosphere at the time of formation. In contrast, models of tectonic deformation at other locations point to a very thin lithosphere and high heat flow (e.g., Smrekar and Solomon 1992; Raitala et al. 1995; Ruiz, 2007). As on Earth, the lithospheric thickness and heat flow are likely variable in time and space on Venus.

Elastic lithospheric thickness can also be estimated from the ratio of gravity and topography, or the admittance, due to the wavelength dependence of the deformation response. This method has been applied at numerous highland regions, typically returning estimates of elastic thickness of ~20–30 km (e.g., Smrekar 1994; Simons et al., 1997)—hence the assumption of ~20 km on average by James et al. (2013). Global admittance maps (Anderson and Smrekar 2006; Jiménez-Díaz et al. 2015) find a range of elastic thicknesses of ~0–94 km. Jiménez-Díaz et al. (2015) derive an average value of ~30–40 km, although this value is likely a bit high based on their approach of limiting the minimum value to 14 km.

Heat flow estimates are a strong constraint on the models of convective state and thermal evolution of a planetary body (see Rolf et al. 2022, this collection). Models of mantle convection estimate thermal lithospheric thickness using geoid and topography data and magmatic activity as constraints (see Rolf et al. 2022, this collection). Using geoid–topography ratios, the average thermal lithospheric thickness may be 200 to 400 km, with reduced thickness below volcanic highlands (Moore & Schubert, 1997; Solomatov and Moresi, 1996). Larger average thicknesses of up to $\sim$600 km may be feasible if geoid and topography can be explained by the thermal isostasy of a stagnant lid (Orth & Solomatov, 2011). The unknown temperature and viscosity of Venus' mantle allow for a very wide range of values from such models.

In contrast, a thinner thermal lithosphere of $\sim$100–150 km is required to permit pressure-release melting at Venus hot spots (e.g., Smrekar & Parmentier 1996; Nimmo and McKenzie 1998), and to agree with estimates of heat flow (Smrekar et al. 2023). These estimates are compatible with mantle temperatures calculated from inferred compositions of basalts at the surface (e.g., Nimmo, 2002; Lee et al. 2009; Filiberto, 2014). Estimates of the mantle potential temperature (T_P) (i.e., the mantle adiabat projected up to the surface) have been derived based on data from Venera 13 ($T_P = 1732 \pm 73$ K) and Venera 14 ($T_P = 1603$ K, 1643 $\pm$ 70 K, 1732 $\pm$ 101 K)—within the range of ambient conditions of Earth (i.e., $T_P = 1623 \pm 50$ K). Of note, estimates based on measurements from Vega 2 ($T_P = 2051 \pm 167$ K) may be substantially higher but could be within the range of Earth's ambient mantle, too (e.g., McKenzie et al. 1992; Lee et al. 2009; Weller et al., 2015; Shellnutt 2016).

On Mars, where surface ages span several Gyr and flexural features can be dated back to the first $\sim$1–3 Ga, it is possible to show a history of cooling with time based on increasing lithospheric thickness with the age of the terrain (e.g., McGovern et al., 2004). The inability to date individual areas on Venus means that different lithospheric thickness on Venus globally could reflect spatial or temporal variations. Given Venus' young surface age, the average lithospheric thickness may not have increased significantly over the last $\sim$500–1000 Myr. The question remains as to whether the estimates of elastic thickness and heat flow reflect the present-day lithosphere or the time of lithospheric loading. In areas of thicker lithosphere, the loading signature from the time of loading may be 'frozen in' even if the lithosphere continues to cool and thicken with time. Smrekar et al. (2022b) argue that some regions with thin lithosphere and high heat flow, such Parga Chasma, are likely to represent areas with high heat flow today. Even given some uncertainties, lithospheric thickness and heat flow can inform models of the present-day lithosphere and interior (e.g., Rolf et al. 2022, this collection).

5 Models for the Formation of Volcanic and Tectonic Features

Understanding the evolution of Venus and planets in general requires modelling of specific geologic processes to explore plausible conditions in the mantle (potential temperature, decompression melting including composition and volatile content, etc.) and the lithosphere (mechanical thickness, strength, thermal gradient, stress state, etc.). The key constraints described above form the recent or present conditions that evolutionary models should match. In this section, we discuss models of the origin of specific features and the conditions implied over the resurfacing age of Venus up to the present. We then use all these inputs to explore the hypothesis that conditions on Venus currently are much like those on early Earth. Rolf et al. 2022 (this collection) examines models of Venus' interior evolution, including the deep parts of the planet, across a billion-year timescale.

5.1 Formation of Volcanic Features

Despite their broad similarities, large volcanoes on Venus appear to have some differences in their formation relative to their Terran and Martian counterparts (Fig. 1). Higher surface temperatures and pressures on Venus today yield smaller edifice heights than would otherwise be the case (e.g., Head and Wilson 1986; Bridges 1997; Head and Wilson 1992). Higher surface temperatures lead to smaller thermal gradients, shallower magma sources, hotter erupting lava, and therefore lava flows that spread farther laterally than on Earth (e.g., Flynn et al. 2023). The capacity of the thick, dense atmosphere to transport heat by radiation and convection also affects the spreading of lava flows (e.g., Snyder 2002; Flynn et al. 2023). The high surface pressure of Venus was also found to inhibit magma volatile exsolution, leading to fewer gas bubbles and thus potentially to less explosive volcanic eruptions than on Earth (e.g., Airey et al. 2015).

5.1.1 Plains Topography and Volcanism

Most of Venus' surface consists of volcanic plains. The plains are defined as regions of low elevation, low relief topography. As described above, there are a wide range of volcanic features in the plains. Studies to date have examined the formation of these plains via investigating the origin of the low lying topography as a possible result of thermal isostasy, and via numerical models that predict either global or regional volcanism. Globally, Venus' hypsometry is largely unimodal, lying within ± 1 km of the mean planetary radius (6051.8 km, although see Smrekar et al. 2018 for a discussion of elevations above 2 km). The hypsometry of the lowlands—and thus of the volcanic plains—is well-fit by a straight line (Rosenblatt et al., 1994), indicating that these regions are dominantly due to thermal isostasy. For an Earth-like mantle temperature, a thermal isostasy model applied to Venus gives a thermal lithospheric thickness of ~ 90 km. Means other than plate tectonic spreading centres can produce thermal isostasy (e.g., Morgan and Phillips 1983; Rosenblatt et al. 1994). For example, Morgan and Phillips (1983) showed that hot spot volcanism could produce the hypsometry of the plains. Ghail (2015) explained the plains topography as lithospheric rejuvenation—proposing that Venus may have a thin crust underlain by a localized asthenosphere activated by the presence of CO_2 in regions where the heat flow is relatively high, such as at mantle upwellings, plumes, or hot spots. Local topographic deviations from the global pattern of up to of several 100 m, such as those in Baltis Vallis, may be due to current convection-driven dynamic compensation of the topography (e.g. Conrad and Nimmo 2023).

Numerous geodynamic models predict widespread volcanism, or plains volcanism, and resurfacing, as discussed in detail in Rolf et al. 2022. These models predict volcanism and crustal thickness as a function of the lithospheric thickness, mantle temperature, and volatile content. We describe some example models here for completeness. In terms of explaining the occurrence of global plain volcanism on Venus, Parmentier and Hess (1992) and Head et al. (1994) presented a model linking the vertical accretion of a basaltic crust and the subsequent delamination linked to the emplacement of widespread volcanic plains. A 'catastrophic' tectonic resurfacing model initiated by delamination of negatively buoyant, depleted mantle was proposed to cause upwelling of warm, fertile mantle and pressure-release melting to produce extensive surface volcanism. Recent geodynamic modelling work by Adams et al. (2022) also focused on lithospheric delamination, in a 'peel-back' tectonic style, to be followed by the emplacement of hot, buoyant asthenosphere beneath the crust that may give rise to regional-scale volcanism. This style of resurfacing is proposed to occur on a more regional scale rather than a 'catastrophic' one.

Some geodynamic models focus on the role of mantle upwellings and decompression melting to understand the origins of plains volcanism. Reese et al. (2007) suggest that the lithosphere has thickened over the last 1 Ga, causing a decrease in melting consistent with the impact crater population and resurfacing history. Smrekar and Sotin (2012) focused on modeling the formation of mantle plumes capable of producing the hotspots. As discussed above, available thermal emissivity data from VIRTIS, as well as Magellan radar emissivity measurements, suggest that recent, large-scale volcanism is confined to hotspot/mantle plume settings. If this observation is borne out by future missions, it may turn out that tiny amounts of volatiles from the lower mantle are required to enable decompression melting in mantle plumes. Localized upwelling confined to the upper mantle may not be capable of crossing the solidus today due to desiccation of the upper mantle by prior, widespread melting (e.g., Smrekar and Sotin, 2012). This hypothesis is potentially consistent with the interpretation of Ar isotope data as indicating that Venus' interior volatiles are $\sim$25% outgassed (e.g., Kaula, 1999; O'Rourke and Korenga, 2015), if the upper mantle represents $\sim$25% of the total mantle volume.

5.1.2 Lava Channels

Models of canali and sinuous rilles mainly focus on explaining their observed morphology, e.g., meander wavelengths, from Magellan radar images. Kargel et al. (1994) applied a scaling law, which describes the relationship between the meander wavelength and discharge at bank-full stage for streams on Earth, to estimate lava volumes required to form canali. These authors also calculated the duration of the lava flows from a simple estimate of how quickly canali meander could migrate, and found that a typical canale could form from $\sim 10^{13}$ to 10^{15} m^3 of lava, which could cover up to $\sim 2.5 \times 10^6$ km^2 with $\sim$500-m thick flows. If the lava contained $\sim$50 vol% carbonate minerals, the degassed CO_2 would have between 0.05% and 5% of the total mass of the present-day atmosphere. Perhaps older "generations" of canali once existed, contributing even more CO_2 to the atmosphere, but are not visible on the modern surface. Currently, Venus' atmosphere contains nearly as much carbon as exists in Earth's lithosphere and atmosphere (e.g., Lécuyer et al. 2000). However, studies have not yet fully elucidated how crustal or mantle differentiation on Venus could have produced huge volumes of carbonatite lava.

Few studies have attempted to estimate the depth profiles of lava channels on Venus because of the difficulty in making these estimates from the $\sim$10–30 km horizontal footprint of Magellan altimetry data (Ford et al. 1993), which is substantially wider than the widths of most channels. However, post-Magellan studies have used radar clinometry at Baltis Vallis (Oshigami and Namiki, 2007) and six sinuous rilles and two valley networks (Oshigami et al. 2009) to obtain depth profiles that were resolved along the length of the channel.

One-dimensional models of channel formation by thermal erosion have been developed for application to Venus and other planets (e.g., Huppert and Sparks, 1985; Williams et al., 2001; Honda, 2005). Briefly, these models solve the heat balance for a lava flow to determine how far it can travel before solidifying given ground slope, the properties of the substrate, and lava composition, rheology, and thickness. For a certain type of lava, the thickness of the flow is adjusted to reproduce the observed length of the channel. The depth of the channel at the source is divided by the predicted vertical erosion rate to determine the effective duration of the lava flow. Oshigami et al. (2009) used this type of model to estimate that basaltic lava flows with initial thicknesses of $\sim$2–6 m and durations of $\sim$6–25 months could explain the lengths and depth profiles of six sinuous rilles, assuming present-day surface temperatures. The total lava volume associated with a typical sinuous rille on Venus is thus $\sim 10^{12}$ m^3, $\sim$1–3 orders of magnitude less than that suggested for a typical canale.

Existing modelling studies of lava channels on Venus have several limitations. First, they assume that Venus' surface temperature has remained constant, but it could have been higher or lower in the past. Second, models of sinuous rilles have not tested lava compositions other than Earth-like basalt. Third, no erosion models have been applied to canali. Since some proposed canali-forming lava compositions (e.g., carbonatite) have a lower melting temperature than the eroded substrate (i.e., basalt), mechanical erosion is likely more important than thermal erosion. Analogous models have been applied to the erosion of bedrock by water in attempts to calculate the discharge rates and volumes required to produce fluvial channels on Mars (e.g., Kleinhans 2005). Developing hybrid mechanical and thermal models of lava on Venus would be very useful for studies of channel formation. Measurements of the depth profiles of all channels, the grain sizes of the flow deposits and surrounding material, and any flow thicknesses via stratigraphy would provide key inputs to these models. Finally, any new information about the composition of surface material would help discriminate between models that feature different types of lavas and substrate materials.

5.1.3 Volcanic Rises

Volcanic rises are one of the few feature types for which there is an almost universally agreed formation mechanism: hotspots. The ten or so large volcanic rises on Venus are characterized by broad topographic swells, large positive gravity anomalies, and the presence of large volcanoes or coronae (e.g., McGill 1994), all of which are characteristic features of hotspots on Earth (see Sect. 2.1.2). Models of the interaction of plumes with the lithosphere seek to produce similar topographic uplift and gravity signatures (Kiefer and Hager, 1991) and decompression melting (Smrekar and Parmentier, 1996) consistent with the presence of large volcanoes. These studies indicate a thermal lithospheric thickness of 100 to 150 km and upwellings from the core–mantle boundary that are consistent with hotspot characteristics. Modelling the initiation of mantle plumes at the core–mantle boundary as well as the observed number of hotspots (e.g., Smrekar and Sotin, 2012) provides important information about mantle convective processes, as discussed in Rolf et al. 2022 (this collection).

As described in Sect. 3.2, more recent modelling work focuses on the lithosphere flexure associated with large volcano emplacement. McGovern and Solomon (1997) noted the lack of flexural induced moats and extensional faults around large (>100 km diameter) volcanoes on Venus. They then used flexural models to calculate the volumes of extrusive lava flows required to fill the expected (but not observed) flexural moat. McGovern and Solomon (1997) concluded that large volcanoes and the filled moat are one structurally coherent volcanic unit, unlike those on Earth and Mars that are partially filled by mass wasting or edifice collapse. The interaction of stresses—related to magma chamber growth and magma ascent, plus lithosphere flexure from the growing edifice—favour lateral subsurface flow as edifices grow, creating oblate magma chambers and promoting dyking (e.g., Galgana et al., 2011; McGovern and Solomon, 1998; McGovern et al. 2013, 2014).

5.1.4 Coronae and Novae

The wide range of corona characteristics (Sect. 2.1.3) have spawned numerous interpretations and models. Most of these modelling studies succeed in recreating at least some of the corona fracture annulus, topographic profiles, and/or gravity signatures, and point to a progression of topographic shapes over time.

Early work on corona formation focused on their generally circular shape, topography, and associated volcanism to conclude that some form of mantle upwelling must be involved

(Barsukov et al. 1986; Basilevsky 1986). A common proposition is that coronae form in response to stresses developed above an upwelling mantle plume or diapir, followed by gravitational relaxation or collapse due to magma withdrawal (e.g., Stofan et al. 1991, 1992; Janes et al. 1992; Squyres et al., 1992; Koch, 1994; Koch and Manga 1996 Grindrod and Hoogenboom, 2006; Gerya, 2014; Gülcher et al. 2020, 2023). Smrekar and Stofan (1997) introduced the first 2D numerical model incorporating plume-induced delamination, wherein lithospheric downwelling occurs at the edges of the laterally spreading plume head. These plume-induced delamination models were better at predicting the observed corona topographies than previous modelling, and placed the varying corona topographies into an evolutionary sequence (Smrekar and Stofan 1997; 1999). Some larger coronae may possess delamination or subduction, described below.

An additional class of models relies on volcanic construction to form coronae. Dombard et al. (2007) argued that partial melting above transient mantle plumes that impinge only on the base of the thermal lithosphere can cause magmatic loading of the crust above. Subsequent lateral crustal flow may cause surface deformation, producing the elevated rim surrounding an interior depression commonly observed at coronae, as first described in a simple model by Stofan et al. (1991). McGovern et al. (2013) suggested that the vertical loads exerted on the lithosphere by large volcanoes could influence magma ascent pathways from the mantle to the surface, which in the case of low elastic thickness may produce annular ridges of volcanic material at the surface, matching the topographic signatures of some coronae. Moreover, for higher values of elastic thickness ($\sim$10–40 km), volcanoes are predicted to form instead of coronae. However, as discussed above, many larger coronae have elastic thickness of $\sim$20–40 km.

The first study to use two-phase flow to model the pressure-release melting process at coronae shows that topographic rims can be produced by uplift above the buoyant melt concentrations (Schools and Smrekar 2024). This model also predicts high strain rate shearing that indicates the location of fracturing at rims. This model shows good agreement with a recent study of the shape of corona rim topography and the relationship to fracture locations (Sabbeth et al. 2024). This model creates rims at an early rather than late stage of evolution, suggesting an alternate evolutionary sequence to prior models.

The first 3D numerical study on corona formation involved thermal mantle plume impingement into warm and thin Venusian lithosphere (Gerya, 2014), assuming a thin elastic lithosphere (Anderson and Smrekar 2006). The results of this study suggest that plume-induced convection in a weak, ductile crust may be a plausible origin for some small-to-moderate sized (<200 km) coronae on Venus. This process may be analogous to those believed responsible for the formation of ancient terrestrial gneiss complexes (Campbell and Hill, 1988). Gerya (2014) noted that the first stage of uplift over a plume predicts radial fracturing, similar to 'novae' that are sometimes associated with coronae. Indeed, some prior studies did suggest that novae represent either an early stage or a "failed" corona (e.g., Janes and Squyres 1995; Stofan et al., 1997; Krassilnikov and Head 2003). Ernst et al. (2003) describe numerous radiating fracture patterns on Venus that they compare to dyke swarms on Earth. Gülcher et al. (2020) expanded on 3D numerical studies of plume-induced corona formation, defining different plume–lithosphere interaction scenarios possible at large coronae, such as lithospheric delamination, subduction (see Sect. 2.1.3), and an underplated plume, dependent on lithospheric and mantle plume properties. Gülcher et al. (2023) model lateral variations in crustal and/or lithospheric thicknesses to create asymmetric coronae with a prolonged tectonic and magmatic lifetime to develop, highlighting the importance of lateral variations in lithospheric properties for geodynamics on Venus.

Alternatively, models involving lithospheric downwelling with no associated mantle plume, such Rayleigh–Taylor instabilities, may also account for some of the topographic characteristics of coronae (e.g., Tackley and Stevenson, 1991; Grindrod and Hoogenboom, 2006; Hoogenboom and Houseman 2006). In these models, the nominal instability is caused by a dense mantle lithospheric layer over a less dense asthenosphere. A hybrid hypothesis involves the interaction between a mantle upwelling associated with large rifts and an adjacent downwelling instability, which may account for the spatial association of coronae with rift zones (Piskorz et al. 2014).

Since many different corona formation models can reproduce at least some of the key corona features, and both up- and downwelling of different sizes are common within the mantle convection regimes that may prevail within Venus (see Rolf et al. 2022, this collection), there may be multiple plausible formation scenarios for the diversity of observed coronae. Additionally, different corona morphologies may also represent different stages in their evolution. In particular, Gülcher et al. (2020) found that the topographic profile displayed by a corona may be completely isostatically inverted when an active plume interacting with the lithosphere cools down and fully crystallizes over time. These authors' analysis reveals that observed coronae on Venus may fall on a spectrum between early-stage ("active") and late-stage ("inactive") structures. However, different evolutionary sequences are possible (e.g., Schools and Smrekar 2024).

5.2 Formation of Tectonic Features

A key aspect of all modelling efforts to date is that subduction under Venusian conditions seems to be short-lived and, in contrast to modelling investigations under Earth-like conditions (e.g. Gerya et al. 2015; Baes et al., 2016), self-sustaining subduction zone does not develop on Venus. The exact conditions and behaviour of slab break-up is model- and condition dependent. For example, in the analogue models of Davaille et al. (2017), the slab separates into segments due to tearing caused by the brittle elasto–plastic behaviour of the lithosphere, while in the numerical models of Gülcher et al. (2020, 2023) the slab can coherently neck and detach, or tear into multiple segments that will caused directional slab retreat and detachment, based on the rheological parameters used. Moreover, the densification of slab material by eclogitization parameterized in the models seems to be a key factor in driving crustal recycling into the deeper mantle (Gülcher et al. 2023).

5.2.1 Rifts and Radiating Fracture Networks

Stoddard and Jurdy (2012) compared topographic profiles of selected rifts and hotspots on Earth and Venus. They calculated correlation coefficients between each topographic profile along a rift and the average of all profiles from that rift. These correlation coefficients were much higher for terrestrial rifts than Venusian rifts—the latter appeared most similar to terrestrial slow spreading rifts. Ghail (2015) found that the topographic profile across Venusian rifts are consistent with different rates of extension below a thermally-induced crustal detachment. Tectonic fractures characterize 'slower' rifts (e.g., Diana Chasma), while 'faster' rifts (e.g., Parga Chasma) host large numbers of coronae.

As discussed in Sect. 3.2, Venus possesses over 100 giant radial fracture systems, with an average radius of ~325 km and a maximum of >2000 km (e.g., Grosfils and Head 1994a, 1994b). Moreover, more such systems are likely to be found with future, higher-resolution radar imagery (Ernst et al. 2003). Early studies noted the similarity of the radial fractures to radial dyke swarms on Earth (McKenzie et al. 1992; Grosfils and Head 1994a, 1994b;

Ernst et al. 1995), although some modelled their formation as faults formed from dome up-lift from a mantle upwelling (Stofan et al., 1991; Janes et al. 1992; Cyr and Melosh 1993), akin to novae formation. Conceptual and analytical models of the dyke formation mechanism (McKenzie et al. 1992; Grosfils and Head 1994a; Koenig and Pollard 1998) used terrestrial field and modelling work to estimate surface stresses and resultant strains on the Venus surface. Grindrod et al. (2005) found that the measured strain at four selected radial fracture centres is too large to be solely uplift related and must be primarily dyke driven, but concluded that the formation process is likely to be a combination of dyking and uplift. The numerical model of Galgana et al. (2013) accounted for both flexural stress from lithosphere uplift and the stress of magma chamber inflation. These authors found that edifice growth creates compressive stresses that eventually halt magma ascent and force existing, ascending dykes to grow laterally, thus becoming radial dykes. As previously discussed with corona models, it is probable that similar features such as radial corona fracture systems may have differing formation mechanisms dependent on emplacement conditions such as lithosphere thickness or strain from regional tectonics, and no single formati

5.2.2 Plume-Induced Subduction

The edges of some large coronae have been proposed to be sites of subduction (e.g., McKenzie et al. 1992; Sandwell and Schubert 1992b; 1995), with the concept of plume-induced subduction probably first described by Sandwell and Schubert (1992b) to explain coronae. Subsequent focus in the terrestrial community on understanding the initiation of subduction as the first step in developing plate tectonics led to a variety of models of plume-induced subduction (e.g., Ueda et al. 2008; Burov and Cloetingh 2010; Gerya et al. 2015). These models show how a long-lived, buoyant mantle plume can overcome the strength of and thus penetrate the lithosphere. The plume head then intrudes between the upper crust and dense mantle lithosphere, pushing the lithosphere downward into the asthenosphere, eventually initiating self-sustained subduction (Ueda et al. 2008; Stern and Gerya 2018).

Numerical explorations of this theory in 3D were undertaken to investigate subduction initiation by a thermal plume in the Archean on Earth (Gerya et al. 2015) and by a thermal-chemical plume on the modern Earth (Baes et al., 2016). Gerya et al. (2015) suggested that plume–lithosphere interactions on Archean Earth initiated subduction zones, which possibly led to the onset of plate tectonics. At a minimum, such complete breaks in the lithosphere are needed to allow individual plates to form. These workers proposed that a combination of three key physical factors is needed to trigger self-sustained, plume-induced subduction: a strong, negatively buoyant lithosphere, focused magmatic weakening leading to thinning of the lithosphere above the plume, and lubrication of the upper slab interface by hydrated crust (Gerya et al. 2015). The first and third factor may be (partially) absent on Venus (e.g. Huang et al. 2013). However, other means of weakening slab interfaces are possible in addition to hydrated crust, such as grain size reduction (Li and Gurnis 2023).

Gülcher et al. (2020) expanded on 3D numerical studies of plume-induced subduction, mapping out the dependency of different plume–lithosphere interaction scenarios on three key factors: plume buoyancy, lithospheric strength, and crustal thickness. Short-lived subduction episodes, leading to slab detachment, were identified in models that featured high plume buoyancy in combination with a low crust–mantle boundary temperature (by a relatively strong lithosphere through a colder lithosphere and/or a thin crust). For higher crust–mantle boundary temperatures (in that study found to be $\sim$1100 K), lithospheric dripping developed instead of retreating slab segments. Follow-up work (Gülcher et al. 2023) on 3D plume-impingement upon laterally changing crustal thicknesses (i.e., lowland transitioning into a plateau) confirmed the occurrence of a short-lived subduction arc on the lowland-side (thinned crust). Several key features of retreating subduction zones were reproduced in these 3D models, such as an outer rise surrounding a deep trench, a topographic feature observed at several Venusian coronae.

Theories of coronae formation involving plume-induced lithospheric subduction are supported by the successful reproduction of some of the tectonic features observed at several coronae margins in 3D analogue experiments (Davaille et al. 2017). They utilized analogue materials to represent the full range of structure and rheology for both the brittle lithosphere and convecting mantle, which is not achieved in numerical studies of corona formation. For example the lithosphere, which forms via drying of surface, has pervasive cracks that allow asymmetry to develop. In particular, this study is able to reproduce characteristic features observed at, for example, Artemis and Quetzalpetlal coronae, such as the arcuate outer rise surrounding a trench and rim. The outer rise of these coronae features extensional deformation, interpreted to result from bending of the subducting plate. Recent models of corona formation predict the locations of faulting where rims are flexed upward above buoyant partially molten regions at the base of the lithosphere (Schools and Smrekar 2024). In these models, topography surface faulting can be attributed to magmatic processes at depth due to plume evolution.

6 Is Venus the Archean Earth, Proterozoic Earth, or Something Completely Different?

Venus' apparent lack of modern Earth-like plate tectonics as well as its basaltic crust, mantle hotspots, and high surface temperature has led to tantalizing comparisons to Archean Earth (Morgan 1983; Hansen 2018; Harris and Bédard 2014). One challenge for assessing whether the Venus of today is a good representation of Earth of the past (i.e., Hadean and/or Archean) is that most of Earth's earliest history (i.e., 4.5-4.0 Ga) has been removed by plate tectonics, leaving primarily geochemical data from limited locations to piece together evidence of past processes. Similarly, very limited chemical data exist for Venus, and then only for a handful of locations. Venus and early Earth offer complementary, though very incomplete, views of the dominant processes on a pre-plate tectonic planet. Venus has been described as having vertical tectonics (Solomon and Head 1982; Phillips and Malin 1983; Morgan and Phillips 1983), dominated by mantle plumes. Similarly, Archean Earth was likely driven by vertical processes of mantle plume-initiated rifting and collision (Smithies et al., 2005a; Condie et al., 2016; Bédard 2018; Brown et al. 2020), and vertical return flow via lithospheric delamination (Johnson et al., 2014). As discussed below, this assumption is further supported by our understanding of Archean cratons and the known rock types and surface features of Venus (Hansen 2007, 2015, 2018; Harris and Bédard 2014, 2015).

6.1 Comparison with Archean Tectonic Systems

Little is known about the tectonic regime and development of the Archean (4.0–2.5 Ga) on Earth because the rock record is not widely exposed or preserved relative to younger terranes and cratons (Brown et al. 2020; Hawkesworth et al. 2020). There is debate on the timing of the initiation of plate tectonics and thus interpretations of a modern plate tectonic system cannot be robustly confirmed for the Archean (Condie and Kröner 2008; Stern 2008; Hamilton 2011, 1998). However, it is expected that the thermal regime under which the Archean crust developed had mantle potential temperatures (T_P) $\sim$300–500 K higher than ambient conditions today (1620 $\pm$ 50 K). There is almost no rock record for the Hadean (>4.0 Ga) as most of the information is inferred from detrital zircons, the Acasta gneiss, or isotopic model ages (Harrison 2009; O'Neil et al., 2012; Roth et al. 2014; O'Neil and Carlson, 2017; Reimink et al. 2020). Earth's Archean cratons are composed of two distinct belts or "terranes". One of these major terranes, the greenstone–granite belts, primarily record surficial rock sequences, whereas the other major terrane, the granulite–gneiss belts, record middle to lower crust metamorphic conditions (Condie 1981; Kröner 1985). Together, it is likely that greenstone–granite and granulite–gneiss belts represent a glimpse into the formation of primitive terrestrial crust or proto-continental crust (Smithies et al., 2005a; Thurston 2015; Bédard 2018).

Greenstone–granite belts are linear to curvilinear rock suites (i.e., $\sim$10–25 km wide, $\sim$100–300 km long, $\sim$5–30 km thick). These dimensions are similar to those of the many tessera inliers distributed around Venus (e.g., Ivanov and Head, 2011). Terrestrial greenstone–granite belts have a characteristic stratigraphy of volcanic and volcaniclastic rocks, and sedimentary rocks accompanied by granitic rocks including tonalite–trondhjemite–granodiorite (TTG) suites (e.g., Condie 1981; Anhaeusser 2014; Thurston 2015). All greenstone belts are metamorphosed, but the degree (i.e., granulite, amphibolite, greenschist, prehnite-pumpellyite facies) and type (i.e. regional, contact, retrograde) of metamorphism are unique to each one. The volcanic suites are divided into an older subaqueous lower komatiite–tholeiitic basalt series and an upper, younger (by 3 to 30 Ma) bimodal sequence composed of tholeiitic basalt and calc-alkaline basalt, andesite, and rhyolite (Anhaeusser 2014; Thurston 2015). Greenstone belt formation and origin (i.e., a plate-tectonic origin or not) has yet to be resolved. However, they may be analogous to oceanic plateaus, volcanic arcs, ophiolites, or flood basalt suites that, at some level, may involve a mantle plume, particularly with respect to the eruption of the lower ultramafic-mafic volcanic series (de Wit and Ashwal, 1995; Smithies et al., 2005a; Bédard et al. 2003, 2013; Bédard 2006; Condie and Benn 2006; Anhaeusser 2014; Thurston 2015). The calc-alkaline nature of the upper silicic volcanic rocks is evidence that favours a volcanic arc-like origin, but similar rocks can be found in extensional tectonic settings since the calc-alkaline signature is a consequence of oxidizing magma conditions (Scott et al. 2002; Wyman et al. 2002; Arculus 2003; Smithies et al. 2005b; Bédard 2018). Modern (Cambrian to present) examples of greenstone-like belts exist and are associated with subduction and extensional tectonic settings but lack komatiitic rocks and banded-iron formations (Turner et al. 2014; Shellnutt and Dostal 2019).

At first glance, the topography of Venus resembles the continental and oceanic crustal dichotomy on Earth, in which tesserae are representative of 'continental' crust and the plains 'oceanic' crust (e.g. Smrekar et al. 2018). Near-infrared mapping spectrometer data suggest that, at least in one location, the plains and tesserae are compositionally different (Hashimoto et al. 2008; Mueller et al. 2008; Gilmore et al. 2015). Crustal thickness estimates indicate that plains are possibly $\sim$10–20 km thick, whereas the tesserae may be $\sim$65 km thick or more (see Sect. 3.1). These estimates are within the range of crustal thicknesses estimated for greenstone–granite belts and granulite–gneiss belts. The inferred compositional differences are consistent with the differences between terrestrial oceanic and continental crust. The major elemental composition of basalt measured at the Venera 14 landing site is very similar to olivine tholeiite of Archean greenstone belts, and the estimated primary melt compositions are ultramafic but not komatiitic (Shellnutt 2016, 2021). In comparison, the estimated composition of the rock at the Venera 8 landing site could be granodiorite/dacite and broadly resemble the calc-alkaline silicic rocks from the bimodal volcanic sequence of greenstone belts. However, uncertainty in its composition means that the Venera 8 rock could be lamprophyric (Nikolayeva 1990; Basilevsky et al. 1992; Kargel et al. 1993; Shellnutt 2019). Mantle potential temperature estimates of Venusian basalts are variable: the rocks at the Venera 13 and Venera 14 sites are probably derived from a thermal regime similar to modern Earth, but estimates of the thermal regime for rocks at the Vega 2 landing site indicate an anomalously hot (2051 ± 167 K) regime (Lee et al. 2009; Weller and Duncan 2015; Shellnutt 2016).

There is debate on the precise composition of the tesserae, possible depositional processes, existence of a hydrosphere, formation of coronae, the mantle thermal regime, and the tectonic regime that was operating on ancient Venus. Recent studies indicate that the tesserae could be composed of mafic volcanic-sedimentary sequences rather than intermediate to silicic volcanic or plutonic rocks (Byrne et al. 2021). No *in situ* geochemical data exist for tesserae, and the high U (2.2 ± 0.7 ppm), Th (6.5 ± 2.2 ppm), and K_2O (4.0 ± 1.2 wt%) contents reported at the Venera 8 landing site could be indicative of a lamprophyre (alkali basalt) rather than a dacitic/syenitic rock (Basilevsky et al. 1992). It is possible that Venus had a hydrosphere during the Archean, but lithified sedimentary rocks (or their metamorphic equivalents) derived by mechanical weathering and chemical precipitation have not been verified (Florensky et al. 1977; Basilevsky et al. 1985). At present, the only evidence for the existence of sulphur-rich evaporites (e.g., gypsum, anhydrite) is the high SO_3 (4.7 ± 1.5 wt%) content measured at the Vega 2 landing site, but sulphates may form by weathering under current Venus conditions (Fegley and Prinn 1989; Fegley et al. 1997; Bullock and Grinspoon 2001; Dyar et al. 2021).

Earth's continental crust began to form during the Archean. One key reason why determining whether or not tesserae are true analogues of continents is that massive volumes of felsic crust require basaltic melt extraction from the mantle, ideally in the presence of water (e.g. Campbell 2002; Bonin 2012). Thus, tesserae may host the geochemical fingerprints of past water. This is a critical question not only for understanding the divergent evolution of Venus and Earth, but also for establishing Venus' potential habitability (Westall et al. 2023). Venus is the only other rocky body in the Solar System with a possible substantial volume of silicic crust (Shellnutt 2013; Wang et al. 2022). The formation and erosion of Earth's continental crust into the oceans is proposed to be the source of the elements needed for life to flourish (Duncan and Dasgupta 2017). In addition to the plume-induced subduction mechanism discussed above, some coronae types may form via delamination (Hoogenboom and Houseman 2006), possibly assisted by mantle flow due to extension (Piskorz et al. 2014). Moreover, Venus' large-scale ridge belts may be a result of compression above mantle downwelling (e.g. Zuber 1990). Thus, a variety of features on Venus may be indicative of crustal recycling and large-scale mantle melting.

Venus' high surface temperature creates a hot lithosphere and provides a thermal/mechanical analogue to conditions on early Earth. Davaille et al. (2017) argued that the apparent plume-induced subduction seen at features such as Artemis and Quetzelpetlatl Coronae are more likely to form under conditions of hot mantle and moderately thin lithosphere, as inferred for Venus today and Earth in the Archean. Hotter mantle temperatures allow for more buoyant plumes able to break a moderately strong lithosphere. The lithosphere cannot be too thin, or it will be too buoyant to subduct (e.g., Gülcher et al. 2020). Although many agree that subduction, which creates a complete break of the lithosphere, is the first step in plate tectonics, the evolution to mobile plates is not clear. Bercovici and Ricard (2014) suggested that the reason that Venus has developed subduction but has not yet developed Earth-like mobile plates is that breaks created by subduction are annealed over time due to the high temperatures, preventing the formation of large plates. This inference is consistent with the presence of many but moderately sized discrete blocks of lowland lithosphere, which appear to have jostled together since, at least in places, the emplacement of the locally youngest plains materials (Byrne et al. 2021).

When and how modern (oceanic) plate tectonics started on Earth is debated. There are advocates for plate tectonics having always operated and those that think it began at $\sim$3.2 Ga, $\sim$2.5 Ga, $\sim$1.0 Ga, or $\sim$0.8 Ga (e.g., Condie and Kröner 2008; Stern 2008; Hamilton 2011, 1998; Kusky et al. 2018; Windley et al., 2021). The precise origin of greenstone–granite and granulite–gneiss belts is still debated, and there are compelling arguments for and against the operation of plate tectonic-related processes in their development (de Wit and Ashwal, 1995; Anhaeusser 2014; Thurston 2015). Nevertheless, it appears that the generation of highly differentiated continental crust was either absent, very slow, or stunted on Venus. Currently, the continental crust represents $\sim$41% of the surface area of the Earth and $\sim$0.7% of its volume. It has taken $\sim$4.5 Ga to create this volume of continental crust; however, although the rate of crustal growth throughout geologic time is uncertain (Honing and Spohn, 2016; Honing et al., 2019; Hawkesworth et al. 2019, 2020), we do know that this rate has not been constant. The three most common models for crustal growth are rapid development followed by steady-state; continuous growth; or episodic growth (Fig. 13). There are a number of uncertainties in these models, but perhaps the most important uncertainty is the timing of the initiation of modern plate tectonics (Honing and Spohn, 2016; Condie 2018; Windley et al. 2021). Assuming Venus is a close analogue of Earth and many fundamental planetary properties (e.g., thermal structure, abundances of heat-producing elements) are proportional to the size difference between the planets, then Venus' surface should have a similar area and volume of continental crust as Earth does today. Under the assumption that tesserae are similar to continental crust, then only $\sim$7.3% of the surface area of Venus is continental crust (Ivanov and Head 2011; James et al. 2013)—that is, only 17% to 20% of the expected continental crust is present on Venus taking into consideration differences in crustal thickness estimates and that no crustal recycling occurred. According to different terrestrial crustal growth models (Fig. 13), Earth produced 17% to 20% of its current continental crust during one of three different time periods (Fig. 5): the Hadean to Eoarchean ($\sim$4.4–3.9 Ga); the Mesoarchean to Neoarchean ($\sim$3.1–2.7 Ga); or the Paleoproterozoic ($\sim$2.4–2.1 Ga). Thus, either the processes of crustal evolution on Venus have operated significantly more slowly than on Earth, or the development of continental crust was arrested relatively early. Alternatively, Ivanov and Head (1996) suggest that felsic crust could underlie as much as $\sim$55% of Venus and that there may be only a surface veneer of basalt ($\sim$2–4 km thick). If this is the case, then Venus would have a larger surface area of continental crust than Earth.

Is Venus therefore analogous to Archean Earth? Possibly. However, our overall understanding of Venus' crustal evolution and overall geodynamics is in its infancy. A possible Venusian hydrosphere is currently only speculation. New geologic, geochemical, and isotopic surface measurements are required to thoroughly investigate the possibility that Venus can help us understand the nature and development of continental crust in the absence of modern plate tectonics, which may be, in turn, analogous to the Archean Earth.

6.2 Comparison with Proterozoic Tectonic Systems

By the end of the Archean, Earth was certainly developing its present-day dynamics. But there were also important tectonic, hydrospheric, atmospheric, and biological shifts that occurred during the Proterozoic that laid the foundation of Earth's present habitability. It is during the Archean–Proterozoic transition that the planetary evolution of Earth and Venus may have diverged. There are suggestions that by the Neoarchean (2.8 to 2.5 Ga), horizontal tectonic processes responsible for collision and accretion of island-arc-like terranes was becoming increasingly dominant (Zegers and van Keken, 2001; O'Neill et al. 2016; Windley et al., 2021). Once Earth transitioned into the Paleoproterozoic (2.5 to 1.6 Ga), some geologic

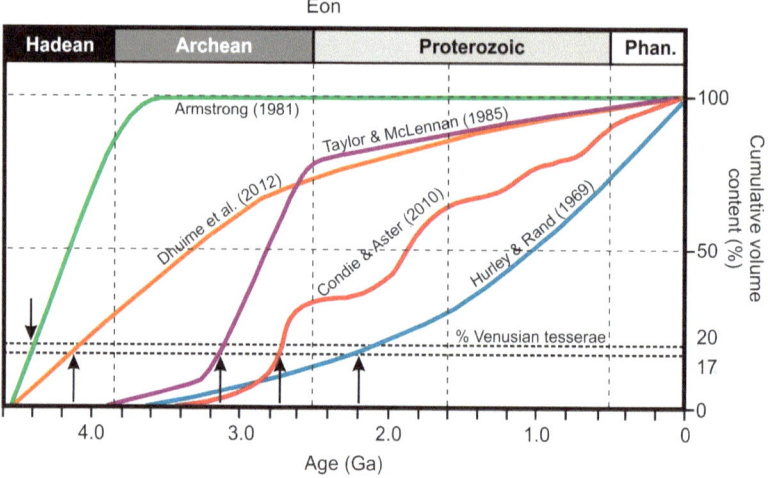

Fig. 13 - Terrestrial crustal growth curves showing the models that constrain the volume of crust in the Earth's past independent of present-day age distributions (modified from Cawood et al. 2013). The intersections (arrows) of the terrestrial crustal growth curves are shown at 17–20 vol.% crust to highlight the time periods when the amount of tesserae on Venus may have reached its present state if it evolved along the same path as Earth

features appeared (e.g., giant radiating mafic dyke swarms, anorthosite massifs, ophiolites) that were either rare or absent during the Archean.

A considerable proportion of the cratons that existed at the end of the Archean (∼2.7–2.5 Ga) apparently amalgamated for the first time to form a supercontinent or multiple large continental blocks (Bleeker 2002; Pesonen et al., 2003). The emplacement of giant radiating mafic dyke swarms (e.g., Matachewan, Mistassini, Dharwar, Fort Frances, Kikkertavak, Black Range) and the possible existence of ophiolites at ∼2.7–2.2 Ga indicates that a continent or continents broke up during the Early Paleoproterozoic (Cadman et al. 1993; Kusky et al. 2001; Ernst and Bleeker 2010; Sarma et al. 2020; Windley et al., 2021). Associated with these dykes are the correspondingly voluminous, and probably mantle plume-derived, flood basalt provinces that are collectively referred to as large igneous provinces (LIPs) (Ernst and Buchan 2001). The volcanic rocks of Paleoproterozoic LIPs are not commonly preserved due to erosion but their Neoproterozoic (∼1.0–0.54 Ga old) and Phanerozoic (∼0.54 Ga to present day) equivalents are better preserved. Archean giant radiating dyke swarms are rare. Their appearance during the latest Archean to earliest Paleoproterozoic marks a major shift in Earth's tectonic regime, becoming a common feature during the break-up of younger supercontinents (e.g., Rodinia, Pangea).

Complementary to the Paleoproterozoic dyke swarms and continental break-up events are well defined ophiolites that are within collisional and accretionary orogenic belts. The Purtuniq, Jormua, and Kandra ophiolites demonstrate that oceanic crust was obducted at active continental margin settings at ∼2.0–1.8 Ga (Scott et al. 1992; Peltonen and Kontinen 2004; Kumar et al. 2010). At ∼1.9–1.8 Ga, a major global orogenic cycle occurred that is evidenced by the number of contemporaneous orogenic belts (e,g., Trans-Hudson, Thelon, Svecofennian), followed by post-collisional granites and anorthosite massifs were emplacement into the roots of the old orogens and another cycle of plate break-up, collisional/accretion orogens, and post-orogenic magmatism (Nance et al. 2014; Mitchell et al.

2021). The break-up and collision of cratons is indicative of the supercontinent cycle, which has become a defining characteristic of modern plate tectonics on Earth.

In comparison to Earth, the corresponding Proterozoic Eon of Venus was probably less geologically eventful. Per our current understanding, Venus appears not to have undergone the transition from a non-plate tectonic regime to a plate tectonic regime. Consequently, accretionary belts, island-arc terranes, ophiolites, and accretionary prisms are not expected to be found on Venus. Although collision-related mountains exist (e.g., Maxwell Montes) and underthrusting of crust probably occurred (Ansan and Vergely 1995; Davaille et al. 2017), the processes of proto-continental crust formation via the development of Earth-like subduction zones, their volcanic systems, and accretionary processes seemingly did not occur (Nimmo and McKenzie 1998). The converse of this is true as well: MORB-like oceanic crust did not develop either. From the Th/U ratios of Venusian basalt, it is likely that the overall internal heat budget generated by radioactive decay is the same as or similar to Earth (e.g., Steinberger et al. 2010; Armann and Tackley 2012; Taylor et al. 2018). Tonalite–trondhjemite–granodiorite suites, anorthosite, and post-collisional granite may be present within the crust of Venus as they do not require special geologic processes in order to form (Gilmore et al. 2015; Shellnutt and Manu Prasanth 2021; Wang et al. 2022). For example, anorthosite can be produced by crystallization and accumulation of plagioclase from a mafic parental magma, whereas post-collisional granite can develop after a period of compressive crustal stress. Since basalt and mountains generated by compressive stress are known to exist on Venus, it stands to reason that anorthosite and post-collisional granite could be present within the crust of Venus. Moreover, large, ore-bearing layered igneous complexes like the Bushveld intrusion or Great Dyke of Zimbabwe should be present as well, because they principally require large volumes of primitive mafic and ultramafic magma to differentiate. It is possible that some lobate or rounded smaller coronae of Venus are analogous to Bushveld-type complexes, but the Sudbury Igneous Complex-type intrusions may be unique to Earth since it was likely derived, in part, by bolide impact-induced melting (Therriault et al. 2002).

The major uncertainties in the Proterozoic evolution of Venus are the development of the atmosphere, hydrosphere, and biosphere. The current CO_2-rich atmosphere of Venus may or may not be primary, but it is likely that Venus lost a substantial amount of H_2 and without becoming O_2-rich like Earth (Donahue et al. 1997; Shaw 2008; Lammer et al. 2018). The evolution of Earth's atmosphere is directly related to the hydrosphere and biosphere, since the evolution of anaerobic cyanobacteria likely increased the amount of oxygen, whereas their subsequent demise due to the conversion of atmospheric CH_4 to CO_2 and H_2O may have assisted, along with lower solar luminosity, in the development of the Paleoproterozoic glaciations and reduced photosynthesis until after glacial retreat (Kopp et al. 2005). Could a similar process have occurred on Venus only for it to revert back to a CO_2-rich atmosphere? It depends on whether there was a hydrosphere and if life (i.e., cyanobacteria) evolved to the point where photosynthesis started to change the composition of the atmosphere. Venus may have lost a huge quantity of water during its early evolution, and surface water might have been stable until as recently as one billion years ago (Donahue et al. 1997; Way et al., 2016; Way and Del Genio, 2020). If Venus did indeed have a vibrant, Earth-like hydrosphere until ~1 Ga, then prokaryotic and possibly early eukaryotic life could have evolved. Furthermore, a hydrosphere would have permitted the chemical precipitation of limestone, banded-iron formations, primary and secondary sulphate rocks, and the formation of hydrothermal and hydro-magmatic mineral deposits (sulphate materials, along with Fe-oxides and Fe-sulfides, might also result from likely weathering of current surface rocks, see Gilmore et al, 2023, this collection). Nevertheless, if there was a global hydrosphere, Venus did not sustain a tropopause 'cold trap', allowing the hydrosphere to evaporate and generating a significant volume of evaporites (i.e., carbonates, sulphates, halites). If all these events took place, how could the atmosphere become CO_2-rich again? The mostly likely explanation would be the devolatilization of carbonate and/or evaporite rocks due to contact metamorphism associated with magmatism (c.f., Aarnes et al. 2011; Pang et al. 2013). The possible implications of a Venusian hydrosphere are profound but purely speculative without evidence for chemically precipitated sedimentary rocks, glaciations, or hydro-magmatic mineral deposits. Although it is possible that Venus and Earth were similar during the Archean, they must have diverged by the Proterozoic. However, it is possible that life on both planets evolved in parallel until the loss of the Venusian hydrosphere.

7 Knowledge Gaps and Measurements Required to Fill Them

Venus is the least geologically understood of the terrestrial planetary bodies of the inner Solar System. Enormous knowledge gaps remain (Treiman 2007; Glaze et al., 2018). The thick, cloudy atmosphere of Venus is a formidable challenge to remote sensing of the surface, motivating the need for surface or near-surface measurements. New missions will greatly advance our understanding and lead to new questions about the evolution of Venus through time (Widemann et al. 2023, this collection, and references therein). Additionally, fundamental geophysical techniques of seismology, electromagnetic sounding, heat flow, and magnetometry have yet to be exploited for interrogating the interior.

7.1 Geochemistry

Venera landers provided in situ data at four sites, providing images, atmospheric composition, wind speed, temperature, and the composition of surface materials. Major elemental compositions were reported at three sites, and Th, U, K contents at two others, with the Vega 2 site providing the only measurements of both major and trace elemental compositions (Surkov et al. 1984, 1986; Kargel et al. 1993).

First-order geochemical measurements of the surface rocks and regolith are needed. Advances in X-ray fluorescence spectrometry, particle X-ray spectrometry, Mössbauer spectrometry, and sample preparation, as well as the development of laser-induced breakdown spectrometry (LIBS) and remote micro-imager (RMI) technology, permit smaller equipment to be included in surface landers (e.g., Gorevan et al. 2003; Treiman 2007; Clegg et al. 2009; Wiens et al. 2012). The measurement accuracy and precision of the various techniques would be a substantial improvement over the data reported from the Venera and Vega probes (c.f., Treiman 2007). The elements that can be measured range from major elements (Si, Ti, Al, Fe, Mg, Ca, Na, K, P) to many trace elements (Li, Sr, Rb, Mn). Depending on the rock type (ultramafic, mafic, felsic) additional elements (e.g., Ni, Cr, V) may be of a suitable concentration for measurements. The key problem with Venus is the harsh surface conditions (93 bars, 740 K), and any spectrometer would have to be properly prepared to overcome these extreme environmental conditions.

Of critical importance are the landing sites. The Venera and Vega probes landed at equatorial to tropical latitudes within 2 km of the mean planetary radius in eastern Aphrodite Terra and eastern to southeastern Beta Regio (Kargel et al. 1993). If possible, suitable landing sites should be considered across the various landscapes of Venus (in tesserae, coronae, and hotspots). The highland regions of Ishtar Terra, Aphrodite Terra, Beta Regio, and Lada Terra should be prioritized as likely preserving the oldest rocks and the greatest diversity of lithologies and compositions (e.g., Hashimoto et al. 2008; Gilmore et al. 2015,

2017; Treiman et al. 2016; Gilmore and Head 2018; Byrne et al. 2021; Resor et al. 2021). From the surface rocks and regolith, a number of important processes and geologic conditions can potentially be investigated, including crustal differentiation, magma crystallization and mineral chemistry, lava viscosity, eruptive temperature, formation of magmatic mineral deposits, mantle source composition, mantle redox conditions, mantle potential temperature, depositional environment, metamorphic facies (pressure, temperature, fO_2), stress and strain rates, as well as investigations of whether any evidence exists in support of ancient hydrological cycles, phreatomagmatic eruptions, hydro-magmatic and hydrothermal mineral deposits, water composition, paleoenvironment, and biosphere development. Clarity on any or all of these facets of Venus geology and geophysics would offer a huge advancement in our understanding of the interior properties and development of the planet.

Of equal importance to basic surface geochemistry, but technologically problematic to measure, is the isotopic systematics of Venusian rocks. Light and heavy stable isotopes (e.g., H, O, C, S, Fe, Zn, Mo) and radiogenic isotopes (e.g., Sr, Nd, Pb, Hf, Os) are invaluable for addressing the interior and surficial development of a planetary body. The stable isotopes offer insight into mass-dependent fractionation, primarily of atmosphere–lithosphere interactions that involve the hydrosphere, biosphere, mantle volatile budget, and climate, and can also be useful for assessing mantle evolution and mineral deposits (e.g., Hoefs 2009). The radiogenic isotopes of Venusian rocks offer insight to planetary accretion, interior differentiation of the core, mantle, and crust, and secular mantle and crust evolution (e.g., Faure 2001). Most of the isotopic systems require the application of relatively intense laboratory preparation prior to measurement and would not be practical for a surface lander. The return of surface samples or the identification of Venusian rocks in the meteorite collection are currently the only viable means to measure some or all of the stable and radiogenic isotopic systems. However, there are breakthroughs in the measurement of some radiogenic isotopes by in situ laser ($\sim$20–50 mm beam size) ablation methods (Bolea-Fernandez et al. 2016; Spencer et al. 2020). Sr and Hf isotopes can be measured with a suitable degree of accuracy by in situ methods with minimal sample preparation but rely on the presence of Rb–Sr- and Lu–Hf-rich silicate minerals such as feldspar ($KAlSi_3O_8$–$NaAlSi_3O_8$–$CaAl_2Si_2O_8$) and zircon ($ZrSiO_4$). If the in-situ measurement of Sr ($^{87}Sr/^{86}Sr$) and Hf ($^{176}Hf/^{177}Hf$, $^{176}Lu/^{177}Hf$) isotopic ratios can be achieved by a lander, such measurements would provide a major advance in the understanding of planetary accretion and internal differentiation of Venus.

Perhaps the most ambitious and desirable measurement that can be achieved is radioisotopic geochronology of rocks and minerals from the surface (e.g., Coleman et al. 2012; Cohen et al., 2019). Crater retention and distribution, and relative geologic relationships, are the only methods for estimating the average surface age (<1 Ga) and general stratigraphy (e.g., Strom et al. 1994; Ivanov and Head 2015), in which, for example, tesserae are usually regarded as older than the plains (e.g., Basilevsky and Head, 2002; Gilmore and Head 2018). However, radiometric dating is the only method that can return a measure of the absolute age of a rock. Over the past twenty years, in situ geochronology has exploded in the geologic literature and, on Earth, is a common, reliable, affordable, and low-intensity sample preparation method of choice (Spencer et al., 2016). Under ideal circumstances, a rock sample from Venus would be necessary for mineral separation and measurement but, it may be feasible to measure the isotopic ratios directly with minimal sample preparation by in situ methods. Specifically, U–Pb dating of zircon, a common accessory mineral of granitic rocks, can provide a robust measurement of the $^{207}Pb/^{206}Pb$, $^{207}Pb/^{235}U$, and $^{206}Pb/^{238}U$ ratios with an uncertainty in the range $\sim$2–4% (Spencer et al., 2016). Another possibility is the application of the Rb/Sr isochron method on minerals (e.g., feldspar, biotite, amphibole, pyroxene) using the measured $^{87}Sr/^{86}Sr$ ratio and Rb and Sr concentrations to estimate the $^{87}Rb/^{86}Sr$

ratio (Coleman et al. 2012; Bolea-Fernandez et al. 2016). Although the uncertainty of the Rb/Sr isochron method makes it less reliable, it would be useful (i.e., better than ±100 Ma) and in situ Rb/Sr geochronology may be more practical because the potential target minerals are common in mafic to silicic rocks. Regardless, the first step in the application of in situ geochronology methods would be the positive identification of differentiated igneous rocks (i.e., intermediate to silicic) by surface geochemical remote sensing.

7.2 Seismology

A better knowledge of Venus' seismicity would majorly improve our understanding of rock rheology, deformation mechanisms, and Venus geotherms, as well as help us constrain estimates of crustal thickness, mantle viscosity, and core properties. As described above, Venus' young surface has a wide range of volcanic and tectonic features capable of producing substantial seismicity. The spatial distribution of seismic events related to tectonic features, like faults and volcanoes, is much more homogeneous on Venus than on other planets, due in part to the extremely limited erosion. Seismicity estimates have assumed an intraplate level of seismic activity because Venus lacks Earth-like plate tectonics. As expected, this approach leads to a lower predicted rate of activity than on Earth (e.g., Stofan et al., 1993; Stevenson et al. 2015). Sources of uncertainty come from the limited constraints on heat flow and rock rheology discussed above. Lognonné and Johnson (2007) predicted >100 quakes >Mw5 per year, with a maximum seismic moment magnitude of ∼6.5 based on analogy to terrestrial intraplate activity. This low maximum quake magnitude calculation assumed a thickness of the seismogenic layer to be 30 km. Knapmeyer et al. (2006) estimated seismicity on Mars based on extensive (albeit hillshade-based) mapping of surface faults and their conversion into quake magnitudes. This approach, which has large error bars, predicts a range of seismicity consistent with seismic data from the InSight mission (Giardini et al. 2020). Sabbeth et al. (2023) applied a similar approach to a specific fault type on Venus, predicting an order of magnitude higher level of seismicity at Venus than InSight measured in Cerberus Fossae on Mars. This estimate was based on fault lengths. Future high-resolution topography would allow fault throw to be measured as well, providing greater fidelity to this approach. Seismic sources due to meteoroid impacts are likely to be rare because the thick atmosphere destroys impactors that produce craters less than a few kilometres in diameter (e.g., Zahnle 1992; Korycansky and Zahnle 2005). Airbursts due to impactor disruption and breakup in the thick Venus atmosphere could help provoke a seismic signal, although an impactor large enough to airburst in the lower atmosphere is statistically unlikely to hit Venus in the next several decades (e.g., McKinnon et al. 1997).

Although long duration, in-situ seismic measurements would be ideal, Venus' surface temperature motivates a range of possible methods, including orbital searches for atmospheric airglow or balloon-born infrasound measurements (Stevenson et al. 2015). If present, aerial platforms could detect infrasound from explosive volcanism (e.g., Brissaud et al. 2021; Garcia et al. 2022; Rossi et al. 2023). Pyroclastic flows, likely from explosive volcanism (e.g., Ganesh et al. 2021), appear to be rare but indicate that such eruptions have occurred on Venus (e.g., Campbell et al. 2017; Ganesh et al. 2022). See also a review of future seismic investigations and concepts in Widemann et al. 2023, Sect. 10 (This collection).

7.3 Electromagnetic Sounding

Future missions could use electromagnetic (EM) sounding to probe properties of the lithosphere that influence volcanic and tectonic processes. Grimm et al. (2012) proposed exploiting Schumann resonances at ∼10–40 Hz to determine the depth profiles of conductivity and temperature within the lithosphere. If the lithosphere is dry as expected, then EM sounding could probe depths of hundreds of kilometres on Venus. If the lithosphere is wet (i.e., with hundreds of ppm H_2O), then sounding would be limited to depths of <20 km—but the discovery of a wet lithosphere would be a major result by itself. Speculatively, EM sounding could also search for molten salt aquifers as proposed by Kargel et al. (1994). An aerial platform could perform EM sounding with instruments such as magnetometers and electrometers at ~55-km altitudes in the clouds (e.g., Wilson et al. 2012; Cutts et al. 2022). However, no previous mission has confirmed the presence of Schumann resonances.

On Earth, frequent lightning discharges (tens each second) dump so much EM energy below the ionosphere that Schumann resonances exist always around the globe (e.g., Grimm et al. 2012). Observations of whistler waves from Venus Express and Pioneer Venus Orbiter have been claimed to reveal that lightning happens on Venus several times more often than on Earth (e.g., Hart et al. 2022). However, these whistlers might have a non-lightning origin (George et al. 2023). Recently, the Parker Solar Probe mission observed whistlers at Venus during a gravity assist and measured their direction of propagation for the first time. Whereas lightning-generated whistlers would propagate outwards from the atmosphere, the observed whistlers travelled towards Venus from its nightside magnetotail (George et al. 2023). Multiple processes can produce whistlers at Venus, but no existing measurement provides proof of lightning. Long-duration observations from an orbiter would provide much better statistical constraints on the rate of lightning-derived whistlers.

Lightning could be absent or extremely rare on Venus (see also Lorenz 2018 for a comprehensive review). The Akatsuki mission (Lorenz et al. 2019; Takahashi et al. 2020) and earlier ground-based observations (Hansell et al., 1995) have identified <10 optical flashes total that are potentially attributable to lightning, yielding a rate of only a few flashes per hour on all of Venus—less than 10^{-5} times the rate of cloud-based lightning on Earth. Other processes may have produced some or all of these flashes, including meteors (e.g., Blaske et al. 2023) and instrument artifacts (Lorenz et al. 2018), meaning that the rate of cloud-level lightning on Venus could be (nearly) zero. In principle, explosive volcanism and/or aeolian processes on the surface could generate lightning, albeit likely not at high rates (e.g., Lorenz et al. 2018). Unlike lightning in the clouds, near-surface lightning probably would create EM signals but not optical flashes detectable via remote observation. Ultimately, our ignorance about the EM environment below the ionosphere is profound and coupled to our struggles to understand and explore volcanism on Venus generally.

7.4 Magnetometry

Crustal remanent magnetism has revealed the history of volcanism and tectonics on other terrestrial planets. Most notably, scientists found "smoking gun" evidence for plate tectonics in the 1950s in the form of "stripes" of magnetized crust painted with alternating polarities outwards from spreading centres in the seafloor. As described in the introductory article of this collection (O'Rourke et al. 2023, this collection), no mission has so far discovered any crustal remanent magnetism on Venus. However, the detection limits are so weak that huge amounts of magnetized crust could await detection (O'Rourke et al. 2019)—for example, equivalent to the magnetic anomalies detected in the crusts of Mercury, Mars, Earth, and the Moon. As with EM sounding, we cannot promise to use crustal magnetometry to study volcanism and tectonics at Venus. Perhaps Venus never had a strong, global magnetic field that would enable lava or magma to acquire remnant magnetization. Maybe surface temperatures were higher in the recent past (e.g., Bullock and Grinspoon 2001; Warren and Kite 2023), which would have demagnetized any magnetized rocks in the crust. Again, as for

EM sounding, an aerial platform could characterize the magnetic environment below the ionosphere of Venus for the first time—and determine if measurements that are so powerful on Earth and other planets are applicable to Venus.

7.5 In Situ Heat Flow

Heat-flow data would provide valuable new information on local thermal gradients and, coupled with crustal thickness and thermal evolution models, the abundance and distribution of radiogenic elements. Heat flow is also a key discriminator between different modes of mantle convection (see Rolf et al. 2022, this collection). One approach is to measure the heat coming out of the interior with a flux plate (e.g., Kremic et al. 2020). However, scientists have inferred small ($\sim$1 K) diurnal surface temperature variations from remote observations of thermal emission (e.g., Mueller et al. 2008) and global climate models (e.g., Lebonnois et al. 2018). Such small changes in surface temperature could be problematic for the flux plate approach, but these estimates could change after future measurements of surface brightness temperature, acquired from orbit. Analyses could compensate for the effects of these variations if a long enough baseline of surface temperature could be made (i.e., over multiple Venus days). Alternatively, subsurface drilling could deploy a more traditional heat flow probe (Widemann et al. 2023, this collection, and references therein). As for surface seismology, the engineering challenge of drilling is considerable. Given the high scientific value of heat flow measurements to understanding the planet's interior evolution, new technologies to enable this kind of science investigation are important to develop.

8 Conclusion

In this article, we have explored how Venus' tectonic and volcanic processes provide essential insights on Venus' evolution, as any rocky body's crustal layer provides a complex archive for the planet's long term evolution. A primary constraint on the interpretation of geologic processes on another planet is the size and spatial distribution of impact craters, the study of tectonic processes that removed older terrains, which remains controversial. Our understanding of current processes and surface structures effectively provides boundary conditions for extrapolating back into Venus' geologic and convective history: the size, type and distribution of volcanic features provide a window into interior processes such as upwelling/downwelling, melting and mantle temperature. We have introduced the fact that a completely stagnant lid does not provide sufficient heat loss, leading to the idea that Venus has either cycled between a stagnant lid and past plate tectonic regimes, or has undergone regional equilibrium resurfacing: these different interpretations address the fundamental role that both the mean surface age and local variability play in framing the long-term geologic evolution of Venus, and in interpreting interior properties.

Tectonic features provide separate constraints on the rheology, thickness of the lithosphere, and the origin and magnitude of driving stress fields. Gravity data provide further information on the thickness of the crust, the lithosphere, and subsurface density variations due to processes such as mantle upwellings. The goal of this review work has been not to provide a single view, but rather to represent a range of possible interpretations, the constraints available from current data, and to outline specific needs for future datasets. Global mapping of the tessera terrain shows that most tessera margins are embayed by plains, thus establishing the tessera as the oldest materials in a proposed global stratigraphic column. Their composition has not yet been measured directly, but that of some tessera exposures can be inferred from both NIR and radar observations. The generation of such large amounts of felsic melt could require both water and a lithospheric recycling mechanism for formation, and thus possibly must have formed during a more habitable era, assuming the tesserae are igneous in origin.

Volcanic and tectonic constraints on the evolution of Venus are therefore particularly important to our understanding of terrestrial planets' habitability, providing a natural laboratory to understand its evolution in time. A renaissance in Venus exploration is underway with ESA's EnVision orbiter mission, NASA's VERITAS orbiter and the DAVINCI in-situ probe missions all going to Venus by the early 2030s. VERITAS's payload is composed of two instruments crafted to study Venus' surface coupled with a radio science investigation to measure the gravity field. The two instruments are an X-band interferometric synthetic aperture radar, VISAR, and a fourteen-band infrared spectrometer, VEM (Smrekar et al., 2022a). DAVINCI's Venus Descent Imager (VenDI), a near-IR descent-imaging system will deliver clear, high contrast, high SNR images, providing the first geologic constraints on the tesserae surface environment at 2–200 m length scales (Garvin et al., 2022). EnVision's suite of investigations will perform targeted surface imaging as well as polarimetric and stereo imaging, radiometry, and altimetry, a subsurface radar that will sound the upper crust in search of material boundaries, and a suite three spectrometers operating in the UV and Near- and Short Wave-IR (Widemann et al. 2023, this collection). Together, the three missions will fundamentally improve our understanding of the planet's long term history, current activity and evolutionary path.

Declarations

Competing Interests The authors have no competing interests to declare that are relevant to the content of this article and no funding was received to assist with the preparation of this manuscript.

References

Aarnes I, Fristad F, Planke S, Svensen H (2011) The impact of host-rock composition on devolatilization of sedimentary rocks during contact metamorphism around mafic sheet intrusions. Geochem Geophys Geosyst 12:Q10019. https://doi.org/10.1029/2011GC003636

Adams AC, Stegman DR, Smrekar SE, Tackley PJ (2022) Regional-scale lithospheric recycling on Venus via peel-back delamination. J Geophys Res, Planets 127:e2022JE007460. https://doi.org/10.1029/2022JE007460

Addington EA (2001) A stratigraphic study of small volcano clusters on Venus. Icarus 149(1):16–36. https://doi.org/10.1006/icar.2000.6529

Airey MW, Mather TA, Pyle DM, Glaze LS, Ghail RC, Wilson CF (2015) Explosive volcanic activity on Venus: the roles of volatile contribution, degassing, and external environment. Planet Space Sci 113–114:33–48. https://doi.org/10.1016/j.pss.2015.01.009

Aittola M, Kostama VP (2002) Chronology of the formation process of Venusian novae and the associated coronae. J Geophys Res, Planets 107(E11):22-1–22-26. https://doi.org/10.1029/2001JE001528

Anderson FS, Smrekar SE (1999) Tectonic effects of climate change on Venus. J Geophys Res 104(E12):30743–30756. https://doi.org/10.1029/1999JE001082

Anderson FS, Smrekar SE (2006) Global mapping of crustal and lithospheric thickness on Venus. J Geophys Res, Planets 111:E08006. https://doi.org/10.1029/2004JE002395

Anhaeusser CR (2014) Archaean greenstone belts and associated granitic rocks – a review. J Afr Earth Sci 100:684–732. https://doi.org/10.1016/j.jafrearsci.2014.07.019

Ansan V, Vergely P (1995) Evidence of vertical and horizontal motions on Venus: Maxwell Montes. Earth Moon Planets 69:285–310. https://doi.org/10.1007/BF00643789

Arculus RJ (2003) Use and abuse of the terms calcalkaline and calcalkalic. J Petrol 44:929–935. https://doi.org/10.1093/petrology/44.5.929

Armann M, Tackley PJ (2012) Simulating the thermochemical magmatic and tectonic evolution of Venus's mantle and lithosphere: Two-dimensional models. J Geophys Res, Planets 117(E12). https://doi.org/10.1029/2012JE004231

Bachmann O, Bergantz GW (2008) Deciphering magma chamber dynamics from styles of compositional zoning in large silicic ash flow sheets. Rev Mineral Geochem 69(1):651–674. https://doi.org/10.2138/rmg.2008.69.17

Baes M, Gerya T, Sobolev SV (2016) 3-D thermo-mechanical modeling of plume-induced subduction initiation. Earth Planet Sci Lett 453:193–203. https://doi.org/10.1016/j.epsl.2016.08.023

Baker VR, Komatsu G, Parker TJ, Gulick VC, Kargel JS, Lewis JS (1992) Channels and valleys on Venus: preliminary analysis of Magellan data. J Geophys Res, Planets 97(E8):13421–13444. https://doi.org/10.1029/92JE00927

Banerdt WB (1986) Support of long-wavelength loads on Venus and implications for internal structure. J Geophys Res, Solid Earth 91(B1):403–419. https://doi.org/10.1029/JB091iB01p00403

Barsukov VL, Basilevsky AT, Burba GA, Bobinna NN, Kryuchkov VP, Kuzmin RO, Nikolaeva OV, Pronin AA, Ronca LB, Chernaya IM, Shashkina VP, Garanin AV, Kushky ER, Markov MS, Sukhanov AL, Kotelnikov VA, Rzhiga ON, Petrov GM, Alexandrov YN, Sidorenko AI, Bogomolov AF, Skrypnik GI, Bergman MY, Kudrin LV, Bokshtein IM, Kronrod MA, Chochia PA, Tyuflin YS, Kadnichansky SA, Akim EL (1986) The geology and geomorphology of the Venus surface as revealed by the radar images obtained by Veneras 15 and 16. J Geophys Res, Solid Earth 91(B4):378–398. https://doi.org/10.1029/JB091iB04p0D378

Basilevsky AT, Head JW (1996) Evidence for rapid and widespread emplacement of volcanic plains on Venus: stratigraphic studies in the Baltis Vallis region. Geophys Res Lett 23(12):1497–1500. https://doi.org/10.1029/96GL00975

Basilevsky AT, Head JW (2002) Venus: timing and rates of geologic activity. Geology 30(11):1015–1018. https://doi.org/10.1130/0091-7613(2002)030<1015:VTAROG>2.0.CO;2

Basilevsky AT, Kuzmin RO, Nikolaeva OV, Pronin AA, Ronca LB, Avduevsky VS, Uspensky GR, Cheremukhina ZP, Semenchenko VV, Ladygin VM (1985) The surface of Venus as revealed by the Venera landings: part II. Geol Soc Am Bull 96(1):137–144. https://doi.org/10.1130/0016-7606(1985)96<137:TSOVAR>2.0.CO;2

Basilevsky AT, Nikolaeva OV, Weitz CM (1992) Geology of the Venera 8 landing site region from Magellan data: morphological and geochemical considerations. J Geophys Res, Planets 97:16315–16335. https://doi.org/10.1029/92JE01557

Bédard JH (2006) A catalytic delamination-driven model for coupled genesis of Archaean crust and subcontinental lithospheric mantle. Geochim Cosmochim Acta 70(5):1188–1214. https://doi.org/10.1016/j.gca.2005.11.008

Bédard JH (2018) Stagnant lids and mantle overturns: implications for Archaean tectonics, magmagenesis, crustal growth, mantle evolution, and the start of plate tectonics. Geosci Front 9(1):19–49. https://doi.org/10.1016/j.gsf.2017.01.005

Bédard JH, Brouillette P, Madore L, Berclaz A (2003) Archaean cratonization and deformation in the northern Superior Province, Canada: an evaluation of plate tectonic versus vertical tectonic models. Precambrian Res 127:61–87. https://doi.org/10.1016/S0301-9268(03)00181-5

Bédard JH, Harris LB, Thurston PC (2013) The hunting of the snArc. Precambrian Res 229:20–48. https://doi.org/10.1016/j.precamres.2012.04.001

Bercovici D, Ricard Y (2014) Generation of plate tectonics with two-phase grain-damage and pinning: source-sink model and toroidal flow. Earth Planet Sci Lett 365:275–288. https://doi.org/10.1016/j.epsl.2013.02.002

Bilotti F, Suppe J (1999) The global distribution of wrinkle ridges on Venus. Icarus 139(1):137–157. https://doi.org/10.1006/icar.1999.6092

Bindschadler DL, Parmentier EM (1990) Mantle flow tectonics: the influence of a ductile lower crust and implications for the formation of topographic uplands on Venus. J Geophys Res, Solid Earth 95(B13):21329–21344. https://doi.org/10.1029/JB095iB13p21329

Bindschadler DL, Schubert G, Kaula WM (1992) Coldspots and hotspots: global tectonics and mantle dynamics of Venus. J Geophys Res, Planets 97(E8):13495–13532. https://doi.org/10.1029/92JE01165

Bjonnes E, Johnson BC, Evans AJ (2021) Estimating Venusian thermal conditions using multiring basin morphology. Nat Astron 5:498–502. https://doi.org/10.1038/s41550-020-01289-6

Blaske CH, O'Rourke JG, Desch SJ, Borrelli ME (2023) Meteors may masquerade as lightning in the atmosphere of Venus. J Geophys Res, Planets 128(9):e2023JE007914. https://doi.org/10.1029/2023JE007914

Bleeker W (2002) The late Archean record: a puzzle in ca. 35 pieces. Lithos 71:99–134. https://doi.org/10.1016/j.lithos.2003.07.003

Bolea-Fernandez E, Van Malderen SJM, Balcaen L, Resano M, Vanhaecke F (2016) Laser ablation-tandem ICP-mass spectrometry (LA-ICP-MS/MS) for direct Sr isotopic analysis of solid samples with high Rb/Sr ratios. J Anal At Spectrom 31:464–472. https://doi.org/10.1039/C5JA00404G

Bondarenko NV, Head JW, Ivanov MA (2010) Present-day volcanism on Venus: evidence from microwave radiometry. Geophys Res Lett 37(23). https://doi.org/10.1029/2010GL045233

Bonin B (2012) Extra-terrestrial igneous granites and related rocks: a review of their occurrence and petrogenesis. Lithos 153:3–24. https://doi.org/10.1016/j.lithos.2012.04.007

Borrelli ME, O'Rourke JG, Smrekar SE, Ostberg CM (2021) A global survey of lithospheric flexure at steep-sided domical volcanoes on Venus reveals intermediate elastic thicknesses. J Geophys Res, Planets 126(7):e2020JE006756. https://doi.org/10.1029/2020JE006756

Brian AW, Stofan ER, Guest JE, Smrekar SE (2004) Laufey Regio: a newly discovered topographic rise on Venus. J Geophys Res, Planets 109(E7). https://doi.org/10.1029/2002JE002010

Bridges NT (1997) Ambient effects on basalt and rhyolite lavas under Venusian, subaerial, and subaqueous conditions. J Geophys Res, Planets 102(E4):9243–9255. https://doi.org/10.1029/97JE00390

Brissaud Q, Krishnamoorthy S, Jackson JM, Bowman DC, Komjathy A, Cutts JA, Zhan Z, Pauken MT, Izraelevitz JS, Walsh GJ (2021) The first detection of an earthquake from a balloon using its acoustic signature. Geophys Res Lett 48(12):e2021GL093013. https://doi.org/10.1029/2021GL093013

Brossier JF, Gilmore MS, Toner K (2020) Low radar emissivity signatures on Venus volcanoes and coronae: new insights on relative composition and age. Icarus 343:113693. https://doi.org/10.1016/j.icarus.2020.113693

Brown CD, Grimm RE (1995) Tectonics of Artemis Chasma: a Venusian "Plate" boundary. Icarus 117(2):219–249. https://doi.org/10.1006/icar.1995.1155

Brown CD, Grimm RE (1996) Lithospheric rheology and flexure at Artemis Chasma, Venus. J Geophys Res 101(E5):12697–12708. https://doi.org/10.1029/96JE00834

Brown CD, Grimm RE (1997) Tessera deformation and the contemporaneous thermal state of the plateau highlands. Venus, Earth and Planetary Science Letters 147(1):1–10. https://doi.org/10.1016/S0012-821X(97)00007-1

Brown CD, Grimm RE (1999) Recent tectonic and lithospheric thermal evolution of Venus. Icarus 139(1):40–48. https://doi.org/10.1006/icar.1999.6083

Brown M, Johnson T, Gardiner NJ (2020) Plate tectonics and the Archean Earth. Annu Rev Earth Planet Sci 48:291–320. https://doi.org/10.1146/annurev-earth-081619-052705

Bullock MA, Grinspoon DH (2001) The recent evolution of climate on Venus. Icarus 150:19–37. https://doi.org/10.1006/icar.2000.6570

Burov E, Cloetingh S (2010) Plume-like upper mantle instabilities drive subduction initiation. Geophys Res Lett 37(3). https://doi.org/10.1029/2009GL041535

Byrne PK, Krishnamoorthy S (2022) Estimates on the frequency of volcanic eruptions on Venus. J Geophys Res, Planets 127:e2021JE007040. https://doi.org/10.1029/2021JE007040

Byrne PK, Ghail RC, Gilmore MS, Şengör AMC, Klimczak C, Senske DA, Whitten JL, Khawja S, Ernst RE, Solomon SC (2021) Venus tesserae feature layered, folded, and eroded rocks. Geology 49:81–85. https://doi.org/10.1130/G47940.1

Cadman AC, Heaman L, Tarney J, Wardle R, Krogh TE (1993) U-Pb geochronology and geochemical variation within two Proterozoic mafic dyke swarms. Labrador Canad J Earth Sci 30:1490–1504. https://doi.org/10.1139/e93-128

Campbell IH (2002) Implications of the Nb/U, Th/U and Sm/Nd in plume magmas for the relationship between continental and oceanic crust formation and the development of depleted mantle. Geochim Cosmochim Acta 66(9):1651–1661. https://doi.org/10.1016/S0016-7037(01)00856-0

Campbell IH, Hill RI (1988) A two-stage model for the formation of the granite-greenstone terrains of the Kalgoorlie-Norseman area, Western Australia. Earth Planet Sci Lett 90(1):11–25. https://doi.org/10.1016/0012-821X(88)90107-0

Campbell IH, Taylor SR (1983) No water, no granites - no oceans, no continents. Geophys Res Lett 10(11):1061–1064. https://doi.org/10.1029/GL010i011p01061

Campbell BA, Campbell DB, Morgan GA, Carter LM, Nolan MC, Chandler JF (2015) Evidence for crater ejecta on Venus tessera terrain from Earth-based radar images. Icarus 250:123–130. https://doi.org/10.1016/j.icarus.2014.11.025

Campbell BA, Morgan GA, Whitten JL, Carter LM, Glaze LS, Campbell DB (2017) Pyroclastic flow deposits on Venus as indicators of renewed magmatic activity. J Geophys Res, Planets 122(7):1580–1596. https://doi.org/10.1002/2017JE005299

Carter LM, Gilmore MS, Ghail RC, Byrne PK, Smrekar SE, Ganey TM, Izenberg N (2023) Sedimentary processes on Venus. Space Sci Rev 219:85. https://doi.org/10.1007/s11214-023-01033-2

Cawood PA, Hawkesworth CJ, Dhuime B (2013) The continental record and the generation of continental crust. Geol Soc Am Bull 125(1–2):14–32. https://doi.org/10.1130/B30722.1

Chantel J, Manthilake G, Andrault D, Novella D, Yu T, Wang YB (2016) Experimental evidence supports mantle partial melting in the asthenosphere, Experimental evidence supports mantle partial melting in the asthenosphere. Sci Adv 2, e1600246. https://doi.org/10.1126/sciadv.1600246. 2016

Christiansen RL (2005) Post-Laramide Tectonomagmatics of the U.S. Cordillera. Geochimica et Cosmochimica Acta Supplement 69(10). Supplement, 2005, A136

Christiansen EH, McCurry M (2008) Contrasting origins of Cenozoic silicic volcanic rocks from the western Cordillera of the United States. Bull Volcanol 70:251–267. https://doi.org/10.1007/s00445-007-0138-1

Clegg SM, Sklute E, Dyar MD, Barefield JE, Wiens RC (2009) Multivariate analysis of remote laser-induced breakdown spectroscopy spectra using partial least squares, principal component analysis, and related techniques. Spectrochim Acta, Part B, At Spectrosc 64(1):79–88. https://doi.org/10.1016/j.sab.2008.10.045

Cohen BA, Malespin CA, Farley KA, Martin PE, Cho Y, Mahaffy PR (2019) In situ geochronology on Mars and the development of future instrumentation. Astrobiology 19(11):1303–1314. https://doi.org/10.1089/ast.2018.1871

Coleman M, Hecht M, Hurowitz J, Neidholdt E, Polk J, Sinha MP, Sturhahn W, Zimmerman W (2012) In situ geochronology as a mission-enabling technology. In: 2012 IEEE aerospace conference, pp 1–8. https://doi.org/10.1109/AERO.2012.6187053

Condie KC (1981) Archean greenstone belts. Dev Precambrian Geol 3:1–434

Condie KC (2018) A planet in transition: the onset of plate tectonics on Earth between 3 and 2 Ga? Geosci Front 9(1):51–60. https://doi.org/10.1016/j.gsf.2016.09.001

Condie KC, Benn K (2006) Archean geodynamics: similar to or different from modern geodynamics? In: Benn K, Mareschal J-C, Condie KC (eds) Archean geodynamics and environments. Geophysical monograph series, vol 164, pp 47–59

Condie KC, Kröner A (2008) When did plate tectonics begin? Evidence from the geologic record. In: Condie KC, Pease V (eds) When did plate tectonics begin on planet Earth? Spec Pap Geol Soc Am, vol 440, pp 281–294

Condie KC, Aster RC van Hunen J (2016) A great thermal divergence in the mantle beginning 2.5 Ga: geochemical constraints from greenstone basalts and komatiites. Geosci Front 7(4):543–553. https://doi.org/10.1016/j.gsf.2016.01.006

Conrad JW, Nimmo F (2023) Constraining characteristic morphological wavelengths for Venus using Baltis Vallis. Geophys Res Lett 50(10):e2022GL101268. https://doi.org/10.1029/2022GL101268

Cordier D, Bonhommeau DA, Port S, Chevrier V, Lebonnois S, García-Sánchez F (2019) The physical origin of the Venus low atmosphere chemical gradient. Astrophys J 880:2. https://doi.org/10.3847/1538-4357/ab27bd

Crumpler LS, Head JW, Aubele JC (1993) Relation of major volcanic center concentration on Venus to global tectonic patterns. Science 261(5121):591–595. https://doi.org/10.1126/science.261.5121.591

Crumpler LS, Aubele JC, Senske DA, Keddie ST, Magee KP, Head JW (1997) Volcanoes and centers of volcanism on Venus. In: Bougher SW, Hunten DM, Phillips RJ (eds) Venus II. University of Arizona Press, Tucson, pp 697–756

Crumpler LS, Aubele JC (2000) Volcanism on Venus. In: Sigurdsson H et al (eds) Encyclopedia of volcanoes. Academic Press, New York, pp 727–770

Cutts J, Baines K, Dorsky L, Frazier W, Izraelevitz J, Krishnamoorthy S et al (2022) Exploring the clouds of Venus: science driven Aerobot missions to our sister planet. In: 2022 IEEE aerospace conference (AERO), pp 1–20. https://doi.org/10.1109/AERO53065.2022.9843740

Cyr KE, Melosh HJ (1993) Tectonic patterns and regional stresses near Venusian coronae. Icarus 102(2):175–184. https://doi.org/10.1006/icar.1993.1042

Davaille A, Smrekar SE, Tomlinson S (2017) Experimental and observational evidence for plume-induced subduction on Venus. Nat Geosci 10(5):349–355. https://doi.org/10.1038/ngeo2928

Dawson JB (1962) Sodium Carbonate Lavas from Oldoinyo Lengai. Tanganyika Nat 195(4846):1075–1076. https://doi.org/10.1038/1951075a0

Dawson JB, Pinkerton H, Norton GE, Pyle DM (1990) Physicochemical properties of alkali carbonatite lavas: data from the 1988 eruption of Oldoinyo Lengai. Tanzania Geol 18(3):260–263. https://doi.org/10.1130/0091-7613(1990)018<0260:PPOACL>2.3.CO;2

de WitMJ, Ashwal LD (1995) Greenstone belts: what are they? S Afr J Geol 98:505–520

D'Incecco P, Müller N, Helbert J, D'Amore M (2017) Idunn Mons on Venus: location and extent of recently active lava flows. Planet Space Sci 136:25–33. https://doi.org/10.1016/j.pss.2016.12.002

Dombard AJ, Johnson CL, Richards MA, Solomon SC (2007) A magmatic loading model for coronae on Venus. J Geophys Res, Planets 112(E4). https://doi.org/10.1029/2006JE002731

Donahue TM, Grinspoon DH, Hartle RE, Hodges RR (1997) Ion/neutral escape of hydrogen and deuterium: evolution of water. In: Bougher SW, Hunten DM, Phillips RJ (eds) Venus II. Univ. Arizona Press, Tucson, pp 385–414

Duncan M, Dasgupta R (2017) Rise of Earth's atmospheric oxygen controlled by efficient subduction of organic carbon. Nat Geosci 10:387–392. https://doi.org/10.1038/ngeo2939

Dyar MD, Helbert J, Marturilli A, Müller NT, Kappel D (2020) Probing Venus surface iron contents with six-band visible near-infrared spectroscopy from orbit. Geophys Res Lett 47(23):e2020GL090497. https://doi.org/10.1029/2020GL090497

Dyar MD, Helbert J, Cooper RC, Sklute EC, Marurilli A, Mueller NT, Kappel D, Smrekar SE (2021) Surface weathering on Venus: constraints from kinetic, spectroscopic, and geochemical data. Icarus 358:114139. https://doi.org/10.1016/j.icarus.2020.114139

Elkins-Tanton LT, Smrekar SE, Hess PC, Parmentier EM (2007) Volcanism and volatile recycling on a one-plate planet: Applications to Venus. J Geophys Res, Planets 112(E4). https://doi.org/10.1029/2006JE002793

Ernst R, Bleeker W (2010) Large igneous provinces (LIPs), giant dyke swarms, and mantle plumes: significance for break-up events within Canada and adjacent regions from 2.5 Ga to the present. Can J Earth Sci 47:695–739. https://doi.org/10.1139/E10-025

Ernst RE, Buchan KL (2001) Large mafic magmatic events through time and links to mantle plume heads. In: Buchan RE, Buchan KL (eds) Mantle plumes: their identification through time. Spec Pap Geol Soc Am, vol 352, pp 483–575. https://doi.org/10.1130/0-8137-2352-3.483

Ernst RE, Head JW, Parfitt E, Grosfils E, Wilson L (1995) Giant radiating dyke swarms on Earth and Venus. Earth-Sci Rev 39(1):1–58. https://doi.org/10.1016/0012-8252(95)00017-5

Ernst RE, Grosfils EB, Mège D (2001) Giant Dike Swarms: Earth, Venus, and Mars. Annu Rev Earth Planet Sci 29:489–534. https://doi.org/10.1146/annurev.earth.29.1.489

Ernst RE, Dosnoyers DW, Head JW, Grosfils EB (2003) Graben–fissure systems in Guinevere Planitia and Beta Regio (264o–312oE, 24o–60oN), Venus, and implications for regional stratigraphy and mantle plumes. Icarus 164:282–316. https://doi.org/10.1016/S0019-1035(03)00126-X

Fagereng Å, Biggs J (2019) New perspectives on 'geological strain rates' calculated from both naturally deformed and actively deforming rocks. J Struct Geol 125:100–110. https://doi.org/10.1016/j.jsg.2018.10.004

Faure G (2001) Origin of igneous rocks: the isotopic evidence. Springer, Heidelberg

Fegley B Jr, Prinn RG (1989) Estimation of the rate of volcanism on Venus from reaction rate measurements. Nature 337:55–58. https://doi.org/10.1038/337055a0

Fegley B (2003) Venus. In: Holland HD, Turekian KK (eds) Treatise on geochemistry, vol 1. Elsevier, Amsterdam, pp 487–507. https://doi.org/10.1016/b0-08-043751-6/01150-6

Fegley B, Klingelhofer G, Lodders K, Widemann T (1997) Geochemistry of surface-atmosphere interactions on Venus. In: Bougher SW, Hunten DM, Phillips RJ (eds) Venus II. University of Arizona Press, Tucson, pp 591–636

Filiberto J (2014) Magmatic diversity on Venus: constraints from terrestrial analog crystallization experiments. Icarus 231:131–136. https://doi.org/10.1016/j.icarus.2013.12.003

Filiberto J, Trang D, Treiman AH, Gilmore MS (2020) Present-day volcanism on Venus as evidenced from weathering rates of olivine. Sci Adv 6(1):eaax7445. https://doi.org/10.1126/sciadv.aax7445

Florensky CP, Ronca LB, Basilevsky AT, Burba GA, Nikolaeva OV, Pronin AA, Trakhtman AM, Volkov VP, Zazetsky VV (1977) The surface of Venus as revealed by Soviet Venera 9 and 10. Geol Soc Am Bull 88(11):1537–1545. https://doi.org/10.1130/0016-7606(1977)88<1537:TSOVAR>2.0.CO;2

Flynn ITW, Chevrel MO, Ramsey MS (2023) Adaptation of a thermorheological lava flow model for Venus conditions. J Geophys Res, Planets 128(7):e2022JE007710. https://doi.org/10.1029/2022JE007710

Ford JP, Plaut JJ, Weitz CM, Farr TG, Senske DA, Stofan ER, Michaels G, Parker TJ (1993) Guide to Magellan image interpretation. JPL Publ 93(24)

Galgana GA, McGovern PJ, Grosfils EB (2011) Evolution of large Venusian volcanoes: Insights from coupled models of lithospheric flexure and magma reservoir pressurization. Journal of Geophysical Research Planets 116(E3). https://doi.org/10.1029/2010JE003654

Galgana GA, Grosfils EB, McGovern PJ (2013) Radial dike formation on Venus: insights from models of uplift, flexure and magmatism. Icarus 225(1):538–547. https://doi.org/10.1016/j.icarus.2013.04.020

Ganesh I, McGuire LA, Carter LM (2021) Modeling the dynamics of dense pyroclastic flows on Venus: insights into pyroclastic eruptions. J Geophys Res, Planets 126(9):e2021JE006943. https://doi.org/10.1029/2021JE006943

Ganesh I, Carter LM, Henz TN (2022) Radar backscatter and emissivity models of proposed pyroclastic density current deposits on Venus. J Geophys Res, Planets 127(10):e2022JE007318. https://doi.org/10.1029/2022JE007318

Garcia RF, Klotz A, Hertzog A, Martin R, Gérier S, Kassarian E, Bordereau J, Venel S, Mimoun D (2022) Infrasound from large earthquakes recorded on a network of balloons in the stratosphere. Geophys Res Lett 49(15):e2022GL098844. https://doi.org/10.1029/2022GL098844

Garvin JB, Getty SA, Arney GN, Johnson NM, Kohler E, Schwer KO, Sekerak M, Bartels A, Saylor RS, Elliott VE, Goodloe CS, Garrison MB, Cottini V, Izenberg N, Lorenz R, Malespin CA, Ravine M, Webster CR, Atkinson DH, Aslam S, Atreya S, Bos BJ, Brinckerhoff WB, Campbell B, Crisp D, Filiberto JR, Forget F, Gilmore M, Gorius N, Grinspoon D, Hofmann AE, Kane SR, Kiefer W, Lebonnois S, Mahaffy PR, Pavlov A, Trainer M, Zahnle KJ, Zolotov M (2022) Revealing the mysteries of Venus: the DAVINCI mission. Planet Sci J 3:117. https://doi.org/10.3847/psj/ac63c2

George H, Malaspina DM, Goodrich K, Ma Y, Ramstad R, Connor D, Bale SD, Curry S (2023) Non-lightning-generated whistler waves in near-Venus space. Geophys Res Lett 50(19):e2023GL105426. https://doi.org/10.1029/2023GL105426

Gerya T (2014) Precambrian geodynamics: concepts and models. Gondwana Res 25:442–463. https://doi.org/10.1016/j.gr.2012.11.008

Gerya TV, Stern RJ, Baes M, Sobolev S, Whattam SA (2015) Plate tectonics on the Earth triggered by plume-induced subduction initiation. Nature 527:221–225. https://doi.org/10.1038/nature15752

Ghail RC (2002) Structure and evolution of southeast Thetis Regio. J Geophys Res, Planets 107(E8):4-1–4-7. https://doi.org/10.1029/2001JE001514

Ghail RC (2015) Rheological and petrological implications for a stagnant lid regime on Venus. Planet Space Sci 113–114:2–9. https://doi.org/10.1016/j.pss.2015.02.005

Ghent RR, Hansen VL (1999) Structural and kinematic analysis of eastern Ovda Regio, Venus: implications for crustal Plateau formation. Icarus 139(1):116–136. https://doi.org/10.1006/icar.1999.6085

Ghent RR, Tibuleac IM (2002) Ribbon spacing in Venusian tessera: implications for layer thickness and thermal state. Geophys Res Lett 29(20):61. https://doi.org/10.1029/2002GL015994

Giardini D, Lognonné P, Banerdt WB et al (2020) The seismicity of Mars. Nat Geosci 13:205–212. https://doi.org/10.1038/s41561-020-0539-8

Gilmore MS, Head JW (2000) Sequential deformation of plains at the margins of Alpha Regio, Venus: implications for tessera formation. Meteorit Planet Sci 35(4):667–687. https://doi.org/10.1111/j.1945-5100.2000.tb01451.x

Gilmore MS, Head JW (2018) Morphology and deformational history of Tellus Regio, Venus: evidence for assembly and collision. Planet Space Sci 154:5–20. https://doi.org/10.1016/j.pss.2018.02.001

Gilmore MS, Ivanov MA, Head JW, Basilevsky AT (1997) Duration of tessera deformation on Venus. J Geophys Res, Planets 102(E6):13357–13368. https://doi.org/10.1029/97JE00965

Gilmore MS, Collins GC, Ivanov MA, Marinangeli L, Head JW (1998) Style and sequence of extensional structures in tessera terrain. Venus J Geophys Res Planets 103(E7):16813–16840. https://doi.org/10.1029/98JE01322

Gilmore MS, Mueller N, Helbert J (2015) VIRTIS emissivity of Alpha Regio, Venus, with implications for tessera composition. Icarus 254:350–361. https://doi.org/10.1016/j.icarus.2015.04.008

Gilmore M, Treiman A, Helbert J, Smrekar S (2017) Venus surface composition constrained by observation and experiment. Space Sci Rev 212(3–4):1511–1540

Gilmore MS, Dyar MD, Mueller N et al (2023) Mineralogy of the Venus Surface. Space Sci Rev 219:52. https://doi.org/10.1007/s11214-023-00988-6

Glaze LS, Stofan ER, Smrekar SE, Baloga SM (2002) Insights into corona formation through statistical analyses. J Geophys Res, Planets 107(E12):18-1–18-12. https://doi.org/10.1029/2002JE001904

Glaze LS, Wilson CF, Zasova LV, Nakamura M, Limaye S (2018) Future of Venus research and exploration. Space Sci Rev 214:89. https://doi.org/10.1007/s11214-018-0528-z

Global Volcanism Program (2023). [Database] Volcanoes of the World (v. 5.1.5; 15 Dec 2023). Distributed by Smithsonian Institution, compiled by Venzke, E. https://doi.org/10.5479/si.GVP.VOTW5-2023.5.1

Gorevan SP, Myrick T, Davis K, Chau JJ, Bartlett P, Mukherjee S, Anderson R, Squyres SW, Arvidson RE, Madsen MB, Bertelsen P, Goetz W, Binau CS, Richter L (2003) Rock Abrasion Tool: Mars Exploration Rover mission. J Geophys Res, Planets 108(E12). https://doi.org/10.1029/2003JE002061

Griffiths RW, Campbell IH (1990) Stirring and structure in mantle starting plumes. Earth Planet Sci Lett 99(1–2):66–78. https://doi.org/10.1016/0012-821X(90)90071-5

Grimm RE, Hess PC (1997) The crust of Venus. In: Venus II. University of Arizona Press, Tucson, pp 1205–1244

Grimm RE, Barr AC, Harrison KP, Stillman DE, Neal KL, Vincent MA, Delory GT (2012) Aerial electromagnetic sounding of the lithosphere of Venus. Icarus 217(2):462–473. https://doi.org/10.1016/j.icarus.2011.07.021

Grindrod PM, Hoogenboom T (2006) Venus: the corona conundrum. Astron Geophys 47(3):3.16–3.21. https://doi.org/10.1111/j.1468-4004.2006.47316.x

Grindrod PM, Nimmo F, Stofan ER, Guest JE (2005) Strain at radially fractured centers on Venus. J Geophys Res, Planets 110(E12):E12002. https://doi.org/10.1029/2005JE002416

Grindrod PM, Stofan ER, Guest JE (2010) Volcanism and resurfacing on Venus at the full resolution of Magellan SAR data. Geophys Res Lett 37(15). https://doi.org/10.1029/2010GL043424

Grosfils E, Head JW (1994a) Emplacement of a radiating dike swarm in western Vinmara Planitia, Venus: interpretation of the regional stress field orientation and subsurface magmatic configuration. Earth Moon Planets 66:153–171. https://doi.org/10.1007/BF00644129

Grosfils E, Head JW (1994b) The global distribution of giant radiating dike swarms on Venus: implications for the global stress state. Geophys Res Lett 21(8):701–704. https://doi.org/10.1029/94GL00592

Grosfils E, Aubele J, Crumpler L, Gregg T, Sakimoto S (2000) Volcanism on Earth's Seafloor and Venus. In: Zimbelman JR, Gregg TKP (eds) Environmental effects on volcanic eruptions: from deep oceans to deep space

Guest JE, Stofan ER (1999) A new view of the stratigraphic history of Venus. Icarus 139(1):55–66. https://doi.org/10.1006/icar.1999.6091

Gülcher AJP, Gerya TV, Montési LGJ, Munch J (2020) Corona structures driven by plume–lithosphere interactions and evidence for ongoing plume activity on Venus. Nat Geosci 13(8):547–554. https://doi.org/10.1038/s41561-020-0606-1

Gülcher AJP, Yu T-Y, Gerya TV (2023) Tectono-magmatic evolution of asymmetric coronae on Venus: topographic classification and 3D thermo-mechanical modeling. J Geophys Res, Planets 123(1):e2023JE007978. https://doi.org/10.1029/2023JE007978

Hahn RM, Byrne PK (2022) Kernel density analysis of volcanoes on Venus at varying spatial scales. Lun Planet Sci Conf 53:2437

Hamilton WB (1998) Archean magmatism and deformation were not products of plate tectonics. Precambrian Res 91(1–2):143–179. https://doi.org/10.1016/S0301-9268(98)00042-4

Hamilton WB (2011) Plate tectonics began in Neoproterozoic time, and plumes from deep mantle have never operated. Lithos 123:1–20. https://doi.org/10.1016/j.lithos.2010.12.007

Hamilton VE, Stofan ER (1996) The geomorphology and evolution of Hecate Chasma, Venus. Icarus 121(1):171–194. https://doi.org/10.1006/icar.1996.0077

Hansell SA, Wells WK, Hunten DH (1995) Optical detection of lightning on Venus. Icarus 117(2):345–351. https://doi.org/10.1006/icar.1995.1160

Hansen VL (2007) Venus: a thin-lithosphere analog for early Earth? In: Van Kranendonk MJ, Smithies RH, Bennett VC (eds) Earth's oldest rocks, developments in Precambrian geology 15, pp 987–1012

Hansen VL (2015) Impact origin of Archean cratons. Lithosphere 7(5):563–578. https://doi.org/10.1130/L371.1

Hansen VL (2018) Global tectonic evolution of Venus, from exogenic to endogenic over time, and implications for early Earth processes. Philos Trans R Soc A376:20170412. https://doi.org/10.1098/rsta.2017.0412

Hansen VL, Phillips RJ (1993) Tectonics and volcanism of eastern Aphrodite Terra, Venus: no subduction, no spreading. Science 260(5107):526–530. https://doi.org/10.1126/science.260.5107.526

Hansen VL, Willis JA (1996) Structural analysis of a sampling of tesserae: implications for Venus geodynamics. Icarus 123(2):296–312. https://doi.org/10.1006/icar.1996.0159

Hansen VL, Banks BK, Ghent RR (1999) Tessera terrain and crustal plateaus, Venus. Geology 27(12):1071–1074. https://doi.org/10.1130/0091-7613(1999)027<1071:TTACPV>2.3.CO;2

Hansen VL, Phillips RJ, Willis JJ, Ghent RR (2000) Structures in tessera terrain, Venus: issues and answers. J Geophys Res, Planets 105(E2):4087–4368. https://doi.org/10.1029/1999JE001137

Harris LB, Bédard JH (2014) Crustal evolution and deformation in a non-plate-tectonic Archaean Earth: comparisons with Venus. In: Dilek Y, Furnes H (eds) Evolution of Archean crust and early life, modern approaches in solid Earth sciences, pp 215–291

Harris LB, Bédard JH (2015) Interactions between continent-like 'drift' rifting and mantle flow on Venus: gravity interpretations and Earth analogues. In: Plattz T, Massironi M, Byrne PK, Hiesinger H (eds) Volcanism and tectonism across the inner solar system. Geologic society of London special publication, vol 401, pp 327–356

Harrison TM (2009) The hadean crust: evidence from >4 Ga zircons. Annu Rev Earth Planet Sci 37:479–505. https://doi.org/10.1146/annurev.earth.031208.100151

Hart RA, Russell CT, Zhang T (2022) Statistical study of lightning-generated whistler-mode waves observed by Venus Express. Icarus 380:114993. https://doi.org/10.1016/j.icarus.2022.114993

Hashimoto GL, Roos-Serote M, Sugita S, Gilmore MS, Kamp LW, Carlson RW, Baines K-H (2008) Felsic highland crust on Venus suggested by Galileo near-infrared mapping spectrometer data. J Geophys Res, Planets 113:E00B24. https://doi.org/10.1029/2008JE003134

Hawkesworth CJ, Cawood PA, Bhuime B (2019) Rates of generation and growth of the continental crust. Geosci Front 10:165–173. https://doi.org/10.1016/j.gsf.2018.02.004

Hawkesworth CJ, Cawood PA, Bhuime B (2020) The evolution of the continental crust and the onset of plate tectonics. Front Earth Sci 8:326. https://doi.org/10.3389/feart.2020.00326

Head JW (1990) Processes of crustal formation and evolution on Venus: an analysis of topography, hypsometry, and crustal thickness variations. Earth Moon Planets 50(1):25–55. https://doi.org/10.1007/BF00142388

Head JW, Wilson L (1986) Volcanic processes and landforms on Venus: theory, predictions, and observations. J Geophys Res, Solid Earth 91(B9):9407–9446. https://doi.org/10.1029/JB091iB09p09407

Head JW, Wilson L (1992) Magma reservoirs and neutral buoyancy zones on Venus: implications for the formation and evolution of volcanic landforms. J Geophys Res, Planets 97(E3):3877–3903. https://doi.org/10.1029/92JE00053

Head JW, Crumpler LS, Aubele JC, Guest JE, Saunders RS (1992) Venus volcanism: classification of volcanic features and structures, associations, and global distribution from Magellan data. J Geophys Res, Planets 97(E8):13153–13197. https://doi.org/10.1029/92JE01273

Head JW, Parmentier EM, Hess PC (1994) Venus: vertical accretion of crust and depleted mantle and implications for geologic history and processes. Planet Space Sci 42(10):803–811. https://doi.org/10.1016/0032-0633(94)90061-2

Heaman LM, Kjarsgaard BA (2000) Timing of eastern North American kimberlite magmatism: continental extension of the Great Meteor hotspot track? Earth Planet Sci Lett 178(3):253–268. https://doi.org/10.1016/S0012-821X(00)00079-0

Helbert J, Müller N, Kostama P, Marinangeli L, Piccioni G, Drossart P (2008) Surface brightness variations seen by VIRTIS on Venus Express and implications for the evolution of the Lada Terra region. Venus Geophys Res Lett 35(11):L11201. https://doi.org/10.1029/2008GL033609

Hensley S, Wallace MS, Martin J, Perkovic-Martin D, Smrekar S, Younis M, Lachaise M, Prats P, Rodriguez M, Zebker H, Campbell B, Mastrogiuseppe M (2022) Planned differential interferometric SAR observations at Venus by the Veritas mission. In: Proceedings of IGARSS 2022, International Geoscience and Remote Sensing Symposium, Kuala Lumpur, Indonesia, 17-22 July, 2022

Herrick RR, Hensley S (2023) Surface changes observed on a Venusian volcano during the Magellan mission. Science 379(6638):1205–1208. https://doi.org/10.1126/science.abm7735

Herrick RR, Phillips RJ (1992) Geological correlations with the interior density structure of Venus. J Geophys Res, Planets 97(E10):16017–16034. https://doi.org/10.1029/92JE01498

Herrick RR, Rumpf ME (2011) Postimpact modification by volcanic or tectonic processes as the rule, not the exception, for Venusian craters. J Geophys Res, Planets 116(E2). https://doi.org/10.1029/2010JE003722

Herrick RR, Stahlke DL, Sharpton VL (2012) Fine-scale Venusian topography from Magellan stereo data. Eos, Trans Am Geophys Union 93(12):125–132. https://doi.org/10.1029/2012EO120002

Herrick RR, Bjonnes ET, Carter LM, Gerya T, Ghail RC, Gillmann C, Gilmore MS, Hensley S, Ivanov MA, Izenberg NR, Mueller NT, O'Rourke JG, Rolf T, Smrekar SE, Weller MB (2023) Resurfacing history and volcanic activity of Venus. Space Sci Rev 219:29. https://doi.org/10.1007/s11214-023-00966-y

Hess PC, Head JW (1990) Derivation of primary magmas and melting of crustal materials on Venus: some preliminary petrogenetic considerations. Earth Moon Planets 50–51(1):57–80. https://doi.org/10.1007/BF00142389

Hoefs J (2009) Stable isotope geochemistry. Springer, Berlin. https://doi.org/10.1007/978-3-540-70708-0

Hoggard MJ, Parnell-Turner R, White N (2020) Hotspots and mantle plumes revisited: towards reconciling the mantle heat transfer discrepancy. Earth Planet Sci Lett 542:116317. https://doi.org/10.1016/j.epsl.2020.116317

Honda C, Fujimura A (2005) Formation process of lunar sinuous rilles by thermal erosion of basaltic lava flow. In: Lunar and Planetary Science Conference XXXVI, Abstract 1562. https://www.lpi.usra.edu/meetings/lpsc2005/pdf/1562.pdf

Höning D, Spohn T (2016) Continental growth and mantle hydration as intertwined feedback cycles in the thermal evolution of Earth. Phys Earth Planet Inter 255:27–49. https://doi.org/10.1016/j.pepi.2016.03.010

Höning D, Tosi N, Spohn T (2019) Carbon cycling and interior evolution of water-covered plate tectonics and stagnant-lid planets. Astron Astrophys 627:A48. https://doi.org/10.1051/0004-6361/201935091

Hoogenboom T, Houseman GA (2006) Rayleigh–Taylor instability as a mechanism for corona formation on Venus. Icarus 180(2):292–307. https://doi.org/10.1016/j.icarus.2005.11.001

Huang J, Yang A, Zhong S (2013) Constraints of the topography, gravity and volcanism on Venusian mantle dynamics and generation of plate tectonics. Earth Planet Sci Lett 362:207–214. https://doi.org/10.1016/j.epsl.2012.11.051

406

Hunt GR, Salisbury JW (1970) Visible and near-infrared spectra of minerals and rocks: I silicate minerals. Mod Geol 1:283–300

Huppert HE, Sparks RSJ (1985) Komatiites I: eruption and flow. J Petrol 26:694–725. https://doi.org/10.1093/petrology/26.3.694

Ivanov MA (2001) Morphology of the Tessera Terrain on Venus: implications for the composition of Tessera material. Sol Syst Res 35:1–17. https://doi.org/10.1023/A:1005289305927

Ivanov MA, Basilevsky AT (1993) Density and morphology of impact craters on Tessera Terrain. Venus Geophys Res Lett 20(23):2579–2582. https://doi.org/10.1029/93GL02692

Ivanov MA, Head JW (1996) Tessera terrain on Venus: a survey of the global distribution, characteristics, and relation to surrounding units from Magellan data. J Geophys Res, Planets 101(E6):14861–14908. https://doi.org/10.1029/96JE01245

Ivanov MA, Head JW (2004) Stratigraphy of small shield volcanoes on Venus: Criteria for determining stratigraphic relationships and assessment of relative age and temporal abundance. J Geophys Res, Planets 109(E10). https://doi.org/10.1029/2004JE002252

Ivanov MA, Head JW (2011) Global geologic map of Venus. Planet Space Sci 59(13):1559–1600. https://doi.org/10.1016/j.pss.2011.07.008

Ivanov MA, Head JW (2013) The history of volcanism on Venus. Planet Space Sci 84:66–92. https://doi.org/10.1016/j.pss.2013.04.018

Ivanov MA, Head JW (2015) Volcanically embayed craters on Venus: testing the catastrophic and equilibrium resurfacing models. Planet Space Sci 106:116–121. https://doi.org/10.1016/j.pss.2014.12.004

James PB, Zuber MT, Phillips RJ (2013) Crustal thickness estimates and support of topography on Venus. J Geophys Res, Planets 118(4):859–875. https://doi.org/10.1029/2012JE004237

Janes DM, Squyres SW (1995) Viscoelastic relaxation of topographic highs on Venus to produce coronae. J Geophys Res, Planets 100(E10):21173–21187. https://doi.org/10.1029/95JE01748

Janes DM, Squyres SW, Bindschadler DL, Baer G, Schubert G, Sharpton VL, Stofan ER (1992) Geophysical models for the formation and evolution of coronae on Venus. J Geophys Res, Planets 97(E10):16055–16067. https://doi.org/10.1029/92JE01689

Janle P, Jannsen D, Basilevsky AT (1988) Tepev Mons on Venus: morphology and elastic bending models. Earth Moon Planets 41(2):127–139. https://doi.org/10.1007/BF00056398

Jeffery AJ, Gertisser R (2018) Peralkaline Felsic Magmatism of the Atlantic Islands. Front Earth Sci 6. https://doi.org/10.3389/feart.2018.00145

Jellinek AM, Manga M (2004) Links between Long-Lived Hot Spots, Mantle Plumes, D″, and Plate Tectonics. Rev Geophys 42(3). https://doi.org/10.1029/2003RG000144

Jiménez-Díaz A, Ruiz J, Kirby JF, Romeo I, Tejero R, Caopte R (2015) Lithospheric structure of Venus from gravity and topography. Icarus 260:215–231. https://doi.org/10.1016/j.icarus.2015.07.020

Johnson CL, Sandwell DT (1992) Joints in Venusian lava flows. J Geophys Res, Planets 97(E8):13601–13610. https://doi.org/10.1029/92JE01212

Johnson CL, Sandwell DT (1994) Lithospheric flexure on Venus. Geophys J Int 119(2):627–647. https://doi.org/10.1111/j.1365-246X.1994.tb00146.x

Johnson TE, Brown M, Kaus BJP, VanTongeren JA (2014) Delamination and recycling of Archaean crust caused by gravitational instabilities. Nat Geosci 7:47–52. https://doi.org/10.1038/ngeo2019

Johnson TE, Brown M, Gardiner NJ, Kirkland CL, Smithies RH (2017) Earth's first stable continents did not form by subduction. Nature 543(7644):239–242. https://doi.org/10.1038/nature21383

Jull MG, Arkani-Hamed J (1995) The implications of basalt in the formation and evolution of mountains on Venus. Phys Earth Planet Inter 89(3–4):163–175. https://doi.org/10.1016/0031-9201(95)03015-O

Kargel JS, Komatsu G, Baker VR, Strom RG (1993) The volcanology of Venera and VEGA landing sites and the geochemistry of Venus. Icarus 103:253–275. https://doi.org/10.1006/icar.1993.1069

Kargel JS, Kirk RL, Fegley Jr B, Treiman AH (1994) Carbonate-sulfate volcanism on Venus? Icarus 112(1):219–252. https://doi.org/10.1006/icar.1994.1179

Kasbohm J, Schoene B (2018) Rapid eruption of the Columbia River flood basalt and correlation with the mid-Miocene climate optimum. Sci Adv 4(9):eaat8223. https://doi.org/10.1126/sciadv.aat8223

Kaula WM (1999) Constraints on Venus evolution from radiogenic argon. Icarus 139(1):32–39. https://doi.org/10.1006/icar.1999.6082

Keller J, Krafft M (1990) Effusive natrocarbonatite activity of Oldoinyo Lengai, June 1988. Bull Volcanol 52(8):629–645. https://doi.org/10.1007/BF00301213

Khawja S, Ernst RE, Samson C, Byrne PK, Ghail RC, MacLellan LM (2020) Tesserae on Venus may preserve evidence of fluivial erosion. Nat Commun 11:5789. https://doi.org/10.1038/s41467-020-19336-1

Kiefer WS, Hager BH (1991) A mantle plume model for the equatorial highlands of Venus. J Geophys Res 96:20947–20966. https://doi.org/10.1029/91JE02221

Kleinhans MG (2005) Flow discharge and sediment transport models for estimating a minimum timescale of hydrological activity and channel and delta formation on Mars. J Geophys Res 110:E1. https://doi.org/10.1029/2005JE002521

Klimczak C, Byrne PK, Şengör AM, Solomon SC (2019) Principles of structural geology on rocky planets. Can J Earth Sci 56(12):1437–1457. https://doi.org/10.1139/cjes-2019-0065

Klose KB, Wood JA, Hashimoto A (1992) Mineral equilibria and the high radar reflectivity of Venus mountaintops. J Geophys Res, Planets 97(E10):16353–16369. https://doi.org/10.1029/92JE01865

Knapmeyer M, Oberst J, Hauber E, Wählisch M, Deuchler C, Wagner R (2006) Working models for spatial distribution and level of Mars' seismicity. J Geophys Res, Planets 111(E11). https://doi.org/10.1029/2006JE002708

Koch DM (1994) A spreading drop model for plumes on Venus. J Geophys Res, Planets 99(E1):2035–2052. https://doi.org/10.1029/93JE03097

Koch DM, Manga M (1996) Neutrally buoyant diapirs: a model for Venus coronae. Geophys Res Lett 23(3):225–228. https://doi.org/10.1029/95GL03776

Koenig E, Pollard DD (1998) Mapping and modeling of radial fracture patterns on Venus. J Geophys Res, Solid Earth 103(B7):15183–15202. https://doi.org/10.1029/98JB00577

Komatsu G, Baker VR (1994) Meander properties of Venusian channels. Geology 22(1):67–70. https://doi.org/10.1130/0091-7613(1994)022<0067:MPOVC>2.3.CO;2

Komatsu G, Baker VR, Gulick VC, Parker TJ (1993) Venusian channels and valleys: distribution and volcanological implications. Icarus 102(1):1–25. https://doi.org/10.1006/icar.1993.1029

Komatsu G, Gulick VC, Baker VR (2001) Valley networks on Venus. Geomorphology 37(3):225–240. https://doi.org/10.1016/S0169-555X(00)00084-2

Konopliv AS, Banerdt WB, Sjogren WL (1999) Venus gravity: 180th degree and order model. Icarus 139(1):3–18. https://doi.org/10.1006/icar.1999.6086

Kopp RE, Kirschvink JL, Hilburn IA, Nash CZ (2005) The Paleoproterozoic snowball Earth: a climate disaster triggered by the evolution of oxygenic photosynthesis. Proc Natl Acad Sci 102(32):11131–11136. https://doi.org/10.1073/pnas.0504878102

Korycansky DG, Zahnle KJ (2005) Modeling crater populations on Venus and Titan. Planet Space Sci 53(7):695–710. https://doi.org/10.1016/j.pss.2005.03.002

Krassilnikov AS, Head JW (2003) Novae on Venus: Geology, classification, and evolution. J Geophys Res, Planets 108(E9). https://doi.org/10.1029/2002JE001983

Kremic T, Ghail R, Gilmore M, Hunter G, Kiefer W, Limaye S, Pauken M, Tolbert C, Wilson C (2020) Long-duration Venus lander for seismic and atmospheric science. Planet Space Sci 190:104961. https://doi.org/10.1016/j.pss.2020.104961

Kröner A (1985) Evolution of the Archean continental crust. Annu Rev Earth Planet Sci 13:49–74. https://doi.org/10.1146/annurev.ea.13.050185.000405

Kumar KV, Ernst WG, Leelandadam C, Wooden JL, Grove MJ (2010) First Paleoproterozoic ophiolite from Gondwana: geochronologic-geochemical documentation of ancient oceanic crust from Kandra, SE India. Tectonophysics 487:22–32. https://doi.org/10.1016/j.tecto.2010.03.005

Kusky TM, Li J-H, Tucker RD (2001) The Archean Dongwanzi ophiolite complex, North China craton: 2.505-billion-year-old oceanic crust and mantle. Science 292(5519):1142–1145. https://doi.org/10.1126/science.1059426

Kusky TM, Windley BF, Polat A (2018) Geologic evidence for the operation of plate tectonics throughout the Archean: records from Archean paleo-plate boundaries. J Earth Sci 29:1291–1303. https://doi.org/10.1007/s12583-018-0999-6

Lammer H, Zerkle AL, Gebauer S, Tost N, Noack L, Scherf M, Pilat-Lohinger E, Güdel M, Grenfell JL, Godolt M, Nikalaou A (2018) Origin and evolution of the atmospheres of early Venus, Earth and Mars. Astron Astrophys Rev 26:2. https://doi.org/10.1007/s00159-018-0108-y

Lancaster MG, Guest JE, Magee KP (1995) Great lava flow fields on Venus. Icarus 118(1):69–86. https://doi.org/10.1006/icar.1995.1178

Lang NP, López I (2015) The magmatic evolution of three Venusian coronae. Geol Soc (Lond) Spec Publ 401(1):77–95. https://doi.org/10.1144/SP401.3

Lebonnois S, Schubert G (2017) The deep atmosphere of Venus and the possible role of density-driven separation of CO_2 and N_2. Nat Geosci 10(7):473–477. https://doi.org/10.1038/ngeo2971

Lebonnois S, Schubert G, Forget F, Spiga A (2018) Planetary boundary layer and slope winds on Venus. Icarus 314:149–158. https://doi.org/10.1016/j.icarus.2018.06.006

Lécuyer C, Simon L, Guyot F (2000) Comparison of carbon, nitrogen and water budgets on Venus and the Earth. Earth Planet Sci Lett 181(1):33–40. https://doi.org/10.1016/S0012-821X(00)00195-3

Lee CT-A, Luffi P, Plank T, Dalton H, Leeman WP (2009) Constraints on the depths and temperatures of basaltic magma generation on Earth and other terrestrial planets using new thermobarometers for mafic magmas. Earth Planet Sci Lett 279:20–33. https://doi.org/10.1016/j.epsl.2008.12.020

Lenardic A, Kaula WM, Bindschadler DL (1991) The tectonic evolution of Western Ishtar Terra, Venus. Geophys Res Lett 18(12):2209–2212. https://doi.org/10.1029/91GL02734

 Springer

Li Y, Gurnis M (2023) Strike slip motion and the triggering of subduction initiation. Front Earth Sci 11:1156034. https://doi.org/10.3389/feart.2023.1156034

Lithgow-Bertelloni C, Richards MA (1995) Cenozoic plate driving forces. Geophys Res Lett 22:1317–1320. https://doi.org/10.1029/95GL01325

Lognonné P, Johnson C (2007) 10.03—Planetary seismology. In: Schubert G (ed) Treatise on geophysics. Elsevier, Amsterdam, pp 69–122. https://doi.org/10.1016/B978-044452748-6.00154-1

Lorenz RD (2018) Lightning detection on Venus: a critical review. Prog Earth Planet Sci 5:34. https://doi.org/10.1186/s40645-018-0181-x

Lorenz RD, Imai M, Takahashi Y, Sato M, Yamazaki A, Imamura T, Satoh T, Nakamura M (2019) Constraints on Venus lightning from Akatsuki's first 3 years in orbit. Geophys Res Lett 46(14):7955–7961. https://doi.org/10.1029/2019GL083311

Lourenço DL, Rozel AB, Ballmer MD, Tackley PJ (2020) Plutonic-squishy lid: a new global tectonic regime generated by intrusive magmatism on Earth-like planets. Geochem Geophys Geosyst 21(4):e2019GC008756. https://doi.org/10.1029/2019GC008756

Magee KP, Head JW (2001) Large flow fields on Venus: implications for plumes, rift associations, and resurfacing. In: Ernst RE, Buchan KL (eds) Mantle plumes: their identification through time. Spec Pap Geol Soc Am, vol 352, pp 81–101. https://doi.org/10.1130/0-8137-2352-3.81

Maia JS, Wieczorek MA (2022) Lithospheric structure of Venusian crustal plateaus. J Geophys Res, Planets 127:e2021JE007004. https://doi.org/10.1029/2021JE007004

Marcq E, Bertaux J-L, Montmessin F, Belyaev D (2013) Variations of sulphur dioxide at the cloud top of Venus's dynamic atmosphere. Nat Geosci 6:25–28. https://doi.org/10.1038/ngeo1650

Martin P, Stofan ER, Glaze LS, Smrekar SE (2007) Corona of Parga Chasma, Venus. J Geophys Res, Planets 112:E04S03. https://doi.org/10.1029/2006JE002758

McGill GE (1994) Hotspot evolution and Venusian tectonic style. J Geophys Res, Planets 99(E11):23149–23161. https://doi.org/10.1029/94JE02319

McGovern PJ, Solomon SC (1997) Filling of flexural moats around large volcanoes on Venus: implications for volcano structure and global magmatic flux. J Geophys Res 102:16303–16318

McGovern PJ, Solomon SE (1998) Growth of large volcanoes on Venus: mechanical models and implications for structural evolution. J Geophys Res, Planets 103(E5):11071–11101. https://doi.org/10.1029/98JE01046

McGovern PJ, Sean C, Solomon SC, Smith DE, Zuber MT, Simons M, Wieczorek MA, Phillips RJ, Neumann GA, Aharonson O, Head JW (2004) Correction to "Localized gravity/topography admittance and correlation spectra on Mars: implications for regional and global evolution" by Patrick J. McGovern, Sean C. Solomon, David E. Smith, Maria T. Zuber, Mark Simons, Mark A. Wieczorek, Roger J. Phillips, Gregory A. Neumann, Oded Aharonson, and James W. Head. J Geophys Res 107(E12):5136. https://doi.org/10.1029/2002JE001854

McGovern PJ, Rumpf ME, Zimbelman JR (2013) The influence of lithospheric flexure on magma ascent at large volcanoes on Venus. J Geophys Res, Planets 118(11):2423–2437. https://doi.org/10.1002/2013JE004455

McGovern PJ, Galgana GA, Verner KR, Herrick RR (2014) New constraints on volcanotectonic evolution of large edifices on Venus from stereo topography-derived strain estimates. Geology 42:59–62. https://doi.org/10.1130/G34919.1

McKenzie DP, Ford PG, Johnson CL, Parsons B, Sandwell DT, Saunders RS, Solomon SC (1992) Features on Venus generated by plate boundary processes. J Geophys Res, Planets 97(E8):13533–13544. https://doi.org/10.1029/92JE01350

McKinnon WB, Zahnle KJ, Ivanov BA, Melosh HJ (1997) Cratering on Venus: models and observations. In: Bougher SW, Hunten DM, Phillips RJ (eds) Venus II. University of Arizona Press, Tucson, pp 969–1014

McNutt MK (1984) Lithospheric flexure and thermal anomalies. J Geophys Res, Solid Earth 89(B13):11180–11194. https://doi.org/10.1029/JB089iB13p11180

Mitchell RN, Zhang N, Salminen J, Liu Y, Spencer CJ, Steinberger B, Murphy JB, Li Z-X (2021) The supercontinent cycle. Nature Rev Earth Environ 2:358–374. https://doi.org/10.1038/s43017-021-00160-0

Molnar N, Cruden A, Betts P (2020) The role of inherited crustal and lithospheric architecture during the evolution of the Red Sea: insights from three dimensional analogue experiments. Earth Planet Sci Lett 544:116377. https://doi.org/10.1016/j.epsl.2020.116377

Moore WB, Schubert G (1997) Venusian crustal and lithospheric properties from nonlinear regressions of highland geoid and topography. Icarus 128:415–428. https://doi.org/10.1006/icar.1997.5750

Moreels P, Smrekar SE (2003) Identification of polygonal patterns on Venus using mathematical morphology. IEEE, Trans Image Proc 1. https://doi.org/10.1109/TIP.2003.814254

Morgan WJ (1972) Deep mantle convection plumes and plate Motions1. AAPG Bull 56(2):203–213. https://doi.org/10.1306/819A3E50-16C5-11D7-8645000102C1865D

409

Morgan P (1983) Hot spot heat loss and tectonic style on Venus and in the Earth's Archean. Lunar and Planetary Science XIV, 515–516

Morgan P, Phillips RJ (1983) Hot spot heat transfer: its application to Venus and implications to Venus and Earth. J Geophys Res, Solid Earth 88(B10):8305–8317. https://doi.org/10.1029/JB088iB10p08305

Moyen JF, Martin H (2012) Forty years of TTG research. Lithos 148:312–336. https://doi.org/10.1016/j.lithos.2012.06.010

Mueller N, Helbert J, Hashimoto GL, Tsang CCC, Erard S, Piccioni G, Drossart P (2008) Venus surface thermal emission at 1 μm in VIRTIS imaging observations: evidence for variation of crust and mantle differentiation conditions. J Geophys Res, Planets 113(E5):E00B17. https://doi.org/10.1029/2008JE003118

Namiki N, Solomon SC (1993) The gabbro-eclogite phase transition and the elevation of mountain belts on Venus. J Geophys Res, Planets 98:15025–15031. https://doi.org/10.1029/93JE01626

Nance RD, Murphy JB, Santosh M (2014) The supercontinent cycle: a retrospective essay. Gondwana Res 25(1):4–29. https://doi.org/10.1016/j.gr.2012.12.026

Nikolayeva OV (1990) Geochemistry of the Venera 8 material demonstrates the presence of continental crust on Venus. Earth Moon Planets 50:329–341. https://doi.org/10.1007/BF00142398

Nimmo F (2002) Why does Venus lack a magnetic field? Geology 30(987). https://doi.org/10.1130/0091-7613(2002)030<0987:WDVLAM>2.0.CO;2

Nimmo F, Mackwell S (2023) Viscous relaxation as a probe of heat flux and crustal Plateau composition on Venus. Proc Natl Acad Sci 120(3):e2216311120. https://doi.org/10.1073/pnas.2216311120

Nimmo F, McKenzie D (1998) Volcanism and tectonics on Venus. Annu Rev Earth Planet Sci 26:23–51. https://doi.org/10.1146/annurev.earth.26.1.23

Nunes DC, Phillips RJ, Brown CD, Dombard AJ (2004) Relaxation of compensated topography and the evolution of crustal plateaus on Venus. J Geophys Res, Planets 109(E1):E01006. https://doi.org/10.1029/2003JE002119

O'Neil J, Carlson RW (2017) Building Archean cratons from Hadean mafic crust. Science 355:1199–1202. https://doi.org/10.1126/science.aah3823

O'Neil J, Carlson RW, Paquette JL, Francis D (2012) Formation age and metamorphic history of the Nuvvuagittuq Greenstone Belt. Precambrian Res 220:23–44. https://doi.org/10.1016/j.precamres.2012.07.009

O'Neill C, Lenardic A, Weller M, Moresi L, Quenette S, Zhang S (2016) A window for plate tectonics in terrestrial planet evolution? Phys Earth Planet Inter 255:80–92. https://doi.org/10.1016/j.pepi.2016.04.002

O'Rourke JG, Korenaga J (2015) Thermal evolution of Venus with argon degassing. Icarus 260:128–140. https://doi.org/10.1016/j.icarus.2015.07.009

O'Rourke JG, Smrekar SE (2018) Signatures of lithospheric flexure and elevated heat flow in stereo topography at coronae on Venus. J Geophys Res, Planets 123(2):369–389. https://doi.org/10.1002/2017JE005358

O'Rourke JG, Buz J, Fu RR, Lillis RJ (2019) Detectability of remanent magnetism in the Crust of Venus. Geophys Res Lett 46(11):5768–5777. https://doi.org/10.1029/2019GL082725

O'Rourke JG, Wilson CF, Borrelli ME, Byrne PK, Dumoulin C, Ghail R, Gülcher AJP, Jacobson SA, Korablev O, Spohn T, Way MJ, Weller M, Westall F (2023) Venus, the planet: introduction to the evolution of Earth's sister planet. Space Sci Rev 219:10. https://doi.org/10.1007/s11214-023-00956-0

Orth CP, Solomatov VS (2011) The isostatic stagnant lid approximation and global variations in the Venusian lithospheric thickness. Geochem Geophys Geosyst 12:Q07018. https://doi.org/10.1029/2011GC003582

Oshigami S, Namiki N (2007) Cross-sectional profiles of Baltis Vallis channel on Venus: reconstructions from Magellan SAR brightness data. Icarus 190(1,Pages):1–14. https://doi.org/10.1016/j.icarus.2007.03.011

Oshigami S, Namiki N, Komatsu G (2009) Depth profiles of venusian sinuous rilles and valley networks. Icarus 199(2):250–263. https://doi.org/10.1016/j.icarus.2008.10.012

Pang K-N, Arnd N, Svensen H, Polozov A, Polteau S, Iizuka Y, Chung S-L (2013) A petrologic, geochemical and Sr-Nd isotopic study on contact metamorphism and degassing of Devonian evaporates in the Norilsk aureoles. Siberia Contrib Mineral Petrol 165(4):683–704

Parfitt EA, Head JW (1993) Buffered and unbuffered dike emplacement on Earth and Venus: implications for magma reservoir size, depth, and rate of magma replenishment. Earth Moon Planets 61(3):249–281. https://doi.org/10.1007/BF00572247

Parmentier EM, Hess PC (1992) Chemical differentiation of a convecting planetary interior: consequences for a one plate planet such as Venus. Geophys Res Lett 19(20):2015–2018. https://doi.org/10.1029/92GL01862

Pauer M, Fleming K, Čadek O (2006) Modeling the dynamic component of the geoid and topography of Venus. J Geophys Res, Planets 111(E11). https://doi.org/10.1029/2005JE002511

Pavri B, Head III JW, Klose KB, Wilson L (1992) Steep-sided domes on Venus: characteristics, geologic setting, and eruption conditions from Magellan data. J Geophys Res, Planets 97(E8):13445–13478. https://doi.org/10.1029/92JE01162

Peltonen P, Kontinen A (2004) The Jormua Ophiolite: a mafic-ultramafic complex from an ancient ocean-continent transition zone. Dev Precambrian Geol 13:35–71. https://doi.org/10.1016/S0166-2635(04)13001-6

Pesonen LJS-A, Elming SÅ, Meranen S, Pisarevsky S, D'Agrella-Filho MS, Meert JG, Schmidt PW, Abrahamsen N, Bylund G (2003) Palaeomagnetic configuration of supercontinents during the Proterozoic. Tectonophysics 375:289–324. https://doi.org/10.1016/S0040-1951(03)00343-3

Phillips RJ, Hansen VL (1998) Geological evolution of Venus: rises, plains, plumes, and plateaus. Science 279(5356):1492–1497. https://doi.org/10.1126/science.279.5356.1492

Phillips RJ, Malin MC (1983) The interior of Venus and tectonic implications. In: Hunten DM, Colin L, Donahue TM, Moroz VI (eds) Venus. Univ. Arizona Press, Tucson, pp 159–214.

Phillips RJ, Raubertas RF, Arvidson RE, Sarkar IC, Herrick RR, Izenberg N, Grimm RE (1992) Impact craters and Venus resurfacing history. J Geophys Res, Planets 97(E10):15293–15948. https://doi.org/10.1029/92JE01696

Piskorz D, Elkins-Tanton LT, Smrekar SE (2014) Coronae formation on Venus via extension and lithospheric instability. J Geophys Res, Planets 119(12):2568–2582

Raitala J, Kauhanen K, Black M, Tokkonen T (1995) Crustal bending at Salme Dorsa on Venus. Planet Space Sci 43(8):1001–1012. https://doi.org/10.1016/0032-0633(95)00004-O

Reese CC, Solomatov VS, Moresi LN (1998) Heat transport efficiency for stagnant lid convection with dislocation viscosity: application to Mars and Venus. J Geophys Res 103(E6):13643–13657. https://doi.org/10.1029/98je01047

Reese CC, Solomatov VS, Orth CP (2007) Mechanisms for cessation of magmatic resurfacing on Venus. J Geophys Res 112:E04S04. https://doi.org/10.1029/2006JE002782

Reimink JR, Davies JHFL, Bauer AM, Chacko T (2020) A comparison between zircons from the Acasta gneiss complex and the Jack Hills region. Earth Planet Sci Lett 561:115975. https://doi.org/10.1016/j.epsl.2019.115975

Resor PG, Gilmore MS, Straley B, Senske DA, Herrick RR (2021) Felsic tesserae on Venus permitted by lithospheric deformation models. J Geophys Res, Planets 126(4):e2020JE006642. https://doi.org/10.1029/2020JE006642

Robinson CA, Wood JA (1993) Recent volcanic activity on Venus: evidence from radiothermal emissivity measurements. Icarus 102(1):26–39. https://doi.org/10.1006/icar.1993.1030

Rolf T, Steinberger B, Sruthi U, Werner SC (2018) Inferences on the mantle viscosity structure and the post-overturn evolutionary state of Venus. Icarus 313:107–123. https://doi.org/10.1016/j.icarus.2018.05.014

Rolf T, Weller M, Gülcher A et al (2022) Dynamics and evolution of Venus' mantle through time. Space Sci Rev 218:70. https://doi.org/10.1007/s11214-022-00937-9

Romeo I, Capote R (2011) Tectonic evolution of Ovda Regio: an example of highly deformed continental crust on Venus? Planet Space Sci 59(13):1428–1445. https://doi.org/10.1016/j.pss.2011.05.013

Romeo I, Turcotte DL (2008) Pulsating continents on Venus: an explanation for crustal plateaus and tessera terrains. Earth Planet Sci Lett 276(1–2):85–97. https://doi.org/10.1016/j.epsl.2008.09.00

Rosenblatt P, Pinet PC, Thouvenot E (1994) Comparative hypsometric analysis of Earth and Venus. Geophys Res Lett 21(6):465–468. https://doi.org/10.1029/94GL00419

Rossi F, Saboia M, Krishnamoorthy S, Vander Hook J (2023) Proximal exploration of Venus volcanism with teams of autonomous buoyancy-controlled balloons. Acta Astronaut 208:389–406. https://doi.org/10.1016/j.actaastro.2023.03.003

Roth ASG, Bourdon B, Mojzsis SJ, Rudge JF, Guitreau M, Blichert-Toft J (2014) Combined ^{147,146}Sm-^{143,142}Nd constraints on the longevity and residence time of early terrestrial crust. Geochem Geophys Geosyst 15(6):2329–2345. https://doi.org/10.1002/2014GC005313

Rubie DC, Jacobson SA, Morbidelli A, O'Brien DP, Young ED, de Vries J, Nimmo F, Palme H, Frost DJ (2015) Accretion and differentiation of the terrestrial planets with implications for the compositions of early-formed Solar System bodies and accretion of water. Icarus 248:89–108. https://doi.org/10.1016/j.icarus.2014.10.015

Ruiz J (2007) The heat flow during the formation of ribbon terrains on Venus. Planet Space Sci 55(14):2063–2070. https://doi.org/10.1016/j.pss.2007.05.003

Russell MB, Johnson CL (2021) Evidence for a locally thinned lithosphere associated with recent Volcanism at Aramaiti Corona, Venus. J Geophys Res, Planets 126(8):e2020JE006783. https://doi.org/10.1029/2020JE006783

Sabbeth L, Smrekar SE, Stock JM (2023) Estimated seismicity of Venusian wrinkle ridges based on fault scaling relationships. Earth Planet Sci Lett 619:118308. https://doi.org/10.1016/j.epsl.2023.118308

Sabbeth L, Carrington MA, Smrekar SE (2024) Constraints on corona formation from an analysis of topographic rims and fracture annuli. Earth Planet Sci Lett 633:118568. https://doi.org/10.1016/j.epsl.2024.118568

Sandwell DT, Schubert G (1992b) Flexural ridges, trenches, and outer rises around coronae on Venus. J Geophys Res, Planets 97(E10):16069–16083. https://doi.org/10.1029/92JE01274

Sandwell DT, Schubert G (1992a) Evidence for retrograde lithospheric subduction on Venus. Science 257(5071):766–770. https://doi.org/10.1126/science.257.5071.766

Sandwell DT, Johnson CL, Bilotti F, Suppe J (1997) Driving Forces for Limited Tectonics on Venus. Icarus 129(1):232–244. https://doi.org/10.1006/icar.1997.5721

Sarma DS, Parachuramulu V, Santosh M, Nagaraju E, Babu NR (2020) Pb-Pb baddeleyite ages of mafic dyke swarms from the Dharwar craton: implications for Paleoproterozoic LIPs and diamond potential of mantle keel. Geosci Front 11:2127–2139

Schaber GG, Strom RG, Moore HJ, Soderblom LA, Kirk RL, Chadwick DJ, Dawson DD, Gaddis LR, Boyce JM, Russell J (1992) Geology and distribution of impact craters on Venus: what are they telling us?. J Geophys Res, Planets 97(E8):13257–13301. https://doi.org/10.1029/92JE01246

Schools J, Smrekar SE (2024) Formation of coronae topography and fractures via plume buoyancy and melting. Earth Planet Sci Lett 633:118643. https://doi.org/10.1016/j.epsl.2024.118643.

Schubert G, Sandwell DT (1995) A global survey of possible subduction sites on Venus. Icarus 117(1):173–196. https://doi.org/10.1006/icar.1995.1150

Scott DJ, Helmstaedt H, Bickle MJ (1992) Purtuniq ophiolite, Cape Smith belt, northern Quebec, Canada: a reconstructed section of Early Proterozoic oceanic crust. Geology 20(2):173–176. https://doi.org/10.1130/0091-7613(1992)020<0173:POCSBN>2.3.CO;2

Scott CR, Mueller WU, Pilote P (2002) Physical volcanology, stratigraphy, and lithogeochemistry of an Archean volcanic arc: evolution from plume-related volcanism to arc rifting of SE Abitibi greenstone belt, Val d'Or, Canada. Precambr Res 115:223–260. https://doi.org/10.1016/S0301-9268(02)00011-6

Senske DA, Schaber GG, Stofan ER (1992) Regional topographic rises on Venus: geology of Western Eistla Regio and comparison to Beta Regio and Atla Regio. J Geophys Res, Planets 97(E8):13395–13420. https://doi.org/10.1029/92JE01167

Shalygin EV, Markiewicz WJ, Basilevsky AT, Titov DV, Ignatiev NI, Head JW (2015) Active volcanism on Venus in the Ganiki Chasma rift zone. Geophys Res Lett 42(12):4762–4769. https://doi.org/10.1002/2015GL064088

Shaw GH (2008) Earth's atmosphere – Hadean to early Proterozoic. Chem Erde 68(3):235–264. https://doi.org/10.1016/j.chemer.2008.05.001

Shellnutt JG (2013) Petrological modeling of basaltic rocks from Venus: a case for the presence of silicic rocks. J Geophys Res, Planets 118(6):1350–1364. https://doi.org/10.1002/jgre.20094

Shellnutt JG (2016) Mantle potential temperature estimates of basalt from the surface of Venus. Icarus 277:98–102. https://doi.org/10.1016/j.icarus.2016.05.014

Shellnutt JG (2018) Derivation of intermediate to silicic magma from the basalt analyzed at the Vega 2 landing site. Venus Public Library of Science PLOS ONE 13(3):e0194155. https://doi.org/10.1371/journal.pone.0194155

Shellnutt JG (2019) The curious case of the rock at Venera 8. Icarus 321:50–61. https://doi.org/10.1016/j.icarus.2018.11.001

Shellnutt JG (2021) Construction of a Venusian greenstone belt: a petrological perspective. Geosci Can 48(3):97–116. https://doi.org/10.12789/geocanj.2021.48.176

Shellnutt JG, Dostal J (2019) Haida Gwaii (British Columbia, Canada): a Phanerozoic analogue of a subduction-unrelated Archean greenstone belt. Sci Rep 9:3251. https://doi.org/10.1038/s41598-019-39818-7

Shellnutt JG, Manu Prasanth MP (2021) Modeling results for the composition and typology of non-primary Venusian anorthosite. Icarus 366:114531. https://doi.org/10.1016/l.icarus.2021.114531

Sifré D, Gardés E, Massuyeau M, Hashim L, Hier-Majumder S, Gaillard F (2014) Electrical conductivity during incipient melting in the oceanic low-velocity zone. Nature 509:81–85. https://doi.org/10.1038/nature13245

Simons M, Solomon SC, Hager BH (1997) Localization of gravity and topography: constraints on the tectonics and mantle dynamics of Venus. Geophys J Int 131(1):24–44. https://doi.org/10.1111/j.1365-246X.1997.tb00593.x

Smithies RH (2000) The Archaean tonalite-trondhjemite-granodiorite (TTG) series is not an analogue of Cenozoic adakite. Earth Planet Sci Lett 182(1):115–125 https://doi.org/10.1016/S0012-821X(00)00236-3

Smithies RH, Champion DC, Van Kranendonk MJ, Howard HM, Hickman AH (2005b) Modern-style subduction processes in the Mesoarchaean: geochemical evidence from the 3.12 Ga Whundo intra-oceanic arc. Earth Planet Sci Lett 231:221–237. https://doi.org/10.1016/j.epsl.2004.12.026

Smithies RH, Van Kranendonk MJ, Champion DC (2005a) It started with a plume – early Archaean basaltic proto-continental crust. Earth Planet Sci Lett 238:284–297. https://doi.org/10.1016/j.epsl.2005.07.023

Smrekar SE (1994) Evidence for active hotspots on Venus from analysis of Magellan Gravity Data. Icarus 112:2–26. https://doi.org/10.1006/icar.1994.1166

Smrekar SE, Parmentier EM (1996) The interaction of mantle plumes with surface thermal and chemical boundary layers: applications to hotspots on Venus. J Geophys Res, Solid Earth 101:5397–5410. https://doi.org/10.1029/95jb02877

Smrekar SE, Phillips RJ (1988) Gravity-driven deformation of the crust on Venus. Geophys Res Lett 15(7):693–696. https://doi.org/10.1029/GL015i007p00693

Smrekar SE, Phillips RJ (1991) Venusian highlands: geoid to topography ratios and their implications. Earth Planet Sci Lett 107(3–4):582–597. https://doi.org/10.1016/0012-821X(91)90103-O

Smrekar SE, Solomon SC (1992) Gravitational spreading of high terrain in Ishtar Terra, Venus. J Geophys Res, Planets 97(E10):16121–16148. https://doi.org/10.1029/92JE01315

Smrekar SE, Sotin C (2012) Constraints on mantle plumes on Venus: implications for volatile history. Icarus 217(2):510–523. https://doi.org/10.1016/j.icarus.2011.09.011

Smrekar SE, Stofan ER (1997) Corona formation and heat loss on Venus by coupled upwelling and delamination. Science 277(5330):1289–1294. https://doi.org/10.1126/science.277.5330.1289

Smrekar SE, Kiefer WS, Stofan ER (1997) Large Volcanic Rises on Venus. In: Bougher SW, Hunten DM, Phillips RJ, Matthews MS, Ruskin AS, Guerrieri ML (eds) Venus II. Geology, geophysics, atmosphere, and solar wind environment. University of Arizona Press, Tucson, pp 845–878

Smrekar SE, Moreels P, Franklin BJ (2002) Characterization and formation of polygonal fractures on Venus. J Geophys Res, Planets 107(E11):1–17. https://doi.org/10.1029/2001JE001808

Smrekar SE, Stofan ER, Mueller NT, Treiman AH, Elkins-Tanton LT, Helbert J, Piccioni G, Drossart P (2010) Recent hotspot volcanism on Venus from VIRTIS emissivity data. Science 328(5978):605–608. https://doi.org/10.1126/science.1186785

Smrekar SE, Davaille A, Sotin C (2018) Venus interior structure and dynamics. Space Sci Rev 214:88. https://doi.org/10.1007/s11214-018-0518-1

Smrekar SE, Hensley S, Nybakken R, Wallace MS, Perkovic-Martin D, You T-H, Nunes D, Brophy J, Ely T, Burst E, Dyar MD, Helbert J, Miller B, Hartley J, Kallemeyn P, Whitte J, Iess L, Mastrogiuseppe M, Younis M, Prts P, Rodriguez M, Mazarico R (2022a) VERITAS (Venus emissivity, radio science, InSAR, topography, and spectroscopy): a discovery mission. In: 2022 institute for electrical and electronics engineers/IEEE Aerospace Conference (AERO), pp 1–20. https://doi.org/10.1109/AERO53065.2022.9843269

Smrekar SE, Ostberg C, O'Rourke JG (2022b) Evidence for active rifting and Earth-like lithospheric thickness and heat flow on Venus. Nat Geosci. https://doi.org/10.1038/s41561-022-01068-0

Smrekar SE, Ostberg C, O'Rourke JG (2023) Earth-like lithospheric thickness and heat flow on Venus consistent with active rifting. Nat Geosci 16:13–18. https://doi.org/10.1038/s41561-022-01068-0

Snyder D (2002) Cooling of lava flows on Venus: the coupling of radiative and convective heat transfer. J Geophys Res, Planets 107(E10):5080. https://doi.org/10.1029/2001JE001501

Solomatov VS, Moresi L-N (1996) Stagnant lid convection on Venus. J Geophys Res, Planets 101:4737–4753. https://doi.org/10.1029/95je03361

Solomon SC, Head JW (1982) Mechanisms for lithospheric heat transport on Venus: implications for tectonic style and volcanism. J Geophys Res, Solid Earth 87(B11):9236–9246. https://doi.org/10.1029/JB087iB11p09236

Solomon SC, Head JW, Kaula WM, McKenzie D, Parsons B, Phillips RJ, Schubert G, Talwani M (1991) Venus tectonics: initial analysis from Magellan. Science 252(5003):297–312. https://doi.org/10.1126/science.252.5003.297

Solomon SC, Smrekar SE, Bindschadler DL, Grimm RE, Kaula WM, McGill GE, Phillips RJ, Saunders RS, Schubert G, Squyres SW, Stofan ER (1992) Venus tectonics: an overview of Magellan observations. J Geophys Res, Planets 97(E8):13199–13255. https://doi.org/10.1029/92JE01418

Solomon SC, Bullock MA, Grinspoon DH (1999) Climate change as a regulator of tectonics on Venus. Science 286(5437):87–90. https://doi.org/10.1126/science.286.5437.87

Spencer CJ, Kirkland CL, Taylor RJM (2016) Strategies towards statistically robust interpretations of in situ U-Pb zircon geochronology. Geosci Front 7(4):581–589. https://doi.org/10.1016/j.gsf.2015.11.006

Spencer CJ, Kirkland CL, Roberts NMW, Evans NJ, Liebmann J (2020) Strategies towards robust interpretations of in situ zircon Lu-Hf isotope analyses. Geosci Front 11(3):843–853. https://doi.org/10.1016/j.gsf.2019.09.004

Squyres SW, Jankowski DG, Simons M, Solomon SC, Hager BH, McGill GE (1992) Plains tectonism on Venus: the deformation belts of Lavinia Planitia. J Geophys Res 97(E8):13579–13599. https://doi.org/10.1029/92JE00481

Steinberger B, Werner SC, Torsvik TH (2010) Deep versus shallow origin of gravity anomalies, topography and volcanism on Earth, Venus and Mars. Icarus 207:564–577. https://doi.org/10.1016/j.icarus.2009.12.025

Stern RJ (2008) Modern-style plate tectonics began in Neoproterozoic time: an alternative interpretation of Earth's tectonic history. In: Condie KC, Pease V (eds) When did plate tectonics begin on planet Earth? Spec Pap Geol Soc Am, vol 440, pp 265–280

Stern RJ, Gerya T (2018) Subduction initiation in nature and models: a review. Tectonophysics 746:173–198. https://doi.org/10.1016/j.tecto.2017.10.014

Stevenson DJ, Cutts J, Mimoun D, Arrowsmith S, Banerdt B, Blom P, Brageot E, Brissaud Q, Chin G, Gao P, Garcia R, Hall J, Hunter G, Jackson J Kerzhanovic V, Kiefer W, Komjathy A, Lee C, Lognonné P, Lorenz R, Majid W, Majorradi M, Nolet G, O'Rourke J, Rolland L, Schubert G, Simons M, Sotin C, Spilker T, Tsai V (2015) Probing the interior structure of Venus. Keck Institute of Space Studies, California Institute of Technology, Pasadena. https://doi.org/10.26206/C1CX-EV12

Stoddard PR, Jurdy DM (2012) Topographic comparisons of uplift features on Venus and Earth: implications for Venus tectonics. Icarus 217(2):524–533. https://doi.org/10.1016/j.icarus.2011.09.003

Stofan ER, Saunders RS (1990) Geologic evidence of hotspot activity on Venus: predictions for Magellan. Geophys Res Lett 17(9):1377–1380. https://doi.org/10.1029/GL017i009p01377

Stofan ER, Smrekar SE (2005) Large topographic rises, coronae, large flow fields, and large volcanoes on Venus: evidence for mantle plumes? In: Foulger GR, Natland JH, Presnall DC, Anderson DL (eds) Plates, plumes and paradigms. Geological Society of America

Stofan ER, Bindschadler D, Parmentier EM, Head J (1991) Corona structures on Venus: models of origin. J Geophys Res 96:20933–20946. https://doi.org/10.1029/91JE02218

Stofan ER, Sharpton VL, Schubert G, Baer G, Bindschadler DL, Janes DM, Squyres SW (1992) Global distribution and characteristics of coronae and related features on Venus: implications for origin and relation to mantle processes. J Geophys Res, Planets 97(E8):13347–13378. https://doi.org/10.1029/92JE01314

Stofan ER, Saunders RS, Senske D et al (1993) Venus interior structure mission (VISM): establishing a seismic network on Venus. In: Workshop on Advanced Technologies for Planetary Instruments, Part 1, SEE N93-28764 11-91. Lunar and Planetary Institute, Houston, pp 23–24

Stofan ER, Smrekar SE, Bindschadler DL, Senske DA (1995) Large topographic rises on Venus: implications for mantle upwelling. J Geophys Res, Planets 100(E11):23317–23327. https://doi.org/10.1029/95JE01834

Stofan ER, Hamilton VE, Janes DM, Smrekar SE (1997) Coronae on Venus: morphology and origin, Venus II: geology, geophysics, atmosphere, and solar wind environment. In Bougher SW, Hunten DM, Philips RJ (eds) University of Arizona Press, Tucson, p 931

Stofan ER, Anderson SW, Crown DA, Plaut JJ (2000) Emplacement and composition of steep-sided domes on Venus. J Geophys Res 105(E11):26,757–26,771. https://doi.org/10.1029/1999JE001206

Stofan ER, Smrekar SE, Tapper SW, Guest JE, Grindrod PM (2001) Preliminary analysis of an expanded corona database for Venus. Geophys Res Lett 28(22):4267–4270. https://doi.org/10.1029/2001GL013307

Stofan ER, Smrekar SE, Mueller N, Helbert J (2016) Themis regio, Venus: evidence for recent (?) volcanism from VIRTIS data. Icarus 271:375–386. https://doi.org/10.1016/j.icarus.2016.01.034

Strom RG, Schaber GG, Dawson DD (1994) The global resurfacing of Venus. J Geophys Res, Planets 99(E5):10899–10926. https://doi.org/10.1029/94JE00388

Suppe J, Connors C (1992) Critical taper wedge mechanics of fold-and-thrust belts on Venus: initial results from Magellan. J Geophys Res, Planets 97(E8):13545–13561. https://doi.org/10.1029/92JE01164

Surkov YA, Barsukov VL, Moskalyeva LP, Kharyukova VP, Kemurdzhian AL (1984) New data on the composition, structure, and properties of Venus rock obtained by Venera 13 and Venera 14. J Geophys Res, Solid Earth 89(S02):B393–B402. https://doi.org/10.1029/JB089iS02p0B393

Surkov YA, Moskalyova LP, Kharyukova VP, Dudin AD, Smirnov GG, Zaitseva SY (1986) Venus rock composition at the Vega 2 Landing Site. J Geophys Res, Solid Earth 91(B13):E215–E218. https://doi.org/10.1029/JB091iB13p0E215

Tackley PJ, Stevenson DJ (1991) The production of small Venusian coronae by Rayleigh-Taylor instabilities in the uppermost mantle. Eos, Trans Am Geophys Union 72:287–287

Taylor SR, McLennan SM (1995) The geochemical evolution of the continental crust. Rev Geophys 33(2):241–265. https://doi.org/10.1029/95RG00262

Taylor FW, Svedhem H, Head JW (2018) Venus: the atmosphere, climate, surface, interior and near-space environment of an Earth-like planet. Space Sci Rev 214. https://doi.org/10.1007/s11214-018-0467-8

Therriault AM, Fowler AD, Grieve RA (2002) The Sudbury Igneous Complex: a differentiated impact melt sheet. Econ Geol 97(7):1521–1540. https://doi.org/10.2113/gsecongeo.97.7.1521

Thordarson T, Garcia MO (2018) Variance of the Flexure Model Predictions with Rejuvenated Volcanism at Kīlauea Point, Kaua'i, Hawai'i. Front Earth Sci 6. https://doi.org/10.3389/feart.2018.00121

Thurston PC (2015) Greenstone belts and granite-greenstone terranes: constraints on the nature of the Archean world. Geosci Can 42(4):437–484. https://doi.org/10.12789/geocanj.2015.42.081

Torsvik TH, van der Voo R, Doubrovine PV, Burke K, Steinberger B, Ashwal LD, Trønnes RG, Webb SJ, Bull AL (2014) Deep mantle structure as a reference frame for movements in and on the Earth. Proc Natl Acad Sci 111(24):8735–8740. https://doi.org/10.1073/pnas.1318135111

Treiman AH (2007) Geochemistry of Venus' surface: current limitations as future opportunities. In: Esposito LW, Stofan ER, Cravens TE (eds) Exploring Venus as a Terrestrial Planet. Geophysical monograph series, vol 176. https://doi.org/10.1029/176GM03

Treiman A, Harrington E, Sharpton V (2016) Venus' radar-bright highlands: different signatures and materials on Ovda Regio and on Maxwell Montes. Icarus 280:172–182. https://doi.org/10.1016/j.icarus.2016.07.001

Tucker WS, Dombard AJ (2023) Evidence of Topographic Change Recorded by Lava Flows at Atete and Aruru Coronae on Venus. Journ Geophys Res Planets 128(11):e2023JE007971. https://doi.org/10.1029/2023JE007971

Tuckwell GW, Ghail RC (2003) A 400-km-scale strike-slip zone near the boundary of Thetis Regio. Venus Earth Planet Sci Lett 211(1–2):45–55. https://doi.org/10.1016/S0012-821X(03)00128-6

Turcotte DL (1993) An episodic hypothesis for Venusian tectonics. J Geophys Res, Planets 98(E9):17061–17068. https://doi.org/10.1029/93JE01775

Turcotte DL, Morein G, Roberts D, Malamud BD (1999) Catastrophic resurfacing and episodic subduction on Venus. Icarus 139(1):49–54. https://doi.org/10.1006/icar.1999.6084

Turner S, Rushmer T, Reagan M, Moyen J-F (2014) Heading down early on? Start of subduction on Earth. Geology 42(2):139–142. https://doi.org/10.1130/G34886.1

Ueda K, Gerya T, Sobolev SV (2008) Subduction initiation by thermal-chemical plumes: numerical studies. Phys Earth Planet Inter 171(1–4):296–312. https://doi.org/10.1016/j.pepi.2008.06.032

Van Kranendonk MJ (2010) Two types of Archean continental crust: plume and plate tectonics on early Earth. Am J Sci 310(10):1187–1209

Wang YJ, Shellnutt JG, Kung J, Iizuka Y, Lai Y-M (2022) The formation of tonalitic and granodiortiic melt from Venusian basalt. Sci Rep 12:1652. https://doi.org/10.1038/s41598-022-05745-3

Warren AO, Kite ES (2023) Narrow range of early habitable Venus scenarios permitted by modeling of oxygen loss and radiogenic argon degassing. Proc Natl Acad Sci 120(11):e2209751120. https://doi.org/10.1073/pnas.2209751120

Way MJ Del Genio AD (2020) Venusian habitable climate scenarios: modeling Venus through time and applications to slowly rotating Venus-like exoplanets. J Geophys Res, Planets 125(5):e2019JE006276. https://doi.org/10.1029/2019JE006276

Way MJ, Del Genio AD, Kiang NY, Sohl LE, Grinspoon DH, Aleinov I, Kelley M, Clune T (2016) Was Venus the first habitable world of our solar system? Geophys Res Lett 43(16):8376–8383. https://doi.org/10.1002/2016GL069790

Wei DY, Yang A, Huang JS (2014) The gravity field and crustal thickness of Venus. Sci China Earth Sci 57(9):2025–2035. https://doi.org/10.1007/s11430-014-4824-5

Wei SS, Shearer PM, Lithgow-Bertelloni C, Stixrude L, Tian D (2020) Oceanic Plateau of the Hawaiian mantle plume head subducted to the uppermost lower mantle. Science 370(6519):983–987. https://doi.org/10.1126/science.abd0312

Weller MB, Duncan MS (2015) Insight into terrestrial planetary evolution via mantle potential temperatures. 46th Lunar Planetary Science Conference. Abstract #2749

Weller MB, Lenardic A, O'Neill C (2015) The effects of internal heating and large-scale climate variations on tectonic bi-stability in terrestrial planets. Earth Planet Sci Lett 420:85–94. https://doi.org/10.1016/j.epsl.2015.03.021

Westall F, Höning D, Avice G, Gentry D, Gerya T, Gillmann C, Izenberg N, Way MJ, Wilson C (2023) The habitability of Venus. Space Sci Rev 219:17. https://doi.org/10.1007/s11214-023-00960-4

Whattam SA, Stern RJ (2015) Late Cretaceous plume-induced subduction initiation along the southern margin of the Caribbean and NW South America: The first documented example with implications for the onset of plate tectonics. Gondwana Research 27(1):38–63. https://doi.org/10.1016/j.gr.2014.07.011

White RS, McKenzie D (1995) Mantle plumes and flood basalts. J Geophys Res, Solid Earth 100(B9):17543–17585. https://doi.org/10.1029/95JB01585

Whitten JL, Campbell BA (2016) Recent volcanic resurfacing of Venusian craters. Geology 44(7):519–522. https://doi.org/10.1130/G37681.1

Widemann T, Smrekar S, Garvin J, Straume-Lindner AG, Ocampo A, Schulte M, Voirin T, Hensley S, Dyar MD, Whitten J, Nunes D, Getty S, Arney G, Johnson N, Kohler E, Spohn T, O'Rourke JG, Wilson C, Way M, Ostberg C, Westall F, Höning D, Jacobson S, Salvador A, Avice G, Breuer D, Carter L, Gilmore M, Ghail R, Helbert J, Byrne P, Santos A, Herrick R, Izenberg N, Marcq E, Rolf T, Weller M, Gillmann C, Korablev O, Zelenyi L, Zasova L, Gorinov D, Seth G, Narasimha Rao CV, Desai N (2023) Venus evolution through time: key science questions, selected mission concepts and future investigations. Space Sci Rev 219:56. https://doi.org/10.1007/s11214-023-00992-w

Wiens RC, Maurice S, Barraclough B et al (2012) The ChemCam instrument suite on the Mars Science Laboratory (MSL) rover: body unit and combined system tests. Space Sci Rev 170:167–227. https://doi.org/10.1007/s11214-012-9902-4

Williams DA, Kerr RC, Lesher CM, Barnes SJ (2001) Analytical/numerical modeling of komatiite lava emplacement and thermal erosion at Perseverance, Western Australia. J Volcanol Geotherm Res 110(1–2):27–55. https://doi.org/10.1016/S0377-0273(01)00206-2

Williams-Jones G, Williams-Jones AE, Stix J (1998) The nature and origin of Venusian canali. J Geophys Res, Planets 103(E4):8545–8555. https://doi.org/10.1029/98JE00243

Wilson CF, Chassefière E, Hinglais E et al (2012) The 2010 European Venus Explorer (EVE) mission proposal. Exp Astron 33:305–335. https://doi.org/10.1007/s10686-011-9259-9

Wilson CF, Marcq E, Gillmann C, Widemann T, Korablev O, Mueller N, Lefevre M, Rimmer P, Robert S, Zolotov M (2024) Magmatic volatiles and effects on the modern atmosphere of Venus. Space Sci Rev this collection, in revision

Windely BF, Kusky T, Polat A (2021) Onset of plate tectonics by the Eoarchean. Precambrian Res 352:105980. https://doi.org/10.1016/j.precamres.2020.105980

Windley BF, Kusky T, Polat A (2021) Onset of plate tectonics by the Eoarchean. Precambrian Res 352:105980. https://doi.org/10.1016/j.precamres.2020.105980

Wroblewski FB, Treiman AH, Bhiravarasu S, Gregg TKP (2019) Ovda Fluctus, the festoon lava flow on Ovda Region, Venus: not silica-rich. J Geophys Res, Planets 124:2233–2245. https://doi.org/10.1029/2019JE006039

Wyman DA, Kerrich R, Polat A (2002) Assembly of Archean cratonic mantle lithosphere and crust: plume-arc interaction in the Abitibi-Wawa subduction-accretion complex. Precambrian Res 115:37–62. https://doi.org/10.1016/S0301-9268(02)00005-0

Zahnle KJ (1992) Airburst origin of dark shadows on Venus. J Geophys Res, Planets 97(E6):10243–10255. https://doi.org/10.1029/92JE00787

Zegers TE, van Keken PE (2001) Middle Archean continent formation by crustal delamination. Geology 29(12):1083–1086. https://doi.org/10.1130/0091-7613(2001)029<1083:MACFBC>2.0.CO;2

Zuber MT (1990) Ridge belts: evidence for regional- and local-scale deformation on the surface of Venus. Geophys Res Lett 17(9):1369–1372. https://doi.org/10.1029/GL017i009p01369

Publisher's Note Springer Nature remains neutral with regard to jurisdictional claims in published maps and institutional affiliations.

Authors and Affiliations

Richard C. Ghail[1] · **Suzanne E. Smrekar[2]** · **Thomas Widemann[3,4]** · **Paul K. Byrne[5]** · **Anna J.P. Gülcher[2]** · **Joseph G. O'Rourke[6]** · **Madison E. Borrelli[6]** · **Martha S. Gilmore[7]** · **Robert R. Herrick[8]** · **Mikhail A. Ivanov[9]** · **Ana-Catalina Plesa[10]** · **Tobias Rolf[11]** · **Leah Sabbeth[2]** · **Joe W. Schools[12]** · **J. Gregory Shellnutt[13]**

✉ R.C. Ghail
richard.ghail@rhul.ac.uk

S.E. Smrekar
suzanne.e.smrekar@jpl.nasa.gov

T. Widemann
thomas.widemann@obspm.fr

P.K. Byrne
paul.byrne@wustl.edu

A.J.P. Gülcher
anna.gulcher@caltech.edu

J.G. O'Rourke
jgorourk@asu.edu

M.E. Borrelli
meborrel@asu.edu

M.S. Gilmore
mgilmore@wesleyan.edu

R.R. Herrick
rrherrick@alaska.edu

M.A. Ivanov
mikhail_ivanov@brown.edu

A.-C. Plesa
ana.plesa@dlr.de

T. Rolf
tobias.rolf@geo.uio.no

L. Sabbeth
leah.sabbeth@jpl.nasa.gov

J.W. Schools
jschools@arizona.edu

J. Gregory Shellnutt
jgshelln@ntnu.edu.tw

[1] Department of Earth Sciences, Royal Holloway, University of London, Egham, Surrey, TW20 0EX, UK

[2] Jet Propulsion Laboratory, California Institute of Technology, 4800 Oak Grove Drive, Pasadena, CA 91109, USA

[3] LESIA, Observatoire de Paris, Université PSL, CNRS, Sorbonne Université, Université Paris Cité, 5 place Jules Janssen, 92195 Meudon, France

[4] Université Paris-Saclay, UVSQ, DYPAC, 78000 Versailles, France

[5] Department of Earth, Environmental, and Planetary Sciences, Washington University, St. Louis, MO 63130, USA

[6] Arizona State University, School of Earth and Space Exploration, Tempe, AZ 85287, USA

[7] Department of Earth and Environmental Sciences, Wesleyan University, 265 Church Street, Middletown, CT 06457, USA

[8] Institute of Northern Engineering, University of Alaska Fairbanks, Fairbanks, AK 99775-5910, USA

[9] Vernadsky Institute of Geochemistry and Analytical Chemistry, RAS, 119991, Moscow, Russia

[10] Institute of Planetary Research, Deutsches Zentrum für Luft- und Raumfahrt, Rutherfordstraße 2, 12489 Berlin, Germany

[11] Centre for Earth Evolution and Dynamics, Dept. of Geosciences, University of Oslo, PO Box 1028 Blindern, 0315 Oslo, Norway

[12] Lunar and Planetary Laboratory, The University of Arizona, PO Box 210092, Tucson, AZ 85721, USA

[13] Department of Earth Sciences, National Taiwan Normal University, 88 Tingzhou Road Sect. 4, Taipei 11677, Taiwan

Space Science Reviews (2023) 219:52
https://doi.org/10.1007/s11214-023-00988-6

Mineralogy of the Venus Surface

Martha S. Gilmore[1] · M. Darby Dyar[2,3] · Nils Mueller[4] · Jérémy Brossier[5] ·
Alison R. Santos[1] · Mikhail Ivanov[6] · Richard Ghail[7] · Justin Filiberto[8] ·
Jörn Helbert[4]

Received: 27 February 2023 / Accepted: 11 July 2023 / Published online: 20 September 2023
© The Author(s) 2023

Abstract

Surface mineralogy records the primary composition, climate history and the geochemical cycling between the surface and atmosphere. We have not yet directly measured mineralogy on the Venus surface in situ, but a variety of independent investigations yield a basic understanding of surface composition and weathering reactions in the present era where rocks react under a supercritical atmosphere dominated by CO_2, N_2 and SO_2 at ~460 °C and 92 bars. The primary composition of the volcanic plains that cover ~80% of the surface is inferred to be basaltic, as measured by the 7 Venera and Vega landers and consistent with morphology. These landers also recorded elevated SO_3 values, low rock densities and spectral signatures of hematite consistent with chemical weathering under an oxidizing environment. Thermodynamic modeling and laboratory experiments under present day atmospheric conditions predict and demonstrate reactions where Fe, Ca, Na in rocks react primarily with S species to form sulfates, sulfides and oxides. Variations in surface emissivity at ~1 μm detected by the VIRTIS instrument on the Venus Express orbiter are spatially correlated to geologic terrains. Laboratory measurements of the near-infrared (NIR) emissivity of geologic materials at Venus surface temperatures confirms theoretical predictions that 1 μm emissivity is directly related to Fe^{2+} content in minerals. These data reveal regions of high emissivity that may indicate unweathered and recently erupted basalts and low emissivity associated with tessera terrain that may indicate felsic materials formed during a more clement era. Magellan radar emissivity also constrain mineralogy as this parameter is inversely related to the type and volume of high dielectric minerals, likely to have formed due to surface/atmosphere reactions. The observation of both viscous and low viscosity volcanic flows in Magellan images may also be related to composition. The global NIR emissivity and high-resolution radar and topography collected by the VERITAS, EnVision and DAVINCI missions will provide a revolutionary advancement of these methods and our understanding of Venus mineralogy. Critically, these datasets must be supported with both laboratory experiments to constrain the style and rate weathering reactions and laboratory measurements of their NIR emissivity and radar characteristics at Venus conditions.

Keywords Venus · Mineralogy · Spectroscopy

Venus: Evolution Through Time
Edited by Colin F. Wilson, Doris Breuer, Cédric Gillmann, Suzanne E. Smrekar, Tilman Spohn and Thomas Widemann

Extended author information available on the last page of the article

1 Introduction

Mineralogy is key to understanding the geologic and climate evolution of all terrestrial planets. The mineralogical analysis of meteorites, returned samples and in situ measurements for the Moon, Mars and some asteroids form the basis of our understanding of the bulk composition of the solar system and the influence of water over the history of these bodies. For airless bodies and those with thinner atmospheres, mineralogy can be assessed from orbit using spectroscopy in the visible to thermal wavelengths, however the massive atmosphere of Venus prohibits hyperspectral imaging from orbit in these wavelengths.

Meadows and Crisp (1996) observed the nightside of Venus from Earth and identified 5 spectral windows between 1.00–1.31 μm in the near infrared (NIR) where radiance from the surface can be detected above the Venus atmosphere. These windows were exploited by the Cassini Visible and Infrared Mapping Spectrometer (VIMS) instrument (Baines et al. 2000), the Galileo Near Infrared Mapping Spectrometer (NIMS) instrument (Hashimoto et al. 2008) and the Parker Solar Probe Wide-Field Imager for Solar Probe (WISPR) instrument (Wood et al. 2022) during Venus flybys. Hashimoto et al. (2008) showed that recognizable surface features were visible from orbit at 1.18 μm and developed a methodology to use other channels to correct for the influence of the atmosphere in the surface emissivity data. Because the ∼1 μm spectral region is sensitive to FeO content in minerals, these data can be used to constrain global surface composition (Dyar et al. 2020). Hashimoto and Sugita (2003) calculated that NIR radiance differences could be related to variations in mineralogy. The Visible and InfRared Thermal Imaging Spectrometer (VIRTIS) instrument on Venus Express (VEx) mapped the NIR radiance of the southern hemisphere of Venus, and demonstrated that radiance anomalies correlate to geological features mapped by Magellan where variations can be plausibly related to mineralogy derived from thermodynamic modeling of the primary composition and weathering reactions at the Venus surface (Mueller et al. 2008; Helbert et al. 2008; Smrekar et al. 2010). These inferences are supported by recent laboratory measurements of rocks and minerals at Venus temperatures that confirm a relationship between FeO content and ∼1 μm emissivity (Dyar et al. 2020; Helbert et al. 2021). These observations are the basis for the design of the Venus Emissivity Mapper (VEM) instrument on NASA's Venus Emissivity, Radio science, InSAR, Topography, And Spectroscopy (VERITAS) mission (Helbert et al. 2020), the Venus Imaging System for Observational Reconnaissance (VISOR) and Venus Descent Imager (VenDI) instruments on NASA's Deep Atmosphere Venus Investigation of Noble gases, Chemistry, and Imaging (DAVINCI) mission (Garvin et al. 2022) and the VenSpec-M imaging system on ESA's EnVision mission (Widemann et al. 2021), which will for the first time provide maps of the global distribution of ∼1 μm emissivity from orbit and at a single site below the clouds.

On the eve of these revolutionary measurements, we summarize the current understanding of the mineralogy of Venus.

2 Theory

2.1 NIR Emissivity

Spectral windows permit the observation of Venus surface thermal emission through the cloud deck. Surface thermal emission has been detected in windows within the wavelength range of 0.7 to 1.2 μm (Lecacheux et al. 1993; Meadows and Crisp 1996; Baines et al. 2000; Wood et al. 2022). Surface thermal emission can be described by the product of

Planck-function of the surface temperature and surface emissivity, which is indicative of surface composition. The radiation from the surface is modified during its transfer through the atmosphere by scattering, absorption, and emission by atmospheric gasses and aerosols. Thus, the derivation of emissivity from orbiter observations requires constraints on the surface temperature as well as on the variables going into the radiative transfer model used for atmospheric correction. The overall uncertainties of these parameters are large, so that an absolute emissivity cannot be derived with a useful uncertainty from single observations (Kappel et al. 2015). Some uncertainties of atmospheric variables such as the continuum absorption are large but affect all locations equally. Other error sources are related to cloud microphysics and can be assumed to be random errors at each time a location is observed. Taken together it is possible to average images of a specific region acquired at different times to reduce the uncertainty of the relative emissivity so that studies of surface composition are feasible (e.g., Dyar et al. 2020). In the following we try to highlight the most important open questions and limitations of deriving emissivity from orbiter observations.

The atmospheric transmittance within the spectral windows is dominated by the effect of scattering of molecules and cloud particles, with little absorption. This has two important consequences. First, the scattering randomizes the direction of photons so that the top-of-atmosphere radiance is a blurred image of the surface thermal emission. Moroz (2002) estimates that the blurring reduces the achievable spatial resolution to about two times the altitude of the cloud base, which is limited by the stability of H_2SO_4 to be above 45 km (Ragent et al. 1985). Hashimoto and Imamura (2001) and Basilevsky et al. (2012) ran Monte Carlo models of photon scattering and describe the effect of scattering as similar to convolution of the surface image with a Gaussian curve with a full width at half maximum of 90 and 100 km, respectively. There is so far no study that confirms this theoretical prediction. Kappel et al. (2015) note that the averaging of several images acquired by orbiters with a certain pointing uncertainty introduces an additional blurring effect. The achievable resolution of observations from below the cloud deck can be significantly higher as the fraction of photons only slightly deviating from the direct path is larger (Knicely and Herrick 2020; Ekonomov 2015), although some reduction of the contrast from diffuse radiation is to be expected (Moroz 2002), and has to be corrected for in surface imaging studies.

The second important effect of scattering is that a significant amount of upwelling radiation is reflected downward, and thus illuminates the surface. Due to Kirchhoff's law this results in a non-linear relation of top-of-atmosphere radiance as function of surface emissivity, which reduces the sensitivity of the measurements to emissivity at high emissivity values (Hashimoto and Sugita 2003). Since the relation is non-linear, and the absolute emissivity cannot be strongly constrained, there is additionally an uncertainty in the amplitude of any observed emissivity differences. Smrekar et al. (2010) used the analytical approximation of this effect from Hashimoto and Sugita (2003) to estimate that the average emissivity of Venus is around 0.6 to maintain relatively high emissivity locations within a plausible range of less than 0.95. However, this approximation neglects the effect of gaseous absorption, and recent analyses of these data with numerical radiative transfer models shows much smaller emissivity differences on the order of a few percent (Kappel et al. 2016; Mueller et al. 2020). As discussed in the following, the atmospheric absorption itself is not well known. This highlights the importance of "ground truth" comparison of orbiter data to in situ data, such as that collected by the spectrophotometers on the Venera landers (Ekonomov et al. 1980; Helbert et al. 2021), which will be an extremely useful synergy for the VERITAS, DAVINCI and EnVision missions.

The gaseous absorption deep in the atmosphere of Venus at wavelengths far from absorption band centers is not well known. The deep atmosphere is far from an ideal gas, molecular collisions are frequent, which results in a significant continuum of absorption (Pollack et al. 1993). Laboratory measurements of this continuum at both Venus near-surface temperature and pressure have not yet been achieved, although laboratory capabilities are approaching these conditions (Snels et al. 2021). The lack of constraints on this collision induced absorption (CIA) coefficient means that the average radiance observed at the cloud tops can correspond to a wide range of surface emissivity (Mueller et al. 2020). The CIA also has an effect on the gradient of radiance with respect to topography, and this has been used to constrain the CIA (Fedorova et al. 2015). This estimate however implies that the relation of surface temperature and topography is well known.

The surface is generally assumed to be very near to thermal equilibrium with the lowest atmosphere, due to the long solar day (243 Earth days), the high heat capacity of the atmosphere and the small insolation at the surface. The temperature of the lowest atmospheric layer is however not well known. Most studies aiming to derive surface emissivity from NIR radiances have relied on the temperature profile defined by the Venus International Reference Atmosphere (VIRA, Seiff et al. 1985). Most of the studies that use VIRA to derive emissivity find that there is a trend of the derived emissivity with topography (Meadows and Crisp 1996; Haus and Arnold 2010; Basilevsky et al. 2012; Kappel et al. 2016; Mueller et al. 2020). The trend is much more gradual than the trend observed in the radiothermal emissivity, and begins at lower elevations, indicating that there is not a common cause for these two phenomena. Meadows and Crisp (1996) instead propose that the lapse rate is different from the VIRA model.

The VIRA model is based mostly on the four Pioneer Venus descent probes, which provide temperature profiles from a wide range of local times and latitudes that converge to similar temperatures with increasing depth in the atmosphere. They however all experienced malfunctions around 18 km altitude. The temperatures in the VIRA model are extrapolated downwards and thus may not include any variation in the temperature lapse rate. The Vega 2 probe observed such variation (Seiff 1987) in the form of a layer between 2 and 7 km altitude, where the lapse rate exceeds that of the VIRA model. The lapse rate also exceeds the calculated adiabatic lapse rate, indicating an unstable stratification that is difficult to explain. The possibility of such a superadiabatic lapse rate is discussed as explanation for the low brightness temperature of Venus highlands observed by Galileo NIMS, but Hashimoto et al. (2008) reject this as impossible and instead interpret it as low emissivity highlands to have a more felsic composition. Lebonnois and Schubert (2017) more recently propose that the Vega profile can be explained by a chemical gradient in the lower atmosphere, i.e., that increasing density stabilizes the apparently superadiabatic layer. It is not clear yet what could cause such a chemical gradient, but it is becoming clearer that the structure and composition of the lower atmosphere may still surprise us when the DAVINCI probe will make its descent. Without better knowledge of the lowest atmosphere, it should be stressed that only robust interpretations are those that compare observed emissivity differences between areas of similar elevation and thus temperature. This approach is utilized by Gilmore et al. (2015) to show that the Alpha Regio tessera terrain has a lower emissivity than the volcanic plains of the adjacent Eve Corona in the same topography range. The close proximity of Alpha and Eve Corona also mitigate potential regional variations in surface temperature, further supporting a correlation between the low emissivity values and surface properties (Sect. 3.2.1).

Mueller et al. (2020) has shown that the apparent trend of derived emissivity with topography varies from region to region. The trend appears to be consistent in regions with scales on the order of 1000 km, so that emissivity variation on smaller scales can be interpreted by correcting for this trend. The causes for these trends are likely the temperatures in the planetary boundary layer and the effect of surface topography on atmospheric dynamics. Global

Circulation Models (GCM) by Lebonnois et al. (2018) show that the atmospheric temperatures show lateral temperature variations that are qualitatively consistent with the variations of the emissivity trend (Mueller et al. 2018). The GCM predicts slope winds associated with the topography of the two regions compared by Mueller et al. (2020): (1) Lavinia Planitia, surrounded on three sides by highlands, and (2) Themis Regio, a highland consisting of a broad swell and volcanic edifices surrounded by lowlands, that may provide a hint as to why there is a temperature difference that is not equilibrated by the expected efficient convection on Venus. This topic requires more research, specifically the systematic comparison of observed brightness temperatures and the surface temperatures predicted by GCM models. It should be stressed that the GCMs still predict that the temperature contrasts at night are small within the same region, so that local emissivity variation can be derived with high accuracy. If emissivity differences are to be derived globally, e.g., when comparing a global emissivity map to an in situ measurement, the lateral temperature differences may be larger on the order of 5 K, corresponding to an emissivity difference of about 10%.

2.2 Radar Emissivity

Radar is largely insensitive to surface mineralogy except for minerals with a high dielectric constant, e.g., pyrite, or metals (which have essentially infinite dielectric constants). Materials with a high dielectric constant are more efficient reflectors and poor emitters and will thus have a higher Fresnel reflectivity and lower emissivity. In active radar, the reflectivity and roughness both contribute to the radar backscatter cross section (Hagfors 1970), which can be measured by both radar imaging and radar altimeter modes, each of which have different sensitivities (e.g., Garvin et al. 1985; Pettengill et al. 1988). However, in many circumstances it can be difficult to disentangle the relative contributions of dielectric constant and roughness in radar brightness measurements.

Passive radiometry, in which the radar sensor simply receives the naturally emitted radio energy from the hot Venus surface, is also sensitive to the dielectric constant, where radiance is a function of emissivity and material temperature as determined from the Stefan-Boltzman equation. This method was first used from orbit by the Pioneer Venus (PV) radar mapper ($\lambda = 17$ cm; Ford and Pettengill 1983). Taking the surface temperatures and atmospheric lapse rates measured by the Venera landers and the PV probes, they recognized that low regions of radio brightness could not be due to temperature differences alone, and required surface materials of low radar emissivity. These anomalously low values were corroborated by observations from the Very Large Array (VLA; $\lambda = 20$ cm; Pettengill et al. 1988). The Magellan radar ($\lambda = 12.6$ cm), made the first global emissivity map of Venus, resampled to 4.5 km, and gives a mean global value of emissivity of 0.845 (Pettengill et al. 1992). Using the relative Magellan SAR backscatter brightness as a proxy for surface roughness to ratio between smooth and rough surfaces, the dielectric permittivity can be calculated from Magellan emissivity (Fig. 1). Campbell (1994) calculated empirically dependencies on Magellan emissivity, SAR (HH and VV) radar cross section and emission angle that allow a more accurate and comparable derivation of emissivity values between smooth (plains) and rough (tesserae, mountain belts, rifts) terrains.

The Magellan data agree with previous observations of regions with anomalously low microwave emissivity and thus high values (up to $\sim$160) for dielectric permittivity of surface materials (Fig. 1). Geologically, these values require that the rocks contain minerals or phases with a high dielectric constant (=real part of the relative dielectric permittivity), where the observed value is a function of the composition, volume, dispersion, permittivity of the matrix, porosity, density, temperature, grain size and shape of the phases. Olhoeft

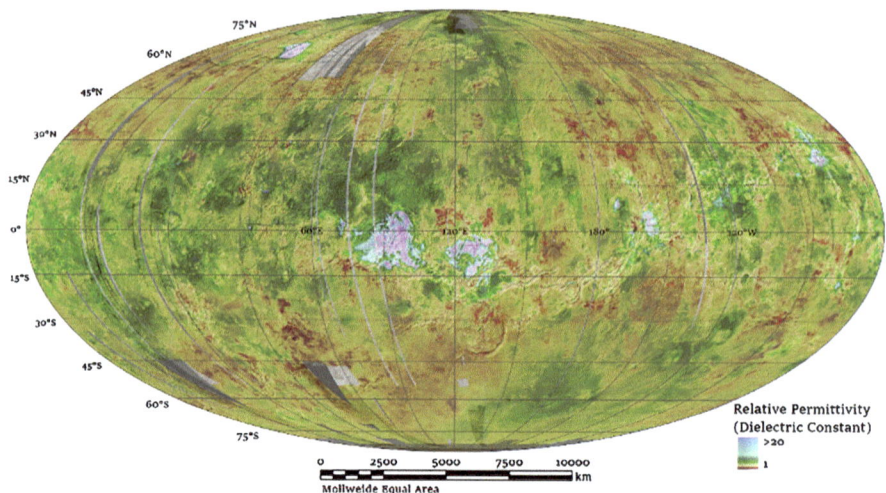

Fig. 1 Relative permittivity (dielectric constant) derived from Magellan emissivity, corrected for surface roughness inferred from SAR images. Red areas are low density unlithified materials, yellow areas are low density lithified materials, and green areas are medium to high density rocks. Blue and purple areas are high dielectric materials of unknown composition. See Sect. 3.3 for further discussion

and Strangway (1975) derived an empirical relationship between density and bulk dielectric constant of rocks and soils and show a dependence on Fe-Ti content of constituent minerals confirmed by others (e.g., Carrier et al. 1991). Shepard et al. (1994) show that the shape of the grains (e.g., needle-like, tabular, or spherical) can change the volume required by 3 orders of magnitude. The change in the crystal lattice with temperature has been shown to affect the dielectric constant (e.g., Havinga 1961; Roberts and Tyburczy 1991). Thus, perhaps most limiting to our ability to derive specific mineralogy from the dielectric permittivity data is the lack of systematic laboratory measurements of well-characterized (composition, density) Venus-relevant minerals at Venus temperatures and relevant radio frequencies.

The derivation of emissivity from radiance also depends strongly on the assumption of physical temperature of the material. As mentioned in the previous section, this depends on a limited number of direct measurements of the surface, the inferred change in lapse rate of atmospheric temperature and its relationship to topography. Also, as described in Pettengill et al. (1992), the brightness temperature as measured by the spacecraft is dependent on a number of factors, including properties of the radar antenna, the absorption and emission of the Venus atmosphere, and the estimate of the surface reflection of thermal energy from the sky. Better knowledge of the properties of the Venus atmosphere and topography by the upcoming missions will refine our measurements of radiothermal emission.

3 Observations

3.1 Venera Landers

To date, the only in situ direct measurements we have of Venus surface materials are from the Venera and Vega landers, which collected data from the surface of the planet in a period from 1972 (Venera 8) to 1985 (Vega 1,2). Chemical measurements of surface materials were made

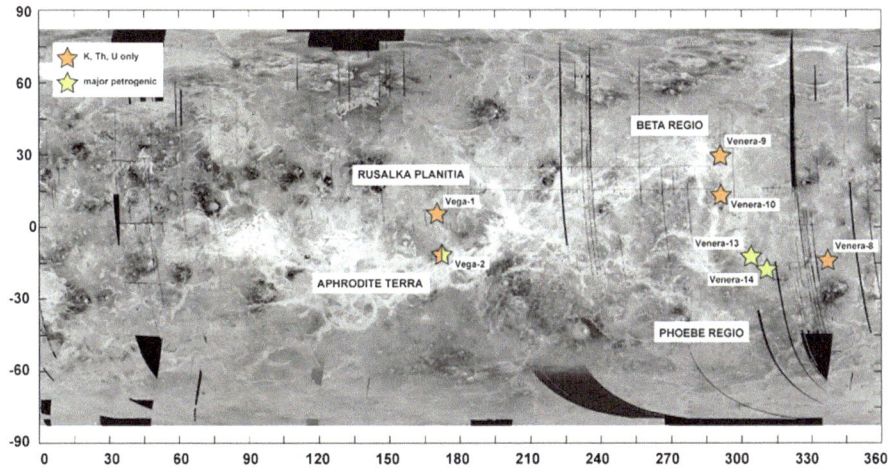

Fig. 2 Landing sites of the Soviet stations of the Venera-Vega series

Table 1 Gamma ray analyses conducted by the Venera and Vega landers. Concentrations are reported with 2σ uncertainty (Surkov et al. 1987; Treiman 2007), compared to averaged values for Mid Ocean Ridge Basalt (MORB) glass

Lander	K (%)	Th (ppm)	U (ppm)
Venera 8	4.0 ± 2.4	6.5 ± 0.4	2.2 ± 1.4
Venera 9	0.5 ± 0.2	3.7 ± 0.8	0.6 ± 0.4
Venera 10	0.3 ± 0.4	0.7 ± 0.6	0.5 ± 0.6
Vega 1	0.45 ± 0.44	1.5 ± 2.4	0.64 ± 0.94
Vega 2	0.40 ± 0.40	2.0 ± 2.0	0.68 ± 0.76
MORB Glass[a]	0.14 ± 0.15	0.4 ± 0.5	0.12 ± 0.14

[a]Average and standard deviation (2σ) of all reported values from 3190 samples in Table S2 of Le Voyer et al. (2019).

at seven locations that are concentrated in the Beta-Phoebe region and in Rusalka Planitia to the north of Aphrodite Terra (Kargel et al. 1993; Abdrakhimov and Basilevsky 2002; Fig. 2). Additionally, color photometry (Venera 9 and 10) and color images (Venera 13 and 14) of the surface yield some information about the surface mineralogy and oxidation state (Pieters et al. 1986). Selection of the landing sites was based purely on the interplanetary ballistic and communication constraints because no knowledge of the surface geology existed when the Venera-Vega missions were planned and implemented. Despite this formidable challenge, ten Soviet automatic landers successfully operated on the surface of Venus for periods of up to two hours, including the first soft landing of a spacecraft on the surface of another planet (Venera 7, 1970) and the first returned images from the surface of another planet (Venera 9, 1975).

The concentrations of K, Th, U were measured at five landing sites (Venera 8, 9, 10, Vega 1 and Vega 2) using gamma-ray spectrometry (Table 1; Surkov et al. 1987). The concentrations of the major petrogenic elements were measured at the sites of Vega 2, Venera 13, and 14 by X-ray fluorescence (XRF) (Table 2; Surkov et al. 1984, 1986). There is general consensus that the major element data are consistent with tholeiitic basalts at the Venera 14

Table 2 X-ray Fluorescence chemical analyses (wt%, (2σ) conducted by the Venera and Vega landers, compared to averaged values for MORB glass

Oxide	Venera 13[a]	Venera 14[a]	Vega 2[b]	MORB Glass[c]
SiO_2	45.1 ± 6.0	48.7 ± 7.2	45.6 ± 6.4	50.42 ± 0.87
TiO_2	1.59 ± 0.9	1.25 ± 0.82	0.2 ± 0.2	1.47 ± 0.38
Al_2O_3	15.8 ± 6.0	17.9 ± 5.2	16 ± 3.6	15.21 ± 0.93
FeO	9.3 ± 4.4	8.8 ± 3.6	7.7 ± 2.2[*]	10.05 ± 1.21
MnO	0.2 ± 0.2	0.16 ± 0.16	0.14 ± 0.24	0.18 ± 0.03
MgO	11.4 ± 12.4	8.1 ± 6.6	11.5 ± 7.4	7.63 ± 0.88
CaO	7.1 ± 1.92	10.3 ± 2.4	7.5 ± 1.4	11.74 ± 0.78
K_2O	4.0 ± 1.26	0.2 ± 0.14	0.1 ± 0.16	0.17 ± 0.17
SO_3	1.62 ± 2.0	0.88 ± 1.54	4.7 ± 3.0	0.37 ± 0.05^
Cl	<3000 ppm	<4000 ppm	<3000 ppm	182 ± 279 ppm
Cu			<3000 ppm	79.2 ± 21.0 ppm
Zn			<2000 ppm	91.8 ± 26.3 ppm
As, Se, Br			<800 ppm	0.3 to 0.4 ppm (As, Se)
Sr, Y, Zr, Nb, Mo			<1000 ppm	0.7 to 125 ppm
Pb			<3000 ppm	0.52 ± 0.29 ppm
Totals	96.1	96.3	94.4	
Estimates of other phases				
Fe^{3+}/FeTotal				0.14 ± 0.01
Na_2O	2.0 ± 0.5[a]	2.4 ± 0.4[a]	2.0[b]	2.61 ± 0.35
	6.0[d]	2.4[d]	2.4[d]	
P_2O_5				0.15 ± 0.07
H_2O				0.30 ± 0.18 ppm
CO_2				194 ± 85 ppm
F				211 ± 118 ppm
S				1109 ± 224 ppm

[a]Surkov et al. (1984), equation specified in Volkov et al. (1986). [b]Surkov et al. (1986), equation specified in Barsukov et al. (1982). Venera and Vega oxides are reported with 2σ uncertainty following Treiman (2007). [c]Average and standard deviation (2σ) of all reported values from 3190 samples in Table S2 of Le Voyer et al. (2019). [d]Semprich et al. (2020). [*]This value is reported as Fe_2O_3 in Barsukov et al. (1986). ^Reported S value recast as SO_3.

and Vega 2 site and an alkaline rock (alkali basalt or leucitite) at the Venera 13 site (Treiman 2007). Felsic compositions are also considered for Venera 8 and 13 by Nikolaeva (1990, 1995, 1997) by compairison to terrestrial arc rocks.

3.1.1 Constraints on Primary Mineralogy

Mineralogy was not directly measured by these landers, but the geochemical data has been used to infer mineralogy using multiple approaches: normative calculations, comparing with terrestrial analogs and experimental results, and modeling calculations using the MELTS software package (Ghiorso and Sack 1995; Asimow and Ghiorso 1998).

The major element abundances provided by XRF instruments on Venera 13, 14 and Vega 2 allow the calculation of normative mineralogy, and thus a rock classification, assuming that

Table 3 CIPW normative compositions for compositions in Table 2 and other terrestrial planet basalts. Norms are calculated on a volatile-free basis and renormalized to 100% using https://minetoshsoft.com/cipw/index. html. Venus norms are calculated using the Na_2O, P_2O_5 are from MORB (average composition of 3190 samples in Table S2 of Le Voyer et al. (2019))

	Ideal Formula	Venera 13	Venera 14	Vega 2	MORB
Quartz	SiO_2	–	–	–	–
Plagioclase		20.1	59.7	58.7	51.6
Anorthite	$CaAl_2Si_2O_8$	*20.1*	*36.4*	*33.7*	*28.6*
Albite	$NaAlSi_3O_8$	–	*23.3*	*25.0*	*23.0*
Orthoclase	$KAlSi_3O_8$	22.6	1.2	0.7	1.0
Nepheline	$(Na,K)AlSiO_4$	12.3	–	–	–
Leucite	$KAlSi_2O_6$	1.3	–	–	–
Diopside	$CaMgSi_2O_6$	12.2	11.3	4.1	23.1
Hypersthene	$(Mg,Fe)SiO_3$	–	12.3	12.3	13.1
Olivine	$(Mg,Fe)_2SiO_4$	27.9	12.7	23.5	8.1
Ilmenite	$FeTiO_3$	3.1	2.4	0.4	2.8
Apatite	$Ca_5(PO_4)_3(Cl,F,OH)$	0.4	0.4	0.4	0.4

these rocks are unweathered, volatile-free and igneous (Table 3). The Cross, Iddings, Pirsson, Washington (CIPW) norm calculation (Cross et al. 1902) nominally includes oxides, particularly Na_2O, that was not measured by the XRF on Venus. To perform the norm calculation, Na_2O has been estimated using different methodologies. For Venera 13 and 14, Na_2O content is calculated from the "adopted relation" [$K_2O/(K_2O + Na_2O)$ vs. $\Sigma FeO/(\Sigma FeO + MgO)$] (Volkov et al. 1986), presumably derived from analysis of 4 Siberian alkali rock types listed in Table 1 of (Barsukov et al. 1982). For Vega 2, Barsukov et al. (1986) plotted the compositions of ~ 50 terrestrial and lunar gabbroic rocks on an "A-S" diagram, where $A = Al_2O_3 + CaO + Na_2O + K_2O$ and $S = SiO_2 - (FeO + MgO + MnO + TiO_2)$, and derived an estimate for the Vega 2 Na_2O content by comparison. Both methods yield values that are similar to Na_2O contents in MORB (Table 2). The measured XRF values and calculated Na_2O were entered into a volatile-free CIPW norm calculation by Barsukov et al. (1982) and Barsukov et al. (1986) (Table 3). Semprich et al. (2020) used 2.4 wt% and 6.0 wt% Na_2O for Venera 14/Vega 2 and Venera 13, respectively, based on comparison to terrestrial analogue tholeiitic and alkali basalts.

Phosphorous was also not measured by the Venus landers, although it is ubiquitous in igneous materials, generally held in the mineral apatite. Apatite is one of the minerals invoked to explain regions with anomalously lower radar emissivity (see Sect. 3.3). Magmatic phosphorous has also been considered as a source for PH_3, which may have been detected in the Venus clouds (Greaves et al. 2020; Truong and Lunine 2021).

We have recalculated the normative mineralogy for the Venus lander XRF data including MORB values for Na2O, P_2O_5 and Fe^{3+}/Fe_{Total} values (Table 2). In agreement with prior normative calculations, Venera 14 and Vega 2 are consistent with olivine-tholeiites, while Venera 13 is consistent with a silica-undersaturated basalt. Apatite is predicted to be a primary phase. Kargel et al. (1993) and Filiberto (2014) compared the analyses with terrestrial analog volcanic rocks and in the case of Filiberto (2014) with experimental crystallization experiments as well. These suggest that Venera 14 and Vega 2 are consistent with terrestrial olivine-tholeiites formed from shallow melting of a potentially hydrous lherzolitic or peridotite source region. Whereas, Venera 13 is consistent with an alkali basalt suggesting

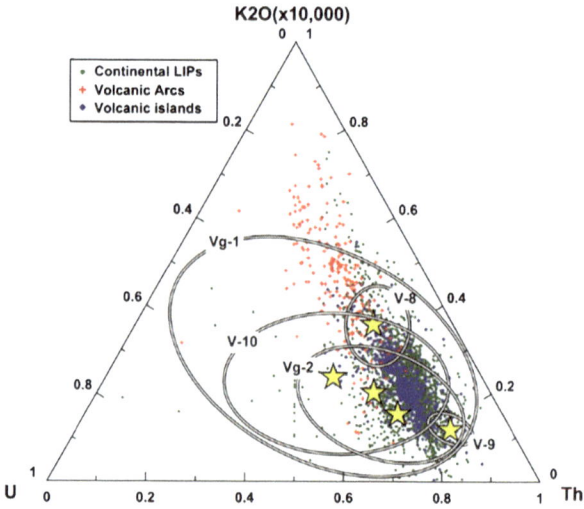

Fig. 3 Ternary plot of U, K, Th values for volcanic rocks from different geodynamic settings on Earth. Yellow stars correspond to the mean values of the measurements by the Venera-Vega landers. Ellipses around the stars correspond to the 1σ measurement errors. Modified from Venera-D JSDT (2017)

a deep partial melting of a potentially carbonated source region. These inferences are also consistent with MELTS modeling where hydrous fractional crystallization is likely needed to produce evolved high alkali rocks (Shellnutt 2013).

Two important factors limit the value of the Venera/Vega data and prevent their robust interpretation. The most important limitation of the Venera/Vega geochemical data is the low precision and unknown accuracy of the measurements (Tables 1, 2) discussed extensively by Treiman (2007). For example, the U, K, Th data Venera 8 and Venera 9 landers have the smallest errors and show that these rocks are geochemically distinct from one another, and correlate with terrestrial rock types, however the error bars of the Vega 1, 2 and Venera 10 measurements are so large they are almost completely unconstrained (Fig. 3).

Secondly, the Venera and Vega landers followed ballistic trajectories, landing within a ~300 km circle within error (Akim and Stepanyantz 1993). Subsequent mapping of the surface using Magellan shows that each landing site included multiple geomorphologic units (e.g., Basilevsky et al. 1992; Weitz and Basilevsky 1993; Basilevsky et al. 2007). For example, the landing circle of Venera 10 includes six various and extensive units Fig. 4). The Venera-10 panorama (Fig. 5, bottom) and inclinometer data indicate that the station is on a flat, sub-horizontal surface. This type of surface is most consistent with the volcanic plains (e.g., the shield plains, psh, and regional plains, rp_1, rp_2) as the hosting units for the lander and disfavors more structurally deformed units such as tessera terrain (t), densely lineated plains (pdl), and groove belts (gb), although no units within the landing circle can be ruled out.

Thus, in the Venera 10 landing circle and in all other landing sites as well, association of the chemical data with the specific terrains can be made on a probabilistic basis only (Abdrakhimov 2005), which limits the ability to use the morphologic data for geochemical context.

All of the above interpretations are based on the assumption that the chemical weathering did not significantly change the surface material on Venus. Such an assumption is potentially consistent with the lack of free water on the surface, but cannot be reasonably constrained by the available chemical data, particularly because of the large error bars for the sulfur measurements (Table 2). All thermodynamic calculations predict that rocks on the Venus surface

Fig. 4 a) Geological map of the Venera 10 (red star) landing area. Units: t - tessera, pdl - densely lineated plains, gb - groove belts, psh - shield plains, rp1 - regional plains, lower unit, rp2 - regional plains, upper unit. North is up. Modified from Basilevsky et al. (2007)

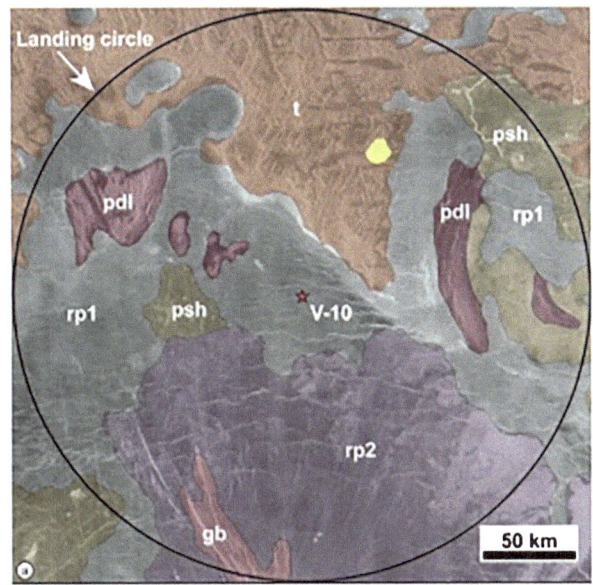

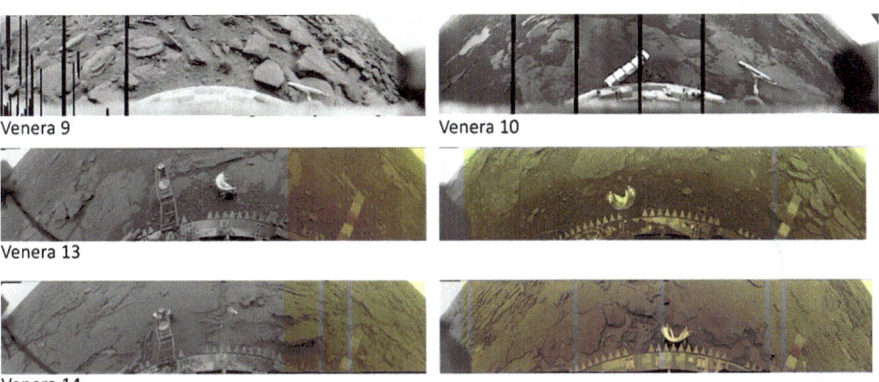

Venera 9   Venera 10

Venera 13

Venera 14

Fig. 5 Images of the Venus surface recorded by Venera Landers (reprocessed images © Don P Mitchell, used with permission)

will be weathered by the present-day atmosphere (see Sect. 4.1.2). The current observations of secondary mineralogy are discussed in the next section.

3.1.2 Constraints on Secondary Mineralogy

The Venera 9 and 10 landers each carried photometers that measured upward and downward radiation in 5 channels with effective wavelengths centered at 0.54, 0.59, 0.67, 0.77 and 0.87 μm (Ekonomov et al. 1980; Golovin et al. 1983). The surface albedo derived from both sites was <0.2, consistent with basalts. The Venera 13 and 14 landers each carried two scanning cameras with filters in the blue, green and red. Pieters et al. (1986) used the green and red channels from Venera 13 to produce a "white light" image and show that the rocks at this site are similarly dark in the visible. The data from Venera 9 and 10 show an increase in reflectance at wavelengths > 0.7 μm, which are similar in shape to the charge transfer and crystal field absorptions of ferric oxides (e.g., Morris et al. 1985). Pieters et al. (1986) show that the shape of the curves is consistent with high temperature reflectance spectra of hematite (Fe_2O_3). This result, the low albedo and measured chemistry at these sites are all consistent with the presence of basalts chemically weathered under oxidizing conditions to produce hematite.

There are several lines of observational evidence that the surface of Venus is physically weathered. The photometers on Venera 9 and 10 both recorded a transient dimming of up-welling and downwelling radiation, interpreted to be due to dust liberated by the lander (Ekonomov et al. 1980). Dust was also deposited on the Venera 13 lander ring (Florensky et al. 1983a) and observed to change appearance in consecutive images, indicating active particle transport by wind during the 85 min surface lifetime (Selivanov et al. 1982).

Regolith is present in each of the Venera panoramas surrounding rocks that have varying degrees of layering, fracturing, rounding, pits, and rock burial (Fig. 5), all consistent with physical degradation of the surface rocks (Florensky et al. 1977a,b, 1983b; Garvin et al. 1984; Basilevsky et al. 1985).

The overall strength of the materials was measured by several methods. Venera 13 and 14 carried a penetrometer instrument that measured the load carrying capacity of the surface. The sediment at the Venera 13 site is 1.3 to 5 MPa, similar to a dense sand or weak rock. Unconfined compressive strengths (UCS) of 31 to 126 MPa at the Venera 14 site are less than half that of basaltic rocks. Such a scenario is consistent with the actual densities (~ 1.5 g cm^{-3}) and porosities ($\sim 50\%$) inferred from mechanical data recorded by the Venera 13 and 14 probes (Florensky et al. 1983b). Strength was also measured by comparing the dynamics of the impact of the probe with the surface to experiments of a Venera model with various materials on Earth (Basilevsky et al. 1985). The bearing strength of Venera 13 is consistent with materials similar to cemented clastic rocks (Table 4).

Although no images were obtained from the Vega 2 site, load carrying capacities measured from the penetrometer and dynamic loads during lander impact (Marov and Grinspoon 1998; Surkov et al. 1984) allow for an estimation of UCS. Vega 2 recorded a lower porosity ($\sim 20\%$) and higher density (~ 2.6 g cm^{-3}) than the Venera landers, but the low UCS (~ 10 MPa) and high sulphate content implies a fully weathered and cemented clastic rock, rather than a fresh basalt.

The samples collected by the landers for geochemistry were derived from the uppermost surface. Moroz (1983) states the drill cores were ~ 1 cm^3 in volume collected from 3 cm depth for both Venera 13 and 14. Barsukov et al. (1986) report a time of 172 sec to acquire and deliver the sample for analysis during Vega 1 and 2. These data both suggest that the samples measured for chemistry by V13, V14 and Vega 2 were from the low-density surface materials and therefore weathered to some degree.

This is corroborated by the abundance of SO_3 in V13, 14 and Vega 2, each of which is elevated above an estimated average MORB composition (Table 4). The source of this SO_3 is most likely due to thermodynamically favored surface - atmosphere interactions which predict atmospheric SO_2 will react with Ca in silicates to produce anhydrite ($CaSO_4$; Zolotov 2007 and references therein). Fegley (2003) calculates if all measured SO_3 in Venera 13, 14, and Vega 2 is contained in anhydrite then that yields 2.8 ± 1.7 mass %, 1.5 ± 1.3 mass %, 8.2 ± 2.6 mass %, for the three landers respectively. Zolotov (2007) points that the S:Ca composition of Venus' rocks indicates incomplete sulfurization of Ca silicates to anhydrite. Dyar et al. (2021) point out that the reported values for S (0.35-1.9 wt%) when considered in light of basalt breakdown reactions modeled by Treiman and Schwenzer (2009) suggest that chemical weathering reactions have gone only 29% to completion at the Vega 2 site,

Table 4 Bearing capacity measurements and inferred surface density and porosity from Venera and Vega landers

Lander	Bearing capacity kg/cm^2	Unconfined compressive strength (MPa)	Inferred density (g/cm^3)	Inferred porosity[d]	Ref	SO$_3$[a,f]
V9	40–300[a]	4–27			b,d	
V10	40–300[a]	4–27			d	
			1.5 or 2.88 ± 0.1[c] (top 5–7 cm)	1–7%	c,d	
V13	2.6–10^1	0.2–0.9	1.4–1.5	50–53%	a	1.62 ± 1.0
	4.0–5.0[b]	0.4				
V14	~2.0 (top 5–10 cm)	0.2	1.15–1.2	60–62%	a,b	0.88 ± 0.77
	4–5 (below 5–10 cm)[a]	0.4	1.4–1.5	50–53%		
	65–250[b]*	6–22			b	
Vega-2	115	10 MPa	2.6	13%	e	4.7 ± 1.5

Measurement Method: [a]Impact Dynamic Loading, assumes entire lander torus is in contact with the ground [b]Penetrometer/Trellis Girder, [c]Gamma Densimeter, [d]Relative to basalt with a density of 3.0 g/cm^3 T *Landed on Lens Cap. [a] Surkov et al. (1984), [b] Basilevsky et al. (1985), [c] Surkov et al. (1977), [d] Basilevsky et al. (2004) and references therein [e] Marov and Grinspoon (1998) and references therein [f] Surkov et al. (1986).

and 5-10% at Venera 14 and 13, respectively. However, Vega 2 also has a low analysis total, suggesting either missing elements or substantially more alteration than rocks at Venera 14 and 13. The lower SO$_3$ amounts measured at Venera 14 relative to the Venera 13 suggest that they are less altered with respect to S consumption. This is consistent with reduced physical weathering relative to Venera 13 rocks (Garvin et al. 1984; Basilevsky et al. 1985), and suggests younger rocks at the Venera 14 landing site.

Thus, the presence of regolith, the relatively low density, the VNIR spectra and the elevated S at the Venera and Vega sites, where measured, are consistent with a weathered basaltic clastic rock that contains hematite and sulfate.

3.2 Near Infrared Imaging

Surface emissivity was derived from the data of several instruments observing the night-side of Venus near 1 µm, using various models. Most of these derivations involved the statistical removal of the correlation with topography, blurred to match the theoretical NIR resolution of 90 to 100 km (Hashimoto and Imamura 2001; Basilevsky et al. 2012). Thus, even if these emissivity derivations may show large differences in average value or amplitude, they are qualitatively comparable, i.e., it can be determined whether a location is high or low emissivity relative to the average. In the following we provide an overview of locations with significant emissivity anomalies.

3.2.1 Low Emissivity Anomalies

Hashimoto et al. (2008) interpret one flyby image of the Galileo Near Infrared Imaging Spectrometer (NIMS) at 1.18 µm wavelength and report that highlands 2 km above mean planetary radius have on average lower emissivity than other locations. The highlands imaged by NIMS are parts of Ishtar and Aphrodite Terra. A more detailed analysis was not possible due to the high noise of the single image. In this study there was no statistical removal of emissivity correlations with topography, the authors argue that this trend cannot be due to a temperature lapse rate different from their model assumption since this would be a dynamically unstable atmospheric stratification. Such an apparently unstable lapse rate was actually observed by Vega 2, although its cause is not yet clear (Seiff 1987; Lebonnois and Schubert 2017). It is unclear whether there are any significant emissivity anomalies in this data after statistical removal of the trend with topography.

VIRTIS was mostly interpreted with some removal of correlation of topography (Helbert et al. 2008; Mueller et al. 2008; Stofan et al. 2016; Kappel et al. 2016; Mueller et al. 2020). Tessera terrain mostly appears to have lower emissivity compared to plains in the same elevation range. Some of this may be attributed to biased errors in the altimetry (artificial pits from off-nadir reflections) that may be more frequent in tessera terrain due its high roughness (Mueller et al. 2008), however Gilmore et al. (2015) show that roughness does not control the observed emissivity variation within the Alpha Regio tesserae, and that the emissivity is instead correlated to differences in precursor material inferred during earlier mapping (Gilmore and Head 2000). The large tessera plateaus observed by VIRTIS repeat imaging are Alpha Regio, Cocomama Tessera, parts of Thetis Regio, Chimon-mana Tessera, Phoebe Regio, and Dolya Tessera. All are associated with some locally low emissivity, although the latter four are only observed by few VIRTIS images and at unfavorable viewing angles so that the uncertainty is large (Mueller et al. 2008). Chimon-mana Tessera was also observed to have relatively low emissivity by the Venus Monitoring Camera (VMC) (Basilevsky et al. 2012).

There are also volcanic features associated with low emissivity in near infrared data. Basilevsky et al. (2012) find that the central part of Tuulikki Mons, which features a steep sided dome at the summit, has relatively low emissivity in VMC data. Stofan et al. (2016) investigated the emissivity of Themis Regio using a statistical removal of the trend with topography that did not include tessera terrain and find several low emissivity anomalies associated with Mielikki Mons, Shiwanokia Corona and Ukemochi Corona, as well as some patches of regional plains. Kappel et al. (2016) also derived the emissivity of Themis Regio, using the more sophisticated multi-spectrum retrieval method (Kappel et al. 2015) and also found low emissivity in the central plateau of Shiwanokia Corona.

The rock emissivity near 1 μm is dominated by Fe bearing minerals (Hashimoto and Sugita 2003; Dyar et al. 2020, 2021; Helbert et al. 2021). The in situ surface spectra by Venera 9 and 10 are consistent with those of un-weathered basalt samples (Helbert et al. 2021; Filiberto et al. 2020; Cutler et al. 2020; Teffeteller et al. 2022; Zhong et al. 2023) which in turn supports the model of chemical weathering by Dyar et al. (2021) that predicts that emissivity is not strongly affected by alteration because the formation of secondary mineral crusts is self-limiting and cannot completely mask the underlying mineralogy due to their high optical transparency. Therefore, it is likely that the low emissivity locations correspond to a primary rock composition with less Fe-bearing minerals than basalts. There are a number of possible rock compositions that would fit this constraint (e.g., discussions in Gilmore et al. 2015, 2017), but since the tessera terrain constitute a relatively large volume (∼8% of the surface) the discussion has been focused on felsic rocks derived from partial melting of hydrated crust (Hashimoto et al. 2008; Gilmore et al. 2015), as only this process can produce such large volumes of melt from basaltic crust (Shellnutt 2018). This however implies the existence of surface water or a hydrated crust and crustal recycling during the formation of tessera terrain. This was likely not the case for the formation of the volcanic

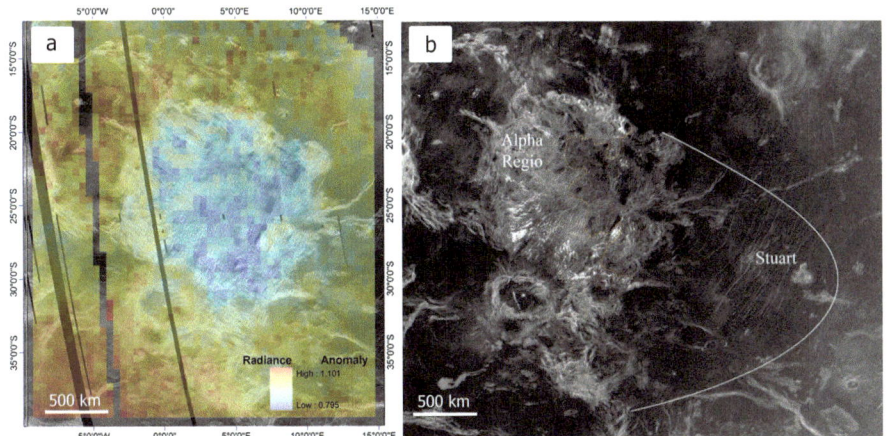

Fig. 6 a) Modified from Gilmore et al. (2015) showing VIRTIS top of atmosphere radiance relative to average radiance versus topography trend, interpreted as low emissivity of Alpha Regio tessera terrain, superposed over Magellan radar image mosaic (2200 m/pixel). b) Modified from Campbell et al. (2015) showing Arecibo circular polarized radar backscatter interpreted as evidence for ejecta from crater Stuart on Alpha Regio

features with low emissivity, however smaller amounts of rhyolite/granite can be produced by anhydrous fractional crystallization of basalt (e.g., Shellnutt 2018).

The presence of particles of micron scale size can also result in lower emissivity. The presence of particles on Venus, in general, can be inferred from the paraboloid impact ejecta blankets (e.g., Campbell et al. 1992), features interpreted as resulting from explosive volcanism (Campbell et al. 2017) and aeolian features (e.g., Greeley et al. 1992). Gilmore et al. (2015) discuss the possibility that the rough tessera terrain collects dust in the topographic lows sheltered from wind. Gilmore et al. (2015) argue that the differences of emissivity within Alpha Regio tessera terrain of comparable roughness is inconsistent with this hypothesis. However, it is also possible that the dust accumulates slowly and that these emissivity differences reflect different ages. The western part of Alpha that does not show low emissivity is one of the few locations where tessera may post-date the plains (Gilmore and Head 2000).

There is however evidence for more recent material on Alpha Regio. Arecibo 12 cm wavelength polarimetric radar observations are interpreted by Campbell et al. (2015) to show ejecta from crater Stuart extending over at least the eastern part of Alpha Regio, however there is no correlation between this same-sense-circular polarized backscatter and the NIR emissivity (Fig. 6). Campbell et al. (2015) estimate the reduction in roughness inferred from the lower radar backscatter to be the equivalent of about 4–7 cm of mantling material. The timescale of settling out of the atmosphere is a function of particle size (Campbell et al. 1992). The presence of ejecta visible in radar data of eastern Alpha Regio therefore strongly suggests some material over the rest of Alpha Regio, with a gradient in particle size decreasing from east to west. It is therefore unlikely that the low NIR emissivity is due to small ejecta particles, since the emissivity is neither correlated with the amount nor with the particle size of ejecta material. The reduction in emissivity must be caused by outcrops that are relatively free of the Stuart ejecta material originating in the presumably basaltic plains, likely by removal of fines by wind.

3.2.2 High Emissivity Anomalies

Locally high emissivity anomalies were reported in several studies of Venus Express VIRTIS data (Helbert et al. 2008; Mueller et al. 2008; Smrekar et al. 2010; Stofan et al. 2016; Kappel et al. 2016; Mueller et al. 2020). The following locations with relatively high emissivity in VIRTIS data coincide with stratigraphically young lava flows: Idunn Mons in Imdr Regio (Smrekar et al. 2010) with recent flows at the summit and on the eastern flank (D'Incecco et al. 2017, 2021; López et al. 2022); Abeona, Chloris, Mertsegger, and Mielikki Montes in Themis Regio (Smrekar et al. 2010; Stofan et al. 2016); Innini and Hathor Montes in Dione Regio (Smrekar et al. 2010), as well as Semiramus, Zywie, Latta, Shulamite, Ukemochi, Shiwanokia Corona, Nzambi, Bibi Patma Coronae in Themis Regio (Stofan et al. 2016). Additionally, there is an area of significantly increased emissivity coinciding with the radar bright inner paraboloid ejecta feature associated with crater Sabin in Themis Regio (Mueller et al. 2020).

As discussed in Dyar et al. (2020), the observed local increase of emissivity is 0.02 to 0.06 according to the most recent radiative transfer models (Kappel et al. 2016; Mueller et al. 2020). This emissivity difference is consistent with the chemical weathering of basalt under Venus surface conditions, where Fe^{2+} in silicates is converted to Fe^{3+} in oxides with higher ~ 1 μm emissivity (e.g., Dyar et al. 2020; Filiberto et al. 2020; Cutler et al. 2020; Teffeteller et al. 2022; Zhong et al. 2023). The high emissivity values and their correspondence with young flows has led to the interpretation that these regions are locations of unweathered and thus recent basaltic volcanism relative to the average plains (Smrekar et al. 2010; Stofan et al. 2016; D'Incecco et al. 2017, 2021; López et al. 2022). Smrekar et al. (2010) estimated the timing of this volcanism is be $< 250,000$ years, based on calculated rates of resurfacing for Venus, however laboratory studies olivine (Fe, $Mg)_2SiO_4$ at Venus temperatures imply much faster rates of oxidation on the order of days to months (Filiberto et al. 2020) or 10s-100s years (Zhong et al. 2023).

These observations highlight the urgent need for both laboratory studies of the weathering of rocks and minerals under Venus conditions and laboratory measurement of their 1 μm emissivity. For example, the NIR emissivity of natural basalts can vary by ~ 0.07 due to FeO content (Helbert et al. 2021), thus emissivity variations must consider variability in the initial composition of the lavas. Emissivity is also dependent on the type, distribution and thickness of secondary minerals. Both thermodynamic calculation and laboratory experiments predict the formation of other secondary minerals such as anhydrite, which is predicted to have a low emissivity. Modeling of the diffuse reflectance and derive emissivity of two-component mixtures of mineral particulates predict that the 10-20% hematite shifts the emissivity signature of basalt by 0.01–0.03 units and 20% anhydrite shifts the emissivity of basalt by 0.04–0.06 unit (Dyar et al. 2021). The effects of grain size in 1 μm emissivity are unclear; limited measurements (n = 4) show that particle size is not a dominant factor (Helbert et al. 2021). Much work remains to be done.

3.3 Magellan

3.3.1 Global Survey of Venus Highlands

Pioneer Venus and Magellan data show that many of Venus' highlands have distinctly elevated values of radar reflectivity relative to the global average (Masursky et al. 1980; Ford and Pettengill 1983) and thus low values of radar emissivity at their summits (Pettengill et al. 1992). The reflectivity values are ascribed to the presence of minerals or compounds

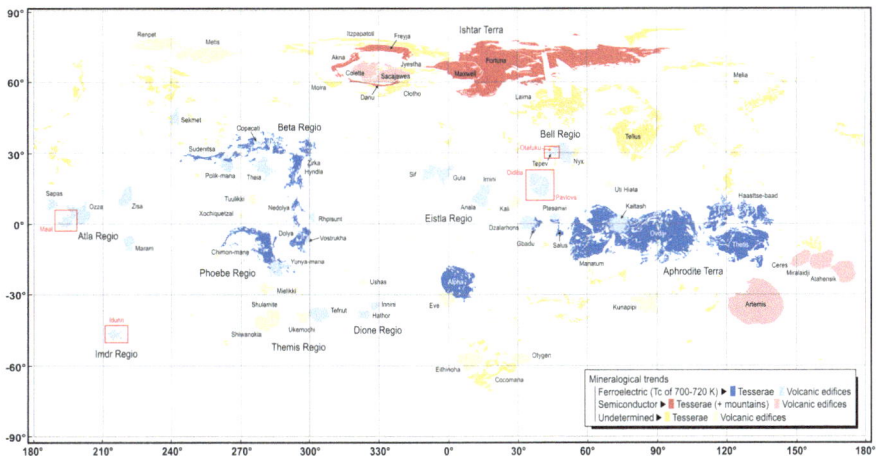

Fig. 7 Mineralogical trends on volcanic edifices (Brossier et al. 2020) and tesserae and mountain belts (Brossier and Gilmore 2021). Red frames indicate sites of possible recent activity (Fig. 8)

with a high dielectric constant as inclusion or coatings on surface rocks, as materials with high dielectric constants will enhance their radar reflectivity and lower their radar emissivity (Pettengill et al. 1992; Campbell 1994). Proposed minerals include: (1) pyrite (Pettengill et al. 1988; Klose et al. 1992; Wood and Brett 1997; Kohler 2016; Port et al. 2016; Berger et al. 2019; Sempich et al. 2020), (2) metallic frosts like lead and bismuth sulfides (Brackett et al. 1995; Pettengill et al. 1996; Schaefer and Fegley 2004; Kohler et al. 2015; Port et al. 2020), and (3) ferroelectric minerals, such as chlorapatite or perovskites that become highly dielectric at certain temperatures (Curie temperatures; Arvidson et al. 1994; Shepard et al. 1994; Treiman et al. 2016). These type and volume of dielectric materials are a function of rock composition, atmospheric composition, temperature and, if due to surface – atmosphere reactions, the length of exposure time at the surface. Because of the temperature dependence of these reactions, the detailed description of the variations in radar emissivity with altitude provide new insights into the relative composition and potential exposure age of the surface materials.

Klose et al. (1992) compared a sample of major volcanoes, mountains and tesserae and recognized differences in the style, magnitude and altitude of emissivity variations between and among these regions. Treiman et al. (2016) also noted similar differences in the radar emissivity signatures between Ovda Regio tessera and Ishtar Terra. More recently, Brossier et al. (2020) and Brossier and Gilmore (2021) measured and compared the variation of radar emissivity with altitude (and thus temperature) to understand the variety of radar emissivity signatures among all major venusian highlands. All of these studies conclude that variations in radar emissivity indicate differences in the composition, volume and type of dielectric materials in these regions.

As shown in Fig. 7, the majority of volcanoes and coronae on Venus (e.g., Maat and Ozza Mons; Brossier et al. 2020) and most tesserae (Aphrodite Terra, Alpha, Beta, and Phoebe regions; Brossier and Gilmore 2021) are compatible with the presence of ferroelectrics. The elevation and shape of the emissivity patterns indicate the presence of at least two ferroelectric compounds with distinct Curie temperatures, varying from 700 K to 720 K (6054–6056 km). This varying "critical altitude" reported in Klose et al. (1992) and seen by Brossier and colleagues could be due to different primary mineralogy, or local differences

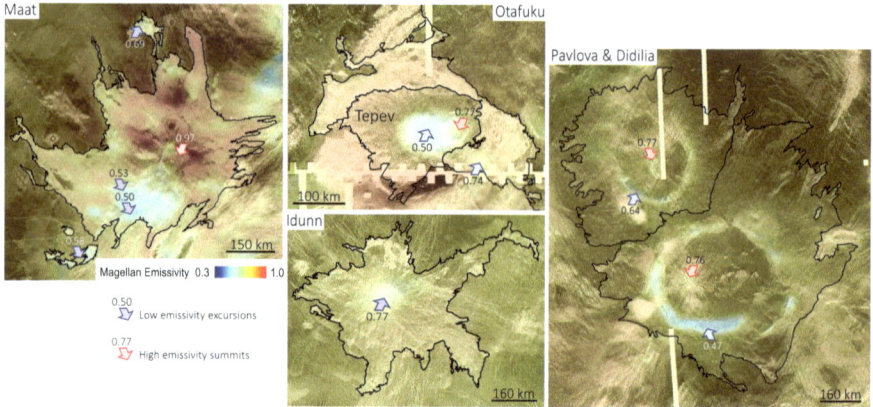

Fig. 8 Radar emissivity maps overlapping SAR images of Maat (194°E, 1°N), Otafuku (45°E, 30°N) and Idunn montes (215°E, 47°S), as well as Pavlova and Didilia coronae (40°E, 16°N). The high radar emissivity values found at their summits suggest the presence of young, unweathered lava flows

in the atmospheric composition or temperature at the time of the formation of the minerals by weathering (Treiman et al. 2016; Strezoski and Treiman 2022). The wide distribution of ferroelectrics shows that these minerals are common to volcanic, tesserae and mountain materials.

Conversely, the emissivity pattern of volcanoes, tesserae and mountain belts in western Ishtar Terra and the coronae in eastern Aphrodite Terra (Diana – Dali chasmata) are more consistent with the presence of semiconductor materials rather than ferroelectrics (Treiman et al. 2016; Brossier et al. 2020; Brossier and Gilmore 2021; Strezoski and Treiman 2022). The different signatures observed on Venus' highlands could be associated with mineralogical variability, perhaps related to differences in geologic settings (mantle processes), or surface temperature gradient.

Additionally, Brossier et al. (2020) reported that emissivity signatures of Maat, Idunn and Otafuku montes and the novae within Pavlova and Didilia coronae are consistent with lower volumes of ferroelectric minerals, consistent with relatively recent and less weathered lava flows (Fig. 8). These volcanic edifices are associated with presumably active hotspots that are among the most likely sites for recent or current volcanic activity on Venus based on geophysical (Smrekar 1994; Stofan et al. 1995) and morphological (Herrick and Hensley 2023) data. A less weathered surface has also been suggested for Idunn based on 1-μm VIRTIS data (Smrekar et al. 2010) (see Sect. 3.2). Thus, the radar emissivity data provide an independent constraint on recent volcanic activity on Venus in agreement with both 1-μm emissivity and gravity signatures.

3.3.2 Local Investigations in Atla Regio

The two largest volcanoes on Venus, Maat and Ozza Montes, display multiple reductions in radar emissivity at different altitudes including, atypically, values at or close to mean planetary radius (MPR = 6051.8 km) (Fig. 9; e.g., Wilt 1992; Robinson and Wood 1993; Brossier et al. 2020). The range of Curie temperatures are derived from the altitude and magnitude of the emissivity excursions, corresponding to values of 693–731 K over a range of elevation of 6052.5 km to 6056.7 km. Low elevation excursions at Maat and Ozza indicate ferroelectric minerals at higher Curie temperatures than any excursions previously observed

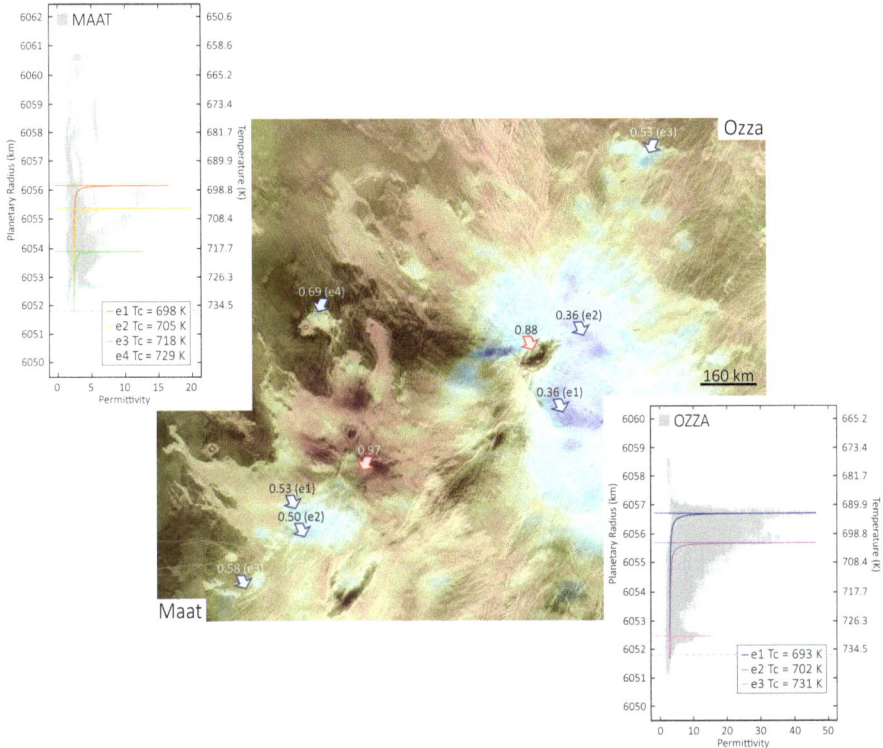

Fig. 9 Radar emissivity map overlapping a SAR image of Maat and Ozza montes located in Atla Regio (198°E, 4°N). Elevation – permittivity plots showing the multiple excursions detected on the two volcanoes (Brossier et al. 2021). Here, permittivity values are derived from radar emissivity values (Campbell 1995), while temperatures are given by the Vega 2 lander data (Seiff 1987; Lorenz et al. 2018). Application of the ferroelectric model (Shepard et al. 1994) to reproduce the low emissivity excursions (color-coded curves)

on Venus. Brossier et al. (2021) showed that specific ferroelectric signatures correlate with individual lava flows, indicating that the ferroelectric minerals are related to rock chemistry as opposed to the regional deposition of atmospheric precipitates.

Laboratory measurements of the Curie temperatures of minerals at Venus temperatures and radar wavelengths are necessary to constrain the candidate minerals to explain the emissivity patterns on Venus observed by Magellan, VERITAS and EnVision. Table 5 contains the list of ferroelectric minerals measure at Venus temperature found in the current literature. From this non-exhaustive list, the derived Curie temperatures from Magellan emissivity are consistent with the mineral chlorapatite ($Ca_5(PO_4)_3Cl$) and perovskite oxides (e.g., $X_2 + TiO_3$) and inconsistent with other substances, including GeTe which has been previously proposed for Venus. Using the method of Shepard et al. 1994, Brossier et al. (2021) calculated minimum estimates for the volumes and types of ferroelectric inclusions responsible for each emissivity excursion and they obtained volumes in the ppm range (6–113 ppm). Experiments on the formation, stability, and Curie temperatures of candidate ferroelectrics at Venus conditions will advance our understanding of the petrology of these volcanoes and the nature of the surface – atmosphere interactions.

Brossier et al. (2022) recently extended their investigation in Ganis Chasma, a rift valley in Atla Regio where recent activity was already suggested based on the superposition of rift

Table 5 List of ferroelectric minerals with Curie temperatures near Venus surface temperature (Brossier et al. 2021). Minerals proposed specifically for Venus are indicated by an asterisk (*). References: (1) Subbarao 1973; (2) Treiman et al. 2016; (3) Brackett et al. 1995; (4) Kadlec et al. 2011; (5) Shepard et al. 1994; (6) Young and Frederikse 1973

Compound	Formula	Curie Temp. Tc (K)	Curie Constant	Refs
Barium Bismuth Tungstate	$Ba(Bi_{0.7}W_{0.3})O_3$	723		1
Cadmium Iron Niobate	$Cd(Fe_{0.5}Nb_{0.5})O_3$	723		1
Cadmium Scandium Niobate	$Cd(Sc_{0.5}Nb_{0.5})O_3$	703		1
Chlorapatite*	$Ca_5(PO_4)_3Cl$	675–775		2
Lead Bismuth Tantalate	$PbBi_2Ta_2O_9$	703		1
Germanium Telluride*	GeTe	623–670	$\sim 1.0 \times 10^5$	1,3,4
Lead Bismuth Tantalate – Niobate*	$Pb_2Bi(Ta,Nb)O_6$	693–748		1,5
Lead Tantalate – Niobate*	$Pb(Ta,Nb)_2O_6$	533–843		5,6
Lead Titanate*	$(Pb(Ba,Sr,Ca))TiO_3$	110–763	1.1×10^5	1,5,6
Sodium Niobate-Tantalate*	$Na(Nb,Ta)O_3$	627–753		1,5
Lutetium Chromite	$LuCrO_3$	713		1
Potassium Lithium Niobate	$K_3Li_2Nb_5O_{15}$	703		1
Potassium Niobate – Tantalate*	$K(Nb,Ta)O_3$	2–708	2.4×10^5	1,5,6
Strontium Bismuth Niobate	$SrBi_2Nb_2O_9$	713		1

structures on young impact deposits (Basilevsky 1993). Ganis Chasma bears several sites with high (1-μm) emission with varying intensity over several days or months (Shalygin et al. 2015). Shalygin and colleagues suggested that these transient high emission sites are possibly associated with short-lived effusive activity, locally causing significant increases of surface temperatures. According to Brossier et al. (2022), the pattern of radar emissivity in these sites indicates the presence of ferroelectrics with subtle differences in the mineral composition, in agreement with the other volcanoes in Atla Regio (Brossier et al. 2020, 2021). They also demonstrated that these sites are consistent with relatively young and unweathered materials, providing independent corroboration of active (rift-associated) volcanism in Ganis Chasma since the 1990's.

3.4 Morphology

The rheology of lava is affected by its composition (e.g., Giordano et al. 2008), therefore the morphology of volcanic features can - in some cases - give evidence of lava composition. Generally, the amount of SiO_2 increases lava viscosity due to the formation of polymer chains in the melt (Baker et al. 1992). The interpretation is complicated by the fact that the rheology of silicate lavas is also highly dependent on temperature, volatile content, and shear forces (Lejeune and Richet 1995), all of which are expected to be different on Venus when compared to more accessible morphological analogues with known compositions (Head and Wilson 1986). In addition, it is not clear that the climate of Venus is stable over long durations, such that formation of any volcanic features at different periods of time might have taken place under very different ambient temperature conditions (Bullock and Grinspoon 1996, 2001, Noack et al. 2012), which influence the cooling rate (Snyder 2002) and thus viscosity contrasts within bodies of lava (Stofan et al. 2000). The high atmospheric pressure inhibits exsolution of volatiles (Garvin et al. 1982; Head and Wilson 1986) which is expected to reduce the viscosity of basalt and rhyolite lavas by 20-30% and an order of magnitude, respectively (Bridges 1997).

Most volcanic features observed by Magellan on Venus are morphologically consistent with basaltic compositions (Head et al. 1992). Features that have been discussed as potentially indicative of non-basaltic composition are steep sided domes (Pavri et al. 1992), festoon flows (Moore et al. 1992), sinuous rilles and canali (Baker et al. 1992; Komatsu et al. 1993).

Steep sided domes are features that are 10 to 10^2 km diameter, hundreds of m high, typically near circular volcanic features, that resemble pancakes (Pavri et al. 1992). The morphology and thickness to diameter ratio suggests slow emplacement of highly viscous lava, similar to dacite or rhyolite domes on Earth (Pavri et al. 1992; Fink et al. 1993). Stofan et al. (2000) on the other hand propose that lava inflates the flow under a cooled crust that fractures and anneals as it expands. Stofan et al. (2000) find that basalt is a more likely composition in the context of this mechanism since the high viscosity of rhyolite would not allow annealing, which however would not be true under a significantly hotter climate.

Festoon flows have been proposed to be composed of more silicic lavas than basalt for similar reasons as the steep sided dome: the large thickness and relatively broad interior structures suggests a high lava viscosity during emplacements (Head et al. 1992; Moore et al. 1992), and resemble rhyolitic coulees on Earth (Fink and Anderson 1999). Only three of these flows are found on Venus. One of them occurs on tessera terrain in Ovda Regio, which has been mentioned in support of the hypothesis tessera are more silicic than basalt, but recently Wroblewski et al. (2019) find that the morphology of this particular flow is more consistent with basalt.

Sinuous rilles on Venus resemble features of the same name on the Moon: single meandering channels that are widest and deepest near the source and appear to incise into pre-existing terrain (Baker et al. 1992; Komatsu et al. 1993; Oshigami et al. 2009). This suggests thermal erosion by a turbulent low viscosity lava as the formation mechanism, where underlying material is partially melted and not just mechanically eroded (Baker et al. 1992; Komatsu et al. 1993). Basalt lava is formed by partially melting ultramafic mantle material and thus erupts close to its solidus temperature ($\sim$1100 °C), which means that its capability to thermally erode underlying basalt is limited (Komatsu et al. 1993; Gregg and Greeley 1993; Fagents and Greeley 2001). Increased ambient temperature increases the substrate's susceptibility to thermal erosion due to the lower amount of heat that needs to be transferred from the lava. Ultramafic magmas formed by higher degrees of partial melting would have lower viscosity – increasing turbulence – and significantly higher eruption temperatures ($\sim$1600 °C), thus have a much higher potential for thermal erosion (Komatsu et al. 1993). Turbulent ultramafic komatiite lava flows likely occurred on Earth during the Precambrian (Kilburn 2000; Williams et al. 2001) and high temperature ultramafic eruptions have likely been observed on Io (McEwen et al. 1998), thus there is some support for the hypothesis that mantle temperature higher than that of recent Earth could result in such lava compositions. Komatiite consists mostly of Mg-rich olivine (Williams et al. 2001) and may be more resistant to chemical weathering compared to tholeitic basalt (see Sect. 4).

Canali type channels (Head et al. 1992; Baker et al. 1992; Komatsu et al. 1993), are very distinct from sinuous rilles and more similar to terrestrial rivers in that they can meander over distances over thousands of km at near constant width and depths (Oshigami and Namiki 2007). They exhibit morphologies suggesting fluvial sediment deposits such as scroll bars, streamlined islands, bird-foot-deltas, and oxbows (Kargel et al. 1994), indicating that some particulate materials were transported and redeposited by a fluid. They also appear to have levees in some locations (Oshigami and Namiki 2007), which is a common feature of lava channels. Assuming that the channels formed by deposition would require tremendous volumes of lava and resurface large parts of the planet within the cooling timescale of the flow (Komatsu et al. 1993). Thermal erosion by a silicate lava is unlikely since the heat loss would increase the viscosity long before traveling the entire distance (Gregg and Greeley 1993; Williams-Jones et al. 1998). Mechanical erosion by a fluid that remains liquid at ambient or very near to ambient temperatures fits the observed morphology better; proposed compositions are carbonates, sulfates, and elemental sulfur (Gregg and Greeley 1993; Kargel et al. 1994). The question whether the proposed compositions remain liquid at ambient (e.g., Treiman 1994) has to be discussed in the context that there is no reason to assume that the climate is stable and may have been 150 K hotter in the past (Bullock and Grinspoon 1996, 2001; Noack et al. 2012).

4 Laboratory Data

4.1 Thermodynamic and Experimental Constraints on Surface Mineralogy

Laboratory experiments and modeling are powerful tools to constrain the mineralogy of surface with acknowledgment of the limitations and assumptions of each approach. Thermodynamic modeling has been extensively used to predict equilibrium alteration mineralogy of the surface; while, more recently, experimental studies have been employed to investigate mineral stability, as well as the rate of mineral alteration. Thermodynamic modeling can calculate and explore a larger parameter space than experimental approaches due to the speed of modeling, but this approach typically assumes equilibrium conditions. Experiments are limited to laboratory timescales, where the amount of reaction is limited by kinetics. Combining modeling and experimental results can therefore cover a larger range of parameter space, alteration processes, and timescales than either approach alone.

Both of these methods rely on a number of variables such as temperature and pressure, oxygen fugacity, sulfur fugacity, and atmospheric composition. Most of these parameters are not well constrained in the near-surface atmosphere on Venus (e.g., Fegley 2003); approaches to mitigate this uncertainty can include using a range in conditions covering the predicted values for Venus. Thermodynamic calculations can more easily accommodate investigations over ranges in these variables, while it takes numerous, often long, experimental runs to accomplish the same feat in the laboratory. A second assumption is the starting composition (i.e., primary mineralogy or chemistry) of rocks on Venus. Most investigations have used minerals likely to be present on Venus based on our understanding of petrology and planetary evolution (e.g., olivine, pyroxene, plagioclase feldspar), as well as the limited rock chemistries that have been measured (e.g., basalt). However, other Venus-specific materials, suggested based on mission observations, have also been investigated (e.g., apatite and pyrite, as possible candidates for the high radar emissivity anomaly; Treiman et al. 2016; Fegley et al. 1995). Due to the number of variables (and unknown values) involved, few studies have utilized the exact same conditions when examining minerals on Venus, but despite this, some overall trends have been determined, which are discussed below.

4.1.1 Thermodynamic Modeling of Venus Weathering

Several studies have performed thermodynamic calculations to assess the stability of a variety of minerals at the surface of Venus. These studies have used different starting compositions and different environmental conditions, ranging from those of the modern Venus

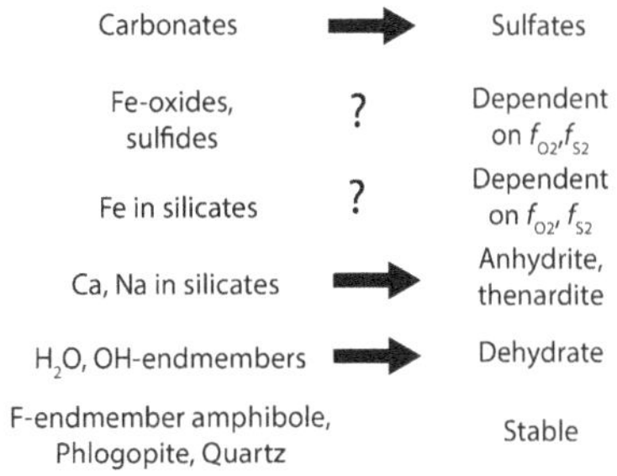

Fig. 10 General trends in Venus weathering from thermodynamics. See Zolotov (2018) for details

Carbonates	→	Sulfates
Fe-oxides, sulfides	?	Dependent on f_{O_2}, f_{S_2}
Fe in silicates	?	Dependent on f_{O_2}, f_{S_2}
Ca, Na in silicates	→	Anhydrite, thenardite
H_2O, OH-endmembers	→	Dehydrate
F-endmember amphibole, Phlogopite, Quartz		Stable

atmosphere, to more water-rich atmospheres that possibly reflect an ancient Venus environment (Semprich et al. 2020). Some studies have focused on specific mineral-gas reactions, while others use more complex modeling that involves reactions between many gases and solids (i.e., Gibbs free energy minimizations). A summary of solid-gas reactions on Venus can be found in Zolotov (2018) and is summarized in Fig. 10. The advantage of thermodynamic modeling over experiments is the ability to easily cover a wide range of parameter space, something that is useful for examining Venus in the current time as we still have poor constraints on the conditions of the near surface environment. This modeling is limited, however, by available databases and their quality, our knowledge of the actual conditions at the surface of Venus, and the fact that thermodynamics determines equilibrium states, which may or may not be achieved on natural planetary surfaces due to rates of reactions.

Gibbs free energy minimization modeling efforts have covered a wide range of conditions, but a broad division can be made between those that included sulfur species as an atmospheric gas species, and those that did not. Two studies that conducted modeling in H_2O-CO_2 atmospheres without sulfur, those of Khodakovsky et al. (1979) and Semprich et al. (2020), determined a number of hydrous phases (e.g., micas, amphiboles, tremolite) were stable products of basalts, alkali basalts, and granite at conditions of the mean planetary radius. The exact phases differ due to differences in the model parameters, updated databases, and the use of solid solutions in the work of Semprich et al. (2020).

Modeling of reactions of basalts and granites that included sulfur species in the atmosphere resulted in stable plagioclase, orthopyroxene, clinopyroxene, K-feldspar, quartz, and anhydrite at different modeled conditions (Barsukov et al. 1982; Klose et al. 1992; Treiman and Schwenzer 2009; Semprich et al. 2020). Differences due to use of solid solutions in some models but not others, as well as differences in databases and input parameters, can account for some discrepancies in the model outputs. For example Semprich et al. (2020), Treiman and Schwenzer (2009), Klose and Zolotov (1992) find cordierite $(Mg,Fe)_2Al_3(AlSi_5O_{18})$ to form as a result of reactions of anorthite and diopside at high oxygen fugacities in basalt and alkali basalt compositions, while studies that do not include solid solutions predict andalusite $Al_2(SiO_4)O$ to be a product (e.g., Klose et al. 1992). The stability of iron-bearing oxides and sulfides is highly dependent on the chosen conditions of the model, but multiple studies show pyrite to only be stable at high elevations (Semprich et al. 2020; Klose et al. 1992) (though Barsukov et al. (1982) calculate either magnetite or pyrite to be present at the mean planetary radius), pyrrhotite is shown to only be stable at reduced conditions that are unlikely to be present on Venus (Klose et al. 1992; Semprich et al. 2020), hematite solid solutions with ilmenite $(FeTiO_3)$ and geikielite $(MgTiO_3)$ are stable to lower oxygen fugacities than pure hematite (Zolotov 1994; Semprich et al. 2020). Calcite was not found to be stable at any elevation in the sulfur-bearing models, but Semprich et al. (2020) did show dolomite and calcite to be stable in conditions of low oxygen fugacity and high sulfur fugacity for limited bulk compositional space (i.e., having low f_{SO_2}/f_{S_2}).

Thermodynamic calculations have also been applied to specific open questions for Venus, particularly to explore possible mineralogical causes of the low radar emissivity regions. Calculations by Schaefer and Fegley (2004) and Port and Chevrier (2021) addressed the likelihood and stability of minerals composed of a number of volatile metals (e.g., tellurium, mercury), and found neither elemental Te nor Hg compounds are likely to be responsible for this phenomenon on Venus, as had been proposed previously by Brackett et al. (1995) and Pettengill et al. (1996) due to their high dielectric constants. Treiman et al. (2016) proposed conversion of fluorapatite, the most common variety of apatite in terrestrial igneous systems, to chloroapatite by reaction with HCl in the Venus atmosphere could be a source for a ferroelectric mineral. Other studies have calculated the stability of different iron oxides or sulfides at high elevations (Fegley et al. 1995; Zolotov 2018). However, the stability of these phases is largely dependent on f_{O2} and f_{S2}, which are not well constrained, and for this reason, thermodynamic calculations cannot resolve this question without further measurements from missions.

4.1.2 Experiments

Crystallization experiments are used to determine the minerals initially formed by the cooling of lava or magma, and also can provide information about mantle mineralogy if conducted at high pressure. These experiments have not been extensively conducted for Venus because of the lack of chemical data from rock units, and due to the quality of the measurements we do have (e.g., large uncertainty in MgO a key parameter for igneous petrology and a lack of data on Na_2O or P_2O_5, both of which influence mineralogy). However, experimental results from studies on terrestrial compositions have been useful in helping interpret the rocks measured on the surface (e.g., Filiberto 2014).

Weathering experiments to determine stable minerals, reaction products, and reaction rates have been the major focus of current Venus experimental work. Different experimental approaches have been applied to study Venus weathering, including heating in air or CO_2 (typically looking at oxidation or dehydration reactions, e.g., Knafelc et al. 2019; Cutler et al. 2020; Filiberto et al. 2020; Johnson and Fegley 2000, 2002, 2003; Teffeteller et al. 2022), heating with sulfur-bearing gases (e.g., Aveline et al. 2011; Fegley and Prinn 1989; Port and Chevrier 2020; Port et al. 2020; Reid 2021), heating under complex atmospheres (Radoman-Shaw et al. 2022; Santos et al. 2023) and observations of basaltic weathering in the field (McCanta et al. 2014). These studies have been done at ambient pressure or at Venus-relevant pressures. Overall, experimental results have been in agreement with predictions of Venus weathering from thermodynamics, with the exception of reactions that may take place over long timescales, and thus are not observable on experimental timescales. Both experimental and modeling methods suggest that primary minerals on the Venus surface (as well as glasses and bulk rocks) will lose Fe, Ca, and Na to weathering reactions that will produce secondary mineral phases such as Fe-oxides, Fe-sulfides, or sulphates (Ca- or Na-) on their surfaces. This will also result in the formation of more Mg-, Al-, and Si-rich phases such as forsterite, aluminosilicates, and quartz (e.g., Zolotov 2018).

A critical aspect of weathering is the rate at which minerals weather. As noted by Johnson and Fegley (2000, 2002, 2003), even if a mineral is not stable at the current surface conditions of Venus, a slow rate of breakdown may allow it to persist on the surface over geologic time. This may allow minerals formed in past climate eras by different weathering styles to still exist today. Determining reaction rates involves a number of time series experiments, as demonstrated by Fegley and Prinn (1989), who studied the rate of calcite reaction, and Johnson and Fegley (2000, 2002, 2003), who studied the rate of tremolite dehydration.

Calcite reaction was show to take place faster than tremolite dehydration in these studies, and calcite was observed to react relatively quickly in mineral suites exposed to complex Venus atmospheres in Radoman-Shaw et al. (2022) and Santos et al. (2023). Additionally, mechanisms of reaction are important to determine, as some reactions may produce protective coatings of secondary minerals that limit further reaction progress, leaving only thin layers of surface minerals changed from their primary composition (discussed in King et al. 2018). Thus far, no experimental studies have reported finding reactions limited by a protective surface layer. Both reaction rate and mechanism are critical to determine to understand the mineralogy of the Venus surface today and in the past, and how mineral-gas reactions influence both the surface and atmospheric chemistry over time.

4.2 High Temperature Spectroscopy

4.2.1 Theory

One of the problems with interpreting Venus surface spectra is the need to understand spectral properties of geologic materials at high temperatures, for which little laboratory data exist. High (and low) surface temperatures affect band positions in mineral spectra as predicted by crystal field theory. In the few cases where experimental data exist, temperature (T) and pressure (P) have opposing effects (i.e., shifts to higher and lower wavelengths, respectively) on band positions, but such conclusions are difficult to generalize because they vary with crystal structures and site geometries of different mineral groups. Existing data are a mishmash of emissivity and reflectance measurements taken under various viewing geometries acquired over a period of more than 60 years. There is a need for clarity in understanding the relationships among high- and low-T spectra of single minerals and the underlying physical characteristics that govern them.

Pioneering work from the 1960's to 1980's led to the use of crystal field theory for understanding spectral feature positions, summarized in the seminal book of Roger Burns (1970, 1993), *Mineralogical Applications of Crystal Field Theory*. It details how spin-allowed crystal field transitions take place when transition metal cations reside in different types of ligand fields, splitting the energy levels of d-orbital electrons in response to asymmetry induced by the crystal structure. An incident photon may be absorbed at certain wavelengths, exciting an electron from a lower to a higher energy state (Burns 1993). If the resultant electronic transition is spin-allowed, it produces a prominent absorption feature, such as the familiar 1 and 2 μm Fe^{2+} bands in pyroxene (Adams 1974; Cloutis and Gaffey 1991), which result from transitions made possible by splitting within the e_g and t_{2g} orbitals around Fe cations in the M1 and M2 crystallographic sites. The relative intensities of such absorptions increase with the asymmetry of the (in this case, octahedral) site (Burns 1993). The energy of crystal field splitting and the resulting band positions are a function of the ligand field, which changes in response to the structure and composition of the mineral studied.

Crystal field splitting (Δ) describes the difference in energy between the lowest and highest orbital in the electron cloud surrounding (generally) a transition metal cation. The energies of Δ vary according to several factors including (1) the symmetry and coordination number of the polyhedral (often the silica tetrahedron in its various forms), (2) the valence state of the cation, (3) the strength of its bond with the surrounding anions, (4) the distance between the cation and the surrounding anions, (5) pressure, and (6) temperature. Of particular interest are the relative magnitudes of these different variables on spectra, and for Venus, particularly the latter two.

In general, the largest factor affecting Δ values is coordination number. The separation (also termed Δ) between energy levels can be expressed as:

$$\Delta_o : \Delta_c : \Delta_d : \Delta_d = 1 : \frac{8}{9} : \frac{-1}{2} : \frac{-4}{9}.$$

These ratios correspond to the magnitude of the splitting of the d-orbitals between t_{2g} orbitals and e_g orbitals. The Δ values also represent the amount of energy (photon) that will be needed to move an electron from the low energy orbitals to the high energy ones, and thus they correspond to the energies of bands seen in spectra.

The Δ values are also a function of valence state and type of ligand, in the general order of $Mn^{2+} < Ni^{2+} < Co^{2+} < Fe^{2+} < V^{2+} < Fe^{3+} < Cr^{3+} < V^{3+} < Co^{3+} < Mn^{4+}$. However, coordination number, valence state, and bond strength rarely change with pressure and temperature. On the other hand, bond length is directly related to Δ, and it changes with both pressure (P) and temperature (T).

Pressure effects on Δ are described by the relation (Burns 1993):

$$\frac{\Delta_P}{\Delta_0} = \left(\frac{R_0}{R_P} \right)^5,$$

where Δ_0 and R_0 are the splitting and typical cation to anion distance under ambient conditions and Δ_P and R_P represent those parameters at a specific pressure. However, such changes in bond length can be difficult to generalize because they depend on the compressibility of each coordination polyhedron, which is in turn a function of bond type/strength. High pressure may increase the degree of covalent bonding, increasing peak intensity. When pressure increases do cause contractions of cation-anion bonds, higher Δ and (lower) blueshifted peak energies relative to ambient conditions are observed. However, it must be noted that such pressure effects have primarily been observed under conditions in the Earth's mantle, where $P \geq 140$ GPa. By comparison, the increased pressure on the surface of Venus is negligible (93 bars = 0.0093 GPa). Thus, under Venus conditions, the effects of the increased P on spectra are trivial, and the primary spectral changes will result from variations in T.

The temperature variation of Δ is expressed by Burns (1993) as:

$$\frac{\Delta_0}{\Delta_T} = \left(\frac{\Delta_0}{\Delta_T} \right)^{-5/3} = [1 - \alpha \, (T - T_0)]^{-5/3},$$

where α is the volume coefficient of thermal expansion, T is temperature, T_0 is ambient, and V and V_0 represent molar volumes at those temperatures. Interatomic distances increase as atoms grow farther apart as the result of increasing temperature. The net effect is that Δ decreases as T increases. This results in an inverse fifth power dependence of crystal field splitting (Δ) on cation-anion distance (R), expressed for the octahedral case as $\Delta \propto R^{-5}$ by Burns (1993). For high temperature spectra, there is also a general increase in the position of peak absorption intensity due to the enhanced thermal vibrations and enhanced vibronic coupling within each crystal site; see Burns (1982) and Burns (1993) for examples of such peak shifts in olivine group minerals.

In contrast, intervalence charge transfer phenomena, which also give rise to many features in the visible region, become more probable at high P because the sharing ions come closer together, but are decreased with increasing temperature, which pushes the ions farther apart. Oxygen-to-metal charge transfer features occur at the high end of the ultraviolet (UV) portion of the spectrum but extend into the visible. These features are highly dependent on the specific coordination polyhedra studied, but they generally red-shift into the visible region at increasing temperatures. These general trends of peak shifts with changing T and P are summarized in Tables 9.2, 9.3, and 9.4 of Burns (1993), which include 42 papers (too numerous to list here) containing high temperature and/or high-pressure optical data published before 1992. A diminishing number of more recent papers have subsequently been published because our understanding of mantle phases has evolved and refocused on other types of spectroscopy.

This effect of temperature on NIR reflectance was discussed in the context of Venus by Pieters et al. (1986) who collected reflectance data of basaltic materials in the laboratory from 20°–500 °C. They found that with increasing temperature, the spectrum of hematite (and oxidized basalts) exhibited both a weakening of the Fe^{3+} crystal field absorption and a $\sim$0.1 μm shift of the Fe-O charge transfer absorption to longer wavelengths. Pieters et al. (1983) concluded that the relative brightness of hematite at 500 °C observed in the lab can explain the high reflectance of the Venus surface recorded by the 0.9 μm channel on Venera 9 and 10, consistent with the presence of ferric oxides on Venus and establishing the critical linkage between laboratory work and Venus observations. With the dawning age of Venus orbital spectroscopy, it is necessary to resume such laboratory measurements to understand spectral signatures as viewed from landers, descent imagers, and orbiters.

4.2.2 Venus Spectral Library Development

To address the need for high-temperature spectra, the Planetary Spectroscopy Laboratory (PSL) at DLR has created a novel facility dedicated to acquiring high-T emissivity data appropriate for interpretation of Venus surface spectra. The facility allows the collection of near to far-infrared emission spectroscopy (0.7-200 μm) of solid and powdered samples to temperatures up to 1000 K under medium vacuum conditions ($\sim$10-100 Pa) (Maturilli et al. 2018; Helbert et al. 2021). High temperature emissivity spectra for a set of natural mafic and felsic igneous rocks are shown in Fig. 11. Mafic materials have higher (by $\sim$30%) emissivity values than felsic materials, regardless of grain size, and can be readily distinguished even when subsampled to bands in the 5 atmospheric windows that permit nightside measurement of radiation emitted from the Venus surface (Fig. 12; Dyar et al. 2020, 2021; Helbert et al. 2021). These data confirm and allow quantification of the inverse relationship between FeO content and high-temperature emissivity (Dyar et al. 2020; Helbert et al. 2021). Helbert et al. (2021) applied this relationship directly to the spectrophotometer data collected by Venera 9 and 10, to show that FeO content of the rocks can be derived from the lander data and are consistent with basaltic compositions.

High temperature laboratory emissivity data, advancements in atmospheric characterization and modeling (Sect. 2.1), and improved knowledge of the surface temperature from topography are the cornerstones that allow derivation of surface composition from the nightside emissivity measurements to be taken by VERITAS, EnVision and DAVINCI orbiters and the dayside surface images to be collected by the DAVINCI probe. The expansion of the Venus spectral library critically enables the revolution in our understanding of surface mineralogy from the upcoming missions.

4.2.3 Spectral Measurements of Venus-Relevant Weathering Experiments

There are currently very limited spectroscopic measurements of weathered Venus analog materials and none, to date, measured spectroscopically at Venus temperatures. What little work has been analyzing VNIR reflectance spectroscopy on samples analyzed at Venus

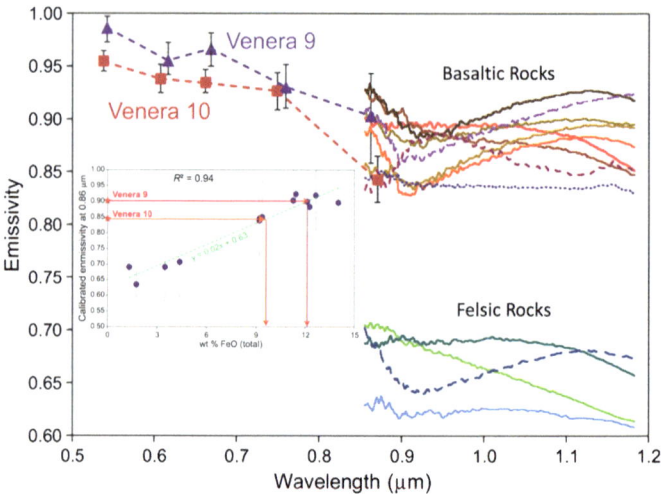

Fig. 11 Laboratory emissivity of geologic samples collected at 440 °C. Solid lines are solid samples and dashed lines are powders. The spectra are compared to reflectance data from the Venera 9 and 10 spectrophotometers (Ekonomov et al. 1980) converted to emissivity via Kirchoff's Law (Helbert et al. 2021). Inset shows the relationship between emissivity and wt% FeO for laboratory and the Venera data. Modified from Helbert et al. (2021)

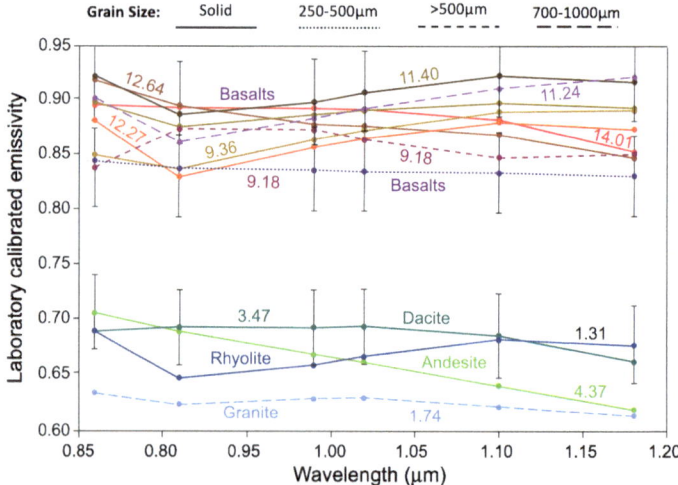

Fig. 12 Laboratory emissivity data of samples in Fig. 11 downsampled to 6 NIR bands sensitive to surface retrievals on Venus. Numbers are wt% FeO values of the samples, plotted in Fig. 11, inset. Error bars correspond to 4% estimated uncertainty of measurement from orbit. Modified from Helbert et al. (2021)

temperatures but under a terrestrial atmosphere (Filiberto et al. 2020; Cutler et al. 2020). Oxidation at Venus surface conditions forms discontinuous coatings of iron oxides mainly on glassy material and olivine crystals. Spectral measurements of these samples are consistent with first a flattening and reddening in the 1um region, consistent with the formation of magnetite first, before an increase in the reflectance and formation of new spectral features consistent with hematite. Limited high-temperature VNIR reflectance of Venus analog materials have also been recently done with a direct comparison of room temperature reflectance measurements (Treiman et al. 2021). This work has shown that at first order, the reflectance of igneous and sedimentary rocks are nearly identical except for the effects of oxidation on some surfaces. Basalt has reflectance near 7.5% and a leucogranite has reflectance near 50% (Treiman et al. 2021). Powdered hematite does shift its absorption edge slightly because it appears darker brown at high temperature (Yamanoi et al. 2009; Treiman et al. 2021).

There is clearly a need for more and diverse types of experiments to elucidate these interpretations, including measurement of altered surfaces at high temperatures. Such work is ongoing.

5 Summary

As with all planets, the mineralogy of the Venus surface provides a critical record of geologic and climatologic history and the current chemical exchanges between the atmosphere and solid body. In the absence of direct measurements of mineralogy, a range of approaches has been utilized to begin to constrain surface composition. Major element data from three Venera and Vega landers are consistent with tholeiitic or alkali basalts whose normative mineralogy can be calculated and serve as a starting point for comparison to other planets. The SO_3 content and density of these rocks indicate they have been altered under the current Venus atmosphere. Spectral reflectance from the Venera 9 and 10 photometers indicate oxidation, which is consistent with the range of f_{O2} constrained by compositional and temperature measurements of the Venus atmosphere at specific altitudes, latitudes and time of day. Despite these limitations, thermodynamic modeling of equilibrium gas/surface chemistry provides a critical baseline for weathering reactions on Venus. These models show reactions driven primarily by sulfur chemistry and oxidation state, where Fe, Ca, Na, in minerals will convert to sulfates or sulfides (and/or oxides) as a function of S and O fugacity. The specifics and kinetics of these reactions are being investigated using experimental apparatus, ranging from low pressure, single gas experiments to the complex environment provided by the Glenn Extreme Environments Rig. These experiments, in concert with thermodynamic modeling, are essential to advancing our understanding of the stability and variety of minerals we can expect on the Venus surface.

The Venus Express mission demonstrated that surface emissivity through near-infrared windows in the CO_2-rich atmosphere correlates with geomorphologic units, and provided, for the first time, a method to evaluate surface composition globally. The development of laboratory facilities to measure high temperature NIR emissivity of geologic materials has confirmed a relationship, predicted from theory, between emissivity and Fe^{2+} content in minerals. These measurements allow the community to develop a Venus spectral library that will prove essential for the interpretation of NIR data to be collected from the upcoming VERITAS, EnVision and DAVINCI missions and the ISRO Shukrayaan-1 (Haider et al. 2018; Sundararajan 2021) and Chinese Space Agency VOICE (Dong et al. 2023) missions in development. Spectral measurements of primary minerals and materials weathered under Venus conditions in the laboratory will strengthen our ability to recognize and infer mineralogy from these missions.

Reevaluation of the Magellan data also show spatial variability in the microwave emissivity of different terranes and locations on Venus. Microwave emissivity is related to the type and volume of high dielectric minerals in the rocks formed via igneous processes or through surface/atmosphere weathering reactions at varying temperatures. Laboratory measurements of the electrical properties of rocks at Venus temperatures and radar wavelengths will facilitate our understanding these data, which provide an independent constraint on surface mineralogy that can be compared to the NIR data.

All of these approaches: theory, modeling, experiment and spectroscopy can and must be brought to bear collaboratively to understand the mineralogy and history of this Earth-sized world. But all of these rely heavily on direct, in situ measurements of the Venus environment: the chemistry and dynamics of the atmosphere via probe, balloon and orbit and the mineralogy of chemistry of surface rocks via landed assets such as the Roscosmos Venera-D mission in development (Senske et al. 2017; Zasova et al. 2019).

Acknowledgements Thanks to organizers of, participants in and conversations fostered by the International Space Science Institute "Venus: Evolution through Time" workshop, and to the patient editors of this volume.

Funding Open Access funding enabled and organized by Projekt DEAL.

Declarations

Competing Interests The authors declare no competing interests.

References

Abdrakhimov AM (2005) Geochemical comparison of volcanic rocks from terrestrial intraplate oceanic hot spots with Venusian surface material. Geochem Int 43:732–747

Abdrakhimov AM, Basilevsky AT (2002) Geology of the Venera and Vega landing-site regions. Sol Syst Res 36:136–159

Adams JB (1974) Visible and near-infrared diffuse reflectance spectra of pyroxenes as applied to remote sensing of solid objects in the solar system. J Geophys Res 79:4829–4836

Akim EL, Stepanyantz VA (1993) Landing sites of automatic interplanetary probes on new map of Venus. Russian, Astron Vestn

Arvidson RE et al (1994) Microwave signatures and surface properties of Ovda Regio and surroundings, Venus. Icarus 112:171–186. https://doi.org/10.1006/icar.1994.1176

Asimow PD, Ghiorso MS (1998) Algorithmic modifications extending MELTS to calculate subsolidus phase relations. Am Mineral 83:1127–1131

Aveline DC, Abbey WJ, Choukroun M, Treiman AH, Dyar MD, Smrekar SE, Feldman SM (2011) Rock and mineral weathering experiments under model Venus conditions. Lunar Planet Sci Conf 42:2165

Baines KH et al (2000) Detection of sub-micron radiation from the surface of Venus by Cassini/VIMS. Icarus 148:307. https://doi.org/10.1006/icar.2000.6519

Baker VR et al (1992) Channels and valleys on Venus: preliminary analysis of Magellan data. J Geophys Res 97:13421. https://doi.org/10.1029/92JE00927

Barsukov VL, YuA S, Moskaleva LP et al (1982) Geochemical studies of the surface of Venus by the Venera 13 and 14 probes. Geochem Int 899–919

Barsukov VL et al (1986) Geochemical studies on Venus with the landers from the Vega-1 and Vega-2 probes. Geochem Int 23:53–65

Basilevsky AT (1993) Age of rifting and associated volcanism in Atla Regio, Venus. Geophys Res Lett 20:883–886. https://doi.org/10.1029/93GL00736

Basilevsky AT et al (1985) The surface of Venus as revealed by the Venera landings: Part II. Geol Soc Am Bull 96:137–144

Basilevsky AT, Nikolaeva OV, Weitz CM (1992) Geology of the Venera 8 landing site region from Magellan data: morphological and geochemical considerations. J Geophys Res 97:16315–16335

Basilevsky AT, Head JW, Abdrakhimov AM (2004) Impact crater air fall deposits on the surface of Venus: areal distribution, estimated thickness, recognition in surface panoramas, and implications for provenance of sampled surface materials. J Geophys Res 109:E12003. https://doi.org/10.1029/2004JE002307

Basilevsky AT et al (2007) Landing on Venus: past and future. Planet Space Sci 55:2097–2112

Basilevsky AT et al (2012) Geologic interpretation of the near-infrared images of the surface taken by the Venus monitoring camera, Venus express. Icarus 217:434. https://doi.org/10.1016/j.icarus.2011.11.003

Berger G et al (2019) Experimental exploration of volcanic rocks-atmosphere interactions under Venus surface conditions. Icarus 329:8–23. https://doi.org/10.1016/j.icarus.2019.03.033

Brackett RA, Fegley B, Arvidson RE (1995) Volatile transport on Venus and implications for surface geochemistry and geology. J Geophys Res 100:1553–1563. https://doi.org/10.1029/94JE02708

Bridges NT (1997) Ambient effects on basalt and rhyolite lavas under Venusian, subaerial, and subaqueous conditions. J Geophys Res 102:9243. https://doi.org/10.1029/97JE00390

Brossier J, Gilmore MS (2021) Variations in the radiophysical properties of tesserae and mountain belts on Venus: classification and mineralogical trends. Icarus 355:114161. https://doi.org/10.1016/j.icarus.2020.114161

Brossier J, Gilmore MS, Toner K (2020) Low radar emissivity signatures on Venus volcanoes and coronae: new insights on relative composition and age. Icarus 343:113693. https://doi.org/10.1016/j.icarus.2020.113693

Brossier J et al (2021) Distinct mineralogy and age of individual lava flows in Atla Regio, Venus derived from Magellan radar emissivity. J Geophys Res 126:e2020JE006722. https://doi.org/10.1029/2020JE006722

Brossier J, Gilmore MS, Head JW (2022) Extended rift-associated volcanism in Ganis Chasma, Venus detected from Magellan radar emissivity. Geophys Res Lett 49:e2022GL099765. https://doi.org/10.1029/2022GL099765

Bullock MA, Grinspoon DH (1996) The stability of climate on Venus. J Geophys Res 101:7521. https://doi.org/10.1029/95JE03862

Bullock MA, Grinspoon DH (2001) The recent evolution of climate on Venus. Icarus 150:19. https://doi.org/10.1006/icar.2000.6570

Burns RG (1970) Mineralogical applications of crystal field theory. Cambridge University Press, Cambridge

Burns RG (1982) Electronic spectra of minerals at high pressures: how the mantle excites electrons. In: Schreyer W (ed) High-pressure research in geosciences. E. Schweizerbart'sche Verlagsbuchhandlung, Stuttgart, pp 223–246

Burns RG (1993) Mineralogical applications of crystal field theory, 2nd edn. Cambridge University Press, Cambridge

Campbell BA (1994) Merging Magellan emissivity and SAR data for analysis of Venus surface dielectric properties. Icarus 112:187–203. https://doi.org/10.1006/icar.1994.1177

Campbell BA (1995) Use and presentation of Magellan quantitative data in Venus mapping. USGS Open-File Report 95-519. https://doi.org/10.3133/ofr95519

Campbell DB, Stacy NJS, Newman WI, Arvidson RE, Jones EM, Musser GS, Roper AY, Schaller C (1992) Magellan observations of extended impact crater related features on the surface of Venus. J Geophys Res 97:16249–16277. https://doi.org/10.1029/92je01634

Campbell BA, Campbell DB, Morgan GA et al (2015) Evidence for crater ejecta on Venus tessera terrain from Earth-based radar images. Icarus 250:123–130. https://doi.org/10.1016/j.icarus.2014.11.025

Campbell BA, Morgan GA, Whitten JL et al (2017) Pyroclastic flow deposits on Venus as indicators of renewed magmatic activity. J Geophys Res 122:1580–1596. https://doi.org/10.1002/2017je005299

Carrier WD et al (1991) Physical properties of the lunar surface. In: Heiken GH, Vaniman DT, French BM (eds) Lunar sourcebook, a user's guide to the moon. Cambridge University Press, Cambridge

Cloutis EA, Gaffey MJ (1991) Pyroxene spectroscopy revisited: spectral-compositional correlations and relationship to geothermometry. J Geophys Res 96:22809–22826

Cross W, Iddings JP, Pirsson LV, Washington HS (1902) A quantitative chemicomineralogical classification and nomenclature of igneous rocks. J Geol 10:555–690

Cutler KS, Filiberto J, Treiman AH, Trang D (2020) Experimental investigation of oxidation of pyroxene and basalt: implications for spectroscopic analyses of the surface of Venus and the ages of lava flows. Planet Space Sci 1:21

D'Incecco P, Müller N, Helbert J, D'Amore M (2017) Idunn Mons on Venus: location and extent of recently active lava flows. Planet Space Sci 136:25–33. https://doi.org/10.1016/j.pss.2016.12.002

D'Incecco P, Filiberto J, López I, Gorinov DA, Komatsu G (2021) Idunn Mons: evidence for ongoing volcano-tectonic activity and atmospheric implications on Venus. Planet Space Sci 2:215

Dong X, Liu Y, He J (2023) VOICE: a Venus Volcano Imaging and Climate Explorer mission. LPI Contributions 2807, 8068. https://ui.adsabs.harvard.edu/abs/2023LPICo2807.8068D

Dyar MD et al (2020) Probing Venus surface iron contents with six-band visible near-infrared spectroscopy from orbit. Geophys Res Lett 47:e2020GL090497. https://doi.org/10.1029/2020GL090497

Dyar MD et al (2021) Surface weathering on Venus: constraints from kinetic, spectroscopic, and geochemical data. Icarus 358:114139. https://doi.org/10.1016/j.icarus.2020.114139

Ekonomov AP (2015) Resolving the surface details on Venus in the balloon- or lander-borne images with a computer modeling method. Sol Syst Res 49:110. https://doi.org/10.1134/S003809461502001X

Ekonomov AP, YuM G, Moshkin BE (1980) Visible radiation observed near the surface of Venus: results and their interpretation. Icarus 41:65–75. https://doi.org/10.1016/0019-1035(80)90159-1

Fagents S, Greeley R (2001) Factors influencing lava-substrate heat transfer and implications for thermomechanical erosion. Bull Volcanol 62:519. https://doi.org/10.1007/s004450000113

Fedorova A et al (2015) The CO_2 continuum absorption in the 1.10- and 1.18-μm windows on Venus from Maxwell Montes transits by SPICAV IR onboard Venus express. Planet Space Sci 113:66. https://doi.org/10.1016/j.pss.2014.08.010

Fegley B Jr, Prinn RG (1989) Estimation of the rate of volcanism on Venus from reaction rate measurements. Nature 337:55–58. https://doi.org/10.1038/337055a0

Fegley B (2003) Venus. In: Holland HD, Turekian KK (eds) Treatise on geochemistry, vol 1. Elsevier, pp 487–507. https://doi.org/10.1016/b0-08-043751-6/01150-6

Fegley BK, Lodders K, Treiman AH, Klingelhöfer G (1995) The rate of pyrite decomposition on the surface of Venus. Icarus 115:159–180. https://doi.org/10.1006/icar.1995.1086

Filiberto J (2014) Magmatic diversity on Venus: constraints from terrestrial analog crystallization experiments. Icarus 231:131–136. https://doi.org/10.1016/j.icarus.2013.12.003

Filiberto J et al (2020) Present-day volcanism on Venus as evidenced from weathering rates of olivine. Sci Adv 6:aax7445. https://doi.org/10.1126/sciadv.aax7445

Fink JH, Bridges NT, Grimm RE (1993) Shapes of Venusian "pancake" domes imply episodic emplacement and silicic composition. Geophys Res Lett 20:261. https://doi.org/10.1029/92GL03010

Fink JH, Anderson SW (1999) Lava domes and Coulées. In: Sigurdsson H et al (eds) Encyclopedia of volcanoes. Academic, New York, pp 307–319

Florensky CP et al (1977b) The surface of Venus as revealed by Soviet Venera 9 and 10. Geol Soc Am Bull 88:1537. https://doi.org/10.1130/0016-7606(1977)88<1537:TSOVAR>2.0.CO;2

Florensky CP, Ronca LB, Basilevsky AT (1977a) Geomorphic degradations on the surface of Venus: an analysis of Venera 9 and Venera 10 data. Science 196:869–871. https://doi.org/10.1126/science.196.4292.869

Florensky CP et al (1983a) The oxidizing-reducing conditions on the surface of Venus according to the data of the "KONTRAST" geochemical indicator on the Venera 13 and Venera 14 spacecraft. Cosm Res 21:278–281

Florensky CP et al (1983b) Venera 13 and Venera 14: sedimentary rocks on Venus? Science 221:57–59. https://doi.org/10.1126/science.221.4605.57

Ford PG, Pettengill GH (1983) Venus: global surface radio emissivity. Science 220:1379–1381. https://doi.org/10.1126/science.220.4604.1379

Garvin JB, Head JW, Wilson L (1982) Magma vesiculation and pyroclastic volcanism on Venus. Icarus 52:365–372. https://doi.org/10.1016/0019-1035(82)90119-1

Garvin JB et al (1984) Venus: the nature of the surface from Venera panoramas. J Geophys Res 89:3381–3399. https://doi.org/10.1029/JB089iB05p03381

Garvin JB, Head JW, Pettengill GH, Zisk SH (1985) Venus global radar reflectivity and correlations with elevation. J Geophys Res 90:6859

Garvin JB et al (2022) Revealing the mysteries of Venus: the DAVINCI mission. Planet Sci J 3:117

Ghiorso MS, Sack RO (1995) Chemical mass transfer in magmatic processes. IV. A revised and internally consistent thermodynamic model for the interpolation and extrapolation of liquid-solid equilibria in magmatic systems at elevated temperatures and pressures. Contrib Mineral Petrol 119:197–212

Gilmore MS, Head JW (2000) Sequential deformation of plains at the margins of Alpha Regio, Venus: implications for tessera formation. Meteorit Planet Sci 35:667–687. https://doi.org/10.1111/j.1945-5100.2000.tb01451.x

Gilmore M, Mueller N, Helbert J (2015) VIRTIS emissivity of Alpha Regio, Venus, with implications for tessera composition. Icarus 254:350–361. https://doi.org/10.1016/j.icarus.2015.04.008

Gilmore MS et al (2017) Venus surface composition constrained by observation and experiment. Space Sci Rev 212:1–30. https://doi.org/10.1007/s11214-017-0370-8

Giordano D, Russell JK, Dingwell DB (2008) Viscosity of magmatic liquids: a model. Earth Planet Sci Lett 27:123. https://doi.org/10.1016/j.epsl.2008.03.038

Golovin YM, Moshkin BY, Ekonomov AP (1983) Some optical properties of the Venus surface. In: Venus, p 131136

Greaves JS et al (2020) Phosphine gas in the cloud decks of Venus. Nat Astron 1–20. https://doi.org/10.1038/s41550-020-1174-4

Greeley R et al (1992) Aeolian features on Venus: preliminary Magellan results. J Geophys Res 97:13319–13345. https://doi.org/10.1029/92JE00980

Gregg TKP, Greeley R (1993) Formation of venusian canali: considerations of lava types and their thermal behaviors. J Geophys Res 98:10873. https://doi.org/10.1029/93JE00692

Hagfors T (1970) Remote probing of the moon by infrared and microwave emissions and by radar. Radio Sci 5:189–227

Haider SA, Bhardwaj A, Shanmugam M, Goyal SK, Sheel V, Pabari J, Prasad Karanam D (2018) Indian Mars and Venus missions: science and exploration. 42nd COSPAR Scientific Assembly 42:B4-1

Hashimoto GL, Imamura T (2001) Elucidating the rate of volcanism on Venus: detection of lava eruptions using near-infrared observations. Icarus 154:239. https://doi.org/10.1006/icar.2001.6713

Hashimoto GL, Sugita S (2003) On observing the compositional variability of the surface of Venus using nightside near-infrared thermal radiation. J Geophys Res 108:5109. https://doi.org/10.1029/2003JE002082

Hashimoto GL et al (2008) Felsic highland crust on Venus suggested by Galileo near-infrared mapping spectrometer data. J Geophys Res 113:E00B24. https://doi.org/10.1029/2008JE003134

Haus R, Arnold G (2010) Radiative transfer in the atmosphere of Venus and application to surface emissivity retrieval from VIRTIS/VEX measurements. Planet Space Sci 5:1578. https://doi.org/10.1016/j.pss.2010.08.001

Havinga EE (1961) The temperature dependence of dielectric constants. J Phys Chem Solids 18:253–255. https://doi.org/10.1016/0022-3697(61)90169-X

Head JW, Wilson L (1986) Volcanic processes and landforms on Venus: theory, predictions, and observations. J Geophys Res 91:9407. https://doi.org/10.1029/JB091iB09p09407

Head JW et al (1992) Venus volcanism: classification of volcanic features and structures, associations, and global distribution from Magellan data. J Geophys Res 97:13153. https://doi.org/10.1029/92JE01273

Helbert J et al (2008) Surface brightness variations seen by VIRTIS on Venus express and implications for the evolution of the Lada Terra region, Venus. Geophys Res Lett 35:L11201. https://doi.org/10.1029/2008GL033609

Helbert J, Säuberlich T, Dyar MD et al (2020) The Venus Emissivity Mapper (VEM): advanced development status and performance evaluation. In: Proc SPIE 11502, infrared remote sensing and instrumentation XXVIII, 20 August 2020, vol 1150208. https://doi.org/10.1117/12.2567634

Helbert J et al (2021) Deriving iron contents from past and future Venus surface spectra with new high-temperature laboratory emissivity data. Sci Adv 7:eaba9428. https://doi.org/10.1126/sciadv.aba9428

Herrick RR, Hensley S (2023) Surface changes observed on a Venusian volcano during the Magellan mission. Science 379:1205–1208. https://doi.org/10.1126/science.abm7735

Johnson NM, Fegley B Jr (2000) Water on Venus: new insights from tremolite decomposition. Icarus 146:301–306

Johnson NM, Fegley B Jr (2002) Tremolite decomposition and Venus. Meteorit Plan Sci Suppl 37:A72

Johnson NM, Fegley B Jr (2003) Tremolite decomposition on Venus II. Products, kinetics, and mechanism. Icarus 164:317–333

Kadlec F et al (2011) Study of the ferroelectric phase transition in germanium telluride using time-domain terahertz spectroscopy. Phys Rev B 84:205. https://doi.org/10.1103/PhysRevB.84.205209

Kappel D, Haus R, Arnold G (2015) Error analysis for retrieval of Venus' IR surface emissivity from VIRTIS/VEX measurements. Planet Space Sci 113–114:49–65. https://doi.org/10.1016/j.pss.2015.01.014

Kappel D, Arnold G, Haus R (2016) Multi-spectrum retrieval of Venus IR surface emissivity maps from VIRTIS/VEX nightside measurements at Themis Regio. Icarus 265:42. https://doi.org/10.1016/j.icarus.2015.10.014

Kargel JS et al (1993) The volcanology of Venera and VEGA landing sites and the geochemistry of Venus. Icarus 103:253–275. https://doi.org/10.1006/icar.1993.1069

Kargel JS et al (1994) Carbonate-sulfate volcanism on Venus? Icarus 112:219. https://doi.org/10.1006/icar.1994.1179

Khodakovsky IL, Volkov VP, Sidorov YI, Borisov MV, Lomonosov MV (1979) Venus: preliminary prediction of the mineral composition of surface rocks. Icarus 39:352–363

Kilburn C (2000) Lava flows. In: Sigurdsson H et al (eds) Encyclopedia of volcanoes. Academic, New York, pp 291–306

King PL, Wheeler VW, Renggli CJ, Palm AB, Wilson S, Harrison AL, Morgan B, Nekvasil H, Troitzsch U, Mernagh T, Yue L (2018) Gas–solid reactions: theory, experiments and case studies relevant to Earth and planetary processes. Rev Mineral Geochem 84:1–56

Klose KB, Zolotov MY (1992) Chemical weathering of evolved igneous rocks on Venus. Lunar Plan Sci Conf 23

Klose KB, Wood JA, Hashimoto A (1992) Mineral equilibria and the high radar reflectivity of Venus mountaintops. J Geophys Res 97:16353–16369. https://doi.org/10.1029/92JE01865

Knafelc J et al (2019) The effect of oxidation on the mineralogy and magnetic properties of olivine. Am Mineral 104:694–702

Knicely J, Herrick RR (2020) Evaluation of the bandwidths and spatial resolutions achievable with near-infrared observations of Venus below the cloud deck. Planet Space Sci 181:104787. https://doi.org/10.1016/j.pss.2019.104787

Kohler E (2016) Investigating mineral stability under Venus conditions: a focus on the Venus radar anomalies. Theses and Dissertations

Kohler E et al (2015) Radar-reflective minerals investigated under Venus near-surface conditions. Lunar Plan Sci Conf 46:2563

Komatsu G et al (1993) Venusian channels and valleys: distribution and volcanological implications. Icarus 102:1. https://doi.org/10.1006/icar.1993.1029

Le Voyer M et al (2019) Carbon fluxes and primary magma CO_2 contents along the global mid-ocean ridge system. Geochem Geophys Geosyst 20:1387–1424. https://doi.org/10.1029/2018GC007630

Lebonnois S, Schubert G (2017) The deep atmosphere of Venus and the possible role of density-driven separation of CO_2 and N_2. Nat Geosci 10:473. https://doi.org/10.1038/ngeo2971

Lebonnois S et al (2018) Planetary boundary layer and slope winds on Venus. Icarus 314:149. https://doi.org/10.1016/j.icarus.2018.06.006

Lecacheux J et al (1993) Detection of the surface of Venus at 1.0 μm from ground-based observations. Planet Space Sci 41:543. https://doi.org/10.1016/0032-0633(93)90035-Z

Lejeune A-M, Richet P (1995) Rheology of crystal-bearing silicate melts: an experimental study at high viscosities. J Geophys Res 100:4215. https://doi.org/10.1029/94JB02985

López I, D'Incecco P, Filiberto J, Komatsu G (2022) The volcanology of Idunn Mons, Venus: the complex evolution of a possible active volcano. J Volcanol Geotherm Res 421:107428

Lorenz RD, Crisp D, Huber L (2018) Venus atmospheric structure and dynamics from the VEGA lander and balloons: new results and PDS archive. Icarus 305:277–283. https://doi.org/10.1016/j.icarus.2017.12.044

Marov MY, Grinspoon DH (1998) The planet Venus. Yale University Press, New Haven

Masursky H et al (1980) Pioneer Venus radar results: geology from images and altimetry. J Geophys Res 85:8232–8260. https://doi.org/10.1029/JA085iA13p08232

Maturilli A et al (2018) The Planetary Spectroscopy Laboratory (PSL): wide spectral range, wider sample temperature range. In: Proc. SPIE 10765, infrared remote sensing and instrumentation XXVI:107650A. https://doi.org/10.1117/12.2319944

McCanta MC, Dyar MD, Treiman AH (2014) Alteration of Hawaiian basalts under sulfur-rich conditions: applications to understanding surface-atmosphere interactions on Mars and Venus. Am Mineral 99:291–302. https://doi.org/10.2138/am.2014.4584

McEwen AS, Keszthelyi L, Geissler P, Simonelli DP, Carr MH, Johnson TV, Klaasen KP, Breneman HH, Jones TJ, Kaufman JM, Magee KP (1998) Active volcanism on Io as seen by Galileo SSI. Icarus 135:181–219

Meadows VS, Crisp D (1996) Ground-based near-infrared observations of the Venus nightside: the thermal structure and water abundance near the surface. J Geophys Res 101:4595. https://doi.org/10.1029/95JE03567

Moore HJ et al (1992) An unusual volcano on Venus. J Geophys Res 97:13479. https://doi.org/10.1029/92JE00957

Moroz VI (1983) Summary of preliminary results of the Venera 13 and Venera 14 missions. In: Venus. University of Arizona Press, Tucson, pp 45–68

Moroz VI (2002) Estimates of visibility of the surface of Venus from descent probes and balloons. Planet Space Sci 50:287. https://doi.org/10.1016/S0032-0633(01)00128-3

Morris RV, Lauer HV, Lawson C, Gibson EK, Nace GA, Stewart C (1985) Spectral and other physicochemical properties of submicron powders of hematite (α-Fe2O3), maghemite (γ-Fe2O3), magnetite (Fe3O4), goethite (α-FeOOH), and lepidocrocite (γ-FeOOH). J Geophys Res 90:3126–3144. https://doi.org/10.1029/jb090ib04p03126

Mueller N et al (2008) Venus surface thermal emission at 1 μm in VIRTIS imaging observations: evidence for variation of crust and mantle differentiation conditions. J Geophys Res 113:E00B17. https://doi.org/10.1029/2008JE003118

Mueller NT et al (2018) Regional Venus surface temperature variations in models and infrared observations. Lunar Plan Sci Conf 49:2400

Mueller NT, Smrekar SE, Tsang CCC (2020) Multispectral surface emissivity from VIRTIS on Venus express. Icarus 335:113400. https://doi.org/10.1016/j.icarus.2019.113400

Nikolaeva OV (1990) Geochemistry of the Venera 8 material demonstrates the presence of continental crust on Venus. Earth Moon Planets 50/51:329–341. https://doi.org/10.1007/BF00142398

Nikolaeva OV (1995) K-U-Th systematics of terrestrial magmatic rocks for planetary comparisons: terrestrial N-MORBs and venusian basaltic material. Geochem Int 33:1–11

Nikolaeva OV (1997) K-U-Th systematics of igneous rocks for planetological comparisons: oceanic island-arc volcanics on Earth versus rocks on the surface of Venus. Geochem Int 35:424–447

Noack L, Breuer D, Spohn T (2012) Coupling the atmosphere with interior dynamics: implications for the resurfacing of Venus. Icarus 217:484. https://doi.org/10.1016/j.icarus.2011.08.026

Olhoeft GR, Strangway DW (1975) Dielectric properties of the first 100 meters of the moon. Earth Planet Sci Lett 24:394–404

Oshigami S, Namiki N (2007) Cross-sectional profiles of Baltis Vallis channel on Venus: reconstructions from Magellan SAR brightness data. Icarus 190:1. https://doi.org/10.1016/j.icarus.2007.03.011

Oshigami S, Namiki N, Komatsu G (2009) Depth profiles of venusian sinuous rilles and valley networks. Icarus 199:250. https://doi.org/10.1016/j.icarus.2008.10.012

Pavri B et al (1992) Steep-sided domes on Venus: characteristics, geologic setting, and eruption conditions from Magellan data. J Geophys Res 97:13445. https://doi.org/10.1029/92JE01162

Pettengill GH, Ford PG, Chapman BD (1988) Venus: electromagnetic properties. J Geophys Res 93:14881–14892. https://doi.org/10.1029/JB093iB12p14881

Pettengill GH, Ford PG, Wilt RJ (1992) Venus surface radiothermal emission as observed by Magellan. J Geophys Res 97:13091–13102. https://doi.org/10.1029/92JE01356

Pettengill GH, Ford PG, Simpson RA (1996) Electrical properties of the Venus surface from bistatic radar observations. Science 272:1628–1631. https://doi.org/10.1126/science.272.5268.1628

Pieters CM et al (1983) Strength of mineral absorption features in the transmitted component of near-infrared reflected light: first results from RELAB, J Geophys Res 88:9534–9544. https://doi.org/10.1029/JB088iB11p09534

Pieters CM et al (1986) The color of the surface of Venus. Science 234:1379–1383. https://doi.org/10.1126/science.234.4782.1379

Pollack JB, Dalton JB, Grinspoon D et al (1993) Near-infrared light from Venus' nightside: a spectroscopic analysis. Icarus 103:1–42. https://doi.org/10.1006/icar.1993.1055

Port ST, Chevrier VF (2020) Stability of pyrrhotite under experimentally simulated Venus conditions. Planet Space Sci 105022. https://doi.org/10.1016/j.pss.2020.105022

Port ST, Chevrier VF (2021) Numerical investigation of Mercury-bearing minerals to address the low emissivity anomaly on the highlands of Venus. Lunar Plan Sci Conf 52:2548

Port ST, Kohler E, Craig PI, Chevrier V (2016) Stability of pyrite under Venusian surface conditions. Lunar Planet Sci 47:2144

Port ST, Chevrier VF, Kohler E (2020) Investigation into the radar anomaly on Venus: the effect of Venus conditions on bismuth, tellurium, and sulfur mixtures. Icarus 336:113432. https://doi.org/10.1016/j.icarus.2019.113432

Radoman-Shaw BG, Harvey RP, Costa G et al (2022) Experiments on the reactivity of basaltic minerals and glasses in Venus surface conditions using the Glenn Extreme Environment Rig. Meteorit Planet Sci. https://doi.org/10.1111/maps.13902

Ragent B et al (1985) Particulate matter in the Venus atmosphere. Adv Space Res 5:85. https://doi.org/10.1016/0273-1177(85)90199-1

Reid RB (2021) Experimental alteration of venusian surface basalts in a hybrid CO_2-SO_2 atmosphere. Masters thesis, Univ. Tennessee

Roberts JJ, Tyburczy JA (1991) Frequency dependent electrical properties of polycrystalline olivine compacts. J Geophys Res 96:16205–16222

Robinson CA, Wood JA (1993) Recent volcanic activity on Venus: evidence from radiothermal emissivity measurements. Icarus 102:26–39. https://doi.org/10.1006/icar.1993.1030

Santos AR, Gilmore MS, Greenwood JP, Nakley LM, Phillips K, Kremic T, Lopez X (2023) Experimental weathering of rocks and minerals at Venus conditions in the Glenn Extreme Environments Rig (GEER). J Geophys Res. https://doi.org/10.1029/2022JE007423

Schaefer L, Fegley B (2004) Heavy metal frost on Venus. Icarus 168:215–219. https://doi.org/10.1016/j.icarus.2003.11.023

Seiff A (1987) Further information on structure of the atmosphere of Venus derived from the VEGA Venus Balloon and Lander mission. Adv Space Res 7:323–328. https://doi.org/10.1016/0273-1177(87)90239-0

Seiff A, Schofield JT, Kliore AJ et al (1985) Models of the structure of the atmosphere of Venus from the surface to 100 kilometers altitude. Adv Space Res 5:3–58. https://doi.org/10.1016/0273-1177(85)90197-8

Selivanov AS, Gektin YM, Naraeva MK, Panfilov AS, Fokin AB (1982) Evolution of the VENERA-13 imagery. Sov Astron Lett 8:235–236

Semprich J, Filiberto J, Treiman AH (2020) Venus: a phase equilibria approach to model surface alteration as a function of rock composition, oxygen- and sulfur fugacities. Icarus 346:113779. https://doi.org/10.1016/j.icarus.2020.113779

Senske DA, Zasova LV, Ignatiev NI et al (2017) Venera-D: Expanding our horizon of terrestrial planet climate and geology through the comprehensive exploration of Venus. Report of the Venera-D Joint Science Definition Team

Shalygin EV et al (2015) Active volcanism on Venus in the Ganiki Chasma rift zone. Geophys Res Lett 42:4762–4769. https://doi.org/10.1002/2015GL064088

Shellnutt JG (2013) Petrological modeling of basaltic rocks from Venus: a case for the presence of silicic rocks. J Geophys Res, Planets 118:1350–1364. https://doi.org/10.1002/jgre.20094

Shellnutt JG (2018) Derivation of intermediate to silicic magma from the basalt analyzed at the Vega 2 landing site, Venus. PLoS ONE 13:e0194155

Shepard MK et al (1994) A ferroelectric model for the low emissivity highlands on Venus. Geophys Res Lett 21:469–472. https://doi.org/10.1029/94GL00392

Smrekar SE (1994) Evidence for active hotspots on Venus from analysis of Magellan gravity data. Icarus 112:2–26. https://doi.org/10.1006/icar.1994.1166

Smrekar SE et al (2010) Recent hotspot volcanism on Venus from VIRTIS emissivity data. Science 328:605–608. https://doi.org/10.1126/science.1186785

Snels M et al (2021) A simulation chamber for absorption spectroscopy in planetary atmospheres. Atmos Meas Tech 14:7187. https://doi.org/10.5194/amt-14-7187-2021

Snyder D (2002) Cooling of lava flows on Venus: the coupling of radiative and convective heat transfer. J Geophys Res 107:5080. https://doi.org/10.1029/2001JE001501

Stofan ER et al (1995) Large topographic rises on Venus: implications for mantle upwellings. J Geophys Res 327:317–323. https://doi.org/10.1029/95JE01834

Stofan ER et al (2000) Emplacement and composition of steep-sided domes on Venus. J Geophys Res 105:26757. https://doi.org/10.1029/1999JE001206

Stofan ER, Smrekar SE, Mueller N, Helbert J (2016) Themis regio, Venus: evidence for recent (?) volcanism from VIRTIS data. Icarus 271:375–386. https://doi.org/10.1016/j.icarus.2016.01.034

Strezoski A, Treiman AH (2022) The "Snow Line" on Venus's Maxwell Montes: varying elevation implies a dynamic atmosphere. Planet Sci J 3:264. https://doi.org/10.3847/PSJ/ac9f3a

Subbarao EC (1973) Ferroelectric and antiferroelectric materials. Ferroelectrics 5:267–280

Sundararajan V (2021) Tradespace exploration of space system architecture and design for India's Shukrayaan-1, Venus orbiter mission. In: ASCEND 2021, p 4103

Surkov YA et al (1977) Density of surface rock on Venus from data obtained by the Venera 10 automatic interplanetary station. Cosm Res 14:612–618

Surkov YA et al (1984) New data on the composition, structure, and properties of Venus rock obtained by Venera 13 and 14. J Geophys Res 89:B393–B402. https://doi.org/10.1029/JB089iS02p0B393

Surkov YA et al (1986) Venus rock composition at the Vega 2 landing site. J Geophys Res 91:E215. https://doi.org/10.1029/JB091iB13p0E215

Surkov YA, Kirnozov FF, Glazov VN (1987) Uranium, thorium, and potassium in the Venusian rocks at the landing sites of Vega 1 and 2. J Geophys Res 92:E537–E540. https://doi.org/10.1029/JB092iB04p0E537

Teffeteller H et al (2022) An experimental study of the alteration of basalt on the surface of Venus. Icarus 384:115085. https://doi.org/10.1016/j.icarus.2022.115085

Treiman AH (1994) Comment on "Formation of Venusian canali: Considerations of lava types and their thermal behaviors" by Gregg, T. K. P. and Greeley, R.. J Geophys Res 99:17163. https://doi.org/10.1029/93JE03576

Treiman AH (2007) Geochemistry of Venus' surface: current limitations as future opportunities. In: Esposito LW, Stofan ER, Cravens TE (eds) Exploring Venus as a terrestrial planet. https://doi.org/10.1029/176GM03

Treiman AH, Schwenzer SP (2009) Basalt–atmosphere interaction on Venus: preliminary results on weathering of minerals and bulk rock. In: Venus geochemistry: progress, prospects, and new missions, 26–27 Feb 2009. Houston, TX, USA

Treiman A, Harrington E, Sharpton V (2016) Venus' radar-bright highlands: different signatures and materials on Ovda Regio and on Maxwell Montes. Icarus 280:172–182. https://doi.org/10.1016/j.icarus.2016.07.001

Treiman AH, Filiberto J, Kaaden KEV (2021) Near-infrared reflectance of rocks at high temperature: preliminary results and implications for near-infrared emissivity of Venus's surface. Plan Sci J 2:43. https://doi.org/10.3847/PSJ/abd546

Truong N, Lunine JI (2021) Volcanically extruded phosphides as an abiotic source of Venusian phosphine. Proc Natl Acad Sci 118:e2021689118. https://doi.org/10.1073/pnas.2021689118

Venera-D Joint Science Definition Team (JSDT) (2017) Venera-D: Expanding our Horizon of Terrestrial Planet Climate and Geology through the Comprehensive Exploration of Venus. https://www.lpi.usra.edu/vexag/meetings/meetings-of-interest/Venera-D-Report.pdf

Volkov VP, Zolotov MY, Khodakovsky IL (1986) Lithospheric-atmospheric interaction on Venus. In: Saxena SK (ed) Chemistry and physics of terrestrial planets. Springer, New York, pp 136–190. https://doi.org/10.1007/978-1-4612-4928-3_4

Weitz CM, Basilevsky AT (1993) Magellan observations of the Venera and Vega landing site regions. J Geophys Res 98:17069

Widemann T, Ghail R, Wilson C, Titov D (2021) EnVision at Venus: Europe's next medium-class science mission. 19th Meeting of the Venus Exploration Analysis Group. Abstract #8068

Williams DA, Kerr RC, Lesher CM, Barnes SJ (2001) Analytical/numerical modeling of komatiite lava emplacement and thermal erosion at Perseverance, Western Australia. J Volcanol Geotherm Res 110:27–55. https://doi.org/10.1016/S0377-0273(01)00206-2

Williams-Jones G, Williams-Jones AE, Stix J (1998) The nature and origin of Venusian canali. J Geophys Res 103:8545–8555. https://doi.org/10.1029/98JE00243

Wilt JR (1992) A study of areas of low radio-thermal emissivity on Venus. PhD thesis, Massachusetts Institute of Technology, Cambridge

Wood JA, Brett R (1997) Comment on "The rate of pyrite decomposition on the surface of Venus". Icarus 128:472–473. https://doi.org/10.1006/icar.1997.5743

Wood BE et al (2022) Parker Solar Probe imaging of the night side of Venus. Geophys Res Lett 49:e96302. https://doi.org/10.1029/2021GL096302

Wroblewski FB et al (2019) Ovda Fluctus, the festoon lava flow on Ovda Regio, Venus: not silica-rich. J Geophys Res 124:2233. https://doi.org/10.1029/2019JE006039

Yamanoi Y, Nakashima S, Katsura M (2009) Temperature dependence of reflectance spectra and color values of hematite by in situ, high-temperature visible micro-spectroscopy. Am Mineral 94:90–97. https://doi.org/10.2138/am.2009.2779

Young KF, Frederikse HPR (1973) Compilation of the static dielectric constant of inorganic solids. J Phys Chem Ref Data 2:313. https://doi.org/10.1063/1.3253121

Zasova LV, Gorinov DA, Eismont NA et al (2019) Venera-D: a design of an automatic space station for Venus exploration. Sol Syst Res 53:506–510. https://doi.org/10.1134/S0038094619070244

Zhong S-S et al (2023) High temperature oxidation of magnesium- and iron-rich olivine under a CO_2 atmosphere: implications for Venus. Remote Sens 15:1959. https://doi.org/10.3390/rs15081959

Zolotov MY (1994) Phase relations in the Fe-Ti-Mg-O oxide system and hematite stability at the condition of Venus'. Surface Lunar Plan Sci Conf 25

Zolotov MY (2007) Solid planet–atmosphere interactions. In: Schubert G (ed) Treatise on Geophysics, vol 10. Elsevier, pp 349–369. https://doi.org/10.1016/B978-044452748-6.00181-4

Zolotov MY (2018) Gas–solid interactions on Venus and other solar system bodies. Rev Mineral Geochem 84:351–392. https://doi.org/10.2138/rmg.2018.84.10

Publisher's Note Springer Nature remains neutral with regard to jurisdictional claims in published maps and institutional affiliations.

Authors and Affiliations

Martha S. Gilmore[1]  **· M. Darby Dyar[2,3]**  **· Nils Mueller[4]**  **· Jérémy Brossier[5]**  **· Alison R. Santos[1]**  **· Mikhail Ivanov[6]**  **· Richard Ghail[7]**  **· Justin Filiberto[8]**  **· Jörn Helbert[4]**

✉ J. Helbert
Joern.Helbert@dlr.de

M.S. Gilmore
mgilmore@wesleyan.edu

[1] Department of Earth and Environmental Sciences, Wesleyan University, Middletown, CT, USA

[2] Planetary Science Institute, Tucson, AZ, USA

3 Department of Astronomy, Mount Holyoke College, South Hadley, MA, USA

4 German Aerospace Center (DLR) Institute for Planetary Research, Berlin, Germany

5 Institute for Space Astrophysics and Planetology IAPS, National Institute of Astrophysics, Rome, Italy

6 Vernadsky Institute of Geochemistry and Analytical Chemistry, Moscow, Russia

7 Earth Sciences, Royal Holloway, University of London, Egham, UK

8 NASA Johnson Space Center, Houston, TX, USA

Space Science Reviews (2023) 219:29
https://doi.org/10.1007/s11214-023-00966-y

Resurfacing History and Volcanic Activity of Venus

Robert R. Herrick[1] · Evan T. Bjonnes[2] · Lynn M. Carter[3] · Taras Gerya[4] ·
Richard C. Ghail[5] · Cédric Gillmann[6,4] · Martha Gilmore[7] · Scott Hensley[8] ·
Mikhail A. Ivanov[9] · Noam R. Izenberg[10] · Nils T. Mueller[11] · Joseph G. O'Rourke[12] ·
Tobias Rolf[13,14] · Suzanne E. Smrekar[8] · Matthew B. Weller[2,15]

Received: 28 June 2022 / Accepted: 1 March 2023 / Published online: 25 April 2023
© The Author(s) 2023

Abstract

Photogeologic principles can be used to suggest possible sequences of events that result in
the present planetary surface. The most common method of evaluating the absolute age of
a planetary surface remotely is to count the number of impact craters that have occurred
after the surface formed, with the assumption that the craters occur in a spatially random
fashion over time. Using additional assumptions, craters that have been partially modified
by later geologic activity can be used to assess the time frames for an interpreted sequence
of events. The total number of craters on Venus is low and the spatial distribution taken by
itself is nearly indistinguishable from random. The overall implication is that the Venusian
surface is much closer to Earth in its youthfulness than the other, smaller inner solar system
bodies. There are differing interpretations of the extent to which volcanism and tectonics
have modified the craters and of the regional and global sequences of geologic events. Consequently, a spectrum of global resurfacing views has emerged. These range from a planet
that has evolved to have limited current volcanism and tectonics concentrated in a few zones
to a planet with Earth-like levels of activity occurring everywhere at similar rates but in
different ways. Analyses of the geologic record have provided observations that are challenging to reconcile with either of the endmember views. The interpretation of a global
evolution with time in the nature of geologic activity relies on assumptions that have been
challenged, but there are other observations of areally extensive short-lived features such
as canali that are challenging to reconcile with a view of different regions evolving independently. Future data, especially high-resolution imaging and topography, can provide the
details to resolve some of the issues. These different global-evolution viewpoints must tie to
assessments of present-day volcanic and tectonic activity levels that can be made with the
data from upcoming missions.

Keywords Venus · Venus volcanism · Venus geology · Venus resurfacing history

Venus: Evolution Through Time
Edited by Colin F. Wilson, Doris Breuer, Cédric Gillmann, Suzanne E. Smrekar, Tilman Spohn and
Thomas Widemann

Extended author information available on the last page of the article

 Springer

1 Introduction

At minimum, it will be a few decades before samples from the Venusian surface can be dated in situ or returned to Earth for analysis. Understanding the volcanic and tectonic history of the planet thus involves primarily two types of activities: determining sequences of events on the surface by applying photogeologic analysis techniques to synthetic aperture radar (SAR) images, and evaluating the age of surfaces by assuming that impact cratering of the surface occurs in a spatially random manner with a known frequency so that craters can be used as a crude clock.

Dating the very last activity that occurred on a surface can be accomplished by counting the number of craters that occurred after that activity ceased. As we discuss below, using modeling and certain assumptions, craters that do not postdate all geologic activity can provide information about the ages of the events that altered the craters (e.g., post-impact faulting or partial volcanic filling). The timing, pattern, and nature of observed geologic events and accompanying crater alteration provide critical clues about the global geodynamic scenarios that have led to the current geomorphology. Other articles in this collection delve into the details of volcanism and deformation and the geodynamic models that have driven resurfacing. In this article we concern ourselves with efforts to calibrate the cratering clock, our understanding of how and where craters have been altered since formation, and attempts to determine the parameter space of acceptable geomorphic models for global resurfacing that are consistent with observations. Unfortunately, the combination of a relatively young surface and atmospheric filtering of small meteoroids means that Venus has far fewer impact craters than Mars, Mercury, or Earth's moon, making the job particularly difficult.

We begin (Sect. 2) by reviewing estimates of the rate at which impact craters form on the surface, the first step in using the cratering record to understand the history of Venus. A remarkable observation from the initial results of the Magellan mission was that the distribution of craters, taken alone, was nearly indistinguishable from a random distribution. We devote a section to a statistical assessment of the pattern of craters and its relationship to the global pattern of geomorphologic units (Sect. 3). We then discuss the general ways in which craters get altered by subsequent geologic processes (Sect. 4). Two- and three-dimensional global models of crater emplacement and volcano/tectonic resurfacing have been developed that attempt to match broad-brush characterizations of global crater modification states (Sect. 5). The overall relationships between craters and different surface features have led to a few end-member models for global geologic surface activity, and we review these unifying concepts (Sect. 6). Some of these concepts involve a present-day level of volcanic and tectonic activity much lower than present-day Earth, while others postulate a level as high or higher. Consequently, we add sub-sections discussing evidence of young and present-day volcanism (Sect. 7). We also include a section (Sect. 8) of miscellaneous geologic observations that provide additional constraints on the global resurfacing history.

The article concludes with an overview of the big-picture models for the planet's geologic history and a summary of how well these ideas mesh with the models and observations presented here. Suggestions for future data collection that can be used to assess these models is provided.

2 Constraints on the Production Rate of Craters

Because we currently have no capability to radioactively date rocks on the surface of Venus, our primary constraint on the absolute age of the Venusian surface comes from using the impact cratering record as a calendar. The starting assumption is that craters form in a spatially random pattern over time with a rate that is constant or slowly changing with time, and a size-frequency distribution that can be fit with a smooth curve. There are three basic components to understanding how frequently craters of a specified size form on Venus: 1) knowledge of the velocities and size-frequency distribution of objects entering the Venusian atmosphere; 2) understanding of how the Venusian atmosphere affects those objects; and 3) for those objects that reach the surface, knowledge of the size of impact crater produced. In general terms, one of two approaches are used to solve for 1) and 3) as if Venus were an airless body, and then craters on "atmosphere-free Venus" are convolved with a model of how the atmosphere filters projectiles and modifies crater formation. Estimates of the three components are used to determine a crater production function, or the size-frequency distribution of craters for a surface of a specified area and age.

The first approach to estimating crater production on an airless Venus is to take the size-frequency distribution of lunar craters, which have been calibrated by dating lunar samples for selected areas, and attempt to translate those curves to Venus. The knowledge required is the rate of impact on Venus relative to the moon and how large a crater on Venus would be relative to the same-sized crater on the moon. Impacts on both bodies come from a mix of asteroids, short-period comets, and long-period comets, and the relative numbers and proportions of these categories differ between bodies. Relative crater size for a given impactor on the two bodies is a function of primarily planetary gravity, impact velocity, and impactor density. Thus, a crater scaling law must be identified and applied, relative impact velocities must be assessed, and impactor densities must be assumed. A problem with using the lunar curve for Venus is that the latter has a much younger surface than the most confidently calibrated lunar curves, which date to formation of the lunar maria 3+ Ga; a constant or slowly decreasing impact rate over the last 3 Gy is often assumed (e.g., Stöffler et al. 2006).

A second approach to estimating crater production on an airless Venus is to use a present-day census of potential Venus meteoroids, assess their probability of impact, and use a crater scaling law to evaluate likely impact crater size. This approach trades the problem of interpolating the lunar curve from 3 Ga for the problem of assuming that the current observable population of asteroids and comets is representative of the past few hundred My. Also, while the asteroid population is relatively well surveyed at the size range of objects capable of making it through the Venusian atmosphere, populations of short- and long-period comets are less so. Finally, the first approach only requires estimating the relative size of a crater formed under lunar gravity compared to Venusian gravity, while the second approach requires that the absolute size scaling is correct.

Of some of the more widely cited papers with production ages estimates, the post-Venera 15/16 paper by Ivanov et al. (1986) used the first approach, the Magellan-era work of Phillips et al. (1992) compared both approaches, and the second approach was used by Korycansky and Zahnle (2005), McKinnon et al. (1997), Schaber et al. (1992), and Strom et al. (1994). Phillips et al. (1992), Schaber et al. (1992), and Strom et al. (1994) all used estimates of the impactor populations (numbers and proportions of asteroid and comets) by Shoemaker et al. (1991), while McKinnon et al. (1997) developed their own functions that were also used in Korycansky and Zahnle (2005).

The final major hurdle in determining a production age is accounting for the effects of the atmosphere on incoming meteoroids. Meteoroid disruption and deceleration prevent small objects from making it to the surface and forming a crater, while the separation of fragments of other meteoroids results in irregular craters or crater fields (Herrick and Phillips 1994a,b). The end result is that the smallest impact crater on Venus is ~1.5 km across, and the incremental size-frequency distribution rolls over and has reductions relative to the

Fig. 1 Size-frequency
distribution of certain and
probable craters from the
database of Herrick et al. (1997).
Craters are binned in $\sqrt{2}$
increments of D and plotted at
the minimum diameter of the bin.
The rollover with decreasing
diameter and the lack of craters
with D < 1.5 km is attributable to
atmospheric filtering of incoming
meteoroids

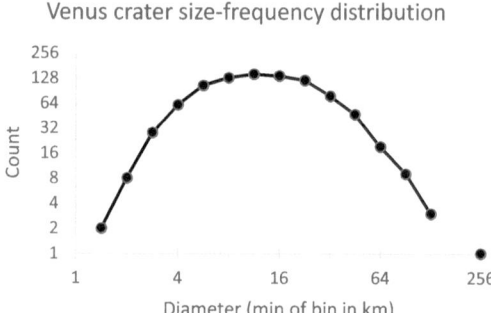

airless body case to crater sizes up to perhaps 30 km in diameter (Herrick and Phillips 1994b; Ivanov et al. 1986; Tauber and Kirk 1976) (Fig. 1). Note that reductions in the size-frequency distribution occur not only from meteoroids not making it to the surface, but also from those that do reach the surface but are slowed and make smaller craters.

The simplest approach to dealing with atmospheric effects is to assume that the straight part of the size-frequency curve, diameter D > ~30 km, is either unaffected by the atmosphere or only affected by a modification of the scaling law to calculate final crater size from an impacting meteoroid; Phillips et al. (1992), and Schaber et al. (1992) consider this approach. Using a small portion of what is already a limited number of craters on Venus to estimate surface age precludes evaluating all but the largest regional differences in surface age. Because different meteoroid types (e.g., iron versus chondritic asteroids, asteroids versus comets) should behave differently in the atmosphere, attempting to match the rollover in the incremental size-frequency distribution not only allows more craters to be used in age calculations, it also constrains the relative proportions of different impactor types. Thus, there have been attempts to match the rollover in the size-frequency curve (Herrick and Phillips 1994b; Ivanov et al. 1986; Korycansky and Zahnle 2005; McKinnon et al. 1997; Phillips et al. 1992) and the number and geometry of craters from clustered impact and crater fields (Herrick and Phillips 1994b; Korycansky and Zahnle 2005). Approaches involve considering meteoroid deceleration that includes flattening (which increases cross-sectional area) and disruption (fragments have larger ratios of surface area to mass), with criteria set for atmospheric explosion (and thus no surface crater) and/or a minimum impact velocity below which a crater is assumed not to form. An important caveat to the curve-fitting studies is an assumption that the shape of the incremental size-frequency distribution is reflective of a production function and that there is no preferential removal of small craters by volcanic or tectonic processes. A never-stated but required assumption of the above studies is that the atmosphere of Venus has not changed in nature over the time of observation; alternatively, the lack of sub-km craters can be seen as evidence that there are no well-preserved areas that formed during a period when Venus had a thin, Earth-like atmosphere.

In short, the range of production ages for the Venusian size-frequency distribution from credible past works spans from ~300-800 Ma, and the spread is mostly due to interpretations of meteoroid flux and crater scaling rather than treatments of atmospheric effects on crater formation. In considering the remainder of this article, it is important to remember that a production age assumes that all of the craters being counted are the last significant thing to have occurred on the surface that they are located. For example, in dating lunar mare deposits, one does not count the craters whose rims poke through the deposit and thus predate it. Furthermore, we cannot use the partially buried craters to date the surface underlying the deposit without some mechanism for determining the number of pre-existing craters that

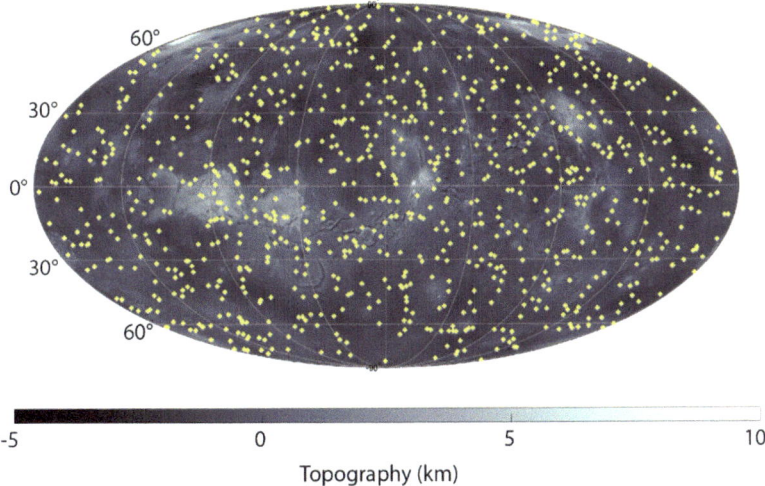

Fig. 2 Global distribution of craters shown by yellow crosses overlaid on Magellan topography. Data shown in Mollweide projection

were completely buried by the deposit. Thus, if the interpretation of Herrick and Rumpf (2011) is correct that 80% or more of craters have experienced post-emplacement partial volcanic modification, then the production age of Venus should be considered to be tens of millions of years, and modeling of surface ages using the entire crater population requires some model of how volcanism buries craters.

3 The Global Pattern of Craters

The distribution and condition of impact craters on a planetary surface can provide insight into how old and geologically active that planet has been through time. Initial results from the Magellan mission indicated that Venus hosts fewer craters than previously expected and these craters are largely unaffected by post-impact erosional, tectonic, or volcanic processes. There are ~900 craters across the surface of Venus with diameters between 1.5 to 270 km (Herrick et al. 1997; Phillips et al. 1992; Schaber et al. 1992). Initial assessments of the Venus crater population showed that all but a few percent of the craters have radar-bright ejecta deposits around most or all of the crater rim, few of the craters have been distorted by tectonic deformation or even have obvious through-going faults, and few of the craters have rims clearly breached by volcanism or volcanically buried central structures (Herrick and Phillips 1994a; Phillips et al. 1992, 1991; Schaber et al. 1992). Based on these observations, it was initially assessed that 80% or more of the craters are pristine (Herrick et al. 1997; Phillips et al. 1992; Schaber et al. 1992), and this was assumed as a constraint for much of the subsequent statistical and modeling work that took place until at least the late 1990s. However, a large portion of these craters have radar-dark floors, suggesting that these craters may have been flooded with lava flows and are not as pristine as previously thought (Herrick and Rumpf 2011; Herrick and Sharpton 2000; O'Rourke et al. 2014).

Maps of impact crater spatial densities can indicate regions of significantly different surface ages because older surfaces have been exposed to impact events for a longer time and, consequently, are predicted to have more craters (Shoemaker and Hackman 1962; Wilhelms

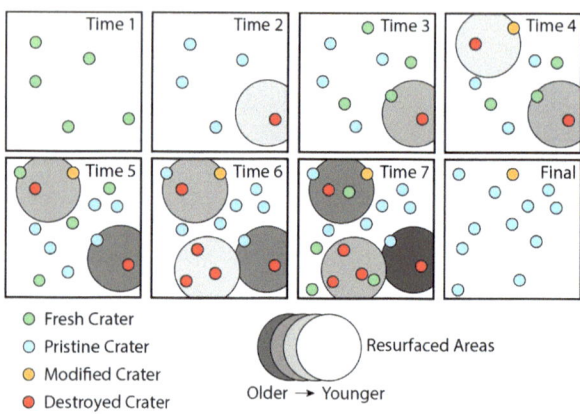

Fig. 3 Schematic illustrating the random crater generation and resurfacing used in Monte Carlo modeling of Venusian crater evolution. Craters which fall on the perimeters of subsequent resurfacing events may be tagged as "Modified" if modification is included as a consideration of the final crater distribution (e.g. Bjonnes et al. 2012), or as "Destroyed" if the analysis does not include modification (e.g. Strom et al. 1994). Because no specific resurfacing mechanism has been identified, final frame shows the modern-day distribution of "pristine" and "modified" craters and omits the resurfaced areas. Figure modified from Bjonnes et al. (2012)

et al. 1987). Crater density maps on Venus reveal that the planet has a distribution of craters across the surface that looks like a set of points placed randomly on a sphere (Fig. 2); in a broad sense, craters are scattered across the surface, but unlike a uniform distribution the inter-crater distance varies and small clusters and gaps occur. This distribution is commonly interpreted as indicating that Venus lacks large areas of notably different ages (Phillips et al. 1992) and instead Venus has a globally-uniform surface age of ∼300-800 Ma (McKinnon et al. 1997). One of the main reasons for this conclusion is the low overall number of impact craters on Venus which make it difficult to determine statistically significant regions of variable surface ages. Mapping the global crater distribution against areas with suspected high volcanic activity such as Eastern Aphrodite Terra shows that there is a higher proportion of tectonically deformed or volcanically embayed craters on the margins of areas with lower overall crater densities (Herrick and Phillips 1994a; Phillips et al. 1992; Price and Suppe 1994; Price et al. 1996), and there are some significant differences in crater spatial density between different geomorphologic map units (Kreslavsky et al. 2015; Price and Suppe 1995). That the distribution of small craters is mostly indistinguishable from that of large craters (Phillips et al. 1992; Schaber et al. 1992) places additional constraints on resurfacing processes and has been used to bolster the position that the current crater population is mostly a production function (Strom et al. 1994).

Several studies have used Monte Carlo models to quantify the spatial randomness of impact craters on Venus. Monte Carlo models are a statistical tool used to determine the probability of a certain outcome by generating large datasets of randomized outcomes and determining the probability distribution of the outcomes. Monte Carlo models simulating impact crater distributions on Venus test the spatial randomness of craters emplaced on its surface with or without accompanying simulated resurfacing events (Fig. 3). With a large number of these Monte Carlo models complete, it is possible to generate statistically robust probability distributions of several metrics of spatial randomness, including intercrater angles and nearest-neighbor distances. Some common comparisons between model results and the observed craters are known as probability-probability (P-P) plot and quantile-quantile (Q-Q) plots. P-P plots are a graphical representation of how the cumulative distributions of two probability functions compare to each other; if two distributions are the same, a P-P plot will be a straight line at 45°. Q-Q plots are similar but instead compare the quantiles of two different datasets, and for identical data, will also plot along a 45° line. The two plots differ in that the centers of probability distributions are more apparent in P-P plots whereas differences in the tails of two distributions are more apparent in Q-Q plots. Observed craters on Venus are statistically indistinguishable from random distributions generated in Monte Carlo models in both P-P and Q-Q plots (Phillips et al. 1992).

Other statistical tests used to compare the existing Venus crater record to simulated distributions are the t-test and chi-squared test. Although similar, a t-test compares two datasets to determine if they are identical and a chi-squared test determines if there is a relationship between two datasets. Strom et al. (1994) tested Venus' crater distribution against Monte Carlo models with and without simulated resurfacing events and determined that the distribution of craters on Venus is indistinguishable from random as determined by both t-tests and chi-squared tests. Before it was realized that most dark floored craters have likely been modified by post-impact volcanism, it was interpreted that the craters on Venus are statistically unlike Monte Carlo simulations with resurfacing patches of 0.001-0.03% of total surface area (too many volcanically modified craters) and 10-50% of total surface area (non-random distribution; Strom et al. 1994). However, impact craters on Venus match both the number of modified craters and their random distribution if resurfaced areas are between 0.1 and 1% of the total surface area if modeled resurfacing events are limited to the first 3 billion years of simulated time (Bjonnes et al. 2012). This conclusion hinges on the timing of volcanic activity; if lava flows between 0.03 – 10% of the total surface area are younger than 3 Ga, then these non-random resurfacing events will remove more craters from the surface than can be recovered by subsequent impact events, resulting in an overall non-random spatial distribution in the present day (Phillips et al. 1992; Romeo and Turcotte 2009).

Although indistinguishable from random when the spatial pattern is considered alone, there is subtle variability within the global distribution of impact craters on Venus. A test of the distance between a crater and its nearest neighbors is more able to determine subtle variability in a global dataset compared to tests which consider the entire probability distribution (Hauck et al. 1998). Testing the mean distance between a crater and its 1st – 4th nearest neighbors reveal that the Venusian plains, which cover 60% of the surface (Ivanov and Head 2011), can be subdivided into up to 4 different units based on morphology, and these morphologic units have statistically significant different mean ages (Hauck et al. 1998). The global crater distribution is also non-random with respect to topography, with a higher-than-expected number of craters located between 6050.9 and 6051.9 km planetary radius (Herrick and Phillips 1994a). Topographic analysis shows a significant crater deficit at planetary radii between 6052.4 and 6052.9 km, an elevation range associated with tectonic rift zones and volcanoes (Herrick and Phillips 1994a).

4 Modification and Degradation of Craters

Most Venusian craters occur in what can generally be called "volcanic plains" (Herrick and Phillips 1994a; Ivanov and Head 2011; Price and Suppe 1995), which is to say that they occur in areas where the terrain is relatively flat and with uniform, low radar backscatter compared to the rough ejecta of an impact crater. The crater population exhibits signs of different degrees of post-impact modification and/or degradation (Herrick and Phillips 1994a; Izenberg et al. 1994; Schaber et al. 1992). Most craters have a circular platform, and continuous ejecta that appear bright in Magellan SAR. "Fresh" appearing craters have bright,

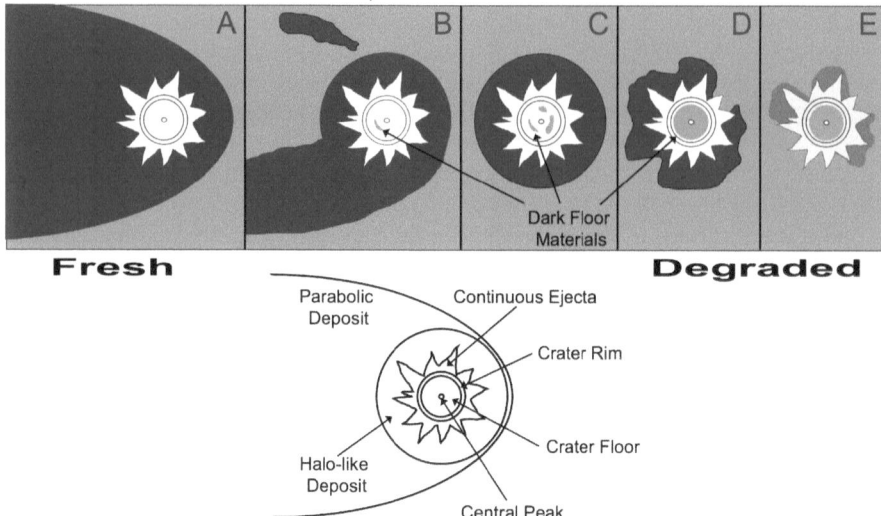

Fig. 4 Model of crater degradation on Venus, from Izenberg et al. (1994). **A**) Crater forms with high backscatter rim, bright floor and continuous ejecta, and low backscatter halo and parabola ejecta deposits with high contrast relative to surrounding plains. **B-E**) Progression of removal of parabola and dark halo, while radar-dark deposits form on the crater floor

rough floors, usually flat. While increasing crater size shows a familiar progression from flat floors to central peaks to peak-rings, there are few simple, bowl-shaped craters on Venus due to atmospheric shielding from small impactors. Many plains craters have fluidized ejecta, and/or dark extended ejecta features or halos that can extend several crater radii farther than the continuous ejecta boundary. The youngest fraction of craters have areally extensive parabolic ejecta features, always opening to the west, resulting from the wind-borne movement of ejecta thrown into the atmosphere's global East-West prevailing winds (Campbell et al. 1992). The parabolas are almost universally superposed on all other surface geologic features, indicating that they are both young and ephemeral. The correlation of the parabolas with dark haloes, radar-bright floors, etc., provides the primary evidence that this is the common appearance of craters immediately after formation (keeping in mind differences as a function of crater size), and that differences from this appearance represent post-impact alteration (Herrick and Phillips 1994a; Izenberg et al. 1994).

There are insufficient numbers of craters to use them as local, or even regional age indicators for the surface, but they are time-stratigraphic markers for their immediate surroundings. Craters, once emplaced, are affected by planetary aeolian, tectonic, and volcanic processes, and the degradation process can be used to better understand local and regional geologic activity (Izenberg et al. 1994) (Fig. 4). For example, the absence of parabolas from most craters, and the existence of both partial parabolas and partial halos infer that surface processes remove or homogenize these features over time. This process could be aeolian removal of the fine ejecta that makes the parabolas and halos appear dark, or volcanic burial, or lithification in place. Crater interiors exhibit a range of potential age effects. While some craters may begin with smooth, dark floors or patches on their floors due to impact melt, some appear to be infilled after formation by volcanic deposits. Bright crater ejecta, on the other hand usually appears pristine unless severely modified or clearly buried by volcanic activity. Crater and continuous ejecta modification by tectonic activity can also be seen in rift zones and in the tessera.

Crater aging degradation on Venus appears morphologically controlled by two vectors: absolute age, and regional activity. A simply aging crater, undisturbed by planetary volcanism or tectonism will evolve from a bright crater with a dark parabola, to a partial parabola, to a halo, to a partial halo, to no halo. The bright interior and ejecta may darken through weathering of rougher blocky materials, and the floor may darken due to interior flooding. Volcanic activity may accelerate the apparent aging process. Low level embayment could very quickly bury/destroy a parabola or halo, while leaving the continuous ejecta apparently untouched. Higher levels of volcanic activity could embay small to large portions of continuous ejecta and potentially breach and flood the crater interior. Significant activity can disrupt crater walls, but lower levels of tectonism may be harder to see in continuous ejecta and crater interiors.

The distinction between crater degradation by volcanism versus mechanical and chemical weathering is important for big-picture resurfacing hypotheses. If degradation proceeds by weathering then the progression shown in Fig. 4 at least loosely represents a defined sequence with time so that the state of a crater would indicate its age of emplacement. If, however, volcanic or tectonic resurfacing play a significant role in degrading craters, then the state of an individual crater conveys only its relative position in the stratigraphic column and little information regarding the absolute age of emplacement. Herrick and Rumpf (2011) have argued that the presence of a radar-dark floor in a crater is not due to weathering but nearly exclusively due to post-impact volcanism, suggesting the latter. Their assessment is based on observations from stereo-derived topography that dark-floored craters, as a family, are shallower than bright-floored craters. That craters with parabolic features all have radar-bright floors (Campbell et al. 1992) argues against craters acquiring dark floors during or shortly after formation through some process like triggered volcanism. They observe no aeolian features such as dune fields in dark-floored craters that would indicate substantial sediment filling, but detailed mapping of several of the largest Venusian craters showed abundant evidence of post-impact volcanism modifying most of these craters (Herrick and Rumpf 2011; Herrick and Sharpton 2000, 1996). Herrick and Rumpf (2011) also found that nearly all dark-floored craters had a portion of the dark halo removed, indicating that if dark-floor formation does indicate volcanic filling that it is probably accompanied by partial volcanic embayment.

On Venus there are a number of near-circular albedo patches with diffuse boundaries that often resemble the dark haloes associated with craters, have no topographic signature or apparent vent associated with them, and are superposed on the terrain with no correlation to local geology (Strom et al. 1994). These "splotches" are thought to be airblast features, similar to the Tunguska event on Earth (Turco et al. 1982), caused by the atmospheric pressure wave created by the disintegration and incineration of intermediate sized impactors (asteroids or comets of $\sim$200 m to 3 km diameter depending on composition and velocity) above the surface due to Venus' thick atmosphere. Splotches may be primarily scour features, but there are indications of ejecta-like materials near some. Stereo imaging may indicate some splotches are slight depressions, however the stereo data lacks enough resolution to form any conclusions. That there seem to be few or no splotches of the $\sim$400 cataloged that are modified by later volcanism has been cited as supporting evidence of a surface that is currently geologically stagnant (Strom et al. 1994).

5 Global Modeling of Resurfacing

Models of resurfacing designed to match some of the general observations can provide constraints and insights on the planet's resurfacing history. The primary observations that the models in this section try to match are statistical measurements of the spatial distribution of craters and the percentage of craters that have experienced post-impact modification. As the discussion below makes clear, changes in our understanding and interpretation of post-impact crater modification have occurred during the time period in which the modeling has been conducted, so that both the nature of the models and the assumed observational constraints have evolved with time. In general terms, the models can be two-dimensional or three-dimensional, with the latter adding a topographic component to how resurfacing occurs.

5.1 Goals of Resurfacing Models

Proposed models for the geologic history of Venus are tested against characteristics of the cratering and stratigraphic record. First, models of resurfacing should produce the correct number of impact craters on Venus and their size-frequency distribution. On other planetary bodies (Mercury, Moon, Mars, etc.), small craters vastly outnumber large ones because larger impactors are rarer. However, the diameters of impact craters on Venus seem to fit a log-normal probability distribution. Testing resurfacing models against this distribution is difficult because volcanism may have less of an effect on the final size-frequency distribution than two factors that are also quite uncertain: the impactor production function and the atmosphere-impactor interactions. Ultimately, models of resurfacing often draw the diameters of modeled craters from the real size-frequency distribution to guarantee that this constraint is satisfied.

Beyond the number of craters, models of resurfacing can attempt to reproduce statistical properties of their locations. Because impacts are stochastic, it would be absurd to seek to match the exact coordinates of every crater. As discussed above, the distribution of impact craters on Venus is nearly indistinguishable from random. When taken with the relatively low number of total craters, there is no statistically reliable assertion of distinct areal provinces based on crater population.

As discussed above, although the global impact crater distribution is sparse, there are indications that crater distribution may not be random with topography (Herrick and Phillips 1994a), geologic units (Hauck et al. 1998), or global stratigraphy-based geologic maps (Ivanov and Head 2011) and buffered crater counts (Kreslavsky et al. 2015). Nearest-neighbor analysis is one method to determine spatial randomness, which has revealed subtle variation in crater densities across the Venusian plains (Hauck et al. 1998) and clustering among volcanically embayed craters (O'Rourke et al. 2014). This approach has an added benefit of analyzing the distance between a crater and its Mth nearest neighbors, not just its closest neighbor. For example, the obviously embayed craters have random-like distances to their nearest neighbors ($M = 1$). However, they are clustered in groups with $M > 3$. The probability of an intercrater distance between a crater in a dataset with N craters and its Mth nearest neighbor is given by the following formula:

$$p(\theta \mid N, M) = \frac{(N-1)!}{2^{N-1}(N-M-1)!(M-1)!} \sin(\theta)[1 - \cos\theta]^{M-1}[1 + \cos\theta]^{N-M-1}$$

where θ varies between $0°$ and $180°$ (i.e., θ is the smallest angle between the two lines that extend from the center of Venus to the craters on the surface) and thus p is normalized to integrate to 1 over this interval. The probabilistic angular distance between points on a sphere can be compared with the angular distance between observed craters using a normalized test statistic z defined by:

$$z = \frac{\mu_{exp} - \mu_{obs}}{\sigma_{exp}}$$

where μ_{exp} and μ_{obs} are the calculated and observed mean angular distance and σ_{exp} is the standard deviation of the probability distribution function. A smaller population (low N) would have σ_{exp} that is relatively large. A perfectly random spatial distribution will have z values close to 0, whereas a surface distribution with either underclustering or overclustering will result in $z < 0$ or $z > 0$, respectively. This test is more sensitive than chi-squared tests on spherical coordinates or intercrater angles, and consequently can provide a more nuanced understanding of the spatial crater distribution.

Finally, successful models of resurfacing should explain the modification states of impact craters. However, as discussed above the underlying observations are controversial. The initial post-Magellan view was that impact craters on Venus are generally pristine with fewer than 10% of total craters showing obvious embayment (Herrick et al. 1997). Unlike craters overall, those obviously embayed craters are spatially clustered (O'Rourke et al. 2014; Romeo 2013; Strom et al. 1994) which could result from volcanic resurfacing concentrated in the Beta-Atla-Themis (BAT) region (Romeo 2013). It is difficult to reconcile the relatively low number of embayed craters with widespread evidence of volcanism across the surface. Consequently, the overall low occurrence of such craters and their spatial clustering remain key observational constraints on any postulated resurfacing history. However, if post-impact volcanism can flood crater floors through cracks and small breaches in crater rims, 750 dark-floored craters would be reclassified as modified. The proportion of volcanically modified craters would thus increase from ~10% to ~80% (Herrick and Rumpf 2011; O'Rourke et al. 2014). Ideally, resurfacing models would consider this ambiguity when determining if a tested resurfacing history is consistent with the observed cratering record on Venus. However, in the extant literature, the number of modified craters is generally considered as a separate question from the spatial randomness of all craters—and no study has fully assessed the spatial distribution of dark-floored craters with respect to topography and geological units.

5.2 Methodology of Resurfacing Models

Modelers face a tradeoff between realism and computational efficiency. Complex models can capture the diversity of volcano-tectonic processes and geologic terrains that is observed on Venus. However, impact cratering is a stochastic process and often volcanism is modeled as a stochastic process, so the outputs of resurfacing models are also stochastic. Statistical comparisons between models and reality require running models repeatedly—thousands of times or more. Specifically, models often use a Monte Carlo approach with a set of random variables that describe how impact craters form and are destroyed over time. Models with more random variables require more model runs (and thus human and computer time) for statistical analyses. Counterintuitively, modelers can thus apply more granular and rigorous statistical tests to resurfacing models that are more abstract and simplified. As in most scientific endeavors, the art of modeling resurfacing on Venus lies in discarding all complexities except those that make a big-picture difference. Monte Carlo models are thus best applied to answering broad questions, such as "was resurfacing catastrophic?"

Resurfacing models must first describe the location, size, and production rate of impact craters. The simplest approach is to assume that impact craters have equal sizes and appear at a constant rate with equal probability at any location on the surface. However, some additional realism is not too expensive to implement. For example, the size-frequency distribution of modeled craters is often generated from the real distribution (e.g., Romeo and Turcotte 2010) — or with a distribution that is skewed slightly large to account for the likelihood that resurfacing processes preferentially remove smaller craters (e.g., Romeo 2013). Treating cratering on Venus as a Poisson process with a single time constant is reasonable because the surface is relatively young. Studies of planetary bodies with older surfaces such as Mars and Moon are more sensitive to how the impactor production function has changed over geologic time. Finally, Venus is indeed massive enough to homogenize impactor trajectories such that crater locations are isotropic (i.e., with degree-2 variations less than a few percent)—in contrast, the spatial density of impacts on Earth's Moon is highly non-uniform at the ±25% level (e.g., Le Feuvre and Wieczorek 2011).

Global resurfacing models on Venus invoke volcanism as the primary means of removing impact craters from the surface. Two end-member scenarios for volcanism—catastrophic and equilibrium resurfacing—are commonly studied. In models with catastrophic resurfacing, the surface is (mostly) wiped clean by a global episode of intensive volcanism at one time (e.g., Strom et al. 1994). Volcanic activity after catastrophic resurfacing is modeled to occur at a reduced rate (e.g., Romeo 2013) and, often, only at certain locations such as the Beta-Atla-Thema region. In equilibrium models, volcanism occurs at smaller scales at a roughly constant rate over time. Resurfaced patches are modeled as circular areas, which have sizes that are scaled to the frequency of resurfacing and cratering events to reproduce the correct number of craters on the surface of Venus today. For example, if one crater is produced every Myr, then a model where resurfacing events occur every Myr and each cover ∼0.1% of the surface will produce ∼1000 craters in equilibrium and a cratering age of ∼1 Gyr. Tectonic processes can also destroy (or at least obscure) craters, especially in tesserae, but are usually ignored in resurfacing models—or tacitly treated as equivalent to volcanism.

In general, existing craters can be unaffected, modified, or destroyed by resurfacing events. Different models take different approaches to quantifying when each outcome occurs. In perhaps the most basic model, the resurfacing event is described with one number: the radius of the circular region covered by a volcanic flow. In this model, craters are unaffected if they lie fully outside the resurfaced region; they are modified if some portion of the crater is touching the resurfaced area; and they are fully destroyed if they lie within the resurfaced area (e.g., Schaber et al. 1992). This simple model is computationally efficient but imposes an unrealistic assumption that fresh crust will always fully cover existing craters in the area. In reality, crater rims are a non-negligible elevation feature—and it is unrealistic to assume that volcanic flows will always be thick enough to bury them. This shortcoming can be addressed in models by adding an annulus around the resurfaced patch and tagging craters within that annulus as "modified craters" (e.g., Bjonnes et al. 2012). Resurfaced patches modeled this way more closely represent volcanic flows which are thicker near their point of origin, presumably the center of the circular resurfaced patch, and get thinner toward the edges to the point where it is not enough to fully cover existing features. The width of the annulus can be adjusted relative to the radius of the inner portion to simulate different lava viscosities. Preexisting craters would then be marked as "completely erased" if they are in the center portion of the resurfaced patch, "modified" if they fall within the defined annulus, or "unaffected" if they are fully outside the resurfaced area. Finally, another type of Monte Carlo model treats volcanic units as three-dimensional cones (Romeo 2013; Romeo and Turcotte 2010). Geological maps provide the edge angles of these cones (e.g., from ∼0.2–2°) and the frequency-area distributions of the units (e.g., from >10^6 km^2 to <20 km^2). In these models, a volcanic unit modifies a crater if the cone is tall enough to cover the crater's rim, using distributions of rim heights and crater diameters derived from bright-floored craters (e.g., Herrick and Sharpton 2000).

Another model of resurfacing was developed in response to the idea that volcanism has modified up to ∼80% of the impact craters on Venus (Herrick and Rumpf 2011; Herrick and Sharpton 2000; Wichman 1999). Like the simplest model, resurfacing patches are circles defined by a single radius. However, in these new models, multiple resurfacing events are required to destroy an impact crater (O'Rourke et al. 2014). Craters that are tagged with fewer than a critical number of resurfacing events are considered modified. If five resurfacing events are required to completely bury crater, then 20% (i.e., 1/5) of the total population will remain pristine in equilibrium. Critically, the horizontal radius of the patch does not affect the final proportion of pristine craters in this type of model. If the average crater has a rim-floor depth of ∼1 km (Herrick and Rumpf 2011), then this ratio of pristine-to-modified craters implies that one resurfacing patch involves extrusive flows that have thicknesses of ∼200 m. In this model, small craters are preferentially erased because they have shallow rim-floor depths. Finally, the radius of the resurfacing patch then controls the spatial clustering of the pristine and modified craters—small patches lead to random-looking distributions.

In the aftermath of the Magellan mission, many scientists claimed that only catastrophic resurfacing was consistent with the cratering record. Immediately after global maps were available, Phillips et al. (1992) showed that either catastrophic or equilibrium models could create a random-looking spatial distribution of all craters. However, others then argued that the ostensibly low fraction of modified impact craters (∼10%) was inconsistent with equilibrium resurfacing. Strawman models were developed that used very small resurfacing patches with proportionally thick annuluses, meaning that volcanic events were almost as likely to modify as destroy craters (Schaber et al. 1992; Strom et al. 1994). Several studies (Bullock et al. 1993; Romeo 2013; Romeo and Turcotte 2010) developed perhaps the most elaborate version of this model—the shapes of volcanic flows and the diameters and rim heights of craters were all drawn from different distributions and modeled in three dimensions. Their catastrophic resurfacing models naturally predicted a low proportion of volcanically modified craters—planet-wide volcanism destroyed rather than modified all extant craters, then volcanic events were concentrated at certain locations (such as the Beta-Atla-Themis region) and occurred at drastically lower rates (e.g., Romeo 2013).

In the last decade or so, equilibrium resurfacing models were refined to show that they can also produce the correct proportion of modified craters. The approach of using a global resurfacing model with variably-sized resurfacing patches and an annular rim wherein craters are modified and not fully destroyed was implemented by Bjonnes et al. (2012). They ran a series of Monte Carlo models testing the effects of resurfacing between 0.01 and 50% of total surface area per resurfacing events, concurrent with impact crater formation throughout Venus' history. Their results show that resurfacing the planet in areas approximately 0.1–1% of total surface area replicates both the seemingly random spatial distribution of craters across the surface as well as the low number of obviously embayed craters. This approach used the more realistic method of incorporating a ring around the resurfacing patch where craters would be embayed rather than destroyed. However, these models randomly selected points for the centers of resurfacing patches, which is inconsistent with observations that some regions on Venus are more geologically active than others. Consequently, the simulated locations of embayed craters were also randomly distributed across the planet. However, nearest-neighbor analysis shows that observed embayed craters are clustered for >3rd nearest-neighbor distances (O'Rourke et al. 2014). Other studies tested equilibrium models with non-uniform spatial distributions and temporal rates of volcanic resurfacing (Romeo 2013; Romeo and Turcotte 2010). Future work could augment these models to allow for even more complex scenarios consistent with new models of Venus's geodynamic evolution (Rolf et al. 2022).

Ultimately, Monte Carlo models that represent either catastrophic or equilibrium resurfacing can reproduce the first-order properties of the cratering record, including the number of impact craters, their size-frequency distribution, their spatial clustering (or lack thereof), and the proportion of craters modified by post-impact volcanism. Claimed inconsistencies between equilibrium models and reality often result from unrealistic assumptions in the models that were made to reduce the parameter space to a manageable size for a single study. For the next generation of Venus exploration, scientists should tie models of cratering and resurfacing to newly available information about the geologic history of the surface—and to increasingly sophisticated simulations of how the deep interior of Venus has evolved over time.

5.3 Resurfacing Models Coupled to Mantle Convection

The techniques outlined in the previous subsections are powerful methods to efficiently map out controls on the surface age distribution and its link to the crater population. However, they do not fully capture the physical processes driving resurfacing, which on Venus are likely a combination of volcanism (lava flows) and tectonism (crustal deformation). Mantle convection is the physical mechanism controlling the thermal and stress state of Venus' crust (Rolf et al. 2022, this collection) and therefore the spatiotemporal evolution of resurfacing. Classically, surface age may be inferred from mantle flow using half-space cooling (see Turcotte and Schubert 2017) in which case surface age is a simple function of heat flux (e.g., Labrosse and Jaupart 2007). However, this approach has several limitations and importantly assumes the lithosphere to be mobile (i.e., to move with rates comparable to that of mantle flow), which is unlikely the case for present Venus. Instead, tracer particles may be used to either track when a respective material patch last melted and rose to the surface (e.g., Noack et al. 2012) or how long a tracer resides in a layer close to the surface (Karlsson et al. 2020; Rolf et al. 2018; Uppalapati et al. 2020). The latter approach captures both tectonic and volcanic resurfacing self-consistently—but has limitations, for instance, regarding the thickness of the layer defined as 'surface'. In addition, existing planetary-scale models typically only track magma ascent to the surface, but not the subsequent lateral lava flow across Venus' surface. The modification of craters by the resurfacing processes is also beyond the capability of current models.

Nevertheless, such physics-based approaches have proven useful to tie resurfacing to the state of the interior. Although sophisticated simulations of mantle convection are too intensive to run tens of thousands of times in a Monte Carlo framework, they can provide guidance about big-picture issues such as the age of the surface and whether resurfacing is catastrophic. For example, a cooler mantle may yield a larger mean surface age, because lower temperatures lead to less mantle melting and thus suppress volcanic resurfacing rates. Catastrophic resurfacing episodes can cool the mantle more efficiently and may promote an older surface, on average. However, this strongly depends also on the eruption efficiency of magma as intrusive magmatism will less efficiently destroy surface craters than extrusive lava flows. Only recently have mantle convection models become capable of realizing low eruption efficiencies (e.g., Lourenço et al. 2020) which are more likely relevant for Venus in its more mature stages of evolution.

In the episodic scenario, surface age also strongly depends on the style of the resurfacing episode. In the classic form, an episode resets surface age globally leading to a uniform surface as inferred for Venus. Extinct (or very limited) volcanism after overturn cessation then allows for uniformly increasing mean age without degenerating the age uniformity. However, models of Venus' mantle convection indicate a more complex process: Lithospheric

Fig. 5 Typical age relationships among tessera (t), the lower unit of regional plains (rp1) and lobate plains (pl). These relationships are repeated in each region of Venus where these units are observed

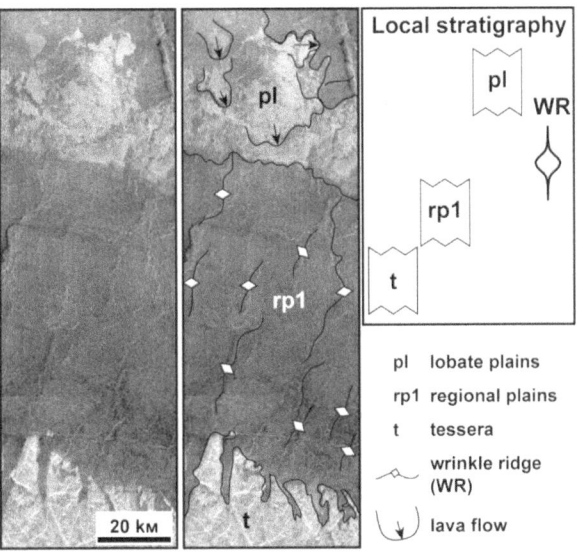

mobilization during the episode may not be global (Gillmann and Tackley 2014; Noack et al. 2012; Romeo and Turcotte 2008) and portions of the surface may not be recycled (e.g., Karlsson et al. 2020; Uppalapati et al. 2020; Weller and Kiefer 2020). Even if global, resurfacing does not occur simultaneously across the surface, but propagates during $\sim$100–200 Myr or more depending on the duration of such events (e.g., Armann and Tackley 2012; King 2018; Rolf et al. 2018). As a consequence, substantial variations in surface age exist during essentially any time. Important steps of future work are thus to identify whether this variability in surface age is due to model limitations or whether it applies to Venus, perhaps on scales that are poorly captured by the sparse density of craters on Venus.

6 Resurfacing Constraints from the Geologic Record

6.1 The Case for a Global Stratigraphic Sequence

The idea of a global stratigraphy for Venus is based on two key interpretations made during a photogeological analysis of 36 randomly selected regions (Basilevsky and Head 1995, 1994): (1) in all cases, sets of morphologically distinct units that make up the surface were relatively small and repetitive, (2) all these units show similar relationships of relative ages and, thus, form similar stratigraphic sequences. Support for the idea of a global stratigraphy came from the results of the USGS program of geological mapping of Venus. These mapping efforts in different regions of Venus have indicated existence of a general progression of units from the oldest strongly tectonized terrain, such as tessera, through a ubiquitous unit of regional plains at middle stratigraphic level, to the youngest plains characterized by prominent internal flow-like features (Fig. 5).

Basilevsky and Head (1995, 1994) have summarized their observations in the form of a hypothesis that the geological record of Venus is made up of globally correlative units and that the accessible geologic history of the planet consists of a series of specific episodes when different volcanic and/or tectonic processes dominated (Basilevsky and Head III 1998;

Herrick 1994). This hypothesis is broadly consistent with the lack of evidence of plate tectonics during the observable part of the Venus' history and suggests a possible alternative mechanism of internal heat loss, primarily conduction through a static lithosphere over a convecting mantle (e.g., Solomon et al. 1991; Solomon and Head 1982). To test the applicability of the observations made in isolated areas (Basilevsky and Head 2000, 1995, 1994), geological mapping in sizable and, most important, contiguous areas was required. Ivanov and Head (2001b) thus attempted such a test by compiling a geologic map consisting of a geotraverse centered at 30°N that extended completely around the Venus globe and connects several isolated regions. The mapping results in the geotraverse revealed that regional plains with uniform morphology can be traced continuously for thousands of kilometers and link remote regions. This extensive and pervasive unit everywhere in the geotraverse embays most of heavily tectonized units (e.g., tessera, ridge belts, and groove belts) and is superposed by flows of lobate plains (Fig. 5). The observations made during mapping of the geotraverse and in several USGS quadrangles (Ivanov et al. 2010; Ivanov and Head 2001a, 2008, 2006, 2005, 2004) have suggested that only a small number of units with clearly different morphology make up the majority of the surface of Venus. These findings allowed a compilation of a global geological map of Venus at 1:10,000,000 scale that portrays the distribution of the defined units/terrains in space and time (Ivanov and Head 2011). The global map has shown that the defined units (1) clearly describe the variety of morphologies at the global scale, and that (2) they form a generalized sequence that consist of three major (composite) units: (1) the older tessera, labeled "t" (and the other heavily tectonized terrains), (2) mildly deformed regional plains, labeled "rp" (and the other units of the plains volcanism), in the middle of the generalized sequence, and (3) mostly non-tectonized lobate plains, labeled "pl", at the top of the sequence (Fig. 5).

The findings made during compilation of the global map suggest that the preserved geologic record of Venus (the last < ~1 Ga) is mostly consistent with a mechanism of the internal heat loss that is radically different from the global plate tectonics mode of heat loss that has characterized Earth through much of its history (Rolf et al. 2022), and that may even have had an episodic character (Parmentier and Hess 1992; Turcotte 1993).

The characteristic age relationships of units shown in the global map suggest that the accessible part of the geologic history of Venus consists of three major episodes, each with a specific style of resurfacing (Fig. 6) (Ivanov and Head 2015): (1) A global tectonic regime, when tectonic resurfacing dominated. Exposed occurrences of units of this regime comprise ~20% of the surface of Venus. (2) A global volcanic regime, when volcanism was the most important process and resurfaced ~60% of the surface of Venus. (3) A network rifting-volcanism regime, when both tectonic and volcanic activity were about equally important. During this regime, ~16% of the surface of Venus was modified.

The global tectonic regime is defined by units labeled t (tessera), pdl (densely lineated plains), pr/RB (ridged plains/ridge belts) and gb (groove belts). Ridges of tesserae and ridge belts suggest that compressional forces dominated during the earlier phases of the global tectonic regime. They were responsible for the formation of the bulk of the tesserae over the sites of mantle downwelling. Ridge belts are due to limited horizontal displacement and warping/buckling of crustal materials at the periphery of the major convective cells where downwelling takes place (Gilmore and Head 2018, 2000). Groove belts characterize the later phase of the global tectonic regime. The scales of the features of the global tectonic regime suggest that they reflect the ancient pattern of mantle convection.

During the global volcanic regime, vast volcanic plains (shield and regional plains, the lower and upper units) were emplaced. These plains have different morphologies that indicate different volcanic styles: (1) eruptions from small, abundant, and broadly distributed

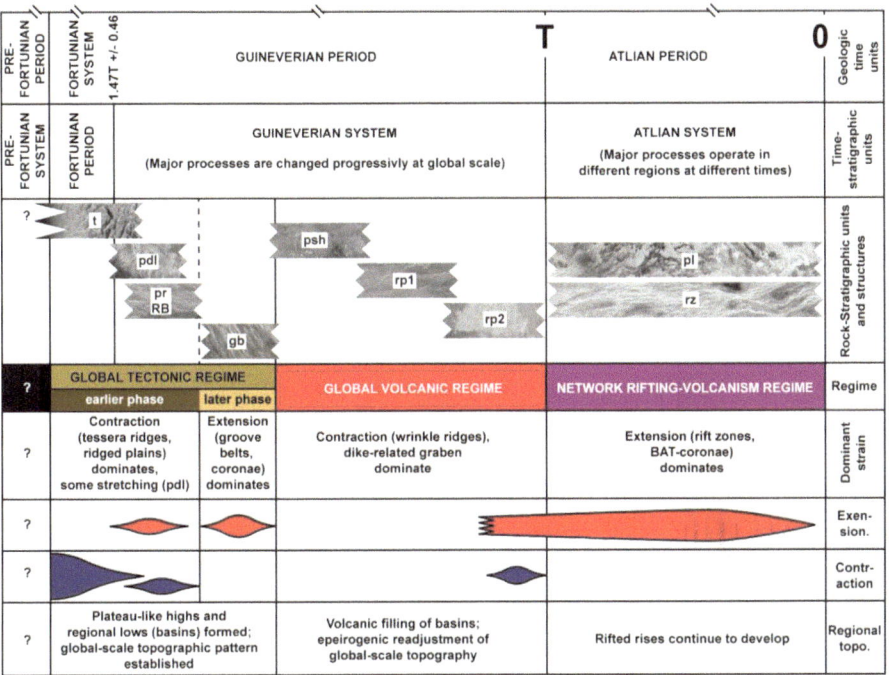

Fig. 6 Three major regimes of resurfacing (see text for discussion). Modified from Ivanov and Head (2015)

sources (psh, shield plains), (2) voluminous eruptions from near globally distributed sources (rp1, lower ridged plains unit), and (3) eruptions that have been concentrated within specific regions and around some large volcanic centers. The major tectonic structures of the global volcanic regime are wrinkle ridges and graben swarms radiating from distinct volcano-tectonic centers (Bilotti and Suppe 1999; Ernst et al. 2003, 2001, 1995; Sandwell et al. 1997). Both types of structures deform the surfaces of the vast plains but are not related directly to specific patterns of mantle circulation.

Approximately equally abundant volcanic and tectonic processes of the formation of lobate plains and rift zones characterize the youngest network rifting-volcanism regime. Rift zones that constitute its tectonic component are very prominent but their total area is about four times smaller than the exposed area of terrains of the global tectonic regime. This means that the tectonic resurfacing diminished with time and evolved from the broadly distributed deformation during the earlier global tectonic regime to the highly concentrated deformation during the later network rifting-volcanism regime. The numerous flows of lobate plains suggest multiple episodes of volcanic activity during the network rifting-volcanism regime. The areal distribution of lobate plains implies that their sources were discrete and that they formed in different areas at different times. The apparently prolonged and broadly synchronous formation of lobate plains and rift zones are fundamentals of the nondirectional model of the geologic history of Venus (Guest and Stofan 1999). This model adequately describes the later stages of the history, whereas the alternative, evolutionary, model (Basilevsky and Head III 1998; Ivanov and Head 2011) embraces the entire assessable portion of the geologic history of Venus.

Impact craters with the surrounding dark mantling materials (parabolas/haloes) are considered as the youngest impact structures on Venus (Campbell et al. 1992). Indeed, the relative abundance of such craters suggests that craters with prominent dark parabolas may have been formed during the last 0.1-0.15T (where "T" is the mean model age of the surface of Venus) and lifetime of craters without parabolas but with distinct haloes is about 0.3T (Basilevsky and Head 2002). These estimates of the dark parabolas/haoles lifetime suggest that the youngest lobate plains of the network rifting-volcanism regime at Beta Regio are younger than about 0.5T and those at Atla Regio may be as young as about 0.1-0.15T (Basilevsky and Head 2002).

6.2 The Case for a "Non-directional" Geologic History

6.2.1 Material Units and Tectonic Structures

Stratigraphic sequences are normally defined on the basis of lithological units (e.g. through sedimentary logging), although biostratigraphy is of course commonly used on Earth because of the abundance of fossils in the Phanerozoic. Mappable units are defined on the basis of common features, such as the 'grey chalk' in distinction to the 'white chalk'. Gaps in succession (unconformities) often indicate episodes of tectonic uplift (orogenies), but these are of course not recorded in the stratigraphic sequence: for example, the Carboniferous Coal Measures do not grade into the Variscan orogeny and then into the Mercia Mudstone, but instead the Mercia Mudstone directly and unconformably overlies the Coal Measures. Geological structures do not feature in the sequence for two reasons: in this example, Variscan structures overprint all older material units, regardless of their age, and so do not correspond with any particular material unit; and additionally, these structures can be reactivated and so appear younger than they really are.

Distinguishing material units separately from structural relationships (Hansen 2000) is therefore helpful for inferring the sequence of tectonic events from kinematic analysis. Structural relationships can often be counter-intuitive, e.g. a straight fracture is more likely the oldest structure, not the youngest. Younger fractures sense the presence of older breaks (because they act as free surfaces) and deviate towards or away from them and, as noted, suitably oriented fractures may be reactivated by later events, sometimes in reverse (i.e. inversion). This can make the extremely fractured surface of Venus difficult to decipher.

6.2.2 Mappable Radar Units

One of the major problems for lithostratigraphic mapping on Venus is that radar is quite insensitive to lithological units, except in the rare instance of a marked difference in relative permittivity (usually in relation to iron content). Instead, radar is sensitive to surface roughness on the scale of the radar wavelength (particularly near the Rayleigh criterion), small-scale slopes, and other surface textures. An example from Earth nicely illustrates this problem (Fig. 7): contrast changes in the radar image only sometimes correlate with geological boundaries, while particularly significant features, such as the Mid-Atlantic ridge, and the edge of the Vatnajökull ice cap, are almost invisible in the radar image. Even fresh lava flows can prove difficult to map, apparently disappearing in the radar image where their flow becomes laminar, and reappearing where a'a or pahoehoe-textured areas. A more serious problem is that very different lithostratigraphic units can appear identical in a radar images: 100 Ma old Upper Greensand Sandstones look very similar to 800 Ma old Torridon Sandstones, but very different to the almost coeval Gault Clay deposited farther offshore. Worse still, a thin sedimentary veneer is sufficient to completely erase contacts between different units. The reason that large parts of Fig. 7 appear similar is that they are covered by alluvial sediments.

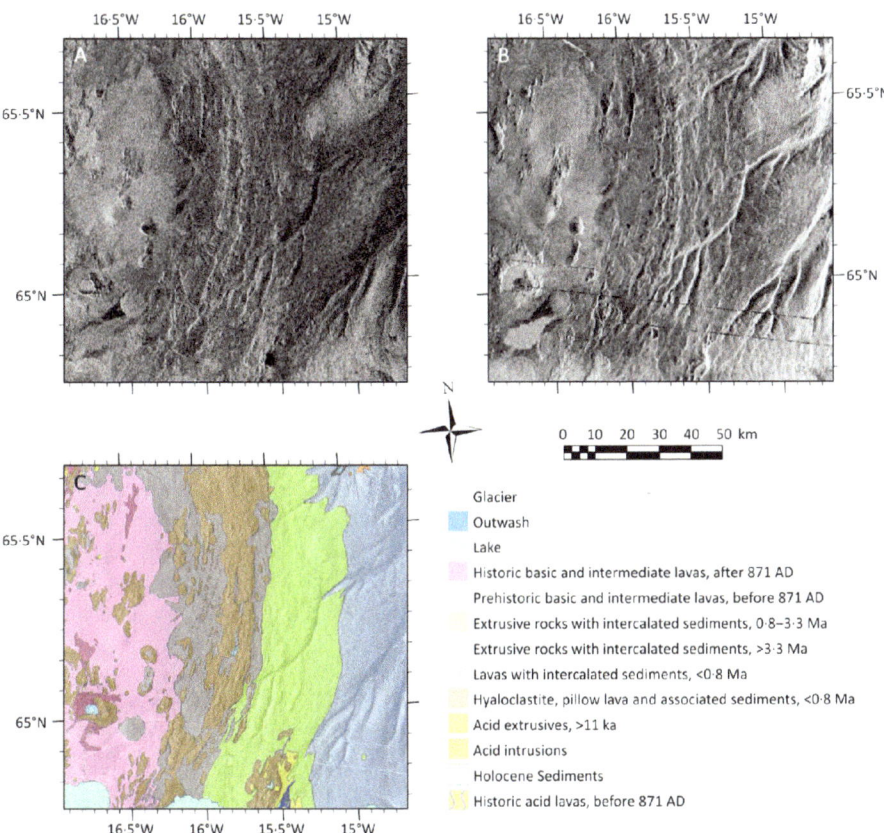

Fig. 7 Simulated Magellan radar images and geological map of the Vatnajökull National Park, Iceland, illustrating the difficulty in identifying geological boundaries from radar imagery. Oskjuvatn volcano is in the lower left (65·0°N, 16·8°W) with the Holuhraun eruption that occurred between the image acquisitions to the south. [A] Sentinel-1A left-look (descending) image acquired on 24 September 2016, degraded to simulate Magellan Cycle 1 image. [B] Sentinel-1A right-look (ascending) image acquired 22 December 2014, degraded to simulate Magellan Cycle 2 image. [C] Shaded relief geological map based on data from The Icelandic Institute of Natural History showing that neither the boundary between the European and North American plates (approximately N-S through the centre of the image) nor the Vatnajökull ice cap (bottom) are obvious in the radar images, despite their geological significance. Note that the growth of the Holuhraun lava flows from 70 to 85 km^2 (and in volume from 1·0 to 1·6 km^3) in the nine months between radar images is not apparent given the significant changes in appearance caused by the changes in viewing geometry

Nearly a third of the Venus surface comprises indistinct plains; not featureless, but lacking traceable boundaries or features, most likely because these plains are covered by a veneer of sediments of the type observed in the Venera lander images (see Carter et al. 2023, this collection). A veneer only tens of centimetres thick is sufficient to obscure boundaries and break down global stratigraphic relationships.

6.2.3 Radiometric and Spatial Resolution

As the average roughness of a surface declines through the Rayleigh criterion, which is close to 2 cm for the Magellan radar, the backscatter brightness changes by ∼12 dB from bright (an average roughness of 4 cm for Magellan, corresponding to a pebbly surface) to dark (at

0.5 cm, typical of gravel). Such a change can occur over a relatively short distance, perhaps only 100 m on debris slopes and fans, which might be just one or two pixels at Magellan resolution (110 m along track and between 101 and 250 m across track) and therefore can appear as an abrupt change that may be misinterpreted as a geological boundary or contact.

Lithostratigraphy from radar images is therefore fraught with difficulty and should be avoided, except in circumstances where different lithologies can be identified with confidence (clearly defined lava flows, for example). More usually, mapping should be geomorphic, to which radar is exquisitely sensitive, which typically means a focus on geological structures and sedimentary processes.

In short, there are two key assumptions that are linked to the global stratigraphic sequence interpretation and its timing based on crater degradation. First, surface geologic units that have been deposited at different times must always be separable, and within a geologic unit there cannot be boundaries that can be confused with stratigraphic boundaries. Second, using crater degradation/alteration to determine the relative timing of this sequence assumes a spatially uniform timing of the degradation process (e.g., parabolas associated with craters are always removed by weathering at a uniform rate). If these assumptions are widely violated, then it becomes impossible to distinguish between a lunar-like directional model (Ivanov and Head 2011) and an Earth-like nondirectional model of Venus stratigraphy using Magellan radar data alone. A key point of (Hansen 2000) was an illustration of how a variety of completely different geologic histories could result in a final landscape that appears identical in Magellan-resolution SAR images.

7 Evidence for Young and Active Volcanism

In this section we discuss three general approaches to searching for evidence of recent or ongoing volcanism. In the first two sections we discuss properties of imaging at radar and infrared wavelengths that may be indicative of young, chemically unweathered surfaces. We then explore radar signatures indicative of mechanical weathering and/or sediments that have been used to infer the presence of young volcanic deposits. Finally, active volcanism can potentially be observed through observation of a thermal signature or changes between images taken at different times.

7.1 Radar Emissivity

The Magellan mission observed the thermal emission of the surface at 12.6 cm wavelength, which together with surface temperature inferred from descent probes and Magellan altimetry provides radiothermal emissivity with a resolution of 20 to 80 km, depending on latitude (Pettengill et al. 1992). Radiothermal emissivity is affected by the dielectric permittivity of the surface, shape of the surface at the scales of the radar wavelength, and emission angle, the latter of which varied with latitude due to the orbit of Magellan. The most striking feature of this map is a steep reduction of the radiothermal emissivity to unexpectedly low values above a critical elevation, which varies from location to location to between 6053.6 to 6056.4 km radius (Pettengill et al. 1992).

The cause of the low emissivity and the complementary high radar reflectivity at high elevations is still a matter of debate (Brackett et al. 1995; Fegley Jr. 1997; Fegley Jr. et al. 1997; Klose et al. 1992; Pettengill et al. 1992; Shepard et al. 1994; Wood 1997), but in general terms the elevation boundary is interpreted to be a temperature boundary across which elements with a high dielectric constant become stable. The temperature of the surface of Venus likely only has small lateral gradients due to the efficient redistribution of heat by convection in combination with small solar heating and radiative cooling rates (Stone 1975), and the temperature is thought to decrease vertically by a little less than 8° C/km (Seiff et al. 1985). Using the highest resolved temperature profile of the lowest 10 km by the VeGa 2 probe (Team and Seiff 1987), this critical elevation corresponds to temperatures between 720 and 700 K (Brossier et al. 2020).

As with formation of a weathering rind on Earth, it is hypothesized that the high-dielectric surface minerals form over a geologically significant time period, perhaps up to millions of years (Robinson and Wood 1993; Wood 1997). Places where there is a significant departure from an expected elevation "snow line" of high reflectivity / low emissivity can be interpreted as where something has occurred recently enough that the normal weathering process has not had time to operate. Robinson and Wood (1993) identified volcanic features and flows with anomalous emissivities and interpreted them to relatively recent volcanism, and speculated that Maat Mons has undergone the most recent episode of volcanic activity. More recent work with Magellan emissivity data (Brossier et al. 2022, 2021, 2020) has affirmed this interpretation, indicated additional sites of young volcanism on larger volcanoes, and suggested recent volcanic activity in Ganis Chasma.

7.2 Near Infrared Emissivity

The Venus atmosphere features atmospheric windows in the near infrared, where surface thermal emission can be observed on the night-side. The VIRTIS instrument on the Venus Express mission (Piccioni et al. 2007) observed this radiance at 1020 nm wavelength, and detected locations that feature thermal emission that was up to 15% higher than other locations with the same surface elevation (Erard et al. 2009; Mueller et al. 2008). VIRTIS observed thermal emission in two additional surface windows at 1100 nm and 1180 nm but in these bands the radiance anomalies are not significant even at the most frequently observed area in Themis Regio (Kappel et al. 2016; Mueller et al. 2020), likely due to higher atmospheric opacity decreasing the surface signal.

The areas with increased emission at 1020 nm are interpreted as higher emissivity because 1) the Venus atmosphere is efficient at redistributing heat via convection, allowing only for small lateral temperature gradients in the lower atmosphere (Seiff et al. 1985; Stone 1975), 2) the anomalies persist over the 2 years of observation and 3) the anomalies are correlated with lava flow features at the flanks of volcanic centers such as Quetzalpetlatl Corona (Helbert et al. 2008), and Themis and Imdr Regios (Mueller et al. 2008).

Smrekar et al. (2010) argue that chemical weathering reduces the emissivity of basaltic surfaces over time by breakdown of high emissivity mafic minerals to lower emissivity minerals such as quartz, anhydrite, and very fine hematite. The high emissivity anomalies correlate with locations where gravity and topography data indicate an active mantle plume (Ivanov and Head 2010; Smrekar 1994; Stofan et al. 2016). Smrekar et al. (2010) therefore conclude that these high emissivity anomalies associated with stratigraphically young lava flows represent unweathered and thus relatively recently resurfaced areas.

It is currently not clear what average rate of resurfacing this interpretation implies. An alternative explanation to chemical weathering for a reduction of emissivity with time, could be the deposition of micrometer thin layers of fines produced by impacts. The timescales at which chemical weathering reduces emissivity has been estimated at widely diverging values, from hours in not entirely realistic laboratory conditions (Filiberto et al. 2020) to 0.5 Ma from theoretical calculations of reaction kinetics (Dyar et al. 2021). A weathering timescale of hours is close to the cooling timescale of lava flow, thus if the estimates of Filiberto et al.

2020 are correct, the radiance anomalies would correspond to a mix of temperature and emissivity contributions.

Atmospheric blurring reduces the spatial resolution to 90 to 100 km (Basilevsky et al. 2012; Hashimoto and Imamura 2001). It is possible to identify large flow fields that could be responsible for the high emissivity (D'Incecco et al. 2017), but with the current data quality it is not possible to state whether the emissivity anomaly is due to a few flow lobes with high thermal emission or a moderate increase of emissivity over the whole flow field. An unweathered area needs to be on the order of 10,000 km^2 for an unambiguous estimate of its emissivity difference.

The observed radiance anomaly of up to 15% does not directly translate into an emissivity difference due to the non-linear relation of top-of-atmosphere radiance and surface emissivity (Hashimoto and Sugita 2003). At high absolute emissivity, a relative anomaly in radiance indicates a higher relative change in emissivity, e.g. a 1% change in emissivity results only in a 0.3% change in observable radiance (Dyar et al. 2020). Uncertainties in the absolute emissivity, i.e. the global average emissivity, are large due to uncertainties of atmospheric opacity and surface temperature (Kappel et al. 2015). In-situ spectrophotometry from the Venera 9 and 10 landers (Ekonomov et al. 1980) indicates an emissivity of $\sim$0.85 to 0.9 at $\sim$900 nm, consistent with unweathered basalt at Venus temperatures (Helbert et al. 2021). This means that either (1) these locations are unweathered, or (2) there is a steep decline in emissivity from 900 nm to 1020 nm in the weathered basalt, or (3) some of the radiance anomalies reported in Helbert et al. (2008), Mueller et al. (2008) are in fact not due to an emissivity anomaly.

Thus, overall it is difficult to use the thermal emission data for dating surfaces. Some progress seems possible in the near future. If DAVINCI happens to descend at a location with a basaltic inter-tessera plains, observations from below the bulk of the atmosphere might constrain the absolute reflectance/emissivity of weathered basalt at 1020 nm. An analysis of the Venera 13 and 14 spectrophotometer data could also provide more insights, since the spectral range extends to 1200 nm. An experimental confirmation of the long timescale of weathering proposed by Dyar et al. (2021) is difficult, as is the production of analog samples weathered under accurate Venus conditions for such timescales.

A way forward could also be an independent estimate of the rate of volcanism specific to the locations with increased emissivity. The volcanic activity of mantle hotspots on Earth, an apparently close analog to the Venus hotspots, is routinely monitored by satellites (Wright et al. 2015). Hotspot eruptions are relatively frequent, and a significant fraction of Earth's eruptions would be detectable through the clouds of Venus (Mueller et al. 2017). If sufficient IR data is gathered by instruments on the coming Venus orbiter missions to detect a robust statistic of active volcanic eruptions, this would allow us to constrain the rate of emissivity decrease with time. The observation of a plains resurfacing event appears unlikely, but the rate of emissivity fading with time constrained from hotspot activity would allow us to date plains with increased emissivity, if any are found. A fading timescale of 0.5 Ma and a resurfacing rate of 1 km^2/yr indicate that some not yet fully weathered plains would exist and be detectable, if plains resurfacing is ongoing in patches with at least 100 km diameter.

Bondarenko et al. (2010) interpret radiothermal data at Bereghynia Planitia as result of emplacement of a 200,000 km^2 lava area within decades preceding the Magellan mission. This area was not observed by VIRTIS but could be by future missions.

7.3 General Radar Properties

Radar backscatter is affected by the cm-scale morphology of the surface. It is possible that this morphology changes over time, and if so this might be used to date surfaces. Campbell and Campbell (1992) studied the incidence angle and polarization dependence of radar backscatter in comparison to Earth analogs and concluded that the surface roughness of most flows is comparable in roughness to pahoehoe flows on Earth, characterized by a crust that has not been disrupted during the emplacement of the flow. Campbell and Campbell (1992) also note that the large lava flows on Venus imply high eruption rates which favor the formation of aa flows, where the flow surface is continuously disrupted during emplacement and is thus covered in rubble. Bruno et al. (1992) studied the outline of Venusian flows and find that their fractal dimensions are more indicative of aa flow. Kratter et al. (2007) studied the huge, radar bright flow field (600,000 km^2) associated with Quetzalpetlatl Corona on the south flank of Lada Terra and found that its radar properties are mostly consistent with pahoehoe flow roughness. Stofan et al. (2001) studied the flow fields of some volcanoes in Magellan radar images and found some evidence of crust disruption during flow emplacement. Some flows have bright margins, interpreted to be blocky levee material; on Earth such flows form when a flow pushes disrupted crust material in front of it.

One possible explanation for the dearth of aa-like surface roughnesses (Campbell and Campbell 1992) is that erosion from physical or chemical weathering reduces the surface roughness over time. Campbell et al. (2017) noted that the bright diffuse deposits they interpret as cm-sized pyroclasts appear to be absent from everywhere except the latest part of the stratigraphic column, and they consider rapid weathering as a possibility. If cm-sized clasts rapidly disappear from the radar signal, the decimeter-sized rubble of a flow with aa texture should disappear soon after.

Potential mechanisms for reducing cm- and decimeter-scale roughness are wind erosion and chemical weathering. Zolotov and Volkov (1992) discuss aspects of chemical weathering on Venus. An alternative explanation could be the reduction of surface roughness by airfall deposits. Basilevsky et al. (2004) suggest that the surfaces in the Venera panoramas that had been interpreted as pahoehoe lava flow (Garvin et al. 1984) are actually lithified aeolian sediments from impact ejecta.

Whatever processes produce a modification of surface roughness over time, it is clear that only surfaces that have some initial roughness are significantly affected by it. To use surface roughness as an age proxy would require one to have a measure of the initial roughness. Future missions with highly resolved images and topography data might be useful in identifying features that could be indicative of flow texture, such as blocky levees for aa flows (Stofan et al. 2001) or inflation features on pahoehoe flows (Voigt et al. 2021). For mostly featureless plains, this approach does not seem promising.

7.4 Radar Properties of Potential Young Pyroclastics

Radar-bright diffuse deposits near the summit regions of some coronae have been proposed to be young pyroclastics, and possible evidence of a renewed epoch of mantle volcanism that taps into deeper volatiles (Campbell et al. 2017). The deposits have radar properties associated with rough terrain, including high backscatter and an increased circular polarization ratio in Arecibo Observatory images compared to the surrounding plains. (Circular polarization ratio is a measure of the same sense circular polarization transmitted to the opposite sense polarization; rougher surfaces generate higher CPR values.). The radar-bright features also have emissivity values that are similar to the plains or only slightly lower, so there is no evidence that they are a high-dielectric coating as is seen in other regions of Venus (Campbell et al. 2017).

The deposits are semi-transparent in places, especially near the edges where it is often possible to detect stratigraphically lower features, such as fractures and lava flows. There is rarely any evidence that the bright deposits are embayed by subsequent flows, and they only appear to have been altered through aeolian processes. There is also little evidence for similar deposits interleaved with other flows. Taken together, these observations suggest that these bright regions are rough mantling pyroclastic flow deposits that are at the top of the stratigraphic column (Campbell et al. 2017). Modeling suggests that the deposits could have been formed through column collapse or perhaps long duration fire-fountaining, with thicknesses ranging from meters to tens-of-meters (Ganesh et al. 2021). Their presence indicates that significant volatiles (CO_2, H_2O) must have been present (Airey et al. 2015), and such volatile-rich eruptions are more typical in the early stages of volcanic eruptions (Campbell et al. 2017). These radar-bright deposits may therefore be evidence of renewed activity in these regions, and a good location to search for ongoing volcanic activity.

7.5 Efforts at Detection of Active Eruptions

Detection of active eruptions is here understood to mean all methods that can generate evidence that an eruption took place at a certain location within a short, well constrained timeframe. One such detection has been made using Magellan data, where Herrick and Hensley (2023) found a volcanic vent near the summit of Maat Mons that changed in shape during the Magellan mission. Because the Magellan mission changed viewing geometry with each imaging cycle, and the resolution of the SAR imaging is 100-200 m, only multi-kilometer shape changes in objects can be confidently identified with Magellan data. Future missions, with higher resolution SAR imaging at a consistent viewing geometry, will be much more capable of searching for changes during their missions, and they will be able to look backward for changes since Magellan. Detection of volcanic gas plumes is treated in Wilson et al. (2023, this collection).

Venus surface thermal emission at radar wavelength has been observed by several spacecraft, most importantly the Pioneer Venus and Magellan orbiters. Radar wavelengths are far from the peak of Venus surface thermal emission at 4 μm wavelength temperatures and are not very sensitive to temperature; however, they have the advantage that lava meters below the surface, where the cooling rate is much lower, also contributes to the observable signal. Bondarenko et al. (2010) investigated the Magellan orbiter radiothermal emission (Pettengill et al. 1992) and found that a $\sim 10^5$ km^2 large lava flow unit in Bereghinia Planitia at 28°E 39°N shows an up to 85 K higher than expected brightness temperatures and interpret this as consistent with lava emplaced within the last few decades. Lorenz et al. (2016) argue that this result is ambiguous since it could be explained by a surface emissivity that is different from that derived by Bondarenko et al. (2010) from radar backscatter under assumption of a smooth lava surface. Lorenz et al. (2016) furthermore point out that such a large eruption would be a very unusual event. Assuming that this is not an unusual event, i.e. a statistical expectation of an area of 10^5 km^2 emplaced within the last 100 years, corresponds to a resurfacing rate of 10^3 km^2 / yr, three orders of magnitude larger than the 1 km^2/yr estimated by Phillips et al. (1992) for resurfacing in equilibrium with cratering. Nevertheless, Lorenz et al. (2016) argue that future observations with radiometry with higher spatial resolution than Magellan, ideally accompanied by radar polarimetry to better understand surface roughness and emissivity, would have a good chance of detecting excess thermal emission from a volcano with an activity similar to Mount Etna. However, MacKenzie and Lorenz (2020) studied the radiometry dataset of the Earth observing Soil Moisture Active Passive (SMAP) instrument and found that the detected thermal emission of known active eruptions was lower than expected.

Thermal emission from the surface of Venus can also be detected in near infrared wavelength between 700 and 1200 nm wavelength (Carlson et al. 1991; Wood et al. 2022). The atmosphere is also mostly transparent in the visible range and therefore an incandescent lava flow (>800 K) could also be detected in this range even if there is otherwise no detectable surface thermal emission. The detection capability in these visible to near-infrared wavelengths is limited by the spatial resolution of $\sim$100 km imposed by scattering in the permanent cloud layer (Basilevsky et al. 2012; Hashimoto and Imamura 2001; Shalygin et al. 2012). This means that the signature of a lava flow small compared to a 100 km diameter, i.e. the vast majority of lava flows, would appear diffused so that its area cannot be resolved and the maximum of thermal emission is close to proportional to the total thermal emission over the flow. The near infrared thermal emission of an active lava flow is dominated by surfaces close to the eruption temperature, which even on Venus cool to temperatures near ambient on a timescale of hours to a day. Based on this fact it is often assumed in Earth volcano monitoring that the total of the surface thermal emission is proportional to the areal rate at which uncooled lava is exposed, which itself can be assumed to be proportional to the lava effusion rate (Harris et al. 1997; Wright et al. 2001).

Several instruments on Venus Orbiters had capabilities that were estimated to be able to detect eruptions that are not completely outside of the range of historically observed eruptions: the Venus Monitoring Camera (VMC) on Venus Express (Shalygin et al. 2012), the Visible and Infrared Thermal Imaging Spectrometer (VIRTIS) on Venus Express (Mueller et al. 2017), and the IR1 camera on Akatsuki (Iwagami et al. 2011). No detections of eruptions were reported for the VIRTIS and IR1 data, but Shalygin et al. (2015) interpret several VMC images as evidence for active lava flows near Ganis Chasma (18°N, 189°E), south of Ganiki Planitia. This interpretation is based on higher brightness temperature than expected for the respective elevation appearing tied to fixed locations for a duration of a few days. Each of the brightness temperature anomalies have dimensions larger than 100 km, thus if volcanic activity is responsible then it must be distributed over at least 100 km of distance. Three of the four anomalies occur within the same 48 hour timespan along a $\sim$$10^3$ km long stretch of the rift. Shalygin et al. (2015) mention that image quality was a limiting factor for detection and that the instrument was able to acquire relatively good data on only 37 days. This suggests that tens of similar anomalies could have been detected with consistently good imaging and the same coverage over a duration of a year. Assuming that the > 100 km long anomalies are caused by 100 km^2 of erupted lava surface, this would indicate a resurfacing rate of >10^3 km^2/yr (in the area observed by VMC).

Mueller et al. (2017) unsuccessfully searched the VIRTIS dataset for eruptions. The dataset comprises imaging covering an area equal to six times Venus surface on separate days, limited to the southern hemisphere. The VIRTIS images which would have been able to detect brightness temperature anomalies the same magnitude as those observed by VMC at Ganis Chasma in the northern hemisphere. Assuming eruption effusion rate and duration statistics based on hotspot volcanism on Earth, Mueller et al. (2017) estimate that only at a rate of volcanism > 300 km^3/yr there would have been a high likelihood of an eruption visible. This constraint is weak to the point of uselessness, mostly because only the eruptions with the highest historically observed effusion rates ($\sim$$10^3$ m^3/s) would have been detected (Fig. 8). The VIRTIS detection capability was mostly limited by the high instrumental noise, which is proportional to the effusion rate that is reliably detectable. The VIRTIS signal to noise ratio was $\sim$15; a 20-fold improvement to 300 is planned with the near-infrared cameras for the VERITAS and EnVision missions. This would allow the detection of flows with 50 m^3/s. This has the further advantage that eruptions or eruption phases with very high effusion rates ($\sim$$10^3$ m^3/s) typically last only hours, while less intense eruptions can be expected to remain active and thus detectable for tens of days.

Present day volcanic resurfacing rates can be constrained by direct observation of new flows from repeat SAR imaging. Although it is dangerous to extrapolate from a data set

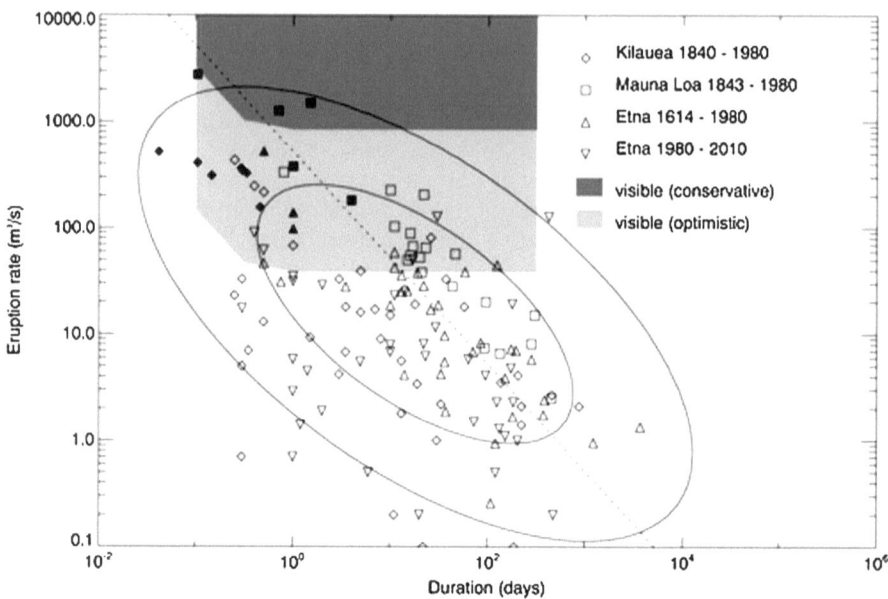

Fig. 8 From Mueller et al. (2017). Historical eruptions of three terrestrial volcanos from the data collated by Wadge [1981] and after 1980 by Harris et al. [2011]. Filled symbols represent only the initial phase of a few of the most intense eruptions. Grey areas indicate combinations of effusion rate and eruption duration that would be detectable by VIRTIS under the optimistic and conservative sets of assumptions. The solid lines correspond to isolines of a fitted probability density function that encompass 95% and 68% of eruptions. The dotted line corresponds to a volume of 0.045 km^3

of one, the observation of change during the Magellan mission by Herrick and Hensley (2023) indicates that it is almost certain that future missions will see changes from volcanism (new flows and structures) since Magellan and during their missions. When combined with various analyses attempting to extrapolate from terrestrial observations (Lorenz 2015; Byrne and Krishnamoorthy 2022a,b; van Zelst 2022), it seems likely that somewhere between a few to several dozen detectable volcanic eruptions occur on Venus in an Earth year.

8 Other Geologic Constraints

Here we consider a few other sets of geologic features whose nature and location bear upon the resurfacing and climate history of Venus. First, we examine the abundance and orientation of wrinkle ridges, a tectonic feature that occurs across most of the planet. We look at the possibility that the timing of formation of wrinkle ridges and polygonal structures may be related to climatic conditions. Then we consider the Venusian canali, a family of extremely long channels, each of which presumably formed over a short time period and can be used as a stratigraphic markers.

The geologic features known as "wrinkle ridges" represent limited contraction on coherent, layered surfaces such as lava plains and are found on multiple planets, including the Moon, Mercury, Venus, and Mars. On Venus, wrinkle ridges are ubiquitous in regional plains units. Efforts at mapping the orientations of wrinkle ridges demonstrated that they are associated predominantly with low elevations and low geoid values. Their strikes are generally perpendicular to both the local topographic slope and the local geoid slope (Bilotti and Suppe 1999; Sandwell et al. 1997), suggesting that convection-driven topography creates sufficient stress to cause wrinkle-ridge formation perpendicular to the direction of flow from mantle upwelling to downwelling. As discussed in more detail in other articles of this collection, the general correlation of the geoid with the current distribution of large volcanoes and rifts makes a compelling argument that the overall mantle convection pattern has not reoriented during the period reflected in the visible geologic record. However, two of the stratigraphically youngest areas on the planet, with high geoid values, Beta and Atla Regiones, are notable for not being ringed by wrinkle ridges that strike perpendicular to their topography (Bilotti and Suppe 1999). An interpretation could be that these areas were uplifted after some internal or surface change (e.g. volcanic resurfacing) occurred that globally inhibits wrinkle ridge formation.

Climate models of a possible massive resurfacing event predict very substantial (several 100 C°) surface temperature excursions (Bullock and Grinspoon 2001), capable of producing tectonic deformation and a global stratigraphic marker. If such major temperature variations occurred, two studies predicted surface extension and contraction due to propagation of thermal waves due to climate change into the interior (Anderson and Smrekar 1999; Solomon et al. 1999). Climate-driven thermal stresses using a continuous plate predict that extension will produce polygons resembling hexagonal cooling of lava flows (Johnson and Sandwell 1992), but on a much larger scale due to the duration of the temperature excursion (Anderson and Smrekar 1999). A continuous plate model does not permit wrinkle ridge formation based on the Bullock and Grinspoon DT estimates. Solomon et al. (1999) assume a broken plate model, similar to what is used to model mid-ocean ridges, and predict both wrinkle ridges and polygons. Polygons occur in over 200 locations (Smrekar et al. 2002). Although wrinkle ridges can be produced by topographic slope, climate-change is the likely explanation for the observed sizes of polygons (up to several kms across). However, the apparent size of polygons observed in Magellan data may be under-resolved (Smrekar et al. 2002). Smaller diameter polygons could be a result of cooling lava flows. Future missions can test between these two hypotheses.

Magellan revealed a variety of channel and valley landforms, all of which likely transported lava and not water (Baker et al. 1992). Of particular interest are the 50 or so "canali", channels that transported lavas hundreds of kilometers; the longest of these, Baltis Vallis, extends 6800 km (Baker et al. 1992; Komatsu et al. 1993). These features were each likely emplaced over relatively short periods of time (Komatsu et al. 1992) and thus represent time-stratigraphic markers that span long distances. Several of these features have topographic profiles that require post-emplacement regional tectonic warping (Komatsu and Baker 1994), and some of them have also experienced limited disruption by tectonic features such as wrinkle ridges. However, there are not clear instances where a canali has been cut by flows from a new shield volcano or other major later volcanism. An extensive mapping study of Baltis Vallis relative to its surroundings has been used as supporting evidence for the idea that the global stratigraphic sequence described above is also globally synchronous (Basilevsky and Head 1996).

9 Summary and Conclusions

The lack of small craters due to atmospheric filtering and the challenges of calibrating the cratering record in the last billion years means that there will likely always be ambiguity about the absolute timeline of the surface record for Venus, but Venus is clearly much closer to Earth than Mars in terms of its overall youthfulness. Upcoming missions (see Widemann et al. 2023, this collection) will help constrain the timeline by 1) determining the level of current volcanic and tectonic activity through various monitoring approaches, 2) elucidating the details of how the impact craters have been altered post-impact with higher resolution imaging and topography, and 3) providing vastly improved imaging that will enhance our ability to separate lava flows and characterize key geologic contacts. Although the evidence suggests that most impact craters have experienced some post-impact volcanic modification, the current state of observations also allows one to interpret that as resulting from the final death knells of a "directional" resurfacing; in that case the near-random appearing spatial distribution of craters would largely represent a production surface. Resurfacing modeling can only produce a highly active present-day Venus with few altered craters in a near-random spatial distribution if resurfacing occurs in physically unrealistic ways that do not match the current geology.

However, to the extent that some or most of the craters are viewed as part of a continuum of features in various stages of being eliminated by volcanic and tectonic processes, then many more models become acceptable for creating a spatially near-random crater distribution. In this scenario, because the distribution of geologic terrain types (e.g., tessera, rifts, plains) is less random than the distribution of craters, there will have to be distinctly different mechanisms for removing craters in different regions that operate at somewhat similar rates.

Tying crater resurfacing to global views of the geologic history, the viewpoint that there is a global stratigraphy that shows a change in the nature of geologic activity and a rapid decline in the level of volcanism relies on a variety of assumptions, some of which will be testable with higher resolution imaging. Conversely, having different regions of Venus evolving independently of each other will require understanding how areally extensive features are maintained; for example, why there is a near-global wrinkle-ridge pattern that follows the current topography or why there are no volcanic flows that superpose canali. The improved resolution from future missions should better constrain how rapidly the vast volcanic plains were emplaced, the size and nature of individual flow events that emplaced those plains, and the timing of volcanism and deformation. Study of key geologic contacts, such as those between tessera units and embaying plains units, will be crucial in understanding the overall global geologic sequence.

Whatever "world view" of the geologic history that one develops from the new data from upcoming missions, its endpoint will have to be consistent with the constraints of where and how volcanism and tectonics are currently occurring.

Funding Note Open access funding provided by University of Oslo (incl Oslo University Hospital).

Declarations

Competing Interests The authors declare no competing interests relevant to this manuscript.

References

Airey MW, Mather TA, Pyle DM, Glaze LS, Ghail RC, Wilson CF (2015) Planet Space Sci 113:33

Anderson FS, Smrekar SE (1999) J Geophys Res, Planets 104:30743

Armann M, Tackley PJ (2012) J Geophys Res, Planets 117:E12003

Baker VR, Komatsu G, Parker TJ, Gulick VC, Kargel JS, Lewis JS (1992) J Geophys Res, Planets 97:13421

Basilevsky AT, Head JW III (1998) J Geophys Res, Planets 103:8531

Basilevsky AT, Head JW (1994) Earth Moon Planets 66:285

Basilevsky AT, Head JW (1995) Planet Space Sci 43:1523

Basilevsky AT, Head JW (1996) Geophys Res Lett 23:1497

Basilevsky AT, Head JW (2000) Planet Space Sci 48:75

Basilevsky AT, Head JW (2002) Geology 30:1015

Basilevsky AT, Head JW, Abdrakhimov AM (2004) J Geophys Res, Planets 109:E12003

Basilevsky AT, Shalygin EV, Titov DV, Markiewicz WJ, Scholten F, Roatsch T, Kreslavsky MA, Moroz LV, Ignatiev NI, Fiethe B (2012) Icarus 217:434

Bilotti F, Suppe J (1999) Icarus 139:137

Bjonnes EE, Hansen VL, James B, Swenson JB (2012) Icarus 217:451

Bondarenko NV, Head JW, Ivanov MA (2010) Geophys Res Lett 37:L23202

Brackett RA, Fegley B Jr, Arvidson RE (1995) J Geophys Res, Planets 100:1553

Brossier JF, Gilmore MS, Toner K (2020) Icarus 343:113693

Brossier J, Gilmore MS, Toner K, Stein AJ (2021) J Geophys Res, Planets 126:e2020JE006722

Brossier J, Gilmore MS, Head JW (2022) Geophys Res Lett 49:e2022GL099765

Bruno BC, Taylor GJ, Rowland SK, Lucey PG, Self S (1992) Geophys Res Lett 19:305

Bullock MA, Grinspoon DH (2001) Icarus 150:19

Bullock MA, Grinspoon DH, Head JW III (1993) Geophys Res Lett 20:2147

Byrne PK, Krishnamoorthy S (2022a) J Geophys Res, Planets 127:e2021JE007040

Byrne PK, Krishnamoorthy S (2022b) J Geophys Res, Planets 127:e2022JE007666

Campbell BA, Campbell DB (1992) J Geophys Res, Planets 97:16293

Campbell DB, Stacy NJS, Newman WI, Arvidson RE, Jones EM, Musser GS, Roper AY, Schaller C (1992) J Geophys Res, Planets 97:16249

Campbell BA, Morgan GA, Whitten JL, Carter LM, Glaze LS, Campbell DB (2017) J Geophys Res, Planets 122:1580

Carlson RW, Baines KH, Encrenaz T, Taylor FW, Drossart P, Kamp LW, Pollack JB, Lellouch E, Collard AD, Calcutt SB (1991) Science 253:1541

Carter et al (2023) Space Sci Rev 219

D'Incecco P, Müller N, Helbert J, D'Amore M (2017) Planet Space Sci 136:25

Dyar MD, Helbert J, Maturilli A, Müller NT, Kappel D (2020) Geophys Res Lett 47:e2020GL090497

Dyar MD, Helbert J, Cooper RF, Sklute EC, Maturilli A, Mueller NT, Kappel D, Smrekar SE (2021) Icarus 358:114139

Ekonomov AP, Golovin YM, Moshkin BE (1980) Icarus 41:65

Erard S, Drossart P, Piccioni G (2009) J Geophys Res, Planets 114:E00B27

Ernst RE, Head JW, Parfitt E, Grosfils E, Wilson L (1995) Earth-Sci Rev 39:1

Ernst RE, Grosfils EB, Mege D (2001) Annu Rev Earth Planet Sci 29:489

Ernst RE, Desnoyers DW, Head JW, Grosfils EB (2003) Icarus 164:282

Fegley B Jr (1997) Icarus 128:474

Fegley B Jr, Klingelhöfer G, Lodders K, Widemann T (1997) In: Bougher SW et al (eds) Venus II: Geology, geophysics, atmosphere, and solar wind environment, p 591

Filiberto J, Trang D, Treiman AH, Gilmore MS (2020) Sci Adv 6:eaax7445

Ganesh I, McGuire LA, Carter LM (2021) J Geophys Res, Planets 126:e2021JE006943

Garvin JB, Head JW, Zuber MT, Helfenstein P (1984) J Geophys Res, Solid Earth 89:3381

Gillmann C, Tackley P (2014) J Geophys Res, Planets 119:1189

Gilmore MS, Head JW (2000) Meteorit Planet Sci 35:667

Gilmore MS, Head JW (2018) Planet Space Sci 154:5

Guest JE, Stofan ER (1999) Icarus 139:55

Hansen VL (2000) Earth Planet Sci Lett 176:527

Harris AJ, Blake S, Rothery DA, Stevens NF (1997) J Geophys Res, Solid Earth 102:7985

Hashimoto GL, Imamura T (2001) Icarus 154:239

Hashimoto GL, Sugita S (2003) J Geophys Res, Planets 108:5109

Hauck SA, Phillips RJ, Price MH (1998) J Geophys Res, Planets 103:13635

Helbert J, Müller N, Kostama P, Marinangeli L, Piccioni G, Drossart P (2008) Geophys Res Lett 35:L11201

Helbert J, Maturilli A, Dyar MD, Alemanno G (2021) Sci Adv 7:eaba9428

Herrick RR (1994) Geology 22:703
Herrick RR, Hensley S (2023) Science 379:1205–1208
Herrick RR, Phillips RJ (1994a) Icarus 111:387
Herrick RR, Phillips RJ (1994b) Icarus 112:253
Herrick RR, Rumpf ME (2011) J Geophys Res, Planets 116:E02004
Herrick RR, Sharpton VL (1996) Geology 24:11
Herrick RR, Sharpton VL (2000) J Geophys Res, Planets 105:20245
Herrick R, Sharpton V, Malin M, Lyons S, Feely K (1997) In: Bougher SW et al (eds) Venus II: Geology, geophysics, atmosphere, and solar wind environment, p 1015
Ivanov M, Head JW (2001a) Geologic map of the Lavinia Planitia quadrangle (V-55), Venus. US Geological Survey Geologic Investigations Series I-2684
Ivanov MA, Head JW (2001b) J Geophys Res, Planets 106:17515
Ivanov MA, Head JW (2010) Planet Space Sci 58:1880
Ivanov MA, Head JW (2011) Planet Space Sci 59:1559
Ivanov MA, Head JW (2015) Planet Space Sci 106:116
Ivanov MA, Head JW (2004) Geologic Map of the Atalanta Planitia Quadrangle (V-4), Venus. US Department of the Interior, US Geological Survey
Ivanov MA, Head JW (2005) Geologic Map of the Nemesis Tesserae Quadrangle, V-13, Venus. US Department of the Interior, US Geological Survey
Ivanov MA, Head JW (2006) Geologic Map of the Mylitta Fluctus Quadrangle (V-61), Venus. US Geological Survey
Ivanov MA, Head JW (2008) Geologic Map of the Meskhent Tessera Quadrangle (V-3), Venus. US Department of the Interior, US Geological Survey
Ivanov BA, Basilevsky AT, Kryuchkov VP, Chernaya IM (1986) J Geophys Res, Solid Earth 91:413
Ivanov M, Head JW, Ryan DA (2010) Geologic map of the Lakshmi Planum quadrangle (V-7), Venus. US Department of the Interior, US Geological Survey
Iwagami N, Takagi S, Ohtsuki S, Ueno M, Uemizu K, Satoh T, Sakanoi T, Hashimoto GL (2011) Earth Planets Space 63:487
Izenberg NR, Arvidson RE, Phillips RJ (1994) Geophys Res Lett 21:289
Johnson CL, Sandwell DT (1992) J Geophys Res, Planets 97:13601
Kappel D, Haus R, Arnold G (2015) Planet Space Sci 113:49
Kappel D, Arnold G, Haus R (2016) Icarus 265:42
Karlsson R, Cheng KW, Crameri F, Rolf T, Uppalapati S, Werner SC (2020) J Geophys Res, Planets 125:e2019JE006340
King SD (2018) J Geophys Res, Planets 123:1041
Klose KB, Wood JA, Hashimoto A (1992) J Geophys Res, Planets 97:16353
Komatsu G, Baker VR (1994) Icarus 110:275
Komatsu G, Kargel JS, Baker VR (1992) Geophys Res Lett 19:1415
Komatsu G, Baker VR, Gulick VC, Parker TJ (1993) Icarus 102:1
Korycansky DG, Zahnle KJ (2005) Planet Space Sci 53:695
Kratter KM, Carter LM, Campbell DB (2007) J Geophys Res, Planets 112:E04008
Kreslavsky MA, Ivanov MA, Head JW (2015) Icarus 250:438
Labrosse S, Jaupart C (2007) Earth Planet Sci Lett 260:465
Le Feuvre M, Wieczorek MA (2011) Icarus 214:1
Lorenz RD (2015) Planet Space Sci 117:356
Lorenz RD, Le Gall A, Janssen MA (2016) Icarus 270:30
Lourenço DL, Rozel AB, Ballmer MD, Tackley PJ (2020) Geochem Geophys Geosyst 21:e2019GC008756
MacKenzie SM, Lorenz RD (2020) Remote Sens 12:2544
McKinnon WB, Zahnle KJ, Ivanov BA, Melosh HJ (1997) In: Bougher SW et al (eds) Venus II: Geology, geophysics, atmosphere, and solar wind environment, p 969
Mueller N, Helbert J, Hashimoto GL, Tsang CCC, Erard S, Piccioni G, Drossart P (2008) J Geophys Res Planets 113:E00B17
Mueller NT, Smrekar S, Helbert J, Stofan E, Piccioni G, Drossart P (2017) J Geophys Res, Planets 122:1021
Mueller NT, Smrekar SE, Tsang CCC (2020) Icarus 335:113400
Noack L, Breuer D, Spohn T (2012) Icarus 217:484
O'Rourke JG, Wolf AS, Ehlmann BL (2014) Geophys Res Lett 41:8252
Parmentier EM, Hess PC (1992) Geophys Res Lett 19:2015
Pettengill GH, Ford PG, Wilt RJ (1992) J Geophys Res, Planets 97:13091
Phillips RJ, Arvidson RE, Boyce JM, Campbell DB, Guest JE, Schaber GG, Soderblom LA (1991) Science 252:288

 Springer

Phillips RJ, Raubertas RF, Arvidson RE, Sarkar IC, Herrick RR, Izenberg N, Grimm RE (1992) J Geophys Res, Planets 97:15923

Piccioni G, Drossart P, Suetta E, Cosi M, Amannito E, Barbis A, Berlin R, Bocaccini A, Bonello G, Bouyé M (2007) ESA Spec Publ, vol 1295. ESA, Noordwijk

Price M, Suppe J (1994) Nature 372:756

Price M, Suppe J (1995) Earth Moon Planets 71:99

Price MH, Watson G, Suppe J, Brankman C (1996) J Geophys Res, Planets 101:4657

Robinson CA, Wood JA (1993) Icarus 102:26

Rolf T, Steinberger B, Sruthi U, Werner SC (2018) Icarus 313:107

Rolf T, Weller M, Gülcher A, Byrne P, O'Rourke JG, Herrick R, Bjonnes E, Davaille A, Ghail R, Gillmann C (2022) Space Sci Rev 218:70

Romeo I (2013) Planet Space Sci 87:157

Romeo I, Turcotte DL (2008) Earth Planet Sci Lett 276:85

Romeo I, Turcotte DL (2009) Icarus 203:13

Romeo I, Turcotte DL (2010) Planet Space Sci 58:1374

Sandwell DT, Johnson CL, Bilotti F, Suppe J (1997) Icarus 129:232

Schaber GG, Strom RG, Moore HJ, Soderblom LA, Kirk RL, Chadwick DJ, Dawson DD, Gaddis LR, Boyce JM, Russell J (1992) J Geophys Res, Planets 97:13257

Seiff A, Schofield JT, Kliore AJ, Taylor FW, Limaye SS, Revercomb HE, Sromovsky LA, Kerzhanovich VV, Moroz VI, Marov MY (1985) Adv Space Res 5:3

Shalygin EV, Basilevsky AT, Markiewicz WJ, Titov DV, Kreslavsky MA, Roatsch T (2012) Planet Space Sci 73:294

Shalygin EV, Markiewicz WJ, Basilevsky AT, Titov DV, Ignatiev NI, Head JW (2015) Geophys Res Lett 42:4762

Shepard MK, Arvidson RE, Brackett RA, Fegley B Jr (1994) Geophys Res Lett 21:469

Shoemaker EM, Hackman RJ (1962) In: Kopal Z, Mikhailov ZK (eds) The Moon. Academic Press, London and New York, pp 289–300

Shoemaker EM, Wolfe RF, Shoemaker CS (1991) In: Abstracts of the Lunar and Planetary Science Conference, vol 22, p 1253

Smrekar SE (1994) Icarus 112:2

Smrekar SE, Moreels P, Franklin BJ (2002) J Geophys Res, Planets 107:8

Smrekar SE, Stofan ER, Mueller N, Treiman A, Elkins-Tanton L, Helbert J, Piccioni G, Drossart P (2010) Science 328(5978):605–608

Solomon SC, Head JW (1982) J Geophys Res, Solid Earth 87:9236

Solomon SC, Head JW, Kaula WM, McKenzie D, Parsons B, Phillips RJ, Schubert G, Talwani M (1991) Science 252:297

Solomon SC, Bullock MA, Grinspoon DH (1999) Science 286:87

Stofan ER, Guest JE, Copp DL (2001) Icarus 152:75

Stofan ER, Smrekar SE, Mueller N, Helbert J (2016) Icarus 271:375

Stöffler D, Ryder G, Ivanov BA, Artemieva NA, Cintala MJ, Grieve RAF (2006) Rev Mineral Geochem 60:519

Stone PH (1975) J Atmos Sci 32:1005

Strom RG, Schaber GG, Dawson DD (1994) J Geophys Res, Planets 99:10899

Tauber ME, Kirk DB (1976) Icarus 28:351

Team VBS, Seiff A (1987) Adv Space Res 7:323

Turco RP, Toon OB, Park C, Whitten RC, Pollack JB, Noerdlinger P (1982) Icarus 50:1

Turcotte DL (1993) J Geophys Res, Planets 98:17061

Turcotte DL, Schubert G (2017) Geodynamics. Cambridge University Press, Cambridge

Uppalapati S, Rolf T, Crameri F, Werner SC (2020) J Geophys Res, Planets 125:e2019JE006258

van Zelst I (2022) J Geophys Res, Planets 127:e2022JE007448

Voigt JR, Hamilton CW, Steinbrügge G, Scheidt SP (2021) Bull Volcanol 83:1

Weller MB, Kiefer WS (2020) J Geophys Res, Planets 125:e2019JE005960

Wichman RW (1999) In: Lunar and Planetary Science XXX, p 1156

Widemann T et al (2023) Space Sci Rev 219

Wilhelms DE, McCauley JF, Trask NJ (1987) The geologic history of the Moon

Wilson et al (2023) Space Sci Rev 219

Wood JA (1997) Bougher SW et al (eds) Venus II: Geology, geophysics, atmosphere, and solar wind environment, p 637

Wood BE, Hess P, Lustig-Yaeger J, Gallagher B, Korwan D, Rich N, Stenborg G, Thernisien A, Qadri SN, Santiago F (2022) Geophys Res Lett 49:e2021GL096302

Wright R, Blake S, Harris AJ, Rothery DA (2001) Earth Planet Sci Lett 192:223

Wright R, Blackett M, Hill-Butler C (2015) Geophys Res Lett 42:282

Zolotov M, Volkov VP (1992) In: Barsukov VL et al (eds) Venus geology, geochemistry and geophysics. University of Arizona Press, Tucson, pp 177–199

Publisher's Note Springer Nature remains neutral with regard to jurisdictional claims in published maps and institutional affiliations.

Authors and Affiliations

Robert R. Herrick[1] [iD] **· Evan T. Bjonnes[2] · Lynn M. Carter[3] · Taras Gerya[4] · Richard C. Ghail[5] · Cédric Gillmann[6,4] · Martha Gilmore[7] · Scott Hensley[8] · Mikhail A. Ivanov[9] · Noam R. Izenberg[10] · Nils T. Mueller[11] · Joseph G. O'Rourke[12] · Tobias Rolf[13,14] · Suzanne E. Smrekar[8] · Matthew B. Weller[2,15]**

[✉] T. Rolf
tobias.rolf@geo.uio.no

R.R. Herrick
rrherrick@alaska.edu

[1] Geophysical Institute, University of Alaska Fairbanks, Fairbanks, 99775-7320, USA

[2] The Lunar and Planetary Institute, 3600 Bay Area Blvd, Houston, TX 77058, USA

[3] Lunar and Planetary Laboratory, University of Arizona, 1629 E. University Blvd, Tucson, AZ 85721, USA

[4] Institute of Geophysics, ETH Zurich, NO H 9.2, Sonneggstrasse 5, 8092 Zürich, Switzerland

[5] Department of Earth Sciences, Royal Holloway, University of London, Egham, Surrey, TW20 0EX, UK

[6] Department of Earth, Environmental and Planetary Sciences, Rice University, MS-126, 6100 Main Street, Houston, TX 77005, USA

[7] Dept. of Earth and Environmental Sciences, Wesleyan University, 265 Church St., Middletown, CT 06459, USA

[8] Jet Propulsion Laboratory, California Institute of Technology, Pasadena, CA 91109, USA

[9] Laboratory of Comparative Planetology, Vernadsky Institute of Geochemistry and Analytical Chemistry, Russian Academy of Science, 119991 Kosygin street, Moscow, Russia

[10] Space Exploration Sector, Atmospheres and Ionospheres Group, Johns Hopkins University Applied Physics Laboratory, 11100 Johns Hopkins Road, Laurel, MD 20723, USA

[11] Institute of Planetary Research – Department of Planetary Physics, German Aerospace Center (DLR e.V.), Rutherfordstr. 2, 12489 Berlin, Germany

[12] School of Earth and Space Exploration, Arizona State University, Tempe, AZ, USA

[13] Centre for Earth Evolution and Dynamics (CEED), University of Oslo, P.O. Box 1028 Blindern, 0315 Oslo, Norway

[14] Institute for Geophysics, University of Münster, Corrensstraße 24, 48149 Münster, Germany

[15] Department of Earth, Environmental and Planetary Sciences, Brown University, 324 Brook Street, Providence, RI 02912, USA

Space Science Reviews (2024) 220:31
https://doi.org/10.1007/s11214-024-01054-5

Possible Effects of Volcanic Eruptions on the Modern Atmosphere of Venus

Colin F. Wilson[1,2] · Emmanuel Marcq[3] · Cédric Gillmann[4] · Thomas Widemann[5,6] ·
Oleg Korablev[7] · Nils T. Mueller[8,9] · Maxence Lefèvre[3] · Paul B. Rimmer[10] ·
Séverine Robert[11] · Mikhail Y. Zolotov[12]

Received: 26 April 2023 / Accepted: 1 February 2024 / Published online: 5 April 2024
© The Author(s) 2024

Abstract

This work reviews possible signatures and potential detectability of present-day volcanically emitted material in the atmosphere of Venus. We first discuss the expected composition of volcanic gases at present time, addressing how this is related to mantle composition and atmospheric pressure. Sulfur dioxide, often used as a marker of volcanic activity in Earth's atmosphere, has been observed since late 1970s to exhibit variability at the Venus' cloud tops at time scales from hours to decades; however, this variability may be associated with solely atmospheric processes. Water vapor is identified as a particularly valuable tracer for volcanic plumes because it can be mapped from orbit at three different tropospheric altitude ranges, and because of its apparent low background variability. We note that volcanic gas plumes could be either enhanced or depleted in water vapor compared to the background atmosphere, depending on magmatic volatile composition. Non-gaseous components of volcanic plumes, such as ash grains and/or cloud aerosol particles, are another investigation target of orbital and *in situ* measurements. We discuss expectations of *in situ* and remote measurements of volcanic plumes in the atmosphere with particular focus on the upcoming DAVINCI, EnVision and VERITAS missions, as well as possible future missions.

Keywords Venus · Venus evolution · Planetary volcanism · Venus exploration · Volcanic plumes · Detecting volcanism

1 Introduction and Overview

In this paper we review how observations of Venus' present-day atmosphere place any constraint on the rate or style of volcanic activity in the present era.

One might first ask whether the very nature of the Venus' atmosphere, in particular its atmospheric pressure ($\sim$47-110 bars depending on elevation) and its thick envelope of sulphuric acid clouds, provides any evidence for volcanism in the present era. In other words,

Venus: Evolution Through Time
Edited by Colin F. Wilson, Doris Breuer, Cédric Gillmann, Suzanne E. Smrekar, Tilman Spohn and Thomas Widemann

Extended author information available on the last page of the article

does the current atmosphere represent an equilibrium state which can exist indefinitely without volcanic outgassing? Notably, Bullock and Grinspoon (2001) calculated that the persistence of the sulphuric acid cloud deck could be maintained only through supply of sulphur dioxide gas (SO_2), presumably through volcanism, within the last 20-50 My. However, such calculations require many assumptions, in particular regarding surface-atmosphere reactions which serve as a sink for atmospheric volatiles (Zolotov 2018). New constraints on surface composition and on volcanic resurfacing styles will provide much needed constraints for assessing the surface-atmosphere exchange, as would the experimental testing of Venus analogue materials, as addressed in companion papers by Gilmore et al. (2023), Ghail et al. (2024), Herrick et al. (2023), and Gillmann et al. (2022). The specific case of using noble gas isotopic ratios such as $^4He/^3He$ to constrain recent volcanism is reviewed by Avice et al. (2022, this collection). There are still too many unknowns today to allow a firm conclusion that the current known atmospheric composition provides evidence of present-day active volcanism, but promising new avenues of investigation are being developed in the coming decade.

Another approach is to search for the transient effects of active volcanism – atmospheric plumes of volcanic gases or particulates – as these may be readily observable. Esposito (1984) suggested that variable SO_2 abundances observed in the Venus' mesosphere could be the result of active volcanism. In the past decade, Marcq et al. (2013) showed that these episodic variations of mesospheric SO_2 continued into the Venus Express years (2006-2014) and proposed that these variations were caused by transient increases of atmospheric mixing between the SO_2-rich troposphere and SO_2-poor mesosphere. These fluctuations in vertical atmospheric mixing could be caused by, for example, vertically propagating atmospheric waves (Kouyama et al. 2017), changes in the chemical composition of the cloud particles (Rimmer et al. 2021), variations in static stability profile caused by variations of solar light absorption, or even buoyancy anomalies triggered by volcanic eruptions (Esposito 1984). This, along with interpretations of surface observations in the near infrared spectral range and the Magellan radar images (Smrekar et al. 2010; Shalygin et al. 2015; D'Incecco et al. 2021; Gülcher et al. 2020; Filiberto et al. 2020; Stofan et al. 2016), was interpreted as possibly providing indirect evidence for a recent or ongoing formation of silicate lava flows. However, many key questions about current-day volcanic emissions are still open. What is the redox state and elemental content of volcanic emissions? What is speciation of magmatic volatiles and volcanic gases compared to the atmosphere of Venus? What is the possible range for volcanic gas emissions rates?

Detectability of volcanic products (gases and particulates) in the atmosphere depends on volcanic gas compositions and fluxes, the dimensions, density and dynamics of volcanic plumes, the altitudes and geographical locations of sampling, and the capabilities of analytical instruments on orbiters, descent probes and landers. In addition, detectability will depend on how volcanically emitted species are transported through the atmosphere and how they interact with gases and aerosols, through coupling of chemical and dynamical processes (Bullock and Grinspoon 2001; Wilson et al. 2008; Titov et al. 2018; Lefèvre et al. 2018, 2020).

In this review we will present the different types of atmospheric compositional observations possible from incoming missions to Venus. For each one of these, we will discuss what species can be observed, and what those observations can tell us about the current and/or geologically recent volcanic activity. In Sect. 2, we address the possible composition breakdown and influx rate of magmatic volatiles entering the atmosphere. In Sect. 3, we discuss how volcanic plumes would be detectable with *in situ* instrumental capabilities, such as the DAVINCI mission or future descent probe or aerial platform concepts (Widemann et al.

2023, this collection). In Sect. 4 we discuss remote sensing capabilities, especially how tropospheric gases could be mapped from orbit, as can be carried out by near-infrared spectroscopy (e.g. EnVision/VenSpec-H, EnVision/VenSpec-M and VERITAS/VEM). In Sect. 5 we turn to mapping of gas species in the upper atmosphere, whether through nadir spectroscopy of reflected sunlight, such as EnVision/VenSpec-U investigation. In Sect. 6 we focus on particulate matter: ash particles, sulfuric acid aerosols, or other materials - which might be measured by future *in situ* missions with cloud-level balloons and/or aerial platforms. Finally, in Sect. 7, we suggest directions for further observational constraints and modeling efforts.

2 Magmatic Volatiles - Which Gas Species Would Venus' Volcanoes Emit Today and Which Can We Measure?

2.1 Composition of Outgassed Magmatic Volatiles

On Earth, chemical footprints of eruptions are often traced by measurements of SO_2 and other S-bearing gases (e.g., Oppenheimer et al. 2011; Henley and Hughes 2016). The relatively high mixing ratio of SO_2 in Venus' lower atmosphere ($\sim$1-2 $\times$ 10^{-4}) limits the probability of recognizing volcanic sources of this gas, and the same evidently applies to the much more abundant CO_2. However, our evaluations demonstrate that the compositions of both Earth-like moderately H-rich volcanic gases and putative H-depleted volcanic gases on Venus (Fig. 1) could be distinguishable from the atmosphere (Fig. 2). Locally elevated concentrations of SO_2, H_2O, CO, OCS, S_2, HCl, and HF may indicate volcanic sources. Ratios between these gases will further help distinguish nominal atmospheric features and signatures related to volcanism. Concentrations of these volcanic plume gases would decrease due to dilution through mixing with the air and through chemical interactions. Several orders of magnitude differences between supposed volcanic and atmospheric concentrations of gases (Fig. 2) suggest gas-gas type chemical reactions that consume volcanic COS, CO, S_2, S_3, S_4, CS_2, SO and exotic S- Cl- F-bearing species that all are out of chemical equilibrium with respect to atmospheric gases. However, reaction kinetics of volcanic and atmospheric gases remains to be assessed and compared with rates of gas mixing and dilution.

Besides the bulk volatile content (e.g., H/C/S ratios) and oxygen fugacity of parent magmas, the outgassed composition is also expected to depend on the vent pressure that reflects ambient atmospheric pressure (Gaillard and Scaillet 2014). The typical breakdown is shown in Fig. 3. Whereas at low to moderate surface pressures ($<$1 bar), SO_2 and H_2O dominate the mix, it is not the case for high vent and atmospheric pressures. On Venus, low-solubility C-bearing species (CO_2 and CO) are expected to dominate owing to suppressed degassing of SO_2 and H_2O due to their high solubility at elevated pressures. This may make direct detection of volcanic outgassing more challenging, since CO_2 is the dominant species ($\sim$97vol%) in the atmosphere, and CO a significant species at a background level of a $\sim$10-20 parts per million by volume (ppmv) in the first scale height. However, direct detection of H_2O-bearing volcanic plumes on Venus may still be possible despite this pressure suppression, just as detection of SO_2 plumes is often achieved on Earth despite some suppression of its degassing at 1 bar pressure. Beyond direct detection, though, one also needs to consider post-eruption perturbations to the atmospheric composition resulting in changes in mixing ratio for a vast array of species, as detailed below.

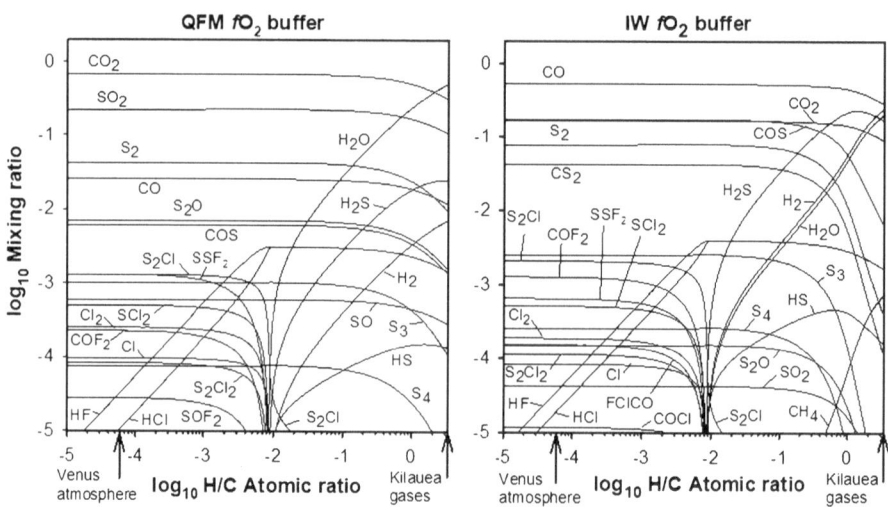

Fig. 1 Effects of H content on speciation of volcanic gases at 95.6 bars and 1500 K (modified after Zolotov and Matsui 2002). The elemental composition of gases corresponds to reconstructed 1918 analysis of Kilauea magma lake gases (Gerlach 1980) with variable C/H ratio. Volcanic gases on Earth are enriched in H relative to C (Symonds et al. 1994; Oppenheimer et al. 2014) with the H/C atomic ratio of $\sim$1 to $\sim$2000. Kilauea emissions are among the most H-depleted gases. Venus' counterparts could be more H-depleted than Kilauea gases owing to both H deficiency in the interior and suppressed degassing of high-solubility H_2O at the ambient pressure of $\sim$95 bars. The QFM fO_2 buffer corresponds to the quartz-fayalite-magnetite equilibrium and roughly represents the oxidation state of the Earth's mantle (Frost and McCammon 2008). The IW buffer stands for a more reduced iron-wüstite mineral assemblage. Venus' basaltic magmas and corresponding magmatic gases could have fO_2 between that of QFM and IW buffers (Schaefer and Fegley 2017). It follows that the figure provides a plausible range of Venus' volcanic gases.

CO_2 As shown in Fig. 3, CO_2 is likely the most abundant gas in Venus' volcanic emissions at the Earth-like oxidation state on parent magma (Gaillard et al. 2021; Zolotov and Matsui 2002) – but plumes of volcanic CO_2 would be indistinguishable from atmospheric CO_2. The speculations about an endogenic cause of supposedly depleted N_2/CO_2 ratio below $\sim$7 km (Lebonnois and Schubert 2017) would require unrealistically high volcanic CO_2 flux (Cordier et al. 2019; Lebonnois et al. 2020). Indeed, the apparent age of volcanic plains (0.2-1 Ga, Basilevsky et al. 1997; Korycansky and Zahnle 2005; Bottke et al. 2016) and unconfirmed signs of only very local present volcanism without much pyroclastic activity (Bondarenko et al. 2010; Smrekar et al. 2010; Shalygin et al. 2015; D'Incecco et al. 2021; Gülcher et al. 2020; Filiberto et al. 2020) are inconsistent with a major current CO_2 flux from the interior. The higher Venus' CO_2 atmospheric abundance than Earth's crust normalized to planetary masses (e.g., Lécuyer et al. 2000) points to degassing of Venus' interior CO_2 in previous epochs via volcanism and/or metamorphic decomposition of carbonates formed in putative aqueous environments.

CO A decrease in CO sub-cloud content toward the surface observed by the Pioneer Venus Large Probe and Venera 12 probe (Oyama et al. 1980; Gel'man et al. 1980), telescopic, and orbital (Venus Express) measurements (Esposito et al. 1997; Mills and Allen 2007; Marcq et al. 2018) is explainable by consumption in gas-gas type reactions (e.g., CO $\to$ OCS, Krasnopolsky 2007, 2013; Yung et al. 2009). The observed latitudinal differences in tropospheric CO profiles are consistent with the CO to OCS conversion in the deep atmosphere and the Hadley cell type circulation that causes lower CO contents at low latitudes where air

492

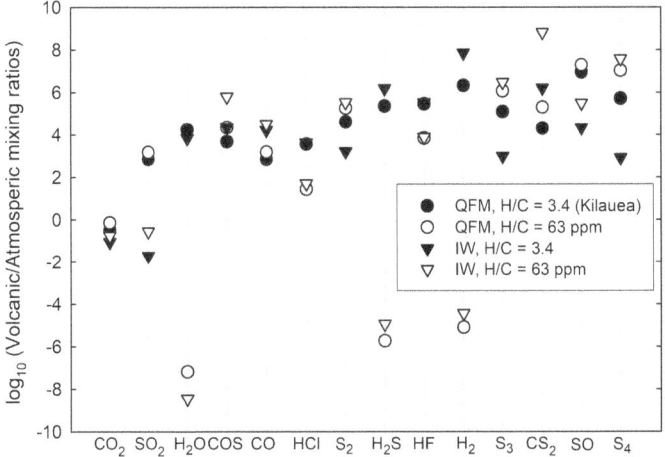

Fig. 2 Volcanic/Venus atmospheric gas ratios for Earth-like volcanic gases (filled symbols) and H-depleted gases (empty symbols). Concentrations of volcanic gases correspond to Kilauea-like compositions and H-depleted composition with the atmospheric H/C atomic ratio from Fig. 1. The atmospheric composition corresponds to observations (CO_2, SO_2, H_2O, CO, HCl, HF) and chemical equilibrium models for the near-surface atmosphere (Zolotov 1996). The figure demonstrates that most volcanic species before dilution and chemical interactions could be distinguished from atmospheric gases.

is upwelling from the near-surface atmosphere (Tsang et al. 2008; Tsang and McGouldrick 2017). Further understanding of CO and OCS chemical cycles may be revealed by improved mapping of their latitudinal and altitudinal variability through remote sensing, as will be discussed below.

H_2O and HDO Tropospheric water vapor can be measured remotely in the infrared range at various wavelengths, more than any other trace species, so that its vertical profile is relatively well constrained and consistent with a vertically uniform abundance of $\sim$30 ppmv at heights of less than 48 km (see Table 1 in Sect. 4.1), pointing to a lack of surface or tropospheric sources or sinks. *In situ* measurements from humidity sensors on board Vega probes (Surkov et al. 1987) exhibited much larger values in the clouds (up to 200 ppmv in the 30-45 km altitude range) - although this measurement may have been contaminated by cloud aerosol particles. In any case, the relatively dry Venusian atmosphere and good remote sensing sensitivity makes tracking of water horizontal, vertical and temporal variations a key observable to constrain any outgassing provided it is water-rich enough. Moreover, gaseous HDO can also be measured, so that D/H ratio is accessible, e.g., through nightside near-IR atmospheric windows between 1 and 2.5 μm, at altitudes 0-45 km (see below, Sect. 4.1) or *in situ* probe measurements (Garvin et al. 2022). Any significantly measured value lower than the background value, $(120 \pm 40) \times$ Standard Mean Ocean Water (De Bergh et al. 1991; Donahue et al. 1997), would also point to ongoing outgassing from an interior source whose D/H ratio is closer to the solar system average than the value in the bulk Venusian atmosphere, often assumed to have been heavily altered by a strong differential H vs. D escape (Avice et al. 2022, this collection; Gillmann et al. 2022, this collection).

SO_2 Sulfur dioxide has been measured *in situ* below the clouds from Pioneer Venus and Venera 12 descent probes using mass spectrometry and gas chromatography (Oyama et al. 1980; Gel'man et al. 1980), but the most extensive vertical profile below the clouds was

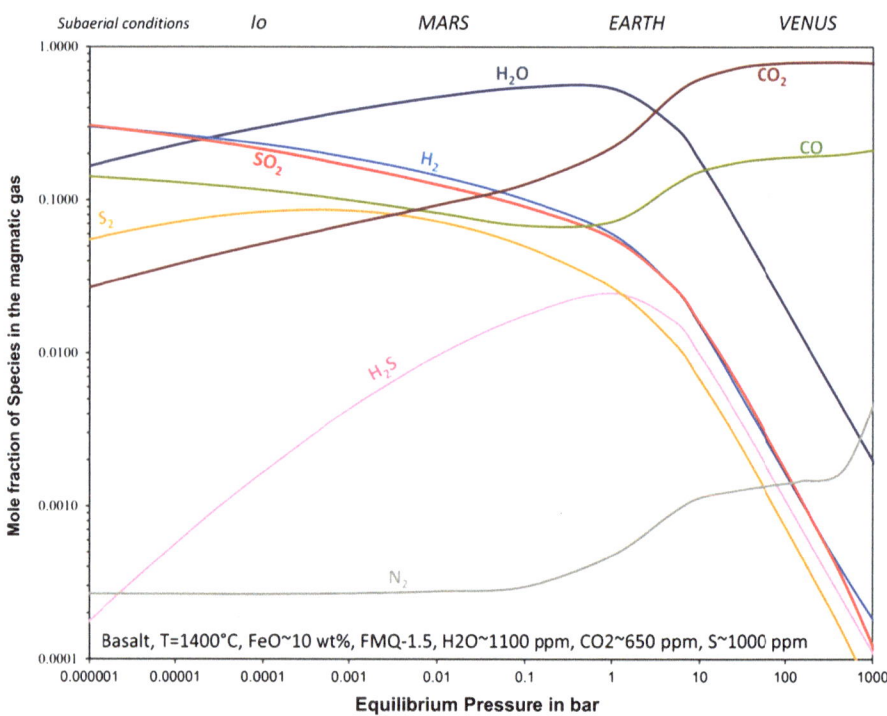

Fig. 3 Modeled composition of C-H-O-S bearing volcanic gases with respect to pressure for a given basalt composition. Note that composition of volatiles at the bottom of the figure represents basaltic melt rather than volcanic gases. From Gaillard et al. (2021).

obtained using UV absorption spectroscopy by the ISAV spectrometers on Vega 1 and 2 entry probes (Bertaux et al. 1996). The ISAV data showed a marked decrease in SO_2 concentration from around 100 ppm at 40 km down to some 25 ppm at 10 km, below which the data become unreliable. Bertaux et al. (1996) suggested that this gradient provided evidence of a strong downward transport of SO_2, with a SO_2 sink in the lower atmosphere and/or surface, and SO_2 source above 40 km, which could be associated with volcanic activity, or recent impacts or cloud processes. However, this gradient and putative sink is difficult to reconcile with known atmospheric processes (Krasnopolsky 2007), and reactions with surface materials are likely too slow to explain such a sink (see the following section). The possible latitudinal variations in SO_2 content below the clouds (at 30-40 km) (see Fig. 6 in Marcq et al. 2021) and in lower clouds (51-54 km; Oschlisniok et al. 2021) reported from remote sensing looks like an atmospheric phenomenon related to the Hadley cell circulation. Constraining SO_2 behavior through further remote sensing is discussed below.

Hydrogen halides HCl and HF in the mesosphere have been measured from ground-based and space-based spectroscopy (e.g. Esposito et al. 1997; Mills and Allen 2007). Recently, extensive monitoring by Venus Express using solar occultation appeared to show some variability at altitudes 70-115 km (Mahieux et al. 2015). Below the clouds at altitudes of 30-40 km, though, no spatial variability has yet been found (Iwagami et al. 2008; Arney et al. 2014). It is likely that the HCl and HF are in equilibrium with the surface minerals held at the observed gas-phase concentrations of HCl (0.4 ppm) and HF (2-8 ppb) (Fegley et al.

1992). The low Cl/S and F/S atmospheric ratios compared to possible volcanic gases could be explained by a more efficient consumption of volcanic halogen-bearing gases to secondary surface minerals, and the low concentrations of HCl and HF indicate past gas-solid reactions that led to gas-solid equilibration (Zolotov 2018). Aside from putative fluctuations in HCl and HF contents just below the main cloud level, any positive compositional anomalies in the sub-cloud atmosphere could be indicative of current volcanism. Detection of such anomalies could be achieved by remote observation in the 1.74 μm spectral window or by multiple atmospheric profiles from entry probes.

PH₃ The claimed detection of phosphine (Greaves et al. 2021), has been debated in recent years and has been proposed as a possible tracer of volcanism (Truong and Lunine 2021; Omran et al. 2021); but this PH_3 purported signature is difficult to reconcile with other non-detections (Encrenaz et al. 2020b; Villanueva et al. 2021) as well as current oxidized (Bains et al. 2021) and dry (Bains et al. 2022) atmospheric composition. Other phosphorus species could constrain mantle composition and volcanic rates, and a combination of these species can provide a loose constraint on mantle redox state, under the assumption that these species are volcanically derived.

2.2 Surface and Atmospheric Chemistry Involving Trace Volcanic Gases

In contrast to atmospheric processes, slow gas-solid type reactions in a permeable upper layer (e.g., Fegley et al. 1992; Fegley et al. 1997a,b; Wood 1997; Zolotov 2018) cannot cause gradients of CO, SO_2, or other atmospheric gases. Atmospheric mixing through eddy diffusion and turbulence (e.g., Krasnopolsky 2007; Morellina and Bellan 2022) in the near surface atmosphere and the Hadley cell circulation are much faster than gas-solid reactions. The thermodynamically favorable oxidation of ferrous minerals (silicates, FeS, $Fe_{1-x}S$) and glasses in basalts produce CO at the expense of CO_2 (Zolotov 2018) as illustrated by simplified reactions that are irrelevant to the CO atmospheric gradient:

$$3FeSiO_3 \text{ (ferrous pyroxene)} + CO_2 \rightarrow Fe_3O_4 \text{ (magnetite)} + 3SiO_2 \text{ (silica)} + CO$$

$$3FeS \text{ (troilite)} + 4CO_2 \rightarrow Fe_3O_4 \text{ (magnetite)} + CO + 3OCS$$

These and other oxidation reactions, especially further oxidation of magnetite to ferric oxide hematite (Fe_2O_3), could be limited because the atmospheric CO/CO_2 ratio is indistinguishable from that controlled by the magnetite-hematite equilibrium (Fegley et al. 1997b; Zolotov 2018). In contrast to CO, atmospheric SO_2 is consumed through reactions with Ca- and Na-bearing phases in basalt leading to formation of sulfates and S_2 gas (Zolotov 2018):

$$CaSiO_3 \text{(Ca silicate)} + 1.5SO_2 \rightarrow CaSO_4 \text{(anhydrite)} + SiO_2 \text{(silica)} + 0.25S_2$$

Feasibility of the latter interactions on Venus has been illustrated by modeling experiments (e.g., Fegley and Prinn 1989; Berger et al. 2019; Radoman-Shaw et al. 2022; Santos et al. 2023). However, trapping of SO_2 would likely be limited by equilibration with sulfate- and/or pyrite-bearing mineral assemblages and because of inefficient or inhibited alteration of plagioclase (Zolotov 2018).

It is possible that volcanic gases would be depleted in hydrogen, for Venus (Zolotov and Matsui 2002), Io (Fegley and Zolotov 2000), and Mercury (Zolotov 2011; Evans et al. 2015). The H-depleted volcanic gases could not have abundant H_2O, H_2S, HCl, and HF (Fig. 1). Chlorine, fluorine and partially sulfur and carbon are presented by OCS, CS_2, S_2,

S_3, S_4, S_2O, S_2Cl, S_2F_2, SO, SCl_2, Cl_2, Cl, COF_2, SOF_2, and S_2Cl_2. With the exception of S_3 and S_4 (Maiorov et al. 2005; Krasnopolsky 2013), and OCS (see Mills and Allen 2007; Marcq et al. 2018 for reviews), none of these gases have been detected in the sub-cloud atmosphere. Any detection of these gases below clouds could therefore be indicative of current volcanism (Fig. 2). However, these gases are thermodynamically unstable in contact with lower atmospheric gases and could be converted into HCl, HF, Cl- and S-bearing gases. Likely consumption of Cl- and F-bearing gases through net hydrolysis reactions

$$SCl_2 + 2H_2O \rightarrow 2HCl + SO_2 + H_2$$

$$Cl_2 + H_2O + CO \rightarrow 2HCl + CO_2$$

will deplete atmospheric water vapor and increase concentration of the reaction products (e.g. HCl) that could be observed in a partially diluted volcanic plume. Reaction kinetics of H-free volcanic gases remain to be evaluated for Venus conditions. Although the consumption lowers chances of detection in diluted plumes away from a vent, near-vent sampling of H-depleted gases from a descent probe or a lander would be more informative than for H-rich gases (Fig. 2). Such a detection will strongly constrain water history and water content in the interior of the planet.

2.3 Possible Volcanic Processes and Outgassing Rates

The volcanic production rate on Venus, and the associated outgassing rate are still poorly constrained. Using an equilibrium resurfacing model based on the age of the surface, Phillips et al. (1992) obtained a rough estimate of $\simeq 1$ km^3/yr, consistent with Earth-like hotspot production rates. Namiki and Solomon (1998) find that ^{40}Ar and ^{4}He abundances in the atmosphere of Venus are consistent with magmatism rates in the order of 1-4 km^3/y, but highlight high uncertainties. Stofan et al. (2005), based their estimate on the volume of lava required to bury impact craters since the emplacement of the present-day surface and found a range of 0.6-46 km^3/yr (we note though that the high end of this range is for an improbably surface age of only 10 Myr). Romeo and Turcotte (2010) propose a post-catastrophe resurfacing rate of ~ 2 km^3/yr and note that equilibrium resurfacing models they tested resulted, by contrast, in global present-day resurfacing rates of > 40 km^3/yr to explain the statistics of the observed crater population. Ivanov and Head (2013) estimated the volume of volcanic units on Venus and derive resurfacing production rates between 0.03 and 0.1 km^3/yr depending on the age of the surface (300 Myr to 750 Myr).

Byrne and Krishnamoorthy (2022) scaled the rates of volcanism from Earth to Venus using the planets' masses and volumes and estimated 120 eruptions per Earth year on Venus. However, van Zelst (2022) took into account tectonic difference associated with volcanism and estimated only 12 per year. Other estimates are based on the outgassing of SO_2 necessary to sustain the current SO_2 abundance in the clouds, and yield values of 0.3-11 km^3/yr (Fegley and Prinn 1989) and 4.6-9.2 km^3/yr (Bullock and Grinspoon 2001). However, two latter evaluations are reliant on poorly known composition of the near-surface atmosphere and may benefit from being revisited based on future data. Both evaluations assume an existence of Ca carbonates in surface materials that is not possible based on SO_2-calcite interaction experiments (e.g., Fegley and Prinn 1989; Santos et al. 2023), evaluations of stability of Ca carbonates with respect to S-bearing atmospheric gasses, and the lack of mechanism for carbonate formation at the current surface conditions (e.g., Zolotov 2018).

The degassing rates will depend on the dynamics of the mantle; and the connection between these two has been explored by e.g., Noack et al. (2012), Gillmann and Tackley

(2014). One should note that a stagnant lid regime limits the efficiency of heat loss from the mantle. It has been calculated that under such conditions, the interior of Venus may heat up and cause widespread melting, where volcanism is used to extract heat to the surface in a heat-pipe regime (e.g. O'Reilly and Davies 1981; Moore et al. 2017). Volcanic production rates consistent with this scenario are in the order of 170-280 km^3/yr, an order of magnitude above Earth's total production (Solomon and Head 1982; Armann and Tackley 2012). Short episodes of intense resurfacing separated by long periods of relative quiescence or a predominantly intrusive emplacement of melt appear to be more consistent with suggested Venus' present-day heat flux and estimations of crustal thickness. The complex links between melt production and outgassing have been briefly mentioned above in Sect. 2.1, and are discussed in greater detail by Gillmann et al. (2022), for example.

Other papers in this collection address the expected styles of volcanism as evidenced by their morphology and occurrence. Ghail et al. (2024, this collection) describe the large diversity of volcanic features with generally no direct analogs in Earth's volcanism. Among volcanic features, large volcanic rises or hotspots may be directly related to mantle plumes, whereas features such as coronae, or extremely long lava channels require more detailed observations to be understood. Coronae have been used to suggest (Ghail et al. 2024, this collection) that most of the magma delivery on Venus would be intrusive (Lourenço et al. 2020), because heat pipe models, in which all mantle heat is extracted through extrusive volcanism, predict crustal production rates that would produce a much younger surface than observed (Moore et al. 2017; Rolf et al. 2022). Geophysical considerations of (Head and Wilson 1986, 1992) suggested abundant magma chambers in a zone of neutral buoyancy in the Venus' crust.

The style of volcanic eruption will affect the detectability of volcanic gas plumes, in that it will affect both how episodic an emission of gases is, and to what altitude the plume of gases will rise. Of particular importance is the volatile content of the erupting magma: not only does this determine whether an eruption will be explosive, injecting gases high into the atmosphere, but would also be an important indicator of the long-term evolution of the interior (Rolf et al. 2022, this collection; Gillmann et al. 2022, this collection). Explosive volcanism would throw not only gas but also ash into the atmosphere, likely expressed on the surface in the form of pyroclastic deposits (Head and Wilson 1986; Airey et al. 2015; Ganesh et al. 2021); a few suspected instances of pyroclastic deposits have been found on Venus (Campbell et al. 2017; Ganesh et al. 2022; Ghail and Wilson 2015; Keddie and Head 1995). The volatile content needed to produce an explosive eruption varies with the barometric pressure and therefore with vent altitude. For a suspected pyroclastic flow feature identified at Scathach Fluctus, it has been calculated that the volatile content needed to produce an explosive eruption here must be relatively high: 4.5 weight percent (wt%) if only H_2O, or 3% H_2O plus 3% CO_2. On the other hand, explosive volcanism at the summit of Ma'at Mons would require only 2% H_2O content (Airey et al. 2015). A thin layering of apparently porous rocks seen as low flat outcrops with darker soil in between them at Venera 13 and 14 landing sites could have formed through airborne deposition of impact-generated particles (Basilevsky et al. 2004, 2019), though a low viscosity lava crystallization phenomenon (Garvin et al. 1984) and volcanic pyroclastic origins (Florensky et al. 1983) are possible. Such deposits could bring evidence regarding the volatile content of lava flows, possibly reflecting an early stage of renewed magmatic activity with volatile-rich, disrupted magma escaping through vents in fractured regions of the upper crust (Campbell et al. 2017). Herrick et al. (2023, this collection) notes that future InSAR missions (see the review in Widemann et al. 2023, this collection) will significantly improve our understanding of the sequence of events on the surface and evaluate whether or not fundamental changes in the nature of geologic activity have occurred over the past several hundred million years.

The above discussion has focused on explosive volcanism largely because it would likely be the easiest to detect, due to the large gas release in a short period of time and possible higher altitude reached by the ensuing volcanic gas plume. Much of the discussion of volcanic gas detection is equally applicable to effusive (non-explosive) eruptions or even from passive outgassing (i.e. involving no coincident eruption of magma). Determining whether the style of eruption is effusive or explosive would help to constrain magmatic volatile content, with implications for the evolution of Venus. This could be achieved either by looking for surface deposits like pyroclastic flows, or by detection of ash plumes in the atmosphere, as discussed below.

2.4 Vertical Transport of Volcanic Gases

In Sect. 2.3 we have presented a discussion of volcanism styles and outgassing rates, which will determine the source characteristics of a volcanic plume. We now discuss the subsequent transport of volcanic gases in the atmosphere. We will first discuss the vertical transport of plumes because this is essential for determination the vertical profiles of volcanic gases in the atmosphere.

The elevation that a volcanic plume can eventually reach is determined by the temperature of the volcanic gases and by the atmospheric temperature-pressure profile: as long as the rising gas plume remains of lower density than the surrounding atmosphere, despite its own adiabatic cooling as its barometric pressure decreases, then it will continue to rise. The canonical view is that volcanic gas plumes typically cannot overcome the convective stability of the current atmosphere of Venus, with the exception of particularly large eruptions (Glaze 1999; Glaze et al. 2011; Airey et al. 2015). However, the above works (in particular the first two) were motivated largely by the question of whether volcanic plumes would rise to the cloud-tops at 70 km or above, to explain mesospheric SO_2 variations from space such as those from Pioneer Venus Orbiter UV measurements (Esposito 1984). A rising plume would require a lot of buoyancy to rise through the convectively stable layers found at altitude ranges of 35-50 km and 60-70 km. However, reaching 30 or 35 km altitude is comparatively much easier: the atmospheric temperature profile appears to be near zero stability at 0-10 km and 20-35 km altitude, indicating that plume rise in these altitude bands could be readily achieved even with a small excess temperature (on the order of only 10 K).

The above discussion is based on the mean atmospheric temperature profile of the Venus International Reference Atmosphere (VIRA) – (Seiff et al. 1985), and therefore comes with two important caveats. The first is that this temperature profile is subject to significant uncertainties, particularly in the deepest 12 km. None of the Pioneer Venus probes returned usable temperature data below 12 km, so the VIRA profiles were based largely on the Venera 10 probe profile, which itself had significant uncertainties. The most reliable temperature profile is from the Vega 2 probe, but this appears to show a superadiabatic (convectively unstable) layer at 0-7 km altitude which is yet unexplained (Lebonnois and Schubert 2017; Lorenz et al. 2018; Morellina et al. 2020; Morellina and Bellan 2022). New measurements of temperature and composition near the surface from DAVINCI descent probe (Garvin et al. 2022) will reduce these uncertainties.

The second caveat about using the VIRA mean profile is that it will vary spatially and diurnally; using a mean profile ignores these variations and also neglects the effects of atmospheric dynamics. Lebonnois et al. (2018) found that the diurnal cycle of the planetary boundary layer convection is expected to be influenced by topography. Above large topographic features, the diurnal cycle of the convective layer is expected to be quite strong, extending up to 7 km above the local surface at noon and dropping below 1 km at night. Such strong vertical mixing would enhance the vertical propagation of the plume material. The low convective stability of the boundary layer in the late afternoon conditions would also enable the propagation of topographically-driven waves from the surface up to the cloud tops, as discussed by Lefèvre et al. 2020. Follow-up observations by Akatsuki LIR camera and modeling lend further support to these findings (Lefèvre 2022; Suzuki et al. 2023).

3 In Situ Measurements of Volcanic Gases in the Troposphere

Having presented an overview of possible volcanic gases of interest, of styles of volcanism, and of vertical transport of volcanic gases in the atmosphere, we now turn to focus on measurements from future missions, starting with *in situ* descent probe measurements in Sect. 3 before turning to orbital remote sensing measurement in Sect. 4. The discussion is based on the DAVINCI probe, as it is, at the time of writing, the only confirmed future mission which will measure gas composition *in situ* in the atmosphere. Other future candidate missions will be discussed in Sect. 5 below.

3.1 Expected Constraints on Volcanism from the DAVINCI Mission

NASA's DAVINCI mission (Garvin et al. 2022; Widemann et al. 2023, this collection) is scheduled to sample chemical and isotopic compositions in the lower atmosphere in early 2030s with the Venus Mass Spectrometer (VMS) and the Venus Tunable Laser Spectrometer (VTLS). Although these instruments have different performance capabilities, they will provide data on mixing ratios of major reactive gases (CO_2, SO_2, H_2O, CO, OCS, H_2S, S_n, H_2SO_4, and HCl) and stable isotope composition of H, O, C, S, He, Ne, Ar, Kr, and Xe. The VMS is capable of measuring gases in the range of 2-535 Da. Measurements by VMS and VTLS will constrain abundances of P-bearing gases that may be present in the atmosphere but may not be volcanic (Krasnopolsky 1989; Baines et al. 2021). Sampling of reactive gases will occur every 0.15-1 km at altitudes ranging from $\sim$11 to 50 km, and more often (50-100 m) below $\sim$11 km. Stable isotope data for H, O, C and S will be obtained through at least 5 VTLS measurements below 40 km. The D/H ratio will be measured with 2% precision up to ten times from 67 km to $\sim$2 km. Abundances of noble gases (He, Ne, Ar, Kr, Xe) and N_2 and their major isotopes will be measured by the VMS at least once below 70 km. The precisions are as follows, ^{4}He (5%), ^{40}Ar (15%), ^{136}Xe (20%), ^{129}Xe (20%). Major atmospheric and purported volcanic gases are certainly detectable with these instruments. Accuracies of measurements (SO_2, 5%; H_2O, 20%, CO, 5%; OCS, 3%), will allow identification of moderately diluted and/or chemically altered plumes. Gases emitted by putative H-depleted melts (CS_2, SCl_2, Cl_2, etc., Fig. 1) are within the mass range of the mass spectrometer, though evaluations and tests are needed to assess their detectability and detection limits.

On Earth, volcanic gases (Symonds et al. 1994; Oppenheimer et al. 2014) are isotopically distinct from the air, and expected measurements from the DAVINCI mission of vertical profiles of ^{16}O/^{18}O, ^{12}C/^{13}C, and ^{32}S/^{33}S/^{34}S would be informative to distinguish a plume. Locally low D/H ratios will indicate current degassing because the high atmospheric D/H ratio is likely resulted from a preferential H escape (e.g., Donahue et al. 1997). Locally elevated ^{40}Ar/^{36}Ar ratio will indicate current degassing because ^{40}Ar accumulates in the interior through decay of ^{40}K. On Earth, the ^{3}He/^{4}He ratio in volcanic gases is used to distinguish interior source regions (e.g. Graham et al. 2002; Hilton et al. 2002), and a local ^{3}He/^{4}He anomaly in the lower atmosphere could place constraints on magmatic sources of volatiles. Note that these constraints could be obtained only with multiple sampling of Ar and He at different altitudes that is not currently planned on DAVINCI for the sub-cloud atmosphere. Overall, several correlated compositional (SO_2, CO, OCS, H_2O, etc.) and isotopic anomalies of H, O, C and S might be sufficient to catch a volcanic plume in the sub-cloud atmosphere.

In addition to putative current volcanism, DAVINCI expected data on bulk atmospheric ^{40}Ar content will constrain both the scale and history of Venus degassing (Kaula 1999; Namiki and Solomon 1998; O'Rourke and Korenaga 2015; Gillmann et al. 2022, this collection; Avice et al. 2022, this collection). Likewise, both early and overall degassing rate could be evaluated from abundances of ^{129}Xe and ^{136}Xe that are products of radioactive decay in the interior (Coltice et al. 2009). The atmospheric escape rate of helium is on the order of hundreds of millions of years (Fedorov et al. 2011; Nordström et al. 2013), so the bulk atmospheric helium isotope ratios measured by DAVINCI will help to constrain volcanic fluxes on these timescales (Namiki and Solomon 1998). Taken together, these constraints on degassing throughout history from isotopic abundances and ratios will place new bounds on the probability of current and future volcanism on various timescales. In particular, these data will help to distinguish between the alternative catastrophic (Schaber et al. 1992; Strom et al. 1994) and equilibrium resurfacing models that imply approximately continuous and current volcanism (Phillips et al. 1992; Herrick and Rumpf 2011; Bjonnes et al. 2012). Further discussion of noble gas isotope geochemistry and its constraints on recent volcanic outgassing is presented by Avice et al. (2022) in this collection.

The planned DAVINCI deep atmosphere probe descent over Alpha Regio, which represents a strongly tectonically-deformed region (tessera) without clear signs of volcanism (e.g., Bindschadler et al. 1992), decreases the probability of direct (localized) sampling of a volcanic plume. Descent imaging of the Alpha surface from the probe in the near-infrared range and at 0.74 to 1.03 μm will provide new geomorphological information about the prospects of volcanic activity at the touchdown region at spatial scales as fine as meters. With the single atmospheric profile, it will be difficult to unequivocally distinguish atmospheric and volcanic causes of any measured compositional phenomenon, if any are detected. Geographical locations of purported current or recent volcanism (Bondarenko et al. 2010; Smrekar et al. 2010; Shalygin et al. 2015; D'Incecco et al. 2017; Stofan et al. 2016; Gülcher et al. 2020; Filiberto et al. 2020) may need to be selected for future sampling of volcanic gases and particulates by descent probes and landers.

4 Orbital Detection of Tropospheric Gas Plumes

In this section we turn to the detection of volcanic gas plumes from orbital remote sounding. We focus at first only on detection of H_2O-rich plumes; this is because water vapor is detectable from both orbit and ground-based telescopes using nightside near-IR spectroscopy and can be mapped at several altitudes from the surface up to 30-40 km altitude, as will be detailed further below. The discussion in Sect. 2.1 showed that the H_2O content of volcanic gases is not known on Venus; volcanic gas from an H-depleted magma might well be depleted in H_2O compared to the atmosphere. Nevertheless, the discussion of plume transport in Sect. 4.1 is applicable to any passive tracer, and can be scaled to any positive or negative water excess, so could even be applied to the transport of H_2O-poor volcanic plumes. In the second half of this section, we will then extend the discussion to consider other volcanic gas species whose tropospheric abundances could be mapped from orbit.

Table 1 Summary of post-Venus Express remote measurements of tropospheric water vapor.

Spectral window	Altitude (km)	H_2O gas abundance (ppmv)	References	Notes
1.10 μm & 1.18 μm	0-15	25-40 25-31 25-30	Chamberlain et al. (2013) Arney et al. (2014) Fedorova et al. (2015)	- retrievals sensitive to assumed surface emissivity - D/H could be measured in principle
1.74 μm	15-30	30-35	Arney et al. (2014)	
2.3 μm	30-45	29-33 30-36	Marcq et al. (2008) Arney et al. (2014)	D/H ≈ 120±40 SMOW (De Bergh et al. 1991)

4.1 H_2O Tropospheric Plumes

The background abundance of water vapor in Venus' atmosphere is close to 30 ppmv below the clouds (e.g., Bézard and de Bergh 2007; Marcq et al. 2018; Ignatiev et al. 1997; Pollack et al. 1993). Remote sensing of tropospheric water is possible on the night side using the thermal emission in a few near infrared windows of relative CO_2 transparency (Taylor et al. 1997), as summarized in Table 1.

Water vapor abundances are expressed in volume mixing ratio (VMR), in units of parts per million by volume (ppmv).

As can be seen in Table 1, the vertical profile of water vapor is uniform in the probed range within the sensitivity of existing measurements. Some hints of horizontal variation of H_2O have been reported in the 2.3 μm band (sensitive to 30-45 km altitude) by Arney et al. 2014, and from Venus Express/VIRTIS observations (Haus et al. 2015), typically at the 10%-20% level; however, it is not clear whether these are associated with cloud 'ghosting' effects in the retrieval as discussed in Arney et al. (2014) and Barstow et al. (2012). No spatial variations in water vapour retrieved in the 0-15 km and 15-30 km altitude windows have yet been found (see e.g., Bézard et al. 2011; Haus et al. 2015). In order to assess the detectability of a plume of volcanic gases, one needs to consider whether it could alter gas abundances by an amount greater than the measurement sensitivity.

Preliminary mesoscale simulations of atmospheric plume advection were performed by Lefèvre et al. (2020). Outgassing was simulated by setting a constant 1.1 multiplicative factor for H_2O (33 ppmv instead of 30 ppmv) in a single 10 km × 10 km (surface) × 10 km (height from surface) atmospheric cell over Idunn Mons in Imdr Regio (this is one of the sites of suspected recent hotspot volcanism identified by Smrekar et al. (2010). The thermochemical lifetime of H_2O below the clouds far exceeds the simulation duration. Results after 10 Earth days are shown in Fig. 4. Thermal buoyancy above the first 10 km is not considered yet, only advection by the atmospheric circulation is. The altitude reached by the plume depends sensitively on the thermal structure of the planetary boundary layer; Lebonnois et al. (2018) showed that this would be highly variable, with high-altitude slopes giving rise to particularly deep convection potentially allowing volcanic gas plumes to reach greater height. Similar modeling is necessary at the equator, for example at the potentially active region found in *Atla Regio* (Shalygin et al. 2015), to study the variability of the deep atmosphere stability and the possible transfer of volatiles between northern and southern hemispheres.

The effective H_2O emission rate in this simulation is about 30 t/s (2.6 Mt/Earth day) during the simulated 10 Earth days. Such an amount is typical of 'mid-sized' eruptions on Earth

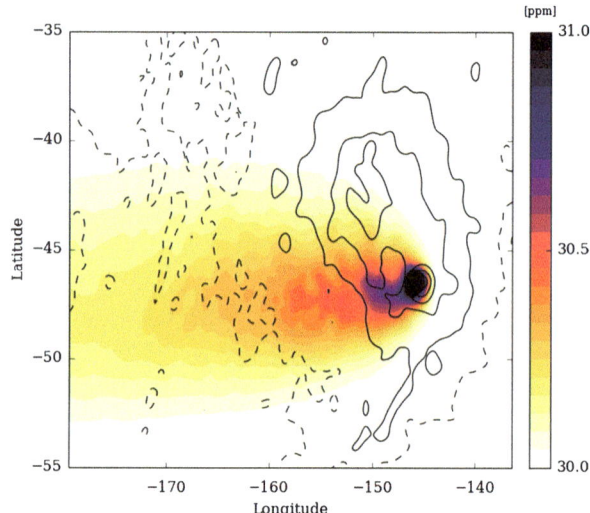

Fig. 4 Simulated dispersion of a volcanic plume released from Idunn Mons in Imdr Regio. Shading indicates water vapor abundance at 13 km altitude, contours indicate surface topography. See text for further details of the simulation.

such as the Bárðarbunga-Holuhraun (Iceland) eruption in 2014 (Schmidt et al. 2015), assuming a $H_2O:SO_2$ mass ratio of 20. The maximal observable enhancement near the source is about 0.8 ppmv, which is marginally below the ppm-level accuracy of water vapor measurements as of 2021. It should however be detectable from future dedicated instruments such as VenSpec-M and -H channels (Helbert et al. 2019) on board EnVision as well as VEM on board VERITAS (Helbert et al. 2018, 2020).

The modelling presented here is preliminary; further investigations of the effects of buoyancy and atmospheric thermal structure, eruptive style and location are needed. However, this approach shows how the detectability of atmospheric plumes from volcanic eruptions of different sizes and styles can be modelled and evaluated. We note that there may also be non-volcanic sources of H_2O variability near the Venus cloud deck, such as strong convective downdrafts as modelled by Baker et al. (1998), or evaporating H_2SO_4 rain (virga) as discussed by Arney et al. (2014) and modelled by Gao et al. (2014); these must be modelled in order to understand potential 'false positives' in the hunt for volcanic plumes.

Radiative transfer numerical simulations show that with a high spectral resolution ($R\sim8000$) and a signal-to-noise ratio of 100, VenSpec-H will determine the abundances of H_2O with an accuracy of 3% in each of the spectral windows centered at 1.18, 1.74 and 2.3 μm (Robert et al. 2021). As shown in Table 1, these three windows allow mapping of water vapor in three different altitude domains, comprising 0-15 km, 15-30 km and 30-45 km respectively.

As well as seeking evidence for plumes (and providing new information about tropospheric dynamics) by mapping the spatial variations of water vapor, VenSpec-H will also search for variations in the HDO/H_2O ratio. For HDO, preliminary spectral modeling shows that it should be possible to reach an accuracy of 5% in HDO retrievals in the 1.18 μm and 2.3 μm windows, assuming only the random error, i.e. the retrieval error directly from the fit (Robert et al. 2021; ESA 2024). Taking into account also the H_2O retrieval accuracy described above, this would result in an accuracy of 8% in mapping spatial variability of the D/H ratio between 0 and 15 km and between 30 and 45 km.

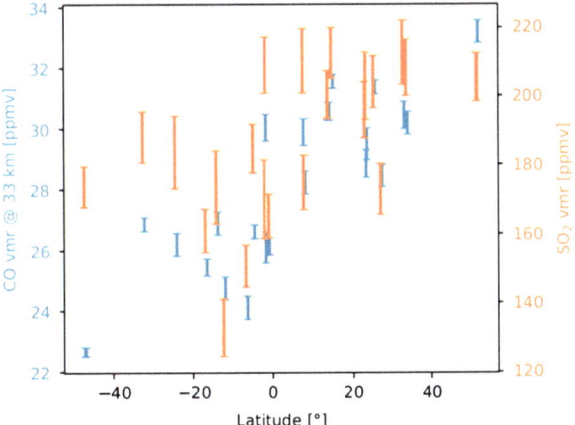

Fig. 5 Latitudinal variations of CO and SO_2 volume mixing ratios at altitude of 33 km are both measurable using remote sounding, as was demonstrated by Marcq et al. (2021) using IRTF/iSHELL. Error bars stand for $\pm 3\sigma$ standard deviations. Systematic uncertainties of the absolute SO_2 mixing ratio are not displayed.

4.2 Other Tropospheric Gas Plumes: SO₂, CO, HCl and OCS

Besides H_2O, several other gases may be related to present volcanic activity as described in above Sect. 2.1. In particular, tropospheric SO_2, CO, HCl and OCS could be detected from the orbit, also using nightside IR-spectroscopy. The 2.3-2.5 μm region is therefore targeted as each of these species are known to absorb in this spectral range. Their detectability is not an easy task as the spectral lines are weak, moreover confirming current volcanic activity requires assessing their background (non-volcanic) variability. In addition, the vertical profiles of OCS and CO are known to have anti-correlated behavior at ∼35 km, which could be explained by reactions (Yung et al. 2009; Pollack et al. 1993; Krasnopolsky 2007), that do not require volcanic gases.

Measurements of tropospheric SO_2 have been performed from Venus Express orbit (Marcq et al. 2008, 2023; Oschlisniok et al. 2021) and from ground-based facilities (Pollack et al. 1993; Arney et al. 2014), leading to an accepted value of 130±50 ppmv in the 30-40 km altitude range (or, in the latest analysis of VIRTIS-H data by Marcq et al. 2023, 190±50 ppmv). The relatively low spectral resolution of the VIRTIS-H spectrometer aboard Venus Express (R∼1800, or more accurately 1500-2000, see Marcq et al. 2008, their Fig. 1) means that SO_2 retrievals from that instrument reach a precision of only ± 20% or so; at this sensitivity, no evidence of spatial variability was found. More recent observations were performed by Marcq et al. (2021) using the high-resolution iSHELL spectrometer at the NASA IRTF facility (R∼20,000); thanks to this higher resolution, the sensitivity to tropospheric SO_2 achieved in the retrieval was typically < 5%. With this higher sensitivity, Marcq et al. (2021) were able to show evidence of latitudinal variations of SO_2 abundances (see Fig. 5). This demonstrates the importance of high spectral resolution for enabling measurement of tropospheric SO_2 variability, as will be performed by VenSpec-H on EnVision.

Carbon monoxide is also absorbing in the 2.3 μm range between 30-40 km in altitude. CO is already known to increase with increasing latitude (Marcq et al. 2008, 2021, 2023; Tsang and McGouldrick 2017), as shown in Fig. 5, resulting from the large-scale meridional circulation and positive vertical gradient due to its photochemical source above the clouds (Yung and Demore 1982). CO is indeed considered as a passive tracer in Venus' atmosphere. The results of the numerical radiative transfer simulations are encouraging as an accuracy of 1.5% will be achieved with VenSpec-H (Robert et al. 2021; ESA 2024).

Robustly assessing the presence of a volcanic gas plume necessitates measuring multiple tracer species simultaneously. Therefore, the VenSpec-H instrument has been designed to

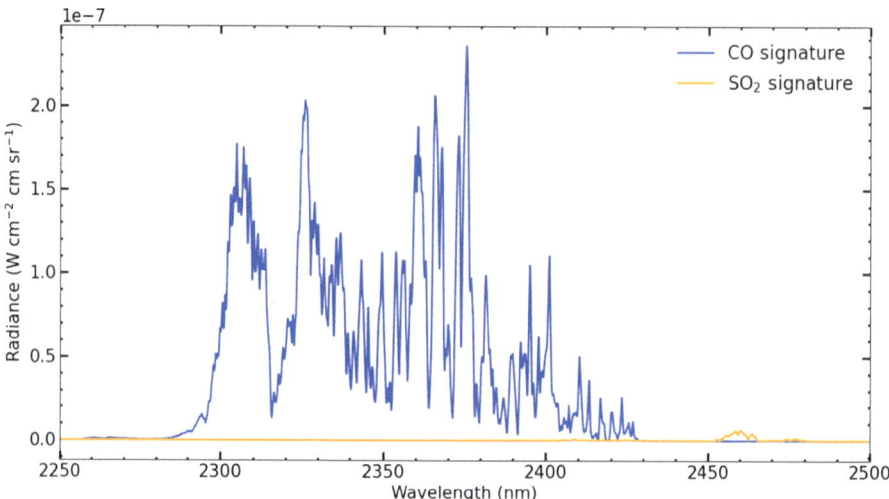

Fig. 6 Spectral signatures of CO and SO_2 in emission from the nightside of Venus in the 2.3–2.4 μm spectral band. The figure shows the difference in spectrum if CO and SO_2, respectively, are removed from the model atmosphere. Simulations are obtained using the ASIMUT-ALVL radiative transfer tool (Vandaele et al. 2006).

allow simultaneous measurement in both of the CO and SO_2 bands as shown in Fig. 6 (as well as H_2O, HDO, HF, CO, COS, which are not shown; Robert et al. 2021).

We have focused here (in Sect. 4.2) on the 2.3 μm spectral window. In addition, the other nightside transparency windows may bring insights on the current volcanic activity of Venus. HCl, for instance, may be measured in the 1.74 μm window together with H_2O, probing the 15-30 km altitude range. Investigating further the information contained in the 2.3 μm window necessitates new mesoscale simulations considering the thermal buoyancy to properly address how these trace species (SO_2, SO, HF) could be used as tracers of present-day outgassing.

5 Detecting Volcanic Gases in the Mesosphere

SO_2 exhibits the most dramatic variations at Venus' cloud top, both spatially and temporally (Esposito 1984; Esposito et al. 1988; Marcq et al. 2013; Vandaele et al. 2017a,b), spanning more than two orders of magnitude on timescales ranging from a few days up to several decades. Esposito (1984) suggested that the observed episodic variations in mesospheric SO_2 might be caused by episodic volcanic activity. The SO_2 injections into the upper atmosphere need not be volcanic SO_2: instead, the thermal energy and dynamical effects of a volcanic eruption could lead to tropospheric (SO_2-rich) air being transported into the mesosphere. However, episodic injections of SO_2-rich tropospheric air into the mesosphere might also be associated with dynamical cycles of the atmosphere, rather than due to volcanism (as argued for example by Marcq et al. 2013). To address this question, EnVision will measure supposed volcanic species both below the clouds (using near-IR nightside spectroscopy as discussed above), and above the clouds (using near-IR and UV spectroscopy on the dayside). If correlations are found between mesospheric and tropospheric gas abundances (or indeed with surface indicators of volcanic activity), that would provide valuable information on how volcanic volatiles are transported through the atmosphere.

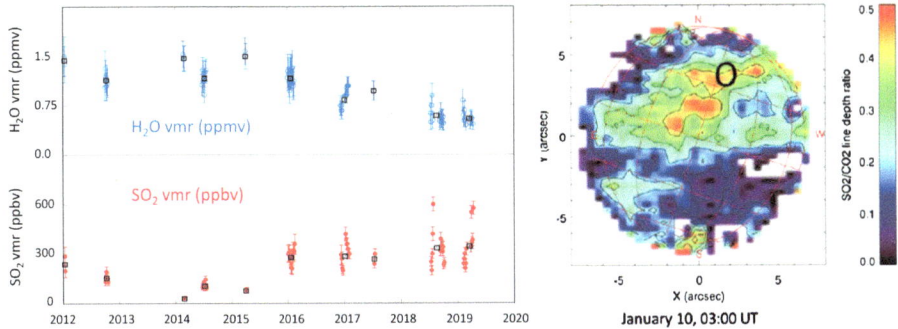

Fig. 7 Left: Long-term variations of the H_2O volume mixing ratio (top, blue points) and the SO_2 volume mixing ratio (bottom, red points), measured at the cloud top from the TEXES data at 7.4 μm. From Encrenaz et al. (2020a, their Fig. 11); Right hand side panel: SO_2:CO_2 line ratio map obtained by TEXES at 7.4 μm by Encrenaz et al. (2012), showing the large instantaneous spatial heterogeneity of SO_2 near cloud top.

Anti-correlated variations of H_2O (or its proxy HDO) in the same altitude range also exist, albeit less spectacular and restricted within one order of magnitude (Encrenaz et al. 2020a) as can be seen in Fig. 7. The greater range of SO_2 variations compared to H_2O is currently explained by the fast photochemical destruction of SO_2 by UV sunlight at cloud tops, making this species a much more sensitive tracer of the atmospheric circulation and vertical mixing between its lower atmospheric source (below the clouds, SO_2 is more abundant by 3-4 orders of magnitude, e.g., Oyama et al. 1980; Gel'man et al. 1980; Bertaux et al. 1996; Pollack et al. 1993; Arney et al. 2014) and its cloud top photochemical sink.

The variations of the vertical mixing are themselves poorly understood and most likely depend on the interplay between atmospheric waves breaking, solar radiative forcing, thus varying with latitude, local solar time and underlying topography. For example, the thickness of the convective layer in the lower cloud layer varies with latitude and local solar time (Tellmann et al. 2009; Imamura et al. 2017), and convective activity near the subsolar point at cloud top level is suggested by observations (Titov et al. 2012) and modeling (Lefèvre et al. 2018). However, the statistical distribution of observed SO_2 "plumes" by TEXES exhibits a minimum near the subsolar point between 10:00 and 14:00 LST (Encrenaz et al. 2019), as does SPICAV-UV (Marcq et al. 2020). The other main sulfur oxide species, namely SO, could be measured using high resolution UV spectroscopy from HST/STIS (Jessup et al. 2015) and appears to be correlated to SO_2 with an average ratio SO:SO_2 of about 10%.

Interestingly, the observed SO_2/H_2O anticorrelation at cloud tops (Fig. 7) was predicted by chemical modelers (Parkinson et al. 2015; Shao et al. 2020), on the grounds that both species can be the limiting progenitor species of the cloud droplets made of mostly sulfuric acid. Unfortunately, no chemical model is currently able to quantitatively reproduce both H_2O and SO_2 profiles below and above the clouds simultaneously (Bierson and Zhang 2020) – most models focus on either below or above the clouds and tune their lower/upper boundary conditions accordingly. This points to unknown processes operating in the cloud region, involving an extra sulfur reservoir beyond SO_2 and the sulfuric acid in the droplets. This unknown reservoir may be linked to the unknown UV absorber (which would be a sulfur-based compound in this hypothesis). The most recent hypothesis (Rimmer et al. 2021) postulates a sulfate salt reservoir in order to solve this long standing issue, and will be further discussed in Sect. 6.

Other possible gaseous species potentially linked with volcanism are harder to monitor above the clouds, since their relatively lower abundance makes them only detectable using

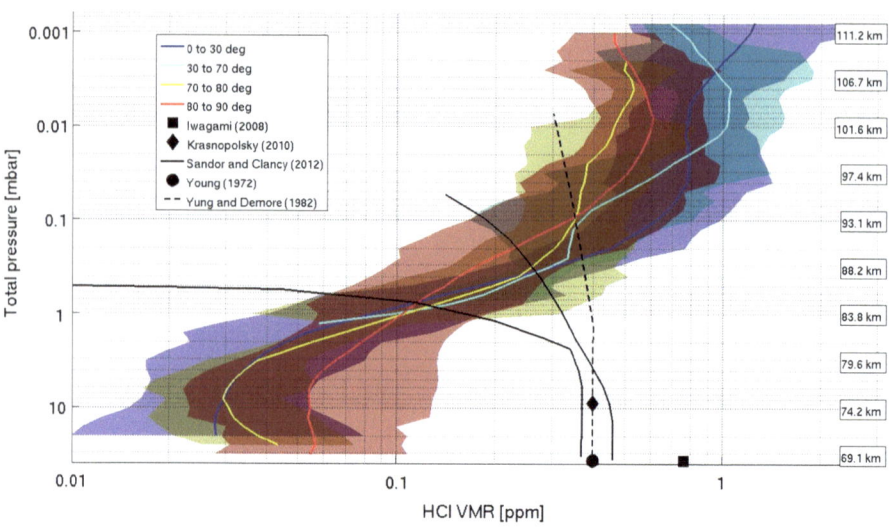

Fig. 8 Mean value of the HCl profile obtained by SOIR for different latitude bins as a function of total pressure. The altitude is also shown in correlation with the pressure scale. The weighted standard deviations are the colored shaded envelopes. The profiles are compared to literature data (models + other observations), hinting at a discrepancy in the vertical gradient. From Mahieux et al. (2015, their Fig. 5).

the very sensitive solar/stellar occultation technique. This technique strongly constrains the vertical profiles of such species, but the drawback is a poor horizontal and temporal coverage compared to nadir measurements (and, in the case of solar occultations, restricted to 6AM/6PM local solar time except at the very highest latitudes). Nevertheless, the SPICAV/SOIR instrument onboard Venus Express was able to record the first vertical profiles of HCl and HF gases above the clouds in the infrared domain (Mahieux et al. 2015), which showed a great spatial and temporal variability as well as yet poorly understood positive vertical gradients — not seen by other observers, e.g. Sato and Sagawa (2023) — hinting at yet unknown sources/sinks (Fig. 8). In the near future as of 2023, using the same observation technique, the VIRAL instrument onboard ISRO's planned Venus Orbiter Mission Shukrayaan-1 will also be able to monitor the vertical profiles of the above mentioned species (hydrogen halides, hydrogen isotopes, sulfur oxides) as well as yet unmeasured other species involved in the sulfur chemistry, including H_2S and OCS at a ppb accuracy level (Patrakeev et al. 2022; Widemann et al. 2023). The expected improvement in measurement sensitivity for mesospheric species from upcoming missions is listed in Table 2.

Carbon monoxide may be a major constituent gas in volcanic gas emission, due to the high Venus surface pressures (Fig. 3). It is readily observable in the mesosphere, both in nadir observations (e.g. Irwin et al. 2008) and in solar occultation (Vandaele et al. 2016). However, CO is not usually considered as related to volcanism above the clouds, since its main source in the upper atmosphere comes from the photolysis of CO_2 (Marcq et al. 2018; Yung and Demore 1982).

In conclusion, variability in many mesospheric trace gas species has already been observed, but it has not yet been possible to attribute any of this variability to volcanism. The EnVision mission will seek to address this by continuing mesospheric and tropospheric composition measurements while simultaneously searching for ongoing volcanic activity on the surface.

Table 2 Expected measurement accuracy for the volume mixing ratios (VMR) of mesospheric trace species with the instrument VIRAL onboard ISRO's Venus Orbiter Mission Shukrayaan-1 (Patrakeev et al. 2022; Widemann et al. 2023) and the instruments VenSpec-U and -H onboard ESA's EnVision spacecraft.

Species	Altitude range (km)	Current VMR measurement accuracy (or upper limit)	Expected VMR measurement accuracy (or upper limit)
H_2O	70-110	~1 ppm (Fedorova et al. 2008)	≤0.3 ppm (VIRAL)
	70	~2 ppm (Cottini et al. 2012)	≤0.2 ppm (VenSpec-H)
HDO	70-95	~0.1 ppm (Fedorova et al. 2008)	≤30 ppb (VIRAL)
	70		≤20 ppb (VenSpec-H)
HCl	70-110	0.1-1 ppm (Mahieux et al. 2015)	≤30 ppb (VIRAL)
		30 ppb (Krasnopolsky 2010)	
	70	0.379 ± 0.013 ppm (Sato and Sagawa 2023)	
HF	70-100	5-50 ppb (Mahieux et al. 2015)	≤1 ppb (VIRAL)
	70	0.3 ppb (Krasnopolsky 2010)	≤1 ppb (VenSpec-H)
SO_2	70-80	50-500 ppb (Belyaev et al. 2008; Mahieux et al. 2015)	≤30 ppb (VIRAL)
		10-1000 ppb (Marcq et al. 2013, 2021)	≤30 ppb (VenSpec-H)
	70		≤5 ppb (VenSpec-U)
SO	70	~1 ppb (Jessup et al. 2015)	≤0.5 ppb (VenSpec-U)
H_2S	~70	<23 ppb (Krasnopolsky 2008)	≤3 ppb (VIRAL)
OCS	~70	~few ppb (Mahieux et al. 2023)	≤0.3 ppb (VIRAL)
			≤1 ppb (VenSpec-H)

6 Clouds and Particulates

We previously have considered gaseous species whose detection, whether by *in situ* or remote sensing, could provide evidence for volcanism. Now we turn instead to non-gaseous species: first we consider whether mapping from orbit could reveal plumes of sulphuric acid cloud droplets formed due to volcanically emitted gases and/or plumes of volcanic ash particles; then we consider whether measurements of cloud droplet/particulate composition from an *in situ* cloud exploration platform could reveal about volcanism.

A large volcanic emission of H_2O or SO_2 could lead to increased formation cloud droplets, and therefore increased optical thickness and lower radiances in nightside observations of near-IR thermal window regions. To a first approximation, an upper bound on this effect can be reached by assuming that all the volcanic volatile species emitted at the volcano are transported to the clouds and are all turned into cloud particulate matter. For example, if we assume the same Bárðarbunga-Holuhraun 2014 eruption as discussed above, in which about 28,000 kg/s of water are released (Schmidt et al. 2015), this could lead to up to 28,000 kg/s of water droplets being produced (or up to 152,000 kg/s of sulfuric acid cloud droplets being produced if there is sufficiently abundant SO_3 to combine with. If this were deposited in a layer at 50 km altitude at the base of the clouds in a plume 100 km wide, and taking into account the mean wind velocity of 60 m/s at this altitude (from VIRA), we can calculate that the additional cloud mass in a column would be in the range 5-30 g/m^2. This is an order of magnitude less than the vertically integrated columnar mass of cloud droplets, of 150 g/m^2 (as reported in Ragent et al. 1985, Table 3-4), and significantly less than the factor of $2\times$ over which the lower cloud optical depth is observed to vary (Titov et al. 2018). This suggests that the direct formation of aerosols from condensing volcanic volatiles could only be detectable for larger ($> 100,000$ km^2) eruptions. We emphasize that this is a very simplistic order-of-magnitude calculation, neglecting dynamical and chemical interactions.

Explosive eruptions may inject not only gases into the atmosphere, but often also ash particulates; for example, the 2010 Eyjafjallajökull eruption is estimated to have ejected at least 7×10^{10} kg of fine ash aerosol (characteristic radius < 28 μm) into the atmosphere (Gudmundsson et al. 2012). Peak emission rates of fine ash in that eruption exceeded 40,000 kg/s (Flanner et al. 2014), so of a similar order of magnitude to the volatile release scenario considered above. Volcanic ash aerosols on Earth may have a single scattering albedo at 2 μm of around 0.9 (Mortier et al. 2013), which is much less than the > 0.99 single scattering albedo typical of the Venus cloud particles, so volcanic ash would be expected to have a much greater extinction effect per columnar unit of mass than sulfuric acid aerosol. On the other hand, the residence time of ash particles in the clouds may be much shorter, depending on their interaction with cloud microphysical cycles. We note that remote sounding in nightside IR channels would not be able to unambiguously distinguish between the spectral signatures of ash or cloud droplets in lower cloud, because of the limited number of spectral windows in which observations can be made and the degeneracy between effects of particle size and composition (Wilson et al. 2008). If volcanically lofted particulates reach the clouds, they can provide nucleation sites for condensation of sulfuric acid, which in turn may dissolve silicate material of volcanic ashes, and therefore could change the gas-phase (Rimmer et al. 2021) and solid/liquid phase (Zolotov et al. 2023) composition in the cloud layer. However, mechanisms of delivery and fluxes of volcanic particulates to the clouds remain unknown.

In situ measurements of particulate composition at cloud level offer a much more capable method of searching for evidence of volcanism. The most sophisticated *in situ* measurement of cloud composition was made by Venera 12 (Petryanov et al. 1981) and then by the Vega 1 and 2 descent probes: they analyzed the composition of collected cloud & aerosol matter using mass spectrometry, gas chromatography and X-ray fluorescence, complementing measurements of particle size and refractive index from nephelometry. The original data from this impressive suite of instrumentation were not archived. The original analyses, published in Russian-language journals, are summarized by Krasnopolsky (1989) and reviewed by Titov et al. (2018). Of particular note, is a tentative measurement by the X-ray fluorescence spectrometers of significant abundances of phosphorus in the lower cloud particles. The molecular form of this phosphorus is not determined by the instrument; Krasnopolsky (1989) proposed it to be in the form of phosphoric acid. Results of X-ray fluorescence and mass spectrometers (Venera 12, Vega 1 and 2) also suggested chlorine in the cloud particles. Iron in cloud aerosols reported based on Venera 12 X-ray fluorescence spectrometry could be in Fe(III)-bearing salts (Krasnopolsky 1985, 1989, 2017; Zolotov et al. 2023). No evidence of any volcanic ash or other surface materials was reported from entry probe mass spectrometers, but this is not conclusive because such materials would dissolve in concentrated sulfuric acid. Minerals that survived the dissolution likely not have been volatilized when the captured aerosol samples were heated to 400 °C for mass spectrometric analysis. A recent reanalysis of Pioneer Venus Large Probe mass spectrometer data suggests thermal decomposition of trapped cloud particles below cloud deck to the surface, and some captured particles could have formed through chemical alteration of surface-delivered grains in the clouds (Zolotov et al. 2023).

For future cloud-level platforms, an aerosol mass spectrometer (AMS) has been proposed which could analyze the composition of both atmospheric gases and cloud/aerosol particulates (Baines et al. 2021). This would be deployed on a balloon-borne gondola which, during a mission duration of two months, be carried by Venus' super-rotating winds to complete at least ten complete circumnavigations of the planet, mapping cloud composition as a function of local time, latitude. Baines et al. (2021) propose to deploy such an instrument on an altitude-controlled balloon which would explore altitudes ranging from 52 km (in the lower convective cloud) up to 62 km (in the convectively stable upper cloud). Just as was described in the proposed DAVINCI investigation (see Sect. 3.1), a Venus aerobot investigation of cloud-level particulate and gas composition would search for species which are out of chemical equilibrium with the environment; it would search for "volcanosignature" gases and also seek to characterize the liquid/solid phases with which they are reacting (Cutts et al. 2021; Byrne et al. 2022). In particular, a cloud-level balloon platform with a lifetime of many weeks offers the opportunity to characterize the background environmental and chemical conditions; marked changes to this baseline composition (such as increases in sulfate aerosols) could be indicative of volcanic activity (see also Widemann et al. 2023, this collection).

Such an instrument could also directly search for volcanic ash aerosol, whether this is directly lofted to altitude by a volcanic explosion or remobilized from the surface by wind. Distinguishing current 'fresh' volcanic and wind-delivered silicates dust should be possible because ancient volcanic materials could be chemically altered through gas-solid type reactions at the surface (Zolotov 2018). Chemical analysis of dust particles could be performed through the X-ray fluorescence method and/or mass spectrometry with laser ablation, for example. Phase composition could be performed via X-ray diffraction method. Even if this were not possible, detection of volcanic ash particles through nephelometry is another viable possibility; volcanic ash particles are typically highly fractal and therefore exhibit very different phase functions compared to liquid (and thus spherical) cloud droplets. Note that glassy silicate ash particles could dissolve in $\sim$75% sulfuric acid in hours to days; in that case, any detection of such particles would imply currently ongoing volcanic ash production. Chemical analysis of liquid aerosols could inform initial composition of dissolved particulates. In particular, relative contributions of dissolved cosmic dust (Zolotov et al. 2023) and surface-sourced volcanic materials in liquid cloud aerosols could be assessed.

The discussion here about measurements of particulate composition to be made from a balloon platform would also apply to a descent probe, although the target species would be somewhat different. Descent probes have reported the existence of hazes below the cloud base (Titov et al. 2018), where temperatures are well above the boiling point of sulfuric acid cloud droplets. These hazes, therefore, cannot be composed of sulfuric acid. Similarly, Grieger et al. (2004) reported that Venera descent probe photometry suggested a discrete layer of particulates at 2 km above the mean planetary radius, whose nature is not known and might also be related to volcanic pyroclastic processes (see Carter et al. 2023, this collection).

7 Summary and Conclusions

7.1 Further Needed Observational Constraints

Since the exact composition of the possible outgassing depends on the poorly constrained physical and chemical conditions of the magmatic phase, we should refrain from any premature assumptions and look for any possible sign of assessed deviation relative to the baseline abundances of minor species. Time-variability of some out-of- background gas abundances could provide evidence of volcanic degassing, even if it is not definitive.

Remote detection of present-day volcanic outgassing should therefore rely on assessed imbalance over as many simultaneous measurements of detectable trace gases species as possible. In the troposphere, high spectral resolution spectroscopy of the night side thermal emission windows in the CO_2 spectrum would provide accurate measurements averaged over a horizontal extent of $\sim$30 km due to multiple scattering in the overlying clouds. HCl, HF, CO, OCS, SO_2, H_2O and HDO can be constrained this way below the clouds despite the very high pressure (and even at multiple altitude levels for H_2O and HDO, with the additional insight provided by the D/H isotopic ratio variations). Incidentally, this same night side thermal infrared spectroscopy is also sensitive to surface mineralogy (through emissivity measurements), and can constrain chemical weathering of surface materials via gas-solid type reactions. Above the clouds, in the mesosphere, evidence for volcanic outgassing would become indirect, and rely on the enhanced sensitivity of some minor species caused by the competition between photochemistry and convective variability (whose origin may be related to ongoing volcanic activity if any). Diagnostic species in this altitude range are therefore photochemically active short-lived species, first and foremost SO_2 and SO, then H_2O and HDO, as well as CO or hydrogen halides (and their deuterated isotopes) to a lesser extent. These species have already been detected or even monitored, either through UV spectroscopy or near infrared spectroscopy over Venus' dayside. These measurements would also benefit from simultaneous observations in the sub-millimeter range of their vertical profiles in the upper mesosphere, providing upper boundary conditions.

Compared to remote sensing observations, *in situ* measurements will be of paramount interest to constrain atmospheric layers that cannot be observed otherwise. Gas chromatographs, mass spectrometers and tunable laser spectrometers on descent probes or more long-lived aerial platforms within the clouds would probe, among other species, volcanogenic volatiles such as sulfur-bearing and halogen-bearing species. Descent probes reaching the surface would be able to measure vertical gradients in the first atmospheric scale height above the ground and could indicate active sources or sinks. On the other hand, dedicated payload on board aerial platforms such as nephelometers will better characterize aerosols and constrain their compositions, including possibly volcanogenic particulate matter (e.g. ashes). Although it is beyond the scope of this paper to discuss it in more detail, we note that infrasound sensors mounted on cloud-level balloons could allow detection of pressure waves from explosive volcanic eruptions, providing a direct measurement on the rate and style of volcanic activity (Rossi et al. 2023). Finally, one of the main objectives of surface landers will be characterization of the surface composition and mineralogy through various means (X-ray fluorescence, X-ray diffraction, Mossbauer spectroscopy, infrared spectroscopy, etc.), which would in return bring constraints on magma composition and/or surface-atmosphere reactions discussed in Sect. 2.

7.2 Further Needed Modeling and Experimental Efforts

Improvements in existing numerical models are already being considered to support future observations described above. Chemical models, including heterogeneous chemistry between the cloud droplets and gases (e.g. Rimmer et al. 2021), are necessary to elucidate the unknown sulfur reservoir issue (Marcq et al. 2018). Coupling these chemical models with dynamical models is also an undergoing effort as of 2023, with projects ranging from general circulation models coupling to mesoscale models (Lefèvre 2022), particularly suited to investigate buoyancy- and dynamically-driven plume dispersal reviewed in this paper's

Sect. 4, along with disequilibrium chemical reactions between the background atmosphere and the volcanogenic volatiles brought by the plume.

Modeling efforts also include laboratory simulations. Some experimental setups are already operational to study alteration of surface minerals in Venusian atmospheric conditions (e.g., Santos et al. 2023), in order to prepare for orbital infrared spectroscopic measurements (Morlok et al. 2019; Berger et al. 2019; Helbert et al. 2021; Treiman et al. 2021; see also the review in Widemann et al. 2023). Another kind of useful simulation would be laboratory analogs of the cloud droplets and surrounding atmosphere (facilitated by the fact that temperature and pressure conditions in the clouds are very close to Earth's) to better assess poorly constrained heterogeneous reactions in particular.

Acknowledgements TW and EM acknowledge France's Centre National d'Études Spatiales (CNES) and ESA for funding support of Venus related studies.

CG acknowledges the support of Rice University and the CLEVER planets group (itself supported by NASA and part of NExSS). This work has been carried out within the framework of the NCCR PlanetS supported by the Swiss National Science Foundation under grants 51NF40_182901 and 51NF40_205606.

ML acknowledges funding from the European Research Council (ERC) under the European Union's Horizon 2020 research and innovation program (grant agreement No. 740963/EXOCONDENSE). ML was granted access to the High-Performance Computing (HPC) resources of Centre Informatique National de l'Enseignement Supérieur (CINES) under the allocations n°A0060110391 and n°A0080110391 made by Grand Équipement National de Calcul Intensif (GENCI). ML acknowledges that this project has received funding from the European Union's Horizon Europe research and innovation program under the Marie Skłodowska-Curie grant agreement 101110489/MuSICA-V.

SR acknowledges funding by the Belgian Science Policy Office (BELSPO) through the FED-tWIN program (Prf-2019-077-RT-MOLEXO) and through financial and contractual support coordinated by the ESA Prodex Office (PEA 4000137943, 4000128137).

MZ was supported by NASA grant 80NSSC19K0787 and by the DAVINCI mission.

Finally, the authors thank the International Space Institute (ISSI) in Berne, Switzerland, for supporting the "Venus: Evolution through Time" workshop and the subsequent book, of which this paper forms a chapter (http://dx.doi.org/10.5281/zenodo.22558).

Funding Open Access funding enabled and organized by Projekt DEAL.

Declarations

Competing Interests CW, TW and CG are Guest Editors of the collection "Venus: Evolution Through Time", but have not been involved in the peer-review process of this article.

References

Airey MW, Mather TA, Pyle DM et al (2015) Explosive volcanic activity on Venus: the roles of volatile contribution, degassing, and external environment. Planet Space Sci 113:33. https://doi.org/10.1016/j.pss.2015.01.009

Armann M, Tackley PJ (2012) Simulating the thermochemical magmatic and tectonic evolution of Venus's mantle and lithosphere: Two-dimensional models. J Geophys Res 117. https://doi.org/10.1029/2012JE004231

Arney G, Meadows V, Crisp D et al (2014) Spatially resolved measurements of H_2O, HCl, CO, OCS, SO_2, cloud opacity, and acid concentration in the Venus near-infrared spectral windows. J Geophys Res, Planets 119:1860. https://doi.org/10.1002/2014JE004662

Avice G, Parai R, Jacobson S et al (2022) Noble gases and stable isotopes track the origin and early evolution of the Venus atmosphere. Space Sci Rev 218:60. https://doi.org/10.1007/s11214-022-00929-9

Baines KH, Nikolić D, Cutts JA et al (2021) Investigation of Venus cloud aerosol and gas composition including potential biogenic materials via an aerosol-sampling instrument package. Astrobiology 21:1316. https://doi.org/10.1089/ast.2021.0001

Bains W, Petkowski JJ, Seager S et al (2021) Phosphine on Venus cannot be explained by conventional processes. Astrobiology 21:1277. https://doi.org/10.1089/ast.2020.2352

Bains W, Shorttle O, Ranjan S et al (2022) Only extraordinary volcanism can explain the presence of parts per billion phosphine on Venus. Proc Natl Acad Sci 119:e2121702119. https://doi.org/10.1073/pnas.2121702119

Baker RD, Schubert G, Jones PW (1998) Cloud-level penetrative compressible convection in the Venus atmosphere. J Atmos Sci 55:1. https://doi.org/10.1175/1520-0469(1998)055<0003:CLPCCI>2.0.CO;2

Barstow JK, Tsang CCC, Wilson CF, Irwin PGJ, Taylor FW, McGouldrick K, Drossart P, Piccioni G, Tellmann S (2012) Models of the global cloud structure on Venus derived from Venus Express observations. Icarus 217(2):542–560. https://doi.org/10.1016/j.icarus.2011.05.018

Basilevsky AT, Head JW, Schaber GG, Strom RG (1997) The resurfacing history of Venus. In: Venus II: geology, geophysics, atmosphere, and solar wind environment, p 1047. https://doi.org/10.2307/j.ctv27tct5m.35

Basilevsky AT, Head JW, Abdrakhimov AM (2004) Impact crater air fall deposits on the surface of Venus: areal distribution, estimated thickness, recognition in surface panoramas, and implications for provenance of sampled surface materials. J Geophys Res, Planets 109:E12003. https://doi.org/10.1029/2004JE002307

Basilevsky AT, Shalygin EV, Bondarenko NV et al (2019) Venus crater-related radar-dark parabolas and neighboring terrains: a comparison of 1-µm emissivity and microwave properties. Icarus 330:103. https://doi.org/10.1016/j.icarus.2019.01.009

Belyaev D, Korablev D, Fedorova D et al (2008) First observations of SO_2 above Venus' clouds by means of Solar Occultation in the Infrared. J Geophys Res Planets 113:E00B25. https://doi.org/10.1029/2008JE003143

Berger G, Cathala A, Fabre S et al (2019) Experimental exploration of volcanic rocks-atmosphere interaction under Venus surface conditions. Icarus 329:8. https://doi.org/10.1016/j.icarus.2019.03.033

Bertaux J-L, Widemann T, Hauchecorne A et al (1996) VEGA 1 and VEGA 2 entry probes: an investigation of local UV absorption (220-400 nm) in the atmosphere of Venus (SO_2, aerosols, cloud structure). J Geophys Res 101:12709. https://doi.org/10.1029/96JE00466

Bézard B, de Bergh C (2007) Composition of the atmosphere of Venus below the clouds. J Geophys Res, Planets 112:E04S07. https://doi.org/10.1029/2006JE002794

Bézard et al (2011) The 1.10- and 1.18-μm nightside windows of Venus observed by SPICAV-IR aboard Venus Express. https://doi.org/10.1016/j.icarus.2011.08.025

Bierson CJ, Zhang X (2020) Chemical cycling in the venusian atmosphere: a full photochemical model from the surface to 110 km. J Geophys Res, Planets 125:e06159. https://doi.org/10.1029/2019JE006159

Bindschadler DL, Decharon A, Beratan KK et al (1992) Magellan observations of Alpha Regio: implications for formation of complex ridged terrains on Venus. J Geophys Res 97:13563. https://doi.org/10.1029/92JE01332

Bjonnes EE, Hansen VL, James B, Swenson JB (2012) Equilibrium resurfacing of Venus: results from new Monte Carlo modeling and implications for Venus surface histories. Icarus 217:451. https://doi.org/10.1016/j.icarus.2011.03.033

Bondarenko NV, Head JW, Ivanov MA (2010) Present-day volcanism on Venus: evidence from microwave radiometry. Geophys Res Lett 37:L23202. https://doi.org/10.1029/2010GL045233

Bottke WF, Vokrouhlicky D, Ghent B et al (2016) On asteroid impacts, crater scaling laws, and a proposed younger surface age for venus. LPI 2036

Bullock MA, Grinspoon DH (2001) The recent evolution of climate on Venus. Icarus 150:19. https://doi.org/10.1006/icar.2000.6570

Byrne PK, Krishnamoorthy S (2022) Estimates on the frequency of volcanic eruptions on Venus. J Geophys Res, Planets 127:e2021JE007040. https://doi.org/10.1029/2021JE007040

Byrne PK et al (2022) Phantom: an aerobot mission to the skies of Venus AGU fall meeting

Campbell BA, Morgan GA, Whitten JL et al (2017) Pyroclastic flow deposits on Venus as indicators of renewed magmatic activity. J Geophys Res, Planets 122:1580. https://doi.org/10.1002/2017JE005299

Carter LM, Gilmore MS, Ghail RC et al (2023) Sedimentary processes on Venus. Space Sci Rev 219:85. https://doi.org/10.1007/s11214-023-01033-2

  Springer

Chamberlain S, Bailey J, Crisp D, Meadows V (2013) Ground-based near-infrared observations of water vapour in the Venus troposphere. Icarus 222:364. https://doi.org/10.1016/j.icarus.2012.11.014

Coltice N, Marty B, Yokochi R (2009) Xenon isotope constraints on the thermal evolution of the early Earth. Chem Geol 266:4. https://doi.org/10.1016/j.chemgeo.2009.04.017

Cordier D, Bonhommeau DA, Port S et al (2019) The physical origin of the Venus low atmosphere chemical gradient. Astrophys J 880:82. https://doi.org/10.3847/1538-4357/ab27bd

Cottini V et al (2012) Water vapor near the cloud tops of Venus from Venus Express/VIRTIS dayside data. Icarus 217:561. https://doi.org/10.1016/j.icarus.2011.06.018

Cutts J et al Scientific Exploration of Venus with Aerial Platforms Bull Am Astron Soc 53(4) (2021). https://doi.org/10.3847/25c2cfeb.29dd4fbb

De Bergh C, Bezard B, Owen T et al (1991) Deuterium on Venus: observations from. Earth Sci 251:547. https://doi.org/10.1126/science.251.4993.547

D'Incecco P, Mueller NT, Helbert J, D'Amore M et al (2017) Idunn Mons on Venus: Location and extent of recently active lava flows. Planet Space Sci 136:25. https://doi.org/10.1016/j.pss.2016.08.004

D'Incecco P, Filiberto J, Lopez I, Gorinov DA, Komatsu G, Martynov A, Pisarenko P (2021) The young volcanic rises on Venus: a key scientific target for future orbital and in-situ measurements on Venus. Sol Syst Res 55:315. https://doi.org/10.1134/S0038094621040031

Donahue TM, Grinspoon DH, Hartle RE, Hodges RR (1997) Ion/neutral escape of hydrogen and deuterium: evolution of water. In: Bougher SW, Hunten DM, Philips RJ (eds) Venus II: geology, geophysics, atmosphere and solar wind environment. University of Arizona Press, Tucson, pp 385–414

Encrenaz T, Greathouse TK, Roe H et al (2012) HDO and SO_2 thermal mapping on Venus: evidence for strong SO_2 variability. Astron Astrophys 543:A153. https://doi.org/10.1051/0004-6361/201219419

Encrenaz T, Greathouse TK, Marcq E et al (2019) HDO and SO_2 thermal mapping on Venus. IV. Statistical analysis of the SO_2 plumes. Astron Astrophys 623:A70. https://doi.org/10.1051/0004-6361/201833511

Encrenaz T, Greathouse TK, Marcq E et al (2020a) HDO and SO_2 thermal mapping on Venus. V. Evidence for a long-term anti-correlation. Astron Astrophys 639:A69. https://doi.org/10.1051/0004-6361/202037741

Encrenaz T, Greathouse TK, Marcq E et al (2020b) A stringent upper limit of the PH_2 abundance at the cloud top of Venus. Astron Astrophys 643:L4. https://doi.org/10.1051/0004-6361/202039559

ESA (2024). EnVision Definition Study Report ('Red Book'), Document ESA-SCI-DIR-RP-003 v. 1.6 released Nov 2023. Publically available at https://www.cosmos.esa.int/web/envision/links

Esposito LW (1984) Sulfur dioxide: episodic injection shows evidence for active Venus volcanism. Science 223:1072. https://doi.org/10.1126/science.223.4640.1072

Esposito LW, Copley M, Eckert R et al (1988) Sulfur dioxide at the Venus cloud tops, 1978-1986. J Geophys Res 93:5267. https://doi.org/10.1029/JD093iD05p05267

Esposito LW, Bertaux J-L, Krasnopolsky V, Moroz VI, Zasova LV (1997) Chemistry of lower atmosphere and clouds. In: Bougher SW, Hunten DM, Phillips RJ (eds) Venus II: geology, geophysics, atmosphere, and solar wind environment. University of Arizona Press, Tucson, pp 415–458

Evans LG, Peplowski PN, McCubbin FM et al (2015) Chlorine on the surface of Mercury: MESSENGER gamma-ray measurements and implications for the planet's formation and evolution. Icarus 257:417. https://doi.org/10.1016/j.icarus.2015.04.039

Fedorov A, Barabash S, Sauvaud J-A et al (2011) Measurements of the ion escape rates from Venus for solar minimum. J Geophys Res Space Phys 116:A07220. https://doi.org/10.1029/2011JA016427

Fedorova A et al (2008) HDO and H2O vertical distributions and isotopic ratio in the Venus mesosphere by Solar Occultation at Infrared spectrometer on board Venus express. J Geophys Res 113. https://doi.org/10.1029/2008JE003146

Fedorova A, Bézard B, Bertaux J-L et al (2015) The CO_2 continuum absorption in the 1.10- and 1.18-μm windows on Venus from Maxwell Montes transits by SPICAV IR onboard Venus express. Planet Space Sci 113:66. https://doi.org/10.1016/j.pss.2014.08.010

Fegley B, Prinn RG (1989) Estimation of the rate of volcanism on Venus from reaction rate measurements. Nature 337:55. https://doi.org/10.1038/337055a0

Fegley B, Zolotov MY (2000) Chemistry of sodium, potassium, and chlorine in Volcanic gases on Io. Icarus 148:193. https://doi.org/10.1006/icar.2000.6490

Fegley B, Treiman AH, Sharpton VL (1992) Venus surface mineralogy: observational and theoretical constraints. Proc LPSC 22:3

Fegley B, Klingelhöfer G, Lodders K, Widemann T (1997a) Geochemistry of surface-atmosphere interactions on Venus. In: Venus II: geology. Geophysics, atmosphere, and solar wind environment. University of Arizona Press, Tucson, p 591

Fegley B, Zolotov MY, Lodders K (1997b) The oxidation state of the lower atmosphere and surface of Venus. Icarus 125:416. https://doi.org/10.1006/icar.1996.5628

Filiberto J, Trang D, Treiman AH, Gilmore MS (2020) Present-day volcanism on Venus as evidenced from weathering rates of olivine. Sci Adv 6:eaax7445. https://doi.org/10.1126/sciadv.aax7445

Flanner MG, Gardner AS, Eckhardt S et al (2014) Aerosol radiative forcing from the 2010 Eyjafjallajökull volcanic eruptions. J Geophys Res, Atmos 119:9481. https://doi.org/10.1002/2014JD021977

Florensky CP, Bazilevakii AT, Kriuchkov VP et al (1983) Venera 13 and Venera 14: sedimentary rocks on Venus? Science 221:57. https://doi.org/10.1126/science.221.4605.57

Frost DJ, McCammon CA (2008) The redox state of Earth's mantle. Annu Rev Earth Planet Sci 36:389–420. https://doi.org/10.1146/annurev.earth.36.031207.124322

Gaillard F, Scaillet B (2014) A theoretical framework for volcanic degassing chemistry in a comparative planetology perspective and implications for planetary atmospheres. Earth Planet Sci Lett 403:307. https://doi.org/10.1016/j.epsl.2014.07.009

Gaillard F, Bouhifd MA, Füri E et al (2021) The diverse planetary ingassing/outgassing paths produced over billions of years of magmatic activity. Space Sci Rev 217:22. https://doi.org/10.1007/s11214-021-00802-1

Ganesh I, McGuire LA, Carter LM (2021) Modeling the dynamics of dense pyroclastic flows on Venus: insights into pyroclastic eruptions. J Geophys Res, Planets 126:e2021JE006943. https://doi.org/10.1029/2021JE006943

Ganesh I, Carter LM, Henz TN (2022) Radar backscatter and emissivity models of proposed pyroclastic density current deposits on Venus. J Geophys Res, Planets 127:e2022JE007318. https://doi.org/10.1029/2022JE00731

Gao P et al (2014) Bimodal distribution of sulfuric acid aerosols in the upper haze of Venus. Icarus 231:83. https://doi.org/10.1016/j.icarus.2013.10.013

Garvin JB, Head JW, Zuber MT, Helfenstein P (1984) Venus: the nature of the surface from Venera panoramas. J Geophys Res 89:3381. https://doi.org/10.1029/JB089iB05p03381

Garvin JB, Getty SA, Arney GN et al (2022) Revealing the mysteries of Venus: the DAVINCI mission. Planet Sci J 3:117. https://doi.org/10.3847/PSJ/ac63c2

Gel'man B, Zolotukhin V, Lamonov N, Levchuk B, Lipatov A, Mukhin L, Nenarokov D, Rotin V, Okhotnikov B (1980) An analysis of the chemical composition of the atmosphere of Venus on an AMS of the Venera-12 using a gas chromatograph. Cosmic Res 17(5):585–589

Gerlach TM (1980) Evaluation of volcanic gas analyses from Kilauea volcano. J Volcanol Geotherm Res 7:295. https://doi.org/10.1016/0377-0273(80)90034-7

Ghail RC, Wilson L (2015) A pyroclastic flow deposit on Venus. Geol Soc (Lond) Spec Publ 401:97. https://doi.org/10.1144/SP401.1

Ghail RG, Smrekar SE, Byrne PK, Gülcher AJP, O'Rourke JG, Borrelli ME, Gilmore MS, Herrick RR, Ivanov MA, Plesa A-C, Rolf T, Sabbeth L, Schools JW, Shellnut JG, Widemann T (2024) Volcanic and tectonic constraints on the evolution of Venus. Space Sci Rev 220

Gillmann C, Tackley P (2014) Atmosphere/mantle coupling and feedbacks on Venus. J Geophys Res, Planets 119(6):1189–1217

Gillmann C, Way MJ, Avice G et al (2022) The long-term evolution of the atmosphere of Venus: processes and feedback mechanisms. Space Sci Rev 218:56. https://doi.org/10.1007/s11214-022-00924-0

Gilmore MS, Darby Dyar M, Mueller N et al (2023) Mineralogy of the Venus Surface. Space Sci Rev 219:52. https://doi.org/10.1007/s11214-023-00988-6

Glaze LS (1999) Transport of SO_2 by explosive volcanism on Venus. J Geophys Res 104:18899. https://doi.org/10.1029/1998JE000619

Glaze LS, Baloga SM, Wimert J (2011) Explosive volcanic eruptions from linear vents on Earth, Venus, and Mars: comparisons with circular vent eruptions. J Geophys Res, Planets 116:E01011. https://doi.org/10.1029/2010JE003577

Graham DW, Connelly DP, German CR et al (2002) Helium-3 and manganese in hydrothermal plumes along the Gakkel Ridge, Arctic Ocean AGUFM. T52, T52E-07

Greaves JS, Richards AMS, Bains W et al (2021) Phosphine gas in the cloud decks of Venus. Nat Astron 5:655–664. https://doi.org/10.1038/s41550-020-1174-4

Grieger B et al (2004) Indication of a near surface cloud layer on Venus from reanalysis of Venera 13/14 spectrophotometer data. In: Wilson A (ed) Proceedings of the International Workshop Planetary Probe Atmospheric Entry and Descent Trajectory Analysis and Science, 6–9 October 2003, Lisbon, Portugal. ESA Special Publications, vol ESA-SP-544. ESA Publications Division, Noordwijk, pp 63

Gudmundsson MT, Thordarson T, Höskuldsson Á et al (2012) Ash generation and distribution from the April-May 2010 eruption of Eyjafjallajökull, Iceland. Sci Rep 2:572. https://doi.org/10.1038/srep00572

Gülcher AJP, Gerya TV, Montési LGJ, Munch J (2020) Corona structures driven by plume-lithosphere interactions and evidence for ongoing plume activity at Venus. Nat Geosci 13:547. https://doi.org/10.1038/s41561-020-0606-1

Haus et al (2015) Lower atmosphere minor gas abundances as retrieved from Venus Express VIRTIS-M-IR data at 2.3 μm. https://doi.org/10.1016/j.pss.2014.11.020

Head JW, Wilson L (1986) Volcanic processes and landforms on Venus: theory, predictions, and observations. J Geophys Res 91:9407. https://doi.org/10.1029/JB091iB09p09407

Head JW, Wilson L (1992) Magma reservoirs and neutral buoyancy zones on Venus: implications for the formation and evolution of volcanic landforms. J Geophys Res 97:3877. https://doi.org/10.1029/92JE00053

Helbert J, Dyar M, Walter I, Wendler D, Widemann T, Marcq E, Guignan G, Ferrari S, Maturilli A, Mueller N, Kappel D (2018) The Venus Emissivity Mapper (VEM): obtaining global mineralogy of Venus from orbit. In: Proc. SPIE 10765, Infrared Remote Sensing and Instrumentation XXVI, 107650D. https://doi.org/10.1117/12.2320112

Helbert J, Vandaele AC, Marcq E et al (2019) The VenSpec suite on the ESA EnVision mission to Venus. SPIE 11128:1112804. https://doi.org/10.1117/12.2529248

Helbert J, Säuberlich T, Darby Dyar M, Ryan C, Walter I, Reess J-M, Rosas-Ortiz Y, Peter G, Maturilli A, Arnold G (2020) The Venus Emissivity Mapper (VEM): advanced development status and performance evaluation. In: Proc. SPIE 11502, Infrared Remote Sensing and Instrumentation XXVIII, p 1150208. https://doi.org/10.1117/12.2567634

Helbert J, Maturilli A, Dyar MD, Alemanno G (2021) Deriving iron contents from past and future Venus surface spectra with new high-temperature laboratory emissivity data. Sci Adv 7:eaba9428. https://doi.org/10.1126/sciadv.aba942

Henley RW, Hughes GO (2016) SO_2 flux and the thermal power of volcanic eruptions. J Volcanol Geotherm Res 324:190–199. https://doi.org/10.1016/j.jvolgeores.2016.04.024

Herrick RR, Rumpf ME (2011) Post-impact modification by volcanic or tectonic processes as the rule, not the exception, for Venusian craters. J Geophys Res, Planets 116:E02004. https://doi.org/10.1029/2010JE003722

Herrick RR, Bjonnes ET, Carter LM et al (2023) Resurfacing history and volcanic activity of Venus. Space Sci Rev 219:29. https://doi.org/10.1007/s11214-023-00966-y

Hilton DR, Fischer TP, Noble MB (2002) Gases and volatile recycling at subduction zones. Rev Mineral Geochem 47:319. https://doi.org/10.2138/rmg.2002.47.9

Ignatiev NI, Moroz VI, Moshkin BE et al (1997) Water vapour in the lower atmosphere of Venus: a new analysis of optical spectra measured by entry probes. Planet Space Sci 45:427. https://doi.org/10.1016/S0032-0633(96)00143-2

Imamura T, Ando H, Tellmann S et al (2017) Initial performance of the radio occultation experiment in the Venus orbiter mission Akatsuki. Earth Planets Space 69:137. https://doi.org/10.1186/s40623-017-0722-3

Irwin PGJ et al (2008) Spatial variability of carbon monoxide in Venus' mesosphere from Venus express/visible and infrared thermal imaging spectrometer measurements. J Geophs Res 113. https://doi.org/10.1029/2008JE003093

Ivanov MA, Head JW (2013) The history of volcanism on Venus. Planet Space Sci 84:66–92

Iwagami N, Ohtsuki S, Tokuda K et al (2008) Hemispheric distributions of HCl above and below the Venus' clouds by ground-based 1.7 μm spectroscopy. Planet Space Sci 56:1424

Jessup KL, Marcq E, Mills F et al (2015) Coordinated Hubble space telescope and Venus express observations of Venus' upper cloud deck. Icarus 258:309. https://doi.org/10.1016/j.icarus.2015.05.027

Kaula WM (1999) Constraints on Venus evolution from radiogenic Argon. Icarus 139:32. https://doi.org/10.1006/icar.1999.6082

Keddie ST, Head JW (1995) Formation and evolution of volcanic edifices on the Dione Regio rise, Venus. J Geophys Res 100:11729. https://doi.org/10.1029/95JE00822

Korycansky DG, Zahnle KJ (2005) Modeling crater populations on Venus and Titan. Planet Space Sci 53:695. https://doi.org/10.1016/j.pss.2005.03.002

Kouyama T, Imamura T, Taguchi M et al (2017) Topographical and local time dependence of large stationary gravity waves observed at the cloud top of Venus. Geophys Res Lett 44:12098. https://doi.org/10.1002/2017GL075792

Krasnopolsky VA (1985) Chemical composition of Venus clouds. Planet Space Sci 33:109. https://doi.org/10.1016/0032-0633(85)90147-3

Krasnopolsky VA (1989) Vega mission results and chemical composition of Venusian clouds. Icarus 80:202. https://doi.org/10.1016/0019-1035(89)90168-1

Krasnopolsky VA (2007) Chemical kinetic model for the lower atmosphere of Venus. Icarus 191:25. https://doi.org/10.1016/j.icarus.2007.04.028

Krasnopolsky VA (2008) High-resolution spectroscopy of Venus: detection of OCS, upper limit to H_2S, and latitudinal variations of CO and HF in the upper cloud layer. Icarus 197:377. https://doi.org/10.1016/j.icarus.2008.05.020

Krasnopolsky VA (2010) Spatially-resolved high-resolution spectroscopy of Venus 1. Variations of CO_2, CO, HF, and HCl at the cloud tops. Icarus 208:539. https://doi.org/10.1016/j.icarus.2010.02.012

Krasnopolsky VA (2013) S_3 and S_4 abundances and improved chemical kinetic model for the lower atmosphere of Venus. Icarus 225:570. https://doi.org/10.1016/j.icarus.2013.04.026

Krasnopolsky VA (2017) On the iron chloride aerosol in the clouds of Venus. Icarus 286:134–137. https://doi.org/10.1016/j.icarus.2016.10.003

Lebonnois S, Schubert G (2017) The deep atmosphere of Venus and the possible role of density-driven separation of CO_2 and N_2. Nat Geosci 10:473. https://doi.org/10.1038/ngeo2971

Lebonnois S, Schubert G, Forget F, Spiga A (2018) Planetary boundary layer and slope winds on Venus. Icarus 314:149. https://doi.org/10.1016/j.icarus.2018.06.006

Lebonnois S, Schubert G, Kremic T et al (2020) An experimental study of the mixing of CO_2 and N_2 under conditions found at the surface of Venus. Icarus 338:113550. https://doi.org/10.1016/j.icarus.2019.113550

Lécuyer C, Simon L, Guyot F (2000) Comparison of carbon, nitrogen and water budgets on Venus and the Earth. Earth Planet Sci Lett 181:33. https://doi.org/10.1016/S0012-821X(00)00195-3

Lefèvre M (2022) Venus boundary layer dynamics; Eolian transport and convective vortex. Icarus 387:115167. https://doi.org/10.1016/j.icarus.2022.115167

Lefèvre M, Lebonnois S, Three-Dimensional SA (2018) Turbulence-resolving modeling of the venusian cloud layer and induced gravity waves: inclusion of complete radiative transfer and wind shear. J Geophys Res, Planets 123:2773. https://doi.org/10.1029/2018JE005679

Lefèvre M, Spiga A, Lebonnois S (2020) Mesoscale modeling of Venus' bow-shape waves. Icarus 335:113376. https://doi.org/10.1016/j.icarus.2019.07.010

Lorenz RD, Crisp D, Huber L (2018) Venus atmospheric structure and dynamics from the VEGA lander and balloons: new results and PDS archive. Icarus 305:277. https://doi.org/10.1016/j.icarus.2017.12.044

Lourenço DL et al (2020) Plutonic-squishy lid: a new global tectonic regime generated by intrusive magmatism on Earth-like planets. Geochem Geophys Geosyst 21:e2019GC008756. https://doi.org/10.1029/2019GC008756

Mahieux A, Wilquet V, Vandaele AC et al (2015) Hydrogen halides measurements in the Venus mesosphere retrieved from SOIR on board Venus express. Planet Space Sci 113:264. https://doi.org/10.1016/j.pss.2014.12.014

Mahieux et al (2023) Update on SO_2, detection of OCS, CS, CS_2, and SO_3, and upper limits of H_2S and HOCl in the Venus mesosphere using SOIR on board Venus Express. Icarus 2023. https://doi.org/10.1016/j.icarus.2023.115556

Maiorov BS, Ignat'ev NI, Moroz VI et al (2005) A new analysis of the spectra obtained by the Venera missions in the venusian atmosphere. I. The analysis of the data received from the Venera 11 probe at altitudes below 37 km in the 0.44–0.66 μm wavelength range. Sol Syst Res 39:267–282. https://doi.org/10.1007/s11208-005-0042-1

Marcq E, Bézard B, Drossart P et al (2008) A latitudinal survey of CO, OCS, H_2O, and SO_2 in the lower atmosphere of Venus: spectroscopic studies using VIRTIS-H. J Geophys Res, Planets 113:E00B07. https://doi.org/10.1029/2008JE003074

Marcq E, Bertaux J-L, Montmessin F, Belyaev D (2013) Variations of sulphur dioxide at the cloud top of Venus's dynamic atmosphere. Nat Geosci 6:25. https://doi.org/10.1038/ngeo1650

Marcq E, Mills FP, Parkinson CD, Vandaele AC (2018) Composition and chemistry of the neutral atmosphere of Venus. Space Sci Rev 214:10. https://doi.org/10.1007/s11214-017-0438-5

Marcq et al (2020) Climatology of SO_2 and UV absorber at Venus' cloud top from SPICAV-UV nadir dataset. Icarus 2020. https://doi.org/10.1016/j.icarus.2019.07.002

Marcq E, Amine I, Duquesnoy M, Bézard B (2021) Evidence for SO_2 latitudinal variations below the clouds of Venus. Astron Astrophys 648:L8. https://doi.org/10.1051/0004-6361/202140837

Marcq E, Bézard B, Reess JM, Henry F, Érard S et al (2023) Minor species in Venus' night side troposphere as observed by VIRTIS-H/Venus express. Icarus 405:115714. https://doi.org/10.1016/j.icarus.2023.115714

Mills FP, Allen M (2007) A review of selected issues concerning the chemistry in Venus' middle atmosphere. Planet Space Sci 55:1729. https://doi.org/10.1016/j.pss.2007.01.012

Moore WB et al (2017) Heat-pipe planets. Earth Planet Sci Lett 474:13. https://doi.org/10.1016/j.epsl.2017.06.015

Morellina S, Bellan J (2022) Turbulent chemical-species mixing in the Venus lower atmosphere at different altitudes: a direct numerical simulation study relevant to understanding species spatial distribution. Icarus 371:114686. https://doi.org/10.1016/j.icarus.2021.114686

Morellina S, Bellan J, Cutts J (2020) Global thermodynamic, transport-property and dynamic characteristics of the Venus lower atmosphere below the cloud layer. Icarus 350:113761. https://doi.org/10.1016/j.icarus.2020.113761

Morlok A, Klemme S, Weber I et al (2019) Mid-infrared spectroscopy of planetary analogs: a database for planetary remote sensing. Icarus 324:86. https://doi.org/10.1016/j.icarus.2019.02.010

Mortier A, Goloub P, Podvin T et al (2013) Detection and characterization of volcanic ash plumes over Lille during the Eyjafjallajökull eruption. Appl Cogn Psychol 13:3705. https://doi.org/10.5194/acp-13-3705-2013

Namiki N, Solomon SC (1998) Volcanic degassing of argon and helium and the history of crustal production on Venus. J Geophys Res 103:3655. https://doi.org/10.1029/97JE03032

Noack L, Breuer D, Spohn T (2012) Coupling the atmosphere with interior dynamics: implications for the resurfacing of Venus. Icarus 217(2):484. https://doi.org/10.1016/j.icarus.2011.08.026

Nordström T, Stenberg G, Nilsson H et al (2013) Venus ion outflow estimates at solar minimum: influence of reference frames and disturbed solar wind conditions. J Geophys Res Space Phys 118:3592. https://doi.org/10.1002/jgra.50305

Omran A, Oze C, Jackson B, Mehta LM, Barge LM, Bada J, Pasek MA (2021) Phosphine generation pathways on rocky planets. Astrobiology 21:10. https://doi.org/10.1089/ast.2021.0034

Oppenheimer C, Scaillet B, Martin RS (2011) Sulfur degassing from volcanoes: source conditions, surveillance, plume chemistry and Earth system impacts. Rev Mineral Geochem 73:363. https://doi.org/10.2138/rmg.2011.73.13

Oppenheimer C, Fischer T, Scaillet B (2014) Volcanic degassing: process and impact. In: Holland HD, Turekian KK (eds) Treatise on geochemistry, 2nd edn. vol 4 (pp. 111–179). https://doi.org/10.1016/B978-0-08-095975-7.00304-1

O'Reilly TC, Davies GF (1981) Magma transport of heat on Io: A mechanism allowing a thick lithosphere. Geophys Res Lett 8. https://doi.org/10.1029/GL008I004p00313

O'Rourke JG, Korenaga J (2015) Thermal evolution of Venus with argon degassing. Icarus 260:128. https://doi.org/10.1016/j.icarus.2015.07.009

Oschlisniok J, Häusler B, Pätzold M, Tellmann S, Bird MK, Peter K, Andert TP (2021) Sulfuric acid vapor and sulfur dioxide in the atmosphere of Venus as observed by the Venus express radio science experiment VeRa. Icarus 362:114405. https://doi.org/10.1016/j.icarus.2021.114405

Oyama VI, Carle GC, Woeller F, Pollack JB, Reynolds RT, Craig RA (1980) Pioneer Venus gas chromatography of the lower atmosphere of Venus. J Geophys Res 85(A13):7861

Parkinson CD, Gao P, Esposito L, Yung Y, Bougher S, Hirtzig M (2015) Photochemical control of the distribution of Venusian water. Planet Space Sci 113–114:226–236. https://doi.org/10.1016/j.pss.2015.02.015

Patrakeev A, Trokhimovskiy A, Korablev O et al (2022) The Venus infrared atmospheric gases linker instrument concept for solar occultation studies of Venus atmosphere composition and structure onboard the Venus Orbiter Mission of the Indian Space Research Organization. In: Proc. SPIE 12138, Optics, Photonics and Digital Technologies for Imaging Applications VII, 1213810. https://doi.org/10.1117/12.2632371

Petryanov IV, Andreichikov BM, Korchuganov BN et al (1981) Application of the FP filter to study aerosol of Venus clouds. Dokl Akad Nauk Ukr SSR 258:57

Phillips RJ, Raubertas RF, Arvidson RE et al (1992) Impact craters and Venus resurfacing history. J Geophys Res 97:15923. https://doi.org/10.1029/92JE01696

Pollack JB, Dalton JB, Grinspoon DH, Wattson RB, Freedman R, Crisp D, Allen DA, Bézard B, de Bergh C, Giver L, Ma Q, Tipping R (1993) Near-infrared light from Venus' nightside: a spectroscopic analysis. Icarus 103:1. https://doi.org/10.1006/icar.1993.1055

Radoman-Shaw BG, Harvey RP, Costa G, Jacobson NS, Avishai A, Nakley LM Vento D (2022) Experiments on the reactivity of basaltic minerals and glasses in Venus surface conditions using the Glenn Extreme Environment Rig. Meteorit Planet Sci 24:1796–1819. https://doi.org/10.1111/maps.13902

Ragent B, Esposito LW, Tomasko MG et al (1985) Particulate matter in the Venus atmosphere. Adv Space Res 5:85. https://doi.org/10.1016/0273-1177(85)90199-1

Rimmer PB, Jordan S, Constantinou T et al (2021) Hydroxide salts in the clouds of Venus: their effect on the sulfur cycle and cloud droplet pH. Planet Sci J 2:133. https://doi.org/10.3847/PSJ/ac0156

Robert S, Macovenco C, Lefèvre M et al (2021) Detecting Venus' volcanic gas plumes with VenSpec-H. In: European planetary science congress EPSC2021-678. https://doi.org/10.5194/epsc2021-678

Rolf T, Weller M, Gülcher A et al (2022) Dynamics and evolution of Venus' mantle through time. Space Sci Rev 218:70. https://doi.org/10.1007/s11214-022-00937-9

Romeo I, Turcotte DL (2010) Resurfacing on Venus. Planet Space Sci 58(10):1374–1380

Rossi F, Saboia M, Krishnamoorthya S, Vander Hook J (2023) Proximal exploration of Venus volcanism with teams of autonomous buoyancy-controlled balloons. Acta Astronaut. https://doi.org/10.1016/j.actaastro.2023.03.003

Santos AR, Gilmore MS, Greenwood JP, Nakley LM, Phillips K, Kremic T, Lopez X (2023) Experimental weathering of rocks and minerals at Venus conditions in the Glenn Extreme Environments Rig (GEER). J Geophys Res 128:e2022JE007423. https://doi.org/10.1029/2022JE007423

Sato TM, Sagawa H (2023) A new constraint on HCl abundance at the cloud top of Venus. Icarus 390:115307. https://doi.org/10.1016/j.icarus.2022.115307

Schaber GG, Strom RG, Moore HJ et al (1992) Geology and distribution of impact craters on Venus: what are they telling us? J Geophys Res 97:13257. https://doi.org/10.1029/92JE01246

Schaefer L, Fegley B Jr (2017) Redox states of initial atmospheres outgassed on rocky planets and planetesimals. Astrophys J 843:120. https://doi.org/10.3847/1538-4357/aa784f

Schmidt A, Leadbetter S, Theys N et al (2015) Satellite detection, long-range transport, and air quality impacts of volcanic sulfur dioxide from the 2014–2015 flood lava eruption at Bárðarbunga (Iceland). J Geophys Res 120:9739. https://doi.org/10.1002/2015JD023638

Seiff A, Schofield JT, Kliore AJ et al (1985) Models of the structure of the atmosphere of Venus from the surface to 100 kilometers altitude. Adv Space Res 5(11):3–58. https://doi.org/10.1016/0273-1177(85)90197-8

Shalygin EV, Markiewicz WJ, Basilevsky AT et al (2015) Active volcanism on Venus in the Ganiki Chasma rift zone. Geophys Res Lett 42:4762. https://doi.org/10.1002/2015GL064088

Shao WD, Zhang X, Bierson CJ, Encrenaz T (2020) Revisiting the sulfur-water chemical system in the middle atmosphere of Venus. J Geophys Res, Planets 125:e06195. https://doi.org/10.1029/2019JE006195

Smrekar SE, Stofan ER, Mueller N et al (2010) Recent hotspot volcanism on Venus from VIRTIS emissivity data. Science 328:605. https://doi.org/10.1126/science.1186785

Solomon SC, Head JW (1982) Mechanisms for lithospheric heat transport on Venus: implications for tectonic style and volcanism. J Geophys Res 87. https://doi.org/10.1029/JB087iB11p09236

Stofan ER et al (2005) Resurfacing styles and rates on Venus: assessment of 18 venusian quadrangles. Icarus 173:312. https://doi.org/10.1016/j.icarus.2004.08.004

Stofan ER, Smrekar SE, Mueller N, Helbert J (2016) Themis Regio, Venus: evidence for recent (?) volcanism from VIRTIS data. Icarus 271:375–386

Strom RG, Schaber GG, Dawsow DD (1994) The global resurfacing of Venus. J Geophys Res 99:10899. https://doi.org/10.1029/94JE00388

Surkov YA, Shcheglov OP, Ryvkin ML et al (1987) Profile of water vapor content in the atmosphere of Venus from the results of experiments aboard VEGA-1 and VEGA-2. Sol Syst Res 20:122

Suzuki A, Ando H, Takagi M, Maejima Y, Sugimoto N, Matsuda Y (2023) Dependency of the vertical propagation of mountain waves on the zonal wind and the static stability in the lower Venusian atmosphere. Icarus 402:115615. https://doi.org/10.1016/j.icarus.2023.115615

Symonds RB, Rose WI, Bluth GJS, Gerlach TM (1994) Volcanic gas studies: methods, results, and applications. In: Carrol MR, Holloway JR (eds) Volatiles in magma. Reviews in Mineralogy, vol 30. Mineral. Soc. Am., Washington, pp 1–66. https://doi.org/10.1515/9781501509674-007

Taylor FW, Crisp D, Bézard B (1997) Near-infrared sounding of the lower atmosphere of Venus. In: Bougher SW, Hunten DM, Philips RJ (eds) Venus II. Geology, geophysics, atmosphere, and solar wind environment. University of Arizona Press, Tucson, p 325

Tellmann S, Haeusler B, Paetzold M, Bird MK, Tyler GL, Andert T, Remus S (2009) The structure of the Venus neutral atmosphere as seen by the radio science experiment VeRa on Venus express. J Geophys Res, Planets 114:E00B36. https://doi.org/10.1029/2008JE003204

Titov DV, Markiewicz WJ, Ignatiev NI et al (2012) Morphology of the cloud tops as observed by the Venus express monitoring camera. Icarus 217:682. https://doi.org/10.1016/j.icarus.2011.06.020

Titov DV, Ignatiev NI, McGouldrick K et al (2018) Clouds and Hazes of Venus. Space Sci Rev 214:126. https://doi.org/10.1007/s11214-018-0552-z

Treiman AH, Justin Filiberto J, Vander KE, Kaaden KE (2021) Near-infrared reflectance of rocks at high temperature: preliminary results and implications for near-infrared emissivity Venus's surface. Planet Sci J 2:43. https://doi.org/10.3847/PSJ/abd546

Truong N, Lunine JI (2021) Volcanically extruded phosphides as an abiotic source of Venusian phosphine. Proc Natl Acad Sci 118:e2021689118. https://doi.org/10.1073/pnas.2021689118

Tsang CCC, McGouldrick K (2017) General circulation of Venus from a long-term synoptic study of tropospheric CO by Venus Express/VIRTIS. Icarus 289:173. https://doi.org/10.1016/j.icarus.2017.02.018

Tsang CCC, Irwin PGJ, Wilson CF et al (2008) Tropospheric carbon monoxide concentrations and variability on Venus from Venus Express/VIRTIS-M observations. J Geophys Res, Planets 113:E00B08. https://doi.org/10.1029/2008JE003089

van Zelst I (2022) Comment on "Estimates on the frequency of volcanic eruptions on Venus" by Byrne and Krishnamoorthy. J Geophys Res, Planets 127:e2022JE007448. https://doi.org/10.1029/2022JE007448

Vandaele AC, Kruglanski M, De Mazière M (2006) Modeling and retrieval of atmospheric spectra using ASIMUT. Proceedings of the 1st EPS/MetOp RAO Workshop, 15-17 May 2006, ESRIN, Frascati, Italy. ESA Special Publications, vol. SP-618. ESA Publications Division, Nordwijk

Vandaele AC et al (2016) Carbon monoxide observed in Venus' atmosphere with SOIR/VEx. Icarus 272:48. https://doi.org/10.1016/j.icarus.2016.02.025

518

 Springer

Vandaele AC, Korablev O, Belyaev D et al (2017a) Sulfur dioxide in the Venus atmosphere: I. Vertical distribution and variability. Icarus 295:16. https://doi.org/10.1016/j.icarus.2017.05.003

Vandaele AC, Korablev O, Belyaev D et al (2017b) Sulfur dioxide in the Venus atmosphere: II. Spatial and temporal variability. Icarus 295:16. https://doi.org/10.1016/j.icarus.2017.05.001

Villanueva GL, Cordiner M, Irwin PGJ, de Pater I, Butler B, Gurwell M, Milam SN, Nixon CA, Luszcz-Cook SH, Wilson CF, Kofman V, Liuzzi G, Faggi S, Fauchez TJ, Lippi M, Cosentino R, Thelen AE, Moullet A, Hartogh P, Molter EM, Charnley S, Arney GN, Mandell AM, Biver N, Vandaele AC, de Kleer KR, Kopparapu R (2021) No evidence of phosphine in the atmosphere of Venus from independent analyses. Nat Astron 5:631–635. https://doi.org/10.1038/s41550-021-01422-z

Widemann T, Smrekar SE, Garvin JB et al (2023) Venus evolution through time: key science questions, selected mission concepts and future investigations. Space Sci Rev 219:56. https://doi.org/10.1007/s11214-023-00992-w

Wilson CF, Guerlet S, Irwin PGJ et al (2008) Evidence for anomalous cloud particles at the poles of Venus. J Geophys Res, Planets 113:E00B13. https://doi.org/10.1029/2008JE003108

Wood JA (1997) Rock weathering on the surface of Venus. In: Bougher SW, Hunten DM, Philips RJ (eds) Venus II. Geology, geophysics, atmosphere, and solar wind environment. University of Arizona Press, Tucson, p 637

Yung YL, Demore WB (1982) Photochemistry of the stratosphere of Venus: implications for atmospheric evolution. Icarus 51:199. https://doi.org/10.1016/0019-1035(82)90080-X

Yung YL, Liang MC, Jiang X et al (2009) Evidence for carbonyl sulfide (OCS) conversion to CO in the lower atmosphere of Venus. J Geophys Res, Planets 114:E00B34. https://doi.org/10.1029/2008JE003094

Zolotov MY (1996) A model of thermochemical equilibrium in the near-surface atmosphere of Venus. Geochem Int 33:80–100

Zolotov MY (2011) On the chemistry of mantle and magmatic volatiles on Mercury. Icarus 212:24–41. https://doi.org/10.1016/j.icarus.2010.12.014

Zolotov MY (2018) Gas-solid interactions on Venus and other Solar system bodies. Rev Mineral Geochem 84:351–392. https://doi.org/10.2138/rmg.2018.84.10

Zolotov MY, Matsui T (2002) Chemical models for volcanic gases on Venus. Lunar Planet Sci Conf 33:1433

Zolotov MY, Mogul R, Limaye SS, Way MJ, Garvin JB (2023) Venus cloud composition suggested from the pioneer Venus large probe neutral mass spectrometer data. Lunar Planet Sci Conf 54:2880

Publisher's Note Springer Nature remains neutral with regard to jurisdictional claims in published maps and institutional affiliations.

Authors and Affiliations

Colin F. Wilson[1,2] ⓘ **· Emmanuel Marcq[3] · Cédric Gillmann[4] · Thomas Widemann[5,6] · Oleg Korablev[7] · Nils T. Mueller[8,9]** ⓘ **· Maxence Lefèvre[3] · Paul B. Rimmer[10] · Séverine Robert[11] · Mikhail Y. Zolotov[12]**

✉ N.T. Mueller
Nils.Mueller@dlr.de

C.F. Wilson
colin.wilson@esa.int

E. Marcq
emmanuel.marcq@latmos.ipsl.fr

C. Gillmann
cgillmann@ethz.ch

T. Widemann
thomas.widemann@obspm.fr

O. Korablev
korab@cosmos.ru

M. Lefèvre
maxence.lefevre@latmos.ipsl.fr

P.B. Rimmer
pbr27@cam.ac.uk

S. Robert
severine.robert@aeronomie.be

M.Y. Zolotov
zolotov@asu.edu

[1] European Space Agency, Keplerlaan 1, 2201, AZ Noordwijk, The Netherlands

[2] Physics Dept, Oxford University, Oxford OX1 3PU, UK

[3] LATMOS/IPSL, UVSQ Sorbonne Université Paris-Saclay, Sorbonne Université, CNRS, Guyancourt, France

[4] ETH Zurich, Institut für Geophysik, Geophysical Fluid Dynamics, Sonneggstraße 5, 8092 Zürich, Switzerland

[5] LESIA, Observatoire de Paris, Université PSL, CNRS, Sorbonne Université, Université Paris Cité, 5 place Jules Janssen, 92195 Meudon, France

[6] Université Paris-Saclay, UVSQ, DYPAC, 78000 Versailles, France

[7] Space Research Institute (IKI), Russian Academy of Sciences, Moscow 117997, Russia

[8] Institute for Planetary Research, DLR, Rutherfordstraße 2, 12489 Berlin, Germany

[9] Institute of Geosciences, Freie Universität Berlin, Malteserstr. 74-100, 12249 Berlin, Germany

[10] Cavendish Laboratory, University of Cambridge, JJ Thomson Avenue, Cambridge CB3 0HE, UK

[11] Royal Belgian Institute for Space Aeronomy, Brussels, Belgium

[12] School of Earth and Space Exploration, Arizona State University, Tempe, AZ 85287-1404, USA

Space Science Reviews (2022) 218:70
https://doi.org/10.1007/s11214-022-00937-9

Dynamics and Evolution of Venus' Mantle Through Time

Tobias Rolf[1,2] · Matt Weller[3,4] · Anna Gülcher[5] · Paul Byrne[6,7] ·
Joseph G. O'Rourke[8] · Robert Herrick[9] · Evan Bjonnes[3,4] · Anne Davaille[10] ·
Richard Ghail[11] · Cedric Gillmann[12] · Ana-Catalina Plesa[13] ·
Suzanne Smrekar[14]

Received: 25 February 2022 / Accepted: 7 November 2022 / Published online: 28 November 2022
© The Author(s) 2022

Abstract
The dynamics and evolution of Venus' mantle are of first-order relevance for the origin and
modification of the tectonic and volcanic structures we observe on Venus today. Solid-state
convection in the mantle induces stresses into the lithosphere and crust that drive deformation leading to tectonic signatures. Thermal coupling of the mantle with the atmosphere and
the core leads to a distinct structure with substantial lateral heterogeneity, thermally and
compositionally. These processes ultimately shape Venus' tectonic regime and provide the
framework to interpret surface observations made on Venus, such as gravity and topography. Tectonic and convective processes are continuously changing through geological time,
largely driven by the long-term thermal and compositional evolution of Venus' mantle. To
date, no consensus has been reached on the geodynamic regime Venus' mantle is presently
in, mostly because observational data remains fragmentary. In contrast to Earth, Venus'
mantle does not support the existence of continuous plate tectonics on its surface. However,
the planet's surface signature substantially deviates from those of tectonically largely inactive bodies, such as Mars, Mercury, or the Moon. This work reviews the current state of
knowledge of Venus' mantle dynamics and evolution through time, focussing on a dynamic
system perspective. Available observations to constrain the deep interior are evaluated and
their insufficiency to pin down Venus' evolutionary path is emphasised. Future missions
will likely revive the discussion of these open issues and boost our current understanding by
filling current data gaps; some promising avenues are discussed in this chapter.

Keywords Venus · Mantle dynamics · Interior evolution · Surface tectonics · Thermal
history

1 Introduction

Early space missions–such as Pioneer, Venera, and Magellan–indicated that Venus' present
tectonic regime differs substantially from the Earth's, but a profound answer for why this is
so remains lacking. One major challenge to resolve this issue is the difficulty of observing

Venus: Evolution Through Time
Edited by Colin F. Wilson, Doris Breuer, Cédric Gillmann, Suzanne E. Smrekar, Tilman Spohn and
Thomas Widemann

Extended author information available on the last page of the article

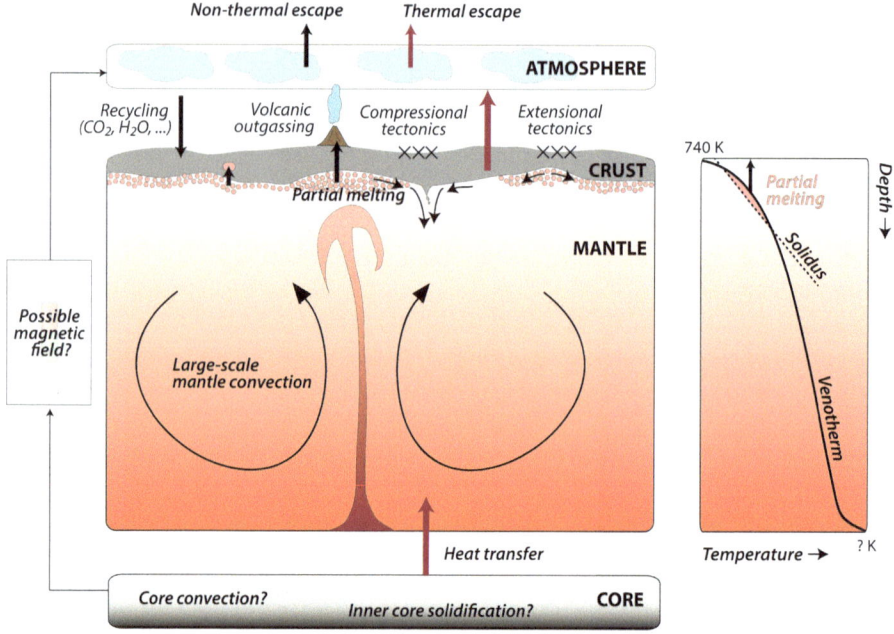

Fig. 1 A schematic of the role of the mantle in planetary evolution. Red arrows denote the exchange of heat; thick black arrows denote other means of transfer such as magma transport from the mantle into the crust and to the surface; thin black arrows indicate flow. The box at right illustrates a hypothetical temperature profile (here named the 'Venotherm'); the dashed line is a hypothetical solidus

robust constraints on Venus, particularly for the deep interior. Nevertheless, those missions provided important insights into Venus' interior structure and dynamics (see e.g., O'Rourke et al. 2023), which are key for understanding the planet's surface tectonics. Bulk density and composition suggest an interior structure that is similar to that of Earth: a massive, iron-rich core overlain by a thick silicate layer (e.g., Margot et al. 2021). The large heat capacity and high viscosity of that silicate mantle cause huge thermal inertia and long dynamic time scales. Over the billions of years of Venus' evolution, however, the mantle is still a highly dynamic system and its evolution determines the state of Venus' interior, the tectonic expressions at the surface, and the interaction with Venus' fluid layers including the atmosphere, the core, and thus potential magnetic field generation and evolution (Fig. 1).

Solid-state convection is the key mechanism of heat transport in the mantle (e.g., Stevenson 2003) and its efficiency determines the temperature in the silicate interior as well as the thickness of Venus' lithosphere and crust. In addition to the thermal impact of convection, upwelling mantle flow moves material from regions of higher pressure to regions of lower pressure (decompression), causing topography at the boundaries of mantle convection cells and controlling the locations and rates of partial melting. Eventually, these partially molten zones form the source regions of intrusive magmatism and extrusive volcanism. Such processes in turn lead to compositional heterogeneity in the interior, determine the thickness and rheological strength of the crust (e.g., Lourenço et al. 2020), and control the outgassing of volatiles into Venus' atmosphere (e.g., O'Rourke and Korenaga 2015). At the bottom of the mantle, thermal exchange determines core cooling and the heat transport efficiency in the planet's central layer. If the mantle allows for sufficient core cooling, an inner core may crystallise at some point during the planet's evolution (e.g., O'Rourke et al. 2018), depending on the core composition (e.g., O'Neill 2021). Core cooling and inner core crystallisation drive convection in the liquid iron-rich core and possibly power a core dynamo. This dynamo can generate a planetary magnetic field, which may interact with the atmosphere in several ways including shielding from stellar winds and volatile loss, but to date this remains strongly debated (see Way et al. 2023; Gillmann et al. 2022).

In summary, the mantle interacts with essentially all other subsystems of a planet over geological time scales. For Venus, these couplings are partly revealed by observational constraints, but many details remain unanswered. This is the case for Venus' present-day state–in principle directly observable–and becomes more challenging for the evolution of Venus' mantle through time. Today, Venus neither features plate tectonics nor generates an Earth-like magnetic field. Evidence for volcanism and tectonics are ubiquitous on the surface, but what caused the divergence between Venus and Earth remains poorly resolved. Were the conditions on Venus less favourable for the development of plate tectonics right from the start, because of differences in the accretion history, or because Venus is closer to the Sun and therefore received more insolation (see Salvador et al. 2023)? Or did the evolution of both planets start off relatively similar before diverging at some later stage because of an endogenic or exogenic trigger?

Previous missions to Venus returned a number of observables useful to constrain interior models, which are summarised in Sect. 3.1 and detailed in Herrick et al. (2023), Ghail et al. (2023), Carter et al. (2023), and Gilmore et al. (2023). Their coverage, resolution, and non-uniqueness leave gaps and uncertainty in Venus' core and mantle dimensions, in how much heat is transferred from the mantle to the atmosphere, and in the thickness, age, and composition of the crust. However, new space missions will launch to Venus within a decade to further fill existing gaps (see Widemann et al. 2023). It is thus timely to review our understanding of Venus' mantle evolution through time and to identify what we know and what we do not yet know.

This work is not the first review on Venus' interior evolution (e.g., Mocquet et al. 2011; Smrekar et al. 2018), but it particularly focuses on how mantle dynamics generate and relate to different regimes of surface tectonics and volcanism, and how this relation may evolve through time. Our goal is to emphasise which possible pathways Venus' mantle may have taken to its present state, and what can be done to reduce the number of feasible scenarios in future. Section 2 provides a background on heat transfer in planetary mantles, the peculiar properties of mantle silicates, and the spectrum of planetary tectonic regimes, driven by mantle convection. Section 3 summarises the observations constraining Venus' contemporary mantle and discusses feasible tectonic regimes relating mantle dynamics, crustal tectonics and volcanism. Section 4 reviews Venus' mantle evolution through time with a focus on the possibility of lateral and particularly temporal variations as well potential triggers. Finally, Sect. 5 gives an overview of possible mantle evolution scenarios for Venus. A perspective is given on how future conceptual understanding and data collection can boost our understanding of Venus' interior and help to resolve unanswered questions specific to the planet's mantle.

2 Planetary Tectonic Regimes

2.1 Basics of Planetary Mantle Convection

The evolution of a planetary body is strongly controlled by its thermal history. Whether the primordial heat accumulated during accretion and differentiation is efficiently kept in the interior or can easily escape determines the planet's cooling rate and the vigour of internal dynamics. Inside a Venus-like planet, heat is transported via thermal conduction or convection (i.e., heat transport via large-scale material flow), where the latter is the much more efficient mechanism for most parts of the mantle. The vigour of convection determines the efficiency of cooling and partial melting of the silicate mantle that leads to crustal production and volcanic outgassing. Thus, interior dynamical processes can be linked to surface expressions such as tectonics and volcanism as well as the evolution of the atmosphere.

A simple analogue to planetary mantle convection is the Rayleigh–Bénard system, where a homogeneous fluid layer of finite thickness D is heated uniformly from below and cooled from above. Heating the fluid leads to unstable density stratification and initiates convective currents once the heating-induced density contrast becomes sufficiently strong. At that point, thermal conduction cannot limit the upward-directed buoyancy anymore and this force overcomes the viscous resistance against the onset of motion inside the fluid layer. Whether convection occurs is determined by the Rayleigh number: $Ra = \alpha g \Delta T \rho D^3 / \kappa \eta$, where α is the thermal expansivity, g the gravitational acceleration, ΔT the superadiabatic temperature contrast across the layer, and ρ, κ, η are the layer density, thermal diffusivity, and viscosity, respectively. For a convective instability to grow and generate large-scale convection, a critical value of Ra must be reached, which depends on the layer geometry and boundary conditions. Typically, this value is $\sim 10^3$ (Turcotte and Schubert 2017). Transferred to Venus' mantle, Ra is $O(10^7$-$10^8)$, using $D = 3000$ km, $\alpha = 3 \times 10^{-5}$ K^{-1}, $\Delta T = 2500$ K, $g = 8.87$ m s^{-2}, $\rho = 3300$ kg m^{-3}, $\kappa = 10^{-6}$ m^2 s^{-1}, $\eta = 10^{20}$-10^{21} Pa s, which implies vigorous, time-dependent convection–despite the huge viscosity of mantle rocks.

A vigorously convecting system develops thin thermal boundary layers (TBLs) near the top and bottom boundary. Across these TBLs, heat is transported via thermal conduction, but this transport cannot accommodate the continuous inflow of heat, so that the TBLs form convective instabilities, expressed as upwellings and downwellings. In comparison to these instabilities, the bulk mantle behaves passively, is essentially stirred around, and adopts a thermal profile that does not actively contribute to driving convection. The larger Ra, the more prone the system is to instabilities and the thinner the TBLs become. On Earth, the TBLs are ~ 100 km thick and the surface TBL is thought to correspond to the (oceanic) lithosphere.

Convective currents self-organise the relative spacings of up- and downwellings depending on the properties of the convecting layer. In the Rayleigh–Bénard system, these spacings are comparable to the layer thickness, but in planetary mantles several peculiarities promote convection cells with large aspect ratios. These include the strength of the lithosphere (e.g., van Heck and Tackley 2008; Yoshida 2008; Rolf et al. 2014, 2018a), pressure-dependence of mantle viscosity (Bunge et al. 1997; Höink and Lenardic 2008, 2010; Höink et al. 2012; Lenardic et al. 2019) and other material properties (e.g., Hansen et al. 1993), as well as the mantle heating mode (McNamara and Zhong 2005). On Earth, this is manifested in the size of the largest tectonic plates, like the Pacific plate. On Venus, however, there seems to be little indication of such long-wavelength flow structures in the mantle.

2.2 Specific Complexities of Planetary Mantle Convection

2.2.1 Planetary Heating Modes

The Rayleigh–Bénard setup oversimplifies planetary mantle convection. For example, the ratio of the Earth's core and mean surface radii of ~ 0.55 implies that the plane-layer approximation is inaccurate, impacting the flow patterns in the mantle (e.g., Weller et al. 2016;

Yanagisawa et al. 2016; Guerrero et al. 2018). Venus' radius ratio is not well understood given the uncertainty in core radius (e.g., Margot et al. 2021), but is typically assumed to be similar to Earth's. Due to planetary curvature, the surface boundary comprises a larger area than the core–mantle boundary. This makes the bottom boundary layer more prone to instabilities compared with the plane-layer geometry, if the same heat flow across the core–mantle boundary is considered.

The mantles of Earth and (probably) Venus contain long-lived radiogenic nuclides (in particular ^{40}K, ^{232}Th, ^{235}U, and ^{238}U) that generate internal heat by radiogenic decay, with rates decreasing through time and possibly varying spatially. Internal heating interplays with basal heating from the hot core established during the planet's accretion and differentiation. The rates of basal and internal heating control the stability of the boundary layers, the spacing of its instabilities, and therefore the planform of mantle flow (Moore 2008; Weller et al. 2016; Korenaga 2017). Internally heated convection exhibits different flow patterns than basally heated convection, because the bottom boundary layer is absent or greatly weakened as the interior adopts the temperature of the bottom boundary and inhibits inflow of heat from below (e.g., Mulyukova and Bercovici 2020). This suppresses active hot instabilities, which become diffuse and passive return flows compared with the active and pronounced downwellings (e.g., McKenzie et al. 1974; Parmentier et al. 1994; Sotin and Labrosse 1999).

Both purely basal or purely internal modes of heating are special cases for planets that either are sufficiently ancient to have depleted all their internal heat sources, or sequestered the bulk of them into a nonrecyclable crust, or have conditions where the temperature of the mantle temporarily becomes equal to that of the outer core. For both Earth and Venus, such special cases are extremely unlikely, suggesting that mixed mode heating dominates heat transport and either planet's thermal evolution. Mixed mode convective heating allows for a strong mechanical interaction between both the upper and lower boundary layers (here the upper boundary layer is analogous to the thermal planetary lithosphere). This exerts a first order control on both the thickness and the heat flux through these layers (Moore 2008; Weller et al. 2016). Critically, the mantle then deviates from the classic definition of an adiabatic interior (Weller et al. 2016; Lenardic et al. 2019). Boundary layer interaction through mixed heating is a stark departure from the regime identified in classical theory (Howard 1966; Fowler 1985), where each boundary layer thermally destabilises on its own upon reaching a critical thickness (Sect. 2.1). This emphasises that the heat loss due to convection in planetary mantles deviates from classical convection experiments, and that extrapolation of real behaviour from classic theory is limited.

For any planet, the ratio of basal to internal heating is debatable as it depends on the style of planetary accretion and differentiation (e.g., McKenzie et al. 1974), and is also a function of the planet's evolution and geodynamic regime. For Earth, traditional estimates of $\sim$90% suggested a mostly internally heated mantle (e.g., Sleep 1990), but more recent estimates indicate a more balanced partitioning of $\sim$60–70% internal heating (e.g., Lay et al. 2008; Leng and Zhong 2008, 2009). On Earth, the main challenge is to estimate the heat flow across the core–mantle boundary, which depends on the poorly understood thermal conductivity of silicates under deep mantle conditions as well as on the temperature of the core. For Venus, not even the heat flow across the surface has been robustly measured, but indirect estimates exist (Sect. 3.1.3); heat flow across Venus' core–mantle boundary is almost wholly unknown. The absence of an intrinsic magnetic field precludes a thermally driven core dynamo, so that the heat flow conducted along the core adiabat may be an upper bound. However, this scenario assumes an Earth-like core for Venus, which is not guaranteed by the available data (Sect. 3.1.6).

2.2.2 Material Properties and Rheology

The mantle's material properties strongly depend on temperature, pressure, and other factors. As a result, viscosity, thermal expansivity, conductivity, diffusivity, and density vary through space (and time) and the system-characterising Rayleigh number of the bulk mantle does not properly determine the dynamics on local scales. Potential consequences of these variations range from different planforms of mantle flow (e.g., Hansen et al. 1993; Tosi et al. 2010), to different surface heat flows (e.g., Ghias and Jarvis 2008), to differences in obtained surface topography and geoid (e.g., Schmeling et al. 2003).

Viscosity in particular is known to vary over many orders of magnitude throughout the mantle as a function of temperature, pressure, composition, mineral phase, water content, grain size, previous deformation history and various other parameters (see Karato 2010). This material property relates the strain rate with the stress that material experiences. Under different strain-rate versus stress conditions, deformation happens through different mechanisms. At relatively low levels of stress, deformation is viscous. In diffusion creep, viscosity does not explicitly depend on stress, but depends strongly on grain size. At higher levels of stress, but still in the viscous regime, dislocation creep becomes dominant and viscosity has a power-law dependence on stress, yet is insensitive to grain size. In Venus' mantle, both creep mechanisms coexist and which one dominates strongly depends on material properties such as water and melt content, mineralogical phase, deformation texture, and oxygen fugacity (e.g., Kohlstedt and Hansen 2015). Although coexisting creep mechanisms are frequently considered in terrestrial studies of crustal to lithospheric scale, these complexities are often ignored on planetary-scale problems (but cf. Rozel et al. 2014; Dannberg et al. 2017; Schulz et al. 2019 and others). This issue could be especially relevant in connection with grain-size evolution, which under Venus' hot surface conditions may affect crustal rheology very differently than under Earth-like conditions (Bercovici and Ricard 2014).

Water content is another key component of Venus' interior rheology. If Venus' present mantle and shallow crust are drier than Earth's, as suggested by some studies (e.g., Namiki 1995), the strength of Venus' crustal rocks may be enhanced compared to Earth's (e.g., Mackwell et al. 1998) with a commensurate impact on the style of tectonics (e.g., Moresi and Solomatov 1998; Turcotte 1996). Holding all else equal, reduced water content would also have a dampening effect on melting and magmatism as water acts to reduce the mantle solidus (e.g., Green et al. 2014; Ohtani 2020). Although abundant water tends to decrease mantle viscosity, the thermo-tectonic feedback between surface motions and cooling of the mantle may actually result in a more viscous mantle if viscosity is very sensitive to water content (Nakagawa et al. 2015). Abundant water in the mantle transition zone could also help to explain variations in geochemical signatures of basaltic lavas across the planetary surface. If water-rich ambient mantle ascends out of the mantle transition zone into a zone of low-water-solubility, it may undergo dehydration-induced partial melting, thereby filtering incompatible elements out of the depleted rising material (Bercovici and Karato 2003). Finally, water transported into the deepest mantle could lead to chemical reactions with iron-rich materials (e.g., Yuan et al. 2018) that may possibly trigger large-scale geodynamic events (Mao et al. 2021).

Depending on the tectonic regime, water transport in Venus' mantle may differ from that inside Earth. Current data for Venus points to relatively low water content in the atmosphere, but constraints on estimates of the water budget of the mantle remain poor (e.g., Zolotov et al. 1997, and references therein), even though Venus' high D/H ratio may imply that substantial parts of Venus' interior water may have been lost, possibly by volcanic outgassing and atmospheric escape (e.g., Grinspoon 1993; Gillmann et al. 2022). The initial water content of the mantle after magma ocean solidification is a major unknown (Salvador et al. 2023). Parameterized convection models coupled with water transport suggest self-regulation effects that make the present water content only weakly sensitive to the initial content (Sandu et al. 2011)—but these models assume a terrestrial tectonic regime that may be inapplicable for Venus, and more complex fully dynamic models have not confirmed such self-regulation effects (Nakagawa et al. 2015).

In the lithosphere and crust, relatively low temperatures lead to high viscosity and stress that can lead to brittle fracturing at the low confining pressure. The consequence is irreversible plastic deformation associated with strong localization that is thought to enable the formation of narrow weak zones, such as faults and, at larger scales, systems of faults such as rifts. The stress threshold that surface rocks can sustain is often called 'yield stress' and is one crucial property of the planetary lithosphere. If the yield stress is small, convection-induced stresses can fragment the lithosphere and may generate tectonic plates such as on Earth (Sect. 2.3). Such a yield stress approach has been used in many numerical studies to map out the feasibility of a plate-like regime (e.g., Moresi and Solomatov 1998; Trompert and Hansen 1998; Tackley 2000; Stein et al. 2004; van Heck and Tackley 2008; Foley and Becker 2009). A typical, only partly resolved problem, however, is that feasible yield stress ranges in large-scale numerical models are substantially smaller than those inferred from laboratory experiments on terrestrial rocks (e.g., Kohlstedt et al. 1995). Non-uniform yield stress across the surface, for instance due to variations in composition (Lenardic et al. 2003; Rolf and Tackley 2011) or inherited weakening from previous deformation (e.g., Fuchs and Becker 2019, 2021; Miyagoshi et al. 2020) may help to facilitate the formation of plates and to keep the lithosphere mobile even at higher yield stress of the bulk and/or undamaged lithosphere. Such structural inheritance may result from various mechanisms inducing damage. Grain-size evolution is a prominent example as it is manifested in terrestrial shear zones, where grain size is strongly reduced during active deformation (e.g., Okudaira et al. 2017). Subsequent grain growth heals the damage at a strongly temperature-dependent rate, which implies that planetary surface temperature could be a key factor in determining the role of structural damage of the crust and lithosphere. Venus' high surface temperature may prevent preservation of lithospheric weakness that could ultimately have led to the formation of the first tectonic plates on the (relatively cold) surface of the Earth (e.g., Bercovici and Ricard 2014; Foley 2018).

2.2.3 Mineralogy and Compositional Variation

Composition and mineralogy are additional factors that determine state and dynamics of the mantle. A detailed review of Venus' surface composition and mineralogy is given in Gilmore et al. (2023). Depending on the ambient temperature–pressure conditions, mantle rocks undergo a series of phase transitions that modify crystallographic structure. Both density and viscosity change, which possibly promotes separation of distinct layers (Weidner and Wang 2000) and affects mantle flow structure and radial heat transport (Tackley 1996; Bunge et al. 1997). Certain components may become abundant in specific regions of the mantle. For instance, subducted basaltic crust is buoyant at the base of the mantle transition zone, but negatively buoyant just above and below as garnet transitions to (Mg-)perovskite at slightly higher pressure than ringwoodite. As a result, basaltic crust may get trapped in the transition zone of the (Venusian) mantle (e.g., Armann and Tackley 2012; Vesterholt et al. 2021). Episodic breakdown may induce mantle avalanches and possibly trigger dramatic periods of volcanism and large-scale tectonic activity on Venus (e.g., Papuc and Davies 2012; Vesterholt et al. 2021), although some geodynamic models have challenged the relevance of this behaviour for Venus (e.g., Huang et al. 2013). Phase transitions also alter mantle flow by either absorbing or releasing latent heat. Secondary plumes (e.g., Yuen et al. 1998; Yoshida 2004) are likely to form when hot upwellings penetrate an endothermic phase change like that from ringwoodite to perovskite at a pressure of 23 GPa or equivalent depth of 730-740 km inside Venus (Ishii et al. 2018; Trønnes et al. 2019).

Moreover, phase changes depend on mineral assemblage and rock composition, which vary temporally and spatially in the mantle. Although the bulk composition is mostly determined by the accretion history, compositional mantle heterogeneity may arise from melting and freezing processes. For example, compositional layering may be a consequence of fractional crystallisation of an early magma ocean (e.g., Labrosse et al. 2007). Later-stage partial melting induced by hot anomalies in a largely solidified mantle leads to small-scale heterogeneity as different rock components melt under different conditions and have different affinity to partition into the molten phase (e.g., Hofmann 1997). Such processes compete with mantle convective stirring that tends to homogenise lateral variation, but the mixing time scales in the mantle are large and compositional heterogeneity occurs likely across multiple scales in the Earth's present mantle (e.g., Tkalčić et al. 2015), as is also supported by modelling studies (Gülcher et al. 2021). However, terrestrial plate tectonics continuously induces new heterogeneity by deep crustal subduction, which may be less relevant to Venus without plate tectonics.

Finally, melting processes likely cause the growth of secondary crust on the planet's surface. Basalt-rich melts in the upper mantle rise to shallow depths where they cool and solidify to form fresh basaltic crust. Such magmatic transport may happen either via magmatic intrusions in the shallow subsurface, or via extrusive volcanism in cases where melt diapirs reach the surface via so-called heat pipes (Fig. 2b, Turcotte 1989; Moore et al. 2017). Upon such melt formation, latent heat is consumed and migrated to the surface, where it can be released to the atmosphere. Therefore, melting, magmatism, and volcanism strongly influence the cooling history of a planet and with that the dominant tectonic regime at its surface (e.g., Ogawa 2000; Ogawa and Yanagisawa 2014; Lourenço et al. 2016, 2018, 2020; Byrne 2019).

2.3 Diversity of Geodynamic Regimes

Material complexity in the mantle allows for different styles of mantle flow. In turn, this flow induces deformation in the crust and thus determines surface tectonics. Present Earth is in a mobile-lid tectonic regime (with plate tectonics being a peculiar subcategory), but most other known terrestrial planets show different regimes, which are introduced in this section and conceptually visualised in Fig. 2. Venus' current and past regime is then specifically discussed in detail in Sects. 3 and 4.

The standard view of convection (like that from Bénard's experiments) is the mobile lid regime. The name implies that the lid (typically defined as the region inside the top thermal boundary layer) is in continuous motion, with a velocity at least as large as the average velocity of the convecting layer. In other words, the surface mobility (M), defined as the ratio of surface-averaged to volume-averaged velocity, is $M \geq 1$ (e.g., Tackley 2000). M is a useful measure to distinguish mobile from immobile (thus, stagnant) surface regimes. Earth's plate tectonics clearly falls into the mobile category given the observed speeds of tectonic plates. One peculiarity of plate tectonics, however, is the high degree of localisation of deformation into narrow plate boundaries (Fig. 2c). This localisation is not a general characteristic of the

Fig. 2 A conceptual visualisation of geodynamic regimes relevant for the terrestrial planets: (**a**) Stagnant Lid; (**b**) Heat Pipe; (**c**) Mobile Lid; (**d**) (Plutonic) Squishy Lid. The regime could conceivably transition between these modes through time (see Sect. 4). The Episodic lid is an often-proposed specific example of such transitional behaviour between mode (**a**) and (**c**). The colour coding qualitatively indicates temperature, as indicated. The white layer at the bottom of each panel indicates the core of the planet, the brown layer at the top is the lithosphere. The zoom-ins are used to highlight characteristic volcanic and tectonic structures

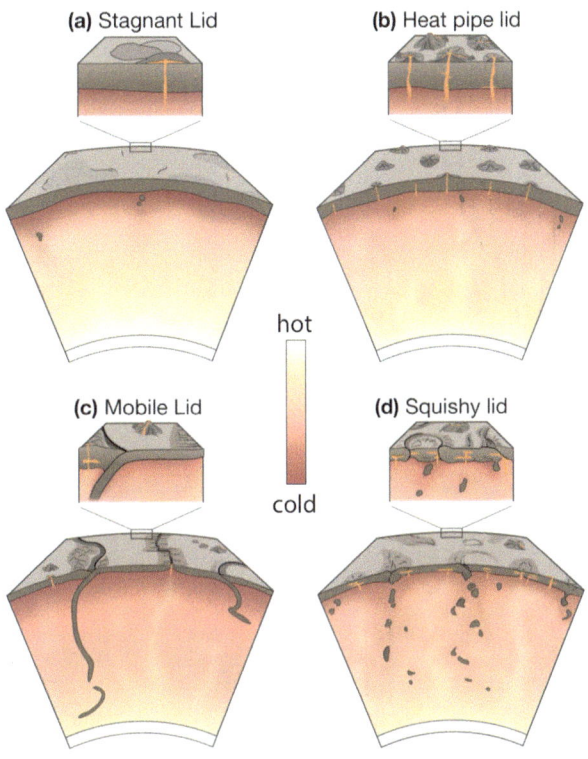

mobile lid regime as deformation may be more diffuse or feature patterns that are absent in Earth's surface motion. In an Earth-like regime, most deformation is confined to a small fractional area of the surface. This can be expressed via the plateness (P), which quantifies how much more localised tectonic deformation occurs in the investigated regime compared to a reference regime, which is typically the simplest possible, thus an isoviscous regime. A value of $P = 1$ indicates an extremely high degree of localisation, while $P = 0$ means that localisation is as poor as in the isoviscous case (Tackley 2000). Therefore, on Earth $P \to 1$, while a wider range of values ($0 \ll P < 1$) is representative of the mobile lid.

The viscosity of mantle rocks is so sensitive to temperature that the surface boundary layer becomes a quasi-rigid lid and decouples from the convecting mantle below (e.g., Solomatov 1995). This regime is often called 'stagnant lid' (Fig. 2a). Small viscosity contrasts between cold lithosphere and hot mantle degrade the decoupling and lead to a mixed regime ('sluggish lid') in which the surface lid is mobile, but at reduced rates ($M < 1$). The strong temperature dependence of mantle rocks makes a stagnant lid seemingly inevitable, unless additional processes lead to localised failure of the stagnant layer. The stagnant lid is thus sometimes seen as the default mode of planetary mantle convection, with present Mars being an archetype of this regime. The stagnant lid is also likely the terminal mode of mantle evolution when the planet has cooled so much that convection is not maintained any longer (e.g., O'Neill et al. 2016; Stern et al. 2018). In the stagnant-lid regime, surface velocity is much smaller than interior velocity ($M \ll 1$). Surface deformation is not particularly focussed into narrow weak zones, so that the characteristic plateness is also small ($P \ll 1$).

The transition between the mobile lid and stagnant lid regimes is determined by the competition between stress induced into the lithosphere and the integrated strength of that

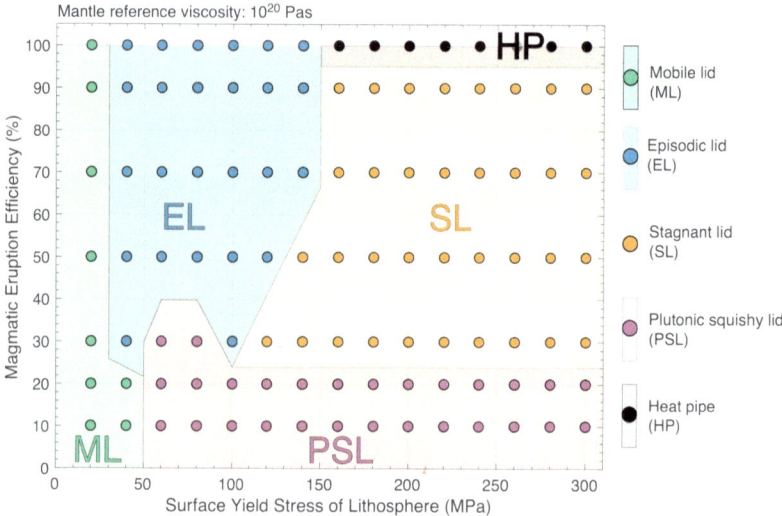

Fig. 3 A geodynamic regime diagram as a function of the surface yield stress of the lithosphere and the magmatic eruption efficiency. Here, the yield stress is defined as the maximum stress material can sustain without deforming plastically; its values cannot be directly compared to laboratory inferred values. Each dot indicates a geodynamic model; its colour denotes the geodynamic regime and is determined on the basis of average surface mobility and plateness–a measure for the localization of surface deformation (Tackley 2000). Background colours illustrate regime fields qualitatively. The regime diagram is also a function of other parameters, such as mantle viscosity (here: 10^{20} Pa s). The plotted data is taken from Lourenço et al. (2020)

lithosphere (Fig. 3). Lid mobility may occur when stresses inside the lithosphere overcome its internal strength locally to induce weakness by failure. The transition to mobile lid is probably not sharp, but includes a transitional range. One flavour of this transition is episodicity. In this context (but not generally), this means that episodes of pronounced surface mobility intersect the evolution in the stagnant-lid mode. Stress may build up gradually during evolution, for instance due to mantle cooling or crustal and lithospheric thickening (e.g., Fowler and O'Brien 1996). At some point the lid is mobilised and tectonic recycling cools the mantle and reduces lid thickness. In turn, the stress in the lithosphere is lowered and active recycling stops again until stress has built up again to initiate another episode of recycling. The scales of such resurfacing events may range from regional (e.g.,Noack et al. 2012; Karlsson et al. 2020; Weller and Kiefer 2020) to global (Turcotte 1993; Armann and Tackley 2012). In its global form, this 'episodic lid' regime has been used as an explanation for Venus' quasi-random distribution of impact craters that imply uniform surface age, but the necessity of such a catastrophic event has been challenged (Sects. 3.2, 4.2). During overturn phases, surface mobility and plateness are similar to the mobile lid characteristics but, between such phases, these diagnostics are representative of stagnant lid behaviour. Temporal averages would thus strongly depend on the time scales of overturn episodes.

The regimes outlined so far do not consider melting and magmatism. As heat transfer across a stagnant lid is much less efficient than across a mobile lid, heat is trapped inside the planet and leads to a hotter mantle. This trapped heat enhances melting and magmatic activity. Latent heat consumption and transport of the hot, buoyant magma via volcanic eruption together facilitate extraction of interior heat, buffer mantle temperatures, and act as a 'mantle thermostat' (e.g., Ogawa and Yanagisawa 2011). This mode has been described as a 'heat pipe' (Fig. 2b, e.g., O'Reilly and Davies 1981; Moore and Webb 2013), as the ascending magma is thought to rise through narrow vertical channels. At present, such a regime may apply to Jupiter's moon Io, which is heated by tidal friction (e.g., Tyler et al. 2015). The 'heat pipe' mode may be particularly relevant for the early phases of terrestrial planet evolution (Stern et al. 2018), such as during the Hadean and Archean epochs on Earth (e.g., Kankanamge and Moore 2016). If Venus featured a stagnant-lid throughout all its history, it could have been in a heat pipe mode during some phases, too (e.g., Turcotte 1989). On modern Venus, there is also evidence for recent and possibly active volcanism (e.g., Smrekar et al. 2010; Bondarenko et al. 2010, D'Incecco et al. 2017, 2021; Campbell et al. 2017; Byrne 2019), but its rate is not well-known.

Heat-piping leads to large volumes of extrusive volcanism, but not to large-scale horizontal motion as in tectonically mobile regimes. The ability of rising magma to reach the surface strongly depends on its buoyancy and overpressure, but also on the strength of the lithosphere and crust. A thick and strong crust prevents magma from reaching the surface and could instead promote the emplacement of magmatic intrusions within the crust. On the other hand, Io features the largest volcanic heat flow in the solar system and may have a thick strong crust, but its lower part may be substantially weakened by magmatic emplacements which could consume up to 80% of the total magma delivered to the crustal base (Spencer et al. 2020). Intrusion could be the dominant mode of terrestrial magmatism as several proposals have been made that only 10–20% of magmatism is extrusive (Crisp 1984; Cawood et al. 2012), but such ratios are spatially highly variable and sensitive to the detailed tectonic setting and structure of the crust (e.g., White et al. 2006). For other planets including Venus, such details are typically unknown, leaving the ratio of extrusive and intrusive magmatism as a major unknown.

If hot crustal intrusions are sufficiently abundant, however, they weaken the surface lid from inside and can lead to yet another tectonic regime, the 'plutonic-squishy lid' (Lourenço et al. 2018, 2020). This regime is characterised by a strong lithosphere that is fragmented into a set of small tectonic units by warm and weak regions caused by plutonism (Fig. 2d). The lithosphere is much thinner and more mobile than in the 'heat pipe' and 'stagnant lid' regime (though not as mobile as in the mobile regime, thus $M < 1$ and $M_{PSL} < M_{ML}$) and deformation is expected to be more localised (thus $P_{PSL} > P_{SL}$). The Archean Earth may have displayed such a regime, but it could also be relevant for Venus given this planet's hot and therefore soft lithosphere (e.g., Gerya 2014; Byrne et al. 2021) and the hints of lateral motion documented in present tectonic features without evidence for Earth-like subduction (Sect. 3.1).

The regimes discussed here may not capture all possible behaviours. Sub- and mixed regimes may exist (Loddoch et al. 2006; Rozel et al. 2015). Moreover, it is challenging to interpret a regime with regards to the state of the mantle: is the displayed regime a snapshot representing the current state of the mantle or an accumulated consequence of planetary evolution of hundreds of millions to billions of years? Describing a planet using the single geodynamic regime observed today seems infeasible in most cases, including Venus (Sect. 4). Determining how and when regime transitions occur is important for interpreting the preserved geological record. One challenge is that evidence from earlier regimes can be (and, on Venus, likely have been) overprinted by signatures from the modern regime. Also, transitions may not be clearly distinct events, but stretch over long time scales (e.g., Weller and Kiefer 2020) and possibly lead to different, coexisting tectonic styles for different parts of a planet's surface (e.g., Robin et al. 2007; Capitanio et al. 2019).

3 Present-Day Mantle Dynamics on Venus

3.1 Observational Constraints

This section briefly reviews the available constraints with the greatest relevance for Venus' deep interior. A more detailed account on these constraints is given in this collection by Ghail et al. (2023), Gilmore et al. (2023), and Herrick et al. (2023).

3.1.1 Gravity and Topography

In the absence of seismic data, gravity and topography are the most powerful constraints on Venus' interior – and at shorter wavelengths – on the strength of the lithosphere and crustal thickness variations. The long–wavelength (i.e., below spherical harmonic degree 40) gravitational and topographic response of a mantle density anomaly depend on its scale, density contrast, depth in the mantle, and on the stress propagation from there through the crust and lithosphere to the surface. Thus, gravity and topography provide key inferences on the viscosity structure of Venus' interior at long wavelengths (e.g., Hager et al. 1985; Kiefer et al. 1986; Zhang and Christensen 1993; Rudolph et al. 2015) and ultimately on the planet's tectonic regime (e.g., Steinberger et al. 2010; Huang et al. 2013).

From the Magellan mission, Venus' topography is available at 10–25 km horizontal resolution and a nominal vertical resolution of 80 m that strongly depends on local topographic gradients (Pettengill et al. 1992). The Magellan gravity field is on average resolved at ~270 km (corresponding to spherical harmonic degree 70, Konopliv et al. 1999), but resolution varies from 170–540 km (spherical harmonic degrees ~35–110) across the surface. Future missions aim to deliver higher resolution (Widemann et al. 2023). As discussed below, gravity and topography analyses can be used to estimate elastic thickness in many areas, but the low resolution of the gravity data results in larger errors (e.g., Anderson and Smrekar 2006). Within these errors, there is evidence for significant variations in crustal and elastic thickness over short spatial scales over much of Venus. Short–wavelength variations may have crustal sources (Fig. 4, Steinberger et al. 2010; Benešová and Čížková 2012).

Lacking seismological data, internal density anomalies are poorly known for Venus, which complicates interpretation of surface gravity with respect to mantle viscosity structure. To overcome this, Steinberger et al. (2010) assumed that Venus' mantle density anomalies are statistically similar to Earth's. Other authors used numerical models to predict synthetic density distributions consistent with Venus' evolution (e.g., Pauer et al. 2006; Armann and Tackley 2012; Orth and Solomatov 2011, 2012; Benešová and Čížková 2012; Huang et al. 2013; King 2018; Rolf et al. 2018b). For these approaches, the spectral representation via a gravity power spectrum is particularly useful (Fig. 4), as models cannot be expected to fit the observed signals from Venus' interior directly, but make predictions with statistically similar amplitudes and scales.

3.1.2 Crater Statistics

Topography and gravity constrain the current state of Venus' interior, but reveal little temporal information. Such insights require reconstruction of the planet's surface chronology, which—lacking dating of Venus' crustal rocks—is mostly based on cratering statistics and the relative age of geologic features. Compared with the stagnant-lid bodies—Mars, Mercury, and the Moon—Venus displays much fewer craters whose distribution taken alone is

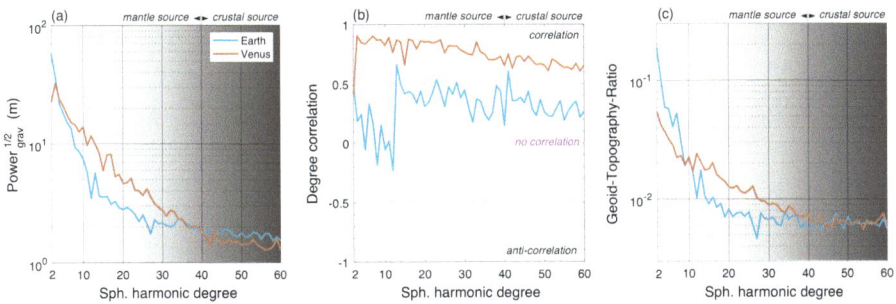

Fig. 4 Power spectra of Venus' present (**a**) surface gravity field, (**b**) degree correlation between gravity and topography, and (**c**) geoid–topography ratio. Power spectra are as defined by Steinberger et al. (2010). The respective spectra for Earth are shown in blue for comparison. In (**a**) and (**c**) the background shading indicates which parts of the spectra may dominantly have deep/mantle (bright) and shallow/crustal sources (dark) according to Steinberger et al. (2010). The gravity model SHGJ180UA01 is used (https://pds-geosciences.wustl.edu/mgn/mgn-v-rss-5-gravity-l2-v1/mg_5201/gravity/). Topography data is obtained from Venus' shape given by Wieczorek (2007) (https://github.com/MarkWieczorek/web/tree/master/spherical-harmonic-models-topography).

not distinguishable from random (Phillips et al. 1992; Strom et al. 1994). The planet's dense atmosphere tends to obliterate small meteoroids, leading to a lack of craters with diameters below ∼2 km (Herrick and Phillips 1994). Considering this atmospheric screening, Venus' craters suggest a young age of ∼150–1000 Myr (Phillips et al. 1992; Strom et al. 1994; McKinnon et al. 1997; Herrick and Rumpf 2011; Le Feuvre and Wieczorek 2011). Overall, there is no consensus on the average surface age, the age of various geological units (e.g., Kreslavsky et al. 2015), the rate of surface weathering or the degree of volcanic embayment of impact craters (Herrick et al. 2023).

In broad terms, accommodating the relatively uniform surface age can be done by the end members of completely wiping the surface clean of craters every few hundred Myr or by having resurfacing occurring everywhere at similar rates but with different mechanisms (e.g., Romeo and Turcotte 2010; Bjonnes et al. 2012). These end members have different implications regarding the long-term evolution of the mantle. The former 'catastrophic' resurfacing—singularly or recurring episodically—implies that mantle temperature and other conditions are not smoothly evolving with time. Global-scale excursions occur, with rapid periods of cooling during a resurfacing episode. The episodes are separated by longer quiescent periods during which the mantle temperature decreases only slowly or even increases due to radiogenic heating (e.g., Turcotte 1993; Nimmo 2002; Armann and Tackley 2012; Rolf et al. 2018b). The latter 'equilibrium' case has a global resurfacing rate (which can be accommodated with different mechanisms) that implies enough surface disruption caused by the interaction between mantle, lithosphere, and crust to eliminate at least several hundred million years of Venus' cratering record, regardless of surface spatial location relative to mantle convection pattern (Herrick et al. 2023). Apart from these end members, the scenario best supported by both the impact crater record (Phillips et al. 1992) and by the removal of extended impact ejecta blankets (Phillips and Izenberg 1995) is regional equilibrium resurfacing. This scenario allows local resurfacing patches on the scale of 100s to ∼1000 km to occur in different locations at different times (O'Rourke et al. 2014). This scenario is consistent with the scale of a variety of volcanic features, including plume-induced subduction (Davaille et al. 2017).

3.1.3 Crustal and Lithospheric Thickness

The thermal state of Venus' interior is tied to the thickness of the lithosphere via conductive heat transfer. It also sets the conditions for the occurrence of magmatism and volcanism (partly) determining crustal thickness. To date, no direct measurements of lithospheric and crustal thickness have been made, but indirect estimates exist (Ghail et al. 2023). Absent seismic data, crustal thickness is typically inferred from inversion of surface gravity and topography (e.g., Maia and Wieczorek 2022). Estimated mean thicknesses vary greatly, from $\sim$10 km (James et al. 2013), to $\sim$20–25 km (Jiménez-Díaz et al. 2015; Maia and Wieczorek 2022) to as much as $\sim$60 km (Steinberger et al. 2010). Lateral variations are substantial, as Anderson and Smrekar (2006) suggested thickness ranges from 0 to $\sim$90 km, whereas James et al. (2013) reported a range of $\sim$5–60 km (Fig. 5a-b). However, such values depend on the non-unique choice of a mean thickness and the assumption that no crust should be thicker than $\sim$70 km–the depth at which Venus' basaltic crust transitions into eclogite. Dense eclogitic crust may be recycled through delamination, unless the crustal root grows too quickly, which could be the case for Maxwell Montes (Namiki and Solomon 1993).

For models of mantle convection, crustal thickness analysis is most easily incorporated in spectral form (Fig. 5c). Wei et al. (2014) combined observed admittance spectra of gravity and topography with convection models to infer which parts of the spectra arise from mantle and from crustal sources and estimated a crustal thickness range of 28–70 km. Thicknesses exceeding 50 km are limited to Ishtar Terra, Ovda Regio, and Thetis Regio, where the gravity and topography signatures suggest isostatic compensation via a thick crustal root (e.g., Smrekar and Phillips 1991; James et al. 2013). In addition, Yang et al. (2016) separated Venus' gravity and topography into dynamic and isostatic components to derive a crustal thickness range of 12–65 km. However, as crustal growth and destruction are strongly tied to volcanism, magmatism and convective processes acting on the crustal base, the long-term evolution of the interior can impose substantial variations of Venus' crustal thickness through time (Sect. 4).

The thickness of the lithosphere, in which the crust is embedded, is defined based on either its elastic strength or temperature profile (Ghail et al. 2023). The thermal thickness is given by the thickness of the conductive thermal boundary layer and depends on the (unknown) temperature and viscosity of the interior. The geoid and topography can be interpreted as indicating average thickness of 200–400 km, possibly less below the volcanic highlands (Solomatov and Moresi 1996; Moore and Schubert 1997). If geoid and topography are explained purely by thermal isostatic adjustment of Venus' stagnant lid, the thermal thickness may be as great as 600 km (Orth and Solomatov 2011). However, a thinner lithosphere (100–150 km) may be more compatible with melt generation rates estimated at Venus' hotspots (Smrekar and Parmentier 1996; Nimmo and McKenzie 1998). The elastic thickness is always less than the thermal thickness and can be estimated from flexural models applied to geological features on Venus (Solomon and Head 1990; Johnson and Sandwell 1994; O'Rourke and Smrekar 2018; Borrelli et al. 2021) and from global admittance maps (Anderson and Smrekar 2006). For most of the planet, the elastic thickness could be as little as 20 km (Anderson and Smrekar 2006), which may indicate a warm lithosphere possibly promoted by intrusive magmatism (Lourenço et al. 2020; Plesa and Breuer 2021) and plume–lithosphere interactions (Gülcher et al. 2020). Admittance for tessera plateaus can be interpreted as evidence of thin elastic lithosphere, reflecting the presumed more ancient time of tessera formation (Maia and Wieczorek 2022), or as simply Airy compensation due to a thick crust (Anderson and Smrekar 2006). Topographic fitting to flexural models can be

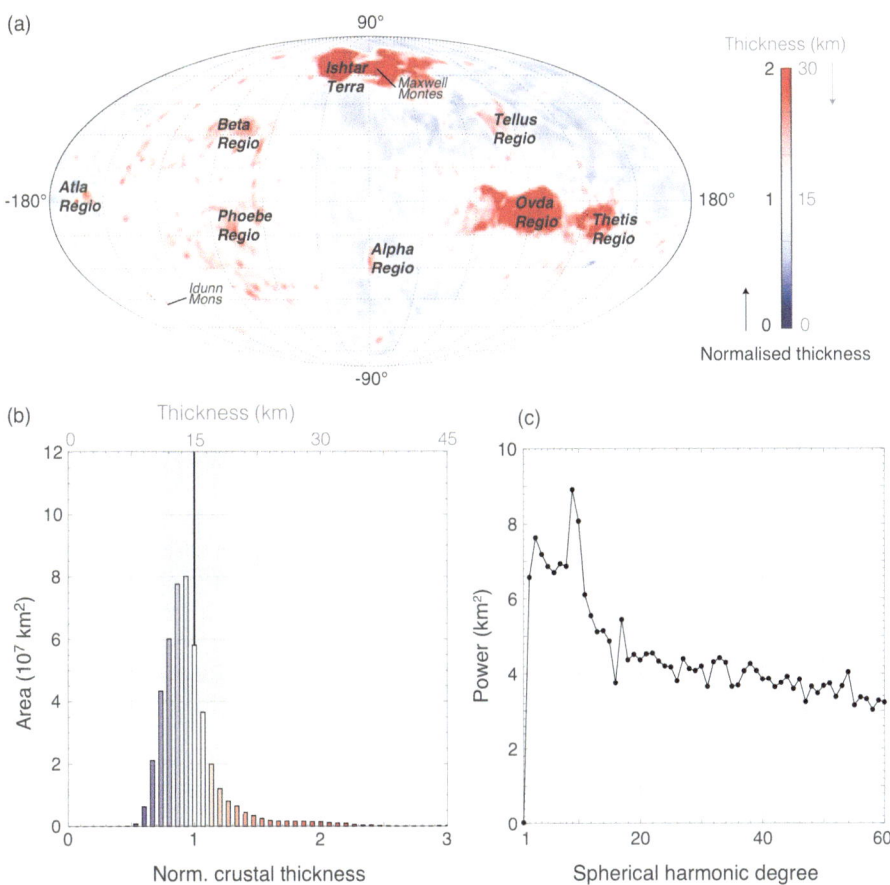

Fig. 5 (**a**) A Mollweide projection of inferred crustal thickness on Venus, centred at 0°E (data from James et al. (2013), assuming a mean thickness of 15 km and a mantle load depth of 250 km). Red and blue indicate higher- and lower-than-average thickness, respectively. (**b**) A histogram representation of panel (**a**). The black vertical line denotes the mean value; the grey-shaded region is bounded by one standard deviation. (**c**) A power spectrum of crustal thickness variations for spherical harmonic degrees 1–60

used to estimate elastic thickness from which heat flux can be inferred (e.g., O'Rourke and Smrekar 2018), based on assuming a particular rheology and strain rate.

3.1.4 Heat Flux and Thermal Emissivity

Venus' average surface heat flux is typically assumed to be smaller than Earth's ($\sim$80 mW/m^2, Davies 2013). Mantle convection simulations in the stagnant lid regime promote low estimates of 10–40 mW/m^2 (Solomatov and Moresi 1996; Gillmann and Tackley 2014; Rolf et al. 2018b; Uppalapati et al. 2020), but episodic overturns may cause a pronounced temporal increase (Armann and Tackley 2012; Gillmann and Tackley 2014; Rolf et al. 2018b; Uppalapati et al. 2020). These average fluxes do not reflect spatial variations. Based on viscoelastic relaxation models, Karimi and Dombard (2017) suggest a higher-than-average flux of 55–90 mW/m^2 at Mead Crater, although the applicability of their model assumptions has been challenged (Ruiz et al. 2019). Impact-crater formation models indicate that Mead's multiring-structure implies much lower flux (<28 mW/m^2, Bjonnes et al. 2021). In contrast, many of Venus' coronae may feature high fluxes (>95 mW/m^2, O'Rourke and Smrekar 2018), whereas domes in the proximity of coronae may display intermediate fluxes (Borrelli et al. 2021). The spatially varying estimates imply a thermally heterogeneous upper mantle and/or lateral thickness variations of the crust and lithosphere, which numerical models of Venus' mantle dynamics can shed further light on (Sect. 4.3).

In addition, thermal emissivity provides constraints on recent volcanic activity and therefore on the thermal conditions in the interior that may make magmatism feasible. Emissivity measurements for Venus' southern hemisphere by Venus Express revealed a number of regions of anomalously high emissivity. These high values do not imply thermal anomalies, but rather are consistent with fresh, unweathered basaltic composition (Smrekar et al. 2010). Each of these features had previously been identified as having the broad topographic rises and major volcanoes analogous to terrestrial hotspot features (McGill 1994). The gravity anomalies for nine such features are interpreted as indicative of active mantles plumes (Smrekar and Phillips 1991; Kiefer and Hager 1991; Smrekar 1994), which offers insight into the planform of Venus' mantle flow (e.g., Huang et al. 2013; Rolf et al. 2018b). The presence of high emissivity and gravity anomalies at both large-scale rift-dominated and corona-dominated features is consistent with current activity at small and large plumes (Smrekar et al. 2010), rather than requiring different convective regimes to allow their formation (Jellinek et al. 2002). The inferred number of hotspots places bounds on mantle viscosity ($\leq 10^{20}$ Pa s, if the mantle is mostly internally heated) and core temperature (≥ 1700 K, Smrekar and Sotin 2012). Future data may reveal evidence of additional locations of recent volcanism and find evidence of additional small-scale plumes at depth, thus providing further insights into mantle temperature and planform.

3.1.5 Surface Tectonic Features Linked to the Deep Interior

Observations of the distributions, types, and spatial and temporal relations of tectonic and volcanic landforms provide important constraints for models of Venus' interior. These constraints include estimates of surface strains, areas of crustal shortening and extension, regions of stratigraphically young volcanism, and the extent and distribution of likely active surface features (see Ghail et al. 2023). A key question for this chapter is to what extent are interior processes manifested at Venus' surface? Gravity–topography admittance ratios obtained during the Magellan mission indicated that some portions of the planet are dynamically supported. For example, the large volcanic rises – such as Beta, Atla and Themis Regiones – and several smaller rises are consistent with their being supported by large mantle upwellings (e.g., Smrekar and Phillips 1991). It is notable that the volcanic flows hosting geochemical evidence of incomplete weathering (e.g., Smrekar et al. 2010; Brossier et al. 2020) are those interpreted to be comparable to hotspots on Earth (Smrekar 1994).

Another issue is whether mantle forces drive large-scale horizontal motion on Venus? There is no morphological evidence for Earth-like oceanic plate movement and convergent (rather than roll-back or retrograde) subduction on Venus today. Nonetheless, because of the high surface temperature, there may be a weak layer within the lower crust or upper mantle, akin to the rheological layering in continental lithosphere on Earth (Buck 1992; Ghail 2015). If so, tractions from mantle flow on this low-strength layer may drive surface deformation (e.g., Leftwich et al. 1999). Rheological data for Venus' lithosphere is sparse, but there is a geological basis for interpreting lateral motions on the planet. For example, Harris and Bédard (2013, 2014) documented evidence for Lakshmi Planum having collided with Ishtar Terra. Their work proposed that, akin to how continents move on

Earth, mantle flow on Venus may have pushed against a deep lithospheric keel of Lakshmi Planum, driving it northwards over perhaps 1000s of kilometres. Similar deep-seated mantle flow may have led to major shear displacements between Ovda and Thetis Regiones. Recent work has suggested more modest horizontal motions, too, again potentially driven by mantle flow. Numerous portions of Venus' lowlands feature intersecting bands of extensional and shortening structures (termed 'groove belts' and 'ridge belts', respectively) that delineate smooth plains-filled lows (Byrne et al. 2021). Many such intersecting belts show evidence of lateral displacements and paint a picture of these plains-filled lows being mechanically coherent crustal blocks that have moved with respect to one another. Calculations from gravity-induced mantle flow show that the tractions arising from mantle motion today may transfer sufficient force to the surface at every location where these blocks have been observed—consistent with this motion having taken place geologically recently, and perhaps even ongoing (Byrne et al. 2021).

Various modelling studies have focused on the formation of Venus' prominent coronae, which are thought to have been formed by mantle upwelling impinging on the lithosphere (Stofan et al. 1992). The observed variety of their morphological forms can be explained by the spectrum of development of individual coronae (e.g., Stofan et al. 1991; Smrekar and Stofan 1997; Koch and Manga 1996; Hoogenboom and Houseman 2006; Gülcher et al. 2020). Models of plume upwelling generally evolve from domes to depressions; the opposite is true for models of corona formation above dripping or delaminating lithosphere (Hoogenboom and Houseman 2006; Piskorz et al. 2014). For coronae that form over small upwellings, a major unresolved question is the origin of the upwelling and their relationship to circulation patterns in the mantle. Are they shallow upwellings or do they rise up from the core–mantle boundary, like classical plumes on Earth (e.g., DePaolo and Manga 2003, French and Romanowicz 2015) and those inferred for large scale (1000-2000 km) features such as Atla Regio (e.g., Smrekar 1994)?

3.1.6 Magnetic Field

Pioneer Venus Orbiter provided the upper limit on the intrinsic magnetic field: any magnetization is $\leq 10^{-5}$ times weaker than Earth's magnetic field today (Phillips and Russell 1987). Other missions have failed to detect any intrinsic magnetism and this apparent dearth of signal has been used to exclude the existence of a magneto-hydrodynamic dynamo in a convecting, electrically conductive core of Venus (see Gillmann et al. 2022) Theory predicts that a dynamo inside Venus would be apparent in available data if it existed (e.g., Stevenson 2003, 2010).

The lack of an intrinsically generated magnetosphere today may inform models of mantle convection. Core convection is primarily affected by core cooling, which is determined by mantle heat transfer. The described (absence of) observation is most useful assuming an Earth-like core for Venus that is partially liquid and its liquid part is chemically homogeneous. Then, the heat flow across the core–mantle boundary today is below the critical value required to drive core convection (e.g., Labrosse 2015; Nimmo 2015). If Venus has no solid inner core, the dynamo has to be sustained by thermal convection alone and the critical value equals the heat flow along the core adiabat. Depending on the core's thermal conductivity, this could range from $\sim$5–15 TW (see Lay et al. 2008 and references therein). With a solid inner core, the latent heat of freezing and the gravitational energy release associated with the partitioning of light elements into the liquid following inner core growth provide additional sources of power and lower the threshold for a dynamo to roughly half the total adiabatic heat flow (Blaske and O'Rourke 2021).

However, using these arguments to constrain Venus' modern core heat flow is problematic, because Venus' core may not be like Earth's. The estimates for the moment of inertia (Spada et al. 1996; Margot et al. 2021) and tidal Love number (Konopliv and Yoder 1996) are too inaccurate even to rule out that Venus' core is fully solid or has a liquid layer too thin to sustain a dynamo (Dumoulin et al. 2017). A fully solidified core still appears unlikely with at least some fraction of sulphur (S) in the Venus core as this would imply core temperatures close to the Fe-FeS eutectic. These are likely too low (e.g., Boehler 1998) to allow for partial melting of the mantle that is prerequisite for volcanic activity. However, core heat flow also depends on heat transport efficiency on the mantle side and if Venus' mantle plumes are merely hot thermals (Jellinek et al. 2002), they may extract as little heat as 3 mW/m^2 from the core. That is an integrated heat flow of less than 0.6 TW for the range of possible core radii (3500 $\pm$ 500 km, Margot et al. 2021).

3.2 Which Geodynamic Regime for Present Venus?

A key question is which tectonic regime matches the observational constraints summarised best? The mobile lid regime implies a comparably thin lithosphere as well as efficient heat transfer through the surface. The typically suggested reduced heat flux on Venus with respect to Earth then points to a thicker lithosphere favoured by less vigorous mantle flow and a more viscous mantle. However, predictions of Venus' modern geoid from mantle flow models (e.g., Steinberger et al. 2010; Rolf et al. 2018b) do not point to systematically higher mantle viscosities. Subduction—the ultimate characteristic of mobile lid convection—cools the mantle efficiently, and a colder mantle triggers less volcanism, confined to regions underlain by hotter-than-average mantle (see e.g., Bondarenko et al. 2010; Smrekar et al. 2010; Smrekar and Sotin 2012). Although active subduction does not occur on Venus today, there are indications for subduction-like processes (e.g., Schubert and Sandwell 1995), either in the form of localised retrograde (Sandwell and Schubert 1992) and/or as plume-induced subduction (Davaille et al. 2017; Zampa et al. 2018). A continuous network of divergent and convergent plate boundaries is still lacking, so that the indication of subduction processes does not imply planet-wide mobile lid tectonics (but see Sect. 4.3). Byrne et al. (2021) argue that in many locations Venus' present lithosphere appears fragmented into dozens of coherently moving crustal blocks – like tectonic plates, albeit the largest identified block is only $\sim 1.9 \times 10^6$ km^2 and thus much smaller than any major tectonic plate on Earth. The small size could be an expression of merely crustal deformation processes rather than fully-developed mobile lid tectonics. This in turn suggests smaller horizontal forces to drive lateral motion of plate-size structures, and has consequences for the rheology of Venus' crust and upper mantle.

On Earth, a low-viscosity asthenosphere is evident in the post-seismic deformation of large earthquakes (e.g., Hu et al. 2016) and known to promote long-wavelength flow in the upper mantle and large tectonic plates (e.g., Höink and Lenardic 2008). On Venus, such evidence is lacking and the asthenosphere may be less pronounced—in particular if its strength is related to water content (Green et al. 2014; Masuti et al. 2016) and Venus' interior is relatively dehydrated. The lacking or weakly pronounced asthenosphere is manifested in the high correlation of Venus' gravity and topography (Fig. 4b), which can also be expressed via the spectral admittance ratio (e.g., Kiefer et al. 1986). At long wavelengths (spherical harmonic degrees < 10), the Earth features positive geoid and negative topography, leading to negative admittance. On Venus, in contrast, the admittance is positive, which requires stronger coupling of the mantle and the lithosphere, hence the absence of a pronounced asthenosphere (e.g., Kiefer et al. 1986; Steinberger et al. 2010; Rolf et al. 2018b). This absence is also supported by the large geoid anomalies associated with Atla and Beta Regiones that require stronger coupling of upwelling plumes to Venus' surface layer (Smrekar and Phillips 1991). In contrast, deformation experiments on crustal plagioclase support a strong rheological contrast between crust and mantle at Venus' Moho conditions. This promotes a decoupling of both layers, which reduces the magnitude of mantle tractions transmitted to the crust and impedes large-scale lateral motion (Azuma et al. 2014; Katayama 2021). Smaller-scale surface mobility is still feasible, especially if facilitated by a weak lower crust (e.g., Arkani-Hamed 1993; Ghail 2015; Byrne et al. 2021). A weaker lower crust may however have difficulty to support some of Venus' highest topography, so that this feature may not be global.

Venus' crater population supports the infeasibility of a large-scale mobile plate-like regime today as it would efficiently renew the surface, deform, and erase craters. Although Venus' cratering record supports a geologically young surface, the mobile-lid framework has difficulty accounting for the apparent random distribution of Venus' craters and the implied age uniformity (Sect. 3.1.2). For example, terrestrial tectonic reconstructions since 200 Ma indicate substantial variations in surface age, even after ignoring the presence of anomalously old continents (e.g., Coltice et al. 2013). Finally, continuous subduction of surface material would foster heat transport from Venus' core into the mantle making dynamo action and inner core nucleation more likely. However, other aspects—such as stable stratification of the core—could explain the lack of dynamo action even if the mantle were in a mobile-lid regime today (Smrekar et al. 2018).

An episodic lid regime with an ongoing resurfacing episode is difficult to distinguish from the mobile lid regime given the (intermittent) surface motions, but peak rates of surface velocities and eruption rate are likely higher during short episodes than in a long-term stable mobile lid. If resurfacing happens via one major zone of convergence as promoted by some numerical models (e.g., Karlsson et al. 2020), the deep mantle density distribution becomes heterogeneous at hemispheric scale and causes a too large offset between Venus' centre of mass and centre of figure (Bindschadler et al. 1994; King 2018). If the resurfacing process operates on localised, regional scales, this issue is less problematic, but still a sudden increase in volcanism would occur due to the lithospheric thinning following the onset of resurfacing. This is challenging to accommodate with the apparent rates of recent volcanism on Venus. An ongoing resurfacing episode on Venus is thus similarly unlikely as a continuous mobile regime. However, resurfacing episodes happen with undetermined frequency (e.g., Uppalapati et al. 2020), and are separated by long stagnant-lid periods especially during mature stages of the planet's evolution (Armann and Tackley 2012). Therefore, a current quasi-stagnant state with long tectonic quiescence between episodes is difficult to rule out based on available data (Rolf et al. 2018b). During a long-lasting stagnant lid state, the mantle becomes mechanically decoupled from the shallow lid because of the high viscosity contrast between mantle and crust. The mantle does not exert enough forces on the lid to coherently move it horizontally in a plate-like fashion. Under this regime, surface heat loss is greatly reduced, but it remains unclear how such a reduction accounts for locally elevated fluxes, such as suggested for coronae structures (Sect. 3.1.4). Convection-induced deformation in a subcrustal lid with a weak lower crust (Ghail 2015; Byrne et al. 2021) and plumes impinging and eroding the lithospheric base (Smrekar and Stofan 1997; O'Rourke and Smrekar 2018; Gülcher et al. 2020) together could provide an explanation.

With inefficient heat loss during a stagnant-lid period, Venus' mantle would have difficulty losing its heat. If such a period is established after previous extended periods of mobile lid tectonics, the mantle even heats up – despite the background trend of decaying radiogenic heat sources – until the mantle temperature has adjusted to the rate of radiogenic heat production. During the heating process, the mantle and core slowly equilibrate thermally, decreasing heat transfer and making a core dynamo less feasible. Moreover, the upper mantle heats up towards its solidus, resulting in substantial magmatism and volcanism. Moderate rates of localised volcanic resurfacing are not at odds with the relatively uniform surface age inferred from Venus' crater distribution (e.g., Kreslavsky et al. 2015). With purely volcanic resurfacing, the uniform surface age requires lava flows to be distributed broadly across the surface, as indeed indicated on geological maps of Venus' surface (e.g., Ivanov and Head 2011). In the absence of cold sinking slabs, lateral variations in mantle temperature may be small in the stagnant-lid regime, so that a global layer beneath Venus' lithosphere could be partially molten and feed spatially random volcanism. Whether this form of equilibrium resurfacing is a feasible scenario to generate a uniform surface age consistent with the crater distribution remains debated (e.g., Romeo and Turcotte 2010; cf. Bjonnes et al. 2012). However, O'Rourke et al. (2014) show that scales of volcanism of several 100 to $\sim$1000 km are consistent with the crater population.

The relevance of the stagnant lid for Venus also depends on this regime's definition. Often defined by small surface mobility, it is undefined what 'small' means. Although less than in the mobile lid, some mobility is permitted, in particular if the main manifestation of the stagnant lid is the dominance of heat transport via conduction through the lithosphere (Byrne et al. 2021). Also, magmatic processes potentially induce mobility (e.g., Noack et al. 2012; Lourenço et al. 2016, 2020), depending on the efficiency of magma eruption. Heat pipes can operate at any non-zero eruption efficiency, but at high efficiency less magma production and therefore a lower temperature is required in the mantle to make heat piping the main mode of planetary resurfacing. The estimated rates of volcanic activity on modern Venus do not reflect a dominance of heat piping. The prospect of a heat-pipe regime is further challenged by the estimates of Venus' crustal thickness, which typically suggest an average crustal thickness of only a few 10s of km (Sect. 3.1.3), whereas evolution models of Venus' mantle with maximum eruption efficiency—resembling the heat-pipe regime—typically predict much larger global crustal thicknesses of $\sim$100 km (Armann and Tackley 2012; Rolf et al. 2018b). The high volcanic fluxes required to form such thick crust are difficult to reconcile with the age of Venus' present crust (e.g., Turcotte 1989). Reduced magma eruption efficiency ($<20\%$) can substantially decrease the thicknesses of the present crust in the stagnant lid regime (Fig. 6). At this point, crustal intrusions control the thickness and strength of the crust so that the plutonic-squishy lid can be entered (Sect. 2.3; Lourenço et al. 2020).

In this regime, Venus' groove and ridge belts serve as the regions where a lot of magma is intruded into the crust, coexisting with relatively coherent crustal blocks (Byrne et al. 2021). Recent volcanism then represents the relatively small extrusive portion of magma that makes it to the surface, but the total volume of magmatism is much higher, so that the rate of volcanism is not directly linked to the thermal state of the upper mantle. Extrusions are more likely where the rising magma has anomalously high buoyancy or where the integrated compressive strength of the crust is low. On Venus, extension in a subcrustal lid around regions of major mantle melting is in line with regional topography variations (Ghail 2015) and may keep the compressive strength low to facilitate the rise of magma to the surface. Idunn Mons may be an example for such a region (D'Incecco et al. 2017). The ubiquitous presence of novae, coronae, and other volcanic features (e.g., Stofan and Smrekar 2005) is another hint to such a spatially heterogeneous regime in which tectonic and volcanic resurfacing interact and widespread volcanism is consistent with a relatively uniform surface age on scales > 1000 km (O'Rourke et al. 2014). Most of the buoyant magma would intrude the crust, keeping it warm and sufficiently weak for crustal flow and surface deformation required to eliminate crater signatures.

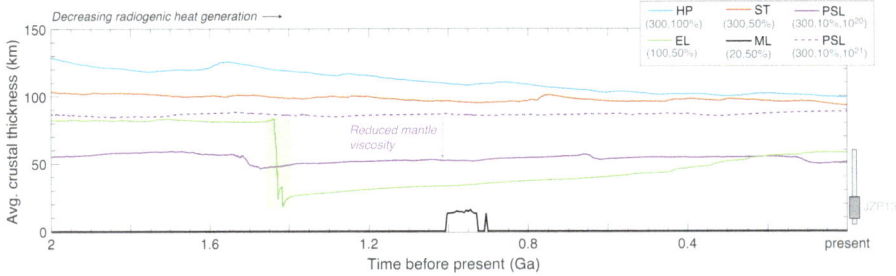

Fig. 6 A time series of mean crustal thickness during 2 Gyr of model evolution in different geodynamic regimes (data from Lourenço et al. 2020). HP: Heat Pipe; ST: Stagnant Lid; PSL: Plutonic Squishy Lid; EL: Episodic Lid; ML: Mobile Lid. Values in parentheses indicate the surface yield stress in MPa, and the magmatic eruption efficiency in %. In the EL model, the shaded area indicates a large-scale overturn episode. The grey bars at the right margin of the plot indicate estimates of Venus' present crustal thickness (James et al. 2013, dark: mean, bright: range). All cases assume a mantle viscosity of 10^{20} Pa s, except one (dashed, 10^{21} Pa s)

In summary, Venus' present dynamic regime remains hard to be defined with the available data. Observations and predictions by numerical models can be combined, but a complete answer to how Venus loses its interior heat and causes deformation of the lithosphere and crust does not yet exist; future missions will provide new data that will help to resolve this issue (Sect. 5.2). For the time being, a regime with a globally mobile lid can be ruled out – this includes, but in a more general way, the Earth's specific regime of plate tectonics – and so can be an ongoing (or recently ceased) global overturn event in the episodic lid regime. However, Venus is also not in the classical stagnant-lid state without any lid mobility such as Mars, Mercury, or the Moon (see e.g., Tosi and Padovan 2021). Improved estimates of strain on Venus' surface could help to distinguish between the different regime evolutions, as characteristic values tend to differ by orders of magnitude (e.g., Grimm 1994). Another promising avenue invokes the interplay of magmatism and tectonics to generate a hybrid regime, such as the plutonic-squishy lid. The interaction between tectonics and magmatism as well as between mantle and crust vary across Venus' surface due to mantle flow patterns and crustal variations resulting from the cooling and deformation history of the planet. Therefore, understanding Venus' current regime also requires insights on past tectonic states.

4 Venus' Mantle Dynamics Through Time

For Earth, the mantle during the Archean (>2.5 Ga) was several 100 K hotter than at present and operated under a different regime, leading to a different style of tectonics (e.g., Moyen and van Hunen 2012). Some aspects of Venus' current geology may be analogous to those present on early Earth (e.g., Ghail et al. 2023). As discussed in this section, Venus may have experienced similar regime transitions, in particular with a more mobile period in its past. However, details on the style of mobility during that epoch and when it occurred still need to be revealed.

4.1 Shaping Venus' Interior Through Time

The thermal evolution of Venus' mantle is determined by the balance of interior heating and the net flux of heat leaving the mantle (e.g., Smrekar et al. 2018). The net flux is mostly the sum of heat loss from the interior to the atmosphere, heat transfer from the core to the mantle, and heat transfer related to melting and magmatism via latent heat consumption (release) upon melting (freezing), as well as extraction of heat through extrusive volcanism. Internal heat generation comprises mostly radiogenic heating, but can also involve shock heating induced by asteroidal impacts and heating by tidal dissipation (negligible for Venus). Even if the present thermal state, fluxes and heating rates were well known, backward integration of the present heat budget is challenging because the different contributions to the heat budget are nonlinearly coupled and depend on the geodynamic regime and its history (see e.g., Korenaga 2008).

4.1.1 Evolution of the Upper Mantle and Crust

The well-known half-lives of the principle radiogenic isotopes expected for Venus suggest a decay of internal heating by a factor of about three over Venus' lifetime (Turcotte and Schubert 2017). Jellinek and Jackson (2015) proposed that ablation from early impactors on Earth may have caused the loss of an isolated geochemical reservoir with a lower $^{142}Nd/^{144}Nd$ ratio lower than ordinary chondrites. Such a loss could imply a 20–45% lower radiogenic heat production on Earth compared to Venus, potentially enough to explain their divergent evolution pathways (Weller and Lenardic 2015). Stronger radiogenic heating increases mantle temperature and decreases viscosity, effectively resulting in weaker coupling to the mantle and lower stress in the lithosphere. Therefore, planets operating with high mantle temperatures are more likely in the stagnant lid regime (e.g., Stein et al. 2013; Weller et al. 2015). However, whether ejected enriched material would not re-accrete to the planet and why impact ablation should happen on Earth, but not on Venus has not been discussed by Jellinek and Jackson (2015). Moreover, measurements on enstatite chondrites (Boyet et al. 2018) suggest a nucleosynthetic origin of the Earth's apparently anomalous ^{142}Nd composition (e.g., Burkhardt et al. 2016); removing this effect leads to almost indistinguishable Nd ratios of chondrites and the accessible Earth. This leaves the proposal of Jellinek and Jackson (2015) as a possibility that cannot be disproven, but it is neither sufficiently supported by available data.

More established is that small variations in surface temperature affect the relationship between radiogenic heating and the geodynamic regime, and may have acted as important triggers for transitions in Venus' tectonic regime (Sect. 4.2). Depending on the thermo-tectonic history, the result can be multiple tectonic states, feasible for the same conditions. To date, the link between radiogenic heat production, mantle thermal state, and tectonic regime has mostly been inferred from theoretical scalings and numerical models employing a pseudo-plastic rheology (Sect. 2.2). As discussed, stress imparted to the lithosphere through convection decreases with increasing mantle temperature. Reaching the yield stress to induce plastic failure is thus more complicated and the feasibility of developed mobile-lid tectonics on a younger, hotter planet is reduced. On the other hand, lithospheric stress increases with increasing slope of the lithospheric base (e.g., Fowler 1985; Wong and Solomatov 2015) making it sensitive to the aspect ratio of convection cells. If Venus mantle established sufficiently wide convection cells at some point, initiation of subduction-like processes would have been facilitated (Wong and Solomatov 2015), in particular when the planet was not in the heat-pipe regime anymore (Kankanamge and Moore 2016). Another possibility to enable lithospheric weakening in an earlier, hotter mantle—and thus a challenge to the notion of the prevalence of early hot stagnant lid mantle convection—is via grain-size-dependent damage evolution as the deformational work driving grain size reduction does not decrease in a hotter mantle (Foley 2018). In contrast, an additional argument for the promotion of stagnant lid behaviour independent of rheological arguments may come from the physics of mantle systems heated from below and within. In such systems, high levels of radiogenic heating tend to inhibit convection, and thus stress imparted via internal convective velocities, due to transition in convective planforms from predominantly sheet-like to plume-like (Weller et al. 2016; Weller and Lenardic 2016; Lenardic et al. 2021).

This discussion typically assumes a uniform radiogenic heat production across the mantle, which is not the case in a planetary mantle with melting-induced differentiation. Incompatible radiogenic elements preferentially partition into the liquid phase upon melting, which leads to an enriched crustal layer, impacts mantle cooling history (e.g., Ogawa 2018), and can alter surface heat flux (e.g., Lourenço et al. 2018; Vilella and Deschamps 2021). However, even with extreme partitioning that forces almost all radiogenic elements into the melt, numerical models of Venus' evolution in the stagnant lid—or rather heat pipe—regime cannot produce a thin crustal layer consistent with other estimates for Venus (Armann and Tackley 2012). As discussed in Sect. 3.2, a regime supporting high volumes of intrusive magmatism—such as the plutonic squishy lid—may be more feasible in this regard as it facilitates remixing of radiogenically enriched material into the mantle. The effects on radiogenic partitioning could then be less pronounced than they would be otherwise (Lourenço et al. 2018).

The melting-induced crustal layer features thickness variations reflecting the lateral variations of temperature and flow in the mantle in a time-integrated sense. Crustal heterogeneity locally increases the stress within the lithosphere and facilitates surface mobilisation (Lourenço et al. 2016). This mobilisation is reinforced when the crustal root transforms into eclogite (at ∼60–70 km depth on Venus) as this denser phase induces additional buoyancy and stress to trigger episodic mobilisation of an otherwise stagnant lid (e.g., Rolf et al. 2018b). Thinning the crustal layer by a mobile episode may let the stress in the lithosphere drop below the yield strength, promoting tectonic quiescence. After shutting down recycling, surface heat loss decreases and the mantle gradually heats up, favouring larger volumes of melting, magmatism and volcanism until the crust and lithosphere eventually experience sufficient melting-induced heterogeneity and stress to reinitiate surface mobility. As radiogenic heat production decreases with time, heating up the mantle and powering magmatism takes progressively more time, possibly increasing the interval between surface recycling events until they may eventually fade (e.g., Armann and Tackley 2012; Vesterholt et al. 2021).

The importance of melting-induced heterogeneity and crustal tectonics for Venus depends on the ability of magma to propagate through the crust and reach the surface. Venus is widely covered in basaltic volcanic rocks, so that the magma eruption efficiency is clearly non-zero. Without knowing the total volume of generated magma in the mantle, refined estimates are difficult to make, but important as different eruption efficiencies can lead to different tectonic regimes (Sect. 2.3, Fig. 3). Reducing the eruption efficiency in the stagnant lid regime tends to increase the average age of Venus' crust, but also creates larger spatial age variations (Uppalapati et al. 2020), which may be difficult to reconcile with Venus' crater distribution. Assuming a continuous stagnant lid, dominantly intrusive volcanism could strongly reduce the mechanical lithospheric thickness and better match present elastic thickness estimates for Venus, especially if intrusions are placed at shallow depths (∼50 km, Plesa and Breuer 2021). Magmatic inclusions cool less efficiently than surface lava so that pockets of high melt fractions may be preserved whose evolution is difficult to address in global models (see Abe 1995; Rozel et al. 2017; Lourenço et al. 2018). Moreover, magma eruption varies with the integrated crustal strength and the buoyancy of rising magma, whose ascent through the crust may also be affected by permeability barriers

(Schools and Montési 2018). For all these complexities, the interaction of magmatism and tectonism remains incompletely understood–even in terrestrial settings–and demands future research.

4.1.2 Evolution of the Deep Mantle and Core–Mantle Coupling

The lack of a magnetosphere is a characteristic that any evolution scenario of Venus' mantle has to match (Sect. 3.1.6). However, without sufficient knowledge of the state and structure of the core, including the existence of an inner core, different evolutionary scenarios remain possible (see Smrekar et al. 2018). Speculatively, a basal magma ocean could exist in Venus' lower mantle, thick enough to suppress core cooling, but too thin to support a dynamo by itself today (O'Rourke 2020). Alternatively, Venus could have experienced a 'gentle' accretion relative to Earth. In the absence of mechanical stirring provided by late energetic impacts, Venus' core could have retained a stable primordial chemical stratification (Jacobson et al. 2017). With a stably stratified or solid core, even rapid core cooling would not produce a dynamo and planetary magnetic field.

Realistic predictive models of planetary dynamo generation are still to be developed (see Wicht and Sanchez 2019), however, simplified thermal evolution models of Venus' interior suggest that the prospects for an internal dynamo were more favourable in the past when Venus' interior was hotter (Nimmo 2002; Driscoll and Bercovici 2013, 2014; O'Rourke et al. 2018; Gillmann et al. 2022). Detecting crustal remanent magnetism could indicate that an internal dynamo once operated on Venus. Although the surface of Venus is hot, modern temperatures are still below both the Curie point and the expected blocking temperatures of magnetite, a common magnetic carrier. The Pioneer Venus Orbiter and Venus Express would have detected large magnetised surface regions northwards of 50°S, at least if magnetization is coherent over horizontal spatial scales comparable to the orbital altitude ($\sim$150 km, Russell et al. 2007). However, crustal magnetization may still be undetected on Venus if located near the south pole or if the spatial scales of preserved crustal magnetization at present day are small.

Recent models of Venus' core–mantle–atmosphere coupling (O'Rourke et al. 2018) point out that an initially hot and chemically homogeneous core should remain at least partially liquid today. If Venus has an Earth-like core, the absence of a dynamo is easiest to explain if the thermal conductivity of core material is at the high end of recent estimates (i.e., >100 $Wm^{-1} K^{-1}$). Internal heating in the core and/or dense insulating layers at the base of the mantle could keep the core fully molten. In particular, a basal magma ocean slows down core cooling, but similarly requires slow mantle cooling to avoid solidification. For this, a continuous stagnant lid regime would be more favourable as it keeps the mantle hotter than one (episodically) cooled by active resurfacing. However, the stagnant lid does not rule out a core dynamo in general, with Mercury being a solar system example for such a planet (e.g., Christensen 2006). If melting and magmatism provide a sufficient heat sink for the mantle, a core dynamo may be active on Venus until $\sim$0.3 Ga in the stagnant lid regime with sufficient extraction ($>50\%$) of magma to cool the mantle, but only until $\sim$3 Ga when magma is inefficiently extruded and/or the magma volume is small (Driscoll and Bercovici 2014).

Thermal insulation of the core may also result from compositional layering of the mantle. When entering the lower mantle, basalt becomes denser than olivine, making recycled crustal material denser than the ambient lower mantle and possibly leading to piles of basaltic material atop the core-mantle boundary. Such ponding can happen to some degree in all regimes discussed in Sect. 2.3, but most pronounced in the episodic lid or mobile lid where sufficiently strong, deeply subducting slabs provide an efficient way of downward transport (Lourenço et al. 2020). The presence of such relatively dense piles may prevent Venus' core from solidifying completely (O'Rourke et al. 2018). If originating from the surface and being transported efficiently across the mantle, the settled material may initially be cold. The temperature contrast would temporarily increase heat transfer from the core (King 2018; Rolf et al. 2018b) and possibly enable intermittent thermally-driven dynamo activity. In the longer term, the accumulated dense layer would heat up, in particular when enriched in radiogenic nuclides, insulating the core and suppressing core cooling. Depending on the competition between the thermal and compositional buoyancy, the dense layer may be permanent or remix back into the mantle if not continuously fed by new settling material.

On Earth, most of the dense material atop the core–mantle boundary is organised in two antipodal provinces centred at the equator (Garnero et al. 2016 and references therein). For Venus, seismological constraints are lacking, but the small offset of $\sim$280 m (Bindschadler et al. 1994) between Venus' centre of mass (CoM) and centre of figure (CoF) rules out strong hemispheric asymmetry in the thickness of the dense layer. This finding seemingly argues against a recent global resurfacing episode, which produces much too large an offset (King 2018). However, the hemispherical-scale variations may distort the offset only for a relatively short time span ($\sim$100 Myr, Fig. 7); afterwards, propagation of active resurfacing zones and the reorganisation of mantle flow may override the hemispherical anomaly in the deep mantle and promote a smaller-scale structure that could have much less impact on the CoM-CoF offset. Clearly, the feasibility of such a scenario depends on Venus' lower mantle properties. Given their uncertainty, the observed small offset may not definitively rule out an episodic resurfacing event on Venus, but at least places bounds on the timing of the latest resurfacing episode: recent cessation less than 150–200 Myr ago seems infeasible. This prediction, in turn, is consistent with the decay of long-wavelength gravity anomalies observed over a time scale of $\sim$100–150 Myr after overturn cessation (Rolf et al. 2018b). Minimum estimates of Venus' mean surface age are similar (e.g., Herrick and Rumpf 2011; Le Feuvre and Wieczorek 2011), but such a young surface can also be generated without lithospheric overturn when magma eruption efficiency is high (Uppalapati et al. 2020). However, mantle overturn and deep recycling wipe out the pattern of mantle plumes established prior to an overturn episode and re-establishing that pattern tends to take much longer after the cessation of active resurfacing (Rolf et al. 2018b).

4.2 Regime Transitions: Triggers and Time Scales

As emphasised previously, the planetary tectonic regime changes through time in response to the thermal and compositional evolution of the mantle. A relevant question for this paper is what could trigger a regime transition and on what time scale?

4.2.1 Surface Temperature Variations

Being closer to the Sun than Earth, Venus would be expected to receive greater solar insolation, but Venus' much higher albedo ultimately leads to smaller absorption of solar energy, at least at present. Nevertheless, Venus' surface is almost 500 K hotter than the Earth's surface, because surface temperature is controlled by Venus' greenhouse atmosphere (Gillmann et al. 2022). Higher surface temperature weakens crustal rocks due to the strong temperature dependence of viscosity. An increase in surface temperature persisting over geological time propagates through the lithosphere into the mantle, acts to reduce the stress imparted

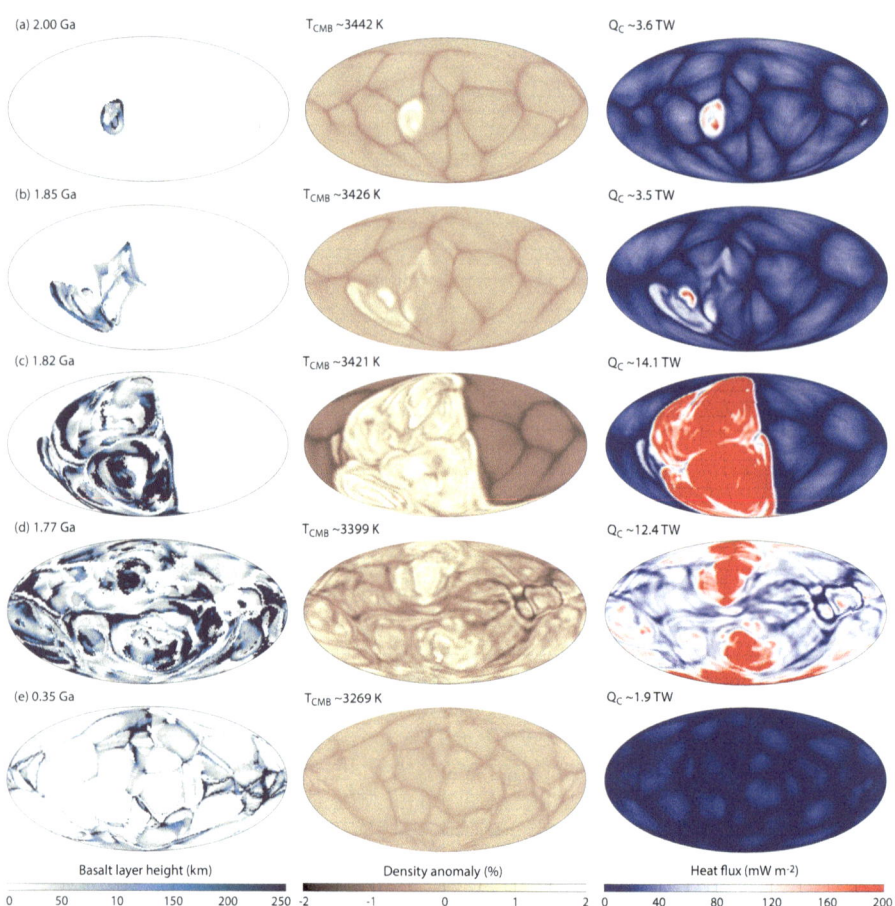

Fig. 7 The evolution of basalt heterogeneity (left column), the deviation from mean density (middle), and heat flux across the core–mantle boundary (Q_C, right) from a 3D thermochemical evolution model of Venus (case 'E50' from Rolf et al. 2018b). Basalt heterogeneity is plotted as the height of the column in which basalt is the dominant composition (i.e., basalt fraction >50%). Each row indicates a different time. Substantial lid mobility is observed between ∼1.85 and ∼1.65 Ga here. The respective temperature at the core–mantle boundary (T_{CMB}) is indicated

to the lithosphere from the convecting mantle, and may induce transition from a mobile to a stagnant lid regime (e.g., Lenardic et al. 2008; Weller et al. 2015). Moreover, higher surface temperature advances healing of previously accumulated damage, since the growth and recovery of mineral grains is faster. As a consequence, reactivation of previously weakened tectonic structures and the formation of plates may be more complicated on Venus than on Earth (e.g., Landuyt and Bercovici 2009; Foley et al. 2012; Bercovici and Ricard 2014).

Both processes inherently require geological time scales to operate. Relaxing this temporal requirement and assuming that the time scale of surface temperature change is less than the mantle mixing time allows mantle temperatures to remain relatively unchanged. Under this condition, an increase in surface temperature reduces the thermal contrast across the lithosphere. If the associated viscosity contrast falls below a critical value ($\sim 10^5$ as suggested by models and scaling analysis), a previously stagnant lithosphere may enter a transitional regime (Moresi and Solomatov 1998). Although the viscosity contrast across Venus' lithosphere is not well determined, comparison to viscosity estimates of the Earth's crust and asthenosphere could make this situation relevant for Venus. If so, the transition could in turn lead to resurfacing and to a comparably young surface, like that of Venus, without the need for global overturn episodes induced by yielding of the lithosphere (Noack et al. 2012). In contrast, Gillmann and Tackley (2014) concluded that lower surface temperatures can trigger transition to mobile lid behaviour. Such relatively lower temperatures foster volcanic activity thereby potentially increasing water content in the atmosphere, which causes surface temperatures to increase again until the interior transitions back into a stagnant and subsequently into an episodic lid regime. Lateral variations in surface temperature can also induce spatial variations in rheology and–in extreme cases of tidally-locked planets–induce hemispherically different tectonics (Meier et al. 2021), but on Venus balancing by the thick atmosphere keeps such lateral variations small.

4.2.2 Stochastic Triggers

A fundamental question is whether a planet's state can always be categorised by a regime that is distinct from others by a number of diagnostics (Sect. 2.3). This depends on the frequency of regime transitions, but also on whether such transitions are reversible. If melting-induced crustal growth triggers overturn events (e.g., Armann and Tackley 2012; Rolf et al. 2018b; Vesterholt et al. 2021), the rate of magmatism may partly determine overturn frequency. Then, overturns may be more frequent in the early phases of evolution when stronger volcanism facilitates crustal growth, and then feature longer intervals until they eventually cease (Armann and Tackley 2012; Vesterholt et al. 2021). With sufficiently long intervals, the mantle may reach a thermal state representative of the stagnant-lid regime in between overturns, but it may still spontaneously enter another isolated resurfacing episode. Although overturn timing may be unpredictable because of the chaotic nature of mantle convection (Wong and Solomatov 2016), the triggers for such spontaneous transitions can include localised lithospheric thinning (Wong and Solomatov 2015), the merging of several upwellings into a stronger one (Loddoch et al. 2006), plume-induced subduction (Crameri and Tackley 2016), or sub-lithospheric, small-scale convection (Solomatov 2003). These mechanisms together add a degree of stochasticity to the evolution of a convecting planet, on top of that arising from different initial states.

Stochasticity in mantle convection is associated with complex feedbacks. Critical system parameters such as surface temperature, internal heating rate, and yield strength couple nonlinearly to the convective system (Crowley and O'Connell 2012; Lenardic and Crowley 2012; Weller and Lenardic 2012, 2018; Weller et al. 2015; Lenardic et al. 2016). This coupling leads to a hysteresis of states in which otherwise identical parameters lead to non-unique tectonic states (Fig. 8). Within the hysteresis window, stagnant, episodic, or mobile states are equally allowable and stable, with none energetically preferred over another. For vigorous mantle convection, as expected for Venus, the region of multistable states may extend over a wide range of lithospheric yield strengths and surface temperatures (Weller and Lenardic 2018). In this framework, the observed tectonic state is inherently controlled by the specific history of the planet. A planet such as early Venus, which may have been in a mobile lid state before transitioning to a stagnant lid through a surface temperature increase for example, may not transition back to a mobile lid state by a reduction in surface temperature alone (e.g., Weller et al. 2015). With all things held equal, either planetary tectonic state represented by Earth and Venus (State A, B in Fig. 8a) is inherently allowable from the same initial condition. Under the surface temperature regime (Fig. 8b), both Venus and Earth are

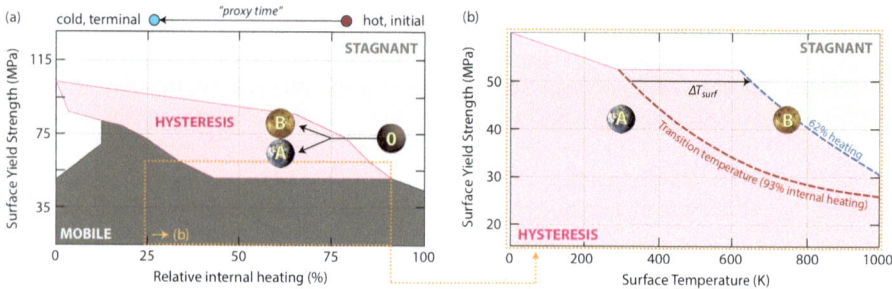

Fig. 8 Tectonic regime diagrams indicating regions of hysteresis (multiple stable tectonic regimes), showing (**a**) the surface yield strength and relative internal heating (the latter is a time proxy: X% indicates that 100-X% of radiogenic heat source are exhausted), and (**b**) yield strength and surface temperature parameter space. The pink threshold lines in (**a**) indicate the yield strength and internal heating combination required to leave an (early) mono-tectonic stagnant lid state 0. With the yield strength held equal, either state (mobile A, stagnant B) is allowable from the same initial state 0. In (**b**), the dashed lines emphasise a widening of the hysteresis window; larger surface temperature changes (ΔT_{surf}) are required to leave the region of multiple states and enter a mono-tectonic state as radiogenics become depleted. Earth and Venus are plotted merely illustratively. These plots are based on Weller and Lenardic (2018)

plotted in the positions of the currently observed surface temperatures, indicating that both planets could currently be within the bistable temperature space. However, this strongly depends on the poorly constrained effective yield strength of the planets' lithospheres. Thus, planets such as Earth and Venus could presently feature different tectonic regimes even under identical present-day conditions, if their surface temperature evolved differently in the past. This finding holds true for other key system parameters such as the yield strength, the global heat budget, radiogenic heating rate, and also for largely stochastic effects occurring in chaotic vigorous mantle convection.

4.2.3 Impact Events

A peculiar class of stochastic events are impacts, which primarily determine the accretion of the terrestrial planets including initial structure, composition, and heat budget. An extremely large, early impact collision could have melted large parts of Venus' early mantle (Davies 2008) and depleted it from most of its water (see Salvador et al. 2023). Water could have been delivered afterwards, but coupled orbital–interior–atmosphere models suggest that such late accretion consisted mostly of relatively dry enstatite chondrites, as otherwise Venus' present atmosphere would be too rich in volatiles (Gillmann et al. 2020).

Apart from water delivery, impacts may trigger changes in mantle dynamics and in interior–atmosphere coupling as reviewed in detail in Gillmann et al. (2022). On Earth, such impacts particularly affected the Hadean and Archean mantle (O'Neill et al. 2017, 2020) which was still hotter and perhaps more comparable to present Venus. Impact energy causes shock heating of the interior and potentially drives magmatic pulses that could trigger resurfacing. The delivered impact energy cannot easily be extracted from the system again, which adds to the discussion of hysteresis above. Under Venus' conditions, a single large impactor could enforce substantial volatile release into the atmosphere, maintaining high surface temperature and promoting a stagnant lid (Gillmann et al. 2016). An otherwise identical evolution lacking such an impact may instead evolve through a period of relatively cold surface conditions that are more prone to lithospheric mobility. On Venus' young surface any detectable relics of such large impacts have been obliterated (e.g., Ruedas and Breuer 2018), but even if preserved, interpreting such anomalies with regards to the triggering of large-scale tectonic events is not unique. As a general problem, the differences between Venus' and Earth's bombardment history are not known well enough to reliably use them as arguments for the diverging pathways of both planets.

4.3 Can We Establish Venus' Geodynamic Regime Evolution?

Tesserae may be among the oldest surfaces preserved on Venus today (e.g., Ivanov and Head 2011; Kreslavsky et al. 2015) and thus possibly important windows into Venus' past. Magellan radar imaging and altimetry reveal the collision of three distinct tessera regions in Tellus Regio (Gilmore and Head 2018), pointing to a phase of surface mobility at some period after the formation of those tesserae. The collision of Lakshmi Planum and Ishtar Terra (Harris and Bédard 2013, 2014; Sect. 3.1.5), and implied 2000–3000 km of crustal convergence may further support such a mobile period (Kiefer 2013). Although ridge belts that possibly formed by crustal convergence over downwelling mantle are ubiquitous on Venus' volcanic plains, they probably have accommodated far less horizontal deformation than tesserae and Ishtar Terra (Moruzzi and Kiefer 2020; Kiefer and Weller 2021), so they may originate from a period with less lithospheric mobility. However, expected regional variations in convergence and deformation rates at the same time would complicate this argument and probably allow for alternative scenarios. In contrast to the evidence of an (undated) earlier mobile epoch, most geophysical evidence points to stagnant lid-like behaviour on Venus at present (Sect. 3.2), which would require a transition in tectonic regime to explain.

Tessera terrains feature characteristic 'ribbons', large-aspect-ratio trough-and-graben structures of debated origin (see Hanmer 2020, and references therein). At least two types of ribbons have been suggested, tensile-fracture and shear-fracture ribbons both of which require a shallow brittle-ductile transition within Venus' crust (<1-2 km, Hansen and Willis 1998). Moreover, the apparent regular spacing and similarity of such ribbons may reflect a thermal control on the thickness of the deformed layer (Ruiz 2007), with which the local heat flux at the time of ribbon formation can be estimated. For a brittle–ductile transition at 1–3 km depth (Ghent and Tibuleac 2002) – in support of prevailing locally hot lithospheric conditions – Ruiz (2007) proposed a heat flux range of 130–780 mW/m^2 assuming present surface temperature, well above all estimates for the present (Sect. 3.1.4). If the surface were 100–150 K hotter when the ribbons formed, heat flux could be much reduced to 20–130 mW/m^2 with a brittle–ductile transition at $\sim$3 km depth. High heat fluxes likely place the solidus inside Venus' crust and support arguments for a weak lower crust (e.g., Ghail 2015) and for intrusive crustal magmatic bodies relevant for the plutonic-squishy lid regime. Additional constraints for Venus' past heat flux come from impact crater morphology, which suggest low heat flux ($\leq$28 mW/m^2) during the formation of Mead (Bjonnes et al. 2021), in line with predictions from mantle convection modelling in the stagnant-lid regime. In contrast, Karimi and Dombard (2017) inferred a substantially higher background heat flow in the vicinity of Mead (55–90 mW/m^2), which simply reflects the uncertainty of heat flux estimates for Venus.

The global relevance of heat flux estimates from a single morphological structure is undetermined, but by analogy to Earth, likely not all that representative. Apart from Mead, several other multi-ring basins are preserved on Venus and are not spatially clustered. If the conditions required to form a multi-ring morphology at these sites are similar as for Mead, the estimated low heat flow may have global relevance (Bjonnes et al. 2021). Recent maps for Earth (e.g., Davies 2013) indicate low heat flux comparable to those derived for Mead crater are common in and near cratonic shields, whereas heat flux is much higher at mid-ocean ridges. Cratons on Earth are the locations of thickest lithosphere; if the tesserae on Venus are similar in nature to cratons, then estimates from these locations are not globally representative.

Independent of the presence of continent-like terrains on Venus, the analogy to heat flux variations on Earth could be challenged as our world is under a regime of plate tectonics whereas Venus is not. Still, heat flux variations are also expected under a stagnant lid regime reflecting variations in crustal thickness and upper mantle temperature linked to the planform of mantle down- and upwellings. Global models of Venus' mantle convection (Rolf et al. 2018b; Uppalapati et al. 2020), however, indicate relatively small variations in surface heat flux, at least if the mantle evolved under stagnant-lid conditions for a sufficiently long time (Fig. 9). In addition, the change over the last billion years is small due to inefficient mantle cooling. Stronger partitioning of radiogenic elements in the crust facilitates surface heat loss, but not greatly, and spatial variations are not more pronounced. Under the assumed model conditions, heat flux lows correlate with regions of thick basaltic crust across which conductive heat transfer is poorer. At the same time, these regions correspond to the zones of strongest magmatic/volcanic activity and thus strong heat transfer by magmatic processes. If magmatic activity is capable of mobilising the surface lid locally, spatial variations in heat flux can be much larger (Noack et al. 2012). On a global scale, this effect is seen during episodic overturn events. During such an event (Fig. 9c–g), heat flux contrasts strongly between recently recycled regions and regions where relatively thick crust is preserved. After cessation of the resurfacing event, the strong variations typically decay on a time scale of 100–200 Myr. At this stage, the variations of predicted heat fluxes across the surface could be within a factor of about two (Fig. 9e–f). If such modelled variations are representative of Venus, this could imply that local estimates of Venus' surface heat flux– such as those made for Mead–may be within a factor of two or less compared to Venus' average, unless the estimate is made for a period during (or shortly after the cessation of) large-scale tectonic lid mobilisation. During those times, the average heat flux would not be related to the present value. High local heat flux estimates such as those derived from ribbon formation (Ruiz 2007) would thus–if confirmed–manifest a period of previous lid mobility on Venus.

Making the assumption that a past mobile epoch did occur on Venus, the question is when and how abruptly the transition to the present state occurred. Even for the Earth – which transitioned from some early regime into the mobile lid regime – neither the timing nor the spatiotemporal evolution of such a transition (gradual or abrupt, regional or global onset) are well established. For Venus conditions, global mantle flow models support long time scales for regime transitions (Weller and Kiefer 2020; Fig. 10). Whereas this time scale is likely sensitive to both mantle structure and convective vigour, tectonic stability is controlled by the system's sensitivity to perturbations (Weller and Lenardic 2018) and by the growth of thermal boundary layers. For a conditionally stable system (i.e., those in the hysteresis windows in Fig. 8), a relatively small perturbation can initiate a transition, such as changes in the global yield strength, water content of the crust, or surface temperature changes of 5–10% (Weller and Kiefer 2020). Once initiated, the perturbation disrupts the established mobile lid pattern on a time scale of ∼500 Myr (dashed line in Fig. 10). As instabilities grow, the system enters the transitory (or episodic-like) state, oscillating between extreme activity and quiescence. Each overturn is marked by plume generation leading to destabilisation of the lithosphere, followed by cessation of yielding and thermal boundary

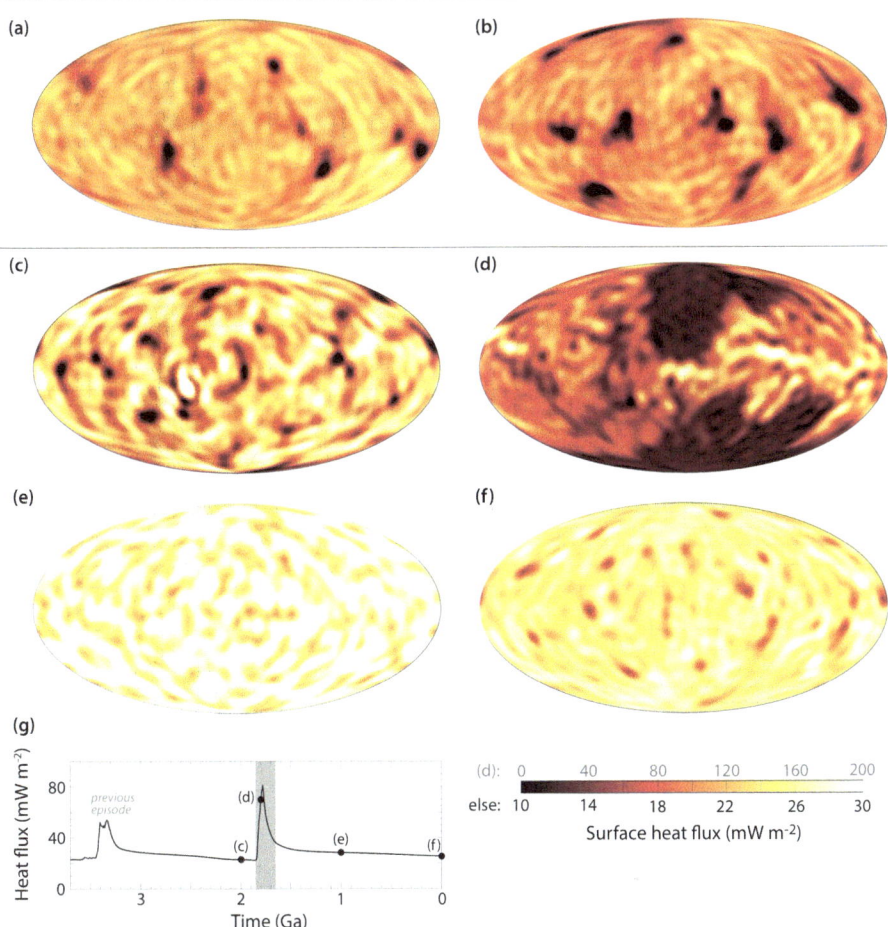

Fig. 9 Maps of surface heat flux (in the spherical harmonic degree range 0-32) for two models from Rolf et al. (2018b). (**a**) Case 'S2' in the stagnant-lid regime at 0 Ga (mean: 22 ± 3 mW/m^2, range: 6–29 mW/m^2) and (**b**) at 1 Ga (20 ± 4 mW/m^2, 4–26 mW/m^2). (**c–f**) Case 'E50' in the episodic regime with a global overturn event ending at ~1.65 Ga at (**c**) 2 Ga (23 ± 6 mW/m^2, 7–58 mW/m^2), (**d**) 1.8 Ga (71 ± 43 mW/m^2, ~0–226 mW/m^2), (**e**) 1 Ga (29 ± 4 mW/m^2, 20–39 mW/m^2), (**f**) 0 Ga (26 ± 3 mW/m^2, 15–33 mW/m^2). Note the anomalous color bar for panel (**d**). (**g**) Evolution of mean surface heat flow for case 'E50', the main resurfacing episode is indicated by the grey box

layer thickening. Although the number and recurrence interval of overturns may be stochastic, each overturn has a minimum operating time scale of 100–300 Myr. Consequently, each different tectonic state likely can operate over 500-1000 Myr time frames, with multiple states requiring several billion years to transition fully.

There are several implications for this behaviour and these time frames of operation. At any given time, a planet that has undergone a tectonic transition would be in a form of dynamic thermal disequilibrium. Properties such as mantle temperature and heat flow would be out of synchronicity with the planet's observed tectonic expressions by perhaps as much as a billion years (Weller and Kiefer 2020). For example, portions of Venus' mantle would be warming at differing rates after a transition (as opposed to assumptions of secular cooling trends), which presents substantial challenges for mission data interpretation. However, the

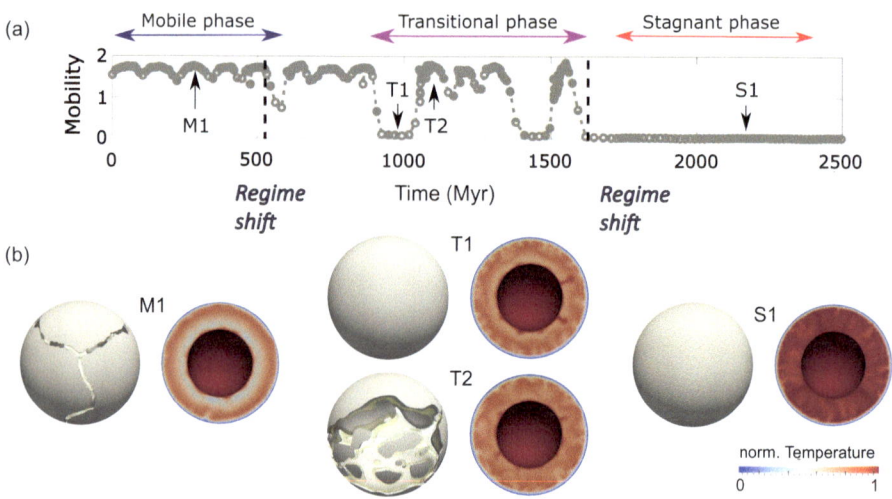

Fig. 10 Oscillation in tectonic states in 3D thermo-tectonic numerical models. The transition follows from a destabilised mobile lid (yield stress increased by 8% at time 0), through an episodic (transitional) state, into a stagnant lid state. (**a**) Time series of surface mobility; time is dimensionalised assuming a mantle overturn time of 100 Myr. Dashed lines indicate regime shifts. (**b**) Snapshots of surface viscosity (left) and temperature (right) that are representative for each regime: mobile (M1), transitional (T1, T2), and stagnant (S1). For viscosity, grey shells indicate high viscosity "plates", yellow bands are regions of active yielding. Temperature is normalised to the temperature drop across the mantle. This figure is modified from Weller and Kiefer (2020)

overall behaviour described may elucidate several outstanding questions regarding Venus' evolution. During a transition, rapid shifts in surface behaviour are predicted that would be discontinuous in nature, often restricted to regional or hemispheric extents. These local scale events may then not be reflective of the global state, in terms of activity, temperature, and heat flux. As a result, the surface of Venus may record differing convection modes and styles of tectonics, with some portions showing extreme activity, yet others reflecting tectonic quiescence. This evolutionary pattern appears consistent with many Venusian geological enigmas, such as inferences of the apparent crustal mobility required to form Ishtar Terra in addition to some tesserae (Gilmore and Head 2018), the hemisphere-scale variation in both volcanic rates and intensity between the BAT region and surrounding areas (Crumpler et al. 1993), the presence of multiple differing and apparently simultaneous styles of mantle upwelling, such as mantle plumes and coronae (Johnson and Richards 2003; Robin et al. 2007; Smrekar et al. 2010), as well as the general and more widespread partial infilling on impact crater floors (Herrick and Rumpf 2011).

Although suggestive, this concept leaves several major questions unanswered. As mentioned above, during a regime transition, a stagnant lid would be effectively indistinguishable from a sufficiently long quiescent period following an overturn. Given the range of proposed surface ages, this disequilibrium potential and the ambiguity regarding stagnant lid convection versus quiescent overturn period could imply that Venus is in an ongoing transition phase today. How long this transition has been ongoing needs to be further resolved. Existing insights are mostly based on numerical simulations, but observable diagnostics need to be defined for future missions (like those in Sect. 5.2) in order to distinguish between the different possibilities predicted by such models.

5 Synthesis and Future Perspectives

5.1 Mantle Dynamics and Evolution on Venus

As discussed throughout this chapter, our understanding of Venus' mantle evolution remains fragmentary to date due to insufficient data for developing robust models of the subsurface. Given the clearly different surface expressions, Venus' mantle likely differs from that of Earth in various aspects. A key difference could be the absence of a weak asthenosphere on Venus as suggested by gravity and topography data (Sects. 3.1.1, 3.2). The asthenosphere is often considered as key for allowing large-scale horizontal surface motions, and thus plate tectonics, on Earth. A reduced water content in Venus' interior is a possible explanation for this discrepancy, but drier conditions need to be confirmed by future observations. Alternatively, different concentrations of carbon dioxide and different degrees of partial melting may also suppress an asthenosphere (e.g., Sifré et al. 2014).

Divergent surface tectonics could also arise from a different crustal structure—like a distinct weak lower crust (e.g., Ghail 2015), possibly analogous to Earth's crust in the Archean (e.g., Ghail et al. 2023). Crustal structure is shaped by the interplay of tectonics and magmatism. Further understanding is needed of this coupling and how mantle and crustal interactions shape the surface. Such insights include establishing the coupling between the interior and the thermo-compositional state of the atmosphere, which is tied to the interior via mantle outgassing and feeds back with surface tectonics by modulating the surface temperature (Gillmann et al. 2022). The coupling between Venus' subsystems (Fig. 1) implies that—as for the Earth—the planet can only be understood as one integrative system: addressing the feedback between the subsystems is and will be essential.

Models are powerful tools with which to shed light on the coupling processes, as we have reviewed here. A major challenge, however, is to evaluate the predictive power of those models for Venus, in particular regarding time scales. Existing studies propose a wide range of possibilities of Venus' present and past interior dynamics (Fig. 11). These options are only converging in the sense that Venus' lithosphere does not feature large-scale coherent horizontal motion in contrast to Earth. Some works suggest that Venus is in the stagnant lid regime, but upon closer inspection this regime is likely different from that of the classical stagnant-lid bodies, such as the Moon or Mercury. The current tectonic state may be transient, with Venus in transition between a past mobile state and a future stagnant state (Weller and Kiefer 2020), or in a quiescent state between episodic mantle overturns (e.g., Rolf et al. 2018b). Moreover, the tectonic state may be linked to the evolution of volcanism, such as in the plutonic-squishy lid regime, where intrusive magmatism may keep the lithosphere hot and more prone to tectonic deformation (Lourenço et al. 2020). The widespread presence of coronae and other volcanic features on Venus (e.g., Stofan and Smrekar 2005), and the possibility of discrete, mobile crustal blocks (Byrne et al. 2021) could be indicative for such a volcano-tectonic regime. From our general understanding of terrestrial planet evolution, the ratio of intrusive to extrusive magmatism increases through time (e.g., Stern et al. 2018), but it is difficult to establish where Venus is currently situated on that trend. However, several inferences of recent volcanic activity (e.g., Smrekar et al. 2010; Brossier et al. 2020) suggest that Venusian magmatism may not (yet) be entirely intrusive (see Byrne 2019).

Observed surface tectonics suggest that Venus' surface has been more mobile in previous epochs, but little is known about the timing, extent, and duration of such mobility. Present geophysical constraints are insufficient to properly distinguish the different models and Venus' sparse and spatially random crater population seems insufficient to reveal these details either, especially on absolute time scales (Herrick et al. 2023; Sect. 3.1.2). On Earth,

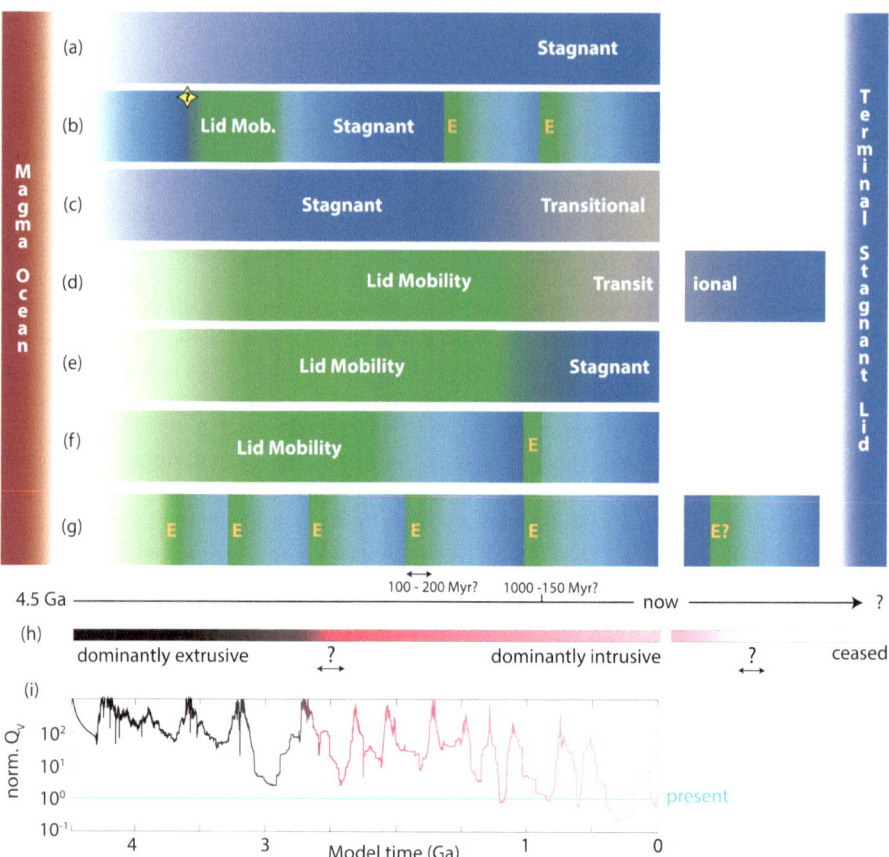

Fig. 11 Proposed evolutionary scenarios for Venus. (**a**) A continuous stagnant lid (O'Rourke and Korenaga 2015); (**b**) regime transitions controlled by surface temperature (Gillmann and Tackley 2014), possibly triggered by an impact (yellow star, Gillmann et al. 2016); (**c**) the transition from early stagnant lid to a transitional regime with localised resurfacing, triggered by surface temperature (Noack et al. 2012); (**d**) the transition from an earlier epoch of lid mobility through a transitional state towards a (future) stagnant lid (Weller and Kiefer 2020); (**e**)+(**f**) an early epoch of mobility motivated by low surface temperature to allow for liquid surface water (Way and Del Genio 2020), transitioning into a stagnant or an episodic lid state; and (**g**) episodic resurfacing with a progressively increasing interval between the episodes (E, Armann and Tackley 2012); future overturns may occur. The time axis is largely unconstrained, except for a few estimates on overturn duration or time since last overturn from modelling studies. All scenarios may start from a magma ocean and terminate in a stagnant lid (Stern et al. 2018). (**h**) Qualitative evolution of magmatism, from dominantly extrusive to more intrusive to intermittent and ultimately faded activity (see Byrne 2019). Panel (**i**) provides a quantitative example of such a scenario; shown is a (smoothed) time series of volcanic heat flux Q_V in a plutonic-squishy lid evolution (normalised to the present-day value) based on Lourenço et al. (2020). This case assumes a mantle viscosity of 10^{21} Pa s, a lithospheric yield stress of 80 MPa and a magmatic eruption efficiency of 10%

dating of tectonic events beyond the limit of seafloor reconstructions relies on geochronology and radiometric dating of preserved rocks, but such an option is infeasible for Venus for the time being. However, potential future boosts for improving our understanding are discussed in Sect. 5.2. Apart from the timing and duration of previous lid mobility, the question arises as to whether this period was similar to mobility on Earth (that is, plate tectonics). On present Earth, mantle plume and plate boundary locations are largely uncorrelated, but laboratory experiments and structural observations on Venus point towards plume-induced subduction, favoured by Venus' hotter and thus weaker lithosphere (Davaille et al. 2017). A similar mechanism may have triggered (proto-)subduction on early Earth (Gerya et al. 2015; Fischer and Gerya 2016; Baes et al. 2020).

If the analogy with early Earth holds, present Venus would be a key observatory for the Earth's past, but why would Earth have evolved from its early tectonic mode into modern plate tectonics, when Venus is still in the early-state mode? Could Venus' evolution have been 'slower', meaning that it has not yet entered the plate tectonics mode? Cooling rate is largely controlled by mantle viscosity, thus a higher viscosity in Venus' mantle could have trapped heat inside more efficiently. Reduced water content could explain such increased viscosity, but would imply that Venus either accreted differently than Earth or was dried out soon after—perhaps by an extreme impact collision (Davies 2008). Also, such a 'slower' evolution would suggest a mobile epoch yet to come, but hints of such a mobile regime are already manifested in the planet's preserved geological record (Sect. 4.3). This could controversially point to a 'faster' evolutionary pace on Venus instead. The role of a hotter lithosphere in this regard is discussed in Ghail et al. (2023). Other possibilities are recurring episodes of mobility or that plate tectonics may develop only under specific circumstances (for water content and surface temperature for example) that the Earth happened to have at some point, but Venus did not (see Stern et al. 2018).

A planet is a highly dynamic system with strongly nonlinear behaviour: small changes can induce large and unpredictable consequences. Coupled atmosphere–interior models (e.g., Gillmann and Tackley 2014) indicate substantial variation of surface temperature through time, more than enough to induce transition from one mode of tectonics to another; bolide impacts could—stochastically—further trigger such transitions (Sect. 4.2). Even if the immediate trigger subsequently vanishes, the transition in tectonic mode may not automatically reverse because of hysteresis. Once pushed off the evolutionary path into a different mode, various processes may establish some of Venus' peculiarities such as lacking an asthenosphere and having high surface temperature, which are both maintained by and help maintain the planet's tectonic regime. In this light, self-organisation could explain the divergent evolutions of two very similar planets. Whether such an explanation applies for the tectonic divergence of Venus and Earth remains difficult to answer. Some avenues to further resolve these issues are given below.

5.2 Future Boosts for Understanding the Venus Mantle?

5.2.1 Expected Insights from Confirmed Future Missions

The upcoming decade(s) will see various space missions targeting Venus. NASA's VERI-TAS and DAVINCI missions, together with ESA's EnVision mission, will look for active volcanism and tectonics, return high-resolution gravity and topography data, provide compositional maps of the surface, and place bounds on estimates of mantle outgassing rates by measuring noble gas concentrations (see Widemann et al. 2023). Some key questions with particular relevance for the mantle that could potentially be resolved with the upcoming mission data are discussed here.

5.2.1.1 How Did Mantle Cooling History Contol the State of Venus' Core? The mantle provides the boundary conditions of the core and controls its cooling history. A better understanding of Venus' core would thus inform our understanding of the mantle evolution. VERITAS will provide much improved measurements of the planet's k_2 tidal Love number,

a characteristic for the planet's tidal deformation from which the size of Venus' core and its state can be estimated, in combination with the moment of inertia factor (MoI, Cascioli et al. 2021). The radius of the core pins the thickness of the mantle and thus places bounds on the temperature and pressure range at and above the core–mantle boundary. Margot et al. (2021) found the first estimate of core size (∼3500 km) using Earth-based radar observations of Venus' spin. However, this method has large error bars (±300 km at best), and cannot determine core state. Scaling core size from that of Earth places Venus' lowermost mantle probably just outside the stability field of post-perovskite, which is known to influence the dynamics of the Earth's lowermost mantle (e.g., Čížková et al. 2010) and outer core (e.g., Amit and Choblet 2009). Assuming an Earth-like mantle density profile and a transition pressure of ∼125 GPa (Murakami et al. 2004; Trønnes et al. 2019), the occurrence of post-perovskite is inconsistent with the range of core sizes estimated by Margot et al. (2021).

The improved data will not only better characterise the size of the core, but also its state, whether it is liquid, solid, or partially both. This will be an important constraint for mantle interior models (Dumoulin et al. 2017), and for telling us how much core cooling should be accounted for in these models. In turn, we will be able to place bounds on core–mantle boundary temperature and the formation of mantle plumes as well as their excess temperature, at least in the deep mantle. For example, if future k_2 and MoI measurements point to a fully solidified core, a mantle evolution in the permanent stagnant lid regime becomes less feasible as not enough heat could have escaped the interior. In contrast, a fully solid core would imply strong core cooling and point to a relatively cold mantle, which was cooled more efficiently like in a mobile lid regime. Such information will also impact the feasibility of a basal magma ocean inside Venus today, and thus our understanding of the planet's magnetic field history (O'Rourke et al. 2019). Mantle cooling rate further strongly depends on the mantle viscosity. To constrain the latter, VERITAS will deliver estimates of Venus' tidal phase lag with a potential accuracy of 0.05° (Cascioli et al. 2021).

5.2.1.2 How Much Heat Does Venus Lose Today? The heat loss from the mantle through the surface of the planet is a cornerstone for every thermal evolution model. VERITAS will measure global topography and gravity at much improved resolution over presently available data. EnVision will also measure topography and gravity (e.g., Rosenblatt et al. 2021). Such refined data is expected to inform models on upper-mantle density anomalies, crustal thickness, and eventually elastic thickness from which thermal gradients and thus surface heat flux can be estimated. The improved resolution could also provide a better picture of lateral heat flux variations, which can be linked to hotter or colder regions of the mantle and possibly to convective structures such as deep mantle plumes. VERITAS will additionally deliver improved measurements of surface thermal emissivity, positive anomalies of which may be linked to plume locations (e.g., Smrekar et al. 2010). Moreover, thermal emissivity is an important observable factor for the detection of recent volcanic activity (e.g., D'Incecco et al. 2017, 2021). Improved estimates of the volcanic eruption rate at Venus could pin the amount of heat leaving the interior via volcanic processes, thus providing constraints on the temperature of the uppermost mantle or the important partitioning of intrusive magmatism and eruptive volcanism.

5.2.1.3 Are Venus' Tesserae Felsic and Older than the Average Crust? High-resolution gravity and topography data will not only allow for improved estimates of elastic thickness and thermal gradients, but also help to further map out the crater population, in particular in terms of searching for buried craters. Subsurface radar measurements from EnVision may support this search. Buried craters may also be apparent in high resolution topography, as on other rocky bodies, or could still be retained in the gravity field if measured at sufficient resolution. Given the small population of preserved craters on Venus, the detection of buried craters, particularly a few large ones (e.g., Karimi et al. 2018), would place new constraints on the absolute model age of Venus' crust and therefore on the rates of resurfacing the internal evolution models must accommodate. This, again, is directly linked to Venus' tectonic regime. In particular, refining the distribution of Venus' modified craters could help to identify whether parts of Venus' surface, specifically tesserae, are older than the average surface age. Deformed craters on tesserae could be seen in high resolution topography, as evidenced in the stereo topography (see Herrick et al. 2023). Higher-resolution mapping of the tesserae's bounding structures can additionally help to determine the structural relationship between the tesserae and their surroundings and thus to constrain the formation mechanism of the tesserae. The advanced mapping of tesserae regions, the search for buried craters and in particular the determination of the processes modifying craters will reveal Venus' resurfacing history in detail.

VERITAS and EnVision will also measure iron content from orbit, providing proxy data for the SiO_2 content of the surface rocks (Dyar et al. 2020; Helbert et al. 2021). Such data for the whole planet will be acquired, so that the composition of tessera regions will be further refined. Alpha Regio—the only tessera region captured by previous data—has a composition that is consistent with, but does not uniquely mean, felsic rock (Gilmore et al. 2015, 2017). Determining whether this finding holds for all Venusian tesserae is important for our understanding of the interior evolution, as a generally felsic composition would require the presence of near-surface water during tessera formation. The DAVINCI mission (Garvin et al. 2022) shall deliver new measurements of Venus' D/H ratio with a precision sufficient to distinguish between the different scenarios proposed for the origin of Venus' water. This will help to constrain how wet the near-surface of Venus may be. If such a wet environment propagates into the deeper interior, it has implications for the viscosity and the degree of partial melting in the upper mantle and therefore volcanic outgassing rates, for which DAVINCI's improved measurement of atmospheric noble gas concentrations will be new constraints. Water in Venus' shallow interior will also affect the strength of the crust. Targeted surface deformation maps—another expected outcome of the VERITAS mission—will further support this line of insight by revealing currently active features for various parts of the planet. Finally, DAVINCI's descending probe (VDI) will take high-resolution images of Alpha Regio when approaching the surface. Such images may offer additional visual information for the formation and evolution of what are possibly Venus' most ancient tectonic features.

5.2.2 Additional Desired Observations Beyond Planned Mission Plans

5.2.2.1 In-Situ Heat Flow The future data to be delivered by DAVINCI, VERITAS, and EnVision will greatly advance our understanding of Venus' interior and mantle, but gaps will naturally remain. For example, surface heat flux will remain an estimate and not be measured in-situ, which would be the ultimate data to pin down Venus' heat loss and the thermal state of the upper mantle. As discussed in Sect. 4.3, however, extrapolating data from single locations to a global characteristic value is difficult when lateral variations are important, so that future heat flow measurements would ideally be taken at various sites.

5.2.2.2 Seismology Seismology would be the most powerful tool to map out the planet's density distribution. With that, the thermo-compositional structure of the interior could be further revealed, including the (spatially varying) thickness of Venus' crust. Such constraints would ultimately inform thermo-magmatic evolution models about the rates of crustal production and destruction and therefore about the volume and timing of volcanic resurfacing. Mapping the relatively shallow crustal boundary could be feasible with relatively low-magnitude seismic events; deeper structures would be more challenging to map, but useful for understanding Venus' deeper mantle structure. Major unknown aspects are for instance the thickness of the mantle transition zone and its spatial variations, which provide information about the thermal state of that zone (e.g., Lawrence and Shearer 2008), the identification of possible heterogeneity in a deep layer comparable to the D" region on Earth (e.g., Cobden et al. 2015), and confirmation of the size and state of the core that are indirectly determined by measurements of k_2 and the MoI. The potential of future seismic investigation is detailed in Widemann et al. (2023); further background and concepts for (future) seismology on Venus are for instance given by Stevenson et al. (2015) and Kremic et al. (2021).

5.2.2.3 Electromagnetics As discussed above, the water content in the crust and upper mantle is a crucial parameter to understand rheology and deformation on Venus.

Electromagnetic methods, for instance magnetotellurics, provide a tool to constrain water, but also melt and carbon dioxide content (e.g., Sifré et al. 2014). Such measurements are best made from the surface, but aerial sounding is a possibility and can still achieve exploration depths exceeding the lithospheric thickness with sufficiently low electromagnetic frequencies (Grimm et al. 2012). If Venus' crust is dry, this method could provide an independent measurement of crustal and lithospheric thickness. If the crust is wet, containing 100s ppm of water, exploration depth is limited to much shallower depth and lithospheric properties can no longer be estimated; however, such a finding would still reveal the wet nature of Venus' upper mantle. Knowing the water content of the upper mantle would further inform our models of its viscosity, jointly with improved gravity and topography measurements by EnVision and VERITAS, and shed light upon whether Venus' interior features an asthenosphere, and how the mantle couples to the lithosphere and crust. An issue for such a measurement is the availability of electromagnetic sources. Lightning in Venus' atmosphere (e.g., Russell et al. 2007)—if found to be present—could provide one such source at frequencies that allow for wave penetration as deep as 100 km into Venus' interior (Grimm et al. 2012).

5.2.3 Future Conceptual Approaches and Modelling

The next generation of missions will deliver new observations with which to improve our understanding of Venus' interior, but a remaining challenge will be to incorporate these observations into a common dynamic framework and ideally to place at least relative time scales on the processes to which these observations can be attributed. To do so, future advances in modelling the evolution of the planet are needed. Of particular importance will be work dedicated to revealing the coupling between subsystems—such as the core and the mantle, the mantle and the atmosphere (including surface processes), and tectonics and volcanism. Some important questions to address with such models are how much water does Venus' interior contain, how is it distributed through time and space, and how does it affect mantle viscosity, melt generation, and the properties of the crust? Improved understanding of how magma migrates through the crust and how much of it makes it to the surface as a function of crustal properties is another target of future modelling efforts. However, such models must also account for the properties and complexities of Venus' crustal rocks, their composition, and mineral assemblage—which, as we know from Earth—can be far from homogeneous even over relatively modest spatial scales. Identifying and analysing Earth materials analogue to Venusian rocks (e.g., Filiberto et al. 2020), as well as detailed understanding of mineral physics of mantle material under high temperature and pressure conditions, are thus also necessary to decipher Venus' still enigmatic mantle structure and evolution.

Acknowledgements The authors thank the International Space Science Institute in Bern/Switzerland for hosting and the editorial board (D. Breuer, C. Gillmann, S. Smrekar, T. Spohn, T. Widemann, C. Wilson) for organising the workshop "Venus Evolution Through Time" from which this work originates. The authors thank W. Kiefer and W. Moore for thoughtful reviews that helped to improve the original manuscript. TR acknowledges funding from the Research Council of Norway (grants 223272, Centre for Earth Evolution and Dynamics, and 276032, PLATONICS).

Funding Note Open access funding provided by University of Oslo (incl Oslo University Hospital).

Declarations

Competing Interests The authors declare no competing interests.

References

Abe Y (1995) Basic equations for evolution of partially molten mantle and core. In: TERRA-PUB, pp 215–230

Amit H, Choblet G (2009) Mantle-driven geodynamo features—effects of post-perovskite phase transition. Earth Planets Space 61:1255–1268. https://doi.org/10.1186/bf03352978

Anderson FS, Smrekar SE (2006) Global mapping of crustal and lithospheric thickness on Venus. J Geophys Res 111:E08006. https://doi.org/10.1029/2004je002395

Arkani-Hamed J (1993) On the tectonics of Venus. Phys Earth Planet Inter 76:75–96. https://doi.org/10.1016/0031-9201(93)90056-f

Armann M, Tackley PJ (2012) Simulating the thermochemical magmatic and tectonic evolution of Venus' mantle and lithosphere: two-dimensional models. J Geophys Res 117:E12003. https://doi.org/10.1029/2012je004231

Azuma S, Katayama I, Nakakuki T (2014) Rheological decoupling at the Moho and implication to Venusian tectonics. Sci Rep 4:4403. https://doi.org/10.1038/srep04403

Baes M, Sobolev S, Gerya T, Brune S (2020) Plume-induced subduction initiation: single- or multi-slab subduction? Geochem Geophys Geosyst 21(2):e2019GC008663. https://doi.org/10.1029/2019GC008663

Benešová N, Čížková H (2012) Geoid and topography of Venus in various thermal convection models. Stud Geophys Geod 56:621–639. https://doi.org/10.1007/s11200-011-0251-7

Bercovici D, Karato S-I (2003) Whole-mantle convection and the transition-zone water filter. Nature 425:39–44. https://doi.org/10.1038/nature01918

Bercovici D, Ricard J. (2014) Plate tectonics, damage and inheritance. Nature 508:513–516. https://doi.org/10.1038/nature13072

Bindschadler DL, Schubert G, Ford P (1994) Venus' center of figure-center of mass offset. Icarus 111:417–432. https://doi.org/10.1006/icar.1994.1153

Bjonnes EE, Hansen VL, James B, Swemson JB (2012) Equilibrium resurfacing of Venus: results from new Monte Carlo modeling and implications for Venus surface histories. Icarus 217:451–461. https://doi.org/10.1016/j.icarus.2011.03.033

Bjonnes EE, Johnson BC, Evans AJ (2021) Estimating Venusian thermal conditions using multiring basin morphology. Nat Astron 5:498–502. https://doi.org/10.1038/s41550-020-01289-6

Blaske CH, O'Rourke JG (2021) Energetic requirements for dynamos in the metallic cores of super-Earth and super-Venus exoplanets. J Geophys Res 126:e2020JE006739. https://doi.org/10.1029/2020je006739

Boehler R (1998) Fe–FeS eutectic temperatures to 620 kbar. Phys Earth Planet Inter 96:181–186. https://doi.org/10.1016/0031-9201(96)03150-0

Bondarenko NV, Head JW, Ivanov MA (2010) Present-day volcanism on Venus: evidence from microwave radiometry. Geophys Res Lett 37:L23202. https://doi.org/10.1029/2010gl045233

Borrelli ME, O'Rourke JG, Smrekar SE, Ostberg CM (2021) A global survey of lithospheric flexure at steep-sided domical volcanoes on Venus reveals intermediate elastic thicknesses. J Geophys Res 126:e2020JE006756. https://doi.org/10.1029/2020je006756

Boyet M, Bouvier A, Frossard P, Hammouda T, Garcon M, Gannoun A (2018) Enstatite chondrites EL3 as building blocks for the Earth: the debate over the 146Sm–142Nd systematics. Earth Planet Sci Lett 488:68–78. https://doi.org/10.1016/j.epsl.2018.02.004

Brossier JF, Gilmore MS, Toner K (2020) Low radar emissivity signatures on Venus volcanoes and coronae: new insights on relative composition and age. Icarus 343:113693. https://doi.org/10.1016/j.icarus.2020.113693

Buck WR (1992) Global decoupling of crust and mantle: implications for topography, geoid and mantle viscosity on Venus. Geophys Res Lett 19:2111–2114. https://doi.org/10.1029/92gl02462

Bunge H-P, Richards MA, Baumgardner JR (1997) A sensitivity study of three-dimensional spherical mantle convection at 10^8 Rayleigh number: effects of depth-dependent viscosity, heating mode, and an endothermic phase change. J Geophys Res 102:11991–12007. https://doi.org/10.1029/96jb03806

Burkhardt C, Borg LE, Brennecka GA, Shollenberger QR, Dauphas N, Kleine T (2016) A nucleosynthetic origin for the Earth's anomalous 142Nd composition. Nature 537:394–398. https://doi.org/10.1038/nature18956

Byrne PK (2019) A comparison of inner solar system volcanism. Nat Astron 4:321–327. https://doi.org/10.1038/s41550-019-0944-3

Byrne PK, Ghail RC, Celâl Sengör AM, James PB, Klimczak C, Solomon SC (2021) A globally fragmented and mobile lithosphere on Venus. Proc Natl Acad Sci 118:e2025919118. https://doi.org/10.1073/pnas.2025919118

Campbell BA, Morgan GA, Whitten JL, Carter LM, Glaze LS, Campbell DB (2017) Pyroclastic flow deposits on Venus as indicators of renewed magmatic activity. J Geophys Res 122:1580–1596. https://doi.org/10.1002/2017je005299

Capitanio FA, Nebel O, Cawood PA, Weinberg RF, Chowdhury P (2019) Reconciling thermal regimes and tectonics of the early Earth. Geology 47:923–927. https://doi.org/10.1130/G46239.1

Carter L, Gilmore M, Ghail R, Byrne P, Izenberg N, Smrekar S (2023) Sedimentary processes on Venus. Space Sci Rev

Cascioli G, Hensley S, De Marchi F, Breuer D, Durante D, Racioppa P, Iess L, Mazarico E, Smrekar SE (2021) The determination of the rotational state and interior structure of Venus with VERITAS. Planet Sci J 2:220. https://doi.org/10.3847/PSJ/ac26c0

Cawood PA, Hawkesworth CJ, Dhuime B (2012) The continental record and the generation of continental crust. Geol Soc Am Bull 125:14–32. https://doi.org/10.1130/b30722.1

Christensen U (2006) A deep dynamo generating Mercury's magnetic field. Nature 444:1056–1058. https://doi.org/10.1038/nature05342

Čížková H, Cadek O, Matiska C, Yuen DA (2010) Implications of post-perovskite transport properties for core–mantle dynamics. Phys Earth Planet Inter 180:235–243. https://doi.org/10.1016/j.pepi.2009.08.008

Cobden L, Thomas C, Trampert J (2015) Seismic detection of post-perovskite inside the Earth. In: Khan A, Deschamps F (eds) The Earth's heterogeneous mantle. Springer, Cham, pp 391–440. https://doi.org/10.1007/978-3-319-15627-9_13

Coltice N, Seton M, Rolf T, Müller RD, Tackley PJ (2013) Convergence of tectonic reconstructions and mantle convection models for significant fluctuations in seafloor spreading. Earth Planet Sci Lett 383:92–100. https://doi.org/10.1016/j.epsl.2013.09.032

Crameri F, Tackley PJ (2016) Subduction initiation from a stagnant lid and global overturn: new insights from numerical models with a free surface. Prog Earth Planet Sci 3:30. https://doi.org/10.1186/s40645-016-0103-8

Crisp JA (1984) Rates of magma emplacement and volcanic output. J Volcanol Geotherm Res 20:177–211. https://doi.org/10.1016/0377-0273(84)90039-8

Crowley JW, O'Connell RJ (2012) An analytic model of convection in a system with layered viscosity and plates. Geophys J Int 188:61–78. https://doi.org/10.1111/j.1365-246x.2011.05254.x

Crumpler LS, Head JW, Aubele JC (1993) Relation of major volcanic center concentration on Venus to global tectonic patterns. Science 261:591–595. https://doi.org/10.1126/science.261.5121.591

560

Springer

Dannberg J, Eilon Z, Faul U, Gassmöller R, Moulik P, Myhill R (2017) The importance of grain size to mantle dynamics and seismological observations. Geochem Geophys Geosyst 18:3034–3061. https://doi.org/10.1002/2017gc006944

Davaille A, Smrekar SE, Tomlinson S (2017) Experimental and observational evidence for plume-induced subduction on Venus. Nat Geosci 10:349–355. https://doi.org/10.1038/ngeo2928

Davies JH (2008) Did a mega-collision dry Venus' interior? Earth Planet Sci Lett 268:376–383. https://doi.org/10.1016/j.epsl.2008.01.031

Davies JH (2013) Global map of solid Earth surface heat flow. Geochem Geophys Geosyst 14:4608–4622. https://doi.org/10.1002/ggge.20271

DePaolo DJ, Manga M (2003) GEOLOGY: deep origin of hotspots–the mantle plume model. Science 300:920–992. https://doi.org/10.1126/science.1083623

D'Incecco P, Müller N, Helbert J, D'Amore M (2017) Idunn Mons on Venus: location and extent of recently active lava flows. Planet Space Sci 136:25–33. https://doi.org/10.1016/j.pss.2016.12.002

D'Incecco P, Filiberto J, López I, Gorinov DA, Komatsu G (2021) Idunn Mons: evidence for ongoing volcano-tectonic activity and atmospheric implications on Venus. Planet Sci J.2(5):215. https://doi.org/10.3847/PSJ/ac2258

Driscoll P, Bercovici D (2013) Divergent evolution of Earth and Venus: influence of degassing, tectonics, and magnetic fields. Icarus 226:1447–1464. https://doi.org/10.1016/j.icarus.2013.07.025

Driscoll P, Bercovici D (2014) On the thermal and magnetic histories of Earth and Venus: influences of melting, radioactivity, and conductivity. Phys Earth Planet Inter 236:36–51. https://doi.org/10.1016/j.pepi.2014.08.004

Dumoulin C, Tobie G, Verhoeven O, Rosenblatt P, Rambaux N (2017) Tidal constraints on the interior of Venus. J Geophys Res, Planets 122:1338–1352. https://doi.org/10.1002/2016je005249

Dyar MD, Helbert J, Maturilli A, Müller NT, Kappel D (2020) Probing Venus surface iron contents with six-band visible near-infrared spectroscopy from orbit. Geophys Res Lett 47:e2020GL090497. https://doi.org/10.1029/2020GL090497

Filiberto J, Trang D, Treiman AH, Gilmore MS (2020) Present-day volcanism on Venus as evidenced from weathering rates of olivine. Sci Adv 6:eaax7445. https://doi.org/10.1126/sciadv.aax7445

Fischer R, Gerya T (2016) Early Earth plume-lid tectonics: a high-resolution 3D numerical modelling approach. J Geodyn 100:198–214. https://doi.org/10.1016/j.jog.2016.03.004

Foley BJ (2018) The dependence of planetary tectonics on mantle thermal state: applications to early Earth evolution. Philos Trans R Soc A 376:20170409. https://doi.org/10.1098/rsta.2017.0409

Foley BJ, Becker TW (2009) Generation of plate-like behavior and mantle heterogeneity from a spherical, viscoplastic convection model. Geochem Geophys Geosyst 10:Q08001. https://doi.org/10.1029/2009gc002378

Foley BJ, Bercovici D, Landuyt W (2012) The conditions for plate tectonics on super-earths: inferences from convection models with damage. Earth Planet Sci Lett 331–332:281–290. https://doi.org/10.1016/j.epsl.2012.03.028

Fowler AC (1985) Fast thermoviscous convection. Stud Appl Math 72:189–219. https://doi.org/10.1002/sapm1985723189

Fowler AC, O'Brien SGB (1996) A mechanism for episodic subduction on Venus. J Geophys Res, Planets 101:4755–4763. https://doi.org/10.1029/95je03261

French SW, Romanowicz B (2015) Broad plumes rooted at the base of the Earth's mantle beneath major hotspots. Nature 525:95–99. https://doi.org/10.1038/nature14876

Fuchs L, Becker T (2019) Role of strain-dependent weakening memory on the style of mantle convection and plate boundary stability. Geophys J Int 218:601–618. https://doi.org/10.1093/gji/ggz167

Fuchs L, Becker TW (2021) Deformation memory in the lithosphere: a comparison of damage-dependent weakening and grain-size sensitive rheologies. J Geophys Res, Solid Earth 126:e2020JB020335. https://doi.org/10.1029/2020jb020335

Garnero EJ, McNamara AK, Shim S-H (2016) Continent-sized anomalous zones with low seismic velocity at the base of Earth's mantle. Nat Geosci 9:481–489. https://doi.org/10.1038/ngeo2733

Garvin JB, Getty SA, Arney GN, Johnson NM, Kohler E, Schwer KO, Sekerak M, Bartels A, Saylor RS, Elliott VE, Goodloe CS, Garrison MB, Cottini V, Izenberg N, Lorenz R, Malespin CA, Ravine M, Webster CR, Atkinson DH, Aslam S, Atreya S, Bos BJ, Brinckerhoff WB, Campbell B, Crisp D, Filiberto JR, Forget F, Gilmore M, Gorius N, Grinspoon D, Hofmann AE, Kane SR, Kiefer W, Lebonnois S, Mahaffy PR, Pavlov A, Trainer M, Zahnle KJ, Zolotov M (2022) Revealing the mysteries of Venus: the DAVINCI mission. Planet Sci J 3:117. https://doi.org/10.3847/PSJ/ac63c2

Gerya TV (2014) Plume-induced crustal convection: 3D thermomechanical model and implications for the origin of novae and coronae on Venus. Earth Planet Sci Lett 391:183–192. https://doi.org/10.1016/j.epsl.2014.02.005

Gerya TV, Stern RJ, Baes M, Sobolev SV, Whattam SA (2015) Plate tectonics on the Earth triggered by plume-induced subduction initiation. Nature 527:221–225. https://doi.org/10.1038/nature15752

Ghail R (2015) Rheological and petrological implications for a stagnant lid regime on Venus. Planet Space Sci 113–114:2–9. https://doi.org/10.1016/j.pss.2015.02.005

Ghail R, Smrekar SE, Byrne PK, Gilmore MS, Herrick RR, Ivanov MA, Plesa AC, Rolf T, Sabbeth L, Schools JW, Shellnut JG (2023) Volcano and tectonic constraints on the evolution of Venus. Space Sci Rev

Ghent RR, Tibuleac IM (2002) Ribbon spacing in Venusian tessera: implications for layer thickness and thermal state. Geophys Res Lett 29:61. https://doi.org/10.1029/2002GL015994

Ghias SR, Jarvis GT (2008) Mantle convection models with temperature- and depth-dependent thermal expansivity. J Geophys Res, Solid Earth 113:B08408. https://doi.org/10.1029/2007jb005355

Gillmann C, Tackley P (2014) Atmosphere/mantle coupling and feedbacks on Venus. J Geophys Res, Planets 119:1189–1217. https://doi.org/10.1002/2013je004505

Gillmann C, Golabek GJ, Tackley PJ (2016) Effect of a single large impact on the coupled atmosphere-interior evolution of Venus. Icarus 268:295–312. https://doi.org/10.1016/j.icarus.2015.12.024

Gillmann C, Golabek GJ, Raymond SN, Schönbächler M, Tackley PJ, Dehant V, Debaille V (2020) Dry late accretion inferred from Venus's coupled atmosphere and internal evolution. Nat Geosci 13:265–269. https://doi.org/10.1038/s41561-020-0561-x

Gillmann C, Way MJ, Avice G, Breuer D, Golabek GJ, Höning D, Krissansen-Totton J, Lammer H, O'Rourke JG, Persson M, Plesa AC, Salvador A, Scherf M, Zolotov MY (2022) The long-term evolution of the atmosphere of Venus: processes and feedback mechanisms. Space Sci Rev 218:56. https://doi.org/10.1007/s11214-022-00924-0

Gilmore MS, Head JW (2018) Morphology and deformational history of Tellus Regio, Venus: evidence for assembly and collision. Planet Space Sci 154:5–20. https://doi.org/10.1016/j.pss.2018.02.001

Gilmore MS, Mueller N, Helbert J (2015) VIRTIS emissivity of Alpha Regio, Venus, with implications for tessera composition. Icarus 254:350–361. https://doi.org/10.1016/j.icarus.2015.04.008

Gilmore MS, Treiman A, Helbert J, Smrekar S (2017) Venus surface composition constrained by observation and experiment. Space Sci Rev 212:1511–1540. https://doi.org/10.1007/s11214-017-0370-8

Gilmore M, Helbert J, Brossier J, Carter L, Darby D, Filiberto F, Gerya T, Ghail R, Ivanov M, Izenberg N, Müller N, Santos A, Smrekar S (2023) Surface composition and mineralogy of the Venus surface. Space Sci Rev

Green DH, Hibberson WO, Rosenthal A, Kovács I, Yaxley GM, Falloon TJ, Brink F (2014) Experimental study of the influence of water on melting and phase assemblages in the upper mantle. J Petrol 55:2067–2096. https://doi.org/10.1093/petrology/egu050

Grimm RE (1994) Recent deformation rates on Venus. J Geophys Res 99:23163–23171. https://doi.org/10.1029/94JE02196

Grimm RE, Barr Milnar A, Harrison K, Stillman D, Neal K, Vincent MA, Delory G (2012) Aerial electromagnetic sounding of the lithosphere of Venus. Icarus 217:462–473. https://doi.org/10.1016/j.icarus.2011.07.021

Grinspoon DH (1993) Implications of the high D/H ratio for the sources of water in Venus' atmosphere. Nature 363:428–431. https://doi.org/10.1038/363428a0

Guerrero JM, Lowman JP, Deschamps F, Tackley PJ (2018) The influence of curvature on convection in a temperature-dependent viscosity fluid: implications for the 2-D and 3-D modeling of moons. J Geophys Res, Planets 123:1863–1880. https://doi.org/10.1029/2017je005497

Gülcher AJP, Gerya TV, Montési L, Munch J (2020) Corona structures driven by plume–lithosphere interactions and evidence for ongoing plume activity on Venus. Nat Geosci 13:547–554. https://doi.org/10.1038/s41561-020-0606-1

Gülcher AJP, Ballmer MD, Tackley PJ (2021) Coupled dynamics and evolution of primordial and recycled heterogeneity in Earth's lower mantle. Solid Earth 12:2087–2107. https://doi.org/10.5194/se-12-2087-2021

Hager BH, Clayton RW, Richards MA, Comer RP, Dziewonski (1985) Lower mantle heterogeneity, dynamic topography and the geoid. Nature 313:541–545. https://doi.org/10.1038/313541a0

Hanmer S (2020) Tessera terrain ribbon fabrics on Venus reviewed: could they be dyke swarms? Earth-Sci Rev 201:103077. https://doi.org/10.1016/j.earscirev.2019.103077. 2020

Hansen VL, Willis JJ (1998) Ribbon terrain formation, southwestern fortuna tessera, Venus: implications for lithosphere evolution. Icarus 132:321–343. https://doi.org/10.1006/icar.1998.5897

Hansen U, Yuen DA, Kroening SE, Larsen TB (1993) Dynamical consequences of depth-dependent thermal expansivity and viscosity on mantle circulations and thermal structure. Phys Earth Planet Inter 77:205–223. https://doi.org/10.1016/0031-9201(93)90099-u

Harris LB, Bédard JH (2013) Crustal evolution and deformation in a non-plate-tectonic Achaean Earth: comparisons with Venus. In: Modern approaches in solid Earth sciences evolution of Archean crust and early life, pp 215–291. https://doi.org/10.1007/978-94-007-7615-9_9

562

Harris LB, Bédard JH (2014) Interactions between continent-like 'drift', rifting and mantle flow on Venus: gravity interpretations and Earth analogues. Geol Soc (Lond) Spec Publ 401:327–356. https://doi.org/10.1144/sp401.9

Helbert J, Maturilli A, Dyar MD, Alemanno G (2021) Deriving iron contents from past and future Venus surface spectra with new high-temperature laboratory emissivity data. Sci Adv 7:eaba9428. https://doi.org/10.1126/sciadv.aba9428

Herrick RR, Phillips RJ (1994) Effects of the Venusian atmosphere on incoming meteoroids and the impact crater population. Icarus 112:543–546. https://doi.org/10.1006/icar.1994.1180

Herrick RR, Rumpf ME (2011) Postimpact modification by volcanic or tectonic processes as the rule, not the exception, for Venusian craters. J Geophys Res 116:E02004. https://doi.org/10.1029/2010je003722

Herrick RR, Bjonnes E, Carter L, Gerya T, Ghail R, Gillmann C, Gilmore M, Hensley S, Ivanov M, Izenberg N, Müller N, O'Rourke JG, Rolf T, Smrekar SE, Weller M (2023) Resurfacing history and volcanic activity of Venus. Space Sci Rev

Hofmann AW (1997) Mantle geochemistry: the message from oceanic volcanism. Nature 385:219–229. https://doi.org/10.1038/385219a0

Höink T, Lenardic A (2008) Three-dimensional mantle convection simulations with a low-viscosity asthenosphere and the relationship between heat flow and the horizontal length scale of convection. Geophys Res Lett 35:L10304. https://doi.org/10.1029/2008gl033854

Höink T, Lenardic A (2010) Long wavelength convection, Poiseuille & Couette flow in the low-viscosity asthenosphere and the strength of plate margins. Geophys J Int 180:23–33. https://doi.org/10.1111/j.1365-246x.2009.04404.x

Höink T, Lenardic A, Richards M (2012) Depth-dependent viscosity and mantle stress amplification: implications for the role of the asthenosphere in maintaining plate tectonics. Geophys J Int 191:30–41. https://doi.org/10.1111/j.1365-246x.2012.05621.x

Hoogenboom T, Houseman GA (2006) Rayleigh-Taylor instability as a mechanism for corona formation on Venus. Icarus 180:292–307. https://doi.org/10.1016/j.icarus.2005.11.001

Howard LN (1966) Convection at high Rayleigh number. In: Görtler H (ed) Applied mechanics. Springer, Berlin, pp 1109–1115. https://doi.org/10.1007/978-3-662-29364-5_147

Hu Y, Bürgmann R, Banerjee P, Feng L, Hill EM, Ito T, Tabei T, Wang K (2016) Asthenosphere rheology inferred from observations of the 2012 Indian Ocean earthquake. Nature 538:368–372. https://doi.org/10.1038/nature19787

Huang J, Yang A, Zhong S (2013) Constraints of the topography, gravity and volcanism on Venusian mantle dynamics and generation of plate tectonics. Earth Planet Sci Lett 362:207–214. https://doi.org/10.1016/j.epsl.2012.11.051

Ishii T et al (2018) Complete agreement of the post-spinel transition with the 660-km seismic discontinuity. Sci Rep 8:6358. https://doi.org/10.1038/s41598-018-24832-y

Ivanov MA, Head JW (2011) Global geological map of Venus. Planet Space Sci 59:1559–1600. https://doi.org/10.1016/j.pss.2011.07.008

Jacobson SA, Rubie DC, Hernlund J, Morbidelli A, Nakajima M (2017) Formation, stratification, and mixing of the cores of Earth and Venus. Earth Planet Sci Lett 474:375–386. https://doi.org/10.1016/j.epsl.2017.06.023

James PB, Zuber MT, Phillips RJ (2013) Crustal thickness and support of topography on Venus. J Geophys Res, Planets 118:859–875. https://doi.org/10.1029/2012je004237

Jellinek AM, Jackson MG (2015) Connections between the bulk composition, geodynamics and habitability of Earth. Nat Geosci 8:587–593. https://doi.org/10.1038/NGEO2488

Jellinek AM, Lenardic A, Manga M (2002) The influence of interior mantle temperature on the structure of plumes: heads for Venus, tails for the Earth. Geophys Res Lett 29:1532. https://doi.org/10.1029/2001gl014624

Jiménez-Díaz A, Ruiz J, Kirby JF, Romeo I, Tejero R, Capote R (2015) Lithospheric structure of Venus from gravity and topography. Icarus 260:215–231. https://doi.org/10.1016/j.icarus.2015.07.020

Johnson CL, Richards MA (2003) A conceptual model for the relationship between coronae and large-scale mantle dynamics on Venus. J Geophys Res 108:5058. https://doi.org/10.1029/2002je001962

Johnson CL, Sandwell DT (1994) Lithospheric flexure on Venus. Geophys J Int 119:627–647. https://doi.org/10.1111/j.1365-246x.1994.tb00146.x

Kankanamge DGJ, Moore WB (2016) Heat transport in the Hadean mantle: from heat pipes to plates. Geophys Res Lett 43:3208–3214. https://doi.org/10.1002/2015gl067411

Karato S-I (2010) Rheology of the deep upper mantle and its implications for the preservation of the continental roots: a review. Tectonophysics 481:82–98. https://doi.org/10.1016/j.tecto.2009.04.011

Karimi S, Dombard AJ (2017) Studying lower crustal flow beneath Mead basin: implications for the thermal history and rheology of Venus. Icarus 282:34–39. https://doi.org/10.1016/j.icarus.2016.09.015

563

Karimi S, Ojha L, Lewis K (2018) Searching for larger buried craters on Venus. In: 48th lunar planet sci conf, #2831

Karlsson RVMK, Cheng KW, Crameri F, Rolf T, Uppalapati S, Werner SC (2020) Implications of anomalous crustal provinces for Venus' resurfacing history. J Geophys Res, Planets 125:e2019JE006340. https://doi.org/10.1029/2019je006340

Katayama I (2021) Strength models of the terrestrial planets and implications for their lithospheric structure and evolution. Prog Earth Planet Sci 8:1. https://doi.org/10.1186/s40645-020-00388-2

Kiefer WS (2013) Making Ishtar Terra, Venus: mobile lid tectonic, continental crust, and implications for liquid water and planetary evolution. In: 44th lunar planet sci conf, #2541

Kiefer WS, Hager BH (1991) A mantle plume model for the equatorial highlands of Venus. J Geophys Res 96:20947–20966. https://doi.org/10.1029/91JE02221

Kiefer WS, Weller MB (2021) Venus, Earth's divergent twin: observations constraining the transition from a mobile lid planet to a stagnant lid planet. In: 52nd lunar planet sci conf, #1792

Kiefer WS, Richards MA, Hager BH (1986) A dynamic model of Venus's gravity field. Geophys Res Lett 13:14–17. https://doi.org/10.1029/GL013i001p00014

King SD (2018) Venus resurfacing constrained by geoid and topography. J Geophys Res, Planets 123:1041–1060. https://doi.org/10.1002/2017je005475

Koch DM, Manga M (1996) Neutrally buoyant diapirs: a model for Venus coronae. Geophys Res Lett 23:225–228. https://doi.org/10.1029/95GL03776

Kohlstedt Dl, Hansen LN (2015) Constitutive equations, rheological behavior, and viscosity of rocks. Treatise Geophys 2:441–472. https://doi.org/10.1016/b978-0-444-53802-4.00042-7

Kohlstedt DL, Evans B, Mackwell SJ (1995) Strength of the lithosphere: constraints imposed by laboratory experiments. J Geophys Res, Solid Earth 100:17587–17602. https://doi.org/10.1029/95jb01460

Konopliv AS, Yoder CF (1996) Venusian k_2 tidal Love number from Magellan and PVO Tracking Data. Geophys Res Lett 23:1857–1860. https://doi.org/10.1029/96gl01589

Konopliv AS, Banerdt WB, Sjogren WL (1999) Venus gravity: 180th degree and order model. Icarus 139:3–18. https://doi.org/10.1006/icar.1999.6086

Korenaga J (2008) Urey ratio and the structure and evolution of Earth's mantle. Rev Geophys 46:2007RG000241. https://doi.org/10.1029/2007rg000241

Korenaga J (2017) Pitfalls in modeling mantle convection with internal heat production. J Geophys Res, Solid Earth 122:4064–4085. https://doi.org/10.1002/2016jb013850

Kremic T, Amato M, Gilmore M, Kiefer W, Johnson N, Sauder J, Hunter G, Thompson T (2021) Venus surface platform study final report. NASA Glen Research Center NP-2021-11-102-GRC, 50 pages. LPI contribution 2660. https://www.lpi.usra.edu/vexag/documents/reports/Venus-Surface-Platform-Study-Final_11-4-21.pdf

Kreslavsky MA, Ivanov MA, Head JW (2015) The resurfacing history of Venus: constraints from buffered crater densities. Icarus 250:438–450. https://doi.org/10.1016/j.icarus.2014.12.024

Labrosse S (2015) Thermal evolution of the core with a high thermal conductivity. Phys Earth Planet Inter 247:36–55. https://doi.org/10.1016/j.pepi.2015.02.002

Labrosse S, Hernlund JW, Coltice N (2007) A crystallizing dense magma ocean at the base of the Earth's mantle. Nature 450:866–869. https://doi.org/10.1038/nature06355

Landuyt W, Bercovici D (2009) Variations in planetary convection via the effect of climate on damage. Earth Planet Sci Lett 277:29–37. https://doi.org/10.1016/j.epsl.2008.09.034

Lawrence JF, Shearer PM (2008) Imaging mantle transition zone thickness with SdS-SS finite-frequency sensitivity kernels. Geophys J Int 174:143–158. https://doi.org/10.1111/j.1365-246x.2007.03673.x

Lay T, Hernlund J, Buffett BA (2008) Core–mantle boundary heat flow. Nat Geosci 1:25–32. https://doi.org/10.1038/ngeo.2007.44

Le Feuvre M, Wieczorek MA (2011) Nonuniform cratering of the Moon and a revised crater chronology of the inner solar system. Icarus 214:1–20. https://doi.org/10.1016/j.icarus.2011.03.010

Leftwich TE, von Frese RRB, Kim HR, Noltimier HC, Potts LV, Roman DR, Tan L (1999) Crustal analysis of Venus from Magellan Satellite Observations at Atalanta Planitia, Beta Regio, and Thetis Regio. J Geophys Res, Planets 104:8441–8462. https://doi.org/10.1029/1999je900007

Lenardic A, Crowley JW (2012) On the notion of well-defined tectonic regimes for terrestrial planets in this solar system and others. Astrophys J 755:132. https://doi.org/10.1088/0004-637x/755/2/132

Lenardic A, Moresi L-N, Mühlhaus H (2003) Longevity and stability of cratonic lithosphere: insights from numerical simulations of coupled mantle convection and continental tectonics. J Geophys Res, Solid Earth 108:2303. https://doi.org/10.1029/2002jb001859

Lenardic A, Jellinek AM, Moresi L-N (2008) A climate induced transition in the tectonic style of a terrestrial planet. Earth Planet Sci Lett 271:34–42. https://doi.org/10.1016/j.epsl.2008.03.031

Lenardic A, Jellinek AM, Foley B, O'Neill C, Moore WB (2016) Climate-tectonic coupling: variations in the mean, variations about the mean, and variations in mode. J Geophys Res 121:1831–1864. https://doi.org/10.1002/2016JE005089

Lenardic A, Weller M, Höink T, Seales J (2019) Toward a boot strap hypothesis of plate tectonics: feedbacks between plates, the asthenosphere, and the wavelength of mantle convection. Phys Earth Planet Inter 296:106299. https://doi.org/10.1016/j.pepi.2019.106299

Lenardic A, Seales J, Moore W, Weller M (2021) Convective and tectonic plate velocities in a mixed heating mantle. Geochem Geophys Geosyst 22:e2020GC009278. https://doi.org/10.1029/2020GC009278

Leng W, Zhong S (2008) Controls on plume heat flux and plume excess temperature. J Geophys Res 113:B04408. https://doi.org/10.1029/2007jb005155

Leng W, Zhong S (2009) More constraints on internal heating rate of the Earth's mantle from plume observations. Geophys Res Lett 36:L02306. https://doi.org/10.1029/2008gl036449

Loddoch A, Stein C, Hansen U (2006) Temporal variations in the convective style of planetary mantles. Earth Planet Sci Lett 251:79–89. https://doi.org/10.1016/j.epsl.2006.08.026

Lourenço DL, Rozel AB, Tackley PJ (2016) Melting-induced crustal production helps plate tectonics on Earth-like planets. Earth Planet Sci Lett 439:18–28. https://doi.org/10.1016/j.epsl.2016.01.024. 2016

Lourenço DL, Rozel AB, Gerya T, Tackley PJ (2018) Efficient cooling of rocky planets by intrusive magmatism. Nat Geosci 11:322–327. https://doi.org/10.1038/s41561-018-0094-8

Lourenço DL, Rozel AB, Ballmer MD, Tackley PJ (2020) Plutonic-squishy lid: a new global tectonic regime generated by intrusive magmatism on Earth-like planets. Geochem Geophys Geosyst 21:e2019GC008756. https://doi.org/10.1029/2019gc008756

Mackwell SJ, Zimmermann ME, Kohlstedt DL (1998) High-temperature deformation of dry diabase with application to tectonics on Venus. J Geophys Res, Solid Earth 103:975–984. https://doi.org/10.1029/97jb02671

Maia JS, Wieczorek MA (2022) Lithospheric structure of Venusian crustal plateaus. J Geophys Res 127:e2021JE007004. https://doi.org/10.1029/2021JE007004

Mao H-K, Hu Q, Yang L, Liu J, Kim DY, Meng Y, Zhang L, Prakapenka VB, Yang W, Mao WL (2021) When water meets iron at Earth's core-mantle boundary. Natl Sci Rev 4:870–878. https://doi.org/10.1093/nsr/nwx109

Margot J-L, Campbell DB, Giorgini JD, Jao JS, Snedeker LG, Ghigo FD, Bonsall A (2021) Spin state and moment of inertia of Venus. Nat Astron 5:676–683. https://doi.org/10.1038/s41550-021-01339-7

Masuti S, Barbot SD, Karato S-I, Feng L, Banerjee P (2016) Upper-mantle water stratification inferred from observations of the 2012 Indian Ocean earthquake. Nature 538:373–377. https://doi.org/10.1038/nature19783

McGill GE (1994) Hotspot evolution and Venusian tectonic style. J Geophys Res 99:23149–23161. https://doi.org/10.1029/94JE02319

McKenzie DP, Roberts JM, Weiss NO (1974) Convection in the Earth's mantle: towards a numerical simulation. J Fluid Mech 62:465. https://doi.org/10.1017/s0022112074000784

McKinnon WB, Zahnle KJ, Ivanov BA, Melosh HJ (1997) Cratering on Venus: models and observations. In: Bougher SW, Hunten DM, Phillips RJ (eds) Venus II: geology, geophysics, atmosphere, and solar wind environment. University of Arizona Press, Tucson, pp 969–1014

McNamara AK, Zhong S (2005) Degree-one mantle convection: dependence on internal heating and temperature-dependent rheology. Geophys Res Lett 32:L01301. https://doi.org/10.1029/2004gl021082

Meier TG, Bower DJ, Lichtenberg T, Tackley PJ, Demory B-O (2021) Interior dynamics of tidally locked super-earths: the case of LHS 3844b. Astrophys J Lett 908:L48. https://doi.org/10.3847/2041-8213/abe400

Miyagoshi T, Kameyama M, Ogawa M (2020) Tectonic plates in 3D mantle convection model with stress-history-dependent rheology. Earth Planets Space 72:70. https://doi.org/10.1186/s40623-020-01195-1

Mocquet A, Rosenblatt P, Dehant V, Verhoeven O (2011) The deep interior of Venus, Mars, and the Earth: a brief review and the need for planetary surface-based measurements. Planet Space Sci 59:1048–1061. https://doi.org/10.1016/j.pss.2010.02.002

Moore WB (2008) Heat transport in a convecting layer heated from within and below. J Geophys Res 113:B11407. https://doi.org/10.1029/2006JB004778

Moore WB, Schubert G (1997) Venusian crustal and lithospheric properties from nonlinear regressions of highland geoid and topography. Icarus 128:415–428. https://doi.org/10.1006/icar.1997.5750

Moore WB, Webb AAG (2013) Heat-pipe Earth. Nature 501:501–505. https://doi.org/10.1038/nature12473

Moore WB, Simon JI, Webb AAG (2017) Heat-pipe planets. Earth Planet Sci Lett 474:13–19. https://doi.org/10.1016/j.epsl.2017.06.015

Moresi L, Solomatov V (1998) Mantle convection with a brittle lithosphere: thoughts on the global tectonic styles of the Earth and Venus. Geophys J Int 133:669–682. https://doi.org/10.1046/j.1365-246x.1998.00521.x

Moruzzi SA, Kiefer WS (2020) Thrust faulting on Venus: tectonic modeling of the Vedma Dorsa Ridge Belt. In: 51st lunar planet sci conf, #1430

Moyen J-F, van Hunen J (2012) Short-term episodicity of Archaean plate tectonics. Geology 40:451–454. https://doi.org/10.1130/g322894.1

Mulyukova E, Bercovici D (2020) Mantle convection in terrestrial planets. In: In Oxford research encyclopedia of planetary science. https://doi.org/10.1093/acrefore/9780190647926.013.109

Murakami M, Hirose K, Kawamura K, Sata N, Ohishi Y (2004) Post-Perovskite phase transition in $MgSiO_3$. Science 304:855–858. https://doi.org/10.1126/science.1095932

Nakagawa T, Nakakuki T, Iwamori H (2015) Water circulation and global mantle dynamics: insight from numerical modeling. Geochem Geophys Geosyst 16:1449–1464. https://doi.org/10.1002/2014GC005701

Namiki N (1995) Tectonics and volcanism on Venus: constraints from topographic relief, impact cratering, and degassing. PhD Thesis, Massachusetts Institute of Technology. 240 p

Namiki N, Solomon SC (1993) The gabbro-eclogite phase transition and the elevation of mountain belts on Venus. J Geophys Res 98:15025. https://doi.org/10.1029/93je01626

Nimmo F (2002) Why does Venus lack a magnetic field? Geology 30:987. https://doi.org/10.1130/0091-7613(2002)030<0987:wdvlam>2.0.co;2

Nimmo F (2015) Thermal and compositional evolution of the core. In: Schubert G (ed) Treatise on geophysics, 2015, vol 9. Elsevier, Amsterdam, pp 201–219. https://doi.org/10.1016/b978-0-444-53802-4.00160-3. Chap 9.08

Nimmo F, McKenzie D (1998) Volcanism and tectonics on Venus. Annu Rev Earth Planet Sci 26:23–51. https://doi.org/10.1146/annurev.earth.26.1.23

Noack L, Breuer D, Spohn T (2012) Coupling the atmosphere with interior dynamics: implications for the resurfacing of Venus. Icarus 217:484–498. https://doi.org/10.1016/j.icarus.2011.08.026

Ogawa M (2000) Coupled magmatism–mantle convection system with variable viscosity. Tectonophysics 322:1–18. https://doi.org/10.1016/s0040-1951(00)00054-8

Ogawa M (2018) The effects of magmatic redistribution of heat producing elements on the lunar mantle evolution inferred from numerical models that start from various initial states. Planet Space Sci 151:43–55. https://doi.org/10.1016/j.pss.2017.10.015

Ogawa M, Yanagisawa T (2011) Numerical models of Martian mantle evolution induced by magmatism and solid-state convection beneath stagnant lithosphere. J Geophys Res 116:E08008. https://doi.org/10.1029/2010je003777

Ogawa M, Yanagisawa T (2014) Mantle evolution in Venus due to magmatism and phase transitions: from punctuated layered convection to whole-mantle convection. J Geophys Res, Planets 119:867–883. https://doi.org/10.1002/2013je004593

Ohtani E (2020) The role of water in Earth's mantle. Nat Sci Rev 7:224–232. https://doi.org/10.1016/j.pss.2017.10.015

Okudaira T, Shigematsu N, Harigane Y, Yoshida K (2017) Grain size reduction due to fracturing and subsequent grain-size-sensitive creep in a lower crustal shear zone in the presence of a CO_2-bearing fluid. J Struct Geol 95:171–187. https://doi.org/10.1016/j.jsg.2016.11.001

O'Neill C (2021) End-member Venusian core scenarios: does Venus have an inner core? Geophys Res Lett 48:e2021GL095499. https://doi.org/10.1029/2021GL095499

O'Neill C, Lenardic A, Weller M, Moresi L, Quenette S, Zhang S (2016) A window for plate tectonics in terrestrial planet evolution? Phys Earth Planet Inter 255:80–92. https://doi.org/10.1016/j.pepi.2016.04.002

O'Neill C, Marchi S, Zhang S, Bottke W (2017) Impact-driven subduction on the Hadean Earth. Nat Geosci 10:793–797. https://doi.org/10.1038/ngeo3029

O'Neill C, Marchi S, Bottke W, Fu R (2020) The role of impacts on Archaean tectonics. Geology 48:174–178. https://doi.org/10.1130/g46533.1

O'Reilly TC, Davies GF (1981) Magma transport of heat on Io: a mechanism allowing a thick lithosphere. Geophys Res Lett 8:313–316. https://doi.org/10.1029/gl008i004p00313

O'Rourke JG (2020) Venus: a thick basal magma ocean may exist today. Geophys Res Lett 47:e2019GL086126. https://doi.org/10.1029/2019gl086126

O'Rourke JG, Korenaga J (2015) Thermal evolution of Venus with argon degassing. Icarus 260:128–140. https://doi.org/10.1016/j.icarus.2015.07.009

O'Rourke JG, Smrekar SE (2018) Signatures of lithospheric flexure and elevated heat flow in stereo topography at coronae on Venus. J Geophys Res, Planets 123:369–389. https://doi.org/10.1002/2017je005358

O'Rourke JG, Wolf AS, Ehlmann BL (2014) Venus: interpreting the spatial distribution of volcanically modified craters. Geophys Res Lett 41:8252–8260. https://doi.org/10.1002/2014GL062121

O'Rourke JG, Gillmann C, Tackley P (2018) Prospects for an ancient dynamo and modern crustal remanent magnetism on Venus. Earth Planet Sci Lett 502:46–56. https://doi.org/10.1016/j.epsl.2018.08.055

O'Rourke JG, Buz J, Fu RR, Lillis RJ (2019) Detectability of remanent magnetism in the crust of Venus. Geophys Res Lett 46:5678–5777. https://doi.org/10.1029/2019GL082725

O'Rourke JG, Wilson C, Borrelli M, Byrne PK, Dumoulin C, Ghail R, Gülcher A, Jacobson S, Korablev O, Spohn T, Way M, Weller M, Westall F (2023) Venus, the planet: introduction to the evolution of Earth's sister planet. Space Sci Rev

Orth CP, Solomatov VS (2011) The isostatic stagnant lid approximation and global variations in the Venusian lithospheric thickness. Geochem Geophys Geosyst 12:Q07018. https://doi.org/10.1029/2011gc003582

Orth CP, Solomatov VS (2012) Constraints on the Venusian crustal thickness variations in the isostatic stagnant lid approximation. Geochem Geophys Geosyst 13:Q11012. https://doi.org/10.1029/2012gc004377

Papuc AM, Davies GF (2012) Transient mantle layering and the episodic behaviour of Venus due to the 'Basalt Barrier' mechanism. Icarus 217:499–509. https://doi.org/10.1016/j.icarus.2011.09.024

Parmentier EM, Sotin C, Travis BJ (1994) Turbulent 3-D thermal convection in an infinite Prandtl number, volumetrically heated fluid: implications for mantle dynamics. Geophys J Int 116:241–251. https://doi.org/10.1111/j.1365-246x.1994.tb01795.x

Pauer M, Fleming K, Cadek O (2006) Modeling the dynamic component of the geoid and topography of Venus. J Geophys Res 111:E11012. https://doi.org/10.1029/2005je002511

Pettengill GH, Ford PG, Wilt RJ (1992) Venus surface radiothermal emission as observed by Magellan. J Geophys Res 97:13091–13102. https://doi.org/10.1029/92je01356

Phillips RJ, Izenberg NR (1995) Ejecta correlations with spatial crater density and Venus resurfacing history. Geophys Res Lett 22:1517–1520. https://doi.org/10.1029/95GL01412

Phillips JL, Russell CT (1987) Upper limit on the intrinsic magnetic field of Venus. J Geophys Res 92:2253. https://doi.org/10.1029/ja092ia03p02253

Phillips RJ, Raubertas RF, Arvidson RE, Sarkar IC, Herrick RR, Izenberg N, Grimm RE (1992) Impact craters and Venus resurfacing history. J Geophys Res 97:15923–15948. https://doi.org/10.1029/92je01696

Piskorz D, Elkins-Tanton LT, Smrekar SE (2014) Coronae formation on Venus via extension and lithospheric instability. J Geophys Res 119:2568–2582. https://doi.org/10.1002/2014JE004636

Plesa A-C, Breuer D (2021) The effects of intrusive magmatism on the mechanical lithosphere thickness of Venus. In: 52nd lunar planetary science conference, p #2130

Robin CMI, Jellinek AM, Thayalan V, Lenardic A (2007) Transient mantle convection on Venus: the paradoxical coexistence of highlands and coronae in the BAT region. Earth Planet Sci Lett 256:100–119. https://doi.org/10.1016/j.epsl.2007.01.016

Rolf T, Tackley PJ (2011) Focussing of stress by continents in 3D spherical mantle convection with self-consistent plate tectonics. Geophys Res Lett 38:L18301. https://doi.org/10.1029/2011gl048677

Rolf T, Coltice N, Tackley PJ (2014) Statistical cyclicity of the supercontinent cycle. Geophys Res Lett 41:2351–2358. https://doi.org/10.1002/2014gl059595

Rolf T, Capitanio FA, Tackley PJ (2018a) Constraints on mantle viscosity structure from continental drift histories in spherical mantle convection models. Tectonophysics 746:339–351. https://doi.org/10.1016/j.tecto.2017.04.031

Rolf T, Steinberger B, Werner SC, Sruthi U (2018b) Inferences on the mantle viscosity structure and the post-overturn evolutionary state of Venus. Icarus 313:107–123. https://doi.org/10.1016/j.icarus.2018.05.014

Romeo I, Turcotte DI (2010) Resurfacing on Venus. Planet Space Sci 58:1374–1380. https://doi.org/10.1016/j.pss.2010.05.022

Rosenblatt P, Dumoulin C, Marty J-C, Genova A (2021) Determination of Venus' interior structure with EnVision. Remote Sens 13:1624. https://doi.org/10.3390/rs13091624

Rozel A, Besserer J, Golabek GJ, Kaplan M, Tackley PJ (2014) Self-consistent generation of single-plume state for Enceladus using non-Newtonian rheology. J Geophys Res, Planets 119:416–439. https://doi.org/10.1002/2013je004473

Rozel AB, Golabek GJ, Näf R, Tackley PJ (2015) Formation of ridges in a stable lithosphere in mantle convection models with a viscoplastic rheology. Geophys Res Lett 42:4770–4777. https://doi.org/10.1002/2015gl063483

Rozel AB, Golabek GJ, Jain C, Tackley PJ, Gerya T (2017) Continental crust formation on early Earth controlled by intrusive magmatism. Nature 545:332–335. https://doi.org/10.1038/nature22042

Rudolph ML, Lekic V, Lithgow-Bertelloni C (2015) Viscosity jump in Earths mid-mantle. Science 350:1349–1352. https://doi.org/10.1126/science.aad1929

Ruedas T, Breuer D (2018) 'Isocrater' impacts: conditions and mantle dynamical responses for different impactor types. Icarus 306:94–115. https://doi.org/10.1016/j.icarus.2018.02.005

Ruiz J (2007) The heat flow during the formation of ribbon terrains on Venus. Planet Spa Sci 55:2063–2070. https://doi.org/10.1016/j.pss.2007.05.003

Ruiz J, Jimenez-Diáz A, Egea-Gonzalez I, Parro LM (2019) Comments on 'Using the viscoelastic relaxation of large impact craters to study the thermal history of Mars' (Karimi et al. (2016) Icarus 272:102–113) and 'Studying lower crustal flow beneath Mead basin: implications for the thermal history and rheology of Venus' (Karimi & Dombard (2017) Icarus 282:34–39). Icarus 322:221–226. https://doi.org/10.1016/j.icarus.2018.10.009

Russell CT, Zhang TL, Delva M, Magnes W, Strangeway WHY (2007) Lightning on Venus inferred from whistler-mode waves in the ionosphere. Nature 450:661–662. https://doi.org/10.1038/nature05930

Salvador A, Avice G, Breuer A, Gillmann C, Jacobson S, Marcq E, Raymond S, Sakuraba H, Scherf M, Way M (2023) Magma ocean, water, and the early atmosphere of Venus. Space Sci Rev

Sandu C, Lenardic A, McGovern P (2011) The effects of deep water cycling on planetary thermal evolution. J Geophys Res 116:B12404. https://doi.org/10.1029/2011jb008405

Sandwell DT, Schubert G (1992) Evidence for retrograde lithospheric subduction on Venus. Science 257:766–770. https://doi.org/10.1126/science.257.5071.766

Schmeling H, Marquart G, Ruedas T (2003) Pressure- and temperature-dependent thermal expansivity and the effect on mantle convection and surface observables. Geophys J Int 154:224–229. https://doi.org/10.1046/j.1365-246x.2003.01949.x

Schools JW, Montési LGJ (2018) The generation of barriers to melt ascent in the Martian lithosphere. J Geophys Res, Planets 123:47–66. https://doi.org/10.1002/2017je005396

Schubert G, Sandwell DT (1995) A global survey of possible subduction sites on Venus. Icarus 117:173–196. https://doi.org/10.1006/icar.1995.1150

Schulz F, Tosi N, Plesa A-C, Breuer D (2019) Stagnant-lid convection with diffusion and dislocation creep rheology: influence of a non-evolving grain size. Geophys J Int 220:18–36. https://doi.org/10.1093/gji/ggz417

Sifré D, Gardés E, Massuyeau M, Hashim L, Hier-Majumder S, Gaillard F (2014) Electrical conductivity during incipient melting in the oceanic low-velocity zone. Nature 509:81–85. https://doi.org/10.1038/nature13245

Sleep NH (1990) Hotspots and mantle plumes: some phenomenology. J Geophys Res 95:6715–6736. https://doi.org/10.1029/jb095ib05p06715

Smrekar SE (1994) Evidence for active hotspots on Venus from analysis of Magellan Gravity Data. Icarus 112:2–26. https://doi.org/10.1006/icar.1994.1166

Smrekar SE, Parmentier EM (1996) The interaction of mantle plumes with surface thermal and chemical boundary layers: applications to hotspots on Venus. J Geophys Res, Solid Earth 101:5397–5410. https://doi.org/10.1029/95jb02877

Smrekar SE, Phillips RJ (1991) Venusian highlands: geoid to topography ratios and their implications. Earth Planet Sci Lett 107:582–597. https://doi.org/10.1016/0012-821X(91)90103-O

Smrekar SE, Sotin C (2012) Constraints on mantle plumes on Venus: implications for volatile history. Icarus 217:510–523. https://doi.org/10.1016/j.icarus.2011.09.011

Smrekar SE, Stofan ER (1997) Corona formation and heat loss on Venus by coupled upwelling and delamination. Science 277:1289–1294. https://doi.org/10.1126/science.277.5330.1289

Smrekar SE, Hoogenboom T, Stofan ER, Martin P (2010) Recent hotspot volcanism on Venus from VIRTIS emissivity data. Science 328:605–608. https://doi.org/10.1126/science.1186785

Smrekar SE, Davaille A, Sotin C (2018) Venus interior structure and dynamics. Space Sci Rev 214:88. https://doi.org/10.1007/s11214-018-0518-1

Solomatov VS (1995) Scaling of temperature- and stress-dependent viscosity convection. Phys Fluids 7:266–274. https://doi.org/10.1063/1.868624

Solomatov VS (2003) Initiation of subduction by small-scale convection. J Geophys Res, Solid Earth 109:B01412. https://doi.org/10.1029/2003jb002628

Solomatov VS, Moresi L-N (1996) Stagnant lid convection on Venus. J Geophys Res, Planets 101:4737–4753. https://doi.org/10.1029/95je03361

Solomon SC, Head JW (1990) Lithospheric flexure beneath the Freya Montes Foredeep, Venus: constraints on lithospheric thermal gradient and heat flow. Geophys Res Lett 17:1393–1396. https://doi.org/10.1029/gl017i009p01393

Sotin C, Labrosse S (1999) Three-dimensional thermal convection in an iso-viscous, infinite Prandtl number fluid heated from within and from below: applications to the transfer of heat through planetary mantles. Phys Earth Planet Inter 112:171–190. https://doi.org/10.1016/s0031-9201(99)00004-7

Spada G, Sabadini R, Boschi E (1996) The spin and inertia of Venus. Geophys Res Lett 23:1997–2000. https://doi.org/10.1029/96gl01765

Spencer DC, Katz RF, Hewitt IJ (2020) Magmatic intrusions control Io's crustal thickness. J Geophys Res 125:e2020JE006443. https://doi.org/10.1029/2020JE006443

Stein C, Schmalzl J, Hansen U (2004) The effect of rheological parameters on plate behaviour in a self-consistent model of mantle convection. Phys Earth Planet Inter 142:225–255. https://doi.org/10.1016/j.pepi.2004.01.006

Stein C, Lowman JP, Hansen U (2013) The influence of mantle internal heating on lithospheric mobility: implications for super-earths. Earth Planet Sci Lett 361:448–459. https://doi.org/10.1016/j.epsl.2012.11.011

Steinberger B, Werner SC, Torsvik TH (2010) Deep versus shallow origin of gravity anomalies, topography and volcanism on Earth, Venus and Mars. Icarus 207:564–577. https://doi.org/10.1016/j.icarus.2009.12.025

Stern RJ, Gerya T, Tackley PJ (2018) Stagnant lid tectonics: perspectives from silicate planets, dwarf planets, large moons, and large asteroids. Geosci Front 9:103–119. https://doi.org/10.1016/j.gsf.2017.06.004

Stevenson DJ (2003) Styles of mantle convection and their influence on planetary evolution. C R Géosci 335:99–111. https://doi.org/10.1016/s1631-0713(03)00009-9

Stevenson DJ (2010) Planetary magnetic fields: achievements and prospects. Space Sci Rev 152:651–664. https://doi.org/10.1007/s11214-009-9572-z

Stevenson DJ, Cutts J, Mimoun D, Arrowsmith S, Banerdt B, Blom P, Brageot E, Brissaud Q, Chin G, Gao P, Garcia R, Hall J, Hunter G, Jackson J, Kerzhanovic V, Kiefer W, Komjathy A, Lee C, Lognonné P, Lorenz R, Majid W, Majorradi M, Nolet G, O'Rourke J, Rolland L, Schubert G, Simons M, Sotin C, Spilker T, Tsai V (2015) Probing the interior structure of Venus. Keck Institute of Space Studies, California Institute of Technology, Pasadena. https://doi.org/10.26206/C1CX-EV12. 85 pages

Stofan ER, Smrekar SE (2005) Large topographic rises, coronae, large flow fields and large volcanoes on Venus: evidence for mantle plumes? In: Foulger GR, Natland JH, Presnall DC, Anderson DL (eds) Plates, plumes, and paradigms. Geological Society of America special papers, vol 388, p 861. https://doi.org/10.1130/SPE388

Stofan ER, Bindschadler D, Parmentier EM, Head J (1991) Corona structures on Venus: models of origin. J Geophys Res 96:20933–20946. https://doi.org/10.1029/91JE02218

Stofan ER, Sharpton VL, Schubert G, Baer G, Bindschadler DL, Janes DM, Squyres SW (1992) Global distribution and characteristics of coronae and related features on Venus: implications for origin and relation to mantle processes. J Geophys Res 97:13347–13378. https://doi.org/10.1029/92je01314

Strom RG, Schaber GG, Dawson DD (1994) The global resurfacing of Venus. J Geophys Res 99:10899–10926. https://doi.org/10.1029/94je00388

Tackley PJ (1996) On the ability of phase transitions and viscosity layering to induce long wavelength heterogeneity in the mantle. Geophys Res Lett 23:1985–1988. https://doi.org/10.1029/96gl01980

Tackley PJ (2000) Self-consistent generation of tectonic plates in time-dependent, three-dimensional mantle convection simulations. Geochem Geophys Geosyst 1:2000GC000036. https://doi.org/10.1029/2000gc000036

Tkalčić H, Young M, Muir JB, Davies DR, Mattesini M (2015) Strong, multi-scale heterogeneity in Earth's lowermost mantle. Sci Rep 5:18416. https://doi.org/10.1038/srep18416

Tosi N, Padovan S (2021) Mercury, Moon, Mars: surface expressions of mantle convection and interior evolution of stagnant-lid bodies. In: Marquardt H, Ballmer MD, Cottaar S, Konter J (eds) Mantle convection and surface expressions. AGU monograph series. Wiley, New York, pp 455–489

Tosi N, Yuen DA, Cadek O (2010) Dynamical consequences in the lower mantle with the post-perovskite phase change and strongly depth-dependent thermodynamic and transport properties. Earth Planet Sci Lett 298:229–243. https://doi.org/10.1016/j.epsl.2010.08.001

Trompert R, Hansen U (1998) Mantle convection simulations with rheologies that generate plate-like behaviour. Nature 395:686–689. https://doi.org/10.1038/27185

Trønnes RG, Baron MA, Eigenmann KR, Guren MG, Heyn BH, Løken A, Mohn CF (2019) Core formation, mantle differentiation and core-mantle interaction within Earth and the terrestrial planets. Tectonophysics 760:165–198. https://doi.org/10.1016/j.tecto.2018.10.021

Turcotte DL (1989) A heat pipe mechanism for volcanism and tectonics on Venus. J Geophys Res, Solid Earth 94:2779–2785. https://doi.org/10.1029/jb094ib03p02779

Turcotte DL (1993) An episodic hypothesis for Venusian tectonics. J Geophys Res 98:17061–17068. https://doi.org/10.1029/93je01775

Turcotte DL (1996) Magellan and comparative planetology. J Geophys Res, Planets 101:4765–4773. https://doi.org/10.1029/95je02295

Turcotte DL, Schubert G (2017) Geodynamics, 3rd edn. Cambridge University Press, Cambridge, 2017

Tyler RH, Henning WG, Hamilton CW (2015) Tidal heating in a magma ocean within Jupiter's moon Io. Astrophys J Suppl Ser 218:22. https://doi.org/10.1088/0067-0049/218/2/22

Uppalapati S, Rolf T, Crameri C, Werner SC (2020) Dynamics of lithospheric overturns and implications for Venus's surface. J Geophys Res, Planets 125:e2019JE006258. https://doi.org/10.1029/2019je006258

van Heck HJ, Tackley PJ (2008) Planforms of self-consistently generated plates in 3D spherical geometry. Geophys Res Lett 35:L19312. https://doi.org/10.1029/2008gl035190

Vesterholt Al, Petersen KD, Nagel TJ (2021) Mantle overturn and thermochemical evolution of a non-plate tectonic mantle. Earth Planet Sci Lett 569:117047. https://doi.org/10.1016/j.epsl.2021.117047

Vilella K, Deschamps F (2021) Heat-blanketed convection and its implications for the continental lithosphere. J Geophys Res, Solid Earth 126:e2020JB020695. https://doi.org/10.1029/2020jb020695

Way M, Del Genio AD (2020) Venusian habitable climate scenarios: modeling Venus through time and applications to slowly rotating Venus-like exoplanets. J Geophys Res, Planets 125:e2019JE006276. https://doi.org/10.1029/2019JE006276

Way M, Ostberg C, Foley BJ, Gillmann C, Höning D, Lammer H, O'Rourke JG, Persson M, Plesa AC, Salvador A, Scherf M, Weller M (2023) Synergies between Venus and exoplanetary observations. Space Sci Rev

Wei D, Yang A, Huang JS (2014) The gravity field and crustal thickness of Venus. Sci China Earth Sci 57:2025–2035. https://doi.org/10.1007/s11430-014-4824-5

Weidner DJ, Wang Y (2000) Phase transformations: implications for mantle structure. In: Karato SI, Forte A, Liebermann R, Masters G, Stixrude L (eds) Earth's deep interior: mineral physics and tomography from the atomic to the global scale. Geophysical monograph series, vol 117, pp 215–235. https://doi.org/10.1029/gm117p0215

Weller MB, Kiefer WS (2020) The physics of changing tectonic regimes: implications for the temporal evolution of mantle convection and the thermal history of Venus. J Geophys Res 125:e2019JE005960. https://doi.org/10.1029/2019je005960

Weller MB, Lenardic A (2012) Hysteresis in mantle convection: plate tectonics systems. Geophys Res Lett 39:L10202. https://doi.org/10.1029/2012gl051232

Weller MB, Lenardic A (2015) Diverging worlds: bi-stability, the evolution of terrestrial planets and its application to Venus and Earth. In: 46th lunar and planetary science conference, p #2670

Weller MB, Lenardic A (2016) The energetics and convective vigor of mixed-mode heating: velocity scalings and implications for the tectonics of exoplanets. Geophys Res Lett 43:9469–9474. https://doi.org/10.1002/2016gl069927

Weller MB, Lenardic A (2018) On the evolution of terrestrial planets: bi-stability, stochastic effects, and the non-uniqueness of tectonic states. Geosci Front 9:91–102. https://doi.org/10.1016/j.gsf.2017.03.001

Weller MB, Lenardic A, O'Neill C (2015) The effects of internal heating and large scale climate variations on tectonic bi-stability in terrestrial planets. Earth Planet Sci Lett 420:85–94. https://doi.org/10.1016/j.epsl.2015.03.021

Weller MB, Lenardic A, Moore WB (2016) Scaling relationships and physics for mixed heating convection in planetary interiors: isoviscous spherical shells. J Geophys Res, Solid Earth 121:7598–7617. https://doi.org/10.1002/2016jb013247

White SM, Crisp JA, Spera FJ (2006) Long-term volumetric eruption rates and magma budgets. Geochem Geophys Geosyst 7:Q03010. https://doi.org/10.1029/2005GC001002

Wicht J, Sanchez S (2019) Advances in geodynamo modelling. Geophys Astrophys Fluid Dyn 113:2–50. https://doi.org/10.1080/03091929.2019.1597074

Widemann T et al (2023) Venus evolution through time: key science questions, selected mission concepts and future investigations. Space Sci Rev

Wieczorek M (2007) Gravity and topography of the terrestrial planets. In: Treatise on geophysics, 2nd edn. Planets and moons, vol 10, pp 165–206. https://doi.org/10.1016/b978-044452748-6/00156-5

Wong T, Solomatov VS (2015) Towards scaling laws for subduction initiation on terrestrial planets: constraints from two-dimensional steady-state convection simulations. Prog Earth Planet Sci 2:18. https://doi.org/10.1186/s40645-015-0041-x

Wong T, Solomatov VS (2016) Variations in timing of lithospheric failure on terrestrial planets due to chaotic nature of mantle convection. Geochem Geophys Geosyst 17:1569–1585. https://doi.org/10.1002/2015gc006158

Yanagisawa T, Kameyama M, Ogawa M (2016) Numerical studies on convective stability and flow pattern in three-dimensional spherical mantle of terrestrial planets. Geophys J Int 206:1526–1538. https://doi.org/10.1093/gji/ggw226

Yang A, Huang JS, Wei D (2016) Separation of dynamic and isostatic components of the Venusian gravity and topography and determination of the crustal thickness of Venus. Planet Space Sci 129:24–31. https://doi.org/10.1016/j.pss.2016.06.001

Yoshida M (2004) Influence of two major phase transitions on mantle convection with moving and subducting plates. Earth Planets Space 56:1019–1033. https://doi.org/10.1186/bf03352544

Yoshida M (2008) Mantle convection with longest-wavelength thermal heterogeneity in a 3-D spherical model: degree one or two? Geophys Res Lett 35:L23302. https://doi.org/10.1029/2008gl036059

Yuan L, Ohtani E, Ikuta D, Kamada S, Tsuchiya J, Naohisa H, Ohishi Y, Suzuki A (2018) Chemical reactions between Fe and H_2O up to megabar pressures and implications for water storage in the Earth's mantle and core. Geophys Res Lett 45:1330–1338. https://doi.org/10.1002/2017GL075720

Yuen DA, Cserepes L, Schroeder BA (1998) Mesoscale structures in the transition zone: dynamical consequences of boundary layer activities. Earth Planets Space 50:1035–1045. https://doi.org/10.1186/bf03352198

Zampa LS, Tenzer R, Eshagh M, Pitoňák M (2018) Evidence of mantle upwelling/downwelling and localized subduction on Venus from the body-force vector analysis. Planet Space Sci 157:48–62. https://doi.org/10.1016/j.pss.2018.03.013

Zhang S, Christensen U (1993) Some effects of lateral viscosity variations on geoid and surface velocities induced by density anomalies in the mantle. Geophys J Int 114:531–547. https://doi.org/10.1111/j.1365-246x.1993.tb06985.x

Zolotov MY, Fegley Jr B, Lodders K (1997) Hydrous silicates and water on Venus. Icarus 130:475–494. https://doi.org/10.1006/icar.1997.5838

Publisher's Note Springer Nature remains neutral with regard to jurisdictional claims in published maps and institutional affiliations.

Authors and Affiliations

Tobias Rolf[1,2] · **Matt Weller[3,4]** · **Anna Gülcher[5]** · **Paul Byrne[6,7]** · **Joseph G. O'Rourke[8]** · **Robert Herrick[9]** · **Evan Bjonnes[3,4]** · **Anne Davaille[10]** · **Richard Ghail[11]** · **Cedric Gillmann[12]** · **Ana-Catalina Plesa[13]** · **Suzanne Smrekar[14]**

✉ T. Rolf
Tobias.Rolf@geo.uio.no

[1] Centre for Earth Evolution and Dynamics (CEED), University of Oslo, Norway

[2] Institute of Geophysics, University of Münster, Münster, Germany

[3] Department of Earth, Environmental & Planetary Sciences, Brown University, Providence, RI, USA

[4] Lunar and Planetary Institute/USRA, Houston, TX, USA

[5] Institute of Geophysics, Department of Earth Sciences, ETH Zürich, Switzerland

[6] Department of Marine, Earth, and Atmospheric Sciences, North Carolina State University, Raleigh, NC, USA

[7] Department of Earth, Environmental, and Planetary Sciences, Washington University in St. Louis, St. Louis, MO, USA

[8] School of Earth and Space Exploration, Arizona State University, Tempe, AZ, USA

[9] Geophysical Institute, University of Alaska Fairbanks, AK, USA

[10] Laboratoire FAST, CNRS and Université Paris-Saclay, Orsay, France

[11] Earth Sciences, Royal Holloway, University of London, Egham, UK

[12] Department of Earth, Environmental & Planetary Sciences, Rice University, Houston, TX, USA

[13] German Aerospace Centre, Institute of Planetary Research, Planetary Physics, Berlin, Germany

[14] Jet Propulsion Laboratory, NASA, Pasadena, CA, US

Space Science Reviews (2022) 218:56
https://doi.org/10.1007/s11214-022-00924-0

The Long-Term Evolution of the Atmosphere of Venus: Processes and Feedback Mechanisms

Interior-Exterior Exchanges

Cedric Gillmann[1] · M.J. Way[2,3] · Guillaume Avice[4] · Doris Breuer[5] · Gregor J. Golabek[6] ·
Dennis Höning[7,8] · Joshua Krissansen-Totton[9] · Helmut Lammer[10] ·
Joseph G. O'Rourke[11] · Moa Persson[12] · Ana-Catalina Plesa[13] · Arnaud Salvador[14,15,16] ·
Manuel Scherf[10,17,18] · Mikhail Y. Zolotov[11]

Received: 13 March 2022 / Accepted: 30 August 2022 / Published online: 7 October 2022
© The Author(s) 2022

Abstract

This work reviews the long-term evolution of the atmosphere of Venus, and modulation of
its composition by interior/exterior cycling. The formation and evolution of Venus's atmosphere, leading to contemporary surface conditions, remain hotly debated topics, and involve questions that tie into many disciplines. We explore these various inter-related mechanisms which shaped the evolution of the atmosphere, starting with the volatile sources and
sinks. Going from the deep interior to the top of the atmosphere, we describe volcanic outgassing, surface-atmosphere interactions, and atmosphere escape. Furthermore, we address
more complex aspects of the history of Venus, including the role of Late Accretion impacts, how magnetic field generation is tied into long-term evolution, and the implications
of geochemical and geodynamical feedback cycles for atmospheric evolution. We highlight
plausible end-member evolutionary pathways that Venus could have followed, from accretion to its present-day state, based on modeling and observations. In a first scenario, the
planet was desiccated by atmospheric escape during the magma ocean phase. In a second
scenario, Venus could have harbored surface liquid water for long periods of time, until its
temperate climate was destabilized and it entered a runaway greenhouse phase. In a third
scenario, Venus's inefficient outgassing could have kept water inside the planet, where hydrogen was trapped in the core and the mantle was oxidized. We discuss existing evidence
and future observations/missions required to refine our understanding of the planet's history
and of the complex feedback cycles between the interior, surface, and atmosphere that have
been operating in the past, present or future of Venus.

Keywords Venus · Atmosphere · Coupled evolution · Feedback cycles · Volatile exchanges

Venus: Evolution Through Time
Edited by Colin F. Wilson, Doris Breuer, Cédric Gillmann, Suzanne E. Smrekar, Tilman Spohn and
Thomas Widemann

Extended author information available on the last page of the article

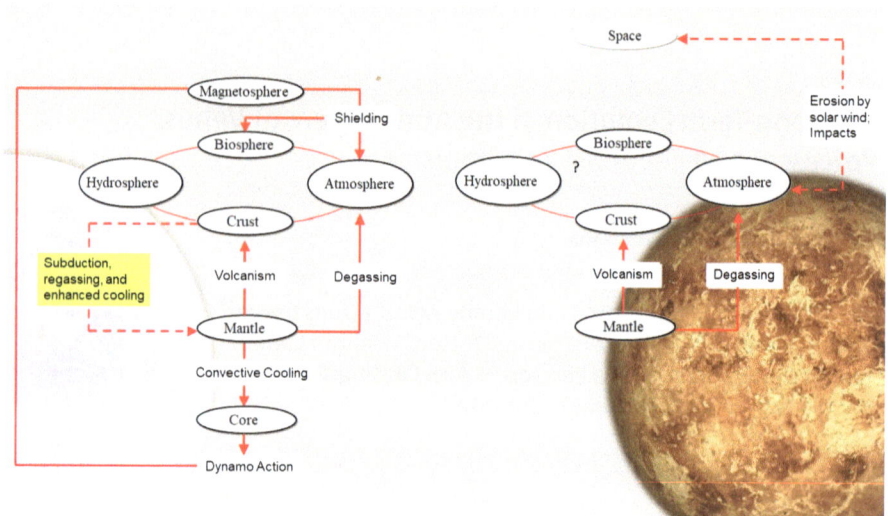

Fig. 1 Sketch of a plate tectonic planet like Earth (left) and a stagnant lid or single plate planet like present Venus (right) showing the interactions between the different reservoirs. Venus shares characteristics with stagnant lid planets but also shows evidence of deformation and horizontal motion. Major difference between plate tectonic planets and stagnant lid planets is the lack of subduction and regassing of volatiles from the surface (yellow rectangle) for the latter. In the plate tectonic scenario, subduction of the cold surface layers (i.e., oceanic crust and lithosphere) leads to efficient cooling of the mantle and core. This can facilitate the generation of a strong magnetic field in the core, which in turn affects atmosphere loss mechanisms caused by the solar wind (see Sect. 3.1 for details). The volatile exchanges in a stagnant lid or a single plate planet are therefore reduced and shows only a one-way path from the mantle into the atmosphere. The reduced volatile circulation may prevent the formation of a biosphere and a hydrosphere over long periods of time. Photos of Earth and Venus in the background, courtesy of NASA/JPL

1 Introduction and Overview

In situ and remote studies of Venus, numerical modelling and experimental work, as well as comparison with the Earth, inform us about the current state of Venus and can provide clues on its past (Fig. 1). Despite past missions and advances in computational models and experimental techniques, many uncertainties remain when it comes to which exact evolutionary pathway Venus has followed. The observation of Venus has challenged many of our assumptions on the evolution and present-day state of terrestrial planets. For example, before observations revealed its surface environment, Venus was imagined to be warm but temperate, while its surface proved to be dry and hot enough to melt lead (see O'Rourke et al. 2022). The planet was thought to be geologically moribund or even dead, but data for the atmosphere and surface from numerous missions indicate that Venus might instead be an active world.

These advances, however, highlight the many gaps in our understanding of Venus. This is even more evident when it comes to Venus' past and its long-term evolution. Whether Venus could have sustained temperate surface conditions at any time during its evolution and for how long this could have lasted, is not known. Whether Venus could have had a magnetic field in the past, and when and why the core dynamo stopped operating is also unknown. What convection regime is most representative for the interior dynamics of Venus, if and how it changed over the planet's history, and what became of Venus's water, are difficult to constrain with current data (see Rolf et al. 2022, for a detailed review). Even

Venus' present-day mantle convection regime is peculiar, lacking the subduction zones and the clear dichotomy between continental and oceanic crust we see on Earth, it also displays more traces of horizontal motion and deformation than the clear stagnant lids operating on Mars and Mercury. For these reasons we label it as single plate planet in this manuscript. In fact, one of the main challenges for reconstructing Venus' evolution is that many observations defy straightforward interpretation. Instead, present-day measurements depend on the cumulative effects of many mechanisms. A good example of this is the high D/H ratio measured for the atmosphere of Venus, indicative of strong fractionation relative to the Earth (Donahue et al. 1982; De Bergh et al. 1991). This is traditionally interpreted as an indication of strong water loss during the evolution of the planet. However, it is actually the result of hydrogen loss combined with at least a possible volcanic source and an external source (impact delivery), both of unknown isotopic composition (Grinspoon 1993). Therefore, timing, mechanism and quantity of water involved remain uncertain.

Recently the selection of new Venus missions (DAVINCI, VERITAS, ENVISION and Shukrayaan-1) recognized both the importance of Venus in our understanding of terrestrial planets and the limited current state of our knowledge. We can only make sense of Venus if we understand how it became the planet we now observe. Despite being Earth's closest neighbor, exploration of Venus by spacecraft has been neglected relative to Mars; we need more data at every level, from its interior structure to its composition, from its surface to its upper atmosphere.

A host of mechanisms have contributed to shape Venus over billions of years into the planet we observe today. Venus, as a whole, is a complex system of interacting processes. While it is important to understand each of them by itself, the evolution of the planet is the result of their joined actions and possible feedback loops. A major goal of this review is to assess the consequences of a wide range of mechanisms through changing planetary conditions. While (Rolf et al. 2022) deals with the interior of Venus, this work is focused on its atmospheric evolution, and how the atmosphere changes over time under the influence of various processes and how, in turn, it affects all other parts of the planet. We first describe the mechanisms involved in volatile evolution that affect atmosphere composition and thickness: sources and sinks of volatiles. Then we discuss peripheral and external mechanisms such as magnetic field generation and the effects of impacts. Finally, we describe the possible evolutionary pathways from the formation of Venus to its present-day state and future, and how the many processes fit into that broad picture. We conclude with highlights of what observations would be needed to distinguish between various evolution scenarios described herein.

2 Volatiles Exchanges

Despite the atmosphere being the most readily observable part of Venus at present-day, its evolution is another matter. It is suspected that the thick ≈ 90 bar CO_2-rich atmosphere has been in place since before the current basalt-dominated surface was created, based on the lack of small craters (smaller meteoroids are destroyed in the current thick atmosphere during entry).

For comparison, despite remaining uncertainties, the evolution of the atmospheres of both Earth and Mars is known to a much finer degree. For Mars, there are estimates for the palaeopressure ≈ 3.6 Gyr ago based on the size distribution of ancient craters (Kite et al. 2014). For Earth, hypotheses exist for the mechanisms that sculpt atmospheric evolution, and these are supported by a variety of geological proxies. For example, despite uncertainties, especially in the Archean and Hadean, we have an idea of the range of pressures at Earth's surface in its distant past, and know about variations of its atmospheric composition (e.g., the rise of O_2, Lyons et al. 2014). Earth has remained habitable for at least the last ≈ 4.3 Gyr (Mojzsis et al. 2001) with probably moderate variations of its atmospheric pressure (Som et al. 2012, 2016; Marty et al. 2013; Avice et al. 2018). In the case of Venus we have no direct constraints on atmospheric pressure through time, although a number of scenarios have been proposed. One end member suggests that the present-day atmosphere was produced in the distant past of the evolution of Venus and was generated by ancient processes (Head et al. 2021), while on the other extreme, a possible geologically more recent origin for the 90 bar worth of CO_2 (Way and Del Genio 2020) has also been proposed.

Additional calculations related to magma ocean lifetime by Hamano et al. (2013) and Lebrun et al. (2013) concern the fate of water, a related crucial question. These works advocate that the divergence between Earth and Venus could have occurred at a primitive epoch due to magma ocean duration, with one planet (Earth) cooling fast and remaining moist, while the other (Venus) would have dried up and stayed hot. Those studies still do not provide a clear way to discriminate between the scenarios, although a 3-D global circulation model (GCM) by Turbet et al. (2021) points to an early and persistent hot-house Venus, based on cloud behaviour. With the lack of definitive answers from currently available data, it is useful to examine mechanisms and try to understand how they interact to produce Venus' modern day atmosphere based on observational clues and the modelling of their respective effects.

Volatile exchanges are at the heart of the interactions between the interior and exterior of Venus. This involves a balance between sources and sinks that governs what the atmosphere looks like, from surface conditions to the edge of space. As such, they define the "face" of the planet: the present-day atmosphere is the outcome of volatile exchange and loss mechanisms throughout its history. Conversely, that also means that it can inform us about what has occurred during the planet's evolution, both inside its mantle and outside. This goes both ways: understanding the mechanisms is necessary for interpreting observations, and data is the basis for our understanding of the mechanisms at play.

Many of these mechanisms of volatile exchange from within and without potentially affect atmosphere composition, as evidenced on Earth. The most straightforward processes can clearly be classified as sources (outgassing from the interior) and sinks (atmospheric escape, chemical interaction with the surface) and are described herein. More complex interactions are left for the next section of this work.

2.1 The Source: Outgassing

A main source of volatile elements on Venus after the initial magma ocean stage and the accretion phase is thought to be volcanic outgassing from the mantle. Avice et al. (2022) offers insights into the initial conditions for the main evolution phase lasting from 4+ Ga to present-day. The outgassing process is complex: to produce outgassing, volatile elements and compounds from the solid mantle must be first (i) extracted during partial melting of the mantle (partitioning into the silicate melt), ascend with the melt to reach the vicinity of the surface and then (ii) be released from magma chambers, volcanic conduits and the lava flows into the atmosphere.

2.1.1 General Principles

Volatile Extraction from the Mantle In a first stage, volatile extraction from the solid mantle depends on melting conditions (pressure, temperature) and mantle composition. For outgassing to occur, the mantle must partially melt: melt is produced when mantle material temperature reaches its solidus temperature (usually due to decompression). Local conditions in the mantle (temperature, pressure, mantle composition) govern how much melt is generated. Melt fraction F is defined as the ratio between the volume of melt that has been generated locally by the partial melting of the mantle and the original volume of mantle affected by said set of physical conditions.

When melt is produced, chemical species can be extracted and enter the liquid phase. However, they distribute unevenly between the melt and the solid phase. The amount of volatiles in the melt depends on their solubility as well as the partitioning behaviour between the melt and solid phase; the smaller value of the two limits the amount of volatiles. For water, the partitioning behaviour is the limiting factor and is represented by the distribution coefficient $D_i = X_i^{sol}/X_i^{liq}$, where X_i denotes the concentration of species i in a given phase at equilibrium. A species with a $D \ll 1$ is said to be incompatible and will transfer easily into the melt, such as water (Katz et al. 2003), while compatible elements will remain in the solid phase. The concentration of a species in the melt will depend on its concentration in the mantle material, on the melt fraction of the material, and on the distribution coefficient for the specific species. The exact relation between these parameters will depend on the type of melting considered: whether it is equilibrium (batch) melting or fractional melting.

The type of melting mechanism is related to the mode of the ascent/extraction of the magma. A plume rising adiabatically would be best described by equilibrium melting: melt and solid stay intermingled as a closed system until melting ceases. Fractional melting rather describes fast-ascending melt in a dike, where melt is extracted as soon as it forms and is only in equilibrium with the solid at the instant it forms before accumulating in a magmatic chamber outside the system.

The amount of H_2O and C in the melt depend on the partition coefficient, the degree of partial melting and their solubility in the melt. Water is much more soluble than CO_2. However the amount of water in magmas (and the mantle) on present-day Venus is still unknown, with no direct measurement, but could be low. Indirect estimates, based on Magellan data (maps of volcanic features) and old calculations of escape rates, or modeling of crustal reservoirs, suggest low water abundances in the lava (about 50 ppm, see Bullock and Grinspoon 2001, and references therein), but are to be taken with caution (see below for the limitation of outgassing by surface pressure).

Solubility of carbon in silicate melts affects the amount of C extracted to the melt in mantle source regions and, ultimately, quantities of C-bearing gases delivered to the atmosphere with volcanism. The solubility depends on oxygen fugacity, melt composition, temperature, water content, and pressure (Holloway 1992; Ni and Keppler 2013). The solubility of C decreases with decreasing fO_2. Reducing mantle conditions are suggested for planets with stagnant lids, but also for the early Earth (e.g., Wadhwa 2008), where substantial recycling of oxidizing agents such as ferric iron, water, or carbonates is/was lacking. At fO_2 roughly below conditions of the iron-wüstite (IW) buffer, C is stable in graphite rather than in melt and C partitions inefficiently in the melt in $Fe(CO)_5$ and minor CH_4 complexes (e.g. Wetzel et al. 2013; Stanley et al. 2014; Armstrong et al. 2015). CO gas will ultimately form through decompression of such melts via decomposition of $Fe(CO)_5$ and/or low-pressure oxidation of graphite grains (e.g. Fogel and Rutherford 1995). Venus's present-day mantle fO_2 is estimated to be between those on Earth and Mars (Schaefer and Bruce 2017; Wadhwa 2008). Then, C in melts is likely present in carbonaceous melt complexes (e.g., CO_3^{2-}) and CO_2 dominates over CO in volcanic gases. C solubility varies with the melt composition (e.g., Dixon 1997, and references therein). Although the SiO_2 content in melts has a minor and unclear effect on the solubility of CO_2, solubility increases with increasing alkali metal content. Tholeiitic basaltic magmas, such as suggested for landing sites of Venera 9, 10,

14, and Vega 2 (Surkov et al. 1984, 1986; Kargel et al. 1993), display the lowest solubility among common silicate melts. Alkali-rich mafic melts, that could have formed Venera 13 and possibly Venera 8 rocks (Surkov et al. 1986; Kargel et al. 1993), show a C solubility that is 2-3 times higher than in tholeiitic melts. Note that formation of K-rich Venera 13 type magmas could be related to the partial melting of carbonated mantle source regions (Kargel et al. 1993; Filiberto 2014). In other words, the presence of K-rich mafic igneous rock may suggest a C-rich mantle that could indicate an heterogeneous mantle, past burial of C-bearing rocks and future CO_2 degassing. For mafic melts, temperature has no definite effect on CO_2 solubility, though solubility in felsic and alkaline melts could be lower or higher at higher temperature (Ni and Keppler 2013). Solubility of CO_2 commonly decreases with increasing dissolved water content in magma (e.g. Ni and Keppler 2013, for a review). The effect of water is not linear and is minor in water-poor melts that could characterize tholeiitic melts on Venus. For all silicate melts, solubility of CO_2 strongly decreases with decreasing pressure (Holloway and Blank 1994; Ni and Keppler 2013).

Volatile Release into the Atmosphere In the second stage, when melt has reached the surface, volatile species can then be released into the atmosphere. The total amount of volatiles released is affected by how much magma reaches the surface (intrusive/extrusive volcanism), and by pre-erupting sub-surface conditions (Berlo et al. 2011; Edmonds and Woods 2018). The outgassing process is further affected by surface conditions, such as atmospheric pressure (Gaillard and Scaillet 2014), since solubility depends strongly on pressure (e.g., Sparks et al. 1994).

For Venus, it is currently unknown how much of the rising magma was extruded at the surface, although modelling attempts have suggested that large crustal magma reservoirs (Head and Wilson 1992) with dikes (Parfitt and Head 1993) could be favoured. Earth shows a wide range of possible extrusive efficiencies based on the volcanic environment. The ratio of intrusive to extrusive magma for Earth ranges from 12:1 (in continental environments) to 5:1 (in oceanic environments) (Crisp 1984; Cawood et al. 2013). It is uncertain whether comparable values are valid for the case of Venus. Direct observation remains limited to estimates of the composition of Venus lava (more basaltic overall; e.g., Surkov et al. 1984, 1986), and surface morphology (Head et al. 1992; Barsukov et al. 1986, indicating the basalt-like morphology of lava in the plains and volcanic centers such as Maat Mons). Additionally, studies of the physics of recent Venus' volcanism and volcanic-tectonic settings indicate no global, or limited subduction-related features. Hot-spots and flood basaltic volcanism dominate recent history (Wilson and Head 1983; Head and Wilson 1986, 1992; Smrekar et al. 2022; Herrick et al. 2022). In the absence of observational data for Venus, Earth-like ranges of the ratio of intrusive to extrusive magma have been assumed. Volcanic outgassing models for Venus usually use a fixed relative extrusion rate in the range of 10:1 (Noack et al. 2012; Gillmann and Tackley 2014), as a simplification of a very complex mechanism. Self-consistent time- and conditions-dependent calculations have not been achieved yet, and a more precise modeling of melt extraction variations will be needed in the future. Since little observational constraint exist for Venus, it is impossible to completely rule out higher values. Higher surface temperatures on Venus might imply a more ductile deformation that translates into relatively low eruption to intrusion ratios. For this reason, it has been suggested that intrusion process may be common on Venus (Gerya 2014). It can have strong implications for the tectonic regime: when hot magma is emplaced in the crust, it can favour a so-called "plutonic squishy lid" regime (a regime where the thin strong lithosphere is divided by warm weak regions generated by plutonism Rozel et al. 2017), which could be favoured on Venus (Lourenço et al. 2020) as it tends to form both a thinner crust and a warmer and thinner lithosphere (Rozel et al. 2017; Lourenço et al. 2018) that could be more in line with more recent lithosphere estimates for Venus (Anderson and Smrekar 2006; James et al. 2013). However, the consequences of plutonic squishy lid convection on outgassing are yet not fully understood.

One should note that knowing the extrusive magma flux is not sufficient to estimate the total fraction of volatiles reaching the atmosphere. It is possible that intrusive magmatism also leads to subsequent outgassing as volatiles find their way through the crust over time. For example, it has been suggested that on Mars 40% of the total volatiles produced by partial melting would ultimately be outgassed into the atmosphere, assuming a fractured upper crust (Grott et al. 2011). The latter number takes into account the mechanical state of the lithosphere, porosity of the crust and calculates the depth at which the pores/fractures would close due to plastic deformation. This number could act as an upper limit for present-day Venus, considering the higher surface temperatures.

An additional uncertainty regarding degassing is a potential change in the tectonic mode associated with resurfacing events (Herrick et al. 2022): the mantle convection mode can change during the history of the planet. Such changes can be as straightforward as a switch from mobile lid regime to stagnant lid, or involve a more complicated pattern (Gillmann and Tackley 2014; Weller and Kiefer 2020). Changes of the convection regime directly affect mantle conditions (for example temperatures), the amount and location of partial melting, the ability of melt to reach the surface and thus outgassing. However, the exact consequences of those changes on outgassing scenarios are uncertain. For stagnant lid planets partial melting occurs below the lid within a region, where the temperature exceeds the solidus temperature (Grott et al. 2011). On planets with plate tectonics, in addition to volcanism related to hotspots and subduction zones, decompression melting below mid-ocean ridges occurs closer to the surface (Kite et al. 2009; Kruijver et al. 2021) which can affect the total amount of melt. As a consequence, if plate tectonics operated on early Venus the degassing rate may have been much higher than today, depending on surface conditions (see Sect. 2.1.2). Issues still remain as the volcanic evolution of Venus is poorly constrained, with no solid data about its ancient history available to date.

2.1.2 Water and the Role of Surface Conditions

Looking at the present-day composition of Venus' atmosphere, it has long been assumed that outgassing of water was limited (e.g., Grinspoon 1993), and that the reason could have been that the planet started off relatively dry (e.g., Namiki and Solomon 1998). A scenario explaining this possible characteristic relied on desiccation of the planet by thermal escape during the slow cooling and solidification of the magma ocean (Hamano et al. 2013, 2015). It was further supported by the observation that the topography of Venus was positively correlated with its geoid (Simons et al. 1994; Johnson and Richards 2003), unlike Earth. This could be explained by Venus having a stiffer interior than Earth, lacking an asthenosphere and possibly being drier, at least in the upper mantle (Kaula 1994; Simons et al. 1994; Kucinskas et al. 1996; Solomatov and Moresi 1996).

However, even a relatively wet Venus mantle could fail to produce significant water outgassing (Holloway 1992). The amount and species of volatiles released can change drastically depending on specific conditions. While surface temperature does not substantially affect the solubility of gaseous species in silicate melt, it can modify some gas speciations (carbon and sulphur being the most sensitive elements; Gaillard et al. 2021). However, lithostatic and surface pressures have been found to be a dominant parameter governing the amount of outgassed species due to their differential effect on volatile solubility. Provided all species are present in sufficient quantities in the mantle that undergoes melting, high surface pressure (100-1000 bar) produces CO_2-rich, N_2-rich, and H_2O-poor volcanic gases, for a high oxyen fugacity (e.g., Venus; Holloway 1992, Zolotov and Matsui 2002), while low pressure ($<10^{-4}$ bar) promotes sulfur-rich gases (e.g., Io). Earth's intermediate surface pressure (1 bar) favors H_2O-rich outgassing. In the case of Venus, with an atmospheric pressure just below 100 bar (and oxygen fugacities above IW+1), outgassing of CO_2 would be favored, even if water degassing may not be completely suppressed ($\sim 1\%$ of the initial lava content, Gaillard et al. (2021), Ortenzi et al. (2020)).

However, the effect of surface pressure should be taken with care given the fact that before reaching and erupting at the surface, volatile exsolution primary occurs during magma ascent and associated lithostatic decompression. Then, magmas are stored within crustal reservoirs and magmatic chambers where volatile exsolution depends on the ambient pressure. The lithostatic pressure of ascending magmas and the pressure within the magmatic chamber is likely higher than the atmospheric pressure and volatile melt-gas partitioning is first ruled by them. Thus, some volatiles exsolution, responsible for the increase of pressure within the chamber, and therefore for the erupting process, already occurred before the magma reaches the surface (e.g., Sparks et al. 1994; Berlo et al. 2011; Edmonds and Woods 2018). Melt composition is also determinant in the process. Thus, an important fraction of volatile species are already outgassed and form gas bubbles before magma erupts so that melt-gas volatile partitioning occurring at depth - and in the magmatic chamber - remains a key aspect of volcanic outgassing (e.g., Gonnermann 2015; Wallace et al. 2015) and needs to be accounted for to reconstruct a consistent outgassing sequence. All outgassed compositions are conditional on the presence of various species in the mantle melt zone. The mantle may be well mixed but not homogeneous (see Salvador et al. 2022), or species may have been depleted earlier during the evolution. For example, it is possible that a large part of N_2 was degassed very early in the evolution of Venus, leaving little N_2 in magmatic gases throughout geological history, especially if no N_2 recycling occurred on Venus (in the case of a full stagnant lid/single plate history, for example). Indeed, the processes affecting the large-scale degassing of the early magma ocean, and likely responsible for most of the atmospheric mass and composition, may significantly differ from the local-scale, volcanic outgassing of the solid-state mantle.

The effect of surface pressure could contribute to understanding the state of the current atmosphere of Venus, but its complete evolution needs to be taken into account. As long as the surface pressure was low, water could have effectively outgassed, but as soon as the surface pressure rose above around 20 bar, mainly CO_2 would have accumulated in the atmosphere. It follows that a CO_2 sink is needed early in the evolution of terrestrial planets to ensure the stability of the low pressure of about 1 bar under which the Earth has been for much of its history. On Earth, this sink is part of the silicate carbon cycle, while recent Venus is unlikely to have been able to sustain it. Conversely, it is likely difficult to return to low pressure conditions that favour water outgassing once a dense and dry CO_2 atmosphere has been established, even through catastrophic scenarios that remove a substantial part of the atmosphere like very large impactors (Schlichting and Mukhopadhyay 2018).

2.2 Outgassing of N_2 and SO_2

Nitrogen is the second most abundant constituent in the atmosphere of Venus after CO_2. Data from the Venera 11, Venera 12 and the Pioneer Venus atmospheric probe indicate a N_2 concentration of $3.5 \pm 0.8\%$ at altitudes lower than 45 km (von Zahn et al. 1983). More recent measurements performed by MESSENGER's Neutron Spectrometer during the second flyby recorded a value of $5.0 \pm 0.4\%$ for altitudes between 60 and 100 km, leading to the conclusion that the atmosphere of Venus might not be as well mixed as previously thought (Peplowski et al. 2020).

Nitrogen isotope data indicate that the nitrogen of the terrestrial planets could originate from carbonaceous chondrites (e.g., Marty et al. 2013), likely from the main accretion phase (Lammer et al. 2018), perhaps as NH_3-ices or organic compounds like HCN, as speculated by Wordsworth (2016). The $^{14}N/^{15}N$ ratio for the atmosphere of Venus is poorly constrained (273 ± 56, Hoffman et al. (1979)) but the measured value remains close to that of Earth's atmosphere, indicating weak escape. This implies that on Venus, the N_2-rich atmosphere might have been shielded from escape. For example, it could either not have been outgassed early on, at a time when escape was still strong enough to remove it from the atmosphere. Alternatively, atmospheric N_2 may have been protected from escape by a thick CO_2 atmosphere efficiently cooling Venus' thermosphere. As a comparison with Earth, atmospheric escape simulations (e.g., Lichtenegger et al. 2010) suggest that an early (older than 3.5 Ga) dense N_2-rich atmosphere could have escaped, in the absence of sufficient CO_2. It has been suggested (Wordsworth 2016) that, during the early stages of planetary evolution, a reducing atmosphere could have favoured reduced N species near the surface. High surface temperatures could have allowed atmospheric N to be dissolved in a reduced (below the IW buffer, since at more oxidized conditions, N_2 has very low solubility in melts) magma ocean and transported into the mantle (Wordsworth 2016). Later during the planetary history, nitrogen could be released into the atmosphere as a consequence of magmatic degassing (Lammer et al. 2018).

In volcanic gases, N_2 remains the dominant species compared to NH_x or NO_x (Gaillard and Scaillet 2014). Calculations that use the C-O-H-S-N system and investigate the abundance of volcanic gas species in equilibrium with basalts indicate that the concentration of N_2 increases for large atmospheric pressures (larger than a few bars), while it becomes diluted in the volcanic gas for low pressures (Gaillard and Scaillet 2014). Present-day Venus surface conditions suppress outgassing of N_2 less than H_2O and SO_2. If the majority of N has not already been outagssed from the interior of Venus (its atmosphere contains more N_2 than Earth's), then this would make nitrogen relatively more abundant in venusian volcanic gases. Another possibility is the exsolution of most of the present-day atmospheric nitrogen inventory at the end of the magma ocean phase (Gaillard et al. 2022a), for oxygen fugacities above IW-3, which would imply limited later volcanic outgassing over Venus' evolution.

For reference, the current amount of N_2 in the Earth's mantle and crust is estimated between 0.32 bar and 5.6 bar (Catling and Zahnle 2020; Wordsworth 2016, and references therein), which still poorly constrains this reservoir. On present Earth (Som et al. 2016), nitrogen is thought to be approximately in balance between sources (half volcanic outgassing and half weathering by O_2) and the sink (burial, with a touch of subducted flux). Some data also suggest large changes in Earth's N_2 partial pressure over geological times (Som et al. 2016), possibly due to changes in volcanic activity or the presence/absence of oxygen (see also Westall et al. 2022). However, the comparison between Venus and Earth is complicated by the fact that the nitrogen cycle is modulated by Earth's biosphere (Jacob 1999; Zerkle and Mikhail 2017).

Sulfur dioxide is the third most abundant gas in the atmosphere of Venus. In contrast to N_2, SO_2 is extremely chemically reactive and is involved in complex reaction chains. The solubility of SO_2 also varies strongly with pressure, which would affect its abundance in released volcanic gases on Venus (possibly reducing its concentration, compared to Earth, Gaillard and Scaillet (2014)). Together with water and CO_2, SO_2 is one of the most important greenhouse gases in the atmosphere of Venus. On Earth and Mars, it is released into the atmosphere during volcanic outgassing. Venus' volcanism is thought to be a probable source for SO_2 that in turn is involved in atmosphere-surface reactions (Zolotov 2018) with calcium-bearing materials to form anhydrite ($CaSO_4$). Since these reactions would deplete the SO_2 in the atmosphere, reduce the production of sulfuric acid (H_2SO_4) and the formation of clouds, an SO_2 source likely in form of volcanic activity needs to be active at present-day. To be able to maintain the measured SO_2 concentration in the atmosphere of Venus, it has been suggested that an eruption rate of about 1 km^3/yr with lava compositions similar to those observed at the Venera 13, 14 and Vega 2 landing sites was necessary to account for the sink by reaction with surface calcium mineral (Fegley and Prinn 1989; Fegley 2009). However, it has since been proposed (see Zolotov 2018) that equilibrium involving pla-gioclase could sustain present-day SO_2 atmospheric concentrations at present-day without recent volcanism (see Sect. 2.4.1).

Models that calculate the gas composition of volcanic gases, indicate that sulfur becomes highly soluble in surface lavas with increasing atmospheric pressure and may be released only in small proportions into a dense atmosphere, such as that of present-day Venus (Gail-lard and Scaillet 2014). For this reason Head et al. (2021) estimate, based on observation of the total surface lava production, that it is unlikely that the present-day SO_2 atmospheric abundance could be solely maintained by extrusive volcanism.

2.2.1 The CO_2 Atmosphere: Role and Origins

Venus' thick present-day atmosphere is dominated by CO_2, containing an amount of car-bon comparable to Earth's combined atmospheric and crustal reservoirs (e.g. Donahue and Pollack 1983; Wedepohl 1995; Lécuyer et al. 2000; Hartmann et al. 2012). Given the un-certainties on the reservoirs, it is often suggested that most of Venus' present-day total CO_2 inventory is now contained in its atmosphere (Lammer et al. 2018) and that its mantle is therefore mostly degassed, but see Sect. 2.2.2 above. Due to the strong greenhouse effect from CO_2 the surface temperature is 737 K at the reference planetary radius of 6052 km, which is approximately 500 K higher than its equilibrium temperature (Lissauer and de Pa-ter 2013). Understanding the evolution of CO_2 in the atmosphere is not only crucial for predicting the potential for the existence of liquid water on early Venus, but also provides insights into the divergent evolution of Earth and Venus. As discussed in Salvador et al. (2022), the solidification of a magma ocean may have outgassed large amounts of CO_2 into the early atmosphere, although it is difficult to constrain (e.g., Bower et al. 2022; Gaillard et al. 2022a). The present day atmosphere could be a combination of an early atmosphere resulting from magma ocean solidification, outside contribution from impactors (Gillmann et al. 2020, both early and late) and a later contribution from subsequent long-term magmatic mantle outgassing (Lammer et al. 2018).

While mantle composition first governs the availability of elements to be released, it also determines the chemical composition of the gases. The oxidation state of the mantle (linked to the oxygen fugacity parameter) controls how much C partitions into the melt during man-tle partial melting (for instance, Ortenzi et al. 2020, find that it is suppressed for oxygen fugacities below IW+2). Then, as the magma rises to the surface, redox conditions also affect surface gas speciation during outgassing. Oxidized conditions (above IW+1) favor oxidized species such as CO_2, more reduced conditions (below IW+1) favor species such as CO (e.g., Ortenzi et al. 2020), or even CH_4 in more extreme situations (Wogan et al. 2020). Due to the lack of data, the oxidation state of Venus' mantle is unconstrained. Only sur-face measurements of the mean FeO/MnO ratio by Venera 13, 14 and Vega 2 (Surkov et al. 1984, 1986), have led to interpretations that its oxidation state lies between Earth and Mars

(Schaefer and Bruce 2017). As discussed before (Sect. 2.1.1), early outgassing may occur under reducing conditions, possibly leading to the formation of a CO-rich atmosphere. If water is present in the atmosphere, or possibly if O accumulates due to hydrodynamic escape (see Sect. 2.3.1), CO may convert efficiently into CO_2, consistent with present-day observation. Primitive evolution scenarios are discussed further in Salvador et al. (2022).

Because of overall low CO_2 solubility, the decompression of ascending magma on Earth causes formation of gas bubbles deep below the surface (Burton et al. 2013). Although Venus' elevated atmospheric pressure slightly affects CO_2 degassing, its effect on degassing of H_2O and SO_2 is much stronger, and CO_2 is likely the most abundant volcanic gas on Venus (Gaillard and Scaillet 2014). Surface pressure, in the thick modern-day CO_2 atmosphere, also affects the species that can be outgassed, as mentioned above. The existence and evolution of the thick CO_2 atmosphere significantly affects the ability of Venus to release water.

The accumulation of CO_2 in the atmosphere after solidification of the magma ocean depends on the tectonic mode and may have occurred either gradually over time or has been substantially enhanced by one or several catastrophic resurfacing events (López et al. 1998; Bullock and Grinspoon 2001). Based on Magellan data (Schaber et al. 1992; Basilevsky et al. 1997), it has been suggested that at least one global resurfacing event may have occurred 200-1000 Myr ago (the catastrophic evolution hypothesis). This was later self-consistently simulated with numerical models and (Armann and Tackley 2012; Rolf et al. 2018, see for instance), and is further discussed in Herrick et al. (2022). If such an event occurred, it would have been accompanied by large-scale melting and therefore outgassing. The amount of CO_2 released by one such resurfacing event depends on the carbonate complexes concentration in the lava and thus the composition (and oxygen fugacity) of the mantle. Assuming Earth-like composition of mafic (basaltic) melts, one such global event could be responsible for a release of 5.6×10^{19} kg of CO_2 (approximately 9 bar) (López et al. 1998) - under more reducing conditions in Venus' mantle the CO_2 release would be much lower. However, long-term continuous activity and outgassing could potentially lead to the same volatile build-up observed at present-day, without requiring any catastrophic event, and still satisfy [40]Ar measurements (Namiki and Solomon 1998). Some modelling efforts propose that it is unlikely that the bulk of the venusian atmosphere could come solely from mantle outgassing (Morschhauser et al. 2009; Gillmann and Tackley 2014; Gillmann et al. 2020, and compare with Weller and Kiefer 2021; Weller et al. 2022; Westall et al. 2022), either due to large initial exsolution from the magma ocean or because models suggest insufficient later CO_2 outgassing. Other sources may thus be required. However, the composition of the mafic melts (possibly water-poor tholeiitic) that could have formed most of Venus' plains could have allowed an efficient CO_2 degassing. Alkaline mafic melts (in the Venera 13 case) could also release substantial amounts of CO_2 upon degassing. A relatively efficient CO_2 outgassing over Venus's history is consistent with the abundances of CO_2 (gram per gram) in the Venus' atmosphere versus Earth's crust (Pollack and Black 1979; Lécuyer et al. 2000).

In addition to the tectonic mode, the melt production and thus magmatic degassing rate of CO_2 depends on the temperature-depth gradient. First, the temperature-depth gradient controls the melting region below Venus' lid. This gradient can depend significantly on the ratio between melt intrusion and melt extrusion (Rozel et al. 2017; Lourenço et al. 2018). On one hand, more extrusion directly translates into more outgassing, but on the other, more intrusion could locally affect the thermal profile in the lid and cooling rates, possibly easing melting. While it is likely that the more straightforward former effect dominates, possible feedback between the two processes has not yet been investigated in depth. Additionally, the temperature-depth profile is also affected by Venus' surface temperature. It is unlikely that Venus' surface temperature has been constant throughout time: solar evolution (Claire et al. 2012) could have contributed to a slow increase, or it could have varied widely with changes in the state and composition of the atmosphere. For example, water concentration and the presence or absence of clouds are major factors (Way and Del Genio 2020; Turbet 2018). In particular a runaway greenhouse would have substantially increased the surface temperature, which would also have an effect on partial melting and degassing as discussed above.

Altogether, in a reasonable but still unproven scenario, outgassing of CO_2 from the mantle into the atmosphere was likely stronger during the early evolution and then gradually decreased with time. This can be explained by the fact that the rate of mantle convection directly affects the degassing rate, as it controls the rate at which new mantle material enters the source region of the partial melt (Kite et al. 2009; O'Neill et al. 2014). After crystallization of the magma ocean, mantle temperatures are high and likely near the solidus (see Salvador et al. 2022), mantle viscosity is low, and thus mantle convection is particularly strong, presumably promoting rapid early degassing (O'Neill et al. 2014; Gillmann and Tackley 2014; Tosi et al. 2017). Rapid outgassing is likely supported by the low ^{40}Ar in the atmosphere, possibly indicating that a large part of ^{40}Ar was not yet formed when outgassing occurred (Kaula 1999), although precise estimations would require knowledge of both initial K content of the mantle and Venus' volcanic history. However, there were possibly one or several massive outgassing events related to catastrophic resurfacing, which eventually contributed to Venus' thick, CO_2-rich atmosphere as observed today. Further, it could be that a substantial CO_2 atmosphere existed prior to the outgassing events and was released during the magma ocean solidification (a few bars to a few dozen bars). A clear understanding of the complete outgassing sequence and timing and the respective contribution of the different processes at play to build-up the current atmosphere is thus still lacking.

2.2.2 Constraining Models of Melt Production with Noble Gases

In principle, noble gases in the atmosphere of Venus are clues to the outgassing history. Over the course of Venus' history several isotopes of noble gases have been produced by extinct (^{129}I, ^{244}Pu) and extant (e.g., ^{238}U, ^{40}K) radioactive nuclides present in the interior of the planet. When a portion of Venus' crust or mantle melted these gaseous daughter nuclides migrated into the magmatic melt and were eventually degassed into the atmosphere. Outgassing produces radiogenic excesses on top of the primordial isotopic composition. While it remains difficult to put constraints on the efficiency of degassing of radiogenic gases from silicate reservoirs, models of outgassing of radiogenic noble gases suggest that, for Earth, degassing efficiency does not play the major role (Hamano and Ozima 1978; Allègre et al. 1987). Because the aforementioned radioactive nuclides have very different half-life times ($t_{1/2}(^{129}I) = 16\,\mathrm{Myr}$, $t_{1/2}(^{244}Pu) = 82\,\mathrm{Myr}$, $t_{1/2}(^{40}K) = 1.25\,\mathrm{Gyr}$ and $t_{1/2}(^{238}U) = 4.47\,\mathrm{Gyr}$), studying the abundances and relative proportions of their daughter products (^4He, ^{40}Ar, ^{86}Kr, $^{129,131-136}Xe$) has the potential to give important constraints on the degassing history of Venus. A simplistic view is that relatively high rates of melt production over time should ultimately lead to higher abundances of radiogenic noble gases in the atmosphere today.

About radiogenic contributions to the atmosphere of Venus, only the atmospheric $^{40}Ar/^{36}Ar$ ratio for Venus is known with reasonable precision. This ratio, combined with estimates of the atmospheric abundance of ^{36}Ar, gives an atmospheric abundance of ^{40}Ar of $3.3 \pm 1.1 \times 10^{-9}$ times the total mass of Venus while ^{40}Ar is $12.7 \pm 1.3 \times 10^{-9}$ times the total mass of Earth (Kaula 1999). The $^{40}Ar/^{36}Ar$ ratio is 1.11 ± 0.02 (Istomin et al. 1983),

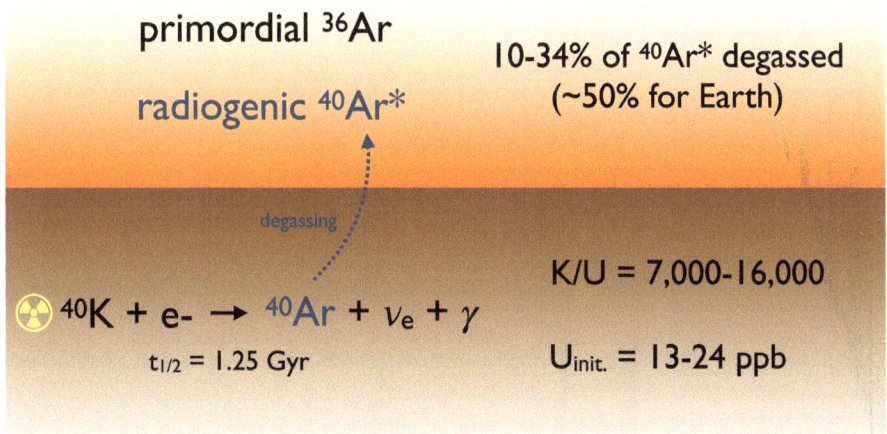

Fig. 2 Schematic explanation of how the degassing state of Venus (and Earth) has been estimated. Radiogenic ^{40}Ar has been produced inside terrestrial planets by the radioactive decay of ^{40}K. Simple mass balance considerations conclude that less radiogenic argon has been degassed from Venus interior to its atmosphere compared to Earth. Results are dependent on the assumed K/U ratio and the abundance of U in bulk silicate Venus, which are poorly constrained. See Kaula (1999), O'Rourke and Korenaga (2015), Allègre et al. (1987) and refs. therein for details on the method to evaluate the degassing state of terrestrial planets

which is much lower than for Earth's atmospheric argon ($\approx$300, Ozima and Podosek 2002). This difference has been interpreted as evidence that Venus experienced less outgassing than Earth through time (Fig. 2).

Four missions (Venera 9 and 10, and Vega 1 and 2) used gamma rays to measure the abundances of K and U in surface material of their landing sites, which ranged from $\sim$0.3–0.5 wt% and $\sim$0.5–0.7 ppm, respectively (Surkov et al. 1987).

The inferred K/U elemental ratio for bulk silicate Venus was $\sim$7000 by mass (see O'Rourke and Korenaga (2015) for details). If this K/U ratio is correct and the bulk abundance of U is Earth-like, then Venus outgassed only about 10-34% of the radiogenic ^{40}Ar that would have been produced by the radioactive decay of ^{40}K ($t_{1/2} = 1.25$ Gyr) over the course of 4.56 Gyrs (Kaula 1999; O'Rourke and Korenaga 2015; Namiki and Solomon 1998b; Volkov and Frenkel 1993). By comparison, Earth degassed $\approx$50% according to this metric (Allègre et al. 1987) although a recent study revisited the meaning of this result and suggested that half of the ^{40}Ar budget has been subducted back into the Earth's mantle (Tucker et al. 2022). If the K/U ratio for Venus is actually higher and closer to Earth-like values (i.e., up to 16,000, Arevalo et al. 2009), then Venus could be only 10–12% degassed, meaning that the vast majority of radiogenic ^{40}Ar is still stored in the interior (O'Rourke and Korenaga 2015). One idea is that less crustal production has occurred at Venus than at Earth.

Speculatively, a basal magma ocean in the mantle could act as a "hidden reservoir" of incompatible elements such as noble gases (Jackson et al. 2021). Such a basal magma ocean may exist today inside Venus (O'Rourke 2020), or could have solidified recently so that its noble gas inventory has not been fully mixed throughout the mantle. On the other hand, though, one should note that the large inventory of N_2 in Venus' present-day atmosphere compared to Earth could be interpreted as evidence of comparatively strong outgassing during Venus' evolution (see Sect. 2.1.2).

Drawing any definite conclusions about the outgassing history of Venus from the present-day abundance of atmospheric ^{40}Ar alone is very difficult. As stated above, all models rely on strong assumptions about the K/U ratio and the abundance of U in bulk silicate Venus (Lammer et al. 2020a). For example, multiplying the bulk abundance of K by two in a model would mean that the best-fit rate of crustal production is halved. Furthermore, argon is assumed not to escape from the atmosphere, meaning that the timing of its outgassing is uncertain. Kaula (1999) showed that a single, catastrophic resurfacing event could outgas the entirety of Venus' atmospheric ^{40}Ar—under a certain set of assumptions. However, several less dramatic outgassing episodes could release the same amount of ^{40}Ar. Namiki and Solomon (1998) demonstrated that steady (i.e., not catastrophic) outgassing, perhaps at different rates before and after a given transition time, is consistent with the present-day atmospheric abundances of both ^{40}Ar and ^{4}He. Building on these studies, O'Neill et al. (2014) and O'Rourke and Korenaga (2015) found that models with continuous stagnant lid convection in the mantle also yield acceptable trajectories of ^{40}Ar outgassing for much or all of Venus' history.

Overall, new measurements of the chemical composition of rocks at the surface of Venus are needed in order to narrow down the range of estimates for the K/U ratio of bulk Venus. Precise knowledge of the bulk U abundance is required to translate the K/U ratio into the bulk K abundance that underpins outgassing models. Additionally, the I-Pu-U-Xe isotope systematics of Venus could be explored if $^{129,131-136}$Xe/^{130}Xe ratios of the Venus atmosphere are determined, which would allow us to put chronological constraints on degassing using only elemental ratios of refractory elements (U, Pu) and not absolute elemental abundances. These new investigations would help eliminate degeneracies between models of outgassing from the interior of Venus and would allow to build a coherent picture of Venus' geodynamics through time (see Widemann et al. (2022) for a more details about future mission projects).

2.3 The Sinks

This section focuses on two types of sinks. The first, atmospheric escape has long been considered a dominant pathway for removing volatile elements from an atmosphere. We detail here the current state of escape-related observations (space & surface based) of Venus including modelling efforts for both the present-day and past history (cumulative effect with time). A brief comparison with Earth and Mars is also included. This includes a review of issues related to extrapolating escape. The lack of solid global direct models of escape leads to uncertainties in escape rates under past conditions possibly different from the present-day observed atmospheric state of Venus (beyond solar energy input variation).

The second type of sink discussed is related to surface interactions as a possible alternative means of volatile loss on Venus. We consider the role of mineral reaction buffering for the atmosphere (for major gases like CO_2 or for minor ones, like SO_2). We next consider the possibility of mechanisms trapping volatile species, such as the oxidation of fresh basaltic material (to sulfates, iron oxides, and pyrite), as a possible way to remove an unknown amount of oxygen from the atmosphere. Further, consequences of crustal recycling are considered. We highlight different styles of recycling, e.g., plate tectonic-like, delamination, vertical advection, and crustal sequestration versus return to mantle.

2.3.1 Hydrogen and Oxygen Escape During Venus' History

Observations by spacecraft and many theoretical studies have found that enhanced solar/stellar X-ray and extreme ultraviolet (XUV) radiation, solar wind plasma and coronal mass ejections (CMEs) result in forcing of the upper atmospheres of planets with no intrinsic magnetic field like Mars and Venus. The short wavelength radiation and the precipitating plasma flux can ionize, chemically modify, heat, expand, and erode upper atmospheres during a planetary lifetime (Jeans 1955; Chamberlain 1963; Chamberlain and Campbell 1967; Öpik 1963; Bauer and Lammer 2004; Lammer 2013).

In addition to impact-related erosion (which we discuss later in this review), one can separate two main categories of atmospheric escape processes: (i) thermal escape and (ii) non-thermal escape. There are generally three thermal escape conditions possible:

- *Jeans escape:* loss of mainly atoms which populate the high energy tail of a Maxwell distribution (Jeans 1955; Chamberlain 1963).
- *Hydrodynamic escape:* very efficient escape of the bulk gas of the upper atmosphere where the atmosphere remains a collisional fluid as it passes through the transonic point (Chamberlain 1963; Chassefière 1996b; Lammer et al. 2016; Owen and Wu 2016; Fossati et al. 2017). In such an extreme condition, nearly all atoms in the upper atmosphere exceed the escape velocity due to heating by XUV radiation and/or the surface temperature of a magma ocean. A special case of hydrodynamic loss is the so-called boil-off where a whole atmosphere is not gravitationally bound to a planetary body and the thermal energy of the atmospheric gas overcomes the gravitational potential. This condition can occur when a hot magmatic low mass planetary embryo or protoplanet dissipates its primordial atmosphere during/after disk evaporation. There are also close-in exoplanets that could have experienced boil-off escape (Owen and Wu 2016; Fossati et al. 2017; Lammer and Blanc 2018).
- *Slow hydrodynamic outflow:* a hybrid condition between Jeans and hydrodynamic escape, means that the lower thermosphere starts to expand hydrodynamically so that the exobase level is moved to high altitudes but the expanding gas cools adiabatically so that the escape can be described with a shifted Maxwellian (Lammer et al. 2008; Tian et al. 2008a,b).

Depending on the atmospheric species and their mixing ratios, the exosphere can expand beyond an atmosphere protecting magnetopause for higher XUV fluxes than that of the present Sun (Tian et al. 2008a,b; Lichtenegger et al. 2010; Lammer et al. 2018), so that non-thermal escape processes become relevant. For present solar activity conditions most non-thermal escape processes are relevant on non-magnetized planets. The most efficient known non-thermal atmospheric escape processes can be separated into two categories, which are ion loss processes and the loss of neutrals (Lammer 2013, and references therein). Ion loss processes are:

- *Ion pick up:* planetary atoms are ionized and accelerated by electric fields within the solar/stellar wind plasma flow around a planetary obstacle.
- *Detached plasma clouds:* at non-magnetic planets plasma instabilities can cause wave structures at plasma boundaries such as the ionopause, where bubbles filled with cool ionospheric plasma can be detached.
- *Cool ion outflow:* At non- or weakly magnetized planets, planetary ions can also be accelerated by electrical fields to escape velocities throughout the tail.
- *Polar outflow/wind:* Ions that are accelerated via electric fields so that they reach escape energy and are lost over magnetospheric cusps of magnetized planets.

Besides these non-thermal ion escape processes, photochemical and particle interaction processes exist, such as:

- *Dissociation:* Dissociative recombination and electron impact dissociation of molecular ions as well as photodissociation of neutral molecules than can yield neutral atoms with excess energies larger than the escape energy so that they can be lost or populate the exosphere with a suprathermal "hot" atom population.

- *Charge exchange:* Charge exchange of solar wind protons with exospheric neutral particles can result in the loss of heavy neutral atoms that are ionized by the process which then produces an energetic neutral hydrogen atom from the former solar wind proton.
- *Atmospheric sputtering:* Precipitating solar wind ions or ionized exospheric atoms that are back-scattered into the upper atmosphere can act as sputter agents for atmospheric particles that are lost from the planet if their energy overcomes the escape energy.

One can expect that hydrodynamic thermal escape processes could have been involved in the loss of Venus' primordial H_2-He-dominated atmosphere at the time when the disk dissipated (Hayashi et al. 1979) and a few 10s to 100 million years after the planet's origin, if proto-Venus accreted primordial gas from the disk (Gillmann et al. 2009; Lammer et al. 2020a,b). Furthermore, hydrodynamic escape of hydrogen should also have played a role if Venus (i) produced water due to interaction of a primordial atmosphere with an underlying magma ocean (Salvador et al. 2022; Lammer et al. 2021); (ii) outgassed a huge amount of hydrogen and/or water vapour before and after its final magma ocean solidified or (iii) if the planet experienced a runaway greenhouse effect that evaporated a huge amount of H_2O (Chassefière 1996a,b; Gillmann et al. 2009; Lammer et al. 2018).

Hydrodynamic escape is most likely related to the escape of light species such as H, H_2 and He. It is also able to cause considerable O escape, directly during the first few hundred million years, and drag neutral O and heavier noble gases, even when they cannot escape directly (for instance Hunten 1973; Zahnle et al. 1990; Gillmann et al. 2009; Odert et al. 2018). However, this escape process depends on a complex interplay between the planet's gravitational potential and the atmospheric mixing ratio related to the availability of potential IR-cooling molecules in the planet's thermosphere (e.g., Kasting and Pollack 1983; Yoshida and Kuramoto 2020).

Here, we focus on the escape processes which have affected atmospheric evolution over most of the past 2.5–3.0 Ga, the time range for which the increase in solar activity can likely be covered by the maximum activity during the present-day solar cycle. As mentioned before, thermal escape from Venus was important during its early history, but became negligible by present-day. Photochemical escape is only relevant for suprathermal H atoms (e.g., Cravens et al. 1980; McElroy et al. 1982a; Rodriguez et al. 1984; Hodges and Tinsley 1981; Hodges 2000). The most important escape process on Venus today is thus non-thermal escape, i.e. ion and energetic neutral escape. The main pathways for the non-thermal escape is either through the induced magnetotail of Venus (the nightside elongation of the induced magnetosphere), through pickup ions in the solar wind or through sputtering of neutral atoms via impinging on the atmosphere of the pickup ions (Lammer et al. 2006, and references therein). Thus far, there are no measurements on how much sputtering contributes to total escape from Venus, but simulations suggest that it is around 25% of the total escape (Lammer et al. 2006). Measurements and simulations indicate that today the largest portion of the non-thermal escape is due to magnetotail escape (e.g., Lammer et al. 2006; Masunaga et al. 2019), which is what we focus on in this section.

Investigating the present non-thermal escape from Venus allows us to investigate the evolution of the D/H ratio and of oxygen back to the estimated time of the average age of surface, i.e., ~ 0.2–1.0 Ga (e.g., Strom et al. 1994; Basilevsky et al. 1997; Kreslavsky et al. 2015; Bottke et al. 2016; McKinnon et al. 1997; Grinspoon 1993). If the observed present-day D/H ratio can be reproduced assuming the surface age as starting point for its initial reservoir - which should have likely been derived from chondrites, comets, the solar nebula or a mixture of these - all of them having significantly lower D/H ratios than present-day

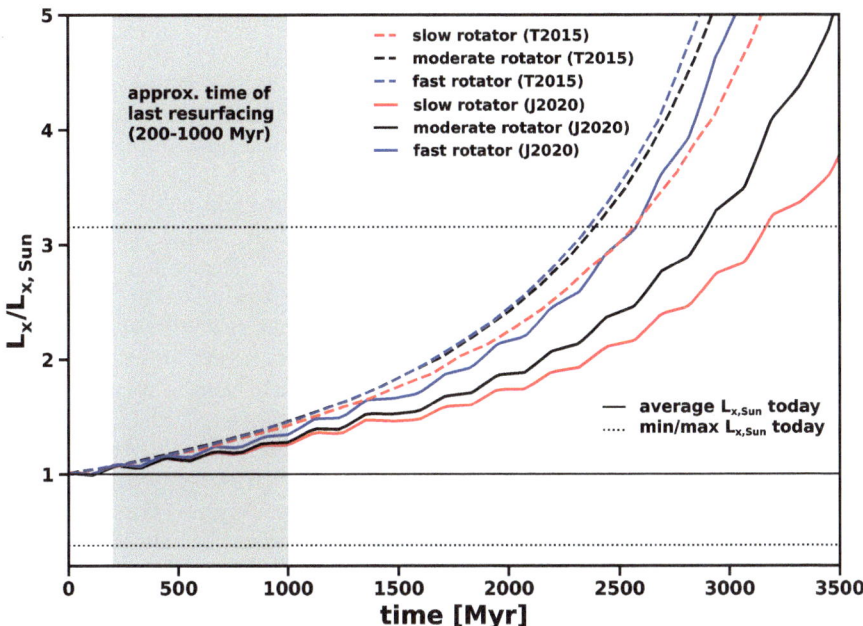

Fig. 3 The evolution of X-ray activity for Sun-like stars normalized to the Sun's present-day value of $10^{27.2}$ erg s^{-1} cm^{-2} according to Tu et al. (2015) (dashed lines) and Johnstone (2020) (solid lines). Red, grey, and blue depict the tracks of initially slow, moderate, and fast rotating Sun-like stars. The evolutionary tracks by Tu et al. (2015) were employed to fit the Sun's average X-ray flux at present-day, while those by Johnstone (2020) show the respective averages of the rotational tracks for all measured Sun-like stars. The grey shaded area shows the approximate average age of Venus' surface. As can be seen, this age falls clearly within a time period that can be covered by the variation of the X-ray flux over the solar cycle (solid line for average value, dashed lines for minimum and maximum)

Venus' water (e.g., Robert et al. 2000), then Venus' present water was obtained during that time frame. In case the D/H ratio cannot be reproduced, it is an indication that the present water inventory is at least partially a remnant from before the average age of the present day surface. However, it neglects any additional unknown sinks for D and H that preferentially remove H from the atmosphere besides atmospheric escape.

The most recent measurements of non-thermal escape of H^{+} and O^{+} from Venus comes from the ASPERA-4 plasma package instrument (Barabash et al. 2007a) on board Venus Express. The Venus Express measurements were performed over almost an entire 11 year long solar cycle (2006–2014), which allows us to reconstruct the escape of these ions during the past ~ 2.5–3.0 Ga assuming an identical atmospheric structure and lack of intrinsic magnetic field. For this we compare the observed solar XUV activity (with X-ray and EUV emissions from 1-10 nm, and 10-118 nm, respectively) with the reconstructed XUV history inferred from Sun-like stars with younger age by Tu et al. (2015) and Johnstone et al. (2021). This can be seen in Fig. 3 where the X-ray part of the spectrum can be directly obtained from observations of other Sun-like stars (e.g., Airapetian et al. 2021), whereas the EUV flux is normally assumed to directly scale with the respective X-ray emissions (Tu et al. 2015; Johnstone et al. 2021). Such reconstructions are typically divided into different rotational evolution tracks (slow, moderate, and fast), since a star's XUV flux is correlated with the rotation rate of the respective star. For G-type stars these rotational tracks typically converge after ~ 1 billion years. This makes it difficult to reconstruct the detailed activity history of the

Sun. However, the XUV flux variation over the present-day Sun's solar cycle only reaches maximum values that can be compared to the XUV flux evolution of G stars for time periods at which the rotational tracks mostly converge. See Fig. 3, which compares the maximum flux of the present solar cycle with the evolution of slow, moderate, and fast rotators.

Several studies have inferred escape rates of H^+ and O^+ using the ASPERA-4 measurements. Although the exact numbers of escape rates found for each study vary, depending on method and time period used for the analysis, in general the results and conclusions agree with each other (Nordström et al. 2013). On average today's H^+ and O^+ ions have escape rates at a ratio close to the stoichiometric ratio of H_2O of 2:1 (Barabash et al. 2007b; Fedorov et al. 2011; Persson et al. 2018), with an average O^+ escape rate of $(3 - 6) \times 10^{24}$ s^{-1} (Futaana et al. 2017). However, the ratio varies with the solar cycle (Persson et al. 2018), and it is yet unclear whether O accumulates in the atmosphere since its actual abundance is not precisely measured (with the mole fraction of O likely being <50 ppm, see Johnson and de Oliveira 2019). In case it does not accumulate in Venus' atmosphere, another oxygen sink has to be envisioned. It may additionally escape along with the hydrogen by a non-mass-dependent non-thermal escape process such as ion escape through moderately XUV-exposed upper atmospheres. On the other hand, some oxygen is consumed through oxidation of atmospheric gases (CO, COS, H_2S, S_2) and surface minerals to produce CO_2, SO_2, Fe_3O_4, Fe_2O_3 and sulfates (Zolotov 2018).

Using all Venus Express measurements, Persson et al. (2018) found that the H^+ escape varied from $\approx 7.6 \times 10^{24}$ H^+ s^{-1} at solar minimum to $\approx 2.1 \times 10^{24}$ H^+ s^{-1} at solar maximum, while the O^+ escape rate exhibited smaller variation with $\approx 2.9 \times 10^{24}$ O^+ s^{-1} at solar minimum to $\approx 2.0 \times 10^{24}$ O^+ s^{-1} at solar maximum. The significant variations in the H^+ escape, as illustrated in Fig. 4b, over the solar cycle provide a change in the ratio between H^+ and O^+ escape rates from 2.6:1 at solar minimum to 1.1:1 at solar maximum. This means that more H^+ ions are lost, but at solar minimum only 1.1 times as much as O, and at maximum about 2.6 times. However, on average the value is in agreement with Barabash et al. (2007b) and is close to 2:1, indicating that no O gets enriched in Venus' atmosphere in relation to H, provided that both O and H originate from H_2O vapor. However, this assumption neglects any other sink for O, as we will discuss below.

The large decrease in H^+ escape from solar minimum to maximum is explained by an apparent change in the average direction of the proton flows in parts of the induced magnetotail. During solar maximum the protons have a larger contribution of ions flowing towards Venus instead of escaping through the magnetotail (Kollmann et al. 2016; Persson et al. 2018). On the other hand, Edberg et al. (2011) showed that in general the escape increases by a factor 1.9 from low solar wind dynamic pressure conditions to high solar wind dynamic pressure conditions. Persson et al. (2020) extended the study by further dividing the data set into ten bins using a combination of solar wind energy flux and solar XUV ranges. In agreement with Edberg et al. (2011), they found a general increase in escape of O^+ as the solar wind energy flux increases, with variations between $\approx 9.0 \times 10^{23}$ O^+ s^{-1} to $\approx 4.9 \times 10^{24}$ O^+ s^{-1}. A small decrease in the O^+ escape rates with an increase in the solar XUV flux was also found. In addition to the escape down the Venusian magnetotail, Masunaga et al. (2019) studied the escape of O^+ in the magnetosheath, and found that approximately 30% of the total O^+ escape from Venus happens through the magnetosheath directly to the solar wind.

Figure 5 shows the O^+ escape rate measured by Venus Express ASPERA-4 (Persson et al. 2020) for five separate ranges of solar wind energy flux using high and low XUV flux which shows that O^+ escape diminishes with high solar activity due to the reasons mentioned above. An earlier mission that also studied the Venus plasma environment was the Pioneer Venus Orbiter mission (PVO), which orbited Venus from 1978 until 1992

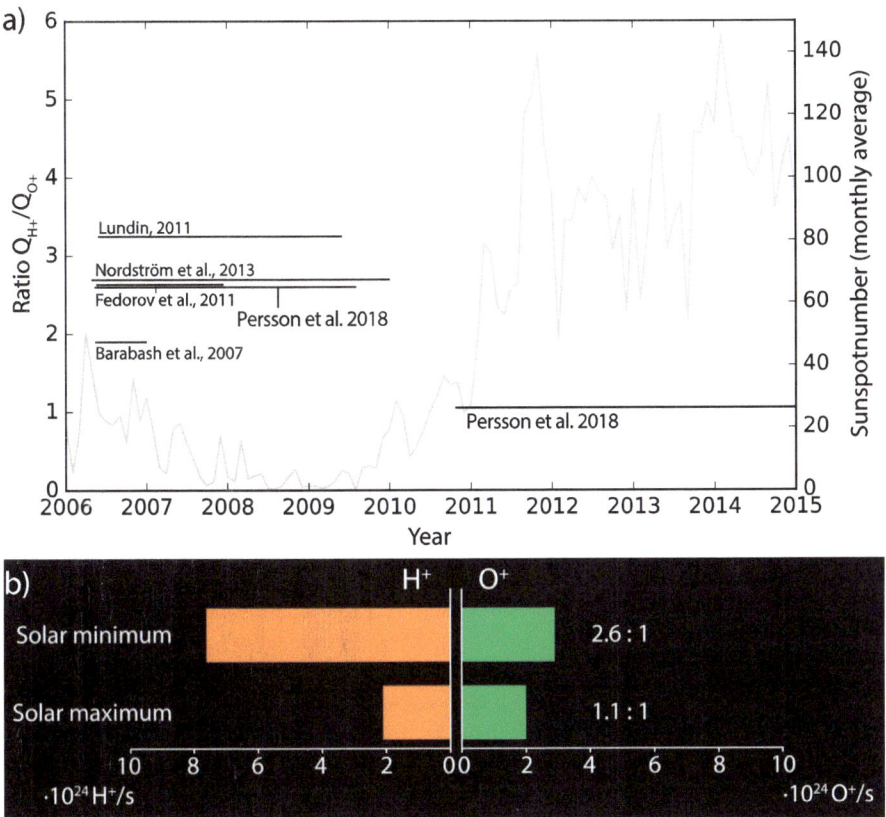

Fig. 4 **a)** Black lines indicate the ratio of H^+ and O^+ escape rates from different studies for the respective time periods in comparison to the results of Persson et al. (2018) based on the analysis of Venus Express ASPERA-4 data over the whole solar cycle. The gray curve depicts the monthly average sunspot number (https://www.sidc.be/silso). Figure adapted from Persson et al. (2018). **b)** Illustration of the H^+ and O^+ ratios and escape ratios for solar minimum and maximum

(Colin 1980). Therefore, a comparison is included as the dotted-red cross, which shows O^+ escape estimates based on the PVO measurements of wave-like plasma irregularities above the Venusian ionosphere (e.g., Brace et al. 1982; Russell et al. 1982). Several studies assumed that such plasma clouds were detached from the ionosphere by plasma instabilities such as the Kelvin-Helmholtz and interchange instabilities and were therefore lost to space (Wolff et al. 1980; Elphic and Ershkovich 1984; Terada et al. 2002; Arshukova et al. 2004). Based on extrapolated occurrence rates of such clouds and assumed model parameters, Brace et al. (1982, 1987) and McComas et al. (1986) indirectly estimated the high O^+ loss rates depicted in Fig. 5, which likely overestimate the escape rates compared to Venus Express measurements. This, together with the higher solar wind energy flux and the stronger solar maximum during the PVO era, can potentially explain the differences found in escape rates between the two missions.

Using the simple relationship between the O^+ escape rates and the solar wind energy flux in Fig. 5, one can make a simple extrapolation backwards in time to estimate the total O^+ escape in time assuming the same atmospheric structure and lack of intrinsic magnetic

Fig. 5 O^+ escape rates from Persson et al. (2020) for five separated ranges of solar wind energy flux using high and low XUV flux including error bars. The dashed lines are best fits for high and low XUV, respectively. The red dashed cross shows the range of estimated escape rates based on PVO measurements (Brace et al. 1987; McComas et al. 1986) and range of solar wind energy flux during the time PVO orbited Venus (McEnulty 2012)

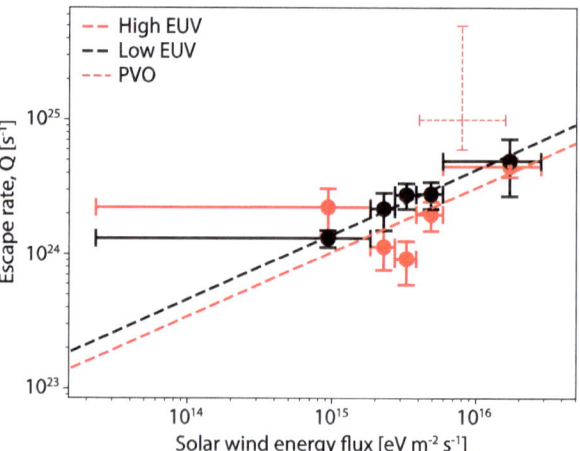

field. This simple exercise gives a total loss of $\approx 3 \times 10^{15}$ kg from now until 1 Ga, and $\approx 1.3 \times 10^{16}$ kg in total from now to 3 Ga. Assuming that the oxygen originates from water, and converting this into a global water depth, the total water loss would be around a few millimeters at 1 Ga, reaching a few centimeters or a few decimeters at 3 Ga (Persson et al. 2020). These results indicate that oxygen originating from water was not lost through escape to space during the later evolution of Venus' atmosphere, but instead either was lost earlier during hydrodynamic escape or by oxidation of surface materials. However, it is important to note that these calculations assume that the atmospheric structure and the interaction between the Venusian ionosphere and the solar wind, and thus the non-thermal escape, remains unchanged over the considered time period.

The Venus Express ASPERA-4 ion escape data obtained during the minimum and maximum solar activity conditions are in disagreement with the work by Hartle and Grebowsky (1993). Hartle and Grebowsky (1993) modeled the H^+ loss rate from Venus' nightside due to acceleration by an outward electric polarization force related to ionospheric holes. These authors estimated this electric polarization force by assuming that electrons are more mobile than ions and therefore they may easily escape from the top of the ionosphere such that a charge separation occurs that leads to the polarization electric field force (Hartle and Grebowsky 1993). By using PVO data and ionospheric modeling they estimated a so-called hydrogen bulge sector which covers an area of $\approx 20\%$ of Venus' sphere that resulted in a planet wide averaged H^+ escape rate by this process of $\approx 7 \times 10^{25}$ s^{-1} (Donahue and Hartle 1992; Hartle and Grebowsky 1993). This is $\approx 9 - 26$ times higher than the total H^+ ion escape rates inferred from the Venus Express ASPERA-4 data during minimum and high solar activity conditions.

This result is very relevant because it has consequences for the time span estimation that is required for escaping H and D to produce the present D/H ratio (HDO/H_2O) in Venus' atmosphere, as well as the characteristics of exogenous and endogenous hydrogen sources. According to Donahue (1999), unless the escape of D is very efficient on Venus, the present H escape rate averaged over a solar cycle should not exceed 5×10^{25} s^{-1}, if the present water vapor in Venus' atmosphere is a remnant of water deposited prior to the estimated average age of the surface (Donahue 1999). If the H escape rate was larger during the past 100 million years then the planet's present atmospheric water would be a remnant of water outgassed only ≈ 500 Myr ago (Grinspoon 1993; Donahue 1999). Since the H^+ ion escape rates inferred from the Venus Express ASPERA-4 data during the solar cycle are much lower than 5×10^{25} s^{-1}, Venus' present atmospheric water is most likely a remnant that is older than the average surface age of Venus. Moreover, the results also indicate that the electric polarization force loss process via ionospheric holes as suggested by Hartle and Grebowsky (1993) overestimates the H$^+$ escape rates.

However, besides ion escape H and D also escape from Venus' upper atmosphere via the photochemical processes shown in the following equations,

$$O^+ + H_2 \rightarrow OH^+ + H^* \qquad \Delta E = 0.6\,\text{eV}, \tag{1}$$

$$OH^+ + e \rightarrow O + H^* \qquad \Delta E = 8\,\text{eV}, \tag{2}$$

$$CO_2^+ + H_2 \rightarrow CO_2H^+H^* \qquad \Delta E = 1\,\text{eV}, \tag{3}$$

$$CO_2H^+ + e \rightarrow CO_2H^* \qquad \Delta E = 8\,\text{eV}, \tag{4}$$

and charge exchange processes,

$$H^+ + O \rightarrow H^* + O^+, \tag{5}$$

$$H^+ + H \rightarrow H^* + H^+. \tag{6}$$

Present-day escape rates of suprathermal H* atoms that originate from the photochemical reactions shown above were studied by Lammer et al. (2006) with a Monte Carlo model. These authors obtained present-day suprathermal H* atom escape rates from the photochemical reactions given in Eq. (1) on Venus' dayside of order $\approx 2.5 \times 10^{24}$ s^{-1}. This is slightly lower than the earlier but simpler estimates of McElroy et al. (1982b) of $\approx 4 \times 10^{24}$ s^{-1}. Regarding the production of H* atoms via the charge exchange reaction shown in Eq. (1), the literature gives a wide range of escape rate values from $\approx 6 \times 10^{25}$ s^{-1} (Hodges and Tinsley 1981; Hodges et al. 1986) to $\approx 1 \times 10^{25}$–2.65×10^{25} s^{-1} (Rodriguez et al. 1984). A more detailed analysis of charge exchange related H* losses by Donahue and Hartle (1992) and Hartle et al. (1996) revealed that the aforementioned studies used model inputs that did not fit the Venus International Reference Atmosphere (VIRA) empirical model that was developed based on the PVO ion mass spectrometer and electron temperature probe data. Donahue and Hartle (1992) used ion profiles for their charge exchange model that were fitted to VIRA and obtained an H* escape rate of $\approx 2.65 \times 10^{25}$ s^{-1}. An analysis of charge exchange related H* for solar minimum to solar maximum conditions by Donahue and Hodges (1992) resulted in escape rates from Venus' dayside of $\approx 1.3 \times 10^{25}$–$2.65 \times 10^{25}$ s^{-1}, or on average $\approx 2.0 \times 10^{25}$ s^{-1} which is higher than the H$^+$ escape rate of $\approx 7.6 \times 10^{24}$ H$^+$ s^{-1} as measured by ASPERA-4 (see above). Although there may also be some H* atoms produced on Venus' nightside, total H* escape rates that reach values $\geq 5 \times 10^{25}$ s^{-1} are very unlikely as long as the solar activity was not higher than at solar maximum. Taking the sum of ion and suprathermal escape for H, this yields a total loss of $\approx 2.8 \times 10^{25}$ s^{-1}.

Suprathermal escape of oxygen atoms, on the other hand, is negligible, since the escape energy of O is almost twice as high as the energy of the photochemically produced hot atoms (Lichtenegger et al. 2009; Gröller et al. 2010). This imbalance between suprathermal H and O escape consequently alters the aforementioned loss ratio of H to O of $\approx 2{:}1$ to a ratio that significantly deviates from 2:1. If one assumes that the majority of hydrogen originated from dissociated water, this would either suggest an accumulation of O in the atmosphere or an additional surface sink. However, H could also originate from other molecules than H$_2$O such as H$_2$SO$_4$, etc.

The findings of lower H$^+$ ion escape rates by Persson et al. (2018) compared to previous estimates by Hartle and Grebowsky (1993) and Donahue (1999) indicate that Venus' present

D/H ratio is most likely a remnant of an ancient water inventory that was mixed with more recently outgassed water vapor. D/H ratio has been modified by the non-thermal atmospheric escape processes discussed above on timescales of ≈ 200–1000 Myr (the average age of the surface) or more.

Even though no study has thus far estimated thermal and non-thermal loss of H and D at Venus even further back in time, it can be expected that at least thermal escape rates of its atmospheric species were significantly higher due to the elevated XUV flux from the young Sun, as different studies on the early Earth and exoplanets might indicate. Thermal losses of water, but also of CO_2 and N_2, were more significant earlier on (e.g., Tian et al. 2008a; Tian 2009; Tian et al. 2018; Johnstone 2020; Johnstone et al. 2021). This can be exemplified by a simple comparison between Earth and Venus. As has been shown by Johnstone et al. (2021), Earth would not have been able to maintain an N_2–CO_2 atmosphere with even 99% CO_2 for an x-ray surface flux of ≥ 10 erg s^{-1} cm^{-2}, i.e., prior to ≈ 3.5 Ga if the Sun would have been a fast rotator. A nitrogen-dominated atmosphere with only 10% CO_2 would have been unstable, i.e., a 1 bar atmosphere would escape to space within less than 10 million years for a surface flux of ≥ 5 erg s^{-1} cm^{-2}, i.e., until ≈ 3.0 Ga for a fast, and ≈ 3.5 Ga for a slow rotator, respectively (Johnstone et al. 2021). Such high fluxes start to dissociate CO_2 (Tian 2009; Johnstone et al. 2021), the main infrared cooler in the upper atmosphere (e.g., Roble 1995), resulting in hotter (e.g., Cnossen 2020) and more extended thermospheres and exobase levels. In extreme cases, the upper thermosphere can even hydrodynamically expand so that the gas flow cools adiabatically with the exobase level reaching several planetary radii, again resulting in massive atmospheric outflows due to the low gravitational potential at the exobase level (Tian et al. 2008a; Tian 2009; Lammer et al. 2008; Johnstone et al. 2021). Oxygen as an additional constituent and heating agent in the upper atmosphere is not needed for an expansion under such high XUV fluxes (Kulikov et al. 2006; Johnstone et al. 2019, 2021)

that are susceptible to strong thermal escape (e.g., Tian et al. 2008a; Johnstone et al. 2021). It is also important to note that besides the incident flux the mixing ratio of CO_2 with other species such as N_2 is the main factor that determines the upper atmosphere structure, while the density of the underlying atmosphere has negligible effects. Since Venus receives approximately twice as much XUV than the Earth and has only about 80% of its mass, such atmospheres would have potentially been susceptible to escape to space at Venus until later times.

Finally, one should note that for such high escape rates, i.e., when the loss becomes hydrodynamic, H will not (or only marginally) be fractionated from D, since both species will escape together with the bulk atmosphere. Only when hydrodynamic escape transitions to Jeans escape D/H fractionation will start to become important (see e.g., Lammer et al. 2020b). Therefore water loss from Venus cannot be simply estimated from the fractionation of D/H. This exercise is even more difficult considering that the starting D/H ratio for Venus' hydrogen is unknown. The extent of water loss would be a ≥ 60 m deep global ocean through non-thermal escape if one assumes carbonaceous chondrites as initial source of H_2O (e.g., Donahue 1999). On the other hand it would be ≥ 420 m in case the Venus' initial D/H ratio was defined by water that was produced due to surface interactions between a magma ocean and a primordial hydrogen-rich atmosphere (Salvador et al. 2022) that had a ≈ 7 times higher solar-like D/H ratio of $21 \pm 5 \times 10^{-6}$ (Geiss and Gloeckler 1998; Robert et al. 2000).

If one assumes comets as a main reservoir, the water layer would have been even thinner, potentially ranging from ≈ 30 m to only ≈ 4 m, depending on the cometary D/H ratio (see, e.g., Robert et al. 2000, for different D/H ratios within comets). Several planned in-situ measurements of the HDO/H_2O ratio by the DAVINCI entry probe will constrain the D/H ratio in the bulk atmosphere along a vertical profile.

2.4 Surface-Gas Reactions, Mineral Buffering and Atmospheric Evolution

In this section, we discuss gas-mineral type reactions that deplete atmospheric volatiles and could control atmospheric composition before and during the last global resurfacing event, at present, and in the future. The mass of chemically active elements in the atmosphere (S, Cl, and F, but not C) is small compared to masses of those elements in permeable near-surface crustal materials, based on surface material compositions data from Venera/Vega landers (Surkov et al. 1984, 1986) and calculated by Zolotov and Volkov (1992). It follows that atmosphere-crust type reactions should have affected atmospheric composition throughout history. Such reactions trapped volcanic, impact-generated, chemically and photochemically produced O-, C-, H-, S-, Cl-, and F-bearing gases, as well as excessive O-bearing compounds (e.g., CO_2, SO_2, O_2) formed via net O accumulation after H escape. In turn, gaseous products of weathering reactions affected the atmosphere. In some cases, the gas-trapping reaction reached chemical equilibria that maintained (buffered) atmospheric abundances of the involved gases.

2.4.1 Atmospheric-Surface Reactions on Ancient Venus

If Venus had an early aqueous history, atmospheric concentrations of chemically active gases (CO_2, H_2O, SO_2, etc.) were likely controlled by aqueous processes such as gas-water partitioning, dissolution and precipitation of solids, as well as erosion, transport and sedimentation. As on Earth, aqueous reactions would have allowed trapping of gases in secondary minerals such as salts (carbonates, chlorides, fluorides, etc.), phyllosilicates, oxides, and hydroxides (Zolotov and Mironenko 2009). These likely occurred in a permeable surface layer on continents and were affected by local temperature, humidity and by water erosion. A geological time-scale carbonate-silicate cycle (Walker et al. 1981) could have regulated the concentration of atmospheric CO_2 and thus surface temperature through the greenhouse effect of CO_2 (see Sect. 2.2.1).

In the case of 'runaway' greenhouse warming without any condensation of surface water (Ingersoll 1969), lower erosion rates and higher temperatures would have favored the formation of secondary mineral assemblages that could have controlled the concentrations of gases. In the absence of any major supply of gases and rocks, concentrations of certain atmospheric gases could be controlled by gas-solid equilibria rather than ongoing reactions. For example, the equilibrium assemblage of ferric and ferrous oxides in surface materials could control the CO/CO_2 ratio in the lower atmosphere (g denotes a gaseous species),

$$2Fe_3O_4 \text{ (magnetite)} + CO_2, g = 3Fe_2O_3 \text{ (hematite)} + CO, g \qquad (7)$$

because the reaction equilibrium constant at a fixed temperature is expressed as $K_1 = (aFe_2O_3)^3 \, fCO/((aFe_3O_4)^2 \, fCO_2) \sim fCO/fCO_2 \sim xCO/xCO_2$ where a, f, and x stand for activity, fugacity, and mixing ratio, correspondingly. A change in gas concentration, e.g., through volcanic degassing, will cause the gas to interact with the corresponding mineral in order to restore its equilibrium concentration. Changes in temperature mean changes in the reaction constant, and therefore would facilitate forward or backward reaction to the new equilibrium concentrations of gases. For reaction (7), CO_2 degassing will cause the partial oxidation of the magnetite in the rocks to restore the CO_2/CO ratio. In contrast, heating will decrease the CO/CO_2 ratio in the atmosphere. As in laboratory gas-mineral buffers, natural buffers in permeable near-surface materials could operate until exhaustion of a reactant. In the case of Eq. (7) on Venus, the CO/CO_2 ratio would be controlled until complete oxidation of the rock's magnetite to hematite. Without liquid water, the massive atmospheric CO_2 may not be controlled by silicate-carbonate equilibria because of the inefficient carbonation of dry near-surface rocks (Tanner et al. 1985) and other factors that are notable for the current epoch (Sect. 2.4.2). However, hot and dense CO_2 is an efficient oxidizer of ferrous minerals and glasses (Berger et al. 2019) and sulfide sulfur S(II). Corresponding gas-mineral reactions (e.g., Eq. (7)) could have affected xCO_2 throughout the anhydrous history.

Volcanic S-, Cl-, and F-bearing gases (SO_2, S_2, H_2S, HCl, etc.) are highly reactive with rocks and atmospheric/surface dust. In other words, they are efficiently trapped in secondary minerals (salts, sulfides) and may not accumulate in the atmosphere without re-supply. If volcanic Cl and F are provided by the degassing of HCl and HF (this is not the case in H-depleted gases; Zolotov and Matsui 2002), rapidly formed secondary minerals would maintain $xHCl$ and xHF at trace levels, that depend on temperature, mineralogy of buffering equilibria, and H_2O,g content (e.g., Lewis 1970; Fegley and Treiman 1992).

Whether atmospheric SO_2-silicate reactions led to buffering sulfate-bearing mineral assemblages (e.g., $CaSO_4$-plagioclase; Zolotov 2018) depended on the relative rates of SO_2 volcanic degassing, fresh rock supply, and gas-solid reactions. Although SO_2 trapping in minerals was unavoidable, slow SO_2-silicate reactions limited by the rates of metal (Ca, Na) diffusion (King et al. 2018; Berger et al. 2019) could have maintained atmospheric xSO_2 above values of corresponding gas-solid equilibria, as at present (Sect. 2.4.2).

Atmospheric xH_2O on early Venus could have been controlled by hydration-dehydration reactions involving salts, hydroxides, and phyllosilicates (serpentine, clay minerals, micas, and amphiboles). The hydrated mineralogy could have affected xH_2O either after the cessation of an aqueous epoch or during a greenhouse evolution without water condensation. At a constant temperature, net consumption of H_2O gas through H escape and oxidation of surface materials could have caused a step-wise decrease in xH_2O buffered by sequentially changing mineral assemblages in surface and subsurface materials. Each dehydration reaction, for example, could have maintained xH_2O until exhaustion of a dehydrating mineral (Eqs. (8), (9)).

$$(Mg,Fe)_3Si_2O_5(OH)_4(\text{serpentine}) \rightarrow$$
$$(Mg,Fe)_2SiO_4(\text{olivine}) + (Mg,Fe)SiO_3(\text{pyroxene}) + 2H_2O,g \tag{8}$$

$$(Mg,Fe)3Si_4O_{10}(OH)_2(\text{saponite}) \rightarrow$$
$$3(Mg,Fe)SiO_3(\text{pyroxene}) + SiO_2(\text{silica}) + H_2O,g \tag{9}$$

Micas and amphiboles could be the last H-bearing phases before the complete decomposition of H-bearing minerals that is expected at present (Zolotov et al. 1997). If surface temperature increased, the dehydration of minerals (e.g., Eqs. (8), (9)) would have caused a positive feedback because the released H_2O is a strong greenhouse gas. The negative feedback, through slowing mineral dehydration in an increasingly H_2O-rich atmosphere, could not have prevented a rapid and thorough dehydration, nor the corresponding temperature spike. It is unclear if subsequent consumption of H_2O,g through H escape and oxidation of surface/atmospheric species could have allowed temporary formation of micas or amphiboles on lower-temperature highlands.

The high D/H ratio in the current atmosphere (Donahue et al. 1982; De Bergh et al. 1991) may imply H escape and net oxidation of rocks by remaining net water's oxygen (e.g., Kasting and Pollack 1983; Lewis and Prinn 1984), but see Sect. 2.3.1 for a more detailed picture. The oxidation of exposed surface materials as an alternative to atmospheric O escape has been considered in recent publications (Gillmann et al. 2009, 2020;

Wordsworth et al. 2018; Krissansen-Totton et al. 2021a; Way and Del Genio 2020; Warren and Kite 2021) though the pathway of post magma ocean O consumption remains to be better quantified. In one pathway, photochemically produced oxygen was consumed through oxidation of atmospheric gases ($CO \rightarrow CO_2$; COS, S_2, $H_2S \rightarrow SO_2$; $H_2 \rightarrow H_2O$) followed by oxidation of rocks (e.g., Eq. (7)) by formed O-bearing gases. In a parallel pathway (e.g., Eqs. (10), (11)), the oxygen was consumed through rock-H_2O,g reactions leading to oxidized solids (Fe(III) oxides, pyrite, sulfates) and subsequent H_2 loss (e.g., Khodakovsky 1982). However, the accumulation of atmospheric O_2 above trace amounts is unlikely because of its reactivity with reduced atmospheric gases and crustal solids, if O_2 ever reached the surface. The amount of oxygen consumed by rocks would have depended on the unknown amount of water that supplies oxygen through photo-dissociation in the upper atmosphere and water-rock reactions followed by H escape.

$$3FeO \text{ (in silicates)} + H_2O,g \rightarrow Fe_3O_4 \text{ (magnetite)} + H_2, g \tag{10}$$

$$2Fe_3O_4 \text{ (magnetite)} + H_2O,g \rightarrow 3Fe_2O_3 \text{ (hematite)} + H_2, g \tag{11}$$

Using Earth-like FeO concentration Lécuyer et al. (2000) proposed that 10^{21} kg of water (roughly one Earth ocean) could be removed by the oxidation of a global layer of ≈ 50 km of basaltic crust into hematite (also see, Kasting and Pollack 1983; Lewis and Prinn 1984). For evolution scenarios, this must be nuanced by actual volcanic production rates required to bring fresh basalt to the surface and bury older material. Earth production (≈ 20 km^3 yr^{-1}; Morgan 1998) is assumed to be a maximum for recent Venus, but no data exist for ≥ 1 Ga Venus. The efficiency and degree of crustal oxidation is also uncertain, but would be limited once the surface is solid. Experimental studies using Venus-like conditions and atmosphere composition (e.g., Berger et al. 2019) showed that the alteration was limited to surface oxidation or a coating at the surface of olivine, but that it occurred rather fast: the Fe content in a 10s to 100s of nanometers thick layer could be oxidized within about 15 days (Teffeteller et al. 2022). Other important factors would include the mineralogy of formed oxidized phases, kinetics of oxidation reactions, grain size and crystallinity of altering materials, and the depth of permeable surface layers. By contrast, oxidation of the liquid material before it solidifies could be much more efficient. This is especially true during the magma ocean phase (Gillmann et al. 2009; Schaefer et al. 2016; Warren and Kite 2021; Salvador et al. 2022). However, some simulations have shown that significant O_2 atmospheres can still remain due to H_2O photolysis once the magma ocean freezes over (Wordsworth et al. 2018; Krissansen-Totton et al. 2021b).

In addition to oxidation of ferrous silicates (Eqs. (7), (10), (11)), formation of pyrite at the expense of ferrous silicates and/or sulfides and sulfatization of Ca-bearing silicates, glasses, and putative ancient water-deposited carbonates by SO_2 (12) could have provided an additional sink of atmospheric O throughout the anhydrous history, including the current epoch (Zolotov 2018).

$$CaO \text{ (in silicates and/or carbonates)} + 1.5SO_2, g \rightarrow$$
$$CaSO_4 \text{ (anhydrite)} + 0.25S_2, g, \tag{12}$$

In the hypothetical case of several global volcanic resurfacing events, the intensity of surface oxidation could have been sluggish between the events. During periods of limited supply of fresh volcanic materials to the atmosphere and surface, the oxidation rate might roughly be comparable to the rate of removing excess oxygen through non-thermal escape

(Persson et al. 2020, see above). Whether gas-solid reactions buffered atmospheric gases between resurfacing events depended on volcanic activity. In the case of minute volcanic supply of rocks and gases, as suggested for the current epoch (Schaber et al. 1992; Basilevsky et al. 1997) and limited physical weathering and erosion, buffering is likely. Secondary minerals (Fe oxides, pyrite, sulfates) in permeable surface materials formed through oxidation by CO_2, H_2O, and SO_2 (Eqs. (7), (10)–(12)), could have controlled an array of redox-dependent atmospheric gases. For example, the magnetite-hematite assemblage is likely a major player in controlling ratios of reduced (COS, CO, S_2, H_2S, H_2) and oxidized (CO_2, H_2O, SO_2) gases at Venus' present and future (Sects. 2.4.2, 2.4.3). Because of gas-solid type redox equilibration in the near-surface atmosphere, further oxidation is only driven by a supply of excess O, which becomes limited for recent history and without endogenic and/or exogenic (comets, carbonaceous chondrites) supplies of H_2O gas to the atmosphere.

Global volcanic resurfacing events, if they occurred, likely eliminated existing buffering reactions due to changes in surface composition, volcanic outgassing, and related greenhouse warming (Solomon et al. 1999). Elevated concentrations of volcanic SO_2, COS, S_2, CO, CO_2, H_2O, HCl, and HF in the atmosphere, increased greenhouse temperatures, and exposure of fresh volcanic materials facilitated trapping of the gases in minerals. HCl and HF gases would have needed to be efficiently trapped and would likely have reached minute gas-solid equilibrium concentrations at the corresponding temperatures. Elevated concentrations of degassed H_2O,g (Berger et al. 2019) could have favored oxidation of solids (e.g., ferrous species to Fe oxides; Warren and Kite 2021) and H_2 formation (Eqs. (10), (11)). High temperatures could have facilitated oxidation of ferrous and sulfide compounds by hot and dense CO_2 (Eq. (7)).

Likewise, a rapid interaction of hot SO_2 with silicate minerals and glasses (Renggli and King 2018) could cause sulfatization of the exposed surfaces of rocks, ash, and dust via sulfur disproportionation reactions exemplified by the sulfatization of Ca in silicates and glasses (e.g., Eq. (12)). As on Earth (Delmelle et al. 2018) alterations of lava surfaces and pyroclastic products could have occurred while the materials were still hot, though they cool faster on Venus than on Earth due to its dense atmosphere (Frenkel and Zabalueva 1983).

Oxidation of surface materials and trapping of atmospheric HCl and HF could have led to the secondary mineralogy on the current surface (Sect. 2.4.2). However, the oxidation potential of the atmosphere could be limited because of overall reduced nature of planetary volcanic gases (Gaillard and Scaillet 2014) and because of the accumulation of reduced products (CO, COS, H_2, S_2, H_2S) of gas-solid reactions (Eqs. (7), (10)-(12)). A low mass (up to a %) of degassed compounds (H_2O, SO_2, HCl, HF), compared to that of erupted rocks, implies that alteration of exposed rocks would be limited. The burying of earlier altered lava flows by fresh ones would remove trapped volcanic volatiles from the atmosphere-surface system, and favor further trapping but not gas-solid equilibration. Consumption of degassed H_2O through rapid oxidation (Eqs. (10), (11)) and H escape would cool the atmosphere. Limited resurfacing during hundreds of Myr could favor co-evolution of trace atmospheric gases with buffering mineral assemblages that partially formed during and shortly after more ancient and intense volcanic activity.

2.4.2 Current Gas-Solid Interactions and Atmospheric Buffering

Current Venus' atmosphere-surface system is a checkpoint to constrain effects of gas-solid reactions and corresponding equilibria on atmospheric composition. Since the early 1960s, atmosphere-surface interactions and buffering reactions have been considered in multiple journal publications (e.g., Mueller 1964; Lewis 1970; Barsukov et al. 1982;

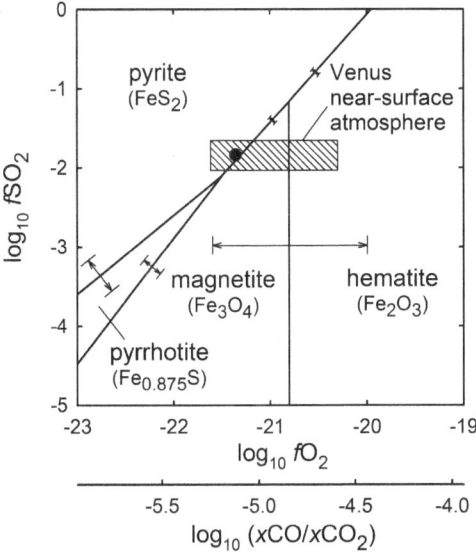

Fig. 6 Stability fields of iron oxides and sulfides at the conditions of modal Venus radius (740 K, 95.6 bars). The Venus box corresponds to mixing ratios of SO_2 (95-228 ppm) and CO (5-23 ppm) based on Pioneer Venus and Venera entry probe data with uncertainties. The lowest CO value reflects an extrapolation of the atmospheric gradient toward surface (Fegley et al. 1997b). The circle symbol corresponds to $xSO_2 = 150$ ppm and $xCO = 17$ ppm. The fO_2 data are for the CO-CO_2 gas equilibrium. The error bars reflect uncertainties of thermodynamic data of minerals. The figure shows that Venus gas composition is close to conditions at iron oxide and oxide-sulfide equilibria. Modified from Zolotov (2019)

Klose et al. 1992; Semprich et al. 2020) and book chapters (e.g., Fegley and Treiman 1992; Fegley et al. 1997a; Wood 1997; Zolotov 2018). This Section provides a summary of the current views.

Although the composition of the sub-cloud atmosphere remains uncertain (Marcq et al. 2018), existing data do not exclude gas-gas type equilibration at Venus' lowlands at ≈740 K (e.g., Fegley et al. 1997a; Zolotov 1996; Krasnopolsky 2007). Partial pressures of equilibrated gases match those at the hematite-magnetite (Hem-Mt) (Eq. (1)), hematite-magnetite-pyrite (Hem-Mt-Py) and magnetite-pyrite (Mt-Py) mineral equilibria (e.g., Fegley et al. 1997a; Zolotov 2018, Fig. 6).

The match does not exclude gas-solid equilibration and these minerals are expected weathering products formed during (Warren and Kite 2021) and after a global volcanic resurfacing event. The presence of hematite is consistent with the near-infrared reflectance and color of surface materials at the landing sites of Venera 9, 10 and 13 (Pieters et al. 1986; Shkuratov et al. 1987; Yamanoi et al. 2009). A hematite-rich layer could be at least a few micrometers thick in order to be optically thick (Gilmore et al. 2017). The current working hypothesis is that Hem-Mt-Py surface mineralogy buffer key ratios of reduced and oxidized gases, CO/CO_2, $SO_2/(COS, H_2S, S_2)$, and H_2/H_2O. If the buffering works, current atmospheric gases are not involved in corresponding weathering reactions, changes in atmospheric composition will be compensated by gas-solid reactions (e.g., Eq. (7)), in one direction or another, and major changes in surface mineralogy (e.g., via volcanism) will affect atmospheric composition and its redox state. In contrast to thermodynamically favorable oxidation of Fe(II) in surface materials to magnetite by atmospheric CO_2, H_2O and/or SO_2, formation of hematite will only be driven by net oxidation of the atmosphere that currently occurs through a slow H escape (see Sect. 2.3.1)

It is unclear whether minerals buffer abundances of S-bearing gases. On the one hand, atmospheric $x SO_2$ is high enough to allow alteration of Ca-bearing pyroxenes to $CaSO_4$ (Eq. (12)), as inferred from calculations of chemical equilibria (e.g., Barsukov et al. 1982; Fegley and Treiman 1992; Zolotov 2018) and supported by modeling experiments (e.g., Berger et al. 2019). The high but variable S/Ca ratio in three Venera and Vega probes of surface basaltic material (Surkov et al. 1984, 1986) indicates the ongoing trapping of atmospheric S. On the other hand, plagioclase, a major mineral in basalt, is more resistant with respect to Venus' SO_2 and a plagioclase-anhydrite assemblage could have a buffering effect on SO_2 (Zolotov 2018). The low current rate of H escape and a comparable rate of O escape (Sect. 2.3.1) imply minute (if any) net supply of oxidants to the atmosphere. This does not contradict the buffering of atmospheric redox conditions by secondary minerals.

The minute abundances of atmospheric HCl and HF are insufficient to alter surface minerals and suggest a major anterior trapping of the gases (Zolotov 2018). Although the mineralogy of secondary Cl- and F-bearing phases remains unclear (Lewis 1970; Fegley and Treiman 1992), buffering of atmospheric HCl and HF by gas-solid equilibria is highly plausible. In contrast, buffering of H_2O gas by H-bearing minerals is unlikely because of their instability at the current surface Zolotov et al. (1997) and a thorough thermal dehydration related to extreme greenhouse heating during volcanic resurfacing events. Likewise, mineral buffering of CO_2 is unlikely because of the instability of carbonates and carbonate-silicate equilibria with respect to greenhouse warming (Hashimoto and Abe 2005), because of inhibited carbonation without aqueous fluid (Tanner et al. 1985), because of the instability of Ca carbonates and silicates (except plagioclase) with respect to Venus' SO_2, and other reasons (Zolotov 2018). A match of Venus' surface $f CO_2$ with $f CO_2$ at the carbonate-silicate equilibrium

$$CaCO_3 \text{ (calcite)} + SiO_2 \text{ (quartz)} = CaSiO_3 \text{ (wollastonite)} + CO_2, g \qquad (13)$$

(e.g., Mueller 1964; Nozette and Lewis 1982; Fegley and Treiman 1992) is probably accidental.

2.4.3 Future Co-Evolution of the Atmosphere-Lithosphere System

Future Venus will likely be characterized by the physical-chemical co-evolution of the atmosphere, surface materials, and shallow interior. It is unclear if the decreasing radiogenic heating in the interior will cause further major volcanic resurfacing and related greenhouse heating. If a global volcanic event occurs, physical-chemical processes would be similar to those during the last global resurfacing at 0.2–1 Ga (Sect. 2.4.1). Intensive gas-solid type reactions would again proceed toward establishing gas-solid equilibria that will control minor chemically active gases. As in the present epoch (Zolotov 2018), Fe oxides and sulfides, Ca sulfate, chlorides and fluorides will be major participants of buffering reactions, though actual gas concentrations will depend on evolving temperature-pressure conditions. Ultimately, exhaustion of radionuclides of U and K will terminate volcanic supply of fresh rocks and gases. That termination will further favor establishing and maintaining buffering gas-solid equilibria in permeable surface materials. Further evolution will be moderately affected by increasing solar luminosity and corresponding warming, gas-solid reactions at the surface, and atmosphere escape. The thermodynamically favorable oxidation of Fe(II) and S(II) in exposed solids by atmospheric CO_2, and traces of SO_2 and H_2O from volcanic

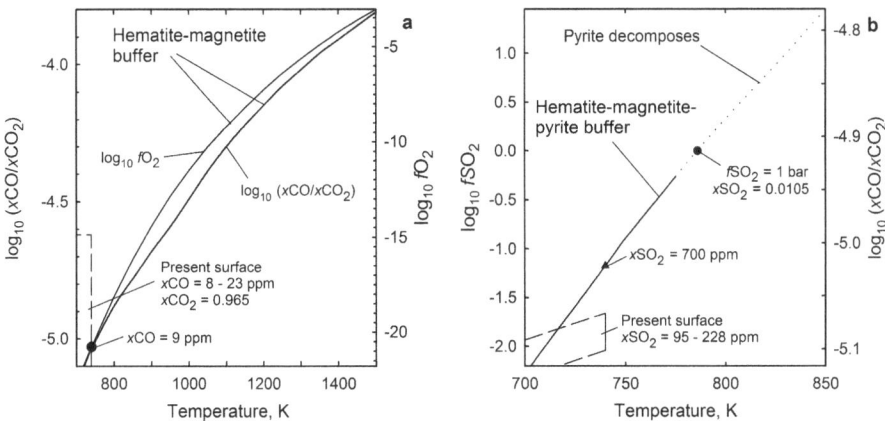

Fig. 7 Possible future changes in concentrations of trace atmospheric gases controlled by mineral-gas equilibria at 95.6 bar. In (**a**), CO/CO_2 mixing ratio and fO_2 are controlled by the hematite-magnetite equilibrium (Eq. (7)) during warming. The area shown by dash lines corresponds to the present near-surface CO/CO_2 ratios (see Fig. 6 for fO_2). The circle symbol shows conditions at which CO extrapolated toward the surface (Fegley et al. 1997b; Zolotov 2019) closely match the equilibrium (Eq. (7)). In (**b**), fugacity of SO_2 is controlled by the hematite-magnetite-pyrite equilibrium at the CO/CO_2 ratio at Eq. (7). The fugacity of SO_2 in the current near-surface atmosphere reflects a range of measured SO_2 concentrations with uncertainties. The triangle symbol shows nominal conditions of the mineral equilibrium at 740 K (see Fig. 6). The circle symbol corresponds to partial pressure of SO_2 of 1 bar at 786 K. The dotted curve shows possible conditions of through decomposition of pyrite in a permeable surface layer. (This figure was developed by M. Zolotov)

gases suggest establishing ferric-ferrous equilibria and equilibria with participation of relatively oxidized S-bearing minerals, such as pyrite, and Ca and Na sulfates. In the likely case of hematite-magnetite and/or hematite-magnetite-pyrite equilibration, one would expect changes in CO_2/CO, SO_2/COS, SO_2/H_2S, and H_2O/H_2 ratios that reflect increasing fO_2 with temperature (Fig. 7). The suppressed gas-solid reaction rates at near-equilibrium conditions would not cause a thorough oxidation of exposed rocky materials to secondary phases. In other words, net reactions at the atmosphere-surface interface

$$2FeO \text{ (in silicates, magnetite)} + O \text{ (in } CO_2, H_2O, SO_2) \rightarrow$$
$$Fe_2O_3 \text{ (hematite)} + \text{reduced gas}(CO, H_2, S_2) \tag{14}$$

and sulfatization reactions (Eq. (12)) will not proceed toward completion, especially given the limited expected O enrichment of the atmosphere by escape of H and O from the atmosphere (Sect. 2.3.1). As in the present epoch, the elevated level of atmospheric SO_2 controlled by equilibria with pyrite and/or sulfates would not allow formation of carbonates of Ca, Na and K that affect atmospheric CO_2. Atmospheric HCl and HF will be controlled by equilibria with corresponding surface Cl- and Fe-bearing phases (chlorides, fluorides, amphiboles, phosphates; Lewis 1970, Fegley and Treiman 1992) and H_2O,g at corresponding temperatures. CO_2 and N_2 will remain the major atmospheric gases and the increasing greenhouse warming could slightly contribute to their abundances through the degassing of rocks. In several Gyr, the Sun will become a red giant and a significant greenhouse heating will cause shallow and then surface salt (e.g. NaCl), sulfide and then silicate melting. These events and corresponding melt-atmosphere interactions will reduce the atmosphere by magma that is rich in Fe(II) and S(II) melt complexes. One would expect a decrease in CO_2/CO and $SO_2/(H_2S, S_2, COS)$ ratios in the atmosphere towards fO_2 expected for Venus'

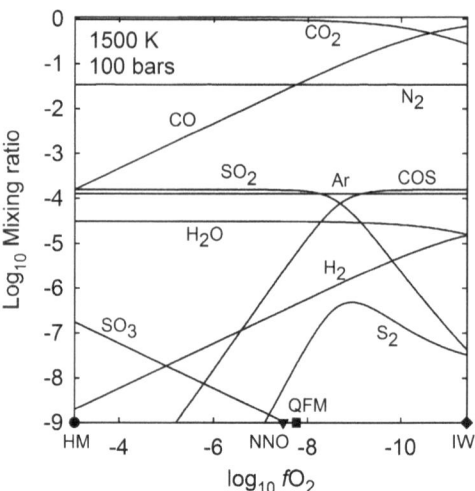

Fig. 8 Mixing ratios of C-O-H-S-N gases with bulk composition of the current atmosphere as functions of fO_2 at 1500 K. The Ar value corresponds to the current atmospheric ^{36}Ar and ^{38}Ar contents and tripled current ^{40}Ar. The figure illustrates possible sequential reduction (fO_2 decrease) of the atmospheric composition after formation of a second magma ocean on Venus. HM, hematite-magnetite buffer; NNO, Ni-NiO buffer; QFM-quartz-fayalite-magnetite buffer; IW, iron-wüstite buffer. (This figure was developed by M. Zolotov)

mantle melts, that is 2 log fO_2 units below the QFM buffer (Wadhwa 2008). In other words, the gas-melt equilibria at the surface of a second magma ocean in Venus history will control the composition of its atmosphere, in which CO_2, CO and N_2 could be the major gases (Fig. 8).

2.4.4 Silicate Weathering and Carbon Cycling

As hydrogen is fused into helium, the density of the Sun increases, and thereby both its fusion rate and luminosity. As a consequence, the incident insolation of Earth has increased by ≈30% throughout its history (Gough 1981). However, liquid water is known to have existed on Earth's surface at least since the early Archean, which requires a substantially higher amount of greenhouse gases in the early atmosphere (e.g., Kasting and Catling 2003). An efficient feedback mechanism that regulates the amount of CO_2 in Earth's atmosphere against long-term changes in the surface temperature is the long-term carbonate-silicate cycle (e.g., Berner and Caldeira 1997): in this cycle, CO_2 reacts with rainwater to form carbonic acid, which dissolves silicate rocks. Weathering products are washed via rivers into the ocean and lakes, where calcium carbonate is precipitated. Burial of calcium carbonates acts as a sink for carbon. In subduction zones, part of the carbon is then thermally decomposed and returned to the atmosphere, while the remainder is subducted into the mantle. Outgassing of CO_2 at mid-ocean ridges completes the cycle. Importantly, the rate of silicate weathering depends on the concentration of CO_2 in the atmosphere and on the surface temperature, which efficiently regulates the climate against fluctuations of atmospheric CO_2 and against increasing incident insolation (Walker et al. 1981). Way et al. (2022b) also offers an in-depth look at weathering and volatile cycles, in the context of exoplanetary studies.

If liquid water existed on early Venus (e.g., Kasting et al. 1984; Kasting 1988; Abe et al. 2011; Way et al. 2016), silicate weathering may have served as a sink for atmospheric CO_2 (e.g., Kasting et al. 1984). This could have involved either dissolution of silicates on land and the subsequent precipitation of carbonates in soils, lakes and ocean basins (so-called continental weathering), or the sub-oceanic circulation of carbon-bearing fluids and precipitation of carbonates from in-situ dissolution products (seafloor weathering) (e.g., Driscoll and Bercovici 2013). In either case, cations produced from rock weathering combine with carbonate ions to remove carbon dioxide from the atmosphere-ocean system (Walker et al. 1981; Brady and Gíslason 1997).

On Earth, the precise dependence of seafloor weathering on dissolved inorganic carbon (and thereby indirectly atmospheric CO_2), bottom ocean temperature, and ocean pH is uncertain (Brady and Gíslason 1997; Sleep and Zahnle 2001; Krissansen-Totton and Catling 2017). In scenarios for early Venus, the functional dependence of continental silicate weathering on atmospheric CO_2 or surface temperature may be important for the climate, since it determines the efficiency at which CO_2 could be removed from the atmosphere and thereby the potential for maintaining habitable conditions.

Recent models considering the thermodynamic, energetic, and lithological limitations to continental weathering suggest the weathering response to changes in climate and atmospheric CO_2 may be strongly dependent on geologic factors. For example, the chemical weathering rate of granite is significantly lower than that of basaltic crust, which becomes particularly relevant for high surface temperature (Hakim et al. 2021). In addition, if weathering is controlled by a thermodynamic limit, climate becomes increasingly sensitive to land fraction and outgassing (Graham and Pierrehumbert 2020). With limited exceptions (Zolotov 2020), the implications of such weathering models for a potentially habitable early Venus have not yet been explored, but point at possible consequences for the CO_2 evolution in the atmosphere of Venus if liquid water was present.

The fate of carbonated crust has important implications for Venus' climate evolution. On Earth, carbonate sediments and carbonatized basalt are subducted with the oceanic lithospheric plates. Depending on the geotherm of subduction zones, subducted carbonates either release CO_2 into the atmosphere via arc volcanism, or remain stable and are recycled into the mantle. It has been suggested that plate tectonics on early Venus, if it ever existed, could have similarly stored carbon in the mantle and thus preserved habitable surface conditions for billions of years (Way and Del Genio 2020). As discussed in (Rolf et al. 2022), the existence of plate tectonics on early Venus is debated. However, plate tectonics may not be a requirement for early Venus habitability (Foley and Smye 2018; Höning et al. 2019; Tosi et al. 2017). In the absence of plate tectonics, silicate weathering may still occur, and carbonate sediments would be buried by new lava flows. Ultimately, these carbonates would heat up as they move downward in the crust and eventually become unstable.

To what extent CO_2 released by metamorphic decarbonation reactions in the buried crust makes its way back into the atmosphere is unclear. At shallow depth, cracks likely form a path for CO_2 to the surface. However, deeper in the crust increasing pressure and temperature may lead to sealing voids, which would cause CO_2 to sink towards the mantle as the crust gets buried by new lava flows. In submerged metamorphosed rocks, CO_2 fugacity could have been buffered by newly established silicate-carbonate equilibria. If rising melt comes into contact with the trapped CO_2, the latter would likely rise with the melt to the surface, similar as at arc volcanoes on Earth. Melting in CO_2-rich source regions could have caused the formation of alkaine silicate melts (e.g., Kargel et al. 1993) that are common on oceanic islands on Earth. Parent melts of K-rich Venera 8 and Venera 13 surface probes could have formed this way, and carbonate-rich lavas (carbonatites) might be responsible for the formation of Venus canali (Kargel et al. 1994). Altogether, part of the released CO_2 from decarbonation reactions of buried carbonates could return to the atmosphere. However, the precise fraction is difficult to assess and depends on mechanical processes in the crust and the path of rising magma. It also depends on the depth and temperature where carbonates become unstable as well as on the crustal burial rate and porosity, which are both uncertain. However, without plate tectonics the rate at which carbonates are recycled into the mantle is certainly limited compared to Earth.

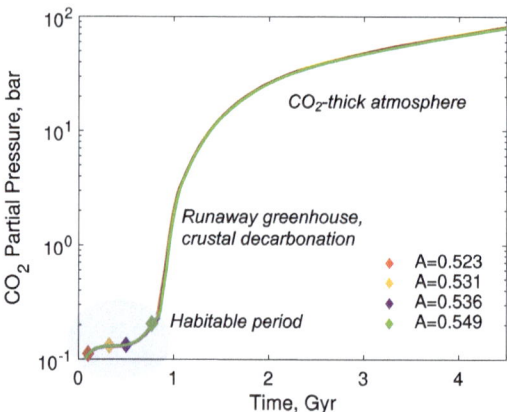

Fig. 9 Modelling results from a coupled interior-atmosphere model of a stagnant lid Venus assuming different planetary albedos, after the solidification of the magma ocean (Höning et al. 2021). The model assumes an initially thin atmosphere containing only the present-day N_2 inventory. Diamonds depict control runs from a 3-D general circulation model (Way et al. 2016; Way and Del Genio 2020) at the given time (corresponding to incident insolation) and atmospheric CO_2 pressure, yielding the indicated albedo

The resulting carbon cycle for a stagnant lid Venus would differ from that of present-day Earth since the mantle would not serve as an efficient inorganic carbon sink (Foley and Smye 2018; Höning et al. 2019). For a stagnant lid Venus, the concentration of carbon in the combined crust-atmosphere reservoir would increase with time, since mantle degassing supplies CO_2 to this reservoir on the long-term. This would ultimately enhance the rate of metamorphic crustal decarbonation and raise the atmospheric carbon budget. In addition, without compensating cloud albedo feedbacks (Way and Del Genio 2020), increasing solar luminosity makes it increasingly difficult to preserve liquid water on the planetary surface. Above a critical absorbed solar radiation threshold, all surface water evaporates and aqueous weathering ceases. Crustal decarbonation would result in a substantial rise of atmospheric CO_2 and surface temperature. Mantle outgassing would then release CO_2 over time, although the exact amount (in the range of 10^{19}-10^{20} kg of CO_2) is highly dependent on mantle composition and oxygen fugacity.

The time up to which aqueous silicate weathering and carbonate formation could keep a stagnant lid Venus habitable depends on the planetary albedo, which in turn depends on the atmospheric composition and the rotation rate of the planet. 3-D general circulation models by Way and Del Genio (2020) indicate that cloud feedbacks of a slowly rotating early Venus would result in a planetary albedo between 0.5 and 0.6, sufficiently high to potentially allow for liquid surface water on early Venus. Coupled atmosphere-interior models that include silicate weathering, carbonate burial and metamorphic decarbonation (Höning et al. 2021) indicate that for planetary albedos in this range an early stagnant lid Venus could have been habitable for up to 1 Gyr, followed by evaporation of water and dramatic rise of atmospheric CO_2 (Fig. 9).

Given the right conditions, the silicate weathering thermostat could have conceivably maintained habitable surface conditions on Venus for several hundreds of millions or even billions of years, even under a stagnant lid regime. Further work, however, is required to address uncertainties in order to better understand a potential habitable period. For example, the fraction of CO_2 released by metamorphic crustal decarbonation that makes it to the surface is a major uncertainty in coupled atmosphere-interior models. In addition, volatile partitioning during magma ocean solidification controls the post-crystallization atmosphere (Elkins-Tanton 2008; Massol et al. 2016; Hier-Majumder and Hirschmann 2017; Salvador et al. 2017; Nikolaou et al. 2019; Bower et al. 2022; Gaillard et al. 2022a,b). The composition of this initial atmosphere may in turn dictate whether the precipitation of water oceans, and therefore CO_2 drawdown and climate regulation via silicate weathering is permissible.

3 It Is Not Only About Volatile Exchanges: Other Important Mechanisms

This section highlights that while important, volatile exchanges are not the only way the atmosphere and the solid planet can affect each other, nor the only complex process that can affect long-term evolution and surface conditions on several scales.

3.1 Magnetic Fields

The links between magnetic field and other processes or layers of terrestrial planets are perhaps the canonical example of complex feedback cycles on a planetary scale. A terrestrial planet's self-generated magnetic field has its origins within a large volume of electrically conductive fluid—typically the liquid metallic core, where a magnetic dynamo effect can be powered by the convection of this electrically conducting fluid. Especially early in the history of Venus-sized planets, liquid silicates in a basal magma ocean may also convect vigorously enough to produce a dynamo (Ziegler and Stegman 2013; Stixrude et al. 2020; Blanc et al. 2020; O'Rourke 2020).

Core convection can be thermally or compositionally driven, e.g., by the exclusion of light elements from an inner core (e.g., Nimmo 2015) or the exsolution of buoyant element components from the liquid core into the mantle (O'Rourke and Stevenson 2016; Badro et al. 2016; Hirose et al. 2017). However, even chemical convection is ultimately caused by the heat loss from the core. Therefore, the existence of a dynamo is governed by the capacity of the mantle to extract and transport heat (see Rolf et al. 2022, for more details). In turn, the core offers an evolving boundary condition that affects mantle evolution, and conversely, evolving lower-mantle conditions may influence core dynamics. Heat sources in the core include radiogenic heating, gravitational energy and latent heat of crystallization of iron. At present, it is still uncertain whether the generation of a strong magnetic field (here we consider the intrinsic magnetic fields of Mercury and Ganymede to be weak) requires plate tectonics to operate in the mantle. In the Solar System at present-day, Venus, Earth and Mars might suggest so, but the number of samples is too small to draw conclusions: only present-day Earth exhibits plate tectonics and incidentally possesses a strong magnetic field. Plate tectonics is indeed more efficient at transporting and removing heat from the planetary interior than stagnant lid or single plate convection. While Venus shows no sign of a strong self-generated magnetic field at present-day, numerical simulations of its evolution suggest that, for similar parameters to Earth's (state of the core, composition, etc.), it is possible that Venus could have sustained a magnetic field until recently, even without long-term plate tectonics (Driscoll and Bercovici 2014; O'Rourke et al. 2019). However, according to thermal evolution models, a solid inner core may or may not exist at present day, because it depends on the current temperature and composition of the core which are unknown.

It is still unknown whether Venus had a magnetic field at any point during its history, but remnants of past magnetization could potentially remain to be discovered. Evolution of the planetary mantle and changes in convection regimes that could have occurred on Venus would directly affect the planet's ability to generate a magnetic field. On the opposite extent of the planetary domain, the magnetic field is known to affect non-thermal atmospheric escape (hydrodynamic escape remains unaffected), see Sect. 2.3.1. It has long been considered as a shield that protects a planet's volatile species from escape mechanisms (i.e., Lundin et al. 2007), which would have explained the differences between Earth, Mars and Venus. Earth would have been protected and remained habitable with plenty of water due to its magnetic field. Venus, lacking a magnetic field would have therefore lost its water to space. Mars, smaller and unprotected after an early dynamo phase of about 700 Myr, would have lost most of its primordial volatile inventory. This hypothesis has been challenged in the past decade, on the basis of the comparison of present-day estimates for escape rates from these three planets (Strangeway et al. 2010): numbers suggest that all three loss rates are comparable and do not directly reflect the alleged larger losses of Venus and Mars. Such an observation has been explained by considering the cross section of the area that interacts with incoming radiation for those three planets (Egan et al. 2019).

In the case of Venus and Mars, it consists of little more than the planetary diameter plus a relatively limited atmospheric extent. In the case of Earth, that would include the magnetosphere, leading to a much larger amount of energy intercepted by the planet, thus fuelling escape, despite the protection provided by the magnetic field to the lower atmosphere. Numerical studies (Kallio and Barabash 2012, in the case of Mars), and semi-analytical work (Gunell et al. 2018, for various planets) have found that some specific magnetic field strengths could be associated with peak escape rates that vary depending on the planet and solar wind pressure (Gronoff et al. 2020). In some cases the magnetic field could enhance atmospheric losses. For example, in the case of Venus with a magnetic field, losses increase by a factor of 2-5 (Gunell et al. 2018).

Other investigators (i.e., Tarduno et al. 2014) have pointed out possible reasons that said observations only partially describe the actual situation. They suggested that a possibly large (Seki et al. 2001) return flux of volatile species (downward) was not taken into account and led to overestimates of the escape rate for magnetized planets. They also raised the question of the response of the atmosphere to high solar wind pressure events that could affect overall escape differently on "shielded" or unmagnetized planets (Wei et al. 2012). Finally, they pointed at the difference between atmospheric loss and the loss of a key species: while Venus' atmosphere remains dense the loss of water by non-thermal mechanisms alone could have had tremendous implications for its climate and geodynamical history. However, water can also be lost neutrally through the hydrodynamic escape of its H atoms: for strong thermal escape, a magnetic field cannot prevent the desiccation of a planet.

The issue has certainly not been resolved yet: it remains unknown whether the presence of a magnetosphere would result in decreased or increased atmospheric losses. It is not a clear-cut choice between shielded planets versus unprotected ones. One certainty is that magnetic fields have a strong and complex influence on loss processes. For example, at Earth the ion escape processes are different from those at Venus and Mars due to Earth's intrinsic magnetic field. The geometry and strength of the field affect both how much of the planetary atmosphere is lost and where it is lost (Tarduno et al. 2014): for internally generated magnetic fields, losses seem to occur at the cusps of the field, near poles, while in the case of non-self-generated magnetic fields, losses could occur in the magnetotail. This seems to be supported by simulations for high-obliquity exoplanets around M-type stars (Dong et al. 2019) exhibiting higher loss when the cusps face the stellar wind. Moreover, atmospheric structure also plays a role in atmosphere escape as discussed in Sect. 2.3.1 and Salvador et al. (2022).

Present-day observation of the remanent magnetization from a past intrinsic magnetic field on Venus would affect our vision of both the interior of the planet and its atmospheric evolution (O'Rourke et al. 2019). No mission has yet conducted a magnetometer survey below the ionosphere, so a large amount of crustal remanent magnetism could await detection. A future mission that, for example, included an aerial platform dwelling in the clouds could detect regions of magnetized crust as thin as $\sim$200 m if the magnetization intensity is comparable to typical values observed on Earth, Mars, and the Moon.

3.2 The Role of Impacts

Meteoritic and cometary impacts are a common feature to all terrestrial bodies in the Solar System and probably beyond: all known planets feature impact craters to some degree (e.g., Zahnle et al. 1999; Neukum et al. 2001). Impacts are also catastrophic and sudden events that affect planetary evolution at many levels: interior, surface, atmosphere. In Earth's history they have been linked to major extinction events (e.g., Keller and Kerr 2014). They have been suggested as possible causes for changes in surface conditions and even interior dynamics. Micrometeorites ($<$2 mm) dominate the extra-terrestrial flux to present-day Earth (40,000 tons/year $\pm$50%, Love and Brownlee 1993; Peucker-Ehrenbrink and Ravizza 2000) and probably on other terrestrial planets (Frankland et al. 2017, for Venus, specifically). However, impacts of km-sized objects have much stronger consequences for planetary evolution (Schlichting et al. 2015).

Due to the young age of most of its surface, few studies have targeted the role and consequences of impacts on Venus. A young surface implies that, unlike the Moon, Mars or even Earth, most traces of impacts that occurred during Venus' history have been erased from the planet's surface. Only impacts occurring in the last $\sim$ 200 Ma–1 Ga are recorded (Bottke et al. 2016; McKinnon et al. 1997), which excludes the largest bodies (10s to 100s km radius) that are most likely to collide with a planet during its early history. Additionally, due to both its young surface and its thick atmosphere, Venus shows no sign of small craters ($<$1 km radius), as the small impacting bodies would vaporize or disintegrate during their atmospheric entry phase (Herrick and Phillips 1994). This shows that at least during the last $\sim$ 200 Ma–1 Ga Venus must have had a thick atmosphere comparable to today's. The current cratering record is mostly used for estimating surface ages via crater counting and to estimate how surface features are affected by surface alteration (erosion, tectonics) or volcanic flows (see Herrick et al. 2022, for a detailed discussion).

Collisions with other solid bodies can have a multitude of consequences depending on the temporal and spatial scale. Short-term effects include ejecta blankets, thermal radiation, blast-wave propagation in the atmosphere, crater excavation and earthquakes (Pierazzo and Artemieva 2012). Here, we will instead focus on long-term, global-scale consequences that affect the evolution of terrestrial planets. Impacts can (i) deliver material to a planet, such as gases (oxidizing and reducing gases, noble gases, etc.); they can (ii) erode its atmosphere, and finally they can (iii) directly affect the solid planet, sometimes down to its core.

All those consequences can vary depending on the specific conditions of the collision (angle, velocity, impactor size, mass and composition). Impactors have also been proposed to explain both Venus' obliquity and rotation rate since the 1960s (McCord 1968). Ward and Reid (1973) pointed out that an impactor less than 1% the mass of the Earth's Moon' can drive Venus retrograde if the planet was initially spinning slowly prograde due to solid body tidal dissipation. See Way and Del Genio (2020, Sect. 5) for a detailed discussion on Venusian tidal dissipation and the possible role of impactors on its rotational history.

3.2.1 Material Deposition

The evolution of a terrestrial planet will first be affected by the mass of material delivered/retained during collision. The impact geometry and physics are essential to determining the repartition (atmosphere, surface, interior, or loss) of material delivery (Golabek et al. 2018; Landeau et al. 2016). A portion of the impactor can be lost back to space as ejecta, depending on the size/mass of the impactor, impact velocity and incident angle, gravity, and atmospheric properties, as evidenced by hydrocode simulations (Shuvalov 2009). The remainder of the repartition of material delivered is complicated by the specific mechanism of the impact (e.g., impact angle). Part of the impactor can remain solid (Pierazzo and Melosh 2000), while part of it can undergo partial melting or be vaporized (Svetsov and Shuvalov 2015).

Mass delivery by impacts plays a major role during the accretion phase, when the planet is forming (Jacobson and Dobson 2022) During later evolution, this effect becomes less important as the incoming mass flux decreases exponentially while the mass of the planet is increasing, leading to only minor consequences for both the solid planet and its atmosphere (Sakuraba et al. 2019; Gillmann et al. 2020). It is possible that large impacts can modify the climate of planets, as has been studied in the case of Mars (Segura et al. 2002, 2008, 2012; Turbet 2018; Haberle et al. 2019). In the case of Venus' post-accretion history, a stray ≈ 10 km sized impactor could still deliver more water to the atmosphere of Venus than its present-day measured H_2O content, with possible consequences on planetary surface conditions/temperatures (Gillmann et al. 2016). Negligible amounts of CO_2 and N_2 would be delivered compared to Venus' present day atmospheric inventory. Additionally, large impacts can deliver enough reducing material to convert H_2O-CO_2 dominated atmospheres to transient H_2-CH_4-CO dominated atmospheres, resulting in enhanced H escape (Genda et al. 2017; Zahnle et al. 2020; Haberle et al. 2017). The effects of such large impacts on Venus' atmospheric evolution remain to be explored.

At a given mass delivered to the planet, the nature of impactors governs the deposition of material on rocky planets during collision processes. On present-day Earth, 80-86% of finds from the influx of meteorites is made of ordinary chondrites (from various collections, Harvey and Cassidy 1989; Krot et al. 2014; Dehant et al. 2019). Carbonaceous chondrites are a rare ($\approx 4\%$ of the finds) type of chondrite (Dehant et al. 2019). Despite their name, only the some subsets of carbonaceous chondrites (namely, CM, CR, and CI) are significantly enriched in carbon and H (up to a few percent mass; Grady and Wright 2003, Marty 2012, Pearson et al. 2006) relative to ordinary chondrites (Gounelle 2011). Even rarer ($\approx 2\%$) are enstatite chondrites (EH, EL), that are reduced specimens. They are comparatively dry, having formed in the inner Solar System. However, studies revealed that enstatite chondrites can contain significant amounts of H (≈ 0.1-0.5 wt.% water equivalent; Pepin 1991, Muenow et al. 1992, Piani et al. 2020).

Beyond simple material delivery, impactor chemistry could prove critical in order to estimate the composition and state of post-impact atmospheres and mantles. It plays a key role in determining what species are released during impact outgassing (depending on how reduced the atmosphere is), and how material (such as iron) delivered by the impactor interacts with the planetary atmosphere. As thermochemical equilibrium modelling has shown (Hashimoto et al. 2007; Schaefer and Fegley 2010), impacts could lead to the outgassing of reduced species (H, CH_4, CO) for all types of chondrites (Hashimoto et al. 2007), except the most wet carbonaceous chondrites (Schaefer and Fegley 2010). However, Lupu et al. (2014) show that post giant impact atmospheres would most likely be dominated by water and CO_2.

Nevertheless, it has been suggested that reaction with iron metal could consume up to several water oceans and produce tens of bars of hydrogen (Genda et al. 2017). However this is debated, as another recent study (Citron and Stewart 2022) suggests that most of the iron delivered by late accretion impactors is deposited in the crust and mantle which implies that only a fraction of the projectile iron can react with water and CO_2 on the planet. Consequently, it is unlikely that this mechanism could trap entire pre-existing Earth-like oceans of water. Therefore, both the chemistry and the physics of the impact should ideally be considered together to assess how they affect the distribution of material between surface and interior.

3.2.2 Atmospheric Erosion

A second effect of collisions is the loss of volatile species from the planetary atmosphere. Three main mechanisms are thought to cause atmosphere erosion during a meteorite impact (Pham et al. 2011):

1. Direct ejection of the atmosphere via atmospheric compression by the impactor upon entry, possibly including an aerial burst if the body disintegrates in an explosion (Shuvalov et al. 2014)
2. Hot vapor plume shock-wave above the impact location involving the vaporized projectile and target body
3. Interaction of high speed ejecta with the atmosphere; particles are ejected into the atmosphere, accelerating and heating atmospheric molecules.
4. For giant collisions a further mechanism has been suggested: pressure waves cause vertical ground motion at the antipode of the impact location, accelerating atmospheric particles above the escape velocity (Genda and Abe 2003; Schlichting and Mukhopadhyay 2018).

At first, simple parameterizations were developed to estimate atmosphere losses, such as energy calculations or simple geometrical models (Cameron 1983; Melosh and Vickery 1989). Early energy-based considerations suggested that large, possibly complete atmospheric losses were possible, if the impact was sufficiently large and/or fast enough. Considering the geometry of the collision, for example by limiting maximum escape to the portion of the atmosphere above a plane tangent to the planet at the impact location (the tangent plane model) reduced estimations to a much lower portion of the planetary atmosphere (Melosh and Vickery 1989; Pham et al. 2011; Schlichting et al. 2015). Other approaches rely on full numerical hydrocode simulations of the entire impact process, from entry to vapor plume, such as those simulations performed with the SOVA hydrocode (Shuvalov 2009). Hydrocode simulations contain limitations for impactor bodies with sizes larger than the thickness of the planetary atmosphere. For that reason, estimates of the loss through giant impacts are mostly done using the tangent plane model and ground motion estimates. This type of collision involves complex material redistribution of the solid bodies involved and can have important consequences for the atmosphere.

Again, most studies are not targeted directly at Venus, but usually cover multiple bodies or are general enough to produce laws including planetary and atmosphere masses as parameters. In general, lower mass planets such as Mars tend to undergo more efficient atmospheric erosion than more massive ones, like Earth or Venus (Pham et al. 2011). Hydrocode results indicate that large impacts ($R > 10$ km, up to a few 100s km) are relatively inefficient at removing planetary atmospheres, especially when volatile delivery is taken into account (see above), and yield results implying lower losses (but of the same order) than simple tangent plane models for Earth and Venus (Shuvalov 2009; Shuvalov et al. 2014; Gillmann et al. 2016). Smaller collisions are even less efficient when only single events are considered, but erosion can then be considerably increased by airburst mechanisms (Shuvalov et al. 2014). Calculations by Schlichting et al. (2015) suggest that the bulk of atmospheric loss by impacts could be attributed to swarms of small impactors (1 km < R < 10 km), as they are sufficiently numerous to have a significant cumulative effect. This is because most of their energy is delivered to the atmosphere and used to power escape. Giant impacts that can theoretically cause very efficient erosion are rare. They require the right set of conditions: large mass relative to the target body (above 40%) and high impact velocity (a few times mutual escape velocity). Such giant impacts could remove 50-90% of a preexisting atmosphere. However, slower, less massive collisions are more likely and would cause the loss of up to 20% of early atmospheres (Schlichting et al. 2015; Genda and Abe 2003). Studies modelling the relative atmospheric loss and delivery by impacts can sometimes differ by orders of magnitude, due to the use of different equations of state and dynamical models (Hamano and Abe 2010; Melosh and Vickery 1989; Newman et al. 1999; Shuvalov 2009; Svetsov 2007; Vickery and Melosh 1990).

3.2.3 Energy Transfer

The third main way for impacts to affect planetary evolution revolves around the large amount of energy they can transfer to the solid part of the target. During the collision event, the kinetic energy of the impactor is transferred to the interior of the planet. A shock wave is generated that propagates hemispherically away from the point of impact, decreasing in amplitude with increasing distance (O'Keefe and Ahrens 1977; Melosh and Kipp 1989). Shock pressure followed by decompression cause heating inside the planet. The temperature increase is proportional to the maximum shock pressure (Pierazzo et al. 1997), possibly leading to melting and vaporization of lithospheric and even mantle material.

Under the impact location, the shock pressure is uniform in a spherical isobaric core, leading to an isothermal central thermal anomaly, where planetary material is heated. The size of the isobaric core is governed by the that of the impactor, its radius being about 1-2 times the impactor radius (Croft 1982; Monteux et al. 2007). Outside the isothermal zone, the thermal anomaly's amplitude decreases rapidly with distance from the isobaric core (Pierazzo et al. 1997; Monteux et al. 2007). The temperature increase depends on the impact velocity, as well as the size of the target (but not the impactor's; Monteux et al. 2007). As a consequence, larger target bodies experience higher impact-generated temperatures. One can expect an amplitude of a few hundred K for a Mars-sized planet, but upward of 1000 K for Venus or Earth (Gillmann et al. 2016; Ruedas 2017). The thermal anomaly geometry is also affected by the impact angle (Bierhaus et al. 2012), and more energy is transferred in a head-on collision than a grazing impact.

As a consequence, fast, head-on and larger impacts on large planets (like Venus) are more likely to have long term consequences (Gillmann et al. 2016); low energy events have mostly local or short-term consequences on the global-scale picture (although they could still affect surface conditions and habitability). Antipodal heating has been suggested to occur due to a convergence of pressure waves on the other side of the planet, but effects have been shown to be limited to displacement and fracturing, with no outright melting, outside of a giant impact scenario (Melosh 2000; Meschede et al. 2011). The models briefly discussed here are inaccurate for larger events (radius above a few 10s to 100s of kilometers; Manske et al. 2018), and should not be used for giant impacts (Nakajima and Stevenson 2014, 2015; Cameron and Ward 1976; Canup 2004; Melosh 1990; Pahlevan and Stevenson 2007). For these larger impacts, new scaling laws have been developed recently (Nakajima et al. 2021).

An additional effect of impacts is the acceleration of target material, by the energy transfer and shock wave, up to velocities sufficient to displace it over long distances or even eject a small portion into space. This causes the excavation of a crater and projection of ejecta (vapor, for the larger planets, solid and molten rocks from both the planet and the impactor). Impactors larger than ≈ 100 km lead to the redistribution of mass several times the impactor mass over the surface of the whole planet (Shuvalov et al. 2012).

These two effects (ejecta and thermal anomaly) can have consequences on the evolution of the planet and its mantle. First, given the low thermal conductivity of ejecta, it has been proposed that layers created by their accumulation from several impacts could insulate the interior of planets and affect their thermal evolution (Rolf et al. 2017, in the case of the Moon). Melting and outgassing of the lithosphere (for smaller impacts, $R < 100$ km) or the mantle (for larger ones, $R > 100$ km) are other common effects of impacts. In that latter case, depletion of the mantle can be expected: large impacts could contribute to removing volatiles from planetary mantles (Davies 2008; Gillmann et al. 2016). In particular, a succession of large impacts ($R \approx 100$s km) could be responsible for the efficient depletion of the upper mantle of Venus, especially in the absence of rehydration mechanisms (Gillmann et al. 2016, 2017). All sizes of impactors, however, contribute to outgassing and the release of volatiles from the mantle or lithosphere in the atmosphere, with potential effects on the climate. The impact-generated melting of crustal carbonate deposits (if they were formed in large quantities at some point of Venus' history) could result in the release of considerable amounts of CO_2, for example (see Sect. 4).

Depending on the size of the impacts, even mantle convection can be affected. The volume of mantle heated by the collision is the governing factor for long term consequences on mantle evolution, therefore, larger thermal anomalies will affect the mantle more strongly. Small impacts that do not penetrate through the lithosphere into the convecting mantle have negligible consequences on the mantle. On the other hand, large impactors can generate a thermal anomaly that reaches the deeper convecting parts of the mantle, and thus affect convection processes. Because it is hotter than the surrounding material, the material forming impact-induced thermal anomalies is positively buoyant and exhibits lower relative viscosity. The thermal anomaly tends to rise and flatten under the lithosphere, causing melting and resurfacing (Watters et al. 2009; Roberts and Arkani-Hamed 2014; Gillmann et al. 2016; O'Neill et al. 2017; Padovan et al. 2017; Ruedas and Breuer 2017; Rolf et al. 2017; Borgeat and Tackley 2022). As the thermal anomaly then spreads laterally over hundreds of thousands to millions of years, the upper mantle is pushed away from the impact location. This effect may be enhanced on larger planets like Venus, due to the larger thermal (and therefore viscosity) contrast between the upper mantle and the anomaly.

Given a sufficiently large impactor (a large thermal anomaly), hot material may accumulate at an antipodal position, where it forms downwellings or subduction events (Gillmann et al. 2016; O'Neill et al. 2017; Borgeat and Tackley 2022). Thus, large impacts, on an otherwise stagnant lid or single plate mantle, can break the lid and shift tectonics toward a mobile lid or plate-like regime. However, the convection regime change is temporary and reverts to the original stagnant lid when the impact flux diminishes or stops (Gillmann et al. 2016; Borgeat and Tackley 2022, based on numerical models). Impact-triggered subduction events also affect the mixing of the mantle (in particular the upper mantle) by recycling crust into the interior. On the long term, especially strong impacts can affect the mantle down to the core-mantle boundary.

The consequences of impacts on the mantle can eventually affect the core of the planet: O'Neill et al. (2017) suggest that vigorous impact-driven convection could strengthen the magnetic dynamo. On the other hand, the thermal anomaly of a sufficiently large collision (impactor radius upward of 800 km) can even reach down to the core-mantle boundary. This would place a high temperature layer on top of part of the core, possibly reducing the heat flux out of the core. In turn, core convection would shut down, and magnetic field generation would cease on long timescales ($\approx$1 Gyr; Arkani-Hamed 2009, Roberts and Arkani-Hamed 2014).

4 The Evolution of Venus: Shaped by a Multitude of Mechanisms

The conditions on a planet are not static, but change with time as evidenced throughout Earth's history. In turn, changes in mantle and atmospheric states affect the mechanisms at work, possibly creating positive or negative feedbacks. Due to its thick atmosphere and seemingly active mantle, Venus is a place of great interest to study the mechanisms and their interactions that have shaped the evolution of the planet and its surface conditions. Changes in the convection regime are one of the prominent examples of feedback loops, due to their scale and possibility for far-ranging consequences. By itself, mantle dynamics change with time. But the convection regime is also affected by external conditions, such as the distribution of water between the atmosphere, the surface and the interior, or surface conditions (pressure, temperature, liquid water availability). Different convection regimes imply variations in the interactions with the atmosphere. For example, it has been suggested that a mobile lid regime favors liquid surface water, while surface temperature variations can lead to a change from stagnant lid to mobile lid convection or vice-versa (Lenardic et al. 2008; Gillmann et al. 2016). Conversely outgassing is affected by the convection regime and vigor. Degassing affects surface temperature via the greenhouse effect and this, in turn, could affect mantle processes such as convection style and velocity, possibly creating a cycle of interactions. Further examples of this include the recycling of surface material into the mantle and the question of formation, stability and recycling of carbonates.

4.1 Possible Evolutionary Pathways

Improved understanding of atmosphere-interior couplings could shed light on the history of water on Venus, and as a result, on the planet's long-term history. Even 35-50 years after pioneering works on Venus' volatile history coupled with solar and/or geological processes (e.g., Ingersoll 1969; Rasool and de Bergh 1970; Walker 1975; Kasting and Pollack 1983; Kasting et al. 1984; Kasting 1988) there is still no consensus on the evolutionary pathways followed by Venus. We still do not know whether Venus was ever habitable in the past or if its surface has always been far too hot. Likewise, little is known about the past history of its mantle convection regime and if it ever supported a mobile lid or even Earth-like plate tectonics before its apparent present-day stagnant lid/single plate state. In our current approach to understand Venus's past, the fate of water particularly unites most of the unknowns. This is because it is intimately linked to fundamental questions such as surface conditions, past atmospheric escape, and interior convection, structure and composition. Based on that, three main end-member scenarios describe what could have led Venus to the hot and dry observable current state, assuming similar building blocks to the Earth's (see Fig. 10).

In a first scenario (the "dry Venus" scenario, top path on Fig. 10), Venus was desiccated at the onset of its evolution, as early as the magma ocean phase. This scenario is described in Salvador et al. (2022), and identifies Venus as a Type-II planet, following the Hamano et al. (2013) classification. In contrast to Type-I Earth's short-lived ($\sim$1 Myr) magma ocean, Venus' vicinity to the Sun would result in a slow ($\sim$100 Myr) magma ocean solidification.

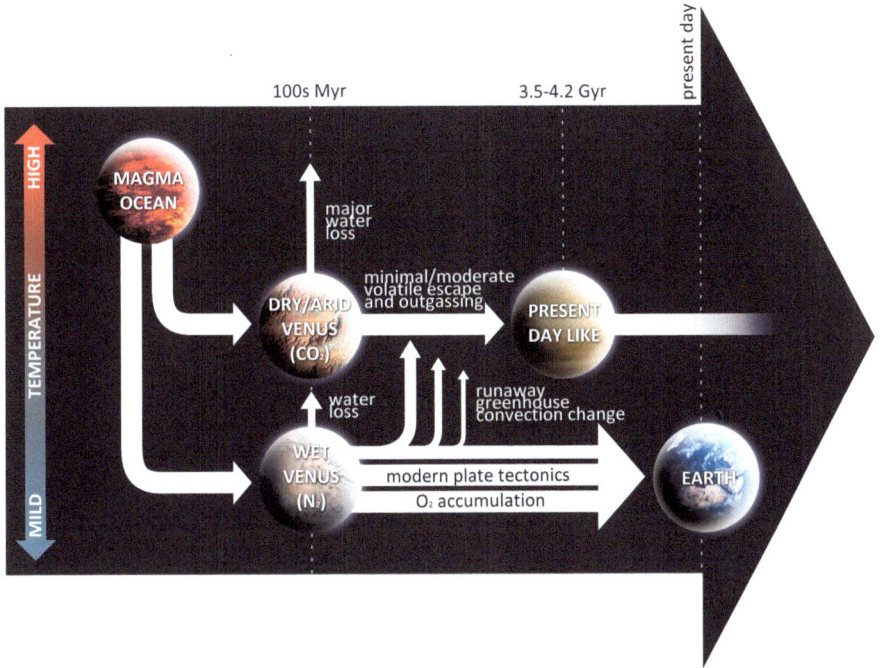

Fig. 10 Current understanding of the extreme tentative scenarios for the evolution of Venus' surface conditions, from its origins to present-day, compared to Earth. On top, Venus lost its surface water early on (desiccated Venus, or stifled outgassing scenarios), while on the bottom evolution, it evolved closer to Earth, retaining a larger portion of its water inventory, until its climate was destabilized. For now, both evolutionary pathways remain consistent with our global knowledge of the planet. Only general evolution trends are represented, Earth-related processes (modern plate tectonics and O_2 accumulation) are not attributed a specific time and only included for comparison with Venus

The associated long-lived steam atmosphere overlying the molten surface would slowly lose its water en route through early intense hydrodynamic escape (Gillmann et al. 2009), and its desiccation would thus preclude water condensation at the surface (Walker 1975; Matsui and Abe 1986; Lebrun et al. 2013; Turbet et al. 2021). Remaining atmospheric oxygen could then be consumed by oxidation of the solid surface or magma ocean. Upper mantle rocks and basalts redox state could help us test the likelihood of this scenario. Overall, it induces a mostly dry planet (inside and outside). After only a few hundred million years, Venus would look rather similar to its present-day state, with a CO_2 and N_2 atmosphere that would slowly grow to its present-day level, through moderate exchanges (moderate outgassing and loss), in a rather straightforward evolution. In such a scenario, Venus' habitability would be ruled out early on and a liquid water ocean would have never existed.

In a second possible scenario, Venus could have followed a Type-I planet, fast cooling magma ocean evolution (Hamano et al. 2013), leading to a temperate Venus (bottom path on Fig. 10). In such a case, Venus would have cooled down fast enough to evolve in a similar way to Earth and allowed condensation of water on its surface early on, resulting in an Earth-like climate for an unspecified period of time (Way and Del Genio 2020). Yet, some cooling mechanisms are required to buffer Venus high incident solar flux and allow for such a rapid cooling. The development of highly reflective clouds during the magma ocean cooling could be an example of such mechanisms (Salvador et al. 2017; Way and Del Genio 2020). While it has been shown that a concentration of clouds in the sub-stellar region could maintain temperate climates at insulation even higher than experienced by early Venus (Yang et al. 2014), thus favoring this temperate Venus scenario, details of 3D atmospheric circulation and resulting clouds distribution might prevent early ocean formation on Venus. To date, the lack of a self-consistent magma ocean-atmosphere coupled model accounting for realistic clouds does not allow one to make a definitive statement. It is discussed in length in Salvador et al. (2022) and Westall et al. (2022). As on Earth, temperate conditions could have favoured the extraction of CO_2 from the atmosphere by the formation of carbonates and possibly a limited carbon cycle. This would have allowed for a N_2 dominated atmosphere to sustain low CO_2 partial-pressures for long periods of time and temperate surface conditions akin to Archean Earth. At a later, but unknown time (between 1-4 Ga), geological events, such as the formation of large igneous provinces (Way and Del Genio 2020; Way et al. 2022a), could have released magmatic CO_2 into the atmosphere and, depending on the planetary albedo response, possibly started a late runaway greenhouse (already suggested by Pollack 1971), ultimately causing the decomposition of crustal carbonates and leading to the present-day conditions, as water is lost. External factors such as the increase of the solar luminosity with time or the occurrence of a large impact on a carbonate-rich crust could also be responsible for triggering a late runaway greenhouse.

In a third scenario (stifled outgassing) the interior of Venus would have retained its water initially due to inefficient outgassing of the magma ocean (Ikoma et al. 2018; Solomatova and Caracas 2021; Salvador and Samuel 2021) and magma ocean redox conditions (an oxidized magma ocean, Gaillard et al. 2022a). These conditions would have been sustained in its later evolution possibly due to high surface pressures (Gaillard and Scaillet 2014) preventing the outgassing of large quantities of H_2O. An initially wet magma ocean could cause part of its H content to partition into the core, leading to oxidized conditions (Ringwood and Anderson 1977; Okuchi 1997; Tagawa et al. 2021) and ultimately a FeO-rich (and water-poor) mantle. In this scenario, Venus would not have lost most of its water to space but could not outgas it, either due to its chemical composition/redox state or due to surface pressure. Additionally, both carbon and nitrogen components of the atmosphere would be a direct result of the early magma ocean outgassing (Gaillard et al. 2022a), with possibly a minor component outgassed later.

4.2 The Role of Water During Evolution

As illustrated by the scenarios mentioned above, the fate of water is decisive for the evolution of Venus (and has been recognized as such since early works, e.g., Ingersoll 1969; Walker 1975). It should be noted that the history of water on Venus is intimately tied to Venus' atmosphere-interior redox evolution. If water has not been lost during the magma ocean solidification and if cloud feedback maintained a high albedo due to Venus' slow rotation (Yang et al. 2014; Way et al. 2016), the longevity of temperate surface conditions is constrained by the time required to lose leftover steam and remaining O after the runaway greenhouse initiation. Very recent habitability may not provide sufficient time for water/hydrogen and oxygen loss. But even for dry Venus scenarios, H loss to space is expected to leave behind substantial oxygen (Chassefière 1996a; Gillmann et al. 2009; Schaefer et al. 2016). Reaching the oxygen-free modern atmosphere may require efficient dry crustal oxidation (any oxidation of the crust that is not attributable to volatile degassing or water-rock reactions, Krissansen-Totton et al. 2021a). As noted in Sect. 2.4.1, one interesting possible mechanism for this surface exchange is oxidation of dust particles from explosive volcanism, which could provide a significant oxygen sink (Warren and Kite 2021). Furthermore, this mechanism could be enhanced if the total atmospheric mass was small during a temperate phase on early Venus, because in such a case, the lowered pressure overburden would potentially increase rates of explosive volcanism.

Another consequence of suppressed water outgassing on Venus would be the formation of a relatively water-rich lithosphere. Such a lithosphere would be rheologically weaker due to the presence of H_2O (Wang et al. 2020), which might be inferred from future observations. In addition, a weak lithosphere might drip or delaminate more easily and result in a thinner crust than would be expected otherwise (Anderson and Smrekar 2006; James et al. 2013). As in terrestrial seafloor basalts (Holloway 2004; Holloway and PA 2000), interaction of remaining H_2O with ferrous Fe in melts could produce magnetite (Fe_3O_4) and H_2 (Eq. (10)), that could be released into the atmosphere through physical or chemical weathering of lavas.

In fact, the fate of water is central to most aspects of the evolution of Venus. Any evidence of the history and state of water through the ages would not only improve considerably our understanding of the evolution of Venus' climate, but of other linked processes as well, both at the surface and in the interior of the planet. Tracking water has thus been proposed as a tool to investigate Venus past and its possible states (Gillmann et al. 2020; Warren and Kite 2021). It has been used to constrain possible accretion scenarios (see Fig. 11), the past climate or more recent changes. Central to this approach is the estimation of volatile species fluxes over time (for non-thermal escape for instance, Persson et al. 2020). For this reason, it depends on our understanding of a wide array of mechanisms. In some aspects we have made progress in the last decades, but the unconstrained parameter space is still large. It is also important to keep in mind that processes evolve with time. Different past conditions could have influenced the way they affect volatile species fluxes. Therefore each step in understanding mechanisms at work in the evolution also informs further attempts to illuminate Venus' history.

4.3 The Link with Origins

The question of very early Venus has long been identified as a major aspect determining the planet's evolution (e.g., Kasting and Pollack 1983; Kasting et al. 1984). Building upon this, recent works have shown that the evolutionary pathway that a planet like Venus can follow can be decided by its "initial" era, or rather the conditions of its transition from the magma ocean (MO) phase to the solid planet phase. The time required for MO solidification (Hamano et al. 2013), and the early loss of water (Gillmann et al. 2009) could be responsible for the distinct evolutionary paths of Earth and Venus. Climate simulations of early Venus have also demonstrated that the initial state of water (liquid vs. vapor) has a strong influence on the post-MO conditions (Way and Del Genio 2020; Turbet et al. 2021). Jacobson and Dobson (2022), Salvador et al. (2022) and Avice et al. (2022) offer a more comprehensive picture of primitive Venus.

Initial distribution of water between the mantle and atmosphere is controlled in part by the crystallization of the post-accretion magma ocean. Given its insolation, early Venus could be located at the boundary between dichotomous magma ocean evolution and lifetimes (Hamano et al. 2013; Salvador et al. 2017; Lebrun et al. 2013). If the insolation absorbed by early Venus was less than the runaway greenhouse limit, then the magma ocean likely persisted for a few million years until Venus had cooled sufficiently for water to condense at the surface as occurred for Earth. However, if the insolation absorbed exceeded the runaway greenhouse threshold, then a magma ocean could have persisted for around 100 Myr or more, until almost all water was photodissociated during the slow solidification process, with most hydrogen lost to space while the oxygen was partially lost to space and partially

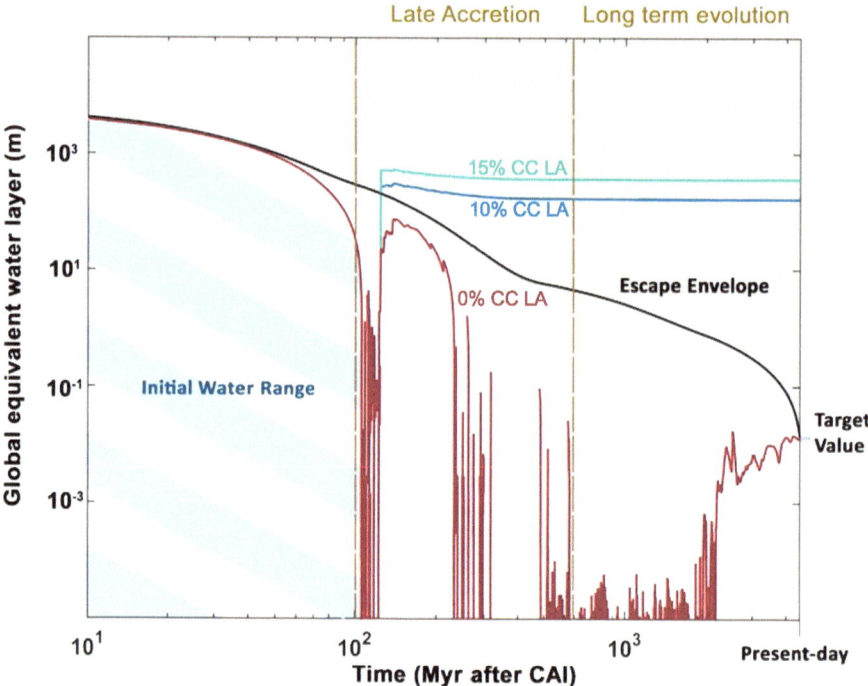

Fig. 11 Evolution of the amount of water at the surface of Venus since its accretion up to present-day, in equivalent global ocean layer depth (adapted from Gillmann et al. 2020). The simulation includes the effects of impacts during late accretion (erosion, delivery, energy transfer), atmospheric escape and volcanism. Three compositions are proposed for the late accretion: fully composed of relatively dry enstatite chondrite-like material (Piani et al. 2020, red line with 0.1% H_2O, 0.4% CO_2 and 0.02% N_2); with 10% wetter material akin to carbonaceous chondrites (blue curve noted CC; by mass, 8% H_2O, 4% CO_2 and 0.2% N_2); and with 15% CC (light blue curve). Only the drier late accretion is consistent with the present-day water inventory in the Venus atmosphere (target value). The escape envelope is the maximum amount of water that could have been present in the atmosphere at a given time and still be consistent with present-day measurements in the absence of all mechanisms other than atmosphere escape. Atmospheric escape is calculated based on Kulikov et al. (2006), and is higher than estimates from Persson et al. (2020). The initial water range illustrates the uncertainty on the water inventory during accretion that is not resolved by the model. Time is measured from the formation of calcium-aluminum rich inclusions (CAIs)

consumed by the MO (e.g., Schaefer et al. 2016) and by reducing atmospheric gases. Which of these scenarios occurred depends on the radiative and cloud feedbacks in the atmosphere.

Some GCM modeling suggests that cloud feedback on a slowly rotating early Venus could have maintained a temperate climate due to high subsolar cloudiness leading to high albedo values (Yang et al. 2014; Way et al. 2016; Way and Del Genio 2020). However, it has been argued that a hot magma ocean on Venus could not plausibly transition to a temperate state because a thick steam atmosphere would inhibit day-side cloud formation whereas night-side clouds would limit radiative cooling, thereby preventing the condensation of surface water (Turbet et al. 2021). This would ultimately maintain a hot climate and lead to a "dry Venus" scenario (Type-II planet in Hamano et al. 2013), when early loss of water is taken into account. The major difference between these two modeling efforts is the initial state of the water: liquid surface layer (Way et al. 2016; Way and Del Genio 2020), or in the atmosphere as steam (Turbet et al. 2021).

In addition to its dynamics and resulting radiative transfer and clouds distribution, the initial composition of the atmosphere is of fundamental importance to decipher whether early Venus was wet or dry. It is influenced by both the magma ocean chemical composition and thermal evolution during the crystallization which control the partitioning of water and other volatile species between the interior and the atmosphere. The cloud feedback invoked to keep early Venus hot and ultimately dry (Turbet et al. 2021) assumes an early H_2O-, H_2O–CO_2-, or H_2O–N_2-dominated atmosphere (or all three species). Such oxidizing, H_2O–CO_2-dominated early atmospheres are generally considered (e.g., Elkins-Tanton 2008; Hamano et al. 2013; Lebrun et al. 2013; Massol et al. 2016; Salvador et al. 2017; Nikolaou et al. 2019; Bower et al. 2019) in analogy to modern volcanic gas composition (e.g., Holland 1984), resulting from the equilibrium with their mantle sources and with hot surfaces of different compositions (Lupu et al. 2014). Yet, despite geological evidence, the origin of life on Earth seems to require a highly reduced (rich in CH_4, H_2, and CO) early atmosphere (e.g., Urey 1952; Miller and Urey 1959; Johnson et al. 2008). Alternative scenarios involving impact degassing, volatiles trapping in solidifying mantle or varying the magma ocean oxygen fugacity have thus emerged and been proposed to match such requirements and enlarged the possible range of early atmosphere compositions (e.g., Hashimoto et al. 2007; Hier-Majumder and Hirschmann 2017; Zahnle et al. 2020; Solomatova and Caracas 2021; Gaillard et al. 2022a; Bower et al. 2022). Extensive studies remain to be done in order to constrain how the dynamics of such atmospheres would affect the heat and clouds redistribution and thus influence the magma ocean cooling and resulting surface conditions.

Incidentally, some early atmosphere composition possibilities also give rise to the "stifled outgassing" scenario for Venus, where the early outgassing of CO_2 is not followed by water release at the end of the magma ocean phase. Instead, in the resulting C-poor, deep magma ocean, H incorporation (due to H being increasingly siderophile at higher pressures, while C is less so) into the core could lead to desiccation of the magma ocean (Gaillard et al. 2022b) and formation of a FeO-rich mantle, through oxidation of remaining Fe-metal in the mantle. The possibly oxidized conditions on Venus (relative to Earth) due to H-loss (Zahnle et al. 2013) may have played a role in the growth of a dry magma ocean atmosphere on this planet compared to an H-bearing one on Earth (Gaillard et al. 2022a).

Furthermore, some dynamical aspects might also favor an incomplete to minimal degassing scenario (Ikoma et al. 2018). Indeed, while the vigorous convection has been thought to promote efficient mantle mixing and thus rapid early degassing, a recent study showed that the convective magma ocean dynamics and detailed patterns may significantly reduce the amount of magma reaching the surface, thereby significantly reducing the outgassing rates (Salvador and Samuel 2021). In addition, extreme magma ocean conditions are still out of the range of the experimental data constraining volatile behavior in magma and large uncertainties remain. It should be noted that in the above scenarios it is usually assumed that Venus had at least one global and deep magma ocean. On Earth, this likely resulted from the moon-forming impact. For Venus, however, it is not clear whether there was a global and deep magma ocean, resulting from the combination of additional heat sources, or rather local magma ponds that contributed to the early atmosphere (Salvador et al.). In the latter case, H incorporation into the core is less likely.

On the other hand, impacts could bring a wide array of volatiles (both possibly reduced and oxidized) to early atmospheres, making the primitive atmospheric conditions even more complex. Additionally, one has to reconcile these results with the case of Earth, where evidence suggests that surface liquid water was present as early as 4.3 Ga ago (Mojzsis et al. 2001). The composition of the early atmosphere will also be determined by the redox state of the terminal magma ocean, which is poorly constrained for both Earth and Venus. Recent experimental work on iron speciation in silicate melts has suggested that the terminal magma ocean atmospheres of Earth and Venus could have been CO-dominated with a comparatively minor H_2-H_2O component (Sossi et al. 2020; Bower et al. 2022). In such a case, one could hypothesise that oxygen remnants from hydrodynamic escape could oxidise CO to produce the more recent CO_2 atmosphere. It is also not clear exactly what types of clouds would be produced in such atmospheres (e.g., Herbort et al. 2021) nor their feedbacks on the climate.

The cloud dynamics of such atmospheres have yet to be explored, but the rapid escape of H to space would presumably result in comparatively cool CO-dominated atmospheres (CO is a poor greenhouse gas). The possible radiative and dynamical effects of photochemical hazes (He et al. 2018) or atmospheric chemical reduction caused by giant impacts (Zahnle et al. 2020) have also yet to be fully investigated.

Figure 12 illustrates the possible dichotomous evolution of water on Venus within a coupled atmosphere-interior model starting with a fully molten mantle (Krissansen-Totton et al. 2021a). The differing evolutionary scenarios are attributable to assumed initial volatile inventories and fixed albedo. The subplots on the left show the long-lasting hot climate and resulting dry surface conditions scenario described above (Turbet et al. 2021). Here, cloud feedback does not allow for buffering the excess (above the runaway greenhouse threshold) of received solar insolation which maintains Venus in a runaway greenhouse state, thus preventing it from cooling down to temperate climates where ocean condensation could occur i.e. a low albedo is assumed. Atmospheric water vapor is gradually lost to space during the slow magma ocean solidification and the atmosphere is mostly desiccated. Once the surface has solidified, it then evolves in a similar way to the dryer models described in Gillmann et al. (2020) (see Fig. 11). The panels on the right illustrate a post-magma ocean wet scenario where liquid water condenses at the surface immediately after magma ocean crystallization, and CO_2 cycle/silicate weathering feedback maintains a temperate surface i.e. a high albedo is assumed. Eventually, increasing solar luminosity triggers a transition to runaway greenhouse, carbon degassing from the mantle returns CO_2 to the atmosphere (crustal decarbonation is neglected), and all remaining water is lost to space. Here, the longevity of surface liquid water is up to 3.5 Gyrs because both plate tectonics and a temperature-dependent silicate weathering feedback are assumed. This differs from the shorter-duration wet scenario illustrated in Fig. 9 which assumes a stagnant-lid early Venus and CO_2-dependent silicate weathering. Both evolutionary scenarios (wet or dry post-magma ocean surface conditions) can explain modern Venus conditions, which is to say they recover the observed bulk atmospheric composition (Fig. 12, bottom panels), but also produce reasonable values for atmospheric ^{40}Ar & ^4He (Sect. 2.2.2), and plausible surface heat flow values (Krissansen-Totton et al. 2021a). This highlights the need for new discriminants (see below).

4.4 Feedback Processes

The role of models based on physics and experimental results becomes especially important when considering the interactions between various mechanisms governing the evolution of terrestrial planets. This has been already highlighted at various points in this chapter, for example with uncertainties related to the interpretation of the atmospheric D/H ratio. Understanding the evolution of Venus (and other terrestrial planets) becomes even more complex if feedback processes are taken into account. Those are interactions between several mechanisms or domains that affect one another and can result in positive or negative control of the given processes. As an example of positive retroaction, we have already described (Sect. 2.4.3) that in a scenario where carbonates were formed, destabilization of the carbonate reservoir by increasing temperatures in the lithosphere could cause carbon outgassing,

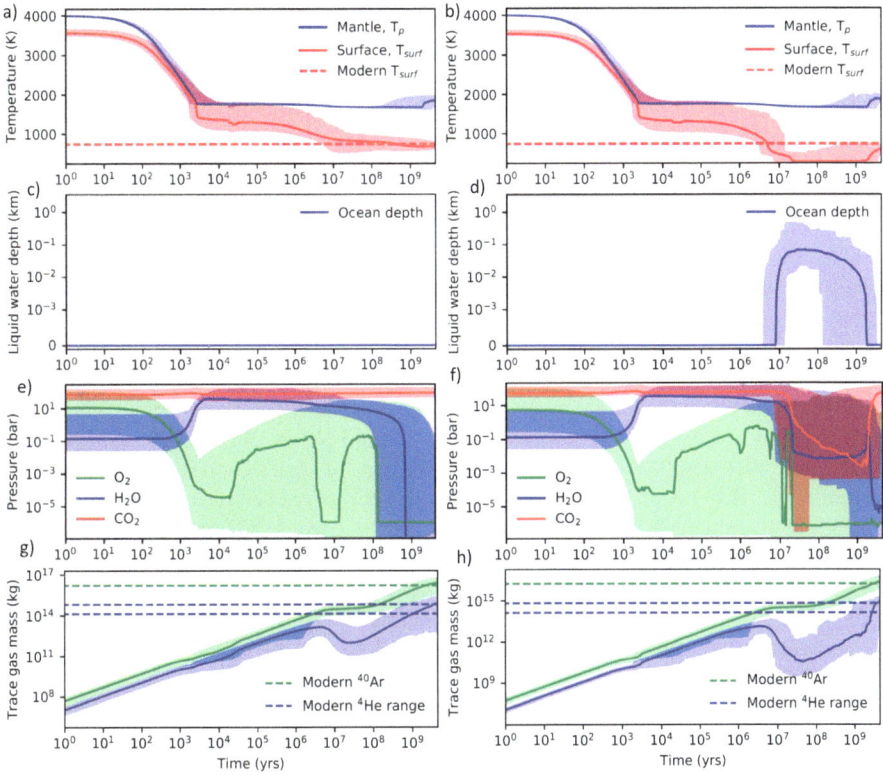

Fig. 12 Two scenarios for the coupled atmosphere-interior evolution of Venus. Solid lines are median values, shaded regions denote 95% confidence intervals, and the time evolution spans from post-accretion magma ocean to the present-day (4.5×10^9 yrs). Left-column plots show low albedo model runs that recover the modern Venus atmosphere and were always dry (no surface water condensation), whereas right-column plots show high albedo model runs that recover modern Venus and were temporarily wet. Subplots (**a**) and (**b**) show the evolution of mantle potential temperature (blue) and surface temperature (red), with the modern mean surface temperature denoted by a dashed red line for reference. Subplots (**c**) and (**d**) show the global average liquid water depth (blue). Subplots (**e**) and (**f**) show the evolution of atmospheric composition, including partial pressure of carbon dioxide (red), oxygen (blue), and steam (green). Subplots (**g**) and (**h**) show the accumulation of atmospheric ^{40}Ar (green) and ^{4}He (blue) from degassing. Dry scenarios show a transition from magma ocean to runaway greenhouse (**a**), and the gradual loss of water to space (**e**). Transiently habitable model runs experience temperate conditions with $\approx$100 m global oceans (**d**) where atmospheric CO_2 is drawn down by silicate weathering (**f**), before Venus reenters a runaway greenhouse and loses remaining water to space. The model successfully recovers modern atmospheric ^{40}Ar abundance, plausible atmospheric ^{4}He abundance (dashed lines), and modern heat flow (not shown). Adapted from Krissansen-Totton et al. (2021a)

followed by further increase of temperature due to enhanced greenhouse effect, until all the reservoir has been depleted. In an Earth-like situation, one could even imagine a more complete carbon cycle (Krissansen-Totton et al. 2021a). The same mechanism would apply to thermal dehydration of crustal materials and water.

One important feedback that has been observed in models directly affects both the climate and the mantle evolution of Venus: the effects of surface temperature variations on mantle convection regime (see Rolf et al. 2022). This is especially relevant in the case of Venus due to its high observed surface temperature, and the possibility that, at some point in its past, surface conditions could have been much cooler, with the potential for large variations ($\approx$400 K) during its history. A simple attempt to model such a feedback was made by Phillips et al. (2001), using a parameterized model. They obtained a positive feedback where higher surface temperatures led to higher partial melting, larger outgassing and in turn even higher surface temperatures because of the release of greenhouse gases. It is now suspected that the situation is more complex. Increases in temperature weaken the lithosphere, leading to a decrease in both lithosphere and mantle viscosity. Convective stresses in the lithosphere would therefore be lower at higher surface temperatures (Lenardic et al. 2008). It is also expected that Venus's present-day lithosphere is more ductile than Earth's and that pores and fractures are healed more efficiently due to its higher temperature and lower viscosity (e.g., Bercovici and Ricard 2014). As a consequence, higher surface temperatures would favor a stagnant-lid like regime, while lower surface temperatures would allow for a mobile lid regime more akin to plate tectonics, where the lithosphere can be fractured by convective stresses (Weller et al. 2015). Therefore the surface temperature could result in a different mantle convection regime on Earth and Venus. While it has not been fully investigated yet, it is likely that the "plutonic squishy lid" convection regime highlighted in recent work (Rozel et al. 2017; Lourenço et al. 2020) would be affected by changes in the surface temperature. One could imagine that this regime would be favoured by temperatures on present-day Venus, as fracture healing and ductile behaviour may prevent a larger part of the magma from reaching the surface. A simple model for changes in intrusive/extrusive volcanism ratio could well suggest a complex interaction between three evolving convection regimes rather than the two end-members studied in past works.

The role of surface temperatures has been observed in long-term modeling of the coupled evolution of Venus (Gillmann and Tackley 2014) as part of a negative feedback cycle. If the greenhouse effect decreases due to the loss of water, a mobile lid regime emerges accompanied by high volcanic production and outgassing. The water reservoir in the atmosphere is replenished, leading to a corresponding increase in surface temperatures that suppresses the mobile lid and favors a return to stagnant lid convection.

On the other hand, Noack et al. (2012) observed a similar stabilizing feedback but with a different mechanism: higher surface temperatures allowed for the mobilization of the surface, by reducing the viscosity contrast between the mantle and lithosphere, which removes heat from the mantle efficiently, and ultimately reduces volatile concentrations in the atmosphere and surface temperature.

As illustrated by these modelling efforts, the consequences of surface temperature variations go beyond changes solely in mantle convection. In turn, different tectonic regimes lead to variations in outgassing efficiency and regulate surface conditions, thus completing a feedback cycle. Beyond this, changes in surface conditions can affect (aqueous) weathering and the state of water, with possibly far reaching consequences. If surface conditions are modified enough to allow for new processes to dominate, large changes in atmosphere composition cannot be ruled out, as suggested previously (Krissansen-Totton et al. 2021a).

In the same way, a change in tectonic regime can also affect how heat is removed from the interior of the planet, with possible consequences on heat fluxes at both the surface and core mantle boundary. Magnetic field generation might therefore be affected by surface changes (O'Rourke et al. 2018). While the exact contribution of the magnetic field to the protection or erosion of the atmosphere is still debated (e.g., Gunell et al. 2018b; Gronoff et al. 2020, as discussed in Sect. 3.1), it is widely accepted that magnetic field modifications would affect atmospheric escape mechanisms and thus, the long-term atmospheric composition. By considering the planet as a coupled system, estimating long-term evolution becomes a complex game that involves identifying the dominant feedback mechanism at specific conditions.

4.5 Reaching Present-Day Venus

All possible scenarios for the past evolution of Venus must reach the planet's present-day state. First, it means that the current bulk atmospheric composition must be met: 11×10^{18} kg of N_2 ($3.5 \pm 0.8\%$, Johnson and Goldblatt 2015), 10^{16} kg of H_2O (30 ± 15 ppm, Kasting and Pollack 1983; Lécuyer et al. 2000), 4.69×10^{20} kg of CO_2 ($96.5 \pm 0.8\%$, Fegley 2014), virtually no O_2 (Oyama et al. 1980; Fegley 2014), and 150 ± 30 ppm SO_2 (Kasting and Pollack 1983; Lécuyer et al. 2000). Note that actual abundances vary with altitude. Other common constraints are given by ^{40}Ar or ^{4}He abundances and isotope ratios of other noble gases, topography, geoid or the age of the surface and volcanic eruption rates. Conversely, understanding late evolution places constraints and gives clues on early scenarios by highlighting what initial conditions can be compatible with present-day. The most drastic changes in surface conditions would take place when branching out of an early temperate state into hot and dry present-day Venus (see also Avice et al. 2022).

If cloud feedback did enable a wet early Venus, then the subsequent evolution and ultimate longevity of such clement surface conditions were also controlled by atmosphere-interior interactions. Indeed, this habitable period could persist for billions of years barring destabilizing outside or internal events (Way et al. 2016; Way and Del Genio 2020). Even under a stagnant lid tectonic regime, the weathering feedback could have maintained temperate surface conditions over long periods of time (Höning et al. 2021). The era of a temperate climate could have ended as buried carbonates became unstable at depth and released CO_2 into the atmosphere. The rising surface temperature would have moved the decarbonation depth even closer to the surface, leading to a catastrophic outgassing of CO_2, thus establishing a strong greenhouse effect (Höning et al. 2021).

The stagnant lid also limits surface heat loss, which may increase interior temperatures and further destabilize crustal carbonates. Against this destabilizing feedback, the drawdown of carbon dioxide by silicate weathering (Sects. 2.4.1 and 2.4.4) could maintain habitable surface climate for a duration on the order of 1 Gyr according to model calculations (Höning et al. 2021) before the transition to runaway greenhouse would occur (Fig. 9). However, the exact timing of this mechanism is extremely model-dependent and is affected by initial mantle temperature, the rate at which inorganic carbon in the lid is transferred to the atmosphere, and the sensitivity of silicate weathering to climate. As discussed in Sect. 2.4.4 the dependence of silicate weathering on atmospheric composition and climate is highly uncertain, even for Earth. A pure CO_2-dependence, as assumed in Fig. 9, is a worst-case scenario for extending habitability since it provides no stabilizing thermostat to offset gradual CO_2 outgassing through time. Early plate tectonics or episodic subduction could also extend the duration of a wet Venus surface via efficient transport of carbonates to the deep mantle, thereby limiting the decarbonation feedback as Venus's surface warms.

It is unlikely that CO_2 accumulation on Venus could solely originate from regular volcanic outgassing, especially in a short period of time. Calculations indicate that between 10 (López et al. 1998) and 100 (Head et al. 2021) times the current total observable crust inventory would be needed for a complete Venus atmosphere build-up. Models (e.g., Gillmann et al. 2020) suggest that around 10 bars of CO_2 could be outgassed during the late evolution (although this depends heavily on composition).

Such late massive outgassing would need to be checked against constraints brought by the various radiogenic noble gas isotopes (e.g., ^{4}He, ^{40}Ar, $^{129,131-136}$Xe). Current estimates for argon suggest less degassing of Venus compared to Earth (Kaula 1999), and are inconsistent with large scale recent volcanism.

The main issue remains the fate of water in this scenario. Atmospheric oxygen loss cannot account for the escape of more than a few tens of centimeters of global equivalent layer of water over the last $\sim$4 Gyr (Persson et al. 2020), assuming present-day atmospheric composition. Solid surface oxidation could account for roughly the same amount of water loss as atmospheric escape, as discussed earlier in this work (Gillmann et al. 2020). Therefore the mechanism could only be significant for a long-lived liquid magma at the surface, such as in a late intense runaway greenhouse event (Zahnle et al. 2007). Additionally, present-day observations suggest that hydrogen and oxygen escape is close to the stoichiometric ratio (2:1) of water molecules (Barabash et al. 2007b). If oxygen and hydrogen originate from water dissociation in Venus' atmosphere, this stoichiometric escape could suggest that no other substantial sink alters the oxygen repartition at present-day. An interesting alternative has recently been proposed by Warren and Kite (2021) where fine ash resulting from explosive volcanism could be fully oxidized due to their larger surface to mass ratio. First results from their parameterized model indicate a large amount of oxygen (stored in H_2O and CO_2) could thus be removed from the atmosphere. However, if this takes place after the modern day thick atmosphere, their models needs to be reconciled with the fast cooling of volcanic ash at Venus' surface conditions due to heat transfer to the dense atmosphere (Frenkel and Zabalueva 1983). Explosive volcanism is thought to require volatile contents >3-5 wt%, several wt% higher than typical Earth magmas (<1 wt%) (Head and Wilson 1986; Head et al. 2021). This aspect also requires further experimental investigation in order to assess the efficiency of this mechanism, which may vary by orders of magnitude, especially compared to solid basalt oxidation. Further surface observation of Venus may yield clues to understand whether a significant amount of ash can be formed under Venus past and present-day conditions. Until now, only very limited pyroclastic activity has been identified on Venus, with most volcanic activity being effusive in nature (Campbell and Clark 2006; Ghail and Wilson 2015; Grosfils et al. 2000, 2011; Keddie and Head 1995; McGill 2000).

In contrast, both the desiccated Venus scenarios (dry Venus and stifled outgassing) have a straightforward evolution from early Venus to present-day, since in those cases, the bulk of the atmosphere was generated early on by oxidized magma ocean outgassing. Competition between moderate/marginal outgassing and loss processes would have kept water and oxygen abundances low and only allowed for a slight increase in CO_2 and N_2 pressure over time. Long-term stagnant lid like (or single plate) convection would have been likely. The main issue could be the low water content of the present-day atmosphere requiring low volcanic outgassing of any remaining water, but that could be explained by high surface pressure impairing water release into the atmosphere (Gaillard and Scaillet 2014). Surface pressures would have remained high because of the CO_2-N_2 atmosphere and the lack of a mechanism to trap them at the surface in the absence of oceans (Sleep and Zahnle 2001). No ocean would be possible because water was never outgassed early on and could not be released later on. The possibly dry nature of the mantle (with H trapped in the core, lost to space, or a low initial water abundance if Venus accreted dry material) in this evolution pathway would also be consistent with geoid-topography correlation data (Kiefer and Hager 1992) and the possible lack of a venusian asthenosphere. If Venus' interior is not dry, the impossibility to outgas water due to surface conditions would also fit this scenario, but not necessarily the geoid-topography correlation data. Further observation of the surface, possibly including estimates for the composition of lava flows may help discriminate between these scenarios.

As a consequence, the temperate Venus scenario requires a way to remove efficiently any surface/atmospheric water it did not lose during its primordial evolution, so that it can reach the present state of Venus (that has been in place since at least the mean age of the present-day surface). The longer the temperate phase lasted, the more difficult it is to remove water while building up the CO_2 atmosphere (unless the latter is primarily inherited from the magma ocean early outgassing) and meeting other observational markers. The dry evolution scenarios are more static, which has to be reconciled with Venus being an active planet. The present-day state of our understanding of mechanisms at work is still unable to distinguish between these evolutionary pathways.

4.6 Discriminating Between Scenarios

Considering the planet as a whole is a complex problem, but it is necessary to understand its evolution. Currently, our knowledge is limited by the lack of data about past (surface, mantle and atmospheric) conditions. This naturally reflects on how well we can assess interactions between interior and atmosphere, and in turn how they could have changed over the history of Venus. The further we step back in time, the wider the range of possible scenarios becomes. Therefore, we search for ways to constrain evolution from observation and assess what could provide valuable landmarks in the evolution of Venus to anchor different scenarios.

Noble gases are powerful tracers of volatile element exchanges between the interior and exterior of terrestrial planets (e.g., Ozima and Podosek 2002; Marty 2020). They are currently the most robust source of information on the distant past of Venus, despite a lack of data, measurement uncertainties and debates about their interpretation. See also Avice et al. (2022) for a review on noble gases in the present-day Venus atmosphere. Many of the physical processes described above potentially altered the starting primordial elemental and isotopic composition of noble gases in the atmosphere of Venus. Depending on its accretion history, Venus could have, similarly to Earth, incorporated in its solid mantle noble gases which were originally present in the Solar nebula (Williams and Mukhopadhyay 2018). Gas partitioning between the melt in the magma ocean and the primitive atmosphere or early outgassing events would have led to the emergence of an early atmosphere of solar composition. However, atmospheric erosion by escape processes or by impacts could have modified the composition of this primordial atmosphere (e.g., Pepin 1991) or could have totally removed it (see Sects. 3.2.2 and 2.3.1).

Impacts of chondritic or cometary bodies at the end of Venus' accretion (O'Brien et al. 2014) would have delivered new noble gases (and other volatile elements) having distinct elemental and isotopic compositions depending on their origin (e.g., Busemann et al. 2000; Füri and Marty 2015; Marty et al. 2017; Rubin et al. 2018). If there were remnants of a primordial solar atmosphere at the time of this late contribution, the mixing between solar and meteoritic and/or cometary gases could be detected in the resulting elemental and isotopic composition of atmospheric noble gases (*e.g.* Marty 2012, 2022). Mixing between components could also result from the long-term degassing of noble gases into the atmosphere. Finally, noble gases could also track long-term escape processes.

On Earth, the isotopic composition of atmospheric xenon evolved through time (see Avice et al. 2018, and refs. therein). This evolution could be due to the long-term escape of xenon ions from Earth's atmosphere to outer space in a photo-ionized hydrogen wind (Zahnle et al. 2019). This escape process would progressively lead to the depletion and isotopic fractionation of atmospheric xenon. The fact that xenon on Earth and Mars share similar features (depletion and isotopic fractionation) means that the selective coupled H^+–Xe^+ escape mechanism could have been operating on both planets. The evolution of the isotopic composition of atmospheric xenon could thus be a tracer of hydrogen escape on terrestrial planets (Avice and Marty 2020). So far the elemental abundance and isotopic composition of atmospheric xenon on Venus remains unknown (see Avice et al. 2022, for a review on existing data). New measurements would certainly help to further understand how hydrogen escape shaped the atmosphere of Venus and may have influenced its geodynamical evolution (Baines et al. 2013).

Over the course of Venus' history, several isotopes of noble gases have been produced by extinct (^{129}I, ^{244}Pu) and extant (e.g., ^{238}U, ^{40}K) radioactive nuclides present in the silicate portions of the planet. When a portion of Venus crust or mantle melted, these gaseous products migrated into the magmatic gas and were eventually degassed into the atmosphere inducing radiogenic/fissiogenic excesses on top of the primordial isotopic composition. Because the aforementioned radioactive nuclides have very different half-life times (e.g., $t_{1/2}(^{129}\text{I}) = 16$ Myr and $t_{1/2}(^{238}\text{U}) = 4.47$ Gyr), studying the abundances and relative proportions of their daughter products has the potential to provide important constraints on the amount and timing of exchanges of volatile elements between the interior and the atmosphere of Venus.

For example, atmospheric Xe on Mars and Earth have very distinct ^{129}Xe/^{130}Xe ratios (Ozima and Podosek 2002). The very high ratio for Mars atmospheric Xe could be due to intense and early episodes of atmospheric escape leaving an atmosphere depleted in primordial noble gases (Swindle 2002). Subsequent degassing of xenon from Mars' interior carrying a strong excess of radiogenic ^{129}Xe would explain the high present-day ^{129}Xe/^{130}Xe ratio of Mars atmospheric Xe. For Earth, recent studies demonstrated that, 3.3 Gyr ago, the atmospheric ^{129}Xe/^{130}Xe ratio was significantly lower than the modern ratio (Avice et al. 2017; Marty et al. 2019). The difference has been used to estimate mantle degassing rates over the past 3.3 Gyr.

Knowing the ^{3}He/^{4}He ratio of Venus' atmospheric helium could also provide important information regarding the present and past interactions between the atmosphere and the interior. However, numerous events with changing magnitudes over time can modify this ratio such as: (i) deposition of ^{3}He-rich extraterrestrial material, (ii) atmospheric escape causing isotopic fractionation of the remaining helium fraction and (iii) degassing of radiogenic ^{4}He produced inside Venus. Only approaches coupling He and Ne-Ar-Xe isotope systematics would allow us to use the ^{3}He/^{4}He ratio as an efficient probe of Venus' ancient geodynamics.

Another look at geodynamics can be obtained from the state of the core, although its composition is still debated, which can affect conclusions. A stagnant lid regime limits core-mantle heat flux (in the common view, all other things equal, the surface heat flux is smaller than for plate tectonics because of the lid, the mantle is hotter and the core cools more slowly), thus under these conditions the core of Venus would be expected to remain in a liquid state until present-day (Nimmo 2002; O'Neill and Debaille 2014; O'Rourke and Korenaga 2015; Smrekar et al. 2018). The size and state of the core can be constrained by measurements of the moment of inertia factor and the tidal Love number k_2 (Dumoulin et al. 2017). Detection of remnant magnetisation of the crust caused by a past magnetic field could help shed light on the state and history of the core (O'Rourke et al. 2018). The absence of a Venusian solid inner core would therefore rule out a scenario involving long-term plate tectonics operating on Venus (O'Neill 2021). In case a solid inner core exists, even the presence of small amounts of hydrogen – as suggested by the stifled outgassing scenario – would increase considerably both the compressional and shear wave velocities in the core (Caracas 2015).

Determining possible scenarios involving water could rely on the study of the very early history of Venus. Coupled models of magma ocean evolution and outgassing (e.g., Elkins-Tanton 2008; Lebrun et al. 2013; Hamano et al. 2013; Salvador et al. 2017; Nikolaou et al. 2019; Bower et al. 2019; Lichtenberg et al. 2021, and Salvador et al. 2022 for a review) and models of atmosphere-interior interactions can provide insights, however determining whether a temporarily habitable (e.g., Fig. 12, right column) or never habitable (e.g., Fig. 12, left column) evolutionary scenario was realized will require more observational constraints.

However, the planet may still hold some clues for the links between the distant past and the present-day state on its surface. Water and its history are a key to understand how Venus reached its present-day state and diverged from Earth. A succession of increasingly difficult questions would ideally need to be answered to form a robust base for evolution scenarios. First, we need to look for any evidence of water (at the surface or in the mantle) over the course of the evolution of the planet, between the magma ocean freezing and the generation of the current state 200-1000 Ma ago. A second step would be to estimate the abundance and state of that water. Then, refining the timing of the observations would clarify whether the changes were catastrophic or progressive. Finally, proof of liquid water would set a standard for Venus' past climate. However, with the caveat that negative results might indicate either a lack of water (either at the surface or in the mantle, depending on the nature of the observation) or that no traces of water were preserved.

It will also be extremely important to gain more knowledge on the mechanisms that define possible evolution scenarios on Venus. This chapter has highlighted the advances made in recent years as well as gaps in our current understanding of the effects of those processes. We hope that a closer examination of both the surface composition and volatile species exchanges will allow to better link the past of Venus to its present state.

The composition of the surface could retain traces of the mechanisms of atmosphere/crust interaction needed to understand the late evolution of Venus. It is important to look for signs of water and chemical, mineralogical and isotopic markers. On one hand, this could mean measuring the product of surface oxidation, or the composition of volcanic gas plumes in the atmosphere of Venus, to understand where the water comes from and where and how it disappeared. That also includes any evidence for past pyroclastic activity.

On the other hand, it can be even much more direct, by searching for actual traces of water on the surface, from surface composition, such as the presence of granite, or aqueous alteration products in the surface mineralogy. A detection of rocks similar to Earth's banded iron formations, silica-rich deposits, salt-rich formations or currently solid salt flows (e.g., in Venus canali) may quantify any past water activity (Zolotov and Mironenko 2009; Zolotov 2019). That also means looking for any evidence of fluvial (e.g., Khawja et al. 2020), lacustrine and/or shoreline processes and deposits in the tesserae and plains, and for evidence in any layered rocks exposed in deformed regions such as tesserae.

In the same way, terrains such as the tesserae, that could correspond to older parts of the surface of Venus (e.g., Ivanov and Head 1996), may inform us about the past mechanisms of their formation, both from the point of view of the dynamics regime and, through their composition, from that of the evolution of the mantle composition. Important information about the water content in the interior of the planet can be obtained from the phase lag of the tidal Love number. The latter depends strongly on the mantle viscosity, which in turn depends on the temperature and volatile content. Four upcoming space missions to Venus (Widemann et al. 2022), DAVINCI, VERITAS, ENVISION and Shukrayaan-1, possibly complemented by the Chinese mission VOICE (VOlcano Imaging and Climate Explorer), will provide information to better discriminate between an aqueous and a dry evolution scenario of the Earth's sister planet.

Acknowledgements The authors thank S. Mojzsis and J. Head for their comments, as well as R. Wordsworth and an anonymous reviewer for their help in improving the manuscript. CG acknowledges the support of Rice University and the CLEVER planets group (itself supported by NASA and part of NExSS) and ET-HoME Excellence of Science programme. GA acknowledges support from the french Centre National d'Etudes Spatiales (CNES) for support of Venus-related studies. MJW acknowledges support from the Goddard Space Flight Center Sellers Exoplanet Environments Collaboration (SEEC), which is funded by the NASA Planetary Science Division's Internal Scientist Funding Model. MJW acknowledges support from NASA's Nexus for Exoplanet System Science (NExSS), the GSFC Sellers Exoplanet Environments Collaboration (SEEC), which is funded by the NASA Planetary Science Division's Internal Scientist Funding Model (ISFM) and the ROCKE-3D Project ISFM jointly funded by a NASA Planetary and Earth Science Divisions. AS acknowledges support from NASA's Habitable Worlds Program (No. 80NSSC20K0226). MYZ acknowledges support from the NASA Solar System Workings program.

Funding Note Open Access funding enabled and organized by Projekt DEAL.

References

Abe Y, Abe-Ouchi A, Sleep NH, Zahnle KJ (2011) Habitable zone limits for dry planets. Astrobiology 11:443–460. https://doi.org/10.1089/ast.2010.0545

Airapetian VS, Jin M, Lüftinger T, Saikia SB, Kochukhov O, Güdel M, Van Der Holst B, Manchester W IV (2021) One year in the life of young suns: data-constrained corona-wind model of k 1 ceti. Astrophys J 916(2):96. https://doi.org/10.3847/1538-4357/ac081e

Allègre CJ, Staudacher T, Sarda P (1987) Rare gas systematics: formation of the atmosphere, evolution and structure of the Earth's mantle. Earth Planet Sci Lett 81(2):127–150. https://doi.org/10.1016/0012-821X(87)90151-8

Anderson FS, Smrekar SE (2006) Global mapping of crustal and lithospheric thickness on Venus. J Geophys Res, Planets 111(E8):E08006

Arevalo R, McDonough WF, Luong M (2009) The K/U ratio of the silicate Earth: insights into mantle composition, structure and thermal evolution. Earth Planet Sci Lett 278(3–4):361–369. https://doi.org/10.1016/j.epsl.2008.12.023

Arkani-Hamed J (2009) Did tidal deformation power the core dynamo of Mars? Icarus 201(1):31–43

Armann M, Tackley PJ (2012) Simulating the thermochemical magmatic and tectonic evolution of Venus's mantle and lithosphere: two-dimensional models. J Geophys Res, Planets 117(E12):E12003

Armstrong LS, Hirschmann MM, Stanley BD, Falksen EG, Jacobsen SD (2015) Speciation and solubility of reduced C–O–H–N volatiles in mafic melt: implications for volcanism, atmospheric evolution, and deep volatile cycles in the terrestrial planets. Geochim Cosmochim Acta 171:283–302

Arshukova IL, Erkaev NV, Biernat HK, Vogl DF (2004) Interchange instability of the venusian ionopause. Adv Space Res 33(2):182–186. https://doi.org/10.1016/j.asr.2003.04.015

Avice G, Marty B (2020) Perspectives on atmospheric evolution from noble gas and nitrogen isotopes on Earth, Mars & Venus. Space Sci Rev 216(3):36. https://doi.org/10.1007/s11214-020-00655-0

Avice G, Marty B, Burgess R (2017) The origin and degassing history of the Earth's atmosphere revealed by Archean xenon. Nat Commun 8:15455. https://doi.org/10.1038/ncomms15455

Avice G, Marty B, Burgess R, Hofmann A, Philippot P, Zahnle K, Zakharov D (2018) Evolution of atmospheric xenon and other noble gases inferred from Archean to Paleoproterozoic rocks. Geochim Cosmochim Acta 232:82–100. https://doi.org/10.1016/j.gca.2018.04.018

Avice G, Parai R, Jacobson S, Labidi J, Trainer MG, Petkov MP (2022) Noble gases and stable isotopes track the origin and early evolution of the Venus atmosphere

Badro J, Siebert J, Nimmo F (2016) An early geodynamo driven by exsolution of mantle components from Earth's core. Nature 536(7616):326–328

Baines KH, Atreya SK, Bullock MA, Grinspoon DH, Mahaffy P, Russell CT, Schubert G, Zahnle K (2013) The atmospheres of the terrestrial planets: clues to the origins and early evolution of Venus, Earth, and Mars. In: Comparative Climatology of Terrestrial Planets. University of Arizona Press, Tucson, pp 1–28. https://doi.org/10.2458/azu_uapress_9780816530595-ch006

Barabash S, Fedorov A, Sauvaud JJ, Lundin R, Russell CT, Futaana Y, Zhang TL, Andersson H, Brinkfeldt K, Grigoriev A, Holmström M, Yamauchi M, Asamura K, Baumjohann W, Lammer H, Coates AJ, Kataria DO, Linder DR, Curtis CC, Hsieh KC, Sandel BR, Grande M, Gunell H, Koskinen HEJ, Kallio E, Riihelä P, Säles T, Schmidt W, Kozyra J, Krupp N, Fränz M, Woch J, Luhmann J, McKenna-Lawlor

S, Mazelle C, Thocaven JJ, Orsini S, Cerulli-Irelli R, Mura M, Milillo M, Maggi M, Roelof E, Brandt P, Szego K, Winningham JD, Frahm RA, Scherrer J, Sharber JR, Wurz P, Bochsler P (2007a) The loss of ions from Venus through the plasma wake. Nature 450(7170):650–653. https://doi.org/10.1038/nature06434

Barabash S, Sauvaud JA, Gunell H, Andersson H, Grigoriev A, Brinkfeldt K, Holmström M, Lundin R, Yamauchi M, Asamura K, Baumjohann W, Zhang TL, Coates AJ, Linder DR, Kataria DO, Curtis CC, Hsieh KC, Sandel BR, Fedorov A, Mazelle C, Thocaven JJ, Grande M, Koskinen HEJ, Kallio E, Säles T, Riihela P, Kozyra J, Krupp N, Woch J, Luhmann J, McKenna-Lawlor S, Orsini S, Cerulli-Irelli R, Mura M, Milillo M, Maggi M, Roelof E, Brandt P, Russell CT, Szego K, Winningham JD, Frahm RA, Scherrer J, Sharber JR, Wurz P, Bochsler P (2007b) The analyser of space plasmas and energetic atoms (ASPERA-4) for the Venus Express mission. Planet Space Sci 55(12):1772–1792. https://doi.org/10.1016/j.pss.2007.01.014

Barsukov V, Volkov V, Khodakovsky I (1982) The crust of Venus: theoretical models of chemical and mineral composition. J Geophys Res, Solid Earth 87(S01):A3–A9

Barsukov VL, Basilevsky AT, Burba GA, Bobinna NN, Kryuchkov V, Kuzmin R, Nikolaeva O, Pronin A, Ronca L, Chernaya I et al (1986) The geology and geomorphology of the Venus surface as revealed by the radar images obtained by Veneras 15 and 16. J Geophys Res, Solid Earth 91(B4):378–398

Basilevsky AT, Head JW, Schaber GG, Strom RG (1997) The resurfacing history of Venus. In: Bougher SW, Hunten DM, Phillips RJ (eds) Venus II: Geology, Geophysics, Atmosphere, and Solar Wind Environment, p 1047

Bauer SJ, Lammer H (2004) Planetary Aeronomy: Atmosphere Environments in Planetary Systems. Springer, Berlin

Bercovici D, Ricard Y (2014) Plate tectonics, damage and inheritance. Nature 508:513. https://doi.org/10.1038/nature13072

Berger G, Cathala A, Fabre S, Borisova AY, Pages A, Aigouy T, Esvan J, Pinet P (2019) Experimental exploration of volcanic rocks-atmosphere interaction under Venus surface conditions. Icarus 329:8–23

Berlo K, Gardner JE, Blundy JD (2011) Timescales of magma degassing. In: Timescales of Magmatic Processes: From Core to Atmosphere. Wiley, New York. https://doi.org/10.1002/9781444328509.ch11. Chap. 11

Berner R, Caldeira K (1997) The need for mass balance and feedback in the geochemical carbon cycle. Geology 25:955–956

Bierhaus EB, Dones L, Alvarellos JL, Zahnle K (2012) The role of ejecta in the small crater populations on the mid-sized saturnian satellites. Icarus 218(1):602–621

Blanc NA, Stegman DR, Ziegler LB (2020) Thermal and magnetic evolution of a crystallizing basal magma ocean in Earth's mantle. Earth Planet Sci Lett 534:116085. https://doi.org/10.1016/j.epsl.2020.116085

Borgeat X, Tackley PJ (2022) Hadean/eoarchean tectonics and mantle mixing induced by impacts: a three-dimensional study. Prog Earth Planet Sci 9:38

Bottke WF, Vokrouhlicky D, Ghent B, Mazrouei S, Robbins S, Marchi S (2016) On asteroid impacts, crater scaling laws, and a proposed younger surface age for Venus. In: Lunar and Planetary Science Conference, p 2036

Bower DJ, Kitzmann D, Wolf AS, Sanan P, Dorn C, Oza AV (2019) Linking the evolution of terrestrial interiors and an early outgassed atmosphere to astrophysical observations. Astron Astrophys 631:A103. https://doi.org/10.1051/0004-6361/201935710

Bower DJ, Hakim K, Sossi PA, Sanan P (2022) Retention of water in terrestrial magma oceans and carbon-rich early atmospheres. Planet Sci J 3(4):93

Brace LH, Theis RF, Hoegy WR (1982) Plasma clouds above the ionopause of Venus and their implications. Planet Space Sci 30(1):29–37. https://doi.org/10.1016/0032-0633(82)90069-1

Brace LH, Kasprzak WT, Taylor HA, Theis RF, Russell CT, Barnes A, Mihalov JD, Hunten DM (1987) The ionotail of Venus: its configuration and evidence for ion escape. J Geophys Res 92(A1):15–26. https://doi.org/10.1029/JA092iA01p00015

Brady PV, Gíslason SR (1997) Seafloor weathering controls on atmospheric CO_2 and global climate. Geochim Cosmochim Acta 61(5):965–973

Bullock MA, Grinspoon DH (2001) The recent evolution of climate on Venus. Icarus 150:19–37. https://doi.org/10.1006/icar.2000.6570

Burton MR, Sawyer GM, Granieri D (2013) Deep carbon emissions from volcanoes. Rev Mineral Geochem 75(1):323–354

Busemann H, Baur H, Wieler R (2000) Primordial noble gases in "phase Q" in carbonaceous and ordinary chondrites studied by closed-system stepped etching. Meteorit Planet Sci 35(5):949–973. https://doi.org/10.1111/j.1945-5100.2000.tb01485.x

Cameron AG (1983) Origin of the atmospheres of the terrestrial planets. Icarus 56(2):195–201

Cameron AG, Ward WR (1976) The origin of the moon. In: Lunar and Planetary Science Conference, vol 7

Campbell BA, Clark DA (2006) Geologic map of the Mead quadrangle (V-21), Venus. US Geological Survey Report, p 2897. https://doi.org/10.3133/sim2897

Canup RM (2004) Simulations of a late lunar-forming impact. Icarus 168(2):433–456. https://doi.org/10.1016/j.icarus.2003.09.028

Caracas R (2015) The influence of hydrogen on the seismic properties of solid iron. Geophys Res Lett 42(10):3780–3785

Catling DC, Zahnle KJ (2020) The archean atmosphere. Sci Adv 6(9):eaax1420

Cawood P, Hawkesworth C, Dhuime B (2013) The continental record and the generation of continental crust. Geol Soc Am Bull 125(1–2):14–32. https://doi.org/10.1130/B30722.1

Chamberlain JW (1963) Planetary coronae and atmospheric evaporation. Planet Space Sci 11(8):901–960. https://doi.org/10.1016/0032-0633(63)90122-3

Chamberlain JW, Campbell FJ (1967) Rate of evaporation of a non-Maxwellian atmosphere. Astrophys J 149:687. https://doi.org/10.1086/149298

Chassefière E (1996a) Hydrodynamic escape of oxygen from primitive atmospheres: applications to the cases of Venus and Mars. Icarus 124(2):537–552

Chassefière E (1996b) Hydrodynamic escape of hydrogen from a hot water-rich atmosphere: the case of Venus. J Geophys Res 101(E11):26039–26056. https://doi.org/10.1029/96JE01951

Citron RI, Stewart ST (2022) Large impacts onto the early Earth: planetary sterilization and iron delivery. Planet Sci J 3:116

Claire MW, Sheets J, Cohen M, Ribas I, Meadows VS, Catling DC (2012) The evolution of solar flux from 0.1 nm to 160 µm: quantitative estimates for planetary studies. Astrophys J 757:95. https://doi.org/10.1088/0004-637X/757/1/95

Cnossen I (2020) Analysis and attribution of climate change in the upper atmosphere from 1950 to 2015 simulated by WACCM-X. J Geophys Res Space Phys 125(12):e28623. https://doi.org/10.1029/2020JA028623

Colin L (1980) The pioneer Venus program. J Geophys Res Space Phys 85(A13):7575–7598. https://doi.org/10.1029/JA085iA13p07575

Cravens TE, Gombosi T, Nagy AF (1980) Hot hydrogen in the exosphere of Venus. Nature 283(5743):178–180. https://doi.org/10.1038/283178a0

Crisp JA (1984) Rates of magma emplacement and volcanic output. J Volcanol Geotherm Res 20(3):177–211. https://doi.org/10.1016/0377-0273(84)90039-8

Croft SK (1982) A first-order estimate of shock heating and vaporization in oceanic impacts. In: Geological Implications of Impacts of Large Asteroids and Comets on Earth, vol 190, pp 143–152

Davies JH (2008) Did a mega-collision dry Venus' interior? Earth Planet Sci Lett 268(3):376–383. https://doi.org/10.1016/j.epsl.2008.01.031

De Bergh C, Bézard B, Owen T, Crisp D, Maillard JP, Lutz BL (1991) Deuterium on Venus: observations from Earth. Science 251(4993):547–549. https://doi.org/10.1126/science.251.4993.547

Dehant V, Debaille V, Dobos V, Gaillard F, Gillmann C, Goderis S, Grenfell JL, Höning D, Javaux EJ, Karatekin Ö, Morbidelli A, Noack L, Rauer H, Scherf M, Spohn T, Tackley P, Van Hoolst T, Wünnemann K (2019) Geoscience for understanding habitability in the solar system and beyond. Space Sci Rev 215(6):42. https://doi.org/10.1007/s11214-019-0608-8

Delmelle P, Wadsworth FB, Maters EC, Ayris PM (2018) High temperature reactions between gases and ash particles in volcanic eruption plumes. Rev Mineral Geochem 84(1):285–308

Dixon JE (1997) Degassing of alkalic basalts. Am Mineral 82(3–4):368–378

Donahue TM (1999) New analysis of hydrogen and deuterium escape from Venus. Icarus 141(2):226–235. https://doi.org/10.1006/icar.1999.6186

Donahue TM, Hartle RE (1992) Solar cycle variations in H^+ and D^+ densities in the Venus ionosphere: implications for escape. Geophys Res Lett 19(24):2449–2452. https://doi.org/10.1029/92GL02927

Donahue TM, Pollack JB (1983) In: Origin and Evolution of the Atmosphere of Venus. University of Arizona Press, Tucson, pp 1003–1036

Donahue TM, Hodges R Jr (1992) Past and present water budget of Venus. J Geophys Res 97(E4):6083–6091. https://doi.org/10.1029/92JE00343

Donahue TM, Hoffman JH, Hodges RR, Watson AJ (1982) Venus was wet - a measurement of the ratio of deuterium to hydrogen. Science 216:630–633. https://doi.org/10.1126/science.216.4546.630

Dong C, Huang Z, Lingam M (2019) Role of planetary obliquity in regulating atmospheric escape: G-dwarf versus M-dwarf Earth-like exoplanets. Astrophys J Lett 882(2):L16. https://doi.org/10.3847/2041-8213/ab372c. 1907.07459

Driscoll P, Bercovici D (2013) Divergent evolution of Earth and Venus: influence of degassing, tectonics, and magnetic fields. Icarus 226(2):1447–1464

Driscoll P, Bercovici D (2014) On the thermal and magnetic histories of Earth and Venus: influences of melting, radioactivity, and conductivity. Phys Earth Planet Inter 236:36–51. https://doi.org/10.1016/j.pepi.2014.08.004

628 Springer

Dumoulin C, Tobie G, Verhoeven O, Rosenblatt P, Rambaux N (2017) Tidal constraints on the interior of Venus. J Geophys Res, Planets 122(6):1338–1352

Edberg NJT, Nilsson H, Futaana Y, Stenberg G, Lester M, Cowley SWH, Luhmann JG, McEnulty TR, Opgenoorth HJ, Fedorov A, Barabash S, Zhang TL (2011) Atmospheric erosion of Venus during stormy space weather. J Geophys Res Space Phys 116(A9):A09308. https://doi.org/10.1029/2011JA016749

Edmonds M, Woods AW (2018) Exsolved volatiles in magma reservoirs. J Volcanol Geotherm Res. https://doi.org/10.1016/j.jvolgeores.2018.10.018

Egan H, Jarvinen R, Ma Y, Brain D (2019) Do Magnetic Fields Prevent Atmospheric Escape? In: AAS/Division for Extreme Solar Systems Abstracts, vol 51, p 329.04

Elkins-Tanton LT (2008) Linked magma ocean solidification and atmospheric growth for Earth and Mars. Earth Planet Sci Lett 271(1–4):181–191. https://doi.org/10.1016/j.epsl.2008.03.062

Elphic RC, Ershkovich AI (1984) On the stability of the ionopause of Venus. J Geophys Res 89(A2):997–1002. https://doi.org/10.1029/JA089iA02p00997

Fedorov A, Barabash S, Sauvaud JA, Futaana Y, Zhang TL, Lundin R, Ferrier C (2011) Measurements of the ion escape rates from Venus for solar minimum. J Geophys Res Space Phys 116(A7):A07220. https://doi.org/10.1029/2011JA016427

Fegley B (2009) In: Gornitzs V (ed) Atmospheric Evolution on Venus. Encyclopedia of Paleoclimatology and Ancient Environments Springer, Dordrecht, pp 75–83

Fegley B (2014) In: Holland HD, Turekian KK (eds) Venus. Treatise on Geochemistry, vol 2. Elsevier, Amsterdam, pp 127–148

Fegley B, Prinn RG (1989) Estimation of the rate of volcanism on Venus from reaction rate measurements. Nature 337(6202):55–58

Fegley JB, Treiman AH (1992) Chemistry of the surface and lower atmosphere of Venus. Sol Syst Res 26(2):97

Fegley JB, Klingelhöfer G, Lodders K, Widemann T (1997a) Geochemistry of surface-atmosphere interactions on Venus. In: Bougher SW, Hunten DM, Phillips RJ (eds) Venus II: Geology, Geophysics, Atmosphere, and Solar Wind Environment, p 591

Fegley JB, Zolotov MY, Lodders K (1997b) The oxidation state of the lower atmosphere and surface of Venus. Icarus 125(2):416–439

Filiberto J (2014) Magmatic diversity on Venus: constraints from terrestrial analog crystallization experiments. Icarus 231:131–136

Fogel RA, Rutherford MJ (1995) Magmatic volatiles in primitive lunar glasses: I. FTIR and EPMA analyses of Apollo 15 green and yellow glasses and revision of the volatile-assisted fire-fountain theory. Geochim Cosmochim Acta 59(1):201–215

Foley BJ, Smye AJ (2018) Carbon cycling and habitability of Earth-sized stagnant lid planets. Astrobiology 18(7):873–896

Fossati L, Erkaev NV, Lammer H, Cubillos PE, Odert P, Juholovan I, Kislyakova KG, Lendl M, Kubyshkina D, Bauer SJ (2017) Aeronomical constraints to the minimum mass and maximum radius of hot low-mass planets. Astron Astrophys 598:A90. https://doi.org/10.1051/0004-6361/201629716. 1612.05624

Frankland VL, James AD, Carrillo-Sánchez JD, Nesvornỳ D, Pokornỳ P, Plane JM (2017) CO oxidation and O_2 removal on meteoric material in Venus' atmosphere. Icarus 296:150–162

Frenkel MY, Zabalueva EV (1983) Solidification of effusive melt on Venus and on the Earth. Geokhimiia 9:1275–1279

Füri E, Marty B (2015) Nitrogen isotope variations in the solar system. Nat Geosci 8(7):1–8. https://doi.org/10.1038/ngeo2451

Futaana Y, Stenberg Wieser G, Barabash S, Luhmann JG (2017) Solar wind interaction and impact on the Venus atmosphere. Space Sci Rev 212(3):1453–1509. https://doi.org/10.1007/s11214-017-0362-8

Gaillard F, Scaillet B (2014) A theoretical framework for volcanic degassing chemistry in a comparative planetology perspective and implications for planetary atmospheres. Earth Planet Sci Lett 403:307–316. https://doi.org/10.1016/j.epsl.2014.07.009

Gaillard F, Bouhifd MA, Füri E, Malavergne V, Marrocchi Y, Noack L, Ortenzi G, Roskosz M, Vulpius S (2021) The diverse planetary ingassing/outgassing paths produced over billions of years of magmatic activity. Space Sci Rev 217(1):22. https://doi.org/10.1007/s11214-021-00802-1

Gaillard F, Bernadou F, Roskosz M, Bouhifd MA, Marrocchi Y, Iacono-Marziano G, Moreira M, Scaillet B, Rogerie G (2022a) Redox controls during magma ocean degassing. Earth Planet Sci Lett 577:117255. https://doi.org/10.1016/j.epsl.2021.117255

Gaillard F, Malavergne V, Bouhifd MA, Rogerie G (2022b) A speciation model linking the fate of carbon and hydrogen during core–magma ocean equilibration. Earth Planet Sci Lett 577:117266

Geiss J, Gloeckler G (1998) Abundances of deuterium and helium-3 in the protosolar cloud. Space Sci Rev 84:239–250

Genda H, Abe Y (2003) Survival of a proto-atmosphere through the stage of giant impacts: the mechanical aspects. Icarus 164(1):149–162. https://doi.org/10.1016/S0019-1035(03)00101-5

Genda H, Brasser R, Mojzsis S (2017) The terrestrial late veneer from core disruption of a lunar-sized impactor. Earth Planet Sci Lett 480:25–32

Gerya TV (2014) Plume-induced crustal convection: 3D thermomechanical model and implications for the origin of novae and coronae on Venus. Earth Planet Sci Lett 391:183–192. https://doi.org/10.1016/j.epsl.2014.02.005

Ghail RC, Wilson L (2015) A pyroclastic flow deposit on Venus. Geol Soc (Lond) Spec Publ 401(1):97–106

Gillmann C, Tackley P (2014) Atmosphere/mantle coupling and feedbacks on Venus. J Geophys Res, Planets 119(6):1189–1217. https://doi.org/10.1002/2013JE004505

Gillmann C, Chassefière E, Lognonné P (2009) A consistent picture of early hydrodynamic escape of Venus atmosphere explaining present Ne and Ar isotopic ratios and low oxygen atmospheric content. Earth Planet Sci Lett 286(3–4):503–513. https://doi.org/10.1016/j.epsl.2009.07.016

Gillmann C, Golabek GJ, Tackley PJ (2016) Effect of a single large impact on the coupled atmosphere-interior evolution of Venus. Icarus 268:295–312. https://doi.org/10.1016/j.icarus.2015.12.024

Gillmann C, Golabek G, Tackley P, Raymond S (2017) The role of late veneer impacts in the evolution of Venus. In: European Planetary Science Congress, EPSC2017-15

Gillmann C, Golabek G, Raymond S, Schönbächler M, Tackley P, Dehant V, Vinciane D (2020) Dry late accretion inferred from Venus's coupled atmosphere and internal evolution. Nat Geosci 13:1–5. https://doi.org/10.1038/s41561-020-0561-x

Gilmore M, Treiman A, Helbert J, Smrekar S (2017) Venus surface composition constrained by observation and experiment. Space Sci Rev 212(3):1511–1540. https://doi.org/10.1007/s11214-017-0370-8

Golabek G, Emsenhuber A, Jutzi M, Asphaug E, Gerya T (2018) Coupling sph and thermochemical models of planets: methodology and example of a Mars-sized body. Icarus 301:235–246. https://doi.org/10.1016/j.icarus.2017.10.003

Gonnermann HM (2015) Magma fragmentation. Annu Rev Earth Planet Sci 43(1):431–458. https://doi.org/10.1146/annurev-earth-060614-105206

Gough DO (1981) Solar interior structure and luminosity variations. Sol Phys 74(1):21–34. https://doi.org/10.1007/BF00151270

Gounelle M (2011) The asteroid–comet continuum: in search of lost primitivity. Elements 7(1):29–34

Grady MM, Wright IP (2003) Elemental and isotopic abundances of carbon and nitrogen in meteorites. Space Sci Rev 106(1):231–248

Graham R, Pierrehumbert R (2020) Thermodynamic and energetic limits on continental silicate weathering strongly impact the climate and habitability of wet, rocky worlds. Astrophys J 896(2):115

Grinspoon DH (1993) Implications of the high D/H ratio for the sources of water in Venus' atmosphere. Nature 363(6428):428–431. https://doi.org/10.1038/363428a0

Gröller H, Shematovich VI, Lichtenegger HIM, Lammer H, Pfleger M, Kulikov YN, Macher W, Amerstorfer UV, Biernat HK (2010) Venus' atomic hot oxygen environment. J Geophys Res, Planets 115:E12017. https://doi.org/10.1029/2010JE003697

Gronoff G, Arras P, Baraka S, Bell JM, Cessateur G, Cohen O, Curry SM, Drake JJ, Elrod M, Erwin J et al (2020) Atmospheric escape processes and planetary atmospheric evolution. J Geophys Res Space Phys 125(8):e2019JA027639

Grosfils EB, Aubele J, Crumpler L, Gregg TK, Sakimoto S (2000) Volcanism on Earth's seafloor and Venus. In: Environmental Effects on Volcanic Eruptions. Springer, Berlin, pp 113–142

Grosfils EB, Long SM, Venechuk EM, Hurwitz DM, Richards JW, Kastl B, Drury DE, Hardin JS (2011) Geologic Map of the Ganiki Planitia Quadrangle (v–14), Venus. US Geological Survey Scientific Investigations Map

Grott M, Morschhauser A, Breuer D, Hauber E (2011) Volcanic outgassing of CO_2 and H_2O on Mars. Earth Planet Sci Lett 308(3):391–400. https://doi.org/10.1016/j.epsl.2011.06.014

Gunell H, Maggiolo R, Nilsson H, Stenberg Wieser G, Slapak R, Lindkvist J, Hamrin M, De Keyser J (2018) Why an intrinsic magnetic field does not protect a planet against atmospheric escape. Astron Astrophys 614:L3. https://doi.org/10.1051/0004-6361/201832934

Gunell H, Maggiolo R, Nilsson H, Stenberg Wieser G, Slapak R, Lindkvist J, Hamrin M, Johan DK (2018b) Why an intrinsic magnetic field does not protect a planet against atmospheric escape. Astron Astrophys 614:L3. https://doi.org/10.1051/0004-6361/201832934

Haberle RM, Clancy RT, Forget F, Smith MD, Zurek RW (2017) The Atmosphere and Climate of Mars. Cambridge University Press, Cambridge

Haberle RM, Zahnle K, Barlow NG, Steakley KE (2019) Impact degassing of H_2 on early Mars and its effect on the climate system. Geophys Res Lett 46(22):13355–13362

Hakim K, Bower DJ, Tian M, Deitrick R, Auclair-Desrotour P, Kitzmann D, Dorn C, Mezger K, Heng K (2021) Lithologic controls on silicate weathering regimes of temperate planets. Planet Sci J 2(2):49

630

Hamano K, Abe Y (2010) Atmospheric loss and supply by an impact-induced vapor cloud: its dependence on atmospheric pressure on a planet. Earth Planets Space 62(7):599–610

Hamano Y, Ozima M (1978) Earth-atmosphere evolution model based on Ar isotopic data. In: Terrestrial Rare Gases, Center for Academic Publications Japan; Japan Scientific Societies Press Edn, Proceedings of the U.S.-Japan Seminar on Rare Gas Abundance and Isotopic Constraints on the Origin and Evolution of the Earth's Atmosphere, Hakone, Kanagawa, Japan, June 28–July 1, 1977. Center for Academic Publications Japan; Japan Scientific Societies Press, Tokyo, pp 155–171

Hamano K, Abe Y, Genda H (2013) Emergence of two types of terrestrial planet on solidification of magma ocean. Nature 497:607–610. https://doi.org/10.1038/nature12163

Hamano K, Kawahara H, Abe Y, Onishi M, Hashimoto GL (2015) Lifetime and spectral evolution of a magma ocean with a steam atmosphere: its detectability by future direct imaging. Astrophys J 806(2):216. https://doi.org/10.1088/0004-637X/806/2/216. 1505.03552

Hartle RE, Grebowsky JM (1993) Light ion flow in the nightside ionosphere of Venus. J Geophys Res 98(E4):7437–7445. https://doi.org/10.1029/93JE00399

Hartle RE, Donahue TM, Grebowsky JM, Mayr HG (1996) Hydrogen and deuterium in the thermosphere of Venus: solar cycle variations and escape. J Geophys Res 101(E2):4525–4538. https://doi.org/10.1029/95JE02978

Hartmann J, Dürr HH, Moosdorf N, Meybeck M, Kempe S (2012) The geochemical composition of the terrestrial surface (without soils) and comparison with the upper continental crust. Int J Earth Sci 101(1):365–376

Harvey R, Cassidy W (1989) A statistical comparison of Antarctic finds and modern falls-mass frequency distributions and relative abundance by type. Meteoritics 24:9–14

Hashimoto GL, Abe Y (2005) Climate control on Venus: comparison of the carbonate and pyrite models. Planet Space Sci 53(8):839–848. https://doi.org/10.1016/j.pss.2005.01.005

Hashimoto GL, Abe Y, Sugita S (2007) The chemical composition of the early terrestrial atmosphere: formation of a reducing atmosphere from CI-like material. J Geophys Res, Planets 112(E5):E05010. https://doi.org/10.1029/2006JE002844

Hayashi C, Nakazawa K, Mizuno H (1979) Earth's melting due to the blanketing effect of the primordial dense atmosphere. Earth Planet Sci Lett 43(1):22–28

He C, Hörst SM, Lewis NK, Yu X, Moses JI, Kempton EMR, Marley MS, McGuiggan P, Morley CV, Valenti JA et al (2018) Photochemical haze formation in the atmospheres of super-earths and mini-neptunes. Astron J 156(1):38

Head JW, Wilson L (1992) Magma reservoirs and neutral buoyancy zones on Venus: implications for the formation and evolution of volcanic landforms. J Geophys Res 97(E3):3877–3903. https://doi.org/10.1029/92JE00053

Head JW, Crumpler L, Aubele JC, Guest JE, Saunders RS (1992) Venus volcanism: classification of volcanic features and structures, associations, and global distribution from Magellan data. J Geophys Res, Planets 97(E8):13153–13197

Head JW, Wilson L, Ivanov MA, Wordsworth R (2021) Contributions of volatiles to the Venus atmosphere from the observed extrusive volcanic record: implications for the history of the Venus atmosphere. In: Lunar and Planetary Science Conference, p 2143

Herbort O, Woitke P, Helling C, Zerkle AL (2021) The Atmospheres of Rocky Exoplanets II. Influence of surface composition on the diversity of cloud condensates. arXiv e-prints 2111.14144

Herrick RR, Phillips RJ (1994) Effects of the venusian atmosphere on incoming meteoroids and the impact crater population. Icarus 112(1):253–281. https://doi.org/10.1006/icar.1994.1180

Herrick RR, Izenberg N, Ghail R, Gülcher A, Weller M, Bjonnes E, O'Rourke J, Rolf T, Smrekar S, Carter L, Mueller N, Davaille A, Gillmann C, Hensley S, Gerya T, Gilmore M, Avice G, Ivanov M (2022) Resurfacing History and Volcanic Activity of Venus. Space Sci Rev

Hier-Majumder S, Hirschmann MM (2017) The origin of volatiles in the Earth's mantle. Geochem Geophys Geosyst 18(8):3078–3092

Hirose K, Morard G, Sinmyo R, Umemoto K, Hernlund J, Helffrich G, Labrosse S (2017) Crystallization of silicon dioxide and compositional evolution of the Earth's core. Nature 543(7643):99–102

Hodges R Jr, Tinsley BA (1981) Charge exchange in the Venus ionosphere as the source of the hot exospheric hydrogen. J Geophys Res 86(A9):7649–7656. https://doi.org/10.1029/JA086iA09p07649

Hodges RR (2000) Distributions of hot oxygen for Venus and Mars. J Geophys Res 105(E3):6971–6982. https://doi.org/10.1029/1999JE001138

Hodges J, Richard R, Tinsley BA (1986) The influence of charge exchange on the velocity distribution of hydrogen in the Venus exosphere. J Geophys Res 91(A12):13649–13658. https://doi.org/10.1029/JA091iA12p13649

Hoffman J, Hodges R Jr, McElroy M, Donahue T, Kolpin M (1979) Composition and structure of the Venus atmosphere: results from pioneer Venus. Science 205(4401):49–52

Holland HD (1984) Chemical Evolution of the Atmosphere and Oceans. Princeton Univ. Press, Princeton

Holloway JR (1992) Volcanic degassing under thick atmospheres: consequences for magmatic volatiles on Venus. In: Lunar and Planetary Science Conference, vol 23, p 545

Holloway JR (2004) Redox reactions in seafloor basalts: possible insights into silicic hydrothermal systems. Chem Geol 210(1–4):225–230

Holloway JR, Blank JG (1994) Application of experimental results to coh species in natural melts. In: Volatiles in Magmas. de Gruyter, Berlin, pp 187–230

Holloway JR, O'Day PA (2000) Production of CO_2 and H_2 by diking-eruptive events at mid-ocean ridges: implications for abiotic organic synthesis and global geochemical cycling. Int Geol Rev 42(8):673–683

Höning D, Tosi N, Spohn T (2019) Carbon cycling and interior evolution of water-covered plate tectonics and stagnant-lid planets. Astron Astrophys 627:A48

Höning D, Baumeister P, Grenfell JL, Tosi N, Way MJ (2021) Early habitability and crustal decarbonation of a stagnant-lid Venus. J Geophys Res, Planets 126(10):e2021JE006895

Hunten DM (1973) The escape of light gases from planetary atmospheres. J Atmos Sci 30(8):1481–1494

Ikoma M, Elkins-Tanton L, Hamano K, Suckale J (2018) Water partitioning in planetary embryos and protoplanets with magma oceans. Space Sci Rev 214(4):76. https://doi.org/10.1007/s11214-018-0508-3. 1804.09294

Ingersoll AP (1969) The runaway greenhouse: a history of water on Venus. J Atmos Sci 26:1191–1198. https://doi.org/10.1175/1520-0469(1969)026<1191:TRGAHO>2.0.CO;2

Istomin VG, Grechnev KV, Kochnev VA (1983) Venera 13 and Venera 14: mass spectrometry of the atmosphere. Kosm Issled 21:410–420

Ivanov MA, Head JW (1996) Tessera terrain on Venus: a survey of the global distribution, characteristics, and relation to surrounding units from Magellan data. J Geophys Res, Planets 101(E6):14861–14908

Jackson CR, Williams CD, Zhixue D, Bennett NR, Mukhopadhyay S, Fei Y (2021) Incompatibility of argon during magma ocean crystallization. Earth Planet Sci Lett 553:116598

Jacob DJ (1999) Introduction to Atmospheric Chemistry. Princeton University Press, Princeton

Jacobson SA, Dobson C (2022) What does it mean to have no moon? Evidence for an early or no giant impact on Venus. LPI Contrib 2680:2030

James PB, Zuber MT, Phillips RJ (2013) Crustal thickness and support of topography on Venus. J Geophys Res, Planets 118(4):859–875

Jeans JH (1955) The Dynamical Theory of Gases. Dover, New York

Johnson B, Goldblatt C (2015) The nitrogen budget of Earth. Earth-Sci Rev 148:150–173

Johnson CL, Richards MA (2003) A conceptual model for the relationship between coronae and large-scale mantle dynamics on Venus. J Geophys Res, Planets 108(E6):5058

Johnson AP, Cleaves HJ, Dworkin JP, Glavin DP, Lazcano A, Bada JL (2008) The Miller volcanic spark discharge experiment. Science 322(5900):404. https://doi.org/10.1126/science.1161527

Johnstone CP (2020) Hydrodynamic escape of water vapor atmospheres near very active stars. Astrophys J 890(1):79. https://doi.org/10.3847/1538-4357/ab6224. 1912.07027

Johnstone CP, Khodachenko ML, Lüftinger T, Kislyakova KG, Lammer H, Güdel M (2019) Extreme hydrodynamic losses of Earth-like atmospheres in the habitable zones of very active stars. Astron Astrophys 624:L10. https://doi.org/10.1051/0004-6361/201935279. 1904.01063

Johnstone CP, Bartel M, Güdel M (2021) The active lives of stars: a complete description of the rotation and XUV evolution of F, G, K, and M dwarfs. Astron Astrophys 649:A96. https://doi.org/10.1051/0004-6361/202038407. 2009.07695

Johnstone CP, Lammmer H, Kislyakova K, Scherf M, Güdel M (2021) The young Sun's XUV-activity as a constraint for lower CO_2-limits in the Earth's Archean atmosphere. Earth Planet Sci

Head JW III, Wilson L (1986) Volcanic processes and landforms on Venus: theory, predictions, and observations. J Geophys Res, Solid Earth 91(B9):9407–9446

Kallio E, Barabash S (2012) Magnetized Mars: spatial distribution of oxygen ions. Earth Planets Space 64(2):149–156. https://doi.org/10.5047/eps.2011.07.008

Kargel J, Komatsu G, Baker V, Strom R (1993) The volcanology of Venera and Vega landing sites and the geochemistry of Venus. Icarus 103(2):253–275

Kargel JS, Kirk RL, Fegley B Jr, Treiman AH (1994) Carbonate-sulfate volcanism on Venus? Icarus 112(1):219–252

Kasting JF (1988) Runaway and moist greenhouse atmospheres and the evolution of Earth and Venus. Icarus 74:472–494. https://doi.org/10.1016/0019-1035(88)90116-9

Kasting JF, Catling D (2003) Evolution of a habitable planet. Annu Rev Astron Astrophys 41(1):429–463

Kasting JF, Pollack JB (1983) Loss of water from Venus. I. Hydrodynamic escape of hydrogen. Icarus 53(3):479–508. https://doi.org/10.1016/0019-1035(83)90212-9

Kasting JF, Pollack JB, Ackerman TP (1984) Response of Earth's atmosphere to increases in solar flux and implications for loss of water from Venus. Icarus 57:335–355. https://doi.org/10.1016/0019-1035(84)90122-2

Katz RF, Spiegelman M, Langmuir CH (2003) A new parameterization of hydrous mantle melting. Geochem Geophys Geosyst 4(9):1073

Kaula WM (1994) The tectonics of Venus. Philos Trans R Soc Lond Ser A, Phys Eng Sci 349(1690):345–355

Kaula WM (1999) Constraints on Venus evolution from radiogenic argon. Icarus 139(1):32–39. https://doi.org/10.1006/icar.1999.6082

Keddie ST, Head JW (1995) Formation and evolution of volcanic edifices on the dione regio rise, Venus. J Geophys Res, Planets 100(E6):11729–11754

Keller G, Kerr AC (2014) Volcanism, Impacts, and Mass Extinctions: Causes and Effects. Geol. Soc. Am., Boulder. https://doi.org/10.1130/SPE505

Khawja S, Ernst R, Samson C, Byrne P, Ghail R, MacLellan L (2020) Tesserae on Venus may preserve evidence of fluvial erosion. Nat Commun 11(1):1–8

Khodakovsky I (1982) Atmosphere-surface interactions on Venus and implications for atmospheric evolution. Planet Space Sci 30(8):803–817

Kiefer WS, Hager BH (1992) Geoid anomalies and dynamic topography from convection in cylindrical geometry: applications to mantle plumes on Earth and Venus. Geophys J Int 108(1):198–214

King P, Fegley B, Seward T et al (2018) High Temperature Gas-Solid Reactions in Earth and Planetary Processes, vol 84. de Gruyter, Berlin.

Kite ES, Manga M, Gaidos E (2009) Geodynamics and rate of volcanism on massive Earth-like planets. Astrophys J 700(2):1732

Kite ES, Williams JP, Lucas A, Aharonson O (2014) Low palaeopressure of the Martian atmosphere estimated from the size distribution of ancient craters. Nat Geosci 7(5):335–339. https://doi.org/10.1038/ngeo2137. 1304.4043

Klose K, Wood J, Hashimoto A (1992) Mineral equilibria and the high radar reflectivity of Venus mountaintops. J Geophys Res, Planets 97(E10):16353–16369

Kollmann P, Brandt PC, Collinson G, Rong ZJ, Futaana Y, Zhang TL (2016) Properties of planetward ion flows in Venus' magnetotail. Icarus 274:73–82. https://doi.org/10.1016/j.icarus.2016.02.053

Krasnopolsky VA (2007) Chemical kinetic model for the lower atmosphere of Venus. Icarus 191(1):25–37

Kreslavsky MA, Ivanov MA, Head JW (2015) The resurfacing history of Venus: constraints from buffered crater densities. Icarus 250:438–450. https://doi.org/10.1016/j.icarus.2014.12.024

Krissansen-Totton J, Catling DC (2017) Constraining climate sensitivity and continental versus seafloor weathering using an inverse geological carbon cycle model. Nat Commun 8(1):1–15

Krissansen-Totton J, Fortney JJ, Nimmo F (2021a) Was Venus ever habitable? Constraints from a coupled interior–atmosphere–redox evolution model. Planet Sci J 2(5):216

Krissansen-Totton J, Fortney JJ, Nimmo F, Wogan N (2021b) Oxygen false positives on habitable zone planets around sun-like stars. AGU Adv 2(2):e2020AV000294

Krot A, Keil K, Scott E, Goodrich C, Weisberg M (2014) Classification of meteorites and their genetic relationships. Meteor Cosmochem Process 1:1–63

Kruijver A, Höning D, van Westrenen W (2021) Carbon cycling and habitability of massive Earth-like exoplanets. Planet Sci J 2(5):208

Kucinskas AB, Turcotte DL, Arkani-Hamed J (1996) Isostatic compensation of Ishtar Terra, Venus. J Geophys Res, Planets 101(E2):4725–4736

Kulikov YN, Lammer H, Lichtenegger HIM, Terada N, Ribas I, Kolb C, Langmayr D, Lundin R, Guinan EF, Barabash S, Biernat HK (2006) Atmospheric and water loss from early Venus. Planet Space Sci 54(13–14):1425–1444. https://doi.org/10.1016/j.pss.2006.04.021

Lammer H (2013) Origin and Evolution of Planetary Atmospheres. Springer, Berlin. https://doi.org/10.1007/978-3-642-32087-3

Lammer H, Blanc M (2018) From disks to planets: the making of planets and their early atmospheres. An introduction. Space Sci Rev 214(2):60. https://doi.org/10.1007/s11214-017-0433-x

Lammer H, Lichtenegger HIM, Biernat HK, Erkaev NV, Arshukova IL, Kolb C, Gunell H, Lukyanov A, Holmstrom M, Barabash S, Zhang TL, Baumjohann W (2006) Loss of hydrogen and oxygen from the upper atmosphere of Venus. Planet Space Sci 54(13–14):1445–1456. https://doi.org/10.1016/j.pss.2006.04.022

Lammer H, Kasting JF, Chassefière E, Johnson RE, Kulikov YN, Tian F (2008) Atmospheric escape and evolution of terrestrial planets and satellites. Space Sci Rev 139(1–4):399–436. https://doi.org/10.1007/s11214-008-9413-5

Lammer H, Erkaev NV, Fossati L, Juvan I, Odert P, Cubillos PE, Guenther E, Kislyakova KG, Johnstone CP, Lüftinger T, Güdel M (2016) Identifying the 'true' radius of the hot sub-Neptune CoRoT-24b by mass-loss modelling. Mon Not R Astron Soc 461(1):L62–L66. https://doi.org/10.1093/mnrasl/slw095. 1605.03595

Lammer H, Zerkle AL, Gebauer S, Tosi N, Noack L, Scherf M, Pilat-Lohinger E, Güdel M, Grenfell JL, Godolt M, Nikolaou A (2018) Origin and evolution of the atmospheres of early Venus, Earth and Mars. Astron Astrophys Rev 26:2. https://doi.org/10.1007/s00159-018-0108-y

Lammer H, Leitzinger M, Scherf M, Odert P, Burger C, Kubyshkina D, Johnstone C, Maindl T, Schäfer C, Güdel M, Tosi N, Nikolaou A, Marcq E, Erkaev N, Noack L, Kislyakova K, Fossati L, Pilat-Lohinger E, Ragossnig F, Dorfi E (2020a) Constraining the early evolution of Venus and Earth through atmospheric Ar, Ne isotope and bulk K/U ratios. Icarus 339:113551. https://doi.org/10.1016/j.icarus.2019.113551

Lammer H, Scherf M, Kurokawa H, Ueno Y, Burger C, Maindl T, Johnstone CP, Leizinger M, Benedikt M, Fossati L, Kislyakova KG, Marty B, Avice G, Fegley B, Odert P (2020b) Loss and fractionation of noble gas isotopes and moderately volatile elements from planetary embryos and early Venus, Earth and Mars. Space Sci Rev 216(4):74. https://doi.org/10.1007/s11214-020-00701-x. 2011.01064

Lammer H, Brasser R, Johansen A, Scherf M, Leitzinger M (2021) Formation of Venus, Earth and Mars: constrained by isotopes. Space Sci Rev 217(1):7. https://doi.org/10.1007/s11214-020-00778-4. 2102.06173

Landeau M, Olson P, Deguen R, Hirsh BH (2016) Core merging and stratification following giant impact. Nat Geosci 9(10):786–789. https://doi.org/10.1038/ngeo2808

Lebrun T, Massol H, ChassefièRe E, Davaille A, Marcq E, Sarda P, Leblanc F, Brandeis G (2013) Thermal evolution of an early magma ocean in interaction with the atmosphere. J Geophys Res, Planets 118:1155–1176. https://doi.org/10.1002/jgre.20068

Lécuyer C, Simon L, Guyot F (2000) Comparison of carbon, nitrogen and water budgets on Venus and the Earth. Earth Planet Sci Lett 181(1):33–40. https://doi.org/10.1016/S0012-821X(00)00195-3

Lenardic A, Jellinek A, Moresi LN (2008) A climate induced transition in the tectonic style of a terrestrial planet. Earth Planet Sci Lett 271(1–4):34–42

Lewis JS (1970) Venus: atmospheric and lithospheric composition. Earth Planet Sci Lett 10(1):73–80. https://doi.org/10.1016/0012-821X(70)90066-X

Lewis JS, Prinn RG (1984) Planets and Their Atmospheres: Origin and Evolution. Orlando FL Academic Press Inc International Geophysics Series, vol 33

Lichtenberg T, Bower DJ, Hammond M, Boukrouche R, Sanan P, Tsai SM, Pierrehumbert RT (2021) Vertically resolved magma ocean–protoatmosphere evolution: H_2, H_2O, CO_2, CH_4, Co, O_2, and N_2 as primary absorbers. J Geophys Res, Planets 126(2):e2020JE006711

Lichtenegger HIM, Gröller H, Lammer H, Kulikov YN, Shematovich VI (2009) On the elusive hot oxygen corona of Venus. Geophys Res Lett 36(10):L10204. https://doi.org/10.1029/2009GL037575

Lichtenegger HIM, Lammer H, Grießmeier JM, von Kulikov YN, Paris P, Hausleitner W, Krauss S, Rauer H (2010) Aeronomical evidence for higher CO_2 levels during Earth's Hadean epoch. Icarus 210(1):1–7. https://doi.org/10.1016/j.icarus.2010.06.042

Lissauer JJ, de Pater I (2013) Fundamental Planetary Science. Cambridge University Press, Cambridge

López I, Oyarzun R, Márquez A, Doblas-Reyes F, Laurrieta A (1998) Progressive build up of CO_2 in the atmosphere of Venus through multiple volcanic resurfacing events. Earth Moon Planets 81:187–192. https://doi.org/10.1023/A:1006369831384

Lourenço DL, Rozel AB, Gerya T, Tackley PJ (2018) Efficient cooling of rocky planets by intrusive magmatism. Nat Geosci 11:322–327

Lourenço DL, Rozel AB, Ballmer MD, Tackley PJ (2020) Plutonic-squishy lid: a new global tectonic regime generated by intrusive magmatism on Earth-like planets. Geochem Geophys Geosyst 21(4):e2019GC008756

Love S, Brownlee D (1993) A direct measurement of the terrestrial mass accretion rate of cosmic dust. Science 262(5133):550–553

Lundin R, Lammer H, Ribas I (2007) Planetary magnetic fields and solar forcing: implications for atmospheric evolution. Space Sci Rev 129(1–3):245–278. https://doi.org/10.1007/s11214-007-9176-4

Lupu RE, Zahnle K, Marley MS, Schaefer L, Fegley B, Morley C, Cahoy K, Freedman R, Fortney JJ (2014) The atmospheres of earthlike planets after giant impact events. Astrophys J 784(1):27. https://doi.org/10.1088/0004-637X/784/1/27

Lyons T, Reinhard C, Planavsky N (2014) The rise of oxygen in Earth's early ocean and atmosphere. Nature 506:307–315. https://doi.org/10.1038/nature13068

Manske L, Wünnemann K, Güldemeister N (2018) Impact-induced melting by giant impact events. In: EGU General Assembly Conference Abstracts, p 15883

Marcq E, Mills FP, Parkinson CD, Vandaele AC (2018) Composition and chemistry of the neutral atmosphere of Venus. Space Sci Rev 214(1):1–55

Marty B (2012) The origins and concentrations of water, carbon, nitrogen and noble gases on Earth. Earth Planet Sci Lett 313:56–66

Marty B (2020) Origins and early evolution of the atmosphere and the oceans. Geochem Perspect 9(2):135–313. https://doi.org/10.7185/geochempersp.9.2

Marty B (2022) Meteoritic noble gas constraints on the origin of terrestrial volatiles. Icarus 381:115020. https://doi.org/10.1016/j.icarus.2022.115020

Marty B, Zimmermann L, Pujol M, Burgess R, Philippot P (2013) Nitrogen isotopic composition and density of the archean atmosphere. Science 342(6154):101–104. https://doi.org/10.1126/science.1240971

Marty B, Altwegg K, Balsiger H, Bar-Nun A, Bekaert DV, Berthelier JJ, Bieler A, Briois C, Calmonte U, Combi M, De Keyser J, Fiethe B, Fuselier SA, Gasc S, Gombosi TI, Hansen KC, Hässig M, Jackel A, Kopp E, Korth A, Le Roy L, Mall U, Mousis O, Owen T, Reme H, Rubin M, Semon T, Tzou CY, Waite JH, Wurz P (2017) Xenon isotopes in 67P/Churyumov-Gerasimenko show that comets contributed to Earth's atmosphere. Science 356(6342):1069–1072. http://science.sciencemag.org/content/356/6342/1069.abstract

Marty B, Bekaert DV, Broadley MW, Jaupart C (2019) Geochemical evidence for high volatile fluxes from the mantle at the end of the archaean. Nature 575(7783):485–488. https://doi.org/10.1038/s41586-019-1745-7

Massol H, Hamano K, Tian F, Ikoma M, Abe Y, Chassefière E, Davaille A, Genda H, Güdel M, Hori Y, Leblanc F, Marcq E, Sarda P, Shematovich VI, Stökl A, Lammer H (2016) Formation and evolution of protoatmospheres. Space Sci Rev pp:1–59. https://doi.org/10.1007/s11214-016-0280-1

Masunaga K, Futaana Y, Persson M, Barabash S, Zhang TL, Rong ZJ, Fedorov A (2019) Effects of the solar wind and the solar EUV flux on O^+ escape rates from Venus. Icarus 321:379–387. https://doi.org/10.1016/j.icarus.2018.11.017

Matsui T, Abe Y (1986) Impact-induced atmospheres and oceans on Earth and Venus. Nature 322:526–528. https://doi.org/10.1038/322526a0

McComas DJ, Spence HE, Russell CT, Saunders MA (1986) The average magnetic field draping and consistent plasma properties of the Venus magnetotail. J Geophys Res 91(A7):7939–7953. https://doi.org/10.1029/JA091iA07p07939

McCord TB (1968) The loss of retrograde satellites in the solar system. J Geophys Res 73(4):1497–1500. https://doi.org/10.1029/JB073i004p01497

McElroy MB, Prather MJ, Rodriguez JM (1982a) Escape of hydrogen from Venus. Science 215(4540):1614–1615. https://doi.org/10.1126/science.215.4540.1614

McElroy MB, Prather MJ, Rodriguez JM (1982b) Loss of oxygen from Venus. Geophys Res Lett 9(6):649–651. https://doi.org/10.1029/0GPRLA000009000006000649000001

McEnulty TR (2012) Oxygen Loss from Venus and the Influence of Extreme Solar Wind Conditions. University of California, Berkeley. https://www.proquest.com/docview/1322039649

McGill GE (2000) Geologic map of the Sappho Patera quadrangle (V-20), Venus. The Survey

McKinnon WB, Zahnle KJ, Ivanov BA, Melosh HJ (1997) Cratering on Venus: models and observations. In: Bougher SW, Hunten DM, Phillips RJ (eds) Venus II: Geology, Geophysics, Atmosphere, and Solar Wind Environment, p 969

Melosh H (1990) Giant impacts and the thermal state of the early Earth. In: Origin of the Earth, pp 69–83

Melosh HJ (2000) A new and improved equation of state for impact computations. In: Lunar and Planetary Science Conference, p 1903

Melosh H, Kipp M (1989) Giant impact theory of the moon's origin: first 3-d hydrocode results. In: Lunar and Planetary Science Conference, vol 20

Melosh H, Vickery A (1989) Impact erosion of the primordial atmosphere of Mars. Nature 338(6215):487–489

Meschede MA, Myhrvold CL, Tromp J (2011) Antipodal focusing of seismic waves due to large meteorite impacts on Earth. Geophys J Int 187(1):529–537

Miller SL, Urey HC (1959) Organic compound synthesis on the primitive Earth. Science 130(3370):245–251. https://doi.org/10.1126/science.130.3370.245

Way MJ, Del Genio AD (2020) Venusian habitable climate scenarios: modeling Venus through time and applications to slowly rotating Venus-like exoplanets. J Geophys Res, Planets 125(5):e2019JE006276

Way MJ, Del Genio AD, Kiang NY, Sohl LE, Grinspoon DH, Aleinov I, Kelley M, Clune T (2016) Was Venus the first habitable world of our solar system? Geophys Res Lett 43(16):8376–8383

Mojzsis SJ, Harrison TM, Pidgeon RT (2001) Oxygen-isotope evidence from ancient zircons for liquid water at the Earth's surface 4,300 Myr ago. Nature 409(6817):178–181

Monteux J, Coltice N, Dubuffet F, Ricard Y (2007) Thermo-mechanical adjustment after impacts during planetary growth. Geophys Res Lett 34(24):L24201

Morgan JP (1998) Thermal and rare gas evolution of the mantle. Chem Geol 145(3–4):431–445

Morschhauser A, Grott M, Breuer D (2009) Mantle degassing and the origin of the Venusian atmosphere. In: European Planetary Science Congress, p 616

Mueller RF (1964) A chemical model for the lower atmosphere of Venus. Icarus 3(4):285–298

Muenow DW, Keil K, Wilson L (1992) High-temperature mass spectrometric degassing of enstatite chondrites: implications for pyroclastic volcanism on the aubrite parent body. Geochim Cosmochim Acta 56(12):4267–4280. https://doi.org/10.1016/0016-7037(92)90267-M

Nakajima M, Stevenson DJ (2014) Investigation of the initial state of the moon-forming disk: Bridging SPH simulations and hydrostatic models. Icarus 233:259–267. https://doi.org/10.1016/j.icarus.2014.01.008

Nakajima M, Stevenson DJ (2015) Melting and mixing states of the Earth's mantle after the moon-forming impact. Earth Planet Sci Lett 427:286–295. https://doi.org/10.1016/j.epsl.2015.06.023

Nakajima M, Golabek GJ, Wünnemann K, Rubie DC, Burger C, Melosh HJ, Jacobson SA, Manske L, Hull SD (2021) Scaling laws for the geometry of an impact-induced magma ocean. Earth Planet Sci Lett 568:116983

Namiki N, Solomon SC (1998) Volcanic degassing of argon and helium and the history of crustal production on Venus. J Geophys Res, Planets 103(E2):3655–3677. https://doi.org/10.1029/97JE03032

Namiki N, Solomon SC (1998b) Volcanic degassing of argon and helium and the history of crustal production on Venus. J Geophys Res. https://doi.org/10.1029/97JE03032

Neukum G, Ivanov BA, Hartmann WK (2001) Cratering records in the inner solar system in relation to the lunar reference system. In: Chronology and Evolution of Mars. Springer, Berlin, pp 55–86

Newman WI, Symbalisty EM, Ahrens TJ, Jones EM (1999) Impact erosion of planetary atmospheres: some surprising results. Icarus 138(2):224–240

Ni H, Keppler H (2013) Carbon in silicate melts. Rev Mineral Geochem 75(1):251–287

Nikolaou A, Katyal N, Tosi N, Godolt M, Grenfell JL, Rauer H (2019) What factors affect the duration and outgassing of the terrestrial magma ocean? Astrophys J 875(1):11. https://doi.org/10.3847/1538-4357/ab08ed. 1903.07436

Nimmo F (2002) Why does Venus lack a magnetic field? Geology 30(11):987–990

Nimmo F (2015) 8.02 - energetics of the core. In: Schubert G (ed) Treatise on Geophysics, 2nd edn. Elsevier, Oxford, p 27–55. https://doi.org/10.1016/B978-0-444-53802-4.00139-1

Johnson NM, de Oliveira MRR (2019) Venus atmospheric composition in situ data: a compilation. Earth Space Sci 6(7):1299–1318. https://doi.org/10.1029/2018EA000536

Noack L, Breuer D, Spohn T (2012) Coupling the atmosphere with interior dynamics: implications for the resurfacing of Venus. Icarus 217(2):484–498. https://doi.org/10.1016/j.icarus.2011.08.026. Advances in Venus Science

Nordström T, Stenberg G, Nilsson H, Barabash S, Zhang TL (2013) Venus ion outflow estimates at solar minimum: influence of reference frames and disturbed solar wind conditions: Venus ion outflow estimates. J Geophys Res Space Phys 118(6):3592–3601. https://doi.org/10.1002/jgra.50305

Nozette S, Lewis JS (1982) Venus: chemical weathering of igneous rocks and buffering of atmospheric composition. Science 216(4542):181–183. https://doi.org/10.1126/science.216.4542.181

O'Brien DP, Walsh KJ, Morbidelli A, Raymond SN, Mandell AM (2014) Water delivery and giant impacts in the 'Grand Tack' scenario. Icarus 239:74–84. https://doi.org/10.1016/j.icarus.2014.05.009

Odert P, Lammer H, Erkaev N, Nikolaou A, Lichtenegger H, Johnstone C, Kislyakova K, Leitzinger M, Tosi N (2018) Escape and fractionation of volatiles and noble gases from Mars-sized planetary embryos and growing protoplanets. Icarus 307:327–346. https://doi.org/10.1016/j.icarus.2017.10.031

O'Keefe JD, Ahrens TJ (1977) Meteorite impact ejecta: dependence of mass and energy lost on planetary escape velocity. Science 198(4323):1249–1251

Okuchi T (1997) Hydrogen partitioning into Molten iron at high pressure: implications for Earth's core. Science 278(5344):1781–1784

O'Neill C (2021) End-member venusian core scenarios: does Venus have an inner core? Geophys Res Lett 48(17):e2021GL095499

O'Neill C, Debaille V (2014) The evolution of hadean–eoarchaean geodynamics. Earth Planet Sci Lett 406:49–58

O'Neill C, Lenardic A, Höink T, Coltice N (2014) Mantle convection and outgassing on terrestrial planets. In: Comparative Climatology of Terrestrial Planets, pp 473–486

O'Neill C, Marchi S, Zhang S, Bottke W (2017) Impact-driven subduction on the Hadean Earth. Nat Geosci 10:793–797. https://doi.org/10.1038/ngeo3029

Öpik EJ (1963) Selective escape of gases? Geophys J 7(4):490–506. https://doi.org/10.1111/j.1365-246X.1963.tb07091.x

O'Rourke JG (2020) Venus: a thick basal magma ocean may exist today. Geophys Res Lett 47(4):e86126. https://doi.org/10.1029/2019GL086126

O'Rourke JG, Korenaga J (2015) Thermal evolution of Venus with argon degassing. Icarus 260:128–140. https://doi.org/10.1016/j.icarus.2015.07.009

O'Rourke JG, Stevenson DJ (2016) Powering Earth's dynamo with magnesium precipitation from the core. Nature 529(7586):387–389

O'Rourke JG, Gillmann C, Tackley P (2018) Prospects for an ancient dynamo and modern crustal remanent magnetism on Venus. Earth Planet Sci Lett 502:46–56

O'Rourke JG, Buz J, Fu RR, Lillis RJ (2019) Detectability of remanent magnetism in the crust of Venus. Geophys Res Lett 46(11):5768–5777. https://doi.org/10.1029/2019GL082725

 Springer

O'Rourke JG, Wilson CF, Borrelli ME, Byrne PK, Dumoulin C, Ghail R, Gülcher AJP, Jacobson SA, Korablev O, Spohn T, Way M, Weller M, Westall F (2022) Venus, the planet: introduction to the evolution of Earth's sister planet. Space Sci Rev

Ortenzi G, Noack L, Sohl F, Guimond C, Dorn C, Schmidt J, Vulpius S, Katyal N, Kitzmann D, Rauer H (2020) Mantle redox state drives outgassing chemistry and atmospheric composition of rocky planets. Sci Rep 10:10907. https://doi.org/10.1038/s41598-020-67751-7

Owen JE, Wu Y (2016) Atmospheres of low-mass planets: the "boil-off". Astrophys J 817(2):107. https://doi.org/10.3847/0004-637X/817/2/107. 1506.02049

Oyama VI, Carle GC, Woeller F, Pollack JB, Reynolds RT, Craig RA (1980) Pioneer Venus gas chromatography of the lower atmosphere of Venus. J Geophys Res 85:7891–7902

Ozima M, Podosek FA (2002) Noble Gas Geochemistry, 2nd edn. Cambridge University Press, Cambridge

Padovan S, Tosi N, Plesa AC, Ruedas T (2017) Impact-induced changes in source depth and volume of magmatism on Mercury and their observational signatures. Nat Commun 8(1):1–10

Pahlevan K, Stevenson DJ (2007) Equilibration in the aftermath of the lunar-forming giant impact. Earth Planet Sci Lett 262(3–4):438–449

Parfitt EA, Head JW (1993) Buffered and unbuffered dike emplacement on Earth and Venus: implications for magma reservoir size, depth, and rate of magma replenishment. Earth Moon Planets 61(3):249–281. https://doi.org/10.1007/BF00572247

Pearson VK, Sephton M, Gilmour I (2006) Molecular and isotopic indicators of alteration in cr chondrites. Meteorit Planet Sci 41(9):1291–1303

Pepin RO (1991) On the origin and early evolution of terrestrial planet atmospheres and meteoritic volatiles. Icarus 92(1):2–79. https://doi.org/10.1016/0019-1035(91)90036-S

Peplowski PN, Lawrence DJ, Wilson JT (2020) Chemically distinct regions of Venus's atmosphere revealed by measured N_2 concentrations. Nat Astron 4(10):947–950

Persson M, Futaana Y, Fedorov A, Nilsson H, Hamrin M, Barabash S (2018) H^+/O^+ escape rate ratio in the Venus magnetotail and its dependence on the solar cycle. Geophys Res Lett 45(20):10805–10811. https://doi.org/10.1029/2018GL079454

Persson M, Futaana Y, Ramstad R, Masunaga K, Nilsson H, Hamrin M, Fedorov A, Barabash S (2020) The venusian atmospheric oxygen ion escape: extrapolation to the early solar system. J Geophys Res, Planets 125(3):e06336. https://doi.org/10.1029/2019JE006336

Peucker-Ehrenbrink B, Ravizza G (2000) The marine osmium isotope record. Terra Nova 12(5):205–219

Pham LBS, Karatekin Ö, Dehant V (2011) Effects of impacts on the atmospheric evolution: comparison between Mars, Earth, and Venus. Planet Space Sci 59:1087–1092. https://doi.org/10.1016/j.pss.2010.11.010

Phillips RJ, Bullock MA, Hauck SA (2001) Climate and interior coupled evolution on Venus. Geophys Res Lett 28(9):1779–1782

Piani L, Marrocchi Y, Rigaudier T, Vacher LG, Thomassin D, Marty B (2020) Earth's water may have been inherited from material similar to enstatite chondrite meteorites. Science 369(6507):1110–1113. https://doi.org/10.1126/science.aba1948

Pierazzo E, Artemieva N (2012) Local and global environmental effects of impacts on Earth. Elements 8:55–60. https://doi.org/10.2113/gselements.8.1.55

Pierazzo E, Melosh H (2000) Hydrocode modeling of oblique impacts: the fate of the projectile. Meteorit Planet Sci 35(1):117–130

Pierazzo E, Vickery A, Melosh H (1997) A reevaluation of impact melt production. Icarus 127(2):408–423

Pieters CM, Head JW, Pratt S, Patterson W, Garvin J, Barsukov VL, Basilevksy AT, Khodakovsky IL, Selivanov AS, Panfilov AS, Gektin YM, Narayeva YM (1986) The color of the surface of Venus. Science 234(4782):1379–1383. https://doi.org/10.1126/science.234.4782.1379

Pollack JB (1971) A nongrey calculation of the runaway greenhouse: implications for Venus' past and present. Icarus 14(3):295–306

Pollack JB, Black DC (1979) Implications of the gas compositional measurements of pioneer Venus for the origin of planetary atmospheres. Science 205(4401):56–59

Renggli CJ, King PL (2018) SO_2 gas reactions with silicate glasses. Rev Mineral Geochem 84(1):229–255

Ringwood A, Anderson DL (1977) Earth and Venus: a comparative study. Icarus 30(2):243–253

Robert F, Gautier D, Dubrulle B (2000) The solar system D/H ratio: observations and theories. Space Sci Rev 92:201–224. https://doi.org/10.1023/A:1005291127595

Roberts JH, Arkani-Hamed J (2014) Impact heating and coupled core cooling and mantle dynamics on Mars. J Geophys Res, Planets 119(4):729–744. https://doi.org/10.1002/2013JE004603

Roble RG (1995) Major greenhouse cooling (yes, cooling): the upper atmosphere response to increased CO_2. Rev Geophys 33(S1):539–546. https://doi.org/10.1029/95RG00118

Rodriguez JM, Prather MJ, McElroy MB (1984) Hydrogen on Venus: exospheric distribution and escape. Planet Space Sci 32(10):1235–1255. https://doi.org/10.1016/0032-0633(84)90067-9

Rolf T, Zhu MH, Wuennemann K, Werner SC (2017) The role of impact bombardment history in lunar evolution. Icarus 286:138–152

Rolf T, Steinberger B, Sruthi U, Werner S (2018) Inferences on the mantle viscosity structure and the post-overturn evolutionary state of Venus. Icarus 313:107–123. https://doi.org/10.1016/j.icarus.2018.05.014

Rolf T, Weller M, Gülcher A, Byrne P, O'Rourke JG, Mulyukova E, Herrick R, Bjonnes E, Davaille A, Ghail R, Gillmann C, Plesa AC, Smrekar S (2022) Venus' mantle dynamics and evolution through time. Space Sci Rev

Rozel AB, Golabek GJ, Jain C, Tackley PJ, Gerya T (2017) Continental crust formation on early Earth controlled by intrusive magmatism. Nature 545(7654):332–335

Rubin M, Altwegg K, Balsiger H, Bar-Nun A, Berthelier JJ, Briois C, Calmonte U, Combi M, Johan DK, Fiethe B, Fuselier SA, Gasc S, Gombosi TI, Hansen KC, Kopp E, Korth A, Laufer D, Le Roy L, Mall U, Marty B, Mousis O, Owen T, Rème H, Sémon T, Tzou CY, Waite JH, Wurz P (2018) Krypton isotopes and noble gas abundances in the coma of comet 67P/Churyumov-Gerasimenko. Sci Adv 4:1–10. https://doi.org/10.1126/sciadv.aar6297

Ruedas T (2017) Globally smooth approximations for shock pressure decay in impacts. Icarus 289:22–33. https://doi.org/10.1016/j.icarus.2017.02.008

Ruedas T, Breuer D (2017) On the relative importance of thermal and chemical buoyancy in regular and impact-induced melting in a Mars-like planet. J Geophys Res, Planets 122(7):1554–1579

Russell CT, Luhmann JG, Elphic RC, Scarf FL, Brace LH (1982) Magnetic field and plasma wave observations in a plasma cloud at Venus. Geophys Res Lett 9(1):45–48. https://doi.org/10.1029/GL009i001p00045

Sakuraba H, Kurokawa H, Genda H (2019) Impact degassing and atmospheric erosion on Venus, Earth, and Mars during the late accretion. Icarus 317:48–58. https://doi.org/10.1016/j.icarus.2018.05.035. 1805.07094

Salvador A, Samuel H (2021, submitted) Convective outgassing efficiency in planetary magma oceans: insights from computational fluid dynamics. Icarus

Salvador A, Massol H, Davaille A, Marcq E, Sarda P, Chassefière E (2017) The relative influence of H_2O and CO_2 on the primitive surface conditions and evolution of rocky planets. J Geophys Res, Planets 122(7):1458–1486

Salvador A, Avice G, Breuer D, Gillmann C, Jacobson S, Lammer H, Marcq E, Raymond SN, Sakuraba H, Scherf M, Way M (2022) Magma ocean, water, and the early atmosphere of Venus. Space Sci Rev

Schaber GG, Strom RG, Moore HJ, Soderblom LA, Kirk RL, Chadwick DJ, Dawson DD, Gaddis LR, Boyce JM, Russell J (1992) Geology and distribution of impact craters on Venus: what are they telling us? J Geophys Res, Planets 97(E8):13257–13301. https://doi.org/10.1029/92JE01246

Schaefer L, Fegley FJ (2017) Redox states of initial atmospheres outgassed on rocky planets and planetesimals. Astrophys J 843(2):120. https://doi.org/10.3847/1538-4357/aa784f

Schaefer L, Fegley B (2010) Chemistry of atmospheres formed during accretion of the Earth and other terrestrial planets. Icarus 208(1):438–448. https://doi.org/10.1016/j.icarus.2010.01.026

Schaefer L, Wordsworth RD, Berta-Thompson Z, Sasselov D (2016) Predictions of the atmospheric composition of GJ 1132b. Astrophys J 829(2):63. https://doi.org/10.3847/0004-637X/829/2/63. 1607.03906

Schlichting HE, Mukhopadhyay S (2018) Atmosphere impact losses. Space Sci Rev 214(1):34. https://doi.org/10.1007/s11214-018-0471-z

Schlichting HE, Sari R, Yalinewich A (2015) Atmospheric mass loss during planet formation: the importance of planetesimal impacts. Icarus 247:81–94. https://doi.org/10.1016/j.icarus.2014.09.053

Segura TL, Toon OB, Colaprete A, Zahnle K (2002) Environmental effects of large impacts on Mars. Science 298(5600):1977–1980

Segura TL, Toon OB, Colaprete A (2008) Modeling the environmental effects of moderate-sized impacts on Mars. J Geophys Res, Planets 113(E11):E11007

Segura TL, McKay CP, Toon OB (2012) An impact-induced, stable, runaway climate on Mars. Icarus 220(1):144–148

Seki K, Elphic RC, Hirahara M, Terasawa T, Mukai T (2001) On atmospheric loss of oxygen ions from Earth through magnetospheric processes. Science 291(5510):1939–1941

Semprich J, Filiberto J, Treiman AH (2020) Venus: a phase equilibria approach to model surface alteration as a function of rock composition, oxygen- and sulfur fugacities. Icarus 346:113779

Shkuratov YG, Kreslavsky MA, Nikolayeva OV (1987) Diagram Albedo-color of Venus surface according to Venera-13 data. In: Lunar and Planetary Science Conference, vol 18, p 914

Shuvalov V (2009) Atmospheric erosion induced by oblique impacts. Meteorit Planet Sci 44(8):1095–1105

Shuvalov V, Artemieva N, Kuz'micheva MY, Losseva T, Svettsov V, Khazins V (2012) Crater ejecta: markers of impact catastrophes. Izv Phys Solid Earth 48(3):241–255

Shuvalov V, Kührt E, de Niem D, Wünnemann K (2014) Impact induced erosion of hot and dense atmospheres. Planet Space Sci 98:120–127

Rasool SI, de Bergh C (1970) The runaway greenhouse and the accumulation of CO_2 in the Venus atmosphere. Nature 226(5250):1037–1039

Simons M, Hager BH, Solomon SC (1994) Global variations in the geoid/topography admittance of Venus. Science 264(5160):798–803

Sleep NH, Zahnle K (2001) Carbon dioxide cycling and implications for climate on ancient Earth. J Geophys Res, Planets 106(E1):1373–1399

Smrekar SE, Davaille A, Sotin C (2018) Venus interior structure and dynamics. Space Sci Rev 214(5):1–34

Smrekar SE, Ghail R, Byrne P, Gülcher A, Garcia RF, Herrick R, Gerya T, O'Rourke J, Davaille A, Mulyukova E, Rolf T, Plesa AC, Shellnutt G, Ivanov M, Borrelli M (2022) Volcano- tectonic processes on Venus. Space Sci Rev pp:1–34

Solomatov V, Moresi LN (1996) Stagnant lid convection on Venus. J Geophys Res, Planets 101(E2):4737–4753

Solomatova NV, Caracas R (2021) Genesis of a CO_2-rich and H_2O-depleted atmosphere from Earth's early global magma ocean. Sci Adv 7(41):eabj0406

Solomon SC, Bullock MA, Grinspoon DH (1999) Climate change as a regulator of tectonics on Venus. Science 286(5437):87–90

Som D, Catling S, Harnmeijer J, Polivka P, Buick R (2012) Air density 2.7 billion years ago limited to less than twice modern levels by fossil raindrop imprints. Nature 484:359. https://doi.org/10.1038/nature10890

Som S, Buick R, Hagadorn J, Blake T, Perreault J, Harnmeijer J, Catling D (2016) Earth's air pressure 2.7 billion years ago constrained to less than half of modern levels. Nat Geosci 9:448. https://doi.org/10.1038/ngeo2713

Sossi PA, Burnham AD, Badro J, Lanzirotti A, Newville M, O'Neill HSC (2020) Redox state of Earth's magma ocean and its Venus-like early atmosphere. Sci Adv 6(48):eabd1387

Sparks RSJ, Barclay J, Jaupart C, Mader HM, Phillips JC (1994) Physical aspects of magma degassing I. Experimental and theoretical constraints on vesiculation. In: Carroll MR, Holloway JR (eds) Volatiles in Magmas. Mineralogical Society of America, Washington, pp 413–446. https://doi.org/10.1515/9781501509674-017. Chapter 11a

Stanley BD, Hirschmann MM, Withers AC (2014) Solubility of COH volatiles in graphite-saturated Martian basalts. Geochim Cosmochim Acta 129:54–76

Stixrude L, Scipioni R, Desjarlais MP (2020) A silicate dynamo in the early Earth. Nat Commun 11(1):1–5

Strangeway R, Russell C, Luhmann J, Moore T, Foster J, Barabash S, Nilsson H (2010) Does a planetary-scale magnetic field enhance or inhibit ionospheric plasma outflows? In: Agu Fall Meeting Abstracts, vol 2010, SM33B-1893

Strom RG, Schaber GG, Dawsow DD (1994) The global resurfacing of Venus. J Geophys Res 99(E5):10899–10926. https://doi.org/10.1029/94JE00388

Surkov YA, Barsukov V, Moskalyeva L, Kharyukova V, Kemurdzhian A (1984) New data on the composition, structure, and properties of Venus rock obtained by Venera 13 and Venera 14. J Geophys Res, Solid Earth 89(S02):B393–B402

Surkov YA, Moskalyova L, Kharyukova V, Dudin A, Smirnov G, Zaitseva SY (1986) Venus rock composition at the Vega 2 landing site. J Geophys Res, Solid Earth 91(B13):E215–E218

Surkov YA, Kirnozov FF, Glazov VN, Dunchenko AG, Tatsy LP, Sobornov OP (1987) Uranium, thorium, and potassium in the venusian rocks at the landing sites of Vega 1 and 2. J Geophys Res, Solid Earth 92(B4):E537–E540. https://doi.org/10.1029/JB092iB04p0E537

Svetsov V (2007) Atmospheric erosion and replenishment induced by impacts of cosmic bodies upon the Earth and Mars. Sol Syst Res 41(1):28–41

Svetsov V, Shuvalov V (2015) Water delivery to the moon by asteroidal and cometary impacts. Planet Space Sci 117:444–452

Swindle TD (2002) Martian noble gases. Rev Mineral Geochem 47(1):171–190. https://doi.org/10.2138/rmg.2002.47.6

Tagawa S, Sakamoto N, Hirose K, Yokoo S, Hernlund J, Ohishi Y, Yurimoto H (2021) Experimental evidence for hydrogen incorporation into Earth's core. Nat Commun 12(1):1–8

Tanner S, Kerrick D, Lasaga A (1985) Experimental kinetic study of the reaction; calcite+ quartz<-> wollastonite+ carbon dioxide, from 1 to 3 kilobars and 500 degrees to 850 degrees C. Am J Sci 285(7):577–620

Tarduno JA, Blackman EG, Mamajek EE (2014) Detecting the oldest geodynamo and attendant shielding from the solar wind: implications for habitability. Phys Earth Planet Inter 233:68–87

Teffeteller H, Filiberto J, McCanta M, Treiman A, Keller L, Cherniak D, Rutherford M, Cooper R (2022) An experimental study of the alteration of basalt on the surface of Venus. Icarus 384:115085

Terada N, Machida S, Shinagawa H (2002) Global hybrid simulation of the Kelvin-Helmholtz instability at the Venus ionopause. J Geophys Res Space Phys 107(A12):1471. https://doi.org/10.1029/2001JA009224

Tian F (2009) Thermal escape from super Earth atmospheres in the habitable zones of M stars. Astrophys J 703(1):905–909. https://doi.org/10.1088/0004-637X/703/1/905

Tian F, Kasting JF, Liu HL, Roble RG (2008a) Hydrodynamic planetary thermosphere model: 1. Response of the Earth's thermosphere to extreme solar EUV conditions and the significance of adiabatic cooling. J Geophys Res, Planets 113(E5):E05008. https://doi.org/10.1029/2007JE002946

Tian F, Solomon SC, Qian L, Lei J, Roble RG (2008b) Hydrodynamic planetary thermosphere model: 2. Coupling of an electron transport/energy deposition model. J Geophys Res, Planets 113(E7):E07005. https://doi.org/10.1029/2007JE003043

Tian F, Güdel M, Johnstone CP, Lammer H, Luger R, Odert P (2018) Water loss from young planets. Space Sci Rev 214(3):65. https://doi.org/10.1007/s11214-018-0490-9

Tosi N, Godolt M, Stracke B, Ruedas T, Grenfell JL, Höning D, Nikolaou A, Plesa AC, Breuer D, Spohn T (2017) The habitability of a stagnant-lid Earth. Astron Astrophys 605:A71

Tu L, Johnstone CP, Güdel M, Lammer H (2015) The extreme ultraviolet and X-ray sun in time: high-energy evolutionary tracks of a solar-like star. Astron Astrophys 577:L3. https://doi.org/10.1051/0004-6361/201526146. 1504.04546

Tucker JM, van Keken PE, Ballentine CJ (2022) Earth's missing argon paradox resolved by recycling of oceanic crust. Nat Geosci. https://doi.org/10.1038/s41561-021-00870-6

Turbet M (2018) Habitability of planets using numerical climate models. Application to extrasolar planets and early Mars. PhD thesis, Sorbonne Université

Turbet M, Bolmont E, Chaverot G, Ehrenreich D, Leconte J, Marcq E (2021) Day–night cloud asymmetry prevents early oceans on Venus but not on Earth. Nature 598(7880):276–280

Urey HC (1952) On the early chemical history of the Earth and the origin of life. Proc Natl Acad Sci 38(4):351–363. https://doi.org/10.1073/pnas.38.4.351

Vickery AM, Melosh HJ (1990) Atmospheric erosion and impactor retention in large impacts, with application to mass extinctions. In: Global Catastrophes in Earth History, vol 247, pp 289–300

Volkov VP, Frenkel MY (1993) The modeling of Venus' degassing in terms of K-Ar system. Earth Moon Planets 62(2):117–129. https://doi.org/10.1007/BF00572140

von Zahn U, Kumar S, Niemann H, Prinn R (1983) In: Composition of the Venus Atmosphere. The University of Arizona Press, Tucson, p 299

Wadhwa M (2008) Redox conditions on small bodies, the moon and Mars. Rev Mineral Geochem 68(1):493–510

Walker JC (1975) Evolution of the atmosphere of Venus. J Atmos Sci 32(6):1248–1256

Walker JC, Hays P, Kasting JF (1981) A negative feedback mechanism for the long-term stabilization of Earth's surface temperature. J Geophys Res, Oceans 86(C10):9776–9782

Wallace PJ, Plank T, Edmonds M, Hauri EH (2015) Chap. 7 – volatiles in magmas. In: Sigurdsson H (ed) The Encyclopedia of Volcanoes, 2nd edn. Academic Press, San Diego, pp 163–183. https://doi.org/10.1016/B978-0-12-385938-9.00007-9

Wang Z, Shi F, Zhang J (2020) Effects of water on the rheology of dominant minerals and rocks in the continental lower crust: a review. J Earth Sci 31(6):1170–1182

Ward WR, Reid MJ (1973) Solar tidal friction and satellite loss. Mon Not R Astron Soc 164:21. https://doi.org/10.1093/mnras/164.1.21

Warren AO, Kite ES (2021) Degassing, decarbonation, and dehydration: investigating the likelihood of a habitable era on Venus. In: 52nd Lunar and Planetary Science Conference, p 1253

Watters WA, Zuber M, Hager BH (2009) Thermal perturbations caused by large impacts and consequences for mantle convection. J Geophys Res, Planets 114(E2):E02001

Way M, Ernst RE, Scargle JD (2022a) Large-scale volcanism and the heat death of terrestrial worlds. Planet Sci J 3(4):92

Way MJ, Ostberg C, Foley BJ, Gillmann C, Höning D, Lammer H, O'Rourke J, Persson M, Plesa AC, Scherf M, Weller M (2022b) Synergies between Venus & exoplanetary observations. Space Sci Rev

Wedepohl KH (1995) The composition of the continental crust. Geochim Cosmochim Acta 59(7):1217–1232

Wei Y, Fraenz M, Dubinin E, Woch J, Lühr H, Wan W, Zong QG, Zhang T, Pu Z, Fu S et al (2012) Enhanced atmospheric oxygen outflow on Earth and Mars driven by a corotating interaction region. J Geophys Res Space Phys 117(A3):A03208

Weller MB, Kiefer WS (2020) The physics of changing tectonic regimes: implications for the temporal evolution of mantle convection and the thermal history of Venus. J Geophys Res, Planets 125(1):e05960. https://doi.org/10.1029/2019JE005960

Weller MB, Kiefer WS (2021) Punctuated evolution of the venusian atmosphere from mantle outgassing. In: 52nd Lunar and Planetary Science Conference, p 1555

Weller M, Lenardic A, O'Neill C (2015) The effects of internal heating and large scale climate variations on tectonic bi-stability in terrestrial planets. Earth Planet Sci Lett 420:85–94

Weller M, Evans A, Ibarra D, Johnson A, Kukla T (2022) Atmospheric evidence of early plate tectonics on Venus. LPI Contrib 2678:2328

Westall F, Höning D, Avice G, Gillmann C, Way M (2022) The Habitability of Venus. Space Sci Rev

Wetzel DT, Rutherford MJ, Jacobsen SD, Hauri EH, Saal AE (2013) Degassing of reduced carbon from planetary basalts. Proc Natl Acad Sci 110(20):8010–8013

Widemann T, Smrekar SE, Garvin JB, Straume-Lindner AG, Ocampo AC, Voirin T, Hensley S, Darby Dyar M, Whitten JL, Nunes DC, Getty SA, Arney GN, Johnson NM, Kohler E, Spohn T, O'Rourke JG, Wilson C, Way MJ, Ostberg C, Westall F, Höning D, Jacobson S, Salvador A, Avice G, Carter L, Gilmore M, Ghail R, Helbert J, Byrne P, Herrick RR, Izenberg N, Marcq E, Rolf T, Weller M, Gillmann C, Korablev O, Zelenyi L, Zasova L, Gorinov D (2022) Venus Evolution Through Time: Key Science Questions, Selected Mission Concepts and Future Investigations. Space Sci Rev

Williams CD, Mukhopadhyay S (2018) Capture of nebular gases during Earth's accretion is preserved in deep-mantle neon. Nature 50:202. https://doi.org/10.1038/s41586-018-0771-1

Wilson L, Head JW (1983) A comparison of volcanic eruption processes on Earth, moon, Mars, io and Venus. Nature 302(5910):663–669

Wogan N, Krissansen-Totton J, Catling DC (2020) Abundant atmospheric methane from volcanism on terrestrial planets is unlikely and strengthens the case for methane as a biosignature. Planet Sci J 1(3):58

Wolff RS, Goldstein BE, Yeates CM (1980) The onset and development of Kelvin-Helmholtz instability at the Venus ionopause. J Geophys Res 85:7697–7707. https://doi.org/10.1029/JA085iA13p07697

Wood JA (1997) Rock weathering on the surface of Venus. In: Bougher SW, Hunten DM, Phillips RJ (eds) Venus II: Geology, Geophysics, Atmosphere, and Solar Wind Environment, p 637

Wordsworth RD (2016) Atmospheric nitrogen evolution on Earth and Venus. Earth Planet Sci Lett 447:103–111

Wordsworth RD, Schaefer LK, Fischer RA (2018) Redox evolution via gravitational differentiation on low-mass planets: implications for abiotic oxygen, water loss, and habitability. Astron J 155(5):195. https://doi.org/10.3847/1538-3881/aab608. 1710.00345

Yamanoi Y, Nakashima S, Katsura M (2009) Temperature dependence of reflectance spectra and color values of hematite by in situ, high-temperature visible micro-spectroscopy. Am Mineral 94(1):90–97

Yang J, Boué G, Fabrycky DC, Abbot DS (2014) Strong dependence of the inner edge of the habitable zone on planetary rotation rate. Astrophys J 787:L2. https://doi.org/10.1088/2041-8205/787/1/L2. 1404.4992

Yoshida T, Kuramoto K (2020) Sluggish hydrodynamic escape of early Martian atmosphere with reduced chemical compositions. Icarus 345:113740. https://doi.org/10.1016/j.icarus.2020.113740

Zahnle K, Kasting JF, Pollack JB (1990) Mass fractionation of noble gases in diffusion-limited hydrodynamic hydrogen escape. Icarus 84(2):502–527. https://doi.org/10.1016/0019-1035(90)90050-J

Zahnle K, Levison H, Dones L, Schenk P (1999) Cratering rates in the outer solar system. In: Bulletin of the American Astronomical Society, vol 31, p 1109

Zahnle K, Arndt N, Cockell C, Halliday A, Nisbet E, Selsis F, Sleep NH (2007) Emergence of a habitable planet. Space Sci Rev 129(1–3):35–78

Zahnle KJ, Catling DC, Claire MW (2013) The rise of oxygen and the hydrogen hourglass. Chem Geol 362:26–34. https://doi.org/10.1016/j.chemgeo.2013.08.004

Zahnle KJ, Gacesa M, Catling DC (2019) Strange messenger: a new history of hydrogen on Earth, as told by xenon. Geochim Cosmochim Acta 244:56–85. https://doi.org/10.1016/j.gca.2018.09.017

Zahnle KJ, Lupu R, Catling DC, Wogan N (2020) Creation and evolution of impact-generated reduced atmospheres of early Earth. Planet Sci J 1(1):11

Zerkle A, Mikhail S (2017) The geobiological nitrogen cycle: from microbes to the mantle. Geobiology 15(3):343–352

Ziegler LB, Stegman DR (2013) Implications of a long-lived basal magma ocean in generating Earth's ancient magnetic field. Geochem Geophys Geosyst 14(11):4735–4742. https://doi.org/10.1002/2013GC005001

Zolotov MY (1996) A model for the thermal equilibrium of the surface venusian atmosphere. Geochem Int 33(10):80–100

Zolotov MY (2018) Gas–solid interactions on Venus and other solar system bodies. Rev Mineral Geochem 84(1):351–392. https://doi.org/10.2138/rmg.2018.84.10

Zolotov M (2019) Chemical weathering on Venus. In: Oxford Research Encyclopedia of Planetary Science. Oxford University Press, London, p 146. https://doi.org/10.1093/acrefore/9780190647926.013.146

Zolotov MY (2020) Water-CO_2-basalt interactions on terrestrial planets and exoplanets. In: Exoplanets in Our Backyard: Solar System and Exoplanet Synergies on Planetary Formation, Evolution, and Habitability, vol 2195, p 3062

Zolotov MY, Matsui T (2002) Chemical models for volcanic gases on Venus. In: Lunar and Planetary Science Conference, p 1433

Zolotov MY, Mironenko MV (2009) On the composition of putative oceans on early Venus. In: Venus Geochemistry: Progress, Prospects, and New Missions, LPI Contributions, vol 1470, pp 53–54

Zolotov M, Volkov V (1992) Chemical processes on the planetary surface. In: Venus Geology, Geochemistry and Geophysics. University of Arizona Press, Tucson, pp 177–199

Zolotov MY, Fegley B Jr, Lodders K (1997) Hydrous silicates and water on Venus. Icarus 130(2):475–494

Publisher's Note Springer Nature remains neutral with regard to jurisdictional claims in published maps and institutional affiliations.

Authors and Affiliations

Cedric Gillmann[1] · M.J. Way[2,3] · Guillaume Avice[4] · Doris Breuer[5] · Gregor J. Golabek[6] · Dennis Höning[7,8] · Joshua Krissansen-Totton[9] · Helmut Lammer[10] · Joseph G. O'Rourke[11] · Moa Persson[12] · Ana-Catalina Plesa[13] · Arnaud Salvador[14,15,16] · Manuel Scherf[10,17,18] · Mikhail Y. Zolotov[11]

✉ D. Breuer
doris.breuer@dlr.de

C. Gillmann
cedric.gillmann@rice.edu

M.J. Way
Michael.J.Way@nasa.gov

G. Avice
avice@ipgp.fr

G.J. Golabek
gregor.golabek@uni-bayreuth.de

D. Höning
dennis.hoening@pik-potsdam.de

J. Krissansen-Totton
jkt@ucsc.edu

H. Lammer
helmut.lammer@oeaw.ac.at

M. Persson
moa.persson@irap.omp.eu

A. Salvador
arnaudsalvador@arizona.edu

M. Scherf
manuel.scherf@oeaw.ac.at

M.Y. Zolotov
zolotov@asu.edu

[1] Department of Earth, Environmental and Planetary Sciences, Rice University, Houston, TX 77005, USA

[2] NASA Goddard Institute for Space Studies, 2880 Broadway, New York, NY 10025, USA

[3] Theoretical Astrophysics, Department of Physics and Astronomy, Uppsala University, Uppsala, Sweden

[4] Université Paris Cité, Institut de physique du globe de Paris, CNRS, 75005 Paris, France

[5] DLR, Institute of Planetary Research, 12489 Berlin, Germany

[6] Bayerisches Geoinstitut, University of Bayreuth, 95440 Bayreuth, Germany

[7] Potsdam Institute for Climate Impact Research, Potsdam, Germany

[8] Department of Earth Sciences, VU, Amsterdam, The Netherlands

[9] Department of Astronomy and Astrophysics, University of California, Santa Cruz, CA, USA

[10] Space Research Institute, Austrian Academy of Sciences, Graz, Austria

[11] School of Earth and Space Exploration, Arizona State University, Tempe, USA

[12] Institut de Recherche en Astrophysique et Planétologie, Centre National de la Recherche Scientifique, Université Paul Sabatier - Toulouse III, Centre National d'Etudes Spatiales, Toulouse, France

[13] Institute of Planetary Research, DLR, Berlin, Germany

[14] Department of Astronomy and Planetary Science, Northern Arizona University, Box 6010, Flagstaff, AZ 86011, USA

[15] Habitability, Atmospheres, and Biosignatures Laboratory, University of Arizona, Tucson, AZ, USA

[16] Lunar and Planetary Laboratory, University of Arizona, Tucson, AZ, USA

[17] Institute of Physics, University of Graz, Graz, Austria

[18] Institute for Geodesy, Technical University, Graz, Austria

Space Science Reviews (2023) 219:56
https://doi.org/10.1007/s11214-023-00992-w

Venus Evolution Through Time: Key Science Questions, Selected Mission Concepts and Future Investigations

**Thomas Widemann · Suzanne E. Smrekar · James B. Garvin ·
Anne Grete Straume-Lindner · Adriana C. Ocampo · Mitchell D. Schulte et al.** *[full
author details at the end of the article]*

Received: 28 September 2022 / Accepted: 7 August 2023 / Published online: 3 October 2023
© The Author(s) 2023

Abstract

In this work we discuss various selected mission concepts addressing Venus evolution
through time. More specifically, we address investigations and payload instrument concepts supporting scientific goals and open questions presented in the companion articles
of this volume. Also included are their related investigations (observations & modeling) and
discussion of which measurements and future data products are needed to better constrain
Venus' atmosphere, climate, surface, interior and habitability evolution through time. A new
fleet of Venus missions has been selected, and new mission concepts will continue to be considered for future selections. Missions under development include radar-equipped ESA-led
EnVision M5 orbiter mission (European Space Agency 2021), NASA-JPL's VERITAS orbiter mission (Smrekar et al. 2022a), NASA-GSFC's DAVINCI entry probe/flyby mission
(Garvin et al. 2022a). The data acquired with the VERITAS, DAVINCI, and EnVision from
the end of this decade will fundamentally improve our understanding of the planet's long
term history, current activity and evolutionary path. We further describe future mission concepts and measurements beyond the current framework of selected missions, as well as the
synergies between these mission concepts, ground-based and space-based observatories and
facilities, laboratory measurements, and future algorithmic or modeling activities that pave
the way for the development of a Venus program that extends into the 2040s (Wilson et al.
2022).

Keywords Venus · Planetary system formation · Geological processes · Atmospheric
dynamics · Atmospheric chemistry · Space instrumentation · Surface processes · Interior
structure · Thermal state · Synthetic aperture radar · Subsurface sounder · Radio-science ·
Multispectral imager · Ground and space-based observatories

1 Introduction

Each of the companion articles in this collection has identified key open questions about
the evolution of Venus' atmosphere, climate, surface, interior and habitability through time,
as well as the measurements or approaches that are needed to address them. To capture the

---

Venus: Evolution Through Time
Edited by Colin F. Wilson, Doris Breuer, Cédric Gillmann, Suzanne E. Smrekar, Tilman Spohn and
Thomas Widemann

A correction to the original article reprinted here can be found online in *Space Sci Rev* **219**, 72 (2023),
https://doi.org/10.1007/s11214-023-01022-5, also reprinted in the Appendix of this chapter.

---

wide variety of scientific domains and fields covered in this collection, and before describing current and future investigations to address these questions, we provide a summary of their conclusions as well as open questions regarding the dynamical properties and various processes of the present-day atmosphere.

VERITAS (Smrekar et al. 2022a), DAVINCI (Garvin et al. 2022a), and EnVision (European Space Agency 2021) will greatly advance our understanding and lead to new questions about the evolution of Venus through time. Key advances will come from new types of data to better constrain the interior, such as improved crustal thickness and structure, mantle viscosity/temperature from seismology, lithospheric thickness from electromagnetic sounding, in-situ heat flow to constrain thermal lithospheric thickness and radiogenic heat budget and distribution. Over the next 15 years, these three missions will work together to answer many of the outstanding questions in Venus science and rocky planet evolution described above (Fig. 2; Table 1).

In addition, several Venus missions are under consideration on in development: Russia's Venera-D orbiter, descent module and lander mission (Zasova et al. 2020); an Indian radar-equipped orbiter, Shukrayaan-1 (Antonita 2022); a Chinese radar-equipped orbiter, VOICE (Dong et al. 2023); Rocket Lab's private, low-cost "Morning Star" concept mission to Venus (Seager et al. 2021). Their science observation strategy is under competitive study or development. Furthermore, various mission concepts, whether from landers, from aerial platforms or from orbit require further technology development beyond the current framework of selected missions to enable long-term surface science (seismic, compositional, heat flow investigations); missions that take advantage of mobility in the surface, near-surface, and atmospheric environments; and collection and return of atmospheric samples to Earth (Wilson et al. 2022; Limaye and Garvin 2023). Therefore, the current definition phase is an ideal time to collate knowledge of Venus long-term evolution scenarios and the observations needed to distinguish between them. These questions are left for future investigators to address through a wide range of research approaches, including Earth-based observations, laboratory and modeling studies based on existing data, and future new spacecraft missions.

Section 2 presents an overview of conclusions and open questions from the companion papers in the following order: (1) Comparison of Venus with exoplanets; (2) Venus initial conditions; (3) Venus surface processes, surface age and evidence for volcanic and tectonic activity; (4) Interior regime through history, water and other volatiles.

Sections 3-8 outline the science objectives of upcoming and future missions, in addition to their observational strategy, including expectations for addressing the conclusions summarized in Sect. 2. In June 2021, NASA selected two missions in its Discovery program: VERITAS (Venus Emissivity, Radio Science, InSAR, Topography, and Spectroscopy) (Smrekar et al. 2022a) and DAVINCI (Deep Atmosphere Venus Investigation of Noble Gases, Chemistry, and Imaging), a descent chemistry/imaging probe coupled with a carrier, relay and imaging spacecraft (Garvin et al. 2022a). EnVision has been selected as ESA's 5th Medium-class mission in the agency's Cosmic Vision plan, and is targeted for launch in the early 2030s. The mission is a partnership between ESA and NASA, with NASA providing the Synthetic Aperture Radar (European Space Agency 2021). In this Section, mission design proposals Venera-D (Zasova et al. 2020) and Shukrayaan-1 (Antonita 2022) are also described in some detail.

Sections 9-12 address future mission concepts and measurements that require further technology development beyond the current framework of selected missions, future mission concepts, the synergies between currently selected missions and future laboratory measurements in different experimental setups, and expected modeling activities to address the evolution of Venus' climate, surface, interior and habitability through time. Complementarity with non-Venus missions (e.g., exoplanet observatories) is also addressed. The review concludes with a Summary and Conclusions (Sect. 13), which discusses key questions about Venus' evolution can be answered convincingly with the current and next-generation mission concepts, and which fundamental questions will remain open for future investigations.

2 Open Science Questions and Required Investigations to Address Venus Evolution Through Time

This section presents an overview of conclusions and open questions from the companion articles, organized along the following science themes:

Comparison of Venus with Exoplanets, summarizes conclusions and open questions about how Venus' ancient evolution can inform exoplanet studies regarding the importance of primordial & basal magma oceans and their evolution toward habitability, and, conversely, how terrestrial exoplanet studies can inform Venus' evolutionary history. We summarize conclusions of Way et al. (2023, this collection) and Westall et al. (2023, this collection) regarding water inventory, early tectonics, and volatile cycling between the interior and atmosphere of Venus, and whether liquid water ever existed on the surface at temperatures conducive to the emergence of life. We explore the longevity of a habitable Venus, the divergent paths for planets in the Venus Zone (Kane et al. 2014), and the conditions of Venus evolution from a habitable to an inhospitable planet.

Initial Conditions, Accretion, and Early Venus, discusses modeling and observational constraints on early Venus based on different accretion scenarios. How did the accretion of Venus and Earth differ? Is Venus a more primordial or primitive body than Earth? Was there a late giant impact and to what extent it could have affected its initial thermal state and differentiation? What are the processes driving the thermal evolution of the latter and the concurrent early atmosphere formation, volatile trapping in the solidified mantle and water distribution? (Salvador et al. 2023, this collection). How does the elemental abundances and isotopic compositions of noble gases (He, Ne, Ar, Kr, Xe) and stable isotopes (H, C, N, O, S) constrain the budget of volatile elements outgassed in the atmosphere, the timing and mechanisms of volatile transport between planetary reservoirs, and the geodynamical history of Venus through time? (Avice et al. 2022, this collection).

Surface Processes, Age of the Surface and Evidence for Current Activity, addresses key open science questions about the resurfacing history and volcanic activity of Venus and their relationship to present-day volcanism and tectonism. Upcoming orbital missions will improve our understanding of the resurfacing history of Venus in crucial ways for a better understanding of the sequence of events that occurred in producing the geologic landscape: how are impact features and their associated deposits (ejecta, haloes, parabolas) altered over time, has the nature of volcanism and tectonics changed over time, and how does this compares with global resurfacing models and constrain the global evolution of Venus through time? (Herrick et al. 2023, this collection; Ghail et al. 2023, this collection). Sediments and sedimentary rocks are also critical to understanding surface modification processes and how Venus works today, but are also extremely important for determining how Venus's climate has changed through time and whether it was once a habitable planet (Carter et al. 2023, this collection). Furthermore, mineralogy of the Venus surface provides a critical record of geologic and climatologic history and the current chemical exchanges between the atmosphere and solid body (Gilmore et al. 2023, this collection).

Interior Regime Throughout History, Water and Other Volatiles, discusses open science questions regarding dynamics and evolution of Venus' mantle: how did mantle cooling

history control the state of Venus' core and the tectonic regime; and constraints on the variability of heat flow through time by Rolf et al. (2022, this collection); what was the evolution of the atmosphere-interior of Venus, including its core (Gillmann et al. 2022, this collection)? What are the signatures and potential detectability of present-day volcanically emitted material in the atmosphere of Venus by incoming Venus missions, is there a non-gaseous component of volcanic plumes? Could these measurements shed light on the compositional history of magmatic volatiles and their reservoirs? (Wilson et al. 2023, this collection).

In addition to open questions along the previous science themes, this section also addresses investigations and open questions based on the recent analysis and exploitation of ESA's Venus Express and JAXA's Venus Climate Orbiter (Akatsuki) - in addition to recent ground-based observations, on how to better constrain the dynamical variability and couplings from surface to cloud tops in present day's atmosphere. Important variability on all time scales, in latitude, in local time of the main dynamical and photochemical tracers at all altitude levels, such as CO, SO or SO_2, variability of the cloud convective layer, atmospheric structure and turbulent processes, and large bow-shaped topography-driven stationary waves above the main equatorial highlands, all contribute to the study of the complex dynamical structure and properties of the Venusian atmosphere, and their relation to the long-term evolutionary path of Venus.

2.1 Comparison of Venus with Exoplanets

2.1.1 Synergies Between Venus and Exoplanetary Observations (Way et al. 2023, this collection)

a) *The importance of magma oceans*: Exoplanetary observations of planets in the Venus Zone (VZ), defined by Kane et al. (2014) as part of the Habitable Zone (HZ) in which an Earth-sized planet is more likely to be a Venus analog than an Earth analog, can help us to constrain the magma ocean (MO) lifetime of Venus. Constraining the magma ocean lifetime prior to solidification is extremely important in understanding the likelihood of water ever condensing on the surface of a Venus-like world. The reason lies in 1-D calculations by Hamano et al. (2013), who demonstrated that Venus may sit at a boundary between a world that receives so much solar insolation that the magma ocean lifetime is long ($\sim$100 Myr), providing ample time for photodissociation and escape of the overlying steam atmosphere, and effectively drying out Venus (classified in Hamano et al. 2013 as a Type II world). On the other hand, Venus may lie on the other side of this boundary with a short-lived magma ocean with a steam atmosphere lifetime of $\sim$1 Myr (i.e., comparable to Earth's, see e.g., Salvador et al. 2017, 2023, this collection) that allows the planet to condense water on the surface (a Type I world). Work by Turbet et al. (2021) suggests that it is more likely that Venus ended up as a Type II world because their 3-D model generates clouds that appear to efficiently trap heat at the poles and night side. However, both the Hamano and Turbet models use CO_2-H_2O or N_2-H_2O atmospheres. It is not clear from recent work by Gaillard et al. (2022) and Bower et al. (2022) that these combinations of gases are adequate. For this reason, exoplanetary observations of young planets around G-stars in the Venus Zone will be critical, in addition to selected Venus missions, to discerning early Venus' history (see Sect. 11.2).

b) *Are there divergent evolutionary paths for exoplanets in the Venus Zone?* As indicated above, the magma ocean lifetime is critical to understanding the likelihood of water ever condensing on the surface of a world in the Venus Zone. At the same time, 3-D General Circulation Modeling (GCM) by Yang et al. (2014), Way et al. (2016) and Way and Del

Genio (2020) has demonstrated that slow rotation is key to keeping a planet temperate within the Venus Zone. Moving present-day Earth into the Venus Zone will rapidly move the planet into a moist and then runaway greenhouse state, as 1-D models have shown that Earth is already at the inner edge of the habitable zone (Kopparapu et al. 2013, Fig. 8). The 3-D GCM modeling studies have shown that if Earth were rotating as slowly as modern Venus does, an efficient, large-scale cloud-albedo feedback at the substellar point would generate high enough albedos for a significant portion of the incident solar radiation to be reflected back to space, keeping the surface of the planet temperate. Most observable habitable-zone exoplanets are terrestrial exoplanets orbiting M dwarfs, with rotation periods of order of 10-30 days; such planets close to their star are expected to become tidally despun into a synchronously rotating state; atmospheres of typical tidally locked terrestrial exoplanets are expected to superrotate (Imamura et al. 2020, and references therein). Examining planets in the Venus Zone of exoplanetary systems will determine whether these 3-D GCMs are correct, and whether Venus-like worlds ever have temperate surface conditions and the role that rotation rate may play. Even if water condenses early on the surface of a Venus-like world, its later evolution may diverge, depending upon its rotation rate.

c) **What is the longevity of habitability of an Earth-sized planet in the Venus Zone?** If Venus had a temperate period, its longevity may be difficult to constrain, but Earth-size worlds in their Venus Zone will help us to bound the problem. Conversely, new data from upcoming Venus missions should give us a constraint on the longevity of water on Venus and encourage the planetary and exoplanetary science communities to search for such worlds in exoplanet databases in the coming decades. Yet the latter is not unambiguous; for instance, key observations that the DAVINCI descent probe analytical instruments within and below the clouds (Sect. 5.3) and the EnVision VenSpec-H spectrometer (Sect. 6.3) are related to the D/H ratio and the heavy noble gas isotopes. Work by Avice et al. 2022 (this collection) demonstrates that the D/H ratio in itself is insufficient to determine when Venus lost its water and the time-scale of that loss, as implied in the published Pioneer Venus D/H measurements by Donahue et al. (1982). The heavy noble gas isotope measurements by DAVINCI will be crucial to understand the epoch and timescale of the loss on Venus (Garvin et al. 2022a; this review, Sect. 5.3). If Venus had a habitable period, what constraints can interior, tectonic, and atmospheric escape models provide to understand the likelihood of long-term volatile cycling? Here again, exoplanetary observations of planets in the Venus Zone will be a unique opportunity to test our models and their application to Venus' long-term evolution.

2.1.2 The Habitability of Venus (Westall et al. 2023, this collection)

a) **What was Venus' water inventory and was there liquid water on its surface at temperatures conducive to the emergence of life?** Did water condense after crystallization of the magma ocean? Given the lack of direct access to the ancient history of the planet, this question is best addressed by refining models and through eventual comparisons with exoplanets exhibiting characteristics such as rotation speed of the planet that may have similarities with early Venus. Important requirements for habitable conditions would be a slow rotation of early Venus and a corresponding weak Coriolis force to allow for a large and reflective cloud cover (e.g., Way and Del Genio 2020). Liquid water on the surface may allow silicate weathering and thereby maintain a low atmospheric pressure of CO_2, protecting a subaqueous habitable environment.

b) *If there was once water on the early planet's surface, how long was the transition from habitable to uninhabitable planet?* This question is closely related to the tectonic state of early Venus. With active plate tectonics, a carbonate-silicate cycle similar to that on Earth could have allowed for a substantial habitable period until the proposed resurfacing event accompanied by catastrophic mantle outgassing some several hundred million years ago (e.g., Way and Del Genio 2020; Krissansen-Totton et al. 2021). In contrast, without plate tectonics but with liquid surface water, recycling of carbonates into the mantle would have been rare, limiting the long-term habitability on Venus (Höning et al. 2021). Knowledge of Venus is currently insufficient to rule out any but the most extreme scenarios, but further observation should yield important evidence to constrain Venus' evolution. In particular, constraints on ancient plate tectonics, which could be derived from future seismic measurements as well as from a more detailed exploration of surface features such as the tesserae and their compositions, would shed light into the early habitable period of Venus.

c) *How habitable is the Venusian cloud environment, and are there signs that it was ever inhabited?* Conditions at today's 55-70 km altitude range are juxtaposed with the observed limits for terrestrial life. By these metrics, hypothetical life in Venusian aerosols may be within required bounds of temperature, pressure, and pH, and energy sources (Grinspoon and Bullock 2007; Nicholson et al. 2010; Limaye et al. 2021; Westall et al. 2023, this collection, Fig. 8); is the purported phosphine signature (Greaves et al. 2021; Encrenaz et al. 2020b; Villanueva et al. 2021) real and is it really a biosignature? The odds against there being life in the clouds of Venus today are high due to extreme conditions in terms of water activity (which takes extreme acidity and aridity into account), and the lack of permanent habitability (for Earth-like life in any case). Venusian life would have to be quite different from that on Earth to not just survive but thrive in its clouds. This does not mean that the hypothesis should be completely disregarded, but it remains just a hypothesis, awaiting further boundary conditions from missions including DAVINCI and EnVision. If Venusian cloud life exists, and if it is indeed very different from terrestrial life, then could it be identified as living and viable? If the answer to these speculations is yes, then this would answer our first question of Venus habitability. This question requires an in-depth characterization of the cloud-level environment: gas and cloud composition, available light levels, cloud droplet microphysics (droplet size, formation / precipitation cycles), UV & ionizing radiation levels. A detailed investigation of how these environmental factors vary with altitude, latitude and local time would require sustained measurements from an aerial platform such as a balloon or powered aircraft (Limaye and Garvin 2023).

2.2 Initial Conditions, Accretion, and Early Venus

2.2.1 The Accretion and Differentiation of Venus

Venus and Earth contain 41% and 51% of the remaining mass of the inner protoplanetary disk vs 3% and 5% for Mercury and Mars; based on this comparison, we might expect the smaller planets to be outliers and Venus and Earth to be similar and represent good averages of the composition of the inner solar system. Yet Earth and Venus appear fundamentally different. The dynamical causes and timing of this globally-important difference are a topic of active work and debate, with broad implications for planet accretion models, early solar system dynamical stability, volatile delivery to the terrestrial planet region, and the early impact rate throughout the solar system (e.g., Bottke et al. 2017).

a) ***When, from where, and how many volatiles were accreted by Venus?*** Venus and Earth are expected to have formed over several million years by accretion of planetesimals and planetary embryos originating from various heliocentric distances, with the majority coming from a narrow annulus near 1 AU (O'Brien et al. 2006). Did Venus accrete as an average body as the protoplanetary disk was rapidly cooling, or did proto-Venus components emerge from distinctive regions? The fundamentally different isotopic compositions of non-carbonaceous (NC) and carbonaceous (CC) meteorites reveal the presence of distinct reservoirs in the solar protoplanetary disk that were likely separated by Jupiter. However, the extent of material exchange between these reservoirs, and how this affected the composition of the inner disk, are not known or strongly underconstrained (e.g., Spitzer et al. 2020; Morbidelli 2020). A variety of different processes such as thermal processing in the primordial atmosphere and atmospheric escape must have had a dramatic impact on the bulk and isotopic compositions of planetary embryos accreting to form Venus, being possibly a major factor in volatile depletion (Lammer et al. 2020).

b) ***Was the accretion of Venus Earth-like with a late giant impact?*** The final stages of planetary accretion involve collisions between the forming planet and leftover bodies such as large planetesimals or planetary embryos. Earth suffered from a final major collision at the end of its accretion and the Moon is the witness of this event. For Venus, it remains unknown if the planet ever suffered from an impact energetic enough to create a moon (Brooks and Jacobson 2019; Jacobson and Dobson 2022). Medium or large impactors on early Venus affect the primordial atmosphere through impact erosion and might trigger further degassing through energy deposition in the mantle and crust. High temperatures generated in the upper mantle and the spreading of the thermal anomaly lead to partial melting and the formation of new basaltic crust. Yet, giant impacts are not the only potential interactions. Alternative scenarios involving smaller and successive multiple impacts have also been proposed to explain the Moon's formation (Rufu et al. 2017). Whether the energy deposition was then sufficient to melt the entire mantle depends primarily on impact parameters (e.g., Nakajima et al. 2021) and on their frequency. Finally, the accretion sequence of the Earth and Venus may significantly differ, with Venus possibly experiencing more hit-an-run collisions (Emsenhuber et al. 2021). To what extent these alternatives apply to Venus and how they affected its initial thermal state and differentiation remains unclear. Is Venus therefore a more primordial or primitive body than Earth, if there was no giant impact in its early history? How might these be related to a plausible water-rich past?

2.2.2 Magma Ocean, Water, and the Early Atmosphere of Venus (Salvador et al. 2023, this collection)

a) ***What proportion of initial mantle volatile inventory is outgassed and escaped?*** Water and its distribution between the different planetary reservoirs are of fundamental importance in controlling the processes and feedbacks at play, from the deep interior to the upper atmosphere, during the entire evolution of the planet (e.g., Crowley et al. 2011; Tikoo and Elkins-Tanton 2017). Furthermore, surface conditions and thus the potential habitability of the planet are direct outcomes of the evolutionary pathways followed (e.g., Hamano et al. 2013).

The earliest stages of planetary evolution, and in particular the so-called magma ocean stage, are crucial in distributing water between the different reservoirs (e.g., Salvador et al. 2017; Nikolaou et al. 2019). During this phase, the surface is molten and the absence of a thick, long-lasting boundary between the molten mantle and the atmosphere allows for free and extremely efficient thermal and chemical exchanges between the interior and the atmosphere. Volatile species initially dissolved within the molten mantle are thought to concentrate readily into the melt during the crystallization of the magma ocean until melt saturation is reached. At that point, volatiles in excess of saturation likely exsolve out of the melt, form gas bubbles, and rise up to burst at the surface and be expelled out, forming the atmosphere. Magma ocean outgassing and therefore (secondary) atmospheric formation have been thought to be efficient because of the vigorous convection at play in the melt, believed to bring the entire melt volume close to the surface, where outgassing occurs rapidly enough.

In part because of the extreme P, T conditions of the magma ocean, which are far out of reach of current numerical models and experimental setups capabilities, many processes affecting the thermo-chemical evolution of the magma ocean, such as the crystallization scenario and the convection regime, remain highly unconstrained. In addition, the initial state of the magma ocean itself is highly uncertain. For instance, the mantle redox state or the initial volatile abundance and their evolution with time are still unsettled. Yet these processes are thought to significantly affect the type and timing of outgassed volatiles and thus their evolving distributions between the interior and the atmosphere. Further investigations considering the range of uncertainties and the interplays between these mechanisms are needed to draw a consistent picture of the early volatiles outgassing and escape processes and reconcile the early evolution with the current state of Venus. Any measurable constraint on the timing and amount of early outgassing, volatile loss, and on the amount and type of remaining volatiles in the present-day mantle could be used in evolution models and would help in choosing among the different scenarios, thereby improving our understanding of the processes at play.

b) *What is the timing of silicate / metal differentiation?* Several parameters of primary importance for the early evolution stages remain highly unconstrained, for instance including the initial water content, and the initial mantle state. These two aspects are tightly linked to the outcomes of the accretion sequence. While it seems reasonable to assume, given their vicinity, that the Earth and Venus experienced similar accretion histories with similar volatiles delivery, the absence of a moon and associated late Moon-forming giant impact challenges the assumptions regarding the timing and extent of large-scale melting episodes on early Venus. Indeed, a moon-forming impact on early Earth (e.g., Canup 2004) is thought to induce a global-scale mantle melting that is generally assumed to be the initial state of coupled magma ocean-atmosphere models. However, it is important to note that this event is not the only heat source that can support large-scale mantle melting. Other plausible scenarios than the single giant moon-forming impact hypothesis, i.e., smaller multiple-impact models, have been proposed and may better explain the compositional similarities of the Earth and Moon (Rufu et al. 2017). These scenarios would not ultimately discard the likelihood of deep and global magma oceans but constraining the timing and timescale of Venus formation/accretion would help clarify the initial state of planetary evolution. This could be achieved using the ^{182}Hf ^{182}W chronometer to constrain the timing of silicate/metal differentiation and thus core formation (e.g., Lee and Halliday 1995; Harper and Jacobsen 1996). This ultimately relates to the molten state of the mantle and thus to the timescale and intensity of the accretion phase (e.g., Zahnle et al. 2007). Knowing the amount of accretionary energy delivered within a constrained time frame would help test the global magma ocean hypothesis and provide clues for the initial state of the early evolution of Venus.

c) *Could the mantellic water content be constrained through the geodynamic regime (and present-day volcanic outgassing)?* Because of the influence of water on mantle melting and viscosity (and thus rheology) (e.g., Lange 1994; Hirschmann 2006; Ohtani 2020), information on current mantle convective regime and dynamics might provide indirect clues about its current and therefore past water content. The mantle present-day out-gassing rate and composition might be another way to sample the planetary interior state and water content and inform models of early and long-term planetary evolution. Esti-mates of the Venusian mantle water content would indeed provide constraints for evo-lutionary models to match with and thus help deciphering which paths and associated mechanisms are most likely to match with these constraints.

d) *Are there hints for the existence of liquid water in the past? Can they be inferred through Venus surface mineralogy, mantle present-day rheology, outgassing and com-position?* Hints for the existence of liquid water in the past would provide significant constraints for the early evolution scenarios. If liquid water ever existed at the surface of Venus, it would be strong evidence that enough water has to be retained, either in the planetary interior or in the atmosphere, for a substantial amount of time and that temper-ate climates were plausible in Venus' past. It would discard all scenarios where water is lost in Venus' early history due to desiccation via a slowly ($\sim$100 Myr) cooling magma ocean (Hamano et al. 2013 type II planets), and can thus never sustain habitable condi-tions (Turbet et al. 2021). Conversely, it would favor scenarios where the early evolution of the magma ocean allows for a rapid ($\sim$1 Myr) solidification (type I planets according to Hamano et al. 2013 classification). For instance, this could either be due to an ineffi-cient magma ocean outgassing (e.g., Ikoma et al. 2018; Salvador and Samuel 2023) and reduced atmospheric greenhouse effect, or due to mechanisms reducing the incoming solar flux and allowing for temperate climates at Venus orbital distance, such as the pres-ence of highly reflecting clouds. In the latter case, temperate surface conditions would directly be inherited from the early magma ocean stage evolution while in the former, habitability would be related to the subsequent long-term evolution of Venus (e.g., Way and Del Genio 2020; see also Gillmann et al. 2022, this collection).

2.2.3 Volatiles and Noble Gas Isotopes (Avice et al. 2022, this collection)

The elemental and isotopic compositions of volatile elements (H, C, N, O, S, P and noble gasses) contained in the Venus atmosphere hold clues to the origin and evolution of the entire planet and can provide decisive answers to three major fundamental questions:

a) *Did Venus acquire its volatile elements from the protoplanetary solar nebula, aster-oids, comets, or a mix of these sources?* Classical views, supported by geochemical constraints and outcomes of models of the formation of terrestrial planets, propose that Earth, and by extension Venus, accreted relatively dry and acquired most volatile ele-ments later by bombardment of volatile-rich material (asteroids/comets) during the final stages of planetary formation (Marty 2012; Halliday 2013). These late events could also have delivered chemical elements and/or compounds conducive to the emergence and development of life. The detection of solar-derived gasses in the interior of Earth (e.g., Williams and Mukhopadhyay 2018) and on Mars (Swindle 2002) also leaves room for an early contribution of gasses from the solar nebula, or from solar wind implanted at the surface of grains (Péron et al. 2018), to the budget of volatile elements on terres-trial planets. Recent investigations propose that the Earth's building blocks did contain significant water (Piani et al. 2020). Estimating the delivery mix of volatile elements to Venus would thus help to constrain the timing of the formation of Venus and the building blocks of the planet, also contributing to placing Venus in the context of the formation of the entire Solar System. For example, the isotopic composition of Venus' atmospheric xenon could carry the signature of a delivery of cometary material to the atmosphere of Venus (Avice et al. 2017; Marty et al. 2017). This cometary contribution is visible in the isotopic composition of Earth atmospheric xenon as a marked depletion in ^{134}Xe and ^{136}Xe isotopes relative to Solar or Meteoritic end-members (Avice et al. 2022, this collection, and references therein). In addition to Ar/Ne and ^{20}Ne/^{22}Ne, detecting (or not) cometary xenon on Venus would thus be a key constraint for models attempting to understand the late delivery of volatile-rich bodies originally formed in the outer Solar System to the terrestrial planets. A list of key measurements of noble gases and of their associated maximal uncertainties required to answer the scientific questions is summarized in Avice et al. 2022, this collection, Table 1. DAVINCI's Descent probe quadrupole mass spectrometer instrument VMS is described in Sect. 5, Sect. 5.3.1 (see also Garvin et al. 2022a, Table 3.2).

b) **How was Venus' atmosphere shaped by early impacts and atmospheric escape?** On Earth, the Moon-forming impact likely removed the primitive, possibly solar-derived, atmosphere and set the stage for the emergence of a secondary atmosphere. Loss of water has been a key driving mechanism for the evolution of Venus (Baines et al. 2013), but the extent of water loss and the history of atmospheric escape, including of other atmospheric species, remain largely unconstrained. Measurements of the isotopic composition of nitrogen and precise determinations of the elemental and isotopic compositions of noble gasses in the atmosphere of Venus would provide constraints on the presence or absence of remnants of primordial solar gasses, on the regime of atmospheric escape (thermal vs. non-thermal) but also on its timing in the planet's history. For example, elevated ^{38}Ar/^{36}Ar and ^{15}N/^{14}N ratios measured in Mars' atmosphere demonstrate that non-thermal escape processes have been active on Mars (Atreya et al. 2013). Knowing the abundance and isotopic composition of xenon in the atmosphere of Venus would also clarify if Venus suffered from joint hydrogen-xenon escape processes (Zahnle et al. 2019; Avice and Marty 2020) like Earth and Mars. Determining the ^{129}Xe/^{132}Xe ratio, which might have recorded contributions of radiogenic ^{129}Xe from the decay of now extinct ^{129}I ($T_{1/2} = 16$ Ma), would also help evaluate the relative timing of atmospheric escape and outgassing processes (see next paragraph). NASA's DAVINCI mission will address these issues directly via *in situ* sampling and measurements of Xe isotopes.

c) **What is the outgassing history of Venus?** Although Venus is currently in a quiet stagnant-lid regime, the planet is not "dead" and there is evidence for recent activity including recent hotspot volcanism (Smrekar et al. 2010a). Several models propose that Venus might have been in a much more active regime in the past and even that plate tectonics was active on ancient Venus. Radioactive decay of extinct (^{129}I, ^{244}Pu) and extant (^{238}U, ^{40}K) nuclides present in silicate portions of Venus have been producing excesses of radiogenic and fissiogenic isotopes of noble gasses (e.g., ^{40}Ar, $^{129,131-136}$Xe) relative to primordial compositions. Given the wide range of half-lives of the parent nuclides (ranging from Ma to Ga), the relative proportions of these excesses measured in a reservoir should vary with time. Magmatic-driven outgassing contributes to the progressive degassing of these radiogenic isotopes from Venus' interior to its atmosphere. Measuring the elemental and isotopic composition of noble gasses in the atmosphere of Venus would thus help to refine current estimates on the relative proportions of radiogenic noble gasses degassed in the atmosphere versus those still retained in the planet's interior (Kaula 1999). Such measurements will also allow a coherent picture to be built of the geodynamical history of Venus through time.

Two broad types of science investigations are envisaged for gathering data on the elemental and isotopic compositions of noble gasses and stable isotopes (H, C, N, O, S) in the atmosphere of Venus. One would be an in-situ mission carrying a scientific payload to sample and measure the abundances and isotope ratios of the chemical elements of interest below the homopause such as DAVINCI (Garvin et al. 2022a, see Sect. 5). Another would be a sample mission during which a portion of the Venus atmosphere would be sampled below the homopause, which corresponds to a pressure level of 10^6 mbar (approximatively 135 km, see Mahieux et al. 2015), such as JPL Cupid's Arrow concept (Sotin et al. 2018a,b; Rabinovitch et al. 2019), with collected sample(s) possibly returned to Earth for characterization with state-of-the-art technologies available in international laboratories (Shibata et al. 2017; see also Sect. 10, Sect. 10.3.2).

2.3 Surface Processes, Age of the Surface and Evidence for Current Activity

2.3.1 Resurfacing History and Volcanic Activity of Venus (Herrick et al. 2023, this collection)

Upcoming orbital missions will improve our understanding of the resurfacing history of Venus in crucial ways. We will have a better understanding of the sequence of events that occurred in producing the geologic landscape. Placement of craters within that sequence will provide a timeline for that sequence, and constraints on the current level of volcanic and tectonic activity will provide a present-day "boundary condition" on that history. Advances in understanding will be achieved by upcoming missions for three critical science questions, including the following:

a) *How are impact features and their associated deposits (ejecta, haloes, parabolas) altered over time?* Key to establishing the absolute timing of geologic events is evaluating whether impact craters largely postdate the volcanic and tectonic activity observed on the Venus surface, or whether they are a population of features that are in various stages of being obliterated. It is expected that improved resolution in imaging and topography, along with SAR polarimetry and imaging at multiple wavelengths (e.g., by the VERITAS and EnVision radar orbiter missions), will enable the processes altering the appearances of impact craters over time on the Venusian surface to be distinguished. If aeolian or chemical weathering processes are the dominant mechanisms for removing the emissivity and backscatter signatures associated with distal impact deposits such as dark haloes and parabolas, and sediment fill is responsible for creating low-backscatter floor deposits, then most of the craters can be viewed as being at the top of the stratigraphic column. In such a case, the surface would have formed from a relatively rapid sequence of events several hundred million years ago. If large portions of the craters have deposits that have been altered by one or tectonic or volcanic events, then the timeline of geologic activity spreads, and much of the surface, are probably younger than 100 My.

b) *Has the nature of volcanism and tectonics changed over time?* Improvements in imaging and topography from the upcoming missions will enable seeing key geologic contacts, individual volcanic flows, fault blocks, and other details of surface geology. Considerable advancements in our knowledge of compositional information will come from both infrared and SAR imaging. This information will allow us to build an understanding of the sequence of events on the surface and evaluate whether or not fundamental changes in the nature of geologic activity have occurred over the past several hundred million years.

c) *What is the current level of volcanic and tectonic activity?* Magellan images compared against changes observed during the upcoming orbital missions (VERITAS and EnVision), will constrain where and how much current geologic activity is occurring on the surface. Most of this work will simply involve change detection among images taken at different times to search for new flows, landslides, new fractures or faults, etc.. Repeat pass SAR interferometry will also provide information regarding cm-scale movement in tectonic zones, caldera inflation and deflation, and other active small-scale deformation. The near-infrared descent imaging by the DAVINCI probe will search for signs of mass wasting in 3D using very high resolution imaging and topography acquired under the clouds for a region within Alpha Regio to complement the orbital SAR observations.

2.3.2 Volcanic and Tectonic Constraints on the Evolution of Venus (Ghail et al. 2023, this collection)

a) ***What stresses and thermal, mechanical and geochemical parameters are responsible for the formation of Venus' extremely diverse volcanic features?*** Venus hosts an enormous diversity of volcanic features: direct analogs and those whose formation mechanisms are extremely challenging to understand, such as narrow lava channels that extend many 1000s of km. Large volcanic rises termed hotspots are directly linked to mantle plumes, probably arising at the core-mantle boundary. But where do smaller plumes that are likely to form at least some coronae originate? Why are coronae arguably unique to Venus? What processes are responsible for Venus' many enigmatic volcanic features? Are differences in features due to spatial or temporal differences in crust/lithosphere/mantle conditions? These questions will be addressed by NASA-JPL's VERITAS orbiter mission (Smrekar et al. 2022a), NASA-GSFC's DAVINCI mission (Garvin et al. 2022a) and by the ESA-led EnVision M5 orbiter mission (European Space Agency 2021), but better understanding of crust/mantle/core composition and rheology will be needed to take the next steps in understanding.

b) ***What are the driving forces and mechanisms for stress accommodation that produce the variable scales and apparent strains seen in Venus complex tectonic terrains?*** Extensional and compressional features on all spatial scales dominate deformation, with limited evidence for strike-slip faulting. In some environments, the origin of stress is clearly linked to mantle plumes or volcanic processes. In most feature types, there are multiple possible origin hypotheses. For example, what causes the ~major 5000-10,000 km rifts? There is no apparent compressional zone of accommodation for the displaced, extended lithosphere. Local scale (<150 square km) studies using fine-scale topography from the DAVINCI probe's sub-cloud imaging will provide boundary conditions for strain within Alpha Regio at scales < 30 m.

c) ***What is the mechanism of tesserae formation?*** Tesserae are characterized by highly elevated topography, small-scale surface roughness and multiple sets of cross-cutting tectonic structures and appear to represent areas of intense, past tectonism. Tessera terrain covers about ~8% of the surface of Venus and is morphologically clearly distinct from the volcanic plains that dominate the remainder of the planet. Detailed study of the type, number, spacing, distribution, and stratigraphic position of tessera structures will yield insight into the geodynamics of Venus before the production of the plains. Are tessera structures compressional or extensional in nature? Does their formation require a different strain rate, heat flow, or composition than in the plains? Is there evidence of lateral accretion of materials to form tessera plateaus? Do the tesserae underlie the plains across Venus? Are the tesserae dynamically compensated? The formational models to explain such high, complex and strained terrain are still the subject of much debate and uncertainty: horizontal convergence, extension, mantle upwelling, sub-crustal flow, crustal underplating, sub-crustal rejuvenation, crustal plateau formation, diapiric intrusion, gravitational sliding and relaxation, or all of these? (Hansen and Willis 1996; Ivanov and Head 2011).

d) ***Absent Earth-like plate tectonics, what is Venus' overall geodynamic system that links mantle convection, surface deformation and volcanism, and volatile history?*** Venus' interior heat engine provides ample energy to the geologic activity that creates Venus' young surface and massive atmosphere. But fundamental questions remain. Why does Venus lack a dynamo? How does Venus lose its heat? Have processes changed with time? What is the extent of lithospheric recycling? Is Venus, with its hot lithosphere, a good analog for Earth's Archean? Are current tectonic processes the precursors of plate tectonics and continent formation? Can up- and downwelling plumes produce all surface features? Numerous hypotheses have been put forward, but additional data are needed to discriminate between them.

2.3.3 Mineralogy of the Venus Surface (Gilmore et al. 2023, this collection)

The Venera, VeGa, and Magellan missions found that Venus is dominated by basaltic rocks associated with widespread volcanism (Basilevsky and Head 2003a). Near-infrared observations from Venus Express and Galileo first detected variations attributed to mineralogy. They suggest that there is diversity in the FeO content of materials, including relatively high FeO content consistent with less weathered rocks, and low FeO content consistent with differentiated, non-basaltic compositions.

The stratigraphically oldest material on Venus are found among the major tessera terrains. They record an extinct geodynamic regime and have a near-IR emissivity signature that is different from the basaltic plains. The nature of the tesserae is critical to our understanding of Venus prior to the emplacement of the plains. Several major questions about Venus history are recorded in tessera terrain. Near-IR observations from VERITAS, EnVision, and DAVINCI (including below the cloud deck at spatial scales < 100 m) will provide the first global assessment of surface composition, which is critical to addressing the following questions:

a) ***What is the composition and diversity of Venus surface materials?*** What are the primary rocks and minerals recorded on the surface of Venus? How do these compositions spatially correspond to morphological units? How do these units vary with time and location? What do these differences tell us about the ancient and modern geologic history of Venus?

b) ***What is the style of weathering recorded in surface rocks over the history of Venus?*** Thermodynamic models of Venus weathering make predictions about the products of surface-atmosphere weathering under current Venus conditions, but there is a lack of consensus over exactly which phases might form from bulk versus diffusion-constrained reactions between surface and atmosphere and the timing of their formation. Weathering reactions depend upon the composition and crystallinity of the surface rocks, which are unknown. Do we detect these predicted phases? Can the presence or absence of these phases be used to constrain the age of surface units? Do we see weathering products that are consistent with weathering under an extinct atmosphere? Can we constrain the composition of the high radar reflectivity materials found across Venus?

c) ***Is there compositional evidence for aqueous or hydrous minerals?*** Is the near-IR signature of the tesserae consistent with Fe-poor magmas, clay minerals, or primary sedimentary phases? Is there evidence for Fe-poor phases in other regions of Venus?

d) ***Are the tesserae felsic?*** The near-IR emissivity of the tesserae may be consistent with Fe-poor materials, which, if igneous, could be consistent with the production of felsic lavas

(Hashimoto et al. 2008; Mueller et al. 2008; Gilmore et al. 2015) or of granitic rocks. If confirmed, these would require a planet with abundant water and a plate-recycling mechanism. Such a discovery would be critical evidence of a once-habitable Venus and elevate the targeting of tessera for future *in situ* study.

e) *Are the tesserae compositionally, morphologically, and stratigraphically heterogeneous?* Is there compositional variation across and within the tesserae? What is the detailed stratigraphic relationship between the tesserae and the plains? Is there evidence of unrecognized craters in the tesserae? Is there evidence of sedimentary materials in the tesserae?

Each of these questions requires laboratory work to examine the near-IR signatures of rocks and minerals and their weathering products expected under Venus conditions (see also Sect. 12.1.2 below).

2.3.4 Sediments, Regolith / Sediment Supply, Evolution (Carter et al. 2023, this collection)

a) *What are the nature, distribution, and range of sedimentary surface modification processes?* How has the surface of Venus been modified since it was formed? In particular, what are the processes that have modified and partially filled impact craters? What are the causes of the lower emissivity material in some highlands, and what do they imply about weathering processes and possible volatile transport? On Venus, sediments are likely produced by impact cratering and weathering (Garvin 1990). Even if these are not volumetrically large, sediments are likely widespread, covering a large fraction of the surface. Thus aeolian erosion and deposition may be important processes at the Venus surface. Local-scale evidence from Venera lander panoramas indicates possible sedimentary processes that would need connection to regional and global scale models to place them in a proper context (Garvin et al. 1984).

b) *Is there evidence of active physical and chemical landscape change?* Landscape evolution refers to processes that modify the morphology of a planet's surface, such as gravity-driven mass-wasting landslides and slumps. Mass-wasting is a ubiquitous geomorphological process operating on any planetary body large enough to have gravity; such features are observed on Earth, the Moon, Mercury, Venus, Mars, icy satellites, comets, and asteroids. Magellan's low-resolution radar imagery provided the first evidence of mass movement on Venus in the form of large-scale slope failures: rock slumps, rock and/or block slides, rock avalanches, debris avalanches, and possibly debris flows are seen in areas of high relief and steep slope gradients (Malin 1992).

c) *Understanding the range and scope of mass-wasting processes.* Although impact cratering is likely the main process behind sediment production on Venus, the planet's hot, dense and highly oxidizing atmospheric conditions may cause chemical weathering of surface materials. In the absence of near-surface water which, on Earth, affects material bulk density, shear strength and pore-pressure, and thus lead to slope instability, the mechanisms of slope instability and failure on Venus are unclear, and it is likely that landslides require triggering by external forces, such as earthquakes. Magellan imagery (Malin 1992) revealed a very strong spatial relationship between the locations of large-scale mass-wasting features and steep slopes related to rift zones and volcanic edifices, which may in turn point to them being geodynamically active in the recent geological past. Wind-streaks and debris-fans (downwind of impact craters) are relatively large-scale features on Venus (kilometer to tens of kilometers in length) and are also commonly observed in Magellan images (Greeley et al. 1992, 1995; Kreslavsky and Bondarenko 2017; Neakrase et al. 2017).

2.4 Interior Regime Throughout History, Water and Other Volatiles

2.4.1 Dynamics and Evolution of Venus' Mantle Through Time (Rolf et al. 2022, this collection)

a) *How did mantle cooling control the state of Venus' core?* The state and size of Venus' core are largely controlled by the cooling efficiency of the mantle and thus provide constraints for the evolution of Venus' mantle convection and surface tectonics. Available observations (MoI, k_2) do not pin down these important characteristics of the core well enough (see Gillmann et al. 2022, this collection), though they will be addressed by the upcoming VERITAS and EnVision missions. A smaller Venus core implies a thicker mantle that experiences high pressures at the core-mantle boundary, possibly high enough for the occurrence of post-perovskite that is known to influence lowermost mantle and core dynamics on Earth. Current estimates of Venus' core size (Margot et al. 2021) seem inconsistent with the occurrence of this high-pressure phase, but further refinement is necessary. Next to core radius, the state of the core remains a fundamental unknown. A completely solidified core would imply efficient cooling through time and a cold state of the mantle. Compared to present-day, such a scenario would point to more efficient cooling during parts of Venus' evolution, possibly expressed as mobile lid tectonics with more large-scale horizontal surface motion than inferred for the present day. The opposing end member of a fully liquid core would in contrast point to smaller core cooling rates and reduced heat loss across the core-mantle boundary. In such a case, remnants of a basal magma ocean may be preserved inside Venus' mantle today, with implications for the planets' magnetic field history (O'Rourke 2020). The explained end members may be extreme scenarios, but they emphasize the importance of further pinning down the properties and state of Venus' core and deep mantle.

b) *How much heat does Venus lose today?* Heat loss from the mantle to the atmosphere and its efficiency through time are crucial for understanding Venus' interior evolution. Venus loses its interior heat via thermal conduction through the lithosphere and via volcanism, but how these fluxes vary through time and across Venus' surface and how they are linked to the various geological and tectonic features on Venus' surface remains to be established. Conductive heat flux has not been measured in-situ, but is indirectly determined from flexural modeling of elastic thickness. Strong lateral variations are indicated, yet insufficiently mapped out. Improving this knowledge gap could provide key information on local differences in lithospheric and crustal thickness as well as in the temperature of the uppermost mantle. The latter is important for the rheology of the uppermost mantle and lower crust, both of which are key aspects for interior-surface coupling on Venus (see Ghail et al. 2023, this collection).

c) *Has Venus' surface preserved anomalously old regions; are these felsic tessera?* The crucial question of Venus' mean surface age and its lateral uniformity links back to the debate of whether Venus' crust is renewed via catastrophic events or by more equilibrium processes. These different options have contrasting implications for the regime of mantle convection inside Venus. Venus' sparse crater distribution provides some constraints (Herrick et al. 2023, this collection), but the degree of age uniformity across the surface has been questioned. Refined surface age characteristics would further pin down the rates and scales with which tectonism and volcanism renew the surface. An important issue is how to link this to underlying mantle patterns, such as up- or downwellings? Tesserae may form some of the oldest regions on Venus, but whether they are substantially older-than-average as proposed by some and perhaps even predate a phase of near-global resurfacing is challenging to answer without knowing their formation mechanism. This relates to their composition, which if felsic as indicated for some tessera (see Gilmore et al. 2023, this collection) would demand a relatively wet environment during formation and thus a tectonomagmatic regime that can maintain enough water in the shallowest interior.

2.4.2 Atmosphere-Interior Evolution of Venus and Evolution of the Core (Gillmann et al. 2022, this collection)

a) *Was there ever liquid water on the surface of Venus?* The presence or absence of water in any form, but especially liquid water on the surface of Venus, is a major unknown in the scenarios for the planet's evolution. Currently, investigated evolutionary pathways range from a Venus that desiccated early and never harbored any substantial water inventory after the magma ocean phase to possibly habitable scenarios until recently, with a full spectrum of intermediate cases. Water in the atmosphere or on the surface could potentially make a huge difference for the evolution of climate, volatile exchanges, and general planetary evolution. It affects weathering, surface reactions, outgassing conditions, atmosphere structure, escape mechanisms, and possibly volatile recycling and interior dynamics. Vague clues to the accretion of Venus and its very early history (about 4 Ga and older) come from comparisons with Earth, modeling, impact hypothesis (Gillmann and Tackley 2014; Gillmann et al. 2016) and noble gas measurements. We also have Venus' present/recent state, because the majority of its surface formed from $\sim$0.3-1 Ga ago. However, no data points lie in between to identify a most likely scenario, define an intermediate state in the planetary evolution (ideally, with time constraints), or help find criteria to refine scenarios. Knowing if liquid water was present on Venus at any point in its history would be an important starting point, followed by estimating the amount of water necessary to explain those observations. Having an idea of the time period for that wet past then defines a basic succession of eras. Definite proof of a liquid water surface could finally constrain a tighter set of evolution scenarios. Lack of proof could mean that Venus was dry or that evidence was just lost. If ancient material is still present at the surface of Venus, it could offer a window into the planet's past. Likewise, the presence and abundance of granite-like rocks (see Gilmore et al. 2023, this collection) could yield information on the conditions during their formation and on the availability of liquid water. VERITAS will distinguish unequivocally the difference between granite and basalt based on their distinctive emissivity signatures near 1 micrometer, resolving this controversy at last. Additionally, new km-scale topography of Alpha Regio developed from analysis of Arecibo polarimetric imaging and radargram-based reanalysis of Magellan altimetry showcases possibilities of stream networks in this region when compared to Earth (Garvin et al. 2022b).

b) *When was the thick CO_2 atmosphere formed?* Another way to look at the situation is to understand when the current state of Venus' atmosphere came to be. The distinctive 90 bar CO_2 atmosphere is in itself an interesting limitation for many processes, from outgassing to climate modeling. Based on the lack of small craters on the surface of Venus, the atmospheric pressure has likely been high at least since the formation of the current surface. Little is known for sure beyond that. Radiogenic argon suggests that Venus is less outgassed than Earth, but re-constructing the entire outgassing history relies on under-constrained models. The timing and means of the CO_2 atmosphere formation can inform evolution models. Is it a recent feature (formed just before the change to present-day conditions, 500-1000 Myr ago), a relic from the magma ocean freezing, or the result of a long term build-up? If it is a recent feature, there would be strong implications for the planet's past climate, under a thin atmosphere. A catastrophic transition could point either to the destabilization of carbonate deposits or mantle-related massive volcanic outgassing. The mineralogy of the surface can help us understand if the CO_2 in the current atmosphere of Venus is buffered by solid-gas reactions.

c) *What is the dominant volatile exchange process on the surface (buffering reactions, outgassing, oxidation...)?* A key to understanding the past is to understand how volatile species are exchanged between the interior and the atmosphere of the planet, and how they can be trapped or recycled. Recent work has shown that quantifying the volatile exchanges could provide invaluable insights into the evolution of the planet, and help understand how various processes and feedback mechanisms have shaped Venus. However, such investigations rely on the ability to constrain volatile flux reasonably well and identify all the processes involved. Because the planet is a complex system that changes with time, the relative importance and even the existence of those processes will also vary, depending on specific conditions such as climate affecting surface reactions for example, or liquid water affecting recycling, thus mantle conditions and finally outgassing. This makes the modeling of fully consistent evolution scenarios a daunting task, with widely different mechanisms operating under different states of the planet on a variety of time scales and domains (from the core to the upper atmosphere). We still do not know if a single exchange process dominates the others at the surface of Venus or if its atmosphere is close to a steady state. Is it slowly evolving due to volcanic outgassing or are surface-gas reactions buffering the atmosphere to a stable level? It has been suspected that water cannot be outgassed beyond marginal concentrations due to the surface pressure (Gaillard and Scaillet 2014): do the observations support this idea? What does it mean for surface lava flow composition? The volatile species may be trapped in the flow during the ascent of the magma and its extrusion. Does this imply high gas fractions and generate explosive volcanism for which there is only very limited evidence on recent Venus? How would it have changed with time? Does it mean that volcanic activity actually is a trap for water due to oxidation processes, rather than being a source?

d) *Where is the oxygen?* This question is perhaps the corollary to all the previous ones. If Venus had water at any time (and it did, at least initially), where did its constituents (two hydrogen and one oxygen atoms) go? Oxygen is especially critical because hydrogen could be lost to space and removed from the atmosphere efficiently, while oxygen mostly remains. Atmospheric escape of oxygen species has been recently suggested to be extremely low, which indicates that either some other process has been involved or that there was very little oxygen in the atmosphere in the first place. Can that be verified with recent instruments that have proved themselves on Mars, with the MAVEN mission? Does it mean the O is in the solid planet instead? If so, is it in the mantle (either not outgassed or recycled), or in the crust? Both confirmation of atmosphere measurements and surface investigation will be needed. NASA's DAVINCI mission will address the oxygen within the clouds and deep atmosphere directly via its VfOx student collaboration experiment (see Sect. 5, Sect. 5.3.5), and the detailed measurements of oxygen-bearing trace gas species by its Venus Mass Spectrometer and Venus Tunable Laser Spectrometer (Garvin et al. 2022a).

2.4.3 Magmatic Volatiles and Effects on the Modern Atmosphere (Wilson et al. 2023, this collection)

a) *What is the rate and style of volcanic outgassing in the present era?* Were the radically different evolutionary paths of Earth and Venus driven solely by distance from the Sun, or did internal dynamics, geological activity, volcanic outgassing, and weathering also played an important part? What types of volcanism may be expected on Venus in light of its unique interior-surface coupling history described in companion articles (Avice et al. 2022; Gillmann et al. 2022; Ghail et al. 2023; Gilmore et al. 2023; Herrick et al. 2023, this collection)? To understand the long-term climate evolution of Venus, we need to establish (1) whether there is any morphological and compositional evidence of an epoch with abundant liquid water on the surface; (2) whether Venus is geologically active now, and whether this is a continuous or episodic style, to constrain interior-atmosphere exchange throughout history; (3) if there is atmospheric evidence of present day volatile sources and sinks at the atmosphere of Venus, including potential active volcanic sources; (4) whether and how sulfur- and water-related volatiles are transported through the atmosphere and how they interact with cloud layers (Bullock and Grinspoon 2001; Wilson et al. 2008; Titov et al. 2018).

b) *How are tropospheric and geological processes coupled on Venus?* Do exchanges take place from direct outgassing of volatiles into the lowermost atmosphere, buffering of atmospheric species with surface reservoirs, or aeolian/chemical alteration of surface minerals? The high surface pressure of Venus is maintained through surface-atmosphere chemical buffering reactions that are, as yet, unidentified. Buffer systems proposed have included calcite-anhydrite (Fegley and Treimann 1992; Fegley et al. 1992, 1997) and pyrite-magnetite (Hashimoto et al. 1997; Hashimoto and Abe 2005) systems, but there is little evidence constraining these claims because several of the relevant minerals including pyrite are unstable in Venus surface conditions (Hashimoto and Abe 2005). Latitudinal variability of minor species in the troposphere is thought to arise either from non-uniform vertical profiles, or from planetary-scale meridional transport through global convection cells, not restricted to cloud layers where most of solar energy is deposited, but extending deeply into the troposphere where latitudinal contrasts are observed. Because latitudinal gradients have already been observed in CO and OCS, these species act as tracers for the meridional circulation and provide glimpses into some of the chemical cycles of the troposphere. The water vapor vertical gradient in the deep atmosphere is not known and may be a steep gradient due to surface-atmosphere reactions (Ignatiev et al. 1997). Studying how trace gas abundances change over terrain of different compositions and/or elevations may yield insights into the surface-atmosphere exchanges and coupling (Zolotov 2018; Zolotov and Garvin 2020; Garvin et al. 2022a).

c) *Does present day atmospheric chemistry involve volcanic trace gasses?* Sulfur dioxide variations in the present day mesosphere have been used as possible evidence of volcanic activity (Esposito 1984). The proximate cause for these variations is related to spatial and temporal fluctuations of the SO_2 supply through vertical mixing within the cloud region. However, the origin of these vertical mixing fluctuations is barely understood. Purely atmospheric phenomena such as momentum deposition from upward propagating atmospheric gravity waves induced by topography (Kouyama et al. 2019; Kitahara et al. 2019) or diurnal variations of cloud top convection through solar absorption certainly play a role. Thermal destabilization of the atmospheric column through hot volcanic outgassing has also been suggested (Esposito 1984). SO_2 exhibits the most dramatic variations at Venus' cloud top, both spatially and temporally (Esposito 1984; Esposito et al. 1988; Marcq et al. 2013; Vandaele et al. 2017a,b, Encrenaz et al. 2016, 2019, 2020a), spanning more than two orders of magnitude on timescales ranging from a few days up to several decades. The greater range of SO_2 variations compared to H_2O is currently explained by the fast photochemical destruction of SO_2 by UV sunlight at cloud top level, making this species a much more sensitive tracer of the atmospheric circulation and vertical mixing between its lower atmospheric source and its cloud top photochemical sink. The variations of the vertical mixing are poorly understood. Conditions at which a buoyant plume could reach an altitude of 40-50 km are very narrow and require volatile-rich eruptions at higher elevation, their signature depending on the flux of volcanic gases, mixing rate with the air through eddy diffusion and turbulence (Lefèvre et al. 2018, 2020; Morellina and Bellan 2022), and wind velocities at different altitudes (Lorenz 2016; Bengtsson et al. 2012). Deep atmosphere gradients (i.e., every 140-200 m) associated with trace gas species involving SO_2 and other sulfur-bearing species will be obtained by the DAVINCI mission together with altitude resolved measurements of pressure, temperature (p, T) to address this issues regionally (Garvin et al. 2022a).

2.5 Global and Mesoscale Atmospheric Processes: Short Term and Long-Term Variability

a) *How to better constrain dynamical variability and couplings from surface to cloud tops?* The present-day surface of Venus is permanently obscured by a layer of optically thick clouds between 45 and 70 km altitude. The atmosphere at cloud level is in retrograde superrotation with a period of 4-7 Earth days in the zonal direction, i.e., parallel to the equator, relative to the solid surfacesurface, a motion driven by its thermal structure (Grassi et al. 2010, 2014). The sidereal day of Venus, by comparison, is 243.02 Earth days, also retrograde. The clouds have been studied *in situ* and are mostly composed of sulfuric acid droplets described as populations with lognormal size distributions with mode diameters of 0.15, 1.0, 1.3, 3.4 mm, called mode 1, 2, 2', and 3 (Knollenberg and Hunten 1980; Ragent et al. 1985). Several structures visible in UV and IR wavelengths at different altitudes indicate a dynamically active atmosphere as well as a significant geodynamic coupling with the surface: huge bow-shaped structures extending from northern to southern latitudes have been detected by the Longwave Infrared Camera (LIR) and the Ultraviolet Imager (UVI) on board JAXA's Akatsuki (Fukuhara et al. 2017, and references therein). The extension of wave trains in both upstream and downstream directions was also observed (Fukuya et al. 2022). Vertical wind oscillations attributed to topographic gravity waves have also been observed by the VeGa balloons over Aphrodite Terra, which has a top height of 3-4 km (Blamont et al. 1986).

One of the remaining questions about the dynamics of the Venusian atmosphere is how the convective cloud layer and topographically generated waves mix momentum, heat, and chemical species (Lefèvre et al. 2022). Radio occultation can be used to monitor the main cloud constituent, H_2SO_4, in both vapor and liquid form, near the cloud base, providing clues to cloud formation and convection processes. Using direct imaging or high frequency (HF) radar, electromagnetic signatures of lightning could be considered (Lorenz 2018), an investigation that would also contribute to the understanding of chemical and microphysical processes at work in the cloud layer. It should also be noted that the complex dynamics of the Venusian atmosphere produce a periodic mass redistribution pattern that generates a time-varying modulation of the Venusian gravitational field (Cascioli et al. 2023).

b) *How variable is the upper atmosphere?* The Venusian mesosphere (65-120 km) is a transition region between the lower atmosphere (from the surface to the cloud layer near 60 km), where the circulation is mainly zonal, and the upper thermosphere (above 120 km), where the wind pattern is mainly driven by diurnal pressure contrasts and flows from the sub-solar to the anti-solar point (the so-called SSAS flow). The zonal component decreases with altitude above the clouds, probably due to the deceleration caused by momentum transport by atmospheric waves (Sánchez-Lavega et al. 2017). Global structure of thermal tides in the upper cloud layer has been studied by the LIR camera on board Akatsuki (Kouyama et al. 2019). Monitoring of thermal profiles and winds in the mesosphere has shown that this transition region, where the two types of circulation coexist, is also characterized by important temporal variability. Large-scale images from Venus orbiters (e.g., VMC/VEx, UVI/Akatsuki) lacked spectroscopic capabilities, while orbiter-borne spectrographs (e.g., VIRTIS-H/VEx, SPICAV-UV/VEx) lacked extensive spatial coverage. Because mesospheric composition varies on time scales ranging from hours to years, measurements over a wide range of latitudes, local times, longitudes, and time scales are needed, including atmospheric airglow at 1.27 μm (Hueso et al. 2008; Soret et al. 2014), measurements of the spatial and temporal variability of albedo and UV absorber (Lee et al. 2019), ionospheric electron density, and temperature, pressure, and density of the neutral atmosphere by radio-occultation (Peter et al. 2023). HF sounding radar can be used to constrain the ionosphere, as routinely performed by MEx/MARSIS (Picardi et al. 2005). The DAVINCI mission will carry a technology demonstration instrument (CUVIS) that will acquire UV (200-400 nm) spectra at 0.2 nm spectral resolution on two Venus dayside flybys in 2030 with up to one million new spectra of the upper clouds, as well as a near UV frame imaging camera (VISOR UV) that will acquire "movies" during these periods to quantify cloud feature motions (Garvin et al. 2022a).

Venus is particularly important to our understanding of terrestrial planets' habitability, providing a natural laboratory to understand its evolution in time. Many significant questions remain on the current state of Venus, suggesting major gaps in our understanding of how Venus's evolutionary pathway diverged from Earth's (Fig. 1). Venus is the only spatially resolvable, Earth-sized world other than the Earth, where a diversity of geophysical envelopes, their interactions and evolution at several time scales and spatial scales, may be monitored from a variety of mission and instrumental concepts and support long-term evolutionary models of Earth-sized planets. Venus exploration offers therefore unique opportunities to answer fundamental questions about the evolution of terrestrial planets and the habitability within our own solar system.

Comparing the interior, surface and atmosphere evolution of Earth and Venus is essential to understanding what processes have shaped our planet, and is particularly relevant in an era where we expect thousands of terrestrial exoplanets to be discovered. Compelling recent insights, and the planet's relevance to exoplanetary systems (Kane 2022), have opened up new questions about the evolution and dynamic nature of Venus and its atmosphere. In Sects. 3-8, we explore how recently selected investigations and payload instrument concepts support the scientific goals and open questions presented in the accompanying papers of this collection. VERITAS (Smrekar et al. 2022a), DAVINCI (Garvin et al. 2022a), and EnVision (European Space Agency 2021) will greatly advance our understanding as well as lead to new questions about geologic evolution. In Sects. 9-12, we also address what key advances would come from new types of data to better constrain the interior, such as improved crustal thickness and structure, mantle viscosity/temperature from seismology, lithospheric thickness from electromagnetic sounding, in-situ heat flow to constrain thermal lithospheric thickness and radiogenic heat budget and distribution. In-situ geochemistry would help constrain rock rheology, mantle conditions, accretion considerations, and the source and history of specific rock types. Information on age of specific features will be very valuable; the detailed composition of weathering products will tightly constrain weathering rates and, if feasible, rock age dating would provide valuable absolute age information. Section 10 addresses, in particular, key areas and investigations at Venus not covered by the fleet of missions under development described in previous sections.

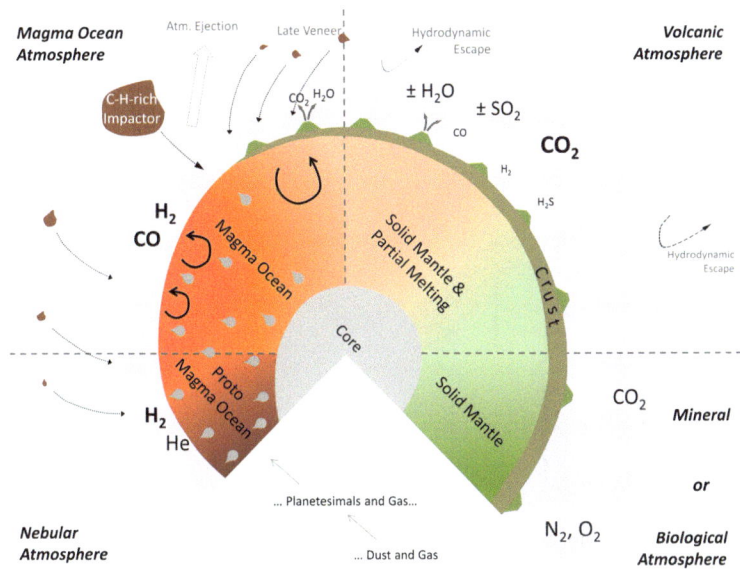

Fig. 1 Venus is essential to our understanding of the habitability of terrestrial planets, providing a natural laboratory for understanding their evolution throughout the history of our solar system. This clockwise helicoidal projection illustrates the parallel growth of an Earth-sized planet's interior, surface, and atmosphere (Gaillard and Scaillet 2014). In companion papers, Way et al. (2023), this collection, and Westall et al. (2023), this collection, discuss ongoing studies of the habitability of Earth-size planets, how they can inform the evolutionary history of Venus, and conversely, how the ancient evolution of Venus can serve as ground truth for studies of the divergent paths for Earth-size planets and the longevity of a habitable Venus-size or Earth-size body. The initial thermal state of the planet, which is closely related to its accretion sequence, determines the amount of energy the planet will dissipate over its history and the initial mantle inventory to be outgassed (Avice et al. 2022; Salvador et al. 2023; this collection), while later stages affect the extent, nature, and distribution of surface modification processes (Herrick et al. 2023, this collection; Ghail et al. 2023, this collection; Gilmore et al. 2023, this collection; Carter et al. 2023, this collection) and interior, surface, and atmosphere exchanges over time (Rolf et al. 2022, this collection; Gillmann et al. 2022, this collection; Wilson et al. 2023, this collection)

3 Overview of Mission Concepts and Observation Strategies to Address Venus Evolution Through Time

3.1 Selected Missions in 2021

On June 2, 2021, NASA announced the selection of two new missions to Venus, VERI-TAS (Venus Emissivity, Radio Science, InSAR, Topography, and Spectroscopy, Sect. 4) and DAVINCI (Deep Atmosphere Venus Investigation of Noble gasses, Chemistry, and Imaging, Sect. 5) as part of the Agency's Discovery 2019 competition. On June 10, 2021, the European Space Agency (ESA) announced the selection of EnVision as its newest Medium-Class science mission in the framework of ESA's Cosmic Vision program (Sect. 6). VERITAS, DAVINCI and EnVision will greatly advance our understanding as well as lead to new questions about Venus evolution through time. Key advances will come from new types of data to better constrain the interior, such as improved crustal thickness and structure, mantle viscosity/temperature from seismology, lithospheric thickness from electromagnetic sounding, in-situ heat flow to constrain thermal lithospheric thickness and radiogenic heat budget and

Fig. 2 A new fleet of elements of a Venus exploration program is now under development. These include two radar-equipped orbiters (ESA-led EnVision orbiter and NASA-JPL's VERITAS orbiter), and NASA-GSFC's DAVINCI entry probe with flyby remote sensing mission. Credit NASA / ESA / JAXA / Paris Observatory / VR2Planets

distribution. Over the next 15 years, these three missions will work together to answer many of the outstanding questions of Venus science and rocky planet evolution described above (Fig. 2; Table 1).

NASA's VERITAS orbiter will map the topography of the planet, create a compositional map, search for surface properties indicative of formation during a water-rich period, search for thermal and chemical signatures of volcanic activity, constrain interior structure, and look for surface change using a combination of radar imagery and differential interferometric techniques (Smrekar et al. 2022a). NASA's DAVINCI probe/flyby mission will measure atmospheric chemistry (noble gas isotopic abundances to constrain Venus' formation and early evolution) and physical properties (including wind speed, pressure and lapse rate) down to the surface and collect multi-scale near-infrared (sub-cloud deck) images of composition and topography at scales from ~200 m down to ~25 cm of the ancient tessera terrain where an oceanic past may be recorded (Garvin et al. 2022a). ESA's EnVision orbital mission will follow with a range of observation modes including multi-resolution radar imaging, radar polarimetry and radiometry, a suite of infrared and UV spectrometers for sensing the atmosphere and surface, and conduct the subsurface sounding (European Space Agency 2021). EnVision's combination of surface and atmospheric measurements will characterize ongoing volcanic processes though an extended-timeline, search for their thermal, morphologic, and gaseous signatures, while also tracing key volatile species from the surface up the mesosphere (Table 1). EnVision is an ESA-led mission in partnership with NASA, providing its Synthetic Aperture Radar instrument, VenSAR and Deep Space Network support for critical mission phases.

Both NASA and ESA orbital missions will provide global context to understand DAVINCI's local, high-resolution near-IR imaging (with topography and band-ratio composition) of its entry ellipse descent corridor within western Alpha Regio (Sect. 5.3.4), which in turn provides 'ground truth' for radar and near-IR VERITAS and EnVision data interpretation (Sects. 5.5 & 6.4). The scientific results of these three synergistic missions will

Table 1 A comparison of missions VERITAS, DAVINCI and EnVision payload instruments as developed in the following Sects. 4-6, their synergies and complementarity; mission design and driving requirements at time of publication (Smrekar et al. 2022a; Garvin et al. 2022a; European Space Agency 2021). VISAR, Venus Interferometric Synthetic Aperture Radar (§4.3.1); VEM: Venus Emissivity Mapper (§4.3.2); VMS: Venus Mass Spectrometer (§5.3.1); VTLS: Venus Tunable Laser Spectrometer (§5.3.2); VenDI: Venus Descent Imager (§5.3.4); VISOR UV: Venus Imaging System for Observational Reconnaissance (§5.4.1-5.4.2); VISOR near-IR: Venus Imaging System for Observational Reconnaissance (§5.4.1); CUVIS: Compact Ultraviolet Imaging System (§5.4.2); VenSAR, Venus Synthetic Aperture Radar (§6.3.1); SRS: Subsurface Radar Sounder (§6.3.2); VenSpec-H, -M, -U: Venus Spectrometer Suite (§6.3.3)

Parameter	VERITAS (Section 4)	DAVINCI (Section 5)	EnVision (Section 6)
Launch date / Duration (at time of publication)	Not earlier than 2031 / 4 cycles (2.7 years beginning ~2 yr after launch) + extended mission opportunities	June 2029: 2 flybys (Jan. 2030, Nov. 2030) + 1 entry, descent, science for chemistry, imaging & physical properties (June 2031)	Dec 2031, May 2032, or Nov 2032 / 6 cycles (4 years beginning ~4 yr after launch) + extended mission opportunites
Orbit / Inclination	180 × 255 km / 85.5 deg	Flybys from 180,000 km to 5,000 km; optional post-probe Venus Orbit Insertion (Jan 2032)	220 x 520 km / ~88 deg
Radar Imagery	VISAR: Global X-band (7.9 GHz) Imaging: 30 m;15 m 25% of surface	N/A	VenSAR: S-band (3.2 GHz) Imaging: 30 m ~30% of surface; 10 m ~2% of surface
Topography at high resolution	VISAR: Single Pass Interferometry Global Topography; 250 m horiz., 5 m vertical	VenDI: Sub-cloud near-IR Structure-from-Motion processed images from 60 m at < 4 m vertical to 25 m at < 3 m vertical; other options include ~90 m at < 7 m vertical & 5 m at 1 m vertical	VenSAR: Stereo, 25% surface, 300 m horiz. 30-50 m vertical Radar Altimetry, 75% surface, 3 km footprint, 2 m vertical precision; SRS Altimetry, 68% of surface
Subsurface boundaries	VISAR: indirect inference from synthetic aperture radar	N/A	SRS: penetration depth up to 1000 m; 20 m resolution
Microwave Radiometer brightness temperature and emissivity maps	N/A	N/A	VenSAR: Dual Polarization HH-HV 7% of surface nadir and off-nadir microwave radiometry, 75% surface. Short-term Temp. var. < 1K
Repeat Pass Interferometry for Deformation	VISAR: Targeted 12-18 200 × 200 km sites; 1.5 cm deformation accuracy	N/A	Not in current baseline; may recover some opportunistic capability depending on ΔV
k₂ tidal Love number and Moment of Inertia Factor (MoIF)	k₂ accuracy 0.2% at 3-sigma, moment of inertia at 0.3% at 3-sigma using two-way Ka-band Doppler tracking and radar tie points	N/A	RSE: k₂ accuracy 1.2% at 3-sigma, moment of inertia at 1.2 % at 3-sigma w/capability to combine Doppler tracking and radar tie points
Gravity field	Two-way Ka/Ka-band and X/X-band tracking; gravity field resolution < 105 km globally with <4 mGal RMS accuracy	N/A	RSE: Two-ways X / X-Ka band tracking; gravity field resolution < 200 km globally and < 170 km on at least 40% of planet with < 10mGal RMS accuracy
Noble gases, trace gases & their isotopic ratios; loss mechanisms and past outgassing	N/A	VMS; VTLS: in-situ probe-based analytical chemistry of S, O species and He isotopes Isotopic ratios for Ne, Ar, Kr, Xe H, S, C, and O & their isotopic ratios including D/H	VenSpec-H: D/H in tropospheric H₂O, HDO
Detect current and recent surface volcanism	VEM: thermal signature 1020 nm, unweathered flows: 6 near IR bands 860 to 1180 nm. VISAR: Surface changes within mission and between Magellan and EnVision	N/A	VEM: thermal signature 1020 nm, unweathered flows: 6 near IR bands 860 to 1180 nm. VenSAR: Surface changes vs Magellan, within mission, vs. VERITAS
Surface composition	VEM: 6 near IR bands 860 to 1180 nm	VISOR: 3 NIR bands from 900 to 1030 nm for Alpha, Ovda, Maat VenDI: sub-cloud near-IR band-ratios at scales from 150 m down to 12 m (740 to 1030 nm)	VenSpec-M: 6 near IR bands 860 to 1180 nm
Current volcanism using water vapor, S-bearing gas detection, and cloud top UV absorber	VEM: water vapor bands at 960 and 1160 nm; cloud bands at 1195, 1310, 1510 nm	CUVIS; hyperspectral UV spectroscopy (0.2 nm resolution) on two dayside flybys; VISOR: near IR emissivity mapping of Maat, Alpha, Ovda on 2 nightside flybys VMS; VTLS: Probe-based analytical chemistry of S, O species and He isotopes	VenSpec-M: water vapor bands 960 & 1160 nm; cloud bands at 1195, 1310, 1510 nm VenSpec-H: 1000-2700 nm: H₂O, HDO, CO and SO₂ VenSpec-U: SO, SO₂ variations 190-380 nm
p, T vertical structure of atmosphere	X/X+Ka/Ka-band radio occultation sounding, SO₂ and H₂SO₄ liquid and gaseous content at 45-55 km, p, T profiles at 45-90 km	p, T, winds, accelerations every 15-50 m from 70 km – surface; for lapse rate (T vs. z). Trace gases every 140 m from 40 km – surface for masses from 8 to 272 Da with gradients in key species (SO₂, OCS, H₂O, CO₂)	RSE: One-way X-Ka band radio occultation sounding, SO₂ and H₂SO₄ liquid and gaseous content at 45-55 km, p, T profiles at 35-90 km, at all local time, lat./long., 4 soundings / day

answer fundamental questions about the early period of Venus history, how the transition to its current forbidding environmental state came about, where volcanic eruptions are re-shaping the surface today, and what the differences from Earth can tell us about possible pathways of extrasolar rocky planet evolution.

3.2 Missions Considered for Selection

In addition to the three selected missions, several international space agencies have developed mission concepts. This collection is based on the papers presented at the Space Science Series of the International Space Science Institute (ISSI) workshop 'Venus: Evolution through Time' held in Bern, Switzerland, on September 13-17, 2021. It therefore captures the discussions that took place among the participants during the workshop. At that time, two large mission concepts were discussed for launch before the end of the current decade and possibly before the planned launch period of the currently selected VERITAS, DAVINCI, and EnVision missions: Venera-D (Zasova et al. 2020) and Shukrayaan-1 (Antonita et al. 2022). We therefore decided to describe extensively these two mission concepts in Sects. 7 and 8.

In conclusion to Sect. 3, we add a brief description of the Chinese radar-equipped VOICE orbiter mission proposal (Dong et al. 2023); and of Venus Life Finder "Morning Star", Rocket Lab's private low-cost concept mission to Venus (Seager et al. 2021; Limaye and Garvin 2023).

Venus Volcano Imaging and Climate Explorer (VOICE) is a Chinese orbiting mission to investigate Venusian geological evolution, atmospheric thermal-chemical processes, surface-atmosphere interactions, and habitable environment and life in the clouds. Three state-of-the-art scientific payloads, the Polarimetric Synthetic Aperture Radar (PolSAR), the Microwave Radiometric Sounder (MWRS) and the Ultraviolet-Visible-Near Infrared Multi-Spectral Imager (UVN-MSI), will be flown on a polar-circular orbit of ~ 350 km. VOICE is currently proposed to Strategic Priority Program (SPP) on Space Science of the Chinese Academy of Sciences (Dong et al. 2023).

Venus Life Finder "Morning Star" Mission, Rocket Lab's low-cost, small entry probe mission to Venus is intended to be the first in a series (Seager et al. 2021). After the cruise phase, the Photon platform, designed for launch on the Electron small launch vehicle, will target an entry interface to deploy a small (~ 20 kg) probe directly into the atmosphere with a flight path angle (EFPA) between -10 and -30 degrees, communicating direct-to-Earth through an S-band antenna, containing up to 1-kg of science payload (French et al. 2022; Seager et al. 2022).

4 Venus Emissivity, Radio Science, InSAR, Topography, and Spectroscopy (VERITAS)

The Venus Emissivity, Radio Science, InSAR, Topography And Spectroscopy (VERITAS) mission will address key science objectives about the geologic and volcanic history of Venus; it will elucidate how Venus' evolution differs from Earth's with three overarching Science Goals: 1) constrain Venus' geologic evolution; 2) determine which geologic processes are active; and 3) search for evidence of past and present water. VERITAS is a partnership between scientists and engineers at NASA/JPL and with the German, Italian and French Space Agencies. DLR provides the Venus Emissivity Mapper (VEM) instrument and VISAR interferometric processing support, ASI, which provides the Integrated Deep Space

Transponder, the lower power RF portion of the VISAR radar and the high gain telecommunications antenna, and CNES, which provides VEM optics and filter array subsystems, and the Ka-band TWTA. Equipped with an interferometric synthetic aperture radar and infrared imaging spectrometer VERITAS will globally map the surface of Venus producing imagery, rock type, topographic, and gravity maps. Following its selection by NASA, and according to the best information available at the time the manuscript is revised, VERITAS is programmed for launch no earlier than 2031 (Table 1).

4.1 VERITAS Science Objectives

4.1.1 Overview

VERITAS is structured to answer three essential science questions about the processes that have shaped and continue to shape the surface today:
1. What processes shape rocky planet evolution?
2. What geologic processes are currently active?
3. Is there evidence of past and present interior water?

These questions are organized into a series of specific investigations that can be addressed with two instruments and a radio science investigation.

VERITAS intends to definitively answer whether volcanism has been steady or catastrophic, why it lacks terrestrial-style plate tectonics, how it loses its heat, and if its plateaus are analogs of Earth's continents. It will also extend our knowledge of Venus with numerous firsts, including constraints on the core size and state (relevant to dynamo formation), high resolution topography and radar imaging, and a search for active surface deformation and active or recent volcanism.

Without an understanding of Venus' geologic evolution through time, we cannot fully test hypotheses for how Earth and other terrestrial planets evolve. Our knowledge of Earth in particular forms the basis for understanding habitability (Way et al. 2023, this collection; Westall et al. 2023, this collection). One of the central issues regarding the potential habitability of extrasolar planets is the extent to which plate tectonics is required to maintain habitability (e.g., Southam et al. 2015). A planet's interior is the engine for geologic activity, which drives climate and atmospheric evolution, thus setting the conditions for its long-term habitability. Because exoplanet surfaces cannot be spatially resolved, only indirect information on planetary radius, interior density, and atmospheric composition is available to predict habitability, in addition to radiative constraints to allow water to condensate in liquid phase (Hays et al. 2015; Meadows et al. 2018). Knowledge of the interior, including the size and state of the core, is critical to understanding how a planet loses its heat and evolves. Habitability models, such as that of Tosi et al. (2017), cannot predict the circumstances leading to habitability of a Venus-like planet, due to a lack of knowledge of Venus' geologic history and evolution.

Plate tectonics is Earth's defining geologic process. For billions of years, distinct lithospheric plates have moved above the upper mantle as they spread apart at undersea volcanic regions (the mid-ocean ridges), sink at subduction zones, and slide past each other at transform faults. Continental crust formed largely via melting of subducting slabs in the presence of water (Campbell and Taylor 1983; Arndt 2013). Earth's continents drift apart and crash together, but are never dragged into the interior because their Si-rich composition is lower in density.

Many new hypotheses linking plate tectonics to Earth's habitability are emerging, including the origin of the great oxygenation event (Duncan and Dasgupta 2017). Processes

Fig. 3 VEM's 6 channel, high signal-to-noise ratio (SNR) spectra will determine definitively if tessera plateaus are felsic (low iron) like Earth's continents. The lower 1000 nm VIRTIS emissivity for Alpha Regio (color overlay) suggests low iron content (Gilmore et al. 2015; Dyar et al. 2020; Helbert et al. 2021). An example of the complex deformation that characterizes tesserae is shown in the Magellan SAR image inset

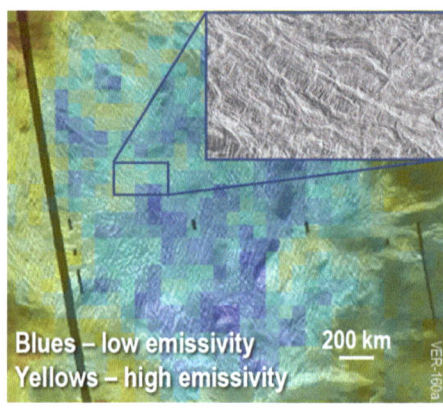

Blues – low emissivity   200 km
Yellows – high emissivity

that link volcanism (that releases volatiles such as H_2O, CO, and SO_2 from the interior to the atmosphere) and subduction (that cycles volatiles back into the mantle) help maintain a stable climate and hydrosphere (Kasting and Catling 2003; Driscoll and Bercovici 2013). In fact, the vast majority of Earth's present-day oceans and atmosphere came from interior degassing (Pearson et al. 2014; Marty et al. 2016, 2017). Interior heat loss maintains our planet's magnetic dynamo, protecting complex organic compounds from radiation damage by magnetically deflecting the solar wind. On Earth, plate tectonics is the dominant heat loss mechanism. Is there evidence of these processes on Venus? VERITAS looks for evidence of "continents," past geologic processes, volcanic history, and subduction—the first step in initiating plate tectonics.

4.1.2 What Processes Shape Rocky Planet Evolution?

Global Rock Type, Terrains, and Tesserae. - Venus surface geology is controversial and geochemistry is largely unconstrained. What little we know about surface composition comes from a handful of Venera and VeGa geochemical analyses (Grimm and Hess 1997; Treiman 2007), laboratory experiments (Shellnutt 2013), thermodynamic modeling (Teffeteller et al. 2019), and conclusions drawn from Magellan radar and limited (single-band) emissivity data from VEx (Mueller et al. 2008; Helbert et al. 2008; Smrekar et al. 2010a; Gilmore et al. 2015; Mueller et al. 2020). Landed measurements suggest geochemically distinct volcanic plains units, but the large uncertainties preclude definitive interpretations (Gilmore et al. 2017). New Magellan emissivity analysis suggests variability among tessera plateaus (Gilmore et al. 2019).

The composition of tessera terrain provides critical constraints on Venus geochemistry, geodynamics and the history of water on the planet. Containing what may be the oldest surface rocks on Venus (Ivanov and Head 1996), they are extremely complex geologic terrains (Figs. 3 and 4). Tesserae occur as both large plateaus with diameters of 1000–4000 km, and as smaller, isolated 100s-km-scale areas embayed by plains volcanism. They have been proposed to be continent-like based on their morphology, gravity signature, and inferred low-Fe, high-Si composition (Hashimoto et al. 2008).

Surface emissivity data from the VIRTIS on VEx, as well as from Galileo, suggest that tesserae are more felsic (lower in iron) than basaltic plains (Hashimoto et al. 2008; Helbert et al. 2008; Gilmore et al. 2015; Helbert et al. 2021) and thus possible analogs of terrestrial continents. Tessera composition and formation are key to assessing the role that volatiles play in shaping Venus' evolution. Large uncertainty in Magellan topographic height (Fig. 4)

Fig. 4 Constraining tesserae composition requires both VEM data and the VERITAS digital elevation model (DEM) to provide the elevation-dependent temperature correction. Tessera heights are not fully resolved in either the spatially limited Magellan stereo DEM data (black) or Magellan global altimetry resolution (blue)

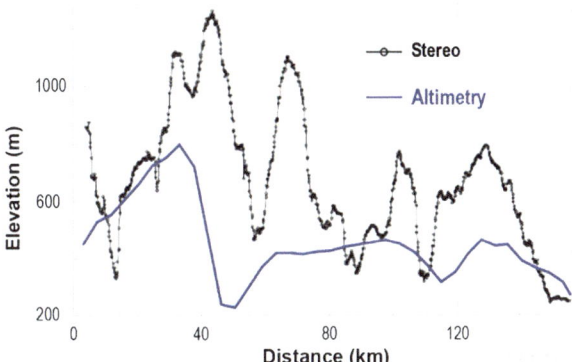

and thus in the altitude-dependent temperature correction for emissivity, as well as lack of spectral data, together preclude an answer to whether Venus has continents.

Two main hypotheses have been proposed for the origin of tesserae: downwelling and upwelling mantle flow. Deciding between these has significant implications for the crustal strength, volatile content, thermal gradient and evolution, and formation timescale (Lenardic and Kaula 1995; Ghent et al. 2005). According to the downwelling scenario, plateaus form as the crust thickens above regions where the cold mantle sinks (Bindschadler and Parmentier 1990; Bindschadler et al. 1992). Downwelling and remelting of basaltic crust could produce more Si-rich compositions, similar to those produced at some of Earth's subduction zones (Elkins-Tanton et al. 2007; Romeo and Turcotte 2008; Gilmore et al. 2017). Alternatively, plateaus may have formed over upwelling mantle plumes, with crustal thickening caused by profuse mantle melting (Phillips and Hansen 1994) that should produce a basaltic composition. Discriminating between these formation models requires distinguishing between Fe-rich, Si-poor basaltic rocks and more Si-rich continental rock types, as well as ascertaining the detailed shape of the small-scale graben that are diagnostic of specific deformation sequences (Bindschadler and Parmentier 1990; Bindschadler et al. 1992; Hansen and Willis 1998). If intermediate rock types are present, they are expected to have intermediate emissivities.

The primary mountain belts on Venus encircle Lakshmi Planum, and may be similar in composition to tesserae. However, there are no VIRTIS emissivity or landed data in that location. If the plains' compositional differences (tholeiitic and highly alkaline basalts) suggested by landed data can be validated by VERITAS' near-global rock type mapping, they would provide evidence of varying depths of melting and fractionation and thus geologic setting (Gilmore et al. 2017).

In addition, gravity data also inform our understanding of rock types and global terrain. Global gravity and topography data can be used to estimate the elastic thickness/thermal gradient, which constrains mechanisms of formation, and can identify processes such as subduction, localized delamination/ upwellings, and fossilized rifts. Other fundamental questions also remain: Did all tesserae form the same way? Are they all the same composition?

Prior Geologic Regimes: Buried Features. - Global high-resolution surface topography, images, and gravity data can provide critical windows into the past. On Mars, high-resolution topography shows subtle signatures of buried impact craters, only revealed by MGS' Mars Orbiter Laser Altimeter (MOLA) instrument (Frey et al. 2002; Buczkowski and McGill 2002; Frey 2006). These demonstrate that an ancient cratered surface was buried

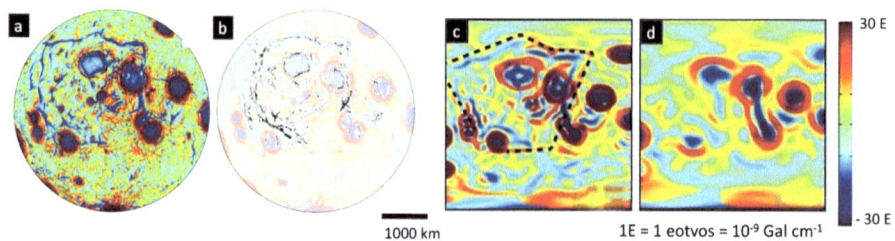

1000 km

1E = 1 eotvos = 10⁻⁹ Gal cm⁻¹

Fig. 5 Bouguer gravity anomalies and gravity gradients reveal a pattern of narrow linear anomalies that border Procellarum associated with buried lunar subsurface structures (a, b; data from the GRAIL mission). VERITAS' high-resolution gravity can reveal such subsurface structures. Gravity gradient maps at VERITAS' resolution (c) would reveal buried rifts (outlined in dots); map at Magellan resolution (d) is insensitive to these features. Figure after Andrews-Hanna et al. 2014; 1 Eotvos (E) $= 10^{-9}$ Gal cm^{-1} $= 10^{-9}$ s^{-2})

beneath plains material, disproving theories of resurfacing by plate tectonics and supportive of lowlands formation via a giant impact (Andrews-Hanna et al. 2008). MESSENGER images of Mercury show volcanically flooded ghost craters (Head et al. 2011). On the Moon, GRAIL gravity data reveal ancient rifts for which no surface signature exists (Fig. 5, Andrews-Hanna et al. 2014), as well as buried impact craters (Evans et al. 2016; Sood et al. 2017).

What lies beneath Venus' volcanic plains? Evidence for buried impact craters would shed new light on the history of volcanism and age of the underlying terrain. Venus may have had plate tectonics earlier in its history, perhaps continuing until the most recent resurfacing event (Armann and Tackley 2012). Interconnected lineations in the gravity or topography could reveal the sites of past spreading centers or subduction zones. Alternatively, tesserae may underlie the plains.

Tesserae inliers are common and could be the gravitationally relaxed remnants of ancient tesserae plateaus (Ivanov and Head 1996; Hansen and López 2010). The topographic or gravity signature of tesserae under the plains would show that tesserae were widespread, and may have formed during a prolonged wet period on Venus.

Impact and Volcanic History. - Venus' impact crater population holds the key to the planet's integrated time history of volcanism. The small number of craters (less than 1000 observed) implies a young surface age, but there is uncertainty in the number of modified craters. The "Catastrophic Resurfacing" hypothesis postulates that a huge pulse of volcanism rapidly covered Venus ~1 Ga ago, followed by limited activity. This theory was based on two observations: 1) The distribution of craters cannot be distinguished from a random one; and 2) few craters appear to be modified by volcanism or tectonism (Schaber et al. 1992; Strom et al. 1994).

Numerous models derive a resurfacing rate by comparing the rates of crater formation and removal or modification. Many models are more consistent with an equilibrium rather than catastrophic resurfacing rate (Basilevsky 1993; Bjonnes et al. 2012; O'Rourke et al. 2014; King 2018). In all cases, the critical parameter is how many craters are modified. The two primary modes of crater modification are volcanic infilling and weathering removal of distal crater ejecta (halos and parabolas). In the catastrophic resurfacing model, major crustal resurfacing is confined to a discrete pulse, and consequently only a small number of craters, preserved from this time, should show volcanic modification. The equilibrium theory suggests resurfacing processes occur continuously, and the majority of craters should display some form of modification, although the degree of modification should vary regionally (Herrick and Rumpf 2011).

Studies investigating crater modification have found evidence of craters with varying levels of modification, however, these results remain controversial. If all dark-floored craters ($\sim$80% of craters) have been partially flooded, then volcanism has persisted over 100s of millions of years (Herrick and Sharpton 2000; Herrick and Rumpf 2011). However, the resolution of the Magellan stereo data ($\sim$100 m) casts the robustness of this interpretation into doubt.

The study of fine-grained distal crater ejecta (halos and parabolas) argues against catastrophic resurfacing. Analysis of variations in halo retention combined with regional crater density, suggest that the surface of Venus is divisible into three major age groupings (Basilevsky et al. 2003b). These include those superposed on wrinkle ridges on regional plains, those superposed on units younger than regional plains, and "other." However, their study relies on the assumption that the processes removing the extended ejecta deposits operate at a uniform rate across the planet, despite no direct observations of the actual mechanisms, and the natural difficulty in mapping the morphology of the dark, diffuse deposits.

Radically different implications follow from the various resurfacing models. The "catastrophic" resurfacing hypothesis suggests that planets may behave episodically, with cycles of plate tectonics or mantle overturn and massive melting (Parmentier and Hess 1992; Reese 1999; Armann and Tackley 2012) separated by periods with no plate motion (a "stagnant lid" regime). These models predict the spike in volcanism implied by catastrophic resurfacing. Volatiles outgassed by volcanism in a catastrophic event could have a significant effect on climate, perhaps changing temperature enough to cause detectable thermal expansion of the surface (Anderson and Smrekar 1999). In contrast, the equilibrium resurfacing model implies more Earth-like rates of volcanism without a need for episodic plate tectonics (O'Rourke et al. 2014).

All resurfacing models are constrained by the number of modified craters and extended ejecta deposits, which remains controversial. Thus, the distribution of volcanism in space and time, the relative age of different locations, and implications for the history of Venus' evolution are uncertain. One method to assess resurfacing history is to model the gradual removal of fine-grained crater ejecta (Phillips and Izenberg 1995; Ghail et al. 2023, this collection). Relatively old regions, without any volcanism for an extended time, would have both high crater density and relatively few halos, since aeolian or chemical weathering can also remove halos and not high-standing crater rims or rocky ejecta. A high-resolution Digital Elevation Model (DEM), gravity measurements, and surface imagery are needed to discriminate between models: to characterize subduction and extension zones, fault patterns, flexed terrains and their margins, elastic thickness anomalies, and the distribution, nature and extent of impact crater modification (Herrick et al. 2023, this collection). VERITAS will provide a new and detailed view of global tectonics, potentially revealing strike-slip faults, extensional zones and other deformation features (Fig. 6).

What Are the Major Tectonic Processes? Is Subduction Currently Active? - Venus does not currently have recognizable Earth-like plates bounded by spreading centers, subduction zones, and transform faults (Solomon et al. 1991). On Earth, plate formation at spreading centers and subduction of old plates dominate heat loss and drive geologic activity. Without global plate tectonics, how does Venus lose its heat?

One clue is that Venus' internal heat engine has created abundant tectonic deformations including >40,000 km of fractured troughs, possible subduction zones, and huge mountain belts. Among the terrestrial planets, only Earth and Venus have such massive deformation zones. Could Venus represent a transitional tectonic state between active (plate tectonic) and stagnant (no plate motion) regimes? Although not globally interconnected, Venus may have key elements of plate tectonics: subduction, major extensional zones, and even transform faults.

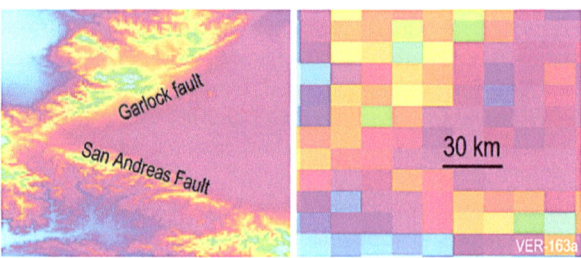

Fig. 6 VERITAS' Digital Elevation Model (DEM) has the needed resolution to discover narrow deformation zones such as strike-slip plate boundaries not apparent in image data. Shuttle Radar Topography Mission (SRTM, Rodriguez et al. 2005; Farr et al. 2007) Earth's topography data (blue = low, purple = high), reduced to VERITAS resolution (left, 240 m horizontal, 5 m vertical noise) clearly show major tectonic boundaries. Such faults are invisible at Magellan resolution (right, 15 km horizontal, 100-m noise)

Subduction is a necessary first step in initiating plate tectonics. Whether it occurs on Venus is vigorously debated (Hansen and Phillips 1993; Sandwell and Schubert 1992a,b; Hansen and Olive 2010; Baes et al. 2016; Crameri and Tackley 2016). Venus' huge trough systems (termed chasmata) are typically interpreted as extensional zones, based on numerous long, linear fractures that can sometimes be identified as graben. However, some chasmata have characteristics of subduction. Specifically, some troughs have the same asymmetric shape, with a ridge on the concave side and a trough on the convex side, as observed at some arcuate terrestrial oceanic subduction zones including the South Sandwich Islands trench (Sandwell and Schubert 1992a; Fig. 7).

Because proposed Venusian subduction zones have estimated elastic thicknesses and bending moments similar to their terrestrial analogs (Sandwell and Schubert 1992a,b), their mechanical behavior should also be similar. If Venus' asymmetric chasmata are produced by subduction, their gravity signature should be asymmetric, in contrast to the symmetric signature of extension (Smrekar et al. 2010b).

A recent theoretical development (Bercovici and Ricard 2014) assesses the likelihood of plate tectonics on exoplanets: subduction evolves into plate tectonics when reductions in lithospheric strength from breakage (due to microcracks and grain-size dependent deformation) dominate over lithospheric healing via grain growth. This suggests that Venus may have a hot lithosphere that anneals too rapidly to allow subduction to develop into full plate tectonics. Global estimates of elastic thickness will allow us to assess whether it indeed has a hot, thin lithosphere, similar to early Earth's. Finding unequivocal evidence for subduction on Venus would elucidate the conditions and mechanisms for subduction initiation on terrestrial planets. Evidence of other major tectonic boundaries (extensional zones like mid-ocean spreading centers or major transform faults) would suggest that Venus could even transition to an Earth-like plate tectonic regime, as predicted by modeling (Armann and Tackley 2012).

Interior Structure and Thermal State. - Our knowledge of the internal structure of Venus is based on limited data for mass, radius, gravity, and topography. One direct existing constraint is provided by the tidal Love number k_2, or gravitational potential modification due to the tidal deformation of the planet.

Lack of a magnetic field at this stage of Venus' evolution does not provide any constraint (Stevenson 2003). Thus, interior structure models use a core size that is simply rescaled from Earth. Published temperature profiles for Venus' interior differ by up to 1000 K and 500 K in the lower and upper mantle, respectively (Steinberger et al. 2010; Armann and

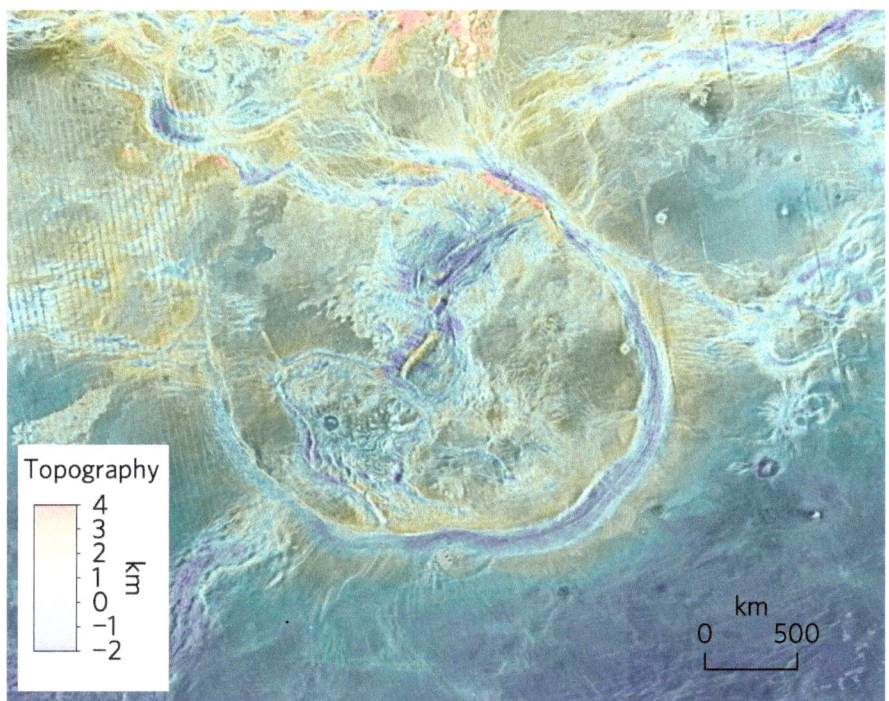

Fig. 7 VERITAS determines whether Venus has subduction. The surface of Venus hosts a variety of different features - volcanoes, rifts, mountain belts - that are typically on the scale of hundreds of kilometers, set within globally extensive regional lowland plains. Many features resemble Earth's arc-shaped oceanic subduction zones, such as the trenches at Artemis Corona, shown in this overlay of Magellan SAR data on altimetry (vertical scale at lower left): troughs have the same asymmetric shape, with a ridge on the concave side and a trough on the convex side (after Davaille et al. 2017)

Tackley 2012) with different implications for its cooling and volcanic history. Gravity and topography (Konopliv and Yoder 1996; Steinberger et al. 2010; Huang et al. 2013; Rolf et al. 2018) are linked at long wavelengths to the structure and dynamics of the sub-lithospheric mantle.

Precise measurements of the Moment of Inertia factor (MOIF) through the pole precession rate, the tidal Love number, k_2 and the tidal phase lag, e, will permit the first useful comparisons between the interior of Venus and other terrestrial planets (see also this Section, Sect. 4.3.3). The state of the core, its size (if liquid), and the tidal phase lag have already been determined by space missions with different accuracy for Mercury, Mars, and Moon. These previous studies demonstrate that knowledge of Venus' interior structure and therefore constraints on its long term evolution, can be derived from the moment of inertia, Love number k_2, and phase lag e.

As an example of results obtained for other Solar System bodies, Moment of Inertia factor constrains the core size of Mercury (Genova et al. 2019) and Mars (Smrekar et al. 2019), and the measured Love number k_2 is indicative of a liquid (or partially liquid) core for the Moon (Williams et al. 2014), Mercury (Margot et al. 2018), and Mars (Yoder et al. 2003). k_2 for Venus has been estimated from Doppler tracking of Magellan and Pioneer Venus Orbiter ($k_2 = 0.295 \pm 0.066$; Konopliv and Yoder 1996), but the large uncertainty does not allow distinguishing between a liquid vs. solid core (Dumoulin et al. 2017). Use of the Martian phase lag (Plesa et al. 2018) constrains core size and volatile components such as sulfur (Rivoldini et al. 2011), as well as mantle viscosity.

4.1.3 What Geological Processes Are Currently Active?

Multiple observations suggest that Venus is geologically active. Magellan topography and gravity data analysis indicate the likely presence of low density, hot mantle plumes under large volcanic rises, or "hotspots," similar to those on Hawaii (Smrekar 1994). Atmospheric sulfur variations imply active volcanic outgassing (Marcq and Lebonnois 2013). The average sulfur content corroborates ongoing volcanism because SO_2 breaks down over time (Fegley and Prinn 1989; Fegley et al. 1997). Campbell et al. (2017) presented evidence for relatively recent pyroclastic volcanic flows based on their radar properties and lack of erosion. Finally, the 2011 Planetary Science Decadal Survey named the case for recent volcanism (Smrekar et al. 2010a) from near-IR emissivity anomalies to be one of the ten major discoveries of the last decade. Enhanced 1020 nm emissivity correlates with stratigraphically young flows at three areas classified as hotspots based on their broad topographic rises and large positive gravity anomalies (interpreted as mantle plumes). The emissivity increase between these areas and the surrounding plains is consistent with that between unweathered and weathered basalt, implying recent volcanism on timescales of years to decades, based on new laboratory data (Knafelc et al. 2019; Berger et al. 2019). But VEx VIRTIS data cover only the southern hemisphere with only a single band.

Are the six hotspots not observed by VIRTIS also active? Is there evidence for recent volcanism in other geologic settings? If present-day volcanism is restricted to hotspots, this may imply a shift from radiogenic internal heating to basal heating at the core-mantle boundary, allowing hotspot formation (Choblet and Parmentier 2009). Alternatively, restriction of volcanism to hotspots could arise from upper mantle heating under an insulating stagnant lid, leading to shutdown of widespread plains volcanism through desiccation of the upper mantle. In this scenario, hotspot volcanism occurs where water-bearing material from the lower mantle melts (Smrekar and Sotin 2012). Identifying the origin of current volcanism will dramatically change our understanding of interior dynamics.

If Venus remains volcanically active, it is also deforming. For the first time on another planet, we will have the ability to detect active tectonic deformation through repeat pass interferometry. Finding active deformation, and identifying the mechanism(s) responsible, would provide unprecedented insights into the geologic evolution of Venus, and by extension to terrestrial planets.

4.1.4 Is There Evidence of Past or Present Water?

Geochemical Fingerprints of Past Water. - Venus may have once had a shallow ocean's worth of water at the surface, based on the D/H ratio of the atmosphere (Donahue et al. 1982; Kumar and Taylor 1985; de Bergh et al. 1991). Its interior may still hold $\sim$75% of its original volatiles (O'Rourke and Korenaga 2015). Its surface composition could retain a fingerprint of past water. Recent VIRTIS emissivity data at 1020 nm from a limited number of tesserae suggest a composition lower in iron than the plains, implying they formed in the presence of water (Mueller et al. 2008; Gilmore et al. 2015; Campbell et al. 2017). With an additional five emissivity bands and more accurate altimetry data to correct for altitude-dependent temperature (as well as detailed T vs z lapse rate boundary conditions from DAVINCI), VERITAS can fully confirm or refute this.

VERITAS uses VEM to map the distribution of FeO contents and relates them to geologic features. If continental-scale low-Fe silicic crustal regions on Venus are confirmed, this will indicate past hydrated source regions. Determining that tesserae are in fact continent-scale silica-rich regions would illuminate the processes that shaped our home planet.

Is Interior Water Being Volcanically Outgassed Today? - Pioneer Venus and VEx (Marcq et al. 2013; Esposito 1984; Esposito et al. 1988) observed strong SO_2 variability at the cloud tops, suggestive of plumes of rising gas from active volcanoes. Recent ground-based observations using Infrared thermal mapping have continuously evidenced such rapid variations (Encrenaz et al. 2013-2020a, see also Sect. 11, Sect. 11.1.3). Plumes must include substantially lighter elements than SO_2, including $\sim$2–5% H_2O, to have sufficient buoyancy to reach heights of tens of kilometers (Airey 2015). Terrestrial volcanoes emit large quantities of H_2O, CO, and SO_2 (e.g., Gerlach 1980). Thus, volcanically outgassed water will only be observed on Venus if there are large, Earth-like concentrations remaining in the interior. Radar-derived evidence for relatively recent pyroclastic flows implies significant outgassing (Campbell et al. 2017). VERITAS observes near-surface water vapor through near-IR atmospheric absorption bands. Unlike higher altitude VEx H_2O measurements, the association of near surface water vapor with volcanism can be confirmed with additional surface observations.

4.2 VERITAS Mission Overview

The VERITAS mission is an orbiter carrying an X-band Interferometric SAR and the VEM instrument to perform global SAR mapping at 30 m resolution; and to acquire topography, gravity, InSAR, and NIR emissivity data. Its indicative mass budget including all margins is 1450 kg (dry mass). After nominal launch VERITAS has a $\sim$7-month cruise to Venus. Following Venus Orbit Insertion (VOI), VERITAS performs a maneuver to reduce the orbit period from 120 hrs to 13 hrs. Aerobraking is then used to place the flight system in its initial and final science orbits. VERITAS has two planned science phases: Science Phase I (SP1) and Science Phase II (SP2). Science Phase I uses a 6.1-hour, highly elliptical orbit, and Science Phase II has a 91-minute orbit period at a mean orbit altitude of 217 km. Both have a nearly an orbit inclination of 85.5°.

The planned aerobraking campaign begins about two weeks after the Period Reduction Maneuver. The aerobraking campaign is divided into two segments: Aerobraking I (AB1, 7 months) and Aerobraking II (AB2, 9 months), with the 4-month SP1 embedded between them. During aerobraking, the spacecraft lowers periapsis into the upper atmosphere, using the additional drag to reduce its apoapsis altitude from over 40,000 km to 400 km. In SP2, which consists of four Venus cycles (roughly equal to a sidereal day of 243 Earth days), all instruments, VEM and VISAR, as well as the gravity science observations are conducting observations. Science Phase II begins with a 60-day VISAR calibration phase after the post-Aerobraking Exit maneuver (post-ABX) 200 × 400 km orbit is reshaped into the final science orbit. Venus rotates so slowly that the ground track moves only 10 km per orbit at the equator. Venus is so spherical, with an equatorial bulge three orders of magnitude smaller than Earth, that the spacecraft's ascending node precesses extremely slowly. These effects create a ground-track repeat cycle slightly longer than the 243-day Venus sidereal day; this causes the orbit to experience the same gravitational perturbations repeatedly over every orbit, which leads to significant eccentricity vector evolution, similar to that experienced by low lunar orbiters such as GRAIL. The mean inclination of 85.5° has been selected to cause this evolution to bend back upon itself, yielding a near-frozen orbit with altitude variation of 182.6 km to 252.5 km that repeats with the topography. A small radial Eccentricity Control Maneuver (ECM) at the end of each cycle corrects the slight mismatch.

Repeat-Pass Interferometry (RPI) enables cm-scale change detection of the surface and requires that the spacecraft fly within 160 m of its previous path over the targeted site. In

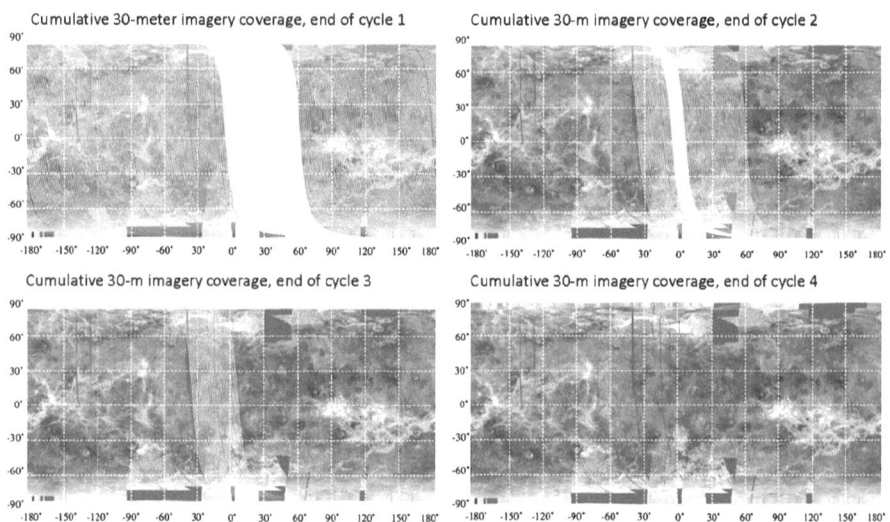

Fig. 8 Map of cumulative 30-meter VISAR radar images over the 4-cycle VERITAS ~3.5 Earth years mission Science Phase II (SP2). SP2 starts after further aerobraking has placed VERITAS in a near-polar, circular, low-altitude orbit. For 57% of each 224-day Venus year, science mapping will cover the entire orbit

addition to maneuver execution errors and differential solar tidal torques, the uncertainty in the rotation period of Venus affects repeatability. To compensate, VERITAS will process radar tie points on the ground, as demonstrated with Magellan data, to improve orbit reconstruction (Chodas et al. 1992, 1993), and modeled for VERITAS (Cascioli et al. 2021).

The VERITAS mission design for SP2 enables a flexible orbit plan that balances 11 science mapping orbits with five downlink orbits each day. For 57% of each 224-day Venus year, science mapping will encompass the entire orbit. In the remainder, eclipses are long enough that additional power management is required. VISAR mapping occurs on a fraction of these orbits selected to optimize coverage within available power balance. Figure 8 shows the build of the 30 m radar imagery over the 4 cycle mission.

4.3 VERITAS Science Payload

VERITAS's payload is composed of two instruments crafted to study Venus' surface coupled with a radio science investigation to measure the gravity field. The two instruments are an X-band interferometric synthetic aperture, VISAR, and a fourteen-band infrared spectrometer, VEM.

4.3.1 Venus Interferometric Synthetic Aperture Radar (VISAR)

Persistent optically opaque cloud cover of Venus necessitates the use of synthetic aperture radar techniques to obtain high resolution imagery and topography of the surface. The Venus Interferometric Synthetic Aperture Radar instrument (VISAR) instrument is an X-band single pass radar interferometer designed to acquire high resolution imagery and topography of Venus as well as to make repeat pass interferometric measurements of surface deformation.

VISAR Specifications. - VERITAS requires a Digital Elevation Model (DEM) with 300 m horizontal postings over 90% of the Venus surface, with height accuracy ≤10 m for

Table 2 Key VISAR radar design and performance parameters

VISAR Parameters	
Instrument Parameters	Value
Platform Altitude Range (km)	182–252
Polarization	VV
Peak RF transmit power (dBW)	26.0
X-band Wavelength (m)	0.038
Antenna azimuth × elevation dimensions (m)	3.9×0.65
Range bandwidth (MHz)	20
Slant Range/Azimuth Resolution (m)	7.5, 2.3
Incidence Angle at swath center (°)	32
Pulse Repetition Frequency (PRF) (Hz)	5500
Pulse length (μs)	35
Baseline length (m), and orientation angle (°)	3.1, 30
Range, Azimuth Ambiguities (dB)	$-36, -25$
Atmospheric Losses (dB)	-9.5 dB

95% of the mapped surface. Global imagery with resolution less than 30 m is also required with radiometric resolution better than 3 dB.

VISAR Design Considerations. - VISAR operates at a center frequency of 7.9 GHz (0.038 m wavelength), which optimizes topographic mapping accuracy by balancing the effects of atmospheric attenuation with baseline and antenna size constraints to fit within the spacecraft fairing. A 20 MHz bandwidth provides ∼15 m ground resolution at the VISAR 30° angle of incidence. The radar must map a swath width greater than 14 km, spanning the 10 km of surface rotation at the equator during an orbital period with 2 km overlap with adjacent orbits. Table 2 lists key radar design and performance parameters.

VISAR Onboard Processing. - VISAR would include an On-board Processing (OBP) element to meet downlink constraints. Key processing steps in the OBP data flow are range compression, motion compensation, azimuth compression, interferogram formation and look averaging of imagery and interferograms.

VISAR Modes of Operation. - The radar has one science-mapping mode with several data downlink options that would accommodate different interferometric and imagery resolutions. For nominal science operations during SP2, we would upload a command table twice weekly specifying, as a function of S/C clock time, the radar parameters needing adjustment.

VISAR Calibration. - Preflight VISAR instrument calibration activities would include measurements of the Solid State Power Amplifier (SSPA) power output, pulse shape, receiver gain, ADC characteristics, Tx-cal-loop phase and amplitude over temperature, and antenna composite waveguides phase and amplitude variation over temperature. For in-flight radar calibration the radar would collect raw data from the two antennas to be downlinked for ground analysis. These data would be used to update calibration parameters needed for the proper collection and onboard processing of the radar data. Primarily, these data would be residual differential time delay between the two radar receive channels any yaw or pitch angle bias adjustments needed for the S/C pointing control to achieve zero-Doppler steering.

VISAR Data Acquisition. - Topography data would be acquired on ascending and descending passes with at least two observations (also called revisits) for 95% of Venus' surface, with more than 80% acquired 3–6 times. Revisits would provide the opportunity to detect surface changes. During descending passes for the VISAR left-looking sensor, matching the dominant East-Looking data acquired by Magellan, data would be acquired to obtain a combination of MedRes (30 m resolution) imagery for nearly 100% coverage and HiRes (15 m resolution) imagery with 27% coverage. VERITAS would be capable of targeting between 12 and 17 200 × 200 km sites with RPI, covering approximately 0.1% of the surface, acquiring each site at least twice to form a repeat pass interferogram. The deformation accuracy including atmospheric variations, mostly due to SO_2 variations, at 50 m posting is about 1.5 cm. These data would be the first deformation SAR interferometry at another planet (Hensley et al. 2022). Potential surface activity areas included those that are likely to have recent volcanism, possible subduction zones and areas of gravitational relaxation; If active regions of Venus experience similar levels of activity as Earth analogs the VERITAS project predicts detection of 3–7 events on a total of 17 RPI acquisitions.

VISAR Expected Performance. - We developed a comprehensive model to evaluate radar performance at Venus including imaging, radar stereo, single and repeat pass interferometric modes (Hensley 2009; Hensley et al. 2018). The radar performance model elements specify the observing geometry and scenario, the instrument configuration and product specification parameters, propagation and scattering parameters that are used to determine radar performance depending on mode, time and location of measurement. Backscatter information is derived from Magellan S-band data using a physical scattering model to convert S-band backscatter measurements to the desired radar frequency and incidence angle. The impact of atmospheric attenuation as a function of terrain height is derived from a model described in Duan et al. (2010). Two-way losses at X-band as a function of elevation (in km) relative to the 6051 km reference sphere is roughly −9.5 dB. In assessing the interferometrically derived height accuracy we have assumed a "bundle adjustment" procedure to remove residual cross-track tilts due to baseline and phase errors. Bundle adjustment uses tie points between adjacent orbits and between crossing ascending and descending passes in a least squares procedure to estimate cross-track tilt and elevation bias between the swaths.

The expected elevation mapping accuracy of the VISAR instrument is shown in Fig. 9. Backscatter contributions to SNR and attenuation losses are factored in the overall performance. Elevation accuracy is computed every 10 km based on the orbital geometry. Phase noise limited elevation accuracy (in green in Fig. 9c) is compared to elevation accuracy before and after bundle adjustment (red and blue in Fig. 9c). The cumulative elevation accuracy is shown in Fig. 9d and shows that 95% of the surface is mapped with elevation accuracy better than 5.9 m.

4.3.2 Venus Emissivity Mapper (VEM)

The permanent cloud cover of Venus prohibits observations of the surface with traditional imaging techniques over much of the electromagnetic (EM) spectral range. Therefore, it was once thought that information about the surface composition of Venus could only be derived from lander missions. Given the harsh environmental conditions on the surface, any type of landed mission will have high complexity and therefore a higher associated risk than orbiting missions. In addition, mission concepts for Venus landers typically focus on one landing site instead of a global reconnaissance, forcing difficult choices to be made between different types of surface units.

The mapping of the southern hemisphere of Venus with VIRTIS instrument on Venus Express using the 1.02-μm thermal emission band can be viewed as a proof-of-concept for an orbital remote sensing approach to surface composition and weathering studies for Venus (Mueller et al. 2008; Helbert et al. 2008; Smrekar et al. 2010a; Gilmore et al. 2015). Thermal emission from the surface is observed on the night side at spatial scales > 50 km.

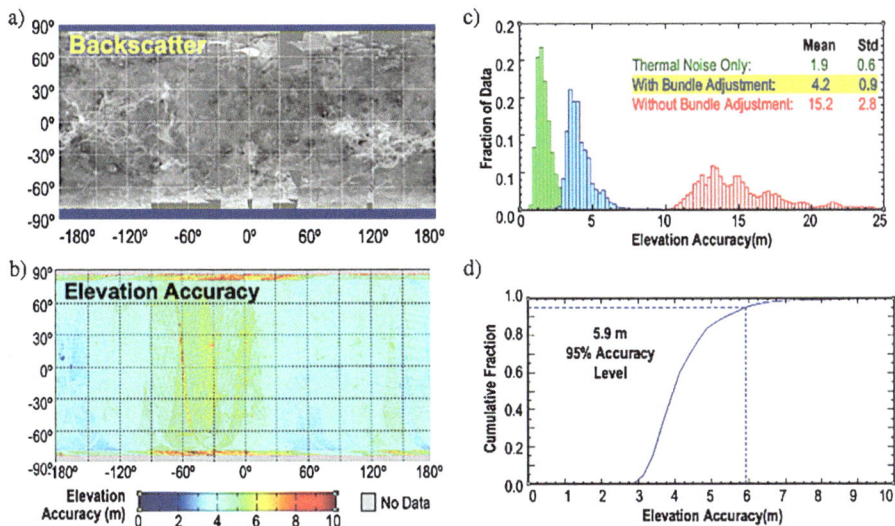

Fig. 9 (a, b) Map of the VISAR elevation mapping backscatter/accuracy, (c) histograms of elevation mapping accuracy with the bundle adjustment and (d) cumulative elevation accuracy showing a 5.9 m 95% accuracy level

Recent advances in high-temperature laboratory spectroscopy at the Planetary Spectroscopy Laboratory at DLR show that the atmospheric windows in the CO_2 clouds of the Venus atmosphere, ranging from 0.86 μm to 1.18 μm, are highly diagnostic for surface composition (Dyar et al. 2020, 2021; Helbert et al. 2021). Night-side observations at shorter wavelengths (<0.80 μm) from the Wide-Field Imager WISPR instrument on board Parker Solar Probe (PSP), allowed to extend measurements of this thermal emission into the optical regime (Wood et al. 2021).

The Venus Emissivity Mapper (Helbert et al. 2016, 2020) builds on these recent advances. It is the first flight instrument specially designed with a focus on mapping the surface of Venus using the narrow atmospheric windows around 1 μm. By observing with six surface bands centered at 0.86 μm, 0.91 μm, 0.99 μm, 1.02 μm, 1.11 μm, 1.18 μm coupled with 8 atmospheric and calibration bands, VEM will provide a global map of surface composition (Dyar et al. 2020, 2021, Table 3 and Fig. 10). Continuous observation of Venus' thermal emission would also provide tight constraints on current day volcanic activity (Smrekar et al. 2010a; Mueller et al. 2017). Measurements of atmospheric water vapor abundance as well as cloud microphysics and dynamics would permit accurate correction of atmospheric interference.

VEM is a pushbroom multispectral imaging system. The telecentric optics images the scene onto a filter array, and the image is relayed by a three-lens objective onto the detector. VEM's optical sub-system sits on top of the electronics compartment and the power supply. A two-stage baffle protects VEM from scattered light. A 45° FOV yields a swath width of 207 km at an altitude of 250 km, providing a thorough sampling of surface emissivity and orbit-orbit repeat coverage.

Scattering at the cloud particles limits the achievable spatial resolution at the surface to approximately 50-100 km (Moroz 2002, 1990; Hashimoto 2003). The VEM optical system has a theoretical on-ground resolution of 300 m from a 250-km orbit. Using digital TDI, the data are reprocessed in the instrument at a spatial resolution of 1 km, providing a significant

Table 3 The 14 bands of the identical VEM instrument on VERITAS and the VenSpec-M instrument on En-Vision, fall into four categories depending on the altitude range from which the near-IR radiation originates (Helbert et al. 2016, 2020, 2021). (1) Radiation for the surface bands at 0.86, 0.91, 0.99, 1.02, 1.11, 1.18 μm originates at the surface. Surface bands are used to determine rock types and to monitor the thermal signature of active volcanism. (2) Radiation in the water vapor bands originates in a layer near the surface (0.96, 1.15 μm) and is sensitive to the abundance of water vapor that may be produced by active volcanic plumes. (3) In the cloud bands at 1.195, 1.31, and 1.51 μm, the radiation originates in an atmospheric layer above the surface but below the clouds. Because the signal in the cloud bands has no surface or water vapor contributions, measurements in these bands can be used to remove cloud-induced contrast variability from the other bands. (4) The background bands at 0.79, 1.06, and 1.37 μm correspond to an atmosphere that is opaque, allowing the removal of background signal on the detector. The spectral widths of the bands, approximately ±10 to ±20 nm, are optimized to cover the full range of atmospheric windows based on radiative transfer modeling while minimizing out-of-band radiation

	Mineralogy & active volcanism	Clouds	Water	Background
Central wavelength (μm)	0.86 0.91 0.99 1.02 1.11 1.18	1.19 1.31 1.51	0.96 1.15	0.79 1.06 1.37

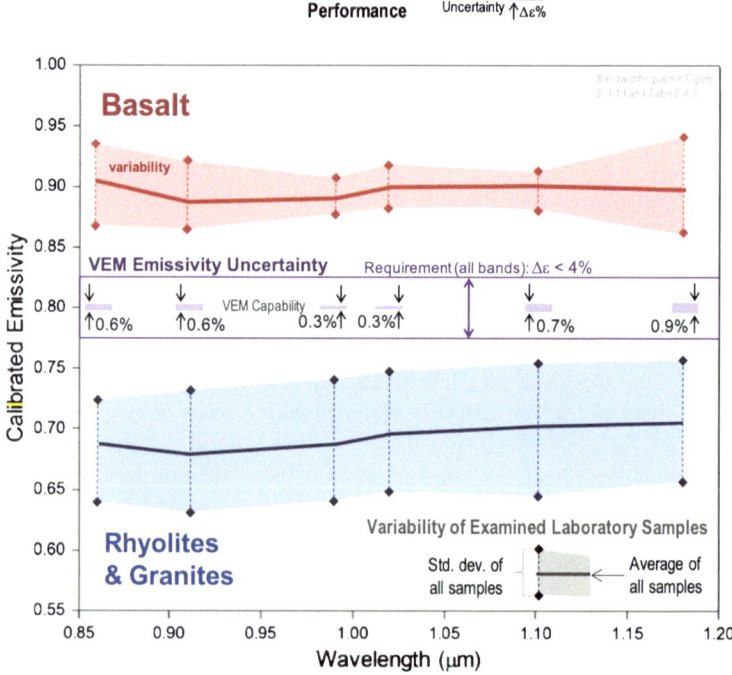

Fig. 10 The projected VEM performance in all surface bands far exceeds the 4% requirement and will enable creation of the first global map of composition on the surface of Venus, and distinguish between felsic and mafic rocks (Dyar et al. 2020, 2021). The 14 bands of VEM fall in four categories depending on the altitude range from which the near-IR radiation is originating. Bands at 0.86, 0.91, 0.99, 1.02, 1.11, 1.18 μm originate at the rocky surface (Helbert et al. 2016, 2020, 2021). See Table 3

gain in signal-to-noise ratio (SNR). Due to the low orbit required for the radar, the wide field of view of the VEM instrument would allow every spot on the surface to be viewed between 5 and 10 times in consecutive orbits. This would allow short-term variability in the atmosphere of Venus to be accounted for. To distinguish between surface and atmospheric contributions, VEM would use an updated version of the extensively tested data pipeline developed to process VIRTIS surface data (Helbert et al. 2008), combined with a radiative transfer model (RTM) (Kappel et al. 2012; Kappel 2014; Kappel et al. 2016). Data would be processed at 10 km spatial resolution and the data from consecutive orbits would be stacked. Both provide an additional increase in the SNR.

Of VEM's total of 14 bands, six would see the surface through all Venus atmospheric windows; three compensate for stray light; three measure cloud transparency; and two measure water abundance. The water vapor and cloud opacity channels would be used as RTM inputs to constrain near-surface water vapor abundance and cloud particle distributions. Observations at 1.16 μm have sufficient accuracy and precision to enable a search for active volcanic outgassing from retrievals of the water concentration in the atmosphere. Multiple observations over the duration of the mission would be used to account for additional unknown atmospheric variability not accounted for in the RTM. This would reduce both atmospheric and instrument noise by averaging image swaths acquired at different times. Applying an updated analysis (Dyar et al. 2020) of atmospheric error for VEM parameters, and taking multiple look averaging into account, our capability for emissivity precision is between 0.3 and 1.2%.

4.3.3 Radio Science / Gravity Experiment

The gravity investigation of VERITAS aims to fill the large knowledge gap on the internal structure of Venus as compared to the other terrestrial planets and the Moon. This investigation addresses Venus evolution science goals involving interior structure of thermal state (Sect. 4.1.2). Geophysical models of the interior are based on limited data for mass, radius, gravity, and topography. Work by Dumoulin et al. (2017) indicates that the state of the core and its size, as well as the viscous response of the interior, can be well constrained by the VERITAS requirements, that is a determination of k_2 to ± 0.01 and of e to $0.25°$ accuracies. These accuracies will be further improved through the use of radar tie points from VISAR (Cascioli et al. 2021).

As introduced in Sect. 4.1.2, precise measurements of the Moment of Inertia factor (MOIF) through the pole precession rate, the tidal Love number, k_2 and the tidal phase lag, e, will permit the first useful comparisons between the interior of Venus and other terrestrial planets. VERITAS will measure the precession rate with an accuracy of 50 arcsec/cycle (1 cycle equals 1 Venus sidereal day or 243.02 Earth days). Information derived from the moment of inertia is crucial to model the thermochemical evolution of Venus' interior including differentiation, as well as key surface processes; e.g., core size which is a key parameter in predicting vigor of mantle convection and the size and number of hot mantle plumes. Recent determinations of the pole precession and the Moment of Inertia factor (MOIF) at 0.337 ± 0.024, using Earth-based observations of radar speckles tied to the rotation of Venus in 2006–2020, find a core radius of approximately 3500 km (58% of the planetary radius) with large (> 500 km) uncertainties due to both model limitations and current uncertainties on normalized moment of inertia C/MR^2 (Margot et al. 2021). VERITAS will determine core to ± 20 km using both the gravity field and radar tie points (Cascioli et al. 2021). Additionally, loading of the surface by the atmospheric thermal tides can be extracted from the gravity field, providing further constraints on both interior structure and atmospheric circulation (Cascioli et al. 2023).

The knowledge of crustal processes, essential to determine why and how Venus and Earth diverged, will greatly benefit from high fidelity mapping of the gravity field. As discussed in

Sect. 4.1, global gravity and topography data are linked at long wavelengths to the structure and dynamics of the sub-lithospheric mantle, and therefore can be used to estimate the elastic thickness/thermal gradient, which constrains mechanisms of formation and how different Venus' evolution processes are from Earth's (Mazarico et al. 2023). Doppler measurements are the primary observables for reconstructing the orbit of the spacecraft and recovering the gravity field of a planet. These measurements are collected by recording the Doppler shift of a radio signal sent from the ground station to the spacecraft, which then coherently retransmits it back to the Earth by means of an onboard transponder. The estimation of Venus' gravity field and VERITAS orbit will rely on the use of advanced orbit determination codes built on accurate mathematical models of the solar system dynamics and of the observables. The 2-way Ka-band radio tracking data are analyzed to reconstruct the VERITAS trajectory and estimate model parameters. The gravity field harmonic coefficients up to an average degree strength of 200 ($\sim$105 km, see Table 1) are generated together with corrections to the spin rate and to the pole right ascension and declination.

By combining global high-resolution gravity and topography VERITAS would look for possible subduction, buried features (e.g., as observed on Mars and the Moon) and unrecognized deformation. In addition, the improved uniformity of the gravity field knowledge would provide precise estimates globally of the elastic thickness, a proxy for heat flow.

4.4 Summary / Outcomes Revealing Venus Evolution

NASA's VERITAS mission is designed to study the geologic evolution of Venus and the processes that affect the habitability of terrestrial planets. Venus most likely had elements essential for habitability because its present conditions can be seen as a geodynamic analog to early Earth, when the lithosphere was hotter and thinner, plate tectonics and continents began to form, and life emerged.

VERITAS will carry two instruments: VISAR and VEM. The Venus Interferometric Synthetic Aperture Radar (VISAR) X-band measurements will provide: 1) a global digital elevation model (DEM) with 250 m postings, 5 m height accuracy, 2) Synthetic aperture radar (SAR) imaging at 30 m horizontal resolution globally, 3) SAR imaging at 15 m resolution for targeted areas, and 4) surface deformation from RPI at 2-centimeter vertical precision for $>12\,200 \times 200$ km potentially active area targets. Community input would be solicited for both RPI and high-resolution imaging targets. VEM will produce surface coverage of most of the surface in 6 near-IR bands located within five atmospheric windows and of eight atmospheric bands for calibration and water vapor measurements. VERITAS will also conduct radio science. Magellan spherical harmonic gravity field has an average resolution of only 550 km. Rigorous modeling shows that VERITAS data, with an average resolution of 155 km, would enable estimation of elastic thickness-a proxy for thermal gradient and resolution of specific geologic processes. Measurements of the moment of inertia factor and k_2 will constrain core size and state (Cascioli et al. 2021). The reader may refer to Smrekar et al. (2022a) for a full mission overview.

The VERITAS mission profile consists of two phases. Science Phase I (SP1) occurs while aerobraking is paused, about 6 months after insertion into a polar elliptical orbit. Science Phase II (SP2) starts after further aerobraking has placed VERITAS in a near-polar, circular, low-altitude orbit that allows global observations throughout the mission. Over $\sim$3.5 Earth years, the mission will return synergistic, global datasets with unprecedented coverage, resolution, and accuracy to meet its science goals: high resolution topography, X-band radar imagery, targeted surface deformation, near-IR spectroscopy and gravity. The VERITAS spacecraft downlinks a total of 20.9 terabits of data to the DSN stations using

CCSDS packets. The total volume of science data products to be archived (raw, reduced, and derived) is estimated to be 134 terabytes. These rich datasets will allow VERITAS to meet key required investigations and science objectives.

VERITAS' rich global datasets will provide an invaluable resource for a new generation of Earth, planetary, and exoplanet scientists, and reveal the truth about how Earthlike Venus really is. These datasets are highly synergistic with DAVINCI (Sect. 5) and EnVision (Sect. 6), providing information about Alpha Regio and identifying key targets for EnVision exploration. These missions also enrich VERITAS contributions by, for example, providing high resolution visual and near-IR images of Alpha Regio (i.e., DAVINCI flyby near-IR emissivity and sub-cloud near-IR imaging and high-resolution local topography before the VERITAS launch) and continuing the search for volcanic activity (EnVision).

5 Deep Atmosphere Venus Investigation of Noble Gasses, Chemistry, and Imaging (DAVINCI)

DAVINCI inherits the legacy of the successes of previous atmospheric probe missions, connecting definitive analytical chemistry of Venus' deepest and bulk atmosphere with new surface compositional constraints linked to the history of water. It provides critical measurements about the evolution of the atmosphere via noble gas isotopes, isotopes of hydrogen (D/H), and other species all in a highly detailed physical context while also imaging the surface in the near-infrared from under the clouds at spatial scales not possible from orbital altitudes.

5.1 DAVINCI Science Objectives

Compelling recent insights (e.g., Garvin et al. 2022a), and the planet's relevance to exoplanetary systems (Kane 2022; Way et al. 2023, this collection), have raised new questions about Venus' atmosphere, climate, and habitability evolution through time. Since 1978, six orbiting missions have comprehensively mapped Venus' surface and upper atmosphere. Despite the success of these reconnaissance missions and prior missions to the surface (Venera and VeGa landers), significant knowledge gaps about Venus' early state and overall evolution remain that can only be addressed through state-of-the-art *in situ* analytical and flyby remote sensing measurements. The international scientific community is now asking key questions about a possible long-lived oceanic state and considering past and present life on Venus, posing new hypotheses that are directly testable by measurements of the local surface and atmosphere.

5.1.1 Overview

The Deep Atmosphere Venus Investigation of Noble gasses, Chemistry, and Imaging (DAVINCI) mission is one of several needed next steps in Venus' exploration, originally suggested by Morrison and Hinners (1983) in their summary of solar system exploration priorities. DAVINCI will be the first mission to Venus to incorporate science-driven Venus dayside and nightside flybys and an instrumented descent sphere (DS) into a unified architecture. The mission will deliver both a deep atmosphere probe and a flyby remote-sensing carrier relay imaging spacecraft (CRIS) to Venus; it will assess the habitability of Venus over time, establishing how the planet evolved from upper atmosphere to the surface. Its complement of *in situ* and remote sensing measurements will reveal the processes that may

Table 4 The DAVINCI mission traces its primary science goals and objectives through the operation of its 7 instruments to specific outcomes related to Venus' evolution and connections to exoplanets like Venus. Each of the Key Science questions (Left column) is linked to Planetary Decadal Survey questions including # 3, 6, 10, and 11 (NASEM 2022) with specific measurement strategies listed (middle column). Color codes map instruments shown at far right (column) to key questions at the far left. The CUVIS technology demonstration instrument will provide 0.2 nm UV (200-400 nm) spectral observations of the Venus dayside at favorable solar phase angles on both flybys (January, November 2030), providing raw spectra as well as AI/Machine Learning analysis on-board (see Sect. 5.4.2)

DAVINCI Goals Address Key Questions		VMS	VTLS	VASI	VfOx	VenDI	VISOR	CUVIS	
Key Science Questions	**Traceability of DAVINCI Measurements**								
What is the origin of the Venus atmosphere and how has it evolved? How and why is Venus different from Earth and Mars, and how does it compare to Earth-sized exoplanets?	**VMS**: noble gas abundance and isotope ratios to test current hypotheses of origin and evolution. **VMS & VTLS**: atmospheric and isotopic composition, search for exotic chemistry, constrain mineralogy by constraining surface-atmosphere exchange. **VfOx**: oxygen abundance near the surface. **VISOR & CUVIS**: UV absorbers in the upper clouds and dynamics from flybys								
Was there an early ocean on Venus? If so, when and where did it go? What is the rate of volcanic activity on Venus?	**VMS, VTLS, & VASI**: D/H and other key trace gases above and below the clouds down to the surface; history of water. **VMS**: radioactive decay products ^{40}Ar and ^{4}He to determine long-term and recent volcanism rate. **VenDI & VISOR**: compositional insights into past water-rock interaction								
What exactly are the tesserae highlands? What is their origin and history? How do they compare with major highlands?	**VenDI**: high-resolution morphology, composition, and role of crustal water in igneous rock formation and erosional processes at Alpha Regio. **VfOx**: constrain surface redox state through oxygen measurements near the surface-atmosphere interface. **VISOR & VenDI**: IR emissivity for composition at scales of 5-200 m (VenDI) and ~70 km (VISOR) to constrain regional composition of mountains on Venus								

have allowed surface water to persist and then dissipate. DAVINCI will further establish new bounds on planetary habitability and enable-improved interpretation of biosignatures in our solar system and beyond (Garvin et al. 2022a; VEXAG 2019; U.S. National Academies of Sciences, Engineering, and Medicine/NASEM 2022).

DAVINCI fully addresses its three overarching science goals (Table 4), which are described in the next three subsections: (1) Atmospheric origin and evolution (Sect. 5.1.2); (2) Atmospheric composition and surface interaction (Sect. 5.1.3); (3) Venus surface properties (Sect. 5.1.4). The results will further catalyze years of productive follow-on scientific analysis that will quantitatively connect Venus to exoplanets and place its evolution into the context of recently selected orbiter missions VERITAS (Sect. 4) and EnVision (Sect. 6).

5.1.2 What Is the Origin of Venus' Atmosphere and How Has It Evolved?

Atmospheric Origin and Evolution. - DAVINCI answers questions about atmospheric formation and evolution of habitable zone planets, including the timing and rate of volcanic outgassing in the past and present that can only be addressed through *in situ* measurements of key noble gasses (including Xe) and nitrogen, never before adequately measured for Venus. DAVINCI unambiguously quantifies atmospheric noble gasses deeply enough (below ~60 km) to avoid strong compositional dependencies on time of day, latitude, and molecular mass prevalent higher in the atmosphere. This information will help us understand similar evolutionary processes for exoplanets of various ages, many of which are expected to be Venus-like.

Enticing Earth-based remote sensing data have suggested the presence of phosphine, a possible biosignature, in the atmosphere of Venus (Greaves et al. 2021), an evidence later questioned (Encrenaz et al. 2020b; Villanueva et al. 2021). Modern life detection science strategies summarized in framework reports by the USA National Academies of Sciences requires that putative biosignatures be evaluated within the systems-level chemical context of the environment – presently poorly constrained at Venus. Investigating its past and present habitability through a definitive analysis of chemical reservoirs and cycles would revolutionize our understanding of Venus, its place in our solar system, and its prospects as a future astrobiology target (Limaye and Garvin 2023).

5.1.3 Was There an Early Ocean on Venus?

Atmospheric Composition and Surface Interaction. - DAVINCI tests hypotheses of when and how Venus lost its putative early oceans, plus chemical processes in the cloud and subcloud atmosphere down to the surface. The descent profile enables vertically resolved (i.e., at altitude scales as fine as 100-200 m) measurements of chemical species across a broad mass range combined with high-precision abundances and isotopes of targeted trace gasses. The highest cadence measurements focus on the deep atmosphere (<16 km), which contains 66% of the atmospheric mass, where no definitive *in situ* data exist and orbiting remote sensing techniques are largely blind. Such precision and broadband analysis is critical to reveal unknown chemical cycles. Cross-calibrated descent and flyby surface emissivity mapping in the near-infrared (~1 μm) will connect the deep atmosphere chemistry measurements to compositional maps of Alpha Regio and other tesserae.

5.1.4 What Are the Tesserae Highlands, Their Origin and History?

Surface Properties. - Analysis of the near-IR radiance of Alpha Regio tessera using VEx VIRTIS data shows that the tessera material differs from the plains materials in a manner that is consistent with a lower FeO (more felsic) content (Gilmore et al. 2015), corroborating earlier measurements by Galileo NIMS during its Venus flyby (Hashimoto et al. 2008). During the descent above western Alpha Regio, which is also the largest (1600 × 1300 km) known exposure of tessera terrain, the DAVINCI probe distinguishes felsic rock (i.e., formed in association with water) from others, such as primitive basalt, at new spatial scales (<100 m) not accessible from orbit via multi-band near-IR descent imaging in 3D context. Alpha Regio is considered an ideal, representative example of tesserae terrain unique to Venus. Compositional constraints from the high-sensitivity 3D views produced by the near-IR descent imaging system will be developed at unprecedented resolution (5-200 m) to connect with ~60-km scale emissivity mapping from the DAVINCI CRIS remote sensing flybys, as well as previous (VEx) and future orbital data (e.g., from Venus emissivity mapping to be accomplished by VERITAS and EnVision in the 2030's). Existing orbital radar (SAR) and near-IR emissivity data have insufficient spatial resolution to characterize such geomorphology definitively without ground-truth from DAVINCI from beneath the clouds, including meter-vertical resolution imaging and derived topography.

5.2 DAVINCI Mission Overview

DAVINCI inherits the legacy of the successes of previous missions, connecting definitive analytical chemistry of Venus' deepest atmosphere with new surface compositional constraints at regional scales from flyby near-infrared remote sensing. DAVINCI probes the composition and physical structure of Venus' atmosphere from an altitude of ~67 km (in the upper clouds) to the surface, addressing terrestrial planet formation, evolution, and the boundaries of habitability (Table 4). During its robust operational phase, including two science-guided Venus flybys, a ~1-hour descent probe *in situ* investigation, and options for extended mission operations (Fig. 11), the DAVINCI mission acquires up to 500 Gbits of uncompressed data about Venus in <2.1 years.

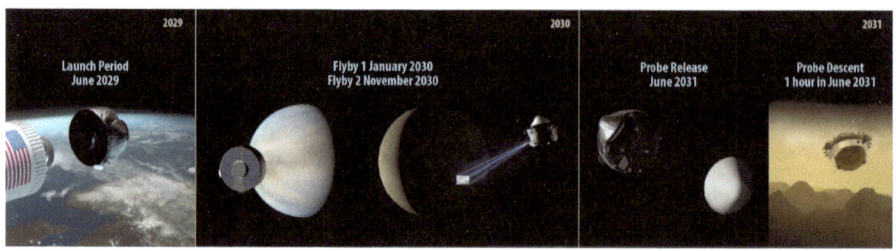

Fig. 11 DAVINCI offers definitive measurements from the top of the Venus cloud deck to the surface, while observing cloud dynamics and regional composition of key highlands including Alpha Regio. The Figure shows the nominal launch in June 2029; after a ~6-month cruise, the spacecraft would fly by Venus in January 2030 for initial remote sensing in the UV and near-IR, then the trajectory returns 9 months later for a second flyby in November 2030. Both flybys will include dayside UV imaging and spectroscopy, as well as night side near-IR surface emissivity mapping of multiple tesserae including Alpha Regio. After an additional 7-month cruise, the flight system will deliver the probe in June 2031 for its entry, descent, and science campaign above western Alpha Regio at very high solar illumination conditions

DAVINCI nominally launches in June 2029 (Garvin et al. 2022a). - After a ~6-month cruise, the Carrier Relay Imaging spacecraft (CRIS) flies by Venus for unique remote sensing science (i.e., near UV cloud motion videos, near-IR surface emissivity of tesserae and volcanic centers), before setting the spacecraft on a trajectory to return for a second science flyby, followed by delivery of the *in situ* probe to Alpha Regio, with favorable solar illumination (Figs. 11, 12). DAVINCI's imaging target area within western Alpha Regio has been comprehensively mapped by prior missions (and Arecibo radiotelescope, see Sect. 11, Sect. 11.1.1) and is large enough to avoid complex controlled descent. DAVINCI's entry-descent-touchdown ellipse (~348 × 160 km) fits within this area with large margin and high-resolution near-IR descent images will assess its relevance to the history of Venus water in association with rock units and geomorphology.

In June 2031, two days before arrival at Venus, the Probe Flight System (PFS) is released. The Carrier-Relay Imaging Spacecraft (CRIS) observes PFS release (via imaging using the VISOR camera system) then conducts a divert maneuver to communicate with the descent sphere (DS) throughout the *in situ* science mission (i.e., by flying overhead with its 2 m HGA for two-way S-band telecommunications). The DAVINCI descent sphere is a hermetically sealed titanium pressure vessel with dimensions (1.1 m × 0.85 m; 250 kg) similar to the Pioneer-Venus Large Probe (PVLP). After DS atmospheric entry and parachute deployment (~70 km altitude), the heat shield is released and the DS-based instruments begin to collect and transmit altitude-resolved, high-fidelity measurements of noble, trace gas, and isotopic abundances; atmospheric temperature, pressure, and winds; and high-resolution broadband and ~1 μm narrow-band images (Fig. 12). Although not required to function following touchdown, the DS has sufficient resources to conduct science and relay data for an additional ~18 minutes from the surface, if it survives the ~13 m/s touchdown. After the CRIS spacecraft has recorded the required probe *in situ* data, it turns toward Earth and transmits those data to the DSN. Via an optional extended science mission, six months after the DS entry-descent-science phase, the CRIS spacecraft conducts a Venus orbit-insertion (VOI) maneuver and enters a 5-day, 60-degree inclined science orbit for most of a Venus year (~6 months), mapping dayside cloud dynamics and nightside near-IR surface emissivity potentially at the same time as VERITAS observes the surface with its powerful payload to evaluate synergies including those associated with different observational times.

DAVINCI delivers definitive atmospheric chemistry measurements, coupled to unprecedented 3D views of ancient tesserae at local scales (5-60 m horizontally) that will transform

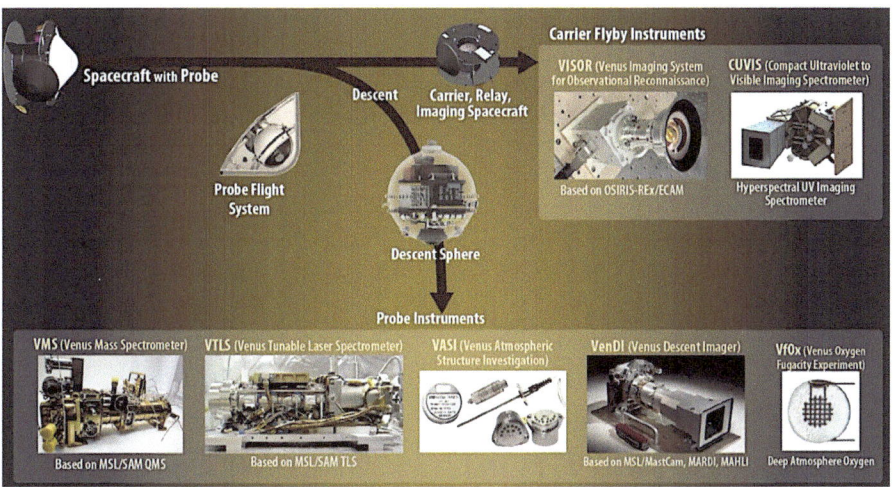

Fig. 12 DAVINCI delivers critical science using five descent sphere (DS)-based instruments, and two remote-sensing instruments on the carrier relay imaging spacecraft (CRIS) (Figure from Garvin et al. 2022a). The probe instruments will operate during its ~59 minute long descent from the upper clouds (~67 km) to the surface over an entry-descent-science corridor with a landing error ellipse located within western Alpha Regio (see Fig. 13)

Fig. 13 DAVINCI entry error ellipse with Magellan S-band SAR mosaic draped over new 1 km scale Digital Elevation Model (DEM) of Alpha Regio based on combined Arecibo polarimetric radar & Magellan radar altimeter datasets (Garvin et al. 2022b, 2023, in preparation)

our understanding of the planet next door and serve as the foundation for future exploration. Given the recent unexpected but contentious discovery of phosphine at Venus (Greaves et al. 2021; Encrenaz et al. 2020b; Villanueva et al. 2021), *in situ* measurements are required to uncover new chemical cycles, including those involving oxygen, sulfur, phosphorus, and others. Enabled by DAVINCI's quantitative investigation of Venus' lower atmosphere and its pathfinding high-resolution near-infrared views of enigmatic tesserae, future missions can follow to accomplish additional 2022 Planetary/Astrobiology Decadal Survey objectives (NASEM 2022). DAVINCI's measurements can also be tied to JWST investigation of Venus analogue exoplanets around M-dwarf stars as "planetary atmosphere ground truth", which is underway at the time of this writing by the JWST observatory (see also Sect. 11, Sect. 11.2).

5.3 DAVINCI Descent Probe Instruments

Five descent probe instruments and two carrier-relay-imaging spacecraft instruments leverage highly successful partnerships between NASA/Goddard Space Flight Center (GSFC), NASA/Caltech Jet Propulsion Laboratory (JPL), NASA/Johns Hopkins University Applied Physics Laboratory (JHU/APL), and Malin Space Science Systems (MSSS), see Table 4; Fig. 12. These include the pairing of a quadrupole mass spectrometer and tunable laser spectrometer evolved from the SAM (Sample Analysis at Mars) suite on MSL/Curiosity rover now in its eleventh year of operation on Mars in Gale Crater. A flyby camera suite based upon imaging systems on the OSIRIS-REx mission provides sensitive nightside NEAR-IR mapping and dayside UV cloud imaging (with movies) from new vantage points over Venus. The *in situ* and flyby observations combine to define the DAVINCI mission baseline.

5.3.1 Venus Mass Spectrometer (VMS)

DAVINCI's Venus Mass Spectrometer (VMS) provides the first comprehensive survey of Venus' noble gasses, as well as detailed analysis of trace gas species – both those expected and those yet to be discovered. Employing mature, tested operational modes, VMS definitively measures isotope ratios for Ne, Ar, Kr, and Xe, and collects hundreds of measurements of each trace species to constrain fine variations with altitude. The VMS is a Quadrupole Mass Spectrometer (QMS) with a gas-enrichment system and pumping system that will provide a comprehensive *in situ* survey of the planet's noble gases. It has significant heritage from the Mars Science Laboratory/MSL (Curiosity) Sample Analysis at Mars/SAM QMS (Webster and Mahaffy 2011; Atreya et al. 2013) and with a broad mass range from 2 to 550 Dalton (Da), VMS has the capability to discover new trace gas species (Garvin et al. 2022a, Table 3.2) within the deep atmosphere where non-equilibrium chemistry is suspected.

VMS measurements will occur every $\sim$200 m or better below 61 km, particularly in the lowest 30 km of the atmosphere (Fig. 12), where it will probe the supercritical CO_2 boundary and properties of the CO_2/N_2 mixture in the temperature and pressure conditions of the deep atmosphere of Venus (Lebonnois & Schubert 2017), and profile new species, including CHNOPS-bearing molecules (including, potentially, P_4O_6 and PH_3) and those trace gases expected to be tied to surface mineralogy and the thermochemical cycle of sulfur-bearing species.

5.3.2 Venus Tunable Laser Spectrometer (VTLS)

DAVINCI's Venus Tunable Laser Spectrometer (VTLS) answers major questions about the Venus atmosphere by providing the first precise abundance and isotopic measurements of key gasses containing hydrogen, sulfur, carbon, oxygen, and potentially phosphorus. VTLS provides a series of ten definitive measurements of the D/H ratio in water vapor throughout the atmosphere (i.e., from 67 km to $\sim$2 km), critical to understanding the longevity, and loss mechanisms of past oceans (Way et al. 2023; Salvador et al. 2023, this collection). The instrument consists of a multipass Herriott cell with three laser channels at 2.64, 4.8, and 7.4 μm, specifically targeting key science questions that discriminate chemical processes in the upper clouds and near-surface environment. VTLS draws heritage from the MSL/SAM tunable laser spectrometer (e.g., Webster and Mahaffy 2011; Pla-Garcia et al. 2019).

VTLS is specifically tailored to answer critical questions about the long-term evolution of Venus' atmosphere by providing the first highly sensitive *in situ* measurements of key gas species containing H, S, C, and O, as well as their high-precision isotope ratios including

D/H (Garvin et al. 2022a, Table 3.2). It should be noted that VTLS offers the possibility to directly measure trace species at different heights in the clouds with a sensitivity of ~1 ppbv, allowing to set new upper limits for PH_3 at the 1 ppbv level discussed in Villanueva et al. (2021) or Encrenaz et al. (2020b).

5.3.3 Venus Atmospheric Structure Investigation (VASI)

DAVINCI's Venus Atmospheric Structure Investigation (VASI) characterizes the fine-scale vertical structure and dynamics of the Venus atmosphere during descent, including wind speed, pressure, and the first detailed profile of the deep atmosphere temperature (e.g., the lapse rate, dT/dz). The instrument consists in a suite of sensors that measure atmospheric pressure, temperature, and dynamics. Internally mounted accelerometers and gyroscopes combined with Doppler tracking via the spacecraft-to-DS communications link enables detailed reconstruction of the descent probe trajectory. VASI provides thermodynamic context for the composition measurements and enables reconstruction of the detailed descent profile and precise landing position, with most measurements every 15-50 m, as well as a final measurement set within ~100 m of the local surface.

5.3.4 Venus Descent Imager (VenDI)

DAVINCI's Venus Descent Imager (VenDI) is a near-IR descent-imaging system with a nadir orientation. It will deliver clear, high contrast, high SNR images (>100:1), providing the first geologic constraints on the highland surface environment at 2-200 m length scales from reflectance imaging below the cloud-deck (and sub-cloud hazes). A narrow-band, near-IR channel delivers 1.02 μm albedo maps with sensitivity to felsic rocks or alteration products when ratioed against broadband images (0.74 to 1.02 μm). Topography can be derived using machine-vision algorithms via Structure-from-Motion (SfM), an expansion of Scale-Invariant Feature Transform (SIFT) algorithm to construct a Digital Elevation Model (DEM) from multiple overlapping images with varying vertical and horizontal baselines (Garvin et al. 2018, 2022a). SfM processing of bundles of descent images produces first 5 - 60 m scale topography of tesserae and establishes boundary conditions for tectonic and erosional models. Final imaging scales from VenDI below ~1.5 km produce unblurred images at scales finer than 1 m, permitting feature-identification resolution of key indicators of sedimentary processes at scales that connect to those observed by prior Venera landers (e.g., Garvin et al. 1984). Evaluation of Earth-based analogue datasets (Pilbara, Zagros mountains) that emulate VenDI near-IR bandpasses and spatial scales have demonstrated discrimination of felsic surfaces at <100 m (down to 5-10 m) providing confidence that sub-cloud descent imaging will complement 50-100 km scale orbital near-IR emissivity mapping by multiple missions (see Sects. 4.3.2; 6.3.3).

5.3.5 Student Collaboration Experiment VfOx; DAVINCI's Engineering Science Investigation (ESI)

The oxygen cycle on Venus, like those involving sulfur, hydrogen, phosphorus, and carbon, is incompletely resolved on the basis of current data and DAVINCI's quadrupole mass spectrometer (VMS) and tunable laser spectrometer (VTLS) measurements of altitude-resolved species will extend beyond extrapolated equilibrium models of likely chemistry to measured abundances across the deep atmosphere all the way to the surface just about the complex ridged terrain in Alpha Regio.

Student Collaboration Experiment VfOx obtains independent measurements of the oxygen fugacity (fO_2) in the lowermost scale height of the Venus atmosphere to compare with indirect (and independent) measurements of oxygen-species retrieved by VTLS. DAVINCI's Student Collaboration Experiment partners with Johns Hopkins University (and others), engaging students to implement an *in situ* sensor that measures oxygen partial pressure, also known as fugacity (VfOx).

DAVINCI's Engineering Science Investigation (ESI) meets high priority NASA measurement objectives for Venus entry with measurements tied to improving future Venus Entry, Descent and Landing (EDL) activities. Measurements that document the entry conditions after Atmospheric-Entry-Interface below 140 km will be obtained via support from NASA's Space Technology Mission Directorate in partnership with the Science Mission Directorate. Final instrument selection is in progress as DAVINCI advances toward its mission confirmation by NASA.

5.4 DAVINCI Carrier/Flyby Instruments

5.4.1 Venus Imaging System for Orbital Reconnaissance (VISOR)

VISOR is an integrated system of four cameras that provides global dayside coverage of Venus in the UV and nightside coverage in the near-IR (0.93–1.03 μm). Each of the VISOR cameras has a field of view of 11.3 degrees by 8.9 degrees which can be converted to a spatial sampling scale as a function of distance to target. Three cameras image night-side Venus in three independent near-IR bands, from 930–938 nm, 947–964 nm, and 990–1030 nm. They will deliver night-side surface emissivity mapping (three near-IR bands to properly characterize clouds and scattered light) to constrain regional composition at $\sim$60 km scales, unveiling new regional patterns associated with highlands during 2030 flybys prior to global mapping by two future orbiters.

DAVINCI first and second Venus gravity-assist flybys, with a closest approach on the night-side hemisphere near Equator at 00:00 LT, are scheduled on January 30 and November 15, 2030 for the planned June 2029 Launch Readiness Date (Table 1). Thousands of images are acquired during the two flybys, potentially identifying felsic regions, "calibrated" by local-scale, sub-cloud VenDI band-ratio mapping at Alpha Regio, as well as with VASI lapse rate information. The fourth VISOR camera will provide global, dayside coverage of Venus in the unknown UV absorber band (355–375 nm). Dayside UV imaging (single band) will measure cloud dynamics as the spacecraft approaches and recedes from pericenter, allowing ultraviolet feature tracking at 355-375 nm at a frequency that exceeds any existing Venus orbital imaging dataset.

5.4.2 CUVIS (Compact Ultraviolet Imaging System)

Technology Demonstration Opportunity (TDO) experiment CUVIS (Compact Ultraviolet Imaging System) acquires 0.2 nm resolution spectra and hyper-cubes of images from 0.2 to 0.4 μm, concurrent with VISOR UV imaging on the dayside flybys of Venus, in a technology demonstration of a new class of small planetary instruments. This UV hyperspectral sensor (CUVIS) will perform upper atmosphere SO_2 and SO chemistry and gather spectral information on the unknown UV absorbing species at a spectral resolution of 0.2 nm in the UV from 0.20 μm to 0.40 μm (Pollack et al. 1980; Wilson et al. 2023, this collection). CUVIS can be accommodated on the CRIS spacecraft and is implemented in a fully separable, *do no harm* fashion. It will further employ Machine Learning to identify key species and test approaches to accommodate effective data transmission for rich datasets such as those delivered by CUVIS. All of its observations will be coupled to VISOR near UV dayside observations which will provide wider field-of-view context (and multi-frame "movies").

5.5 Summary / Outcomes Revealing Venus Evolution

The overall DAVINCI mission, scheduled for launch in June 2029, will provide up to 500 Gbits (uncompressed) new data about the atmosphere and near surface, as well as the first unique characterization of the deep atmospheric environment and chemistry. DAVINCI returns to the Venus atmosphere at a time of heightened interest in understanding terrestrial planet evolution, habitability, and astrobiology in our solar system and beyond (NASEM 2022; Garvin et al. 2022a). Understanding the evolutionary pathways of Venus necessarily involves the interplay between the time-variable atmosphere-climate system, the lithosphere, and the interior. The NASA DAVINCI mission addresses several questions about such components of evolution as they relate to five of the key priorities (as questions) recently published in the US National Academies Planetary and Astrobiology Decadal Survey including for example numbers 3, 4, 5, 6, 10, and 12 (NASEM 2022).

DAVINCI's *in situ* analytical chemistry measurements of noble gasses in the bulk atmosphere will distinguish between current models by resolving the isotopic ratios of xenon, which is currently unmeasured. As for Mars, the full suite of noble gasses and their isotopes will provide chemistry boundary conditions for models that range from early impact blow-off of an initial atmosphere to the consequences of catastrophic volcanic resurfacing on the evolved atmosphere, as well as others. Coupled to these noble gas measurements are ten altitude-resolved observations of D/H in water from as high as 67 km down to the near surface at ~ 2 km altitude.

A suite of 10 such measurements by means of DAVINCI's tunable laser spectrometer will expand upon Pioneer Venus Large Probe based measurements in the cloud deck (above 38 km) and those from remote sensing retrievals from ESA's Venus Express in the upper atmosphere, and connect to the state of D/H in the deep atmosphere, where surface-atmosphere interactions may have affected the history of water. Near surface quantification of the oxygen fugacity by means of multiple experiments (i.e., including the DAVINCI Student Collaboration experiment "VfOx", as well as VTLS and VMS) will resolve the state of oxygen species in the atmosphere and mineral stability near the surface in the tesserae highlands where DAVINCI will come to rest after its atmospheric transect. The oxygen cycle on Venus, like those involving sulfur, hydrogen, phosphorus, and carbon, is incompletely resolved on the basis of current data and DAVINCI's quadrupole mass spectrometer (VMS) and tunable laser spectrometer (VTLS) measurements of altitude-resolved species will extend beyond extrapolated equilibrium models of likely chemistry to measured abundances across the deep atmosphere all the way to the surface just above the complex ridged terrain in Alpha Regio.

Mineral stability assessments associated with rocks containing Fe, S, and other elements will be conducted to infer possible weathering pathways within ~ 2 km of the surface, with direct connections to near-infrared band-ratio descent imaging in the 740 to 1200 nm spectral region at scales as fine as a few meters for potential identification of water-related rock units. Connections between the near-IR descent imaging of possible felsic rock compositions at scales from 100 m down to a few meters and the analytical chemistry of trace gas species potentially relevant to rock formation or modification processes will provide local ground-truth for global assessments of rock unit compositional patterns at 50-100 km scales across Venus. DAVINCI will further address aspects of evolution of the Venus deep atmosphere over time by directly measuring gradients in key species involving S, O, H, P, and

C as often as every 150-200 m in the deepest atmosphere where connections to local rock compositions can be made.

By the time DAVINCI completes its *in situ* transect of the atmosphere (late June 2031), new information of the vertical stratification of the atmosphere from 67 km to the surface as well as a resolved lapse rate (temperature vs altitude at 0.1 K precision every 15-50 m down to the surface) will enable systems-level modeling of Venus evolution that connect the history of the atmosphere to that of the lithosphere, with linkages to Venus tectonic evolution via connections with VERITAS observations. Ultimately the possible evolutionary signatures of water in the Venus system over time will be resolved, preparing the way for future landed experiments that make use of mineralogical signatures in the context of the massive Venus atmosphere that is clearly a major factor in Venus evolutionary divergence from Earth (i.e., see Kane 2022). For further details, see the mission overview in Garvin et al. (2022a).

6 EnVision: Understanding Why Earth's Closest Neighbor Is so Different

On June 10, 2021, the European Space Agency (ESA) announced the selection of EnVision as its 5th Medium-class science mission, targeting a launch in the early 2030s. EnVision is an ESA-led mission in partnership with NASA, providing its Synthetic Aperture Radar instrument, VenSAR and Deep Space Network support for critical mission phases. EnVision will use an array of payload instruments to perform holistic observations of Venus from its inner core to upper atmosphere to better understand how Earth's closest neighbor in the Solar System evolved so differently (European Space Agency 2021).

EnVision's overarching science questions are to explore the full range of geoscientific processes operating on Venus. It will investigate Venus from its inner core to its atmosphere at high resolution, characterizing the interior, signs of past geologic processes, and looking for evidence of past liquid water. As developed in companion articles, recent modeling studies strongly suggest that the evolution of the atmosphere and interior of Venus are coupled at all stages of the planet's long-term evolution (Way and Del Genio 2020; Weller and Kiefer 2020), emphasizing the need to study the atmosphere, surface, and interior of Venus as a system. EnVision's combination of surface and atmospheric measurements will characterize ongoing volcanic processes through an extended-timeline, search for their thermal, morphologic, and gaseous signatures, while also tracing key volatile species from the surface up to the mesosphere.

The mission is scheduled for launch in the fourth quarter of 2031 (see Table 1); the final schedule will be agreed between ESA and NASA at Mission Adoption in January 2024, with back-up launch readiness dates every 6 months in 2032 and 2033, on Ariane 62. Following orbit insertion and periapsis walk-down, orbit circularization will be achieved by aerobraking over a period of several months, followed by a nominal science phase lasting at least 6 Venus sidereal days (4 Earth years). The EnVision payload consists of five instruments provided by European and US institutions (Fig. 14). The five instruments comprise a comprehensive measurement suite spanning infrared, ultraviolet-visible, microwave and high frequency wavelengths. This suite is complemented by the Radio Science investigation exploiting the spacecraft Telemetry, Tracking and Command (TT&C) system. All instruments in the payload have substantial heritage and robust margins relative to the requirements with designs suitable for operation in the Venus environment. This suite of instruments has been selected to meet the wide range of measurement requirements in support of EnVision science investigations.

Fig. 14 Rendering of the generic concept EnVision spacecraft orbiting Venus, with the SRS, VenSAR feeder and reflectarray antennas deployed. Credit ESA / NASA / Paris Observatory / VR2Planets

6.1 EnVision Science Objectives

EnVision will deliver new insights into geological history through complementary imagery, polarimetry, radiometry and spectroscopy of the surface coupled with subsurface sounding and gravity mapping. It will search for thermal, morphological, and gaseous signs of volcanic and other geological activity; and it will trace the development and transport of key volatile species from their sources and sinks at the surface through the clouds up to the mesosphere. Following the same approach through which our understanding of Earth and Mars has been developed, EnVision will combine global observations at low or moderate spatial resolution (e.g., surface emissivity and atmosphere composition) with regionally targeted observations of higher spatial resolutions from a dual polarization S-band synthetic aperture radar (SAR) and subsurface sounding radar profiles.

6.1.1 Overview

EnVision will investigate both present and past geological activity on Venus, and how its atmospheric, surface and interior processes are linked. The background for EnVision's scientific investigations and strategic knowledge gaps is presented in this Section following the lines of three top-level science questions:

1. History - How have the surface and interior of Venus evolved?
2. Activity - How geologically active is Venus?
3. Climate - How are Venus' atmosphere & climate shaped by geological processes?

6.1.2 How Have the Surface and Interior of Venus Evolved?

Geologic Mapping of Volcanic Features and Their Surface Morphology. - Geologic mapping of volcanic features and their surface morphology and dielectric constant is a cornerstone of Magellan data interpretation (Campbell and Campbell 1992; Campbell 1994). There is a need to carry this work to finer spatial scales and into the subsurface to answer fundamental questions of localized stratigraphy (from subsurface profiles and geologic mapping from images), magma composition (from morphology, roughness, and dielectric properties), surface mineralogy, order-of-magnitude eruption rates and volumes (from morphologic features and subsurface profiles), and post-emplacement weathering (from morphologic features and dielectric properties). In a complementary approach to VERITAS, EnVision will accomplish this objective in part through SAR imaging at 10 m resolution and polarimetric imaging at 30 m resolution (Fig. 15), along with VenSpec-M surface investigations. EnVision 30-m SAR imagery will dramatically enhance our understanding of volcanic surface

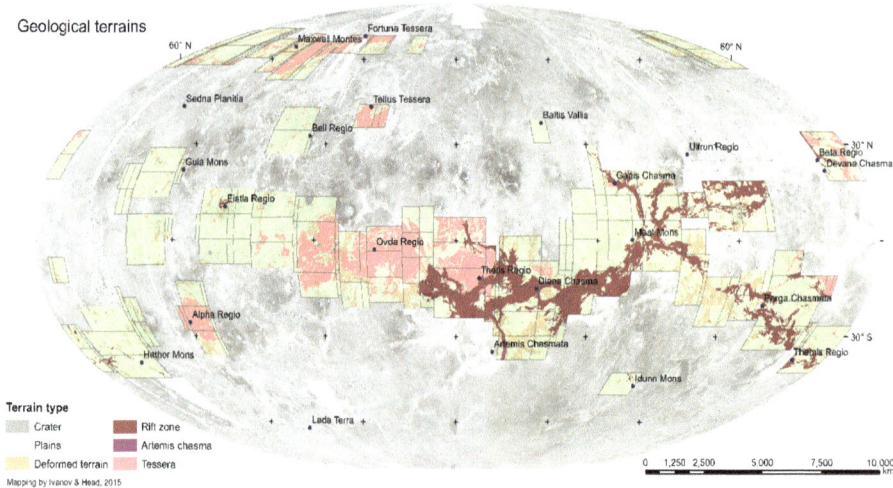

Fig. 15 Map of geological terrains and named landmarks covered by the Regions of Interest (RoIs) defined in EnVision's Science Operations Reference Scenario. The RoIs are chosen to include representative samples of all major geological terrain and feature types. These different features are not distributed at random but are in specific, known locations. EnVision's approach is to define roughly thousand-kilometer square RoIs covering most of the highlands and a representative selection of the lowland features. This strategy allows to progressively build up along the 6 cycles the required global and targeted measurements dataset, in particular over all pre-selected regions of interest, which represent a fraction of about 30% of Venus surface. Definitions of geological terrain types are as mapped by Ivanov and Head (2015)

features. At the >120 m resolution of Magellan (120 m azimuth resolution and 93 m best case range resolution), features like flow channels are visible only where they are at the highest end of those typically seen in terrestrial flow fields, vent locations and associated ash or rugged clinkers are too small to observe, and collapsed tubes or skylights are unseen. Within any single major shield volcano, there are often a wide range of features indicative of magma storage beneath calderas, rapid eruptions that form rugged, channelized flows, fine-grained pyroclastic ash from volatile-rich eruptions, and steep-sided constructs linked with higher-viscosity magma (Campbell and Rogers 1994). Targeted observations at 10 m resolution will bring out crucial details in the stratigraphic relationship between flows, their likely thickness, and the range of scales in flow fields (i.e., short high-volume eruptions or long-term, tube-fed complexes).

Variations in Morphologic Characteristics, Stratigraphic Relationships, and Dielectric Properties of Plains. - The volcanic plains cover around 80% of Venus. Far from being uniform, they exhibit signs of extensive geological activity, from volcanic and tectonic to aeolian and weathering processes. Did the plains form rapidly, with few flow boundaries (like lunar mare) or are they constantly reformed by small-scale volcanism, below the resolution of Magellan? Understanding and mapping stratigraphic boundaries is important in distinguishing geologically old and young units, and between directional and equilibrium surface histories.

The Subsurface Radar Sounder (SRS) will be used to look for layering in the plains and elsewhere on Venus, as has been successfully done on both the Moon and Mars (Fig. 16, and Sect. 6.3.2 below). Analyses of this type enable a far better understanding of Venus's recent geological past and reveal vital information about the character, thickness and mode of resurfacing on Venus. For example, catastrophic resurfacing models for Venus (Strom

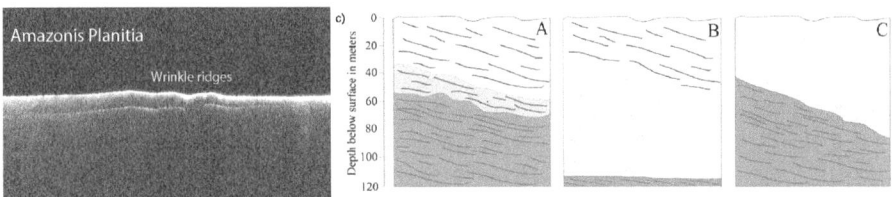

Fig. 16 (a, left): A SHARAD radargram showing layering about 100 meters thick in Amazonis Planitia warped by wrinkle ridges. The image is 400 km across (Campbell et al. 2008). (b, right): Schematic representation of the three potential scenarios of subsurface stratigraphy interpreted from SHARAD radar sounding of volcanic layering in the Arsia Mons caldera (A - stacked lava flows with vesiculated and less dense flows overlying very dense lava, B - less dense lava-flow and a thick tephra deposit overlying denser bedrock, C - pyroclastic or other low-density material deposited over dense lava- flows in the southern part of the caldera, adjacent to the wall (Ganesh et al. 2020; Watters et al. 2006)

et al. 1994) predict that the plains were resurfaced in a brief epoch several hundred million years ago. In such a model, there might not be sufficient time between lava flows to develop thick weathering layers that would produce discrete layered returns in SRS data. If SRS does detect clear layering in the plains, it would tend to favor more gradual resurfacing models for Venus.

Mapping of Tectonic Structures. - Magellan observations provide a valuable overview of tectonic processes on Venus (Solomon et al. 1991), but are limited by the resolution of the radar images and especially the topography (10-30 km). Complementing observations by VERITAS, EnVision's much higher horizontal resolution: 10-30 m imagery, 300 m horizontal resolution of the SAR stereo Digital Elevation Model (DEM) will enable much clearer definition of the styles of tectonic deformation and of the superposition and cross-cutting relationships used by geologists to map the sequence of deformation in a given region.

High resolution radar data are particularly essential in understanding the tesserae of Venus, which contain fine-scale, complex patterns of deformation. We expect that tesserae represent the oldest terrain, locally, but they may not have all formed at the same time; better understanding of their structure and arrangement, their relationship with volcanic terrains and their correlation from one place to another would help to unravel these temporal and structural conundrums. Magellan imagery revealed very varied tesserae interiors often with complex arrangement of solid and deformed rocks, blanketed by finer grained or smoother materials (Hansen and Willis 1996; Ivanov and Head 2011) but without greater spatial resolution and better topographic detail, the nature of the materials and their origins could not be resolved. Multi-polarimetry observations (HH and HV) are needed to better understand their surface textures and physical structures, to reveal emissivity variations of solid lithologies and to discriminate them from unconsolidated materials.

Impact Crater Modification. - The only method for determining the absolute age of a surface, in the absence of measurement of radioactive isotopes, is through the use of crater counts. Because Venus has so few craters it is difficult, or impossible, to distinguish the age of different geological units using craters alone. However, craters on Venus are modified to varying degrees, first by loss of radar-dark halo, and then by infilling, causing dark floors. Some are also modified volcanically or tectonically (Izenberg et al. 1994). Because initial crater depth depends on crater diameter, the extent to which a particular crater deviates from the expected depth-diameter relationship provides a guide to post-impact infilling by lava or sediments at that crater. The height of the crater's rim above the surrounding terrain similarly provides a guide to the thickness of post-impact fill in the crater's ejecta blanket. Initial estimates of crater fill with Magellan data (Herrick and Rumpf 2011) were limited by the accuracy of the available stereo topography digital elevation model. In contrast, Ven-SAR observations will be optimized to produce high resolution DEMs and nadir altimetry profiling will provide global topographic data. Possible direct measurements of crater in-filling with sounding radar will be complementary to topography-based estimates of crater fill thickness. Craters are globally distributed, so such measurements can provide important new information about the global resurfacing history of Venus.

The Heat Engine in Venus' Interior: What Are the Driving Forces for Volcanism and Tectonism? - While Venus and Earth have similar bulk geophysical properties, they have clearly followed divergent geodynamic paths – the former apparently characterized by a strong continuous lithosphere and stagnant lid convection, and the latter characterized by plate tectonic recycling of the lithosphere (Rolf et al. 2022, this collection; Gillmann et al. 2022, this collection; Herrick et al. 2023, this collection). At the root of these distinctions, it is interior dynamics that essentially governs the cooling of a planet. Stagnant lid convection represents a heat transport mechanism much less effective than mobile lid convection. It shows different tectonic characteristics than the plate tectonic regime on Earth. Internal dynamics can therefore cause surface stresses and thus tectonic structures on the planetary surface. Different convection regimes will lead to different tectonic characteristics. As a result, it is expected that planetary surfaces reflect their inner dynamics: a number of features on Venus are tantalizingly similar to structures on Earth, including continent-like tesserae plateaus, chasmata interpreted as rift zones, and some coronae that are surrounded by troughs resembling subduction zones.

6.1.3 How Geologically Active Is Venus?

Detecting Volcanic Activity in Repeated SAR Images. - Detecting and characterizing of relatively large eruptions over the past 40 years will come from three sources in the SAR image data: i) any new, large lava flows (>200 m wide and 100s m long) erupted since the Magellan mission and within EnVision's mapped area will be revealed in the imaging cycles of the EnVision mission; ii) any large scale changes in the morphology of volcanic edifices will also be revealed within EnVision cycles; and iii) any new, small lava flows (>60 m wide and at least a few hundred meters long) erupted in the 4-year duration of the EnVision mission. Detected changes (or non-detection) will be used to place bounds on the volcanic activity rate as described in Lorenz (2015).

Searching for Surface and Near-Surface Temperature Changes. - In addition to SAR imaging, temperature signatures associated with volcanic activity from both hot lava and hot volatile gasses will be detected and monitored in the infrared (IR) and microwave domains. Temperatures associated with volcanic eruptions can range from only 500 °C for low viscosity carbonatite lava to well over 1000 °C for ultramafic lavas. Such young, hot lavas will be directly detectable by their signature in IR emissivity data, (provided lava outflows cover an area of at least 0.1 km^3). Cooling rates at the surface are estimated to be on the order of hours (Mueller et al. 2017), but microwaves offer the prospect of sensing the shallow subsurface and thus may detect warmth from old lava flows, i.e., lava flows which have cooled at the surface possibly years ago and thus have no more IR emission signature but are still hundreds of K above ambient at depth (Lorenz et al. 2016). Polarimetric radiometry measurements (used to determine whether candidate areas have anomalous emissivity rather than high physical temperature) and a better knowledge of the topography (and therefore of the altitude-dependence of the surface physical temperature) will greatly enhance the reliability of the volcanic detection and monitoring.

Understanding the Range and Scope of Mass-Wasting Processes (Landslides). - Though Magellan imagery showed us evidence of mass-wasting and aeolian features, it was not able to reveal their temporal changes during the mission's lifetime, so their geomorphological and temporal properties remain unknown, and we have almost no information about weathering, surface alteration or other aeolian processes. Since there is currently no constraint on the mechanisms and rates at which these processes might be occurring, better topography and nested imaging at multiple resolutions, and repeated imaging during the mission, are needed.

Landscape evolution refers to processes that modify the morphology of a planet's surface, in particular gravity-driven mass-wasting processes such as landslides and slumps. Mass-wasting is a ubiquitous geomorphological process operating on any planetary body with gravity (such as those observed on Earth, the Moon, Mercury, Venus, Mars, icy satellites, comets and asteroids). Malin (1992) provided the first evidence of mass movement on Venus in the form of large-scale slope failures. Magellan's imagery also provided evidence for two dune fields (Greeley et al. 1992, 1995) and indirect evidence for putative 'micro-dunes' (Weitz et al. 1994) that were not resolved by the 100 - 200 m spatial resolution of Magellan's imagery. The surface winds evidenced by these dune fields and by wind streaks and debris fans (downwind of impact craters) are likely to be important agents of aeolian geomorphological change, but data of higher spatial and temporal resolution, and the ability to distinguish loose from consolidated surface materials, are needed to characterize them. Higher resolution, VenSAR observations, with consistent geometry, should reveal many smaller features and better resolve the morphology of features that were not resolved by Magellan. Repeated observations of regions expected to be active, e.g., along rifts, will help to characterize processes operating at decadal (Magellan-VERITAS-EnVision comparison over 40 yrs) and yearly (EnVision inter-cycle comparison) time scales. Local scale DEM's at 5-60 m spatial sampling within the Alpha Regio tesserae will complement the EnVision measurements at scales as wide at 150 km^2.

In the absence of near-surface water which, on Earth, affects material bulk density, shear strength and pore-pressure, and thus leads to slope instability, the mechanisms of slope instability and failure on Venus are unclear, and it is likely that landslides require triggering by external forces, such as earthquakes. Magellan imagery revealed a very strong spatial relationships between the locations of large-scale mass-wasting features and steep slopes related to rift zones and volcanic edifices, which may in turn point to them being geodynamically active in the recent geological past. EnVision's proposed Regions of Interest and higher resolution imaging offer excellent coverage of known mass-wasting features and increase the likelihood of imaging new or previously undetected smaller features. The planned VenSAR investigations will include detailed characterization of mass-wasting geomorphological properties and features with stereo imagery, and of their surface conditions with multi-polarimetry.

6.1.4 How Are Venus' Atmosphere & Climate Shaped by Geological Processes?

Detection of Volcanogenic Gas and Particulate Plumes. - Sulfur dioxide variations in the mesosphere have been attributed as possible evidence of volcanic activity (Esposito 1984), but they also could be due to intrinsic dynamic variability of the atmosphere, associated with temporal changes in transport of SO_2 from troposphere (where it is highly abundant) to mesosphere (where it is detected). On the other hand, volcanic gas plumes in the troposphere (below the clouds) would have quite a distinct signature, with distinct plumes advecting with the prevailing East-to-West winds (Fig. 17). Water vapor is likely to be a better tracer of volcanic activity than sulfur dioxide, because it is less abundant in the Venus atmosphere than SO_2, and because it can be mapped at three different altitudes in the troposphere using different spectral bands on the nightside. Analyses of Venus Express data found no evidence of

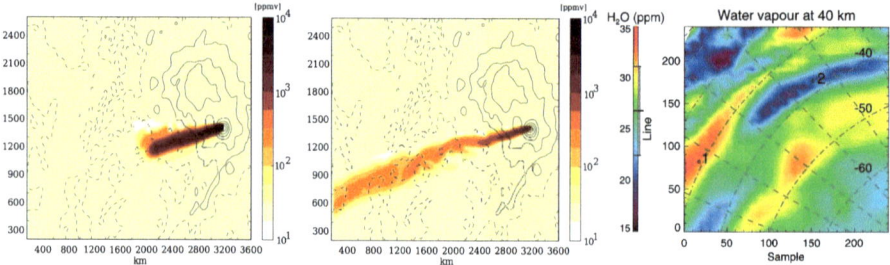

Fig. 17 Simulated advection of a volatile gas plume emitted from Imdr Regio. Black contours represent topography. Colors show excess water vapor (in arbitrary units) after 72 hours of outgassing, at (a, left:) 10 km altitude and (b, center:) 35 km altitude. (Wilson and Lefèvre 2020). (c, right): Variations of water vapor at 40 km altitude (Tsang et al. 2010). This result was later found to be potentially attributed to degeneracies between cloud and water vapor retrieval. The higher spectral resolution of VenSpec-H, compared to VIRTIS-M, will enable unambiguous disentangling of these signals

tropospheric water vapor variations (Bézard et al. 2009, 2011), but these analyses represent data only from a few days and, due to low spectral resolution, could only determine water vapor to a fairly wide range of 25 – 40 ppmv.

The nominal column mass of volcanic gasses in the Venus atmosphere, integrated from surface to space, is $\sim$200 kg m^{-2} for SO_2, $\sim$10 kg m^{-2} for H_2O and $\sim$0.1 kg m^{-2} for HDO. If the composition of Venus volcanic gasses is the same as on Earth - provided that plume dispersion does not exceed 10 km, the limiting spatial resolution induced by cloud scattering - then a large, Pinatubo-size eruption would change H_2O abundance, D/H ratio, and SO_2 abundance, respectively, by $\sim$ +30%, −30%, and +1%. The latter effect may be underestimated with respect to the others, both because the Venusian interior may be much drier than Earth's, and because the outgassed SO_2/H_2O ratio is expected to be higher for a given magma volatile content due to Venus' high atmospheric pressure (Gaillard and Scaillet 2014). The frequency of occurrence, and the ratio of gasses and particulates in any volcanic plumes detected would provide constraints on the upper mantle properties.

Explore the Main Constituent of the Cloud, H_2SO_4, in Both Vapor and Liquid Form. - The main constituent of the clouds, H_2SO_4, in both vapor and liquid form, can be monitored near the cloud base altitude, yielding clues as to cloud formation and convection processes. Geological activity can affect clouds in several ways: (1) volcanic ash can contribute to cloud and haze layers; (2) volcanic sulfur dioxide emissions can contribute to formation of sulfate cloud & haze layers and to the as-yet unidentified UV absorber seen at cloud-tops; (3) volcanically emitted volatiles can form condensate layers; (4) heat from volcanic activity can cause changes in atmospheric circulation (Esposito 1984); (5) near-surface winds in Venus' dense atmosphere can lift dust & other particulates from the surface into airborne suspension. Understanding the dependence of the cloud layer on outgassed mantle volatiles is critical for understanding the long-term climate evolution of the planet. All of these effects can be studied by monitoring the spatial and temporal variations of clouds and hazes. Characteristic timescales of cloud formation and dissipation are expected of the order of hours to days, therefore observations on such timescales are properly addressed from a Venus low polar orbit.

6.2 EnVision Mission Overview

EnVision will be launched on an Ariane 62 in the fourth quarter of 2031 (current working assumption, the final schedule will be agreed together with NASA at Mission Adoption), with back-up launch dates every 6 months until mid 2033. Indicative mass budget including all margins is 1350 kg (dry mass); estimated total wet mass including launch adapter at the time of mission selection is 2500 kg. An interplanetary cruise of 15 months (to be confirmed and pending final launch date) is followed by orbit insertion and then circularization by aerobraking over a period of about 16 months to achieve the nominal science orbit, a low quasi-polar Venus orbit with inclination between 87 and 89 deg, altitudes varying from 220 to 510 km and orbital period of about 92 min. The nominal science phase of the mission will last six Venus sidereal days (four Earth years). The choice of science orbit around Venus is mostly driven by a need for global VenSpec, SRS, and VenSAR altimeter and radiometer coverage, stereo topography, polarimetric and repeated VenSAR imaging, and for high-resolution gravity mapping. The spacecraft is approximately rectangular, 3 m in height × 2 m in depth and width in stowed configuration, with chemical propulsion and powered by two deployable solar arrays. EnVision will downlink ∼210 Tbits of science data, using a Ka-/X-band comms system with a fixed high-gain antenna (HGA) of diameter > 2.5 m.

The communication subsystems relies on X-band uplink for simultaneous telecommand and ranging reception, on X-band downlink for simultaneous spacecraft telemetry and ranging transmission, and Ka-band (32 GHz) downlink for high data rate transmission of science data or alternatively for ranging. The HGA is the primary antenna used for spacecraft communication in X and Ka-band, and is completed by several Low Gain Antennas (LGA) used for X-band communications only, during Launch and Early Operations Phase (LEOP) and spacecraft safe modes. To maximize the data return, the Ka-band communications subsystem relies on a powerful Travel Waveguide Tube Amplifier (TWTA) with a radio frequency power output of 120 W. This architecture, together with daily communication passes with 35 m Deep Space Antennas of 9.3 hours in average, allow to downlink the required science data return whatever the Earth to Venus distance.

6.3 EnVision Science Payload

EnVision's science payload consists of VenSAR, a dual polarization S-band radar also operating as microwave radiometer, three spectrometers VenSpec-M, VenSpec-U and VenSpec-H designed to observe the surface and atmosphere of Venus, and the Subsurface Radar Sounder (SRS), a High Frequency (HF) sounding radar to probe the subsurface. These are complemented by a radio science investigation which achieves gravity mapping and radio occultation of the atmosphere, for a comprehensive investigation of the Venusian surface, interior and atmosphere and their interactions. This extensive suite of instrumentation and experiments work together to comprehensively assess surface and subsurface geological processes, interior geophysics and geodynamics, and atmospheric pathways of key volcanogenic gasses, which together illuminate how and why Venus turned out so differently to Earth. The synergistic and holistic way in which the payload instruments collaborate to investigate processes at different altitudes, depths and spatial scales is characteristic of the EnVision mission (European Space Agency 2021).

6.3.1 VenSAR on EnVision

A **Synthetic Aperture Radar, VenSAR**, will image pre-selected regions of interest at a resolution of 30 m/pixel, and subregions at 10 m/pixel. An order of magnitude better than Magellan and with a better sensitivity, these images are the key to understanding geological processes from local to global scale, discriminating relationships between units of different age, and identifying the changes caused by geological activity. Topographic information at 300 m spatial and 20 m vertical resolution across these regions, derived from stereo imaging at two different incidence angles, is complemented by a global network of altimetry mode tracks with a vertical resolution of 2.5 m. This enables to map the surface at a better resolution than any previous dataset, essential for resolving the geometry of faults, folds and other features, and enabling the quantitative analysis of geological processes. Surface properties such as roughness will be derived from active imaging in both HH and HV polarizations – a first for a Venus orbiter - and passive radiometry at a range of angles, which also permits the detection of surface temperature anomalies. Repeated observations and comparisons with Magellan imagery allow for the detection of volcanic, tectonic and geomorphic changes over periods of months, years and decades.

EnVision will acquire dual-polarization SAR imagery at 30 m resolution for about 7% of the surface after 6 cycles, which aids surface characterization by exploiting the polarimetric reflection properties of the surface. SAR polarimetry is essential for differentiation of surface types and properties, because it is sensitive to surface roughness and structure (e.g., consolidated vs fine-grained material). VenSAR employs a dual polarization mode (transmitting H and recording H and V polarizations) to enable differentiation between terrain types and first-order surface properties characterization. Dual polarization was chosen for data rate and swath width considerations and H polarization to match the Magellan data enhancing change detection studies. Passive radiometry will be carried out in a near-nadir (with an incidence angle of 14°) or nadir viewing geometry, in parallel with other EnVision instruments. The surface microwave brightness temperature will be recorded globally ($>75\%$ of the surface) with repeated observations (at least 2 times) and a final resolution likely better than 10 km when using all overlapping near-nadir observations.

Surface Topography. - Surface topography is integral to many of the EnVision science investigations, either as the primary data source for inferring the type and magnitude of geologic processes that shape the surface, or as ancillary data necessary for proper interpretation of other data. The resolution and vertical accuracy required depends on the investigation and varies from several kilometer scale resolution to roughly quarter kilometer with vertical accuracy of 10s of meters. Magellan global topographic data with its 15-20 km resolution and vertical accuracy of 50-100 m is insufficient to support these investigations (Ford et al. 1992). Quantitative modeling of faulting and folding requires knowledge of topography with a vertical resolution of 25-50 m. Such models can constrain the physical processes that produce the observed tectonic landforms, the magnitude of the deformation, and the mechanical structure of the crust and lithosphere in the vicinity of the tectonic feature.

Topography from SAR stereo data for impact craters, at horizontal resolutions less than a quarter to a third of a crater diameter, i.e., less than 10 km, and vertical resolutions better than 20 m, will enable the measurement of the thickness of post-impact crater fill. Still finer spatial- and vertical-accuracy topography measurements will reduce the uncertainty in crater depth-diameter measurements and more accurate crater fill thickness estimates. Moreover, the plains of Venus are under-represented in the RoIs, and a globally distributed set of topographic measurements will be particularly important for understanding the plains resurfacing history.

Topography data are also needed for investigations other than those of the SAR. The Subsurface Radar Sounder (SRS) requires topographic information to identify likely off-nadir echoes ("clutter") that may confuse subsurface feature identification. Knowledge of the absolute surface temperature is needed for calculation of the absolute surface emissivity from near-IR nightside observations. The variation of surface temperature is primarily dependent on surface altitude; reducing the accuracy of the surface altitude determination to ≤ 10 m also reduces the uncertainty in the absolute determination of surface emissivity.

Surface Properties: Nadir and Near-Nadir Radiometry, Surface Polarimetry, Microwave Emissivity. - Used in passive radiometry mode, EnVision SAR will map the thermal emission emanating from Venus surface with significantly better precision and accuracy than the Magellan radar (0.7 K against 1-2 K and 1.7 K against 15 K, respectively). Emission maps, in the form of surface brightness temperature maps, will then be used to search for thermal anomalies or, if the surface temperature is known, to map the emissivity of the surface which, in turn, provides insight into its composition (through the dielectric constant) and physical properties (roughness, density).

Passive nadir and near-nadir radiometry SAR modes are primarily designed for the search of thermal anomalies but will also be used, based on assumptions on the physical temperature, to build a mosaic of the surface emissivity at 9.5-cm by dividing the measured brightness temperatures by an estimate of the surface temperature. At nadir or near-nadir the microwave emissivity of a surface is largely controlled by its dielectric constant and the surface roughness only has a second order effect. In turn, the dielectric constant is related to the bulk composition and density of the surface material and the dielectric map inferred from radiometry measurements will be used to distinguish surface units. More specifically, for dry materials, the relationship between dielectric constant and the density is generally well described by a power-law function and, with some assumptions, the dielectric map can be readily converted into a global near-surface density map (Campbell and Campbell 1992; Campbell and Rogers 1994).

In addition to near-nadir and nadir observations, polarized radiometry measurements will be acquired in an off-nadir geometry (with a viewing angle of 25-30°) in selected regions. As aforementioned, the main advantage of nadir radiometry is to be less sensitive to roughness than off-nadir radiometry. However, the average of two orthogonally polarized emissivity values (or the polarization ratio) is also less sensitive to roughness than either individual component and can be used to provide an even more reliable estimate of the dielectric constant, requiring no assumption on the physical temperature. Such measurements will be primarily performed in Venus highlands to confirm or inform their unusually high dielectric constant and put new constraints on their composition candidates. Recording of both H and V polarization in an off-nadir geometry will distinguish between the effects of dielectric constant and roughness/volume scattering, thus offering an additional powerful tool for surface characterization (European Space Agency 2021).

By collecting microwave emissivity data at a higher resolution than the radar of Magellan, with better precision and especially accuracy (by a factor ~ 10) and geometries (targeted off-nadir polarized measurements) relevant to the science objectives, the EnVision radar operating as a radiometer combined with the instrument high-resolution topography and polarimetric imaging will refine the mapping of Venus surface in terms of composition and physical properties. It will thus provide key information to retrieve the geological history and age of its terrains. In particular, it will help unravel the nature and rate of alteration in Venus high-altitude low-emissivity regions, investigate impact modification in crater ejectas and maybe unveil deeply weathered regions, thick sedimentary layers or signatures of recent resurfacing. By the end of the EnVision mission (6 cycles) we should be able to produce a radiometry map of >90% of the surface, with a resolution of about 10 km using all overlapping measurements.

6.3.2 EnVision Subsurface Radar (SRS)

A **Subsurface Sounder, SRS**, will characterize the vertical structure and stratigraphy of geological units including volcanic flows. Geological inferences from Magellan data point to

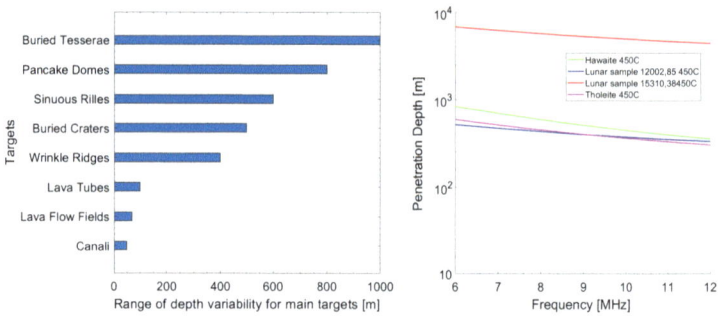

Fig. 18 (a, left): Typical range of subsurface sounding depths in which different geological targets of the EnVision Subsurface Radar (SRS) can be identified; (b, right:) SRS average penetration depth calculated for different Venus-like samples (from measurements on Moon and Earth analogue materials at Venus temperature) in the SRS bandwidth.

a range of subsurface structures and geometries that are as yet unquantified. The SRS provides a unique opportunity to sound the great variety in geologic and geomorphic units. It will also provide unprecedented information on the surface in terms of roughness, composition and permittivity (dielectric) properties at wavelengths completely different from those of VenSAR, thus allowing a better understanding of the surface properties (Fig. 18). SRS observation will also result in altimetry measurements by providing low-resolution profiles of the topography that can be integrated with the altimetric data of VenSAR.

EnVision is the first mission to Venus with a confirmed sounding instrument (ISRO's proposed Venus mission is also considering a sounder, see Sect. 8.3.2) that will allow for the direct measurement of subsurface features. Despite some geological surface investigations that provide hints about possible existence and nature of subsurface structures, no direct measures exist. In this context the Subsurface Radar Sounder (SRS) onboard EnVision mission represents a unique opportunity to sound the great variety of geologic and geomorphic units. SRS will investigate stratigraphic and structural patterns, to test hypotheses related to the origin of structures at the surface and in the shallow subsurface and their relationships. This will enable investigation of interaction processes between surface and subsurface structures as well as subsurface structures not directly linked with surface ones.

There are many geological investigations for which the detection of subsurface boundaries may provide invaluable constraints. They include impact craters and their infilling, buried craters, tesserae and their edges, plains, lava flows and their edges, and tectonic as well as volcanic features. For those features subsurface characteristics are crucial for: the relative dating of surfaces by the analysis of stratigraphic relationships, the modeling of three-dimensional structure, the identification of boundaries between units/edges. The subsurface material boundary delineation by sounding will improve the understanding of Venus resurfacing history and geologic evolution.

These investigations will be performed Venus wide (with an average observation density of 2 per degree of longitude at Equator) and on selected RoIs which include the mentioned features (with an average observation density of 10 per degree of longitude at Equator). The scientific investigations call for a penetration down to a few hundreds of meters (up to 1000 m) and about 20 meters of vertical resolution. The typical depth needed for sounding of different subsurface feature types is shown in Fig. 18a. Calculations of SRS penetration depth, shown in Fig. 18b, determine that the SRS will be able to investigate a wide variety of geological targets. The SRS penetration depth has been calculated using a large variety

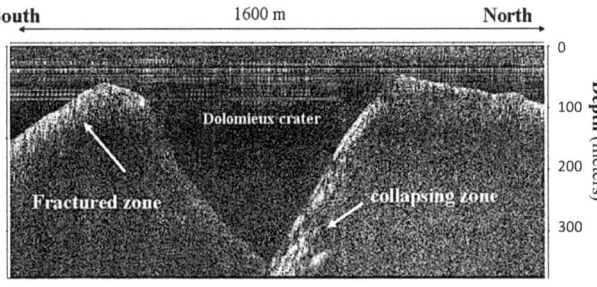

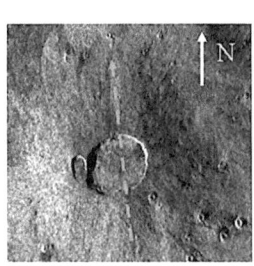

Fig. 19 Airborne radar sounder profile at 40 MHz central frequency (more than four times higher than the SRS one) over the Dolomieu Crater on the top of the Fournaise Volcano in the reunion Island in the Indian Ocean. Fournaise is a hot spot effusive volcano with geomorphological features and magma dynamic very similar to several Venusian volcanoes (Anderson 2005). The radargram crossing from South to North the main crater on the top of the volcano show the fractured areas (white areas before and after the crater) that are materialized by the strong signal scattering resulting from the fractures. Inside the crater the radargrams shows the layering that is on the crater northern wall arising from the succession of debris flowing from the collapsing northern part. The crater depth is approximately 100 m and its width 1 km. The lava temperature ranges from 40 C at the surface to 600 C beyond the 10 m level, demonstrating the viability of HF sounding through rocks at these elevated temperatures

of different rock types and surface topologies; for a quick demonstration of the viability of HF subsurface sounding through rocks at Venus temperatures, Fig. 19 shows an example of sounding through lava at $>600\,°C$ of a volcanic crater floor on Earth.

6.3.3 EnVision Spectrometer Suite (VenSpec)

A **Spectrometer suite, VenSpec**, will obtain global maps of surface emissivity in six wavelength bands using five near-infrared spectral transparency windows in the nightside atmosphere, to constrain surface mineralogy and inform evolutionary scenarios; and measure variations of SO_2, SO and linked gasses in the mesosphere on the dayside, to link these variations to tropospheric variations and volcanism. In combination of its three instruments, detailed below, VenSpec will provide unprecedented insights into the current state of Venus and its past evolution. VenSpec will perform a comprehensive search for volcanic activity by targeting atmospheric signatures, thermal signatures and compositional signatures, as well as a global map of surface composition.

VenSpec-M, like the identical VEM instrument on-board VERITAS (Sect. 4.3.2), is a pushbroom multispectral imager optimized to map thermal emission from Venus' surface using six narrow bands ranging from 0.86 to 1.18 μm, and three bands to study cloud microphysics and dynamics. VenSpec-M will provide near-global compositional data on rock types, weathering, and crustal evolution by mapping the Venus surface in five atmospheric windows. VenSpec-M will use the methodology pioneered by VIRTIS on Venus Express but with more and wider spectral bands, the NASA VERITAS VISAR and Envision VenSAR-derived Digital Elevation Models (DEMs) and EnVision's lower orbit compared to Venus Express to deliver near-global multichannel spectroscopy with wider spectral coverage and an order of magnitude improvement in sensitivity. It will obtain repeated imagery of surface thermal emission, constraining current rates of volcanic activity following earlier observations from Venus Express (Smrekar et al. 2010a; Mueller et al. 2017). In combination with the observations provided by the identical VEM instrument on the NASA VERITAS mission VenSpec-M will provide more than a decade of monitoring for volcanic activity, as well as search for surface changes (Fig. 10).

VenSpec-M uses the same 14 bands filter array as VEM on board VERITAS (Helbert et al. 2016, 2020, see also Table 3 and Fig. 10). Those 14 bands fall in four categories depending on where the radiation is originating. The radiation for the six surface bands at 0.86, 0.91, 0.99, 1.02, 1.11, 1.18 μm originates at the surface. Surface bands are used to determine rock types (Dyar et al. 2020, 2021; Helbert et al. 2021) as well as monitor for the thermal signature of active volcanism. The radiation in the two water vapor bands originates in a layer close to the surface and is sensitive to the abundance of water vapor which may see changes due to volcanic exhalations, complementing the H_2O and HDO measurements by VenSpec-H in the middle atmosphere. In the three cloud bands, radiation originates at an atmospheric layer above the surface but below the clouds. Because the signal in the cloud bands has no surface or water vapor contributions, the measurements in these bands can be used to remove cloud-induced contrast variability from the other bands. Finally, the three background bands are sensitive in spectral regions where the atmosphere is opaque, thus allowing the removal of background signal on the detector. The high density of cloud particles results in multiple scattering of the radiation, reducing the spatial resolution to 50–100 km.

VenSpec-H is dedicated to high spectral resolution atmospheric measurements in the near-infrared. It will focus on the volcanic and cloud forming gases and search for composition anomalies potentially related to the volcanic activity. The instrument, designed to measure H_2O, HDO, CO, OCS, and SO_2 on both the night and day side, is a nadir-pointing, high-resolution (R$\sim$8000) infrared spectrometer that will perform observations in different near-IR spectral windows between 1 and 2.5 μm. Spectra in these bands will be recorded sequentially as the EnVision spacecraft moves along its quasi-polar orbit, and will allow the sounding of different layers in the Venusian atmosphere: close to the surface (1.17 μm), 15-30 km (1.7 μm), 30-40 km (2.4 μm) and above the clouds (1.38 & 2.4 μm). Two additional polarization filters will be used during dayside observations to better characterize the clouds' properties.

The instrument will include a total of four spectral bands: 1.165 - 1.180 μm (B#1), 2.34 - 2.48 μm (B#2), 1.72 - 1.75 μm (B#3) and 1.37 - 1.39 μm (B#4). B#2 is further divided into two ranges: 2.34 - 2.42 μm (2a) and 2.45 - 2.48 μm (2b). Bands 1, 2a, 2b and 3 are observed on the night side, bands 2a, 2b and 4 on the day side. In this near-IR region, the high spectral resolution combined with the high sensitivity of the instrument will allow to clearly identify the absorption features of the targeted species. Spectral band selection is performed in part by a filter wheel mechanism with stringent lifetime requirements and a filter-slit-assembly that allows sequential measurements in the 4 spectral bands of interest. Design measures are taken to make the VenSpec-H observations insensitive to polarization, while exploiting the polarization information contained in the light reflected from Venus.

VenSpec-U, a dual-channel ultraviolet spectrometer, will monitor minor sulfur species (mainly SO and SO_2) and investigate the complex and highly variable upper atmosphere and its relationship with the lower atmosphere. VenSpec-U will search for atmospheric effects of geological activity, in order to determine how much outgassing is occurring, and how the atmospheric chemistry is coupled with surface/subsurface geochemistry and weathering cycles; study how mesospheric gas variations are linked to volcanism, in order to identify the causes of variability in the mesospheric sulfured gases (SO, SO_2); and finally how cloud and particulate variability is linked to volcanism, in order to detect plumes of volcanic ash or sulphate clouds caused by volcanism, and to understand any link between the Venus sulfuric acid clouds and volcanism.

Observations can be conducted in a strict nadir geometry (null emission angle), or in near-nadir geometry (emission angle $< 30°$) thanks to a UV imaging spectrometer operating in the 190 – 380 nm spectral range. Spectral resolutions shall be better than 0.3 nm in the 205 – 235 nm range (typical SNR of 100 at 220 nm) in order to distinguish SO and SO_2 spectral lines, and better than 2 nm in the 190 – 380 nm range (typical SNR of 200 at 220 nm) which encompasses the unknown UV absorber peaking near 365 nm. Spatial sampling shall range from 3 km to 24 km, depending on spectral resolution and orbiter altitude. The narrow-slit axis of the instrument contains the spectral information, whereas the long-slit axis contains the spatial information along the 20° field of view. The remaining spatial direction is provided through orbital scrolling.

6.3.4 EnVision Radio Science / Gravity Experiment (RSE)

A **Radio Science Experiment** uses the spacecraft-Earth radio link for gravity mapping and atmospheric profiling. Measurements of the lateral variations in the strength of a planet's gravity field is an important tool in probing the subsurface structure of a planet. Regional differences in elevation can be supported by differences in crustal thickness, by flexure of the elastic lithosphere, or by convective flow in the mantle. These mechanisms can in turn be distinguished by their expected gravity signatures, resulting in estimates of the thickness of the crust and lithosphere in different regions of Venus. As discussed previously (Ghail et al. 2023; Herrick et al. 2023; Gilmore et al. 2023, this collection), we need to understand whether the tesserae, which contain fine-scale, complex patterns of deformation, represent thick, ancient remnants of deformed and deep-rooted continental crust. The tesserae may also hold clues to the nature of past resurfacing; particularly whether there have been periods of enhanced crustal mobility, or whether Venus has been in its current state for most of its history.

Crustal thickness affects the stratification of mechanical strength in the lithosphere and thus can also affect the style of tectonic deformation (Dumoulin et al. 2017). Magellan gravity data are consistent with an organized pattern of mantle convection broadly similar to Earth; but it lacks the resolution necessary to understand its connection with geological-scale features, such as individual coronae or mountain belts. EnVision can measure the integrated amount of volcanism over time and thus provide tests of thermo-chemical evolution models, but can also sometimes be the product of extensional or compressional tectonism. Determining lithospheric thickness with gravity data is particularly sensitive to data with wavelengths less than 500 km. Higher spatial resolution than the Magellan solution of the Venus gravity field is required to better constrain the crustal and lithospheric structure variations. Combining Magellan and EnVision gravity data will allow determination of the gravity field over at least 95% of the planet, with an average spatial resolution better than 200 km, and an accuracy better than 20 mGal (Rosenblatt et al. 2021; European Space Agency 2021).

Venus' moment of inertia, Love number, and tide-induced phase lag also characterizing the signature of the internal structure in the gravity field will be extensively constrained during the six cycles of the EnVision mission (one cycle equals one Venus sidereal day or 243.02 Earth days). EnVision will constrain the size of the main internal layers crust, lithosphere, mantle and core, and whether or not the core is fluid, will help to understand fundamental differences or even possible similarities between Venus and Earth. Indeed, the overall size of the chemical reservoirs (crust, mantle, core) gives information about the composition of Venus; the average thickness of the crust about the rate of magmatism; the average thickness of the lithosphere about the mechanisms of heat transfer at the surface; and the state of the core about the long-term cooling rates. The EnVision spacecraft-Earth radio link will measure the gravity field of the planet with spatial resolution better than 270 km, and accuracy of <0.2 mm/s^2 globally, with an improved higher spatial resolution of <200 km and accuracy of <0.1 mm/s^2 in most of the Southern hemisphere (40% of the planet) and k_2-Love number with an accuracy of ± 0.01. As discussed before, the potential Love number helps to determine the state of the core and, in the case of a liquid core, also its size. A Love number k_2 lower than 0.27 would indicate the presence of a fully solid iron core, while for larger values, solutions with an entirely or partially liquid core are possible (Dumoulin et al. 2017).

Furthermore, the EnVision radio-occultation experiment aims at sounding of the temperature structure of the Venus atmosphere in the altitude range 90-35 km and abundance of sulfuric acid in gaseous and particulate phases. The experiment relies on the observation of the radio-link propagation (frequency and amplitude) through the atmosphere of Venus during radio-occultation. The radio ray path changes in the ionosphere and neutral atmosphere are induced by a change in the refractivity profile. This leads to a shift in the measured frequency at the ground station. These frequency changes can be used to retrieve the neutral number density, temperature and pressure profiles as a function of the planetary radius at a high vertical resolution. Thanks to the use of the dual X-Ka band, the cloud contents in both gaseous and liquid phase of sulfuric acid, and its spatial and temporal variability, will be estimated for the first time at 35-55 km, with an accuracy of 1 mg/m^3 (liquid) and 1 ppm (gaseous) on time scales from hours to years, with vertical resolution of $\sim$100 m. (European Space Agency 2021). In addition to H_2SO_4 content in both gaseous and liquid phase within and below the clouds, static stability profiles retrieved from temperature, pressure and number density profiles (35-90 km), provide valuable information about small-scale fluctuations in the thermal profiles and the latitudinal dependence of gravity wave activity. Understanding the dependence of the cloud layer on outgassed mantle volatiles is critical for understanding the long-term climate evolution of the planet, and Venus would be indeed the first planet beyond the Earth where we could relate the dynamics of gravity waves and small-scale turbulence and temperature fluctuations, and the cloud composition.

Both VERITAS and EnVision VISAR and VenSAR will produce repeated imaging of surface features throughout their mapping cycles, allowing to create radar tie points, thus tying the inertial position of the probe to the planetary body-fixed frame. Leveraging on the combination of tracking data and radar tie points, both VERITAS and EnVision will be able to measure the precession and monitor the variable spin rate of the planet with a much-improved precision.

6.4 Summary / Outcomes Revealing Venus Evolution

EnVision was selected as ESA's 5th M-class mission, targeting a launch in the early 2030s. The mission is a partnership between ESA and NASA, where NASA provides the Synthetic Aperture Radar payload. The scientific objective of EnVision is to provide a holistic view of the planet from its inner core to its upper atmosphere.

The mission is scheduled for launch on an Ariane 62 in the fourth quarter of 2031 (current working assumption, the final schedule will be agreed together with NASA at mission adoption), with backup launch dates every 6 months until mid-2033. It will provide new insights into geologic history through complementary imaging, polarimetry, radiometry, and spectroscopy of the surface, coupled with subsurface sounding and gravity mapping; search for thermal, morphological, and gaseous signs of volcanic and other geologic activity; and follow the fate of key volatile species from their sources and sinks at the surface through the clouds to the mesosphere. Following the same approach that has advanced our understanding of Earth and Mars, EnVision will combine global observations at low or medium spatial resolution (e.g., surface emissivity & atmospheric composition) with regionally focused observations at higher spatial resolution.

VenSAR, a dual-polarization S-band radar that also operates as a microwave radiometer, builds on NASA-JPL's experience with planetary radars since the Magellan mission; the Subsurface Radar Sounder (SRS), a high-frequency (HF) sounding radar to probe the subsurface, inherits from the RIME instrument on JUICE. These will be complemented by a radio science investigation that will provide gravity mapping and radio occultation of the atmosphere. The three VenSpec spectrometers build on the heritage of ESA's suite of planetary missions, in particular ESA's Venus Express from 2006 to 2014, will also highly benefit from complementarity with DAVINCI and VERITAS measurements.

To achieve its science objectives, EnVision must return 210 Tbits (26.25 Terabytes) of science data to Earth, using a Ka/X-band comms system with a fixed diameter hight-gain antenna, with a large dynamic range of distance to Earth (from 0.3 to 1.7 AU), from a low Venus polar orbit, in the hot Venus environment (exacerbated by the operation of highly dissipative units), while operating three spectrometers in a near cryogenic environment. Achieving the science objectives under these multiple constraints without oversizing the spacecraft requires careful planning of the science operations, making the science planning strategy a critical driver in the overall mission design against which the spacecraft and ground segment are then sized (Sect. 6.2 and Fig. 15).

EnVision science operations strategy is to obtain the widest range of data types that enables us to put the highest resolution datasets into regional and global context. characterize the sequence of events that generated the regional and global surface features of Venus, determine crustal support mechanisms, mantle and core properties, and the geodynamics framework that controls the release of internal heat over Venus history, by determining the styles of volcanic processes which have occurred on Venus, studying the sources, emplacement styles, magma properties and relative ages of different volcanic flows; assessing the styles of tectonic deformation that have operated on Venus by studying their surface expression and gravity signatures, and determining their role in planetary heat loss. It will also characterize surface modification processes - such as impact crater modification, low emissivity/radar bright highlands - to improve our understanding of Venus geochronology and constrain Venus' internal structure, through measurements of gravity field and tidal response, to study the properties and thicknesses of Venus' crust, mantle and core. It will better assess whether Venus once had condensed liquid water on its surface and was thus perhaps hospitable for life in its early history, and therefore fully support the scientific goals and open questions presented in the companion papers of this collection.

7 Venera-D: A Comprehensive Exploration of Venus' Atmosphere, Surface, Interior and Plasma Environment

Since the discovery of Venus' atmosphere by Mikhail Lomonosov in 1761 (Marov 2005), and further observations of the Venus transits, Venus has always been a celestial object of interest among Russian astronomers. It is not surprising that the multistage studies of Venus became the central and most successful part of the Soviet robotic space program. was the site of the first entry probe in any solar system atmosphere in 1967 (Venera-4), first soft landing in Dec. 1970 (Venera-7), first image from the surface of another planet in 1975 (Venera-9). The Soviet series of Venera & VeGa missions were phenomenally successful, not only in their technologically advanced landers which returned color pictures from Venus and successfully analyzed drill samples, but also successfully deployed balloons in the atmosphere in 1985. The Venera-D concept is the logical next step in the highly successful series of Venera and VeGa missions of the 1970s and 1980s (Marov et al. 1973; Marov 1978; Avduevskii et al. 1977; Florensky et al. 1977; Barsukov et al. 1982, 1986; Garvin et al. 1984; Surkov et al. 1984; Moroz 1990; Moroz et al. 1985, 1996; Sagdeev et al. 1986a, 1992).

The Russian Venera-D (Венера-Д) flagship mission concept has been under development with the goal of advancing the investigation of Venus' atmosphere, surface, and interior and the processes that link them as a system. Intense discussions about Venus began in 2013 as part of the Joint Science Definition Team (JSDT) established by NASA and Roscosmos with the goal of shaping a collaborative project. The JSDT developed a full Venera-D mission scenario (Venera-D Joint Science Definition Team 2019; Venera-D Venus Modeling Workshop proceedings 2018; Glaze et al. 2018; Zasova et al. 2020), but as of 2021, for a variety of (mostly non-scientific) reasons, Venera-D has been developed as a national program for Venus. Science objectives of the Venera-D mission concept currently address key questions about atmospheric dynamics, emphasizing atmospheric superrotation and radiative balance; the processes that have formed and modified Venus' surface, highlighting the mineralogical and elemental composition of surface materials; and the chemical processes occurring at the interface of the surface and the atmosphere. The Venera-D lander would not only perform descent phase measurements but would also analyze surface composition. This would of course provide invaluable "ground truth" for VERITAS, DAVINCI and EnVision's surface composition mapping, as well as contributing to understanding of geophysical evolution (Venera-D Joint Science Definition Team 2019).

7.1 Venera-D Science Objectives

Venera-D is designed to study the atmosphere, surface, internal structure and properties of plasma surrounding Venus at new scientific and technological levels. As we discussed in Sect. 2, Venus is an Earth-sized terrestrial planet that has taken a different evolutionary and habitability path. To examine the reasons for this difference is very important to understand the divergent Earth and Venus evolutionary pathways. This is of particular relevance for the study of exoplanets and conditions for their habitability (O'Rourke et al. 2023; Way et al. 2023; Westall et al. 2023; Gillmann et al. 2022, this collection). Among these key aspects and objectives of the Venera-D mission, we can formulate those high-level objectives that directly address the long-term history and evolution of Venus through time, as well as those that are not addressed by currently selected missions:

Coupling Between Geologic and Climate History. - Current and past rates of volcanic outgassing are unknown, as is an understanding of how volcanoes have affected the atmosphere and climate. More fundamentally, the role of water in geodynamics and petrogenesis must be constrained. As on Earth, the geology and climate of Venus are linked (Bullock and Grinspoon 2001). The causes and effects of rapid changes in geologic expression can be studied in detail with a capable surface payload and remote sensing techniques (Helbert et al. 2008; Mueller et al. 2008; Gilmore et al. 2015). To address key geologic questions, it is necessary to characterize the geochemistry, mineralogy, emplacement, sediment supply, and petrology of surface features and terrains (Herrick et al. 2023; Gilmore et al. 2023; Carter et al. 2023, this collection); obtaining this information for the tesserae would provide insight into the oldest exposed rocks. These data will allow us to constrain the history of volatiles, especially water, on Venus and provide a basis for direct comparison of crustal evolution on Earth and Mars. In addition, isotopic measurements of the composition of the Venusian atmosphere and an improved understanding of atmosphere-surface interactions will help constrain the outgassing history, in particular the current and past volcanic outgassing rates (Avice et al. 2022, this collection).

Abundance of Light Elements, Rare Elements and Their Isotopes. - To characterize Venus' origin and evolution through time, accurate assessment of the composition of the atmosphere's composition is essential. Like Earth and Mars, the atmosphere of Venus seems to have substantially evolved from its original composition. Whether the major processes that shaped the atmospheres of Earth and Mars—such as impacts of large bolides and significant solar wind erosion—also occurred on Venus is largely unknown. Detailed chemical measurements of the composition of the atmosphere—in particular, the noble gases and their isotopes along with light elements and isotopes—will aid in understanding if the modern (secondary) atmosphere is a result of degassing from the interior or if it formed from comet or asteroid impacts (Avice et al. 2022; Salvador et al. 2023, this collection). Likewise, it is imperative to determine how the atmospheric abundances of water, sulfur dioxide, and carbon dioxide change under the influence of the exospheric escape of hydrogen, outgassing from the interior, and heterogeneous reactions with surface minerals.

Venus Surface Geochemistry. - The only means by which the geochemical data from Venus' surface can be obtained are the landers. Several landers visited the planet in a period from 1972 (Venera-8) to 1985 (VeGa-1 and 2) and reported the only data on the chemical composition of soils on the surface of Venus. Chemical measurements were made at seven sites that are concentrated in the Beta-Phoebe region and in Rusalka Planitia to the north of Aphrodite Terra. Selection of the landing sites were based purely on the interplanetary ballistic constraints because no knowledge on the surface geology existed when the Venera-VeGa missions were implemented. At four landing sites (Venera-8, -9, -10, and VeGa-1), concentrations of the three major thermal- generating components, K, Th, and U, were determined by gamma spectrometry (Surkov 1997). The mean values of their concentrations on Venus are well within the range that is typical of terrestrial basalts (Kargel et al. 1993; Nikolaeva 1995, 1997). However, enhanced concentrations of K, Th, and U in soils at the Venera-8 landing site raises the possibility for the presence of a non- basaltic material on Venus (Nikolaeva 1990). In two landing sites (Venera-13, and -14), the concentrations of major oxides (without Na_2O) were measured by the X-ray fluorescence (XRF) method. At the VeGa-2 site, both methods (gamma spectrometry and XRF) were used separately and the concentrations of the thermal-generating elements and major oxides were measured (Surkov 1997). The XRF data also suggested that rocks of basaltic composition make up the landing sites (Surkov et al. 1984, 1986; Kargel et al. 1993). Two important factors, unfortunately, strongly limit the value of the Venera and VeGa data and prevent their robust interpretation: (1) we do not know the exact position of the landers. All stations landed somewhere within their own landing circle, which is ~300 km in diameter and usually embraces terrains of different origin and age; (2) past accuracy of measurements. A new generation of lander instruments (Sect. 11.1.4) will determine the mineralogy and chemistry of terrain to ascertain rock type, and look for evidence of past water.

Role of Solar Absorbers and Near-IR Opacity Sources in Venusian Clouds. - The Venusian disk in reflected light is practically featureless in the visible and near-IR spectral regions (contrasts maximum 2 to 3%), but in the UV they reach or exceed 30% at 365 nm. The albedo of Venus decreases from a value of ~0.8 at wavelengths >550 nm to as low as 0.3 at UV wavelengths. Cloud contrast peaks at 365 nm. UV contrasts observed between 0.33 μm and 0.5 μm are the result of absorption by a species of unknown origin. UV absorption at 0.32 to 0.5 μm was observed to disappear below 58 to 60 km by the Pioneer Venus spacecraft (Tomasko et al. 1985). Thus, absorption by the UV-absorbing species of Venus was primarily associated with the upper clouds. However, measurements taken by the VeGa lander during descent show that absorption of UV radiation (220 to 400 nm) occurs down to 47 km altitude, indicating the presence of absorbers whose identities are still unknown (Bertaux et al. 1996). Spatial variations in this absorption produce contrasts in daytime images and are a means of inferring bulk motions in the cloud top atmosphere. Measurements of small-scale feature motions over latitude and longitude provide information about the superrotation of the Venusian atmosphere at the level of cloud contrasts. Both the vertical distribution and the composition of UV-absorbing species are poorly known. The few available profiles of Venus' UV flux obtained between the cloud top and the surface indicate that the UV absorber is present in the middle and upper clouds (between $\sim$47 and 72 $\pm$ 2 km), but may be occasionally found in the upper haze ($\sim$70 and 80 km) (Lee et al. 2015). It is currently unclear whether the cloud-level abundance of the absorber is solely the result of material upwelling from below, or whether it depends on chemical reactions between upwelling and downwelling species (see also Sect. 7.3.2).

Solar Wind-Venus Interaction and Venus Magnetosphere. - Plasma and magnetic field experiments on Venera-9 and Venera-10 in the 1970s provided the first data on the solar wind interaction and magnetosphere formation of Venus (Vaisberg et al. 1976). Important subsequent studies of the magnetic barrier and tail were performed by the Pioneer Venus Orbiter (Russell and Vaisberg 1983). Based on these experimental data, a model of the induced magnetosphere was developed (Vaisberg and Zelenyi 1984; Zelenyi and Vaisberg 1985). To further advance the field, Venus Express also performed investigations of the solar wind interaction with Venus (Barabash et al. 2007; Futaana et al. 2017). The discovery of the comet-like planetary plasma interaction and the processes leading to atmospheric losses allowed us to estimate how these losses vary with solar and interplanetary conditions, and their potential to cause significant changes in the chemical composition of the Venusian atmosphere over time. Despite the significant progress made by previous Venus missions, there are still outstanding problems in the study of the solar wind-Venus interaction and Venusian escape processes on recent and geological time scales.

Many of outstanding questions in Venus exploration are therefore both synergistic and complementary to the new generation of missions discussed above. Venera-D consists of a VeGa-like lander targeting the plains and an orbiter observing the atmosphere at several wavelengths, including near-IR. Beyond its surface science capabilities, the descent module is synergistic and complementary to the currently selected missions EnVision, DAVINCI and VERITAS because it will be conducting similar investigations on a different terrain type—the Venusian plains (Venera-D Joint Science Definition Team 2019). The combination of chemical and mineralogical data from both the plains and tesserae would significantly advance our understanding of the Venusian crust, mantle and igneous processes, the evolution of volcanism with time and the range of surface-atmosphere interactions on modern Venus. Atmospheric measurements from the Venera-D orbiter would complement the volatile mapping carried out by EnVision, in complement to the geophysical studies and atmospheric descent probe measurements which are at the heart of EnVision, DAVINCI and VERITAS science questions (Sects. 4-6).

7.2 Venera-D Mission Overview

The Venera-D mission architecture is composed of orbiting, landing and atmospheric modules. This mission structure provides the opportunity to make measurements in the Venus-induced magnetosphere, in the planetary atmosphere, and at the planetary surface. Figure 20 shows a general view of the mission.

7.2.1 Mission Requirements and Design Drivers

Mass budget of the spacecraft includes: 4800 kg total mass, which includes 1920 kg orbital module (OM), 2660 kg descent module (DM) and 50 kg OM/DM adapter. In addition, the

Fig. 20 Venera-D composite spacecraft consists of orbiting, landing and atmospheric modules. The orbiter module (OM) is designed for Earth-Venus transit, delivery of the lander and payload equipment (PE) to Venus, functioning while in orbit, collecting data, transmitting collected data to Earth and data from the lander (Sect. 7.2.2); The descent module (DM) is designed for performing scientific measurements during the descent and on Venus' surface (Sect. 7.2.3)

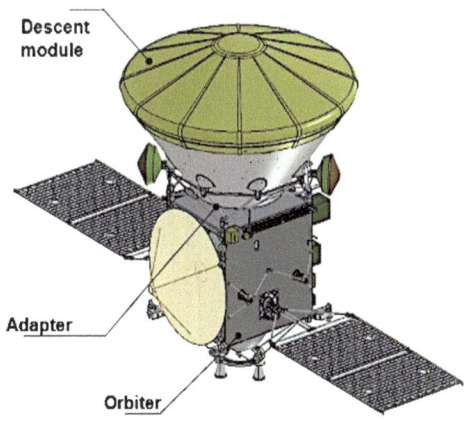

mass of the OM/upper stage adapter is 70 kg. The launch vehicle currently planned for the Venera-D mission is the new Angara-A5 LV launch vehicle with the DM-03 upper stage.

7.2.2 Orbiter Module (OM)

The Orbiter Module (OM) will carry out its science program from a highly elliptical polar orbit with a pericenter of 500 km, located above the Southern Pole, and an apocenter of 69,000 km, with an orbital period of 24 hours. Compared to Venus Express' pericenter of 250 km above the Northern Pole, and apocenter of 60,000 km, the Venera-D Orbiter Module is, essentially, symmetrical to Venus Express' orbit. The expected lifetime of the OM is 7 to 8 years. Figure 21 shows a general view of the orbiter module with a few selected instruments and subsystems. The OM will carry a complex set of science instruments to study the atmosphere of Venus from the surface to the ionosphere. Orbital studies will clarify the climate history of Venus and hopefully reveal the hidden mechanisms of water escape and the extreme greenhouse effect.

The set of imagers operating from UV to longwave IR will monitor dynamics of the atmosphere in the cloud layer on various levels. A combination of a Fourier spectrometer, a mm-radiometer, a long-IR imager and a radio science experiment will allow the construction of a three-dimensional thermal map of Venus atmosphere and the monitoring of its temporal variations. The suite of UV-to-IR spectrometers will provide new insights into minor species of the atmosphere and cloud aerosol properties.

The highly elliptical orbit of OM provides very good opportunities for plasma science. Venus has an extended magnetotail produced by the trapping of planetary ions at interplanetary magnetic field lines. The details of the interactions between the solar wind and the planetary ionosphere, and the corresponding plasma wave excitations, will hopefully be resolved by the charged particle and electromagnetic wave instruments of OM.

7.2.3 Descent Module (DM)

The Descent Module includes

- a ~800-kg Lander Module (LM)
- a ~420 kg Aerial Platform (AP).

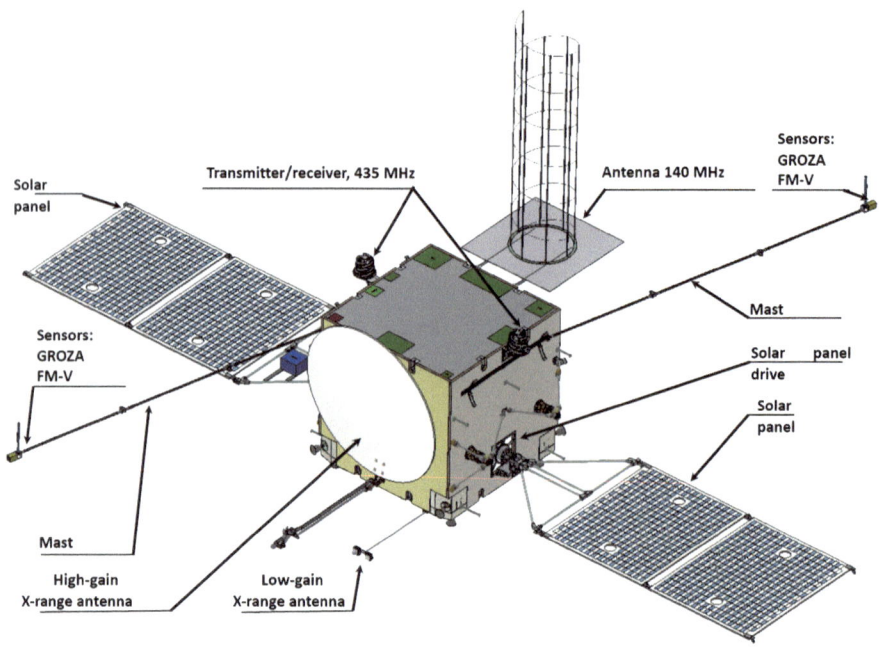

Fig. 21 Venera-D Orbiter Module (OM). OM will carry out its science program from a highly elliptical polar orbit with a pericenter of 500 km and an apocenter of 69,000 km. The orbital period is 24 hours

Landers have always been a key component of Russian missions to Venus. The Venera-D lander will be the first since the successful descent and landing of the Soviet VeGa spacecraft in 1986. The fleet of recently selected Venus missions, described in Sects. 4-6, do not include specific surface components.

Lander Module (LM). - The Venera-D Lander Module (LM) concept resembles in shape the Soviet landers that successfully operated on the surface for 1.5-2 hours, equipped with a modern scientific payload and a much more powerful data transfer system. The design of the lander includes a titanium structure developed for accommodation of onboard avionics equipment (OE); instrument container with OE, including a temperature-resistant cover that allows the LM to operate on Venus' surface for not less than 3 hr; the landing device with the damper, designed to absorb vibrations of the LM during atmospheric entry and to land on the surface; and a separable structure with a parachute system, designed to aerobrake the lander.

High resolution camera system TVS-VD (Table 6) will provide panoramic stereo imaging of the surface around the landing site with the best possible quality. The cameras will use a visible range color imaging system consisting of one landing, four to five panoramic and one close range cameras, providing detailed stereo imaging of the surface with the spatial resolution better than 0.2 mm. See also Sect. 7.3.2 - Lander science, and Fig. 24.

Harsh conditions at the surface of Venus require significant resources to extend the expected lifetime of spacecraft after landing. This lifetime should be sufficient to conduct all planned experiments and upload the obtained data to the Orbiter. As our analysis has shown, the optimum duration of the surface operations in this case is at least 2, maximum 3 hours. Lander descent time in the atmosphere: $\sim$50 min (from 125 km to the surface).

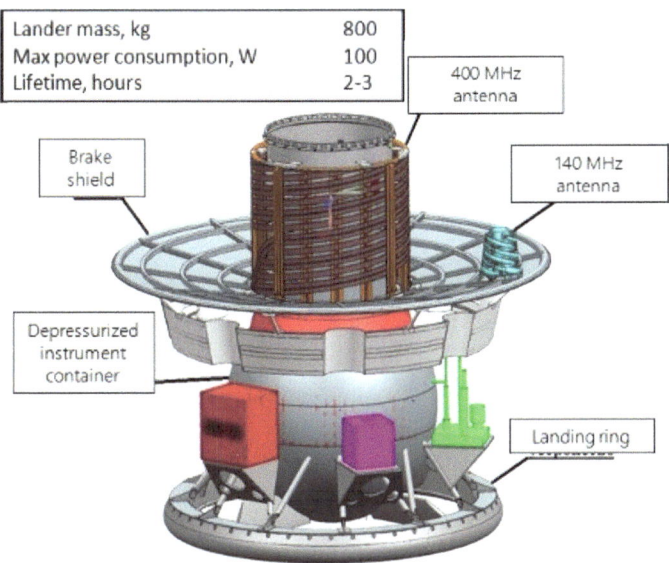

Lander mass, kg	800
Max power consumption, W	100
Lifetime, hours	2-3

400 MHz
antenna

Brake
shield

140 MHz
antenna

Depressurized
instrument
container

Landing ring

Fig. 22 Venera-D Lander Module conceptual design

Lander instrumentation will be operative during the descent and will collect both meteorological data and information about composition of atmospheric gasses and cloud aerosols. The ISKRA-V spectrometer (see Sect. 7.3.2, and Table 6) will study chemical composition of the atmosphere including abundances of gases SO_2, CO, COS, H_2O, NO_2, HCl, and HF, and their isotopologues and isotopic ratios D/H, $^{13}C/^{12}C$, $^{18}O/^{17}O/^{16}O$, and $^{34}S/^{33}S/^{32}S$ during descent from 65 km and after landing; a suite of sensors (MTK-V) will determine temperature, pressure, wind speed, temperature gradient, acceleration from 120 km altitude to the surface and at the surface; at the surface, the package measures chemical composition of rocky sample (which must be delivered inside the lander) and continues measurements of chemical composition of the atmosphere: abundance and isotopic ratio of noble gases in the atmosphere (Sect. 7.3.2, and Table 6). General view of the Venera-D Lander is shown in Fig. 22.

An important event for the planning of the Venera-D mission was the International Landing Site Selection Workshop held in Moscow in October 2019. After intense debate, the collective wisdom of 43 participants prioritized terrain types based on scientific importance and landing safety (see also Ivanov et al. 2017a,b). The results of this analysis are presented in Fig. 23. Regional plains (rp1) were identified as the currently preferred terrain type, as they are considered representative (~30% of the Venus surface), morphologically uniform, and may represent a good sample of the fertile/depleted upper mantle. Last, but not least, is that the regional plains are relatively smooth, which increases the chances of a safe landing. The baseline mission scenario (Eismont et al. 2020), with direct DM insertion, offers a limited choice of accessible landing sites. Alternatively, landing can be achieved virtually anywhere, using gravity assist and resonant trajectories (Eismont et al. 2021a,b), at the cost of a longer transfer.

Aerial Platform (AP). - Another sub-system of the Venera-D descent module is an aerial platform, the design of which is currently less advanced than the other mission elements. The aerial platform (balloon) will have operational altitudes of 54-58 km and will offer very

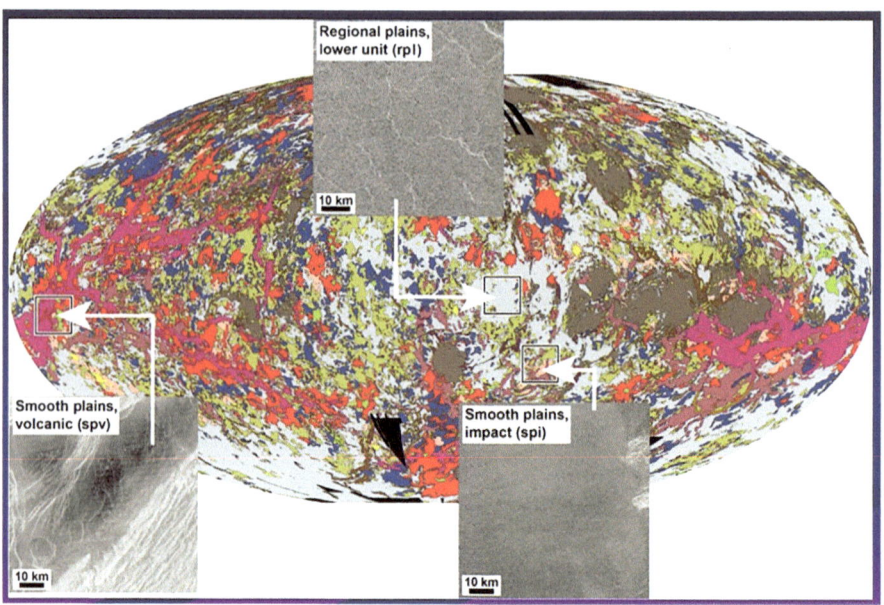

Fig. 23 Potential landing sites for the Venera-D Lander

significant opportunities to make long-term continuous measurements in the planetary atmosphere and in particular in the cloud layer. The preliminary estimate of the mass allocated to the scientific payload is about 30 kg and the set of instruments for the balloon is currently under study.

7.3 Venera-D Science Payload

7.3.1 Orbiter Science

The Orbital Module (OM) investigations aim to clarify the climate history of Venus, the mechanisms of water escape and the characteristics of the greenhouse effect. The potential suite of instruments under study includes 16 scientific instruments to perform studies of Venus atmosphere, ionosphere and magnetosphere, among which a dedicated suite of 7 instruments to analyse plasma spectra and electromagnetic field fluctuations (Table 5).

A set of monitoring cameras (VMC, LIR, IR2R) will allow to follow the dynamic trends of the Venus atmosphere and to improve the general atmospheric circulation models. The synergy of the long-wavelength IR camera (LIR), the Fourier IR thermal spectrometer (SVET) and the radiometer (MM-R) will provide an unprecedented 3D model of the thermal structure of the atmosphere. The heterodyne spectrometer IVOLGA will measure the mesosphere structure and winds, resolving CO_2 lines. UV and IR spectrometers (VOLNA, SVET, VENIS and VIKA) operating both in nadir and solar occultation modes will explore the Venus mesosphere, the cloud layer and possibly identify as yet unknown minor constituents. VIKA includes a nadir channel for the 1.05-1.65 μm range to access the atmosphere below the clouds.

The orbital experiments target to answer many outstanding questions of Venus climate evolution including the role of Solar wind direct interactions with its atmosphere. The OM instrument list is developed in Table 5.

Table 5 Table of Venera-D Orbital Module (OM) instruments under study. Notes: [1] instrument provisionally provided by INAF/ASI; [2] instruments provisionally provided by ISAS/JAXA. A dedicated suite of 7 instruments for measurements of particles (neutral and charged) and electromagnetic fields

Venera-D Orbiter Module (OM)			
Instrument		**Description**	**Scientific goals**
SVET		Thermal IR Fourier transform spectrometer 5-40 µm (2000 – 250 cm–1, Δv=1 cm–1)	3-D thermal structure 55-100 km, SO2 and H20 and clouds composition 55-75 km, dynamics, thermal tides, thermal balance
VOLNA		UV mapping nadir and limb spectrometer (190 – 590 nm, Δλ =0.2-0.3 nm)	UV absorber, SO, SO2, cloud composition, dynamics, night glow emissions
VENIS (1)		IR grating spectrometer 2 – 5 µm and Imager, 2 channels, Δλ=13 nm	Atmospheric structure , composition, dynamics
VIKA		IR spectrometer suite 1.05 – 1.65 µm (λ/Δλ = 20 000, nadir and occultations), 2.3 – 4.3 µm (λ/Δλ = 40 000, occultations)	Atmospheric composition, thermal structure (vertical), aerosols
IVOLGA		Hi-res heterodyne spectrometer in 1.58 and 2.05 µm CO2 lines (solar occultation)	Dynamics above clouds, meso- and thermosphere structure up to 180 km
MM-radiometer		5-channel radiometer (90, 50, 30, 20, 10 GHz); scanning antennae	Thermal structure 0-50 km, and H2SO4, SO2 etc. below the clouds
VMC		4 coaxial monitoring cameras: 365, 513, 965, 1000 nm	Dynamics at day side at different altitudes in upper clouds, UV absorber distribution, aerosol, surface emission at 1 µm
LIR (2)		Thermal IR camera in 8-12 µm	Thermal structure and dynamics in upper cloud layer
IR2R (2)		IR camera in 1.74, 2.02, 2.26 and 2.35 µm	Dynamics and morphologies of middle-to-lower atmosphere, variations of CO
X-range radio occultation		Radio occultation atmospheric experiment	Thermal structure and dynamics of the atmosphere, structure of the ionosphere
Plasma & EM Field complex		7 instruments to analyze plasma spectra and EM field fluctuations	Magnetosphere, ionosphere, escape rate, D/H, lightning
	ARIES-V	Wide angle ion spectrometer 10 eV – 30 k	Composition and intensity of escape ions; ion capture by solar wind; ionospheric clouds;
	ELSPEC	Electron spectrometer 10 eV – 10 keV	influence of solar activity on Venus plasma; transfer of momentum and energy from solar
	NPD	Neutral particles detector 50 eV – 3 keV	wind to ionosphere; interplanetary ion influx
	ASPECT-V	Ion and electron energy spectrometer	Magnetospheric structure; influence of solar activity on magnetosphere
	BMSV-V	Plasma spectrometer 100 eV – 5 keV	Monitoring of solar wind, its interaction with magnetosphere; turbulence of interplanetary medium at high frequencies
	FM-V	Magnetometer ±1000 nT, up to 32 Hz	Interaction of solar wind with magnetosphere; non-thermal atmospheric dissipation; magnetic properties of ionosphere; lightning activity
	GROZA	Radiowave analyzer 15-50 MHz	Detection of radiowave events from lightning activity

Operating orbit of the Venera-D orbiter was designed to satisfy interests and requirements expressed by the atmospheric and plasma science communities. A set of instruments (ARIES-V, ASPECT-V, ELSPEC) for charged particle measurements together with the neutral particle detector (NPD) will be capable to determine the effects of solar activity on Venus' atmosphere and ionosphere, characteristics of the loading of Solar wind plasma stream by planetary ions and parameters of multiscale ionospheric structures. Upstream information on SW parameters will be provided by the BMSV-V instrument and magnetic field variations, produced by various plasma interactions, will be analyzed by the FM-V

Table 6 Table of Venera –D Lander Module (LM) instruments under study. Table of scientific payload (phase A stage). (*): S– instrument requires surface sample; A– instrument requires atmospheric sample; [1]– instrument provisionally provided by Germany

Venera-D Lander Module (LM)			
Instrument	Description	Scientific goals	Sample (S = Surface; A = Atm.)
TVS-VD	Panoramic (5), descent (1) and microscopic (1) cameras	Characterization of local landforms	None
Surface Sampling System	Drill and vacuum pump system to acquire and deliver surface samples to XRD/XRF, MIMOS II, APXS-V,LMS		None
XRD/XRF	X-Ray diffraction and fluorescence spectrometer	Elemental and mineral composition of surface materials; possible detection of bounded water in the surface minerals	S
MIMOS II *(1)*	Moessbauer spectrometer	Surface mineralogical composition, iron phases	S
APXS-V	Analysis of scattering of α-particles from Cm-224 radioisotopic source on atmospheric molecules and surface material	Elemental composition of the atmosphere and surface	S
LMS	Laser mass spectrometer for surface samples	Surface elemental and isotope composition, local geological processes, surface-atmosphere interaction	S
AGNESSA	Active gamma and neutron spectrometer	Radioisotopes (K, U, Th), surface elemental composition to 0.5 m depth (Al, Mg, Fe, O, Na, Si, C)	
VCS	Gas chromatograph & mass spectrometer for atmospheric samples	Vertical profile (70-0 km) of atmospheric composition, trace & noble gasses, aerosol composition	A
ISKRA-V	IR multi-channel tunable laser absorption spectrometer, resolution 10-3 cm-1	Vertical profile (70-0 km) of major atmospheric constituents [SO2 CO CO2 OCS H2O PH3] and their isotopes [C, O, S, D/H]	A
DAVUS	Descent UV spectrometer 250 – 400 nm (Δλ=0.15 nm) for in situ atmosphere analysis	Vertical structure of SO2, SO from 70 to 0 km; unknown UV absorbers; ClO; aerosol extinction in clouds/hazes	None
MTK-V	Suite of sensors for key atmospheric parameters	Vertical profile (70-0 km) of main meteo parameters	None
VERBA	Actinometric spectroradiometer in VIS and IR (0.4-1.1, 1-1.8, 2.4, 3.7, 6.2, 8.7 μm)	Thermal balance below clouds, optical properties of aerosol, profile of H2O	None

magnetometer. Venus is expected to display potential lightning activity (e.g., Lorenz 2018) which will be studied on board by the GROZA radio wave analyzer.

7.3.2 Lander Science

A list of Lander Module science instruments under study is presented in Table 6. The key element of this package is the surface sampling system that will drill the surface to a certain depth and then, using the vacuum pump device, deliver the acquired samples to four analytical instruments:

- X-ray diffraction and fluorescence spectrometer (XRD/XRF) and a Mössbauer spectrometer (MIMOS II) for structural analysis;
- laser mass spectrometer (LMS) and an active APXS-V alpha-particle experiment for elemental analysis.

Gamma and neutron spectrometer with neutron activation AGNESSA will measure the naturally radioactive elements (K, U, Th) and the main rock-forming elements (Al, Mg, Fe, O, Na, Si, Ca) in the activation mode down to ∼0.5 m depth without requiring sampling.

Gas chromatograph (VCS) and Laser absorption spectrometer (ISKRA-V) will study atmospheric samples acquired during the Lander descent (∼50 min) and at the surface, already mentioned in Sect. 7.2.3. These two instruments feature their own sample preparation system and will use the pre-vacuumized Lander as a dump volume. The UV-spectrometer (DAVUS) will continuously study the composition of the atmosphere during the descent down to ∼10 km using an optical cell open to the atmosphere.

Fig. 24 Concept of the Venera-D Lander Module (LM) imaging system (5 + 2 units): FOV: 90° × 90°, frame size: 2048 × 2048 pix, angular resolution-2.5', Linear resolution @ 2.5 m is 2 mm (see Table 6)

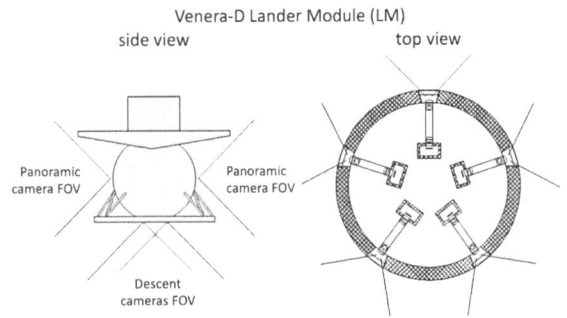

The camera system (TVS-VD) of the Venera-D Lander will provide a series of synoptic images from the descent camera (resolution from a few tens to meters per pixel), complete 360° surface panoramas (Fig. 24). A close-up microscopic camera imaging with resolution of ∼100 microns per pixel aims at observing the sampling spot at sub-millimeter scale to characterize its color, texture, grain sizes, traces of weathering, before and after its brushing/sawing/drilling.

7.4 Summary and Conclusions

The Venera-D mission is currently at the Phase A development stage and some mission parameters as well as characteristics of scientific instruments are to be clarified at the subsequent stages. The Aerial Platform will definitely be an important part of the Venera-D mission, currently its scientific payload as well as operation scenarios are being discussed. At the time of publication, the Venera-D launch is planned for the 2029 launch window, although the funding situation may result in a shift to the early part of the next decade.

The Venera-D science team looks forward to productive cooperation (including coordination between the different missions) with its VERITAS, DAVINCI and EnVision colleagues. Certainly, such cooperation could begin even earlier for the Indian Shukrayaan-1 mission to Venus (see Sect. 8), which will carry Russian instruments. The VeSCoor activity in development by ESA, NASA and other international partners could enhance the synergies between Venera-D and the other missions mentioned above.

8 Shukrayaan-1

Capitalizing on the successes of the Moon and Mars missions, with scientific payloads and instruments such as Synthetic Aperture Radar aboard Chandrayaan-1, 2, Mangalyaan-1 and numerous Earth orbiters, ISRO is considering to take a step towards exploring Venus (Antonita et al. 2022). ISRO has been a leader in the development of Synthetic Aperture Radar (SAR) instruments capable of polarimetric measurements. The hybrid-polarization mode in C-band SAR, RISAT-1 and full-polarimetric mode in Dual Frequency SAR in Chandrayaan-2 were implemented for the first time in any mission for observing Earth and Moon respectively.

The main goal of the planned Venus Orbiter Mission or Shukrayaan-1 is to undertake global mapping of the Venusian surface with a polarimetric SAR and to conduct the first penetration radar experiment to access shallow subsurface stratigraphy. The mission also plans to carry multiple instruments targeting the structure, composition, and dynamics of the atmosphere, and for investigating solar-wind interaction with the Venusian ionosphere. The final schedule and the science payload of the mission will be announced following approval by the Union Government of India.

A non-exhaustive list of Shukrayaan-1 contributions to Venus' evolution study includes the global characterization of current geologic activity, insights to the past by assessing vertical structure and stratigraphy of geological units including buried lava flows, global distribution of craters at better spatial resolution and including buried craters. The mission promises long-term monitoring of the atmosphere and clouds, potentially documenting their response to volcanic events, and detailed characterization of the contemporary escape.

8.1 Shukrayaan-1 Science Objectives

8.1.1 Investigation of the Surface Processes and Shallow Subsurface Stratigraphy

The surface and subsurface observations by Shukrayaan-1 will help in understanding geologic and resurfacing history, aeolian features on the surface, impact processes including detection of buried impact craters, vertical structure and stratigraphy of geological units including active volcanic hotspots and lava flows. The proposed VSAR instrument will carry out full polarimetric SAR observations of the surface of Venus which will provide highly accurate estimates of dielectric permittivity (Fung et al. 2010) and surface feature classifications (Ainsworth et al. 2009; Xie et al. 2015).

The first science objective aims to answer several outstanding questions by characterizing the contemporary and past Venus. The global characterization of volcanic and tectonic landforms will help in understanding how active Venus is at present, and assess its current geologic activity. The global mapping at better surface resolution will deliver a refined distribution of impact craters, quantifying their floor, rim morphology, and the parabola deposits. Characterization of prominent features like tessera terrain, unique geologic landforms such as Coronae and anomalous radar-bright regions found from Magellan using polarimetric surface decomposition techniques will be done.

Potentially, active volcanic hotspots and lava flows will be detected via observations of thermal emission in the near-IR atmospheric windows and brightness temperature measurements by SAR instrument in the radiometer mode. Also, variations of emissivity can be used to assess broad-scale surface composition of tessera terrains.

Venus' past will be addressed by stratigraphy of geological units, potential detection of buried impact craters and larger impact basins, particularly within the older tessera terrains. The subsurface observations will allow estimation of buried lava flows thickness and volume at different times.

8.1.2 Studying the Structure, Composition and Dynamics of the Atmosphere

This goal includes cloud dynamics and morphology, variability of SO_2, studying the superrotation, detection and understanding the role of lightning in the Venusian atmosphere.

By observing the clouds from UV to thermal IR, the mission will address the correlation of the cloud-top altitude with particles' microphysical properties, monitor the spatial and temporal variations of the planet's albedo, particularly in the UV-blue domain, use the cloud opacity to detect changes over various time scales. In the mesosphere, the aerosol will be characterized spectrally, in vertical and horizontal dimensions, including thin clouds and detached layers.

The SO_2 cycle, closely linked to primarily H_2SO_4 clouds and hazes, will be studied via IR occultation spectrometry and the cloud monitoring UV camera. The measurements will provide important insight into the ongoing chemical evolution, atmospheric dynamics, and as an indicator of possible geological activity.

The super-rotation, its long-term trends, the role of wave dynamics, associated planetary-scale cloud structures (the Y-shape) will be studied via the cloud features tracking in the UV, and characterizing the 3D thermal structure and wind field using finely-resolved near-IR CO_2 lines.

Assessing the lightning in Venus' atmosphere is important to understand its rate, strength and spread as well as its possible role in the cloud or atmosphere chemistry. Can lightning be hazardous for atmospheric probes? The detection is planned by analyzing the Hz - kHz electromagnetic spectrum, using the magnetometer and visible imaging data.

8.1.3 Investigation of Solar Wind Interaction with Venusian Ionosphere

This goal includes studying ionospheric dynamics and plasma waves, interaction with the Solar wind, detection of oxygen airglow and understanding its role, to study the upper atmosphere/ionosphere dynamics. Specific questions to address within this goal are: What makes the Venus ionosphere (V-1 and V-3) layers so elusive? What causes ion-holes in the Venus nighttime ionosphere? The measurements are to decipher how the ionospheric parameters depend on the Solar cycle and Solar Zenith Angle (SZA), detect the occasional meteor layer, determine spatio-temporal variation of the neutral composition and density of the upper thermosphere and exosphere, detect the ionosphere plasma waves and characterize escape rates of the Venusian upper atmosphere.

Detecting and observing the oxygen green (557.7 nm) and red (630 nm) lines would help to assess the day-to-night transport in the upper atmosphere, detect signatures of the atmospheric gravity waves. It would probe the atmosphere and ionosphere response to solar disturbances, characterize the sustenance of the nightside ionosphere, and detect large-scale ionospheric structures.

The outstanding questions to address for the Solar wind interaction goal are the loss of the upper atmosphere/ionosphere, the role of different escape mechanisms, characterization of plasma boundaries, assessing the variability of magnetosphere domains and processes responsible for atmospheric sputtering using Energetic Neutral Atom (ENA) measurements.

8.1.4 Astrobiology

The goal of indirect detection of possible bio molecules and life forms in the cloud deck is recognized. The measurement available is the potential PH_3 detection using the IR solar occultation spectrometer.

8.2 Shukrayaan-1 Mission Overview

Shukrayaan-1 is a highly-sophisticated Venus orbiter carrying up to 100 kg of science instruments and delivering up to 500 W of electric power. Shukrayaan-I will be launched on either GSLV Mk II or GSLV Mk III. ISRO has backup launch dates in 2026 and 2028 should it miss the 2024 opportunity. At the time of this publication, ISRO has not yet received approval from the Indian government for the Venus mission and that, as a result, the mission may be delayed until 2031.

Table 7 Key VSAR radar design and performance parameters

Shukrayaan-1 VSAR Parameters	
Instrument Parameters	Value
Platform Altitude Range (km)	200-600
Frequency (GHz)	3.2
Antenna azimuth × elevation dimensions (m)	5.8 × 2
Atmospheric Losses (dB)	−1.6
Polarization modes	Tri-Pol (HH, VV, VH) & Hybrid Pol
Mapping modes RFBW	7.5 MHz, 40 m with 15-20 km swath
Number of looks	6, 13
Radiometric resolution at 10 dB, SNR (dB)	1.2, 1.6
Peak power (W)	300
NES0 (dB)	≤ -25
Swath (km)	15-20

After the Earth-to-Venus cruise and the heliocentric phase, the spacecraft will perform trajectory maneuvers and final braking to reach intermediate Venus $500 \times 60{,}000$ km near-polar orbit. A lower orbit, suitable for SAR mapping, with the pericenter at 200 km and the apocenter at 500–600 km, will be achieved by aerobraking. The aerobraking phase will take 6–8 months with controlled pericenter altitude down to 130–140 km in the atmosphere.

For optimized science operations, the spacecraft features 2D orientable ($0°–360°$; $\pm 90°$) solar panels and a 2D gimbal high-gain antenna (Nigar 2022).

8.3 Shukrayaan-1 Science Payload

The Shukrayaan-1 orbiter will carry the VSAR instrument, for mapping the surface. Up to seventeen further potential science instruments have been shortlisted, including a subsurface sounder and a number of instruments targeting the atmosphere, clouds and the plasma environment (Nigar 2019).

8.3.1 VSAR on Shukrayaan-1

Global mapping by Magellan at resolution of 100–200 m was performed in single polarization (HH). Complete polarimetric data of selected sites are only available from earth-based radar experiments. The main instrument onboard, the S-Band Synthetic Aperture Radar (VSAR) targeting high-resolution (40 m), fully-polarimetric global mapping, would provide a major step in Venus surface studies, complementary to the VISAR and VenSAR operations and performance (Table 7). VSAR will be capable of polarimetric decomposition by retrieval of the complete scattering matrix (all four Stokes parameters). It is well known that the cross-pol channels (HV or VH) are more sensitive towards surface roughness and volumetric inhomogeneities within the medium and are not optimally used for dielectric constant estimation (Fung et al. 2010). Thus, co-polarization channels (HH&VV) will estimate surface dielectric properties with the highest accuracy.

Full polarimetric mode is challenging in terms of handling the radar ambiguities, power requirements and has double penalty in terms of radar coverage due to high data rate and reduced PRI. The radar ambiguities are controlled by partially processing azimuth bandwidth against a trade-off of a number of looks (Shah and Seth 2022). The wide antenna area of 5.8 m × 2 m is optimized to reduce the transmit power requirement which aids in reducing the volume and thermal management requirements of the payload. A trade-off of having antenna width of 2 m is that maximum possible swath would be 15-20 km, which either way was restricted due to the doubled PRF in full pol mode. However due to the slow rotation of Venus consecutive orbit passes are 11 km apart at the equator and will ensure continuity in coverage.

The possibility of cycle-to-cycle interferometry is foreseen. In the radiometer mode, SAR will estimate the brightness temperature with an accuracy better than 2 K. The science objectives of Shukrayaan-1 significantly differ and complement those of EnVision and VER-ITAS, and therefore the overall mission operations and VSAR capabilities, operations and performances are different when compared to VISAR and VenSAR. For each of the three orbiter missions, payload instruments were chosen to meet the broad spectrum of measurement requirements needed to support mission science investigations. Combining the data from all three missions would provide us with deep scientific insights into the geology and evolution of Venus with topographic, polarimetric and global information available at a high resolution.

8.3.2 Subsurface/Surface IR Emissivity

A low frequency (9–30 MHz) radar VARTISS will provide direct measurement of subsurface features. It will provide a vertical resolution of 10–25 m down to a depth of ∼1 km, in a similar technique as the subsurface radar (SRS) currently developed for the EnVision mission (see Sect. 6.3.2 and European Space Agency 2021). VARTISS utilizes three-wire 25-m dipole antennas. VSEAM (Surface Emissivity) instrument to provide planetary temperature and radiation emission data in the 1-μm atmospheric transparency window is also considered.

8.3.3 Atmosphere/Clouds Observations

The atmospheric package includes the Venus Thermal Camera (VTC), UV and visible camera (VCMC) for cloud Monitoring. Together with VSEAM, the imagers will provide cloud tracking at different altitude levels to measure the zonal and meridional wind fields, estimate the horizontal scale of the atmospheric waves.

The near-IR spectro-polarimeter (VASP) aims at cloud microphysics and measuring the cloud-top altitude.

The solar occultation package includes an occultation photometer (SPAV) in the near-IR and visible ranges targeting sub-micron aerosol vertical distribution in the upper clouds. Solar occultation IR spectrometers target detecting trace gasses and studying the mesosphere dynamics through vertical profiling of atmospheric composition and wind speed. Selected as two separate spectrometers, the high-resolution echelle (2.3–4.4 μm) Venus Infrared Atmospheric gas Linker (VIRAL) and a laser heterodyne (1.6 μm) IVOLGA, they are combined into a single two-channel Roscosmos-provided VIRAL. A very high-resolution IR spectrograph ($\lambda/\Delta\lambda > 20{,}000$), VIRAL will measure trace gases at the limb during solar occultations, improving performances of the VEx/SOIR spectrometer by a factor of 10. It will therefore be possible to gain 1 to 2 atmospheric scale heights in depth to explore the upper cloud region and contribute to study the trace species photochemistry at all latitudes at 6 AM and 6 PM local time.

The lightning sensor analyzing the Hz-kHz spectrum (LIVE) will detect lightning activity in coordination with VCMC camera and magnetometer observations.

8.3.4 Measurements Related to Solar Wind/Ionosphere Interactions

The measurements dedicated to Venus' plasma environment will target ionospheric parameters measurements, characterization of neutral and plasma particles, monitoring of radiation environment and solar X-rays. The planned instruments are VEDA (electron density) and RPA (Retarding Potential Analyser), to measure electron and ion density, composition and temperature, VISWAS for mass spectrometry of energetic neutral atoms and ions, VIPER to measure plasma-wave parameters using Langmuir probe, electric field analyzer and magnetometer.

To study the region connecting the neural and ionized atmosphere, a radio-occultation experiment RAVI with German participation and a high-throughput airglow imager NAVA to detect oxygen green and red lines are planned.

To characterize the environment and solar wind, two instruments, VREM to monitor the charged particles (100 keV–100 MeV) and SSXS spectrometer to measure Solar soft X-rays (1–15 keV) are foreseen.

Finally, VODEX, Venus Orbiter Dust EXperiment, to detect interplanetary dust during the cruise phase and at Venus orbits, is being considered.

8.4 Summary and Conclusions

Solar system studies have seen a remarkable growth in the last few decades, due to advances in space technology, observational capabilities and computational technologies. This has enhanced our knowledge and understanding of the diversity of complex processes across the Solar system. It is quite interesting to find clues as to how the planetary systems might have originated and evolved, and how they are different and similar to each other. In this context, the first ever planet explored by humankind for the search of life in the solar system is Venus, our nearest neighbor. Despite many missions including flybys, orbiter and lander, large gaps remain today in our understanding of Venus on its formation, spin, surface evolution, runaway greenhouse phenomenon, super rotation of its atmosphere, its evolution and interaction with solar wind, etc. In this context an orbiter mission to Venus is being considered by ISRO, with the above science objectives, to further improve our understanding of resurfacing processes, neutral atmospheric and ionospheric processes, influence of Sun on Venus atmosphere and ionosphere in particular to produce highest resolution topography, to map sub-surface of Venus, to detect lightning and airglow with better techniques and improved resolutions.

9 Overview of Future Mission Concepts and Science Investigations Addressing Venus Evolution Through Time

In the following Sects. 10-12, we address future mission concepts and measurements that require further technology development beyond the current framework of selected missions, as well as the synergies between these mission concepts, ground-based and space-based observatories and facilities, laboratory measurements, and future algorithmic or modeling activities that pave the way for the development of a Venus program that extends into the 2040s (Wilson et al. 2022; Limaye and Garvin 2023). Continued development of planetary radar technologies, while applicable to many planetary missions, is particularly important for Venus because of the opacity of Venus' cloudy atmosphere to optical imagers. We also discuss complementarities between Venus mission concepts that target specific outcomes that reveal Venus evolution through time, as summarized in Sect. 2, and with non-Venus missions (e.g., exoplanet observatories).

Section 10, Future Mission Concepts and Measurements, addresses key areas and investigations at Venus not covered by the fleet of missions under development described in Sects. 4-8. They include:

Future Surface Investigations. - Since all previous probes have landed in basaltic plains, a new lander in the tesserae highlands, thought to represent the oldest terrains on Venus, is needed to perform a detailed, comprehensive analysis of Venusian surface materials to understand Venus' geologic history, tectonic style, mineralogy, and composition (Sect. 10.1). Meteorology and seismometry, on the other hand, require measurements over months or years. Rather than try to use silicon electronics with associated cooling and power systems, such long-duration lander measurements can be implemented using high- temperature electronics of silicon carbide, gallium nitride or other wide band-gap materials (Wilson et al. 2022). As well as their own scientific investigations, such long-term landers would serve as technology demonstrators developing technology in preparation for mobile surface exploration, which is a post- 2050 Venus exploration goal. Note that the DAVINCI probe will touch-down in tesserae highlands providing ground-truth to enable future landings in complex ridged terrains.

Aerial Platforms; Long-Term Atmospheric **in-situ** *Measurements*. - Longer duration measurements in the cloud layer could be achieved by using balloons, whether these are constant- or variable-altitude balloons (Sect. 10.2). Balloons offer a lifetime of weeks in arguably the most habitable environment found outside Earth, where temperature is around 20 degrees C, pressure is 0.5 bar, and there is ample sunlight and liquid water in the clouds (albeit mixed with sulfuric acid). Such a mission would explore the coupled chemical, dynamical, and radiative processes at work in this critical part of the long-term evolution of the Venusian atmosphere, as well as further investigations of past and present habitability and astrobiological aspects of the cloud layers. As summarized in Sect. 2 (Gillmann et al. 2022, this collection; Avice et al. 2022, this collection; Westall et al. 2023, this collection) volatile elements have a strong influence on the evolutionary pathways of rocky bodies and are critical for understanding the evolution of the Solar System. Because Venus experienced a different volatile element history than Earth, it provides the only accessible example of an end state for habitable Earth-size planets.

Future Orbital or Suborbital Investigations. - Future orbital investigations include longer-term observations of variable phenomena such as the atmospheric signature of surface events (e.g., large Venusian quakes and their potential airglow signature) and the collection of data on the sources, propagation, and dissipation of gravity waves in the Venusian atmosphere, which will require high-frequency imaging of the entire disk (Sect. 10.3). In addition, various concepts for skimmer probes of the upper atmosphere, diving below the Venusian homopause, and returning samples to Earth have been proposed recently (Shibata et al. 2017; Sotin et al. 2018a). Extremely high RF bandwidth radars operating at S or L-bands, could achieve meter-scale polarimetric SAR imaging of key regions after VERITAS and EnVision, building off of planned L-band systems such as NASA-ISRO's NISAR mission and planned Mars-orbital L-band measurements (JAXA, CSA, ASI, and others).

Sections 11 and 12 address the importance of ground-based, Earth orbit-based, observations and laboratory and modeling activities in support of Venus' evolution through time. They include:

Ground-Based and Space-Based Observatories. - Ground-based observations, or observations from Earth orbit by space-based observatories, complement the space-based instrumentation at Venus by providing long temporal baselines of variable phenomena of interest to the long-term evolution of Venus, such as ground-based radar images of the Venusian surface (Campbell et al. 2017), thermal infrared imaging spectroscopy to monitor SO_2 and H_2O variability possibly related to present-day volcanic activity (Encrenaz et al. 2019), millimeter and submillimeter wave observations, observations of key trace gases, or extended series of ground-based dynamical observations of wind speeds and gravity waves (generated by both convection and topography) that support Venusian general circulation models and present-day dynamical coupling between the lower atmosphere and the surface (Sect. 11.1). Section 11.2 focuses on space-based platforms and the synergies between Venus and exoplanetary observations such as TESS, JWST, CHEOPS, and PLATO, as well as next-generation ground-based facilities such as the ELT (Way et al. 2023, this collection).

Laboratory and Modeling Efforts. - Experimental facilities on Venus are critical to advancing our understanding of this planet in the coming decades (Glaze et al. 2018; Santos et al. 2021, see Sect. 12). Several scientific investigations that address the previously discussed scientific themes and knowledge gaps related to how and when Venus' evolutionary path diverged from Earth are enabled by experimental facilities. Section 12.1 highlights a selection of current facilities and experiments that are particularly relevant to understanding ancient Venus and the long-term evolution of the planet, such as geochemical experiments (Sect. 12.1.1), mission calibration libraries (Sect. 12.1.2), rheology experiments (Sect. 12.1.3). Section 12.2 focuses more on progress and future directions in modeling efforts, particularly in the rapidly expanding capabilities for characterizing the chemical and dynamical interactions between the surface and the atmosphere (which are key to understanding the processes driving the long-term evolution of the spin rate); radiative and radiative-convective models for determining the long-term climate evolution. Newly proposed laboratory capabilities in the USA and Europe (2023 and beyond) could facilitate studies that refine the upcoming observations to improve interpretations.

Finally, Sect. 13, Summary and Conclusions, summarizes the key findings of Sects. 3-12, highlighting which key questions about Venusian evolution through time will be answered convincingly by current and next-generation mission concepts and future investigations; how current limitations in our knowledge of Venus affect current and future exoplanetary science and modeling of the evolution of rocky, Earth-sized exoplanets; and how the study of Venus' evolutionary history informs its habitability history through time.

10 Future Mission Concepts and Measurements

The data acquired with the VERITAS, DAVINCI, and EnVision missions from the end of this decade, in addition to other missions concepts described in Sects. 7-8, will fundamentally improve our understanding of the planet's long term history, current activity and evolutionary path. Yet even with the discoveries that await those missions, compelling science questions will likely remain—such as those regarding the chemical and physical cycles of the atmosphere, the interactions between that atmosphere and the surface, and the make-up and structure of the planet itself. Although new missions may be proposed in the near-term using capabilities currently in development, investments in future technologies and new techniques will enable both long-term surface science and missions that take advantage of mobility in the surface, near-surface, and atmospheric environments (Fig. 25).

10.1 Future Surface Investigations

One of the fundamental measurements that is needed is the bulk elemental composition and mineralogy of the surface from key locations, especially tesserae. Tesserae are thought to be

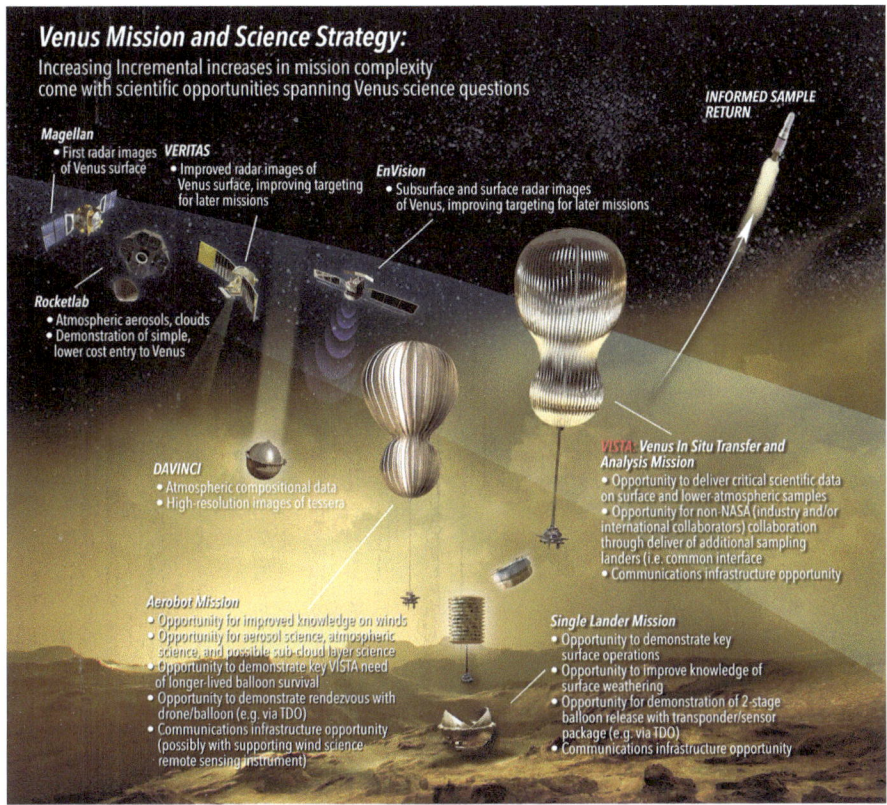

Venus Mission and Science Strategy:
Increasing Incremental increases in mission complexity
come with scientific opportunities spanning Venus science questions

INFORMED SAMPLE RETURN

Magellan
• First radar images of Venus surface

VERITAS
• Improved radar images of Venus surface, improving targeting for later missions

EnVision
• Subsurface and surface radar images of Venus, improving targeting for later missions

Rocketlab
• Atmospheric aerosols, clouds
• Demonstration of simple, lower cost entry to Venus

DAVINCI
• Atmospheric compositional data
• High-resolution images of tessera

VISTA: *Venus In Situ Transfer and Analysis Mission*
• Opportunity to deliver critical scientific data on surface and lower-atmospheric samples
• Opportunity for non-NASA (industry and/or international collaborators) collaboration through deliver of additional sampling landers (i.e. common interface)
• Communications infrastructure opportunity

Aerobot Mission
• Opportunity for improved knowledge on winds
• Opportunity for aerosol science, atmospheric science, and possible sub-cloud layer science
• Opportunity to demonstrate key VISTA need of longer-lived balloon survival
• Opportunity to demonstrate rendezvous with drone/balloon (e.g. via TDO)
• Communications infrastructure opportunity (possibly with supporting wind science remote sensing instrument)

Single Lander Mission
• Opportunity to demonstrate key surface operations
• Opportunity to improve knowledge of surface weathering
• Opportunity for demonstration of 2-stage balloon release with transponder/sensor package (e.g. via TDO)
• Communications infrastructure opportunity

Fig. 25 The Venus *In Situ* Transfer and Analysis mission concept (VISTA) provides an opportunity to obtain measurements that cannot be obtained by a simple, short-term mission to Venus (Izenberg et al. 2023; see also Sect. 10.2.1). This illustration is provided as an example. Data acquired with the VERITAS, DAVINCI, and EnVision missions from the end of this decade will fundamentally improve our understanding of the planet's long-term history, current activity and evolutionary path (top left). Although new missions may be proposed in the near-term using capabilities currently in development, investments in future technologies and new techniques will enable both long-term surface science and missions that take advantage of mobility in the surface, near-surface, and atmospheric environments. Additional future Venus exploration vehicles may include: (1) orbital platforms such as geophysics orbiters for atmospheric remote sensing, and in-situ orbital sensing; aerial or cloud-level platforms; and (3) surface platforms including short-lived landers, long-lived landers, and, ultimately, mobile surface assets. credit: Keck Institute for Space Studies/Noam Izenberg/Chuck Carter

older than the regional plains, and as such may retain evidence of an earlier epoch prior to volcanic resurfacing, including evidence of different climate and weathering (Gilmore et al. 2023, this collection). The ability to perform detailed, comprehensive analysis of Venus surface materials would be game changing. Of course, landers (Fig. 25) must cope with the incredibly harsh conditions of $\sim$470 °C at the surface of Venus. The Venera landers used only thermal inertia and thermal insulation to keep a central electronics compartment cool that, together with limited radio relay capabilities, only allowed for operation times of order an hour (see e.g., Vorontsov et al. 2011). Modern landers using the same, brute-force thermal approach could last for several hours but certainly not more than, say, one Earth day. Even so, a huge amount of surface mineralogical and petrological science could be achieved in that time.

10.1.1 Surface Images and Mineralogy

The principal science payload of such a lander would include surface imagers and non-contact chemical and mineralogical sensors such as a gamma-ray spectrometer with neutron activation, capable of measuring elemental abundances of U, Th, K, Si, Fe, Al, Ca, Mg, Mn, Cl (Mitrofanov et al. 2010) and/or Raman/LIBS (Clegg et al. 2011). Although the extreme Venus surface temperature (740 K) and atmospheric pressure (93 atm) provide a challenging environment for surface geochemical and mineralogical investigations, laboratory Raman experiments have been conducted under supercritical CO_2 Venus surface conditions, involving single-mineral and mixed-mineral samples (Clegg et al. 2016). The inclusion of surface sample ingestion via a drill/grinder/scoop would enable additional analysis techniques such as mass spectroscopy and X-ray fluorescence (XRF) spectroscopy, but would equally require important, continued technology development and verification.

Gamma- and XRF-based spectroscopic measurements were acquired by the Venera and VeGa landers, but modern equivalents of these instruments would provide much improved measurement accuracy. Moreover, carrying out compositional analyses in a tessera region (none of which has been sampled before) would reveal whether these terrains are chemically distinct from the lava plains where previous analyses have been conducted (e.g., Gilmore et al. 2023, this collection).

Collecting imaging data from the surface is not straightforward (e.g., Garvin et al. 1984), as silicon-based detectors typically used in virtually all modern imagers will not work at Venus surface temperatures. For short-lived landers, one can use silicon imaging chips inside insulated housings (e.g., Kremic et al. 2020). For prolonged operation at the surface, however, imagers capable of sustained operations at high temperatures are needed. Such imaging capability would be essential not only for surface science but for eventual mobile platforms (e.g., rover missions).

10.1.2 Surface Geophysical and Meteorological Data

Long-lived landers would prove valuable in numerous other ways, as well. For instance, long-duration surface operations would allow acquiring essential seismological, heat flow, and meteorological data. But the technological barriers to such long-lived surface missions are twofold, and formidable: (1) the high-temperature environment of the Venus surface, which is too hot for silicon electronics; and (2) the lack of sunlight at the surface, making solar-based power systems difficult of even unviable. As has been discussed by Kremic et al. (2020), Kremic and Hunter (2021), and Wilson et al. (2022), recent advances in high-temperature electronics have made long-duration uncooled landers an exciting possibility that can be explored in the coming decades—but doing so requires continued investment in the development of the electronics, in their packaging, and in their environmental qualification under Venus conditions. Power for long-duration landers on the surface could come from primary molten salt batteries, for at least a first generation of such long-lived landers. Second-generation long-life landers could be powered by RTGs or even wind power—again, both of which would require technology development. Yet, long-lived landers would provide not only essential meteorological and geophysical measurements but also would serve as essential precursors for post-2050 surface missions including more capable seismometry stations but also for eventual surface rovers—a technology that we envisage as being tractable beyond 2050.

Of these measurements, meteorological measurements are the most straightforward to make—indeed, Soviet Venera & VeGa landers made measurements of pressure, temperature, and even wind speed during their brief operation at the surface. However, long-duration operations, i.e., those lasting for at least one Venus solar day (118 Earth days) would be needed in order to understand the temporal evolution of surface winds. Recent modeling suggests that the deep atmosphere of Venus, far from being still like the bottom of a pond, exhibits complex diurnal patterns, mesoscale circulations, and slope winds, all driven by the small day-night temperature difference estimated to be of order 1–2 K (Lebonnois et al. 2018; Lefèvre 2022; Lefèvre et al. 2022). Investigating this deep atmosphere circulation from the surface is important for understanding the evolution of Venus; to give two examples, deep circulation governs the transport of magmatic volatiles (or volcanic ash) into the atmosphere (Wilson et al. 2022, this collection) as well as the aeolian and sedimentation processes that shape the surface of Venus (Carter et al. 2023, this collection).

Long-term meteorological measurements on the surface would require sensors capable of operating at ambient temperatures, such as those proposed by Kremic et al. (2020), Kremic and Hunter (2021). As well as the direct relevance to Venus evolution as described here, such meteorological measurement would provide vital information for the next generation of long-lived surface stations: the viability of wind power for surface stations depends, for example, on knowing the reliability of wind speed and direction. The viability of seismometry at the surface also requires characterization of atmospheric turbulence at the surface so that atmospheric perturbations can be removed from seismic data (Spiga et al. 2018).

10.1.3 Surface Heat Flow Measurements

Surface heat flow—that is, the energy flow from the interior to and across the lithosphere—is an important geophysical parameter that can be used to place constraints on estimates of the chemistry of the interior as well as models of the global tectonics (compare Rolf et al. 2022; Gillmann et al. 2022, this collection; and e.g., Breuer et al. 2022). If we assume that Venus formed with the same concentration of heat producing elements as Earth, and further that it has lost heat at a similar rate, an estimate can be made using the Earth's value which is found to be 45–47 TW (Davies and Davies 2010; Jaupart et al. 2015). Given the mass ratio of Venus and Earth, the corresponding global value range for Venus is 37–38 TW or 82 mW/m^2. However, this scaling argument is overly simplified: numerous estimates of heat flow from surface deformation suggest that the global average is higher than this value; O'Rourke and Smrekar (2018) derived elastic lithosphere thicknesses from Magellan data at coronae to estimate heat flow values of up to 95 mW/m^2 or greater Ruiz (2007); and more recently, Maia and Wieczorek (2022) estimate several 100 mW/m^2 for tessera terrains at the time of formation. New elastic thickness estimates from flexural bending (Smrekar et al. 2022b) correlate well with global elastic thickness estimates from admittance (Anderson and Smrekar 2006). This correlation indicates that $\sim$40% of Venus has high heat flow today, with more low-moderate values elsewhere. Very high values are consistent with active regions (Smrekar et al. 2022b). Since such estimates have significant errors, and require assumed or under-constrained parameters, in-situ heat flow would be a very valuable ground truth.

However, measuring the heat flow *in situ* remains a long-term challenge. On Earth and Mars, daily and seasonal temperature perturbations penetrate for meters into the near-surface and disturb the diurnal / seasonal thermal gradient from which the heat flow is usually calculated (if, in addition, the thermal conductivity is measured, see Grott et al. 2021). On Venus, diurnal temperature variations are expected to be small in amplitude (of order 1 K: Lebonnois et al. 2018) but even this small diurnal temperature difference can lead to heat flow disturbances of the order of 1 W/m^2, an order of magnitude larger than the $\sim$100 mW/m^2 internal heat fluxes when assuming the thermal properties of basalt. The thermal skin depth through which the diurnal perturbation decreases by 1/e can then still be a few meters, depending on the thermal diffusivity. At 60 mW/m^2, the temperature difference per unit value of thermal conductivity (W/m K) per m is 60 mK, and thus within the accessible range of space-qualified temperature sensors, assuming that such technology is otherwise inured to the harsh thermal environment at Venus' surface. Although there is a case for sediments on Venus (Carter et al. 2023, this collection), current radar data indicate rock surfaces over most of the planet.

Heat flow is typically measured in a borehole, as done on Earth; and as was done using holes drilled by Apollo astronauts on the Moon's landing sites. Technologies to install sensors at depth include drilling (e.g., Vago et al. 2017), or penetrators or hammering mechanisms such as on the Rosetta or InSight missions (Spohn et al. 2007, 2018, 2022); or pneumatically, as has been studied for the Moon (Zacny et al. 2013). However, drilling into basalt for constraining the heat flow value at Venus is very challenging. Alternatively, the temperature gradient and thermal conductivity could be measured with the "thermal blanket" technique used for heat flow studies at ocean bottoms (e.g., Johnson et al. 2010), which has been considered for Venus by Smrekar et al. (2014) and Kremic et al. (2020). This technique would see a low-thermal-conductivity blanket or plate, equipped with temperature sensors to measure the warming due to the heat flow from below, placed directly onto the ground. The device would need to be large enough to avoid edge effects from lateral heat conduction, and would be soft or flexible enough to smoothly adapt to uneven ground. On Earth, those blankets typically cover a square meter, and are particularly useful in rocky areas lacking a regolith coverage where a drill or penetrator could not be relied upon to work.

As a further alternative to measuring the thermal conductivity and temperature gradient a miniaturized heat flow probe has recently been proposed (Dominguez et al. 2020). It measures the heat flow across the cm-scale probe after thermal equilibrium is attained. The probe still needs to be emplaced below the thermal skin depth but that might be less demanding than installing a suit of temperature sensors.

10.1.4 Seismic Investigations

Seismology is the preeminent methodology for studying the structure and composition of a planet's interior. The behavior of seismic waves traveling through planetary bodies provides constraints on interior composition, compositional boundaries and transitions, and the state of interior materials (e.g., fluid versus solid, hot versus cold, porous or not). Seismology enables the study of several high-scientific priority questions of Venus. The similar size of Venus and Earth, their similar overall surface ages, and recent evidence of ongoing volcanic activity on Venus (Herrick and Hensley 2023) suggest that it should have a level of seismic activity comparable to Earth's, or at least much higher than the Moon or Mars. The nature of Venusian seismic activity, its level, and where earthquakes are occurring on Venus could enable us to distinguish between big-picture geodynamic hypotheses that predict differing versions of current geological activity. Constraints on models for ancient plate tectonics, which could be derived from future seismic measurements in combination with a more detailed exploration of surface features such as the tesserae and their compositions, would shed light into whether Venus had an early habitable period.

A workshop in 2014 (Stevenson et al. 2015) brought together several seismologists to discuss Venus seismicity. Some participants felt that the overall higher temperatures of the uppermost crust on Venus might make most fault movement aseismic, although the consensus at the workshop was that these elevated surface temperatures probably play a minimal role in affecting overall seismicity. The absence of water in the near surface of Venus could also potentially affect the nature of seismicity, as could the higher surface air pressure, but at present these are mostly unstudied and poorly understand aspect of Venus seismicity.

Because of those high surface temperatures on Venus and the limited solar energy reaching the surface relative to the power needed to transmit data, the option of placing solar-powered seismometers on the surface for an extended period—as has been done on Mars and the Moon—is not currently a viable option for Venus. The longest that a surface lander with conventional electronics has lasted on the Venusian surface is about two hours. However, to accomplish even the most basic goal of determining seismicity levels requires the ability to operate over a period of at least several Earth days. Although nuclear power might be viable for either active cooling or simply as a long-lived power source, regulatory and cost considerations are such that they are not likely to be used for seismology on the Venus surface for the foreseeable future (Venus Exploration Analysis Group/VEXAG 2019).

Over the past several years, research and development of high-temperature electronics has advanced to the point where a seismometer that can operate under Venus ambient conditions using battery power has become technically feasible (Kremic et al. 2020). Even so, the constraints on operation of a first-generation seismometer on Venus will be severe. The battery will likely be capable of enabling the seismometer to operate for a period of a handful of months (Glass et al. 2020; Kremic et al. 2020), but transmission of data from the surface to a orbiting relay spacecraft will be power intensive, such that less than ten hours of data will likely be able to be transmitted from the surface. Because data transmission rate depends in part on transmitter power, it may be the case that the frequency and dynamic range of the instrument will be limited to less than its intrinsic capability. Furthermore, although limited computer memory is being developed for high-temperature devices, power consumption will be high, such that storing even small amounts of data for later transmission may not be possible for such first-generation long-endurance landers. A primary mitigation strategy will involve designing a low-memory algorithm that triggers transmission during earthquakes and avoids transmission during wind and other noise events (Tian et al. 2023).

10.2 Future Aerial Investigations

Venus' clouds represent an important exploration target for understanding Venus evolution, whether for characterizing past and present habitability (Westall et al. 2023, this collection), for placing bounds on tectonic and volcanic activity through infrasound measurements (Ghail et al. 2023, this collection), for searching for atmospheric signs of volcanic activity (Wilson et al. 2023, this collection), for conducting long-duration measurements of noble gas isotopic abundances (Avice et al. 2022, this collection), or for mapping felsic surface composition at higher resolution than is possible from orbit (Gilmore et al. 2023, this collection). Some of these goals can be addressed by descent probe missions: notably, the last three of the above five points will be addressed by DAVINCI (2029 launch) and Venera-D during their descent phases, as has been discussed in Sects. 5 and 7 above. The Venus Life Finder Mission "Morning Star" (Seager et al. 2021) considers a small entry probe that would use an ultraviolet autofluorescence backscatter nephelometer to characterize cloud particles and search for biosignatures as it passes through the clouds.

10.2.1 Aerial Platform Technologies and Operations

Descent probes offer an atmospheric investigation time of, at most, 10-100 of minutes. This short duration can be extended to weeks or months, and to a wide range of latitude, longitudes, altitudes and solar times, by using balloons. The deployment of two small balloons at 55 km altitude, in the heart of the main convective cloud layer, was successfully demonstrated by the Soviet VeGa mission in 1985 (Sagdeev et al. 1986a,b; Linkin et al. 1986; Blamont et al. 1986; Preston et al. 1986). At this altitude, the ambient temperature is a comfortable 20 °C and the pressure is 0.5 atm. The main environmental hazard is the concentrated sulfuric acid that makes up the cloud particles (O'Rourke et al. 2023, this collection; Wilson et al. 2023, this collection); however, this environmental threat can be mitigated by choosing appropriate acid-resistant materials for external surfaces. Balloon-borne platforms at this altitude can take advantage of the fast, super-rotating winds that will carry the spacecraft all the way around the planet in a week or less (depending on latitude and altitude), negating the need for horizontal propulsion with motors. A cloud-level aerobot is an ideal platform for studying interlinked dynamical chemical and radiative cloud-level processes (Fig. 25). It also offers a thermally stable, long-lived platform from which measurements of noble gas abundances and isotopic ratios can be carefully carried out and repeated if necessary—in contrast to a descent probe, which offers one chance for making this measurement, in a rapidly changing thermal environment, at a single location, and at a single time.

By the use of a pumping system to alter a balloon's internal pressure, aerial platforms can explore a range of altitudes: (1) One key altitude range to target is the so-called 'habitable zone' of 54–58 km, corresponding approximately to temperatures of 0–40 °C. This region is not only of greatest astrobiological interest but is also at the heart of the 50–60 km convective cloud zone, and also offers the most benign conditions from the standpoint of safely operating a Venus aerial platform (see Sect. 10.2.2). (2) Operation above 60 km, in the convectively stable upper clouds, would be optimal for photochemical processes and identification of the UV absorber, but the low atmospheric density leads to a relatively small mass fraction for scientific payload. (3) Operation below the main cloud deck has been proposed by Japanese researchers with a primary goal of establishing wind fields below the clouds, but temperatures here exceed 100 °C. But operating an aerial platform between 52 and 62 km would enable detailed investigations of the physical and chemical cycles operating in both the convective cloud and the upper (convectively stable) clouds, as well as how volatiles (and even perhaps ash from volcanic eruptions) are transported to, and through, these parts of the atmosphere.

Balloon-borne aerial platforms could also plausibly be used to image the surface, if they are able to descend—or deploy an imaging system—to below the global cloud layer, at an altitude of about 38 km. This capability would be restricted to near IR imagery, since the atmosphere would be optically thick with respect to Rayleigh scattering at shorter wavelengths. Once again, this environment poses considerable thermal challenges to an aerial platform or instrument, and requires continued work to mature technologies capable of functioning at those sub-cloud altitudes. An intriguing means for revealing winds in the lower atmosphere is to use passive balloons, reflective to radio waves, which could be tracked by radar from orbit. It also bears noting that balloons are not the only type of aerial-based platform that could be utilized at Venus: studies of powered-flight vehicles such as Northrop Grumman's Venus Atmospheric Maneuverable Platform (VAMP) concept is one such example of a non-balloon-based Venus aerial platform (Lee et al. 2015; Warwick et al. 2017).

10.2.2 Aeronomy and Surface-Atmosphere Interactions

By being carried passively along with the prevailing winds at a given altitude, a balloon-borne instrument suite could traverse the entire planet longitudinally in less than one Earth week. There is, at present, no prospect for any remotely comparable level of mobility for a surface-based platform.

As discussed in Sect. 10.2.1, balloons are ideally suited for exploring Venus because they can operate at altitudes where pressures and temperatures are far more benign than at the surface. An airborne instrumental payload operating at ~55 km can take advantage of the fast super-rotating winds which will carry the balloon all the way around the planet in a week or less, depending on latitude and altitude (Limaye and Garvin 2023). A cloud-level balloon is an ideal platform for studying interlinked dynamical chemical and radiative cloud-level processes. It also offers a thermally stable long-lived platform from which measurements of noble gas abundances and isotopic ratios can be carefully carried out and repeated if necessary.

The mechanical couplings between the solid and atmosphere parts of the planet are sixty times better than on Earth (Garcia et al. 2005), with almost 6% of the energy of a quake radiated in the atmosphere (Lognonné et al. 2015). It has previously been hypothesized that ground motion on Venus could be detected and characterized using infrasonic waves (or infrasound, pressure waves with a frequency less than 20 Hz) generated by quakes and volcanic activity through coupling between the solid planet and the atmosphere. Infrasound is known to travel large distances from the originating event and could be characterized using barometers suspended from balloons at approximately 60-km altitude on Venus, where the temperature and pressure are more Earthlike, and much longer mission lifetimes compared to surface missions can be achieved. Work has already been undertaken to explore the prospect of recording acoustic infrasound with balloons that is generated by ground movements. For example, various low-altitude experiments have been conducted with pressure and accelerometer sensors using active sources (Krishnamoorthy et al. 2018, 2019; Garcia et al. 2021), demonstrating that pressure sensors and accelerometers are capable of detecting the acoustic waves generated by ground movements above the seismic source, but also those generated by seismic surface waves propagating below the balloon.

Further, it was demonstrated recently that pressure records in the Earth's stratosphere can record seismic surface waves from small- and large-magnitude quakes (Brissaud et al. 2021; Garcia et al. 2022). These studies clearly show that the dispersion of seismic surface waves can be observed in pressure records, and that key quake parameters (magnitude, distance, etc.) can be recovered. These data also show that the response of the balloon system to acoustic forcing has an imprint on the pressure records that should be better modeled and understood for application to Venus (Bowman and Krishnamoorthy 2021), as well as Earth and even other atmosphere-bearing worlds such as Mars and Titan. Besides the detection of seismic events, the measurements of acoustic waves in Venus atmosphere can also be used to investigate volcanic eruptions (Byrne and Krishnamoorthy 2020), bolide events, infrasound from interactions between wind and topography (Hupe 2018; Poler et al. 2020), and even potential thunder signals.

The discrete investigations described above could be combined on a single aerial platform, given that the middle atmosphere offers a unique vantage point for exploring both the surface and the upper atmosphere and its interaction with the space environment. Indeed, a combined focus on atmospheric chemistry and physics, aeronomy, and surface–atmosphere interactions forms the basis of Phantom, a mission concept under development for the NASA New Frontiers 5 competition (Byrne 2022).

10.3 Future Orbital or Sub-Orbital Investigations

10.3.1 High-Frequency Imaging of the Atmosphere and Thermosphere

Atmospheric response to geological and atmospheric events (e.g., volcano eruption, quake, storm) can be of short duration and require a large field of view and a high sampling rate to be measured accurately. The VAMOS (Venus Airglow Measurements and Orbiter for Seismicity) mission concept was designed to cover this need (Didion et al. 2018; Sutin et al. 2018). External events created by solar energy injection in the Venus system also have a short duration. A mission capable of performing high-frequency imaging of the thermosphere would allow recovery of these events, which may play an important part in the atmospheric escape of Venus. In addition, the detection of meteor entry tracks from such high rates of imaging would allow us to infer the meteor flux in the inner Solar System, and the seeding of Venus by external sources. Finally, high-frequency imaging of the atmosphere/thermosphere is required to better understand the sources, propagation, and dissipation of gravity waves in the Venus atmosphere. These waves may be key to explaining the long-term evolution of Venus' rotation dynamics by providing a way to transfer mechanical energy from the solid surface to the atmosphere.

All these science objectives require an imaging of the planet full disk with a horizontal resolution on the order of 5–10 km and a sampling rate around 1 second. Due to the large amount of data generated, onboard data processing and selection must be implemented, possibly requiring machine-learning methods trained on the ground but applied onboard. Other orbital observations potentially involving atmospheric lidar remote sensing (e.g., as in CALIPSO/Cloud-Aerosol Lidar and Infrared Pathfinder Satellite Observations in Earth orbit: e.g., Winker et al. 2008, 2010) could add to these measurements with both spectroscopy but also in terms of vertical structure.

10.3.2 Collection and Return to Earth of an Atmospheric Sample

The elemental and isotopic compositions of noble gasses and stable isotopes in the atmosphere of Venus hold clues to the origin and evolution of the entire planet. Past space missions managed to get a first glance of the composition of volatile elements in the atmosphere, and planned investigations (DAVINCI, Venera-D) will greatly improve our knowledge (see this review, Sect. 2, Sect. 2.2.3; Sect. 5, Sect. 5.1; Sect. 7, Sect. 7.1; see also review by Avice et al. 2022, this collection). However, some rare isotopes of noble gasses are important for identifying the source of volatile elements to Venus but are extremely challenging to detect and measure (e.g., ^{78,80}Kr or ^{124,126}Xe). Measurements of other, more abundant isotopes are also greatly complicated by the presence of isobaric interferences (e.g., CO_2^{++} ions interfering with $^{22}Ne^+$ ions). A direct way to draw up the inventory of volatile elements in the atmosphere of Venus would be to send a probe, skimming through the atmosphere of Venus below the homopause < 120 km (Mahieux et al. 2012; Sotin et al. 2018a,b; Rabinovitch et al. 2019) to collect atmospheric samples, before possibly returning them back to Earth (e.g., Shibata et al. 2017). Such samples would be measured with state-of-the-art instruments in laboratories on Earth, enabling us to characterize the composition of the Venus atmosphere to unprecedented accuracy and precision. Although such investigation will likely allow comparative planetology of Earth, Venus, and Mars on a similar basis, sampling and returning to Earth an atmospheric sample of Venus remains challenging. In the configuration of a skimmer probe, gas would be sampled at high velocity (> 10 km/s), behind a shock wave and the extent of modifications of the molecular, elemental, and isotopic composition of the atmospheric sample remains to be fully studied. After returning the sample to Earth, innovative curating techniques will have to be developed to preserve those invaluable samples for extended periods of time, in the manner of samples from the lunar Apollo program that continue to be assessed to this day.

10.3.3 Venus Lagrange Points Mission

A Venus Lagrange Points (LP) Mission, in which two light spacecraft using heritage, technological readiness & development of Akatsuki, Hayabusa-2 and Destiny+ spacecraft systems are placed in orbit around the Sun-Venus Lagrange points, is under consideration by the Japanese Space Agency's Institute of Space and Astronautical Science (ISAS/JAXA) as one of possible follow-ups of the Akatsuki Mission (Yamashiro et al. 2022). A Venus Lagrange Points Mission, able to continuously and simultaneously monitor the entire planet's atmospheric activity from a distance of about 1 million kilometers, is able to capture phenomena such as atmospheric dissipation and lightning on a global scale and continuous basis.

10.3.4 Cubesat Concepts

Several future projects involve deployment of cubesats in Venus' orbit to monitor atmospheric processes and their evolution: CUVE (CubeSat UV Experiment), a mapping spectrometer concept carrying a multi-spectral UV imager (320-570 nm) complemented by an imaging UV camera (see CUVIS on DAVINCI, Sect. 5.4.2) to study Venus cloud chemical, dynamical properties and radiative balance (Cottini et al. 2018) and TERACUBE, an instrument concept in the Terahertz frequency range. The high spectral resolution (e.g., 100 KHz) and sensitivity of such a heterodyne receiver will allow the spatial and temporal mapping of Doppler lineshifts winds, abundance of minor species (e.g., CO (5-4) at 576 GHz, H_2O 110-101 at 557 GHz) down to a few ppb, and atmospheric temperature, in the altitude range of 70-120 km (Moreno et al. 2020).

10.4 Future Directions

The technology for orbital, aerial, and short-duration surface missions to Venus exists today (Limaye and Garvin 2023). Continued work will naturally lead to ever longer-duration spacecraft, such that—with continued investment in and testing of a variety of technologies from materials to electronics to navigation software—we might reasonably see weeks-long surface operations and perhaps even years-long aerial platform missions in the coming decades.

For the foreseeable future, covering vast distances on Venus will be the exclusive purview of aerial platforms (excluding orbiters, of course); cloud-deck level balloons are to Venus as rovers are to Mars. But the scientific motivation for conducting *in situ*, detailed chemical and geophysical analyses at Venus is at least as compelling as it is at Mars. And so future technology development should take, as its long-term objective, the ability to move along the Venus surface itself (or relocated via hops)—that is, the design and flight of Venus rovers. Leveraging high-temperature electronics designed for long-life landers, rovers (Sect. 10.1.2-10.1.4) could be deployed to numerous terrain types including sites otherwise deemed unsafe for landers such as tesserae. This capability is because of the thick Venus atmosphere: using a delivery system similar in concept to the Skycrane powered descent vehicle that safely placed the Curiosity and Perseverance rovers onto Mars, a Venus rover delivery vehicle could take advantage of the planet's thick atmosphere to target high-standing lower-temperature (<715 K) terrains, in contrast to the Red Planet where low elevations and high surface atmospheric pressures are preferred for safe landing, to slow from hypersonic atmospheric entry speeds.

There are likely no technological showstoppers to exploring, at least with robots, the surface of the second planet, such as the Hybrid Automaton Rover-Venus (HAR-V) studied at the Jet Propulsion Laboratory (Sauder et al. 2019, and references therein). But if we are to see rovers traverse Venus' hostile surface later this century, the foundations for that technology must be laid now.

11 Ground-Based and Space-Based Observatories

Ground-based observations of Venus are complementary to spacecraft observations in a number of fields, from geology to atmospheric composition and dynamics. Venus is distant by a fraction of an Astronomical Unit near its inferior conjunction every 584 terrestrial days (synodic period), and periodically very well suited to observation from Earth (or from near-Earth observatories in space).

The highest resolution ground-based radar images of Venus, from the Arecibo observatory, reached spatial resolutions of 1–2 km; while order of magnitude poorer than Magellan radar images (Campbell et al. 2017) the long temporal baseline offered by decades of observation allows a search for temporal changes on these timescales. In addition, ground-based images include polarimetric information not captured by Magellan allowing constraints on surface properties. Earth-based radar can also be used for monitoring of Venus' spin (Margot et al. 2021, Sect. 11.1.1).

Ground-based observatory facilities and their instrumentation may obtain simultaneous measurements sampling a large range of altitudes, using wavelengths and/or spectral resolutions not available among spaceborne / onboard instruments – for studying the atmosphere, ground based telescopes have the advantage that they can be equipped sophisticated spectrometers far too massive and complex to deploy on a spacecraft; they can thus be sensitive to trace constituents, or faint motions of the atmosphere in a way which is complementary to spacecraft observations; and improve the latitude, longitude and local time coverage, and temporal baseline of rapidly variable phenomena. They can also provide monitoring of properties over decade-long periods of time, particularly times when no spacecraft were at Venus, and expand temporal baseline of rapidly variable phenomena (Lellouch et al. 2007; Lellouch and Witasse 2008; Widemann et al. 2008).

Sulfur dioxide and water vapor are key species in the complex photochemical cycles taking place in the troposphere and mesosphere of Venus, and have been extensively observed from Earth in the past decades (Mills et al. 2021, and references therein). They are also the most variable species in the atmosphere of Venus, and can be observed in millimeter wave (Sect. 11.1.2) or in the near-IR and mid-IR (Sect. 11.1.3). Both play a crucial role in determining climate on Venus, as key volatile species to constrain the rate and style of volcanic outgassing in the present era (see Sect. 2.4.3, Sects. 4.1.3-4.1.4, Sects. 5.3.1-5.3.3, Sects. 6.1.4, 6.3.3-6.3.4, and references therein). Amongst all trace constituents, SO_2 exhibits the most dramatic variations at Venus' cloud top, both spatially and temporally (Esposito 1984; Esposito et al. 1988; Marcq et al. 2013, 2020; Vandaele et al. 2017a,b, Encrenaz et al. 2012, 2016, 2019, 2020a, 2023) and so require continuous observations and long temporal baselines to support our understanding of long term climate evolution. Ground-based observations will continue over the decades ahead as newly selected or proposed missions described in Sects. 3-8 prepare a new era of Venus exploration.

The past years have seen an extraordinary growth in our knowledge of planetary systems around other stars, as well as the remarkable diversity and abundance of exoplanets (Sect. 11.2). These advances have been supported by several ground-based and orbital facilities, such as TESS and JWST, and a progression from planet detection to detailed study and characterization of individual planets. The prevalence of Venus analogs will continue to be relevant to Venus. Such observations capable of identifying key atmospheric abundances for terrestrial planets will face the challenge of distinguishing between possible Venus and Earth-like surface conditions.

11.1 Venus Observations from Ground-Based and Earth Orbiting Facilities

11.1.1 Earth-Based Radar Mapping and Speckle Tracking

Geologic mapping of volcanic features, their surface morphology and dielectric permittivity is a cornerstone of Magellan data interpretation (Campbell and Campbell 1992; Campbell 1994). Earth-based radar mapping can achieve 1–2 km spatial resolution and measure echoes in both the opposite-sense (OC) circular polarization and the same-sense circular (SC) mode. The OC echoes are very similar to Magellan measurements and are strongly modulated by slopes that face toward the radar. The SC echoes are much more sensitive to small-scale surface roughness than to topographic slopes. We can also form the circular polarization ratio (CPR = SC/OC), which allows for simple comparisons with rough surfaces on the Earth. The utility of these data was demonstrated in mapping of fine debris in the Venus highlands (Campbell and Rogers 1994; Whitten and Campbell 2016). Earth-based polarimetric mapping using the Arecibo radar shows that information on small-scale roughness correlates Venus lava flows with those in terrestrial settings (Campbell and Campbell 1992), and may reveal deposits formed during recent, volatile-rich eruptions (Campbell et al. 2017).

New methods for combining Earth-based hybrid astronomical radar polarimetry with refined Magellan radar measurements can extend the scientific value of existing datasets while we await the arrival of the next wave of orbital radar mapping missions, VERITAS and EnVision, as well as other radar missions considered for selection described in Sect. 3. Recent work by Garvin et al. (2022b, 2023 in prep.) demonstrated how multiple viewpoint Arecibo polarimetric radar data could be used via Shape-from-Shading together with re-analyzed Magellan radar altimeter radagrams to produce km-scale topography of Alpha Regio as a test-case. Improved methods building on these efforts could map regions on Venus in advance of the upcoming orbiters with advanced computational tools.

Earth-based radar can also be used to monitor Venus' spin state and moment of inertia (Margot et al. 2021). High-precision measurements of the instantaneous spin state of Venus may be obtained with a radar speckle tracking technique that requires two telescopes and does not involve imaging. Margot et al. used the 70 m antenna (DSS-14) at Goldstone, California (35.24°N, 116.89°E) and transmitted a circularly polarized, monochromatic signal at a frequency of 8560 MHz ($\lambda = 3.5$ cm) and power of $\sim$200-400 kW. Radar echoes were recorded at DSS-14 and also at the 100 m Green Bank Telescope (GBT) in West Virginia (38.24°N, −79.84°E) with fast sampling systems. Despite results not yet sufficient to rule out certain classes of interior models (Dumoulin et al. 2017), the best-fit value of the moment of inertia factor combined with knowledge of the bulk density enable a crude estimate of the size of the core of Venus (Margot et al. 2021). Improved determinations of the spin axis orientation, precession rate, and spin period form the basis of a recommended orientation model for Venus (see also Sects. 4.3.3 and 6.3.4).

11.1.2 Earth-Based Millimeter Wave Observations

Observing Venus at millimeter and submillimeter wavelengths with heterodyne spectroscopy provides unique means to probe the upper mesosphere of Venus (70-120 km). Heterodyne spectroscopy measurements have been performed with single dish antennas for decades in the millimeter range for CO, HDO and H_2O (e.g., Encrenaz et al. 1991, 1995), and in the submillimeter range for CO, SO_2, SO, H_2O and HDO (e.g., Sandor et al. 2010, 2012) using the James Clerk Maxwell Telescope (JCMT) on Maunakea, Hawaii. Altitude resolution is derived from the shape of pressure-broadened spectroscopic lines and the exponential variation of pressure with altitude. This region is dynamically a transition region between the retrograde super-rotation characterizing the lower atmosphere and the day-to-night flow regime prevailing in the thermosphere. Rotational lines of minor species such as carbon monoxide CO and isotopic ^{13}CO are formed at altitudes ranging from 70 km to 110 km, depending upon the strength of the transition used. Millimeter and Submillimeter observations complement the altitudes probed with ground-based IR observations, which investigate atmosphere levels within and below the clouds. in the millimeter-wave range. The Atacama Large Millimeter Array (ALMA) also offered a unique opportunity to probe the upper mesosphere (60–120 km) and monitoring minor species, winds and the thermal structure, targeting CO, SO, HDO and SO_2 transitions in the submillimeter range to derive 3D maps of mesospheric temperatures and minor species in the altitude range 70–105 km (Encrenaz et al. 2015; Piccialli et al. 2017).

11.1.3 Mid-IR, Near-IR and Visible Observations: Chemistry and Dynamics

CO, CO_2, H_2O and SO_2 in the Mid-Infrared 4.3–19 μm Range. - At thermal wavelengths (∼4–50 μm), the spectrum of the planet is close to that of a blackbody at the cloud top temperatures with spectral features mainly belonging to mesospheric CO_2, H_2O, SO_2, and other gasses that absorb at levels within and above the clouds as well as broad signatures of sulfuric acid aerosols. Encrenaz et al. (2012, 2013, 2016, 2019, 2020a, 2023) have performed study of the SO_2 over nearly a decade, using the TEXES high-resolution imaging spectrometer at the NASA InfraRed Telescope Facility (IRTF), also on Maunakea, Hawaii. Maps recorded around 1345 cm^{-1} (7.4 microns, z = 62 km), where SO_2, CO_2 and HDO are observed, and around 530 cm^{-1} (19 μm, z = 57 km) where SO_2 and CO_2 are observed, as well as around 1162 cm^{-1} (8.6 μm, z = 66 km) where CO_2 lines are observed. Mixing ratios are estimated from HDO/CO_2 and SO_2/CO_2 line depth ratios, using weak neighboring transitions of comparable depths. An anti-correlation has been evidenced in the long-term variations of H_2O and SO_2 at the cloud top, a long-term decrease of H_2O associated with a long-term increase of SO_2, as well as a planetocentric distribution of the SO_2 volume mixing ratio enhancement between 120 and 200 East longitude at Venus (Encrenaz et al. 2020a). High-resolution spectroscopic observations of both day and night sides of Venus were also acquired using the CSHELL spectrometer at NASA IRTF between 4.53 and 4.54 μm, to investigate the effect of the decrease of SO_2 (from 2007 onward) at the cloud top level on the spatial distribution of CO, since both species are involved in the mesospheric photochemical cycles (Marcq et al. 2015).

Near-Infrared Windows: H_2O, HCl, CO, OCS, SO_2 in the 1.18, 1.74, 2.32 and 2.46 μm Windows. - Infrared windows at shorter wavelengths in the near-IR probe deeper regions in the atmosphere (Allen and Crawford 1984). Several key gasses can be mapped below the cloud deck, at 0-50 km altitude: water vapor (H_2O and HDO) (Bézard et al. 2009; Arney et al. 2014), sulfur compounds (SO_2, OCS) and carbon monoxide (CO) (Marcq et al. 2006, 2021; Iwagami et al. 2010; Arney et al. 2014) - all potential volcanic volatile gasses or involved in long-term surface-atmosphere exchanges. In particular, discovering spatial variability of the D/H ratio – whether associated with volcanic plumes or other fractionating processes – would be fundamental for understanding the history of the water on Venus. The atmosphere is known to be variable on a range of time scales from minutes to years, so measurements over a wide range of timescales are still required. A plan to monitor Venus' atmosphere using an Earth-orbiting cubesat, CLOVE (Chasing the Long-term Variability of Our Nearest Neighbor Planet Venus), is currently under study by the Institute for Basic Science (IBS) of South Korea to perform observations from Earth's orbit between 320 nm and the near-infrared (Lee et al. 2022).

Atmospheric Circulation in the Visible and Near-IR. - The measurement of wind regimes in support of Venus General Circulation Models is achieved by two means: (1) directly, by Doppler or image correlation velocimetry using cloud features at different altitudes within the cloud layer, and (2) indirectly, using thermal and cyclostrophic wind balance relations to calculate equilibrium wind fields from measured temperature and pressure fields. In complement to cloud tracking in images taken at different wavelengths, which has proved an invaluable tool for extracting wind speeds at distinct vertical levels in the cloud region (Sánchez-Lavega et al. 2008, 2016; Hueso et al. 2012, 2015; Khatuntsev et al. 2013; Titov et al. 2018; Horinouchi et al. 2018; Fujisawa et al. 2022), wind speeds measured using Doppler-shifted spectroscopy, provided signal-to-noise limited precision of ~5 m/s using scattered solar Fraunhofer and CO_2 lines in the visible (dayside), sounding cloud tops (70 km) and a few kilometers above, using ESPaDOnS high-resolution spectrograph at Canada-France-Hawaii telescope (Widemann et al. 2007, 2008; Machado et al. 2012, 2014, 2017, 2021). Cloud-tracked winds may also be observed in the near-IR 2.26 μm window (Peralta et al. 2016, and references therein).

Solar Transits of Venus in 2004 and 2012. - A rare 2004 Venus transit imaging observing campaign with NASA's Transition Region and Coronal Explorer (TRACE) demonstrated the ability of Earth-orbiting observatories to constrain the properties of the upper atmosphere of Venus as a model for a transiting exoplanetary atmosphere (Ehrenreich et al. 2011; Pasachoff et al. 2011; Tanga et al. 2012). A follow-up ground-based campaign was organized in 2012 in coordination with Venus Express/SOIR and the HMI instrument aboard the Solar Dynamic Observatory (SDO) at the time of the 2012 transit to observe the refracted sunlight light curve as a probe of the thermal structure and composition of the upper atmosphere near the terminator (Widemann et al. 2012; Pere et al. 2016; Machado et al. 2023).

11.1.4 Ultraviolet Observations: Albedo Variations

Hubble Space Telescope Imaging Spectrograph (HST/STIS) UV observations of Venus' upper cloud tops have been used in coordination with VMC on board ESA's Venus Express, JAXA's Akatsuki UVI images, and NASA MESSENGER/MASCS UV spectral data to monitor the sulfur cycle and long-term UV albedo variations (Jessup et al. 2015; Lee et al. 2019). Lee et al. discuss the decadal variation of Venus's 365 nm albedo between 2006 and 2017; Solar EUV radiation might affect photochemical reactions involving SO_2 that are necessary for aerosol formation on Venus (Mills et al. 2007; Parkinson et al. 2015). Further studies are required to explore the role of the solar activity cycle on the Venusian upper atmosphere, as many intervening factors that may act in combination to produce the observed albedo variations: the chemical composition and reaction rate of the unknown absorber, its interaction with or dependency on the chemical state of other atmospheric constituents, such as sulfur species SO and SO_2, and the variability of the cloud and haze structure as a function of time (Lee et al. 2019).

11.2 Exoplanets Detection and Characterization

The planetary systems that have been discovered - as exemplified by the 7-planet system around TRAPPIST-1 (Gillon et al. 2016, 2017) only about 12 pc away - are extremely diverse, and study of the demographics of large numbers of exoplanets has led to several advances in understanding, including the recognition that many small-mass planets possess hydrogen atmospheres. Progress in observational technology will enable discovery of an even greater number of systems, and much expanded characterization of individual exoplanets by, e.g., the James Webb Space Telescope. Overall, the ability to image exoplanets both when they are forming and in their mature stage, the ability to characterize these exoplanets and their atmospheres, is providing us with new opportunities to understand planetary systems in the universe and to compare them with the long-term evolution of solar system planets.

Before the discovery of exoplanets, planet formation theories were limited to explaining the Solar system and thus, were unintentionally biased (e.g., Scora et al. 2020). Now, with thousands of extrasolar planetary systems, there is a diverse set of data to test against formation theories. Since planets at the inner edge of can sustain several very different possible atmospheres, depending on the initial water inventory and the water loss time-scales (see e.g., Turbet et al. 2020; Fauchez et al. 2022; Kaltenegger et al. 2023; Barrientos et al. 2023), exoplanetary observations of young planets around G-stars in the Venus Zone will be critical to understanding Venus' long term evolution through time. Determining that a planet resides in the Venus Zone provides a first-order estimate about the potential environment on that planet and criteria for its long term evolution.

Three next generation ground-based (>20 m in diameter) observatories are currently under construction or likely to be built in the near future: the European led Extremely Large Telescope (ELT), the Giant Magellan Telescope (GMT) and the Thirty Meter Telescope (TMT). The former two are currently under construction in Chile while the TMT is proposed for the northern hemisphere, although the exact location remains uncertain. With increasing advances in adaptive optics, they will afford new opportunities to explore the atmospheres of nearby exo-Venuses, as they are discovered by space observatories devoted to detecting such systems via the transit method (e.g., Kepler, TESS, CHEOPS, PLATO).

In space, JWST has already demonstrated how it can detect atmospheres around a few nearby terrestrial planets in systems such as Trappist-1 (Fig. 26), although such observations continue to be challenging, as discussed by (Way et al. 2023, this collection and references therein). The ARIEL mission (Tinetti et al. 2021), led by ESA, is also scheduled to be launched by the end of the decade with the ambition to measure the chemical fingerprints of ∼1000 exoplanetary atmospheres.

12 Laboratory and Modeling Efforts

12.1 Laboratory Experiments and Measurements

There are many areas where new laboratory work is needed to support our understanding of the Venus system. Experimental facilities can replicate the pressure, temperature, and chemical conditions of various layers of the planet and their interfaces; they also can be used to develop, test, and prove technologies to explore Venus. Thus, Venus experimental facilities are critical to moving forward our understanding of this planet in the next decades (Glaze et al. 2018; Santos et al. 2021). In this section we highlight a small selection of current facilities and experiments which the authors consider particularly relevant to the theme of understanding ancient Venus and the long-term evolution of the planet.

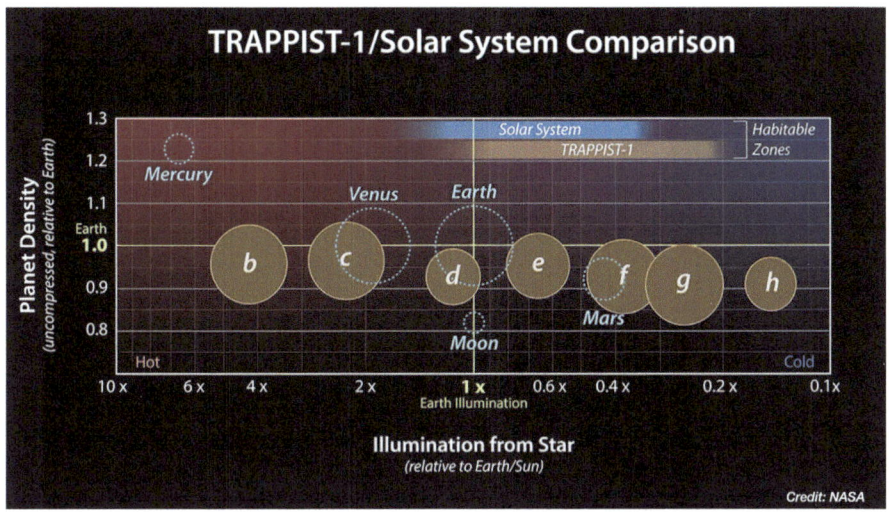

Fig. 26 The Trappist-1 system planets (Gillon et al. 2016, 2017) are among an abundance of Venus Zone planets which are promising candidates for follow-up ground- and space-based observatories, such as JWST and the TESS missions. Of these candidates, the TRAPPIST-1 planets in the Venus Zone are especially intriguing, and observational constraints on their atmospheres will provide an opportunity to compare the differences between Earth and Venus to planets receiving similar insolation flux (Way et al. 2023, this collection)

12.1.1 Geochemical Experiments

Geochemical experiments focus on the stability of minerals in the Venus surface environment and the transfer of elements or isotopes that take place during mineral-mineral or mineral-gas chemical reactions. These studies are important to ancient Venus and long-term planetary evolution because the preservation of mineralogical evidence of ancient planetary processes depends on mineral stability/weathering over geologic time. Additionally, secondary minerals produced by reactions of the atmosphere with the surface of a planet are a major sink of atmospheric gasses and can therefore strongly impact its long-term climate. Many aspects of mineral stability can be assessed using thermodynamic calculations, however major outstanding questions remain in the field of reaction kinetics and mechanisms that are crucial to resolve to understand how mineral-gas reactions unfold over time. Experimental studies that provide relevant reaction kinetic information will also aid in constructing models of surface weathering that include factors such as weathering rates (Gilmore et al. 2017; Santos et al. 2021; Reid 2021; Gilmore et al. 2023, this collection). There are many experimental approaches that can be taken to address these questions, several of them available in typical high temperature lab setups, but here we will highlight one facility that can accommodate both geochemical experiments and exploration technology testing. The reader is referred to many of the other current Venus weathering studies, and references therein, to understand the breadth of approaches in this field (e.g., Berger et al. 2019; Port et al. 2020; Port and Chevrier 2020; Radoman-Shaw 2019; Reid 2021; Teffeteller 2020).

The Glenn Extreme Environments Rig (GEER), located at NASA's Glenn Research Center, is capable of reproducing Venusian temperature, pressure, and complex atmospheric chemistry (CO_2, N_2, traces of SO_2, H_2O, CO, OCS, HCl, HF, and H_2S) for long durations (see experiment description in Radoman-Shaw 2019). The GEER pressure vessel is an

~800 L cylinder made of 304 stainless steel, and this volume allows the accommodation of mission hardware for development and testing. Along with its size, one of the key capabilities of this facility is its precision gas mixing and control system that is capable of making compositional corrections at high pressure and temperature. The gas composition is monitored using an external gas chromatograph. The gas boosting capability is useful for mineral reaction experiments because it can keep the gas composition within a specified range, as opposed to a typical batch reactor where the gas composition is permanently changed by reaction with the samples. This facility has been used for mission and technology development (e.g., Neudeck et al. 2018), materials science investigations (e.g., Costa et al. 2018; Lukco et al. 2018, 2020), and mineral weathering experiments (Radoman-Shaw 2019; Santos et al. 2023).

12.1.2 Mission Calibration Libraries: Planetary Spectroscopy Laboratory (PSL)

The unique environmental conditions on Venus can interfere with some of our traditional remote sensing techniques, for example, the thick atmosphere prevents visible light imaging or mapping of the planet's surface from orbit. Venus-specific exploration methodologies have to be developed and tested as a result. It was determined from data returned by the Galileo mission that there are windows in the CO_2 spectrum around 1 μm that allow us to see surface near-IR emissivity (Carlson et al. 1991), however spectral libraries built from analyses of a large number of geologic materials need to be developed to maximize the return on this type of data (Hashimoto et al. 2008; Mueller et al. 2008). In response to this challenge, the Planetary Spectroscopy Laboratory (PSL) at the German Aerospace Center (DLR) has designed a chamber to demonstrate and calibrate near-IR spectroscopy for Venus (Helbert et al. 2016, 2018). This data can be used as a reference to compare with surface emissivity spectra obtained by a future Venus orbiter. Furthermore, the emissivity chamber has an near-IR transparent window allowing mounting of near-IR spectrometers built for future Venus orbiter missions to take measurements at Venus conditions, for instrument calibration and performance study. A number of sample preparation and analysis tools and experiment sub-systems are available to the facility: a collection of hundreds of rocks and minerals, synthetic minerals, an Apollo 16 lunar sample, several meteorites, set of sample holders for reflectance (plastic, aluminum or stainless steel), various sets of sieves, grinders, mortars, saw, scales, microscope, an oven (20° to 300 °C), ultra-pure water, wet chemistry materials, a second ovens (30° to 3000 °C) for sample treatments, a press to produce pellets (10-mm or 20-mm diameter), purge gas generator for water and CO_2 free air, liquid-nitrogen tank, an ultrasonic cleaning unit and 2 microscopes. When enough sample material is available, the typical grain size separates that are produced for spectral measurements are <25 μm, 25-63 μm, 63-125 μm, 125-250 μm. Larger separates as well as slabs are also routinely measured. Models of anhydrite and hematite coatings on basalt mixtures suggest that changes in emissivity spectra due to chemical weathering can result in shifts in total emissivity, usefully constrain rock types and surface composition based on transition metal contents, but also provide local scale assessments of fresh versus mature lava flows (Dyar et al. 2021).

New High-temperature Dielectric Permittivity Laboratory Measurements relevant to future Venus radar mapping are underway at NASA's JPL (Barmatz et al. 2022) to further extend the value of SAR-based observations and enhance retrievals of surface electrical properties for the upcoming era of Venus radar mapping by VERITAS and EnVision. Measuring the complex dielectric permittivity of Venus analogue rocks and fines, as well as their intrinsic dielectric anisotropy, is important as new radars and radar sounders measure Venus from orbit, and eventually from cloud-deck altitudes (e.g., with balloon-born micro-SAR instruments).

12.1.3 Rheology Experiments: Plume-Induced Subduction on Venus

Venus relatively young surface points towards either a quite recent catastrophic renewal of the whole planet surface, or the continuous renewal of small areas. Geophysical observations suggest that the large coronae are due to the impact of hot plumes rising in the Venusian mantle. Sandwell and Schubert (1992a) had proposed that these plumes could induce subduction, but this hypothesis had remained in the schematic state since then. One challenge for evaluating the plume-induced subduction mechanism is the difficulty of simulating the brittle viscosity, and history-dependent lithospheric rheology in three-dimensional (3D) numerical models, which still cannot fully model deformation on a wide range of scales. Laboratory experiments using complex rheology fluids such as colloidal dispersions provide a means to bridge this gap (Davaille et al. 2017).

A more detailed analysis of Magellan data (radar, topography and gravimetry), where the resolution is sufficient, confirms the existence of the plume-subduction association on the largest corona, Artemis (2300 km diameter), and on Quetzalpetlatl (800 km diameter). In both cases, the proven subduction does not describe a complete circle but only an arc. Furthermore, high emissivities have been measured on Quetzalpetlatl (Smrekar et al. 2010a), and interpreted as a signature of recent volcanism. This suggests that the plume beneath Quetzalpetlatl is still hot and active, which implies that the subduction around it must also be currently active. Larger and probably at a later stage of its evolution, Artemis also shows a large oceanic ridge-like structure inside the corona, where new plate is created by upwelling of hot magma. Different corona structures may represent not only different styles of plume–lithosphere interactions but also different stages in evolution (Smrekar and Stofan 1997; O'Rourke and Smrekar 2018; Smrekar et al. 2018; Gülcher et al. 2020).

Laboratory experiment mechanisms predict the asymmetric, arcuate trenches, and the extensional fractures that radiate outward from the trench, observed at Artemis and Quetzalpetlatl coronae, as well as at other coronae on Venus. Davaille et al. (2017) compare laboratory experiments of plume-induced subduction in a colloidal solution of nanoparticles to observations of proposed subduction sites on Venus. The experimental fluids are heated from below to produce upwelling plumes, which in turn produce tensile fractures in the lithosphere-like skin that forms on the upper surface. Plume material upwells through the fractures and spreads above the skin, analogous to volcanic flooding, and leads to bending and eventual subduction of the skin along arcuate segments. In this unique experiment, the tank is dried from above and uniformly heated from below, allowing for the development of both a gravitationally unstable skin, the experimental 'lithosphere', and several hot upwelling plumes below this skin (Fig. 27). Both processes are due to convection, either solutal or thermal, respectively. In both cases, the intensity of convection is in the range of a planetary mantle. The laboratory experiment mechanisms predict the asymmetric, arcuate trenches, and the extensional fractures that radiate outward from the trench, observed at Artemis and Quetzalpetlatl coronae, as well as at other coronae on Venus. The gravity data are also consistent with the thickness, lengths and dips of those observed in experiments.

Further laboratory experiments are needed to bridge the gap between interior evolution models and surface observations of deformation structures, topography, and volcanism.

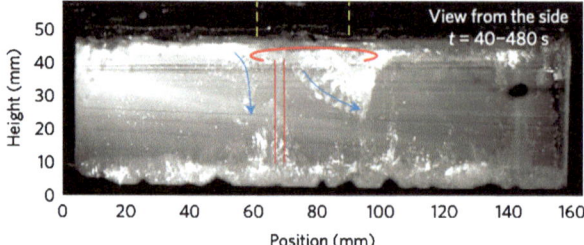

Fig. 27 Experimental facilities can replicate the pressure, temperature, and chemical conditions of different layers of the planet and their interfaces. Here, a laboratory experiment uses an aqueous colloidal dispersion dried from above and heated from below to simulate the plume-induced subduction mechanism in the Venusian mantle (Davaille et al. 2017; see Sect. 12.1.3)

12.2 Numerical Modeling

Modeling studies strongly suggest that the evolution of the atmosphere and interior of Venus are coupled (Way and Del Genio 2020; Weller and Kiefer 2020), emphasizing the need to study the atmosphere, surface, and interior of Venus as a system.

The interaction between the surface and the atmosphere is a key to understanding the processes driving the dynamics of both the atmosphere and the solid planet. The exchanges of heat and angular momentum drive the temperature and wind structure in the deepest layers of Venus's atmosphere. Chemical and dynamical modeling of interactions between the lower atmosphere and the surface, at the inner edge of the habitable zone, must take into account the variety of properties, trace atmospheric compounds and their reaction rates, local circulation and energy balance at the surface (e.g., Leconte et al. 2013; Lebonnois et al. 2018).

1-D radiative and radiative-convective models for the determination of climate are suddenly widespread as researchers worldwide attempt to determine the likely climate of exoplanets (Turbet et al. 2019, 2020; Fauchez et al. 2019, 2022; Way and Del Genio 2020; Bower et al. 2022; Wolf et al. 2019, 2022;Kaltenegger et al. 2023; Barrientos et al. 2023; Way et al. 2023, this collection, and references therein). Venus offers a proving ground for these models much closer to home, one where the conditions are much better known than on exoplanets. Radiative transfer calculations on Venus are difficult: uncertainties in the radiative transfer properties of carbon dioxide at high temperatures and pressures are the main unknown, particularly in the middle- and far- infrared where there are no spectral window regions to allow empirical correction. As on Earth, clouds play an important role, reflecting away sunlight but also trapping upwelling infrared radiation. The state-of-the art Venus radiative balance are still mainly 1-D models representing an average over the whole planet. However, we now know that the clouds are very variable; the vertically integrated optical thickness (as measured at 0.63 μm) can vary by up to 100% (Barstow et al. 2012) and the vertical structure of clouds varies strongly with latitude. In-situ measurements of cloud properties with co-located radiative flux measurements are also needed to determine the diversity of cloud effects on the global radiative balance.

12.3 Summary / Outcomes Revealing Venus Evolution

Future geochemical modeling and experiments will greatly benefit from better constraints on near-surface atmospheric composition and the composition and mineralogy of the solid surface. In fact, the composition of the solid planet (in terms of chemistry, mineralogy, and isotopic values) is a major knowledge gap that is relevant to almost all areas of Venus science. Obtaining better data in this area will also enable other kinds of experiments to be conducted, such as those in the field of experimental petrology, which have provided significant insight into the evolution of other planetary bodies, but cannot be as rigorously applied to Venus due to the current state of our petrologic data from the planet.

13 Summary and Conclusions

The discoveries of many exoplanets, including terrestrial exoplanets perhaps similar to Venus, due to increasingly sensitive methods of discovery and characterization, make exchange between exoplanetary and planetary scientific communities increasingly necessary. The search for exoplanets is largely motivated by the answers to the questions: Is our solar system common and is there life outside our solar system? Answering these questions requires also understanding the habitability of a planet, i.e., the potential of a planet to develop and maintain a living environment. Venus and Earth formed under very similar conditions and were probably supplied with water in the same way. At some point in their history, the evolution of their surfaces and atmospheres diverged dramatically. Venus could be the type of planet that has changed from a habitable and Earth-like state to an uninhabitable one (Way and Del Genio 2020).

Thus, Venus is particularly important to our understanding of terrestrial planets' habitability, providing a natural laboratory to understand its evolution in time. Venus exploration offers therefore unique opportunities to answer fundamental questions about the evolution of terrestrial planets and the habitability within our own solar system. Venus' enhanced D/H ratio suggests that it has lost large amounts (possibly several terrestrial oceans) of water, but it is not clear whether it condensed (as happened on Earth) or whether this water was lost in the steam atmosphere phase; if it had a liquid water ocean phase, Venus may have been habitable for billions of years. There is no consensus on how much water there is in Venus' interior, and how much of this water has been outgassed, a question which has important implications for Venus' atmospheric water and in turn for its habitability through time. Exoplanet transit detection surveys have a bias to detecting exoplanets close to their parent stars: the growing number of such Venus-like exoplanets discoveries emphasizes the relevance of Venus in the search for habitable exoplanets.

Venus has been an object of fascination throughout the space age. It was the site of the first planetary flyby in 1962 (Mariner-2), first entry probe in 1967 (Venera-4), first soft landing in Dec. 1970 (Venera-7), first image from the surface of another planet in 1975 (Venera-9), first orbiter and radar in 1978 (Pioneer). The Soviet series of Venera & VeGa missions were phenomenally successful, not only in their technologically advanced landers which returned colour pictures from Venus and analyzed drill samples despite 450 °C heat, but also successfully deployed balloons in the atmosphere in 1985 (Sagdeev et al. 1986a, 1986b; Linkin et al. 1986; Blamont et al. 1986; Pieters et al. 1986).

Now is a pivotal time in Venus exploration. Since NASA-JPL's Magellan orbiter provided initial global radar imaging and altimetry (1989–1994), and USSR's Venera landers measured major and heat- producing elements in several locations (1975–1985), there have been considerable advances relevant to understanding Venus' evolution. ESA's Venus Express (2006-2014) and JAXA's Akatsuki (2010-present) orbiters focused primarily on atmospheric science. Both revealed many secrets of Venus' atmosphere, but have also left many questions unanswered. There have been new ground-based observations of surface and atmospheric properties e.g., from Arecibo and NASA's Infrared Telescope Facility; new analyses of existing data, in particular surface emissivity from Venus Express; new hypotheses for the origin of plate tectonics; advances in the numerical tools and laboratory simulation of interior convection; and new modeling of the evolution of rocky earth-sized exoplanets, in need of observational constraint using Venus as a reference point. The difficulties in modeling the atmospheric superrotation of Venus meet new developments as the atmospheres of typical tidally locked terrestrial exoplanets are expected to superrotate (Imamura et al. 2020, and references therein).

Many important questions about the current state of Venus remain unanswered, suggesting that there are major gaps in our understanding of how and when Venus's evolutionary pathway diverged from Earth's (Morrison and Hinners 1983). As we developed in this final review, a new fleet of Venus missions is currently in development. These include radar-equipped orbiters (such as the ESA-led EnVision M5 orbiter and NASA-JPL's VERITAS orbiter missions), entry probes / landers/flybys (such as NASA GSFC's DAVINCI mission). Further Venus missions are also considered with Russia's Venera-D orbiter, aerial platform and lander mission and India's ISRO/Shukrayaan-1 orbiter mission. Japan and China have also announced a likely orbiter proposed for launch before the end of this decade. Furthermore, various concepts to detect seismic activity, whether from landers, from balloons or from orbit are also under study (Limaye and Garvin 2023).

The science strategy for all of these missions is in development now and in the coming few years; therefore, now is an ideal time to collate knowledge of Venus evolution scenarios and the observations needed to distinguish between them. Sects. 4-6 captured the considerable advances that the three newly selected missions VERITAS (Sect. 4), DAVINCI (Sect. 5) and EnVision (Sect. 6) will bring to these science goals. These rich, highly synergistic datasets will provide an invaluable resource to assess the long-term history, stability of water reservoirs in the mantle and atmosphere, current levels of activity, divergent pathways and evolution toward habitability. Together they will reveal whether Venus-like and Earth-like planets can potentially transition into one another over time, which would imply that Earth-like exoplanets may be common among Earth-sized exoplanets.

Appendix: Correction to this Chapter

In the original publication, there has been a discrepancy between values mentioned in the text of Sect. 4.3.3 and in Table 1.

The corrected version is provided here:

Wrong: The gravity field harmonic coefficients up to an average degree strength of 130 ($\sim$145 km) are generated together with corrections to the spin rate and to the pole right ascension and declination.

Correct: The gravity field harmonic coefficients up to an average degree strength of 200 ($\sim$105 km, see Table 1) are generated together with corrections to the spin rate and to the pole right ascension and declination.

The original correction can be found online in *Space Sci Rev* **219**, 72 (2023). https://doi.org/10.1007/s11214-023-01022-5

List of Acronyms and Glossary

ABX	Aerobraking Exit Maneuver
ADC	Analog to Digital Converter
AFN	Autofluorescing Nephelometer (Venus Life Finder Science Payload)
AP	Aerial Platform (Venera-D mission)
AU	Astronomical Unit
CC	Carbonaceous
CLOVE	Chasing the Long-term Variability of Our Nearest Neighbor Planet Venus, an Institute for Basic Science (IBS, South Korea) cubesat project to perform observations from 320 nm to the near-infrared
CNES	Centre National d'Études Spatiales
CPR	Circular Polarization Ratio
CRIS	Carrier Relay Imaging Spacecraft (DAVINCI mission)
CUVIS	Compact Ultraviolet Imaging System (DAVINCI science payload)
D/H	Deuterium to Hydrogen isotopic ratio
DAVINCI	Deep Atmosphere Venus Investigation of Noble Gases, Chemistry, and Imaging
DM	Descent Module (Venera-D mission)
DS	Descent sphere (DAVINCI mission)
DSN	Deep Space Network
DV	Delta Velocity
ECM	Eccentricity Control Maneuver
EDL	Entry, Descent and Landing
ELT	39-m Extremely Large Telescope, European Southern Observatory
EM	Electromagnetic
ENA	Energetic Neutral Atom
Eotvos	A non-SI unit of acceleration divied by distance; 1 Eotvos (E) $= 10^{-9}$ galileos cm^{-1}; in SI, 1 E $= 10^{-9}$ s^{-2}; after Loránd Eötvös (1848-1919)
ESA	European Space Agency
ESI	Engineering Science Investigation (DAVINCI)
Ga	Gigayear, one billion years
GBT	Green Bank Telescope, West Virginia
GCM	General Circulation Model
GMT	Greenwich Mean Time
GMT	25-m Giant Magellan Telescope, Las Campanas Observatory, Chile
GRAIL	Gravity Recovery and Interior Laboratory
GROZA	Radio Wave Analyzer 15-50 Mhz (Venera-D Science Payload)
GSFC	Goddard Space Flight Center
GSLV	Geosynchronous Satellite Launch Vehicle
Hadean	Geologic eon extending -4.6 to -4 Ga preceding earliest known minerals on Earth
HF	High Frequency, a range of radio frequencies extending from 3 MHz to 30 MHz i.e., from 10 to 100 m in wavelength.
HGA	High Gain Antenna
HH and HV	Horizontal and Vertical Polarization (conventional imaging radar systems)

HZ	Habitable Zone, a range of distances around a star within which a planetary surface can support liquid water given sufficient atmospheric pressure, and thus provide conditions for the emergence of life, or its precursors. By extension, range of altitudes within the atmosphere or the liquid layers of a planetary or natural satellite interior with similar properties.
InSAR	Interferometric Synthetic Aperture Radar
ISRO	Indian Space Reseach Organisation
IVOLGA	Infrared Heterodyne Fiber Analyzer/Spectrometer (Venera-D Science Payload)
JAXA	Japan Aerospace Exploration Agency
JCMT	James Clerk Maxwell Telescope, Maunakea, Hawaii, a ground-based millimeter- submillimeter wave telescope facility
JPL	Jet Propulsion Laboratory
JSDT	Venera-D Joint Science Definition Team
k_2	Tidal Love number, gravitational potential modification due to the tidal deformation of a planet. After Augustus E. H. Love (1863-1840)
Ka-band	a nominal frequency range, from 26 to 40 GHz (0.8-1.1 cm in wavelength) within the microwave portion of the electromagnetic spectrum
LGA	Low Gain Antenna
LIDAR	LIght Detection And Ranging
LIR	Longwave Infrared Camera (Venera-D Science Payload)
LIVE	Lightning Sensor (Shukrayaan-1 Science Payload)
LM	Lander Module (Venera-D mission)
LOD	Lengh of Day
LOS	Line of Sight
LT	Local (solar) time, hour angle of the Sun as observed from a given point on Venus
LWIR (or Thermal IR)	Long Wavelength Infrared radiation, 8 – 15 μm in wavelength, within the infrared portion of the electromagnetic spectrum
Magellan	NASA Venus Orbiter Mission 1990-1994
MARSIS	Mars Advanced Radar for Subsurface and Ionosphere Sounding
MAVEN	Mars Atmosphere and Volatile EvolutioN
Meridian	Any great circle joining the North and South poles of a planet
MERTIS	Mercury Radiometer and Thermal Infrared Spectrometer (BepiColombo science payload)
MESSENGER	Mercury Surface, Space Environment, Geochemistry, and Ranging
Mid IR or MWIR	Mid Infrared (Wavelength) radiation, 3 – 8 μm in wavelength, within the infrared portion of the electromagnetic spectrum
Millimeter wave	Range of electromagnetic spectrum between 10 millimeters (30 GHz) and 1 millimeter (300 GHz), also known as the extremely high frequency (EHF) band.
(volume) Mixing ratio	Amount of an atmospheric constituent (in moles) divided by the total (in moles) of all other atmospheric constituents. For minor species, it is usually expressed in parts per million (ppm) or parts per billion (ppb)
MGS	Mars Global Surveyor

MM-R	Millimeter Wave Radiometer
MO	Magma Ocean
MoI	Moment of Inertia
MOIF	Moment of Inertia Factor
MOLA	Mars Orbiter Laser Altimeter (MGS science payload)
MSL	Mars Science Laboratory / Curiosity
MWRS	Microwave Radiometric Sounder (VOICE Science Payload)
Myr	Megayear = Million years
N/A	Not applicable
Nadir	Direction pointing directly below a particular location. The radar nadir refers to the downward-facing viewing geometry of an orbiting radar.
NASA	National Aeronautics and Space Agency
NASEM	National Academies of Sciences and Engineering
NC	Non-carbonaceous
Near IR or nIR	Near Infrared radiation, $0.75 - 1.4$ μm in wavelength, within the infrared portion of the electromagnetic spectrum (from the approximate end of the response of the human eye to that of silicon)
NES0	Noise Equivalent Sigma Zero, a measure of SAR sensitivity, usually expressed in dB
OBP	On-board Processing
OC	Opposite Sense, circular polarization
SC	Same Sense, circular polarization
OM	Orbiter Module (Venera-D mission)
p, T	Pressure, temperature
PFS	Probe Flight System (DAVINCI mission)
Polarization	orientation of the electric field vector in an electromagnetic wave, "horizontal" (H) or "vertical" (V) in conventional imaging radar systems.
PolSAR	Polarimetric Synthetic Aperture Radar (VOICE Science Payload)
QMS	Quadrupole Mass Spectrometer
Radar	RAdio Detection And Ranging
R-LIBS	Raman-Laser Induced Breakdown Spectroscopy
RAVI	(Shukrayaan-1 Science Payload)
RPI	Repeat-Pass Interferometry
RTM	Radiative transfer model
S-band	a nominal frequency range, from 2 to 4 GHz (7.5-15 cm in wavelength) within the microwave portion of the electromagnetic spectrum
S/C	Spacecraft
SAM	Sample Analysis at Mars suite (MSL science payload)
SAM	Sample Analysis at Mars
SEP	Solar Electric Propulsion
SfM	Structure-from-Motion processing of descent images (DAVINCI, Sect. 5.3.4)
SHARAD	Mars SHAllow RADar sounder (Mars Reconnaissance Orbiter Payload)
SNR	Signal-to-noise ratio

SP1	Science Phase 1 (VERITAS mission)
SP2	Science Phase 2 (VERITAS mission)
SPICAV	Spectroscopy for Investigation of Characteristics of the Atmosphere of Venus (Venus Express Science Payload)
SRS	Subsurface Radar Sounder (EnVision science payload)
SRTM	Shuttle Radar Topography Mission
SSAS or SS-AS	Subsolar to antisolar (wind circulation)
SSPA	Solid State Power Amplifier
SVET	Fourier Infrared Thermal Spectrometer (Venera-D Science Payload)
SWIR	Short Wavelength Infrared radiation, $1.4 - 3$ µm in wavelength, within the infrared portion of the electromagnetic spectrum
SZA	Solar zenith angle, the angular distance between the vertical direction and the direction of the Sun from a specific location
TanDEM-X	TerraSAR-X satellite add-on for Digital Elevation Measurement
Tb	1 Tb = 1 terabit = 10^{12} bits; 1 terabyte = 1 TB = 8 Tb
TB	1 TB = 1 terabyte = 10^{12} bytes
TDI	Time delay and Integration
TESS	Transiting Exoplanets Survey Satellite
Thermal IR (or LWIR)	Long Wavelength Infrared radiation, $8 - 15$ µm in wavelength, within the infrared portion of the electromagnetic spectrum
TMT	30-m Thirty-Meter Telescope, TMT International Observatory
TRAPPIST	TRAnsiting Planets and Planetesimals Small Telescope, a ground-based observatory on two sites: TRAPPIST-S in La Silla Observatory, Chile; TRAPPIST-N in Oukaïmeden Observatory, Morocco.
TRAPPIST-1	A cold dwarf star in Aquarius constellation, 40.55 ± 0.04 light-years away from Earth, with a planetary system of seven known exoplanets TRAPPIST-1b/h
TT&C	Telemetry, Tracking and Command
TW	1 TW = 1 terawatt = 10^{12} watts
TWTA	Travel Waveguide Tube Amplifier
USO	Ultra-Stable Oscillator
UVN-MSI	Ultraviolet-Visible-Near Infrared Multi-Spectral Imager (VOICE Science Payload)
VARTISS	(Shukrayaan-1 Science Payload)
VASI	Venus Atmospheric Structure Investigation (DAVINCI Science Payload)
VCMC	Visible Camera for Cloud Monitoring (Shukrayaan-1 Science Payload)
VEDA	Electron Density (Shukrayaan-1 Science Payload)
VEM	Venus Emissivity Mapper (VERITAS science payload)
VenDI	Venus Descent Imager (DAVINCI Science Payload)
VENIS	IR grating spectrometer and imager, 2-5 µm (Venera-D Science Payload)
VenSpec	Venus Spectroscopy suite (EnVision science payload)
VenSpec-H	Venus Spectroscopy High Resolution (EnVision science payload)
VenSpec-M	Venus Spectroscopy Mapper (EnVision science payload)
VenSpec-U	Venus Spectroscopy Ultraviolet (EnVision science payload)

VERITAS	Venus Emissivity, Radio Science, InSAR, Topography, and Spectroscopy
VEx	ESA Venus Express orbiter mission 2007-2014
VfOx	Venus Oxygen Fugacity Experiment (DAVINCI Student Collaboration Experiment)
VIKA	near-IR spectrometer suite 1.05 – 1.65 μm (Venera-D Science Payload)
VIRAL	Venus Infrared Atmospheric gas Linker, a high-resolution echelle spectrograph 2.3–4.4 μm (Shukrayaan-1 Science Payload)
VIRTIS	Visible and Infrared Thermal Imaging Spectrometer (Venus Express science payload)
VIRTIS-H	The high spectral resolution channel of the Venus Express VIRTIS IR spectrometer, aboard Venus Express
VIS	Visible spectral range (0.38 – 0.75 μm, or 380 – 750 nm)
VISAR	Venus Interferometric Synthetic Aperture Radar (VERITAS science payload)
VISOR	Venus Imaging System for Orbital Reconnaissance (DAVINCI Science Payload)
VISWAS	(Shukrayaan-1 Science Payload)
VMC	Venus Monitoring Camera (Venus Express Science Payload)
VMS	Venus Mass Spectrometer (DAVINCI science payload)
VODEX	Venus Orbiter Dust EXperiment (Shukrayaan-1 Science Payload)
VOI	Venus Orbit Insertion
VOLNA	(Venera-D Science Payload)
VSAR	(Shukrayaan-1 Science Payload)
VSEAM	(Shukrayaan-1 Science Payload)
VTC	Venus Thermal Camera (Shukrayaan-1 Science Payload)
VTLS	Venus Tunable Laser Spectrometer (DAVINCI Science Payload)
VZ	Venus Zone, defined by Kane et al. (2014) as part of the habitable zone (HZ) in which an Earth-sized planet is more likely to be a Venus analog than an Earth analog
X-band	a nominal frequency range, from 8 to 12 GHz (2.5-3.8 cm in wavelength) within the microwave portion of the electromagnetic spectrum
XRF	X-ray fluorescence

Acknowledgements T.W. and G.A. acknowledge France's Centre National d'Études Spatiales (CNES) for funding support of Venus related studies. J..B.G. (as well as S.G. G.A. N.J, and E. K) acknowledges NASA's Discovery program for support of Venus related studies and the DAVINCI mission. Work by S.E.S., S.H., and D.N was carried out at the Jet Propulsion Laboratory, California Institute of Technology, under a contract with the National Aeronautics and Space Administration. Members of the VERITAS team are grateful for funding by NASA's Discovery Program and JPL's proposal support. C.G. acknowledges the support of Rice University and the CLEVER planets group (itself supported by NASA and part of NExSS) and ET-HoME Excellence of Science programme. A.S. acknowledges support from NASA's Habitable Worlds Program (No. 80NSSC20K0226). Section 7 was prepared with the assistance of Oleg Sedykh (Venera-D). Finally, the authors thank the International Space Institute (ISSI) in Bern, Switzerland, for supporting the "Venus: Evolution through Time" workshop and the subsequent book, of which this paper forms a chapter.

Funding Open access funding provided by University of Oslo (incl Oslo University Hospital).

Declarations

Competing Interests The authors declare that they have no conflict of interest.

References

Ainsworth T, Kelly J, Lee J (2009) Classification comparisons between dual-pol, compact polarimetric and quad-pol sar imagery. ISPRS J Photogramm Remote Sens 64:464–471. https://doi.org/10.1016/j.isprsjprs.2008.12.008

Airey MW (2015) Explosive volcanic activity on Venus: the roles of volatile contribution, degassing, and external environment. Planet Space Sci 113–114:33–48. https://doi.org/10.1016/j.pss.2015.01.009

Allen DA, Crawford JW (1984) Cloud structure on the dark side of Venus. Nature 307:222–224

Anderson DL (2005) Scoring hotspots: the plume and plate paradigms. In: Foulger GR, Natland JH, Presnall DC, Anderson DL (eds) Plates, plumes, and paradigms. Geological Society of America special paper, vol 388, pp 31–54. https://doi.org/10.1130/0-8137-2388-4.31

Anderson FS, Smrekar SE (1999) Tectonic effects of climate change on Venus. J Geophys Res, Planets 104(E12):30743–30756

Anderson FS, Smrekar SE (2006) Global mapping of crustal and lithospheric thickness on Venus. J Geophys Res 111:E08006. https://doi.org/10.1029/2004JE002395

Andrews-Hanna JC, Zuber MT, Banerdt WB (2008) The Borealis basin and the origin of the Martian crustal dichotomy. Nature 453(7199):1212–1215. https://doi.org/10.1038/nature07011

Andrews-Hanna JC, Besserer J, Head JW III, Howett CJA, Kiefer WS, Lucey PJ, McGovern PJ, Melosh HJ, Neumann GA, Phillips RJ (2014) Structure and evolution of the lunar Procellarum region as revealed by GRAIL gravity data. Nature 514(7520):68

Antonita MT (2022) Outstanding science questions of Venus and the proposed Venus Orbiter Mission. Presentation given at National Meet on Venus Science 05/04/2022, ISRO. HQ, Bengaluru, India. https://www.youtube.com/watch?v=yUp6DplyPJk

Antonita MT, Das PTD, Kumar PK (2022) Overview of ISRO's future Venus orbiter mission. In: COSPAR 2022 44th scientific assembly, 16-24 July 2022, Athens, session B4.1 Venus science and exploration

Armann M, Tackley PJ (2012) Simulating the thermochemical magmatic and tectonic evolution of Venus's mantle and lithosphere: two-dimensional models. J Geophys Res, Planets 117(E12):E12003

Arndt NT (2013) The formation and evolution of the continental crust. Geochem Perspect 2(3):405–533

Arney G, Meadows V, Crisp D, Schmidt SJ, Bailey J, Robinson T (2014) Spatially resolved measurements of H_2O, CO, OCS, SO_2, cloud opacity, and acid concentration in the Venus near-infrared spectral windows,J. Geophys Res Planets 119:1860–1891 https://doi.org/10.1002/2014JE004662

Atreya SK, Trainer MG, Franz HB, Wong MH, Manning HLK, Malespin CA, Mahaffy PR, Conrad PG, Brunner AE, Leshin LA, Jones JH, Webster CR, Owen TC, Pepin RO, Navarro-Gonzalez R (2013) Primordial argon isotope fractionation in the atmosphere of Mars measured by the SAM instrument on Curiosity and implications for atmospheric loss. Geophys Res Lett 40:5605–5609. https://doi.org/10.1002/2013GL057763

Avduevskii VS et al (1977) Measurement of wind velocity on the surface of Venus during the operation of stations Venera 9 and Venera 10. Cosm Res 14(5):622–625

Avice G, Marty B (2020) Perspectives on atmospheric evolution from noble gas and nitrogen isotopes on Earth, Mars & Venus. Space Sci Rev 216:36. https://doi.org/10.1007/s11214-020-00655-0

Avice G, Marty B, Burgess R (2017) The origin and degassing history of the Earth's atmosphere revealed by Archean xenon. Nat Commun 8:15455. https://doi.org/10.1038/ncomms15455

Avice G, Parai R, Jacobson SA, Labidi J, Petkov MP, Trainer MG (2022) Noble gases and stable isotopes track the origin and early evolution of the Venus atmosphere. Space Sci Rev 218:60. https://doi.org/10.1007/s11214-022-00929-9

Baes M, Gerya T, Sobolev SV (2016) 3-D thermo-mechanical modeling of plume-induced subduction initiation. Earth Planet Sci Lett 453:193–203

Baines KH, Atreya SK, Bullock MA, Grinspoon DH, Mahaffy P, Russell CT, Schubert G, Zahnle K (2013) The atmospheres of the terrestrial planets: clues to the origins and early evolution of Venus, Earth, and Mars. In: Comparative climatology of terrestrial planets. University of Arizona Press, Tucson, pp 1–28. https://doi.org/10.2458/azu_uapress_9780816530595-ch006

Barabash S, Fedorov A, Sauvaud J, et al (2007) The loss of ions from Venus through the plasma wake. Nature 450:650–653. https://doi.org/10.1038/nature06434

Barmatz MB et al (2022) High-temperature dielectric permittivity laboratory measurements relevant to future Venus radar mapping, in preparation

Barrientos JG, Kaltenegger L, MacDonald RJ (2023) A Venus in the making? Predictions for JWST observations of the ultracool M-dwarf planet LP 890-9 c. Mon Not R Astron Soc Lett 524(1):L5–L9. https://doi.org/10.1093/mnrasl/slad056

Barstow JK, Tsang CCC, Wilson CF, Irwin PGJ, Taylor FW, McGouldrick K, Drossart P, Piccioni G, Tellmann S (2012) Models of the global cloud structure on Venus derived from Venus Express observations. Icarus 217(2):542–560. https://doi.org/10.1016/j.icarus.2011.05.018

Barsukov VL et al (1982) Geochemical studies of Venus surface by Venera 13 and Venera 14 spacecrafts. Geohimiâ 7:899–919. (in Russian)

Barsukov VL et al (1986) The geology and geomorphology of the Venus surface as revealed by radar images obtained by Veneras 15 and 16. J Geophys Res 91:378–398

Basilevsky AT (1993) Age of rifting and associated volcanism in Atla Regio, Venus. Geophys Res Lett 20(10):883–886. https://doi.org/10.1029/93GL00736

Basilevsky AT, Head JW (2003a) The surface of Venus. Rep Prog Phys 66:1699–1734. https://doi.org/10.1088/0034-4885/66/10/R04

Basilevsky AT, Head JW, Setyaeva IV (2003b) Venus: estimation of age of impact craters on the basis of degree of preservation of associated radar-dark deposits. Geophys Res Lett 30(18):1950. https://doi.org/10.1029/2003GL017504

Bengtsson L, Bonnet RM, Grinspoon D, Koumoutsaris S, Lebonnois S, Titov D (eds) (2012) Towards understanding the climate of Venus: applications of terrestrial models to our sister planet, vol 11. Springer. https://doi.org/10.1007/978-1-4614-5064-1

Bercovici D, Ricard Y (2014) Plate tectonics, damage and inheritance. Nature 508(7497):513

Berger G, Cathala A, Fabre S, Borisova AY, Pages A, Aigouy T, Esvan J, Pinet P (2019) Experimental exploration of volcanic rocks-atmosphere interaction under Venus surface conditions. Icarus 329:8–23. https://doi.org/10.1016/j.icarus.2019.03.033

Bertaux JL, Widemann T, Hauchecorne A, Moroz VI, Ekonomov AP (1996) Vega-1 and Vega-2 entry probes: an investigation of local UV absorption (220–400 nm) in the atmosphere of Venus (SO_2, aerosols, cloud structure). Journ Geophys Research 101(E5):12709–12745

Bézard B, Tsang CCC, Carlson RW, Piccioni G, Marcq E, Drossart P (2009) Water vapor abundance near the surface of Venus from Venus Express/VIRTIS observations. J Geophys Res 114:E00B39. https://doi.org/10.1029/2008JE003251

Bézard B, Fedorova A, Bertaux J-L et al (2011) The 1.10- and 1.18-μm nightside windows of Venus observed by SPICAV-IR aboard Venus Express. Icarus 216:173–183

Bindschadler DL, Parmentier EM (1990) Mantle flow tectonics: the influence of a ductile lower crust and implications for the formation of topographic uplands on Venus. J Geophys Res, Solid Earth 95(B13):21329–21344

Bindschadler DL, DeCharon A, Beratan KK, Smrekar SE, Head JW (1992) Magellan observations of Alpha Regio: implications for formation of complex ridged terrains on Venus. J Geophys Res, Planets 97(E8):13563–13577. https://doi.org/10.1029/92JE01332

Bjonnes EE, Hansen VL, James B, Swenson JB (2012) Equilibrium resurfacing of Venus: results from new Monte Carlo modeling and implications for Venus surface histories. Icarus 217(2):451–461

Blamont JE, Young RE, Seiff A, Ragent B, Sagdeev R, Linkin VM, Kerzhanovich VV, Ingersoll AP, Crisp D, Elson LS, Preston RA, Golitsyn GS, Ivanov VN (1986) Implications of the VEGA balloon results for Venus atmospheric dynamics. Science 231(4744):1422–1425. https://doi.org/10.1126/science.231.4744.1422

Bottke WF, Nesvorny D, Marchi S, Levison H, Canup R (2017) Exploring planet migration and early solar system bombardment. In: Planetary science vision 2050 workshop 2017. LPI contrib., vol 1989

Bower D, Hakim K, Sossi P, Sanan P (2022) Retention of water in terrestrial magma oceans and carbon-rich early atmospheres. Planet Sci J 3(4):93

Bowman DC, Krishnamoorthy S (2021) Infrasound from a buried chemical explosion recorded on a balloon in the lower stratosphere. Geophys Res Lett 48(21):e2021GL094861

Breuer D et al (2022) Interiors of Earth-like planets and satellites of the solar system. Surv Geophys 43(1):177–226

Brissaud Q, Krishnamoorthy S, Jackson JM, Bowman DC, Komjathy A, Cutts JA, et al, Walsh GJ (2021) The first detection of an earthquake from a balloon using its acoustic signature. Geophys Res Lett 48(12):e2021GL093013

Brooks J, Jacobson SA (2019) Losing moons: the gravitational influence of close encounters on satellite orbits, AAS Division on Dynamical Astronomy meeting #50, id. 302.05. Bull Am Astron Soc 51:5

Buczkowski DL, McGill GE (2002) Topography within circular grabens: implications for polygon origin, Utopia Planitia, Mars. Geophys Res Lett 29(7):59-1–59-4

Bullock MA, Grinspoon DH (2001) The recent evolution of climate on Venus. Icarus 150:19. https://doi.org/10.1006/icar.2000.6570

Byrne PK (2022) Phantom, an aerobot mission to the skies of Venus. In: 19th international planetary probe workshop (IPPW), Santa Clara/Silicon Valley, Aug. 29-Sep 2, 2022

Byrne PK, Krishnamoorthy S (2020) Estimates on the frequency of volcanic eruptions on Venus. J Geophys Res, Planets 127:e2021JE007040

Campbell BA (1994) Merging Magellan emissivity and SAR data for analysis of Venus surface dielectric properties. Icarus 112:187–203

Campbell BA, Campbell DB (1992) Analysis of volcanic surface morphology on Venus from comparison of Arecibo, Magellan, and terrestrial airborne radar data. J Geophys Res 97:16293–16314

Campbell BA, Rogers PG (1994) Bell Regio, Venus: integration of remote sensing data and terrestrial analogs for geologic analysis. J Geophys Res 99:21,153–21,171

Campbell IH, Taylor SR (1983) No water, no granites-no oceans, no continents. Geophys Res Lett 10(11):1061–1064

Campbell B, Carter L, Phillips R, Plaut J, Putzig N, Safaeinili A, Seu R, Biccari D, Egan A, Orosei R (2008) SHARAD radar sounding of the Vastitas Borealis Formation in Amazonis Planitia. J Geophys Res 113:E12010. https://doi.org/10.1029/2008JE003177

Campbell BA, Morgan GA, Whitten JL, Carter LM, Glaze LS, Campbell DB (2017) Pyroclastic flow deposits on Venus as indicators of renewed magmatic activity. J Geophys Res, Planets 122(7):1580–1596

Canup RM (2004) Simulations of a late lunar-forming impact. Icarus 168(2):433–456. https://doi.org/10.1016/j.icarus.2003.09.028

Carlson RW, Baines KH, Encrenaz Th, Taylor FW, Drossart P, Kamp LW, Pollack JB, Lellouch E, Collard AD, Calcutt SB, Grinspoon DH, Weissman PR, Smythe WD, Ocampo AC, Danielson GE, Fanale FP, Johnson TV, Kieffer HH, Matson DL, McCord TB, Soderblom LA (1991) Galileo infrared imaging spectrometer measurements at Venus. Science 253:1541–1548

Carter LM, Gilmore MS, Ghail RC, Byrne PK, Smrekar SE, Ganey TM, Izenberg N (2023) Sedimentary processes on Venus. Space Sci Rev, this collection, in revision

Cascioli G, De Marchi F, Racioppa P, Durante D, Iess L, Hensley S, Mazarico E, Smrekar SE (2021) The determination of the rotational state and interior structure of Venus with VERITAS. Planet Sci J 2:220. https://doi.org/10.3847/PSJ/ac26c0

Cascioli G, Renaud JP, Mazarico E, Durante E, Iess L, Gossen S, Smrekar S (2023) Constraining the Venus interior structure with future VERITAS measurements of the gravitational atmospheric loading. Planet Sci J 4:65. https://doi.org/10.3847/PSJ/acc73c

Choblet G, Parmentier EM (2009) Thermal convection heated both volumetrically and from below: implications for predictions of planetary evolution. Phys Earth Planet Inter 173(3):290–296. https://doi.org/10.1016/j.pepi.2009.01.005

Chodas PW, Wang TC, Sjogren WL, Ekelund JE (1992) Magellan ephemeris improvement using synthetic aperture radar landmark measurements. In: Astrodynamics 1991; proceedings of the AAS/AIAA astrodynamics conference, Durango, CO, aug. 19-22, 1991. Pt. 2 (A92-43251 18-13). Advances in the astronautical sciences. Univelt, San Diego, pp 875–889

Chodas PW, Lewicki SA, Hensley S, Masters WC (1993) High precision Magellan orbit determination for stereo image processing. In: Astrodynamics 1993. Advances in the astronautical sciences, vol 85. Univelt, San Diego, pp 279–296

Clegg SM, Sharma SK, Misra AK, Dyar MD, Hecht MH, Lambert J, Feldman S, Dallmann N, Wiens RC, Humphries SD, Vaniman DT, Speicher EA, Carmosino ML, Smrekar SE, Treiman A, Wang A, Maurice S, Esposito L (2011) Remote Raman-laser induced breakdown spectroscopy (LIBS) geochemical investigation under Venus atmospheric conditions. In: 42nd lunar and planetary science conference. Abstract #1568. https://www.lpi.usra.edu/meetings/lpsc2011/pdf/1568.pdf

Clegg SM, Wiens RC, Newell RT, DeCroix DS, Sharma SK, Dyar MD, Anderson RB, Angel SM, Martinez R, McInroy R (2016) Remote geochemical and mineralogical analyses under Venus atmospheric conditions by Raman - laser induced breakdown spectroscopy (LIBS). In: American geophysical union, fall general assembly 2016, abstract id.P41B-2068

Costa GC, Jacobson NS, Lukco D, Hunter GW, Nakley L, Radoman-Shaw BG, Harvey RP (2018) Oxidation behavior of stainless steels 304 and 316 under the Venus atmospheric surface conditions. Corros Sci 132:260–271

Cottini V, Aslam S, Gorius N, Hewagama T, Ignatiev N, Piccioni G, D'Aversa E (2018) Cuve - cubesat UV experiment. In: European planetary science congress 2018, held 16-21 September 2018 at TU Berlin, Berlin, Germany, id. EPSC2018-1156

Crameri F, Tackley PJ (2016) Subduction initiation from a stagnant lid and global overturn: new insights from numerical models with a free surface. Prog Earth Planet Sci 3(1):30. https://doi.org/10.1186/s40645-016-0103-8

Crowley JW, Gérault M, O'Connell RJ (2011) On the relative influence of heat and water transport on planetary dynamics. Earth Planet Sci Lett 310:380–388. https://doi.org/10.1016/j.epsl.2011.08.035

Davaille A, Smrekar SE, Tomlinson S (2017) Experimental and observational evidence for plume-induced subduction on Venus. Nat Geosci 10:349–355. https://doi.org/10.1038/ngeo2928

Davies JH, Davies DR (2010) Earth's surface heat flux. Solid Earth 1:5–24. https://doi.org/10.5194/se-1-5-2010

de Bergh C, Bézard B, Owen T, Crisp D, Maillard J-P, Lutz BL (1991) Deuterium on Venus: observations from Earth. Science 251(4993):547–549. https://doi.org/10.1126/science.251.4993.547

Didion A, Komjathy A, Sutin B, Nakazono B, Karp A, Wallace M, Lantoine G, Krishnamoorthy S, Rud M, Cutts J (2018) Remote sensing of venusian seismic activity with a small spacecraft, the VAMOS mission concept. In: 2018 IEEE aerospace conference. IEEE, pp 1–14

Dominguez MD, Rodriguez-Manfredi J-A, Jiménez V, Bermejo S, Pons-Nin J (2020) A miniaturized 3d heat flux sensor to characterize heat transfer in regolith of planets and small bodies. Sens Actuators 20:4135:1–4135:17. https://doi.org/10.3390/s20154135

Donahue TM, Hoffman JH, Hodges RR, Watson AJ (1982) Venus was wet - a measurement of the ratio of deuterium to hydrogen. Science 216:630–633. https://doi.org/10.1126/science.216.4546.630

Dong X, Liu Y, He J, the mission team (2023) VOICE: a Venus volcano imaging and climate explorer mission. In: Venus surface and atmosphere 30 jan-1st feb 2023, Lunar and Planetary Institute, Houston, Abstract #8068. https://www.hou.usra.edu/meetings/venussurface2023/pdf/8068.pdf

Driscoll P, Bercovici D (2013) Divergent evolution of Earth and Venus: influence of degassing, tectonics, and magnetic fields. Icarus 226(2):1447–1464. https://doi.org/10.1016/j.icarus.2013.07.025

Duan X, Moghaddam M, Wenkert D, Jordan RL, Smrekar SE (2010) X band and model of Venus atmosphere permittivity. Radio Sci 45:1–19

Dumoulin C, Tobie G, Verhoeven O, Rosenblatt P, Rambaux N (2017) Tidal constraints on the interior of Venus. J Geophys Res, Planets 122:1338–1352. https://doi.org/10.1002/2016JE005249

Duncan MS, Dasgupta R (2017) Great oxygenation event: rise of Earth's atmospheric oxygen control by ancient subduction of organic carbon. Nat Geosci 10:387–392

Dyar MD, Helbert J, Maturilli A, Mueller N, Kappel D (2020) Probing Venus surface iron contents with six-band VNIR spectroscopy from orbit. Geophys Res Lett 47:e2020GL090497. https://doi.org/10.1029/2020GL090497

Dyar MD, Helbert J, Cooper RD, Skulte EC, Maturilli A, Mueller NT, Kappel D, Smrekar SS (2021) Surface weathering on Venus: constraints from kinetic, spectroscopic, and geochemical data. Icarus 358:114139. https://doi.org/10.1016/j.icarus.2020.114139

Ehrenreich D, Vidal-Madjar A, Widemann T, Gronoff G, Tanga P, Barthélemy M, Lilensten J, des Lecavelier EA, Arnold L (2011) Transmission spectrum of Venus as a transiting exoplanet. Astron Astrophys 527:L2. https://doi.org/10.1051/0004-6361/201118400

Eismont NA, Zasova LV, Simonov AV, Kovalenko ID, Gorinov DA, Abbakumov AS, Bober SA (2020) Venera-D mission scenario and trajectory. Sol Syst Res 53:578–585. https://doi.org/10.1134/S0038094619070062

Eismont NA, Nazirov RR, Fedyaev KS, Zubko VA, Belyaev AA, Zasova LV, Gorinov DA, Simonov AV (2021a) Resonant orbits in the problem of expanding the reachable landing areas on the surface of Venus. Astron Lett 47:316–330. https://doi.org/10.1134/S1063773721050042

Eismont NA, Zubko VA, Belyaev AA, Zasova LV, Gorinov DA, Simonov AV, Nazirov RR, Fedyaev KS (2021b) Gravity assists maneuver in the problem of extension accessible landing areas on the Venus surface. Open Astron J 30:103–109. https://doi.org/10.1515/astro-2021-0013

Elkins-Tanton LT, Smrekar SE, Hess PC, Parmentier EM (2007) Volcanism and volatile recycling on a one-plate planet: applications to Venus. J Geophys Res, Planets 112(E4):E04S06. https://doi.org/10.1029/2006JE002793

Emsenhuber A, Asphaug E, Cambioni S, Gabriel TSJ, Schwartz SR (2021) Collision chains among the terrestrial planets. II. An asymmetry between Earth and Venus. Planet Sci J 2(5):199. https://doi.org/10.3847/psj/ac19b1

Encrenaz T, Lellouch E, Paubert G, Gulkis S (1991) First detection of HDO in the atmosphere of Venus at radio wavelengths: an estimate of the H_2O vertical distribution. Astron Astrophys 246:L63–L66

Encrenaz T, Lellouch E, Cernicharo J, Paubert G, Gulkis S, Spilker T (1995) The thermal profile and water abundance in the Venus mesosphere from H_2O and HDO millimeter observations. Icarus 117(1):162–172

Encrenaz T, Greathouse TK, Roe H, Richter M, Lacy J, Bézard B, Fouchet T, Widemann T (2012) HDO and SO_2 thermal mapping on Venus: evidence for strong SO_2 variability. Astron Astrophys 543:A153

Encrenaz T, Greathouse TK, Richter MJ, Lacy J, Widemann T, Bézard B, Fouchet T, deWitt C, Atreya SK (2013) HDO and SO_2 thermal mapping on Venus. II. The SO_2 spatial distribution above and within the clouds. Astron Astrophys 559:A65, 9p

Encrenaz T, Moreno R, Moullet A, Lellouch E, Fouchet T (2015) Submillimeter mapping of mesospheric minor species on Venus with ALMA (2015). Planet Space Sci 113–114:275–291

Encrenaz T, Greathouse TK, Richter MJ, DeWitt C, Widemann T, Bézard B, Fouchet T, Atreya SK, Sagawa H (2016) HDO and SO_2 thermal mapping on Venus. III. Short-term and long-term variations between 2012 and 2016. Astron Astrophys 595:A74, 15 pp

Encrenaz T, Greathouse TK, Marcq E, Sagawa H, Widemann T, Bézard B, Fouchet T, Lefèvre F, Lebonnois S, Atreya SK, Lee YJ, Giles R, Watanabe S (2019) HDO and SO_2 thermal mapping on Venus. IV. Statistical analysis of the SO_2 plumes. Astron Astrophys 623:A70, 11 pp

Encrenaz T, Greathouse TK, Marcq E, Sagawa H, Widemann T, Bézard B, Fouchet T, Lefèvre F, Lebonnois S, Atreya SK, Lee YJ, Giles R, Watanabe S, Shao W, Zhang X, Bierson CJ (2020a) HDO and SO_2 thermal mapping on Venus. V. Evidence for a long-term anti-correlation. Astron Astrophys 639:A69. https://doi.org/10.1051/0004-6361/202037741

Encrenaz T, Greathouse TK, Marcq E, Widemann T, Bézard B, Fouchet T, Giles R, Sagawa H, Greaves J, Sousa-Silva C (2020b) A stringent upper limit of the PH_2 abundance at the cloud top of Venus. Astron Astrophys 643:L4. https://doi.org/10.1051/0004-6361/202039559

Encrenaz T, Greathouse TK, Giles R, Widemann T, Bézard B, Lefèvre M, Shao W (2023) HDO and SO_2 thermal mapping on Venus: VI. Anomalous SO_2 behavior during late 2021. Astron Astrophys 674:A199. https://doi.org/10.1051/0004-6361/202245831

Esposito LW (1984) Sulfur dioxide: episodic injection shows evidence for active Venus volcanism. Science 223(4640):1072–1074. https://doi.org/10.1126/science.223.4640.1072

Esposito LW, Copley M, Eckert R, Gates L, Stewart AIF, Worden H (1988) Sulfur dioxide at the Venus cloud tops, 1978–1986. J Geophys Res, Atmos 93(D5):5267–5276. https://doi.org/10.1029/JD093iD05p05267

European Space Agency (ESA) (2021), EnVision assessment study report. Yellow book, ESA/SCI(2021)1, pages 1-111. https://sci.esa.int/documents/34375/36249/EnVision_YB_final.pdf

Evans AJ, Soderblom JM, Andrews-Hanna JC, Solomon SC, Zuber MT (2016) Identification of buried lunar impact craters from GRAIL data and implications for the nearside Maria. Geophys Res Lett 43(6):2445–2455

Farr TG, Rosen PA, Caro E, Crippen R, Duren R, Hensley S, Kobrick M, Paller M, Rodriguez E, Roth L, Seal D, Shaffer S, Shimada J, Umland J, Werner M, Oskin M, Burbank D, Alsdorf D (2007) The shuttle radar topography mission. Rev Geophys 45:RG2004. https://doi.org/10.1029/2005RG000183

Fauchez TJ, Turbet M, Villanueva GL, Wolf ET, Arney G, Kopparapu RK, Lincowski A, Mandell A, de Wit J, Pidhorodetska D, Domagal-Goldman SD, Stevenson KB (2019) Impact of clouds and hazes on the simulated JWST transmission spectra of habitable zone planets in the TRAPPIST-1 system. Astrophys J 887(2):194. https://doi.org/10.3847/1538-4357/ab5862

Fauchez TJ, Villanueva GL, Sergeev DE, Turbet M, Boutle IA, Tsigaridis K, Way MJ, Wolf ET, Domagal-Goldman SD, Forget F, Jacob Haqq-Misra J, Kopparapu RK, Manners J, Mayne NJ (2022) The TRAPPIST-1 habitable atmosphere intercomparison (THAI). III. Simulated observables—the return of the spectrum. Planet Sci J 3:213. https://doi.org/10.3847/PSJ/ac6cf1

Fegley B Jr, Prinn RG (1989) Estimation of the rate of volcanism on Venus from reaction rate measurements. Nature 337:55. https://doi.org/10.1038/337055a0

Fegley B Jr, Treiman AH (1992) Chemistry of atmosphere-surface interactions on Venus and Mars. In: Luhmann JG, Tatrallyay M, Pepin RO (eds) Venus and Mars: atmospheres, ionospheres, and solar wind interactions. AGU, Washington, pp 7–71

Fegley B Jr, Klingelhofer G, Lodders K, Widemann T (1997) Geochemistry of surface-atmosphere interactions on Venus. In: Bougher SW, Hunten DM, Phillips RJ (Eds) Venus II. University of Arizona Press, Tucson, pp 591–636

Fegley B, Treiman AH, Sharpton VL (1992) Venus surface mineralogy: observational and theoretical constraints. LPSC 22, 3

Florensky KP et al (1977) The surface of Venus as revealed by Soviet Venera 9 and 10. Geol Soc Am Bull 88:1537–1545

Ford P et al (1992) J Geophys Res 97(E8):13103–13114. https://doi.org/10.1029/92JE01085

French R, Mandy C, Hunter R, Mosleh E, Sinclair D, Beck P, Seager S, Petkowski JJ, Carr CE, Grinspoon DH, et al (2022) Rocket lab mission to Venus. Aerospace 9:445. https://doi.org/10.3390/aerospace9080445

Frey HV (2006) Impact constraints on, and a chronology for, major events in early Mars history. J Geophys Res, Planets 111(E8:E08S91). https://doi.org/10.1029/2005JE002449

Frey HV, Roark JH, Shockey KM, Frey EL, Sakimoto SEH (2002) Ancient lowlands on Mars. Geophys Res Lett 29(10):22-1–22-4. https://doi.org/10.1029/2001GL013832

Fujisawa Y, Murakami S, Sugimoto N, Takagi M, Imamura T, Horinouchi T, Hashimoto GL, Ishiwatari M, Enomoto T, Miyoshi T, Kashimura H, Hayashi Y-Y (2022) The first assimilation of Akatsuki single-layer winds and its validation with Venusian atmospheric waves excited by solar heating. Sci Rep 12:14577. https://doi.org/10.1038/s41598-022-18634-6

Fukuhara T, Futaguchi M, Hashimoto G, Horinouchi T, Imamura T, Iwagami N, Kouyama T, Murakami S, Nakamura M, Ogohara K, Sato M, Sato TM, Suzuki M, Tagushi M, Takagi S, Ueno M, Watanabe S, Yamada M, Yamazaki A (2017) Large stationary gravity wave in the atmosphere of Venus. Nat Geosci 10:85–88. https://doi.org/10.1038/ngeo2873

Fukuya K, Imamura T, Taguchi M, Kouyama T (2022) Horizontal structures of bow-shaped mountain wave trains seen in thermal infrared images of Venusian clouds taken by Akatsuki LIR. Icarus 378:114936. https://doi.org/10.1016/j.icarus.2022.114936

Fung AK, Chen K-S, Chen K (2010) Microwave scattering and emission models for users. Artech House remote sensing library. Artech House, Norwood

Futaana Y et al (2017) Solar wind interaction and impact on the Venus atmosphere. Space Sci Rev 212(3–4):1453–1509. https://doi.org/10.1007/s11214-017-0362-8

Gaillard F, Scaillet B (2014) A theoretical framework for volcanic degassing chemistry in a comparative planetology perspective and implications for planetary atmospheres. Earth Planet Sci Lett 403:307–316

Gaillard F, Bernadou F, Roskosz M, Bouhifd MA, Marrocchi Y, Iacono-Marziano G, Moreira M, Scaillet B, Rogerie G (2022) Redox controls during magma ocean degassing. Earth Planet Sci Lett 577:117255

Ganesh ILM, Carter LM, Smith IB (2020) J Volcanol Geotherm Res 390:106748. https://doi.org/10.1016/j.jvolgeores.2019.106748

Garcia R, Lognonné P, Bonnin X (2005) Detecting atmospheric perturbations produced by Venus quakes. Geophys Res Lett 32(16):L16205

Garcia RF, Martire L, Chaigneau Y, Cadu A, Mimoun D, Bassas Portus M, et al, Martin R (2021) An active source seismo-acoustic experiment using tethered balloons to validate instrument concepts and modeling tools for atmospheric seismology. Geophys J Int 225(1):186–199

Garcia RF, Klotz A, Hertzog A, Martin R, Gérier S, Kassarian E et al (2022) Infrasound from large earthquakes recorded on a network of balloons in the stratosphere. Geophys Res Lett 49:e2022GL098844. https://doi.org/10.1029/2022GL098844

Garvin JB (1990) The global budget of impact-derived sediments on Venus. Earth Moon Planets 50:175–190. https://doi.org/10.1007/BF00142394

Garvin JB, Head JW, Zuber MR, Helfenstein P (1984) Venus: the nature of the surface from Venera panoramas. J Geophys Res 89(B5):3381–3399. https://doi.org/10.1029/JB089iB05p03381

Garvin JB, Glaze LS, Ravine MA et al (2018) Venus descent imaging for surface topography and geomorphology. In: 49th lunar and planetary science conference 2018. LPI contrib., vol 2083, LPSC, 49, 2287

Garvin JB, Campbell B, Pimentel E, Dotson R, Gilmore M, Arney G, Getty S, Slayback D (2022b) Km-scale topography of Alpha Regio: DAVINCI entry corridor for descent imaging science. In: American geophysical union fall meeting, Chicago, Il., Dec. 2022, Abstract #1445

Garvin JB, Getty SA, Arney GN, Johnson NM, Kohler E, Schwer KO, Sekerak M, Bartels A, Saylor RS, Elliott VE, Goodloe CS, Garrison MB, Cottini V, Izenberg N, Lorenz R, Malespin CA, Ravine M, Webster CR, Atkinson DH, Aslam S, Atreya S, Bos BJ, Brinckerhoff WB, Campbell B, Crisp D, Filiberto JR, Forget F, Gilmore M, Gorius N, Grinspoon D, Hofmann AE, Kane SR, Kiefer W, Lebonnois S, Mahaffy PR, Pavlov A, Trainer M, Zahnle KJ, Zolotov M (2022a) Revealing the mysteries of Venus: the DAVINCI mission. Planet Sci J 3:117. https://doi.org/10.3847/psj/ac63c2

Garvin JB, Campbell B, Gilmore M, Arney GN, Getty S et al (2023). AAS/PSJ, in preparation

Genova A, Goossens S, Mazarico E, Lemoine FG, Neumann GA, Kuang W, Sabaka TJ, Hauck I, Steven A, Smith DE, Solomon SC (2019) Geodetic evidence that Mercury has a solid inner core. Geophys Res Lett 46:3625–3633

Gerlach TM (1980) Evaluation of volcanic gas analyses from Kilauea volcano. J Volcanol Geotherm Res 7(3–4):295–317

Ghail R, Smrekar SE, Borrelli ME, Byrne PK, Gilmore MS, Herrick RR, Ivanov MA, O'Rourke JG, Plesa A-C, Rolf T, Sabbeth L, Schools JW, Shellnutt G (2023) Volcanic and tectonic constraints on the evolution of Venus. Space Sci Rev, this collection, in revision

Ghent RR, Phillips RJ, Hansen VL, Nunes DC (2005) Finite element modeling of short-wavelength folding on Venus: implications for the plume hypothesis for crustal plateau formation. J Geophys Res, Planets 110(E11):E11006. https://doi.org/10.1029/2005JE002522

Gillmann C, Tackley P (2014) Atmosphere/mantle coupling and feedbacks on Venus. J Geophys Res, Planets 119(6):1189–1217

Gillmann C, Golabek GJ, Tackley PJ (2016) Effect of a single large impact on the coupled atmosphere-interior evolution of Venus. Icarus 268:295–312. https://doi.org/10.1016/j.icarus.2015.12.024

Gillmann C, Way MJ, Avice G, Breuer D, Golabek GJ, Höning D, Krissansen-Totton J, Lammer H, O'Rourke JG, Persson M, Plesa A-C, Salvador A, Scherf M, Zolotov M (2022) The long-term evolution of the atmosphere of Venus: processes and feedback mechanisms. Space Sci Rev 218:56. https://doi.org/10.1007/s11214-022-00924-0

Gillon M, Jehin E, Lederer SM, Delrez L, de Wit J, Burdanov A, Grootel VV, Burgasser AJ, Triaud AHMJ, Opitom C, Demory BO, Sahu DK, Gagliuffi DCB, Magain P, Queloz D (2016) Temperate Earth-sized planets transiting a nearby ultracool dwarf star. Nature 533:221–224. https://doi.org/10.1038/nature17448

Gillon M, Triaud A, Demory BO, Jehion E, Agol E, Deck KM, Lederer SM, de Wit J, Burdanov A, Ingalls JG, Bolmont E, Leconte J, Raymond SN, Selsis F, Turbet M, Barkaoui K, Burgasser A, Burleigh M, Carey SJ, Chaushev A, Copperwheat CM, Delrez L, Fernandes CS, Holdsworth DL, Kotze EJ, Van Grootel V, Almeaky Y, Benkhaldoun Z, Magain P, Queloz D (2017) Seven temperate terrestrial planets around the nearby ultracool dwarf star TRAPPIST-1. Nature 542:456–460. https://doi.org/10.1038/nature21360

Gilmore MS, Mueller N, Helbert J (2015) VIRTIS emissivity of Alpha Regio, Venus, with implications for tessera composition. Icarus 254:350–361. https://doi.org/10.1016/j.icarus.2015.04.008

Gilmore M, Treiman A, Helbert J, Smrekar S (2017) Venus surface composition constrained by observation and experiment. Space Sci Rev 212(3–4):1511–1540

Gilmore MS, Brossier JF, Zalewski N, Stein AJ (2019) Contrasts between low emissivity tessera and plains materials on Venus mountaintops. In: International Venus conference. https://www.cps-jp.org/~akatsuki/venus2019/program/IVC2019_Program.pdf

Gilmore MS, Dyar MD, Mueller N, Brossier J, Santos A, Filiberto J, Ivanov MA, Ghail R, Helbert J (2023) Mineralogy of the Venus surface. Space Sci Rev 219:5. https://doi.org/10.1007/s11214-023-00988-6

Glass DE, Jones J-P, Shevade AV, Bhakta D, Raub E, Sim R, Bugga RV (2020) High temperature primary battery for Venus surface missions. J Power Sources 449:227492. https://doi.org/10.1016/j.jpowsour.2019.227492

Glaze LS, Wilson CF, Zasova LV, Nakamura M, Limaye S (2018) Future of Venus research and exploration. Space Sci Rev 214:89. https://doi.org/10.1007/s11214-018-0528-z

Grassi D, Migliorini A, Montabone L, Lebonnois S, Cardesìn-Moinelo A, Piccioni G, Drossart P, Zasova LV (2010) Thermal structure of Venusian nighttime mesosphere as observed by VIRTIS-Venus Express. J Geophys Res 115:E09007. https://doi.org/10.1029/2009JE003553

Grassi D, Politi R, Ignatiev NI, Plainaki C, Lebonnois S, Wolkenberg P, Montabone L, Migliorini A, Piccioni G, Drossart P (2014) The Venus nighttime atmosphere as observed by the VIRTIS-M instrument. Average fields from the complete infrared data set. J Geophys Res, Planets 119:837–849. https://doi.org/10.1002/2013JE004586

Greaves JS, Richards AMS, Bains W, Rimmer PB, Sagawa H et al (2021) Phosphine gas in the cloud decks of Venus. Nat Astron 5:655–664. https://doi.org/10.1038/s41550-020-1174-4

Greeley R, Arvidson RE, Elachi C, Geringer MA, Plaut JJ, Saunders RS, Schubert G, Stofan ER, Thouvenot EJP, Wall SD, Weitz CM (1992) Aeolian features on Venus: preliminary Magellan results. J Geophys Res 97:13319–13345. https://doi.org/10.1029/92JE00980

Greeley R, Bender K, Thomas PE, Schubert G, Limonadi D, Weitz C (1995) Wind-related features and processes on Venus: summary of Magellan results. Icarus 115:399

Grimm RE, Hess PC (1997) The crust of Venus. In: Venus II. University of Arizona Press, Tucson, pp 1205–1244

Grinspoon DH, Bullock MA (2007) Astrobiology and Venus exploration. In: Esposito LW, Stofan ER, Cravens TE (eds) Exploring Venus as a terrestrial planet. https://doi.org/10.1029/176GM12

Grott M, Spohn T, Knollenberg J, Krause C, Hudson TL, Piqueux S, et al, Banerdt WB (2021) Thermal conductivity of the Martian soil at the InSight landing site from HP3 active heating experiments. J Geophys Res, Planets 126(7):e2021JE006861

Gülcher AJP, Gerya TV, Montési LGJ et al (2020) Corona structures driven by plume–lithosphere interactions and evidence for ongoing plume activity on Venus. Nat Geosci 13:547–554. https://doi.org/10.1038/s41561-020-0606-1

Halliday AN (2013) The origins of volatiles in the terrestrial planets. Geochim Cosmochim Acta 105:146–171. https://doi.org/10.1016/j.gca.2012.11.015

Hamano K, Abe Y, Genda H (2013) Emergence of two types of terrestrial planet on solidification of magma ocean. Nature 497:607–610. https://doi.org/10.1038/nature12163

Hansen VL, López I (2010) Venus records a rich early history. Geology 38(4):311–314

Hansen VL, Olive A (2010) Artemis, Venus: the largest tectonomagmatic feature in the solar system? Geology 38(5):467–470

Hansen VL, Phillips RJ (1993) Tectonics and volcanism of eastern Aphrodite Terra, Venus: no subduction, no spreading. Science 260(5107):526–530

Hansen VL, Willis JJ (1996) Structural analysis of sampling of tesserae: implications for Venus geodynamics. Icarus 123:296–312

Hansen VL, Willis JJ (1998) Ribbon terrain formation, southwestern Fortuna Tessera, Venus: Implications for lithosphere evolution. Icarus 132(2):321–343

Harper CL, Jacobsen SB (1996) Evidence for ^{182}Hf in the early Solar System and constraints on the timescale of terrestrial accretion and core formation. Geochim Cosmochim Acta 60(7):1131–1153. https://doi.org/10.1016/0016-7037(96)00027-0

Hashimoto GL (2003) On observing the compositional variability of the surface of Venus using nightside near-infrared thermal radiation. J Geophys Res 108(E9):5109

Hashimoto GL, Abe Y (2005) Climate control on Venus: comparison of the carbonate and pyrite models. Planet Space Sci 53(8):839–848. https://doi.org/10.1016/j.pss.2005.01.005

Hashimoto GL, Abe Y, Sasaki S (1997) CO_2 amount on Venus constrained by a criterion of topographic-greenhouse instability. Geophys Res Lett 24:289. https://doi.org/10.1029/96GL04006

Hashimoto GL, Roos-Serote M, Sugita S, Gilmore MS, Kamp LW, Carlson RW, Baines KH (2008) Felsic highland crust on Venus suggested by Galileo near-infrared mapping spectrometer data. J Geophys Res, Planets 113(E5):E00B24. https://doi.org/10.1029/2008JE003134

Hays L, Archenbach L, Bailey J, Barnes R, Barros J, Bertka C, Boston P (2015) NASA astrobiology strategy, NASA, Washington

Head JW, Chapman CR, Strom RG, Fassett CI, Denevi BW, Blewett DT, Ernst CM, Watters TR, Solomon SC, Murchie SL (2011) Flood volcanism in the northern high latitudes of Mercury revealed by MESSENGER. Science 333(6051):1853–1856

Helbert J, Müller N, Kostama P, Marinangeli L, Piccioni G, Drossart P (2008) Surface brightness variations seen by VIRTIS on Venus Express and implications for the evolution of the Lada Terra region, Venus. Geophys Res Lett 35(11):L11201. https://doi.org/10.1029/2008GL033609

Helbert J, Wendler D, Walter I, Widemann T, Marcq E, Ferrari S, Maturilli A, Mueller N, Jaenchen J, Kappel D, Boerner A, d'Amore M, Dyar MD, Arnold GE, Smrekar SE (2016) The Venus emissivity mapper (VEM) concept. In: Infrared remote sensing and instrumentation XXIV. Proceedings SPIE, San Diego, Aug 2016, Paper 9973-26

Helbert J, Dyar M, Walter I, Wendler D, Widemann T, Marcq E, Guignan G, Ferrari S, Maturilli A, Mueller N, Kappel D (2018) The Venus emissivity mapper (VEM): obtaining global mineralogy of Venus from orbit. In: Infrared remote sensing and instrumentation XXVI, San Diego, United States, Aug 2018, 107650D. https://doi.org/10.1117/12.2320112

Helbert J, Säuberlich T, Darby Dyar M, Ryan C, Walter I, Reess J-M, Rosas-Ortiz Y, Peter G, Maturilli A, Arnold G (2020) The Venus emissivity mapper (VEM): advanced development status and performance evaluation. In: Proc. SPIE 11502, Infrared remote sensing and instrumentation XXVIII, 20 August 2020, 1150208

Helbert J, Maturilli A, Dyar MD et al (2021) Deriving iron contents from past and future Venus surface spectra with new high-temperature laboratory emissivity data. Sci Adv. https://doi.org/10.1126/sciadv.aba9428

Hensley S (2009) A combined methodology for SAR interferometric and stereometric error modeling. In: Radar conference, 2009 IEEE. IEEE

Hensley S, Martin J, Oveisgsharan S, Duan X, Campbell BA (2018) Radar performance modeling for Venus. In: VEXAG meeting, Applied Physics Laboratory. https://www.lpi.usra.edu/vexag/meetings/archive/vexag-16/presentations/Hensley.pdf

Hensley S, Wallace MS, Martin J, Perkovic-Martin D, Smrekar S, Younis M, Lachaise M, Prats P, Rodriguez M, Zebker H, Campbell B, Mastrogiuseppe M (2022) Planned differential interferometric SAR observations at Venus by the Veritas mission. In: Proceedings of IGARSS 2022, international geoscience and remote sensing symposium, Kuala Lumpur, Indonesia, 17-22 July, 2022

Herrick RR, Hensley S (2023) Surface changes observed on a Venusian volcano during the Magellan mission. Science 379(6638):1205–1208. https://doi.org/10.1126/science.abm7735

Herrick RR, Rumpf ME (2011) Postimpact modification by volcanic or tectonic processes as the rule, not the exception, for Venusian craters. J Geophys Res, Planets 116(E2):E02004

Herrick RR, Sharpton VL (2000) Implications from stereo-derived topography of Venusian impact craters. J Geophys Res, Planets 105(E8):20245–20262

Herrick RR, Bjonnes EE, Carter L, Ghail R, Gillmann C, Gilmore MS, Hensley S, Ivanov MA, Izenberg NR, Mueller N, O'Rourke JG, Rolf T, Smrekar SE, Weller MB (2023) Resurfacing history and volcanic activity of Venus. Space Sci Rev 219:29. https://doi.org/10.1007/s11214-023-00966-y

Hirschmann MM (2006) Water, melting, and the deep Earth H_2O cycle. Annu Rev Earth Planet Sci 34:629–653. https://doi.org/10.1146/annurev.earth.34.031405.125211

Höning D, Baumeister P, Grenfell JL, Tosi N, Way MJ (2021) Early habitability and crustal decarbonation of a stagnant-lid Venus. J Geophys Res, Planets 126(10):e2021JE006895

Horinouchi T, Kouyama T, Lee YJ, Murakami S, Ogohara K, Takagi M, Imamura T, Nakajima K, Peralta J, Yamazaki A, Yamada M, Watanabe S (2018) Mean winds at the cloud top of Venus obtained from two-wavelength UV imaging by Akatsuki. Earth Planets Space 70:10. https://doi.org/10.1186/s40623-017-0775-3

Huang J, Yang A, Zhong S (2013) Constraints of the topography, gravity and volcanism on Venusian mantle dynamics and generation of plate tectonics. Earth Planet Sci Lett 362:207–214

Hueso R, Sánchez-Lavega A, Piccioni G, Drossart P, Gérard JC, Khatuntsev I, Zasova L, Migliorini A (2008) Morphology and dynamics of Venus oxygen airglow from Venus Express/visible and infrared thermal imaging spectrometer observations. J Geophys Res 113:E00B02. https://doi.org/10.1029/2008JE003081

Hueso R, Peralta J, Sánchez-Lavega A (2012) Assessing the long-term variability of Venus winds at cloud level from VIRTIS-Venus Express. Icarus 217:585–598. https://doi.org/10.1016/j.icarus.2011.04.020

Hueso R, Peralta J, Garate-Lopez I, Bandos TV, Sánchez-Lavega A (2015) Six years of Venus winds at the upper cloud level from UV, visible and near infrared observations from VIRTIS on Venus Express. Planet Space Sci 113:78–99. https://doi.org/10.1016/j.pss.2014.12.010

Hupe P (2018) Global infrasound observations and their relation to atmospheric tides and mountain waves. Ph.D. Thesis, Faculty of Physics. https://edoc.ub.uni-muenchen.de/23790/

Ignatiev I, Moroz VI, Moshkin BE, Ekonomov AP, Gnedykh VI, Grigoriev AV, Khatuntsev IV (1997) Water vapour in the lower atmosphere of Venus: a new analysis of optical spectra measured by entry probes. Planet Space Sci 45:427–438. https://doi.org/10.1016/S0032-0633(96)00143-2

Ikoma M, Elkins-Tanton L, Hamano K, Suckale J (2018) Water partitioning in planetary embryos and proto-planets with magma oceans. Space Sci Rev 214:76. https://doi.org/10.1007/s11214-018-0508-3

Imamura T, Mitchell J, Lebonnois S, Kaspi Y, Showman AP, Korablev O (2020) Superrotation in planetary atmospheres. Space Sci Rev 216:87. https://doi.org/10.1007/s11214-020-00703-9

Ivanov MA, Head JW III (2011) Planet Space Sci 59:1559–1600. https://doi.org/10.1016/j.pss.2011.07.008

Ivanov MA, Head JW III (2015) Planet Space Sci 113–114:10–32. https://doi.org/10.1016/j.pss.2015.03.016

Ivanov MA, Head JW (1996) Tessera terrain on Venus: a survey of the global distribution, characteristics, and relation to surrounding units from Magellan data. J Geophys Res, Planets 101(E6):14861–14908

Ivanov MA, Zasova LV, Gerasimov MV, Korablev OI, Marov MY, Zelenyi LM, Ignatiev NI, Tuchin AG (2017a) The nature of terrains of different types on the surface of Venus and selection of potential landing sites for a descent probe of the Venera-D mission. Sol Syst Res 51:1–19

Ivanov MA, Zasova LV, Zeleny LM, Gerasimov MV, Ignatiev NI, Korablev OI, Marov MY (2017b) Estimates of abundance of the short-baseline (1-3 meters) slopes for different Venusian terrains using terrestrial analogues. Sol Syst Res 51:87–103

Iwagami N, Yamaji T, Ohtsuki S, Hashimoto GL (2010) Hemispherical distribution of CO above the Venus' clouds by ground-based 2.3 μm spectroscopy. Icarus 207:558–563

Izenberg et al (1994) Geophys Res Lett 21:289–292

Izenberg N, Scott V, Fultz B (2023) VISTA: Venus in situ transfer and analysis mission concept. In: 2023 IEEE aerospace conference, Big Sky, MT, USA, pp 1–17. https://doi.org/10.1109/AERO55745.2023.10115688

Jacobson SA, Dobson C (2022) What does it mean to have no moon? Evidence for an early or no giant impact on Venus. In: Ancient Venus conference, held virtually 25-27 July, 2022. LPI contribution, vol 2680, id.2030

Jaupart C, Labrosse S, Lucazeau F, Mareschal J-C (2015) Temperatures, heat, and energy in the mantle of the Earth. In: Bercovici D, Schubert G (eds) Treatise on geophysics, 2nd ed., vol 7. Elsevier, New York, pp 253–303. https://doi.org/10.1016/B978-0-444-53802-4.00126-3

Jessup KL, Marcq E, Mills F, Mahieux A, Limaye S, Wilson C, Allen M, Bertaux J-L, Markiewicz W, Roman T, Vandaele AC, Wilquet V, Yung Y (2015) Coordinated Hubble space telescope and Venus Express observations of Venus' upper cloud deck. Icarus 258:309–336

Johnson HP, Tivey MA, Bjorklund TA, Salmi MS (2010) Hydrothermal circulation within the Endeavor Segment, Juan de Fuca Ridge. Geochem Geophys Geosyst 11:Q05002. https://doi.org/10.1029/2009GC002957

Kaltenegger L, Payne RC, Lin Z, Kasting J, Delrez L (2023) Hot Earth or young Venus? A nearby transiting rocky planet mystery. Mon Not R Astron Soc Lett 524(1):L10–L14. https://doi.org/10.1093/mnrasl/slad064

Kane SR (2022) Atmospheric dynamics of a near tidally locked Earth-sized planet. Nat Astron 6:420–427. https://doi.org/10.1038/s41550-022-01626-x

Kane SR, Kopparapu RK, Domagal-Goldman SD (2014) On the frequency of potential Venus analogs from Kepler data. Astrophys J Lett 794:L5. https://doi.org/10.1088/2041-8205/794/1/L5

Kappel D (2014) MSR, a multi-spectrum retrieval technique for spatially-temporally correlated or common Venus surface and atmosphere parameters. J Quant Spectrosc Radiat Transf 133:153–176

760

Kappel D, Arnold G, Haus R et al (2012) Refinements in the data analysis of VIRTIS-M-IR Venus nightside spectra. Adv Space Res 50(2):228–255

Kappel D, Arnold G, Haus R (2016) Multi-spectrum retrieval of Venus IR surface emissivity maps from VIRTIS/VEx nightside measurements at Themis Regio. Icarus 265:42–62

Kargel JS, Komatsu G, Baker VR, Strom RG (1993) The volcanology of Venera and VEGA landing sites and the geochemistry of Venus. Icarus 103:253–275

Kasting JF, Catling D (2003) Evolution of a habitable planet. Annu Rev Astron Astrophys 41(1):429–463

Kaula WM (1999) Constraints on Venus evolution from radiogenic argon. Icarus 139:32–39. https://doi.org/10.1006/icar.1999.6082

Khatuntsev IV, Patsaeva MV, Titov DV, Ignatiev NI, Turin AV, Limaye SS, Markiewicz WJ, Almeida M, Roatsch T, Moissl R (2013) Cloud level winds from the Venus Express monitoring camera imaging. Icarus 226:140–158. https://doi.org/10.1016/j.icarus.2013.05.018

King SD (2018) Venus resurfacing constrained by geoid and topography. J Geophys Res, Planets 123:1041–1060. https://doi.org/10.1002/2017JE005475

Kitahara T, Imamura T, Sato TM, Yamazaki A, Lee Y-J, Yamada M, Watanabe S, Taguchi M, Fukuhara T, Kouyama T, Murakami S, Hashimoto GL, Ogohara K, Kashimura H, Horinouchi T, Takagi M (2019) Stationary features at the cloud top of Venus observed by Ultraviolet Imager onboard Akatsuki. J Geophys Res, Planets 124:1266–1281. https://doi.org/10.1029/2018JE005842

Knafelc J, Filiberto J, Ferré EC, Conder JA, Costello L, Crandall JR, Dyar MD, Friedman SA, Hummer DR, Schwenzer SP (2019) The effect of oxidation on the mineralogy and magnetic properties of olivine. Am Mineral 104(5):694–702

Knollenberg R, Hunten D (1980) The microphysics of the clouds of Venus: results of the pioneer Venus particle size spectrometer experiment. J Geophys Res Space Phys 85(A13):8039–8058

Konopliv AS, Yoder CF (1996) Venusian k_2 tidal love number from Magellan and PVO tracking data. Geophys Res 23(14):1857–1860

Kopparapu RK, Ramirez R, Kasting JF, Eymet V, Robinson TD, Mahadevan S, Terrien RC, Domagal-Goldman S, Meadows V, Deshpande R (2013) Habitable zones around main-sequence stars: new estimates. Astrophys J 765(2):131

Kouyama T, Taguchi M, Fukuhara T, Imamura T, Horinouchi T, Sato TM, Murakami S, Hashimoto GL, Lee Y-J, Futaguchi M, Yamada T, Akiba M, Satoh T, Nakamura M (2019) Global structure of thermal tides in the upper cloud layer of Venus revealed by LIR onboard Akatsuki. Geophys Res Lett 46:9457–9465. https://doi.org/10.1029/2019GL083820

Kremic T, Hunter G (2021) Long-lived in-situ solar system explorer (LLISSE) potential contributions to solar system exploration. Bull Am Astron Soc 53. https://doi.org/10.3847/25c2cfeb.cb6775e1

Kremic T, Ghail R, Gilmore M, Hunter G, Kiefer W, Limaye S, Pauken M, Tolbert C, Wilson C (2020) Long-duration Venus lander for seismic and atmospheric science. Planet Space Sci 190:104961. https://doi.org/10.1016/j.pss.2020.104961

Kreslavsky MA, Bondarenko NV (2017) Aeolian sand transport and aeolian deposits on Venus: a review. Aeolian Res 26:29–46. https://doi.org/10.1016/j.aeolia.2016.06.001

Krishnamoorthy S, Komjathy A, Pauken MT, Cutts JA, Garcia RF, Mimoun D, et al, Bowman DC (2018) Detection of artificially generated seismic signals using balloon-borne infrasound sensors. Geophys Res Lett 45(8):3393–3403

Krishnamoorthy S, Lai VH, Komjathy A, Pauken MT, Cutts JA, Garcia RF, et al, Cadu A (2019) Aerial seismology using balloon-based barometers. IEEE Trans Geosci Remote Sens 57(12):10191–10201

Krissansen-Totton J, Fortney JJ, Nimmo F (2021) Was Venus ever habitable? Constraints from a coupled interior–atmosphere–redox evolution model. Planet Sci J 2(5):216

Kumar S, Taylor HA (1985) Deuterium on Venus: model comparisons with pioneer Venus observations of the predawn bulge ionosphere. Icarus 62(3):494–504. https://doi.org/10.1016/0019-1035(85)90189-7

Lammer H, Leitzinger M, Scherf M, Odert P, Burger C, Kubyshkina D, Johnstone C, Maindl T, Schäfer CM, Güdel M, Tosi N, Nikolaou A, Marcq E, Erkaev NV, Noack L, Kislyakova KG, Fossati L, Pilat-Lohinger E, Ragossnig F, Dorfi EA (2020) Constraining the early evolution of Venus and Earth through atmospheric Ar, Ne isotope and bulk K/U ratios. Icarus. https://doi.org/10.1016/j.icarus.2019.113551

Lange RA (1994) The effect of H_2O, CO_2 and F on the density and viscosity of silicate melts. In: Carroll MR, Holloway JR (eds) Volatiles in magmas. Mineralogical Society of America, Washington, pp 331–370. https://doi.org/10.1515/9781501509674-015. Chap. 9

Lebonnois S, Schubert G (2017) The deep atmosphere of Venus and the possible role of density-driven separation of CO_2 and N_2. Nat Geosci 10:473–477

Lebonnois S, Schubert G, Forget F, Spiga A (2018) Planetary boundary layer and slope winds on Venus. Icarus 314:149–158

Leconte J, Forget F, Charnay B, Wordsworth R, Selsis F, Millour E, Spiga A (2013) 3D climate modeling of close-in land planets: circulation patterns, climate moist bistability, and habitability. Astron Astrophys 554:A69. https://doi.org/10.1051/0004-6361/201321042. arXiv:1303.7079

Lee D-C, Halliday AN (1995) Hafnium–tungsten chronometry and the timing of terrestrial core formation. Nature 378(6559):771–774. https://doi.org/10.1038/378771a0

Lee G, Polidan RS, Ross F (2015) Venus atmospheric maneuverable platform (VAMP) - a low cost Venus exploration concept. In: American geophysical union, fall meeting 2015, abstract id. P23A-2109

Lee YJ, Jessup KL, Perez-Hoyos S, Titov DV, Lebonnois S, Peralta J, Horinoushi T, Imamura T, Limaye S, Marcq E, Takagi M, Yamazaki A, Yamada M, Watanabe S, Murakami S, Ogohara K, McClintock WM, Holsclaw G, Roman A (2019) Long-term variations of Venus' 365 nm albedo observed by Venus Express, Akatsuki, MESSENGER, and the Hubble space telescope. Astron J 158:126

Lee YJ, Garcia Muñoz A, Choi YJ, Rauer H, Michaelis H, Cabrera J, Marcq E, Granzer T, Young E, Lebonnois S, Imamura T (2022) Long-term plan to monitor Venus using Earth-orbiting CubeSats: chasing the long-term variability of our nearest neighbor planet Venus (CLOVE). In: 44th COSPAR scientific assembly. Held 16-24 July, 2022. Online at https://www.cosparathens2022.org/. Abstract B4.1-0032-22

Lefèvre M (2022) Venus boundary layer dynamics: eolian transport and convective vortex. Icarus 387:115167. https://doi.org/10.1016/j.icarus.2022.115167

Lefèvre M, Lebonnois S, Spiga A (2018) Three-dimensional turbulence-resolving modeling of the Venusian cloud layer and induced gravity waves: inclusion of complete radiative transfer and wind shear. J Geophys Res, Planets 123:2773. https://doi.org/10.1029/2018JE005679

Lefèvre M, Spiga A, Lebonnois S (2020) Mesoscale modeling of Venus' bow-shape waves. Icarus 335:113376. https://doi.org/10.1016/j.icarus.2019.07.010

Lefèvre M, Marcq E, Lefèvre F (2022) The impact of turbulent vertical mixing in the Venus clouds on chemical tracers. Icarus 386:115148. https://doi.org/10.1016/j.icarus.2022.115148

Lellouch E, Witasse O (2008) A coordinated campaign of Venus ground-based observations and Venus Express measurements. Planet Space Sci 56(10):1317–1319. https://doi.org/10.1016/j.pss.2008.07.001

Lellouch E, Widemann T, Luz D, Moreno R (2007) ESA Support Investigation to the Venus Express Mission, European Space Agency

Lenardic A, Kaula WM (1995) More thoughts on convergent crustal plateau formation and mantle dynamics with regard to Tibet. J Geophys Res, Solid Earth 100(B8):15193–15203

Limaye S, Garvin JB (2023) Exploring Venus: next generation missions beyond those currently planned. Front Astron Space Sci 10:1188096. https://doi.org/10.3389/fspas.2023.1188096

Limaye SS, Mogul R, Baines KH, Bullock MA, Cockell C, Cutts JA, Gentry DM, Grinspoon DH, Head JW, Jessup KL, Kompanichenko V, Lee YJ, Mathies R, Milojevic T, Pertzborn RA, Rothschild L, Sasaki S, Schulze-Makuch D, Smith DJ, Way MJ (2021) Venus, an astrobiology target. Astrobiology 21:1163–1185. https://doi.org/10.1089/ast.2020.2268

Linkin VM, Kerzhanovich VV, Lipatov AN, Pichkadze KM, Shurupov AA, Terterashvili AV, Ingersoll AP, Crisp D, Grossman AW, Young RE, Seiff A, Ragent B, Blamont JE, Elson LS, Preston RA (1986) VEGA balloon dynamics and vertical winds in the Venus middle cloud region. Science 231(4744):1417–1419. https://doi.org/10.1126/science.231.4744.1417

Lognonné P, Johnson CL, Schubert G (2015) 10.03—planetary seismology. In: Treatise on geophysics, vol 2, pp 65–120.

Lorenz RD (2015) Probabilistic constraints from existing and future radar imaging on volcanic activity on Venus. Planet Space Sci. https://doi.org/10.1016/j.pss.2015.07.009i

Lorenz RD (2016) Surface winds on Venus: probability distribution from in-situ measurements. Icarus 264:311. https://doi.org/10.1016/j.icarus.2015.09.036

Lorenz RD (2018) Lightning detection on Venus: a critical review. Prog Earth Planet Sci 5:34. https://doi.org/10.1186/s40645-018-0181-x

Lorenz RD, Le Gall A, Janssen MA (2016) Detecting volcanism on Titan and Venus with microwave radiometry. Icarus 270:30–36

Lukco D, Spry DJ, Harvey RP, Costa GCC, Okojie RS, Avishai A et al (2018) Chemical analysis of materials exposed to Venus temperature and surface atmosphere. Earth Space Sci 5:270–284. https://doi.org/10.1029/2017EA000355

Lukco D, Spry DJ, Neudeck PG, Nakley LM, Phillips KG, Okojie RS, Hunter GW (2020) Experimental study of structural materials for prolonged Venus surface exploration missions. J Spacecr Rockets 57:1118–1128

Machado P, Luz D, Widemann T, Lellouch E, Witasse O (2012) Characterizing the atmospheric dynamics of Venus from ground-based Doppler velocimetry. Icarus 221:248–261. https://doi.org/10.1016/j.icarus.2012.07.012

Machado P, Widemann T, Luz D, Peralta J (2014) Wind circulation regimes at Venus' cloud tops: ground-based Doppler velocimetry using CFHT/ESPaDOnS and comparison with simultaneous cloud tracking measurements using VEx/VIRTIS in February 2011. Icarus 243:249–263. https://doi.org/10.1016/j.icarus.2014.08.030

Machado P, Widemann T, Peralta J, Gonçalves R, Donati J, Luz D (2017) Venus cloud-tracked and Doppler velocimetry winds from CFHT/ESPaDOnS and Venus Express/VIRTIS in April 2014. Icarus 285:8–26. https://doi.org/10.1016/j.icarus.2016.12.017

Machado P, Widemann T, Peralta J, Gilli G, Espadinha D, Silva JE, Brasil F, Ribeiro J, Gonçalves R (2021) Venus atmospheric dynamics at two altitudes: Akatsuki and Venus Express cloud tracking, ground-based Doppler observations and comparison with modeling. Atmosphere 12:506. https://doi.org/10.3390/atmos12040506

Machado P, Silva T, Branco A, Jaeggli S, Tanga P, Widemann T (2023) Transmission spectroscopy along the transit of Venus used for probing the atmosphere's upper layers and as a proxy for exoplanets atmosphere characterization. In: DPS-EPSC meeting, San Antonio, TX, Oct. 1-6 2023. American Astronomical Society

Mahieux A, Vandaele AC, Robert S, Wilquet V, Drummond R, Montmessin F, Bertaux JL (2012) Densities and temperatures in the Venus mesosphere and lower thermosphere retrieved from SOIR on board Venus Express: carbon dioxide measurements at the Venus terminator. J Geophys Res 117:E07001. https://doi.org/10.1029/2012JE004058

Mahieux A, Vandaele AC, Bougher SW, Drummond R, Robert S, Wilquet V, Chamberlain S, Piccialli A, Montmessin F, Tellmann S, Pätzold M, Häusler B, Bertaux JL (2015) Update of the Venus density and temperature profiles at high altitude measured by SOIR on board Venus Express. Planet Space Sci 113–114:309–320. https://doi.org/10.1016/j.pss.2015.02.002

Maia JS, Wieczorek MA (2022) Lithospheric structure of Venusian crustal plateaus. J Geophys Res, Planets 127:e2021JE007004. https://doi.org/10.1029/2021JE007004

Malin MC (1992) Mass movements on Venus: preliminary results from Magellan cycle 1 observations. J Geophys Res, Planets 97:16337–16352. https://doi.org/10.1029/92je01343

Marcq E, Lebonnois S (2013) Simulations of the latitudinal variability of CO-like and OCS-like passive tracers below the clouds of Venus using the Laboratoire de Météorologie Dynamique GCM. J Geophys Res, Planets 118(10):1983–1990. https://doi.org/10.1002/jgre.20146

Marcq E, Encrenaz T, Bézard B, Birlan M (2006) Remote sensing of Venus' lower atmosphere from ground-based IR spectroscopy: latitudinal and vertical distribution of minor species. Planet Space Sci 54:1360–1370

Marcq E, Bertaux J-L, Montmessin F, Belyaev D (2013) Variations of sulphur dioxide at the cloud top of Venus's dynamic atmosphere. Nat Geosci 6:25. https://doi.org/10.1038/ngeo1650

Marcq E, Lellouch E, Encrenaz T, Widemann T, Birlan M, Bertaux JL (2015) Search for horizontal and vertical variations of CO in the day and night side lower mesosphere of Venus from CSHELL/IRTF 4.53 µm observations. Planet Space Sci 113–114:256–263. https://doi.org/10.1016/j.pss.2014.12.013

Marcq E, Jessup KL, Baggio L, Encrenaz T, Lee YJ, Montmessin F, Belyaev D, Korablev O, Bertaux J-L (2020) Icarus 335:113368. https://doi.org/10.1016/j.icarus.2019.07.002

Marcq E, Amine I, Duquesnoy M, Bézard B (2021) Astron Astrophys 648:L8. https://doi.org/10.1051/0004-6361/202140837

Margot J-L, Hauck SA, Mazarico E, Padovan S, Peale SJ (2018) Mercury's internal structure. In: Salomon SC, Nittler LR, Anderson BJ (eds) Mercury - the view after MESSENGER. Cambridge University Press, Cambridge. https://doi.org/10.1017/9781316650684

Margot JL, Campbell DB, Giorgini JD, Jao JS, Snedeker LG, Ghigo FD, Bonsall A (2021) Spin state and moment of inertia of Venus. Nat Astron 5:676–683. https://doi.org/10.1038/s41550-021-01339-7

Marov MYA (1978) Results of Venus missions. Annu Rev Astron Astrophys 16:141–169. https://doi.org/10.1146/annurev.aa.16.090178.001041

Marov M (2005) Mikhail Lomonosov and the discovery of the atmosphere of Venus during the 1761 transit. In: Kurtz DW, Bromage GE (eds) Transits of Venus: new views of the solar system and galaxy. IAU colloq., vol 196. Cambridge University Press, Cambridge, pp 209–219. https://doi.org/10.1017/S1743921305001390

Marov MYA, Avduevsky VS, Kerzhanovich VV, Rozhdestevensky MK, Borodin NF, Ryabov OL (1973) Venera 8: measurements of the temperature, pressure and wind velocity on the illuminated side of Venus. J Atmos Sci 30:1210–1214

Marty B (2012) The origins and concentrations of water, carbon, nitrogen and noble gases on Earth. Earth Planet Sci Lett 313–314:56–66. https://doi.org/10.1016/j.epsl.2011.10.040

Marty B, Avice G, Sano Y, Altwegg K, Balsiger H, Hässig M, Morbidelli A, Mousis O, Rubin M (2016) Origins of volatile elements (H, C, N, noble gases) on Earth and Mars in light of recent results from the ROSETTA cometary mission. Earth Planet Sci Lett 441:91–102

Marty B, Altwegg K, Balsiger H, Bar-Nun A, Bekaert DV, Berthelier J-J, Bieler A, Briois C, Calmonte U, Combi M (2017) Xenon isotopes in 67P/Churyumov-Gerasimenko show that comets contributed to Earth's atmosphere. Science 356(6342):1069–1072

Mazarico E, Iess L, Cascioli G, Durante D, De Marchi F, Hensely S, Smrekar S (2023) The Venus gravity field from VERITAS. In: International EnVision Venus science workshop, Berlin-Adlershof, 9-11 May 2023. https://atpi.eventsair.com/2023-envision-workshop/programme

Meadows VS, Reinhard CT, Arney GN, Parenteau MN, Schweiteman EW, Domagal-Goldman SD, Lincowski AP, Stapelfeldt KR, Rauer H, DasSarma S et al (2018) Exoplanet biosignatures: understanding oxygen as a biosignature in the context of its environment. Astrobiology 18(6):630–662

Mills FP, Esposito LW, Yung YL (2007) Atmospheric composition, chemistry, and clouds. In: Esposito L, Stofan ER, Cravens TE (eds) Exploring Venus as a terrestrial planet. American Geophysical Union, Washington, pp 73–blpage100. https://doi.org/10.1029/176GM06

Mills F, Jessup KL, Brecht AS (2021) Atmospheric chemistry on Venus — new observations and laboratory studies to progress significant unresolved issues, White Paper for NASA 2021 Decadal Survey. Bull Am Astron Soc 53(4). https://doi.org/10.3847/25c2cfeb.7a0b2f82

Mitrofanov I, Jun I, the SAGE NAGRS Team (2010) Neutron –activated gamma ray spectrometer (NA-GRS) for the Venus surface and atmosphere geochemical explorer (SAGE) mission. In: European planetary science congress 2010, EPSC abstracts vol. 5, EPSC2010-264, 2010. https://meetingorganizer.copernicus.org/EPSC2010/EPSC2010-264.pdf

Morbidelli A (2020) Planet formation by pebble accretion in ringed disks. Astron Astrophys 638:A1. https://doi.org/10.1051/0004-6361/202037983

Morellina S, Bellan J (2022) Turbulent chemical-species mixing in the Venus lower atmosphere at different altitudes: a direct numerical simulation study relevant to understanding species spatial distribution. Icarus 371:114686. https://doi.org/10.1016/j.icarus.2021.114686

Moreno R, Treuttel J, González-Ovejero D, Gatilova L, Segret B, Lellouch E (2020) TERACUBE: THz instrument concept for CubeSat. In: Europlanet science congress 2020, online, 21 September–9 Oct 2020, EPSC2020-350, 2020. https://doi.org/10.5194/epsc2020-350

Moroz VI (1990) Atmospheric structure of Venus according to optical measurements by VENERA–11 VENERA–13 and VENERA–14. Sol Syst Res 23(4):206

Moroz VI (2002) Estimates of visibility of the surface of Venus from descent probes and balloons. Planet Space Sci 50(3):287–297

Moroz VI et al (1985) Solar and thermal radiation in the Venus atmosphere. Adv Space Res 5:197–232

Moroz VI, Zasova LV, Linkin VM (1996) Venera-15, 16 and VEGA mission results as sources for improvements of the Venus reference atmosphere. Adv Space Res 17(11):171–180

Morrison D, Hinners N (1983) A program for planetary exploration. Science 220(4597):561–567. https://www.jstor.org/stable/1690000

Mueller N, Helbert J, Hashimoto GL, Tsang CCC, Erard S, Piccioni G, Drossart P (2008) Venus surface thermal emission at 1 μm in VIRTIS imaging observations: evidence for variation of crust and mantle differentiation conditions. J Geophys Res, Planets 113(E5):E00B17. https://doi.org/10.1029/2008JE003118

Mueller NT, Smrekar SE, Helbert J, Stofan E, Piccioni G, Drossart P (2017) Search for active lava flows with VIRTIS on Venus Express. J Geophys Res, Planets 122(5):1021–1045

Mueller NT, Smrekar SE, Tsang CCC (2020) Multispectral surface emissivity from VIRTIS on Venus Express. Icarus 335:113400

Nakajima M, Golabek GJ, Wünnemann K, Rubie DC, Burger C, Melosh HJ, Jacobson SA, Manske L, Hull SD (2021) Scaling laws for the geometry of an impact-induced magma ocean. Earth Planet Sci Lett 568:116983. https://doi.org/10.1016/j.epsl.2021.116983

National Academies of Sciences, Engineering & Medicine (NASEM) (2022) Origins, worlds, and life: a decadal strategy for planetary science and astrobiology 2023-2032. https://doi.org/10.17226/26522

Neakrase LDV, Klose M, Titus TN (2017) Terrestrial subaqueous seafloor dunes: possible analogs for Venus. Aeolian Res 26:47–56. https://doi.org/10.1016/j.aeolia.2017.03.002

Neudeck PG, Chen L, Meredith RD, Lukco D, Spry DJ, Nakley LM, Hunter GW (2018) Operational testing of 4H-SiC JFET ICs for 60 days directly exposed to Venus surface atmospheric conditions. IEEE J Electron Dev Soc 7:100–110

Nicholson WL, Fajardo-Cavazos P, Fedenko J, Ortiz-Lugo JL, Rivas-Castillo A, Waters SM, Schuerger AC (2010) Exploring the low-pressure growth limit: evolution of bacillus subtilis in the laboratory to enhanced growth at 5 kilopascals. Appl Environ Microbiol 76(22):7559–7565. https://doi.org/10.1128/AEM.01126-10

Nigar S (2019) Venus Orbiter Mission to study surface, atmosphere and plasma environment. Talk given at 17th Venus Exploration Analysis Group (VEXAG) 11/06/2019, LASP, Boulder, USA. https://www.lpi.usra.edu/vexag/meetings/archive/vexag-17/presentations/Nigar.pdf

Nigar S (2022) Mission to Venus. Challenges and opportunities. Talk given at National Meet on Venus Science 05/04/2022, ISRO HQ, Bengaluru, India. https://www.youtube.com/watch?v=yUp6DplyPJk

Nikolaeva OV (1990) Geochemistry of the Venera 8 material demonstrates the presence of continental crust on Venus. Earth Moon Planets 50/51:329–341

Nikolaeva OV (1995) K-U-Th systematics of terrestrial magmatic rocks for planetary comparisons: terrestrial N-MORBs and Venusian basaltic material. Geochem Int 33:1–11

Nikolaeva OV (1997) K-U-Th systematics of igneous rocks for planetological comparisons: oceanic island-arc volcanics on Earth versus rocks on the surface of Venus. Geochem Int 35:424–447

Nikolaou A, Katyal N, Tosi N, Godolt M, Grenfell JL, Rauer H (2019) What factors affect the duration and outgassing of the terrestrial magma ocean? Astrophys J 875:11. https://doi.org/10.3847/1538-4357/ab08ed

O'Brien DP, Morbidelli A, Levison HF (2006) Terrestrial planet formation with strong dynamical friction. Icarus 184:39–58. https://doi.org/10.1016/j.icarus.2006.04.005

Ohtani E (2020) The role of water in Earth's mantle. Nat Sci Rev 7:224–232. https://doi.org/10.1093/nsr/nwz071

O'Rourke JG (2020) Venus: a thick basal magma ocean may exist today. Geophys Res Lett 47:1–11. https://doi.org/10.1029/2019GL086126

O'Rourke JG, Korenaga J (2015) Thermal evolution of Venus with argon degassing. Icarus 260:128–140

O'Rourke JG, Smrekar SE (2018) Signatures of lithospheric flexure and elevated heat flow in stereo topography at coronae on Venus. J Geophys Res, Planets 123(2):369–389

O'Rourke JG, Wolf AS, Ehlmann BL (2014) Venus: interpreting the spatial distribution of volcanically modified craters. Geophys Res Lett 41(23):8252–8260. https://doi.org/10.1002/2014GL062121

O'Rourke JG, Wilson CF, Borelli ME, Byrne PK, Dumoulin C, Ghail R, Gülcher AJP, Jacobson SA, Korablev O, Spohn T, Way MJ, Weller M, Westall F (2023) Venus, the planet: introduction to the evolution of Earth's sister planet. Space Sci Rev 219:10. https://doi.org/10.1007/s11214-023-00956-0

Parkinson CD, Gao P, Esposito L, Yung Y, Bouguer S, Hirtzig M (2015) Photochemical control of the distribution of Venusian water. Planet Space Sci 113–114:226–236. https://doi.org/10.1016/j.pss.2015.02.015

Parmentier EM, Hess PC (1992) Chemical differentiation of a convecting planetary interior: consequences for a one plate planet such as Venus. Geophys Res Lett 19(20):2015–2018

Pasachoff JM, Schneider G, Widemann T (2011) High-resolution satellite imaging of the 2004 transit of Venus and asymmetries in the cytherean atmosphere. Astron J 141:112. https://doi.org/10.1088/0004-6256/141/4/112

Pearson DG, Brenker FE, Nestola F, McNeill J, Nasdala L, Hutchison MT, Matveev S, Mather K, Silversmit G, Schmitz S (2014) Hydrous mantle transition zone indicated by ringwoodite included within diamond. Nature 507(7491):221

Peralta J, Muto K, Hueso R, Horinouchi T, Sánchez-Lavega A, Murakami S, Machado P, Young EF, Lee YJ, Kouyama T, Sagawa H, McGouldrick K, Satoh T, Imamura T, Limaye SS, Sato TM, Ogohara K, Nakamura M, Luz D (2016) Nightside winds at the lower clouds of Venus with Akatsuki/IR2: longitudinal, local time, and decadal variations from comparison with previous measurements. Astrophys J Suppl Ser 239:29. https://doi.org/10.3847/1538-4365/aae844

Pere C, Tanga P, Widemann T, Bendjoya P, Mahieux A, Wilquet V, Vandaele AC (2016) A multilayer modeling of the aureole photometry during the Venus transit: comparison between SDO/HMI and VEx/SOIR data. Astron Astrophys 595:A115, 9 pp. https://doi.org/10.1051/0004-6361/201628528

Péron S, Moreira M, Agranier A (2018) Origin of light noble gases (He, Ne, and Ar) on Earth: a review. Geochem Geophys Geosyst 461:1227. https://doi.org/10.1002/2017GC007388

Peter K, Tellmann S, Pätzold M, Fränz M, Oschlisniok J, Imamura T, Häusler B (2023) Potential exploration of the Venus ionosphere with EnVision radio science. In: EnVision international Venus science workshop, DLR Berlin, May 9-11, 2023

Phillips RJ, Hansen VL (1994) Tectonic and magmatic evolution of Venus. Annu Rev Earth Planet Sci 22(1):597–656

Phillips RJ, Izenberg N (1995) Ejecta correlations with spatial crater density and Venus resurfacing history. Geophys Res Lett 22(12):1517–1520. https://doi.org/10.1029/95gl01412

Piani L, Marrocchi Y, Rigaudier T, Vacher LG, Thomassin D, Marty B (2020) Earth's water may have been inherited from material similar to enstatite chondrite meteorites. Science 369:1110–1113. https://doi.org/10.1126/science.aba1948

Picardi G, Plaut JJ, Biccari D, Calabrese D, Cartacci M, Cichetti A, Clifford SM, Edenhofer P, Farrell WM, Federico C, Frigeri A, Gurnett DA, Hagfors T, Heggy E, Herique A, Huff RL, Ivanov AB, Johnson WTK, Jordan RL, Kirchner DL, Kofman W, Leuschen CJ, Nielsen E, Orosei R, Pettinelli E, Phillips RJ, Plettemeier D, Safaeinili A, Seu R, Stofan ER, Vannaroni G, Watters TR, Zampolini E (2005) Radar soundings of the subsurface of Mars. Science 310:1925–1928. https://doi.org/10.1126/science.112216

Piccialli A, Moreno R, Encrenaz T, Fouchet T, Lellouch E, Widemann T (2017) Mapping the thermal structure and minor species of Venus mesosphere with ALMA submillimeter observations. Astron Astrophys 606:A53. https://doi.org/10.1051/0004-6361/201730923

Pieters CM, Head JW, Pratt S, Patterson W, Garvin J, Barsukov VL, Basilevsky AT, Khodakovsky IL, Selivanov AS, Panfilov AS, Gektin YM, Narayeva YM (1986) The color of the surface of Venus. Science 234:1379–1383. https://doi.org/10.1126/Science.234.4782.1379

Pla-Garcia J, Rafkin SCR, Karatekin Ö, Gloesener E (2019) Comparing MSL Curiosity rover TLS-SAM methane measurements with Mars regional atmospheric modeling system atmospheric transport experiments. J Geophys Res, Planets 124:2141–2167. https://doi.org/10.1029/2018JE005824

Plesa A-C, Padovan S, Tosi N et al (2018) The thermal state and interior structure of Mars. Geophys Res Lett 45(22):12198–12209

Poler G, Garcia RF, Bowman DC, Martire L (2020) Infrasound and gravity waves over the Andes observed by a pressure sensor on board a stratospheric balloon. J Geophys Res, Atmos 125(6):e2019JD031565

Pollack JB, Toon OB, Whitten RC, Boese R, Ragent B, Tomasko M et al (1980) Distribution and source of the UV absorption in Venus' atmosphere. J Geophys Res 85:8141–8150. https://doi.org/10.1029/JA085iA13p08141

Port ST, Chevrier VF (2020) Stability of pyrrhotite under experimentally simulated Venus conditions. Planet Space Sci 193:11

Port ST, Chevrier VF, Kohler E (2020) Investigation into the radar anomaly on Venus: the effect of Venus conditions on bismuth, tellurium, and sulfur mixtures. Icarus 336:113432. https://doi.org/10.1016/j.icarus.2019.113432

Preston RA, Hildebrand CE, Purcell GH Jr, Ellis J, Stelzried CT, Finley SG, Sagdeev RZ, Linkin VM, Kerzhanovich VV, Altunin VI, Kogan LR, Kostenko VI, Matveenko LI, Pogrebenko SV, Strukov IA, Akim EL, Alexandrov YN, Armand NA, Bakitko RN, Vyshlov AS, Bogomolov AF, Gorchankov YN, Selivanov AS, Ivanov NM, Tichonov VF, Blamont JE, Boloh L, Laurans G, Boischot A, Biraud F, Ortega-Molina A, Rosolen C, Petit G (1986) Determination of Venus winds by ground-based radio tracking of the VEGA balloons. Science 231(4744):1414–1416. https://doi.org/10.1126/science.231.4744.1414

Rabinovitch J, Borner A, Gallis MA, Sotin C (2019) Hypervelocity noble gas sampling in the upper atmosphere of Venus. In: AIAA aviation 2019 forum, Dallas, TX, 17-21 Jun 2019. https://doi.org/10.2514/6.2019-3223

Radoman-Shaw BG (2019) Exposure of basaltic materials to Venus surface conditions using the Glenn extreme environment rig (GEER). Case Western Reserve University

Ragent B, Esposito LW, Tomasko MG, Marov MY, Shari VP, Lebedev VN (1985) Particulate matter in the Venus atmosphere. Adv Space Res 5(11):85–115

Reese CC (1999) Stagnant lid convection and magmatic resurfacing on Venus. Icarus 139:67–80

Reid RB (2021) Experimental alteration of Venusian surface basalts in a hybrid CO_2-SO_2 atmosphere. University of Tennessee, Knoxville

Rivoldini A, Van Hoolst T, Verhoeven O, Mocquet A, Dehant V (2011) Geodesy constraints on the interior structure and composition of Mars. Icarus 213(2):451–472

Rodriguez E, Morris CS, Belz JE, Chapin EC, Martin JM, Daffer W, Hensley S (2005) An assessment of the SRTM topographic products. Technical Report JPL D-31639, Jet Propulsion Laboratory, Pasadena, California. 143 pp

Rolf T, Steinberger B, Sruthi U, Werner SC (2018) Inferences on the mantle viscosity structure and the post-overturn evolutionary state of Venus. Icarus 313:107–123. https://doi.org/10.1016/j.icarus.2018.05.014

Rolf T, Weller MB, Ghail R, Byrne PK, Gulcher A, Gillmann C, Davaille A, Bjonnes EE, O'Rourke JG, Smrekar SE, Herrick RR, Plesa A-C (2022) Dynamics and evolution of Venus' mantle through time. Space Sci Rev 218:70. https://doi.org/10.1007/s11214-022-00937-9

Romeo I, Turcotte DL (2008) Pulsating continents on Venus: an explanation for crustal plateaus and tessera terrains. Earth Planet Sci Lett 276(1):85–97. https://doi.org/10.1016/j.epsl.2008.09.009

Rosenblatt P, Dumoulin C, Marty J-C, Genova A (2021) Determination of Venus' interior structure with EnVision. Remote Sens 13(9):1624. https://doi.org/10.3390/rs13091624

Rufu R, Aharonson O, Perets HB (2017) A multiple-impact origin for the Moon. Nat Geosci 10:89–95. https://doi.org/10.1038/ngeo2866

Ruiz J (2007) The heat flow during the formation of ribbon terrains on Venus. Planet Space Sci 55(14):2063–2070

Russell CT, Vaisberg OL (1983) The interaction of the solar wind with Venus. In: Hunten DM, Colin L, Donahue TM, Moroz VI (eds) Venus. University of Arizona Press, Tuscon, pp 873–940

Sagdeev RZ, Linkin VM, Blamont JE, Preston RA (1986a) The VEGA Venus balloon experiment. Science 231(4744):1407–1408. https://doi.org/10.1126/science.231.4744.1407

Sagdeev RZ, Linkin VM, Kerzhanovich VV, Lipatov AN, Shurupov AA, Blamont JE, Crisp D, Ingersoll AP, Elson LS, Preston RA, Hildebrand CE, Ragent B, Seiff A, Young RE, Petit G, Boloh L, Alexandrov YN, Armand NA, Bakitko RV, Selivanov AS (1986b) Overview of VEGA Venus balloon in situ meteorological measurements. Science 231(4744):1411–1414. https://doi.org/10.1126/science.231.4744.1411

Sagdeev RZ et al (1992) Differential VLBI measurements of the Venus atmosphere dynamics by balloons: VEGA project. Astron Astrophys 254:387–392

Salvador A, Samuel H (2023) Convective outgassing efficiency in planetary magma oceans: insights from computational fluid dynamics. Icarus 390:115265. https://doi.org/10.1016/j.icarus.2022.115265

Salvador A, Massol H, Davaille A, Marcq E, Sarda P, Chassefière E (2017) The relative influence of H_2O and CO_2 on the primitive surface conditions and evolution of rocky planets. J Geophys Res, Planets 122(7):1458–1486. https://doi.org/10.1002/2017JE005286

Salvador A, Avice G, Breuer D, Gillmann C, Jacobson S, Lammer H, Marcq E, Raymond SN, Sakuraba H, Scherf M, Way MJ (2023) Magma ocean, water, and the early atmosphere of Venus. Space Sci Rev 219:51. https://doi.org/10.1007/s11214-023-00995-7

Sánchez-Lavega A, Hueso R, Piccioni G, Drossart P, Peralta J, Pérez-Hoyos S, Wilson CF, Taylor FW, Baines KH, Luz D, Erard S, Lebonnois S (2008) Variable winds on Venus mapped in three dimensions. Geophys Res Lett 35:L13204. https://doi.org/10.1029/2008GL033817

Sánchez-Lavega A, Peralta J, Gomez-Forrellad JM, Hueso R, Pérez-Hoyos S, Mendikoa I, Rojas JF, Horinouchi T, Lee YJ, Watanabe S (2016) Venus cloud morphology and motions from ground-based images at the time of the Akatsuki orbit insertion. Astrophys J Lett 833:L7

Sánchez-Lavega A, Lebonnois S, Imamura T, Read P, Luz D (2017) The atmospheric dynamics of Venus. Space Sci Rev 212:1541–1616. https://doi.org/10.1007/s11214-017-0389-x

Sandor BJ, Clancy RT, Moriarty-Schieven G, Mills FP (2010) Sulfur chemistry in the Venus mesosphere from SO_2 and SO microwave spectra. Icarus 208:49–60

Sandor BJ, Clancy RT, Moriarty-Schieven G (2012) Upper limits for H_2SO_4 in the mesosphere of Venus. Icarus 217(2):839–844

Sandwell DT, Schubert G (1992a) Evidence for retrograde lithospheric subduction on Venus. Science 257(5071):766–770. https://doi.org/10.1126/science.257.5071.766

Sandwell DT, Schubert G (1992b) Flexural ridges, trenches, and outer rises around coronae on Venus. J Geophys Res, Planets 97(E10):16069–16083. https://doi.org/10.1029/92JE01274

Santos AR, Gilmore MS, Greenwood JP, Nakley LM, Phillips K, Kremic T, Lopez X (2023) Experimental weathering of rocks and minerals at Venus conditions in the Glenn extreme environments rig (GEER). J Geophys Res, Planets 128:e2022JE007423

Santos A, Balcerski J, Burr DM, Helbert J, Hunter G, Izenberg N, Johnson N, Kohler E, Port S (2021) The importance of Venus experimental facilities. Bull Am Astron Soc 53(4). https://doi.org/10.3847/25c2cfeb.19b48da4

Sauder J, Hilgemann E, Stack K, Kawata J, Parness A, Johnson M (2019) Hybrid automaton rover-Venus. In: 17th meeting of the Venus exploration group (VEXAG), held 6-8 November, 2019 in Boulder, Colorado. LPI contribution, vol 2193, id.8030

Schaber GG, Strom RG et al (1992) Geology and distribution of impact craters on Venus: what are they telling us? J Geophys Res 97:13257–13301. https://doi.org/10.1029/92JE01246

Scora J, Valencia D, Morbidelli A, Jacobson S (2020) Chemical diversity of super-earths as a consequence of formation. Mon Not R Astron Soc 493:4910–4924

Seager S, Petkowski JJ, Carr CE, Grinspoon D, Ehlmann B, Saikia SJ, Agrawal R, Buchanan W, Weber MU, French R, Kluper P, Worde SP (2021) Venus life finder mission study. https://doi.org/10.48550/arxiv.2112.05153

Seager S, Petkowski JJ, Carr CE, Grinspoon DH, Ehlmann BL, Saikia SJ, Agrawal R, Buchanan WP, Weber MU, French R, et al (2022) Venus life finder missions motivation and summary. Aerospace, 9:385. https://doi.org/10.3390/aerospace9070385

Shah H, Seth G (2022) System design of polarimetric synthetic aperture radar for Venus: a case study. Adv Space Res, under review

Shellnutt JG (2013) Petrological modeling of basaltic rocks from Venus: a case for the presence of silicic rocks. J Geophys Res, Planets 118(6):1350–1364. https://doi.org/10.1002/jgre.20094

Shibata E, Lu Y, Pradeepkumar A, Cutts JA, Saikia SJ (2017) A Venus atmospheric sample return mission concept: feasibility and technology requirements. In: Planetary science vision 2050 workshop 2017. LPI contrib., vol 1989

Smrekar SE (1994) Evidence for active hotspots on Venus from analysis of Magellan gravity data. Icarus 112:2–26

Smrekar SE, Sotin C (2012) Constraints on mantle plumes on Venus: implications for volatile history. Icarus 217(2):510–523

Smrekar SE, Stofan ER (1997) Coupled upwelling and delamination: a new mechanism for coronae formation and heat loss on Venus. Science 277:1289–1294. https://doi.org/10.1126/science.277.5330.1289

Smrekar SE, Hoogenboom T, Stofan ER, Martin P (2010b) Gravity analysis of Parga and Hecate chasmata: implications for rift and corona formation. J Geophys Res, Planets 115(E7):E07010. https://doi.org/10.1029/2009JE003435

Smrekar SE, Stofan ER, Mueller N, Treiman A, Elkins-Tanton L, Helbert J, Piccioni G, Drossart P (2010a) Recent hotspot volcanism on Venus from VIRTIS emissivity data. Science 328:605–608. https://doi.org/10.1126/science.1186785

Smrekar SE, Pauken M, Morgan P, Chase J, Fleurial J-P (2014) Measuring heat flow on Venus: instrumentation and rationale. 45th LPSC, 2825.pdf

Smrekar SE, Davaille A, Sotin C (2018) Venus interior structure and dynamics. Space Sci Rev 214:88. https://doi.org/10.1007/s11214-018-0518-1

Smrekar SE, Lognonné P, Spohn T, Banerdt WB, Breuer D, Christensen U, Dehant V, Drilleau M, Folkner W, Fuji N (2019) Pre-mission InSights on the interior of Mars. Space Sci Rev 215(1):3

Smrekar SE, Hensley S, Nybakken R, Wallace MS, Perkovic-Martin D, You T-H, Nunes D, Brophy J, Ely T, Burst E, Dyar MD, Helbert J, Miller B, Hartley J, Kallemeyn P, Whitte J, Iess L, Mastrogiuseppe M, Younis M, Prts P, Rodriguez M, Mazarico R (2022a) VERITAS (Venus emissivity, radio science, InSAR, topography, and spectroscopy): a discovery mission. In: 2022 institute for electrical and electronics engineers/IEEE aerospace conference (AERO), pp 1–20. https://doi.org/10.1109/AERO53065.2022.9843269

Smrekar SE, Ostberg C, O'Rourke JG (2022b) Evidence for active rifting and Earth-like lithospheric thickness and heat flow on Venus. Nat Geosci. https://doi.org/10.1038/s41561-022-01068-0

Solomon SC, Head JW, Kaula WM, McKenzie D, Parsons B, Phillips RJ, Schubert G, Talwani M (1991) Venus tectonics: initial analysis from Magellan. Science 252(5003):297–312. https://doi.org/10.1126/science.252.5003.297

Sood R, Chappaz L, Melosh HJ, Howell KC, Milbury C, Blair DM, Zuber MT (2017) Detection and characterization of buried lunar craters with GRAIL data. Icarus 289:157–172

Soret L, Gérard J-C, Piccioni G, Drossart P (2014) Time variations of $O_2(a^1\Delta)$ nightglow spots on the Venus nightside and dynamics of the upper mesosphere. Icarus 237:306–314. https://doi.org/10.1016/j.icarus.2014.03.034

Sotin C, Avice G, Baker J, Freeman A, Madzunkov S, Stevenson T, Arora N, Darrach M, Lightsey G, Marty B (2018a) Cupid's arrow: a small satellite concept to measure noble gases in Venus' atmosphere. In: 49th lunar and planetary science conference, Abstract #1763

Sotin C, Borner AP, Gallis MA, Rabinovitch J, Avice G, Darrach M, Madzunkov S, Marty B, Baker J, Mansour NN (2018b) Sampling Venus' atmosphere to measure noble gases and their isotope ratios. In: AGU fall meeting, Washington, DC, 10–14 December 2018

Southam G, Westall F, Spohn T (2015) Geology, life and habitability. In: Spohn T (ed) Treatise on geophysics, vol 10. Elsevier, Amsterdam, p 473

Spiga A, Banfield D, Teanby NA, Forget F, Lucas A, Kenda B, Rodriguez Manfredi JA, Widmer-Schnidrig R, Murdoch N, Lemmon MT, Garcia RF, Martire L, Karatekin Ö, Le Maistre S, Van Hove B, Dehant V, Lognonné P, Mueller N, Lorenz R, Mimoun D, Rodriguez S, Beucler É, Daubar I, Golombek MP, Bertrand T, Nishikawa Y, Millour E, Rolland L, Brissaud Q, Kawamura T, Mocquet A, Martin R, Clinton J, Stutzmann É, Spohn T, Smrekar S, Banerdt WB (2018) Atmospheric science with InSight. Space Sci Rev 214(7):1–64. https://doi.org/10.1007/s11214-018-0543-0

Spitzer F, Burkhardt C, Budde G, Kruijer TS, Morbidelli A, Kleine T (2020) Isotopic evolution of the inner solar system inferred from molybdenum isotopes in meteorites. Astrophys J Lett 898:L2. https://doi.org/10.3847/2041-8213/ab9e6a

Spohn T, Seiferlin K, Hagermann A, Knollenberg J, Ball AJ, Banaszkiewicz M, et al, Zarnecki JC (2007) MUPUS–a thermal and mechanical properties probe for the Rosetta lander Philae. Space Sci Rev 128(1):339–362

Spohn T, Grott M, Smrekar SE, Knollenberg J, Hudson TL, Krause C, et al, Banerdt WB (2018) The heat flow and physical properties package (HP3) for the InSight mission. Space Sci Rev 214(5):1–33

Spohn T, Hudson TL, Witte L, Wippermann T, Wisniewski L, Kedziora B, et al, Grygorczuk J (2022) The InSight-HP3 Mole on Mars: lessons learned from attempts to penetrate to depth in the Martian soil. Adv Space Res 69(8):3140–3163

Steinberger B, Werner S, Torsvik TH (2010) Deep versus shallow origin of gravity anomalies, topography and volcanism on Earth, Venus and Mars. Icarus 207(2):564–577

Stevenson DJ (2003) Planetary magnetic fields. Earth Planet Sci Lett 208(1–2):1–11

Stevenson DJ, Cutts JA, Mimoun D, Arrowsmith S, Banerdt WB, Blom P, Brageot E, Brissaud Q, Chin G, Gao P, Tsai VC (2015) Probing the interior structure of Venus. Keck Institute for Space Studies

Strom RG, Schaber GG, Dawson DD (1994) The global resurfacing of Venus. J Geophys Res 99(E5):10899–10926. https://doi.org/10.1029/94JE00388

Surkov YA (1997) Exploration of terrestrial planets from spacecraft: instrumentation, investigation, interpretation, 2nd edn. Praxis pub., vol 446. Wiley, New York

Surkov YA, Barsukov VL, Moskalyova VP, Kharyukova AD, Kemurdzhian AL (1984) New data on the composition, structure, and properties of Venus rock obtained by Venera 13 and 14. J Geophys Res 896(suppl):B393–B402. Proc. Lunar Planet. Sci. Conf. 14th, Part 2

Surkov YA, Moskalyova VP, Kharyukova AD, Dudin AD, Smirnov GG, Zaitseva SE (1986) Venus rock composition at the Vega 2 landing site. J Geophys Res 17(suppl):E215–E218. Proc. Lunar Planet. Sci. Conf. 17, Part 1

Sutin BM, Cutts J, Didion AM, Drilleau M, Grawe M, Helbert J, et al, Wallace M (2018) VAMOS: a smallsat mission concept for remote sensing of Venusian seismic activity from orbit. In: Space telescopes and instrumentation 2018: optical, infrared, and millimeter wave, vol 10698. International Society for Optics and Photonics, p 106985T

Swindle TD (2002) Martian noble gases. Rev Mineral Geochem 47:171–190. https://doi.org/10.2138/rmg.2002.47.6

Tanga P, Widemann T, Sicardy B, Pasachoff J, Arnaud J, Comolli L, Rondi A, Rondi S, Suetterlin P (2012) Sunlight refraction in the mesosphere of Venus during the transit on June 8th, 2004 (2012). Icarus 218:207–219. https://doi.org/10.1016/j.icarus.2011.12.004

Teffeteller H (2020) Experimental study of the alteration of basalt on the surface of Venus. University of Tennessee

Teffeteller H, McCanta M, Cherniak D, Treiman A, Filiberto J, Rutherford M (2019) Experimental study of the alteration of basalt on the surface of Venus. LPSC 5(1858)

Tian Y, Herrick RR, West M, Kremic T (2023) Mitigating power and memory constraints on a Venusian seismometer. Seismol Res Lett 94(1):159–171. https://doi.org/10.1785/0220220085

Tikoo SM, Elkins-Tanton LT (2017) The fate of water within Earth and super-earths and implications for plate tectonics. Philos Trans R Soc Lond A 375:20150394. https://doi.org/10.1098/rsta.2015.0394

Tinetti G, Eccleston P, Haswell C, Lagage PO, Leconte J, Lüftinger T, Micela G, Min M, Pilbratt G, Puig L et al (2021) Ariel: enabling planetary science across light-years. ArXiv:e-prints arXiv:2104.04824

Titov DV, Ignatiev NI, McGouldrick K, Wilquet V, Wilson CW (2018) Clouds and hazes of Venus. Space Sci Rev 214:126. https://doi.org/10.1007/s11214-018-0552-z

Tomasko MG, Doose LR, Smith PH (1985) The absorption of solar energy and the heating rate in the atmosphere of Venus. Adv Space Res 5:71–79

Tosi N, Godolt M, Stracke B, Ruedas T, Grenfel JL, Höning D, Nikolaou A, Plesa A-C, Breuer D, Spohn T (2017) The habitability of a stagnant-lid Earth. Astron Astrophys 605:A71. https://doi.org/10.1051/0004-6361/201730728

Treiman AH (2007) Geochemistry of Venus' surface: current limitations as future opportunities. Geophysical monograph, vol 176

Tsang CCC, Wilson CF, Barstow JK, Irwin PGJ, Taylor FW, McGouldrick K, Piccioni G, Drossart P, Svedhem H (2010) Correlations between cloud thickness and sub-cloud water abundance on Venus. Geophys Res Lett 37:2202. https://doi.org/10.1029/2009GL041770

Turbet M, Ehrenreich D, Lovis C, Bolmont E, Fauchez T (2019) The runaway greenhouse radius inflation effect – an observational diagnostic to probe water on Earth-sized planets and test the habitable zone concept. Astron Astrophys 628:A12. https://doi.org/10.1051/0004-6361/201935585

Turbet M, Bolmont E, Ehrenreich D, Gratier P, Leconte J, Selsis F, Hara N, Lovis C (2020) Revised mass-radius relationships for water-rich rocky planets more irradiated than the runaway greenhouse limit. Astron Astrophys 638:A41. https://doi.org/10.1051/0004-6361/201937151

Turbet M, Bolmont E, Chaverot G, Ehrenreich D, Leconte J, Marcq E (2021) Day-night cloud asymmetry prevents early oceans on Venus but not on Earth. Nature 598:276–280. https://doi.org/10.1038/s41586-021-03873-w

Vago JL, Westall F, Pasteur Instrument Teams, Landing Site Selection Team et al(2017) Habitability on early Mars and the search for biosignatures with the ExoMars rover. Astrobiology 17:471–510. https://doi.org/10.1089/ast.2016.1533

Vaisberg OL, Zelenyi LM (1984) Formation of the plasma mantle in the Venusian magnetosphere. Icarus 58(6):412–430

Vaisberg OL, Romanov SA, Smirnov VN, Karpinsky IP, Khazanov BI, Polenov BV, Bogdanov AV, Antonov NM (1976) Ion flux parameters in the solar wind-Venus interaction region. In: Williams DJ (ed) Physics of solar planetary environment. AGU, Boulder, pp 904–917

Vandaele AC, Korablev O, Belyaev D, Chamberlain S, Evdokimova D, Encrenaz T, Esposito L, Jessup KL, Lefèvre F, Limaye S, Mahieux A, Marcq E, Mills FP, Montmessin F, Parkinson CD, Robert S, Roman T, Sandor B, Stolzenbach A, Wilson C, Wilquet V (2017a) Sulfur dioxide in the Venus atmosphere: I. Vertical distribution and variability. Icarus 295:1–15. https://doi.org/10.1016/j.icarus.2017.05.001

Vandaele AC, Korablev O, Belyaev D, Chamberlain S, Evdokimova D, Encrenaz T, Esposito L, Jessup KL, Lefèvre F, Limaye S, Mahieux A, Marcq E, Mills FP, Montmessin F, Parkinson CD, Robert S, Roman T, Sandor B, Stolzenbach A, Wilson C, Wilquet V (2017b) Sulfur dioxide in the Venus atmosphere: II. Spatial and temporal variability. Icarus 295:16–33. https://doi.org/10.1016/j.icarus.2017.05.003

Venera-D Joint Science Definition Team (2019) Phase II report http://www.iki.rssi.ru/events/2019/Venera-DPhaseIIFinalReport.pdf (accessed 7.31.22)

Venera-D Venus Modeling Workshop proceedings (2018), held in Moscow, Russia October 5-7 2017, L.M. Zelenyi Editor, with L.V. Zasova and D.A. Gorinov. http://venera-d.cosmos.ru/fileadmin/user_upload/documents/Workshop2017_Proceedings.pdf (accessed 7.31.22)

VEXAG (2019) Venus goals objectives and investigations [WWW Document]. URL https://www.lpi.usra.edu/vexag/documents/reports/VEXAG_Venus_GOI_2019.pdf (accessed 4.26.22)

Villanueva GL, Cordiner M, Irwin PGJ, de Pater I, Butler B, Gurwell M, Milam SN, Nixon CA, Luszcz-Cook SH, Wilson CF, Kofman V, Liuzzi G, Faggi S, Fauchez TJ, Lippi M, Cosentino R, Thelen AE, Moullet A, Hartogh P, Molter EM, Charnley S, Arney GN, Mandell AM, Biver N, Vandaele AC, de Kleer KR, Kopparapu R (2021) No evidence of phosphine in the atmosphere of Venus from independent analyses. Nat Astron 5:631–635. https://doi.org/10.1038/s41550-021-01422-z

Vorontsov VA, Lokhmatova MG, Martynov MB, Pichkhadze KM, Simonov AV, Khartov VV, Zasova L, Zelenyi LM, Korablev OI (2011) Prospective spacecraft for Venus research: Venera-D design. Sol Syst Res 45:710–714. https://doi.org/10.1134/S0038094611070288

Warwick S, Ross F, Sokol D (2017) Venus atmospheric maneuverable platform (VAMP) future work and scaling for a mission. In: 15th meeting of the Venus exploration analysis group (VEXAG), abstract #8029

Watters T, Leuschen C, Plaut J et al (2006) MARSIS radar sounder evidence of buried basins in the northern lowlands of Mars. Nature 444:905–908. https://doi.org/10.1038/nature05356

Way MJ, Del Genio AD (2020) Venusian habitable climate scenarios: modeling Venus through time and applications to slowly rotating Venus-like exoplanets. J Geophys Res, Planets 125(5):e2019JE006276. https://doi.org/10.1029/2019je006276

Way MJ, Del Genio AD, Kiang NY, Sohl LE, Grinspoon DH, Aleinov I, Kelley M, Clune T (2016) Was Venus the first habitable world of our solar system? Geophys Res Lett 43:8376–8383. https://doi.org/10.1002/2016GL069790. arXiv:1608.00706

Way MJ, Ostberg CM, Foley BJ, Gillmann C, Höning D, Lammer H, O'Rourke JG, Persson M, Plesa A-C, Salvador A, Scherf M, Weller MB (2023) Synergies between Venus & exoplanetary observations. Space Sci Rev 219:13. https://doi.org/10.1007/s11214-023-00953-3

Webster CR, Mahaffy PR (2011) Determining the local abundance of Martian methane and its' 13C/12C and D/H isotopic ratios for comparison with related gas and soil analysis on the 2011 Mars science laboratory (MSL) mission. Planet Space Sci 59:271–283

Weitz CM, Plaut JJ, Greeley R, Saunders RS (1994) Dunes and microdunes on Venus: why were so few found in the Magellan data? Icarus 112:282–295

Weller MB, Kiefer WS (2020) The physics of changing tectonic regimes: implications for the temporal evolution of mantle convection and the thermal history of Venus. J Geophys Res, Planets 125:1. https://doi.org/10.1029/2019JE005960

Westall F, Höning D, Avice G, Gentry D, Gerya T, Gillmann C, Isenberg NR, Way MJ, Wilson CF (2023) The habitability of Venus. Space Sci Rev 219:17. https://doi.org/10.1007/s11214-023-00960-4

Whitten JL, Campbell BA (2016) Recent volcanic resurfacing of Venusian craters. Geology 44:519–522. https://doi.org/10.1130/G37681.1

Widemann T, Lellouch E, Campargue A (2007) New wind measurements in Venus lower mesosphere from visible spectroscopy. Planet Space Sci 55:1741–1756. https://doi.org/10.1016/j.pss.2007.01.005

Widemann T, Lellouch E, Donati JF (2008) Venus Doppler winds at cloud tops observed with ESPaDOnS at CFHT. Planet Space Sci 56:1320–1334. https://doi.org/10.1016/j.pss.2008.07.005

Widemann T, Tanga P, Reardon KP, Limaye S, Wilson C, Vandaele A, Wilquet V, Mahieux A, Robert S, Pasachoff JM, Schneider G (2012) Asymmetry in the polar mesosphere revealed by the 2012 Venus transit aureole. In: DPS meeting #44, Reno, NV, Oct. 14-19 2012, American Astronomical Society, #508.08

Williams CD, Mukhopadhyay S (2018) Capture of nebular gases during Earth's accretion is preserved in deep-mantle neon. Nature 50:202. https://doi.org/10.1038/s41586-018-0771-1

Williams JG, Konopliv AS, Boggs DH, Park RS, Yuan DN, Lemoine FG, Goossens S, Mazarico E, Nimmo F, Weber RC (2014) Lunar interior properties from the GRAIL mission. J Geophys Res, Planets 119(7):1546–1578

Wilson C, Lefèvre F (2020) EnVision Science Conference, Feb. 2020, CNES, Paris. Abstract #3.03 https://sites.lesia.obspm.fr/envision/conference-program/

Wilson CF, Guerlet S, Irwin PGJ et al (2008) Evidence for anomalous cloud particles at the poles of Venus. J Geophys Res 113:E00B13. https://doi.org/10.1029/2008JE003108

Wilson CF, Widemann T, Ghail R (2022) Venus: key to understanding the evolution of terrestrial planets. Exp Astron 54:575–595. https://doi.org/10.1007/s10686-021-09766-0

Wilson CF, Marcq E, Gillmann C, Widemann T, Korablev O, Mueller N, Lefevre M, Rimmer P, Robert S, Zolotov M (2023) Possible effects of volcanic eruptions on the modern atmosphere of Venus. Space Sci Rev, this collection, in revision

Springer

Winker DM, Hunt W, Weimer C (2008) The on-orbit performance of the CALIOP lidar on CALIPSO. In: Proceedings of the 7th ICSO (international conference on space optics) 2008, Toulouse, France, Oct. 14–17, 2008

Winker DM, Pelon J, Coakley JA Jr, Ackerman SA, Charlson RJ, Colarco PR, Flamant P, Fu Q, Hoff RM, Kittaka C, Kubar TL, Le Treut H, McCormick MP, Mégie G, Poole L, Powell K, Trepte C, Vaughan MA, Wielicki BA (2010) The CALIPSO mission - a global 3D view of aerosols and clouds. Bull Am Meteorol Soc 91(9):1211–1229

Wolf ET, Kopparapu R, Airapetian V, Fauchez T, Guzewich SD, Kane SR, Pidhorodetska D, Way MJ, Abbot DS, Checlair JH et al (2019) The importance of 3D general circulation models for characterizing the climate and habitability of terrestrial extrasolar planets. ArXiv preprint. arXiv:1903.05012

Wolf ET, Kopparapu R, Haqq-Misra J, Fauchez T (2022) ExoCAM: a 3D climate model for exoplanet atmospheres. Planet Sci J 3:7. https://doi.org/10.3847/PSJ/ac3f3d

Wood BE, Hess P, Lustig-Yaeger J, Gallagher B, Korwan D, Rich N et al (2021) Parker solar probe imaging of the night side of Venus. Geophys Res Lett 48:e2021GL096302. https://doi.org/10.1029/2021GL096302

Xie L, Zhang H, Li H, Wang C (2015) A unified framework for crop classification in southern China using fully polarimetric, dual polarimetric, and compact polarimetric SAR data. Int J Remote Sens 36:3798–3818. https://doi.org/10.1080/01431161.2015.1070319

Yamashiro Y, Imamura T, Nakamura M, Ikari T, Kawabata Y, Sato T, Kouyama T, Imai M, Ando H, Sagawa H, Harada Y, Yamazaki A, Sato T, Aoki S, Funasa R, Hashimoto GL, Hirashima Y, Karyu H, Kashimura H, Nakagawa H, Horinouchi T, Kasaba Y, Huixin L, Maezawa H, Masunaga K, Sato M, Murakami S, Noguchi S, Sugimoto N, Ogawa H, Saito H, Sakai S, Sato N, Sugiyama K, Taguchi M, Takagi M, Terada N, Yamamoto M, Fujisawa Y, Futaana Y, Ishii N, Hirose C, Nakamura R, Matsumoto T, Akiyama Y, Nakatsuka J, Goto K, Toyota H, Toda T (2022) Mission study status of Venus Explorer succeeding Akatsuki. In: COSPAR 2022 44th scientific assembly, 16-24 July 2022, Athens, session B4.1 Venus science and exploration

Yang J, Boué G, Fabrycky DC, Abbot DS (2014) Strong dependence of the inner edge of the habitable zone on planetary rotation rate. Astrophys J 787:L2. https://doi.org/10.1088/2041-8205/787/1/L2. arXiv:1404.4992

Yoder CF, Konopliv AS, Yuan DN et al (2003) Fluid core size of Mars from detection of the solar tide. Science 300(5617):299–303

Zacny K, Nagihara S, Hedlund M et al (2013) Pneumatic and percussive penetration approaches for heat flow probe emplacement on robotic lunar missions. Earth Moon Planets 111:47–77. https://doi.org/10.1007/s11038-013-9423-5

Zahnle K, Arndt N, Cockell C, Halliday A, Nisbet E, Selsis F, Sleep NH (2007) Emergence of a habitable planet. Space Sci Rev. https://doi.org/10.1007/s11214-007-9225-z

Zahnle KJ, Gacesa M, Catling DC (2019) Strange messenger: a new history of hydrogen on Earth, as told by Xenon. Geochim Cosmochim Acta 244:56–85. https://doi.org/10.1016/j.gca.2018.09.017

Zasova LV, Gorinov DA, Eismont NA, Kovalenko ID, Abbakumov AS, Bober SA (2020) Venera-d: a design of an automatic space station for Venus exploration. Sol Syst Res 53:506–510. https://doi.org/10.1134/S0038094619070244

Zelenyi LM, Vaisberg OL (1985) Venus interaction with the solar wind plasma as a limiting case of the cometary type interaction. In: Advances in space plasma physics, p 59

Zolotov MY (2018) Gas-solid interactions on Venus and other solar system bodies. Rev Mineral Geochem 84:351. https://doi.org/10.2138/rmg.2018.84.10

Zolotov MY, Garvin JB (2020) Phosphorous-bearing compounds and atmosphere-surface chemical interactions on Venus, AGU FM P091 2020, P091-0004

Publisher's Note Springer Nature remains neutral with regard to jurisdictional claims in published maps and institutional affiliations.

Authors and Affiliations

**Thomas Widemann[1,2] · Suzanne E. Smrekar[3] · James B. Garvin[4] ·
Anne Grete Straume-Lindner[5] · Adriana C. Ocampo[6] · Mitchell D. Schulte[6] ·
Thomas Voirin[7] · Scott Hensley[3] · M. Darby Dyar[8] · Jennifer L. Whitten[9] ·
Daniel C. Nunes[3] · Stephanie A. Getty[4] · Giada N. Arney[4] · Natasha M. Johnson[4] ·
Erika Kohler[4] · Tilman Spohn[10,11] · Joseph G. O'Rourke[12] · Colin F. Wilson[5,13] ·**

Michael J. Way[14,15] · Colby Ostberg[16] · Frances Westall[17] · Dennis Höning[18,19] ·
Seth Jacobson[20] · Arnaud Salvador[21,22,23] · Guillaume Avice[24] ·
Doris Breuer[25] · Lynn Carter[23] · Martha S. Gilmore[26] · Richard Ghail[27] ·
Jörn Helbert[25] · Paul Byrne[28] · Alison R. Santos[26] · Robert R. Herrick[29] ·
Noam Izenberg[30] · Emmanuel Marcq[31] · Tobias Rolf[32] · Matt Weller[33] ·
Cedric Gillmann[34] · Oleg Korablev[35] · Lev Zelenyi[35] · Ludmila Zasova[35] ·
Dmitry Gorinov[35] · Gaurav Seth[36] · C. V. Narasimha Rao[36] · Nilesh Desai[36]

✉ T. Widemann
thomas.widemann@obspm.fr

✉ T. Rolf
tobias.rolf@geo.uio.no

S.E. Smrekar
suzanne.e.smrekar@jpl.nasa.gov

J.B. Garvin
james.b.garvin@nasa.gov

A.G. Straume-Lindner
anne.straume@esa.int

A.C. Ocampo
acouria@gmail.com

M.D. Schulte
mitchell.d.schulte@nasa.gov

T. Voirin
thomas.voirin@esa.int

S. Hensley
scott.hensley@jpl.nasa.gov

M.D. Dyar
mdyar@psi.edu

J.L. Whitten
jwhitten1@tulane.edu

D.C. Nunes
daniel.nunes@jpl.nasa.gov

S.A. Getty
stephanie.a.getty@nasa.gov

G.N. Arney
giada.n.arney@nasa.gov

N.M. Johnson
natasha.m.johnson@nasa.gov

E. Kohler
erika.kohler@nasa.gov

T. Spohn
tilman.spohn@issibern.ch

J.G. O'Rourke
jgorourk@asu.edu

C.F. Wilson
colin.wilson@physics.ox.ac.uk

M.J. Way
michael.way@nasa.gov

C. Ostberg
costb001@ucr.edu

F. Westall
frances.westall@cnrs.fr

D. Höning
dennis.hoening@pik-potsdam.de

S. Jacobson
seth@msu.edu

A. Salvador
arnaudsalvador@arizona.edu

G. Avice
avice@ipgp.fr

D. Breuer
doris.breuer@dlr.de

L. Carter
lmcarter@lpl.arizona.edu

M.S. Gilmore
mgilmore@wesleyan.edu

R. Ghail
richard.ghail@rhul.ac.uk

J. Helbert
joern.helbert@dlr.de

P. Byrne
paul.byrne@wustl.edu

A.R. Santos
asantos@wesleyan.edu

R.R. Herrick
rrherrick@alaska.edu

N. Izenberg
noam.izenberg@jhuapl.edu

E. Marcq
emmanuel.marcq@latmos.ipsl.fr

M. Weller
mweller@lpi.usra.edu

C. Gillmann
cgillmann@ethz.ch

O. Korablev
korab@iki.rssi.ru

L. Zelenyi
lzelenyi@iki.rssi.ru

L. Zasova
zasova@iki.rssi.ru

D. Gorinov
dmitry_gorinov@rssi.ru

G. Seth
gauravseth@sac.isro.gov.in

C.V.N. Rao
cvnrao@sac.isro.gov.in

N. Desai
nmdesai@sac.isro.gov.in

1   LESIA, Observatoire de Paris, Université PSL, CNRS, Sorbonne Université, Université Paris Cité, 5 place Jules Janssen, 92195 Meudon, France

2   Université Paris-Saclay, UVSQ, DYPAC, 78000 Versailles, France

3   Jet Propulsion Laboratory, California Institute of Technology, 4800 Oak Grove Drive, Pasadena, CA 91109, USA

4   NASA Goddard Space Flight Center, 8800 Greenbelt Road, Greenbelt, MD 20771, USA

5   Directorate of Science, Solar System Section, ESA's European Space Research and Technology Centre, Keplerlaan 1, 2201 AZ Noordwijk, Netherlands

6   NASA Science Mission Directorate, Mary W. Jackson NASA Headquarters, Washington DC 20546, USA

7   ESA's European Space Research and Technology Centre, Keplerlaan 1, 2201 AZ Noordwijk, Netherlands

8   Planetary Science Institute, Tucson, AZ 85719, USA

9   Dept. Earth and Environmental Sciences, Tulane University, 101 Blessey Hall, New Orleans, LA 70118, USA

10   International Space Science Institute, Hallerstrasse 6, 3012 Bern, Switzerland

11   Institute of Space Research, Deutsches Zentrum für Luft- und Raumfahrt, Rutherfordstraße 2, 12489 Berlin, Germany

12   School of Earth and Space Exploration, Arizona State University, Tempe, AZ 85287, USA

13   Department of Atmospheric, Oceanic and Planetary Physics, Oxford University, Oxford OX1 3PU, UK

14   NASA Goddard Institute for Space Studies, 2880 Broadway, New York, NY 10025, USA

15   Theoretical Astrophysics, Department of Physics and Astronomy, Uppsala University, Uppsala, Sweden

16   Department of Earth and Planetary Sciences, University of California, Riverside, CA 92521, USA

17   CNRS - Centre de Biophysique Moléculaire, rue Charles Sadron, 45071 Orléans, France

18   Potsdam Institute for Climate Impact Research, 14473 Potsdam, Germany

19   Department of Earth Sciences, VU Amsterdam, Amsterdam, Netherlands

20   Michigan State University, Natural Science Building, 288 Farm Lane, East Lansing, MI 48824, USA

21   Department of Astronomy and Planetary Science, Northern Arizona University, Flagstaff, AZ 86011, USA

22   Habitability, Atmospheres, and Biosignatures Laboratory, University of Arizona, Tucson, AZ 85721, USA

23   Lunar and Planetary Laboratory, University of Arizona, Tucson, AZ 85721, USA

24   Institut de physique du globe de Paris, CNRS, Université de Paris, 75005 Paris, France

25   Institute of Planetary Research, Deutsches Zentrum für Luft- und Raumfahrt, Rutherfordstraße 2, 12489 Berlin, Germany

774    Springer

[26] Dept. of Earth and Environmental Sciences, Wesleyan University, Middletown, CT 06459, USA

[27] Department of Earth Sciences, Royal Holloway, University of London, Egham, Surrey, TW20 0EX, UK

[28] Department of Earth and Planetary Sciences, Washington University, St. Louis, MO 63130, USA

[29] Geophysical Institute, University of Alaska Fairbanks, 1731 South Chandalar Drive, Fairbanks, AK 99775, USA

[30] Applied Physics Laboratory, Johns Hopkins University, Laurel, MD 20723, USA

[31] LATMOS/CNRS/Sorbonne Université/UVSQ, 11 boulevard d'Alembert, 78280 Guyancourt, France

[32] Centre for Earth Evolution and Dynamics, Dept. of Geosciences, University of Oslo, Blindern, 0316 Oslo, Norway

[33] Lunar and Planetary Institute, 3600 Bay Area Boulevard, Houston, TX 77058, USA

[34] Institut für Geophysik, Geophysical Fluids Dynamics, ETH Zurich, Sonneggstrasse 5, 8092 Zürich, Switzerland

[35] Space Research Institute (IKI), Russian Academy of Sciences, Moscow 117997, Russia

[36] Space Applications Centre, ISRO, Ahmedabad-380015, India